Environmental Remediation Cost Data – Unit Price

11th Annual Edition

2005

Senior Editor
Scott Martin, E.I.T.

Technical Editors
Jacqueline Rast, P.E.
Richard Rast

Reed Construction Data

Senior Vice President & General Manager
John Ware

Product Management
Jane Crudden

Vice President of Sales
John M. Shea

Senior Engineering Operations Manager
John H. Ferguson, P.E.

Production Manager
Michael Kokernak

Production Coordinator
Marion E. Schofield

Contributing Editors
Jeannene D. Murphy
Phillip R. Waier, P.E.
Gary Hoitt

ECHOS
ENVIRONMENTAL COST HANDLING OPTIONS AND SOLUTIONS

Preface

THE BOOK

The **ECHOS** ***Environmental Cost Data - Unit Price Book*** is "the cost source book" for environmental restoration activities beginning with initial site investigation and continuing through studies, design, remediation, and long-term monitoring and operation. Containing over 5,000 assembly cost items, this publication is updated and expanded annually to reflect the latest cost and technology information in the rapidly changing environmental field.

The **ECHOS** ***Environmental Cost Data - Unit Price Book*** provides you with the detailed line-items, component costs, forms, instructions, and guidelines needed to prepare or verify cost estimates for almost any type of environmental restoration project, ranging from simple underground storage tank removals to complex multimedia/multi-contaminant hazardous waste sites listed on the US EPA's Superfund - National Priority List.

The assembly numbering follows the US Government Interagency Code of Accounts, an evolving standard for organizing environmental restoration costs. Cost information is provided for labor, equipment, and materials with guidelines for adjusting costs to reflect work performed at various OSHA-dictated safety levels. Location factors are supplied by zip code, allowing you to adjust your estimate to local conditions.

THE DATA

The **ECHOS** research staff is constantly gathering, monitoring, and developing construction and environmental restoration cost information throughout the US. In so doing, the **ECHOS** database reflects the most current trends in both procedures and unit costs for environmental restoration activities. This book is the result of over eleven years of research in environmental restoration cost, and the cost information in this book has been used successfully on more than 10,000 environmental restoration projects throughout the U.S.

This data is received by us from sources we believe to be reliable, but no warranty, guaranty or representation is made by **ECHOS** as to the correctness or sufficiency of any information, prices, or representation contained in the **ECHOS** ***Environmental Cost Data - Unit Price Book***, and **ECHOS** assumes no responsibility or liability in connection therewith.

Please note that the costs contained in **ECHOS** do not include contractor overhead or profit.

Material costs are determined through contact with product manufacturers, dealers, supply houses, distributors, and contractors. Labor costs are based on crews and productivity factors determined by **ECHOS** environmental engineering and construction experts. Equipment costs are based on either rental rates or purchase and annual cost of ownership.

REGULATORY ENVIRONMENT

There are numerous state and federal laws and regulations that govern the practice of environmental restoration activities. The two primary laws that set the standards for this book are the Comprehensive Environmental Response Compensation and Liability Act (CERCLA) and its amendments (commonly referred to as the SUPERFUND Law), and the Resource Conservation and Recovery Act (RCRA). The environmental restoration technologies and processes used in this book are primarily designed to be used on projects that are regulated by these laws, but the data can be used for other unregulated projects.

ABOUT ECHOS

ECHOS is a joint venture between Azimuth Group, Ltd., experts in environmental restoration cost estimating and technology application, and Reed Construction Data, the leading publisher of construction cost information in North America. Through this collaboration, exhaustive cost research, and seasoned technical experts, all of the practical tools necessary to assemble or analyze restoration costs have been established.

TABLE OF CONTENTS

TABLE OF CONTENTS

TABLE OF CONTENTS

TABLE OF CONTENTS

TABLE OF CONTENTS

TABLE OF CONTENTS

Estimating Forms and Instructions

Creating a cost estimate for environmental restoration, whether an RI/FS, RFI/CMS, or Remedial Action, is a multi-step process, as depicted below. This section includes blank forms that can be used when developing an estimate, instructions for using those forms, and an explanation of how to locate cost data in the ***Environmental Remediation Cost Data – Unit Price Book***.

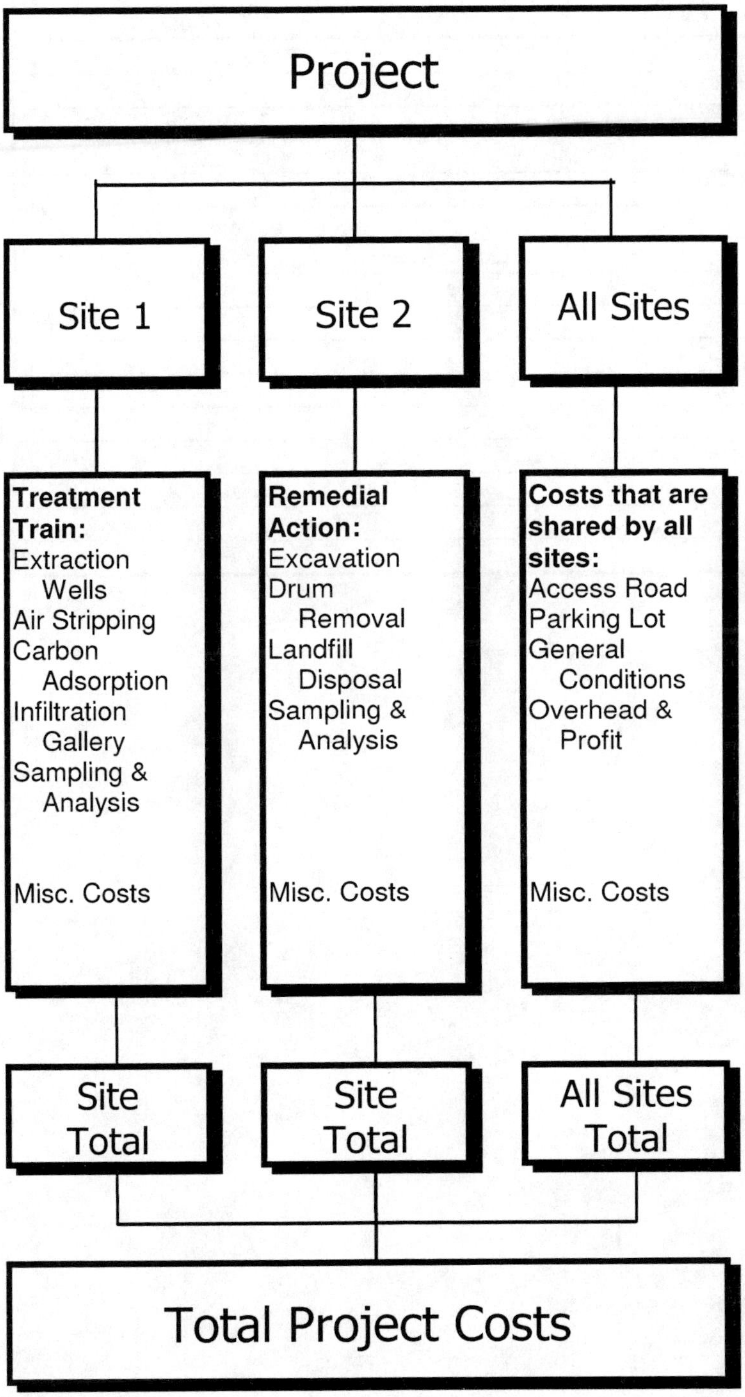

Step #1 -- Estimate Overview

Describe project, site(s), and any assumptions that were made for estimating purposes.

Projects may involve only one site or there may be many sites. For multiple-site projects, separate the site-specific costs from the project costs, shown here as "All Sites," for costs that pertain to all sites in the project.

Finally, identify the technologies, treatments, processes, etc. that are to be included in the estimate for each site.

Step #2 -- Cost Estimates

Estimate the cost of each technology, treatment, process, etc. for each site. The estimates are prepared by selecting, quantifying, and pricing assemblies and line items from this book. ("Pricing" in this context refers to adjusting the costs to account for reduced productivity at higher safety levels.)

Finally, identify and estimate costs for any miscellaneous items that were not included in the technology/process estimates.

Step #3 -- Site Summary

Compute the total costs for each site (technologies, treatments, processes, and miscellaneous costs).

Step #4 -- Project Summary

Sum the costs for all sites and adjust for location and escalation. Add design costs (if applicable), overhead and profit, and contingencies.

Project	Project Name		
Location	Number of Sites	Estimate Prepared By	Preparation Date

Project Description:

Form P1

Site	Site Description	Type of Site

Site Number /Name	Period of Performance: ___ / ___ / ___ through ___ / ___ / ___	Depth to Groundwater feet	Type of Work: ____ RI/FS or RFI/CMS ____ Remedial Design ____ Remedial Action	Start-up Period weeks	O&M Duration months

Media/Contaminants of Concern:	Treatment:	Other Cost Items:

Form S1

Cost Estimate	Site			Technology/Process				Page ____ of ____	
Assembly/Line Item	Quantity & Unit	Safety Level	Unit Price at Safety Level			Total Costs			
			Labor	Equip.	Matrl.	Labor	Equip.	Matrl.	Total
					Total				
Notes									

Form E1

Cost Estimate	Site			Technology/Process				Page ____ of ____	
Assembly/Line Item	Quantity & Unit	Safety Level	Unit Price at Safety Level			Total Costs			
			Labor	Equip.	Matrl.	Labor	Equip.	Matrl.	Total
					Total				
Notes									

Form E2

Site Summary	Site			
Process	Labor	Equipment	Materials	Total
Total				

Form S2

Project Summary	Project Name			
Site	Labor	Equipment	Materials	Total
			Total for All Sites	
			Location Multiplier	
			Site Total, Adjusted for Location	
			Escalation to ___ / ___ /___	
			Escalated Costs	
			Design (____ %)	
			Overhead and Profit (____ %)	
			Contingencies (___ %)	
			Project Total	

Form P2

Step #1 -- Estimate Overview

Project	Project Name *Use this space for the project name, proposal number, contract number, etc.*		
Location *Refer to the "Localization Information" section of this book.*	Number of Sites *Enter the number of sites included in the estimate*	Estimate Prepared By *Enter the name of the estimator(s), company, etc.*	Preparation Date *Enter the proposal due date or actual date of preparation.*
Project Description : *Use this space to describe the project and estimate. Include key assumptions that were made in the estimating process, cost sources, sources of project information (RFP, drawings, etc.).*			

Form P1

Step #1 -- Estimate Overview (continued)

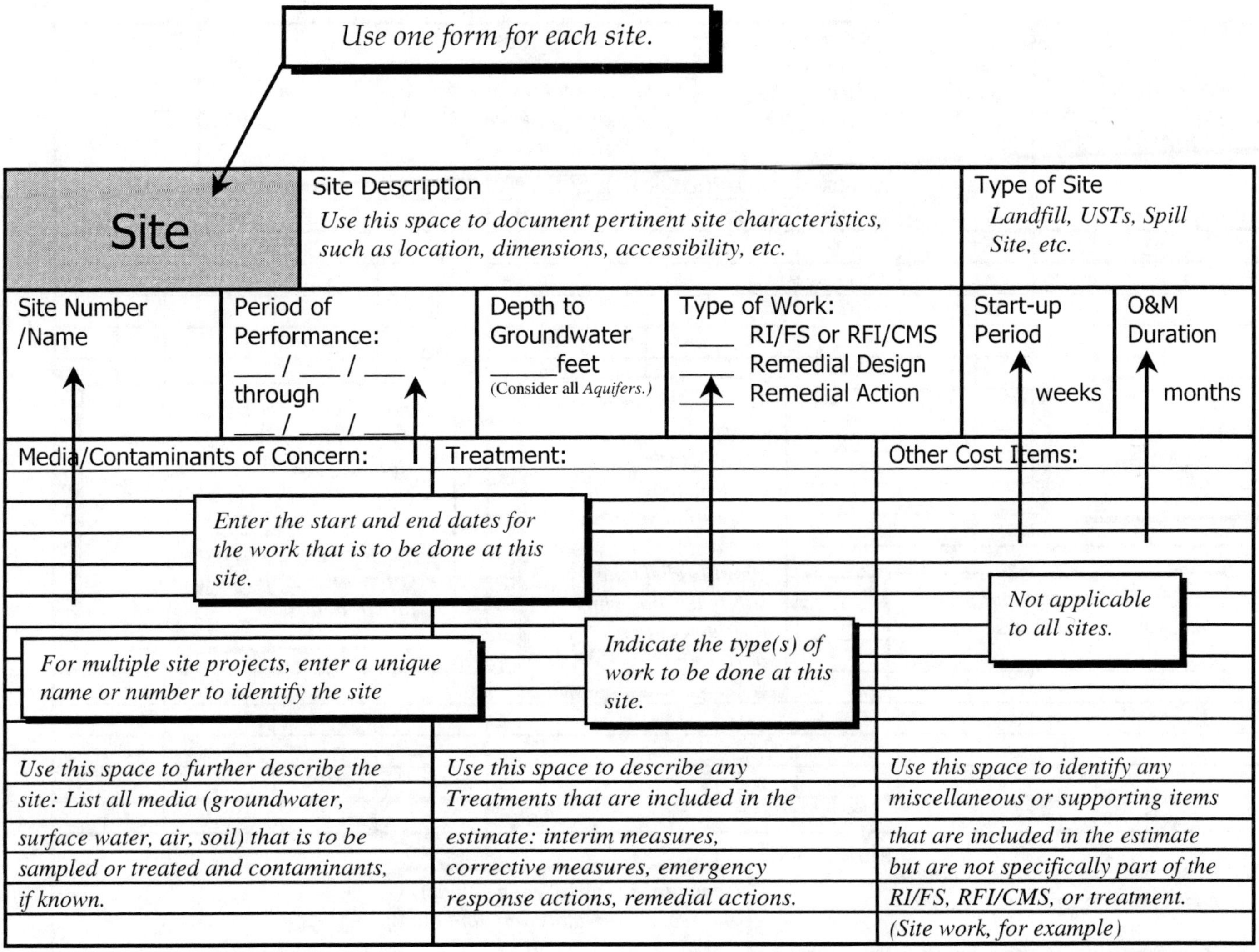

Form S1

Step #2 -- Cost Estimates

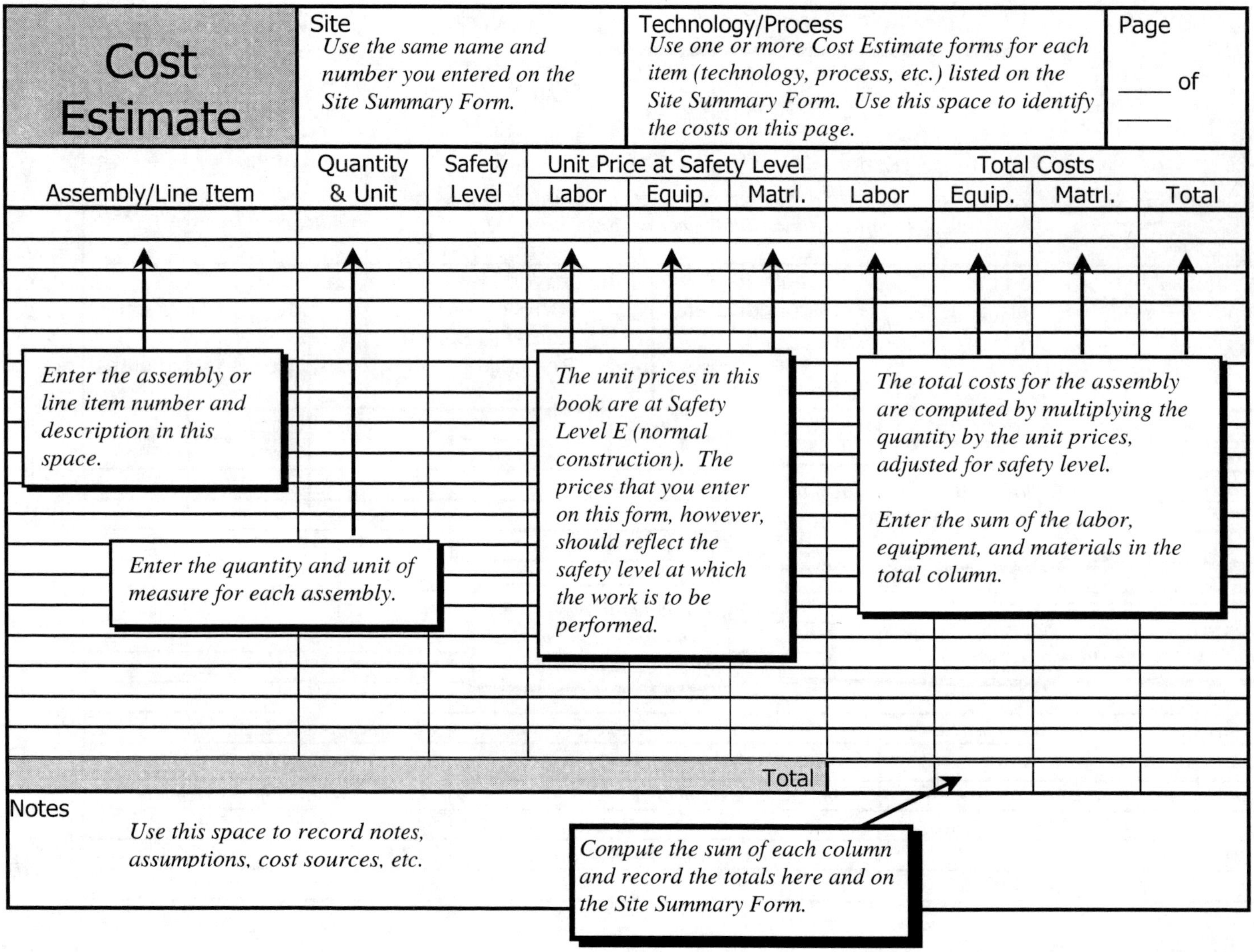

Cost Estimate

Site
Use the same name and number you entered on the Site Summary Form.

Technology/Process
Use one or more Cost Estimate forms for each item (technology, process, etc.) listed on the Site Summary Form. Use this space to identify the costs on this page.

Page ____ of ____

Assembly/Line Item	Quantity & Unit	Safety Level	Unit Price at Safety Level			Total Costs			
			Labor	Equip.	Matrl.	Labor	Equip.	Matrl.	Total
					Total				

Notes
Use this space to record notes, assumptions, cost sources, etc.

Enter the assembly or line item number and description in this space.

Enter the quantity and unit of measure for each assembly.

The unit prices in this book are at Safety Level E (normal construction). The prices that you enter on this form, however, should reflect the safety level at which the work is to be performed.

The total costs for the assembly are computed by multiplying the quantity by the unit prices, adjusted for safety level.

Enter the sum of the labor, equipment, and materials in the total column.

Compute the sum of each column and record the totals here and on the Site Summary Form.

Form E1

Step #2 -- Cost Estimates (continued)

Safety level requirements affect the productivity of labor and equipment. Although the unit prices in this book are for normal conditions (Safety Level E), the prices that you enter on the Cost Estimate form should reflect the safety level at which the work is to be performed. The conversion process is as follows:

$$\text{Labor\$ at Safety Level X} = \frac{\text{Labor\$ at Safety Level E}}{\text{Labor Productivity at Safety Level X}}$$

$$\text{Equip\$ at Safety Level X} = \frac{\text{Equip\$ at Safety Level E}}{\text{Equip Productivity at Safety Level X}}$$

Safety Level	Labor Productivity	Equipment Productivity
A	37% (10-40% range)	50%
B	48% (25-60% range)	60%
C	55% (25-70% range)	75%
D	82% (50-90% range)	100%

Step #2 -- Cost Estimates (continued)

Understanding the assembly numbering system will help you find cost items in this book . . .

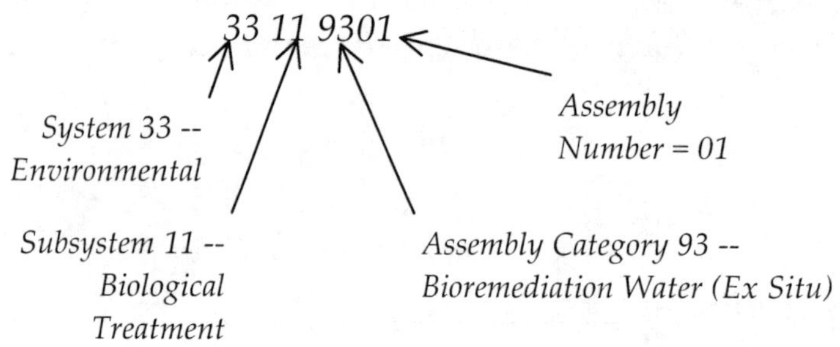

When you are looking for a particular cost item, begin with the ***Table of Contents.*** *Start at the highest and most general level (system). Select the system that seems most likely to contain the data that you need, then look through the list of that system's subsystems. Select a subsystem, then look through the list of assembly categories for that subsystem. When you have selected an assembly category, turn to the page indicated. All of the assemblies for that system-subsystem-assembly category will be listed. If you do not find what you need, go "up" one level (i.e., from subsystem to system) and make another selection.*

Step #3 -- Site Summary

Site Summary	Site *Use the same name or number entered on the Site Summary Form.*			
Process	**Labor**	**Equipment**	**Materials**	**Total**
Total				

This form provides a summary page for each site in your estimate. Use one line for each component (technology, process, etc.) of your site estimate. Enter a description of each cost component (Step #1) and the estimated costs (Step #2).

Form S2

Compute totals for labor, equipment, materials, and total costs. Enter the information here and on the Project Summary form.

Step #4 -- Project Summary

Project Summary	Project Name *Use the same project name or number entered on the Site form.*			
Site	Labor	Equipment	Materials	Total
			Total for All Sites	
			Location Multiplier	
			Site Total, Adjusted for Location	
			Escalation to ___ / ___ /___	
			Escalated Costs	
			Design (____ %)	
			Overhead and Profit (____ %)	
			Contingencies (___ %)	
			Project Total	

This form provides a summary page for your project. Using one line for each site, enter the site name and the estimated costs from Step #3.

Calculate Total Project Costs:

Location Multiplier: Refer to the localization information section.
Escalation: Enter the escalation date (usually the midpoint of construction) and the dollar amount.
Design and Overhead & Profit: If you estimated these costs, enter the dollar amounts and calculate the percentages. Otherwise, enter the percentages and calculate the dollar amounts.
Contingencies: Enter a percentage and calculate the dollar amount.

Form P2

How to Interpret the Cost Pages

The assembly information (assembly number, description, unit of measure, cost) is shown in **bold-face** *type, on the first line of each group.*

This field represents the estimated hourly output per the unit of measure associated with the line item. In this example, the crew CODEU can output 112.8750 CY of line item #02 in one hour.

The labor, equipment, and materials costs for the assembly are the sum of the factored costs for the line items in the assembly, as shown below:

$$\text{Labor Cost}_{Assm} = (\text{Labor}_{\text{Item 1}} * \text{Factor}_{\text{Item 1}}) + (\text{Labor}_{\text{Item 2}} * \text{Factor}_{\text{Item 2}}) \ldots + (\text{Labor}_{\text{Item N}} * \text{Factor}_{\text{Item N}})$$

Assembly/ Line Item	Description	Unit	Crew	Hourly Output	Factor	Unit Costs (No Safety Level Adjustment)			
						Labor	Equipment	Materials	Total
33 06 xxxx	**Assembly Description**	**CY**				**2.00**	**3.00**	**1.50**	**6.50**
44445 5555	Line Item #01 Description	SY	COFCV	106.3800	2.00000	0.50	1.50	0.75	5.50
99999 9999	Line Item #02 Description	CY	CODEU	112.8750	0.00001	10,000.00	0.00	0.00	1.00
33 06 xxxx	**Assembly Description**	**EA**				**50.00**	**100.00**	**15.00**	**165.00**
22224 4444	Line Item Description (Single Line Item)	EA	CODTU	145.7400	1.00000	50.00	100.00	15.00	165.00

The line item information is just below the assembly information and is shown in regular *type.*

The crew is a 5-digit code assigned to a group of cost items representing all of the labor and equipment costs required to accomplish a specific kind of work.

Labor costs are bare, or non-marked up, hourly rates.

Equipment costs represent the total cost of ownership for each piece of equipment. The total cost of ownership includes purchase price as well as life-cycle maintenance costs for the equipment. These costs are then allocated into an hourly rate.

Refer to the Crew Detail chapter for the composition of each crew.

The line item factor tells you how much of a given item is included in one unit of an assembly. In this example, there are two square yards of line item #01 in one cubic yard of assembly 33 06 XXXX and there are 0.00001 cubic yards of line item #02.

Total cost is the sum of the costs for labor, equipment, and materials.

The assembly and line item costs throughout this book are shown at Safety Level E (normal conditions). The costs for labor and equipment will typically be adjusted to account for the reduced productivity related to safety level requirements, as shown below:

Safety Level	Labor Productivity	Equipment Productivity
A	37% (typically 10-40%)	50%
B	48% (typically 25-60%)	60%
C	55% (typically 25-70%)	75%
D	82% (typically 50-90%)	100%

$$\text{Labor Cost}_P = \text{Labor Cost}_E \;/\; \text{Labor Productivity}_P$$

$$\text{Equipment Cost}_P = \text{Equipment Cost}_E \;/\; \text{Equipment Productivity}_P$$

where P = Safety Level for the project

E = Safety Level E

Project Markups

All costs in the ECHOS books are shown as Direct Costs (sometimes referred to as "Bare Costs"). To compute total project costs, markups must be applied. Total project cost is the sum of direct costs plus project markups. Project markups may be referenced by many names but can generally be referred to as contractor general conditions, overhead and profit.

Direct Cost Components

Direct costs (sometimes referred to as "bare costs") include all of the costs that can be directly attributed to a particular item of work or activity. Direct costs generally include:

1. **Professional Labor** – generally comprised of engineers, scientists, site health and safety officials and other professional disciplines required to support the work on the site. Professional Labor is distinguished from Craft Labor because it generally includes a significantly higher labor overhead than craft labor.

2. **Craft Labor** – generally comprised of the direct pay to the laborer plus any required fringe benefits, but not including overhead or profit items, for performing the particular items of work.

3. **Material Items** – as required for the work including permanent equipment items such as groundwater extraction pumps, air stripping equipment, chemical treatment equipment, etc.

4. **Construction Equipment** – includes items such as earthmoving equipment, trucks, drill rigs, etc., as required to perform the work.

The prime contractor's direct costs also include the total subcontractor's price, including subcontractor's overhead and profit.

General Conditions Cost Components

General conditions costs are field-related tasks required to execute a contract, but not specific to a technology or restoration task. For multiple-site projects, general conditions items are typically estimated for the overall project rather than for individual sites. Examples of general conditions items include:

1. Supervision - Labor
2. Temporary Facilities including:
    - Job trailers
    - Storage trailers
    - Portable toilets
3. Temporary plants
4. Personal protective equipment (PPE)
5. Travel and per diem
6. Permits, taxes, insurance, bonds
7. Decontamination facilities (These costs are included in a separate section of this book.)
8. Professional labor personnel, such as project managers, engineers, hydrogeologists, industrial hygienists, and health and safety engineers.

General Conditions costs can be estimated on a percentage of direct project cost basis. General Conditions cost percentages tend to be higher for small projects and lower for large projects. However, every project is different and the estimator should carefully consider the details of any project to estimate these costs. The following table provides guidelines for estimating general conditions cost.

Phase	*Total* Project Phase Direct Cost Range	Professional Labor	Craft Labor	Material	Construction Equipment
Site Assessment and Long Term Monitoring	< $10,000	25.00%	25.00%	20.00%	50.00%
	10,000 - 25,000	15.00%	20.00%	15.00%	35.00%
	25,000 - 50,000	10.00%	17.50%	12.00%	25.00%
	50,000 - 100,000	7.50%	15.00%	10.00%	20.00%
	100,000 - 250,000	6.00%	13.50%	8.00%	15.00%
	250,000 - 500,000	5.00%	12.00%	6.50%	10.00%
	> 500,000	5.00%	11.00%	5.50%	8.50%
Remedial Action Construction	< $10,000	25.00%	25.00%	17.00%	40.00%
	10,000 - 25,000	15.00%	20.00%	12.00%	30.00%
	25,000 - 50,000	10.00%	17.50%	10.00%	20.00%
	50,000 - 100,000	7.50%	15.00%	8.00%	15.00%
	100,000 - 250,000	5.00%	12.00%	6.50%	10.00%
	250,000 - 500,000	5.00%	10.00%	5.00%	8.00%
	> 500,000	5.00%	8.00%	5.00%	6.00%
Operations and Maintenance	< $10,000	10.00%	15.00%	15.00%	20.00%
(direct cost range - annual costs)	10,000 - 25,000	7.50%	11.50%	10.00%	15.00%
	25,000 - 50,000	5.00%	8.50%	6.50%	10.00%
	50,000 - 100,000	4.00%	7.00%	5.00%	8.00%
	100,000 - 250,000	4.00%	6.00%	5.00%	7.00%
	250,000 - 500,000	4.00%	6.00%	5.00%	6.00%
	> 500,000	4.00%	6.00%	5.00%	6.00%

Overhead Cost Components

Overhead costs are items required to support the contractor's permanent staff and offices that provide indirect support to the contractor's field operations. These costs are typically allocated to project expenses as a percentage of the project direct cost for professional labor, craft labor, materials and equipment. Overhead costs may be applied in different ways but generally can be grouped into the following categories:

1. Professional services overhead
2. Home office expense
3. Operations and maintenance service contract overhead

Overhead Guidelines	
Description	**Guidelines**
Office Overhead	Can vary widely depending on project size and the organization of the contractor performing the work. Ranges from 10 – 200% of Direct Professional Labor Cost.
General and Administration (G&A)	Can vary widely depending on project size and the organization of the contractor performing the work. Ranges from 5 – 75% of Direct Professional Labor Cost.
Home Office Expense	Typically 2.5 – 6%; the percentage tends to be smaller for larger projects and vice versa.
O&M Service Contract Overhead	
Administrative Cost	Typically 6 - 10% of Direct Project Cost
Insurance	Typically 5% (varies from 2 - 10%)
Fees and Permits	Project/technology specific
Sales Tax	Applied to materials and equipment only; refer to table in preceding section.
Profit on O&M Work	Typically 5 - 7%

Profit

Profit is the fee earned by the contractor on top of direct costs, general conditions, and overhead items. Profit is generally applied as a percentage of the sum of direct costs, general conditions and overhead. Different profit percentages may be applied to prime contractor costs and subcontractor costs. The following table provides guidelines for profit.

Profit Guidelines	
Description	**Guidelines**
Subcontractor's Profit On Own Work	Typically 5 – 12% of Direct Project Cost plus General Conditions paid by the subcontractor
Prime Contractor's Profit On Own Work	Typically 4 – 15% of Direct Project Cost plus General Conditions for that portion of the project performed by the Prime Contractor
Prime Contractor's Profit On Sub's Work	Typically 4 - 8% of the Total Project Cost for that portion of work performed by subcontractors

ECHOS© Cost Data

All cost data in this book are based on U.S. national average prices and are given in U.S. dollars. The localization factors can be used to adjust these national averages to local costs for a specific location. The localization factors for Canada can also be used to adjust the cost data to local costs in Canadian dollars.

The components of the location adjustment factors include a weighted average index that consists of 66 commonly used construction materials, specific labor hours for 21 construction trades, and specific days of equipment rental for 6 types of construction equipment typically used to install the 66 construction material items by the 21 trades.

Since each section of the book contains many different items, any particular item multiplied by the particular location index may give incorrect results. However, when all the book costs for a particular division are summarized and then factored, the result should be very close to the actual costs for that particular location.

The location adjustment tables are organized by U.S. state and then by the first three digits of the postal zip codes within each state. Canadian factors are organized for key locations by province.

How to Adjust the Cost Data to Local Dollars

All costs in this book, including professional labor and analysis rates, are bare national average costs. Material, labor, and equipment costs can be adjusted based on their area of origin. The following equation can be used to adjust any cost to reflect a project's location:

[ECHOS Unit Price] x [Localization Factor] = Localized Cost

Detailed Description of Safety Levels in ECHOS

The ECHOS cost books use five safety levels, defined as A through E. Safety levels A, B, C, and D are based on the Occupational Safety and Health Administration (OSHA) regulations in 29 CFR Part 1910. Safety Level E (Modified level D) corresponds to the EPA "No Hazard" designation. Safety levels are assumed to impact labor and equipment productivity according to the table outlined, below.

Safety Level	Labor Productivity	Equipment Productivity
A	37% (typically 10-40%)	50%
B	48% (typically 25-60%)	60%
C	55% (typically 25-70%)	75%
D	82% (typically 50-90%)	100%
E (Non-Hazardous)	100% (basis)	100% (basis)

List of Abbreviations

ABS	Asbestos Bonded Steel
ACI	American Concrete Institute
ACRE	per Acre
ASME	American Society of Mechanical Engineers
ASTM	American Society of Testing and Materials
Avg	Average
AWWA	American Water Works Association
BCY	Bank Cubic Yards
BF	Board Feet
BOD	Biochemical Oxygen Demand
BTEX	Benzene, Toluene, Ethylbenzene, Xylene
BTU	British Thermal Unit
BTUH	British Thermal Unit per Hour
CaCl2	Calcium Chloride
cc	Cubic Centimeters
CEC	Cation Exchange Capacity
CERCLA	Comprehensive Environmental Response Compensation & Liability Act
CF	Cubic Feet
CFM	Cubic Feet per Minute
CI	Cast Iron/Curies
CIP	Cast-In-Place
Cl	Chlorine
CLF	Hundred Linear Feet
CLO2/TCO	Chloride Oxygen Content/Total Carbon Dioxide
CLP	Certified Laboratory Program
cm	Centimeter
CMP	Corrugated Metal Pipe
Cmpct	Compaction
Cntnr	Container(s)
CO2	Carbon Dioxide
COD	Chemical Oxygen Demand
Cpr	Hundred Pair
Cprsr	Compressor(s)
CPT	Cone Penetrometer Test
CPVC	Chlorinated Polyvinyl Chloride
CS	Carbon Steel
CSF	Hundred Square Feet
CSF*W	Hundred Square Feet multiplied by Wall Height (Volume)
CT	Cooling Tower
CWT	Hundred Pounds
CY	Cubic Yard(s)
CY/hr	Cubic Yard(s) per hour
D&D	Decontamination & Decommissioning
DAY	per Day
Dbl	Double
DD	Double Drum
Decom	decommissioning

List of Abbreviations

Decon	Decontamination
Demob	Demobilization
Desorp	desorption
Dia	Diameter
DIP	Ductile-Iron Pipe
DOT	Department of Transportation
DRM	Drums
EA	Each
EDTA	Ethylene Diamine Tetraacetic Acid
Encl	Enclosure
EOD	Explosives/Ordnance Waste Disposal
EPA	Environmental Protection Agency
ESVCP	Extra Strength Vitrified Clay Pipe
FML	Fabric Membrane Liner
FNPT	female national pipe thread
FPM	Feet per Minute
FRP	Fiberglass Reinforced Plastic
FT	Feet
ft-lbs	Foot-pound(s)
GAL	Gallons
Gndwtr	Groundwater
GPD	Gallons per Day
GPH	Gallons per Hour
GPM	Gallons per Minute
HD	Header
HDPE	High-Density Polyethylene
Hdr	Header
HP	Horse Power
HPLC	High Performance Liquid Chromatography
HR	Hour(s)
HS	Hollow Stem
HTRW	Hazardous, Toxic & Radioactive Waste
HW	Hazardous Waste
Hydr	Hydraulic
Hydr Exc	Hydraulic Excavator
Hz	Hertz (cycles)
I/O	input/output
ID	Inside Diameter
Incl	Including
Ind	Indirect
Inj	Injection
IP	Iron Pipe
IPT	Iron Pipe Threaded
ISV	In Situ Vitrification
JOB	per Job
kg	Kilogram
KGAL	Thousand Gallons

List of Abbreviations

kHz	Kilohertz
KVA	Kilovolt Ampere
KW	kilowatt
KWH	Kilowatt Hours
LB	Pounds
LB/HR	Pounds per Hour
LB/LF	Pounds per Linear Feet
LF	Linear Feet
LFHD	Linear Feet of Header
LGLCS	Landfill Gas & Leachate Control System
LLRW	Low Level Radioactive Waste
LPG	Liquid Petroleum Gas
LS	Lump Sum
MBH	Thousand BTUs per Hour
MCF	Thousand Cubic Feet
MCFM	Thousand Cubic Feet per Minute
MF	Thousand Feet
MGD	Million Gallons per Day
MH	Manhole/Man-Hour
MHz	Megahertz
MI	Miles
MLF	Thousand Linear Feet
MO	Month(s)
Mob	Mobilization
MPH	Miles per Hour
MSY	Thousand Square Yards
MTBE	Methyl Tertiary Butyl Ether
MWK	Man-weeks
NAPL	Nonaqueous Phase Liquid
Non-Haz	Non-Hazardous
NPL	National Priority List
NPT	National Pipe Thread
O & M	Operations and Maintenance
O & P	overhead and profit
OD	Outside Diameter
ODCs	Other Direct Costs
OH	Overhead
OSHA	Occupational Safety and Health Administration
OVA	Organic Vapor Analyzer
PAH	Polynuclear Aromatic Hydrocarbons
PAIR	per Pair
PCB	Polychlorinated Biphenyl
PCT	Percentage
PE	Polyethylene
pH	Hydrogen-Ion Concentration
Pmp	pump(s)
PNT	per Point

List of Abbreviations

POTW	Publicly-Owned Treatment Works
PPE	Personal Protective Equipment
PPHPR	Parts per Hundred Pair
PPM	Parts per Million
PR	Pair
PSF	Pounds per Square Foot
PSI	Pounds per Square Inch
PSIG	Pounds per Square Inch Gauge
PT	Packed Tower
PTTU	Packed Tower Treatment Unit
PVC	Polyvinyl Chloride
PWR	Power
QA	Quality Assurance
QC	Quality Control
RA	Remedial Action
RCP	Reinforced Concrete Pipe
RCRA	Resource Conservation and Recovery Act
RFI/CMS	RCRA Facility Investigation / Corrective Measures Study
RI/FS	Remedial Investigation / Feasibility Study
RIS	per Riser/Step
RW	Radioactive Waste
S&SA	Soil & Sediment Analysis
SCBASCF	Self-contained Breathing Apparatus Standard Cubic Feet
SCFM	Standard Cubic Feet per Minute
SET	per Set
SF	Square Feet
SQ	100 Square Feet
Sq In	Square Inch
STOP	per Stop/Floor
SVE	Soil Vapor Extraction
SY	Square Yards
TAL	Target Analyte List
TAT	Turn Around Time
TCLP	Toxicity Characteristic Leach Procedure
Thd	Threaded
Thk	Thick
TN	Ton(s)
TOC	Total Organic Carbon
TON	per Ton(s)
TPH	Total Petroleum Hydrocarbons
UST	Underground Storage Tank
UXO	Unexploded Ordnance
VEP	Vapor Extraction Point
Vitr	Vitrification
VLDPE	Very Low Density Polyethylene
VLF	Vertical Linear Foot
VOC	Volatile Organic Compound

List of Abbreviations

w/	With
w/o	Without
WEEK	per Week(s)
WK	per Week(s)
YD	Yard(s)
YR	Year(s)

Localization Factors: Canadian (by City) and U.S. (by Zip)

Canada, Alberta	
Calgary	1.05
Edmonton	1.06
Fort Mcmurray	1.01
Lethbridge	1.01
Lloydminster	1.01
Medicine Hat	1.01
Red Deer	1.01

Canada, British Columbia	
Kamloops	1.03
Prince George	1.03
Vancouver	1.09
Victoria	1.04

Canada, Manitoba	
Brandon	0.96
Portage La Prairie	0.96
Winnipeg	1.00

Canada, New Brunswick	
Bathurst	0.91
Dalhousie	0.91
Fredericton	0.94
Moncton	0.92
Newcastle	0.91
Saint John	0.95

Canada, Newfoundland	
Corner Brook	0.94
St Johns	0.94

Canada, Northwest Territories	
Yellowknife	0.99

Canada, Nova Scotia	
Dartmouth	0.96
Halifax	0.97
New Glasgow	0.95
Sydney	0.93
Yarmouth	0.94

Canada, Ontario	
Barrie	1.05
Brantford	1.06
Cornwall	1.04
Hamilton	1.09

Canada, Ontario	
Kingston	1.05
Kitchener	1.02
London	1.07
North Bay	1.04
Oshawa	1.05
Ottawa	1.06
Owen Sound	1.04
Peterborough	1.04
Sarnia	1.06
St Catharines	1.01
Sudbury	1.01
Thunder Bay	1.01
Toronto	1.11
Windsor	1.03

Canada, Prince Edward Islan	
Charlottetown	0.91
Summerside	0.91

Canada, Quebec	
Cap-De-La-Madele	1.00
Charlesbourg	1.00
Chicoutimi	1.00
Gatineau	1.00
Laval	1.00
Montreal	1.02
Quebec	1.02
Sherbrooke	1.00
Trois Rivieres	1.00

Canada, Saskatchewan	
Moose Jaw	0.92
Prince Albert	0.91
Regina	0.93
Saskatoon	0.92

Canada, Yukon	
Whitehorse	0.91

United States, AK	
995	1.25
996	1.25
997	1.24
998	1.25
999	1.30

United States, AL	
350	0.87
351	0.87
352	0.87
354	0.78
355	0.77
356	0.79
357	0.86
358	0.86
359	0.80
360	0.80
361	0.80
362	0.74
363	0.76
364	0.76
365	0.82
366	0.82
367	0.76
368	0.79
369	0.75

United States, AR	
716	0.80
717	0.69
718	0.72
719	0.69
720	0.81
721	0.81
722	0.81
723	0.77
724	0.77
725	0.74
726	0.74
727	0.72
728	0.73
729	0.78

United States, AZ	
850	0.88
852	0.84
853	0.88
855	0.82
856	0.85
857	0.85
859	0.84
860	0.86
863	0.83

Localization Factors: Canadian (by City) and U.S. (by Zip), Cont.

United States, AZ	
864	0.83
865	0.81

United States, CA	
900	1.07
901	1.07
902	1.07
903	1.03
904	1.03
905	1.03
906	1.04
907	1.04
908	1.04
910	1.04
911	1.04
912	1.04
913	1.06
914	1.06
915	1.06
916	1.06
917	1.05
918	1.05
919	1.04
920	1.04
921	1.04
922	1.03
923	1.02
924	1.02
925	1.06
926	1.03
927	1.03
928	1.07
930	1.06
931	1.06
932	1.04
933	1.04
934	1.05
935	1.02
936	1.08
937	1.08
938	1.08
939	1.10
940	1.22
941	1.22
942	1.10

United States, CA	
943	1.15
944	1.16
945	1.12
946	1.17
947	1.16
948	1.16
949	1.16
950	1.13
951	1.17
952	1.07
953	1.08
954	1.12
955	1.05
956	1.10
957	1.10
958	1.10
959	1.09
960	1.08
961	1.08

United States, CO	
800	0.96
801	0.96
802	0.96
803	0.93
804	0.94
805	0.93
806	0.86
807	0.93
808	0.94
809	0.94
810	0.93
811	0.92
812	0.92
813	0.92
814	0.91
815	0.92
816	0.94

United States, CT	
060	1.06
061	1.06
062	1.06
063	1.04
064	1.05
065	1.07

United States, CT	
066	1.06
067	1.06
068	1.06
069	1.08

United States, DC	
200	0.97
201	0.97
202	0.97
203	0.97
204	0.97
205	0.97

United States, DE	
197	1.03
198	1.02
199	1.03

United States, FL	
320	0.81
321	0.88
322	0.81
323	0.76
324	0.71
325	0.81
326	0.81
327	0.86
328	0.86
329	0.91
330	0.87
331	0.87
332	0.87
333	0.86
334	0.84
335	0.87
336	0.87
337	0.81
338	0.87
339	0.83
340	0.87
341	0.83
342	0.85
344	0.81
346	0.87
347	0.86
349	0.84

Localization Factors: Canadian (by City) and U.S. (by Zip), Cont.

United States, GA

300	0.90
301	0.90
302	0.90
303	0.90
304	0.73
305	0.78
306	0.80
307	0.76
308	0.79
309	0.79
310	0.79
311	0.79
312	0.79
313	0.80
314	0.80
315	0.76
316	0.75
317	0.79
318	0.81
319	0.81
399	0.90

United States, HI

967	1.20
968	1.23
969	1.30

United States, IA

500	0.91
501	0.91
502	0.91
503	0.91
504	0.81
505	0.80
506	0.81
507	0.81
508	0.83
509	0.91
510	0.87
511	0.87
512	0.77
513	0.77
514	0.78
515	0.88
516	0.76
520	0.88

United States, IA

521	0.78
522	0.92
523	0.92
524	0.92
525	0.85
526	0.86
527	0.96
528	0.96

United States, ID

832	0.90
833	0.76
834	0.79
835	0.97
836	0.91
837	0.91
838	0.84

United States, IL

600	1.07
601	1.07
602	1.07
603	1.07
604	1.07
605	1.07
606	1.12
609	1.00
610	1.03
611	1.03
612	0.96
613	0.99
614	0.98
615	1.00
616	1.00
617	0.98
618	1.00
619	1.00
620	0.97
621	0.97
622	0.97
623	0.95
624	0.96
625	0.97
626	0.98
627	0.98
628	0.95

United States, IL

629	0.95

United States, IN

460	0.89
461	0.94
462	0.94
463	0.97
464	0.97
465	0.89
466	0.89
467	0.88
468	0.88
469	0.88
470	0.87
471	0.85
472	0.89
473	0.89
474	0.90
475	0.90
476	0.91
477	0.91
478	0.91
479	0.88

United States, KS

660	0.95
661	0.95
662	0.95
664	0.85
665	0.85
666	0.85
667	0.84
668	0.80
669	0.79
670	0.86
671	0.86
672	0.86
673	0.78
674	0.80
675	0.75
676	0.79
677	0.79
678	0.80
679	0.74

United States, KY

United States, KY	
400	0.92
401	0.92
402	0.92
403	0.83
404	0.83
405	0.83
406	0.84
407	0.71
408	0.71
409	0.71
410	0.94
411	0.95
412	0.95
413	0.72
414	0.72
415	0.80
416	0.80
417	0.71
418	0.71
420	0.91
421	0.89
422	0.89
423	0.85
424	0.91
425	0.71
426	0.71
427	0.87

United States, LA	
700	0.86
701	0.86
703	0.83
704	0.81
705	0.78
706	0.81
707	0.79
708	0.79
710	0.79
711	0.79
712	0.78
713	0.77
714	0.77

United States, MA	
010	1.03
011	1.03
012	1.01
013	1.00
014	1.06
015	1.09
016	1.09
017	1.08
018	1.11
019	1.10
020	1.15
021	1.15
022	1.15
023	1.08
024	1.15
025	1.05
026	1.07
027	1.08

United States, MD	
206	0.87
207	0.90
208	0.90
209	0.89
210	0.93
211	0.93
212	0.93
214	0.90
215	0.89
216	0.74
217	0.89
218	0.78
219	0.82

United States, ME	
039	0.85
040	0.89
041	0.89
042	0.89
043	0.86
044	0.88
045	0.86
046	0.85
047	0.87
048	0.84
049	0.83

United States, MI	
480	1.00
481	1.01
482	1.07
483	1.00
484	0.96
485	0.96
486	0.95
487	0.95
488	0.96
489	0.96
490	0.92
491	0.91
492	0.94
493	0.84
494	0.89
495	0.84
496	0.85
497	0.87
498	0.92
499	0.92

United States, MN	
550	1.10
551	1.10
553	1.13
554	1.13
555	1.13
556	1.04
557	1.04
558	1.04
559	1.02
560	0.99
561	0.89
562	0.91
563	1.07
564	0.98
565	0.99
566	0.98
567	0.95

United States, MO	
630	1.02
631	1.02
633	0.93
634	0.90
635	0.88

Localization Factors: Canadian (by City) and U.S. (by Zip), Cont.

United States, MO	
636	0.95
637	0.93
638	0.90
639	0.90
640	1.03
641	1.03
644	0.95
645	0.95
646	0.85
647	0.97
648	0.85
650	0.92
651	0.92
652	0.93
653	0.91
654	0.87
655	0.87
656	0.87
657	0.87
658	0.87

United States, MS	
386	0.67
387	0.75
388	0.71
389	0.69
390	0.75
391	0.75
392	0.75
393	0.74
394	0.69
395	0.79
396	0.76
397	0.70

United States, MT	
590	0.89
591	0.89
592	0.87
593	0.86
594	0.90
595	0.87
596	0.89
597	0.87
598	0.86
599	0.85

United States, NC	
270	0.76
271	0.76
272	0.76
273	0.76
274	0.76
275	0.76
276	0.76
277	0.76
278	0.69
279	0.70
280	0.75
281	0.75
282	0.75
283	0.76
284	0.75
285	0.68
286	0.68
287	0.75
288	0.75
289	0.68

United States, ND	
580	0.85
581	0.85
582	0.82
583	0.82
584	0.79
585	0.85
586	0.83
587	0.86
588	0.82

United States, NE	
680	0.89
681	0.89
683	0.84
684	0.84
685	0.84
686	0.76
687	0.82
688	0.83
689	0.80
690	0.75
691	0.80
692	0.72
693	0.71

United States, NH	
030	0.93
031	0.93
032	0.91
033	0.91
034	0.77
035	0.81
036	0.75
037	0.75
038	0.89

United States, NJ	
070	1.11
071	1.11
072	1.10
073	1.10
074	1.11
075	1.11
076	1.10
077	1.09
078	1.10
079	1.09
080	1.07
081	1.07
082	1.06
083	1.07
084	1.06
085	1.09
086	1.09
087	1.08
088	1.09
089	1.09

United States, NM	
870	0.89
871	0.89
872	0.89
873	0.89
874	0.89
875	0.88
877	0.88
878	0.88
879	0.86
880	0.84
881	0.87
882	0.88
883	0.89

Localization Factors: Canadian (by City) and U.S. (by Zip), Cont.

United States, NM	
884	0.88

United States, NV	
889	1.01
890	1.01
891	1.01
893	0.91
894	0.97
895	0.97
897	0.97
898	0.93

United States, NY	
100	1.32
101	1.32
102	1.32
103	1.26
104	1.25
105	1.16
106	1.16
107	1.19
108	1.16
109	1.11
110	1.26
111	1.27
112	1.27
113	1.27
114	1.26
115	1.21
116	1.27
117	1.21
118	1.21
119	1.21
120	0.96
121	0.96
122	0.96
123	0.96
124	1.07
125	1.09
126	1.09
127	1.06
128	0.92
129	0.91
130	0.95
131	0.95
132	0.95

United States, NY	
133	0.93
134	0.93
135	0.93
136	0.94
137	0.92
138	0.92
139	0.92
140	1.02
141	1.02
142	1.02
143	1.00
144	0.98
145	0.98
146	0.98
147	0.92
148	0.91
149	0.91

United States, OH	
430	0.94
431	0.94
432	0.94
433	0.92
434	0.99
435	0.99
436	0.99
437	0.90
438	0.90
439	0.95
440	0.97
441	1.01
442	0.98
443	0.98
444	0.96
445	0.96
446	0.93
447	0.93
448	0.94
449	0.94
450	0.93
451	0.94
452	0.94
453	0.90
454	0.90
455	0.91

United States, OH	
456	0.93
457	0.90
458	0.93

United States, OK	
730	0.82
731	0.82
734	0.80
735	0.82
736	0.80
737	0.80
738	0.80
739	0.68
740	0.80
741	0.80
743	0.81
744	0.72
745	0.76
746	0.79
747	0.79
748	0.79
749	0.80

United States, OR	
970	1.04
971	1.04
972	1.04
973	1.02
974	1.02
975	1.02
976	1.02
977	1.02
978	0.99
979	0.94

United States, PA	
150	1.00
151	1.00
152	1.00
153	0.98
154	0.96
155	0.94
156	0.97
157	0.95
158	0.95
159	0.95

Localization Factors: Canadian (by City) and U.S. (by Zip), Cont.

United States, PA	
160	0.96
161	0.96
162	0.98
163	0.94
164	0.95
165	0.95
166	0.94
167	0.95
168	0.95
169	0.92
170	0.95
171	0.95
172	0.92
173	0.93
174	0.93
175	0.92
176	0.92
177	0.88
178	0.92
179	0.92
180	0.99
181	0.99
182	0.94
183	0.95
184	0.96
185	0.96
186	0.94
187	0.94
188	0.94
189	1.05
190	1.14
191	1.14
193	1.06
194	1.06
195	0.99
196	0.99

United States, PR	
009	0.87

United States, RI	
028	1.06
029	1.06

United States, SC	
290	0.75
291	0.75
292	0.75
293	0.73
294	0.75
295	0.73
296	0.73
297	0.67
298	0.84
299	0.70

United States, SD	
570	0.80
571	0.80
572	0.77
573	0.76
574	0.79
575	0.78
576	0.77
577	0.78

United States, TN	
370	0.87
371	0.87
372	0.87
373	0.79
374	0.79
375	0.88
376	0.80
377	0.79
378	0.79
379	0.79
380	0.88
381	0.88
382	0.75
383	0.78
384	0.78
385	0.76

United States, TX	
750	0.80
751	0.80
752	0.84
753	0.84
754	0.73
755	0.77
756	0.73
757	0.79
758	0.71
759	0.74
760	0.81
761	0.81
762	0.77
763	0.79
764	0.72
765	0.75
766	0.79
767	0.79
768	0.71
769	0.75
770	0.87
771	0.87
772	0.87
773	0.72
774	0.75
775	0.85
776	0.82
777	0.82
778	0.81
779	0.76
780	0.77
781	0.82
782	0.82
783	0.78
784	0.78
785	0.75
786	0.80
787	0.80
788	0.69
789	0.71
790	0.80
791	0.80
792	0.77
793	0.79
794	0.79
795	0.78
796	0.78
797	0.77
798	0.77
799	0.77
885	0.77

Localization Factors: Canadian (by City) and U.S. (by Zip), Cont.

United States, UT	
840	0.88
841	0.88
842	0.86
843	0.87
844	0.86
845	0.78
846	0.87
847	0.87

United States, VA	
220	0.91
221	0.91
222	0.91
223	0.93
224	0.84
225	0.84
226	0.79
227	0.79
228	0.76
229	0.81
230	0.85
231	0.85
232	0.85
233	0.84
234	0.84
235	0.84
236	0.83
237	0.83
238	0.84
239	0.73
240	0.76
241	0.76
242	0.75
243	0.73
244	0.75
245	0.77
246	0.74

United States, VT	
050	0.76
051	0.76
052	0.76
053	0.77
054	0.85
056	0.83
057	0.85

United States, VT	
058	0.77
059	0.76

United States, WA	
980	1.05
981	1.05
982	1.01
983	1.02
984	1.02
985	1.01
986	1.02
987	1.05
988	0.95
989	0.98
990	0.96
991	0.96
992	0.96
993	0.96
994	0.95

United States, WI	
530	1.00
531	0.98
532	1.00
534	0.98
535	0.97
537	0.96
538	0.94
539	0.94
540	0.95
541	0.96
542	0.96
543	0.96
544	0.93
545	0.95
546	0.94
547	0.96
548	0.95
549	0.93

United States, WV	
247	0.89
248	0.89
249	0.92
250	0.96
251	0.96

United States, WV	
252	0.96
253	0.96
254	0.90
255	0.95
256	0.95
257	0.95
258	0.93
259	0.93
260	0.97
261	0.95
262	0.96
263	0.95
264	0.95
265	0.96
266	0.95
267	0.93
268	0.94

United States, WY	
820	0.78
821	0.75
822	0.75
823	0.74
824	0.74
825	0.75
826	0.81
827	0.73
828	0.77
829	0.76
830	0.76
831	0.76

Crew Detail

Crew: AASBA

Quantity	Description	Hourly Rate
1	Asbestos Workers	37.10
0.25	Asbestos Workers (Foreman)	39.10
Total Hourly Rate for AASBA		$46.88

Crew: AASBB

Quantity	Description	Hourly Rate
0.5	Asbestos Workers (Foreman)	39.10
2	Asbestos Workers	37.10
Total Hourly Rate for AASBB		$93.75

Crew: AASBC

Quantity	Description	Hourly Rate
0.5	Asbestos Workers (Foreman)	39.10
2	Asbestos Workers	37.10
Total Hourly Rate for AASBC		$93.75

Crew: AASBE

Quantity	Description	Hourly Rate
2	HEPA Vacuum 2 Hp Electric	2.07
1	Asbestos Workers (Foreman)	39.10
2	Asbestos Workers	37.10
Total Hourly Rate for AASBE		$117.44

Crew: AASBF

Quantity	Description	Hourly Rate
2	HEPA Vacuum 2 Hp Electric	2.07
1	Asbestos Workers (Foreman)	39.10
3	Asbestos Workers	37.10
Total Hourly Rate for AASBF		$154.54

Crew: AASBG

Quantity	Description	Hourly Rate
1	Asbestos Workers (Foreman)	39.10
2	Asbestos Workers	37.10
1	Flatbed, 8' x 16.0' (Truck Body Only, add truck)	22.03
1	Truck, Highway, 24,500 GVW, 4x2, 2 Axle	21.63
Total Hourly Rate for AASBG		$156.95

Crew: AASBH

Quantity	Description	Hourly Rate
1	Asbestos Workers (Foreman)	39.10
1	Equipment Operator, Crane/Shovel	35.90
1	Hydraulic Crane, 4WD, 40T/84' Boom	76.60
2	Asbestos Workers	37.10
Total Hourly Rate for AASBH		$225.80

Crew: ACAMC

Quantity	Description	Hourly Rate
2	Millwrights	35.50
Total Hourly Rate for ACAMC		$71.00

Crew: ACARA

Quantity	Description	Hourly Rate
1	Carpenters	34.25
0.25	Carpenters (Foreman)	36.25
Total Hourly Rate for ACARA		$43.31

Crew: ACARB

Quantity	Description	Hourly Rate
0.5	Carpenters (Foreman)	36.25
2	Carpenters	34.25
Total Hourly Rate for ACARB		$86.63

Crew: ACARC

Quantity	Description	Hourly Rate
0.25	Carpenters (Foreman)	36.25
1	Carpenters	34.25
Total Hourly Rate for ACARC		$43.31

Crew: ACARD

Quantity	Description	Hourly Rate
0.25	Carpenters (Foreman)	36.25
2	Carpenters	34.25
Total Hourly Rate for ACARD		$77.56

Crew: ACARF

Quantity	Description	Hourly Rate
2	Carpenters	34.25
1	Carpenters (Apprentice)	27.40
1	Carpenters (Foreman)	36.25
Total Hourly Rate for ACARF		$132.15

Crew: ACARI

Quantity	Description	Hourly Rate
0.25	Carpenters (Foreman)	36.25
2	Carpenters	34.25
1	Laborer (Semi-Skilled)	26.70
Total Hourly Rate for ACARI		$104.26

Crew: ACARJ

Quantity	Description	Hourly Rate
1	Carpenters (Foreman)	36.25
1	Laborer (Semi-Skilled)	26.70
2	Carpenters	34.25
Total Hourly Rate for ACARJ		$131.45

Crew: ACARL

Quantity	Description	Hourly Rate
1	Carpenters	34.25
1	Carpenters (Apprentice)	27.40
1	Laborer (Semi-Skilled)	26.70
2	Carpenters (Foreman)	36.25
Total Hourly Rate for ACARL		$160.85

Crew: ACARS

Quantity	Description	Hourly Rate
1	Carpenters (Foreman)	36.25
1	Hydraulic Crane, 4WD, 20T/70' Boom	71.88
1	Carpenters (Apprentice)	27.40
1	Equipment Operator, Medium	34.65
2	Carpenters	34.25
Total Hourly Rate for ACARS		$238.68

Crew: ACART

Quantity	Description	Hourly Rate
2	Carpenters	34.25
1	Equipment Operator, Crane/Shovel	35.90
1	Carpenters (Apprentice)	27.40
1	Carpenters (Foreman)	36.25
1	Hydraulic Crane, Truck Mounted, 60T /115' Boom	111.93
1	Equipment Operator, Oilers	30.10
Total Hourly Rate for ACART		$310.07

Crew: ACMAA

Quantity	Description	Hourly Rate
1	Cement Finishers	32.85
0.25	Cement Finishers (Foreman)	34.85
Total Hourly Rate for ACMAA		$41.56

Crew: ACMAC

Quantity	Description	Hourly Rate
1	Cement Finishers	32.85
0.1	Cement Finishers	32.85
Total Hourly Rate for ACMAC		$36.14

Crew: ACMAE

Quantity	Description	Hourly Rate
1	Laborer (Semi-Skilled)	26.70
1	Cement Finishers	32.85
1	Carpenters	34.25
Total Hourly Rate for ACMAE		$93.80

Crew: ACMAF

Quantity	Description	Hourly Rate
2	Cement Finishers	32.85
1	Equipment Operator, Light	33.05
0.25	Cement Finishers (Foreman)	34.85
1	Laborer (Semi-Skilled)	26.70
1	Air Hose, 1.00", 100'	0.91
1	Air Compressor, 375 CFM, 100 PSI	18.35
Total Hourly Rate for ACMAF		$153.43

Crew: AGLAB

Quantity	Description	Hourly Rate
2	Glaziers	33.00
0.5	Glaziers (Foreman)	35.00
Total Hourly Rate for AGLAB		$83.50

Crew: ALABE

Quantity	Description	Hourly Rate
1	Cement Finishers	32.85
1	Generator, 5.5 KW, 120/240V, Portable	4.03
2	Concrete Vibrator, 2.50D, High Frequency	1.17
1	Laborer (Semi-Skilled - Foreman)	28.70
4	Laborer (Semi-Skilled)	26.70
Total Hourly Rate for ALABE		$174.71

Crew Detail (Cont.)

Crew: ALABG

Quantity	Description	Hourly Rate
1	Laborer (Semi-Skilled - Foreman)	28.70
1	Equipment Operator, Crane/Shovel	35.90
1	Generator, 5.5 KW, 120/240V, Portable	4.03
5	Laborer (Semi-Skilled)	26.70
1	Equipment Operator, Oilers	30.10
1	Hydraulic Crane, Truck Mounted, 60T /115' Boom	111.93
2	Concrete Vibrator, 2.50D, High Frequency	1.17
1	Cement Finishers	32.85
Total Hourly Rate for ALABG		$379.34

Crew: ALABI

Quantity	Description	Hourly Rate
1	Concrete Pump, 117CY/HR, 75' Boom, with Truck	88.03
1	Equipment Operator, Medium	34.65
2	Concrete Vibrator, 2.50D, High Frequency	1.17
1	Laborer (Semi-Skilled - Foreman)	28.70
1	Cement Finishers	32.85
1	Generator, 5.5 KW, 120/240V, Portable	4.03
5	Laborer (Semi-Skilled)	26.70
Total Hourly Rate for ALABI		$324.09

Crew: ALABM

Quantity	Description	Hourly Rate
1	Carpenters (Foreman)	36.25
0.5	Rodmen, (Reinforcing)	37.95
3	Laborer (Semi-Skilled)	26.70
0.5	Cement Finishers	32.85
1	Carpenters	34.25
Total Hourly Rate for ALABM		$186.00

Crew: ALABN

Quantity	Description	Hourly Rate
0.3	Loader, Front End, Wheeled, 2.5 CY, Articulated, 936E	35.08
0.9	Equipment Operator, Medium	34.65
1.8	Carpenters	34.25
0.1	Hydraulic Excavator, Crawler, 1.00 CY Bucket	69.73
1	Truck, Highway, 46,000 GVW, 6x4, 3 Axle	42.47
2.7	Laborer (Semi-Skilled)	26.70
0.9	Hydraulic Crane, 4WD, 20T/70' Boom	71.88
0.3	Cement Finishers	32.85
1.4	Equipment Operator, Medium	34.65
1	Carpenters (Foreman)	36.25
0.4	Structural Steel Workers	38.15
1	Rear Dump Body, 8 CY, 30,000 GVW	3.54
Total Hourly Rate for ALABN		$403.00

Crew: AMABD

Quantity	Description	Hourly Rate
0.25	Bricklayers (Foreman)	37.25
1	Bricklayers (Semi-Skilled)	26.90
1	Bricklayers	35.25
Total Hourly Rate for AMABD		$71.46

Crew: AMABH

Quantity	Description	Hourly Rate
3	Bricklayers (Semi-Skilled)	26.90
2	Bricklayers	35.25
1	Bricklayers (Foreman)	37.25
0.25	Carpenters	34.25
Total Hourly Rate for AMABH		$197.01

Crew: APLHB

Quantity	Description	Hourly Rate
0.5	Plasters, (Semi-Skilled - Foreman)	27.60
2	Plasters, (Semi-Skilled)	27.10
Total Hourly Rate for APLHB		$68.00

Crew: APTRA

Quantity	Description	Hourly Rate
0.25	Painters, Ordinary (Foreman)	32.60
1	Painters, Ordinary	30.60
Total Hourly Rate for APTRA		$38.75

Crew: APTRB

Quantity	Description	Hourly Rate
0.5	Painters, Ordinary (Foreman)	32.60
2	Painters, Ordinary	30.60
Total Hourly Rate for APTRB		$77.50

Crew: ARFCB

Quantity	Description	Hourly Rate
2	Roofers, Composition	29.45
0.5	Roofers, Composition (Foreman)	31.45
Total Hourly Rate for ARFCB		$74.63

Crew: ARFCD

Quantity	Description	Hourly Rate
0.25	Roofers, Composition (Foreman)	31.45
1	Roofers, Composition	29.45
Total Hourly Rate for ARFCD		$37.31

Crew: CDRGA

Quantity	Description	Hourly Rate
1	170 HP 60 CY/HR (1200 GPM) Dredging Unit	87.77
1	Equipment Operator, Crane/Shovel	35.90
1	Outside Laborer (Semi-Skilled)	26.09
1	Integral Floating Discharge 8" Pipe	6.45
Total Hourly Rate for CDRGA		$156.22

Crew: CDRGB

Quantity	Description	Hourly Rate
1	Outside Laborer (Semi-Skilled)	26.09
1	Integral Floating Discharge 8" Pipe	6.45
1	In-Line Booster Pump	19.80
1	Equipment Operator, Crane/Shovel	35.90
1	1000' On-Shore Discharge Pipe	22.25
1	170 HP 60 CY/HR (1200 GPM) Dredging Unit	87.77
Total Hourly Rate for CDRGB		$198.27

Crew Detail (Cont.)

Crew: CLABB

Quantity	Description	Hourly Rate
1	Air Compressor, 250 CFM, 100 PSI	14.05
3	Laborer (Semi-Skilled)	26.70
1	Air Hose, 1.00", 100'	0.91
1	Laborer (Semi-Skilled - Foreman)	28.70
1	Carpenters	34.25
Total Hourly Rate for CLABB		$158.01

Crew: CLABE

Quantity	Description	Hourly Rate
1	Asphalt Kettle/Pavement,110 gallon,Trailer	5.12
1	Concrete Saw, 6.63" Depth, Walk-Behind	6.23
1	Laborer (Semi-Skilled - Foreman)	28.70
1	Air Compressor, 375 CFM, 100 PSI	18.35
2	Laborer (Semi-Skilled)	26.70
1	Equipment Operator, Light	33.05
1	Air Hose, 1.00", 100'	0.91
Total Hourly Rate for CLABE		$145.76

Crew: CLACA

Quantity	Description	Hourly Rate
4	Laborer (Semi-Skilled)	26.70
1	Laborer (Semi-Skilled - Foreman)	28.70
2	Compactor, Rammer, 13"X11" Shoe	4.35
Total Hourly Rate for CLACA		$144.20

Crew: CLACC

Quantity	Description	Hourly Rate
1	Compactor, Rammer, 13"X11" Shoe	4.35
1	Laborer (Semi-Skilled - Foreman)	28.70
2	Laborer (Semi-Skilled)	26.70
Total Hourly Rate for CLACC		$86.45

Crew: CLACD

Quantity	Description	Hourly Rate
1	Roller, Walk Behind, Vibratory Double, 0.8T, 29.5"W	73.95
0.5	Laborer (Semi-Skilled)	26.70
1	Equipment Operator, Light	33.05
Total Hourly Rate for CLACD		$120.35

Crew: CLADA

Quantity	Description	Hourly Rate
0.25	Laborer (Semi-Skilled - Foreman)	28.70
1	Paving Breaker, 86 LB (Add Compressor)	1.23
0.5	Air Compressor, 375 CFM, 100 PSI	18.35
1	Laborer (Semi-Skilled)	26.70
0.5	Equipment Operator, Light	33.05
1	Air Hose, 1.00", 100'	0.91
Total Hourly Rate for CLADA		$61.71

Crew: CLADB

Quantity	Description	Hourly Rate
2	Paving Breaker, 86 LB (Add Compressor)	1.23
2	Air Hose, 1.00", 100'	0.91
1	Laborer (Semi-Skilled - Foreman)	28.70
4	Laborer (Semi-Skilled)	26.70
1	Air Compressor, 375 CFM, 100 PSI	18.35
	Total Hourly Rate for CLADB	$158.13

Crew: CLADD

Quantity	Description	Hourly Rate
1	Loader, Front End, Crawler, 3.75 CY, 973	136.75
1	Air Compressor, 375 CFM, 100 PSI	18.35
2	Paving Breaker, 86 LB (Add Compressor)	1.23
4	Laborer (Semi-Skilled)	26.70
1	Laborer (Semi-Skilled - Foreman)	28.70
2	Air Hose, 1.00", 100'	0.91
1	Equipment Operator, Medium	34.65
	Total Hourly Rate for CLADD	$329.52

Crew: CLADF

Quantity	Description	Hourly Rate
0.25	Powdermen (Foreman)	40.65
1	Flatbed, 8'x 16.0' (Add Truck)	22.03
0.25	Truck, Highway, 4x2, 3/4T, 8600 GVW	9.75
2	Laborer (Semi-Skilled)	26.70
1	Truck, Highway, 24,500 GVW, 4x2, 2 Axle	21.63
2	Powdermen	38.13
	Total Hourly Rate for CLADF	$185.92

Crew: CLADH

Quantity	Description	Hourly Rate
0.25	Truck, Highway, 4x2, 3/4T, 8600 GVW	9.75
0.5	Laborer (Semi-Skilled)	26.70
1.5	Equipment Operator, Medium	34.65
0.25	Powdermen	38.13
2	Air Hose, 3.00", 100'	4.42
1	Drill, Air Track, 2.5-4" Diameter, 12' FD	83.05
1	Air Compressor, 750 CFM, 100 PSI	39.68
	Total Hourly Rate for CLADH	$208.87

Crew: CLADI

Quantity	Description	Hourly Rate
2	Laborer (Semi-Skilled)	26.70
1	Flatbed, 8'x 16.0' (Add Truck)	22.03
1	Drill, Core, Skid Mounted, NX 400' Depth	75.58
1	Laborer (Semi-Skilled - Foreman)	28.70
1	Truck, Highway, 4x2, 3/4T, 8600 GVW	9.75
1	Truck, Highway, 24,500 GVW, 4x2, 2 Axle	21.63
	Total Hourly Rate for CLADI	$211.08

Crew: CLADM

Quantity	Description	Hourly Rate
1	Drill, R-Blast, 6.75"-9.88",Truck	208.25
4	Laborer (Semi-Skilled)	26.70
1	Laborer (Semi-Skilled - Foreman)	28.70
	Total Hourly Rate for CLADM	$343.75

Crew: CLADQ

Quantity	Description	Hourly Rate
0.5	Truck, Highway, 4x4, 3/4T, 8800 GVW	10.65
1	Laborer (Semi-Skilled)	26.70
2	Drill, Air Track, 4.0" Hole, 12' Feed (Add Compressor)	102.83
1	Equipment Operator, Light	33.05
0.5	Laborer (Semi-Skilled - Foreman)	28.70
1	Air Compressor, 750 CFM, 100 PSI	39.68
Total Hourly Rate for CLADQ		$324.75

Crew: COBBC

Quantity	Description	Hourly Rate
1	Bobcat With Backhoe	24.73
1	Equipment Operator, Light	33.05
Total Hourly Rate for COBBC		$57.77

Crew: CODBB

Quantity	Description	Hourly Rate
0.25	Laborer (Semi-Skilled)	26.70
1	Equipment Operator, Medium	34.65
1	Trencher, Chain, 60" Deep, 6-18" W	6.00
Total Hourly Rate for CODBB		$47.32

Crew: CODBC

Quantity	Description	Hourly Rate
1	Plumbers	40.85
3	Laborer (Semi-Skilled)	26.70
1	Trencher, Chain, 6-18" Wide x 60" Deep	6.00
Total Hourly Rate for CODBC		$126.95

Crew: CODBG

Quantity	Description	Hourly Rate
1	Trencher, Wheel, 90" Deep, 48"W, 22" R-Bucket	182.50
0.17	Laborer (Semi-Skilled)	26.70
1	Equipment Operator, Medium	34.65
Total Hourly Rate for CODBG		$221.69

Crew: CODE1

Quantity	Description	Hourly Rate
1	Hydraulic Excavator, Crawler, 2.00 CY Bucket	113.73
1	Equipment Operator, Crane/Shovel	35.90
Total Hourly Rate for CODE1		$149.63

Crew: CODEA

Quantity	Description	Hourly Rate
1	Equipment Operator, Crane/Shovel	35.90
1	Laborer (Semi-Skilled)	26.70
1	Hydraulic Excavator, Crawler, 1.00 CY Bucket, Long Carriage	69.73
Total Hourly Rate for CODEA		$132.32

Crew: CODEE

Quantity	Description	Hourly Rate
1	Hydraulic Excavator, Crawler, 3.125 CY Bucket	192.88
1	Equipment Operator, Crane/Shovel	35.90
Total Hourly Rate for CODEE		$228.78

Crew Detail (Cont.)

Crew: CODEF

Quantity	Description	Hourly Rate
1	Loader, Backhoe, Wheeled, 1.38 CY Front End Bucket, 30" Dip	32.05
1	Hydraulic RA-PAK Tamper, 21"W, 30"L, 27"H	2.49
2	Equipment Operator, Medium	34.65
2	Laborer (Semi-Skilled)	26.70
0.25	Laborer (Semi-Skilled - Foreman)	28.70
1	Loader, Backhoe, Wheeled, 1.38 CY Front End Bucket, 30" Dip	32.05
Total Hourly Rate for CODEF		$196.47

Crew: CODEG

Quantity	Description	Hourly Rate
0.5	Laborer (Semi-Skilled)	26.70
1	Loader, Backhoe, Wheeled, 1.38 CY Front End Bucket, 30" Dip	32.05
1	Equipment Operator, Medium	34.65
Total Hourly Rate for CODEG		$80.05

Crew: CODEJ

Quantity	Description	Hourly Rate
2	Laborer (Semi-Skilled)	26.70
0.25	Laborer (Semi-Skilled - Foreman)	28.70
1	Equipment Operator, Medium	34.65
1	Loader, Backhoe, Wheeled, 1.38 CY Front End Bucket, 30" Dip	32.05
Total Hourly Rate for CODEJ		$127.27

Crew: CODEK

Quantity	Description	Hourly Rate
1	Equipment Operator, Medium	34.65
4	Laborer (Semi-Skilled)	26.70
1	Laborer (Semi-Skilled - Foreman)	28.70
1	Loader, Backhoe, Wheeled, 1.38 CY Front End Bucket, 30" Dip	32.05
Total Hourly Rate for CODEK		$202.20

Crew: CODEL

Quantity	Description	Hourly Rate
1	Hydraulic Excavator, Crawler, 1.00 CY Bucket, Long Carriage	69.73
1	Equipment Operator, Crane/Shovel	35.90
Total Hourly Rate for CODEL		$105.63

Crew: CODEO

Quantity	Description	Hourly Rate
1	Laborer (Semi-Skilled)	26.70
1	Hydraulic Excavator, Crawler, 1.50 CY Bucket	90.05
1	Equipment Operator, Crane/Shovel	35.90
Total Hourly Rate for CODEO		$152.65

Crew: CODES

Quantity	Description	Hourly Rate
1	Equipment Operator, Crane/Shovel	35.90
1	Hydraulic Excavator, Crawler, 1.50 CY Bucket	90.05
1	Laborer (Semi-Skilled)	26.70
Total Hourly Rate for CODES		$152.65

Crew: CODET

Quantity	Description	Hourly Rate
1	Laborer (Semi-Skilled)	26.70
1	Hydraulic Excavator, Crawler, 2.00 CY Bucket	113.73
1	Equipment Operator, Crane/Shovel	35.90
Total Hourly Rate for CODET		**$176.32**

Crew: CODEU

Quantity	Description	Hourly Rate
1	Hydraulic Excavator, Crawler, 3.125 CY Bucket	192.88
1	Equipment Operator, Crane/Shovel	35.90
1	Equipment Operator, Oilers	30.10
1	Laborer (Semi-Skilled)	26.70
Total Hourly Rate for CODEU		**$285.58**

Crew: CODEV

Quantity	Description	Hourly Rate
1	Equipment Operator, Crane/Shovel	35.90
1	Equipment Operator, Oilers	30.10
1	Laborer (Semi-Skilled)	26.70
1	Hydraulic Excavator, Crawler, 4.50 CY Bucket	408.61
Total Hourly Rate for CODEV		**$501.31**

Crew: CODEX

Quantity	Description	Hourly Rate
1	Laborer (Semi-Skilled - Foreman)	28.70
3	Laborer (Semi-Skilled)	26.70
1	Hydraulic Excavator, Crawler, 1.50 CY Bucket	90.05
1	Equipment Operator, Crane/Shovel	35.90
Total Hourly Rate for CODEX		**$234.75**

Crew: CODEY

Quantity	Description	Hourly Rate
1	Hydraulic Excavator, Crawler, 1.25 CY Bucket	73.77
1	Equipment Operator, Crane/Shovel	35.90
Total Hourly Rate for CODEY		**$109.67**

Crew: CODEZ

Quantity	Description	Hourly Rate
1	Equipment Operator, Medium	34.65
1	Hydraulic Excavator, Crawler, 1.50 CY Bucket	90.05
1	Laborer (Semi-Skilled - Foreman)	28.70
2	Laborer (Semi-Skilled)	26.70
Total Hourly Rate for CODEZ		**$206.80**

Crew: CODFB

Quantity	Description	Hourly Rate
1	Loader, Front End, Wheeled, 5.25 CY, Articulated, 980C	87.55
0.5	Laborer (Semi-Skilled)	26.70
1	Equipment Operator, Crane/Shovel	35.90
Total Hourly Rate for CODFB		**$136.80**

Crew: CODGD

Quantity	Description	Hourly Rate
1	Laborer (Semi-Skilled)	26.70
1	Hydraulic Excavator, Truck Mounted, 1.40 CY Bucket, 4x4	121.38
1	Equipment Operator, Crane/Shovel	35.90
Total Hourly Rate for CODGD		$183.97

Crew: CODKG

Quantity	Description	Hourly Rate
1	Equipment Operator, Light	33.05
1	Loader, Front End, Crawler, 2.0 CY, 953	68.98
Total Hourly Rate for CODKG		$102.03

Crew: CODL1

Quantity	Description	Hourly Rate
1	Equipment Operator, Medium	34.65
1	Loader, Front End, Wheeled, 4.0 CY, Articulated, 966E	55.70
Total Hourly Rate for CODL1		$90.35

Crew: CODL2

Quantity	Description	Hourly Rate
4	Laborer (Semi-Skilled)	26.70
0.25	Laborer (Semi-Skilled - Foreman)	28.70
1	Loader, Front End, Wheeled, 3.0 CY, Articulated, 950E	38.13
1	Equipment Operator, Light	33.05
Total Hourly Rate for CODL2		$185.15

Crew: CODL3

Quantity	Description	Hourly Rate
1	Equipment Operator, Light	33.05
1	Loader, Front End, Crawler, 1.3 CY, 935C	39.08
Total Hourly Rate for CODL3		$72.13

Crew: CODL4

Quantity	Description	Hourly Rate
1	Equipment Operator, Light	33.05
1	Loader, Front End, Wheeled, 2.5 CY, Articulated, 936E	35.08
Total Hourly Rate for CODL4		$68.13

Crew: CODLA

Quantity	Description	Hourly Rate
1	Equipment Operator, Medium	34.65
1	Loader, Front End, Wheeled, 2.5 CY, Articulated, 936E	35.08
0.5	Laborer (Semi-Skilled)	26.70
Total Hourly Rate for CODLA		$83.08

Crew: CODLB

Quantity	Description	Hourly Rate
1	Loader, Front End, Crawler, 1.5 CY, 943	46.10
1	Equipment Operator, Medium	34.65
0.5	Laborer (Semi-Skilled)	26.70
Total Hourly Rate for CODLB		$94.10

Crew: CODLC

Quantity	Description	Hourly Rate
1	Equipment Operator, Medium	34.65
1	Loader, Front End, Wheeled, 5.25 CY, Articulated, 980C	87.55
0.5	Laborer (Semi-Skilled)	26.70
	Total Hourly Rate for CODLC	$135.55

Crew: CODLD

Quantity	Description	Hourly Rate
0.5	Laborer (Semi-Skilled)	26.70
1	Equipment Operator, Medium	34.65
1	Loader, Front End, Crawler, 3.75 CY, 973	136.75
	Total Hourly Rate for CODLD	$184.75

Crew: CODLE

Quantity	Description	Hourly Rate
1	Loader, Front End, Crawler, 1.0 CY, 931C	35.71
1	Equipment Operator, Light	33.05
	Total Hourly Rate for CODLE	$68.76

Crew: CODLF

Quantity	Description	Hourly Rate
1	Loader, Front End, Crawler, 3.75 CY, 973	136.75
1	Laborer (Semi-Skilled)	26.70
1	Equipment Operator, Medium	34.65
	Total Hourly Rate for CODLF	$198.10

Crew: CODLG

Quantity	Description	Hourly Rate
1	Loader, Front End, Crawler, 2.0 CY, 953	68.98
1	Equipment Operator, Light	33.05
	Total Hourly Rate for CODLG	$102.03

Crew: CODLH

Quantity	Description	Hourly Rate
1	Loader, Front End, Crawler, 2.5 CY, 963	98.50
1	Equipment Operator, Light	33.05
	Total Hourly Rate for CODLH	$131.55

Crew: CODLI

Quantity	Description	Hourly Rate
1	Equipment Operator, Medium	34.65
0.5	Laborer (Semi-Skilled)	26.70
1	Loader, Front End, Wheeled, 4.0 CY, Articulated, 966E	55.70
	Total Hourly Rate for CODLI	$103.70

Crew: CODLJ

Quantity	Description	Hourly Rate
1	Loader, Front End, Wheeled, 3.0 CY, Articulated, 950E	38.13
1	Equipment Operator, Light	33.05
	Total Hourly Rate for CODLJ	$71.17

Crew: CODLK

Quantity	Description	Hourly Rate
1	Loader, Front End, Wheeled, 5.25 CY, Articulated, 980C	87.55
1	Equipment Operator, Medium	34.65
	Total Hourly Rate for CODLK	$122.20

Crew: CODLP

Quantity	Description	Hourly Rate
1	Dozer, Crawler, D-6D, Powershift, (Add Blade)	71.50
2	Equipment Operator, Medium	34.65
0.5	Equipment Operator, Medium	34.65
3	Laborer (Semi-Skilled)	26.70
1	Blade, Angle, Hydraulic, for D6	5.38
1	Wheeled Loader with Backhoe,1.00 CY Bucket	39.08
Total Hourly Rate for CODLP		$282.68

Crew: CODLQ

Quantity	Description	Hourly Rate
1	Loader, Front End, Wheeled, 1.25 CY Articulated	24.23
1	Equipment Operator, Light	33.05
Total Hourly Rate for CODLQ		$57.27

Crew: CODLR

Quantity	Description	Hourly Rate
1	Equipment Operator, Light	33.05
1	Loader, Front End, Wheeled, 2.0 CY, Articulated, 926E	35.08
Total Hourly Rate for CODLR		$68.13

Crew: CODLT

Quantity	Description	Hourly Rate
3	Equipment Operator, Light	33.05
3	Loader, Front End, Wheeled, 3.0 CY, Articulated, 950E	38.13
6	Laborer (Semi-Skilled)	26.70
0.5	Truck, Highway, 4x4, 3/4T, 8800 GVW	10.65
Total Hourly Rate for CODLT		$379.05

Crew: CODLU

Quantity	Description	Hourly Rate
1	Loader, Front End, Wheeled, 13.5 CY, Articulated, 992C	228.75
1	Equipment Operator, Medium	34.65
Total Hourly Rate for CODLU		$263.40

Crew: CODLV

Quantity	Description	Hourly Rate
1	Equipment Operator, Medium	34.65
1	Loader, Front End, Crawler, 1.5 CY, 943	46.10
Total Hourly Rate for CODLV		$80.75

Crew: CODLW

Quantity	Description	Hourly Rate
1	Loader, Front End, Crawler, 3.75 CY, 973	136.75
1	Equipment Operator, Medium	34.65
Total Hourly Rate for CODLW		$171.40

Crew: CODLX

Quantity	Description	Hourly Rate
1	Equipment Operator, Light	33.05
1	Loader, Front End, Wheeled, 1.5 CY, Articulated, 916E	30.13
Total Hourly Rate for CODLX		$63.17

Crew: CODLY

Quantity	Description	Hourly Rate
1	Equipment Operator, Light	33.05
1	Loader, Front End, Wheeled, 4.0 CY, Articulated, 966E	55.70
Total Hourly Rate for CODLY		$88.75

Crew: CODLZ

Quantity	Description	Hourly Rate
1	Equipment Operator, Medium	34.65
1	Loader, Front End, Wheeled, 7.0 CY, Articulated, 988B	152.88
Total Hourly Rate for CODLZ		$187.53

Crew: CODMA

Quantity	Description	Hourly Rate
0.5	Equipment Operator, Medium	34.65
0.5	Winch, Dozer	6.02
4	Scraper, 22 CY, 25T, Powershift	206.00
0.5	Dozer, Crawler, D-8L, (Add Blade)	149.38
0.5	Laborer (Semi-Skilled)	26.70
0.5	Blade, Universal, Hydraulic, for D8	9.78
0.5	Truck, Highway, 4x4, 3/4T, 8800 GVW	10.65
4	Equipment Operator, Crane/Shovel	35.90
Total Hourly Rate for CODMA		$1,086.19

Crew: CODSC

Quantity	Description	Hourly Rate
0.5	Dozer, Crawler, D-8L, (Add Blade)	149.38
0.5	Truck, Highway, 4x4, 3/4T, 8800 GVW	10.65
0.5	Equipment Operator, Medium	34.65
4	Scraper, 11 CY, 13T, Powershift	102.43
4	Equipment Operator, Crane/Shovel	35.90
0.5	Blade, Universal, Hydraulic, for D8	9.78
0.5	Winch, (without Cable), for D6	6.02
0.5	Laborer (Semi-Skilled)	26.70
Total Hourly Rate for CODSC		$671.89

Crew: CODSD

Quantity	Description	Hourly Rate
0.5	Laborer (Semi-Skilled)	26.70
4	Scraper, Self Propelled, 14-20 CY, 24T, Powershift	194.00
0.5	Equipment Operator, Medium	34.65
4	Equipment Operator, Crane/Shovel	35.90
0.5	Dozer, Crawler, D-8L, Powershift, (Add Blade)	149.38
0.5	Winch, (without Cable), for D6	6.02
0.5	Truck, Highway, 4x4, 3/4T, 8800 GVW	10.65
0.5	Blade, Universal, Hydraulic, for D8	9.78
Total Hourly Rate for CODSD		$1,038.19

Crew: CODSI

Quantity	Description	Hourly Rate
1	Equipment Operator, Crane/Shovel	35.90
1	Scraper, 22 CY, 25T, Powershift	206.00
Total Hourly Rate for CODSI		$241.90

Crew: CODSJ

Quantity	Description	Hourly Rate
0.5	Blade, Universal, Hydraulic, for D9	12.22
0.5	Equipment Operator, Medium	34.65
0.5	Dozer, Crawler, D-9N, Powershift, (Add Blade)	190.50
1	Scraper, Self Propelled, 21-31 CY, 37.5T, Powershift	289.63
1	Equipment Operator, Crane/Shovel	35.90
Total Hourly Rate for CODSJ		$444.21

Crew: CODSM

Quantity	Description	Hourly Rate
0.25	Equipment Operator, Medium	34.65
1	Scraper, Tandem, 14-20 CY, 24T, Powershift	135.31
0.25	Blade, Universal, Hydraulic, for D8	9.78
0.25	Dozer, Crawler, D-8L, Powershift, (Add Blade)	149.38
1	Equipment Operator, Crane/Shovel	35.90
Total Hourly Rate for CODSM		$219.66

Crew: CODSN

Quantity	Description	Hourly Rate
0.25	Blade, Universal, Hydraulic, for D9	12.22
1	Scraper, Tandem, 37.5 T, 21-31CY, Powershift	159.99
0.25	Dozer, Crawler, D-9N, Powershift, (Add Blade)	190.50
0.25	Equipment Operator, Medium	34.65
1	Equipment Operator, Crane/Shovel	35.90
Total Hourly Rate for CODSN		$255.24

Crew: CODSO

Quantity	Description	Hourly Rate
0.5	Equipment Operator, Medium	34.65
0.5	Blade, Universal, Hydraulic, for D9	12.22
0.5	Dozer, Crawler, D-9N, Powershift, (Add Blade)	190.50
1	Equipment Operator, Crane/Shovel	35.90
1	Scraper, Tandem, 52T, 32-44CY, Powershift	216.28
Total Hourly Rate for CODSO		$370.86

Crew: CODSP

Quantity	Description	Hourly Rate
1	Equipment Operator, Crane/Shovel	35.90
0.5	Laborer (Semi-Skilled)	26.70
1	Scraper, 11 CY, 13T, Powershift	102.43
0.5	Truck, Highway, 4x4, 3/4T, 8800 GVW	10.65
Total Hourly Rate for CODSP		$157.00

Crew Detail (Cont.)

Crew: CODSQ

Quantity	Description	Hourly Rate
1	Laborer (Semi-Skilled)	26.70
0.25	Dozer, Crawler, D-6H, Powershift, (Add Blade)	71.50
0.25	Loader, Front End, Crawler, 3.75 CY, 973	136.75
11	Truck, Off-Highway, 6x6, Rear Dump, 19T, 12-16CY	90.12
1.5	Equipment Operator, Medium	34.65
11	Truck Drivers, Heavy	27.55
0.25	Blade, Angle, Hydraulic, for D6	5.38
1	Ripper, Multishank Beam, Hydraulic, D-9	13.81
1	Dozer, Crawler, D-9N, Powershift, (Add Blade)	190.50
1	Blade, Universal, Hydraulic, for D9	12.22
3	Ripper, Shanks Each, D-9	0.86
1	Loader, Front End, Crawler, 2.5 CY, 963	98.50

Total Hourly Rate for CODSQ $1,744.05

Crew: CODSR

Quantity	Description	Hourly Rate
1	Dozer, Crawler, D-9N, Powershift, (Add Blade)	190.50
2	Equipment Operator, Medium	34.65
1	Laborer (Semi-Skilled)	26.70
1	Ripper, Shanks Each, D-9	0.86
1	Blade, Universal, Hydraulic, for D9	12.22
6	Truck, Off-Highway, 6x6, Rear Dump, 23T,13-17 CY	98.13
1	Ripper, Multishank Beam, Hydraulic, D-9	13.81
1	Loader, Front End, Crawler, 2.5 CY, 963	98.50
6	Truck Drivers, Heavy	27.55

Total Hourly Rate for CODSR $1,165.94

Crew: CODSS

Quantity	Description	Hourly Rate
0.5	Blade, Universal, Hydraulic, for D9	12.22
1	Scraper, Self Propelled, 32-44 CY, 52T, Powershift	319.99
0.5	Equipment Operator, Medium	34.65
0.5	Dozer, Crawler, D-9N, Powershift, (Add Blade)	190.50
1	Equipment Operator, Crane/Shovel	35.90

Total Hourly Rate for CODSS $474.57

Crew: CODST

Quantity	Description	Hourly Rate
1	Equipment Operator, Crane/Shovel	35.90
1	Scraper, Self Propelled, 14-20 CY, 24T, Powershift	194.00
0.25	Blade, Universal, Hydraulic, for D8	9.78
0.25	Dozer, Crawler, D-8L, Powershift, (Add Blade)	149.38
0.25	Equipment Operator, Medium	34.65
Total Hourly Rate for CODST		$278.35

Crew: CODSU

Quantity	Description	Hourly Rate
1	Equipment Operator, Crane/Shovel	35.90
1	Scraper, 11 CY, 13T, Powershift	102.43
Total Hourly Rate for CODSU		$138.32

Crew: CODSV

Quantity	Description	Hourly Rate
0.5	Blade, Universal, Hydraulic, for D9	12.22
0.25	Equipment Operator, Medium	34.65
0.5	Dozer, Crawler, D-9N, Powershift, (Add Blade)	190.50
1	Scraper, Tandem, 37.5 CY, Push-Pull	159.99
1	Equipment Operator, Crane/Shovel	35.90
Total Hourly Rate for CODSV		$305.92

Crew: CODSW

Quantity	Description	Hourly Rate
1	Equipment Operator, Crane/Shovel	35.90
1	Scraper, 11 CY, 13T, Powershift	102.43
Total Hourly Rate for CODSW		$138.32

Crew: CODT1

Quantity	Description	Hourly Rate
1	Equipment Operator, Medium	34.65
1	Dozer, Crawler, D-3B PS, (Add Blade) Powershif, (Add Blade)	39.35
1	Blade, Power Angle Tilt, for D3	2.47
0.5	Laborer (Semi-Skilled)	26.70
1	Outside Laborer (Semi-Skilled)	26.09
Total Hourly Rate for CODT1		$115.92

Crew: CODT2

Quantity	Description	Hourly Rate
1	Outside Laborer (Semi-Skilled)	26.09
1	Blade, Angle, Hydraulic, for D4	3.34
1	Dozer, Crawler, D-4H, Powershift, (Add Blade)	39.35
0.25	Equipment Operator, Medium	34.65
1	Equipment Operator, Medium	34.65
Total Hourly Rate for CODT2		$112.10

Crew Detail (Cont.)

Crew: CODT3

Quantity	Description	Hourly Rate
0.25	Equipment Operator, Medium	34.65
1	Outside Laborer (Semi-Skilled)	26.09
1	Equipment Operator, Medium	34.65
1	Blade, Angle, Hydraulic, for D6	5.38
1	Dozer, Crawler, D-6H, Powershift, (Add Blade)	71.50
Total Hourly Rate for CODT3		$146.29

Crew: CODT4

Quantity	Description	Hourly Rate
1	Outside Laborer (Semi-Skilled)	26.09
1	Equipment Operator, Medium	34.65
0.25	Equipment Operator, Medium	34.65
1	Blade, Universal, Hydraulic, for D7	7.39
1	Dozer, Crawler, D-7H, (Add Blade)	114.95
Total Hourly Rate for CODT4		$191.74

Crew: CODT5

Quantity	Description	Hourly Rate
1	Blade, Universal, Hydraulic, for D9	12.22
0.25	Equipment Operator, Crane/Shovel	35.90
1	Dozer, Crawler, D-9N, Powershift, (Add Blade)	190.50
1	Outside Laborer (Semi-Skilled)	26.09
1	Equipment Operator, Crane/Shovel	35.90
Total Hourly Rate for CODT5		$273.69

Crew: CODTB

Quantity	Description	Hourly Rate
1	Blade, Angle, Hydraulic, for D4	3.34
1	Equipment Operator, Light	33.05
1	Dozer, Crawler, D-4H, Powershift, (Add Blade)	39.35
Total Hourly Rate for CODTB		$75.74

Crew: CODTC

Quantity	Description	Hourly Rate
1	Dozer, Crawler, D-4H, Powershift, (Add Blade)	39.35
0.25	Equipment Operator, Medium (Foreman)	36.65
1	Equipment Operator, Medium	34.65
1	Blade, Angle, Hydraulic, for D4	3.34
Total Hourly Rate for CODTC		$86.51

Crew: CODTD

Quantity	Description	Hourly Rate
1	Dozer, Crawler, D-5H, Powershift, (Add Blade)	56.73
1	Equipment Operator, Light	33.05
1	Blade, Angle, Hydraulic, for D5	3.71
Total Hourly Rate for CODTD		$93.49

Crew: CODTE

Quantity	Description	Hourly Rate
1	Dozer, Crawler, D-6H, Powershift, (Add Blade)	71.50
1	Blade, Angle, Hydraulic, for D6	5.38
1	Equipment Operator, Medium	34.65
0.25	Equipment Operator, Medium (Foreman)	36.65
	Total Hourly Rate for CODTE	$120.70

Crew: CODTF

Quantity	Description	Hourly Rate
1	Equipment Operator, Crane/Shovel	35.90
	Total Hourly Rate for CODTF	$35.90

Crew: CODTG

Quantity	Description	Hourly Rate
0.25	Equipment Operator, Medium (Foreman)	36.65
1	Equipment Operator, Medium	34.65
1	Blade, Universal, Hydraulic, for D7	7.39
1	Dozer, Crawler, D-7H, (Add Blade)	114.95
	Total Hourly Rate for CODTG	$166.15

Crew: CODTJ

Quantity	Description	Hourly Rate
1	Blade, Universal, Hydraulic, for D7	7.39
1	Dozer, Crawler, D-7G, Powershift, (Add Blade)	114.95
1	Equipment Operator, Medium	34.65
	Total Hourly Rate for CODTJ	$156.99

Crew: CODTK

Quantity	Description	Hourly Rate
1	Dozer, Crawler, D-8L, (Add Blade)	149.38
1	Equipment Operator, Crane/Shovel	35.90
1	Blade, Universal, Hydraulic, for D8	9.78
1	Roller, Vibratory, Towed, Sheepsfoot, 25.5 T, 72" Wide	76.28
0.25	Equipment Operator, Crane/Shovel (Foreman)	37.90
	Total Hourly Rate for CODTK	$280.81

Crew: CODTL

Quantity	Description	Hourly Rate
1	Dozer, Crawler, D-8L, (Add Blade)	149.38
1	Equipment Operator, Crane/Shovel	35.90
0.25	Equipment Operator, Crane/Shovel	35.90
1	Ripper, Shanks Each, D-8	0.80
1	Ripper, 1-Shank Beam, Hydraulic, D-8	13.43
1	Blade, Universal, Hydraulic, for D8	9.78
	Total Hourly Rate for CODTL	$218.27

Crew: CODTM

Quantity	Description	Hourly Rate
1	Ripper, Single Shank Beam, Hydraulic, D-8	12.19
1	Equipment Operator, Medium	34.65
1	Ripper, Shanks Each, D-8	0.80
1	Blade, Universal, Hydraulic, for D8	9.78
1	Dozer, Crawler, D-8L, Powershift, (Add Blade)	149.38
	Total Hourly Rate for CODTM	$206.80

Crew: CODTO

Quantity	Description	Hourly Rate
1	Ripper, Shanks Each, D-9	0.86
0.25	Equipment Operator, Crane/Shovel	35.90
1	Ripper, Multishank Beam, Hydraulic, D-9	13.81
1	Equipment Operator, Crane/Shovel	35.90
1	Blade, Universal, Hydraulic, for D9	12.22
1	Dozer, Crawler, D-9N, (Add Blade)	190.50
	Total Hourly Rate for CODTO	$262.26

Crew: CODTP

Quantity	Description	Hourly Rate
1	Ripper, Single-Shank Beam, D-10	19.36
1	Ripper, Shanks Each, D-10	1.36
1	Equipment Operator, Medium	34.65
1	Dozer, Crawler, D-10N, Powershift, (Add Blade)	190.50
1	Blade, Universal, Hydraulic, for D10	17.42
	Total Hourly Rate for CODTP	$263.29

Crew: CODTQ

Quantity	Description	Hourly Rate
1	Ripper, Multishank Beam, Hydraulic, D-9	13.81
1	Dozer, Crawler, D-9N, Powershift, (Add Blade)	190.50
1	Ripper, Shanks Each, D-9	0.86
1	Blade, Universal, Hydraulic, for D9	12.22
1	Equipment Operator, Medium	34.65
	Total Hourly Rate for CODTQ	$252.04

Crew: CODTR

Quantity	Description	Hourly Rate
1	Dozer, Crawler, D-5H, Powershift, (Add Blade)	56.73
1	Equipment Operator, Medium	34.65
1	Blade, Angle, Hydraulic, for D5	3.71
	Total Hourly Rate for CODTR	$95.09

Crew: CODTS

Quantity	Description	Hourly Rate
1	Dozer, Crawler, D-7H, Low Ground Pressure, (Add Blade)	114.95
1	Equipment Operator, Medium	34.65
0.25	Equipment Operator, Oilers	30.10
1	Winch, (without Cable), for D7	8.44
1	Blade, Universal, Hydraulic, for D7	7.39
	Total Hourly Rate for CODTS	$172.95

Crew: CODTT

Quantity	Description	Hourly Rate
1	Dozer, Crawler, D-7H, Low Ground Pressure, (Add Blade)	114.95
1	Blade, Universal, Hydraulic, for D7	7.39
0.5	Laborer (Semi-Skilled - Foreman)	28.70
2	Laborer (Semi-Skilled)	26.70
1	Equipment Operator, Medium	34.65
	Total Hourly Rate for CODTT	$224.74

Crew Detail (Cont.)

Crew: CODTU

Quantity	Description	Hourly Rate
1	Loader, Front End, Wheeled, 4.0 CY, Articulated, 966E	55.70
3	Equipment Operator, Medium	34.65
2	Dozer, Crawler, D-7H, Low Ground Pressure, (Add Blade)	114.95
2	Blade, Universal, Hydraulic, for D7	7.39
0.5	Laborer (Semi-Skilled)	26.70
	Total Hourly Rate for CODTU	$417.67

Crew: CODTV

Quantity	Description	Hourly Rate
1	Tiller, 5x7 V-Shaped, Deep	4.89
1	Blade, Power Angle Tilt, for D3	2.47
1	Dozer, Crawler, D-3C, Powershift, (Add Blade)	39.35
1	Equipment Operator, Medium	34.65
	Total Hourly Rate for CODTV	$81.36

Crew: CODTW

Quantity	Description	Hourly Rate
1	Blade, Straight, Hydraulic, for D4	3.00
1	Dozer, Crawler, D-4H, Powershift, (Add Blade)	39.35
1	Equipment Operator, Medium	34.65
1	Tiller, 5x7 V-Shaped, Deep	4.89
	Total Hourly Rate for CODTW	$81.89

Crew: CODTX

Quantity	Description	Hourly Rate
1	Equipment Operator, Medium	34.65
1	Tiller, 5x7 V-Shaped, Deep	4.89
1	Dozer, Crawler, D-7G, Powershift, (Add Blade)	114.95
1	Blade, Straight, Hydraulic, for D7	6.73
	Total Hourly Rate for CODTX	$161.22

Crew: CODTY

Quantity	Description	Hourly Rate
1	Dozer, Crawler, D-9N, Powershift, (Add Blade)	190.50
1	Tiller, 5x7 V-Shaped, Deep	4.89
1	Equipment Operator, Medium	34.65
1	Blade, Universal, Hydraulic, for D9	12.22
	Total Hourly Rate for CODTY	$242.26

Crew: CODTZ

Quantity	Description	Hourly Rate
1	Blade, Straight, Hydraulic, for D6	4.74
1	Equipment Operator, Medium	34.65
1	Tiller, 5x7 V-Shaped, Deep	4.89
1	Dozer, Crawler, D-6D, Powershift, (Add Blade)	71.50
	Total Hourly Rate for CODTZ	$115.78

Crew: COEIA

Quantity	Description	Hourly Rate
1	Truck, Highway, 43,000 GVW, 6x4, 3 Axle	42.47
1	Truck Drivers, Heavy	27.55
1	Truck, Rear Dump Body, 12 CY, 36,000 and up GVW	3.93
	Total Hourly Rate for COEIA	$73.95

Crew: COEIB

Quantity	Description	Hourly Rate
1	Truck Drivers, Heavy	27.55
1	Truck, Off-Highway, Rear Dump, 23T, 13-17CY, 6X6	98.13
Total Hourly Rate for COEIB		$125.68

Crew: COEIC

Quantity	Description	Hourly Rate
1	Truck Drivers, Heavy	27.55
1	Truck Trailer, End Dump, 20 CY, 24T	18.26
1	Truck, Highway, 78,000 GVW, 6x4, Dump/Mixer	48.06
Total Hourly Rate for COEIC		$93.87

Crew: COEID

Quantity	Description	Hourly Rate
1	Truck Drivers, Heavy	27.55
1	Rear Dump Body,12 CY, 36,000 GVW	3.64
1	Truck, Highway, 46,000 GVW, 6x4, 3 Axle	42.47
Total Hourly Rate for COEID		$73.66

Crew: COEIE

Quantity	Description	Hourly Rate
4	Truck, Rear Dump Body, 12 CY, 36,000 and up GVW	3.93
1	Equipment Operator, Light	33.05
4	Truck, Highway, 43,000 GVW, 6x4, 3 Axle	42.47
4	Truck Drivers, Heavy	27.55
1	Loader, Front End, Wheeled, 3.0 CY, Articulated, 950E	38.13
Total Hourly Rate for COEIE		$366.98

Crew: COEIF

Quantity	Description	Hourly Rate
4	Truck, Rear Dump Body, 12 CY, 36,000 and up GVW	3.93
4	Truck Drivers, Heavy	27.55
1	Equipment Operator, Light	33.05
4	Truck, Highway, 43,000 GVW, 6x4, 3 Axle	42.47
1	Loader, Front End, Wheeled, 4.0 CY, Articulated, 966E	55.70
Total Hourly Rate for COEIF		$384.56

Crew: COELC

Quantity	Description	Hourly Rate
1	Bushhog, Rotary Cutter, 10.5' Wide	28.60
2	Equipment Operator, Medium	34.65
1	Tractor, Wheeled, Farm, JD-2355	8.65
Total Hourly Rate for COELC		$106.55

Crew: COELD

Quantity	Description	Hourly Rate
1	Truck, Highway, 24,700 GVW, 2 Axle	21.63
1	Power Mulcher	12.90
1	Laborer (Semi-Skilled)	26.70
1	Truck Drivers, Light	26.80
Total Hourly Rate for COELD		$88.03

Crew: COEMA

Quantity	Description	Hourly Rate
0.25	Equipment Operator, Medium (Foreman)	36.65
1	Equipment Operator, Medium	34.65
Total Hourly Rate for COEMA		$43.81

Crew: COEMB

Quantity	Description	Hourly Rate
0.5	Equipment Operator, Medium (Foreman)	36.65
2	Equipment Operator, Medium	34.65
Total Hourly Rate for COEMB		$87.63

Crew: COET1

Quantity	Description	Hourly Rate
1	Truck Drivers, Heavy	27.55
1	Truck Trailer, Bottom Dump, 18 CY, 27T	12.88
1	Truck, Highway, 33,000 GVW, 76,800 GCW, 2 Axle	42.06
Total Hourly Rate for COET1		$82.48

Crew: COET2

Quantity	Description	Hourly Rate
1	Truck Drivers, Heavy	27.55
1	Truck Trailer, Bottom Dump, 30 CY, 30T	23.71
1	Truck, Highway, 33,000 GVW, 76,800 GCW, 2 Axle	42.06
Total Hourly Rate for COET2		$93.31

Crew: COETA

Quantity	Description	Hourly Rate
1	Truck, Highway, 46,000 GVW, 6x4, 3 Axle	42.47
1	Truck Drivers, Heavy	27.55
1	Rear Dump Body, 8 CY, 30,000 GVW	3.54
Total Hourly Rate for COETA		$73.57

Crew: COETC

Quantity	Description	Hourly Rate
1	Truck Drivers, Heavy	27.55
1	26 CY Semi Dump, Tractor, & Trailer	67.04
Total Hourly Rate for COETC		$94.59

Crew: COETD

Quantity	Description	Hourly Rate
1	Truck Drivers, Heavy	27.55
1	32 CY Semi Dump, Tractor, & Trailer	70.39
Total Hourly Rate for COETD		$97.94

Crew: COETF

Quantity	Description	Hourly Rate
1	Equipment Operator, Medium	34.65
1	Loader, Backhoe, Wheeled, 1.38 CY Front End Bucket, 30" Dip	32.05
1	Truck, Highway, 46,000 GVW, 6x4, 3 Axle	42.47
2	Laborer (Semi-Skilled)	26.70
1	Truck Drivers, Heavy	27.55
1	Rear Dump Body, 8 CY, 30,000 GVW	3.54
Total Hourly Rate for COETF		$193.67

Crew: COETG

Quantity	Description	Hourly Rate
1	36 CY Semi Dump, Tractor, & Trailer	74.61
1	Truck Drivers, Heavy	27.55
Total Hourly Rate for COETG		$102.16

Crew Detail (Cont.)

Crew: COETH

Quantity	Description	Hourly Rate
1	Truck, Off-Highway, Rear Dump, 14-19 CY, 25 T	114.03
1	Truck Drivers, Heavy	27.55
	Total Hourly Rate for COETH	$141.58

Crew: COETI

Quantity	Description	Hourly Rate
1	Truck, Highway, 46,000 GVW, 6x4, 3 Axle	42.47
1	Trailer, Bottom Dump, 18 CY, 27T (Add Truck)	17.08
1	Truck Drivers, Heavy	27.55
	Total Hourly Rate for COETI	$87.10

Crew: COETL

Quantity	Description	Hourly Rate
1	Laborer (Semi-Skilled - Foreman)	28.70
1	Equipment Operator, Medium	34.65
2	Laborer (Semi-Skilled)	26.70
1	Loader, Front End, Crawler, 3.75 CY, 973	136.75
2	Truck, Highway, 46,000 GVW, 6x4, 3 Axle	42.47
2	Rear Dump Body,12 CY, 36,000 GVW	3.64
2	Truck Drivers, Heavy	27.55
	Total Hourly Rate for COETL	$400.82

Crew: COETO

Quantity	Description	Hourly Rate
1	Truck Drivers, Heavy	27.55
1	20 CY Bottom Dump Truck & Trailer	60.80
	Total Hourly Rate for COETO	$88.35

Crew: COETQ

Quantity	Description	Hourly Rate
1	Truck Drivers, Heavy	27.55
1	28 CY Bottom Dump Truck & Trailer	62.37
	Total Hourly Rate for COETQ	$89.92

Crew: COETV

Quantity	Description	Hourly Rate
2	Truck, Highway, 46,000 GVW, 6x4, 3 Axle	42.47
2	Rear Dump Body, 8 CY, 30,000 GVW	3.54
1	Hydraulic Excavator, Crawler, 1.50 CY Bucket	90.05
2	Truck Drivers, Heavy	27.55
1	Equipment Operator, Medium	34.65
	Total Hourly Rate for COETV	$271.83

Crew: COETY

Quantity	Description	Hourly Rate
1	Hydraulic Crane, 4WD, 14T/49' Boom	63.23
1	Concrete Bucket, 2 CY	3.08
1	Roller, Static, 3WHL, S/P, 12T,84"W	30.23
1	Equipment Operator, Oilers	30.10
1	Equipment Operator, Crane/Shovel	35.90
	Total Hourly Rate for COETY	$162.53

Crew: COETZ

Quantity	Description	Hourly Rate
2	Laborer (Semi-Skilled)	26.70
1	Laborer (Semi-Skilled - Foreman)	28.70
1	Hydraulic Crane, Self Propelled, 22T, Rough Terrain, 4WD	71.88
1	Equipment Operator, Medium	34.65
Total Hourly Rate for COETZ		$188.63

Crew: COFCB

Quantity	Description	Hourly Rate
1	Equipment Operator, Light	33.05
4	Laborer (Semi-Skilled)	26.70
1	Laborer (Semi-Skilled - Foreman)	28.70
1	Roller, Static, 5T, 42" Wide	13.43
Total Hourly Rate for COFCB		$181.98

Crew: COFCC

Quantity	Description	Hourly Rate
1	Equipment Operator, Medium	34.65
1	Grader, Motor, 12-G, Articulated	85.25
Total Hourly Rate for COFCC		$119.90

Crew: COFCD

Quantity	Description	Hourly Rate
1	Roller, Static, 3WHL, S/P, 12T,84"W	30.23
0.5	Laborer (Semi-Skilled)	26.70
1	Equipment Operator, Medium	34.65
Total Hourly Rate for COFCD		$78.22

Crew: COFCK

Quantity	Description	Hourly Rate
0.5	Truck, Highway, 4x4, 3/4T, 8800 GVW	10.65
0.5	Laborer (Semi-Skilled)	26.70
1	Dozer, 105HP, Cat D-5, w/Angle Blade	56.73
1	Equipment Operator, Light	33.05
1	Scarifier-Ripper for D5	7.13
Total Hourly Rate for COFCK		$115.58

Crew: COFCN

Quantity	Description	Hourly Rate
1	Ripper, Single Shank Beam, Hydraulic, D-5	3.70
1	Loader, Front End, Crawler, 2.5 CY, 963	98.50
0.5	Laborer (Semi-Skilled)	26.70
0.5	Truck, Highway, 4x4, 3/4T, 8800 GVW	10.65
1	Equipment Operator, Medium	34.65
Total Hourly Rate for COFCN		$155.52

Crew: COFCO

Quantity	Description	Hourly Rate
1	Truck Drivers, Heavy	27.55
1	Truck, Highway, 46,000 GVW, 6x4, 3 Axle	42.47
1	Water Tank, 3000 Gallon (Add Truck)	11.55
1	Roller, Vibratory, Towed, Sheepsfoot, 25.5 T, 72" Wide	76.28
0.5	Laborer (Semi-Skilled)	26.70
0.25	Equipment Operator, Medium (Foreman)	36.65
1	Blade, Universal, Hydraulic, for D7	7.39
1	Dozer, Crawler, D-7H, (Add Blade)	114.95
1	Equipment Operator, Medium	34.65
Total Hourly Rate for COFCO		$337.35

Crew: COFCP

Quantity	Description	Hourly Rate
0.25	Equipment Operator, Medium (Foreman)	36.65
1	Roller, Static, 13T, 84"W, 11 Tire	30.23
0.5	Laborer (Semi-Skilled)	26.70
2	Equipment Operator, Medium	34.65
1	Truck, Water, Off-Highway, 6000 Gallon	31.21
Total Hourly Rate for COFCP		$153.25

Crew: COFCQ

Quantity	Description	Hourly Rate
0.25	Equipment Operator, Medium (Foreman)	36.65
1	Roller, Vibratory, Sheepsfoot, 13 Ton, 66" Wide	140.63
0.5	Laborer (Semi-Skilled)	26.70
2	Equipment Operator, Medium	34.65
Total Hourly Rate for COFCQ		$232.44

Crew: COFCR

Quantity	Description	Hourly Rate
1	Equipment Operator, Medium	34.65
1	Equipment Operator, Medium (Foreman)	36.65
5	Equipment Operator, Light	33.05
1	Blade, Angle, Hydraulic, for D6	5.38
1	Truck, Water, Off-Highway, 5000 Gallon	14.60
2	Landfill Compactor, 58" Diameter x 38.5" Long	54.15
1	Truck, Highway, 4x4, 3/4T, 8800 GVW	10.65
1	Dozer, Crawler, D-6H, Powershift, (Add Blade)	71.50
1	Truck Drivers, Heavy	27.55
2	Roller, Static, 30T, 96" Wide, 11 Tire	51.48
1	Grader, Motor, 14-G, Articulated	117.25
Total Hourly Rate for COFCR		$694.73

Crew: COFCS

Quantity	Description	Hourly Rate
1	Dozer, Crawler, D-4H, Powershift, (Add Blade)	39.35
1	Loader, Front End, Wheeled, 3.0 CY, Articulated, 950E	38.13
1	Truck, Highway, 4x4, 3/4T, 8800 GVW	10.65
1	Roller, Vibratory, Self Propelled, 11.4 T, 84" W	43.10
2	Equipment Operator, Light	33.05
6	Laborer (Semi-Skilled)	26.70
1	Equipment Operator, Medium	34.65
1	Laborer (Semi-Skilled - Foreman)	28.70
1	Blade, Angle, Hydraulic, for D4	3.34
1	Roller, Vibratory, Self Propelled, 2.7 T, 47.2" W	73.95
1	Grader, Motor, 120-G, Articulated	48.33
1	Roller, Self Propelled, 1T, 34" Wide, Tandem	13.43
Total Hourly Rate for COFCS		$559.92

Crew: COFCT

Quantity	Description	Hourly Rate
1	Equipment Operator, Light	33.05
1	Truck Drivers, Heavy	27.55
1	Truck, Water, Off-Highway, 6000 Gallon	31.21
1	Equipment Operator, Medium	34.65
1	Roller, Static, 30T, 96" Wide, 11 Tire	51.48
1	Grader, Motor, 14-G, Articulated	117.25
1	Equipment Operator, Medium (Foreman)	36.65
1	Truck, Highway, 4x4, 3/4T, 8800 GVW	10.65
Total Hourly Rate for COFCT		$342.49

Crew Detail (Cont.)

Crew: COFCU

Quantity	Description	Hourly Rate
1	Roller, Vibratory, Self Propelled, 84" W x 48" Dia., 18 T	54.15
0.6	Truck Drivers, Heavy	27.55
0.6	Equipment Operator, Light	33.05
1.2	Equipment Operator, Medium	34.65
1	Grader, Motor, 14-G, Articulated	117.25
1	Dozer, Crawler, D-6H, Powershift, (Add Blade)	71.50
0.5	Truck, Highway, 4x4, 3/4T, 8800 GVW	10.65
1	Truck, Water, Off-Highway, 5000 Gallon	14.60
1	Blade, Angle, Hydraulic, for D6	5.38
1	Transit Mixer, 7 CY (Add Truck)	76.20
0.3	Laborer (Semi-Skilled)	26.70
1	Truck, Highway, 45,000 GVW, 3 Axle	38.93
1	Roller, Static, 30T, 96" Wide, 11 Tire	51.48

Total Hourly Rate for COFCU $520.77

Crew: COFCV

Quantity	Description	Hourly Rate
1	Grader, Motor, 14-G, Articulated	117.25
1	Blade, Angle, Hydraulic, for D6	5.38
1	Truck Drivers, Heavy	27.55
1	Truck, Highway, 4x4, 3/4T, 8800 GVW	10.65
1	Transit Mixer, 7 CY (Add Truck)	76.20
2	Equipment Operator, Medium	34.65
1	Roller, Static, 30T, 96" Wide, 11 Tire	51.48
1	Laborer (Semi-Skilled)	26.70
1	Dozer, Crawler, D-6H, Powershift, (Add Blade)	71.50
1	Truck, Highway, 45,000 GVW, 3 Axle	38.93
2	Equipment Operator, Light	33.05
1	Roller, Vibratory, Self Propelled, 84" W x 48" Dia., 18 T	54.15
1	Truck, Water, Off-Highway, 5000 Gallon	14.60

Total Hourly Rate for COFCV $629.79

Crew: COFCX

Quantity	Description	Hourly Rate
1	Loader, Front End, Crawler, 3.75 CY, 973	136.75
1	Laborer (Semi-Skilled)	26.70
1	Equipment Operator, Medium	34.65
2	Laborer (Semi-Skilled)	26.70

Total Hourly Rate for COFCX $251.50

Crew Detail (Cont.)

Crew: COFCY

Quantity	Description	Hourly Rate
3	Ripper, Shanks Each, D-7	0.42
0.5	Laborer (Semi-Skilled)	26.70
0.5	Truck, Highway, 4x4, 3/4T, 8800 GVW	10.65
1	Ripper, Multishank Beam, Hydraulic. D-7	7.65
1	Dozer, Crawler, D-7H, Low Ground Pressure, (Add Blade)	114.95
1	Blade, Universal, Hydraulic, for D7	7.39
1	Equipment Operator, Medium	34.65
	Total Hourly Rate for COFCY	$184.56

Crew: COFCZ

Quantity	Description	Hourly Rate
0.5	Truck, Highway, 4x4, 3/4T, 8800 GVW	10.65
1	Ripper, Single Shank Beam, Hydraulic, D-5	3.70
1	Equipment Operator, Light	33.05
1	Blade, Angle, Hydraulic, for D5	3.71
1	Dozer, Crawler, D-5H, Low Ground Pressure, (Add Blade)	56.73
3	Ripper, Shanks Each, D-5	0.24
0.5	Laborer (Semi-Skilled)	26.70
	Total Hourly Rate for COFCZ	$116.59

Crew: COFGA

Quantity	Description	Hourly Rate
1	Grader, Motor, 12-G, Articulated	85.25
1	Laborer (Semi-Skilled)	26.70
1	Equipment Operator, Medium (Apprentice)	27.72
	Total Hourly Rate for COFGA	$139.67

Crew: COFGB

Quantity	Description	Hourly Rate
1	Equipment Operator, Medium	34.65
1	Grader, Motor, 130-G, Articulated	57.08
	Total Hourly Rate for COFGB	$91.72

Crew: COFGC

Quantity	Description	Hourly Rate
1	Equipment Operator, Medium	34.65
1	Grader, Motor, 120-G, Articulated	48.33
	Total Hourly Rate for COFGC	$82.97

Crew: COFGD

Quantity	Description	Hourly Rate
0.5	Outside Equipment Operator, Oiler	30.10
1	Grader, Motor, 14-G, Articulated	117.25
1	Equipment Operator, Medium	34.65
0.5	Outside Equipment Operator, Oiler (Foreman)	32.10
	Total Hourly Rate for COFGD	$183.00

Crew: COFTR

Quantity	Description	Hourly Rate
1	Roller, Self Propelled, 3 Wheel 10T, 85 HP (7775)	30.23
0.5	Laborer (Semi-Skilled)	26.70
1	Equipment Operator, Medium	34.65
	Total Hourly Rate for COFTR	$78.22

Crew: COFWB

Quantity	Description	Hourly Rate
1	Pump, Centrifugal, 4" Dia, 465 GPM/40' Head	6.70
1	Equipment Operator, Light	33.05
0.5	Laborer (Semi-Skilled)	26.70
Total Hourly Rate for COFWB		$53.10

Crew: COFWI

Quantity	Description	Hourly Rate
1	Water Tank, 3000 Gallon (Add Truck)	11.55
0.25	Equipment Operator, Medium	34.65
1	Truck Drivers, Heavy	27.55
0.5	Equipment Operator, Light	33.05
1	Truck, Highway, 46,000 GVW, 6x4, 3 Axle	42.47
0.5	Pump, Submersible, 6" Dia., 1950 GPM/40' Head	24.27
Total Hourly Rate for COFWI		$118.89

Crew: COFWK

Quantity	Description	Hourly Rate
1	Truck, Water, Off-Highway, 6000 Gallon	31.21
0.25	Equipment Operator, Light	33.05
1	Truck Drivers, Heavy	27.55
0.5	Equipment Operator, Light	33.05
1	Pump, Submersible, 6" Dia., 1950 GPM/40' Head	24.27
Total Hourly Rate for COFWK		$107.82

Crew: COFWL

Quantity	Description	Hourly Rate
1	Truck Drivers, Heavy	27.55
0.25	Pump, Submersible, 6" Dia., 1950 GPM/40' Head	24.27
1	Truck, Water, Off-Highway, 6000 Gallon	31.21
2	Laborer (Semi-Skilled)	26.70
Total Hourly Rate for COFWL		$118.23

Crew: COKBD

Quantity	Description	Hourly Rate
2	Laborer (Semi-Skilled)	26.70
1	Truck, Highway, 46,000 GVW, 6x4, 3 Axle	42.47
1	Truck, Highway, 24,500 GVW, 4x2, 2 Axle	21.63
1	Street Sweeper, 7' Wide with Sprinkler, Towed	5.41
1	Truck Drivers, Light	26.80
1	Truck Drivers, Heavy	27.55
1	Asphalt Distributor, 3000 GAL, (Add Truck)	29.23
Total Hourly Rate for COKBD		$206.48

Crew: COKBM

Quantity	Description	Hourly Rate
1	Street Sweeper, 7' Wide with Sprinkler, Towed	5.41
2	Outside Truck Driver, Light	26.80
1	Truck, Highway, 4x2 Pick-up, 5200 GVW	14.49
Total Hourly Rate for COKBM		$73.49

Crew: COKCA

Quantity	Description	Hourly Rate
1	Laborer (Semi-Skilled)	26.70
1	Asphalt Finisher, 10' with Screed, Pneumatic	198.75
3	Laborer (Semi-Skilled)	26.70
Total Hourly Rate for COKCA		$305.55

Crew: COKCC

Quantity	Description	Hourly Rate
1	Asphalt Finisher, 10' with Screed, Pneumatic	198.75
1	Laborer (Semi-Skilled - Foreman)	28.70
1	Roller, Static, 3 Wheeled, 12T, 84" Wide	30.23
2	Equipment Operator, Medium	34.65
7	Laborer (Semi-Skilled)	26.70
Total Hourly Rate for COKCC		$513.88

Crew: COKSA

Quantity	Description	Hourly Rate
1	Truck Drivers, Light	26.80
1	Pavement Line Striper Truck	68.65
1	Painters, Ordinary	30.60
Total Hourly Rate for COKSA		$126.05

Crew: COMAC

Quantity	Description	Hourly Rate
1	Equipment Operator, Oilers	30.10
1	Loader, Belt, Single Deck Vibrating Grizzle, 72" Wide	7.82
6	Loader, Belt, 24"X50', 355 Ton/HR	19.10
0.5	Laborer (Semi-Skilled)	26.70
1	Laborer (Semi-Skilled)	26.70
1	Loader, Belt, Triple Deck Screen, 12' x 60"	7.09
1	Loader, Front End, Wheeled, 4.0 CY, Articulated, 966E	55.70
1	Truck, Highway, 4x4, 3/4T, 8800 GVW	10.65
1.5	Equipment Operator, Medium	34.65
1	Skilled Worker	34.85
Total Hourly Rate for COMAC		$352.83

Crew: COMAD

Quantity	Description	Hourly Rate
1	Truck, Highway, 4x4, 3/4T, 8800 GVW	10.65
1	Loader, Front End, Wheeled, 3.0 CY, Articulated, 950E	38.13
1	Equipment Operator, Medium	34.65
1	Portable Jaw Plant with 15x36" Jaw Crusher	70.44
1	Skilled Worker	34.85
2	Loader, Belt, 24"X50', 355 Ton/HR	19.10
1.5	Equipment Operator, Oilers	30.10
1	Laborer (Semi-Skilled)	26.70
0.5	Laborer (Semi-Skilled - Foreman)	28.70
1	Loader, Front End, Crawler, 2.0 CY, 953	68.98
1	Loader, Belt Feeder, Dozer Trap	2.71

Total Hourly Rate for COMAD $384.80

Crew: COMCA

Quantity	Description	Hourly Rate
1	Brush Chipper, 16" Dia. Log Capacity, Trailer-Mounted	25.28
1	Laborer (Semi-Skilled - Foreman)	28.70
1	Equipment Operator, Medium	34.65
4	Laborer (Semi-Skilled)	26.70
1	Loader, Front End, Crawler, 3.75 CY, 973	136.75

Total Hourly Rate for COMCA $332.18

Crew: COMCB

Quantity	Description	Hourly Rate
1	Laborer (Semi-Skilled - Foreman)	28.70
1	Dozer, Crawler, D-6D, Powershift, (Add Blade)	71.50
1	Winch, (without Cable), for D6	6.02
1	Equipment Operator, Medium	34.65
1	Multi-application Rake for D8L	6.93
1	Brush Chipper, Trailer 6" Dia. Log Capacity	12.90
1	Equipment Operator, Light	33.05
1	Blade, Angle, Hydraulic, for D6	5.38
3	Laborer (Semi-Skilled)	26.70
1	Truck, Highway, 4x4, 3/4T, 8800 GVW	10.65

Total Hourly Rate for COMCB $289.89

Crew: COMCC

Quantity	Description	Hourly Rate
1	Blade, Universal, Hydraulic, for D8	9.78
2	Equipment Operator, Medium	34.65
1	Blade, Universal, Hydraulic, for D9	12.22
2	Multi-application Rake for D8L	6.93
1	Winch, (without Cable), for D8	8.44
1	Bushhog, Rotary Cutter, 15' Wide	30.05
1	Ripper, Multishank Beam, Hydraulic, D-9	13.81
1	Dozer, Crawler, D-9N, Powershift, (Add Blade)	190.50
1	Laborer (Semi-Skilled - Foreman)	28.70
1	Dozer, Crawler, D-8L, Powershift, (Add Blade)	149.38
1	Equipment Operator, Light	33.05
3	Ripper, Shanks Each, D-9	0.86
2	Chain Saw, 31" Bar	4.17
4	Laborer (Semi-Skilled)	26.70
1	Truck, Highway, 4x4, 3/4T, 8800 GVW	10.65

Total Hourly Rate for COMCC $687.47

Crew: COMCD

Quantity	Description	Hourly Rate
1	Truck, Highway, 4x4, 3/4T, 8800 GVW	10.65
2	Multi-application Rake for D8L	6.93
1	Blade, Universal, Hydraulic, for D8	9.78
2	Blade, Universal, Hydraulic, for D9	12.22
1	Bushhog, Rotary Cutter, 15' Wide	30.05
1	Laborer (Semi-Skilled - Foreman)	28.70
2	Dozer, Crawler, D-9N, Powershift, (Add Blade)	190.50
1	Dozer, Crawler, D-8L, Powershift, (Add Blade)	149.38
3	Chain Saw, 31" Bar	4.17
6	Laborer (Semi-Skilled)	26.70
1	Ripper, Multishank Beam, Hydraulic, D-9	13.81
2	Equipment Operator, Light	33.05
2	Equipment Operator, Medium	34.65
3	Ripper, Shanks Each, D-9	0.86
1	Winch, (without Cable), for D8	8.44

Total Hourly Rate for COMCD $980.82

Crew: COMCE

Quantity	Description	Hourly Rate
1	Blade, Angle, Hydraulic, for D6	5.38
2	Multi-application Rake for D8L	6.93
1	Laborer (Semi-Skilled - Foreman)	28.70
1	Truck, Highway, 4x4, 3/4T, 8800 GVW	10.65
2	Brush Chipper, 6" Diameter Log Capacity, Trailer	12.90
4	Laborer (Semi-Skilled)	26.70
1	Winch, (without Cable), for D6	6.02
1	Blade, Universal, Hydraulic, for D8	9.78
1	Ripper, Shanks Each, D-8	0.80
2	Equipment Operator, Light	33.05
1	Dozer, Crawler, D-6D, Powershift, (Add Blade)	71.50
1	Dozer, Crawler, D-8L, Powershift, (Add Blade)	149.38
2	Equipment Operator, Medium	34.65
1	Ripper, Single Shank Beam, Hydraulic, D-8	12.19
Total Hourly Rate for COMCE		$576.27

Crew: COMCF

Quantity	Description	Hourly Rate
1	Blade, Universal, Hydraulic, for D9	12.22
3	Equipment Operator, Medium	34.65
2	Dozer, Crawler, D-8L, Powershift, (Add Blade)	149.38
2	Blade, Universal, Hydraulic, for D8	9.78
4	Laborer (Semi-Skilled)	26.70
2	Chain Saw, 31" Bar	4.17
1	Dozer, Crawler, D-9N, Powershift, (Add Blade)	190.50
1	Brush Chipper, 6" Diameter Log Capacity, Trailer	12.90
3	Multi-application Rake for D8L	6.93
1	Winch, (without Cable), for D9	9.68
1	Laborer (Semi-Skilled - Foreman)	28.70
1	Equipment Operator, Light	33.05
Total Hourly Rate for COMCF		$845.26

Crew: COMCG

Quantity	Description	Hourly Rate
1	Truck, Highway, 4x4, 3/4T, 8800 GVW	10.65
1	Blade, Angle, Hydraulic, for D5	3.71
1	V-Tree Cutter for D8L Tractor	10.67
1	Multi-application Rake for D8L	6.93
6	Laborer (Semi-Skilled)	26.70
2	Dozer, Crawler, D-5H, Low Ground Pressure, (Add Blade)	56.73
1	Winch, (without Cable), for D5	4.78
2	Equipment Operator, Light	33.05
1	Laborer (Semi-Skilled - Foreman)	28.70
3	Chain Saw, 31" Bar	4.17

Total Hourly Rate for COMCG $417.72

Crew: COMCH

Quantity	Description	Hourly Rate
1	Multi-application Rake for D8L	6.93
1	Truck, Highway, 4x4, 3/4T, 8800 GVW	10.65
1	Winch, (without Cable), for D6	6.02
1	Equipment Operator, Medium	34.65
6	Laborer (Semi-Skilled)	26.70
1	Dozer, Crawler, D-7H, Low Ground Pressure, (Add Blade)	114.95
3	Chain Saw, 31" Bar	4.17
1	Equipment Operator, Light	33.05
1	V-Tree Cutter for D8L Tractor	10.67
1	Dozer, Crawler, D-5H, Low Ground Pressure, (Add Blade)	56.73
1	Laborer (Semi-Skilled - Foreman)	28.70

Total Hourly Rate for COMCH $475.07

Crew: COMCI

Quantity	Description	Hourly Rate
1	Multi-application Rake for D8L	6.93
1	Truck, Highway, 4x4, 3/4T, 8800 GVW	10.65
2	Equipment Operator, Medium	34.65
1	Winch, (without Cable), for D6	6.02
1	Laborer (Semi-Skilled - Foreman)	28.70
3	Chain Saw, 31" Bar	4.17
1	V-Tree Cutter for D8L Tractor	10.67
6	Laborer (Semi-Skilled)	26.70
2	Dozer, Crawler, D-7H, Low Ground Pressure, (Add Blade)	114.95

Total Hourly Rate for COMCI $534.90

Crew: COMCJ

Quantity	Description	Hourly Rate
1	Winch, (without Cable), for D6	6.02
1	Bushhog, Rotary Cutter, 15' Wide	30.05
1	Multi-application Rake for D8L	6.93
1	Equipment Operator, Light	33.05
0.5	Laborer (Semi-Skilled - Foreman)	28.70
1	Dozer, Crawler, D-5H, Powershift, (Add Blade)	56.73
1	Skilled Worker	34.85
0.5	Truck, Highway, 4x4, 3/4T, 8800 GVW	10.65
3	Chain Saw, 31" Bar	4.17
5	Laborer (Semi-Skilled)	26.70
Total Hourly Rate for COMCJ		$333.33

Crew: CPIDC

Quantity	Description	Hourly Rate
1	Air Compressor, 1,600 CFM, 100 PSI	64.16
1	Pile Hammer, Single, 19,500 Ft-Lbs	44.98
2	Equipment Operator, Crane/Shovel	35.90
1	Equipment Operator, Oilers	30.10
2	Pile Drivers (Apprentice)	26.64
2	Pile Drivers	33.30
2	Air Hose, 3.00", 100'	4.42
1	Pile Leads, 8"X 26", 60' L	27.29
1	Crane, Crawler 85T/160' Boom	146.25
1	Pile Drivers (Foreman)	35.30
Total Hourly Rate for CPIDC		$548.61

Crew: CPIDV

Quantity	Description	Hourly Rate
1	Pile Hammer,Vibratory,116T Force Drive	175.38
1	Crane, Crawler 85T/160' Boom	146.25
2	Pile Drivers	33.30
1	Pile Drivers (Foreman)	35.30
2	Pile Drivers (Apprentice)	26.64
2	Equipment Operator, Crane/Shovel	35.90
1	Equipment Operator, Oilers	30.10
Total Hourly Rate for CPIDV		$578.70

Crew: CTDHH

Quantity	Description	Hourly Rate
1	Off-Highway Dump, 50T, 773	147.63
1	Truck Drivers, Heavy	27.55
Total Hourly Rate for CTDHH		$175.18

Crew: CTDHI

Quantity	Description	Hourly Rate
1	Off-Highway Dump, 85T, 777	159.50
1	Truck Drivers, Heavy	27.55
Total Hourly Rate for CTDHI		$187.05

Crew: EELEA

Quantity	Description	Hourly Rate
0.25	Electricians (Foreman)	42.75
1	Electricians	40.75
Total Hourly Rate for EELEA		$51.44

Crew Detail (Cont.)

Crew: EELEB

Quantity	Description	Hourly Rate
2	Electricians	40.75
0.5	Electricians (Foreman)	42.75
Total Hourly Rate for EELEB		$102.88

Crew: EELEC

Quantity	Description	Hourly Rate
0.25	Electricians (Foreman)	42.75
2	Laborer (Semi-Skilled)	26.70
1	Electricians	40.75
Total Hourly Rate for EELEC		$104.84

Crew: EELED

Quantity	Description	Hourly Rate
1	Electricians (Foreman)	42.75
1	Electricians	40.75
0.13	Hydraulic Crane, 4WD, 20T/70' Boom	71.88
0.13	Equipment Operator, Medium	34.65
Total Hourly Rate for EELED		$97.35

Crew: EELEE

Quantity	Description	Hourly Rate
0.1	Equipment Operator, Medium	34.65
3	Electricians	40.75
1	Electricians (Foreman)	42.75
0.1	Hydraulic Crane, 4WD, 20T/70' Boom	71.88
Total Hourly Rate for EELEE		$175.65

Crew: EELEF

Quantity	Description	Hourly Rate
2	Electricians	40.75
0.66	Electricians (Foreman)	42.75
2	Electricians	40.75
Total Hourly Rate for EELEF		$191.22

Crew: EELEG

Quantity	Description	Hourly Rate
1	Electricians (Foreman)	42.75
2	Laborer (Semi-Skilled)	26.70
3	Electricians	40.75
Total Hourly Rate for EELEG		$218.40

Crew: EELEH

Quantity	Description	Hourly Rate
0.13	Hydraulic Crane, 4WD, 40T/84' Boom	76.60
3	Electricians	40.75
0.13	Equipment Operator, Crane/Shovel	35.90
1	Electricians (Foreman)	42.75
Total Hourly Rate for EELEH		$179.63

Crew: EELEJ

Quantity	Description	Hourly Rate
2	Electricians	40.75
2	Electricians (Apprentice)	32.60
1	Electricians (Foreman)	42.75
1	Highway, Line-Truck, with Aerial Platform	31.57
Total Hourly Rate for EELEJ		$221.02

Crew Detail (Cont.)

Crew: EELEK

Quantity	Description	Hourly Rate
3	Electricians	40.75
1	Electricians	40.75
1	Highway, Line-Truck, with Aerial Platform	31.57
3	Electricians	40.75

Total Hourly Rate for EELEK $316.82

Crew: ESELA

Quantity	Description	Hourly Rate
2	Laborer (Semi-Skilled)	26.70
3	Elevator Constructors	45.05
3	Elevator Constructors (Apprentice)	36.04
1	Electricians	40.75

Total Hourly Rate for ESELA $337.42

Crew: ESELB

Quantity	Description	Hourly Rate
1	Elevator Constructors (Apprentice)	36.04
2	Elevator Constructors	45.05
1	Laborer (Semi-Skilled)	26.70
0.17	Elevator Constructors (Foreman)	47.05

Total Hourly Rate for ESELB $160.84

Crew: ESELC

Quantity	Description	Hourly Rate
0.17	Elevator Constructors (Foreman)	47.05
1	Elevator Constructors	45.05
1	Elevator Constructors (Apprentice)	36.04

Total Hourly Rate for ESELC $89.09

Crew: ESELH

Quantity	Description	Hourly Rate
1	Elevator Constructors	45.05
1	Electricians	40.75
1	Laborer (Semi-Skilled)	26.70
0.25	Equipment Operator, Crane/Shovel	35.90
0.25	Hydraulic Crane, 4WD, 40T/84' Boom	76.60
1	Elevator Constructors (Apprentice)	36.04

Total Hourly Rate for ESELH $176.66

Crew: HBLDA

Quantity	Description	Hourly Rate
5	Laborer (Semi-Skilled)	26.70
2	Lift, Hydraulic, Platform Scissors	9.88

Total Hourly Rate for HBLDA $153.25

Crew: HLABD

Quantity	Description	Hourly Rate
2	Laborer (Semi-Skilled)	26.70
0.25	Equipment Operator, Light	33.05
0.25	Fork Lift 8000 lb.	14.05
0.25	Laborer (Semi-Skilled - Foreman)	28.70

Total Hourly Rate for HLABD $72.35

Crew: HLABK

Quantity	Description	Hourly Rate
1	Laborer (Semi-Skilled)	26.70
1	Drum Crusher, 30 T, Mobile	28.85
1	Equipment Operator, Medium	34.65

Total Hourly Rate for HLABK $90.20

Crew: HLABN

Quantity	Description	Hourly Rate
0.25	Equipment Operator, Medium	34.65
1	Laborer (Semi-Skilled)	26.70
1	Hydraulic Excavator, Crawler, 1.13 CY Bucket, Long Carriage	69.73
1	Equipment Operator, Medium	34.65
1	2 CY Rotating Grapple	16.01
Total Hourly Rate for HLABN		$155.75

Crew: HLABO

Quantity	Description	Hourly Rate
1	Equipment Operator, Light	33.05
1	Truck, Highway, 3 Axle, 50,000 GVW, 85,000 GCW	44.46
0.5	Laborer (Semi-Skilled - Foreman)	28.70
1	Trailer, Water Tanker, 5000 Gallon	22.14
2	Laborer (Semi-Skilled)	26.70
1	Fork Lift 8000 lb.	14.05
1	Pump, Centrifugal, 6" Dia, 100 GPM/40' Head	27.13
Total Hourly Rate for HLABO		$208.57

Crew: HLABP

Quantity	Description	Hourly Rate
0.5	Laborer (Semi-Skilled - Foreman)	28.70
1	Fork Lift 8000 lb.	14.05
1	Trailer, End Dump, 17 CY	15.95
1	Truck, Highway, 45,000 GVW, 3 Axle	38.93
2	Laborer (Semi-Skilled)	26.70
1	Equipment Operator, Light	33.05
Total Hourly Rate for HLABP		$169.74

Crew: HLABQ

Quantity	Description	Hourly Rate
0.5	Laborer (Semi-Skilled)	26.70
1	Laborer (Semi-Skilled)	26.70
1	Equipment Operator, Light	33.05
Total Hourly Rate for HLABQ		$73.10

Crew: HLABR

Quantity	Description	Hourly Rate
1	Trailer, Lowboy, 35 Ton, 2 Axle	15.96
1	Flatbed, 8' x 16.0' (Truck Body Only, add truck)	22.03
1	Equipment Operator, Light	33.05
1	Truck, Highway, 24,500 GVW, 4x2, 2 Axle	21.63
1	Fork Lift 2500 lb.	26.95
1	Truck Drivers, Heavy	27.55
0.25	Equipment Operator, Light	33.05
1	Outside Laborer (Semi-Skilled)	26.09
0.25	Outside Laborer (Semi-Skilled - Foreman)	28.32
1	Flatbead Hoist, 2 Ton	1.51
Total Hourly Rate for HLABR		$190.11

Crew: HLABS

Quantity	Description	Hourly Rate
0.33	Field Superintendent, Maximum	43.13
2	Outside Equipment Operator, Heavy	35.90
0.33	Safety Engineer, Maximum	38.62
Total Hourly Rate for HLABS		$98.78

Crew: HLABT

Quantity	Description	Hourly Rate
0.67	Field Superintendent	43.13
0.33	Safety Engineer	38.62
4	Outside Equipment Operator, Heavy	35.90
Total Hourly Rate for HLABT		$185.24

Crew: HLABU

Quantity	Description	Hourly Rate
5	Laborer (Semi-Skilled)	26.70
2	Lift, Hydraulic, Platform Scissors	9.88
Total Hourly Rate for HLABU		$153.25

Crew: HLABV

Quantity	Description	Hourly Rate
1	Decontamination Technician Supervisor	33.37
Total Hourly Rate for HLABV		$33.37

Crew: HLABW

Quantity	Description	Hourly Rate
1	Decontamination Technician Supervisor	33.37
Total Hourly Rate for HLABW		$33.37

Crew: HLABX

Quantity	Description	Hourly Rate
1	Outside Equipment Operator, Oiler (Foreman)	32.10
1	Outside Electrician	40.75
4	Outside Laborer (Semi-Skilled)	26.09
1	Decontamination Technician	26.89
2	Outside Equipment Operator, Medium	34.65
1	Outside Equipment Operator, Heavy	35.90
1	Outside Equipment Operator, Oiler	30.10
1	Decontamination Technician Supervisor	33.37
Total Hourly Rate for HLABX		$372.77

Crew: HLABY

Quantity	Description	Hourly Rate
1	Hoist, 2.5 Ton Air Tugger	1.58
Total Hourly Rate for HLABY		$1.58

Crew: HLACA

Quantity	Description	Hourly Rate
1	Fork Lift 8000 lb.	14.05
0.5	Laborer (Semi-Skilled)	26.70
1	Truck Drivers, Heavy	27.55
0.25	Water Tanker Trailer, 6000 Gallon	11.55
1	Laborer (Semi-Skilled)	26.70
0.18	Truck, Highway, 45,000 GVW, 3 Axle	38.93
Total Hourly Rate for HLACA		$91.55

Crew: HLACB

Quantity	Description	Hourly Rate
1	Fork Lift 8000 lb.	14.05
0.18	Truck, Highway, 45,000 GVW, 3 Axle	38.93
1	Truck Drivers, Heavy	27.55
0.5	Laborer (Semi-Skilled)	26.70
1	Laborer (Semi-Skilled)	26.70
0.33	Truck Trailer, Bottom Dump, 20 CY, 30T	14.48
Total Hourly Rate for HLACB		$93.43

Crew: HLACC

Quantity	Description	Hourly Rate
0.18	Truck, Highway, 35,000 GVW, 4x2	36.05
0.5	Outside Laborer (Semi-Skilled)	26.09
1	Outside Laborer (Semi-Skilled)	26.09
1	Fork Lift 2500 lb.	26.95
0.35	Truck, Flatbed, 8' x 24.0'	22.03
1	Fork Lift 8000 lb.	14.05
1	Truck Drivers, Heavy	27.55
Total Hourly Rate for HLACC		$121.89

Crew: HLACD

Quantity	Description	Hourly Rate
1	Water Tanker Trailer, 4000 Gallon	11.55
1	Truck, Highway, 3 Axle, 50,000 GVW, 85,000 GCW	44.46
Total Hourly Rate for HLACD		$56.01

Crew: HORDA

Quantity	Description	Hourly Rate
1	EOD Technician	33.04
Total Hourly Rate for HORDA		$33.04

Crew: HORDB

Quantity	Description	Hourly Rate
1	Master EOD Technician	41.59
Total Hourly Rate for HORDB		$41.59

Crew: HORDC

Quantity	Description	Hourly Rate
1	Hydraulic Crane, Self Propelled, 22T, Rough Terrain, 4WD	71.88
1	Laborer (Semi-Skilled)	26.70
1	Laborer (Semi-Skilled)	26.70
1	Equipment Operator, Medium	34.65
Total Hourly Rate for HORDC		$159.92

Crew: HORDD

Quantity	Description	Hourly Rate
1	Magnetometer Mapping System Operator	36.74
1	Surface Towed Ordnance Locator System	1356.75
1	Surface Towed Magnetometer Operator	36.74
1	Assistant Towed Magnetometer Operator	39.85
Total Hourly Rate for HORDD		$1,470.07

Crew: HORDE

Quantity	Description	Hourly Rate
1	Emergency Medical Technician	32.97
1	Paramedic	34.99
1	Ambulance, F350, 1T, 10,000 GVW	44.89
Total Hourly Rate for HORDE		$112.85

Crew: HRADB

Quantity	Description	Hourly Rate
2	Radiation Protection Technicians	54.88
0.25	Hot Cell Technicians, Senior	42.77
4	Hot Cell Technicians	39.61
Total Hourly Rate for HRADB		$278.90

Crew: HSHMA

Quantity	Description	Hourly Rate
2	Sheet Metal Workers	40.20
1	Sheet Metal Workers (Apprentice)	32.16
0.75	Equipment Operator, Medium	34.65
0.75	Hydraulic Crane, Self Propelled, 22T, Rough Terrain, 4WD	71.88
1	Sheet Metal Workers (Foreman)	42.20
Total Hourly Rate for HSHMA		$234.65

Crew: HSHMB

Quantity	Description	Hourly Rate
1	Sheet Metal Workers (Foreman)	42.20
2	Sheet Metal Workers	40.20
0.75	Hydraulic Crane, Self Propelled, 22T, Rough Terrain, 4WD	71.88
1	Sheet Metal Workers (Apprentice)	32.16
0.75	Equipment Operator, Medium	34.65
1	Laborer (Semi-Skilled)	26.70
Total Hourly Rate for HSHMB		$261.35

Crew: HSHMC

Quantity	Description	Hourly Rate
0.75	Equipment Operator, Medium	34.65
0.75	Hydraulic Crane, 4WD, 20T/70' Boom	71.88
1	Laborer (Semi-Skilled)	26.70
1	Sheet Metal Workers (Foreman)	42.20
2	Sheet Metal Workers	40.20
1	Sheet Metal Workers (Apprentice)	32.16
Total Hourly Rate for HSHMC		$261.35

Crew: MPLUA

Quantity	Description	Hourly Rate
1	Plumbers	40.85
0.25	Plumbers (Foreman)	42.85
Total Hourly Rate for MPLUA		$51.56

Crew: MPLUB

Quantity	Description	Hourly Rate
0.5	Plumbers (Foreman)	42.85
1	Plumbers (Apprentice)	32.68
1	Plumbers	40.85
Total Hourly Rate for MPLUB		$94.95

Crew: MPLUC

Quantity	Description	Hourly Rate
0.1	Plumbers (Foreman)	42.85
1	Plumbers	40.85
Total Hourly Rate for MPLUC		$45.14

Crew: MPLUD

Quantity	Description	Hourly Rate
1	Plumbers	40.85
0.25	Plumbers (Foreman)	42.85
Total Hourly Rate for MPLUD		$51.56

Crew: MPLUE

Quantity	Description	Hourly Rate
0.5	Plumbers (Foreman)	42.85
1	Plumbers	40.85
1	Laborer (Semi-Skilled)	26.70
Total Hourly Rate for MPLUE		$88.97

Crew: MPLUF

Quantity	Description	Hourly Rate
1	Laborer (Semi-Skilled)	26.70
1	Plumbers (Apprentice)	32.68
0.25	Plumbers (Foreman)	42.85
1	Plumbers	40.85
Total Hourly Rate for MPLUF		$110.94

Crew: MPLUG

Quantity	Description	Hourly Rate
0.1	Hydraulic Crane, 4WD, 20T/70' Boom	71.88
1	Plumbers	40.85
0.25	Plumbers (Foreman)	42.85
1	Plumbers (Apprentice)	32.68
0.1	Equipment Operator, Medium	34.65
Total Hourly Rate for MPLUG		$94.90

Crew: MPLUH

Quantity	Description	Hourly Rate
1	Plumbers (Apprentice)	32.68
0.5	Plumbers (Foreman)	42.85
1	Welder, 300 AMP, Skid, Electric Drive	10.08
1	Plumbers	40.85
Total Hourly Rate for MPLUH		$105.04

Crew: MPLUI

Quantity	Description	Hourly Rate
2	Laborer (Semi-Skilled)	26.70
1	Plumbers	40.85
0.25	Laborer (Semi-Skilled)	26.70
1	Plumbers (Apprentice)	32.68
Total Hourly Rate for MPLUI		$133.60

Crew: MPLUJ

Quantity	Description	Hourly Rate
1	Plumbers (Outside)	40.85
0.5	Plumbers (Outside - Foreman)	42.85
1	Laborer (Semi-Skilled)	26.70
1	Plumbers (Outside - Apprentice)	32.68
Total Hourly Rate for MPLUJ		$121.66

Crew: MPLUK

Quantity	Description	Hourly Rate
1	Welder, 300 AMP, Skid, Electric Drive	10.08
1	Laborer (Semi-Skilled)	26.70
0.5	Plumbers (Outside - Foreman)	42.85
1	Plumbers (Outside)	40.85
1	Outside Plumber (Apprentice)	32.68
Total Hourly Rate for MPLUK		$131.74

Crew: MPLUM

Quantity	Description	Hourly Rate
1	Laborer (Semi-Skilled)	26.70
1	Plumbers (Outside - Foreman)	42.85
1	Plumbers (Outside - Apprentice)	32.68
1	Plumbers (Outside)	40.85
Total Hourly Rate for MPLUM		$143.08

Crew: MPLUO

Quantity	Description	Hourly Rate
1	Welder, 300 AMP, Skid, Electric Drive	10.08
1	Plumbers (Outside - Apprentice)	32.68
1	Plumbers (Outside - Foreman)	42.85
1	Laborer (Semi-Skilled)	26.70
1	Plumbers (Outside)	40.85
Total Hourly Rate for MPLUO		$153.16

Crew: MPLUQ

Quantity	Description	Hourly Rate
1	Plumbers (Outside - Foreman)	42.85
1	Plumbers (Outside)	40.85
0.4	Hydraulic Crane, 4WD, 20T/70' Boom	71.88
1	Welder, 300 AMP, Skid, Electric Drive	10.08
1	Laborer (Semi-Skilled)	26.70
1	Plumbers (Outside - Apprentice)	32.68
0.4	Equipment Operator, Medium	34.65
Total Hourly Rate for MPLUQ		$195.77

Crew: MPLUQ1

Quantity	Description	Hourly Rate
1	Plumbers (Apprentice)	32.68
1	Plumbers	40.85
Total Hourly Rate for MPLUQ1		$73.53

Crew: MPLUS

Quantity	Description	Hourly Rate
1	Laborer (Semi-Skilled)	26.70
1	Plumbers (Outside)	40.85
0.5	Equipment Operator, Medium	34.65
0.5	Hydraulic Crane, 4WD, 20T/70' Boom	71.88
1	Plumbers (Outside - Foreman)	42.85
Total Hourly Rate for MPLUS		$163.66

Crew: MPLUT

Quantity	Description	Hourly Rate
1	Plumbers (Outside)	40.85
1	Plumbers (Outside - Foreman)	42.85
0.75	Equipment Operator, Medium	34.65
1	Laborer (Semi-Skilled)	26.70
0.75	Hydraulic Crane, 4WD, 20T/70' Boom	71.88
Total Hourly Rate for MPLUT		$190.29

Crew: MPLUU

Quantity	Description	Hourly Rate
1	Outside Electrician	40.75
1	Laborer (Semi-Skilled)	26.70
1	Outside Plumber	40.85
Total Hourly Rate for MPLUU		$108.30

Crew: MSHMA

Quantity	Description	Hourly Rate
0.25	Sheet Metal Workers	40.20
1	Sheet Metal Workers	40.20
Total Hourly Rate for MSHMA		$50.25

Crew: MSHMB

Quantity	Description	Hourly Rate
1	Sheet Metal Workers (Apprentice)	32.16
1	Sheet Metal Workers	40.20
0.5	Sheet Metal Workers (Foreman)	42.20
Total Hourly Rate for MSHMB		$93.46

Crew: MSHMC

Quantity	Description	Hourly Rate
2	Sheet Metal Workers	40.20
0.25	Sheet Metal Workers (Foreman)	42.20
Total Hourly Rate for MSHMC		$90.95

Crew: MSHMD

Quantity	Description	Hourly Rate
1	Sheet Metal Workers	40.20
1	Sheet Metal Workers (Apprentice)	32.16
0.5	Sheet Metal Workers (Foreman)	42.20
Total Hourly Rate for MSHMD		$93.46

Crew: MSHME

Quantity	Description	Hourly Rate
0.2	Hydraulic Crane, 4WD, 20T/70' Boom	71.88
0.5	Sheet Metal Workers (Foreman)	42.20
1	Sheet Metal Workers	40.20
0.2	Equipment Operator, Medium	34.65
1	Sheet Metal Workers (Apprentice)	32.16
Total Hourly Rate for MSHME		$114.77

Crew: MSHMF

Quantity	Description	Hourly Rate
1	Sheet Metal Workers	40.20
1	Laborer (Semi-Skilled)	26.70
0.5	Sheet Metal Workers (Foreman)	42.20
1	Sheet Metal Workers (Apprentice)	32.16
Total Hourly Rate for MSHMF		$120.16

Crew: MSHMG

Quantity	Description	Hourly Rate
1	Sheet Metal Workers	40.20
1	Sheet Metal Workers (Foreman)	42.20
1	Sheet Metal Workers (Apprentice)	32.16
1	Laborer (Semi-Skilled)	26.70
Total Hourly Rate for MSHMG		$141.26

Crew: MSPFA

Quantity	Description	Hourly Rate
0.25	Steam/Pipefitters (Foreman)	42.95
1	Steam/Pipefitters	40.95
Total Hourly Rate for MSPFA		$51.69

Crew: MSPFB

Quantity	Description	Hourly Rate
1	Steam/Pipefitters (Apprentice)	32.76
1	Steam/Pipefitters	40.95
0.5	Steam/Pipefitters (Foreman)	42.95
Total Hourly Rate for MSPFB		$95.19

Crew: MSPFC

Quantity	Description	Hourly Rate
0.5	Steam/Pipefitters (Foreman)	42.95
2	Steam/Pipefitters	40.95
Total Hourly Rate for MSPFC		$103.38

Crew: MSPFD

Quantity	Description	Hourly Rate
0.5	Steam/Pipefitters (Foreman)	42.95
1	Steam/Pipefitters (Apprentice)	32.76
1	Steam/Pipefitters	40.95
Total Hourly Rate for MSPFD		$95.19

Crew: MSPFE

Quantity	Description	Hourly Rate
1	Steam/Pipefitters (Apprentice)	32.76
0.5	Steam/Pipefitters (Foreman)	42.95
1	Welder, 300 AMP, Skid, Electric Drive	10.08
1	Steam/Pipefitters	40.95
Total Hourly Rate for MSPFE		$105.27

Crew: MSPFF

Quantity	Description	Hourly Rate
1	Steam/Pipefitters	40.95
0.2	Hydraulic Crane, 4WD, 20T/70' Boom	71.88
0.5	Steam/Pipefitters (Foreman)	42.95
1	Steam/Pipefitters (Apprentice)	32.76
0.2	Equipment Operator, Medium	34.65
Total Hourly Rate for MSPFF		$116.49

Crew: MSPFG

Quantity	Description	Hourly Rate
0.5	Steam/Pipefitters (Foreman)	42.95
0.3	Equipment Operator, Medium	34.65
1	Steam/Pipefitters	40.95
1	Steam/Pipefitters (Apprentice)	32.76
0.3	Hydraulic Crane, 4WD, 20T/70' Boom	71.88
Total Hourly Rate for MSPFG		$127.14

Crew Detail (Cont.)

Crew: MSPFH

Quantity	Description	Hourly Rate
1	Laborer (Semi-Skilled)	26.70
1	Steam/Pipefitters	40.95
1	Steam/Pipefitters (Apprentice)	32.76
1	Welder, 200 AMP, with 1 Axle Trailer	8.31
0.25	Steam/Pipefitters (Foreman)	42.95

Total Hourly Rate for MSPFH $119.46

Crew: MSPFI

Quantity	Description	Hourly Rate
0.5	Steam/Pipefitters (Foreman)	42.95
1	Steam/Pipefitters (Apprentice)	32.76
1	Steam/Pipefitters	40.95
1	Laborer (Semi-Skilled)	26.70

Total Hourly Rate for MSPFI $121.88

Crew: MSPFJ

Quantity	Description	Hourly Rate
1	Laborer (Semi-Skilled)	26.70
1	Steam/Pipefitters (Apprentice)	32.76
0.5	Steam/Pipefitters (Foreman)	42.95
1	Welder, 200 AMP, with 1 Axle Trailer	8.31
1	Steam/Pipefitters	40.95

Total Hourly Rate for MSPFJ $130.20

Crew: MSPFL

Quantity	Description	Hourly Rate
1	Steam/Pipefitters	40.95
1	Steam/Pipefitters (Foreman)	42.95
1	Laborer (Semi-Skilled)	26.70
1	Steam/Pipefitters (Apprentice)	32.76

Total Hourly Rate for MSPFL $143.36

Crew: MSPFM

Quantity	Description	Hourly Rate
1	Steam/Pipefitters (Foreman)	42.95
1	Steam/Pipefitters (Apprentice)	32.76
1	Laborer (Semi-Skilled)	26.70
1	Welder, 300 AMP, Skid, Electric Drive	10.08
1	Steam/Pipefitters	40.95

Total Hourly Rate for MSPFM $153.44

Crew: MSPFN

Quantity	Description	Hourly Rate
1	Steam/Pipefitters	40.95
1	Laborer (Semi-Skilled)	26.70
0.2	Hydraulic Crane, 4WD, 20T/70' Boom	71.88
1	Steam/Pipefitters (Foreman)	42.95
0.2	Equipment Operator, Medium	34.65
1	Steam/Pipefitters (Apprentice)	32.76

Total Hourly Rate for MSPFN $164.67

Crew: MSPFO

Quantity	Description	Hourly Rate
1	Steam/Pipefitters	40.95
1	Steam/Pipefitters (Foreman)	42.95
0.2	Hydraulic Crane, 4WD, 20T/70' Boom	71.88
1	Laborer (Semi-Skilled)	26.70
0.2	Equipment Operator, Medium	34.65
1	Welder, 300 AMP, Skid, Electric Drive	10.08
1	Steam/Pipefitters (Apprentice)	32.76

Total Hourly Rate for MSPFO $174.75

Crew: MSPFP

Quantity	Description	Hourly Rate
0.3	Equipment Operator, Medium	34.65
0.3	Hydraulic Crane, 4WD, 20T/70' Boom	71.88
1	Steam/Pipefitters (Apprentice)	32.76
1	Steam/Pipefitters (Foreman)	42.95
1	Steam/Pipefitters	40.95
1	Laborer (Semi-Skilled)	26.70
Total Hourly Rate for MSPFP		$175.32

Crew: MSPFQ

Quantity	Description	Hourly Rate
0.3	Equipment Operator, Medium	34.65
0.3	Hydraulic Crane, 4WD, 20T/70' Boom	71.88
1	Steam/Pipefitters	40.95
1	Steam/Pipefitters (Foreman)	42.95
1	Steam/Pipefitters (Apprentice)	32.76
1	Laborer (Semi-Skilled)	26.70
Total Hourly Rate for MSPFQ		$175.32

Crew: MSPFR

Quantity	Description	Hourly Rate
0.5	Equipment Operator, Medium	34.65
1	Steam/Pipefitters (Apprentice)	32.76
1	Laborer (Semi-Skilled)	26.70
1	Steam/Pipefitters (Foreman)	42.95
1	Steam/Pipefitters	40.95
0.5	Hydraulic Crane, 4WD, 20T/70' Boom	71.88
Total Hourly Rate for MSPFR		$196.62

Crew: MSPFS

Quantity	Description	Hourly Rate
1	Steam/Pipefitters (Foreman)	42.95
1	Steam/Pipefitters	40.95
1	Laborer (Semi-Skilled)	26.70
0.5	Hydraulic Crane, 4WD, 20T/70' Boom	71.88
1	Steam/Pipefitters (Apprentice)	32.76
0.5	Equipment Operator, Medium	34.65
Total Hourly Rate for MSPFS		$196.62

Crew: MSPFT

Quantity	Description	Hourly Rate
0.3	Hydraulic Crane, 4WD, 40T/84' Boom	76.60
1	Steam/Pipefitters (Foreman)	42.95
0.3	Equipment Operator, Crane/Shovel	35.90
1	Laborer (Semi-Skilled)	26.70
1	Steam/Pipefitters	40.95
1	Steam/Pipefitters (Apprentice)	32.76
1	Welder, 300 AMP, Skid, Electric Drive	10.08
Total Hourly Rate for MSPFT		$187.19

Crew Detail (Cont.)

Crew: MSPFU

Quantity	Description	Hourly Rate
1	Welder, 300 AMP, Skid, Electric Drive	10.08
1	Steam/Pipefitters (Foreman)	42.95
1.8	Laborer (Semi-Skilled)	26.70
0.2	Hydraulic Crane, 4WD, 40T/84' Boom	76.60
1	Steam/Pipefitters (Apprentice)	32.76
1	Steam/Pipefitters	40.95
0.2	Equipment Operator, Crane/Shovel	35.90
Total Hourly Rate for MSPFU		$197.30

Crew: MSPFW

Quantity	Description	Hourly Rate
1	Steam/Pipefitters (Apprentice)	32.76
1	Steam/Pipefitters (Foreman)	42.95
0.5	Hydraulic Crane, 4WD, 20T/70' Boom	71.88
0.5	Equipment Operator, Medium	34.65
2	Laborer (Semi-Skilled)	26.70
1	Steam/Pipefitters	40.95
Total Hourly Rate for MSPFW		$223.32

Crew: MSPFX

Quantity	Description	Hourly Rate
0.5	Hydraulic Crane, 4WD, 40T/84' Boom	76.60
1	Steam/Pipefitters (Foreman)	42.95
1	Steam/Pipefitters (Apprentice)	32.76
3	Laborer (Semi-Skilled)	26.70
1	Steam/Pipefitters	40.95
0.5	Equipment Operator, Crane/Shovel	35.90
Total Hourly Rate for MSPFX		$253.01

Crew: MSPFY

Quantity	Description	Hourly Rate
1	Equipment Operator, Medium	34.65
1	Trailer, Lowboy, 75T, 3 Axle (Add Truck)	21.28
1	Truck Drivers, Heavy	27.55
1	Hydraulic Crane, 4WD, 20T/70' Boom	71.88
0.25	Steam/Pipefitters (Foreman)	42.95
1	Laborer (Semi-Skilled)	26.70
1	Steam/Pipefitters	40.95
1	Steam/Pipefitters (Apprentice)	32.76
1	Truck, Highway, 50,000 GVW, 6x4, 3 Axle	43.26
1	Welder, 200 AMP, with 1 Axle Trailer	8.31
Total Hourly Rate for MSPFY		$318.07

Crew: MSPFZ

Quantity	Description	Hourly Rate
1	Laborer (Semi-Skilled)	26.70
0.75	Welder, 200 AMP, with 1 Axle Trailer	8.31
0.25	Steam/Pipefitters (Foreman)	42.95
1	Steam/Pipefitters	40.95
Total Hourly Rate for MSPFZ		$84.62

Crew: MSPIA

Quantity	Description	Hourly Rate
0.25	Sprinkler Installers (Foreman)	42.75
1	Sprinkler Installers	40.75
Total Hourly Rate for MSPIA		$51.44

Crew: MSPIB

Quantity	Description	Hourly Rate
0.5	Sprinkler Installers (Foreman)	42.75
2	Sprinkler Installers	40.75
Total Hourly Rate for MSPIB		$102.88

Crew: MSPIC

Quantity	Description	Hourly Rate
1	Steam/Pipefitters (Apprentice)	32.76
1	Sprinkler Installers	40.75
0.5	Sprinkler Installers (Foreman)	42.75
Total Hourly Rate for MSPIC		$94.88

Crew: MSPID

Quantity	Description	Hourly Rate
1	Sprinkler Installers	40.75
1	Sprinkler Installers (Apprentice)	32.60
0.5	Sprinkler Installers (Foreman)	42.75
0.1	Hydraulic Crane, 4WD, 20T/70' Boom	71.88
0.1	Equipment Operator, Medium	34.65
Total Hourly Rate for MSPID		$105.38

Crew: SIWRB

Quantity	Description	Hourly Rate
2	Rodmen, (Reinforcing)	37.95
0.5	Rodmen, (Reinforcing - Foreman)	39.95
Total Hourly Rate for SIWRB		$95.88

Crew: SIWRC

Quantity	Description	Hourly Rate
3	Rodmen, (Reinforcing)	37.95
1	Rodmen, (Reinforcing - Foreman)	39.95
Total Hourly Rate for SIWRC		$153.80

Crew: SIWRF

Quantity	Description	Hourly Rate
1	Equipment Operator, Medium	34.65
1	Hydraulic Crane, 4WD, 20T/70' Boom	71.88
2	Rodmen, (Reinforcing)	37.95
1	Rodmen, (Reinforcing - Foreman)	39.95
1	Rodmen, (Reinforcing - Apprentice)	30.36
Total Hourly Rate for SIWRF		$252.73

Crew: SIWSA

Quantity	Description	Hourly Rate
0.25	Structural Steel Workers (Foreman)	40.15
1	Structural Steel Workers	38.15
Total Hourly Rate for SIWSA		$48.19

Crew: SIWSB

Quantity	Description	Hourly Rate
0.5	Structural Steel Workers (Foreman)	40.15
2	Structural Steel Workers	38.15
Total Hourly Rate for SIWSB		$96.38

Crew: SIWSC

Quantity	Description	Hourly Rate
1	Welder, 200 AMP, with 1 Axle Trailer	8.31
1	Structural Steel Workers (Apprentice)	30.52
2	Structural Steel Workers (Foreman)	40.15
1	Structural Steel Workers	38.15
Total Hourly Rate for SIWSC		$157.28

Crew: SIWSE

Quantity	Description	Hourly Rate
1	Equipment Operator, Medium	34.65
2	Structural Steel Workers	38.15
1	Hydraulic Crane, 4WD, 20T/70' Boom	71.88
0.5	Structural Steel Workers (Foreman)	40.15
Total Hourly Rate for SIWSE		$202.90

Crew: SIWSF

Quantity	Description	Hourly Rate
1	Hydraulic Crane, 4WD, 20T/70' Boom	71.88
1	Equipment Operator, Medium	34.65
1	Structural Steel Workers (Foreman)	40.15
2	Structural Steel Workers	38.15
Total Hourly Rate for SIWSF		$222.97

Crew: SIWSG

Quantity	Description	Hourly Rate
1	Structural Steel Workers (Apprentice)	30.52
1	Structural Steel Workers (Foreman)	40.15
1	Hydraulic Crane, 4WD, 20T/70' Boom	71.88
2	Structural Steel Workers	38.15
1	Equipment Operator, Medium	34.65
Total Hourly Rate for SIWSG		$253.50

Crew: SIWSH

Quantity	Description	Hourly Rate
1	Structural Steel Workers (Foreman)	40.15
1	Equipment Operator, Medium	34.65
1	Hydraulic Crane, 4WD, 20T/70' Boom	71.88
2	Structural Steel Workers (Apprentice)	30.52
2	Structural Steel Workers	38.15
Total Hourly Rate for SIWSH		$284.01

Crew Detail (Cont.)

Crew: SIWSK

Quantity	Description	Hourly Rate
1	Hydraulic Crane, 4WD, 40T/84' Boom	76.60
2	Welder, 200 AMP, with 1 Axle Trailer	8.31
1	Equipment Operator, Crane/Shovel	35.90
1	Structural Steel Workers (Foreman)	40.15
2	Welders, Structural Steel	38.15
2	Equipment Operator, Light	33.05
2	Structural Steel Workers	38.15
2	Structural Steel Workers (Apprentice)	30.52
Total Hourly Rate for SIWSK		$449.02

Crew: SIWSM

Quantity	Description	Hourly Rate
1	Hydraulic Crane, Truck Mounted, 60T /115' Boom	111.93
1	Equipment Operator, Oilers	30.10
1	Structural Steel Workers (Foreman)	40.15
1	Equipment Operator, Crane/Shovel	35.90
3	Structural Steel Workers	38.15
2	Structural Steel Workers (Apprentice)	30.52
Total Hourly Rate for SIWSM		$393.57

Crew: SIWSN

Quantity	Description	Hourly Rate
1	Welder, 200 AMP, with 1 Axle Trailer	8.31
1	Welders, Structural Steel	38.15
1	Hydraulic Crane, Truck Mounted, 60T /115' Boom	111.93
1	Structural Steel Workers (Foreman)	40.15
1	Equipment Operator, Crane/Shovel	35.90
2	Structural Steel Workers	38.15
1	Equipment Operator, Oilers	30.10
2	Structural Steel Workers (Apprentice)	30.52
Total Hourly Rate for SIWSN		$401.88

Crew: SIWSO

Quantity	Description	Hourly Rate
2	Structural Steel Workers (Apprentice)	30.52
2	Welders, Structural Steel	38.15
1	Equipment Operator, Oilers	30.10
2	Welder, 200 AMP, with 1 Axle Trailer	8.31
1	Hydraulic Crane, Truck Mounted, 60T /115' Boom	111.93
1	Structural Steel Workers (Foreman)	40.15
1	Equipment Operator, Light	33.05
3	Structural Steel Workers	38.15
1	Equipment Operator, Crane/Shovel	35.90
Total Hourly Rate for SIWSO		$519.54

Crew: SIWSP

Quantity	Description	Hourly Rate
0.5	Equipment Operator, Crane/Shovel	35.90
0.5	Equipment Operator, Oilers	30.10
6	Structural Steel Workers	38.15
0.5	Hydraulic Crane, Truck Mounted, 60T /115' Boom	111.93
4.5	Structural Steel Workers (Apprentice)	30.52
1.5	Structural Steel Workers (Foreman)	40.15
1	Welder, 200 AMP, with 1 Axle Trailer	8.31
Total Hourly Rate for SIWSP		$523.74

Crew: SIWST

Quantity	Description	Hourly Rate
1	Equipment Operator, Crane/Shovel	35.90
2	Structural Steel Workers (Apprentice)	30.52
1	Equipment Operator, Oilers	30.10
1	Welders, Structural Steel	38.15
1	Welder, 200 AMP, with 1 Axle Trailer	8.31
1	Structural Steel Workers (Foreman)	40.15
1	Crane, Crawler 240T/260' Boom	273.50
4	Structural Steel Workers	38.15
Total Hourly Rate for SIWST		$639.75

Crew: SIWSV

Quantity	Description	Hourly Rate
1	Laborer (Semi-Skilled)	26.70
1	Steam/Pipefitters	40.95
1	Electricians	40.75
1	Laborer (Semi-Skilled - Foreman)	28.70
Total Hourly Rate for SIWSV		$137.10

Crew: SIWSW

Quantity	Description	Hourly Rate
1	Equipment Operator, Oilers	30.10
1	Crane, Crawler 40T/50' Boom	111.03
1	Laborer (Semi-Skilled)	26.70
1	Equipment Operator, Medium	34.65
1	Steam/Pipefitters	40.95
1	Electricians	40.75
2	Laborer (Semi-Skilled - Foreman)	28.70
Total Hourly Rate for SIWSW		$341.57

Crew: SIWWC

Quantity	Description	Hourly Rate
1	Welders, Structural Steel	38.15
1	Welder, 200 AMP, with 1 Axle Trailer	8.31
1	Structural Steel Workers	38.15
1	Equipment Operator, Light	33.05
Total Hourly Rate for SIWWC		$117.66

Crew: UEMCD

Quantity	Description	Hourly Rate
1	Steam/Pipefitters	40.95
0.2	Steam/Pipefitters (Foreman)	42.95
1	Steam/Pipefitters (Apprentice)	32.76
1	Electricians	40.75
Total Hourly Rate for UEMCD		$123.05

Crew: ULABA

Quantity	Description	Hourly Rate
1	Laborer (Semi-Skilled)	26.70
0.25	Laborer (Semi-Skilled - Foreman)	28.70
Total Hourly Rate for ULABA		$33.88

Crew: ULABB

Quantity	Description	Hourly Rate
2	Laborer (Semi-Skilled)	26.70
0.5	Laborer (Semi-Skilled - Foreman)	28.70
Total Hourly Rate for ULABB		$67.75

Crew: ULABC

Quantity	Description	Hourly Rate
3	Laborer (Semi-Skilled)	26.70
0.5	Laborer (Semi-Skilled - Foreman)	28.70
Total Hourly Rate for ULABC		$94.45

Crew: ULABD

Quantity	Description	Hourly Rate
2	Skilled Worker	34.85
1	Laborer (Semi-Skilled)	26.70
0.25	Skilled Worker (Foreman)	36.85
Total Hourly Rate for ULABD		$105.61

Crew: ULABE

Quantity	Description	Hourly Rate
1	Laborer (Semi-Skilled)	26.70
0.25	Laborer (Semi-Skilled - Foreman)	28.70
Total Hourly Rate for ULABE		$33.88

Crew: ULABF

Quantity	Description	Hourly Rate
3	Laborer (Semi-Skilled)	26.70
0.5	Laborer (Semi-Skilled - Foreman)	28.70
1	Hydraulic Crane, 4WD, 20T/70' Boom	71.88
Total Hourly Rate for ULABF		$166.32

Crew: ULABG

Quantity	Description	Hourly Rate
1	Laborer (Semi-Skilled - Foreman)	28.70
2	Laborer (Semi-Skilled)	26.70
Total Hourly Rate for ULABG		$82.10

Crew: ULABH

Quantity	Description	Hourly Rate
4	Laborer (Semi-Skilled)	26.70
1	Laborer (Semi-Skilled - Foreman)	28.70
Total Hourly Rate for ULABH		$135.50

Crew: ULABI

Quantity	Description	Hourly Rate
1	Laborer (Semi-Skilled)	26.70
1	Laborer (Semi-Skilled - Foreman)	28.70
1	Truck Drivers, Light	26.80
1	Truck, Highway, 4x2, 3/4T, 8600 GVW	9.75
Total Hourly Rate for ULABI		$91.95

Crew: ULABJ

Quantity	Description	Hourly Rate
0.4	Truck, Highway, 4x2, 3/4T, 8600 GVW	9.75
1	Laborer (Semi-Skilled - Foreman)	28.70
2	Laborer (Semi-Skilled)	26.70
Total Hourly Rate for ULABJ		$86.00

Crew: ULABK

Quantity	Description	Hourly Rate
0.2	Truck, Highway, 24,500 GVW, 4x2, 2 Axle	21.63
2	Laborer (Semi-Skilled)	26.70
1	Laborer (Semi-Skilled - Foreman)	28.70
0.2	Flatbed, 8'x 16.0' (Add Truck)	22.03
Total Hourly Rate for ULABK		$90.83

Crew: ULABL

Quantity	Description	Hourly Rate
1	Laborer (Semi-Skilled - Foreman)	28.70
1	Truck, Highway, 24,500 GVW, 4x2, 2 Axle	21.63
1	Flatbed, 8'x 16.0' (Add Truck)	22.03
2	Laborer (Semi-Skilled)	26.70
Total Hourly Rate for ULABL		$125.75

Crew: ULABN

Quantity	Description	Hourly Rate
1	Flatbed, 8'x 16.0' (Add Truck)	22.03
1	Hydraulic Crane, 8T/85' Boom	75.33
1	Truck, Highway, 46,000 GVW, 6x4, 3 Axle	42.47
1	Laborer (Semi-Skilled - Foreman)	28.70
1	Equipment Operator, Light	33.05
2	Laborer (Semi-Skilled)	26.70
Total Hourly Rate for ULABN		$254.97

Crew: ULABR

Quantity	Description	Hourly Rate
1	Grout Pump, 1-20 GPM, 100 PSI	13.47
0.5	Laborer (Semi-Skilled - Foreman)	28.70
1	Equipment Operator, Medium	34.65
2	Laborer (Semi-Skilled)	26.70
Total Hourly Rate for ULABR		$115.87

Crew: ULABW8

Quantity	Description	Hourly Rate
1	Drill, Rotary, 16" Water Well, Truck	397.38
1	Equipment Operator, Medium	34.65
1	Trailer, Platform, 40 Ton	31.67
1	Laborer (Semi-Skilled)	26.70
1	Laborer (Semi-Skilled - Foreman)	28.70
Total Hourly Rate for ULABW8		$519.09

Crew: ULADA

Quantity	Description	Hourly Rate
0.5	Laborer (Semi-Skilled - Foreman)	28.70
1	Drill, Rotary, 16" Water Well, Truck	397.38
1	Equipment Operator, Medium	34.65
1	Laborer (Semi-Skilled)	26.70
Total Hourly Rate for ULADA		$473.07

Crew: ULADB

Quantity	Description	Hourly Rate
1	Laborer (Semi-Skilled)	26.70
1	Laborer (Semi-Skilled - Foreman)	28.70
1	Drill, Rotary, 16" Water Well, Truck	397.38
1	Equipment Operator, Medium	34.65
Total Hourly Rate for ULADB		$487.42

Crew: ULADC

Quantity	Description	Hourly Rate
1	Laborer (Semi-Skilled)	26.70
0.26	Equipment Operator, Medium	34.65
0.26	Drill, Rotary, 16" Water Well, Truck	397.38
1	Laborer (Semi-Skilled - Foreman)	28.70
Total Hourly Rate for ULADC		$167.73

Crew: ULADD

Quantity	Description	Hourly Rate
1	Pump, Centrifugal, 6" Dia, 100 GPM/40' Head	27.13
0.5	Equipment Operator, Medium	34.65
1	Laborer (Semi-Skilled - Foreman)	28.70
1	Laborer (Semi-Skilled)	26.70
0.5	Drill, Rotary, 16" Water Well, Truck	397.38
Total Hourly Rate for ULADD		$298.54

Crew: UOEDW

Quantity	Description	Hourly Rate
1	Equipment Operator, Crane/Shovel	35.90
1	Laborer (Semi-Skilled)	26.70
1	Equipment Operator, Oilers	30.10
1	1.5CY Dragline, 80T Crane with 80' Boom	189.88
Total Hourly Rate for UOEDW		$282.57

Crew: UOEHA

Quantity	Description	Hourly Rate
1	Equipment Operator, Crane/Shovel	35.90
0.25	Equipment Operator, Crane/Shovel (Foreman)	37.90
Total Hourly Rate for UOEHA		$45.38

Crew: UOEHB

Quantity	Description	Hourly Rate
2	Equipment Operator, Crane/Shovel	35.90
0.5	Equipment Operator, Crane/Shovel (Foreman)	37.90
Total Hourly Rate for UOEHB		$90.75

Crew: UOEHC

Quantity	Description	Hourly Rate
4	Laborer (Semi-Skilled)	26.70
1	Hydraulic Crane, 4WD, 20T/70' Boom	71.88
1	Laborer (Semi-Skilled - Foreman)	28.70
1	Equipment Operator, Medium	34.65
Total Hourly Rate for UOEHC		$242.02

Crew: UOEHD

Quantity	Description	Hourly Rate
0.25	Hydraulic Crane, 4WD, 40T/84' Boom	76.60
1	Laborer (Semi-Skilled - Foreman)	28.70
0.25	Equipment Operator, Oilers	30.10
4	Laborer (Semi-Skilled)	26.70
0.25	Equipment Operator, Crane/Shovel	35.90
Total Hourly Rate for UOEHD		$171.15

Crew: UOEHE

Quantity	Description	Hourly Rate
1	Equipment Operator, Crane/Shovel	35.90
1	Hydraulic Crane, 4WD, 40T/84' Boom	76.60
1	Equipment Operator, Oilers	30.10
Total Hourly Rate for UOEHE		$142.60

Crew: UOEHH

Quantity	Description	Hourly Rate
1	Equipment Operator, Oilers	30.10
1	Laborer (Semi-Skilled - Foreman)	28.70
4	Laborer (Semi-Skilled)	26.70
1	Equipment Operator, Crane/Shovel	35.90
1	Crane, Crawler 100T/210' Boom	189.88
Total Hourly Rate for UOEHH		$391.38

Crew: UOEHI

Quantity	Description	Hourly Rate
1	Crane, Dragline/Clamshell, 2.5 CY, 50' Boom, (Add Bucket)	146.25
1	Bucket, Clamshell, 2.375 CY, General Purpose/Square Nose	8.53
1	Equipment Operator, Crane/Shovel	35.90
1	Equipment Operator, Oilers	30.10
Total Hourly Rate for UOEHI		$220.78

Crew: UOEHM

Quantity	Description	Hourly Rate
1	Equipment Operator, Crane/Shovel	35.90
1	1 CY Dragline, 50T Crane w/60' Boom	150.75
Total Hourly Rate for UOEHM		$186.65

Crew Detail (Cont.)

Crew: UOEHN

Quantity	Description	Hourly Rate
1	Crane, Dragline, Clamshell, 2 CY, 60' Bucket	150.75
1	Equipment Operator, Oilers	30.10
1	Bucket, Dragline, Medium Weight, 1.5 CY	3.83
1	Equipment Operator, Crane/Shovel	35.90
	Total Hourly Rate for UOEHN	$220.57

Crew: UOEHO

Quantity	Description	Hourly Rate
1	Bucket, Dragline, Medium Weight, 2.5 CY	4.30
1	Crane, Dragline/Clamshell, 2.5 CY, 50' Boom, (Add Bucket)	146.25
1	Equipment Operator, Oilers	30.10
1	Equipment Operator, Crane/Shovel	35.90
	Total Hourly Rate for UOEHO	$216.55

Crew: UOEHQ

Quantity	Description	Hourly Rate
1	Crane, Dragline/Clamshell, 1.0 CY, 40' Boom, (Add Bucket)	104.15
1	Bucket, Clamshell, 1.125 CY, General Purpose, Square Nose	5.42
1	Equipment Operator, Crane/Shovel	35.90
1	Equipment Operator, Oilers	30.10
	Total Hourly Rate for UOEHQ	$175.57

Crew: UOEHV

Quantity	Description	Hourly Rate
1	Crane, Dragline/Clamshell, 2 CY, 60' Boom, (Add Bucket)	150.75
1	Equipment Operator, Crane/Shovel	35.90
1	Bucket, Clamshell, 1.5 CY, General Purpose/Square Nose	7.67
1	Equipment Operator, Oilers	30.10
	Total Hourly Rate for UOEHV	$224.43

Crew: UOEHW

Quantity	Description	Hourly Rate
1	Crane, Dragline, Clamshell, 2 CY, 60' Bucket	150.75
1	Bucket, Dragline, Medium Weight, 1.5 CY	3.83
1	Equipment Operator, Oilers	30.10
1	Equipment Operator, Crane/Shovel	35.90
1	Laborer (Semi-Skilled - Foreman)	28.70
	Total Hourly Rate for UOEHW	$249.27

Crew: UOEHX

Quantity	Description	Hourly Rate
1	Bucket, Dragline, Medium Weight, 1.0 CY	2.55
1	Equipment Operator, Oilers	30.10
1	Equipment Operator, Crane/Shovel	35.90
1	Crane, Dragline/Clamshell, 1.0 CY, 40' Boom, (Add Bucket)	104.15
	Total Hourly Rate for UOEHX	$172.70

Crew: UOEHY

Quantity	Description	Hourly Rate
1	Crane, Dragline/Clamshell, 4.5 CY, 95' Boom, (Add Bucket)	189.88
1	Bucket, Dragline, Medium Weight, 4.0 CY	5.92
1	Equipment Operator, Crane/Shovel	35.90
1	Equipment Operator, Oilers	30.10
Total Hourly Rate for UOEHY		$261.80

Crew: UOEHZ

Quantity	Description	Hourly Rate
1	Air Compressor, 600 CFM, 100 PSI	37.23
1	Laborer (Semi-Skilled)	26.70
3	Equipment Operator, Light	33.05
1	Crane, Crawler 25T/50' Boom	70.92
Total Hourly Rate for UOEHZ		$234.00

Crew: USKBK

Quantity	Description	Hourly Rate
1	36" Barricade With Flashing Light	0.21
0.05	Laborer (Semi-Skilled)	26.70
Total Hourly Rate for USKBK		$1.55

Crew: USKCA

Quantity	Description	Hourly Rate
0.25	Skilled Worker (Foreman)	36.85
1	Skilled Worker	34.85
Total Hourly Rate for USKCA		$44.06

Crew: USKCB

Quantity	Description	Hourly Rate
0.5	Skilled Worker (Foreman)	36.85
2	Skilled Worker	34.85
Total Hourly Rate for USKCB		$88.13

Crew: USKCD

Quantity	Description	Hourly Rate
0.5	Skilled Worker (Foreman)	36.85
2	Skilled Worker	34.85
Total Hourly Rate for USKCD		$88.13

Crew: USKCE

Quantity	Description	Hourly Rate
2	Flatbed, 8'x 16.0' (Add Truck)	22.03
12	Laborer (Semi-Skilled)	26.70
1	Skilled Worker (Foreman)	36.85
2	Truck, Highway, 24,500 GVW, 4x2, 2 Axle	21.63
2	Skilled Worker	34.85
Total Hourly Rate for USKCE		$514.26

Crew: USKCF

Quantity	Description	Hourly Rate
1	Truck, Highway, 24,500 GVW, 4x2, 2 Axle	21.63
1	Skilled Worker	34.85
1	Skilled Worker (Foreman)	36.85
6	Laborer (Semi-Skilled)	26.70
1	Flatbed, 8'x 16.0' (Add Truck)	22.03
Total Hourly Rate for USKCF		$275.55

Crew Detail (Cont.)

Crew: USKCG

Quantity	Description	Hourly Rate
0.5	Hydraulic Crane, 4WD, 20T/70' Boom	71.88
2	Skilled Worker	34.85
1	Laborer (Semi-Skilled)	26.70
0.5	Outside Equipment Operator, Light	33.05
0.25	Skilled Worker (Foreman)	36.85
Total Hourly Rate for USKCG		$158.08

Crew: USKCI

Quantity	Description	Hourly Rate
1	Rotary Hammer, Electric Hand Held	2.23
Total Hourly Rate for USKCI		$2.23

Crew: XASPA

Quantity	Description	Hourly Rate
1	Asphalt Finisher, 10' with Screed, Pneumatic	198.75
1	Truck, Highway, 24,500 GVW, 4x2, 2 Axle	21.63
7	Outside Laborer (Semi-Skilled)	26.09
1	Truck, Highway, 24,500 GVW, 4x2, 2 Axle	21.63
2	Roller, Static, 3 Wheeled, 12T, 84" Wide	30.23
1	Street Sweeper, 7' Wide with Sprinkler, Towed	5.41
3	Outside Equipment Operator, Medium	34.65
1	Outside Truck Driver, Heavy	27.55
1	Flatbed, 8'x 16.0' (Add Truck)	22.03
1	Outside Equipment Operator, Medium (Foreman)	36.65
1	Roller, Static, 13T, 84" Wide, 11 Tire	30.23
Total Hourly Rate for XASPA		$710.91

Crew: XASPB

Quantity	Description	Hourly Rate
1	Dozer, Crawler, D-4H, Powershift, (Add Blade)	39.35
1	Roller, Self Propelled, 12T, 3 Wheel, 3" Overlap	30.23
1	Blade, Angle, Hydraulic, for D4	3.34
0.5	Laborer (Semi-Skilled)	26.70
2	Equipment Operator, Medium	34.65
1	Loader, Front End, Wheeled, 3.0 CY, Articulated, 950E	38.13
1	Grader, Motor, 14-G, Articulated	117.25
0.5	Truck, Highway, 4x4, 3/4T, 8800 GVW	10.65
1	Equipment Operator, Light	33.05
Total Hourly Rate for XASPB		$349.32

Crew: XASPC

Quantity	Description	Hourly Rate
0.7	Laborer (Semi-Skilled)	26.70
2.73	Laborer (Semi-Skilled)	26.70
0.7	Truck, Highway, 2 Axle, 24,000 GVW	21.63
0.91	Equipment Operator, Medium	34.65
0.7	Truck, Highway, 4x4, 3/4T, 8800 GVW	10.65
0.7	Truck Drivers, Heavy	27.55
0.91	Concrete Paver, 25'W, Paver, Crawler	238.75
2.73	Outside Cement Finisher	32.85
0.7	Truck, Flatbed, 8' x 16.0'	22.03
Total Hourly Rate for XASPC		$487.35

Crew: XASPD

Quantity	Description	Hourly Rate
0.7	Truck, Highway, 2 Axle, 24,000 GVW	21.63
0.7	Laborer (Semi-Skilled)	26.70
0.7	Truck Drivers, Heavy	27.55
0.7	Truck, Flatbed, 8' x 16.0'	22.03
2.73	Laborer (Semi-Skilled)	26.70
0.5	Outside Carpenter	34.25
0.7	Truck, Highway, 4x4, 3/4T, 8800 GVW	10.65
0.5	Equipment Operator, Crane/Shovel	35.90
0.5	Hydraulic Crane, 4WD, 9T, Rough Terrain	39.63
2	Skilled Worker	34.85
Total Hourly Rate for XASPD		$273.47

Crew: XASPE

Quantity	Description	Hourly Rate
1.8	Equipment Operator, Medium	34.65
0.75	Truck, Flatbed, 8' x 16.0'	22.03
0.75	Laborer (Semi-Skilled - Foreman)	28.70
0.75	Truck, Highway, 2 Axle, 24,000 GVW	21.63
3.5	Outside Cement Finisher	32.85
0.75	Truck, Highway, 4x4, 3/4T, 8800 GVW	10.65
3.66	Laborer (Semi-Skilled)	26.70
0.75	Truck Drivers, Heavy	27.55
Total Hourly Rate for XASPE		$357.98

Crew: XASPF

Quantity	Description	Hourly Rate
0.68	Hydraulic Crane, 4WD, 9T, Rough Terrain	39.63
0.75	Laborer (Semi-Skilled - Foreman)	28.70
0.75	Truck, Highway, 4x4, 3/4T, 8800 GVW	10.65
0.75	Truck, Flatbed, 8' x 16.0'	22.03
0.68	Equipment Operator, Crane/Shovel	35.90
2.7	Skilled Worker	34.85
1.8	Concrete Mixer, 6 CF, Portable	12.88
0.75	Truck, Highway, 2 Axle, 24,000 GVW	21.63
0.75	Truck Drivers, Heavy	27.55
2.44	Laborer (Semi-Skilled)	26.70
Total Hourly Rate for XASPF		$316.69

Crew: XCBGA

Quantity	Description	Hourly Rate
1	Outside Cement Finisher	32.85
1	Outside Carpenter	34.25
1	Outside Carpenter (Foreman)	36.25
1	Outside Laborer (Semi-Skilled)	26.09
Total Hourly Rate for XCBGA		$129.44

Crew: XCMAD

Quantity	Description	Hourly Rate
1	Outside Cement Finisher	32.85
1	Outside Cement Finisher (Apprentice)	26.28
1	Concrete Vibrator, 2.50D, High Frequency	1.17
1	Outside Laborer (Semi-Skilled)	26.09
0.17	Outside Cement Finisher (Foreman)	34.85
Total Hourly Rate for XCMAD		$92.32

Crew: XCMAF

Quantity	Description	Hourly Rate
0.82	Equipment Operator, Light	33.05
1.22	Laborer (Semi-Skilled)	26.70
1.74	Outside Cement Finisher	32.85
Total Hourly Rate for XCMAF		$116.83

Crew: XLABA

Quantity	Description	Hourly Rate
1	Outside Laborer (Semi-Skilled)	26.09
Total Hourly Rate for XLABA		$26.09

Crew: XLABB

Quantity	Description	Hourly Rate
2	Outside Laborer (Semi-Skilled)	26.09
0.34	Outside Laborer (Semi-Skilled - Foreman)	28.32
Total Hourly Rate for XLABB		$61.81

Crew: XLABC

Quantity	Description	Hourly Rate
0.34	Outside Laborer (Semi-Skilled - Foreman)	28.32
3	Outside Laborer (Semi-Skilled)	26.09
Total Hourly Rate for XLABC		$87.90

Crew: XLABK

Quantity	Description	Hourly Rate
1	Outside Equipment Operator, Medium	34.65
1	Loader, Backhoe, Wheeled, 1.0 CY Front End Bucket, 24" Dip	26.98
Total Hourly Rate for XLABK		$61.63

Crew: XPLUB

Quantity	Description	Hourly Rate
0.17	Outside Plumber (Foreman)	42.85
1	Outside Laborer (Semi-Skilled)	26.09
1	Outside Plumber	40.85
Total Hourly Rate for XPLUB		$74.23

Crew: XPLUC

Quantity	Description	Hourly Rate
0.17	Outside Plumber (Foreman)	42.85
2	Outside Laborer (Semi-Skilled)	26.09
1	Outside Plumber	40.85
Total Hourly Rate for XPLUC		$100.32

Crew: XPLUD

Quantity	Description	Hourly Rate
1	Outside Plumber (Apprentice)	32.68
1	Outside Plumber	40.85
0.17	Outside Plumber (Foreman)	42.85
1	Outside Laborer (Semi-Skilled)	26.09
Total Hourly Rate for XPLUD		$106.91

Crew: XSABA

Quantity	Description	Hourly Rate
1	Outside Equipment Operator, Medium (Foreman)	36.65
1	Truck, Water, Off-Highway, 6000 Gallon	31.21
1	Chip Spreader, 10' W, Mechanical	24.23
2	Outside Equipment Operator, Medium	34.65
1	Blade, Straight, Hydraulic, for D8	8.96
1	Loader, Front End, Crawler, 1.5 CY, 943	46.10
2	Outside Truck Driver, Light	26.80
1	Grader, Motor, 12-G, Articulated	85.25
1	Dozer, Crawler, D-8L, (Add Blade)	149.38
3	Outside Laborer (Semi-Skilled)	26.09
Total Hourly Rate for XSABA		$582.95

Crew: XSABB

Quantity	Description	Hourly Rate
20	Truck, Off-Highway, 6x6, Rear Dump, 23T,13-17 CY	98.13
1	Equipment Operator, Crane/Shovel	35.90
1	Loader, Front End, Crawler, 3.75 CY, 973	136.75
20	Truck Drivers, Heavy	27.55
1	Hydraulic Excavator, Crawler, 3.4 CY Bucket	252.50
1	Equipment Operator, Medium	34.65
Total Hourly Rate for XSABB		$2,973.31

Crew: XSABC

Quantity	Description	Hourly Rate
1	Loader, Front End, Wheeled, 3.0 CY, Articulated, 950E	38.13
1	Outside Equipment Operator, Light	33.05
Total Hourly Rate for XSABC		$71.17

Crew: XSABD

Quantity	Description	Hourly Rate
3	Outside Equipment Operator, Medium	34.65
1	Blade, Angle, Hydraulic, for D6	5.38
1	Dozer, Crawler, D-6H, Powershift, (Add Blade)	71.50
1	Outside Equipment Operator, Oiler	30.10
1	Loader, Front End, Crawler, 1.5 CY, 943	46.10
1	Outside Equipment Operator, Oiler (Foreman)	32.10
1	Roller, Self Propelled, 12T, 3 Wheel, 3" Overlap	30.23
Total Hourly Rate for XSABD		$319.36

Crew: XTRHC

Quantity	Description	Hourly Rate
1	Truck, Highway, 46,000 GVW, 6x4, 3 Axle	42.47
1	Outside Truck Driver, Heavy	27.55
1	Water Tank, 3000 Gallon (Add Truck)	11.55
Total Hourly Rate for XTRHC		$81.57

Crew: XTRLC

Quantity	Description	Hourly Rate
1	Truck, Highway, 24,500 GVW, 4x2, 2 Axle	21.63
1	Outside Truck Driver, Light	26.80
1	Flatbed, 8'x 16.0' (Add Truck)	22.03
Total Hourly Rate for XTRLC		$70.45

Crew: XXBBW

Quantity	Description	Hourly Rate
0.5	Equipment Operator, Light	33.05
2.5	Laborer (Semi-Skilled)	26.70
Total Hourly Rate for XXBBW		$83.28

Crew: XXBBX

Quantity	Description	Hourly Rate
1	Truck, Flatbed, 8' x 9.0'	16.10
1	Equipment Operator, Light	33.05
1	Laborer (Semi-Skilled)	26.70
1	Truck, Highway, 2 Axle, 24,000 GVW	21.63
Total Hourly Rate for XXBBX		$97.48

Crew: XXBBZ

Quantity	Description	Hourly Rate
2	Laborer (Semi-Skilled)	26.70
Total Hourly Rate for XXBBZ		$53.40

Crew: XXPLA

Quantity	Description	Hourly Rate
1	Outside Equipment Operator, Medium	34.65
0.5	Outside Plumber (Foreman)	42.85
1	Outside Plumber	40.85
1	Hydraulic Crane, 4WD, 20T/70' Boom	71.88
3	Outside Laborer (Semi-Skilled)	26.09
Total Hourly Rate for XXPLA		$247.08

Crew: XXPLB

Quantity	Description	Hourly Rate
1	Outside Plumber (Apprentice)	32.68
1	Hydraulic Crane, 4WD, 20T/70' Boom	71.88
1	Outside Equipment Operator, Medium	34.65
4	Outside Laborer (Semi-Skilled)	26.09
0.5	Outside Plumber (Foreman)	42.85
1	Outside Plumber	40.85
Total Hourly Rate for XXPLB		$305.85

Crew: XXPLD

Quantity	Description	Hourly Rate
1	Outside Plumber (Foreman)	42.85
1	Air Hose, 2.00", 100'	2.80
1	Trailer, Lowboy, 75T, 3 Axle (Add Truck)	21.28
1	Outside Equipment Operator, Oiler	30.10
1	Outside Equipment Operator, Heavy	35.90
1	Air Compressor, 450 CFM, 100 PSI	29.97
2	Welder, 300 AMP, Skid, Electric Drive	10.08
6	Outside Plumber	40.85
1	Crane, Crawler 100T/210' Boom	189.88
1	Outside Equipment Operator, Oiler (Foreman)	32.10
1	Outside Truck Driver, Heavy	27.55
1	Truck, Highway, 4x2, 3/4T, 8600 GVW	9.75
1	Truck, Highway, 50,000 GVW, 6x4, 3 Axle	43.26
1	Outside Plumber (Apprentice)	32.68

Total Hourly Rate for XXPLD $763.36

Crew: XXPLE

Quantity	Description	Hourly Rate
1	Trailer, Lowboy, 75T, 3 Axle (Add Truck)	21.28
1	Outside Truck Driver, Heavy	27.55
1	Outside Equipment Operator, Oiler (Foreman)	32.10
1	Air Compressor, 450 CFM, 100 PSI	29.97
1	Truck, Highway, 4x2, 3/4T, 8600 GVW	9.75
7	Outside Plumber	40.85
1	Air Hose, 2.00", 100'	2.80
3	Welder, 300 AMP, Skid, Electric Drive	10.08
1	Outside Plumber (Apprentice)	32.68
1	Truck, Highway, 50,000 GVW, 6x4, 3 Axle	43.26
1	Outside Plumber (Foreman)	42.85
1	Outside Equipment Operator, Heavy	35.90
1	Crane, Crawler 100T/210' Boom	189.88
1	Outside Equipment Operator, Oiler	30.10

Total Hourly Rate for XXPLE $814.30

Crew: XXPLF

Quantity	Description	Hourly Rate
1	Outside Equipment Operator, Oiler (Foreman)	32.10
1	Truck, Highway, 50,000 GVW, 6x4, 3 Axle	43.26
1	Outside Equipment Operator, Heavy	35.90
1	Crane, Crawler 100T/210' Boom	189.88
1	Outside Plumber (Apprentice)	32.68
4	Welder, 300 AMP, Skid, Electric Drive	10.08
1	Outside Truck Driver, Heavy	27.55
8	Outside Plumber	40.85
1	Air Compressor, 450 CFM, 100 PSI	29.97
1	Outside Plumber (Foreman)	42.85
1	Truck, Highway, 4x2, 3/4T, 8600 GVW	9.75
1	Trailer, Lowboy, 75T, 3 Axle (Add Truck)	21.28
1	Outside Equipment Operator, Oiler	30.10
Total Hourly Rate for XXPLF		$862.43

Crew: XXPLZ

Quantity	Description	Hourly Rate
0.5	Laborer (Semi-Skilled - Foreman)	28.70
1.5	Laborer (Semi-Skilled)	26.70
Total Hourly Rate for XXPLZ		$54.40

Crew: XXQEW

Quantity	Description	Hourly Rate
1	Outside Equipment Operator, Light	33.05
1	Forklift 6000 lb., Rough Terrain	29.13
Total Hourly Rate for XXQEW		$62.17

Crew: XXQHK

Quantity	Description	Hourly Rate
1	Outside Laborer (Semi-Skilled)	26.09
1	Hydraulic Excavator, Crawler, 3.125 CY Bucket	192.88
1	Outside Equipment Operator, Heavy	35.90
1	Outside Equipment Operator, Oiler	30.10
1	Outside Equipment Operator, Oiler (Foreman)	32.10
Total Hourly Rate for XXQHK		$317.08

Crew: XXQHL

Quantity	Description	Hourly Rate
1	Hydraulic Excavator, Crawler, 4.50 CY Bucket	408.61
1	Outside Equipment Operator, Heavy	35.90
1	Outside Laborer (Semi-Skilled)	26.09
1	Outside Equipment Operator, Oiler	30.10
1	Outside Equipment Operator, Oiler (Foreman)	32.10
Total Hourly Rate for XXQHL		$532.80

Crew: XXQHO

Quantity	Description	Hourly Rate
1	Equipment Operator, Crane/Shovel	35.90
1	Hydraulic Excavator, Crawler, 3.125 CY Bucket	192.88
Total Hourly Rate for XXQHO		$228.78

Crew: XXQHP

Quantity	Description	Hourly Rate
1	Hydraulic Excavator, Crawler, 5.5 CY	479.57
1	Equipment Operator, Crane/Shovel	35.90
Total Hourly Rate for XXQHP		**$515.47**

Crew: XXQHQ

Quantity	Description	Hourly Rate
1	Equipment Operator, Light	33.05
1	Loader, Front End, Wheeled, 3.0 CY, Articulated, 950E	38.13
Total Hourly Rate for XXQHQ		**$71.17**

Crew: XXQHR

Quantity	Description	Hourly Rate
1	Loader, Front End, Wheeled, 5.25 CY, Articulated, 980C	87.55
1	Equipment Operator, Medium	34.65
Total Hourly Rate for XXQHR		**$122.20**

Crew: XXQLK

Quantity	Description	Hourly Rate
1	Blade, Power Angle Tilt, for D3	2.47
1	Equipment Operator, Light	33.05
1	Dozer, Crawler, D-3C, Powershift, (Add Blade)	39.35
Total Hourly Rate for XXQLK		**$74.87**

Crew: XXQMT

Quantity	Description	Hourly Rate
1	Hydraulic Excavator, Crawler, 1.00 CY Bucket, Long Carriage	69.73
1	Outside Laborer (Semi-Skilled)	26.09
1	Outside Equipment Operator, Medium	34.65
Total Hourly Rate for XXQMT		**$130.47**

Crew: XXQMX

Quantity	Description	Hourly Rate
1	Blade, Angle, Hydraulic, for D6	5.38
1	Dozer, Crawler, D-6D, Powershift, (Add Blade)	71.50
1	Equipment Operator, Medium	34.65
Total Hourly Rate for XXQMX		**$111.53**

Crew: XXYBH

Quantity	Description	Hourly Rate
1	Outside Equipment Operator, Heavy	35.90
0.5	Outside Equipment Operator, Oiler (Foreman)	32.10
1	Crane With 1 CY Backhoe (3035) w/4.00/Hr Backhoe	62.67
0.5	Outside Equipment Operator, Oiler	30.10
Total Hourly Rate for XXYBH		**$129.67**

Crew: XXYRP

Quantity	Description	Hourly Rate
1	Dozer D-60 Powershift 140 HP w/Multi-Shk Beam & 4 Rips	71.50
2	Outside Laborer (Semi-Skilled)	26.09
2	Outside Equipment Operator, Medium	34.65
	Total Hourly Rate for XXYRP	$192.98

Crew: XYABC

Quantity	Description	Hourly Rate
1	Survey Rodman	33.00
1	Survey Party Chief	42.90
0.25	Field Engineer, Average	24.88
1	Truck, Suburban 4x4 (9735)	10.65
1	Survey Equipment	16.57
	Total Hourly Rate for XYABC	$109.34

Crew: XYABD

Quantity	Description	Hourly Rate
1	Survey Party Chief	42.90
2	Survey Rodman	33.00
0.25	Field Engineer, Average	24.88
1	Survey Equipment	16.57
1	Truck, Suburban 4x4 (9735)	10.65
	Total Hourly Rate for XYABD	$142.34

Crew: XYABE

Quantity	Description	Hourly Rate
0.25	Field Engineer, Average	24.88
1	Truck, Suburban 4x4 (9735)	10.65
3	Survey Rodman	33.00
1	Survey Equipment	16.57
1	Survey Party Chief	42.90
	Total Hourly Rate for XYABE	$175.34

Crew: XYBBA

Quantity	Description	Hourly Rate
1	Field Office Clerk	8.00
	Total Hourly Rate for XYBBA	$8.00

Crew: XYBBC

Quantity	Description	Hourly Rate
1	Field Engineer, Minimum	19.13
	Total Hourly Rate for XYBBC	$19.13

Crew: XYBBD

Quantity	Description	Hourly Rate
1	Field Engineer, Average	24.88
	Total Hourly Rate for XYBBD	$24.88

Crew: XYBBE

Quantity	Description	Hourly Rate
1	Field Engineer, Maximum	28.75
	Total Hourly Rate for XYBBE	$28.75

Crew: XYBBF

Quantity	Description	Hourly Rate
1	General Purpose Laborer	26.70
	Total Hourly Rate for XYBBF	$26.70

Crew: XYBBG

Quantity	Description	Hourly Rate
1	Project Manager, Minimum	35.61
Total Hourly Rate for XYBBG		**$35.61**

Crew: XYBBH

Quantity	Description	Hourly Rate
1	Project Manager, Average	39.89
Total Hourly Rate for XYBBH		**$39.89**

Crew: XYBBI

Quantity	Description	Hourly Rate
1	Project Manager, Maximum	45.06
Total Hourly Rate for XYBBI		**$45.06**

Crew: XYBBJ

Quantity	Description	Hourly Rate
1	Field Superintendent, Minimum	34.38
Total Hourly Rate for XYBBJ		**$34.38**

Crew: XYBBK

Quantity	Description	Hourly Rate
1	Field Superintendent, Average	37.50
Total Hourly Rate for XYBBK		**$37.50**

Crew: XYBBL

Quantity	Description	Hourly Rate
1	Field Superintendent, Maximum	43.13
Total Hourly Rate for XYBBL		**$43.13**

Crew: XYBBM

Quantity	Description	Hourly Rate
1	Timekeeper, Average	22.25
Total Hourly Rate for XYBBM		**$22.25**

Crew: XYBBN

Quantity	Description	Hourly Rate
1	Watchman Uniformed, Minimum	8.90
Total Hourly Rate for XYBBN		**$8.90**

Crew: XYBBO

Quantity	Description	Hourly Rate
1	Watchman Uniformed, Maximum	16.13
Total Hourly Rate for XYBBO		**$16.13**

Crew: XYBBP

Quantity	Description	Hourly Rate
1	Master Mechanic	36.50
Total Hourly Rate for XYBBP		**$36.50**

Crew: XYBBQ

Quantity	Description	Hourly Rate
1	Safety Engineer, Minimum	38.62
Total Hourly Rate for XYBBQ		**$38.62**

Crew: XYBBR

Quantity	Description	Hourly Rate
1	Safety Engineer, Average	38.62
Total Hourly Rate for XYBBR		**$38.62**

Crew: XYBBS

Quantity	Description	Hourly Rate
1	Safety Engineer, Maximum	38.62
Total Hourly Rate for XYBBS		**$38.62**

Crew: XYBBT

Quantity	Description	Hourly Rate
1	Quality Control, Minimum	30.60
Total Hourly Rate for XYBBT		**$30.60**

Crew: XYBBU

Quantity	Description	Hourly Rate
1	Quality Control, Average	35.05
Total Hourly Rate for XYBBU		**$35.05**

Crew: XYBBV

Quantity	Description	Hourly Rate
1	Quality Control, Maximum	32.27
Total Hourly Rate for XYBBV		**$32.27**

Crew: XYBBW

Quantity	Description	Hourly Rate
2.5	Laborer (Semi-Skilled)	26.70
0.5	Equipment Operator, Light	33.05
Total Hourly Rate for XYBBW		**$83.28**

Crew: XYBBX

Quantity	Description	Hourly Rate
1	Laborer (Semi-Skilled)	26.70
1	Truck, Flatbed, 8' x 9.0'	16.10
1	Truck, Highway, 2 Axle, 24,000 GVW	21.63
1	Equipment Operator, Light	33.05
Total Hourly Rate for XYBBX		**$97.48**

Crew: XYBBZ

Quantity	Description	Hourly Rate
2	Laborer (Semi-Skilled)	26.70
Total Hourly Rate for XYBBZ		**$53.40**

Crew: XYCBA

Quantity	Description	Hourly Rate
0.2	Field Engineer, Average	24.88
1	Laboratory Equipment Materials Testing	27.05
1	Lab. Technician Testing	13.47
Total Hourly Rate for XYCBA		**$45.50**

Crew: XYCBB

Quantity	Description	Hourly Rate
0.5	Field Engineer, Average	24.88
1	Laboratory Equipment Materials Testing	27.05
1	Lab. Technician Testing	13.47
Total Hourly Rate for XYCBB		**$52.96**

Crew: XYCBC

Quantity	Description	Hourly Rate
2	Air Hose, 1.00", 100'	0.91
1	Paving Breaker, 86 LB (Add Compressor)	1.23
1	Air Compressor, 375 CFM, 100 PSI	18.35
Total Hourly Rate for XYCBC		**$21.40**

Crew: XYCBD

Quantity	Description	Hourly Rate
1	Drill, Core, Skid Mounted, NC 1000' Max. NX, 1000 Max. Drill Depth	75.58
Total Hourly Rate for XYCBD		**$75.58**

Crew Detail (Cont.)

Crew: XYCDA

Quantity	Description	Hourly Rate
1	Dozer, Crawler, D-3C, Powershift, (Add Blade)	39.35
1	Blade, Power Angle Tilt, for D3	2.47
Total Hourly Rate for XYCDA		$41.82

Crew: XYCDB

Quantity	Description	Hourly Rate
1	Dozer, Crawler, D-4H, Powershift, (Add Blade)	39.35
1	Blade, Angle, Hydraulic, for D4	3.34
Total Hourly Rate for XYCDB		$42.69

Crew: XYCDC

Quantity	Description	Hourly Rate
1	Blade, Angle, Hydraulic, for D5	3.71
1	Dozer, Crawler, D-5H, Powershift, (Add Blade)	56.73
Total Hourly Rate for XYCDC		$60.44

Crew: XYCDD

Quantity	Description	Hourly Rate
1	Blade, Angle, Hydraulic, for D6	5.38
1	Dozer, Crawler, D-6D, Powershift, (Add Blade)	71.50
Total Hourly Rate for XYCDD		$76.88

Crew: XYCDE

Quantity	Description	Hourly Rate
1	Blade, Universal, Hydraulic, for D7	7.39
1	Dozer, Crawler, D-7G, Powershift, (Add Blade)	114.95
Total Hourly Rate for XYCDE		$122.34

Crew: XYCDF

Quantity	Description	Hourly Rate
1	Dozer, Crawler, D-8L, Powershift, (Add Blade)	149.38
1	Ripper, Single Shank Beam, Hydraulic, D-8	12.19
1	Blade, Universal, Hydraulic, for D8	9.78
1	Ripper, Shanks Each, D-8	0.80
Total Hourly Rate for XYCDF		$172.15

Crew: XYCDG

Quantity	Description	Hourly Rate
1	Dozer, Crawler, D-9N, Powershift, (Add Blade)	190.50
1	Ripper, Shanks Each, D-9	0.86
1	Blade, Universal, Hydraulic, for D9	12.22
1	Ripper, Multishank Beam, Hydraulic, D-9	13.81
Total Hourly Rate for XYCDG		$217.39

Crew: XYCDH

Quantity	Description	Hourly Rate
1	Ripper, Single-Shank Beam, D-10	19.36
1	Blade, Universal, Hydraulic, for D10	17.42
1	Dozer, Crawler, D-10N, Powershift, (Add Blade)	190.50
1	Ripper, Shanks Each, D-10	1.36
Total Hourly Rate for XYCDH		$228.64

Crew: XYCDI

Quantity	Description	Hourly Rate
1	Dozer, Crawler, D-4H, Powershift, (Add Blade)	39.35
1	Tiller, 5x7 V-Shaped, Deep	4.89
Total Hourly Rate for XYCDI		$44.24

Crew: XYCDJ

Quantity	Description	Hourly Rate
1	Tiller, 5x7 V-Shaped, Deep	4.89
1	Dozer, Crawler, D-7G, Powershift, (Add Blade)	114.95
Total Hourly Rate for XYCDJ		$119.84

Crew: XYCDK

Quantity	Description	Hourly Rate
1	Dozer, Crawler, D-9N, Powershift, (Add Blade)	190.50
1	Tiller, 5x7 V-Shaped, Deep	4.89
Total Hourly Rate for XYCDK		$195.39

Crew: XYCDL

Quantity	Description	Hourly Rate
1	Dozer, Crawler, D-3C, Powershift, (Add Blade)	39.35
1	Tiller, 5x7 V-Shaped, Deep	4.89
Total Hourly Rate for XYCDL		$44.24

Crew: XYCEA

Quantity	Description	Hourly Rate
1	Hydraulic Excavator, Crawler, 1.00 CY Bucket, Long Carriage	69.73
Total Hourly Rate for XYCEA		$69.73

Crew: XYCEB

Quantity	Description	Hourly Rate
1	Hydraulic Excavator, Crawler, 1.25 CY Bucket	73.77
Total Hourly Rate for XYCEB		$73.77

Crew: XYCEC

Quantity	Description	Hourly Rate
1	Hydraulic Excavator, Crawler, 2.00 CY Bucket	113.73
Total Hourly Rate for XYCEC		$113.73

Crew: XYCED

Quantity	Description	Hourly Rate
1	Hydraulic Excavator, Crawler, 3.125 CY Bucket	192.88
Total Hourly Rate for XYCED		$192.88

Crew: XYCEE

Quantity	Description	Hourly Rate
1	Hydraulic Excavator, Crawler, 3.125 CY Bucket	192.88
Total Hourly Rate for XYCEE		$192.88

Crew: XYCEF

Quantity	Description	Hourly Rate
1	Hydraulic Excavator, Crawler, 5.5 CY	479.57
Total Hourly Rate for XYCEF		$479.57

Crew: XYCSA

Quantity	Description	Hourly Rate
0.25	Blade, Universal, Hydraulic, for D8	9.78
1	Scraper, Self Propelled, 14-20 CY, 24T, Powershift	194.00
0.25	Dozer, Crawler, D-8L, Powershift, (Add Blade)	149.38
Total Hourly Rate for XYCSA		$233.79

Crew: XYCSB

Quantity	Description	Hourly Rate
0.5	Blade, Universal, Hydraulic, for D9	12.22
0.5	Dozer, Crawler, D-9N, Powershift, (Add Blade)	190.50
1	Scraper, Self Propelled, 21-31 CY, 37.5T, Powershift	289.63
Total Hourly Rate for XYCSB		$390.99

Crew: XYCSC

Quantity	Description	Hourly Rate
1	Scraper, Self Propelled, 32-44 CY, 52T, Powershift	319.99
0.5	Dozer, Crawler, D-9N, Powershift, (Add Blade)	190.50
0.5	Blade, Universal, Hydraulic, for D9	12.22
Total Hourly Rate for XYCSC		$421.35

Crew: XYCSD

Quantity	Description	Hourly Rate
1	Scraper, 11 CY, 13T, Powershift	102.43
Total Hourly Rate for XYCSD		$102.43

Crew: XYCSE

Quantity	Description	Hourly Rate
1	Scraper, 22 CY, 25T, Powershift	206.00
Total Hourly Rate for XYCSE		$206.00

Crew: XYCSF

Quantity	Description	Hourly Rate
1	Scraper, Self Propelled, 14-20 CY, 24T, Powershift	194.00
0.25	Dozer, Crawler, D-8L, Powershift, (Add Blade)	149.38
0.25	Blade, Universal, Hydraulic, for D8	9.78
Total Hourly Rate for XYCSF		$233.79

Crew: XYCSG

Quantity	Description	Hourly Rate
1	Scraper, Self Propelled, 21-31 CY, 37.5T, Powershift	289.63
0.25	Dozer, Crawler, D-9N, Powershift, (Add Blade)	190.50
0.25	Blade, Universal, Hydraulic, for D9	12.22
Total Hourly Rate for XYCSG		$340.31

Crew: XYCSH

Quantity	Description	Hourly Rate
0.5	Dozer, Crawler, D-9N, Powershift, (Add Blade)	190.50
0.5	Blade, Universal, Hydraulic, for D9	12.22
1	Scraper, Self Propelled, 32-44 CY, 52T, Powershift	319.99
Total Hourly Rate for XYCSH		$421.35

Crew: XYCTA

Quantity	Description	Hourly Rate
1	Loader, Front End, Crawler, 1.0 CY, 931C	35.71
Total Hourly Rate for XYCTA		$35.71

Crew: XYCTB

Quantity	Description	Hourly Rate
1	Loader, Front End, Crawler, 1.5 CY, 943	46.10
Total Hourly Rate for XYCTB		$46.10

Crew: XYCTC

Quantity	Description	Hourly Rate
1	Loader, Front End, Crawler, 2.0 CY, 953	68.98
Total Hourly Rate for XYCTC		$68.98

Crew: XYCTD

Quantity	Description	Hourly Rate
1	Loader, Front End, Crawler, 2.5 CY, 963	98.50
Total Hourly Rate for XYCTD		$98.50

Crew: XYCTE

Quantity	Description	Hourly Rate
1	Loader, Front End, Crawler, 3.75 CY, 973	136.75
Total Hourly Rate for XYCTE		$136.75

Crew: XYCWA

Quantity	Description	Hourly Rate
1	Loader, Front End, Wheeled, 1.25 CY Articulated	24.23
Total Hourly Rate for XYCWA		$24.23

Crew: XYCWB

Quantity	Description	Hourly Rate
1	Loader, Front End, Wheeled, 1.5 CY, Articulated, 916E	30.13
Total Hourly Rate for XYCWB		$30.13

Crew: XYCWC

Quantity	Description	Hourly Rate
1	Loader, Front End, Wheeled, 2.0 CY, Articulated, 926E	35.08
Total Hourly Rate for XYCWC		$35.08

Crew: XYCWD

Quantity	Description	Hourly Rate
1	Loader, Front End, Wheeled, 3.0 CY, Articulated, 950E	38.13
Total Hourly Rate for XYCWD		$38.13

Crew: XYCWE

Quantity	Description	Hourly Rate
1	Loader, Front End, Wheeled, 4.0 CY, Articulated, 966E	55.70
Total Hourly Rate for XYCWE		$55.70

Crew: XYCWF

Quantity	Description	Hourly Rate
1	Loader, Front End, Wheeled, 5.25 CY, Articulated, 980C	87.55
	Total Hourly Rate for XYCWF	$87.55

Crew: XYCWG

Quantity	Description	Hourly Rate
1	Loader, Front End, Wheeled, 7.0 CY, Articulated, 988B	152.88
	Total Hourly Rate for XYCWG	$152.88

Crew: XYCWH

Quantity	Description	Hourly Rate
1	Loader, Front End, Wheeled, 13.5 CY, Articulated, 992C	228.75
	Total Hourly Rate for XYCWH	$228.75

Crew: XYSEA

Quantity	Description	Hourly Rate
1	1,500 CY/Month Solvent Extraction Process Unit	254.04
	Total Hourly Rate for XYSEA	$254.04

Crew: XYSEB

Quantity	Description	Hourly Rate
1	6,000 CY/Month Solvent Extraction Process Unit	560.75
	Total Hourly Rate for XYSEB	$560.75

Crew: XYSEC

Quantity	Description	Hourly Rate
1	18,000 CY/Month Solvent Extraction Process Unit	1155.47
	Total Hourly Rate for XYSEC	$1,155.47

Crew: XYSEO

Quantity	Description	Hourly Rate
1	Solidification/Stabilization Equipment	121.12
	Total Hourly Rate for XYSEO	$121.12

Crew: YAABB

Quantity	Description	Hourly Rate
1	Ripper, Shanks Each, D-8	0.80
1	Ripper, Single Shank Beam, Hydraulic, D-8	12.19
1	Dozer, Crawler, D-8L, Powershift, (Add Blade)	149.38
1	Blade, Universal, Hydraulic, for D8	9.78
	Total Hourly Rate for YAABB	$172.15

Crew: YAABC

Quantity	Description	Hourly Rate
1	Loader, Backhoe, Wheeled, 1.0 CY Front End Bucket, 24" Dip	26.98
	Total Hourly Rate for YAABC	$26.98

Crew: YAABD

Quantity	Description	Hourly Rate
1	Hydraulic Crane, Self Propelled, 22T, Rough Terrain, 4WD	71.88
	Total Hourly Rate for YAABD	$71.88

16 01 Nonhazardous Selective Building Demolition

16 01 01 Substructure and Superstructure		Unit	Hourly Output	Crew	Factor	Labor	Equip.	Matl.	Total
16 01 0102	**Remove Concrete Footing**	**CF**				**2.02**	**1.89**	**0.00**	**3.91**
02049 0610	Demolish Reinforced Concrete Building Foundation	CF	84.38	CLADD	1.000	2.02	1.89	0.00	3.91
16 01 0104	**Excavation to Remove Foundation**	**CY**				**0.99**	**1.10**	**0.00**	**2.08**
02228 0302	Trench, 1/2 CY Hydraulic Excavator, Medium Soil, 75 CY/Hour (58 m3)/Hour	CY	63.5	CODEA	1.000	0.99	1.10	0.00	2.08
16 01 0106	**Backfill & Compaction @ Foundation Removal**	**CY**				**6.25**	**0.95**	**0.00**	**7.21**
02215 1220	Backfill Trench with Small Front-end Loader, without Compaction	CY	70	CODLB	1.000	0.69	0.66	0.00	1.34
02220 7260	Compact Backfill with Vibrating Plate Around Structures and Trenches	CY	14.75	CLACC	1.000	5.57	0.29	0.00	5.86
16 01 0108	**Select Backfill**	**CY**				**0.00**	**0.00**	**7.90**	**7.90**
02081 2242	Select Granular Fill (Sand)	CY	0	N/A	1.000	0.00	0.00	7.90	7.90
16 01 0110	**Remove Masonry Foundation Wall**	**CF**				**1.51**	**1.42**	**0.00**	**2.93**
02049 0602	Demolish Brick/Stone Building Foundation	CF	112.5	CLADD	1.000	1.51	1.42	0.00	2.93
16 01 0112	**Remove Concrete Foundation Wall**	**CF**				**2.02**	**1.89**	**0.00**	**3.91**
02049 0610	Demolish Reinforced Concrete Building Foundation	CF	84.38	CLADD	1.000	2.02	1.89	0.00	3.91
16 01 0114	**Remove Concrete Stair on Grade**	**CF**				**0.79**	**0.27**	**0.00**	**1.06**
02112 1223	Remove Concrete Stair on Grade	SF	120	CODEJ	1.000	0.79	0.27	0.00	1.06
16 01 0116	**Remove Metal Stair Railing**	**LF**				**3.39**	**0.00**	**0.00**	**3.39**
02112 1224	Remove Metal Stair Rail	LF	20	ULABB	1.000	3.39	0.00	0.00	3.39
16 01 0118	**Remove Pilings - Concrete**	**VLF**				**5.14**	**5.83**	**0.00**	**10.97**
02112 2006	Remove Concrete Piles	VLF	50	CPIDC	1.000	5.14	5.83	0.00	10.97
16 01 0120	**Remove Pilings - Steel**	**VLF**				**5.14**	**5.83**	**0.00**	**10.97**
02112 2007	Remove H-Piles	VLF	50	CPIDC	1.000	5.14	5.83	0.00	10.97
16 01 0122	**Remove Pilings - Wood**	**VLF**				**5.14**	**5.83**	**0.00**	**10.97**
02112 2008	Remove Wood Piles	VLF	50	CPIDC	1.000	5.14	5.83	0.00	10.97
16 01 0123	**Remove Slab on Grade - Selective**	**CF**				**21.68**	**3.62**	**0.00**	**25.30**
02056 1000	Concrete Cutouts, < 6 CF with Bar Reinforced Interior Elevated Slabs (<.17 m3)	CF	6.25	CLADB	1.000	21.68	3.62	0.00	25.30
16 01 0124	**Remove Slab on Grade (4" - 6")**	**SF**				**0.56**	**0.19**	**0.00**	**0.75**
02112 1220	Remove Slab on Grade (4"-6")	SF	170	CODEJ	1.000	0.56	0.19	0.00	0.75
16 01 0126	**Remove Slab on Grade (7" - 8")**	**SF**				**0.63**	**0.21**	**0.00**	**0.85**
02112 1221	Remove Slab on Grade (7"-8")	SF	150	CODEJ	1.000	0.63	0.21	0.00	0.85

16 01 Nonhazardous Selective Building Demolition

16 01 01 Substructure and Superstructure		Unit	Hourly Output	Crew	Factor	Labor	Equip.	Matl.	Total
16 01 0128	**Remove Slab on Grade (9" - 12")**	**SF**				**0.79**	**0.27**	**0.00**	**1.06**
02112 1222	Remove Slab on Grade (9"-12")	SF	120	CODEJ	1.000	0.79	0.27	0.00	1.06
16 01 0130	**Remove Concrete Beams**	**CF**				**12.55**	**2.09**	**0.00**	**14.64**
02111 2512	Remove Concrete Beams	CF	10.8	CLADB	1.000	12.55	2.09	0.00	14.64
16 01 0132	**Remove Concrete Columns**	**CF**				**11.24**	**1.88**	**0.00**	**13.12**
02111 2514	Remove Concrete Columns	CF	12.05	CLADB	1.000	11.24	1.88	0.00	13.12
16 01 0134	**Remove Concrete Elevated Slab**	**CF**				**2.60**	**0.43**	**0.00**	**3.03**
02111 2511	Remove Concrete Floor	CF	52.2	CLADB	1.000	2.60	0.43	0.00	3.03
16 01 0136	**Remove Concrete Fill (No Deck Demolition)**	**SF**				**1.30**	**0.22**	**0.00**	**1.51**
02111 2513	Remove Concrete Fill (No Deck Demolition)	SF	104.4	CLADB	1.000	1.30	0.22	0.00	1.51
16 01 0138	**Remove Steel Beams and Columns**	**TON**				**297.94**	**16.63**	**0.00**	**314.57**
02111 2522	Remove Steel Beams & Columns	TN	0.5	SIWSC	1.000	297.94	16.63	0.00	314.57
16 01 0140	**Remove Metal Joist**	**TON**				**248.28**	**13.85**	**0.00**	**262.14**
02111 2523	Remove Metal Joist	TN	0.6	SIWSC	1.000	248.28	13.85	0.00	262.14
16 01 0142	**Remove Metal Deck**	**SF**				**0.45**	**0.03**	**0.00**	**0.48**
02111 2521	Remove Steel Deck	SF	330	SIWSC	1.000	0.45	0.03	0.00	0.48
16 01 0144	**Remove Metal Stair**	**RIS**				**37.24**	**2.08**	**0.00**	**39.32**
02111 2524	Remove Metal Stair	RIS	4	SIWSC	1.000	37.24	2.08	0.00	39.32
16 01 0146	**Remove Wood Floor Structure**	**SF**				**0.53**	**0.00**	**0.00**	**0.53**
02111 2534	Remove Wood Floor Structure	SF	250	ACARF	1.000	0.53	0.00	0.00	0.53
16 01 0148	**Remove Wood Roof Truss Structure**	**SF**				**0.67**	**0.29**	**0.00**	**0.95**
02111 2535	Remove Wood Roof Truss Structure	SF	250	ACARS	1.000	0.67	0.29	0.00	0.95
16 01 0150	**Remove Plywood - Floor**	**SF**				**0.58**	**0.00**	**0.00**	**0.58**
02111 2531	Remove Plywood - Floor	SF	133.5	ACARD	1.000	0.58	0.00	0.00	0.58
16 01 0152	**Remove Plywood - Roof**	**SF**				**0.41**	**0.00**	**0.00**	**0.41**
02111 2532	Remove Plywood - Roof	SF	186.9	ACARD	1.000	0.41	0.00	0.00	0.41
16 01 0190	**Remove Misc. (2 Laborers)**	**HR**				**67.75**	**0.00**	**0.00**	**67.75**
02111 4052	Miscellaneous Removal (2 Laborers)	HR	1	ULABB	1.000	67.75	0.00	0.00	67.75

16 01 02 Exterior Closure		Unit	Hourly Output	Crew	Factor	Labor	Equip.	Matl.	Total
16 01 0202	**Remove Concrete Wall**	**CF**				**4.93**	**0.82**	**0.00**	**5.75**
02071 0020	Demolish Reinforced Concrete Partition	CF	13.75	CLADB	0.500	9.85	1.65	0.00	5.75
16 01 0204	**Remove Wood/Metal Stud Wall**	**SF**				**0.73**	**0.55**	**0.00**	**1.28**
02111 3506	Remove Metal/Wood Stud Partition	SF	130	ULABF	1.000	0.73	0.55	0.00	1.28

16 01 02 Exterior Closure		Unit	Hourly Output	Crew	Factor	Labor	Equip.	Matl.	Total
16 01 0206	**Remove Masonry Wall - 4" - 6"**	**SF**				**1.08**	**0.18**	**0.00**	**1.27**
02071 0200	Demolish 4" Concrete Block Partition (10 cm) Thick	SF	125	CLADB	1.000	1.08	0.18	0.00	1.27
16 01 0208	**Remove Masonry Wall - 8"**	**SF**				**1.34**	**0.22**	**0.00**	**1.56**
02071 0280	Demolish 8" Concrete Block Partition (20 cm) Thick	SF	101.3	CLADB	1.000	1.34	0.22	0.00	1.56
16 01 0210	**Remove Masonry Wall - 12"**	**SF**				**1.50**	**0.25**	**0.00**	**1.74**
02074 2100	Demolish 12" Concrete Block Partition (31 cm) Thick	SF	90.63	CLADB	1.000	1.50	0.25	0.00	1.74
16 01 0212	**Remove Brick Veneer**	**SF**				**1.36**	**0.23**	**0.00**	**1.58**
02111 3202	Remove Brick Veneer	SF	100	CLADB	1.000	1.36	0.23	0.00	1.58
16 01 0214	**Remove Tilt-up/Precast Concrete**	**SF**				**1.13**	**0.48**	**0.00**	**1.61**
02111 3801	Remove Precast Cladding	SF	250	SIWSN	1.000	1.13	0.48	0.00	1.61
16 01 0216	**Remove Metal Siding**	**SF**				**0.71**	**0.00**	**0.00**	**0.71**
02111 3802	Remove Metal Siding	SF	200	MSHMG	1.000	0.71	0.00	0.00	0.71
16 01 0218	**Remove Cedar Siding**	**SF**				**0.68**	**0.00**	**0.00**	**0.68**
02111 3803	Remove Cedar Siding	SF	100	ULABB	1.000	0.68	0.00	0.00	0.68
16 01 0220	**Remove Vinyl Siding**	**SF**				**0.68**	**0.00**	**0.00**	**0.68**
02111 3804	Remove Vinyl Siding	SF	100	ULABB	1.000	0.68	0.00	0.00	0.68
16 01 0222	**Remove Stucco/Plaster**	**SF**				**0.90**	**0.00**	**0.00**	**0.90**
02111 3805	Remove Stucco	SF	75	ULABB	1.000	0.90	0.00	0.00	0.90
16 01 0224	**Remove E.I.F.S.**	**SF**				**0.68**	**0.00**	**0.00**	**0.68**
02111 3806	Remove Exterior Insulation & Finish System	SF	100	ULABB	1.000	0.68	0.00	0.00	0.68
16 01 0226	**Remove Curtain Wall**	**SF**				**5.57**	**0.00**	**0.00**	**5.57**
02111 3807	Remove Curtain Wall	SF	15	AGLAB	1.000	5.57	0.00	0.00	5.57
16 01 0227	**Remove Door Only (No Frame)**	**EA**				**16.94**	**0.00**	**0.00**	**16.94**
02112 1450	Remove Door Only (No Frame)	EA	4	ULABB	1.000	16.94	0.00	0.00	16.94
16 01 0228	**Remove Door - Solid**	**EA**				**33.88**	**0.00**	**0.00**	**33.88**
02112 1441	Remove Solid Door	EA	2	ULABB	1.000	33.88	0.00	0.00	33.88
16 01 0230	**Remove Door - Glazed**	**EA**				**55.67**	**0.00**	**0.00**	**55.67**
02112 1442	Remove Glazed Door	EA	1.5	AGLAB	1.000	55.67	0.00	0.00	55.67
16 01 0232	**Remove Door - Overhead**	**SF**				**2.48**	**0.00**	**0.00**	**2.48**
02112 1443	Remove Overhead Door	SF	35	ACARB	1.000	2.48	0.00	0.00	2.48

16 01 02	Exterior Closure	Unit	Hourly Output	Crew	Factor	Labor	Equip.	Matl.	Total
16 01 0234	**Remove Door - Hangar**	**SF**				**2.01**	**0.96**	**0.00**	**2.97**
02112 1444	Remove Hangar Door	SF	75	SIWSF	1.000	2.01	0.96	0.00	2.97
16 01 0236	**Remove Door - Revolving**	**EA**				**792.60**	**447.70**	**0.00**	**1,240**
02112 1445	Remove Revolving Door	EA	0.25	ACART	1.000	792.60	447.70	0.00	1,240
16 01 0238	**Remove Door - Freezer**	**EA**				**206.75**	**0.00**	**0.00**	**206.75**
02112 1446	Remove Freezer Door	EA	0.5	MSPFC	1.000	206.75	0.00	0.00	206.75
16 01 0240	**Remove Door Hardware**	**EA**				**8.47**	**0.00**	**0.00**	**8.47**
02112 1447	Remove Door Hardware	EA	4	ULABE	1.000	8.47	0.00	0.00	8.47
16 01 0242	**Remove Window**	**SF**				**2.71**	**0.00**	**0.00**	**2.71**
02072 3020	Demolish Windows >31 SF (2.9 m2), Wood/Metal Frame	SF	25	ULABB	1.000	2.71	0.00	0.00	2.71
16 01 0244	**Remove Storefront**	**SF**				**4.17**	**0.00**	**0.00**	**4.17**
02112 1815	Remove Storefront	SF	20	AGLAB	1.000	4.17	0.00	0.00	4.17
16 01 0246	**Remove Misc. Metals**	**TON**				**595.88**	**33.25**	**0.00**	**629.13**
02111 2525	Remove Miscellaneous Metals	TN	0.25	SIWSC	1.000	595.88	33.25	0.00	629.13
16 01 0248	**Remove Plywood - Wall**	**SF**				**0.52**	**0.00**	**0.00**	**0.52**
02111 2533	Remove Plywood - Wall	SF	150	ACARD	1.000	0.52	0.00	0.00	0.52
16 01 0250	**Remove Insulation - Rigid**	**SF**				**0.21**	**0.00**	**0.00**	**0.21**
02111 4053	Remove Rigid Insulation	SF	325	ULABB	1.000	0.21	0.00	0.00	0.21
16 01 0252	**Remove Insulation - Batt**	**SF**				**0.14**	**0.00**	**0.00**	**0.14**
02111 4054	Remove Batt Insulation	SF	500	ULABB	1.000	0.14	0.00	0.00	0.14
16 01 0254	**Remove Louver**	**SF**				**1.94**	**0.00**	**0.00**	**1.94**
02111 4055	Remove Louver	SF	35	ULABB	1.000	1.94	0.00	0.00	1.94
16 01 0256	**Remove Chain-link Fence**	**LF**				**1.75**	**0.00**	**0.00**	**1.75**
02046 0752	Remove & Reuse Chain Link Fence	LF	38.75	ULABB	1.000	1.75	0.00	0.00	1.75
16 01 0290	**Remove Misc. (2 Laborers)**	**HR**				**67.75**	**0.00**	**0.00**	**67.75**
02111 4052	Miscellaneous Removal (2 Laborers)	HR	1	ULABB	1.000	67.75	0.00	0.00	67.75

16 01 03	Roofing	Unit	Hourly Output	Crew	Factor	Labor	Equip.	Matl.	Total
16 01 0302	**Remove Roofing - Single Ply**	**SF**				**0.19**	**0.00**	**0.00**	**0.19**
02112 1713	Remove Single Ply Roofing	SF	350	ULABB	1.000	0.19	0.00	0.00	0.19
16 01 0304	**Remove Roofing - Built-up**	**SF**				**0.27**	**0.00**	**0.00**	**0.27**
02068 2800	Demolish Built-up Roof on Metal Deck Steel Frame Roof	SF	250	ULABB	1.000	0.27	0.00	0.00	0.27

16 01 03 Roofing		Unit	Hourly Output	Crew	Factor	Labor	Equip.	Matl.	Total
16 01 0305	**Remove Roofing Gravel**	**SF**				**0.23**	**0.00**	**0.00**	**0.23**
02112 1715	Remove Roofing Gravel	SF	300	ULABB	1.000	0.23	0.00	0.00	0.23
16 01 0306	**Remove Roofing - Shingle**	**SF**				**0.27**	**0.00**	**0.00**	**0.27**
02068 4000	Demolish Roofing Shingles, Wood Frame Roof	SF	250	ULABB	1.000	0.27	0.00	0.00	0.27
16 01 0308	**Remove Metal Roofing**	**SF**				**0.54**	**0.00**	**0.00**	**0.54**
02068 1150	Demolish Corrugated Metal Roofing On Steel Frame	SF	125	ULABB	1.000	0.54	0.00	0.00	0.54
16 01 0310	**Remove Roofing - Clay Tile**	**SF**				**0.27**	**0.00**	**0.00**	**0.27**
02068 4460	Demolish Roof Tiles, Wood Frame Roof	SF	250	ULABB	1.000	0.27	0.00	0.00	0.27
16 01 0312	**Remove Roof Walkway**	**SF**				**0.31**	**0.00**	**0.00**	**0.31**
02112 1731	Remove Walkway	SF	220	ULABB	1.000	0.31	0.00	0.00	0.31
16 01 0314	**Remove Roof Insulation**	**SF**				**0.38**	**0.00**	**0.00**	**0.38**
02112 1732	Remove Insulation	SF	180	ULABB	1.000	0.38	0.00	0.00	0.38
16 01 0316	**Remove Roof Hatch**	**EA**				**104.23**	**0.00**	**0.00**	**104.23**
02112 1733	Remove Roof Hatch	EA	0.65	ULABB	1.000	104.23	0.00	0.00	104.23
16 01 0318	**Remove Skylight**	**SF**				**1.13**	**0.00**	**0.00**	**1.13**
02112 1734	Remove Skylight	SF	60	ULABB	1.000	1.13	0.00	0.00	1.13
16 01 0320	**Remove Gutters/Downspouts**	**LF**				**1.13**	**0.00**	**0.00**	**1.13**
02112 1735	Remove Gutters/Downspouts	LF	60	ULABB	1.000	1.13	0.00	0.00	1.13
16 01 0322	**Remove Flashing**	**SF**				**1.36**	**0.00**	**0.00**	**1.36**
02112 1736	Remove Flashing	SF	50	ULABB	1.000	1.36	0.00	0.00	1.36
16 01 0324	**Remove Blocking**	**SF**				**1.36**	**0.00**	**0.00**	**1.36**
02112 1737	Remove Blocking	SF	50	ULABB	1.000	1.36	0.00	0.00	1.36

16 01 04 Interior Construction and Finishes		Unit	Hourly Output	Crew	Factor	Labor	Equip.	Matl.	Total
16 01 0402	**Remove Concrete Wall**	**CF**				**9.85**	**1.65**	**0.00**	**11.50**
02071 0020	Demolish Reinforced Concrete Partition	CF	13.75	CLADB	1.000	9.85	1.65	0.00	11.50
16 01 0404	**Remove Wood/Metal Stud Wall**	**SF**				**0.73**	**0.55**	**0.00**	**1.28**
02111 3506	Remove Metal/Wood Stud Partition	SF	130	ULABF	1.000	0.73	0.55	0.00	1.28
16 01 0406	**Remove Masonry Wall - 4" to 6"**	**SF**				**1.08**	**0.18**	**0.00**	**1.27**
02071 0200	Demolish 4" Concrete Block Partition (10 cm) Thick	SF	125	CLADB	1.000	1.08	0.18	0.00	1.27
16 01 0408	**Remove Masonry Wall - 8"**	**SF**				**1.34**	**0.22**	**0.00**	**1.56**
02071 0280	Demolish 8" Concrete Block Partition (20 cm) Thick	SF	101.3	CLADB	1.000	1.34	0.22	0.00	1.56

16 01 04	Interior Construction and Finishes	Unit	Hourly Output	Crew	Factor	Labor	Equip.	Matl.	Total
16 01 0410	**Remove Masonry Wall - 12"**	**SF**				**1.50**	**0.25**	**0.00**	**1.74**
02074 2100	Demolish 12" Concrete Block Partition (31 cm) Thick	SF	90.63	CLADB	1.000	1.50	0.25	0.00	1.74
16 01 0412	**Remove Wood/Metal Furring**	**SF**				**0.94**	**0.72**	**0.00**	**1.66**
02111 3507	Remove Metal/Wood Furring	SF	100	ULABF	1.000	0.94	0.72	0.00	1.66
16 01 0413	**Remove Door Only (No Frame)**	**EA**				**16.94**	**0.00**	**0.00**	**16.94**
02112 1450	Remove Door Only (No Frame)	EA	4	ULABB	1.000	16.94	0.00	0.00	16.94
16 01 0414	**Remove Door - Solid**	**EA**				**33.88**	**0.00**	**0.00**	**33.88**
02112 1441	Remove Solid Door	EA	2	ULABB	1.000	33.88	0.00	0.00	33.88
16 01 0416	**Remove Door - Glazed**	**EA**				**55.67**	**0.00**	**0.00**	**55.67**
02112 1442	Remove Glazed Door	EA	1.5	AGLAB	1.000	55.67	0.00	0.00	55.67
16 01 0418	**Remove Door - Overhead**	**SF**				**2.48**	**0.00**	**0.00**	**2.48**
02112 1443	Remove Overhead Door	SF	35	ACARB	1.000	2.48	0.00	0.00	2.48
16 01 0420	**Remove Door - Freezer**	**EA**				**206.75**	**0.00**	**0.00**	**206.75**
02112 1446	Remove Freezer Door	EA	0.5	MSPFC	1.000	206.75	0.00	0.00	206.75
16 01 0422	**Remove Door - Vault**	**EA**				**605.40**	**239.58**	**0.00**	**844.98**
02112 1448	Remove Vault Door	EA	0.3	SIWSG	1.000	605.40	239.58	0.00	844.98
16 01 0424	**Remove Door - Revolving**	**EA**				**792.60**	**447.70**	**0.00**	**1,240**
02112 1445	Remove Revolving Door	EA	0.25	ACART	1.000	792.60	447.70	0.00	1,240
16 01 0425	**Remove Door - Revolving Darkroom**	**EA**				**198.15**	**111.93**	**0.00**	**310.07**
02112 1451	Remove Revolving Darkroom Door	EA	1	ACART	1.000	198.15	111.93	0.00	310.07
16 01 0426	**Remove Door Hardware**	**EA**				**8.47**	**0.00**	**0.00**	**8.47**
02112 1447	Remove Door Hardware	EA	4	ULABE	1.000	8.47	0.00	0.00	8.47
16 01 0428	**Remove Window**	**SF**				**2.71**	**0.00**	**0.00**	**2.71**
02072 3020	Demolish Windows >31 SF (2.9 m2), Wood/Metal Frame	SF	25	ULABB	1.000	2.71	0.00	0.00	2.71
16 01 0430	**Remove Gypsum - Bonus to Wall Demolition**	**SF**				**0.29**	**0.00**	**0.00**	**0.29**
02111 3503	Drywall Removal, Nailed	SF	115	ULABE	1.000	0.29	0.00	0.00	0.29
16 01 0432	**Remove Plaster - Bonus to Wall Demolition**	**SF**				**0.68**	**0.00**	**0.00**	**0.68**
02111 3505	Plaster Removal with Lathe	SF	50	ULABE	1.000	0.68	0.00	0.00	0.68
16 01 0434	**Remove Ceramic - Wall**	**SF**				**0.68**	**0.00**	**0.00**	**0.68**
02111 4201	Remove Ceramic Wall Finish	SF	100	ULABB	1.000	0.68	0.00	0.00	0.68
16 01 0436	**Remove Vinyl Wall Covering**	**SF**				**0.90**	**0.00**	**0.16**	**1.07**
02111 4202	Remove Vinyl Wall Covering	SF	75	ULABB	1.000	0.90	0.00	0.16	1.07

16 01 04 Interior Construction and Finishes		Unit	Hourly Output	Crew	Factor	Labor	Equip.	Matl.	Total
16 01 0438	**Remove Insulation Liner Panel**	**SF**				**0.68**	**0.00**	**0.00**	**0.68**
02111 4203	Remove Insulation Liner Panel	SF	100	ULABB	1.000	0.68	0.00	0.00	0.68
16 01 0440	**Remove Wood Paneling**	**SF**				**0.68**	**0.00**	**0.00**	**0.68**
02111 4204	Remove Wood Paneling	SF	100	ULABB	1.000	0.68	0.00	0.00	0.68
16 01 0441	**Remove Ceiling Suspension System**	**SF**				**0.21**	**0.00**	**0.00**	**0.21**
02112 1625	Remove Suspension System	SF	325	ULABB	1.000	0.21	0.00	0.00	0.21
16 01 0442	**Remove Gypsum - Ceiling**	**SF**				**0.54**	**0.00**	**0.00**	**0.54**
02055 0210	Demolish Drywall Ceiling, Nailed to Frame	SF	125	ULABB	1.000	0.54	0.00	0.00	0.54
16 01 0443	**Remove Gypsum - Ceiling - Selective**	**SF**				**0.90**	**0.00**	**0.00**	**0.90**
02112 1623	Remove Drywall Ceiling, Selective	SF	75	ULABB	1.000	0.90	0.00	0.00	0.90
16 01 0444	**Remove Plaster - Ceiling**	**SF**				**0.54**	**0.00**	**0.00**	**0.54**
02055 1020	Demolish Suspended Plastered Ceiling	SF	125	ULABB	1.000	0.54	0.00	0.00	0.54
16 01 0446	**Remove Acoustic Ceiling Tile - 12" x 12"**	**SF**				**0.36**	**0.00**	**0.00**	**0.36**
02055 1564	Demolish Acoustical Tile Ceiling, Furred and Glued	SF	187.5	ULABB	1.000	0.36	0.00	0.00	0.36
16 01 0448	**Remove Acoustic Ceiling Tile - 2' x 2' / 2' x 4'**	**SF**				**0.10**	**0.00**	**0.00**	**0.10**
02112 1614	Remove Acoustical Ceiling Tile	SF	650	ULABB	1.000	0.10	0.00	0.00	0.10
16 01 0452	**Remove Acoustic Ceiling Grid - 2' x 2' / 2' x 4'**	**SF**				**0.21**	**0.00**	**0.00**	**0.21**
02112 1615	Remove Acoustical Ceiling Grid	SF	325	ULABB	1.000	0.21	0.00	0.00	0.21
16 01 0454	**Remove Plywood Ceiling**	**SF**				**0.54**	**0.00**	**0.00**	**0.54**
02112 1641	Remove Plywood Ceiling	SF	125	ULABB	1.000	0.54	0.00	0.00	0.54
16 01 0456	**Remove Ceramic - Floor**	**SF**				**0.91**	**0.00**	**0.00**	**0.91**
02059 2020	Demolish Ceramic/Quarry Tile Floor	SF	148.8	ULABH	1.000	0.91	0.00	0.00	0.91
16 01 0458	**Remove Terrazzo Floor**	**SF**				**1.39**	**0.00**	**0.00**	**1.39**
02059 2600	Demolish Terrazzo Flooring	SF	97.5	ULABH	1.000	1.39	0.00	0.00	1.39
16 01 0460	**Remove Wood Floor**	**SF**				**0.90**	**0.00**	**0.00**	**0.90**
02111 6004	Remove Wood Flooring	SF	150	ULABH	1.000	0.90	0.00	0.00	0.90
16 01 0462	**Remove Vinyl Composition Tile**	**SF**				**0.59**	**0.00**	**0.00**	**0.59**
02111 6005	Remove Vinyl Composition Tile	SF	230	ULABH	1.000	0.59	0.00	0.00	0.59
16 01 0464	**Remove Sheet Vinyl**	**SF**				**0.41**	**0.00**	**0.00**	**0.41**
02111 6006	Remove Sheet Vinyl Flooring	SF	330	ULABH	1.000	0.41	0.00	0.00	0.41
16 01 0466	**Remove Carpet**	**SF**				**0.25**	**0.00**	**0.00**	**0.25**
02111 6007	Remove Carpet	SF	550	ULABH	1.000	0.25	0.00	0.00	0.25

16 01 04	Interior Construction and Finishes	Unit	Hourly Output	Crew	Factor	Labor	Equip.	Matl.	Total
16 01 0468	**Remove Access Floor**	**SF**				**1.23**	**0.00**	**0.00**	**1.23**
02111 6008	Remove Access Flooring	SF	110	ULABH	1.000	1.23	0.00	0.00	1.23
16 01 0470	**Remove Wire Mesh Partition**	**SF**				**1.36**	**0.00**	**0.00**	**1.36**
02111 3601	Remove Wire Mesh Partition	SF	50	ULABB	1.000	1.36	0.00	0.00	1.36
16 01 0472	**Remove Retractable Partition**	**SF**				**4.52**	**0.00**	**0.00**	**4.52**
02111 3602	Remove Retractable Partition	SF	15	ULABB	1.000	4.52	0.00	0.00	4.52
16 01 0474	**Remove Roll-up Shutter**	**SF**				**6.30**	**0.00**	**0.00**	**6.30**
02112 1449	Remove Roll-Up Shutter	SF	10.75	ULABB	1.000	6.30	0.00	0.00	6.30
16 01 0476	**Remove Urinal Screen**	**EA**				**16.94**	**0.00**	**0.00**	**16.94**
02111 4101	Remove Urinal Screen	EA	2	ULABE	1.000	16.94	0.00	0.00	16.94
16 01 0477	**Remove Toilet Partition**	**EA**				**67.75**	**0.00**	**0.00**	**67.75**
02111 4102	Remove Toilet Partition	EA	1	ULABB	1.000	67.75	0.00	0.00	67.75
16 01 0478	**Remove Bath Accessories**	**EA**				**3.39**	**0.00**	**0.00**	**3.39**
02111 4103	Remove Bath Accessories	EA	10	ULABE	1.000	3.39	0.00	0.00	3.39
16 01 0479	**Remove Patient Cubicle Track**	**LF**				**0.68**	**0.00**	**0.00**	**0.68**
02111 4104	Remove Patient Cubicle Track	LF	100	ULABB	1.000	0.68	0.00	0.00	0.68
16 01 0480	**Remove Chalkboard/Bulletin Board**	**SF**				**0.26**	**0.00**	**0.00**	**0.26**
02111 4105	Remove Chalkboard/Bulletin Board	SF	128	ULABE	1.000	0.26	0.00	0.00	0.26
16 01 0481	**Remove Projection Screen**	**SF**				**0.34**	**0.00**	**0.00**	**0.34**
02111 4106	Remove Projection Screen	SF	100	ULABE	1.000	0.34	0.00	0.00	0.34
16 01 0482	**Remove Directory**	**EA**				**8.47**	**0.00**	**0.00**	**8.47**
02111 4107	Remove Directory	EA	4	ULABE	1.000	8.47	0.00	0.00	8.47
16 01 0483	**Remove Lockers - Single Tier**	**EA**				**6.78**	**0.00**	**0.00**	**6.78**
02111 4108	Remove Lockers - Single Tier	EA	10	ULABB	1.000	6.78	0.00	0.00	6.78
16 01 0484	**Remove Lockers - Double Tier**	**EA**				**9.03**	**0.00**	**0.00**	**9.03**
02111 4109	Remove Lockers - Double Tier	EA	7.5	ULABB	1.000	9.03	0.00	0.00	9.03
16 01 0485	**Remove Shelving**	**LF**				**0.68**	**0.00**	**0.00**	**0.68**
02111 4110	Remove Shelving	LF	50	ULABE	1.000	0.68	0.00	0.00	0.68
16 01 0486	**Remove Fire Extinguisher Cabinet**	**LF**				**16.94**	**0.00**	**0.00**	**16.94**
02111 4111	Remove Fire Extinguisher Cabinet	EA	2	ULABE	1.000	16.94	0.00	0.00	16.94
16 01 0487	**Remove Wall Corner Guard/Hand Rail**	**LF**				**0.34**	**0.00**	**0.00**	**0.34**
02111 4112	Remove Wall Guard/Hand Rail	LF	100	ULABE	1.000	0.34	0.00	0.00	0.34

16 01 04	Interior Construction and Finishes	Unit	Hourly Output	Crew	Factor	Labor	Equip.	Matl.	Total
16 01 0488	**Remove Countertop**	**LF**				**4.52**	**0.00**	**0.00**	**4.52**
02111 4113	Remove Counter Top	LF	15	ULABB	1.000	4.52	0.00	0.00	4.52
16 01 0489	**Remove Cabinet - Base**	**LF**				**6.78**	**0.00**	**0.00**	**6.78**
02111 4114	Remove Base Cabinets	LF	10	ULABB	1.000	6.78	0.00	0.00	6.78
16 01 0490	**Remove Cabinet - Wall**	**LF**				**6.78**	**0.00**	**0.00**	**6.78**
02111 4115	Remove Wall Cabinets	LF	10	ULABB	1.000	6.78	0.00	0.00	6.78
16 01 0491	**Remove Cabinet - Full Height**	**LF**				**13.55**	**0.00**	**0.00**	**13.55**
02111 4116	Remove Cabinet - Full Height	LF	5	ULABB	1.000	13.55	0.00	0.00	13.55
16 01 0492	**Remove Floor Mounted Benches**	**LF**				**8.47**	**0.00**	**0.00**	**8.47**
02111 4117	Remove Floor Mounted Benches	LF	8	ULABB	1.000	8.47	0.00	0.00	8.47

16 01 05	Conveying Systems	Unit	Hourly Output	Crew	Factor	Labor	Equip.	Matl.	Total
16 01 0502	**Remove Elevator**	**STOP**				**887.95**	**0.00**	**0.00**	**887.95**
02112 9001	Remove Elevator	STOP	0.38	ESELA	1.000	887.95	0.00	0.00	887.95
16 01 0504	**Remove Monorail**	**LB**				**0.15**	**0.07**	**0.00**	**0.22**
02112 9002	Remove Monorail	LB	1000	SIWSF	1.000	0.15	0.07	0.00	0.22
16 01 0506	**Remove Monorail Crane**	**EA**				**318.17**	**0.00**	**0.00**	**318.17**
02112 9003	Remove Monorail Crane	EA	0.28	ESELC	1.000	318.17	0.00	0.00	318.17
16 01 0507	**Remove Overhead Crane**	**EA**				**1,212**	**147.31**	**0.00**	**1,359**
02112 9005	Remove Overhead Crane	EA	0.13	ESELH	1.000	1,212	147.31	0.00	1,359
16 01 0508	**Remove Auto Lift**	**EA**				**643.35**	**0.00**	**0.00**	**643.35**
02112 9004	Remove Auto Lift	EA	0.25	ESELB	1.000	643.35	0.00	0.00	643.35

16 01 06	Mechanical Systems	Unit	Hourly Output	Crew	Factor	Labor	Equip.	Matl.	Total
16 01 0610	**Remove Lavatory**	**EA**				**45.83**	**0.00**	**0.00**	**45.83**
02067 1200	Demolish Wall Hung Lavatory	EA	1.125	MPLUA	1.000	45.83	0.00	0.00	45.83
16 01 0611	**Remove Water Closet**	**EA**				**58.93**	**0.00**	**0.00**	**58.93**
02067 1400	Demolish Floor Mounted Water Closet	EA	0.875	MPLUA	1.000	58.93	0.00	0.00	58.93
16 01 0612	**Remove Urinal**	**EA**				**68.75**	**0.00**	**0.00**	**68.75**
02067 1520	Demolish Wall Mounted Urinal	EA	0.75	MPLUA	1.000	68.75	0.00	0.00	68.75
16 01 0613	**Remove Water Fountain**	**EA**				**68.75**	**0.00**	**0.00**	**68.75**
02067 1620	Demolish Recessed Water Fountain	EA	0.75	MPLUA	1.000	68.75	0.00	0.00	68.75
16 01 0614	**Remove Tub and Shower Unit**	**EA**				**147.92**	**0.00**	**0.00**	**147.92**
02111 9105	Remove Tub And Shower Unit	EA	0.75	MPLUF	1.000	147.92	0.00	0.00	147.92

16 01 06	Mechanical Systems	Unit	Hourly Output	Crew	Factor	Labor	Equip.	Matl.	Total
16 01 0615	**Remove Misc. Fixtures**	**EA**				**51.56**	**0.00**	**0.00**	**51.56**
02111 9106	Remove Miscellaneous Fixtures	EA	1	MPLUA	1.000	51.56	0.00	0.00	51.56
16 01 0616	**Remove Double Bowl Sink/Janitorial Sink**	**EA**				**75.66**	**0.00**	**0.00**	**75.66**
02111 9107	Remove Double Bowl Sink/Janitorial Sink	EA	1.176	MPLUE	1.000	75.66	0.00	0.00	75.66
16 01 0617	**Remove Emergency Shower with Eye Wash**	**EA**				**114.81**	**0.00**	**0.00**	**114.81**
02111 9108	Remove Emergency Shower with Eye Wash	EA	0.775	MPLUE	1.000	114.81	0.00	0.00	114.81
16 01 0618	**Remove Electric Hot Water Heater to 50 Gallons**	**EA**				**118.63**	**0.00**	**0.00**	**118.63**
02111 9109	Remove Electric Hot Water Heater to 50 Gallon	EA	0.75	MPLUE	1.000	118.63	0.00	0.00	118.63
16 01 0619	**Remove Gas Hot Water Heater to 100 Gallons**	**EA**				**237.27**	**0.00**	**0.00**	**237.27**
02111 9110	Remove Gas Hot Water Heater to 100 Gallon	EA	0.375	MPLUE	1.000	237.27	0.00	0.00	237.27
16 01 0620	**Remove Oil-fired Hot Water Heater to 50 Gallons**	**EA**				**118.63**	**0.00**	**0.00**	**118.63**
02111 9111	Remove Oil Fired Hot Water Heater to 50 Gallon	EA	0.75	MPLUE	1.000	118.63	0.00	0.00	118.63
16 01 0621	**Remove Metal Pipe to 2"**	**LF**				**2.06**	**0.00**	**0.00**	**2.06**
02067 2000	Demolish Metal Pipe to 2" (5 cm) Diameter	LF	25	MPLUA	1.000	2.06	0.00	0.00	2.06
16 01 0622	**Remove Metal Pipe to 4"**	**LF**				**3.44**	**0.00**	**0.00**	**3.44**
02067 2050	Demolish Metal Pipe to 4" (10 cm) Diameter	LF	15	MPLUA	1.000	3.44	0.00	0.00	3.44
16 01 0623	**Remove Metal Pipe to 8"**	**LF**				**7.60**	**0.00**	**0.00**	**7.60**
02067 2100	Demolish Metal Pipe to 8" (20 cm) Diameter	LF	12.5	MPLUB	1.000	7.60	0.00	0.00	7.60
16 01 0624	**Remove Metal Pipe to 16"**	**LF**				**15.19**	**0.00**	**0.00**	**15.19**
02067 2150	Demolish Metal Pipe to 16" (41 cm) Diameter	LF	6.25	MPLUB	1.000	15.19	0.00	0.00	15.19
16 01 0625	**Remove Plastic Gas Pipe to 6" Diameter**	**LF**				**10.31**	**0.00**	**0.00**	**10.31**
02111 9221	Remove Plastic Gas Pipe to 6" Diameter	LF	5	MPLUA	1.000	10.31	0.00	0.00	10.31
16 01 0626	**Remove PVC Pipe to 4" Aboveground**	**LF**				**1.35**	**0.00**	**0.00**	**1.35**
02111 9222	Remove PVC Pipe to 4" Above Ground	LF	66	MPLUE	1.000	1.35	0.00	0.00	1.35
16 01 0627	**Remove Copper Pipe to 4"**	**LF**				**3.79**	**0.00**	**0.00**	**3.79**
02111 9231	Remove Copper Pipe to 4"	LF	23.5	MPLUE	1.000	3.79	0.00	0.00	3.79
16 01 0628	**Remove Gas Meter & Regulator**	**EA**				**51.56**	**0.00**	**0.00**	**51.56**
02111 9241	Remove Gas Meter & Regulator	EA	1	MPLUA	1.000	51.56	0.00	0.00	51.56
16 01 0629	**Remove Valve to 4"**	**EA**				**63.30**	**0.00**	**0.00**	**63.30**
02111 9242	Remove Valve to 4"	EA	1.5	MPLUB	1.000	63.30	0.00	0.00	63.30
16 01 0630	**Remove Backflow Preventor to 2"**	**EA**				**68.44**	**0.00**	**0.00**	**68.44**
02111 9243	Remove Back Flow Preventor to 2"	EA	1.3	MPLUE	1.000	68.44	0.00	0.00	68.44

16 01 06 Mechanical Systems		Unit	Hourly Output	Crew	Factor	Labor	Equip.	Matl.	Total
16 01 0631	**Remove Backflow Preventor to 8"**	**EA**				**121.66**	**0.00**	**0.00**	**121.66**
02111 9244	Remove Back Flow Preventor to 8"	EA	1	MPLUJ	1.000	121.66	0.00	0.00	121.66
16 01 0632	**Remove Floor Drain to 4"**	**EA**				**52.34**	**0.00**	**0.00**	**52.34**
02111 9245	Remove Floor Drain to 4"	EA	1.7	MPLUE	1.000	52.34	0.00	0.00	52.34
16 01 0633	**Remove Roof Drain to 4"**	**EA**				**52.34**	**0.00**	**0.00**	**52.34**
02111 9246	Remove Roof Drain to 4"	EA	1.7	MPLUE	1.000	52.34	0.00	0.00	52.34
16 01 0634	**Remove Pump to 1 HP**	**EA**				**81.26**	**0.00**	**0.00**	**81.26**
02111 9251	Remove Pump to 1 HP	EA	1.5	MSPFI	1.000	81.26	0.00	0.00	81.26
16 01 0635	**Remove Pump to 5 HP**	**EA**				**81.26**	**0.00**	**0.00**	**81.26**
02111 9252	Remove Pump to 5 HP	EA	1.5	MSPFI	1.000	81.26	0.00	0.00	81.26
16 01 0636	**Remove Pump to 25 HP**	**EA**				**319.11**	**44.92**	**0.00**	**364.03**
02111 9253	Remove Pump to 25 HP	EA	0.32	MSPFF	1.000	319.11	44.92	0.00	364.03
16 01 0637	**Remove Pump to 50 HP**	**EA**				**351.93**	**71.88**	**0.00**	**423.81**
02111 9254	Remove Pump to 50 HP	EA	0.3	MSPFG	1.000	351.93	71.88	0.00	423.81
16 01 0638	**Remove Pump to 100 HP**	**EA**				**351.93**	**71.88**	**0.00**	**423.81**
02111 9255	Remove Pump to 100 HP	EA	0.3	MSPFG	1.000	351.93	71.88	0.00	423.81
16 01 0639	**Remove Pump to 200 HP**	**EA**				**422.32**	**86.25**	**0.00**	**508.57**
02111 9256	Remove Pump to 200 HP	EA	0.25	MSPFG	1.000	422.32	86.25	0.00	508.57
16 01 0640	**Remove Diesel Fire Pump to 2,500 GPM**	**EA**				**392.76**	**28.75**	**0.00**	**421.51**
02111 9257	Remove Diesel Fire Pump to 2,500 GPM	EA	0.25	MSPID	1.000	392.76	28.75	0.00	421.51
16 01 0641	**Remove Hydraulic Pump to 75 HP**	**EA**				**844.64**	**172.50**	**0.00**	**1,017**
02111 9258	Remove Hydraulic Pump to 75 HP	EA	0.125	MSPFG	1.000	844.64	172.50	0.00	1,017
16 01 0642	**Remove Boiler to 500 MBH**	**EA**				**2.51**	**0.00**	**0.00**	**2.51**
02111 9261	Remove Boiler 0 to 500 MBH	MBH	48.51	MSPFI	1.000	2.51	0.00	0.00	2.51
16 01 0643	**Remove Boiler to 1,000 MBH**	**EA**				**2.01**	**0.14**	**0.00**	**2.15**
02111 9262	Remove Boiler 501 to 5,000 MBH	MBH	60.64	MSPFJ	1.000	2.01	0.14	0.00	2.15
16 01 0644	**Remove Boiler to 5,000 MBH**	**EA**				**1.66**	**0.11**	**0.00**	**1.78**
02111 9263	Remove Boiler 5,001 to 10,000 MBH	MBH	73.29	MSPFJ	1.000	1.66	0.11	0.00	1.78
16 01 0645	**Remove Boiler to 10,000 MBH**	**EA**				**1.34**	**0.24**	**0.00**	**1.58**
02111 9264	Remove Boiler over 10,000 MBH	MBH	160.3	MSPFX	1.000	1.34	0.24	0.00	1.58
16 01 0647	**Remove Chiller to 50 Ton**	**EA**				**84.95**	**11.91**	**0.00**	**96.87**
02111 9411	Remove Chiller 0 to 50 Ton	TON	1.81	MSPFP	1.000	84.95	11.91	0.00	96.87

16 01 06 Mechanical Systems		Unit	Hourly Output	Crew	Factor	Labor	Equip.	Matl.	Total
16 01 0648	**Remove Chiller to 100 Ton**	**EA**				**89.79**	**17.22**	**0.00**	**107.01**
02111 9412	Remove Chiller 51 to 200 Ton	TON	2.087	MSPFW	1.000	89.79	17.22	0.00	107.01
16 01 0649	**Remove Chiller to 150 Ton**	**EA**				**51.34**	**9.16**	**0.00**	**60.50**
02111 9413	Remove Chiller 201 to 500 Ton	TON	4.182	MSPFX	1.000	51.34	9.16	0.00	60.50
16 01 0650	**Remove Chiller to 200 Ton**	**EA**				**38.69**	**6.90**	**0.00**	**45.59**
02111 9414	Remove Chiller 501 to 1,000 Ton	TON	5.549	MSPFX	1.000	38.69	6.90	0.00	45.59
16 01 0651	**Remove Chiller to 500 Ton**	**EA**				**26.70**	**4.76**	**0.00**	**31.46**
02111 9415	Remove Chiller over 1,000 Ton	TON	8.042	MSPFX	1.000	26.70	4.76	0.00	31.46
16 01 0654	**Remove Cooling Tower to 200 Ton**	**EA**				**1,260**	**176.74**	**0.00**	**1,437**
02111 9451	Remove Cooling Tower to 200 Ton	EA	0.122	MSPFQ	1.000	1,260	176.74	0.00	1,437
16 01 0655	**Remove Cooling Tower to 300 Ton**	**EA**				**1,373**	**192.52**	**0.00**	**1,565**
02111 9452	Remove Cooling Tower to 300 Ton	EA	0.112	MSPFQ	1.000	1,373	192.52	0.00	1,565
16 01 0656	**Remove Cooling Tower to 500 Ton**	**EA**				**2,460**	**345.00**	**0.00**	**2,805**
02111 9453	Remove Cooling Tower to 500 Ton	EA	0.063	MSPFQ	1.000	2,460	345.00	0.00	2,805
16 01 0657	**Remove Cooling Tower to 1,000 Ton**	**EA**				**3,696**	**792.84**	**0.00**	**4,489**
02111 9454	Remove Cooling Tower to 1,000 Ton	EA	0.042	MSPFT	1.000	3,696	792.84	0.00	4,489
16 01 0658	**Remove Condensing Unit to 10 Ton**	**EA**				**145.88**	**20.54**	**0.00**	**166.41**
02111 9471	Remove Condensing Unit to 10 Ton	EA	0.7	MSPFF	1.000	145.88	20.54	0.00	166.41
16 01 0659	**Remove Condensing Unit to 20 Ton**	**EA**				**408.46**	**57.50**	**0.00**	**465.96**
02111 9472	Remove Condensing Unit to 20 Ton	EA	0.25	MSPFF	1.000	408.46	57.50	0.00	465.96
16 01 0660	**Remove Condensing Unit to 30 Ton**	**EA**				**1,021**	**143.75**	**0.00**	**1,165**
02111 9473	Remove Condensing Unit to 30 Ton	EA	0.1	MSPFF	1.000	1,021	143.75	0.00	1,165
16 01 0661	**Remove Condensing Unit to 150 Ton**	**EA**				**1,929**	**431.42**	**0.00**	**2,360**
02111 9474	Remove Condensing Unit to 150 Ton	EA	0.083	MSPFR	1.000	1,929	431.42	0.00	2,360
16 01 0662	**Remove Computer Room Air Conditioner to 10 Ton**	**EA**				**560.65**	**0.00**	**0.00**	**560.65**
02111 9481	Remove Computer Room Air Conditioning to 10 Ton	EA	0.167	MSHMD	1.000	560.65	0.00	0.00	560.65
16 01 0663	**Remove Computer Room Air Conditioner to 25 Ton**	**EA**				**803.12**	**115.00**	**0.00**	**918.12**
02111 9482	Remove Computer Room Air Conditioning to 25 Ton	EA	0.125	MSHME	1.000	803.12	115.00	0.00	918.12
16 01 0664	**Remove Gas Furnace to 500 MBH**	**EA**				**1,499**	**287.50**	**0.00**	**1,787**
02111 9511	Remove Gas Furnace to 500 MBH	EA	0.125	MSPFW	1.000	1,499	287.50	0.00	1,787

16 01 06 Mechanical Systems		Unit	Hourly Output	Crew	Factor	Labor	Equip.	Matl.	Total
16 01 0665	**Remove Unit Heater to 50 MBH**	**EA**				**166.73**	**0.00**	**0.00**	**166.73**
02111 9515	Remove Unit Heater to 50 MBH	EA	0.62	MSPFC	1.000	166.73	0.00	0.00	166.73
16 01 0666	**Remove Unit Heater to 500 MBH**	**EA**				**408.46**	**57.50**	**0.00**	**465.96**
02111 9516	Remove Unit Heater to 500 MBH	EA	0.25	MSPFF	1.000	408.46	57.50	0.00	465.96
16 01 0667	**Remove Heating Coils to 10,000 CFM**	**EA**				**196.90**	**0.00**	**0.00**	**196.90**
02111 9517	Remove Heating Coils to 10,000 CFM	EA	0.525	MSPFC	1.000	196.90	0.00	0.00	196.90
16 01 0668	**Remove Heat Exchanger to 500 GPM**	**EA**				**68.92**	**0.00**	**0.00**	**68.92**
02111 9521	Remove Heat Exchanger to 500 GPM	EA	1.5	MSPFC	1.000	68.92	0.00	0.00	68.92
16 01 0669	**Remove Heat Exchanger to 5,000 GPM**	**EA**				**408.46**	**57.50**	**0.00**	**465.96**
02111 9522	Remove Heat Exchanger to 5,000 GPM	EA	0.25	MSPFF	1.000	408.46	57.50	0.00	465.96
16 01 0670	**Remove Heat Pump to 5 Ton**	**EA**				**199.97**	**0.00**	**0.00**	**199.97**
02111 9551	Remove Heat Pump to 5 Ton	EA	0.476	MSPFB	1.000	199.97	0.00	0.00	199.97
16 01 0671	**Remove Heat Pump to 15 Ton**	**EA**				**816.92**	**115.00**	**0.00**	**931.92**
02111 9552	Remove Heat Pump to 15 Ton	EA	0.125	MSPFF	1.000	816.92	115.00	0.00	931.92
16 01 0672	**Remove Fan to 2,000 CFM**	**EA**				**186.92**	**0.00**	**0.00**	**186.92**
02111 9611	Remove Fan to 2,000 CFM	EA	0.5	MSHMD	1.000	186.92	0.00	0.00	186.92
16 01 0673	**Remove Fan to 5,000 CFM**	**EA**				**207.69**	**0.00**	**0.00**	**207.69**
02111 9612	Remove Fan to 5,000 CFM	EA	0.45	MSHMD	1.000	207.69	0.00	0.00	207.69
16 01 0674	**Remove Fan to 10,000 CFM**	**EA**				**233.65**	**0.00**	**0.00**	**233.65**
02111 9613	Remove Fan to 10,000 CFM	EA	0.4	MSHMD	1.000	233.65	0.00	0.00	233.65
16 01 0675	**Remove Fan to 50,000 CFM**	**EA**				**467.30**	**0.00**	**0.00**	**467.30**
02111 9614	Remove Fan to 50,000 CFM	EA	0.2	MSHMD	1.000	467.30	0.00	0.00	467.30
16 01 0676	**Remove Roof Fan to 60"**	**EA**				**186.92**	**0.00**	**0.00**	**186.92**
02111 9615	Remove Roof Fan to 60"	EA	0.5	MSHMD	1.000	186.92	0.00	0.00	186.92
16 01 0677	**Remove Low/Medium Pressure Duct**	**LB**				**1.72**	**0.00**	**0.00**	**1.72**
02111 9711	Remove Low/Medium Pressure Duct	LB	70	MSHMF	1.000	1.72	0.00	0.00	1.72
16 01 0678	**Remove Galvanized Duct to 16" Diameter**	**LF**				**3.12**	**0.00**	**0.00**	**3.12**
02111 9715	Remove Galvanized Duct to 16" Diameter	LF	30	MSHMB	1.000	3.12	0.00	0.00	3.12
16 01 0679	**Remove Galvanized Duct to 24" Diameter**	**LF**				**4.67**	**0.00**	**0.00**	**4.67**
02111 9716	Remove Galvanized Duct to 24" Diameter	LF	20	MSHMB	1.000	4.67	0.00	0.00	4.67
16 01 0680	**Remove Flexible Duct to 20" Diameter**	**LF**				**4.01**	**0.00**	**0.00**	**4.01**
02111 9717	Remove Flexible Duct to 20" Diameter	LF	30	MSHMF	1.000	4.01	0.00	0.00	4.01

16 01 06 Mechanical Systems		Unit	Hourly Output	Crew	Factor	Labor	Equip.	Matl.	Total
16 01 0681	**Remove Air-handling Unit to 2.0 MCFM**	**EA**				**342.53**	**0.00**	**0.00**	**342.53**
02111 9721	Remove Air Handling Unit to 2 MCFM	EA	0.412	MSHMG	1.000	342.53	0.00	0.00	342.53
16 01 0682	**Remove Air-handling Unit to 8.0 MCFM**	**EA**				**706.30**	**0.00**	**0.00**	**706.30**
02111 9722	Remove Air Handling Unit to 8 MCFM	EA	0.2	MSHMG	1.000	706.30	0.00	0.00	706.30
16 01 0683	**Remove Air-handling Unit to 16.5 MCFM**	**EA**				**1,413**	**0.00**	**0.00**	**1,413**
02111 9723	Remove Air Handling Unit to 16.5 MCFM	EA	0.1	MSHMG	1.000	1,413	0.00	0.00	1,413
16 01 0684	**Remove Air-handling Unit to 28.0 MCFM**	**EA**				**3,214**	**718.75**	**0.00**	**3,932**
02111 9724	Remove Air Handling Unit to 28 MCFM	EA	0.05	MSPFS	1.000	3,214	718.75	0.00	3,932
16 01 0685	**Remove Air-handling Unit to 40.0 MCFM**	**EA**				**5,150**	**1,152**	**0.00**	**6,302**
02111 9725	Remove Air Handling Unit to 40 MCFM	EA	0.031	MSPFS	1.000	5,150	1,152	0.00	6,302
16 01 0686	**Remove Rooftop Air-handling Unit/Air Conditioner/Heating Unit to 2 MCFM**	**EA**				**332.38**	**0.00**	**0.00**	**332.38**
02111 9726	Remove Rooftop Air Handling Unit/Air Conditioner/Heating Unit to 2 MCFM	EA	0.425	MSHMG	1.000	332.38	0.00	0.00	332.38
16 01 0687	**Remove Rooftop Air-handling Unit/Air Conditioner/Heating Unit to 8 MCFM**	**EA**				**706.30**	**0.00**	**0.00**	**706.30**
02111 9727	Remove Rooftop Air Handling Unit/Air Conditioner/Heating Unit to 8 MCFM	EA	0.2	MSHMG	1.000	706.30	0.00	0.00	706.30
16 01 0688	**Remove Rooftop Air-handling Unit/Air Conditioner/Heating Unit to 16 MCFM**	**EA**				**1,413**	**0.00**	**0.00**	**1,413**
02111 9728	Remove Rooftop Air Handling Unit/Air Conditioner/Heating Unit to 16 MCFM	EA	0.1	MSHMG	1.000	1,413	0.00	0.00	1,413
16 01 0689	**Remove Rooftop Air-handling Unit/Air Conditioner/Heating Unit to 28 MCFM**	**EA**				**3,214**	**718.75**	**0.00**	**3,932**
02111 9729	Remove Rooftop Air Handling Unit/Air Conditioner/Heating Unit to 28 MCFM	EA	0.05	MSPFS	1.000	3,214	718.75	0.00	3,932
16 01 0690	**Remove Rooftop Air-handling Unit/Air Conditioner/Heating Unit to 47 MCFM**	**EA**				**6,427**	**1,438**	**0.00**	**7,865**
02111 9730	Remove Rooftop Air Handling Unit/Air Conditioner/Heating Unit to 47 MCFM	EA	0.025	MSPFS	1.000	6,427	1,438	0.00	7,865
16 01 0691	**Remove Fan Coil Unit to .8 MCFM**	**EA**				**106.03**	**0.00**	**0.00**	**106.03**
02111 9731	Remove Fan Coil Unit to .8 MCFM	EA	0.975	MSPFC	1.000	106.03	0.00	0.00	106.03
16 01 0692	**Remove VAV to 1,400 CFM**	**EA**				**363.80**	**0.00**	**0.00**	**363.80**
02111 9741	Remove VAV to 1,400 CFM	EA	0.25	MSHMC	1.000	363.80	0.00	0.00	363.80
16 01 0693	**Remove VAV to 2.8 MCFM**	**EA**				**727.60**	**0.00**	**0.00**	**727.60**
02111 9742	Remove VAV to 2.8 MCFM	EA	0.125	MSHMC	1.000	727.60	0.00	0.00	727.60
16 01 0694	**Remove Register/Diffuser**	**EA**				**19.27**	**0.00**	**0.00**	**19.27**
02111 9751	Remove Diffuser - 24" x 24"	EA	4.72	MSHMC	1.000	19.27	0.00	0.00	19.27

16 01 06	Mechanical Systems	Unit	Hourly Output	Crew	Factor	Labor	Equip.	Matl.	Total
16 01 0695	**Remove Roof Vent to 60" Square**	**EA**				**93.46**	**0.00**	**0.00**	**93.46**
02111 9755	Remove Roof Vent to 60" Square	EA	1	MSHMD	1.000	93.46	0.00	0.00	93.46
16 01 0696	**Remove Louver to 60" Square**	**SF**				**5.29**	**0.00**	**0.00**	**5.29**
02111 9756	Remove Louver to 60" Square	SF	9.5	MSHMA	1.000	5.29	0.00	0.00	5.29
16 01 0697	**Remove HEPA Filter to 24" Square**	**EA**				**77.88**	**0.00**	**0.00**	**77.88**
02111 9761	Remove HEPA Filter to 24" Square	EA	1.2	MSHMB	1.000	77.88	0.00	0.00	77.88

16 01 07	Electrical Systems	Unit	Hourly Output	Crew	Factor	Labor	Equip.	Matl.	Total
16 01 0701	**Remove 4' Fluorescent Light Fixtures 1 - 4 Lamp**	**EA**				**40.02**	**0.00**	**0.00**	**40.02**
02111 8103	Remove 4' Fluorescent Light Fixtures 1 - 4 Lamp	EA	2.619	EELEC	1.000	40.02	0.00	0.00	40.02
16 01 0702	**Remove 8' Fluorescent Light Fixtures 1 - 4 Lamp**	**EA**				**47.95**	**0.00**	**0.00**	**47.95**
02111 8104	Remove 8' Fluorescent Light Fixtures 1 - 4 Lamp	EA	2.186	EELEC	1.000	47.95	0.00	0.00	47.95
16 01 0703	**Remove Recessed Can Light Fixture**	**EA**				**28.31**	**0.00**	**0.00**	**28.31**
02111 8105	Remove Recessed Can Light Fixture	EA	3.703	EELEC	1.000	28.31	0.00	0.00	28.31
16 01 0704	**Remove Floodlight Fixture to 1,000 W**	**EA**				**69.89**	**0.00**	**0.00**	**69.89**
02111 8106	Remove Flood Light Fixture to 1,000W	EA	1.5	EELEC	1.000	69.89	0.00	0.00	69.89
16 01 0705	**Remove Wall-mount Fixture to 1,000 W**	**EA**				**69.89**	**0.00**	**0.00**	**69.89**
02111 8107	Remove Wall Mount Fixture to 1,000W	EA	1.5	EELEC	1.000	69.89	0.00	0.00	69.89
16 01 0706	**Remove Ceiling/Pendant-Mount Fixture to 300 W**	**EA**				**35.17**	**0.00**	**0.00**	**35.17**
02111 8108	Remove Ceiling/Pendant Mount Fixture to 300W	EA	2.981	EELEC	1.000	35.17	0.00	0.00	35.17
16 01 0707	**Remove Low Bay, High-intensity Discharge Fixture to 400 W**	**EA**				**65.94**	**0.00**	**0.00**	**65.94**
02111 8109	Remove Low Bay, HID Fixture to 400W	EA	1.59	EELEC	1.000	65.94	0.00	0.00	65.94
16 01 0708	**Remove High Bay, High-intensity Discharge Fixture to 400 W**	**EA**				**69.89**	**0.00**	**0.00**	**69.89**
02111 8110	Remove High Bay HID Fixture to 400W	EA	1.5	EELEC	1.000	69.89	0.00	0.00	69.89
16 01 0709	**Remove High Bay, High-intensity Discharge Fixture 400 - 1,000 W**	**EA**				**78.98**	**0.00**	**0.00**	**78.98**
02111 8111	Remove High Bay HID Fixture 400W - 1,000W	EA	1.327	EELEC	1.000	78.98	0.00	0.00	78.98
16 01 0710	**Remove Hazardous/Explosionproof Fixture**	**EA**				**77.96**	**0.00**	**0.00**	**77.96**
02111 8112	Remove Hazardous/Explosion Proof Fixture	EA	1.345	EELEC	1.000	77.96	0.00	0.00	77.96
16 01 0711	**Remove Hazardous/Explosionproof Floor Fixture**	**EA**				**114.82**	**0.00**	**0.00**	**114.82**
02111 8113	Remove Hazardous/Explosion Proof Floor Fixture	EA	0.896	EELEB	1.000	114.82	0.00	0.00	114.82

16 01 07	Electrical Systems	Unit	Hourly Output	Crew	Factor	Labor	Equip.	Matl.	Total
16 01 0712	**Remove Incandescent Fixture to 200 W**	**EA**				**31.97**	**0.00**	**0.00**	**31.97**
02111 8114	Remove Incandescent Fixture to 200W	EA	3.279	EELEC	1.000	31.97	0.00	0.00	31.97
16 01 0713	**Remove Light Fixture Track 4' - 8'**	**EA**				**34.95**	**0.00**	**0.00**	**34.95**
02111 8115	Remove Light Fixture Track 4' - 8'	EA	3	EELEC	1.000	34.95	0.00	0.00	34.95
16 01 0714	**Clean Light Fixture**	**EA**				**34.95**	**0.00**	**0.00**	**34.95**
02111 8116	Clean Light Fixture	EA	3	EELEC	1.000	34.95	0.00	0.00	34.95
16 01 0715	**Remove Main Switchboard 200 - 600 Amps**	**EA**				**523.74**	**0.00**	**0.00**	**523.74**
02111 8211	Remove Main Switchboard 200-600 Amps	EA	0.417	EELEG	1.000	523.74	0.00	0.00	523.74
16 01 0716	**Remove Main Switchboard 800 - 1,200 Amps**	**EA**				**1,050**	**0.00**	**0.00**	**1,050**
02111 8212	Remove Main Switchboard 800-1,200 Amps	EA	0.208	EELEG	1.000	1,050	0.00	0.00	1,050
16 01 0717	**Remove Main Switchboard 1,600 - 3,000 Amps**	**EA**				**1,706**	**0.00**	**0.00**	**1,706**
02111 8213	Remove Main Switchboard 1,600-3,000 Amps	EA	0.128	EELEG	1.000	1,706	0.00	0.00	1,706
16 01 0718	**Remove Main Switchboard 4,000 - 6,000 Amps**	**EA**				**4,282**	**0.00**	**0.00**	**4,282**
02111 8214	Remove Main Switchboard 4,000-6,000 Amps	EA	0.051	EELEG	1.000	4,282	0.00	0.00	4,282
16 01 0719	**Remove Main Switchboard 7,000 - 8,000 Amps**	**EA**				**6,825**	**0.00**	**0.00**	**6,825**
02111 8215	Remove Main Switchboard 7,000-8,000 Amps	EA	0.032	EELEG	1.000	6,825	0.00	0.00	6,825
16 01 0720	**Remove Main Switchboard 9,000 - 10,000 Amps**	**EA**				**8,400**	**0.00**	**0.00**	**8,400**
02111 8216	Remove Main Switchboard 9,000-10,000 Amps	EA	0.026	EELEG	1.000	8,400	0.00	0.00	8,400
16 01 0721	**Remove Panel 50 - 125 Amps**	**EA**				**419.35**	**0.00**	**0.00**	**419.35**
02111 8221	Remove Panel 50 - 125 Amps	EA	0.25	EELEC	1.000	419.35	0.00	0.00	419.35
16 01 0722	**Remove Panel 150 - 225 Amps**	**EA**				**524.19**	**0.00**	**0.00**	**524.19**
02111 8222	Remove Panel 150 - 225 Amps	EA	0.2	EELEC	1.000	524.19	0.00	0.00	524.19
16 01 0723	**Remove Panel 250 - 400 Amps**	**EA**				**628.90**	**0.00**	**0.00**	**628.90**
02111 8223	Remove Panel 250 - 400 Amps	EA	0.167	EELEC	1.000	628.90	0.00	0.00	628.90
16 01 0724	**Remove Panel 450 - 600 Amps**	**EA**				**733.64**	**0.00**	**0.00**	**733.64**
02111 8224	Remove Panel 450 - 600 Amps	EA	0.143	EELEC	1.000	733.64	0.00	0.00	733.64
16 01 0725	**Remove Enclosed Circuit Breakers 30 - 60 Amps**	**EA**				**111.82**	**0.00**	**0.00**	**111.82**
02111 8231	Remove Enclosed Circuit Breakers30 Amps - 60 Amps	EA	0.92	EELEB	1.000	111.82	0.00	0.00	111.82
16 01 0726	**Remove Enclosed Circuit Breakers 100 - 225 Amps**	**EA**				**205.75**	**0.00**	**0.00**	**205.75**
02111 8232	Remove Enclosed Circuit Breakers100 Amps - 225 Amps	EA	0.5	EELEB	1.000	205.75	0.00	0.00	205.75

16 01 07	Electrical Systems	Unit	Hourly Output	Crew	Factor	Labor	Equip.	Matl.	Total
16 01 0727	**Remove Combo Motor Starter Size 0 - 2**	**EA**				**65.53**	**0.00**	**0.00**	**65.53**
02111 8241	Remove Combo Motor Starter Size 0 - 2	EA	1.57	EELEB	1.000	65.53	0.00	0.00	65.53
16 01 0728	**Remove Combo Motor Starter Size 3 - 5**	**EA**				**102.88**	**0.00**	**0.00**	**102.88**
02111 8242	Remove Combo Motor Starter Size 3 - 5	EA	1	EELEB	1.000	102.88	0.00	0.00	102.88
16 01 0729	**Remove MCC Structure Per Section**	**EA**				**109.38**	**0.00**	**0.00**	**109.38**
02111 8243	Remove MCC Structure Per Section	EA	0.941	EELEB	1.000	109.38	0.00	0.00	109.38
16 01 0730	**Remove Secondary Air Circuit Breakers, 3 Phase, 400 - 800 Amps**	**EA**				**205.75**	**0.00**	**0.00**	**205.75**
02111 8246	Remove Secondary Air Circuit Breakers, 3-Phase, 400 - 800 Amps	EA	0.5	EELEB	1.000	205.75	0.00	0.00	205.75
16 01 0731	**Remove Safety Switch 30 Amps**	**EA**				**73.48**	**0.00**	**0.00**	**73.48**
02111 8251	Remove Safety Switch 30 Amps	EA	1.4	EELEB	1.000	73.48	0.00	0.00	73.48
16 01 0732	**Remove Safety Switch 60 Amps**	**EA**				**85.73**	**0.00**	**0.00**	**85.73**
02111 8252	Remove Safety Switch 60 Amps	EA	1.2	EELEB	1.000	85.73	0.00	0.00	85.73
16 01 0733	**Remove Safety Switch 100 Amps**	**EA**				**102.88**	**0.00**	**0.00**	**102.88**
02111 8253	Remove Safety Switch 100 Amps	EA	1	EELEB	1.000	102.88	0.00	0.00	102.88
16 01 0734	**Remove Safety Switch 200 Amps**	**EA**				**128.59**	**0.00**	**0.00**	**128.59**
02111 8254	Remove Safety Switch 200 Amps	EA	0.8	EELEB	1.000	128.59	0.00	0.00	128.59
16 01 0735	**Remove Explosionproof Safety Switch 30 - 60 Amps**	**EA**				**171.46**	**0.00**	**0.00**	**171.46**
02111 8256	Remove Explosion Proof Safety Switch 30 - 60 Amps	EA	0.6	EELEB	1.000	171.46	0.00	0.00	171.46
16 01 0736	**Remove Explosionproof Safety Switch 100 Amps**	**EA**				**187.05**	**0.00**	**0.00**	**187.05**
02111 8257	Remove Explosion Proof Safety Switch100 Amps	EA	0.55	EELEB	1.000	187.05	0.00	0.00	187.05
16 01 0737	**Remove Explosionproof Safety Switch 200 Amps**	**EA**				**205.75**	**0.00**	**0.00**	**205.75**
02111 8258	Remove Explosion Proof Safety Switch 200 Amps	EA	0.5	EELEB	1.000	205.75	0.00	0.00	205.75
16 01 0738	**Remove 25.0 KVA Dry-type Transformer**	**EA**				**88.00**	**9.34**	**0.00**	**97.35**
02111 8270	Remove 25 KVA Dry-type Transformer	EA	1	EELED	1.000	88.00	9.34	0.00	97.35
16 01 0739	**Remove 37.5 KVA Dry-type Transformer**	**EA**				**118.93**	**12.63**	**0.00**	**131.55**
02111 8271	Remove 37.5 KVA Dry-type Transformer	EA	0.74	EELED	1.000	118.93	12.63	0.00	131.55
16 01 0740	**Remove 75.0 KVA Dry-type Transformer**	**EA**				**154.88**	**16.44**	**0.00**	**171.33**
02111 8272	Remove 75 KVA Dry-type Transformer	EA	0.568	EELED	1.000	154.88	16.44	0.00	171.33

16 01 07	Electrical Systems	Unit	Hourly Output	Crew	Factor	Labor	Equip.	Matl.	Total
16 01 0741	**Remove 112.5 KVA Dry-type Transformer**	**EA**				**176.01**	**18.69**	**0.00**	**194.70**
02111 8273	Remove 112.5 KVA Dry-type Transformer	EA	0.5	EELED	1.000	176.01	18.69	0.00	194.70
16 01 0742	**Remove Bus/Feeder Duct 200 - 500 Amps**	**LF**				**9.56**	**0.00**	**0.00**	**9.56**
02111 8275	Remove Bus/Feeder Duct 200 - 500 Amps	LF	20	EELEF	1.000	9.56	0.00	0.00	9.56
16 01 0743	**Remove Bus/Feeder Duct 1,000 - 2,000 Amps**	**LF**				**19.12**	**0.00**	**0.00**	**19.12**
02111 8276	Remove Bus/Feeder Duct 1,000 - 2,000 Amps	LF	10	EELEF	1.000	19.12	0.00	0.00	19.12
16 01 0744	**Remove Bus/Feeder Duct 3,000 - 4,000 Amps**	**LF**				**38.24**	**0.00**	**0.00**	**38.24**
02111 8277	Remove Bus/Feeder Duct 3,000 - 4,000 Amps	LF	5	EELEF	1.000	38.24	0.00	0.00	38.24
16 01 0745	**Remove Electric Metallic Tubing 1/2" to 1"**	**LF**				**1.03**	**0.00**	**0.00**	**1.03**
02111 8311	Remove Electric Metallic Tubing, 1/2" to 1"	LF	100	EELEB	1.000	1.03	0.00	0.00	1.03
16 01 0746	**Remove Electric Metallic Tubing 1 1/4" to 2"**	**LF**				**1.29**	**0.00**	**0.00**	**1.29**
02111 8312	Remove Electric Metallic Tubing, 1-1/4" to 2"	LF	80	EELEB	1.000	1.29	0.00	0.00	1.29
16 01 0747	**Remove Electric Metallic Tubing 2 1/2" to 3 1/2"**	**LF**				**2.57**	**0.00**	**0.00**	**2.57**
02111 8313	Remove Electric Metallic Tubing, 2-1/2" to 3-1/2"	LF	40	EELEB	1.000	2.57	0.00	0.00	2.57
16 01 0748	**Remove Electric Metallic Tubing 4" to 6"**	**LF**				**6.86**	**0.00**	**0.00**	**6.86**
02111 8314	Remove Electric Metallic Tubing, 4" to 6"	LF	15	EELEB	1.000	6.86	0.00	0.00	6.86
16 01 0749	**Remove Rigid Galvanized Steel/PVC Conduit 1/2" to 1"**	**LF**				**1.71**	**0.00**	**0.00**	**1.71**
02111 8315	Remove Rigid Galvanized Steel/PVC Conduit, 1/2" to 1"	LF	60	EELEB	1.000	1.71	0.00	0.00	1.71
16 01 0750	**Remove Rigid Galvanized Steel/PVC Conduit 1 1/4" to 2"**	**LF**				**2.06**	**0.00**	**0.00**	**2.06**
02111 8316	Remove Rigid Galvanized Steel/PVC Conduit, 1-1/4" to 2"	LF	50	EELEB	1.000	2.06	0.00	0.00	2.06
16 01 0751	**Remove Rigid Galvanized Steel/PVC Conduit 2 1/2" to 3 1/2"**	**LF**				**2.71**	**0.00**	**0.00**	**2.71**
02111 8317	Remove Rigid Galvanized Steel/PVC Conduit, 2-1/2" to 3-1/2"	LF	38	EELEB	1.000	2.71	0.00	0.00	2.71
16 01 0752	**Remove Rigid Galvanized Steel/PVC Conduit 4" to 6"**	**LF**				**7.35**	**0.00**	**0.00**	**7.35**
02111 8318	Remove Rigid Galvanized Steel/PVC Conduit, 4" to 6"	LF	14	EELEB	1.000	7.35	0.00	0.00	7.35
16 01 0753	**Remove Flexible Conduit 1/2" - 1" with Connectors**	**LF**				**0.51**	**0.00**	**0.00**	**0.51**
02111 8320	Remove Flexible Conduit, 1/2" - 1" with Connectors	LF	100	EELEA	1.000	0.51	0.00	0.00	0.51

16 01 07 Electrical Systems		Unit	Hourly Output	Crew	Factor	Labor	Equip.	Matl.	Total
16 01 0754	**Remove Flexible Conduit 1 1/4" - 2" with Connectors**	**LF**				**0.64**	**0.00**	**0.00**	**0.64**
02111 8321	Remove Flexible Conduit, 1-1/4" - 2" with Connectors	LF	80	EELEA	1.000	0.64	0.00	0.00	0.64
16 01 0755	**Remove Wire from Conduit #18 - #10**	**MLF**				**63.11**	**0.00**	**0.00**	**63.11**
02111 8331	Remove Wire from Conduit #18 - #10	MLF	1.63	EELEB	1.000	63.11	0.00	0.00	63.11
16 01 0756	**Remove Wire from Conduit #8 - #4**	**MLF**				**102.88**	**0.00**	**0.00**	**102.88**
02111 8332	Remove Wire from Conduit #8 - #4	MLF	1	EELEB	1.000	102.88	0.00	0.00	102.88
16 01 0757	**Remove Wire from Conduit #3 - #1/0**	**MLF**				**171.46**	**0.00**	**0.00**	**171.46**
02111 8333	Remove Wire from Conduit #3 - #1/0	MLF	0.6	EELEB	1.000	171.46	0.00	0.00	171.46
16 01 0758	**Remove Wire from Conduit #2/0 - #4/0**	**MLF**				**228.61**	**0.00**	**0.00**	**228.61**
02111 8334	Remove Wire from Conduit #2/0 - #4/0	MLF	0.45	EELEB	1.000	228.61	0.00	0.00	228.61
16 01 0759	**Remove Wire from Conduit #250 MCM - #350 MCM**	**MLF**				**478.04**	**0.00**	**0.00**	**478.04**
02111 8335	Remove Wire from Conduit #250 MCM - #350 MCM	MLF	0.4	EELEF	1.000	478.04	0.00	0.00	478.04
16 01 0760	**Remove Wire from Conduit #400 MCM - #500 MCM**	**MLF**				**503.20**	**0.00**	**0.00**	**503.20**
02111 8336	Remove Wire from Conduit #400 MCM - #500 MCM	MLF	0.38	EELEF	1.000	503.20	0.00	0.00	503.20
16 01 0761	**Remove Wire #24 - #18, to 20 Pair**	**MLF**				**102.88**	**0.00**	**0.00**	**102.88**
02111 8337	Remove Wire #24 - #18, to 20 Pair	MLF	1	EELEB	1.000	102.88	0.00	0.00	102.88
16 01 0762	**Remove Wire #24 - #18, 20 - 50 Pair**	**MLF**				**171.46**	**0.00**	**0.00**	**171.46**
02111 8338	Remove Wire #24 - #18, 20 - 50 Pair	MLF	0.6	EELEB	1.000	171.46	0.00	0.00	171.46
16 01 0763	**Remove 6" - 10" Cable Tray**	**LF**				**1.07**	**0.00**	**0.00**	**1.07**
02111 8340	Remove 6" - 10" Cable Tray	LF	96	EELEB	1.000	1.07	0.00	0.00	1.07
16 01 0764	**Remove 12" - 18" Cable Tray**	**LF**				**1.47**	**0.00**	**0.00**	**1.47**
02111 8341	Remove 12" - 18" Cable Tray	LF	70	EELEB	1.000	1.47	0.00	0.00	1.47
16 01 0765	**Remove 6" - 10" Cable Tray Fittings**	**EA**				**14.70**	**0.00**	**0.00**	**14.70**
02111 8342	Remove 6" - 10" Cable Tray Fittings	EA	7	EELEB	1.000	14.70	0.00	0.00	14.70
16 01 0766	**Remove 12" - 18" Cable Tray Fittings**	**EA**				**17.74**	**0.00**	**0.00**	**17.74**
02111 8343	Remove 12" - 18" Cable Tray Fittings	EA	5.8	EELEB	1.000	17.74	0.00	0.00	17.74
16 01 0767	**Remove Underfloor Raceway with Fittings**	**SF**				**1.71**	**0.00**	**0.00**	**1.71**
02111 8344	Remove Underfloor Raceway with Fittings	SF	60	EELEB	1.000	1.71	0.00	0.00	1.71
16 01 0768	**Remove 120 - 277 V Receptacle 15 - 60 Amp**	**EA**				**4.29**	**0.00**	**0.00**	**4.29**
02111 8411	Remove 120V - 277V Receptacle 15 - 60 Amp	EA	12	EELEA	1.000	4.29	0.00	0.00	4.29

16 01 07 Electrical Systems		Unit	Hourly Output	Crew	Factor	Labor	Equip.	Matl.	Total
16 01 0769	**Remove 208 - 480 V Receptacle 15 - 60 Amp**	**EA**				**5.14**	**0.00**	**0.00**	**5.14**
02111 8412	Remove 208V - 480V Receptacle 15 - 60 Amp	EA	10	EELEA	1.000	5.14	0.00	0.00	5.14
16 01 0770	**Remove 208 - 480 V Receptacle 70 - 100 Amp**	**EA**				**8.57**	**0.00**	**0.00**	**8.57**
02111 8413	Remove 208V - 480V Receptacle 70 - 100 Amp	EA	6	EELEA	1.000	8.57	0.00	0.00	8.57
16 01 0771	**Remove Plug Mold Receptacle**	**LF**				**1.71**	**0.00**	**0.00**	**1.71**
02111 8414	Remove Plug Mold Receptacle	LF	30	EELEA	1.000	1.71	0.00	0.00	1.71
16 01 0772	**Remove 400 Amp Generator Receptacle**	**EA**				**28.58**	**0.00**	**0.00**	**28.58**
02111 8415	Remove 400 Amp, Generator Receptacle	EA	1.8	EELEA	1.000	28.58	0.00	0.00	28.58
16 01 0773	**Remove 120 V Explosionproof Receptacle to 30 Amps**	**EA**				**19.26**	**0.00**	**0.00**	**19.26**
02111 8420	Remove 120V Explosion Proof Receptacle to 30 Amps	EA	2.67	EELEA	1.000	19.26	0.00	0.00	19.26
16 01 0774	**Remove 120 V, 20 Amp Switch**	**EA**				**3.43**	**0.00**	**0.00**	**3.43**
02111 8431	Remove 120V, 20 Amp Switch	EA	15	EELEA	1.000	3.43	0.00	0.00	3.43
16 01 0775	**Remove Dimmer 2,000 W or Less**	**EA**				**5.14**	**0.00**	**0.00**	**5.14**
02111 8432	Remove Dimmer, 2,000W or Less	EA	10	EELEA	1.000	5.14	0.00	0.00	5.14
16 01 0776	**Remove Switch/Receptacle Plate**	**EA**				**1.71**	**0.00**	**0.00**	**1.71**
02111 8451	Remove Device Plate	EA	30	EELEA	1.000	1.71	0.00	0.00	1.71
16 01 0777	**Remove Outlet Box Up to 4" Square/Octagonal**	**EA**				**5.14**	**0.00**	**0.00**	**5.14**
02111 8452	Remove Outlet Box, Up to 4" Square/Octagonal	EA	10	EELEA	1.000	5.14	0.00	0.00	5.14
16 01 0778	**Remove Explosionproof Outlet Box to 4" x 4"**	**EA**				**10.29**	**0.00**	**0.00**	**10.29**
02111 8453	Remove Explosion Proof Outlet Box to 4" x 4"	EA	5	EELEA	1.000	10.29	0.00	0.00	10.29
16 01 0779	**Remove Explosionproof Outlet Box to 10' x 18"**	**EA**				**12.86**	**0.00**	**0.00**	**12.86**
02111 8454	Remove Explosion Proof Outlet Box to 10' x 18"	EA	4	EELEA	1.000	12.86	0.00	0.00	12.86
16 01 0780	**Remove Plaster Ring/Blank Cover**	**EA**				**2.57**	**0.00**	**0.00**	**2.57**
02111 8455	Remove Plaster Ring/Blank Cover	EA	20	EELEA	1.000	2.57	0.00	0.00	2.57
16 01 0781	**Remove Fixture Whip**	**EA**				**6.43**	**0.00**	**0.00**	**6.43**
02111 8456	Remove Fixture Whip	EA	8	EELEA	1.000	6.43	0.00	0.00	6.43
16 01 0782	**Remove Floor Box & Flange**	**EA**				**34.29**	**0.00**	**0.00**	**34.29**
02111 8457	Remove Floor Box & Flange	EA	3	EELEB	1.000	34.29	0.00	0.00	34.29
16 01 0783	**Remove 6" to 12" Square Junction Box**	**EA**				**25.72**	**0.00**	**0.00**	**25.72**
02111 8458	Remove 6" to 12" Square Junction Box	EA	2	EELEA	1.000	25.72	0.00	0.00	25.72

16 01 Nonhazardous Selective Building Demolition

16 01 07 Electrical Systems		Unit	Hourly Output	Crew	Factor	Labor	Equip.	Matl.	Total
16 01 0784	**Remove Residential Light Fixture/Fan**	**EA**				**25.72**	**0.00**	**0.00**	**25.72**
02058 2500	Remove Residential Light Fixture/Fan	EA	4	EELEB	1.000	25.72	0.00	0.00	25.72
16 01 0785	**Remove Residential 110 V Outlet & Wire**	**EA**				**17.15**	**0.00**	**0.00**	**17.15**
02058 1762	Remove Residential 110V Outlet and Wire	EA	6	EELEB	1.000	17.15	0.00	0.00	17.15
16 01 0786	**Remove Residential 220 V Outlet & Wire**	**EA**				**34.29**	**0.00**	**0.00**	**34.29**
02058 1764	Remove Residential 220V Outlet and Wire	EA	3	EELEB	1.000	34.29	0.00	0.00	34.29
16 01 0787	**Remove Residential Panel & Wire 200 Amp or Less**	**EA**				**411.50**	**0.00**	**0.00**	**411.50**
02058 1240	Remove Residential 200 Amp or Less Load Center with Breakers and Wire	EA	0.25	EELEB	1.000	411.50	0.00	0.00	411.50
16 01 0788	**Remove Residential TV/Telephone Outlet & Wire**	**EA**				**17.15**	**0.00**	**0.00**	**17.15**
02058 1766	Remove Residential TV/Telephone Outlet and Wire	EA	6	EELEB	1.000	17.15	0.00	0.00	17.15
16 01 0789	**Remove Bath Fan Motor**	**EA**				**25.72**	**0.00**	**0.00**	**25.72**
02111 8117	Remove Bath Fan Motor	EA	4	EELEB	1.000	25.72	0.00	0.00	25.72
16 01 0790	**Remove Light Fixture Globe**	**EA**				**20.57**	**0.00**	**0.00**	**20.57**
02111 8118	Remove Light Fixture Globe	EA	5	EELEB	1.000	20.57	0.00	0.00	20.57

16 01 08 Equipment and Furnishings		Unit	Hourly Output	Crew	Factor	Labor	Equip.	Matl.	Total
16 01 0802	**Remove Dock Leveler**	**EA**				**422.45**	**0.00**	**0.00**	**422.45**
02112 9502	Remove Dock Leveler	EA	0.25	ULABD	1.000	422.45	0.00	0.00	422.45
16 01 0804	**Remove Kitchen Equipment/Appliance - Small**	**EA**				**33.88**	**0.00**	**0.00**	**33.88**
02112 9503	Remove Appliance - Small	EA	2	ULABB	1.000	33.88	0.00	0.00	33.88
16 01 0806	**Remove Kitchen Equipment/Appliance - Medium**	**EA**				**67.75**	**0.00**	**0.00**	**67.75**
02112 9504	Remove Appliance - Medium	EA	1	ULABB	1.000	67.75	0.00	0.00	67.75
16 01 0808	**Remove Kitchen Equipment/Appliance - Large**	**EA**				**135.50**	**0.00**	**0.00**	**135.50**
02112 9505	Remove Appliance - Large	EA	0.5	ULABB	1.000	135.50	0.00	0.00	135.50
16 01 0810	**Remove Kitchen Equipment/Appliance - Extra Large**	**EA**				**271.00**	**0.00**	**0.00**	**271.00**
02112 9506	Remove Appliance - Extra Large	EA	0.25	ULABB	1.000	271.00	0.00	0.00	271.00
16 01 0812	**Remove Workstation**	**EA**				**173.25**	**0.00**	**0.00**	**173.25**
02112 9507	Remove Workstation	EA	0.5	ACARB	1.000	173.25	0.00	0.00	173.25
16 01 0814	**Remove Blinds**	**SF**				**0.69**	**0.00**	**0.00**	**0.69**
02112 9508	Remove Blinds	SF	62.5	ACARA	1.000	0.69	0.00	0.00	0.69

16 01 Nonhazardous Selective Building Demolition

16 01 08 Equipment and Furnishings		Unit	Hourly Output	Crew	Factor	Labor	Equip.	Matl.	Total
16 01 0816	**Remove Miscellaneous Portable Fixture (Equipment) Small**	**EA**				**33.88**	**0.00**	**0.00**	**33.88**
02111 4052	Miscellaneous Removal (2 Laborers)	HR	1	ULABB	0.500	67.75	0.00	0.00	33.88
16 01 0818	**Remove Miscellaneous Portable Fixture (Equipment) Large**	**EA**				**67.75**	**0.00**	**0.00**	**67.75**
02111 4052	Miscellaneous Removal (2 Laborers)	HR	1	ULABB	1.000	67.75	0.00	0.00	67.75
16 01 0820	**Remove Miscellaneous Fixed Fixture (Equipment) Small**	**EA**				**101.63**	**0.00**	**0.00**	**101.63**
02111 4052	Miscellaneous Removal (2 Laborers)	HR	1	ULABB	1.500	67.75	0.00	0.00	101.63
16 01 0822	**Remove Miscellaneous Fixed Fixture (Equipment) Large**	**EA**				**203.25**	**0.00**	**0.00**	**203.25**
02111 4052	Miscellaneous Removal (2 Laborers)	HR	1	ULABB	3.000	67.75	0.00	0.00	203.25
16 01 0824	**Remove Stage Curtain and Track**	**SF**				**2.89**	**0.00**	**0.00**	**2.89**
02111 4119	Remove Stage Curtain and Track	SF	30	ACARB	1.000	2.89	0.00	0.00	2.89

16 01 90 Other Nonhazardous Selective Building Demolition		Unit	Hourly Output	Crew	Factor	Labor	Equip.	Matl.	Total
16 01 9020	**Excavation to Remove Mechanical**	**CY**				**0.99**	**1.10**	**0.00**	**2.08**
02228 0302	Trench, 1/2 CY Hydraulic Excavator, Medium Soil, 75 CY/Hour (58 m3)/Hour	CY	63.5	CODEA	1.000	0.99	1.10	0.00	2.08
16 01 9021	**Backfill & Compaction, Mechanical Removal**	**CY**				**0.69**	**0.66**	**0.00**	**1.34**
02215 1220	Backfill Trench with Small Front-end Loader, without Compaction	CY	70	CODLB	1.000	0.69	0.66	0.00	1.34
16 01 9022	**Select Backfill**	**CY**				**0.00**	**0.00**	**7.90**	**7.90**
02081 2242	Select Granular Fill (Sand)	CY	0	N/A	1.000	0.00	0.00	7.90	7.90
16 01 9023	**Pump Out Tank**	**GAL**				**2.37**	**0.00**	**0.00**	**2.37**
01957 3125	Pump Out Tank	GAL	40	MPLUB	1.000	2.37	0.00	0.00	2.37
16 01 9024	**Purge Tank**	**KGAL**				**26.27**	**1.25**	**0.02**	**27.54**
01957 3115	Purge Tank with Dry Ice	KGAL	3.125	ULABJ	1.000	26.27	1.25	0.02	27.54
16 01 9025	**Test Soils @ Tank Removal**	**EA**				**0.00**	**0.00**	**78.50**	**78.50**
01954 5191	Soil Gas Investigation and Analysis	DAY	0	N/A	1.000	0.00	0.00	78.50	78.50
16 01 9030	**Remove Tank to 100 Gallons**	**EA**				**51.69**	**0.00**	**0.00**	**51.69**
02111 9271	Remove Tank to 100 Gallon	EA	2	MSPFC	1.000	51.69	0.00	0.00	51.69
16 01 9031	**Remove Tank to 300 Gallons**	**EA**				**103.38**	**0.00**	**0.00**	**103.38**
02111 9272	Remove Tank to 300 Gallon	EA	1	MSPFC	1.000	103.38	0.00	0.00	103.38
16 01 9032	**Remove Tank to 1,000 Gallons**	**EA**				**286.72**	**0.00**	**0.00**	**286.72**
02111 9273	Remove Tank to 1,000 Gallon	EA	0.5	MSPFL	1.000	286.72	0.00	0.00	286.72

16 01 90	Other Nonhazardous Selective Building Demolition	Unit	Hourly Output	Crew	Factor	Labor	Equip.	Matl.	Total
16 01 9033	**Remove Tank to 5,000 Gallons**	**EA**				**615.02**	**86.25**	**0.00**	**701.27**
02111 9274	Remove Tank to 5,000 Gallon	EA	0.25	MSPFP	1.000	615.02	86.25	0.00	701.27
16 01 9034	**Remove Tank to 10,000 Gallons**	**EA**				**1,281**	**179.69**	**0.00**	**1,461**
02111 9275	Remove Tank to 10,000 Gallon	EA	0.12	MSPFP	1.000	1,281	179.69	0.00	1,461
16 01 9035	**Remove Tank to 50,000 Gallons**	**EA**				**2,563**	**359.38**	**0.00**	**2,922**
02111 9276	Remove Tank to 50,000 Gallon	EA	0.06	MSPFQ	1.000	2,563	359.38	0.00	2,922
16 01 9036	**Remove Acid Neutralization Tank to 100 Gallons**	**EA**				**189.91**	**0.00**	**0.00**	**189.91**
02111 9281	Remove Acid Neutralization Tank to 100 Gallon	EA	0.5	MPLUB	1.000	189.91	0.00	0.00	189.91
16 01 9037	**Remove Oil Interceptor to 100 GPM**	**EA**				**576.03**	**0.00**	**0.00**	**576.03**
02111 9282	Remove Oil Interceptor to 100 GPM	EA	0.193	MPLUF	1.000	576.03	0.00	0.00	576.03
16 01 9038	**Remove Grease Interceptor to 2,000 Gallons**	**EA**				**715.76**	**0.00**	**0.00**	**715.76**
02111 9283	Remove Grease Interceptor to 2,000 Gallon	EA	0.155	MPLUF	1.000	715.76	0.00	0.00	715.76
16 01 9039	**Remove Water Softener to 58 GPM**	**EA**				**108.52**	**0.00**	**0.00**	**108.52**
02111 9284	Remove Water Softener to 58 GPM	EA	0.875	MPLUB	1.000	108.52	0.00	0.00	108.52
16 01 9040	**Remove Cooler Unit to 110 MBH**	**EA**				**249.23**	**0.00**	**0.00**	**249.23**
02111 9811	Remove Cooler Unit to 110 MBH	EA	0.375	MSHMD	1.000	249.23	0.00	0.00	249.23
16 01 9041	**Remove Fire Hose Station to 100'**	**EA**				**75.91**	**0.00**	**0.00**	**75.91**
02111 9821	Remove Fire Hose Station to 100'	EA	1.25	MSPIC	1.000	75.91	0.00	0.00	75.91
16 01 9042	**Remove Dust Collector to 4,500 CFM**	**EA**				**747.68**	**0.00**	**0.00**	**747.68**
02111 9831	Remove Dust Collector to 4,500 CFM	EA	0.125	MSHMD	1.000	747.68	0.00	0.00	747.68
16 01 9043	**Remove Oil/Grease Pumping Unit**	**EA**				**255.45**	**71.88**	**0.00**	**327.32**
02111 9832	Remove Oil/Grease Pumping Unit	EA	0.5	MPLUS	1.000	255.45	71.88	0.00	327.32
16 01 9044	**Remove Oil/Grease Hose Reel**	**EA**				**60.83**	**0.00**	**0.00**	**60.83**
02111 9833	Remove Oil/Grease Hose Reel	EA	2	MPLUJ	1.000	60.83	0.00	0.00	60.83
16 01 9045	**Remove Control Valve Oil/Grease**	**EA**				**121.66**	**0.00**	**0.00**	**121.66**
02111 9834	Remove Control Valve Oil/Grease	EA	1	MPLUJ	1.000	121.66	0.00	0.00	121.66
16 01 9046	**Remove Liquid Oxygen Tank to 6,000 Gallons**	**EA**				**2,042**	**287.50**	**0.00**	**2,330**
02111 9835	Remove Liquid Oxygen Tank to 6,000 Gallon	EA	0.05	MSPFF	1.000	2,042	287.50	0.00	2,330
16 01 9047	**Remove Liquid Vaporizer**	**EA**				**951.85**	**0.00**	**0.00**	**951.85**
02111 9836	Remove Liquid Vaporizer	EA	0.1	MSPFB	1.000	951.85	0.00	0.00	951.85
16 01 9048	**Remove Liquid Nitrous Oxide Tank**	**EA**				**761.48**	**0.00**	**0.00**	**761.48**
02111 9837	Remove Liquid Nitrous Oxide Tank	EA	0.125	MSPFB	1.000	761.48	0.00	0.00	761.48

16 01 90	Other Nonhazardous Selective Building Demolition	Unit	Hourly Output	Crew	Factor	Labor	Equip.	Matl.	Total
16 01 9049	**Remove Air Compressor/Vacuum Pump to 10 HP**	**EA**				**609.42**	**0.00**	**0.00**	**609.42**
02111 9838	Remove Air Compressor/Vacuum Pump to 10 HP	EA	0.2	MSPFI	1.000	609.42	0.00	0.00	609.42
16 01 9050	**Remove Air Compressor/Vacuum Pump to 75 HP**	**EA**				**1,625**	**0.00**	**0.00**	**1,625**
02111 9839	Remove Air Compressor/Vacuum Pump to 75 HP	EA	0.075	MSPFI	1.000	1,625	0.00	0.00	1,625
16 01 9052	**Remove Duct Board**	**SF**				**0.62**	**0.00**	**0.00**	**0.62**
02111 9718	Remove Duct Board	SF	150	MSHMD	1.000	0.62	0.00	0.00	0.62
16 01 9053	**Remove Baseboard Heater to 8' Long**	**EA**				**59.07**	**0.00**	**0.00**	**59.07**
02111 9518	Remove Baseboard Heater to 8' Long	EA	1.75	MSPFC	1.000	59.07	0.00	0.00	59.07
16 01 9054	**Remove Exhaust Hood**	**EA**				**14.26**	**0.00**	**0.00**	**14.26**
02111 9757	Remove Exhaust Hood	SF	8.425	MSHMF	1.000	14.26	0.00	0.00	14.26
16 01 9055	**Remove Refrigerant Dryer to 100 SCFM**	**EA**				**135.43**	**0.00**	**0.00**	**135.43**
02111 9814	Remove Refrigerant Air Dryer to 100 SCFM	EA	0.9	MSPFI	1.000	135.43	0.00	0.00	135.43
16 01 9056	**Remove Oil/Water Filter**	**EA**				**34.82**	**0.00**	**0.00**	**34.82**
02111 9285	Remove Oil/Water Filter	EA	3.5	MSPFI	1.000	34.82	0.00	0.00	34.82
16 01 9057	**Remove Unit Cooler to 60 MBH**	**EA**				**93.46**	**0.00**	**0.00**	**93.46**
02111 9812	Remove Unit Cooler to 60 MBH	EA	1	MSHMD	1.000	93.46	0.00	0.00	93.46
16 01 9058	**Remove Unit Cooler to 120 MBH**	**EA**				**124.61**	**0.00**	**0.00**	**124.61**
02111 9813	Remove Unit Cooler to 120 MBH	EA	0.75	MSHMD	1.000	124.61	0.00	0.00	124.61
16 01 9059	**Remove Vacuum Cleaning System**	**EA**				**220.31**	**0.00**	**0.00**	**220.31**
02111 9840	Remove Vacuum Cleaning System	EA	0.2	USKCA	1.000	220.31	0.00	0.00	220.31
16 01 9060	**Remove Steam Humidifier**	**EA**				**129.22**	**0.00**	**0.00**	**129.22**
02111 9841	Remove Steam Humidifier	EA	0.8	MSPFC	1.000	129.22	0.00	0.00	129.22
16 01 9061	**Remove High-efficiency Particulate Air Filter Housing**	**EA**				**116.83**	**0.00**	**0.00**	**116.83**
02111 9762	Remove High-efficiency Particulate Air Filter Housing	EA	0.8	MSHMB	1.000	116.83	0.00	0.00	116.83
16 01 9062	**Remove Sound Attentuation 36 x 36 x 60**	**EA**				**72.76**	**0.00**	**0.00**	**72.76**
02111 9763	Remove Sound Attentuation 36 x 36 x 60	EA	1.25	MSHMC	1.000	72.76	0.00	0.00	72.76
16 01 9063	**Remove Sound Attentuation 48 x 60 x 60**	**EA**				**90.95**	**0.00**	**0.00**	**90.95**
02111 9764	Remove Sound Attentuation 48 x 60 x 60	EA	1	MSHMC	1.000	90.95	0.00	0.00	90.95
16 01 9064	**Remove Fire Extinguisher**	**EA**				**17.15**	**0.00**	**0.00**	**17.15**
02111 9822	Remove Fire Extinguisher	EA	3	MSPIA	1.000	17.15	0.00	0.00	17.15

16 01 90	Other Nonhazardous Selective Building Demolition	Unit	Hourly Output	Crew	Factor	Labor	Equip.	Matl.	Total
16 01 9065	**Remove Sprinklerhead**	**EA**				**25.72**	**0.00**	**0.00**	**25.72**
02111 9823	Remove Sprinkler Head	EA	4	MSPIB	1.000	25.72	0.00	0.00	25.72
16 01 9066	**Remove Halon System Complete**	**CF**				**1.03**	**0.00**	**0.00**	**1.03**
02111 9824	Remove Halon System Complete	CF	100	MSPIB	1.000	1.03	0.00	0.00	1.03
16 01 9067	**Remove Wash Fountain**	**EA**				**138.68**	**0.00**	**0.00**	**138.68**
02111 9112	Remove Wash Fountain	EA	0.8	MPLUF	1.000	138.68	0.00	0.00	138.68
16 01 9068	**Remove Commercial Dishwasher**	**EA**				**103.13**	**0.00**	**0.00**	**103.13**
02112 9510	Remove Commercial Dishwasher	EA	0.5	MPLUD	1.000	103.13	0.00	0.00	103.13
16 01 9069	**Remove Surgeon Scrub Station**	**EA**				**68.75**	**0.00**	**0.00**	**68.75**
02111 9113	Remove Surgeon Scrub Station	EA	0.75	MPLUD	1.000	68.75	0.00	0.00	68.75
16 01 9070	**Remove Fire Hydrant**	**EA**				**139.83**	**0.00**	**0.00**	**139.83**
02111 9825	Remove Fire Hydrant	EA	0.87	MPLUJ	1.000	139.83	0.00	0.00	139.83
16 01 9071	**Remove Chlorinator**	**EA**				**187.50**	**0.00**	**0.00**	**187.50**
02111 9286	Remove Chlorinator	EA	0.275	MPLUA	1.000	187.50	0.00	0.00	187.50
16 01 9072	**Remove Gas Manifold**	**EA**				**275.00**	**0.00**	**0.00**	**275.00**
02111 9287	Remove Gas Manifold	EA	0.188	MPLUA	1.000	275.00	0.00	0.00	275.00
16 01 9073	**Remove Hose Reel & Support**	**EA**				**32.09**	**0.00**	**0.00**	**32.09**
02111 9288	Remove Hose Reel & Support	EA	1.406	MPLUC	1.000	32.09	0.00	0.00	32.09
16 01 9074	**Remove Gas Panel**	**EA**				**51.56**	**0.00**	**0.00**	**51.56**
02111 9289	Remove Gas Panel	EA	1	MPLUA	1.000	51.56	0.00	0.00	51.56

16 01 95	Other Electrical Systems	Unit	Hourly Output	Crew	Factor	Labor	Equip.	Matl.	Total
16 01 9501	**Remove 120 V/208 V, Cable Reel with Receptacle**	**EA**				**34.29**	**0.00**	**0.00**	**34.29**
02111 8460	Remove 120V/208V, Cable Reel with Receptacle	EA	1.5	EELEA	1.000	34.29	0.00	0.00	34.29
16 01 9502	**Remove 7-day Time Clock with Battery**	**EA**				**25.72**	**0.00**	**0.00**	**25.72**
02111 8461	Remove 7-day Time Clock with Battery	EA	2	EELEA	1.000	25.72	0.00	0.00	25.72
16 01 9503	**Remove Lighting Contactor 30 - 100 Amps**	**EA**				**104.84**	**0.00**	**0.00**	**104.84**
02111 8462	Remove Lighting Contactor 30 - 100 Amps	EA	1	EELEC	1.000	104.84	0.00	0.00	104.84
16 01 9504	**Remove 120 V, 10' Tele-Power Pole**	**EA**				**41.94**	**0.00**	**0.00**	**41.94**
02111 8463	Remove 120V, 10' Tele-Power Pole	EA	2.5	EELEC	1.000	41.94	0.00	0.00	41.94
16 01 9505	**Remove to 10 HP Equipment Connection**	**EA**				**12.86**	**0.00**	**0.00**	**12.86**
02111 8464	Remove to 10 HP Equipment Connection	EA	4	EELEA	1.000	12.86	0.00	0.00	12.86

16 01 95 Other Electrical Systems		Unit	Hourly Output	Crew	Factor	Labor	Equip.	Matl.	Total
16 01 9506	**Remove 15 - 30 HP Equipment Connection**	**EA**				**17.15**	**0.00**	**0.00**	**17.15**
02111 8465	Remove 15 - 30 HP Equipment Connection	EA	3	EELEA	1.000	17.15	0.00	0.00	17.15
16 01 9507	**Remove 40 - 60 HP Equipment Connection**	**EA**				**20.57**	**0.00**	**0.00**	**20.57**
02111 8466	Remove 40 - 60 HP Equipment Connection	EA	2.5	EELEA	1.000	20.57	0.00	0.00	20.57
16 01 9508	**Remove 75 - 100 HP Equipment Connection**	**EA**				**25.72**	**0.00**	**0.00**	**25.72**
02111 8467	Remove 75 - 100 HP Equipment Connection	EA	2	EELEA	1.000	25.72	0.00	0.00	25.72
16 01 9509	**Remove 30 - 75 KW Generator**	**EA**				**1,305**	**76.60**	**0.00**	**1,382**
02111 8471	Remove 30 KW - 75 KW Generator	EA	0.13	EELEH	1.000	1,305	76.60	0.00	1,382
16 01 9510	**Remove 100 - 200 KW Generator**	**EA**				**1,885**	**110.64**	**0.00**	**1,996**
02111 8472	Remove 100 KW - 200 KW Generator	EA	0.09	EELEH	1.000	1,885	110.64	0.00	1,996
16 01 9511	**Remove 225 - 500 KW Generator**	**EA**				**2,828**	**165.97**	**0.00**	**2,994**
02111 8473	Remove 225 KW - 500 KW Generator	EA	0.06	EELEH	1.000	2,828	165.97	0.00	2,994
16 01 9512	**Remove 600 - 1,000 KW Generator**	**EA**				**4,242**	**248.95**	**0.00**	**4,491**
02111 8474	Remove 600 KW - 1000 KW Generator	EA	0.04	EELEH	1.000	4,242	248.95	0.00	4,491
16 01 9513	**Remove Uninterruptible Power Supply with Test Switch**	**KVA**				**79.67**	**0.00**	**0.00**	**79.67**
02111 8475	Remove Uninterruptible Power Supply with Test Switch	KVA	2.4	EELEF	1.000	79.67	0.00	0.00	79.67
16 01 9514	**Remove Battery Charger**	**EA**				**64.30**	**0.00**	**0.00**	**64.30**
02111 8476	Remove Battery Charger	EA	1.6	EELEB	1.000	64.30	0.00	0.00	64.30
16 01 9515	**Remove Clock with Buzzer**	**EA**				**25.72**	**0.00**	**0.00**	**25.72**
02111 8477	Remove Clock with Buzzer	EA	4	EELEB	1.000	25.72	0.00	0.00	25.72
16 01 9516	**Remove Master Clock & Control Panel**	**EA**				**51.44**	**0.00**	**0.00**	**51.44**
02111 8478	Remove Master Clock & Control Panel	EA	2	EELEB	1.000	51.44	0.00	0.00	51.44
16 01 9517	**Remove Time Clock & Program System**	**EA**				**51.44**	**0.00**	**0.00**	**51.44**
02111 8479	Remove Time Clock & Program System	EA	2	EELEB	1.000	51.44	0.00	0.00	51.44
16 01 9518	**Remove Thermal/Smoke Detector**	**EA**				**12.86**	**0.00**	**0.00**	**12.86**
02111 8511	Remove Thermal/Smoke Detector	EA	4	EELEA	1.000	12.86	0.00	0.00	12.86
16 01 9519	**Remove Pull Station**	**EA**				**10.29**	**0.00**	**0.00**	**10.29**
02111 8512	Remove Pull Station	EA	5	EELEA	1.000	10.29	0.00	0.00	10.29
16 01 9520	**Remove Horn/Light**	**EA**				**19.78**	**0.00**	**0.00**	**19.78**
02111 8513	Remove Horn/Light	EA	2.6	EELEA	1.000	19.78	0.00	0.00	19.78
16 01 9521	**Remove Flame Detector**	**EA**				**68.58**	**0.00**	**0.00**	**68.58**
02111 8514	Remove Flame Detector	EA	1.5	EELEB	1.000	68.58	0.00	0.00	68.58

16 01 95	Other Electrical Systems	Unit	Hourly Output	Crew	Factor	Labor	Equip.	Matl.	Total
16 01 9522	**Remove Duct Detector**	**EA**				**102.88**	**0.00**	**0.00**	**102.88**
02111 8515	Remove Duct Detector	EA	1	EELEB	1.000	102.88	0.00	0.00	102.88
16 01 9523	**Remove 4 - 8 Zone Fire Control Panel**	**EA**				**205.75**	**0.00**	**0.00**	**205.75**
02111 8516	Remove 4 - 8 Zone Fire Control Panel	EA	0.5	EELEB	1.000	205.75	0.00	0.00	205.75
16 01 9524	**Remove 12 - 16 Zone Fire Control Panel**	**EA**				**428.65**	**0.00**	**0.00**	**428.65**
02111 8517	Remove 12 - 16 Zone Fire Control Panel	EA	0.24	EELEB	1.000	428.65	0.00	0.00	428.65
16 01 9525	**Remove 4 - 8 Zone Annunciation Panel**	**EA**				**154.30**	**0.00**	**0.00**	**154.30**
02111 8518	Remove 4 - 8 Zone Annunciation Panel	EA	0.667	EELEB	1.000	154.30	0.00	0.00	154.30
16 01 9526	**Remove 12 - 16 Zone Annunciation Panel**	**EA**				**220.15**	**0.00**	**0.00**	**220.15**
02111 8519	Remove 12 - 16 Zone Annunciation Panel	EA	0.467	EELEB	1.000	220.15	0.00	0.00	220.15
16 01 9527	**Remove Transmitter & Antenna**	**EA**				**257.19**	**0.00**	**0.00**	**257.19**
02111 8520	Remove Transmitter & Antenna	EA	0.4	EELEB	1.000	257.19	0.00	0.00	257.19
16 01 9528	**Remove Telephone/Data Receptacle**	**EA**				**5.14**	**0.00**	**0.00**	**5.14**
02111 8531	Remove Telephone/Data Receptacle	EA	10	EELEA	1.000	5.14	0.00	0.00	5.14
16 01 9529	**Remove Terminal Cabinet**	**EA**				**137.17**	**0.00**	**0.00**	**137.17**
02111 8532	Remove Terminal Cabinet	EA	0.75	EELEB	1.000	137.17	0.00	0.00	137.17
16 01 9530	**Remove TV Outlet**	**EA**				**5.14**	**0.00**	**0.00**	**5.14**
02111 8533	Remove TV Outlet	EA	10	EELEA	1.000	5.14	0.00	0.00	5.14
16 01 9531	**Remove TV Antenna**	**EA**				**17.15**	**0.00**	**0.00**	**17.15**
02111 8534	Remove TV Antenna	EA	6	EELEB	1.000	17.15	0.00	0.00	17.15
16 01 9532	**Remove Nurse Call Bedside Station**	**EA**				**26.21**	**0.00**	**0.00**	**26.21**
02111 8551	Remove Nurse Call Bedside Station	EA	4	EELEC	1.000	26.21	0.00	0.00	26.21
16 01 9533	**Remove Nurse Call Dome Light**	**EA**				**13.10**	**0.00**	**0.00**	**13.10**
02111 8552	Remove Nurse Call Dome Light	EA	8	EELEC	1.000	13.10	0.00	0.00	13.10
16 01 9534	**Remove Nurse Call Duty Station**	**EA**				**52.42**	**0.00**	**0.00**	**52.42**
02111 8553	Remove Nurse Call Duty Station	EA	2	EELEC	1.000	52.42	0.00	0.00	52.42
16 01 9535	**Remove Nurse Call Emergency Call Button**	**EA**				**20.97**	**0.00**	**0.00**	**20.97**
02111 8554	Remove Nurse Call Emergency Call Button	EA	5	EELEC	1.000	20.97	0.00	0.00	20.97
16 01 9536	**Remove Nurse Call Master Station**	**EA**				**209.68**	**0.00**	**0.00**	**209.68**
02111 8555	Remove Nurse Call Master Station	EA	0.5	EELEC	1.000	209.68	0.00	0.00	209.68
16 01 9537	**Remove Nurse Call Main Equipment Panel**	**EA**				**102.88**	**0.00**	**0.00**	**102.88**
02111 8556	Remove Nurse Call Main Equipment Panel	EA	1	EELEB	1.000	102.88	0.00	0.00	102.88

16 01 95	Other Electrical Systems	Unit	Hourly Output	Crew	Factor	Labor	Equip.	Matl.	Total
16 01 9538	**Remove Speaker**	**EA**				**12.86**	**0.00**	**0.00**	**12.86**
02111 8557	Remove Speaker	EA	4	EELEA	1.000	12.86	0.00	0.00	12.86
16 01 9539	**Remove Outlet Protector**	**EA**				**12.86**	**0.00**	**0.00**	**12.86**
02111 8558	Remove Outlet Protector	EA	4	EELEA	1.000	12.86	0.00	0.00	12.86
16 01 9540	**Remove Microphone/Outlet**	**EA**				**25.72**	**0.00**	**0.00**	**25.72**
02111 8559	Remove Microphone/Outlet	EA	2	EELEA	1.000	25.72	0.00	0.00	25.72
16 01 9541	**Remove Public Address System Cabinet/Panel**	**EA**				**121.03**	**0.00**	**0.00**	**121.03**
02111 8560	Remove Public Address System Cabinet/Panel	EA	0.85	EELEB	1.000	121.03	0.00	0.00	121.03
16 01 9542	**Remove Tape Deck**	**EA**				**32.56**	**0.00**	**0.00**	**32.56**
02111 8561	Remove Tape Deck	EA	1.58	EELEA	1.000	32.56	0.00	0.00	32.56
16 01 9543	**Remove Volume Control**	**EA**				**12.86**	**0.00**	**0.00**	**12.86**
02111 8562	Remove Volume Control	EA	4	EELEA	1.000	12.86	0.00	0.00	12.86
16 01 9544	**Remove Amplifier**	**EA**				**68.58**	**0.00**	**0.00**	**68.58**
02111 8563	Remove Amplifier	EA	0.75	EELEA	1.000	68.58	0.00	0.00	68.58
16 01 9545	**Remove Fingerprint/Card Reader**	**EA**				**32.15**	**0.00**	**0.00**	**32.15**
02111 8571	Remove Finger Print/Card Reader	EA	3.2	EELEB	1.000	32.15	0.00	0.00	32.15
16 01 9546	**Remove Exit Horn/Light**	**EA**				**6.43**	**0.00**	**0.00**	**6.43**
02111 8572	Remove Exit Horn/Light	EA	8	EELEA	1.000	6.43	0.00	0.00	6.43
16 01 9547	**Remove Control Panel 10 to 16 Zone**	**EA**				**102.88**	**0.00**	**0.00**	**102.88**
02111 8573	Remove Control Panel 10 to 16 Zone	EA	1	EELEB	1.000	102.88	0.00	0.00	102.88
16 01 9548	**Remove Video Camera**	**EA**				**34.29**	**0.00**	**0.00**	**34.29**
02111 8574	Remove Video Camera	EA	3	EELEB	1.000	34.29	0.00	0.00	34.29
16 01 9549	**Remove Motion Detector Multichannel**	**EA**				**51.44**	**0.00**	**0.00**	**51.44**
02111 8575	Remove Motion Detector Multichannel	EA	2	EELEB	1.000	51.44	0.00	0.00	51.44
16 01 9550	**Remove Microprocessor with Cameras & Matrix Switch**	**EA**				**128.59**	**0.00**	**0.00**	**128.59**
02111 8576	Remove Microprocessor with Cameras & Matrix Switch	EA	0.8	EELEB	1.000	128.59	0.00	0.00	128.59
16 01 9551	**Remove Electric Door Strike**	**EA**				**19.41**	**0.00**	**0.00**	**19.41**
02111 8577	Remove Electric Door Strike	EA	5.3	EELEB	1.000	19.41	0.00	0.00	19.41
16 01 9552	**Remove Infrared Detector**	**EA**				**51.44**	**0.00**	**0.00**	**51.44**
02111 8578	Remove Infrared Detector	EA	2	EELEB	1.000	51.44	0.00	0.00	51.44
16 01 9553	**Remove Signal Booster**	**EA**				**128.59**	**0.00**	**0.00**	**128.59**
02111 8579	Remove Signal Booster	EA	0.8	EELEB	1.000	128.59	0.00	0.00	128.59

16 01 95 Other Electrical Systems		Unit	Hourly Output	Crew	Factor	Labor	Equip.	Matl.	Total
16 01 9554	**Remove Personal Computer with Line Printer**	**EA**				**51.44**	**0.00**	**0.00**	**51.44**
02111 8580	Remove Personal Computer with Line Printer	EA	2	EELEB	1.000	51.44	0.00	0.00	51.44
16 01 9555	**Remove Motion Sensor**	**EA**				**51.44**	**0.00**	**0.00**	**51.44**
02111 8581	Remove Motion Sensor	EA	2	EELEB	1.000	51.44	0.00	0.00	51.44
16 01 9556	**Remove Intrusion Software System**	**EA**				**51.44**	**0.00**	**0.00**	**51.44**
02111 8582	Remove Intrusion Software System	EA	2	EELEB	1.000	51.44	0.00	0.00	51.44
16 01 9557	**Remove 19" Video Monitor**	**EA**				**19.05**	**0.00**	**0.00**	**19.05**
02111 8583	Remove 19" Video Monitor	EA	2.7	EELEA	1.000	19.05	0.00	0.00	19.05
16 01 9558	**Remove Ground Rod 8' - 10'**	**EA**				**52.42**	**0.00**	**0.00**	**52.42**
02111 8601	Remove Ground Rod 8' - 10'	EA	2	EELEC	1.000	52.42	0.00	0.00	52.42
16 01 9559	**Remove Ground Rod Receptacle**	**EA**				**54.87**	**0.00**	**0.00**	**54.87**
02111 8602	Remove Ground Rod Receptacle	EA	1.875	EELEB	1.000	54.87	0.00	0.00	54.87
16 01 9560	**Remove Ground Reel**	**EA**				**34.29**	**0.00**	**0.00**	**34.29**
02111 8603	Remove Ground Reel	EA	3	EELEB	1.000	34.29	0.00	0.00	34.29
16 01 9561	**Remove Copper Bar to 1/4" x 2" with Insulation**	**LF**				**5.24**	**0.00**	**0.00**	**5.24**
02111 8604	Remove Copper Bar to 1/4" x 2" with Insulation	LF	20	EELEC	1.000	5.24	0.00	0.00	5.24
16 01 9562	**Remove Air Terminal with Base & Connector**	**EA**				**10.29**	**0.00**	**0.00**	**10.29**
02111 8605	Remove Air Terminal with Base & Connector	EA	5	EELEA	1.000	10.29	0.00	0.00	10.29
16 01 9563	**Remove Signal Filter to 200 Conductors**	**EA**				**79.13**	**0.00**	**0.00**	**79.13**
02111 8621	Remove Signal Filter to 200 Conductors	EA	1.3	EELEB	1.000	79.13	0.00	0.00	79.13
16 01 9564	**Remove Signal Filter 201 - 500 Conductors**	**EA**				**93.52**	**0.00**	**0.00**	**93.52**
02111 8622	Remove Signal Filter 201 - 500 Conductors	EA	1.1	EELEB	1.000	93.52	0.00	0.00	93.52
16 01 9565	**Remove Power Filter 200 Amp**	**EA**				**51.44**	**0.00**	**0.00**	**51.44**
02111 8623	Remove Power Filter 200 Amp	EA	2	EELEB	1.000	51.44	0.00	0.00	51.44
16 01 9566	**Remove Power Filter 600 Amp**	**EA**				**68.58**	**0.00**	**0.00**	**68.58**
02111 8624	Remove Power Filter 600 Amp	EA	1.5	EELEB	1.000	68.58	0.00	0.00	68.58
16 01 9567	**Remove Master Light Control Panel**	**EA**				**411.50**	**0.00**	**0.00**	**411.50**
02111 8625	Remove Master Light Control Panel	EA	0.25	EELEB	1.000	411.50	0.00	0.00	411.50
16 01 9568	**Remove Light Transceiver Panel with Relays**	**EA**				**205.75**	**0.00**	**0.00**	**205.75**
02111 8626	Remove Light Transceiver Panel with Relays	EA	0.5	EELEB	1.000	205.75	0.00	0.00	205.75
16 01 9569	**Remove EMCS Start/Stop Control Relay**	**EA**				**17.15**	**0.00**	**0.00**	**17.15**
02111 8641	Remove EMCS Start/Stop Control Relay	EA	3	EELEA	1.000	17.15	0.00	0.00	17.15

16 01 95	Other Electrical Systems	Unit	Hourly Output	Crew	Factor	Labor	Equip.	Matl.	Total
16 01 9570	**Remove EMCS Pressure Switch/Actuators**	**EA**				**26.72**	**0.00**	**0.00**	**26.72**
02111 8642	Remove EMCS Pressure Switch/Actuators	EA	3.85	EELEB	1.000	26.72	0.00	0.00	26.72
16 01 9571	**Remove EMCS Electrical Meter 3P <600V**	**EA**				**137.17**	**0.00**	**0.00**	**137.17**
02111 8643	Remove EMCS Electrical Meter 3P < 600V	EA	0.75	EELEB	1.000	137.17	0.00	0.00	137.17
16 01 9572	**Remove EMCS Status/Costs Point**	**EA**				**68.58**	**0.00**	**0.00**	**68.58**
02111 8644	Remove EMCS Status/Costs Point	EA	1.5	EELEB	1.000	68.58	0.00	0.00	68.58
16 01 9573	**Remove EMCS Field Interface Device**	**EA**				**102.88**	**0.00**	**0.00**	**102.88**
02111 8645	Remove EMCS Field Interface Device	EA	1	EELEB	1.000	102.88	0.00	0.00	102.88
16 01 9574	**Remove EMCS Multiplexer Field Processing Device**	**EA**				**51.44**	**0.00**	**0.00**	**51.44**
02111 8646	Remove EMCS Multiplexer Field Processing Device	EA	2	EELEB	1.000	51.44	0.00	0.00	51.44
16 01 9575	**Remove EMCS Thermowell Misc. Device**	**EA**				**82.30**	**0.00**	**0.00**	**82.30**
02111 8647	Remove EMCS Thermowell Miscellaneous Device	EA	1.25	EELEB	1.000	82.30	0.00	0.00	82.30
16 01 9576	**Remove 1/2" to 1" Rigid Galvanized Steel Conduit with Couplings**	**LF**				**2.33**	**0.00**	**0.00**	**2.33**
02112 1933	Remove 1/2" to 1" Rigid Galvanized Steel Conduit with Couplings	LF	45	EELEC	1.000	2.33	0.00	0.00	2.33
16 01 9577	**Remove 1 1/4" to 2" Rigid Galvanized Steel Conduit with Couplings**	**LF**				**3.15**	**0.00**	**0.00**	**3.15**
02112 1934	Remove 1-1/4" to 2" Rigid Galvanized Steel Conduit with Couplings	LF	33.33	EELEC	1.000	3.15	0.00	0.00	3.15
16 01 9578	**Remove 2 1/2" to 3 1/2" Rigid Galvanized Steel Conduit with Couplings**	**LF**				**4.99**	**0.00**	**0.00**	**4.99**
02112 1935	Remove 2-1/2" to 3-1/2" Rigid Galvanized Steel Conduit with Couplings	LF	21	EELEC	1.000	4.99	0.00	0.00	4.99
16 01 9579	**Remove 4"-6" Rigid Galvanized Steel Conduit w/Couplings**	**LF**				**6.29**	**0.00**	**0.00**	**6.29**
02112 1936	Remove 4" - 6" Rigid Galvanized Steel Conduit with Couplings	LF	16.67	EELEC	1.000	6.29	0.00	0.00	6.29
16 01 9580	**Remove 1/2" to 3" Weatherhead**	**EA**				**7.49**	**0.00**	**0.00**	**7.49**
02112 1937	Remove 1/2" to 3" Weatherhead	EA	6.868	EELEA	1.000	7.49	0.00	0.00	7.49
16 01 9581	**Remove #18 AWG to #4/0 AWG Wire**	**MLF**				**514.38**	**0.00**	**0.00**	**514.38**
02112 1938	Remove #18 AWG to #4/0 AWG Wire	MLF	0.2	EELEB	1.000	514.38	0.00	0.00	514.38
16 01 9582	**Remove 250 MCM to 500 MCM Wire**	**MLF**				**1,324**	**0.00**	**0.00**	**1,324**
02112 1939	Remove 250 MCM to 500 MCM Wire	MLF	0.144	EELEF	1.000	1,324	0.00	0.00	1,324

16 01 95	Other Electrical Systems	Unit	Hourly Output	Crew	Factor	Labor	Equip.	Matl.	Total
16 01 9583	**Remove 600 MCM to 1,000 MCM Wire**	**MLF**				**1,530**	**0.00**	**0.00**	**1,530**
02112 1940	Remove 600 MCM to 1,000 MCM Wire	MLF	0.125	EELEF	1.000	1,530	0.00	0.00	1,530
16 01 9584	**Remove Pad-mounted Transformer, Oil Filled, 150 - 300 KVA**	**EA**				**758.85**	**32.38**	**0.00**	**791.23**
02112 1953	Remove Pad-mounted Transformer, Oil Filled, 150 - 300 KVA	EA	0.222	EELEE	1.000	758.85	32.38	0.00	791.23
16 01 9585	**Remove Pad-mounted Transformer, Oil Filled, 500 - 1,000 KVA**	**EA**				**842.32**	**35.94**	**0.00**	**878.26**
02112 1954	Remove Pad-mounted Transformer, Oil Filled, 500 - 1,000 KVA	EA	0.2	EELEE	1.000	842.32	35.94	0.00	878.26
16 01 9586	**Remove Pad-mounted Transformer, Oil Filled, 1,500 - 3,750 KVA**	**EA**				**1,685**	**71.88**	**0.00**	**1,757**
02112 1955	Remove Pad-mounted Transformer, Oil Filled, 1,500 - 3,750 KVA	EA	0.1	EELEE	1.000	1,685	71.88	0.00	1,757

16 01 99	Other Nonhazardous Building Demolition - General	Unit	Hourly Output	Crew	Factor	Labor	Equip.	Matl.	Total
16 01 9901	**Concrete Dump Charge**	**CY**				**0.00**	**0.00**	**11.74**	**11.74**
02114 3021	Dump Charges	CY	0	N/A	1.000	0.00	0.00	11.74	11.74
16 01 9902	**Masonry Dump Charge**	**CY**				**0.00**	**0.00**	**11.74**	**11.74**
02114 3021	Dump Charges	CY	0	N/A	1.000	0.00	0.00	11.74	11.74
16 01 9903	**Roofing Dump Charge**	**CY**				**0.00**	**0.00**	**11.74**	**11.74**
02114 3021	Dump Charges	CY	0	N/A	1.000	0.00	0.00	11.74	11.74
16 01 9904	**Tanks Dump Charge**	**CY**				**0.00**	**0.00**	**11.74**	**11.74**
02114 3021	Dump Charges	CY	0	N/A	1.000	0.00	0.00	11.74	11.74
16 01 9905	**Equipment Dump Charge**	**CY**				**0.00**	**0.00**	**11.74**	**11.74**
02114 3021	Dump Charges	CY	0	N/A	1.000	0.00	0.00	11.74	11.74
16 01 9906	**Metals Dump Charge**	**CY**				**0.00**	**0.00**	**11.74**	**11.74**
02114 3021	Dump Charges	CY	0	N/A	1.000	0.00	0.00	11.74	11.74
16 01 9907	**Plastics Dump Charge**	**CY**				**0.00**	**0.00**	**11.74**	**11.74**
02114 3021	Dump Charges	CY	0	N/A	1.000	0.00	0.00	11.74	11.74
16 01 9908	**Wood Dump Charge**	**CY**				**0.00**	**0.00**	**11.74**	**11.74**
02114 3021	Dump Charges	CY	0	N/A	1.000	0.00	0.00	11.74	11.74
16 01 9909	**Miscellaneous Dump Charge**	**CY**				**0.00**	**0.00**	**11.74**	**11.74**
02114 3021	Dump Charges	CY	0	N/A	1.000	0.00	0.00	11.74	11.74
16 01 9910	**Security Labor Adjustment**	**LS**				**0.00**	**0.00**	**1.49**	**1.49**
01985 1002	Labor Adjustment	LS	0	N/A	1.000	0.00	0.00	1.49	1.49

16 01 99	Other Nonhazardous Building Demolition - General	Unit	Hourly Output	Crew	Factor	Labor	Equip.	Matl.	Total
16 01 9911	**Phasing Labor Adjustment**	**LS**				**0.00**	**0.00**	**1.49**	**1.49**
01985 1002	Labor Adjustment	LS	0	N/A	1.000	0.00	0.00	1.49	1.49
16 01 9912	**20 CY, Semi Dump**	**HR**				**27.55**	**66.32**	**0.00**	**93.87**
02225 3139	20 CY, Dump, Highway	HR	1	COEIC	1.000	27.55	66.32	0.00	93.87
16 01 9913	**950, 3.0 CY Wheel Loader**	**HR**				**33.05**	**38.13**	**0.00**	**71.17**
02225 2509	950, 3.0 CY, By Hour	HR	1	XXQHQ	1.000	33.05	38.13	0.00	71.17
16 01 9915	**Dust Partition, 2" x 4" Frame, 6 Mil Polyethylene**	**SF**				**1.04**	**0.00**	**0.74**	**1.79**
06128 6020	Sills & Plates, 2x4 Southern Yellow Pine	BF	100.1	ACARD	0.294	0.77	0.00	0.68	0.43
06128 6140	Stud Framing, 2x3 - 2x4 Southern Yellow Pine for Partition Walls	BF	87.38	ACARD	0.705	0.89	0.00	0.68	1.10
07192 0900	6 Mil Polyethylene Vapor Barrier	CSF	4.625	ACARA	0.020	9.36	0.00	3.30	0.25
16 01 9916	**Dust Partition, 2"x4" Frame, 1/2" Plywood**	**SF**				**1.40**	**0.00**	**1.29**	**2.69**
06128 6020	Sills & Plates, 2x4 Southern Yellow Pine	BF	100.1	ACARD	0.294	0.77	0.00	0.68	0.43
06128 6140	Stud Framing, 2x3 - 2x4 Southern Yellow Pine for Partition Walls	BF	87.38	ACARD	0.705	0.89	0.00	0.68	1.10
06154 0600	Wall Sheathing,1/2" Interior Plywood Standard Grade Applied to Wood Studs	SF	141.3	ACARD	1.000	0.55	0.00	0.61	1.16
16 01 9917	**Dumpster Rental**	**DAY**				**0.00**	**0.00**	**1.68**	**1.68**
02111 1306	Dumpster Rental 30 CY	DAY	0	N/A	1.000	0.00	0.00	1.68	1.68
16 01 9918	**Dumpster Pull Charge**	**EA**				**0.00**	**0.00**	**386.77**	**386.77**
02111 1307	Dumpster Pull Charge	EA	0	N/A	1.000	0.00	0.00	386.77	386.77
16 01 9919	**36" Diameter Steel Chute**	**LF**				**25.19**	**19.17**	**35.00**	**79.35**
02053 0460	Install 36"Diameter (91 cm) Prefabricated Steel Chute	LF	3.75	ULABF	1.000	25.19	19.17	35.00	79.35

16 02 01	Substructure and Superstructure	Unit	Hourly Output	Crew	Factor	Labor	Equip.	Matl.	Total
16 02 0102	**Remove Asbestos Fireproofing - Ceilings**	**SF**				**0.75**	**0.02**	**0.06**	**0.83**
02090 4000	Remove Acoustic/Fireproofing, Ceiling/Wall Scrape Plaster, Flat Ceiling/Wall	SF	200	AASBF	1.000	0.75	0.02	0.06	0.83
16 02 0104	**Remove Asbestos Fireproofing - Beams/Columns**	**SF**				**0.96**	**0.03**	**0.06**	**1.05**
02090 4115	Remove Acoustic/Fireproofing, Metal Beams Scrape, Metal Beams	SF	156.3	AASBF	1.000	0.96	0.03	0.06	1.05

16 02 02	Exterior Closure	Unit	Hourly Output	Crew	Factor	Labor	Equip.	Matl.	Total
16 02 0202	**Remove Asbestos Transite Shingle Siding**	**SF**				**1.59**	**0.82**	**0.10**	**2.51**
02090 8100	Remove Transite Shingle Siding	SF	93.75	AASBH	1.000	1.59	0.82	0.10	2.51

16 02 03	Roofing	Unit	Hourly Output	Crew	Factor	Labor	Equip.	Matl.	Total

16 02 Hazardous Selective Building Demolition

16 02 03	Roofing	Unit	Hourly Output	Crew	Factor	Labor	Equip.	Matl.	Total
16 02 0302	**Remove Asbestos Shingle Roofing**	**SF**				**0.80**	**0.02**	**0.07**	**0.89**
02090 8200	Remove Asbestos Shingle Roofing	SF	187.5	AASBF	1.000	0.80	0.02	0.07	0.89

16 02 04	Interior Construction and Finishes	Unit	Hourly Output	Crew	Factor	Labor	Equip.	Matl.	Total
16 02 0402	**Remove Asbestos Plaster Walls**	**SF**				**1.71**	**0.05**	**0.81**	**2.56**
02089 0410	Remove Interior Asbestos Plaster Walls	SF	88.06	AASBF	1.000	1.71	0.05	0.81	2.56
16 02 0403	**Remove Gypsum Board Containing Asbestos Joint Compound**	**SF**				**0.16**	**0.00**	**0.40**	**0.56**
02090 0060	Demolish Asbestos from Ceiling & Walls, Maximum Thickness	SF	416.6	ULABB	1.000	0.16	0.00	0.40	0.56
16 02 0404	**Remove Lead Paint - Walls**	**SF**				**1.25**	**0.00**	**0.34**	**1.59**
02113 7015	Remove Lead Paint from Walls	SF	75	AASBB	1.000	1.25	0.00	0.34	1.59
16 02 0406	**Remove Vinyl Asbestos Flooring**	**SF**				**1.25**	**0.00**	**0.04**	**1.29**
02090 5000	Remove Vinyl Asbestos Flooring	SF	75	AASBB	1.000	1.25	0.00	0.04	1.29

16 02 06	Mechanical Systems	Unit	Hourly Output	Crew	Factor	Labor	Equip.	Matl.	Total
16 02 0602	**Remove Asbestos Pipe Insulation 1/2" - 4"**	**LF**				**2.41**	**0.07**	**0.06**	**2.53**
02090 0665	Remove 1/2" to 4" Diameter Asbestos Pipe Insulation Magnesium Block Type, Isolated Area	LF	62.5	AASBF	1.000	2.41	0.07	0.06	2.53
16 02 0604	**Remove Asbestos Pipe Insulation 4" - 6"**	**LF**				**3.01**	**0.08**	**0.10**	**3.19**
02090 0670	Remove 4" to 6" Diameter Asbestos Pipe Insulation Magnesium Block Type, Isolated Area	LF	50	AASBF	1.000	3.01	0.08	0.10	3.19
16 02 0606	**Remove Asbestos Pipe Insulation 7" - 12"**	**LF**				**4.01**	**0.11**	**0.11**	**4.23**
02090 0680	Remove 7" to 12" Diameter Asbestos Pipe Insulation Magnesium Block Type, Isolated Area	LF	37.5	AASBF	1.000	4.01	0.11	0.11	4.23
16 02 0608	**Remove Asbestos Plaster Fittings 1/2" - 4"**	**EA**				**14.16**	**0.39**	**0.08**	**14.62**
02090 1000	Remove 1/2" to 4" Diameter Asbestos Pipe Insulation Hand Packed Plastic Fittings, Isolated	EA	10.63	AASBF	1.000	14.16	0.39	0.08	14.62
16 02 0610	**Remove Asbestos Plaster Fittings 4" - 6"**	**EA**				**26.74**	**0.74**	**0.12**	**27.59**
02090 1100	Remove 4" to 6" Diameter Asbestos Pipe Insulation Hand Packed Plastic Fittings, Isolated	EA	5.625	AASBF	1.000	26.74	0.74	0.12	27.59
16 02 0612	**Remove Asbestos Plaster Fittings 7" - 12"**	**EA**				**48.13**	**1.32**	**0.16**	**49.61**
02090 1110	Remove 7" to 12" Diameter Asbestos Pipe Insulation Hand Packed Plastic Fittings, Isolated	EA	3.125	AASBF	1.000	48.13	1.32	0.16	49.61
16 02 0614	**Remove Asbestos Boiler Insulation**	**SF**				**8.59**	**0.24**	**0.07**	**8.90**
02090 0200	Remove Up to 3-1/2" Boiler Insulation	SF	17.5	AASBF	1.000	8.59	0.24	0.07	8.90
16 02 0616	**Remove Asbestos Plaster from Ductwork**	**SF**				**4.06**	**0.00**	**0.06**	**4.12**
02090 5400	Remove Hard Coat Plaster	SF	23.08	AASBB	1.000	4.06	0.00	0.06	4.12

16 02 06	Mechanical Systems	Unit	Hourly Output	Crew	Factor	Labor	Equip.	Matl.	Total
16 02 0620	**Remove and Dispose of Refrigerant from System**	**LB**				**10.31**	**0.00**	**5.86**	**16.18**
02111 9815	Remove & Dispose of Refrigerant from System	LB	5	MPLUA	1.000	10.31	0.00	5.86	16.18

16 02 07	Electrical Systems	Unit	Hourly Output	Crew	Factor	Labor	Equip.	Matl.	Total
16 02 0701	**Removal and Incineration of PCB Light Ballast**	**LB**				**0.00**	**0.00**	**1.95**	**1.95**
13277 2714	Drummed PCB Light Ballasts	LB	0	N/A	1.000	0.00	0.00	1.95	1.95
16 02 0702	**Removal & Incineration of PCB Transformer 500 PPM & Over**	**LB**				**0.00**	**0.00**	**1.75**	**1.75**
13277 2715	Drummed PCB Transformer, 55 Gallon Drum	EA	0	N/A	0.002	0.00	0.00	805.00	1.75

16 02 90	Other Hazardous Selective Building Demolition	Unit	Hourly Output	Crew	Factor	Labor	Equip.	Matl.	Total
16 02 9002	**Mobilization & Fee**	**JOB**				**0.00**	**0.00**	**800.00**	**800.00**
02088 0020	Asbestos Abatement Notification Fees	JOB	0	N/A	1.000	0.00	0.00	300.00	300.00
02088 0030	Asbestos Decontamination of Trucks & Equipment Mobilization Fees	JOB	0	N/A	1.000	0.00	0.00	500.00	500.00
16 02 9004	**Pre-Clean, HEPA Vacuum and Wet Clean**	**SF**				**0.15**	**0.01**	**0.05**	**0.21**
02088 0105	HEPA Vacuum & Wet Clean, 6-11' Ceiling, SF of Floor, Wall, Ceiling	SF	750	AASBE	1.000	0.15	0.01	0.05	0.21
16 02 9006	**Containment Barrier**	**SF**				**0.55**	**0.02**	**2.38**	**2.94**
02088 0490	Setup Containment Barrier for Asbestos Removal, Wood Frame Wall with Visqueen Layer	SF	275	AASBF	1.000	0.55	0.02	2.38	2.94
16 02 9008	**Portable Decontamination Facility**	**EA**				**188.83**	**72.76**	**157.50**	**419.09**
02088 0460	Setup Portable Asbestos Decontamination Facility, Cost Amortized Over 10 Jobs	EA	1.5	AASBG	1.000	75.53	29.10	145.00	249.64
02092 4100	Remove Portable Decontamination Facility Post Abatement Cleanup	EA	1	AASBG	1.000	113.30	43.65	12.50	169.45
16 02 9010	**Negative Air Machine**	**EA**				**75.00**	**0.00**	**289.31**	**364.31**
02088 0900	Setup Negative Air Machine for Asbestos Removal 1,000-2,000 CFM Unit/25 KCF Room Volume	EA	0.625	AASBA	1.000	75.00	0.00	289.31	364.31
16 02 9012	**Air Monitoring**	**DAY**				**0.00**	**0.00**	**625.00**	**625.00**
02087 0270	Asbestos Air Monitor, 4 - 6 Hours/Day, Bulk Sampling, Technician Time Only	DAY	0	N/A	1.000	0.00	0.00	475.00	475.00
02087 0280	Rent 5 Pump Asbestos Air Monitoring Set	DAY	0	N/A	1.000	0.00	0.00	150.00	150.00
16 02 9013	**PCM Air Sample Analysis**	**EA**				**0.00**	**0.00**	**30.00**	**30.00**
02095 1100	PCM Air Sample Analysis for Asbestos	EA	0	N/A	1.000	0.00	0.00	30.00	30.00
16 02 9014	**Air Testing - Post Abatement**	**JOB**				**0.00**	**0.00**	**33.50**	**33.50**
02095 1300	Post Asbestos Abatement Air Testing	JOB	0	N/A	1.000	0.00	0.00	33.50	33.50

16 02 90	Other Hazardous Selective Building Demolition	Unit	Hourly Output	Crew	Factor	Labor	Equip.	Matl.	Total
16 02 9015	**TEM Air Sample Analysis**	**EA**				**0.00**	**0.00**	**350.00**	**350.00**
02095 1220	TEM Air Sample Analysis for Asbestos, 24 to 48 Hour Turnaround Cost	EA	0	N/A	1.000	0.00	0.00	350.00	350.00
16 02 9016	**Personnel Protection**	**EA**				**0.00**	**0.00**	**36.36**	**36.36**
02087 2000	Personnel Equipment, Worker/Day Coveralls, Respirator, Gloves	EA	0	N/A	1.000	0.00	0.00	36.36	36.36
16 02 9018	**Post Abatement Inspection**	**EA**				**0.00**	**0.00**	**257.17**	**257.17**
02092 6000	Post Abatement Visual Inspection	EA	0	N/A	1.000	0.00	0.00	257.17	257.17
16 02 9020	**Post-Clean, High Efficiency Particulate Air Vacuum**	**SF**				**0.55**	**0.02**	**0.09**	**0.66**
02089 0100	Initial Asbestos Abatement Cleanup, Isolated Work Area	SF	285	AASBE	1.000	0.40	0.01	0.05	0.46
02092 9000	Final Cleanup, HEPA Vacuuming, Remove Final Sheeting Layer	SF	750	AASBE	1.000	0.15	0.01	0.04	0.20
16 02 9022	**Disposal Bag Handling**	**EA**				**1.76**	**0.00**	**0.00**	**1.76**
02091 3100	Handling of Asbestos Disposal Bags from Work Site to Haul Truck	EA	53.13	AASBB	1.000	1.76	0.00	0.00	1.76
16 02 9024	**Hazardous Landfill Disposal**	**CY**				**0.00**	**0.00**	**11.74**	**11.74**
02114 3021	Dump Charges	CY	0	N/A	1.000	0.00	0.00	11.74	11.74

17 01 01	Clearing	Unit	Hourly Output	Crew	Factor	Labor	Equip.	Matl.	Total
17 01 0101	**Light Brush without Grub, Clearing**	**ACRE**				**41.53**	**26.68**	**0.00**	**68.21**
02110 0500	Clearing - Light Brush without Grub	ACRE	4.25	COMCB	1.000	41.53	26.68	0.00	68.21
17 01 0102	**Medium Brush without Grub, Clearing**	**ACRE**				**77.40**	**87.25**	**0.00**	**164.65**
02110 0550	Clearing - Medium Brush without Grub	ACRE	3.5	COMCE	1.000	77.40	87.25	0.00	164.65
17 01 0103	**Medium Brush with Average Grub & Some Trees, Clearing**	**ACRE**				**223.86**	**423.17**	**0.00**	**647.03**
02110 1060	Clearing - Medium Brush, Average Grub & Trees	ACRE	1.063	COMCC	1.000	223.86	423.17	0.00	647.03
17 01 0104	**Heavy Brush without Grub, Clearing**	**ACRE**				**109.00**	**229.10**	**0.00**	**338.10**
02110 0600	Clearing - Heavy Brush without Grub	ACRE	2.5	COMCF	1.000	109.00	229.10	0.00	338.10
17 01 0105	**Heavy Brush with Average Grub, Medium - Heavy Trees, Clearing**	**ACRE**				**305.22**	**617.90**	**0.00**	**923.12**
02110 1070	Clearing - Heavy Brush, Average Grub, Medium Trees	ACRE	1.063	COMCD	1.000	305.22	617.90	0.00	923.12
17 01 0106	**Heavy Brush, Light Trees, Clear, Grub, Haul**	**ACRE**				**2,806**	**2,589**	**0.00**	**5,395**
02109 0010	Clear and Grub, Light Trees to 6" (15 cm) Diameter, Cut and Chip	ACRE	0.088	COMCA	1.000	1,945	1,852	0.00	3,796
02110 0600	Clearing - Heavy Brush without Grub	ACRE	2.5	COMCF	1.000	109.00	229.10	0.00	338.10
02110 2000	Machine Load Spoils, 2 Mile Haul (3.2 km), Haul to Dump	CY	5.375	COETF	35.000	21.51	14.52	0.00	1,261
17 01 0107	**Medium Brush, Medium Trees, Clear, Grub, Haul**	**ACRE**				**3,421**	**2,974**	**0.00**	**6,395**
02109 0152	Clear and Grub, Medium Trees to 10" (25 cm) Diameter, Cut and Chip	ACRE	0.075	COMCA	1.000	2,269	2,160	0.00	4,429
02110 0550	Clearing - Medium Brush without Grub	ACRE	3.5	COMCE	1.000	77.40	87.25	0.00	164.65
02110 2000	Machine Load Spoils, 2 Mile Haul (3.2 km), Haul to Dump	CY	5.375	COETF	50.000	21.51	14.52	0.00	1,802
17 01 0108	**Light Brush, Heavy Trees, Clear, Grub, Haul**	**ACRE**				**4,651**	**3,850**	**0.00**	**8,501**
02109 0154	Clear and Grub, Heavy Trees to 16" (41 cm) Diameter, Cut and Chip	ACRE	0.066	COMCA	1.000	2,566	2,444	0.00	5,010
02110 0500	Clearing - Light Brush without Grub	ACRE	4.25	COMCB	1.000	41.53	26.68	0.00	68.21
02110 2000	Machine Load Spoils, 2 Mile Haul (3.2 km), Haul to Dump	CY	5.375	COETF	95.000	21.51	14.52	0.00	3,423
17 01 0110	**Light, without Grub D5 LGP, Wet Clearing**	**ACRE**				**370.91**	**236.68**	**0.00**	**607.59**
02110 8010	Wet Clearing - Light, without Grub, D5LGP	ACRE	0.688	COMCG	1.000	370.91	236.68	0.00	607.59
17 01 0111	**Medium, without Grub D7 LGP, Wet Clearing**	**ACRE**				**315.82**	**268.89**	**0.00**	**584.71**
02110 8020	Wet Clearing - Medium, without Grub, D7LGP	ACRE	0.813	COMCH	1.000	315.82	268.89	0.00	584.71
17 01 0112	**Heavy, without Grub D7 LGP, Wet Clearing**	**ACRE**				**590.17**	**632.45**	**0.00**	**1,223**
02110 8030	Wet Clearing - Heavy, without Grub, D7LGP	ACRE	0.438	COMCI	1.000	590.17	632.45	0.00	1,223
17 01 0113	**Heavy Brush & Trees, Wet Clearing**	**ACRE**				**590.17**	**632.45**	**0.00**	**1,223**
02110 1050	Wet Clearing - Heavy Brush & Trees	ACRE	0.438	COMCI	1.000	590.17	632.45	0.00	1,223

17 01 Site Clearing

17 01 02	Tree Removal	Unit	Hourly Output	Crew	Factor	Labor	Equip.	Matl.	Total
17 01 0201	**<= 6" Tree Removal, Hand**	**EA**				**56.72**	**54.01**	**0.00**	**110.72**
02109 7300	Tree Removal to 6" (15 cm) Diameter	EA	3	COMCA	1.000	56.72	54.01	0.00	110.72
17 01 0202	**> 6" and <= 12" Tree Removal, Hand**	**EA**				**85.07**	**81.01**	**0.00**	**166.09**
02109 7310	Tree Removal, 6" to 12" (15 cm to 30 cm) Diameter"	EA	2	COMCA	1.000	85.07	81.01	0.00	166.09
17 01 0203	**> 12" and <= 24" Tree Removal, Hand**	**EA**				**113.43**	**108.02**	**0.00**	**221.45**
02109 7320	Tree Removal, 12" to 24" (30 cm to 61 cm) Diameter"	EA	1.5	COMCA	1.000	113.43	108.02	0.00	221.45
17 01 0204	**> 24" and <= 36" Tree Removal, Hand**	**EA**				**170.15**	**162.03**	**0.00**	**332.17**
02109 7330	Tree Removal, 24" to 36" (61 cm to 91 cm) Diameter"	EA	1	COMCA	1.000	170.15	162.03	0.00	332.17
17 01 0205	**Brush Removal Equipment**	**EA**				**0.00**	**0.00**	**124.55**	**124.55**
02102 1601	Brush Removal Equipment	EA	0	N/A	1.000	0.00	0.00	124.55	124.55
17 01 0210	**Clear Trees to 6" Diameter with D8 Cat**	**EA**				**1.28**	**4.95**	**0.00**	**6.24**
02109 9000	Site Clearing, Clear Trees to 6" Diameter with 335 HP Dozer	EA	35	CODTL	1.000	1.28	4.95	0.00	6.24
17 01 0211	**Clear Trees to 12" Diameter with D8 Cat**	**EA**				**2.39**	**9.25**	**0.00**	**11.64**
02109 9010	Site Clearing, Clear Trees to 12" Diameter with 335 HP Dozer	EA	18.75	CODTL	1.000	2.39	9.25	0.00	11.64
17 01 0212	**Clear Trees to 24" Diameter with D8 Cat**	**EA**				**3.59**	**13.87**	**0.00**	**17.46**
02109 9020	Site Clearing, Clear Trees to 24" Diameter with 335 HP Dozer	EA	12.5	CODTL	1.000	3.59	13.87	0.00	17.46
17 01 0213	**Clear Trees to 36" Diameter with D8 Cat**	**EA**				**7.18**	**27.74**	**0.00**	**34.92**
02109 9030	Site Clearing, Clear Trees to 36" Diameter with 335 HP Dozer	EA	6.25	CODTL	1.000	7.18	27.74	0.00	34.92

17 01 03	Stump Removal	Unit	Hourly Output	Crew	Factor	Labor	Equip.	Matl.	Total
17 01 0301	**<= 6" Stump Removal, with Excavator**	**EA**				**17.95**	**36.42**	**0.00**	**54.37**
02109 0740	Stump Removal, to 6" (15 cm) Diameter	EA	5	COETV	1.000	17.95	36.42	0.00	54.37
17 01 0302	**> 6" and <= 12" Stump Removal, with Excavator**	**EA**				**22.44**	**45.52**	**0.00**	**67.96**
02109 0750	Stump Removal, 6" to 12" (15 cm to 30 cm) Diameter	EA	4	COETV	1.000	22.44	45.52	0.00	67.96
17 01 0303	**> 12" and <= 24" Stump Removal, with Excavator**	**EA**				**29.92**	**60.69**	**0.00**	**90.61**
02109 0760	Stump Removal, 12" to 24" (30 cm to 61 cm) Diameter	EA	3	COETV	1.000	29.92	60.69	0.00	90.61

17 01 Site Clearing

17 01 03	Stump Removal	Unit	Hourly Output	Crew	Factor	Labor	Equip.	Matl.	Total
17 01 0304	**> 24" and <= 36" Stump Removal, with Excavator**	**EA**				**44.88**	**91.04**	**0.00**	**135.92**
02109 0770	Stump Removal, 24" to 36" (61 cm to 91 cm) Diameter	EA	2	COETV	1.000	44.88	91.04	0.00	135.92
17 01 0310	**< 6" Wet Stump Removal, with LGP D7**	**EA**				**16.87**	**52.31**	**0.00**	**69.18**
02109 0700	Wet Stump Removal <= 6"	EA	2.5	CODTS	1.000	16.87	52.31	0.00	69.18
17 01 0311	**> 6" < 12" Wet Stump Removal, with LGP D7**	**EA**				**21.09**	**65.39**	**0.00**	**86.48**
02109 0710	Wet Stump Removal >= 6" <= 12"	EA	2	CODTS	1.000	21.09	65.39	0.00	86.48
17 01 0312	**> 12" < 24" Wet Stump Removal, with LGP D7**	**EA**				**28.12**	**87.19**	**0.00**	**115.30**
02109 0720	Wet Stump Removal >= 12" < 24"	EA	1.5	CODTS	1.000	28.12	87.19	0.00	115.30
17 01 0313	**> 24" < 36" Wet Stump Removal, with LGP D7**	**EA**				**42.17**	**130.78**	**0.00**	**172.95**
02109 0730	Wet Stump Removal >= 24" <= 36"	EA	1	CODTS	1.000	42.17	130.78	0.00	172.95
17 01 0314	**<= 6" Stump Removal, with D8**	**EA**				**0.90**	**3.47**	**0.00**	**4.37**
02109 9100	Site Clearing, Grub Stumps to 6" Diameter with 335 HP Dozer	EA	50	CODTL	1.000	0.90	3.47	0.00	4.37
17 01 0315	**> 6" and <= 12" Stump Removal, with D8**	**EA**				**1.44**	**5.55**	**0.00**	**6.98**
02109 9110	Site Clearing, Grub Stumps to 12" Diameter with 335 HP Dozer	EA	31.25	CODTL	1.000	1.44	5.55	0.00	6.98
17 01 0316	**> 12" and <= 24" Stump Removal, with D8**	**EA**				**14.36**	**55.49**	**0.00**	**69.85**
02109 9120	Site Clearing, Grub Stumps to 24" Diameter with 335 HP Dozer	EA	3.125	CODTL	1.000	14.36	55.49	0.00	69.85
17 01 0317	**> 24" and <= 36" Stump Removal, with D8**	**EA**				**35.90**	**138.71**	**0.00**	**174.61**
02109 9130	Site Clearing, Grub Stumps to 36" Diameter with 335 HP Dozer	EA	1.25	CODTL	1.000	35.90	138.71	0.00	174.61

17 01 04	Chipping	Unit	Hourly Output	Crew	Factor	Labor	Equip.	Matl.	Total
17 01 0401	**Light Brush without Grub, Chipping**	**ACRE**				**605.09**	**576.19**	**0.00**	**1,181**
02112 0600	Chipping - Light Brush	ACRE	0.281	COMCA	1.000	605.09	576.19	0.00	1,181
17 01 0402	**Medium Brush without Grub, Chipping**	**ACRE**				**778.01**	**740.86**	**0.00**	**1,519**
02112 0650	Chipping - Medium Brush	ACRE	0.219	COMCA	1.000	778.01	740.86	0.00	1,519
17 01 0403	**Heavy Brush without Grub, Chipping**	**ACRE**				**1,089**	**1,037**	**0.00**	**2,127**
02112 0700	Chipping - Heavy Brush	ACRE	0.156	COMCA	1.000	1,089	1,037	0.00	2,127

17 01 05	Grubbing	Unit	Hourly Output	Crew	Factor	Labor	Equip.	Matl.	Total
17 01 0501	**Dozer 105 HP D5, Grubbing & Stacking**	**CY**				**2.65**	**4.01**	**0.00**	**6.66**
02109 0360	Grub & Stack, D5 Dozer	CY	17.5	COFCZ	1.000	2.65	4.01	0.00	6.66

17 01 Site Clearing

17 01 05 Grubbing		Unit	Hourly Output	Crew	Factor	Labor	Equip.	Matl.	Total
17 01 0502	**Dozer 200 HP D7, Grubbing & Stacking**	**CY**				**0.78**	**2.23**	**0.00**	**3.01**
02109 0365	Grub & Stack, D7 Dozer	CY	61.25	COFCY	1.000	0.78	2.23	0.00	3.01
17 01 0503	**Dozer 460 HP D9, Grubbing & Stacking**	**CY**				**0.11**	**0.53**	**0.00**	**0.64**
02109 0370	Grub & Stack, D9 Dozer	CY	411.3	CODTO	1.000	0.11	0.53	0.00	0.64
17 01 0505	**Track Loader, 2.50 CY 963, Grubbing & Stacking**	**CY**				**1.85**	**4.14**	**0.00**	**5.98**
02109 0375	Grub & Stack, 963 Track Loader	CY	26	COFCN	1.000	1.85	4.14	0.00	5.98
17 01 0506	**Track Loader, 3.75 CY 973, Grubbing & Stacking**	**CY**				**1.57**	**3.51**	**0.00**	**5.08**
02109 0380	Grub & Stack, 973 Track Loader	CY	39	CODLF	1.000	1.57	3.51	0.00	5.08
17 01 0510	**Dozer 105 HP D5 LGP, Wet Grubbing & Stacking**	**CY**				**2.65**	**4.01**	**0.00**	**6.66**
02109 0360	Grub & Stack, D5 Dozer	CY	17.5	COFCZ	1.000	2.65	4.01	0.00	6.66
17 01 0511	**Dozer 200 HP D7 LGP, Wet Grubbing & Stacking**	**CY**				**0.78**	**2.23**	**0.00**	**3.01**
02109 0365	Grub & Stack, D7 Dozer	CY	61.25	COFCY	1.000	0.78	2.23	0.00	3.01
17 01 0515	**Scraper, 11 CY, 613, Grubbing, Haul 1 Mile & Spoil**	**CY**				**0.76**	**2.18**	**0.00**	**2.95**
02109 1000	Grubbing, 11 CY, Scraper	CY	228	CODSC	1.000	0.76	2.18	0.00	2.95
17 01 0516	**Scraper, 15 CY, 621, Grubbing, Haul 1 Mile & Spoil**	**CY**				**0.50**	**2.50**	**0.00**	**3.00**
02109 1100	Grubbing, 15 CY, Scraper	CY	345.6	CODSD	1.000	0.50	2.50	0.00	3.00
17 01 0517	**Scraper, 22 CY, 623, Grubbing, Haul 1 Mile & Spoil**	**CY**				**0.32**	**1.68**	**0.00**	**2.00**
02109 1200	Grubbing, 22 CY, Scraper	CY	544	CODMA	1.000	0.32	1.68	0.00	2.00
17 01 0520	**Scarifying, 12G - 2 Pass**	**SY**				**0.09**	**0.14**	**0.00**	**0.22**
02245 2206	Semi-Grade, 12G, Double Pass	SY	625	COFGA	1.000	0.09	0.14	0.00	0.22
17 01 0521	**Scarifying, 14G - 2 Pass**	**SY**				**0.06**	**0.14**	**0.00**	**0.19**
02245 2205	Semi-Grade, 14G, Double Pass	SY	625	COFCC	1.000	0.06	0.14	0.00	0.19

17 01 06 Selective Thinning		Unit	Hourly Output	Crew	Factor	Labor	Equip.	Matl.	Total
17 01 0601	**Selective Thinning**	**ACRE**				**287.67**	**156.78**	**0.00**	**444.44**
02116 1500	Selective Thinning, 8"-12" Diameter	ACRE	0.75	COMCJ	1.000	287.67	156.78	0.00	444.44

17 01 07 Debris Disposal		Unit	Hourly Output	Crew	Factor	Labor	Equip.	Matl.	Total
17 01 0701	**Light Burning**	**ACRE**				**447.55**	**384.60**	**3,150**	**3,982**
02109 0420	Burning - Light (Incinerator)	ACRE	0.281	UOEHZ	1.000	447.55	384.60	3,150	3,982

17 01 07	Debris Disposal	Unit	Hourly Output	Crew	Factor	Labor	Equip.	Matl.	Total
17 01 0702	**Medium Burning**	**ACRE**				**575.45**	**494.51**	**4,629**	**5,699**
02109 0430	Burning - Medium (Incinerator)	ACRE	0.219	UOEHZ	1.000	575.45	494.51	4,629	5,699
17 01 0703	**Heavy Burning**	**ACRE**				**805.70**	**692.38**	**5,786**	**7,284**
02109 0440	Burning - Heavy (Incinerator)	ACRE	0.156	UOEHZ	1.000	805.70	692.38	5,786	7,284
17 01 0705	**5 Miles, Dump Truck, Load & Haul Debris**	**CY**				**1.12**	**1.88**	**0.00**	**3.01**
02234 0545	Haul, 12 CY (91 m3) Truck, 6 Miles (10 km), 40 MPH (60 km/Hour), 2.1 Cycles/Hour	CY	24.5	COEID	1.000	1.12	1.88	0.00	3.01
17 01 0706	**5 Miles, Semi Dump, Load & Haul Debris**	**CY**				**0.80**	**3.33**	**0.00**	**4.13**
02234 1115	Haul, 16.5 CY (12.6 m3) Truck, 6 Miles (10 km), 40 MPH (60 km/Hour), 2.1 Cycles/Hour	CY	34.25	COETH	1.000	0.80	3.33	0.00	4.13

17 02 01	Building Mass Demolition	Unit	Hourly Output	Crew	Factor	Labor	Equip.	Matl.	Total
17 02 0101	**Multilevel, Steel, Nonexplosive, Building Demolition**	**CF**				**0.04**	**0.03**	**0.00**	**0.07**
02049 8110	Steel, No Disposal, Multi-Level Building, Non-Explosive Demolition	CF	2688	COETZ	1.000	0.04	0.03	0.00	0.07
17 02 0102	**Multilevel, Concrete, Nonexplosive, Building Demolition**	**CF**				**0.06**	**0.04**	**0.00**	**0.10**
02049 8120	Concrete, No Disposal, Multi-Level Building, Non-Explosive Demolition	CF	1913	COETZ	1.000	0.06	0.04	0.00	0.10
17 02 0103	**Multilevel, Masonry, Nonexplosive, Building Demolition**	**CF**				**0.05**	**0.03**	**0.00**	**0.08**
02049 8130	Masonry, No Disposal, Multi-Level Building, Non-Explosive Demolition	CF	2513	COETZ	1.000	0.05	0.03	0.00	0.08
17 02 0105	**Single-level, Steel, Nonexplosive, Building Demolition**	**CF**				**0.06**	**0.07**	**0.00**	**0.14**
02049 8000	Steel, No Disposal, Single-Level Building, Non-Explosive Demolition	CF	1850	COFCX	1.000	0.06	0.07	0.00	0.14
17 02 0106	**Single-level, Concrete, Nonexplosive, Building Demolition**	**CF**				**0.08**	**0.10**	**0.00**	**0.18**
02049 8010	Concrete, No Disposal, Single-Level Building, Non-Explosive Demolition	CF	1413	COFCX	1.000	0.08	0.10	0.00	0.18
17 02 0107	**Single-level, Masonry, Nonexplosive, Building Demolition**	**CF**				**0.06**	**0.07**	**0.00**	**0.14**
02049 8020	Masonry, No Disposal, Single-Level Building, Non-Explosive Demolition	CF	1850	COFCX	1.000	0.06	0.07	0.00	0.14
17 02 0108	**Single-level, Wood, Nonexplosive, Building Demolition**	**CF**				**0.06**	**0.07**	**0.00**	**0.14**
02049 8030	Wood, No Disposal, Single-Level Building, Non-Explosive Demolition	CF	1850	COFCX	1.000	0.06	0.07	0.00	0.14

17 02 02	Aboveground Site Demolition	Unit	Hourly Output	Crew	Factor	Labor	Equip.	Matl.	Total

17 02 02 Aboveground Site Demolition		Unit	Hourly Output	Crew	Factor	Labor	Equip.	Matl.	Total
17 02 0201	**Demolish Bituminous Road with Power Equipment**	**CY**				**14.17**	**7.26**	**0.00**	**21.43**
02046 1710	Demolish Bituminous Roads, Remove Paving and Base Course	SY	55	CODEF	6.000	2.36	1.21	0.00	21.43
17 02 0202	**Demolish Bituminous Driveway with Power Equipment**	**CY**				**15.39**	**7.89**	**0.00**	**23.29**
02046 1800	Demolish Bituminous Driveways, Remove Paving and Base Course	SY	50.63	CODEF	6.000	2.57	1.32	0.00	23.29
17 02 0203	**Demolish Bituminous Pavement with Air Equipment**	**CY**				**26.18**	**6.96**	**0.00**	**33.14**
02046 1760	Demolish Bituminous Pavement with Air Equipment	SF	237.5	CLABE	54.000	0.48	0.13	0.00	33.14
17 02 0205	**Demolish Unreinforced Concrete to 6" Thick with Air Equipment**	**CY**				**41.45**	**11.02**	**0.00**	**52.47**
02046 2325	Demolish Unreinforced Concrete to 6" with Air Equipment	SF	150	CLABE	54.000	0.77	0.20	0.00	52.47
17 02 0206	**Demolish Mesh Reinforced Concrete to 6" Thick with Air Equipment**	**CY**				**59.93**	**15.93**	**0.00**	**75.86**
02046 2340	Demolish Mesh Reinforced Concrete to 6" with Air Equipment	SF	103.8	CLABE	54.000	1.11	0.30	0.00	75.86
17 02 0207	**Demolish Rod Reinforced Concrete to 6" Thick with Air Equipment**	**CY**				**65.03**	**17.28**	**0.00**	**82.31**
02046 2345	Demolish Rod Reinforced Concrete to 6" with Air Equipment	SF	95.63	CLABE	54.000	1.20	0.32	0.00	82.31
17 02 0208	**Demolish Mesh Reinforced Concrete to 6" Thick with Power Equipment**	**CY**				**29.69**	**15.22**	**0.00**	**44.91**
02046 1900	Demolish 6" (15 cm) Thick Concrete Pavement with Wire Mesh	SY	26.25	CODEF	6.000	4.95	2.54	0.00	44.91
17 02 0209	**Demolish Rod Reinforced Concrete to 6" Thick with Power Equipment**	**CY**				**32.81**	**16.82**	**0.00**	**49.63**
02046 2000	Demolish 6" (15 cm) Thick Concrete Pavement with Reinforcing Rod	SY	23.75	CODEF	6.000	5.47	2.80	0.00	49.63
17 02 0210	**Demolish Unreinforced Concrete 7" to 24" Thick with Power Equipment**	**CY**				**79.31**	**40.67**	**0.00**	**119.98**
02046 2110	Demolish Unreinforced Concrete, 7" to 24", with Power Equipment	CY	1.638	CODEF	1.000	79.31	40.67	0.00	119.98
17 02 0211	**Demolish Reinforced Concrete 7" to 24" Thick with Power Equipment**	**CY**				**109.37**	**56.08**	**0.00**	**165.45**
02046 2115	Demolish 7" to 24" Thick Concrete Pavement (17.8 cm to 61 cm) , Reinforced	CY	1.188	CODEF	1.000	109.37	56.08	0.00	165.45
17 02 0212	**Demolish Unreinforced Concrete to 6" Thick with Power Equipment**	**CY**				**20.72**	**10.63**	**0.00**	**31.35**
02112 1212	Demolish 6" (15 cm) Thick Concrete Pavement with No Reinforcement	SY	37.6	CODEF	6.000	3.45	1.77	0.00	31.35

17 02 02 Aboveground Site Demolition		Unit	Hourly Output	Crew	Factor	Labor	Equip.	Matl.	Total
17 02 0215	**Demolish Bituminous Sidewalk**	**CY**				**54.01**	**18.18**	**0.00**	**72.18**
02046 1770	Bituminous Sidewalk	SY	25.39	CODEJ	14.400	3.75	1.26	0.00	72.18
17 02 0216	**Demolish Unreinforced Concrete Sidewalk**	**CY**				**42.85**	**14.42**	**0.00**	**57.27**
02112 1209	Unreinforced Concrete Sidewalk	SY	20	CODEJ	9.000	4.76	1.60	0.00	57.27
17 02 0217	**Demolish Mesh Reinforced Concrete Sidewalk**	**CY**				**45.71**	**15.38**	**0.00**	**61.09**
02046 2350	Mesh Reinforced Concrete Sidewalk	SY	18.75	CODEJ	9.000	5.08	1.71	0.00	61.09
17 02 0218	**Demolish Brick-Set-in-Mortar Sidewalk**	**CY**				**37.06**	**12.47**	**0.00**	**49.53**
02046 4050	Brick Set In Mortar Sidewalk	SY	23.13	CODEJ	9.000	4.12	1.39	0.00	49.53
17 02 0220	**Demolish Bituminous Curbs**	**LF**				**0.93**	**0.31**	**0.00**	**1.24**
02046 2700	Demolish and Load Bituminous Curbing	LF	102.5	CODEJ	1.000	0.93	0.31	0.00	1.24
17 02 0221	**Demolish Unreinforced Concrete Curbs**	**LF**				**2.54**	**0.85**	**0.00**	**3.39**
02046 2400	Demolish and Load Plain Concrete Curbing	LF	37.5	CODEJ	1.000	2.54	0.85	0.00	3.39
17 02 0222	**Demolish Reinforced Concrete Curbs**	**LF**				**3.46**	**1.17**	**0.00**	**4.63**
02046 2500	Reinforced Concrete Curbs	LF	27.5	CODEJ	1.000	3.46	1.17	0.00	4.63
17 02 0223	**Demolish Granite Curbs**	**LF**				**2.31**	**0.78**	**0.00**	**3.09**
02046 2600	Demolish and Load Granite Curbing	LF	41.25	CODEJ	1.000	2.31	0.78	0.00	3.09
17 02 0225	**Remove Chain-link Fence**	**LF**				**1.75**	**0.00**	**0.00**	**1.75**
02046 0752	Remove & Reuse Chain Link Fence	LF	38.75	ULABB	1.000	1.75	0.00	0.00	1.75
17 02 0228	**Remove 3-strand Barbed-wire Fence**	**LF**				**1.31**	**0.14**	**0.00**	**1.45**
02046 0600	Demolish 3-Strand Barbed Wire Fence	LF	62.5	ULABK	1.000	1.31	0.14	0.00	1.45
17 02 0229	**Remove 5-strand Barbed-wire Fence**	**LF**				**1.88**	**0.20**	**0.00**	**2.08**
02046 0650	Demolish 5-Strand Barbed Wire Fence	LF	43.75	ULABK	1.000	1.88	0.20	0.00	2.08
17 02 0230	**Remove Wood Fence (To 6' High), Minimum**	**LF**				**0.99**	**0.00**	**0.00**	**0.99**
02046 0760	Wood to 6', Minimum	LF	62.5	XLABB	1.000	0.99	0.00	0.00	0.99
17 02 0231	**Remove Wood Fence (To 6' High), Maximum**	**LF**				**1.98**	**0.00**	**0.00**	**1.98**
02046 0770	Wood to 6', Maximum	LF	31.25	XLABB	1.000	1.98	0.00	0.00	1.98

17 02 03 Underground Site Demolition		Unit	Hourly Output	Crew	Factor	Labor	Equip.	Matl.	Total
17 02 0301	**Remove 24" Diameter Concrete Pipe, No Excavation**	**LF**				**7.01**	**2.88**	**0.00**	**9.88**
02046 2960	Demolish Concrete Pipe, 21" to 24" Diameter, Sewer/Water Pipe, No Excavation	LF	25	XXPLA	1.000	7.01	2.88	0.00	9.88
17 02 0305	**Abandon Catch Basin/Manhole**	**CY**				**18.76**	**6.32**	**0.00**	**25.08**
02046 0010	Demolish Catch Basin or Manhole, Masonry or Concrete, Abandon	EA	0.875	CODEJ	0.172	108.83	36.63	0.00	25.08

17 02 03 Underground Site Demolition		Unit	Hourly Output	Crew	Factor	Labor	Equip.	Matl.	Total
17 02 0306	**Remove Catch Basin/Manhole**	**CY**				**32.84**	**11.05**	**0.00**	**43.89**
02046 0020	Demolish Masonry Catch Basin/Manhole, 5' (1.5 m) Deep by 4' (1.2 m) Diameter	EA	0.5	CODEJ	0.172	190.45	64.10	0.00	43.89

17 02 04 Debris Disposal		Unit	Hourly Output	Crew	Factor	Labor	Equip.	Matl.	Total
17 02 0401	**Dump Charges (General, Non-Hazardous)**	**CY**				**0.00**	**0.00**	**11.74**	**11.74**
02114 3021	Dump Charges	CY	0	N/A	1.000	0.00	0.00	11.74	11.74
17 02 0402	**CY Dump Charges for Building Construction Material**	**CY**				**0.00**	**0.00**	**21.94**	**21.94**
02114 3011	Dump Charges for Construction Material	CY	0	N/A	1.000	0.00	0.00	21.94	21.94
17 02 0403	**CY Dump Charges for Building Construction Material**	**CY**				**0.00**	**0.00**	**21.94**	**21.94**
02114 3009	Dump Charges for Construction Material	CY	0	N/A	1.000	0.00	0.00	21.94	21.94
17 02 0404	**CY Dump Charges for Building Construction Material**	**CY**				**0.00**	**0.00**	**21.94**	**21.94**
02114 3008	Dump Charges for Construction Material	CY	0	N/A	1.000	0.00	0.00	21.94	21.94
17 02 0405	**CY Dump Charges for Building Construction Material**	**CY**				**0.00**	**0.00**	**21.94**	**21.94**
02114 3007	Dump Charges for Construction Material	CY	0	N/A	1.000	0.00	0.00	21.94	21.94
17 02 0406	**Disposal, Steel Frame, Including Loading**	**CY**				**3.20**	**4.26**	**0.00**	**7.46**
02050 0200	Steel Building	CY	53.75	COETL	1.000	3.20	4.26	0.00	7.46
17 02 0407	**Disposal, Concrete Frame, Including Loading**	**CY**				**3.77**	**5.02**	**0.00**	**8.79**
02050 0300	Concrete Building	CY	45.63	COETL	1.000	3.77	5.02	0.00	8.79
17 02 0408	**Disposal, Masonry Construction, Including Loading**	**CY**				**3.09**	**4.12**	**0.00**	**7.21**
02050 0400	Masonry Building	CY	55.63	COETL	1.000	3.09	4.12	0.00	7.21
17 02 0409	**Disposal, Wood Frame, Including Loading**	**CY**				**5.57**	**7.42**	**0.00**	**12.98**
02050 0500	Wood Building	CY	30.88	COETL	1.000	5.57	7.42	0.00	12.98
17 02 0415	**2 CY Loader (Load Only)**	**CY**				**0.15**	**0.32**	**0.00**	**0.48**
02225 2516	2 CY, Track Loader, Load & Haul	CY	214.4	CODKG	1.000	0.15	0.32	0.00	0.48
17 02 0416	**12 CY Dump Truck Haul/Hour**	**HR**				**27.55**	**46.40**	**0.00**	**73.95**
02225 3129	12 CY, Dump Truck, Haul by Hour	HR	1	COEIA	1.000	27.55	46.40	0.00	73.95

17 02 06 Utility Relocation		Unit	Hourly Output	Crew	Factor	Labor	Equip.	Matl.	Total
17 02 0601	**Remove 4" Diameter Steel-welded Connections, Not Including Excavation**	**LF**				**5.39**	**2.21**	**0.00**	**7.60**
02046 3200	Demolish Welded Steel Pipe, 4" Diameter, Sewer/Water Pipe, No Excavation	LF	32.5	XXPLA	1.000	5.39	2.21	0.00	7.60

17 02 Site Demolition and Relocation

17 02 06 Utility Relocation		Unit	Hourly Output	Crew	Factor	Labor	Equip.	Matl.	Total
17 02 0602	**Remove 10" Diameter Steel-welded Connections, Not Including Excavation**	**LF**				**10.01**	**4.11**	**0.00**	**14.12**
02046 3310	Demolish Welded Steel Pipe, 6" to 12" Diameter, Sewer/Water Pipe, No Excavation	LF	17.5	XXPLA	1.000	10.01	4.11	0.00	14.12
17 02 0603	**Remove 12" Diameter Concrete Pipe, Not Including Excavation**	**LF**				**4.67**	**1.92**	**0.00**	**6.59**
02046 2900	Demolish Concrete Pipe to 12" Diameter, Sewer/Water Pipe, No Excavation	LF	37.5	XXPLA	1.000	4.67	1.92	0.00	6.59
17 02 0604	**Remove 15" Diameter Concrete Pipe, Not Including Excavation**	**LF**				**5.61**	**2.30**	**0.00**	**7.91**
02046 2930	Demolish Concrete Pipe, 15" to 18" Diameter, Sewer/Water Pipe, No Excavation	LF	31.25	XXPLA	1.000	5.61	2.30	0.00	7.91
17 02 0605	**Remove 24" Diameter Concrete Pipe, Not Including Excavation**	**LF**				**7.01**	**2.88**	**0.00**	**9.88**
02046 2960	Demolish Concrete Pipe, 21" to 24" Diameter, Sewer/Water Pipe, No Excavation	LF	25	XXPLA	1.000	7.01	2.88	0.00	9.88
17 02 0606	**Remove 36" Diameter Concrete Pipe, Not Including Excavation**	**LF**				**9.34**	**3.83**	**0.00**	**13.18**
02046 3000	Demolish Concrete Pipe, 27" to 36" Diameter, Sewer/Water Pipe, No Excavation	LF	18.75	XXPLA	1.000	9.34	3.83	0.00	13.18
17 02 0610	**Piping Demolition**	**LF**				**5.39**	**2.21**	**0.00**	**7.60**
02046 3200	Demolish Welded Steel Pipe, 4" Diameter, Sewer/Water Pipe, No Excavation	LF	32.5	XXPLA	1.000	5.39	2.21	0.00	7.60

17 02 07 Fencing Relocation		Unit	Hourly Output	Crew	Factor	Labor	Equip.	Matl.	Total
17 02 0701	**Remove and Reset Chain-link Fence**	**LF**				**10.84**	**0.00**	**0.00**	**10.84**
02046 0750	Remove & Reset Chain Link Fence	LF	6.25	ULABB	1.000	10.84	0.00	0.00	10.84

17 03 01 Grading		Unit	Hourly Output	Crew	Factor	Labor	Equip.	Matl.	Total
17 03 0101	**Rough Grading, D6 Dozer**	**SY**				**0.97**	**2.72**	**0.00**	**3.69**
02224 7000	Rough Grade with Dozer	CSY	0.45	CODTG	0.010	97.36	271.86	0.00	3.69
17 03 0102	**Rough Grading, 12G, 1 Pass**	**SY**				**0.18**	**0.45**	**0.00**	**0.64**
02245 2203	Rough Grader, 12G, Single Pass	SY	187.5	COFCC	1.000	0.18	0.45	0.00	0.64
17 03 0103	**Rough Grading, 14G, 1 Pass**	**SY**				**0.35**	**0.63**	**0.00**	**0.98**
02245 2204	Rough Grade, 14G, Single Pass	SY	187.5	COFGD	1.000	0.35	0.63	0.00	0.98
17 03 0105	**Fine Grading, Hand**	**SY**				**2.03**	**0.00**	**0.00**	**2.03**
02227 4020	Trim & Shape Machine Excavated Area by Hand - Finish Grade	SF	150	ULABA	9.000	0.23	0.00	0.00	2.03
17 03 0106	**Fine Grading, 12G, 2 Passes**	**SY**				**0.12**	**0.19**	**0.00**	**0.31**
02245 2209	Fine Grade, 12G, Double Pass	SY	450	COFGA	1.000	0.12	0.19	0.00	0.31

17 03 Site Earthwork

17 03 01 Grading		Unit	Hourly Output	Crew	Factor	Labor	Equip.	Matl.	Total
17 03 0107	**Fine Grading, 120G, 2 Passes**	**SY**				**0.08**	**0.11**	**0.00**	**0.18**
02245 2208	Fine Grade, 120G, Double Pass	SY	450	COFGC	1.000	0.08	0.11	0.00	0.18
17 03 0108	**Fine Grading, 130G, 2 Passes**	**SY**				**0.08**	**0.13**	**0.00**	**0.20**
02245 2207	Fine Grade, 130G, Double Pass	SY	450	COFGB	1.000	0.08	0.13	0.00	0.20
17 03 0109	**Pad Subgrade Preparation**	**CY**				**1.57**	**2.57**	**0.00**	**4.13**
02220 7070	Subgrade, 8" Lifts	CY	1670	COFCR	1.000	0.16	0.26	0.00	0.42
02227 5000	Rough Grade Structures and Slabs, Push to Stockpile with 140 HP Dozer	CY	81.38	CODTE	1.000	0.54	0.94	0.00	1.48
02512 1100	Fine Grade Structure & Slabs with Grader	SY	125	COFGA	2.000	0.44	0.68	0.00	2.23
17 03 0110	**Construction Labor**	**HR**				**26.09**	**0.00**	**0.00**	**26.09**
02245 1005	Construction Labor	HR	1	XLABA	1.000	26.09	0.00	0.00	26.09

17 03 02 Common Excavation and Disposal		Unit	Hourly Output	Crew	Factor	Labor	Equip.	Matl.	Total
17 03 0201	**Excavation, Spoil to Side**	**CY**				**0.45**	**0.43**	**0.00**	**0.88**
02215 2500	Waste Excess, Excavated Material On Site	CY	107.5	CODLB	1.000	0.45	0.43	0.00	0.88
17 03 0202	**Ditch Excavation, Normal Soil, Haul Spoil 1 Mile**	**CY**				**1.78**	**3.15**	**0.00**	**4.92**
02228 0200	Trenching, 1 CY Gradall, Light Soil, 95 CY/Hour (73 m3/Hour), Continuous Footing Excavation	CY	96	CODGD	1.000	0.65	1.26	0.00	1.92
02234 0340	Haul, 12 CY (91 m3) Truck, 1 Mile (2 km), 20 MPH (30 km/Hour), 4.2 Cycles/Hour	CY	49	COEID	2.000	0.56	0.94	0.00	3.01
17 03 0203	**Roadway Soil Excavation, with Scraper, Load & Haul Spoil**	**CY**				**2.13**	**3.70**	**0.00**	**5.83**
02226 3320	Roadway Excavation Scraper, Soil	CY	80.63	CODSP	1.000	0.61	1.34	0.00	1.95
02234 4000	Load & Haul Soil, 12 CY, 5 Miles	CY	94.5	COEIE	1.000	1.52	2.37	0.00	3.88
17 03 0204	**Roadway Sand Excavation, with Scraper, Load & Haul Spoil**	**CY**				**3.76**	**8.96**	**0.00**	**12.72**
02226 3310	Roadway Excavation Scraper, Sand	CY	15	CODSU	1.000	2.39	6.83	0.00	9.22
02234 4100	Load & Haul Sand, 12 CY, 5 Miles	CY	104.9	COEIE	1.000	1.37	2.13	0.00	3.50
17 03 0205	**Curb/Sidewalk Excavation & Backfill, 27% Haul Spoil, 1 Mile**	**CY**				**10.80**	**7.70**	**0.00**	**18.50**
02220 8800	Compact Backfill, by Hand with 1-Ton Roller, 6" (15 cm) Lift	CY	49.13	CLACD	1.000	0.94	1.51	0.00	2.45
02221 5006	Backfill with Medium Front-end Loader	CY	30	CODLT	1.000	8.64	3.99	0.00	12.64
02228 0200	Trenching, 1 CY Gradall, Light Soil, 95 CY/Hour (73 m3/Hour), Continuous Footing Excavation	CY	96	CODGD	1.000	0.65	1.26	0.00	1.92
02234 0340	Haul, 12 CY (91 m3) Truck, 1 Mile (2 km), 20 MPH (30 km/Hour), 4.2 Cycles/Hour	CY	49	COEID	1.000	0.56	0.94	0.00	1.50
17 03 0211	**Hand Excavation, Normal Soil**	**CY**				**38.71**	**0.00**	**0.00**	**38.71**
02228 1650	Trench, Hand Excavation, Normal Soil, 2'-6' Deep, Piled Only	CY	0.875	ULABA	1.000	38.71	0.00	0.00	38.71

17 03 02 Common Excavation and Disposal		Unit	Hourly Output	Crew	Factor	Labor	Equip.	Matl.	Total
17 03 0212	**Hand Excavation, Sand/Gravel**	**CY**				**33.88**	**0.00**	**0.00**	**33.88**
02228 1655	Trench, Hand Excavation, Sand/Gravel, 2'-6' Deep, Piled Only	CY	1	ULABA	1.000	33.88	0.00	0.00	33.88
17 03 0213	**Hand Excavation, Medium Clay**	**CY**				**45.17**	**0.00**	**0.00**	**45.17**
02228 1660	Trench, Hand Excavation, Medium Clay, 2'-6' Deep, Piled Only	CY	0.75	ULABA	1.000	45.17	0.00	0.00	45.17
17 03 0214	**935, 1.3 CY, Track Loader**	**HR**				**33.05**	**39.08**	**0.00**	**72.13**
02225 2517	935, 1.3 CY, Track Loader	HR	1	CODL3	1.000	33.05	39.08	0.00	72.13
17 03 0215	**931, 1.0 CY, Track Loader**	**HR**				**33.05**	**35.71**	**0.00**	**68.76**
02225 2501	931, 1 CY, by Hour	HR	1	CODLE	1.000	33.05	35.71	0.00	68.76
17 03 0216	**943, 1.5 CY, Track Loader**	**HR**				**34.65**	**46.10**	**0.00**	**80.75**
02225 2502	943, 1.5 CY, by Hour	HR	1	CODLV	1.000	34.65	46.10	0.00	80.75
17 03 0217	**953, 2.0 CY, Track Loader**	**HR**				**33.05**	**68.98**	**0.00**	**102.03**
02225 2503	953, 2 CY, by Hour	HR	1	CODLG	1.000	33.05	68.98	0.00	102.03
17 03 0218	**963, 2.5 CY, Track Loader**	**HR**				**33.05**	**98.50**	**0.00**	**131.55**
02225 2504	963, 2.5 CY, by Hour	HR	1	CODLH	1.000	33.05	98.50	0.00	131.55
17 03 0219	**973, 3.75 CY, Track Loader**	**HR**				**34.65**	**136.75**	**0.00**	**171.40**
02225 2505	973, 3.75 CY, by Hour	HR	1	CODLW	1.000	34.65	136.75	0.00	171.40
17 03 0220	**910, 1.25 CY, Wheel Loader**	**HR**				**33.05**	**24.23**	**0.00**	**57.27**
02225 2506	910, 1.25 CY, by Hour	HR	1	CODLQ	1.000	33.05	24.23	0.00	57.27
17 03 0221	**916, 1.5 CY, Wheel Loader**	**HR**				**33.05**	**30.13**	**0.00**	**63.17**
02225 2507	916, 1.5 CY, by Hour	HR	1	CODLX	1.000	33.05	30.13	0.00	63.17
17 03 0222	**926, 2.0 CY, Wheel Loader**	**HR**				**33.05**	**35.08**	**0.00**	**68.13**
02225 2508	926, 2.0 CY, by Hour	HR	1	CODLR	1.000	33.05	35.08	0.00	68.13
17 03 0223	**950, 3.0 CY, Wheel Loader**	**HR**				**33.05**	**38.13**	**0.00**	**71.17**
02225 2509	950, 3.0 CY, by Hour	HR	1	XXQHQ	1.000	33.05	38.13	0.00	71.17
17 03 0224	**966, 4.0 CY, Wheel Loader**	**HR**				**33.05**	**55.70**	**0.00**	**88.75**
02225 2511	966, 4.0 CY, by Hour	HR	1	CODLY	1.000	33.05	55.70	0.00	88.75
17 03 0225	**980, 5.25 CY, Wheel Loader**	**HR**				**34.65**	**87.55**	**0.00**	**122.20**
02225 2512	980, 5.25 CY, by Hour	HR	1	XXQHR	1.000	34.65	87.55	0.00	122.20
17 03 0226	**988, 7.0 CY, Wheel Loader**	**HR**				**34.65**	**152.88**	**0.00**	**187.53**
02225 2513	988, 7 CY, by Hour	HR	1	CODLZ	1.000	34.65	152.88	0.00	187.53
17 03 0227	**992, 13.5 CY, Wheel Loader**	**HR**				**34.65**	**228.75**	**0.00**	**263.40**
02225 2514	992, 13.5 CY, by Hour	HR	1	CODLU	1.000	34.65	228.75	0.00	263.40

17 03 02 Common Excavation and Disposal		Unit	Hourly Output	Crew	Factor	Labor	Equip.	Matl.	Total
17 03 0263	**Cat 235, 2.5 CY, Soil/Sand, 10' - 20' Deep Trench Box, Trench**	**CY**				**0.95**	**1.97**	**0.49**	**3.41**
02228 0384	Trench, 2-1/2 CY Hydraulic Excavator, Loose Rock, 196 CY/Hour (150 m3/Hour)	CY	196	CODEU	2.000	0.47	0.98	0.00	2.91
02228 3120	Trench, 5 CY Hydraulic Excavator, Daily Rental, Trench Box, 10' x 20'	EA	0	N/A	0.002	0.00	0.00	222.00	0.49
17 03 0264	**Cat 235, 2.5 CY, Soil/Sand with Boulders, Trenching**	**CY**				**0.82**	**1.71**	**0.00**	**2.53**
02228 0388	Trench, 2-1/2 CY Hydraulic Excavator, 113 CY/Hour (86 m3)/Hour)	CY	112.9	CODEU	1.000	0.82	1.71	0.00	2.53
17 03 0265	**Cat 245, 3.0 CY, Soil/Sand, Trenching**	**CY**				**0.64**	**0.99**	**0.00**	**1.63**
02228 0392	Trench, 3 CY Hydraulic Excavator, Medium Soil, 14'-20' Deep, 194 CY/Hour	CY	194	XXQHK	1.000	0.64	0.99	0.00	1.63
17 03 0266	**Cat 245, 3.0 CY, Soil/Sand, 10' - 20' Deep Trench Box, Trench**	**CY**				**2.39**	**3.71**	**0.53**	**6.63**
02228 0396	Trench, 3 CY Hydraulic Excavator, Loose Rock, 14'-20' Deep, 104 CY/Hour	CY	104	XXQHK	2.000	1.19	1.85	0.00	6.10
02228 3120	Trench, 5 CY Hydraulic Excavator, Daily Rental, Trench Box, 10' x 20'	EA	0	N/A	0.002	0.00	0.00	222.00	0.53
17 03 0267	**Cat 245, 3.0 CY, Soil/Sand with Boulders, Trenching**	**CY**				**1.19**	**1.85**	**0.00**	**3.05**
02228 0396	Trench, 3 CY Hydraulic Excavator, Loose Rock, 14'-20' Deep, 104 CY/Hour	CY	104	XXQHK	1.000	1.19	1.85	0.00	3.05
17 03 0268	**Koehring 1166, 4.0 CY, Soil/Sand, Trenching**	**CY**				**0.48**	**1.58**	**0.00**	**2.06**
02228 0401	Trench, 4 CY Hydraulic Excavator, Medium Soil, 20'-25' Deep, 259 CY/Hour	CY	259	XXQHL	1.000	0.48	1.58	0.00	2.06
17 03 0269	**Koehring 1166, 4.0 CY, Soil/Sand,10' - 20' Deep Trench Box, Trench**	**CY**				**1.27**	**4.17**	**0.28**	**5.72**
02228 0402	Trench, 4 CY Hydraulic Excavator, Heavy Soil, 20'-25' Deep, 196 CY/Hour	CY	196	XXQHL	2.000	0.63	2.08	0.00	5.44
02228 3120	Trench, 5 CY Hydraulic Excavator, Daily Rental, Trench Box, 10' x 20'	EA	0	N/A	0.001	0.00	0.00	222.00	0.28
17 03 0270	**Koehring 1166, 4.0 CY, Soil/Sand with Boulders, Trenching**	**CY**				**0.63**	**2.08**	**0.00**	**2.72**
02228 0402	Trench, 4 CY Hydraulic Excavator, Heavy Soil, 20'-25' Deep, 196 CY/Hour	CY	196	XXQHL	1.000	0.63	2.08	0.00	2.72
17 03 0272	**Pull Trench Box, Cat 225, 1.5 CY, Soil/Sand, Trenching**	**CY**				**1.24**	**1.78**	**0.35**	**3.37**
02228 0346	Trench, 1-1/2 CY Hydraulic Excavator, Loose Rock, 101 CY/Hour (77 m3/Hour)	CY	101.3	CODES	2.000	0.62	0.89	0.00	3.02
02228 3120	Trench, 5 CY Hydraulic Excavator, Daily Rental, Trench Box, 10' x 20'	EA	0	N/A	0.002	0.00	0.00	222.00	0.35

17 03 02 Common Excavation and Disposal		Unit	Hourly Output	Crew	Factor	Labor	Equip.	Matl.	Total
17 03 0251	**1 CY Clamshell, with 40' Boom**	**HR**				**66.00**	**109.58**	**0.00**	**175.57**
02225 1416	1 CY, Clamshell	HR	1	UOEHQ	1.000	66.00	109.58	0.00	175.57
17 03 0252	**1.5 CY Clamshell, with 60' Boom**	**HR**				**66.00**	**158.43**	**0.00**	**224.43**
02225 1417	1.5 CY, Clamshell	HR	1	UOEHV	1.000	66.00	158.43	0.00	224.43
17 03 0253	**2.5 CY Clamshell, with 50' Boom**	**HR**				**66.00**	**154.78**	**0.00**	**220.78**
02225 1436	2.5 CY, Clamshell	HR	1	UOEHI	1.000	66.00	154.78	0.00	220.78
17 03 0254	**Trench, 5 CY Hydraulic Excavator, Daily Rental Trench Box, 1 - 12'**	**CY**				**0.00**	**0.00**	**0.58**	**0.58**
02228 3120	Trench, 5 CY Hydraulic Excavator, Daily Rental, Trench Box, 10' x 20'	EA	0	N/A	0.003	0.00	0.00	222.00	0.58
17 03 0255	**Trenching to 48" Deep, Including Backfill & Compaction**	**CY**				**3.92**	**0.94**	**0.00**	**4.86**
02215 1220	Backfill Trench with Small Front-end Loader, without Compaction	CY	70	CODLB	1.000	0.69	0.66	0.00	1.34
02228 1900	Compaction, 6" Layers, Vibrating Plate (15 cm) Layers	CY	40.5	CLACC	1.000	2.03	0.11	0.00	2.13
02230 0820	Trench, Chain Trencher, Medium Soil, 1' Wide x 4' Deep	CY	34.38	CODBB	1.000	1.20	0.17	0.00	1.38
17 03 0256	**Cat 215, 1.0 CY, Sand, Shallow, Trenching**	**CY**				**0.40**	**0.58**	**0.00**	**0.98**
02228 0320	Trench, 1 CY Hydraulic Excavator, Light Soil, 4' to 6' Deep, 155 CY/Hour	CY	156	CODEO	1.000	0.40	0.58	0.00	0.98
17 03 0257	**Cat 215, 1.0 CY, Soil, Shallow, Trenching**	**CY**				**0.49**	**0.70**	**0.00**	**1.19**
02228 0322	Trench, 1 CY Hydraulic Excavator, Medium Soil, 4' to 6' Deep, 128 CY/Hour	CY	128	CODEO	1.000	0.49	0.70	0.00	1.19
17 03 0259	**Cat 225, 1.5 CY, Soil/Sand, Trenching**	**CY**				**0.37**	**0.54**	**0.00**	**0.91**
02228 0342	Trench, 1-1/2 CY Hydraulic Excavator, Medium Soil, 168 CY/Hour (128 m3/Hour)	CY	168.1	CODES	1.000	0.37	0.54	0.00	0.91
17 03 0260	**Cat 225, 1.5 CY, Soil/Sand, 10' - 20' Deep Trench Box, Trench**	**CY**				**1.24**	**1.78**	**0.70**	**3.72**
02228 0346	Trench, 1-1/2 CY Hydraulic Excavator, Loose Rock, 101 CY/Hour (77 m3/Hour)	CY	101.3	CODES	2.000	0.62	0.89	0.00	3.02
02228 3120	Trench, 5 CY Hydraulic Excavator, Daily Rental, Trench Box, 10' x 20'	EA	0	N/A	0.003	0.00	0.00	222.00	0.70
17 03 0261	**Cat 225, 1.5 CY, Soil/Sand with Boulders, Trenching**	**CY**				**0.79**	**1.14**	**0.00**	**1.93**
02228 0348	Trench, 1-1/2 CY Hydraulic Excavator, Blast Rock, 79 CY/Hour (61 m3)/Hour)	CY	79.25	CODES	1.000	0.79	1.14	0.00	1.93
17 03 0262	**Cat 235, 2.5 CY, Soil/Sand, Trenching**	**CY**				**0.39**	**0.80**	**0.00**	**1.19**
02228 0382	Trench, 2-1/2 CY Hydraulic Excavator, Medium Soil, 240 CY/Hour (183 m3/Hour)	CY	240	CODEU	1.000	0.39	0.80	0.00	1.19

17 03 02	Common Excavation and Disposal	Unit	Hourly Output	Crew	Factor	Labor	Equip.	Matl.	Total
17 03 0228	**936, 2.5 CY, Wheel Loader**	**HR**				**33.05**	**35.08**	**0.00**	**68.13**
02225 2518	936, 2.5 CY, Wheel Loader	HR	1	CODL4	1.000	33.05	35.08	0.00	68.13
17 03 0229	**Hand Excavation, Heavy Clay**	**CY**				**54.20**	**0.00**	**0.00**	**54.20**
02228 1665	Trench, Hand Excavation, Heavy Clay, 2'-6' Deep, Piled Only	CY	0.625	ULABA	1.000	54.20	0.00	0.00	54.20
17 03 0230	**Crawler-mounted, 1.0 CY, 215 Hydraulic Excavator**	**HR**				**35.90**	**69.73**	**0.00**	**105.63**
02225 2115	215 Hydraulic Excavator 1 CY, by Hour	HR	1	CODEL	1.000	35.90	69.73	0.00	105.63
17 03 0231	**Crawler-mounted, 1.25 CY, 225 Hydraulic Excavator**	**HR**				**35.90**	**69.73**	**0.00**	**105.63**
02225 2125	225 Hydraulic Excavator 1.25 CY, by Hour	HR	1	CODEL	1.000	35.90	69.73	0.00	105.63
17 03 0232	**Crawler-mounted, 2.0 CY, 235 Hydraulic Excavator**	**HR**				**35.90**	**113.73**	**0.00**	**149.63**
02225 2135	235 Hydraulic Excavator 2 CY, by Hour	HR	1	CODE1	1.000	35.90	113.73	0.00	149.63
17 03 0233	**Crawler-mounted, 3.125 CY, 245 Hydraulic Excavator**	**HR**				**35.90**	**192.88**	**0.00**	**228.78**
02225 2145	245 Hydraulic Excavator 3.125 CY, by Hour	HR	1	CODEE	1.000	35.90	192.88	0.00	228.78
17 03 0234	**Crawler-mounted, 4.0 CY, Koehring 1166 Hydraulic Excavator**	**HR**				**35.90**	**192.88**	**0.00**	**228.78**
02225 2155	Koehring 1166, 4 CY, by Hour	HR	1	XXQHO	1.000	35.90	192.88	0.00	228.78
17 03 0235	**Crawler-mounted, 5.5 CY, Koehring 1266, Hydraulic Excavator**	**HR**				**35.90**	**479.57**	**0.00**	**515.47**
02225 2165	Koehring 1266, 5.5 CY, by Hour	HR	1	XXQHP	1.000	35.90	479.57	0.00	515.47
17 03 0240	**Scraper - Elevating, 11 CY - 613**	**HR**				**35.90**	**102.43**	**0.00**	**138.32**
02226 2305	Scraper, Elevating 11 CY, 613 by Hour	HR	1	CODSW	1.000	35.90	102.43	0.00	138.32
17 03 0241	**Scraper - Elevating, 22 CY - 623**	**HR**				**35.90**	**206.00**	**0.00**	**241.90**
02226 2415	Scraper, Elevating 22 CY, 623, by Hour	HR	1	CODSI	1.000	35.90	206.00	0.00	241.90
17 03 0247	**1.0 CY Dragline, with 40' Boom**	**HR**				**66.00**	**106.70**	**0.00**	**172.70**
02225 1415	1 CY, Dragline	HR	1	UOEHX	1.000	66.00	106.70	0.00	172.70
17 03 0248	**1.5 CY Dragline, with 60' Boom**	**HR**				**66.00**	**154.57**	**0.00**	**220.57**
02225 1425	1.5 CY, Dragline	HR	1	UOEHN	1.000	66.00	154.57	0.00	220.57
17 03 0249	**2.5 CY Dragline, with 50' Boom**	**HR**				**66.00**	**150.55**	**0.00**	**216.55**
02225 1435	2.5 CY, Dragline	HR	1	UOEHO	1.000	66.00	150.55	0.00	216.55
17 03 0250	**4 CY Dragline, with 95' Boom**	**HR**				**66.00**	**195.80**	**0.00**	**261.80**
02225 1455	4 CY, Dragline	HR	1	UOEHY	1.000	66.00	195.80	0.00	261.80

17 03 02 Common Excavation and Disposal		Unit	Hourly Output	Crew	Factor	Labor	Equip.	Matl.	Total
17 03 0273	**Pull Trench Box, Cat 235, 2.5 CY, Soil/Sand, Trenching**	**CY**				**0.95**	**1.97**	**0.25**	**3.16**
02228 0384	Trench, 2-1/2 CY Hydraulic Excavator, Loose Rock, 196 CY/Hour (150 m3/Hour)	CY	196	CODEU	2.000	0.47	0.98	0.00	2.91
02228 3120	Trench, 5 CY Hydraulic Excavator, Daily Rental, Trench Box, 10' x 20'	EA	0	N/A	0.001	0.00	0.00	222.00	0.25
17 03 0274	**Pull Trench Box, Cat 245, 3.0 CY, Soil/Sand, Trenching**	**CY**				**2.39**	**3.71**	**0.27**	**6.36**
02228 0396	Trench, 3 CY Hydraulic Excavator, Loose Rock, 14'-20' Deep, 104 CY/Hour	CY	104	XXQHK	2.000	1.19	1.85	0.00	6.10
02228 3120	Trench, 5 CY Hydraulic Excavator, Daily Rental, Trench Box, 10' x 20'	EA	0	N/A	0.001	0.00	0.00	222.00	0.27
17 03 0275	**Pull Trench Box, Koehring 1166, 4.0 CY, Soil/Sand, Trench**	**CY**				**1.27**	**4.17**	**0.14**	**5.58**
02228 0402	Trench, 4 CY Hydraulic Excavator, Heavy Soil, 20'-25' Deep, 196 CY/Hour	CY	196	XXQHL	2.000	0.63	2.08	0.00	5.44
02228 3120	Trench, 5 CY Hydraulic Excavator, Daily Rental, Trench Box, 10' x 20'	EA	0	N/A	0.001	0.00	0.00	222.00	0.14
17 03 0276	**1 CY, Crawler-mounted, Hydraulic Excavator**	**CY**				**1.56**	**2.25**	**0.00**	**3.82**
02232 0100	Excavate & Load, 1 CY Hydraulic Excavator, Medium Material, 40 CY/Hour (31 m3)	CY	40	CODEO	1.000	1.56	2.25	0.00	3.82
17 03 0277	**2 CY, Crawler-mounted, Hydraulic Excavator**	**CY**				**0.83**	**1.52**	**0.00**	**2.35**
02232 0140	Excavate & Load, 2 CY Hydraulic Excavator, Medium Material, 75 CY/Hour (57 m3)	CY	75	CODET	1.000	0.83	1.52	0.00	2.35
17 03 0278	**3 CY, Crawler-mounted, Hydraulic Excavator**	**CY**				**0.84**	**1.31**	**0.00**	**2.16**
02232 0160	Excavate & Load, 3 CY Hydraulic Excavator, Medium Material, 147 CY/Hour (112 m3)	CY	147	XXQHK	1.000	0.84	1.31	0.00	2.16
17 03 0279	**4 CY, Crawler-mounted, Hydraulic Excavator**	**CY**				**0.63**	**2.08**	**0.00**	**2.72**
02225 2152	Excavate & Load, 4 CY Hydraulic Excavator, Medium Material, 196 CY/Hour (150 m3)	CY	196	XXQHL	1.000	0.63	2.08	0.00	2.72
17 03 0280	**Borrow Sand, Load & Haul 5 Miles, No Spreading**	**CY**				**1.64**	**2.66**	**0.00**	**4.30**
02225 3301	Cost for each additional mile (50.89% of total Labor & Equipment costs)	MI	0	N/A	0.000	0.00	0.00	2.27	0.00
02228 0392	Trench, 3 CY Hydraulic Excavator, Medium Soil, 14'-20' Deep, 194 CY/Hour	CY	194	XXQHK	1.000	0.64	0.99	0.00	1.63
02234 0470	Hauling 12 LCY 4 Miles (Base Wide Rate)	CY	27.63	COEID	1.000	1.00	1.67	0.00	2.67

17 03 02 Common Excavation and Disposal		Unit	Hourly Output	Crew	Factor	Labor	Equip.	Matl.	Total
17 03 0281	**Borrow Subgrade, Load & Haul 5 Miles, Spreading**	**CY**				**2.21**	**3.87**	**0.00**	**6.08**
02225 2505	973, 3.75 CY, by Hour	HR	1	CODLW	0.003	34.65	136.75	0.00	0.44
02225 3301	Cost for each additional mile (50.89% of total Labor & Equipment costs)	MI	0	N/A	0.000	0.00	0.00	2.27	0.00
02228 0392	Trench, 3 CY Hydraulic Excavator, Medium Soil, 14'-20' Deep, 194 CY/Hour	CY	194	XXQHK	1.000	0.64	0.99	0.00	1.63
02234 0470	Hauling 12 LCY 4 Miles (Base Wide Rate)	CY	27.63	COEID	1.000	1.00	1.67	0.00	2.67
02240 0030	Spread Borrow with Dozer	CY	90	CODTE	1.000	0.49	0.85	0.00	1.34
17 03 0282	**Soil, 5 Miles, Dump Truck, Load/Haul Spoil From Trench**	**CY**				**1.52**	**2.37**	**0.00**	**3.88**
02234 4000	Load & Haul Soil, 12 CY, 5 Miles	CY	94.5	COEIE	1.000	1.52	2.37	0.00	3.88
17 03 0283	**Sand, 5 Mile, Dump Truck, Load/Haul Spoil From Trench**	**CY**				**1.37**	**2.13**	**0.00**	**3.50**
02234 4100	Load & Haul Sand, 12 CY, 5 Miles	CY	104.9	COEIE	1.000	1.37	2.13	0.00	3.50
17 03 0284	**8 CY, Dump Truck**	**HR**				**27.55**	**46.02**	**0.00**	**73.57**
02225 3146	8 CY, Dump Truck	HR	1	COETA	1.000	27.55	46.02	0.00	73.57
17 03 0285	**12 CY, Dump Truck**	**HR**				**27.55**	**46.40**	**0.00**	**73.95**
02225 3137	12 CY Dump, Highway	HR	1	COEIA	1.000	27.55	46.40	0.00	73.95
17 03 0286	**16.5 CY, Dump Truck**	**HR**				**27.55**	**98.13**	**0.00**	**125.68**
02225 3138	16.5 CY Dump, Highway	HR	1	COEIB	1.000	27.55	98.13	0.00	125.68
17 03 0287	**20 CY, Semi Dump**	**HR**				**27.55**	**66.32**	**0.00**	**93.87**
02225 3139	20 CY, Dump, Highway	HR	1	COEIC	1.000	27.55	66.32	0.00	93.87
17 03 0288	**26 CY, Semi Dump**	**HR**				**27.55**	**67.04**	**0.00**	**94.59**
02225 3141	26 CY, Dump, Highway	HR	1	COETC	1.000	27.55	67.04	0.00	94.59
17 03 0289	**32 CY, Semi Dump**	**HR**				**27.55**	**70.39**	**0.00**	**97.94**
02225 3142	32 CY, Dump, Highway	HR	1	COETD	1.000	27.55	70.39	0.00	97.94
17 03 0290	**36 CY, Semi Dump**	**HR**				**27.55**	**74.61**	**0.00**	**102.16**
02225 3143	36 CY, Dump, Highway	HR	1	COETG	1.000	27.55	74.61	0.00	102.16
17 03 0291	**20 CY, Bottom Dump**	**HR**				**27.55**	**60.80**	**0.00**	**88.35**
02225 3144	20 CY, Bottom Dump	HR	1	COETO	1.000	27.55	60.80	0.00	88.35
17 03 0292	**28 CY, Bottom Dump**	**HR**				**27.55**	**62.37**	**0.00**	**89.92**
02225 3145	28 CY, Bottom Dump	HR	1	COETQ	1.000	27.55	62.37	0.00	89.92
17 03 0293	**18 CY, Bottom Dump, 27 Ton**	**HR**				**27.55**	**54.93**	**0.00**	**82.48**
02225 3147	18 CY, Bottom Dump, 27 Ton	HR	1	COET1	1.000	27.55	54.93	0.00	82.48

17 03 Site Earthwork

17 03 02 Common Excavation and Disposal		Unit	Hourly Output	Crew	Factor	Labor	Equip.	Matl.	Total
17 03 0294	**30 CY, Bottom Dump, 30 Ton**	**HR**				**27.55**	**65.76**	**0.00**	**93.31**
02225 3148	30 CY, Bottom Dump, 30 Ton	HR	1	COET2	1.000	27.55	65.76	0.00	93.31
17 03 0295	**35 Ton, 769, Off-highway Truck**	**HR**				**20.66**	**110.72**	**0.00**	**131.38**
02225 3134	35 Ton, 769, Off Highway Truck	HR	1	CTDHH	0.750	27.55	147.63	0.00	131.38
17 03 0296	**50 Ton, 773, Off-highway Truck**	**HR**				**27.55**	**147.63**	**0.00**	**175.18**
02225 3135	50 Ton, 773, Off Highway Truck	HR	1	CTDHH	1.000	27.55	147.63	0.00	175.18
17 03 0297	**85 Ton, 777, Off-highway Truck**	**HR**				**27.55**	**159.50**	**0.00**	**187.05**
02225 3136	85 Ton, 777, Off Highway Truck	HR	1	CTDHI	1.000	27.55	159.50	0.00	187.05
17 03 0298	**1/2 CY Crawler-mounted, Hydraulic Excavator**	**CY**				**1.21**	**1.35**	**0.00**	**2.56**
02228 0304	Trench, 1/2 CY Hydraulic Excavator, Heavy Soil, 51 CY/Hour (40 m3/Hour)	CY	51.63	CODEA	1.000	1.21	1.35	0.00	2.56
17 03 0299	**Trench, 5 CY Hydraulic Excavator, Daily Rental Trench Box, 12.1' - 24'**	**CY**				**0.00**	**0.00**	**1.17**	**1.17**
02228 3120	Trench, 5 CY Hydraulic Excavator, Daily Rental, Trench Box, 10' x 20'	EA	0	N/A	0.005	0.00	0.00	222.00	1.17

17 03 03 Rock Excavation and Disposal		Unit	Hourly Output	Crew	Factor	Labor	Equip.	Matl.	Total
17 03 0301	**2 1/2" Bore @ 25' with Track Drill**	**LF**				**0.74**	**2.51**	**0.00**	**3.25**
02221 2000	2-1/2" Bore @ 25', 64'/Hour 80%	LF	100	CLADQ	1.000	0.74	2.51	0.00	3.25
17 03 0302	**5 1/2" Bore @ 25' with Track Drill**	**LF**				**5.08**	**9.09**	**0.33**	**14.49**
02221 1500	Quarry Production Drill, 30 LF/Hour (9 m/Hour)	LF	29.5	CLADH	2.000	2.54	4.54	0.16	14.49
17 03 0303	**Ammonium Nitrate Explosive, Loaded, Packed & Blown**	**LB**				**2.39**	**0.11**	**6.37**	**8.87**
02211 2105	Blasting Crew Time Equivalent/Lb	LB	34.38	ULABJ	1.000	2.39	0.11	0.00	2.50
02223 3700	Ammonium Nitrite Explosives	LB	0	N/A	1.000	0.00	0.00	1.99	1.99
02223 3720	Detonator Wire	LF	0	N/A	0.800	0.00	0.00	0.03	0.02
02223 3740	Primacord (50 Grain)	MLF	0	N/A	0.010	0.00	0.00	390.00	3.90
02223 3762	Fuel Oil for Blasting	GAL	0	N/A	0.446	0.00	0.00	1.03	0.46
17 03 0305	**Cat 215, 1.0 CY, Rock, No Haul or Borrow, Trenching**	**BCY**				**50.59**	**34.44**	**54.51**	**139.54**
02221 0800	Rock Removal Drill for Pre-Splitting	LF	39.38	CLADH	4.412	1.90	3.40	0.16	24.12
02221 1605	Quarry Production Drill, 75 LF/Hour, (23m/Hour)	LF	75.5	CLADH	4.412	0.99	1.77	0.00	12.21
02223 3000	Blasting for Pre-Splitting, Load and Detonate	LF	59	CLADF	4.412	2.37	0.78	5.96	40.21
02223 3700	Ammonium Nitrite Explosives	LB	0	N/A	10.300	0.00	0.00	1.99	20.47
02223 3720	Detonator Wire	LF	0	N/A	8.250	0.00	0.00	0.03	0.25
02223 3740	Primacord (50 Grain)	MLF	0	N/A	0.005	0.00	0.00	390.00	2.03
02223 3762	Fuel Oil for Blasting	GAL	0	N/A	4.594	0.00	0.00	1.03	4.73
02232 0110	Excavate & Load, 1 CY Hydraulic Excavator, Blast Rock, 28 CY/Hour, (21 m3)	CY	28	XXQMT	1.000	2.17	2.49	0.00	4.66
02238 0600	Break Rock with Air Hammer, Dense Rock	CY	2	CLADA	1.000	25.20	5.66	0.00	30.86

17 03 03 Rock Excavation and Disposal		Unit	Hourly Output	Crew	Factor	Labor	Equip.	Matl.	Total
17 03 0306	**Cat 225, 1.5 CY, Rock, No Haul or Borrow, Trenching**	**BCY**				**44.13**	**27.52**	**40.97**	**112.63**
02221 0800	Rock Removal Drill for Pre-Splitting	LF	39.38	CLADH	3.333	1.90	3.40	0.16	18.23
02221 1605	Quarry Production Drill, 75 LF/Hour, (23m/Hour)	LF	75.5	CLADH	3.333	0.99	1.77	0.00	9.22
02223 3000	Blasting for Pre-Splitting, Load and Detonate	LF	59	CLADF	3.333	2.37	0.78	5.96	30.38
02223 3700	Ammonium Nitrite Explosives	LB	0	N/A	7.700	0.00	0.00	1.99	15.31
02223 3720	Detonator Wire	LF	0	N/A	6.200	0.00	0.00	0.03	0.19
02223 3740	Primacord (50 Grain)	MLF	0	N/A	0.004	0.00	0.00	390.00	1.52
02223 3762	Fuel Oil for Blasting	GAL	0	N/A	3.434	0.00	0.00	1.03	3.54
02232 0130	Excavate & Load, 1.5 CY Hydraulic Excavator, Blast Rock, 45 CY/Hour (35 m3)	CY	45	CODES	1.000	1.39	2.00	0.00	3.39
02238 0600	Break Rock with Air Hammer, Dense Rock	CY	2	CLADA	1.000	25.20	5.66	0.00	30.86
17 03 0307	**Cat 235, 2.0 CY, Rock, No Haul or Borrow, Trenching**	**BCY**				**40.26**	**23.37**	**32.94**	**96.57**
02221 0800	Rock Removal Drill for Pre-Splitting	LF	39.38	CLADH	2.679	1.90	3.40	0.16	14.65
02221 1605	Quarry Production Drill, 75 LF/Hour, (23m/Hour)	LF	75.5	CLADH	2.679	0.99	1.77	0.00	7.41
02223 3000	Blasting for Pre-Splitting, Load and Detonate	LF	59	CLADF	2.679	2.37	0.78	5.96	24.41
02223 3700	Ammonium Nitrite Explosives	LB	0	N/A	6.200	0.00	0.00	1.99	12.32
02223 3720	Detonator Wire	LF	0	N/A	5.000	0.00	0.00	0.03	0.15
02223 3740	Primacord (50 Grain)	MLF	0	N/A	0.003	0.00	0.00	390.00	1.21
02223 3762	Fuel Oil for Blasting	GAL	0	N/A	2.765	0.00	0.00	1.03	2.85
02232 0150	Excavate & Load, 2 CY Hydraulic Excavator, Blast Rock, 65 CY/Hour (49 m3)	CY	65	CODET	1.000	0.96	1.75	0.00	2.71
02238 0600	Break Rock with Air Hammer, Dense Rock	CY	2	CLADA	1.000	25.20	5.66	0.00	30.86
17 03 0308	**Cat 245, 3.0 CY, Rock, No Haul or Borrow, Trenching**	**BCY**				**38.83**	**21.88**	**27.12**	**87.84**
02221 0800	Rock Removal Drill for Pre-Splitting	LF	39.38	CLADH	2.239	1.90	3.40	0.16	12.24
02221 1605	Quarry Production Drill, 75 LF/Hour, (23m/Hour)	LF	75.5	CLADH	2.239	0.99	1.77	0.00	6.19
02223 3000	Blasting for Pre-Splitting, Load and Detonate	LF	59	CLADF	2.239	2.37	0.78	5.96	20.41
02223 3700	Ammonium Nitrite Explosives	LB	0	N/A	5.200	0.00	0.00	1.99	10.34
02223 3720	Detonator Wire	LF	0	N/A	4.200	0.00	0.00	0.03	0.13
02223 3740	Primacord (50 Grain)	MLF	0	N/A	0.003	0.00	0.00	390.00	1.01
02223 3762	Fuel Oil for Blasting	GAL	0	N/A	1.873	0.00	0.00	1.03	1.93
02232 0170	Excavate & Load, 3 CY Hydraulic Excavator, Blast Rock, 67 CY/Hour (51 m3)	CY	67	XXQHK	1.000	1.85	2.88	0.00	4.73
02238 0600	Break Rock with Air Hammer, Dense Rock	CY	2	CLADA	1.000	25.20	5.66	0.00	30.86

17 03 03 Rock Excavation and Disposal		Unit	Hourly Output	Crew	Factor	Labor	Equip.	Matl.	Total
17 03 0309	**Koehring 1166, 4.0 CY, Rock, No Haul or Borrow, Trench**	**BCY**				**36.72**	**21.71**	**23.80**	**82.22**
02221 0800	Rock Removal Drill for Pre-Splitting	LF	39.38	CLADH	1.923	1.90	3.40	0.16	10.51
02221 1605	Quarry Production Drill, 75 LF/Hour, (23m/Hour)	LF	75.5	CLADH	1.923	0.99	1.77	0.00	5.32
02223 3000	Blasting for Pre-Splitting, Load and Detonate	LF	59	CLADF	1.923	2.37	0.78	5.96	17.53
02223 3700	Ammonium Nitrite Explosives	LB	0	N/A	4.500	0.00	0.00	1.99	8.95
02223 3720	Detonator Wire	LF	0	N/A	3.600	0.00	0.00	0.03	0.11
02223 3740	Primacord (50 Grain)	MLF	0	N/A	0.002	0.00	0.00	390.00	0.90
02223 3762	Fuel Oil for Blasting	GAL	0	N/A	2.007	0.00	0.00	1.03	2.07
02225 2154	Excavate & Load, 4 CY Hydraulic Excavator, Blast Rock, 89 CY/Hour (68 m3)	CY	89	XXQHL	1.000	1.40	4.59	0.00	5.99
02238 0600	Break Rock with Air Hammer, Dense Rock	CY	2	CLADA	1.000	25.20	5.66	0.00	30.86
17 03 0310	**D8 with U-blade & Single-shank Ripper, Bulldozer**	**HR**				**34.65**	**172.15**	**0.00**	**206.80**
02211 4203	D8 with Single Ripper/Hour	HR	1	CODTM	1.000	34.65	172.15	0.00	206.80
17 03 0311	**D9 with U-blade & Single-shank Ripper, Bulldozer**	**HR**				**34.65**	**217.39**	**0.00**	**252.04**
02211 4204	D9 with Double Ripper/Hour	HR	1	CODTQ	1.000	34.65	217.39	0.00	252.04
17 03 0312	**D10 with U-blade & Single-shank Ripper, Bulldozer**	**HR**				**34.65**	**228.64**	**0.00**	**263.29**
02211 4205	D10 with Single Ripper/Hour	HR	1	CODTP	1.000	34.65	228.64	0.00	263.29
17 03 0315	**Roadway Rock Excavator Borrow Subgrade (5 Miles) & Spread**	**CY**				**2.44**	**8.72**	**0.00**	**11.16**
02226 3330	Roadway Excavation Scraper, Rock Borrow	CY	156.3	CODSQ	1.000	2.44	8.72	0.00	11.16
17 03 0316	**Roadway Rock Excavation & Haul (5 Miles), No Borrow**	**CY**				**1.67**	**5.79**	**0.00**	**7.46**
02226 3340	Roadway Excavation Scraper, Rock No Borrow	CY	156.3	CODSR	1.000	1.67	5.79	0.00	7.46
17 03 0317	**Rock, 5 Miles, Dump Truck, Load/Haul Spoil from Trench**	**CY**				**2.14**	**3.61**	**0.00**	**5.75**
02234 4200	Load & Haul Rock, 12 CY, 5 Miles	CY	66.88	COEIF	1.000	2.14	3.61	0.00	5.75
17 03 0323	**Standby, D8 with U-blade & Single-shank Ripper**	**HR**				**0.00**	**43.98**	**0.00**	**43.98**
02211 4206	Standby, D8 with U-Blade & Single Shank Ripper	HR	1	XYCDF	0.255	0.00	172.15	0.00	43.98
17 03 0324	**Standby, D9 with U-blade & Single-shank Ripper**	**HR**				**0.00**	**56.26**	**0.00**	**56.26**
02211 4207	Standby, D9 with U-Blade & Single Shank Ripper	HR	1	XYCDG	0.259	0.00	217.39	0.00	56.26
17 03 0325	**Standby, D10 with U-blade & Single-shank Ripper**	**HR**				**0.00**	**58.42**	**0.00**	**58.42**
02211 4208	Standby, D10 with U-Blade & Single Shank Ripper	HR	1	XYCDH	0.256	0.00	228.64	0.00	58.42

17 03 03 Rock Excavation and Disposal		Unit	Hourly Output	Crew	Factor	Labor	Equip.	Matl.	Total
17 03 0329	**Standby, Scraper - Elevating, 11 CY 613**	**HR**				**0.00**	**25.92**	**0.00**	**25.92**
02226 2313	Standby, Scraper - Elevating, 11 CY 613	HR	1	XYCSD	0.253	0.00	102.43	0.00	25.92
17 03 0330	**Standby, Scraper - Elevating, 22 CY 623**	**HR**				**0.00**	**54.08**	**0.00**	**54.08**
02226 2418	Standby, Scraper - Elevating, 22 CY 623	HR	1	XYCSE	0.263	0.00	206.00	0.00	54.08
17 03 0334	**Standby, Crawler-mounted, 1.0 CY, 215 Hydraulic Excavator**	**HR**				**0.00**	**20.88**	**0.00**	**20.88**
02221 2116	Standby, Crawler Mounted, 1 CY, 215 Hydraulic Excavator	HR	1	XYCEA	0.299	0.00	69.73	0.00	20.88
17 03 0335	**Standby, Crawler-mounted, 1.25 CY, 225 Hydraulic Excavator**	**HR**				**0.00**	**21.42**	**0.00**	**21.42**
02221 2126	Standby, Crawler Mounted, 1.25 CY, 225 Hydraulic Excavator	HR	1	XYCEB	0.290	0.00	73.77	0.00	21.42
17 03 0336	**Standby, Crawler-mounted, 2.0 CY, 235 Hydraulic Excavator**	**HR**				**0.00**	**33.39**	**0.00**	**33.39**
02221 2136	Standby, Crawler Mounted, 2 CY, 235 Hydraulic Excavator	HR	1	XYCEC	0.294	0.00	113.73	0.00	33.39
17 03 0337	**Standby, Crawler-mounted, 3.125 CY, 245 Hydraulic Excavator**	**HR**				**0.00**	**51.94**	**0.00**	**51.94**
02221 2146	Standby, Crawler Mounted, 3.125 CY, 245 Hydraulic Excavator	HR	1	XYCED	0.269	0.00	192.88	0.00	51.94
17 03 0338	**Standby, Crawler-mounted, 4.0 CY, Koehring 1166 Hydraulic Excavator**	**HR**				**0.00**	**49.84**	**0.00**	**49.84**
02221 2156	Standby, Crawler Mounted, 4 CY, Koehring 1166 Hydraulic Excavator	HR	1	XYCEE	0.258	0.00	192.88	0.00	49.84
17 03 0339	**Standby, Crawler-mounted, 5.5 CY, Koehring 1266 Hydraulic Excavator**	**HR**				**0.00**	**123.43**	**0.00**	**123.43**
02221 2166	Standby, Crawler Mounted, 5.5 CY, Koehring 1266 Hydraulic Excavator	HR	1	XYCEF	0.257	0.00	479.57	0.00	123.43
17 03 0340	**Standby, 931, 1.0 CY Track Loader**	**HR**				**0.00**	**7.47**	**0.00**	**7.47**
02225 2519	Standby, 931, 1 CY Track Loader	HR	1	XYCTA	0.209	0.00	35.71	0.00	7.47
17 03 0341	**Standby, 943, 1.5 CY Track Loader**	**HR**				**0.00**	**10.31**	**0.00**	**10.31**
02225 2521	Standby, 943, 1.5 CY Track Loader	HR	1	XYCTB	0.224	0.00	46.10	0.00	10.31
17 03 0342	**Standby, 953, 2.0 CY Track Loader**	**HR**				**0.00**	**15.21**	**0.00**	**15.21**
02225 2522	Standby, 953, 2 CY Track Loader	HR	1	XYCTC	0.221	0.00	68.98	0.00	15.21
17 03 0343	**Standby, 963, 2.5 CY Track Loader**	**HR**				**0.00**	**21.63**	**0.00**	**21.63**
02225 2523	Standby, 963, 2.5 CY Track Loader	HR	1	XYCTD	0.220	0.00	98.50	0.00	21.63
17 03 0344	**Standby, 973, 3.75 CY Track Loader**	**HR**				**0.00**	**30.26**	**0.00**	**30.26**
02225 2524	Standby, 973, 3.75 CY Track Loader	HR	1	XYCTE	0.221	0.00	136.75	0.00	30.26

17 03 Site Earthwork

17 03 03 Rock Excavation and Disposal		Unit	Hourly Output	Crew	Factor	Labor	Equip.	Matl.	Total
17 03 0345	**Standby, 910, 1.25 CY Wheel Loader**	**HR**				**0.00**	**6.20**	**0.00**	**6.20**
02225 2525	Standby, 910, 1.25 CY Wheel Loader	HR	1	XYCWA	0.256	0.00	24.23	0.00	6.20
17 03 0346	**Standby, 916, 1.5 CY Wheel Loader**	**HR**				**0.00**	**7.89**	**0.00**	**7.89**
02225 2526	Standby, 916, 1.5 CY Wheel Loader	HR	1	XYCWB	0.262	0.00	30.13	0.00	7.89
17 03 0347	**Standby, 926, 2.0 CY Wheel Loader**	**HR**				**0.00**	**9.02**	**0.00**	**9.02**
02225 2527	Standby, 926, 2.0 CY Wheel Loader	HR	1	XYCWC	0.257	0.00	35.08	0.00	9.02
17 03 0348	**Standby, 950, 3.0 CY Wheel Loader**	**HR**				**0.00**	**9.53**	**0.00**	**9.53**
02225 2528	Standby, 950, 3.0 CY Wheel Loader	HR	1	XYCWD	0.250	0.00	38.13	0.00	9.53
17 03 0349	**Standby, 966, 4.0 CY Wheel Loader**	**HR**				**0.00**	**13.59**	**0.00**	**13.59**
02225 2529	Standby, 966, 4 CY Wheel Loader	HR	1	XYCWE	0.244	0.00	55.70	0.00	13.59
17 03 0350	**Standby, 980, 5.25 CY Wheel Loader**	**HR**				**0.00**	**22.45**	**0.00**	**22.45**
02225 2531	Standby, 980, 5.25 CY Wheel Loader	HR	1	XYCWF	0.256	0.00	87.55	0.00	22.45
17 03 0351	**Standby, 988, 7.0 CY Wheel Loader**	**HR**				**0.00**	**40.23**	**0.00**	**40.23**
02225 2532	Standby, 988, 7 CY Wheel Loader	HR	1	XYCWG	0.263	0.00	152.88	0.00	40.23
17 03 0352	**Standby, 992, 13.5 CY Wheel Loader**	**HR**				**0.00**	**59.58**	**0.00**	**59.58**
02225 2533	Standby, 992, 13.5 CY Wheel Loader	HR	1	XYCWH	0.260	0.00	228.75	0.00	59.58
17 03 0353	**D8 w/Single Shank Ripper, Bulldozer, No Operator**	**HR**				**0.00**	**172.15**	**0.00**	**172.15**
01001 2109	D8 w/Single Shank Ripper, Bulldozer, No Operator	HR	1	YAABB	1.000	0.00	172.15	0.00	172.15
17 03 04 Fill and Borrow		**Unit**	**Hourly Output**	**Crew**	**Factor**	**Labor**	**Equip.**	**Matl.**	**Total**
17 03 0401	**950, 3.00 CY, Backfill with Excavated Material**	**CY**				**0.47**	**0.54**	**0.00**	**1.00**
02221 5007	Trench Backfill, 3 CY, 950	CY	70.88	CODLJ	1.000	0.47	0.54	0.00	1.00
17 03 0402	**966, 4.00 CY, Backfill with Excavated Material**	**CY**				**0.27**	**0.44**	**0.00**	**0.71**
02221 5008	Trench Backfill, 4 CY, 966	CY	126.8	CODL1	1.000	0.27	0.44	0.00	0.71
17 03 0403	**980, 5.25 CY, Backfill with Excavated Material**	**CY**				**0.22**	**0.55**	**0.00**	**0.77**
02221 5009	Trench Backfill, 5.25 CY, 980	CY	158.5	CODLK	1.000	0.22	0.55	0.00	0.77
17 03 0405	**950, 3.00 CY, Delivered & Dumped, Backfill with Sand**	**CY**				**4.28**	**3.05**	**26.50**	**33.82**
02221 5011	Sand Backfill, 950 Dumped	CY	83.25	CODLJ	1.000	0.40	0.46	0.00	0.85
02244 1505	Washed & Graded Sand, Loaded At Pit	CY	12.38	CODEG	1.000	3.88	2.59	26.50	32.97
17 03 0406	**966, 4.00 CY, Delivered & Dumped, Backfill with Sand**	**CY**				**4.20**	**2.97**	**26.50**	**33.67**
02221 5013	Sand Backfill, 966 Dumped	CY	147.9	CODLI	1.000	0.32	0.38	0.00	0.70
02244 1505	Washed & Graded Sand, Loaded At Pit	CY	12.38	CODEG	1.000	3.88	2.59	26.50	32.97

17 03 04 Fill and Borrow		Unit	Hourly Output	Crew	Factor	Labor	Equip.	Matl.	Total
17 03 0407	**980, 5.25 CY, Delivered & Dumped, Backfill with Sand**	**CY**				**4.07**	**3.06**	**26.50**	**33.63**
02221 5014	Sand Backfill, 980 Dumped	CY	184.9	CODLK	1.000	0.19	0.47	0.00	0.66
02244 1505	Washed & Graded Sand, Loaded At Pit	CY	12.38	CODEG	1.000	3.88	2.59	26.50	32.97
17 03 0410	**950, Delivered & Dumped, Backfill with Cement Stabilized Base Material**	**CY**				**4.28**	**3.05**	**40.42**	**47.74**
02221 5011	Sand Backfill, 950 Dumped	CY	83.25	CODLJ	1.000	0.40	0.46	0.00	0.85
02244 1505	Washed & Graded Sand, Loaded At Pit	CY	12.38	CODEG	1.000	3.88	2.59	26.50	32.97
03314 0350	Cement, Portland, Plain, BulkTruckload	CF	0	N/A	1.500	0.00	0.00	9.28	13.92
17 03 0411	**966, Delivered & Dumped, Backfill with Cement Stabilized Sand**	**CY**				**4.20**	**2.97**	**40.42**	**47.59**
02221 5013	Sand Backfill, 966 Dumped	CY	147.9	CODLI	1.000	0.32	0.38	0.00	0.70
02244 1505	Washed & Graded Sand, Loaded At Pit	CY	12.38	CODEG	1.000	3.88	2.59	26.50	32.97
03314 0350	Cement, Portland, Plain, BulkTruckload	CF	0	N/A	1.500	0.00	0.00	9.28	13.92
17 03 0412	**980, Delivered & Dumped, Backfill with Cement Stabilized Sand**	**CY**				**4.07**	**3.06**	**40.42**	**47.55**
02221 5014	Sand Backfill, 980 Dumped	CY	184.9	CODLK	1.000	0.19	0.47	0.00	0.66
02244 1505	Washed & Graded Sand, Loaded At Pit	CY	12.38	CODEG	1.000	3.88	2.59	26.50	32.97
03314 0350	Cement, Portland, Plain, BulkTruckload	CF	0	N/A	1.500	0.00	0.00	9.28	13.92
17 03 0415	**Backfill with Excavated Material**	**CY**				**2.50**	**0.86**	**0.35**	**3.71**
01418 4736	Soil Density Test, Nuclear Method ASTM D2922-71	EA	2	USKCB	0.000	44.06	0.00	40.00	0.03
02215 1215	Backfill Trench with 200HP Dozer without Compaction	CY	284	CODTG	1.000	0.15	0.43	0.00	0.59
02223 1001	Compaction Water Price $.005/Gallon ($0.454/100 Liters)	CY	174.3	COFWK	1.000	0.30	0.32	0.34	0.96
02228 1900	Compaction, 6" Layers, Vibrating Plate (15 cm) Layers	CY	40.5	CLACC	1.000	2.03	0.11	0.00	2.13
17 03 0416	**Backfill, Large Spread Footing Excavated Material, 950, 3.0 CY**	**CY**				**1.03**	**1.19**	**0.00**	**2.22**
02221 5015	Backfill Large Footing, 950, Limited Access, 40%	CY	32	CODLJ	1.000	1.03	1.19	0.00	2.22
17 03 0417	**Delivered & Dumped - Hand, Backfill with Sand**	**CY**				**19.96**	**7.11**	**26.50**	**53.57**
02221 5011	Sand Backfill, 950 Dumped	CY	83.25	CODLJ	1.000	0.40	0.46	0.00	0.85
02221 5012	Backfill Sand by Hand, 4 Laborers, 25 CY/EA, 75%	CY	9.375	CODL2	1.000	15.68	4.07	0.00	19.75
02244 1505	Washed & Graded Sand, Loaded At Pit	CY	12.38	CODEG	1.000	3.88	2.59	26.50	32.97
17 03 0418	**Delivered & Dumped, Backfill with Stone**	**BCY**				**0.55**	**0.63**	**27.33**	**28.51**
02611 2206	Crushed Stone	CY	72.38	XSABC	1.200	0.46	0.53	22.77	28.51
17 03 0419	**Crushed Stone, 1/2" to 3/4"**	**CY**				**2.93**	**0.99**	**20.71**	**24.63**
02704 0300	3/4" Gravel Fill for Underdrain In Trench, Crusher or Bank Run	CY	32.5	CODEJ	1.000	2.93	0.99	20.71	24.63

17 03 04 Fill and Borrow		Unit	Hourly Output	Crew	Factor	Labor	Equip.	Matl.	Total
17 03 0420	**Backfill Trench, Borrow Material, Delivered & Dumped Only**	**CY**				**1.12**	**1.88**	**6.00**	**9.01**
02081 2243	Borrow Material, Unclassified Fill	CY	0	N/A	1.000	0.00	0.00	6.00	6.00
02234 0545	Haul, 12 CY (91 m3) Truck, 6 Miles (10 km), 40 MPH (60 km/Hour), 2.1 Cycles/Hour	CY	24.5	COEID	1.000	1.12	1.88	0.00	3.01
17 03 0421	**950, 3.0 CY, Backfill with Borrow Material**	**CY**				**1.96**	**2.85**	**0.00**	**4.81**
02221 5007	Trench Backfill, 3 CY, 950	CY	70.88	CODLJ	1.000	0.47	0.54	0.00	1.00
02232 0245	Excavate & Load, 3-1/2 CY Wheel Loader, Medium Material, 103 CY/Hour (99 m3)	CY	130	CODLI	1.000	0.37	0.43	0.00	0.80
02234 0545	Haul, 12 CY (91 m3) Truck, 6 Miles (10 km), 40 MPH (60 km/Hour), 2.1 Cycles/Hour	CY	24.5	COEID	1.000	1.12	1.88	0.00	3.01
17 03 0422	**Unclassified Fill, 6" Lifts, On-Site, Includes Spreading and Compaction**	**CY**				**1.91**	**5.37**	**0.35**	**7.63**
01418 4736	Soil Density Test, Nuclear Method ASTM D2922-71	EA	2	USKCB	0.000	44.06	0.00	40.00	0.03
02215 2360	Spread Dumped Fill/Gravel, 6" Layers, without Compaction	SY	448.5	CODTG	6.000	0.10	0.27	0.00	2.22
02223 1001	Compaction Water Price $.005/Gallon ($0.454/100 Liters)	CY	174.3	COFWK	1.000	0.30	0.32	0.34	0.96
02232 0325	Excavate & Load, 2-3/4 CY Track Loader, Medium Material, 100 CY/Hour (76 m3)	CY	100	CODLD	1.000	0.48	1.37	0.00	1.85
02234 0610	Haul, 16.5 CY (12.6 m3) Truck, 1 Mile (2 km), 20 MPH (30 km/Hour), 4.2 Cycles/Hour	CY	68.75	COETH	1.000	0.40	1.66	0.00	2.06
02239 0120	Spread/Compact with Sheepsfoot Roller, 8" (20 cm) Lift, 654 CY/Hour	CY	654	COFCO	1.000	0.13	0.39	0.00	0.52
17 03 0423	**Unclassified Fill, 6" Lifts, Off-Site, Includes Delivery, Spreading, and Compaction**	**CY**				**1.03**	**2.34**	**6.35**	**9.73**
01418 4736	Soil Density Test, Nuclear Method ASTM D2922-71	EA	2	USKCB	0.000	44.06	0.00	40.00	0.03
02081 2243	Borrow Material, Unclassified Fill	CY	0	N/A	1.000	0.00	0.00	6.00	6.00
02215 2360	Spread Dumped Fill/Gravel, 6" Layers, without Compaction	SY	448.5	CODTG	6.000	0.10	0.27	0.00	2.22
02223 1001	Compaction Water Price $.005/Gallon ($0.454/100 Liters)	CY	174.3	COFWK	1.000	0.30	0.32	0.34	0.96
02239 0120	Spread/Compact with Sheepsfoot Roller, 8" (20 cm) Lift, 654 CY/Hour	CY	654	COFCO	1.000	0.13	0.39	0.00	0.52
17 03 0424	**Unclassified Fill, Delivered, Off-Site**	**CY**				**0.00**	**0.00**	**6.00**	**6.00**
02081 2243	Borrow Material, Unclassified Fill	CY	0	N/A	1.000	0.00	0.00	6.00	6.00

17 03 04 Fill and Borrow		Unit	Hourly Output	Crew	Factor	Labor	Equip.	Matl.	Total
17 03 0425	**Sand, 6" Lifts, On-Site**	**CY**				**1.92**	**5.04**	**0.39**	**7.35**
01418 4605	Grain Size (No. 200 Sieve) ASTM D421	EA	0.64	USKCA	0.001	68.85	0.00	59.00	0.09
01418 4736	Soil Density Test, Nuclear Method ASTM D2922-71	EA	2	USKCB	0.000	44.06	0.00	40.00	0.03
02215 2360	Spread Dumped Fill/Gravel, 6" Layers, without Compaction	SY	448.5	CODTG	6.000	0.10	0.27	0.00	2.22
02223 1001	Compaction Water Price $.005/Gallon ($0.454/100 Liters)	CY	174.3	COFWK	1.000	0.30	0.32	0.34	0.96
02232 0325	Excavate & Load, 2-3/4 CY Track Loader, Medium Material, 100 CY/Hour (76 m3)	CY	100	CODLD	1.000	0.48	1.37	0.00	1.85
02234 0610	Haul, 16.5 CY (12.6 m3) Truck, 1 Mile (2 km), 20 MPH (30 km/Hour), 4.2 Cycles/Hour	CY	68.75	COETH	1.000	0.40	1.66	0.00	2.06
02239 0125	Spread/Compact with Sheepsfoot Roller, 6" (15 cm) Lift, 1,027 CY/Hour	CY	1027	COFCP	1.000	0.09	0.06	0.00	0.15
17 03 0426	**Sand, 6" Lifts, Off-Site**	**CY**				**1.04**	**2.01**	**8.29**	**11.35**
01418 4605	Grain Size (No. 200 Sieve) ASTM D421	EA	0.64	USKCA	0.001	68.85	0.00	59.00	0.09
01418 4736	Soil Density Test, Nuclear Method ASTM D2922-71	EA	2	USKCB	0.000	44.06	0.00	40.00	0.03
02081 2242	Select Granular Fill (Sand)	CY	0	N/A	1.000	0.00	0.00	7.90	7.90
02215 2360	Spread Dumped Fill/Gravel, 6" Layers, without Compaction	SY	448.5	CODTG	6.000	0.10	0.27	0.00	2.22
02223 1001	Compaction Water Price $.005/Gallon ($0.454/100 Liters)	CY	174.3	COFWK	1.000	0.30	0.32	0.34	0.96
02239 0125	Spread/Compact with Sheepsfoot Roller, 6" (15 cm) Lift, 1,027 CY/Hour	CY	1027	COFCP	1.000	0.09	0.06	0.00	0.15
17 03 0427	**Sand Bags**	**EA**				**0.00**	**0.00**	**0.50**	**0.50**
02081 2245	Sand Bags	EA	0	N/A	1.000	0.00	0.00	0.50	0.50
17 03 0428	**Clay, 8" Lifts, Off-Site**	**CY**				**3.08**	**5.80**	**7.26**	**16.15**
01418 4405	Atterberg Limits	EA	0.429	USKCA	0.001	102.76	0.00	60.00	0.11
01418 4736	Soil Density Test, Nuclear Method ASTM D2922-71	EA	2	USKCB	0.004	44.06	0.00	40.00	0.34
01418 4751	Moisture Content	EA	0.471	USKCA	0.004	93.65	0.00	5.49	0.40
02081 2241	Clay Backfill	CY	0	N/A	1.000	0.00	0.00	6.70	6.70
02215 2360	Spread Dumped Fill/Gravel, 6" Layers, without Compaction	SY	448.5	CODTG	6.000	0.10	0.27	0.00	2.22
02220 5660	Compaction, SF, 8" Lifts	CY	118.8	COFCP	1.000	0.77	0.52	0.00	1.29
02223 1001	Compaction Water Price $.005/Gallon ($0.454/100 Liters)	CY	174.3	COFWK	1.000	0.30	0.32	0.34	0.96
02234 1115	Haul, 16.5 CY (12.6 m3) Truck, 6 Miles (10 km), 40 MPH (60 km/Hour), 2.1 Cycles/Hour	CY	34.25	COETH	1.000	0.80	3.33	0.00	4.13

17 03 04 Fill and Borrow		Unit	Hourly Output	Crew	Factor	Labor	Equip.	Matl.	Total
17 03 0429	**Clay, 8" Lifts, On-Site**	**CY**				**3.65**	**7.17**	**0.63**	**11.46**
01418 4405	Atterberg Limits	EA	0.429	USKCA	0.001	102.76	0.00	60.00	0.14
01418 4605	Grain Size (No. 200 Sieve) ASTM D421	EA	0.64	USKCA	0.001	68.85	0.00	59.00	0.13
01418 4736	Soil Density Test, Nuclear Method ASTM D2922-71	EA	2	USKCB	0.004	44.06	0.00	40.00	0.34
01418 4751	Moisture Content	EA	0.471	USKCA	0.004	93.65	0.00	5.49	0.40
02215 2360	Spread Dumped Fill/Gravel, 6" Layers, without Compaction	SY	448.5	CODTG	6.000	0.10	0.27	0.00	2.22
02220 5660	Compaction, SF, 8" Lifts	CY	118.8	COFCP	1.000	0.77	0.52	0.00	1.29
02223 1001	Compaction Water Price $.005/Gallon ($0.454/100 Liters)	CY	174.3	COFWK	1.000	0.30	0.32	0.34	0.96
02232 0325	Excavate & Load, 2-3/4 CY Track Loader, Medium Material, 100 CY/Hour (76 m3)	CY	100	CODLD	1.000	0.48	1.37	0.00	1.85
02234 1115	Haul, 16.5 CY (12.6 m3) Truck, 6 Miles (10 km), 40 MPH (60 km/Hour), 2.1 Cycles/Hour	CY	34.25	COETH	1.000	0.80	3.33	0.00	4.13
17 03 0430	**Gravel, 6" Lifts**	**CY**				**2.93**	**2.06**	**6.64**	**11.63**
01418 4736	Soil Density Test, Nuclear Method ASTM D2922-71	EA	2	USKCB	0.000	44.06	0.00	40.00	0.03
02081 2244	Gravel Backfill	CY	0	N/A	1.000	0.00	0.00	6.28	6.28
02215 2360	Spread Dumped Fill/Gravel, 6" Layers, without Compaction	SY	448.5	CODTG	6.000	0.10	0.27	0.00	2.22
02223 1001	Compaction Water Price $.005/Gallon ($0.454/100 Liters)	CY	174.3	COFWK	1.000	0.30	0.32	0.34	0.96
02228 1900	Compaction, 6" Layers, Vibrating Plate (15 cm) Layers	CY	40.5	CLACC	1.000	2.03	0.11	0.00	2.13
17 03 0431	**580K, 1.0 CY, Backhoe with Front-end Loader**	**HR**				**34.65**	**26.98**	**0.00**	**61.63**
02225 2534	580K, 1 CY, Backhoe with Front-end Loader	HR	1	XLABK	1.000	34.65	26.98	0.00	61.63
17 03 0432	**Bobcat with Backhoe**	**MO**				**0.00**	**0.00**	**1,742**	**1,742**
01500 2001	Skidloader with Bucket	MO	0	N/A	1.000	0.00	0.00	1,742	1,742
17 03 0433	**Crawler Mounted 0.5 CY Hydraulic Excavator**	**HR**				**62.60**	**69.73**	**0.00**	**132.32**
02225 2117	Crawler Mounted 0.5 CY Hydraulic Excavator	HR	1	CODEA	1.000	62.60	69.73	0.00	132.32
17 03 0434	**Wheel Mounted 0.5 CY 206 Hydraulic Excavator**	**HR**				**65.21**	**72.63**	**0.00**	**137.84**
02225 2116	Wheel Mounted 0.5 CY 206 Hydraulic Excavator	HR	0.96	CODEA	1.000	65.21	72.63	0.00	137.84
17 03 0436	**0.75 CY Wheel Loader**	**HR**				**48.00**	**32.05**	**0.00**	**80.05**
02225 2536	0.75 CY Wheel Loader	HR	1	CODEG	1.000	48.00	32.05	0.00	80.05
17 03 0437	**0.75 CY Backhoe with Front End Loader**	**HR**				**50.00**	**33.39**	**0.00**	**83.39**
02225 2537	0.75 CY Backhoe with Front End Loader	HR	0.96	CODEG	1.000	50.00	33.39	0.00	83.39
17 03 0438	**Standby Crawler Mounted 0.5 CY Hydraulic Excavator**	**HR**				**0.00**	**94.57**	**0.00**	**94.57**
02221 2117	Standby Crawler Mounted 0.5 CY Hydraulic Excavator	HR	0.78	XYCEB	1.000	0.00	94.57	0.00	94.57

17 03 04 Fill and Borrow		Unit	Hourly Output	Crew	Factor	Labor	Equip.	Matl.	Total
17 03 0439	**Standby Wheel Mounted 0.5 CY Hydraulic Excavator**	**HR**				**0.00**	**82.03**	**0.00**	**82.03**
02221 2118	Standby Wheel Mounted 0.5 CY Hydraulic Excavator	HR	0.85	XYCEA	1.000	0.00	82.03	0.00	82.03
17 03 0441	**Standby 0.75 CY Wheel Loader**	**HR**				**0.00**	**69.73**	**0.00**	**69.73**
02225 2538	Standby 0.75 CY Wheel Loader	HR	1	XYCEA	1.000	0.00	69.73	0.00	69.73
17 03 0442	**Standby 0.75 CY Backhoe with Front End Loader**	**HR**				**0.00**	**73.39**	**0.00**	**73.39**
02225 2539	Standby 0.75 CY Backhoe with Front End Loader	HR	0.95	XYCEA	1.000	0.00	73.39	0.00	73.39
17 03 0443	**Backhoe w/Front End Loader, No Operator**	**HR**				**0.00**	**26.98**	**0.00**	**26.98**
01001 2110	Backhoe w/Front End Loader, No Operator	HR	1	YAABC	1.000	0.00	26.98	0.00	26.98
17 03 0451	**Horizontal Boring Under Road, 24' Diameter**	**LF**				**197.19**	**219.63**	**0.00**	**416.82**
01001 0000	Horizontal Boring Under Road, 24' Diameter	LF	1	CODEA	3.150	62.60	69.73	0.00	416.82
17 03 0460	**Skidloader Tractor with Bucket**	**MO**				**34.65**	**172.15**	**0.00**	**206.80**
01001 2111	Skidloader Tractor with Bucket	MO	1	CODTM	1.000	34.65	172.15	0.00	206.80

17 03 05 Compaction		Unit	Hourly Output	Crew	Factor	Labor	Equip.	Matl.	Total
17 03 0501	**Compact Subgrade, 2 Lifts**	**CY**				**0.16**	**0.26**	**0.00**	**0.42**
02220 7070	Subgrade, 8" Lifts	CY	1670	COFCR	1.000	0.16	0.26	0.00	0.42
17 03 0504	**Compact Borrowed Subgrade, 4 Lifts**	**BCY**				**0.16**	**0.26**	**6.02**	**6.44**
02220 7070	Subgrade, 8" Lifts	CY	1670	COFCR	1.000	0.16	0.26	0.00	0.42
02244 1002	Borrow, Purchased & Dumped	CY	0	N/A	1.200	0.00	0.00	5.02	6.02
17 03 0506	**Compact Sand Subgrade (Wet & 2 Passes)**	**SY**				**0.16**	**0.26**	**0.00**	**0.43**
02212 1024	1.75 Passes/8" Lifts, Roller	SY	804.4	COFCT	1.000	0.16	0.26	0.00	0.43
17 03 0508	**Compact, Footing Excavator, Excavated Material Backfill**	**CY**				**2.03**	**0.11**	**0.00**	**2.13**
02228 1900	Compaction, 6" Layers, Vibrating Plate (15 cm) Layers	CY	40.5	CLACC	1.000	2.03	0.11	0.00	2.13
17 03 0510	**Dry Roll Gravel, Steel Roller**	**SY**				**0.43**	**0.27**	**0.00**	**0.70**
02212 2103	Gravel, Dry - 12 Ton Roller	SY	112.5	COFCD	1.000	0.43	0.27	0.00	0.70
17 03 0511	**Compact Soil with Vibrating Plate**	**CY**				**3.35**	**0.18**	**0.00**	**3.53**
02220 7000	Compact Backfill, by Hand with Vibrating Plate, 6" (15 cm) Lift	CY	24.5	CLACC	1.000	3.35	0.18	0.00	3.53
17 03 0512	**Compact Roadside Fill, Hand & Machine**	**CY**				**9.65**	**9.01**	**0.00**	**18.66**
02212 3105	by Hand & Machine	CY	30	COFCS	1.000	9.65	9.01	0.00	18.66

17 03 Site Earthwork

17 03 05	Compaction	Unit	Hourly Output	Crew	Factor	Labor	Equip.	Matl.	Total
17 03 0513	**Spread Dumped Borrow & Compact with Roller**	**CY**				**0.16**	**0.26**	**0.00**	**0.42**
02220 7070	Subgrade, 8" Lifts	CY	1670	COFCR	1.000	0.16	0.26	0.00	0.42
17 03 0514	**Compact Soil by Machine with Roller**	**CY**				**0.41**	**0.26**	**0.00**	**0.67**
02220 5900	Compact Backfill, by Machine, Sheepsfoot Roller, 6" (15 cm) Lift	CY	117.1	COFCD	1.000	0.41	0.26	0.00	0.67
17 03 0515	**Compact With Pogosticks**	**CY**				**2.71**	**0.17**	**0.00**	**2.88**
02215 0500	Compaction, 6" Layers with Air Tamper by Hand, (15 cm) Layers	CY	50	CLACA	1.000	2.71	0.17	0.00	2.88
17 03 0516	**Compact with 50% Pogosticks, 50% Hand Roller**	**CY**				**1.83**	**0.84**	**0.00**	**2.67**
02215 0500	Compaction, 6" Layers with Air Tamper by Hand, (15 cm) Layers	CY	50	CLACA	0.500	2.71	0.17	0.00	1.44
02220 8800	Compact Backfill, by Hand with 1-Ton Roller, 6" (15 cm) Lift	CY	49.13	CLACD	0.500	0.94	1.51	0.00	1.22
17 03 0517	**Spread/Compact Large Areas, 6" Lifts, D8 & Towed Sheepsfoot**	**CY**				**0.23**	**0.69**	**0.00**	**0.92**
02239 0110	Spread/Compact with Sheepsfoot Roller, 6" (15 cm) Lift, 366 CY/Hour	CY	366.3	COFCO	1.000	0.23	0.69	0.00	0.92
17 03 0518	**Spread/Compact Large Areas, 8" Lifts, D8 & Towed Sheepsfoot**	**CY**				**0.13**	**0.39**	**0.00**	**0.52**
02239 0120	Spread/Compact with Sheepsfoot Roller, 8" (20 cm) Lift, 654 CY/Hour	CY	654	COFCO	1.000	0.13	0.39	0.00	0.52

17 03 06	Soil Stabilization	Unit	Hourly Output	Crew	Factor	Labor	Equip.	Matl.	Total
17 03 0601	**Lime Stabilization, Dry**	**BCY**				**2.11**	**10.67**	**196.35**	**209.13**
02250 0910	Lime For Soil Stabilization	TON	0	N/A	0.850	0.00	0.00	231.00	196.35
02250 0920	Roller, Grader - Lime Stabilization	CY	40.75	COFCU	1.000	2.11	10.67	0.00	12.78
17 03 0602	**Cement Stabilization, 6%**	**BCY**				**0.73**	**1.69**	**13.92**	**16.33**
02250 1390	Roller, Grader - Cement Stabilization	CY	261	COFCV	1.000	0.73	1.69	0.00	2.41
03314 0350	Cement, Portland, Plain, BulkTruckload	CF	0	N/A	1.500	0.00	0.00	9.28	13.92

17 03 07	Dozers	Unit	Hourly Output	Crew	Factor	Labor	Equip.	Matl.	Total
17 03 0701	**D3 with A-blade Bulldozer**	**HR**				**33.05**	**41.82**	**0.00**	**74.87**
02225 4114	D3 by Hour	HR	1	XXQLK	1.000	33.05	41.82	0.00	74.87
17 03 0702	**D4 with A-blade Bulldozer**	**HR**				**33.05**	**42.69**	**0.00**	**75.74**
02225 4124	D4 by Hour	HR	1	CODTB	1.000	33.05	42.69	0.00	75.74
17 03 0703	**D5 with A-blade Bulldozer**	**HR**				**33.05**	**60.44**	**0.00**	**93.49**
02225 4134	D5 by Hour	HR	1	CODTD	1.000	33.05	60.44	0.00	93.49

17 03 07	Dozers	Unit	Hourly Output	Crew	Factor	Labor	Equip.	Matl.	Total
17 03 0704	**D6 with A-blade Bulldozer**	**HR**				**34.65**	**76.88**	**0.00**	**111.53**
02225 4144	D6 by Hour	HR	1	XXQMX	1.000	34.65	76.88	0.00	111.53
17 03 0705	**D7 with U-blade Bulldozer**	**HR**				**34.65**	**122.34**	**0.00**	**156.99**
02225 4155	D7 by Hour	HR	1	CODTJ	1.000	34.65	122.34	0.00	156.99
17 03 0706	**D8 with U-Blade Bulldozer**	**HR**				**0.00**	**0.00**	**175.79**	**175.79**
01001 0001	D8 w/U Blade Bulldozer	HR	0	N/A	1.000	0.00	0.00	175.79	175.79
17 03 0707	**D9 with U-Blade Bulldozer**	**HR**				**0.00**	**0.00**	**218.80**	**218.80**
01001 0002	D9 w/U Blade Bulldozer	HR	0	N/A	1.000	0.00	0.00	218.80	218.80
17 03 0708	**D10 with U-Blade Bulldozer**	**HR**				**0.00**	**0.00**	**269.29**	**269.29**
01001 0003	D10 w/U Blade Bulldozer	HR	0	N/A	1.000	0.00	0.00	269.29	269.29
17 03 0709	**Standby, D3 with A-blade Bulldozer**	**HR**				**0.00**	**9.26**	**0.00**	**9.26**
02225 4115	Standby, D3 with A-Blade Bulldozer	HR	1	XYCDA	0.221	0.00	41.82	0.00	9.26
17 03 0710	**Standby, D4 with A-blade Bulldozer**	**HR**				**0.00**	**9.52**	**0.00**	**9.52**
02225 4125	Standby, D4 with A-Blade Bulldozer	HR	1	XYCDB	0.223	0.00	42.69	0.00	9.52
17 03 0711	**Standby, D5 with A-blade Bulldozer**	**HR**				**0.00**	**13.79**	**0.00**	**13.79**
02225 4135	Standby, D5 with A-Blade Bulldozer	HR	1	XYCDC	0.228	0.00	60.44	0.00	13.79
17 03 0712	**Standby, D6 with A-blade Bulldozer**	**HR**				**0.00**	**17.65**	**0.00**	**17.65**
02225 4145	Standby, D6 with A-Blade Bulldozer	HR	1	XYCDD	0.230	0.00	76.88	0.00	17.65
17 03 0713	**Standby, D7 with U-blade Bulldozer**	**HR**				**0.00**	**28.42**	**0.00**	**28.42**
02225 4156	Standby, D7 with U-Blade Bulldozer	HR	1	XYCDE	0.232	0.00	122.34	0.00	28.42
17 03 0714	**Standby, D8 with U-Blade Bulldozer**	**HR**				**0.00**	**0.00**	**129.50**	**129.50**
01001 0004	Standby, D8 w/U Blade Bulldozer	HR	0	N/A	1.000	0.00	0.00	129.50	129.50
17 03 0715	**Standby, D9 with U-Blade Bulldozer**	**HR**				**0.00**	**0.00**	**137.29**	**137.29**
01001 0005	Standby, D9 w/U Blade Bulldozer	HR	0	N/A	1.000	0.00	0.00	137.29	137.29
17 03 0716	**Standby, D10 with U-Blade Bulldozer**	**HR**				**0.00**	**0.00**	**139.48**	**139.48**
01001 0006	Standby, D10 w/U Blade Bulldozer	HR	0	N/A	1.000	0.00	0.00	139.48	139.48
17 03 0730	**Scraper - Standard, 15 CY - 621 with D8L Dozer**	**HR**				**44.56**	**233.79**	**0.00**	**278.35**
02226 2304	Scraper, Standard 15 CY, 621 with D8, by Hour	HR	1	CODST	1.000	44.56	233.79	0.00	278.35
17 03 0731	**Scraper - Standard, 22 CY - 631 with D9L Dozers**	**HR**				**53.22**	**390.99**	**0.00**	**444.21**
02226 2414	Scraper, Standard 22 CY, with D9s, by Hour	HR	1	CODSJ	1.000	53.22	390.99	0.00	444.21

17 03 07	Dozers	Unit	Hourly Output	Crew	Factor	Labor	Equip.	Matl.	Total
17 03 0732	**Scraper - Standard, 34 CY - 651 with D9L Dozers**	**HR**				**53.22**	**421.35**	**0.00**	**474.57**
02226 2514	Scraper, Standard 34 CY, 651 with D9s, by Hour	HR	1	CODSS	1.000	53.22	421.35	0.00	474.57
17 03 0733	**Scraper - Tandem, 14 CY - 627 with D8L Dozer**	**HR**				**44.56**	**175.10**	**0.00**	**219.66**
02226 2306	Scraper, Tandem 14 CY, 627 with D8, by Hour	HR	1	CODSM	1.000	44.56	175.10	0.00	219.66
17 03 0734	**Scraper - Tandem, 21 CY - 637 with D9L Dozer**	**HR**				**44.56**	**210.67**	**0.00**	**255.24**
02226 2416	Scraper, Tandem 21 CY, with D9, by Hour	HR	1	CODSN	1.000	44.56	210.67	0.00	255.24
17 03 0735	**Scraper - Tandem, 32 CY - 657 with D9L Dozers**	**HR**				**53.22**	**317.64**	**0.00**	**370.86**
02226 2515	Scraper, Tandem 32 CY, 657 with D9s, by Hour	HR	1	CODSO	1.000	53.22	317.64	0.00	370.86
17 03 0736	**Scraper - Tandem Elevating, 34 CY - with D9L Dozer**	**HR**				**44.56**	**261.35**	**0.00**	**305.92**
02226 2516	Scraper, Tandem & Elevating, 34 CY, with D9s, by Hour	HR	1	CODSV	1.000	44.56	261.35	0.00	305.92
17 03 0737	**Standby, Scraper - Standard, 15 CY 621 with D8L Dozer**	**HR**				**0.00**	**61.02**	**0.00**	**61.02**
02226 2312	Standby, Scraper - Standard, 15 CY 621 with D8L Dozers	HR	1	XYCSA	0.261	0.00	233.79	0.00	61.02
17 03 0738	**Standby, Scraper - Standard, 22 CY 631 with D9L Dozers**	**HR**				**0.00**	**104.10**	**0.00**	**104.10**
02226 2417	Standby, Scraper - Standard, 22 CY 631 with D9L Dozers	HR	1	XYCSB	0.266	0.00	390.99	0.00	104.10
17 03 0739	**Standby, Scraper - Standard, 34 CY 651 with D9L Dozers**	**HR**				**0.00**	**114.06**	**0.00**	**114.06**
02226 2517	Standby, Scraper - Standard, 34 CY 651 with D9L Dozers	HR	1	XYCSC	0.271	0.00	421.35	0.00	114.06
17 03 0740	**Standby, Scraper - Tandem, 14 CY 627 with D8L Dozer**	**HR**				**0.00**	**58.31**	**0.00**	**58.31**
02226 2314	Standby, Scraper - Tandem, 14 CY 627 with D8L Dozer	HR	1	XYCSF	0.249	0.00	233.79	0.00	58.31
17 03 0741	**Standby, Scraper - Tandem 21 CY 637 with D9L Dozer**	**HR**				**0.00**	**88.50**	**0.00**	**88.50**
02226 2419	Standby, Scraper - Tandem 21 CY 637 with D9L Dozer	HR	1	XYCSG	0.260	0.00	340.31	0.00	88.50
17 03 0742	**Standby, Scraper - Tandem 32 CY 657 with D9L Dozers**	**HR**				**0.00**	**108.76**	**0.00**	**108.76**
02226 2518	Standby, Scraper - Tandem 32 CY 657 with D9L Dozers	HR	1	XYCSH	0.258	0.00	421.35	0.00	108.76

17 03 09	Shoring	Unit	Hourly Output	Crew	Factor	Labor	Equip.	Matl.	Total

17 03 09 Shoring		Unit	Hourly Output	Crew	Factor	Labor	Equip.	Matl.	Total
17 03 0901	**Steel Sheeting, Install, Pull and Salvage, to 15 ft**	**SF**				**2.90**	**3.62**	**2.06**	**8.58**
02161 0100	Shoring for 15' (5 m) Excavation, 22 PSF Steel Sheeting, Pull and Salvage	TON	0.475	CPIDV	0.005	541.22	677.11	385.00	8.58
17 03 0902	**Steel Sheeting, Install, Pull and Salvage, to 20 ft**	**SF**				**2.97**	**3.71**	**2.50**	**9.18**
02161 0400	Shoring for 20' (6 m) Excavation, 27 PSF Steel Sheeting, Pull and Salvage	TON	0.563	CPIDV	0.006	457.03	571.78	385.00	9.18
17 03 0903	**Steel Sheeting, Install, Pull and Salvage, to 25 ft**	**SF**				**2.51**	**3.15**	**3.11**	**8.77**
02161 0700	Shoring for 25' (8 m) Excavation, 38 PSF Steel Sheeting, Pull and Salvage	TON	0.825	CPIDV	0.008	311.61	389.85	385.00	8.77
17 03 0904	**Steel Sheeting, Install, Pull and Salvage, to 40 ft**	**SF**				**2.94**	**3.68**	**3.85**	**10.47**
02161 1000	Shoring for 40' (13 m) Excavation, 38 PSF Steel Sheeting, Pull and Salvage	TON	0.875	CPIDV	0.010	293.81	367.57	385.00	10.47
17 03 0905	**Wood Sheeting, 8' Deep Excavation**	**SF**				**3.47**	**0.36**	**1.72**	**5.55**
02161 3910	Wood Shoring, 8' (2.4 m) Deep Excavation, 55 SF/Hour In and 60 SF/Hour Out	SF	41.25	CLABB	1.000	3.47	0.36	1.72	5.55
17 03 0906	**Wood Sheeting, 10' Deep Excavation**	**SF**				**3.81**	**0.40**	**1.77**	**5.98**
02161 4000	Wood Shoring, 10' (3 m) Deep Excavation, 50 SF/Hour In and 150 SF/Hour Out	SF	37.5	CLABB	1.000	3.81	0.40	1.77	5.98
17 03 0907	**Wood Sheeting, 12' Deep Excavation**	**SF**				**4.24**	**0.44**	**1.82**	**6.50**
02161 4100	Wood Shoring, 12' (3.7 m) Deep Excavation, 45 SF/Hour In and 135 SF/Hour Out	SF	33.75	CLABB	1.000	4.24	0.44	1.82	6.50
17 03 0908	**Wood Sheeting, 14' Deep Excavation**	**SF**				**4.58**	**0.48**	**1.88**	**6.94**
02161 4200	Wood Shoring, 14' (4.3 m) Deep Excavation, 42 SF/Hour In and 126 SF/Hour Out	SF	31.25	CLABB	1.000	4.58	0.48	1.88	6.94
17 03 0909	**Wood Sheeting, 16' Deep Excavation**	**SF**				**4.77**	**0.50**	**1.94**	**7.21**
02161 4300	Wood Shoring, 16' (4.9 m) Deep Excavation, 40 SF/Hour In and 120 SF/Hour Out	SF	30	CLABB	1.000	4.77	0.50	1.94	7.21
17 03 0910	**Wood Sheeting, 18' Deep Excavation**	**SF**				**4.98**	**0.52**	**2.00**	**7.50**
02161 4400	Wood Shoring, 18' (5.5 m) Deep Excavation, 38 SF/Hour In and 114 SF/Hour Out	SF	28.75	CLABB	1.000	4.98	0.52	2.00	7.50
17 03 0911	**Wood Sheeting, 20' Deep Excavation**	**SF**				**5.45**	**0.57**	**2.07**	**8.09**
02161 4500	Wood Shoring, 20' (6 m) Deep Excavation, 35 SF/Hour In and 105 SF/Hour Out	SF	26.25	CLABB	1.000	5.45	0.57	2.07	8.09
17 03 0912	**Cat 235, 2.5 CY, Soil/Sand, < 15' Trench in Sheeting**	**CY**				**0.77**	**1.61**	**0.00**	**2.38**
02228 0382	Trench, 2-1/2 CY Hydraulic Excavator, Medium Soil, 240 CY/Hour (183 m3/Hour)	CY	240	CODEU	2.000	0.39	0.80	0.00	2.38

17 03 Site Earthwork

17 03 09 Shoring		Unit	Hourly Output	Crew	Factor	Labor	Equip.	Matl.	Total
17 03 0913	**Cat 235, 2.5 CY, Soil/Sand, 15' - 25' Trench in Sheeting**	**CY**				**1.16**	**2.41**	**0.00**	**3.57**
02228 0382	Trench, 2-1/2 CY Hydraulic Excavator, Medium Soil, 240 CY/Hour (183 m3/Hour)	CY	240	CODEU	3.000	0.39	0.80	0.00	3.57
17 03 10 Temporary Dewatering		**Unit**	**Hourly Output**	**Crew**	**Factor**	**Labor**	**Equip.**	**Matl.**	**Total**
17 03 1001	**Wellpoint for Trench, Install & Remove <500', 1 Month**	**LFHD**				**0.00**	**0.00**	**67.94**	**67.94**
02144 2000	Complete System 2" Wellpoints & 5' OC 200'	LF	0	N/A	1.000	0.00	0.00	67.94	67.94
17 03 1002	**2" Diameter Contractor's Trash Pump, 75 GPM**	**DAY**				**6.68**	**0.00**	**45.91**	**52.59**
15292 2020	75 GPM Cast Iron Sump Pump, 2" Discharge Single Stage, 40' Head, 1-1/2 HP	EA	0.238	MSPFB	0.017	400.78	0.00	2,754	52.59
17 03 1003	**3" Diameter Contractor's Trash Pump, 150 GPM**	**DAY**				**9.76**	**0.00**	**44.33**	**54.09**
15292 2040	150 GPM Cast Iron Sump Pump, 3" Discharge Single Stage, 40' Head, 3 HP	EA	0.163	MSPFB	0.017	585.75	0.00	2,659	54.09
17 03 1004	**4" Diameter Contractor's Trash Pump, 300 GPM**	**DAY**				**12.69**	**0.00**	**68.14**	**80.83**
15292 2060	300 GPM Cast Iron Sump Pump, 4" Discharge Single Stage, 70' Head, 10 HP	EA	0.125	MSPFB	0.017	761.48	0.00	4,087	80.83
17 03 99 Other Site Earthwork		**Unit**	**Hourly Output**	**Crew**	**Factor**	**Labor**	**Equip.**	**Matl.**	**Total**
17 03 9901	**Washing/Screening Plant (125 Tons/Hour) Including 966**	**HR**				**156.97**	**195.86**	**0.00**	**352.83**
02211 5002	125 Tons/Hour Screen/Wash Plant	HR	1	COMAC	1.000	156.97	195.86	0.00	352.83
17 03 9902	**Crusher (200 Tons/Hour) Including 950 & 953 Loaders**	**HR**				**155.70**	**229.10**	**0.00**	**384.80**
02211 5001	200 Tons/Hour Rock Crusher	HR	1	COMAD	1.000	155.70	229.10	0.00	384.80
17 03 9903	**Hand Place Small Earth Fill Berm**	**CY**				**54.20**	**0.00**	**0.00**	**54.20**
02228 1665	Trench, Hand Excavation, Heavy Clay, 2'-6' Deep, Piled Only	CY	0.625	ULABA	1.000	54.20	0.00	0.00	54.20
17 03 9904	**Mulch Hay, 50 Lb Average Bale, 36" x 24" x 18"**	**EA**				**0.00**	**0.00**	**5.02**	**5.02**
02830 2006	Mulch Hay, 50 Lb Average Bale, 36" x 24" x 18"	EA	0	N/A	1.000	0.00	0.00	5.02	5.02
17 03 9911	**2' High Earthen Berm**	**CY**				**0.27**	**0.81**	**6.00**	**7.08**
02081 2243	Borrow Material, Unclassified Fill	CY	0	N/A	1.000	0.00	0.00	6.00	6.00
02239 0012	Spread & compact, 8" lift,roadway embankment, 300 HP	CY	312.5	COFCO	1.000	0.27	0.81	0.00	1.08
17 04 01 Site Cleanup		**Unit**	**Hourly Output**	**Crew**	**Factor**	**Labor**	**Equip.**	**Matl.**	**Total**
17 04 0101	**General Area Cleanup**	**ACRE**				**184.96**	**124.90**	**0.00**	**309.86**
01714 0300	Site Debris Clean-Up & Removal	ACRE	0.625	COETF	1.000	184.96	124.90	0.00	309.86

17 04 01 Site Cleanup		Unit	Hourly Output	Crew	Factor	Labor	Equip.	Matl.	Total
17 04 0102	**Pavement Sweeping, Machine**	**SY**				**0.01**	**0.00**	**0.00**	**0.01**
02971 6420	Pavement Sweeper	SY	7072	COKBM	1.000	0.01	0.00	0.00	0.01
17 04 0103	**Load & Haul Debris, 5 Miles, Dumptruck**	**CY**				**1.56**	**2.61**	**0.00**	**4.17**
02225 3301	Cost for each additional mile (50.89% of total Labor & Equipment costs)	MI	0	N/A	0.000	0.00	0.00	2.27	0.00
02234 0340	Haul, 12 CY (91 m3) Truck, 1 Mile (2 km), 20 MPH (30 km/Hour), 4.2 Cycles/Hour	CY	49	COEID	1.000	0.56	0.94	0.00	1.50
02234 0470	Hauling 12 LCY 4 Miles (Base Wide Rate)	CY	27.63	COEID	1.000	1.00	1.67	0.00	2.67

18 01 Roadways

18 01 01 Bases and Subbases

18 01 01 Bases and Subbases		Unit	Hourly Output	Crew	Factor	Labor	Equip.	Matl.	Total
18 01 0101	**Cement-stabilized Base**	**CY**				**3.33**	**6.73**	**22.95**	**33.01**
02250 1380	Cement Stabilized Base Material	TN	46.88	XASPB	1.350	2.47	4.98	17.00	33.01
18 01 0102	**Gravel, Delivered & Dumped**	**CY**				**1.83**	**1.69**	**22.77**	**26.29**
02611 2205	Graded Crushed Gravel Base	CY	90.88	XSABD	1.000	1.83	1.69	22.77	26.29
18 01 0103	**Gravel (90%) & Sand Base (10%), with Calcium Chloride 3/4 - 1 Lb/CY**	**CY**				**2.03**	**1.78**	**23.39**	**27.20**
02244 1505	Washed & Graded Sand, Loaded at Pit	CY	12.38	CODEG	0.100	3.88	2.59	26.50	3.30
02611 2205	Graded Crushed Gravel Base	CY	90.88	XSABD	0.900	1.83	1.69	22.77	23.66
03318 0420	Admixture Calcium Chloride 100 Lb Bag	BAG	0	N/A	0.010	0.00	0.00	24.66	0.25
18 01 0104	**Asphalt, Intermediate Course (Line Item Includes 5% Waste)**	**TON**				**14.73**	**15.12**	**36.23**	**66.08**
02505 0813	Bituminous Hot Mix Interim Course, Central Plant	TN	25	XASPA	1.050	14.03	14.40	34.50	66.08
18 01 0105	**Asphalt, Stabilized Base Course**	**CY**				**0.62**	**1.26**	**34.92**	**36.81**
02612 1303	Asphalt (Stabilized Base)	TN	250	XASPB	1.350	0.46	0.93	25.87	36.81

18 01 02 Drains, Inlets, Curbs and Gutters

18 01 02 Drains, Inlets, Curbs and Gutters		Unit	Hourly Output	Crew	Factor	Labor	Equip.	Matl.	Total
18 01 0201	**Concrete Curb, 6" x 6"**	**LF**				**0.82**	**0.00**	**1.22**	**2.05**
03170 4660	Radius Metal Formwork, 6" High (10 cm), with Stakes, 10' Length, 100 Uses	LF	52.5	ACARI	0.200	1.99	0.00	0.90	0.58
03217 0600	Grade 60 Rebar, Footings & Slabs, #3-#6	TON	0.313	SIWRC	0.000	492.16	0.00	609.86	0.36
03372 4600	Pour Slab on Grade, >= 6", Direct Chute >= (15 cm) Place 3,000 PSI Concrete	CY	21.88	ALABE	0.010	7.70	0.29	84.22	0.92
03396 0010	Concrete Floor Finishes, Screed	SF	112.5	ACMAA	0.500	0.37	0.00	0.00	0.18
18 01 0202	**Concrete Curb & Gutter, 6" x 24", Formed**	**LF**				**8.63**	**0.00**	**13.50**	**22.13**
02525 0430	6" x 24" Concrete Gutter with 6" Curb Formwork, Rebar, and Finishing	LF	15	XCBGA	1.000	8.63	0.00	13.50	22.13
18 01 0203	**Asphalt Curb 8" W x 6" H**	**LF**				**2.14**	**3.98**	**1.61**	**7.72**
02620 1415	8" (20 cm) Wide x 6" (15 cm) High Asphalt Curb 50 LF/Ton, Bituminous	LF	50	COKCA	1.000	2.14	3.98	1.61	7.72
18 01 0204	**Asphalt Curb 8" W x 8" H**	**LF**				**2.14**	**3.98**	**2.09**	**8.20**
02620 1416	8" (20 cm) Wide x 8" (20 cm) High Asphalt Curb 44 LF/Ton, Bituminous	LF	50	COKCA	1.000	2.14	3.98	2.09	8.20
18 01 0205	**Curb Inlet Frame Grate & Box, 525 Lb**	**EA**				**2,071**	**65.19**	**1,985**	**4,121**
02520 1117	Catch Basin 12' Deep	EA	0.1	ALABM	1.000	1,860	0.00	1,320	3,180
02520 3201	Curb Inlet Opening 48"	EA	0.5	CODEJ	1.000	190.45	64.10	645.05	899.60
03354 3005	Concrete Redimix 2,500 PSI	CY	1.238	CLACC	0.310	66.34	3.52	63.50	41.34

18 01 02 Drains, Inlets, Curbs and Gutters

18 01 02 Drains, Inlets, Curbs and Gutters		Unit	Hourly Output	Crew	Factor	Labor	Equip.	Matl.	Total
18 01 0206	**Concrete Curb, 8" x 8"**	**LF**				**2.82**	**0.00**	**1.95**	**4.77**
03170 4680	Radius Metal Formwork, 8" High (10 cm), with Stakes, 10' Length,100 Uses	LF	47.5	ACARI	0.200	2.19	0.00	1.11	0.66
03217 0600	Grade 60 Rebar, Footings & Slabs, #3-#6	TON	0.313	SIWRC	0.001	492.16	0.00	609.86	0.55
03372 4600	Pour Slab on Grade, >= 6", Direct Chute >= (15 cm) Place 3,000 PSI Concrete	CY	21.88	ALABE	0.016	7.70	0.29	84.22	1.52
03396 0010	Concrete Floor Finishes, Screed	SF	112.5	ACMAA	2.000	0.37	0.00	0.00	0.74
03398 0150	Wall Finish, Carborundum Wet Rub	SF	21.88	ACMAA	0.667	1.90	0.00	0.06	1.31
18 01 0207	**Concrete Curb, 12" High x 8" Deep**	**LF**				**3.81**	**0.01**	**2.72**	**6.54**
03170 4690	Radius Metal Formwork, 12" High (10 cm), with Stakes, 10' Length,100 Uses	LF	42.5	ACARI	0.200	2.45	0.00	1.40	0.77
03217 0600	Grade 60 Rebar, Footings & Slabs, #3-#6	TON	0.313	SIWRC	0.001	492.16	0.00	609.86	0.55
03372 4600	Pour Slab on Grade, >= 6", Direct Chute >= (15 cm) Place 3,000 PSI Concrete	CY	21.88	ALABE	0.025	7.70	0.29	84.22	2.28
03396 0010	Concrete Floor Finishes, Screed	SF	112.5	ACMAA	2.660	0.37	0.00	0.00	0.98
03398 0150	Wall Finish, Carborundum Wet Rub	SF	21.88	ACMAA	1.000	1.90	0.00	0.06	1.96

18 01 03 Paved Surfaces

18 01 03 Paved Surfaces		Unit	Hourly Output	Crew	Factor	Labor	Equip.	Matl.	Total
18 01 0301	**6" Concrete Pavement, Formed, 2 Lanes**	**SY**				**6.83**	**0.26**	**19.21**	**26.29**
01500 2101	Paving Forms, 10,000 LF 6"	HR	1	XASPF	0.002	225.84	90.85	96.80	0.66
01500 2105	Curing Bridge with Spray Forms	HR	1	XCMAF	0.001	116.83	0.00	59.98	0.26
01500 2107	Finishing Screed - Forms	HR	1	XASPE	0.002	317.25	40.73	45.55	0.65
03110 3801	Longitudinal (Construction) Joint	CY	0	N/A	0.170	0.00	0.00	1.03	0.17
03110 3802	Transverse (Expansion) Joint	CY	0	N/A	0.170	0.00	0.00	1.42	0.24
03132 0500	Liquid Neoprene Expansion Joint, 1/2" x 1" Poured, Cold Applied	LF	56.25	ULABA	0.170	0.60	0.00	1.81	0.41
03217 0700	Grade 60 Rebar, Beam, Column, Wall, #3-#6	TON	0.313	SIWRC	0.010	492.16	0.00	609.86	11.21
03334 0205	Curing Paper	CSF	112.5	XASPD	0.090	1.92	0.51	5.63	0.73
03334 0300	Concrete Curing, Sprayed Membrane Curing Compound	CSF	11.88	ULABB	0.090	5.71	0.00	4.05	0.88
03354 3007	Concrete Redimix 3,500 PSI	CY	0	N/A	0.167	0.00	0.00	66.56	11.09
18 01 0302	**8" Concrete Pavement, Formed, 2 Lanes**	**SY**				**7.05**	**0.27**	**23.17**	**30.50**
01500 2101	Paving Forms, 10,000 LF 6"	HR	1	XASPF	0.002	225.84	90.85	96.80	0.66
01500 2105	Curing Bridge with Spray Forms	HR	1	XCMAF	0.002	116.83	0.00	59.98	0.35
01500 2107	Finishing Screed - Forms	HR	1	XASPE	0.002	317.25	40.73	45.55	0.81
03110 3801	Longitudinal (Construction) Joint	CY	0	N/A	0.220	0.00	0.00	1.03	0.23
03110 3802	Transverse (Expansion) Joint	CY	0	N/A	0.220	0.00	0.00	1.42	0.31
03132 0500	Liquid Neoprene Expansion Joint, 1/2" x 1" Poured, Cold Applied	LF	56.25	ULABA	0.220	0.60	0.00	1.81	0.53
03217 0700	Grade 60 Rebar, Beam, Column, Wall, #3-#6	TON	0.313	SIWRC	0.010	492.16	0.00	609.86	11.21
03334 0205	Curing Paper	CSF	112.5	XASPD	0.090	1.92	0.51	5.63	0.73
03334 0300	Concrete Curing, Sprayed Membrane Curing Compound	CSF	11.88	ULABB	0.090	5.71	0.00	4.05	0.88
03354 3007	Concrete Redimix 3,500 PSI	CY	0	N/A	0.222	0.00	0.00	66.56	14.80

18 01 03	Paved Surfaces	Unit	Hourly Output	Crew	Factor	Labor	Equip.	Matl.	Total
18 01 0305	**6" Concrete Pavement, Slipform, Integral Curb, 2 Lanes**	**SY**				**5.75**	**0.42**	**18.72**	**24.89**
03110 3801	Longitudinal (Construction) Joint	CY	0	N/A	0.170	0.00	0.00	1.03	0.17
03110 3802	Transverse (Expansion) Joint	CY	0	N/A	0.170	0.00	0.00	1.42	0.24
03132 0500	Liquid Neoprene Expansion Joint, 1/2" x 1" Poured, Cold Applied	LF	56.25	ULABA	0.170	0.60	0.00	1.81	0.41
03217 0700	Grade 60 Rebar, Beam, Column, Wall, #3-#6	TON	0.313	SIWRC	0.010	492.16	0.00	609.86	11.21
03305 1006	Curing Compound, 5,500 SY/Day	CSF	61.88	XCMAD	0.090	1.47	0.02	2.18	0.33
03334 0205	Curing Paper	CSF	112.5	XASPD	0.090	1.92	0.51	5.63	0.73
03354 3007	Concrete Redimix 3,500 PSI	CY	0	N/A	0.167	0.00	0.00	66.56	11.09
03375 1001	6" Forms	SY	687.5	XASPC	1.000	0.34	0.37	0.00	0.71
18 01 0306	**8" Concrete Pavement, Slipform, Integral Curb, 2 Lanes**	**SY**				**5.88**	**0.50**	**22.81**	**29.19**
03110 3801	Longitudinal (Construction) Joint	CY	0	N/A	0.220	0.00	0.00	1.03	0.23
03110 3802	Transverse (Expansion) Joint	CY	0	N/A	0.220	0.00	0.00	1.42	0.31
03132 0500	Liquid Neoprene Expansion Joint, 1/2" x 1" Poured, Cold Applied	LF	56.25	ULABA	0.220	0.60	0.00	1.81	0.53
03217 0700	Grade 60 Rebar, Beam, Column, Wall, #3-#6	TON	0.313	SIWRC	0.010	492.16	0.00	609.86	11.21
03334 0205	Curing Paper	CSF	112.5	XASPD	0.090	1.92	0.51	5.63	0.73
03334 0305	Curing Compound 4,500 SY/Day	CSF	50.63	XCMAD	0.090	1.80	0.02	4.05	0.53
03354 3007	Concrete Redimix 3,500 PSI	CY	0	N/A	0.222	0.00	0.00	66.56	14.80
03375 1002	8" Forms	SY	562.5	XASPC	1.000	0.41	0.45	0.00	0.87
18 01 0307	**10" Concrete Pavement, Slipform, Integral Curb, 2 Lanes**	**SY**				**6.13**	**0.67**	**26.59**	**33.38**
03110 3801	Longitudinal (Construction) Joint	CY	0	N/A	0.280	0.00	0.00	1.03	0.29
03110 3802	Transverse (Expansion) Joint	CY	0	N/A	0.280	0.00	0.00	1.42	0.40
03132 0500	Liquid Neoprene Expansion Joint, 1/2" x 1" Poured, Cold Applied	LF	56.25	ULABA	0.280	0.60	0.00	1.81	0.67
03217 0700	Grade 60 Rebar, Beam, Column, Wall, #3-#6	TON	0.313	SIWRC	0.010	492.16	0.00	609.86	11.21
03305 1008	Curing Compound 3,300 SY/Day	CSF	37.13	XCMAD	0.090	2.46	0.03	2.18	0.42
03334 0205	Curing Paper	CSF	112.5	XASPD	0.090	1.92	0.51	5.63	0.73
03354 3007	Concrete Redimix 3,500 PSI	CY	0	N/A	0.278	0.00	0.00	66.56	18.49
03375 1003	10" Forms	SY	412.5	XASPC	1.000	0.56	0.62	0.00	1.18
18 01 0310	**Prime Coat**	**SY**				**0.03**	**0.03**	**0.31**	**0.37**
02545 3190	Surface Prime Coat, .28 Gal/SY (1.25 liters/m2)	CSF	345.5	COKBD	0.090	0.31	0.29	3.47	0.37
18 01 0311	**Tack Coat**	**SY**				**0.08**	**0.07**	**0.13**	**0.28**
02545 3260	Tack Coat, .1 Gal/SY (.5 liter/m2)	CSF	123.4	COKBD	0.090	0.87	0.80	1.43	0.28
18 01 0312	**Asphalt Wearing Course, 1 Pass (Line Item Includes 5% Waste)**	**TON**				**14.73**	**15.12**	**36.23**	**66.08**
02505 0854	Bituminous Hot Mix Surf Course 3,774 Lb/CY (2,242 kg/m3), Central Plant	TN	25	XASPA	1.050	14.03	14.40	34.50	66.08

18 01 03	Paved Surfaces	Unit	Hourly Output	Crew	Factor	Labor	Equip.	Matl.	Total
18 01 0313	**Jointed Concrete, Place & Finish, 4" - 12"**	**SY**				**3.64**	**0.29**	**4.10**	**8.02**
01500 2101	Paving Forms, 10,000 LF 6"	HR	1	XASPF	0.002	225.84	90.85	96.80	0.95
01500 2105	Curing Bridge with Spray Forms	HR	1	XCMAF	0.002	116.83	0.00	59.98	0.35
01500 2107	Finishing Screed - Forms	HR	1	XASPE	0.002	317.25	40.73	45.55	0.81
03110 3801	Longitudinal (Construction) Joint	CY	0	N/A	0.220	0.00	0.00	1.03	0.23
03110 3802	Transverse (Expansion) Joint	CY	0	N/A	0.220	0.00	0.00	1.42	0.31
03132 0500	Liquid Neoprene Expansion Joint, 1/2" x 1" Poured, Cold Applied	LF	56.25	ULABA	0.909	0.60	0.00	1.81	2.19
03217 0700	Grade 60 Rebar, Beam, Column, Wall, #3-#6	TON	0.313	SIWRC	0.001	492.16	0.00	609.86	1.10
03334 0200	Concrete Curing, Waterproof Paper, 2 Ply Reinforced	CSF	8.75	ULABB	0.090	7.74	0.00	5.63	1.20
03334 0300	Concrete Curing, Sprayed Membrane Curing Compound	CSF	11.88	ULABB	0.090	5.71	0.00	4.05	0.88
18 01 0314	**Jointed Mesh Reinforced Concrete, Place & Finish, 4" - 12"**	**SY**				**6.02**	**0.29**	**5.65**	**11.96**
01500 2101	Paving Forms, 10,000 LF 6"	HR	1	XASPF	0.002	225.84	90.85	96.80	0.95
01500 2105	Curing Bridge with Spray Forms	HR	1	XCMAF	0.002	116.83	0.00	59.98	0.35
01500 2107	Finishing Screed - Forms	HR	1	XASPE	0.002	317.25	40.73	45.55	0.81
03110 3801	Longitudinal (Construction) Joint	CY	0	N/A	0.220	0.00	0.00	1.03	0.23
03110 3802	Transverse (Expansion) Joint	CY	0	N/A	0.220	0.00	0.00	1.42	0.31
03132 0500	Liquid Neoprene Expansion Joint, 1/2" x 1" Poured, Cold Applied	LF	56.25	ULABA	0.909	0.60	0.00	1.81	2.19
03217 0700	Grade 60 Rebar, Beam, Column, Wall, #3-#6	TON	0.313	SIWRC	0.001	492.16	0.00	609.86	1.10
03227 0300	Welded Wire Fabric in Slabs, 6 x 6 x #6 42 Lb/CSF, (W 2.9 x W 2.9), < 5	SF	362.5	SIWRB	9.000	0.26	0.00	0.17	3.93
03334 0200	Concrete Curing, Waterproof Paper, 2 Ply Reinforced	CSF	8.75	ULABB	0.090	7.74	0.00	5.63	1.20
03334 0300	Concrete Curing, Sprayed Membrane Curing Compound	CSF	11.88	ULABB	0.090	5.71	0.00	4.05	0.88
18 01 0315	**Jointed Reinforced Concrete, Place & Finish, 4" - 12"**	**SY**				**8.64**	**0.29**	**10.30**	**19.23**
01500 2101	Paving Forms, 10,000 LF 6"	HR	1	XASPF	0.002	225.84	90.85	96.80	0.95
01500 2105	Curing Bridge with Spray Forms	HR	1	XCMAF	0.002	116.83	0.00	59.98	0.35
01500 2107	Finishing Screed - Forms	HR	1	XASPE	0.002	317.25	40.73	45.55	0.81
03110 3801	Longitudinal (Construction) Joint	CY	0	N/A	0.220	0.00	0.00	1.03	0.23
03110 3802	Transverse (Expansion) Joint	CY	0	N/A	0.220	0.00	0.00	1.42	0.31
03132 0500	Liquid Neoprene Expansion Joint, 1/2" x 1" Poured, Cold Applied	LF	56.25	ULABA	0.909	0.60	0.00	1.81	2.19
03217 0700	Grade 60 Rebar, Beam, Column, Wall, #3-#6	TON	0.313	SIWRC	0.011	492.16	0.00	609.86	12.31
03334 0200	Concrete Curing, Waterproof Paper, 2 Ply Reinforced	CSF	8.75	ULABB	0.090	7.74	0.00	5.63	1.20
03334 0300	Concrete Curing, Sprayed Membrane Curing Compound	CSF	11.88	ULABB	0.090	5.71	0.00	4.05	0.88
18 01 0316	**Field Mix Concrete**	**CY**				**9.16**	**0.35**	**85.90**	**95.41**
03372 3200	Pour Grade Beam, Regular, Direct Chute, Place 3,000 PSI Concrete Foundations	CY	18.75	ALABE	1.020	8.98	0.34	84.22	95.41

18 01 03 Paved Surfaces		Unit	Hourly Output	Crew	Factor	Labor	Equip.	Matl.	Total
18 01 0330	**Hydraulic Asphalt Concrete**	**SY**				**0.00**	**0.00**	**16.48**	**16.48**
01001 0007	Hydraulic Asphalt Concrete	SY	0	N/A	1.000	0.00	0.00	16.48	16.48

18 01 04 Markings and Signage		Unit	Hourly Output	Crew	Factor	Labor	Equip.	Matl.	Total
18 01 0401	**Crosswalk, Stop Lines, Per Lane, Intersection Painting**	**EA**				**45.73**	**0.00**	**15.10**	**60.83**
02580 4010	Thermoplastic Pavement Marking, 8" Stripe (20 cm) Wide, by Hand	LF	59.38	APTRB	20.000	1.31	0.00	0.43	34.71
02580 4020	Thermoplastic Pavement Marking, 12" Stripe (31 cm) Wide, by Hand	LF	39.5	APTRB	10.000	1.96	0.00	0.65	26.12
18 01 0402	**Turn Lane, Per Lane, Intersection Painting**	**EA**				**13.78**	**16.48**	**2.75**	**33.01**
02618 1011	White, Turn Lane, 6" Wide	LF	125	COKSA	30.000	0.46	0.55	0.09	33.01
18 01 0403	**Arrows, Per Lane, Intersection Painting**	**EA**				**9.89**	**0.00**	**8.39**	**18.28**
02618 4104	Arrows and Letters	SF	31.25	XLABB	5.000	1.98	0.00	1.68	18.28
18 01 0404	**Railroad Crossing Markings, Stop Line, Per Lane**	**EA**				**29.43**	**0.00**	**9.75**	**39.18**
02580 4020	Thermoplastic Pavement Marking, 12" Stripe (31 cm) Wide, by Hand	LF	39.5	APTRB	15.000	1.96	0.00	0.65	39.18
18 01 0405	**No Pass Stripe, Yellow**	**LF**				**0.04**	**0.04**	**0.10**	**0.18**
02618 1009	Yellow Reflecting , 4" Wide	LF	1625	COKSA	1.000	0.04	0.04	0.10	0.18
18 01 0406	**Centerline Stripe, White**	**LF**				**0.08**	**0.09**	**3.36**	**3.52**
02618 1008	White Reflecting, 6" Wide	LF	1500	COKSA	2.000	0.04	0.05	1.68	3.52
18 01 0407	**Edge Stripe, Yellow**	**LF**				**0.04**	**0.04**	**0.10**	**0.18**
02618 1009	Yellow Reflecting , 4" Wide	LF	1625	COKSA	1.000	0.04	0.04	0.10	0.18
18 01 0410	**Street Signs, Average**	**EA**				**33.88**	**0.00**	**19.79**	**53.66**
02842 0590	Directional Sign, 12"x 18" with Post Reflectorized, OSHA Standard	EA	2	ULABB	1.000	33.88	0.00	19.79	53.66
18 01 0411	**Traffic Signs & Posts, Average**	**EA**				**33.88**	**0.00**	**30.66**	**64.53**
02842 0500	Stop Sign, 24" x 24" with Post Reflectorized, OSHA Standard	EA	2	ULABB	1.000	33.88	0.00	30.66	64.53

18 01 05 Guardrails and Barriers		Unit	Hourly Output	Crew	Factor	Labor	Equip.	Matl.	Total
18 01 0501	**Guardrail, Single Rail, Wood Posts**	**LF**				**6.19**	**9.52**	**7.64**	**23.35**
02840 0160	Corrugated Steel, Wood Post	LF	21.88	CLADM	1.000	6.19	9.52	7.64	23.35
18 01 0502	**Guardrail, Single Rail, Wood Posts, Ends**	**EA**				**34.97**	**53.74**	**11.35**	**100.06**
02840 0380	Wraparound Ends	LF	3.875	CLADM	1.000	34.97	53.74	11.35	100.06

18 02 01 Bases and Subbases		Unit	Hourly Output	Crew	Factor	Labor	Equip.	Matl.	Total

18 02 Parking Lots

18 02 01	Bases and Subbases	Unit	Hourly Output	Crew	Factor	Labor	Equip.	Matl.	Total
18 02 0101	**Gravel, Delivered & Dumped**	**CY**				**1.83**	**1.69**	**22.77**	**26.29**
02611 2205	Graded Crushed Gravel Base	CY	90.88	XSABD	1.000	1.83	1.69	22.77	26.29

18 02 02	Drains, Curbs and Gutters	Unit	Hourly Output	Crew	Factor	Labor	Equip.	Matl.	Total
18 02 0201	**Area Inlets, Precast, 4' Deep**	**EA**				**352.57**	**63.63**	**455.04**	**871.23**
02520 1118	Precast 4' Area Inlet	EA	0.513	CODEK	1.000	332.00	62.54	435.35	829.89
03354 3005	Concrete Redimix 2,500 PSI	CY	1.238	CLACC	0.310	66.34	3.52	63.50	41.34
18 02 0202	**Area Drains with Grates, 6' Deep**	**EA**				**1,089**	**21.37**	**1,669**	**2,779**
02520 1115	3'-8" x 3'-8" x 6' Deep CIP Grate Inlet 6" Thick Wall, Storm Drainage	EA	0.181	ALABM	1.000	1,026	0.00	1,314	2,340
02752 4506	24" Diameter Cast Iron Catch Basin Frame & Cover NEENAH R2533	EA	1.5	CODEJ	1.000	63.48	21.37	355.00	439.85
18 02 0203	**26" x 26", 5' Deep Area Drain with Grate**	**EA**				**951.47**	**25.64**	**1,738**	**2,715**
02520 1114	3'-8" x 3'-8" x 5' Deep CIP Grate Inlet 6" Thick Wall, Storm Drainage	EA	0.213	ALABM	1.000	875.29	0.00	1,208	2,083
02752 4530	26"x 26" Cast Iron Catch Basin Frame & Cover NEENAH R3339B	EA	1.25	CODEJ	1.000	76.18	25.64	530.00	631.82
18 02 0204	**27" x 20", 5' Deep Area Drain with Grate**	**EA**				**951.47**	**25.64**	**1,733**	**2,710**
02520 1114	3'-8" x 3'-8" x 5' Deep CIP Grate Inlet 6" Thick Wall, Storm Drainage	EA	0.213	ALABM	1.000	875.29	0.00	1,208	2,083
02752 4510	27"x 20" Cast Iron Catch Basin Frame & Cover NEENAH R3339	EA	1.25	CODEJ	1.000	76.18	25.64	525.00	626.82
18 02 0205	**24" Diameter, 5' Deep Area Drain with Grate**	**EA**				**938.78**	**21.37**	**1,563**	**2,523**
02520 1114	3'-8" x 3'-8" x 5' Deep CIP Grate Inlet 6" Thick Wall, Storm Drainage	EA	0.213	ALABM	1.000	875.29	0.00	1,208	2,083
02752 4506	24" Diameter Cast Iron Catch Basin Frame & Cover NEENAH R2533	EA	1.5	CODEJ	1.000	63.48	21.37	355.00	439.85
18 02 0210	**6" Concrete Irrigation Pipe (Drainage)**	**LF**				**1.75**	**0.33**	**4.46**	**6.53**
02451 1009	6" Extra Strength Non-Reinforced Concrete	LF	97.5	CODEK	1.000	1.75	0.33	4.46	6.53
18 02 0211	**8" Concrete Irrigation Pipe (Drainage)**	**LF**				**4.54**	**0.85**	**4.90**	**10.30**
02451 1002	8" (21 cm) Diameter Extra Strength Concrete Pipe Bell and Spigot/Tongue & Groove	LF	37.5	CODEK	1.000	4.54	0.85	4.90	10.30
18 02 0212	**6" Extra-strength Nonreinforced Concrete Pipe**	**LF**				**5.14**	**0.97**	**4.13**	**10.23**
02762 1010	6" (15 cm) Diameter Extra Strength Concrete Pipe Bell and Spigot/Tongue & Groove	LF	33.13	CODEK	1.000	5.14	0.97	4.13	10.23
18 02 0213	**8" Extra-strength Nonreinforced Concrete Pipe**	**LF**				**1.75**	**0.33**	**4.90**	**6.98**
02451 1011	8" Extra Strength Non-Reinforced Concrete	LF	97.5	CODEK	1.000	1.75	0.33	4.90	6.98

18 02 03	Paved Surfaces	Unit	Hourly Output	Crew	Factor	Labor	Equip.	Matl.	Total

18 02 03 Paved Surfaces		Unit	Hourly Output	Crew	Factor	Labor	Equip.	Matl.	Total
18 02 0301	**Asphalt Pavement - 10" Subgrade, 9" Base, 1 1/2" Topping**	**SY**				**6.98**	**10.94**	**5.91**	**23.83**
02224 7000	Rough Grade with Dozer	CSY	0.45	CODTG	0.010	97.36	271.86	0.00	3.69
02239 0125	Spread/Compact with Sheepsfoot Roller, 6" (15 cm) Lift, 1,027 CY/Hour	CY	1027	COFCP	0.267	0.09	0.06	0.00	0.04
02244 1510	Graded Crushed Aggregate Roadway Base Course	CY	5.202	XSABA	0.100	45.72	66.34	27.50	13.96
02505 0854	Bituminous Hot Mix Surf Course 3,774 Lb/CY (2,242 kg/m3), Central Plant	TN	25	XASPA	0.083	14.03	14.40	34.50	5.20
02512 0200	Grade Roadway Subbase Courses	MSY	0.24	COFGA	0.001	226.75	355.21	0.00	0.58
02545 3190	Surface Prime Coat, .28 Gal/SY (1.25 liters/m2)	CSF	345.5	COKBD	0.090	0.31	0.29	3.47	0.37
18 02 0320	**4" Structural Slab on Grade**	**SF**				**1.95**	**0.17**	**2.18**	**4.30**
02227 4000	Trim & Shape for Slab on Grade by Hand Finish Grade	SF	75.38	ULABA	1.100	0.45	0.00	0.00	0.49
02611 2201	Graded Crushed Aggregate Base	CY	25	XSABA	0.012	9.51	13.80	22.77	0.57
03114 2400	PVC Chamfer Strips, 1" Radius	CLF	0.625	ACARB	0.000	138.60	0.00	51.80	0.05
03170 4500	Straight Edge Metal Formwork, 4" High (10 cm), with Stakes, 10' Length, 100 Uses	LF	65	ACARI	0.003	1.60	0.00	0.58	0.01
03217 0600	Grade 60 Rebar, Footings & Slabs, #3-#6	TON	0.313	SIWRC	0.001	492.16	0.00	609.86	1.38
03334 0300	Concrete Curing, Sprayed Membrane Curing Compound	CSF	11.88	ULABB	0.010	5.71	0.00	4.05	0.10
03372 4600	Pour Slab on Grade, >= 6", Direct Chute >= (15 cm) Place 3,000 PSI Concrete	CY	21.88	ALABE	0.012	7.70	0.29	84.22	1.14
03396 0150	Concrete Floor Finishes, Broom	SF	84.38	ACMAC	1.000	0.43	0.00	0.00	0.43
07192 0900	6 Mil Polyethylene Vapor Barrier	CSF	4.625	ACARA	0.011	9.36	0.00	3.30	0.14
18 02 0321	**6" Structural Slab on Grade**	**SF**				**2.00**	**0.18**	**2.70**	**4.87**
02227 4000	Trim & Shape for Slab on Grade by Hand Finish Grade	SF	75.38	ULABA	1.100	0.45	0.00	0.00	0.49
02611 2201	Graded Crushed Aggregate Base	CY	25	XSABA	0.012	9.51	13.80	22.77	0.57
03114 2400	PVC Chamfer Strips, 1" Radius	CLF	0.625	ACARB	0.000	138.60	0.00	51.80	0.05
03170 4560	Straight Edge Metal Formwork, 6" High (10 cm), with Stakes, 10' Length, 100 Uses	LF	65	ACARI	0.003	1.60	0.00	0.67	0.01
03217 0600	Grade 60 Rebar, Footings & Slabs, #3-#6	TON	0.313	SIWRC	0.001	492.16	0.00	609.86	1.38
03334 0300	Concrete Curing, Sprayed Membrane Curing Compound	CSF	11.88	ULABB	0.010	5.71	0.00	4.05	0.10
03372 4600	Pour Slab on Grade, >= 6", Direct Chute >= (15 cm) Place 3,000 PSI Concrete	CY	21.88	ALABE	0.019	7.70	0.29	84.22	1.71
03396 0150	Concrete Floor Finishes, Broom	SF	84.38	ACMAC	1.000	0.43	0.00	0.00	0.43
07192 0900	6 Mil Polyethylene Vapor Barrier	CSF	4.625	ACARA	0.011	9.36	0.00	3.30	0.14

18 02 03	Paved Surfaces	Unit	Hourly Output	Crew	Factor	Labor	Equip.	Matl.	Total
18 02 0322	**8" Structural Slab on Grade**	**SF**				**2.54**	**0.39**	**4.00**	**6.93**
02227 4000	Trim & Shape for Slab on Grade by Hand Finish Grade	SF	75.38	ULABA	1.100	0.45	0.00	0.00	0.49
02611 2201	Graded Crushed Aggregate Base	CY	25	XSABA	0.028	9.51	13.80	22.77	1.27
03114 2400	PVC Chamfer Strips, 1" Radius	CLF	0.625	ACARB	0.000	138.60	0.00	51.80	0.05
03170 4570	Straight Edge Metal Formwork, 8" High (10 cm), with Stakes, 10' Length, 100 Uses	LF	60	ACARI	0.003	1.74	0.00	0.82	0.01
03217 0600	Grade 60 Rebar, Footings & Slabs, #3-#6	TON	0.313	SIWRC	0.002	492.16	0.00	609.86	2.16
03334 0300	Concrete Curing, Sprayed Membrane Curing Compound	CSF	11.88	ULABB	0.010	5.71	0.00	4.05	0.10
03372 4600	Pour Slab on Grade, >= 6", Direct Chute >= (15 cm) Place 3,000 PSI Concrete	CY	21.88	ALABE	0.025	7.70	0.29	84.22	2.29
03396 0150	Concrete Floor Finishes, Broom	SF	84.38	ACMAC	1.000	0.43	0.00	0.00	0.43
07192 0900	6 Mil Polyethylene Vapor Barrier	CSF	4.625	ACARA	0.011	9.36	0.00	3.30	0.14
18 02 0323	**10" Structural Slab on Grade**	**SF**				**2.67**	**0.39**	**4.62**	**7.68**
02227 4000	Trim & Shape for Slab on Grade by Hand Finish Grade	SF	75.38	ULABA	1.100	0.45	0.00	0.00	0.49
02611 2201	Graded Crushed Aggregate Base	CY	25	XSABA	0.028	9.51	13.80	22.77	1.27
03114 2400	PVC Chamfer Strips, 1" Radius	CLF	0.625	ACARB	0.000	138.60	0.00	51.80	0.05
03170 4580	Straight Edge Metal Formwork, 12" High (10 cm), with Stakes, 10' Length, 100 Uses	LF	52.5	ACARI	0.003	1.99	0.00	1.04	0.01
03217 0600	Grade 60 Rebar, Footings & Slabs, #3-#6	TON	0.313	SIWRC	0.002	492.16	0.00	609.86	2.35
03334 0300	Concrete Curing, Sprayed Membrane Curing Compound	CSF	11.88	ULABB	0.010	5.71	0.00	4.05	0.10
03372 4600	Pour Slab on Grade, >= 6", Direct Chute >= (15 cm) Place 3,000 PSI Concrete	CY	21.88	ALABE	0.031	7.70	0.29	84.22	2.85
03396 0150	Concrete Floor Finishes, Broom	SF	84.38	ACMAC	1.000	0.43	0.00	0.00	0.43
07192 0900	6 Mil Polyethylene Vapor Barrier	CSF	4.625	ACARA	0.011	9.36	0.00	3.30	0.14
18 02 0324	**12" Structural Slab on Grade**	**SF**				**2.84**	**0.42**	**5.31**	**8.57**
02227 4000	Trim & Shape for Slab on Grade by Hand Finish Grade	SF	75.38	ULABA	1.100	0.45	0.00	0.00	0.49
02611 2201	Graded Crushed Aggregate Base	CY	25	XSABA	0.030	9.51	13.80	22.77	1.37
03114 2400	PVC Chamfer Strips, 1" Radius	CLF	0.625	ACARB	0.000	138.60	0.00	51.80	0.05
03170 4580	Straight Edge Metal Formwork, 12" High (10 cm), with Stakes, 10' Length, 100 Uses	LF	52.5	ACARI	0.003	1.99	0.00	1.04	0.01
03217 0600	Grade 60 Rebar, Footings & Slabs, #3-#6	TON	0.313	SIWRC	0.002	492.16	0.00	609.86	2.57
03334 0300	Concrete Curing, Sprayed Membrane Curing Compound	CSF	11.88	ULABB	0.010	5.71	0.00	4.05	0.10
03372 4600	Pour Slab on Grade, >= 6", Direct Chute >= (15 cm) Place 3,000 PSI Concrete	CY	21.88	ALABE	0.037	7.70	0.29	84.22	3.42
03396 0150	Concrete Floor Finishes, Broom	SF	84.38	ACMAC	1.000	0.43	0.00	0.00	0.43
07192 0900	6 Mil Polyethylene Vapor Barrier	CSF	4.625	ACARA	0.011	9.36	0.00	3.30	0.14

18 02 03	Paved Surfaces	Unit	Hourly Output	Crew	Factor	Labor	Equip.	Matl.	Total
18 02 0330	**4" Mesh Reinforced Slab on Grade**	**SF**				**1.71**	**0.17**	**1.79**	**3.67**
02227 4000	Trim & Shape for Slab on Grade by Hand Finish Grade	SF	75.38	ULABA	1.100	0.45	0.00	0.00	0.49
02611 2201	Graded Crushed Aggregate Base	CY	25	XSABA	0.012	9.51	13.80	22.77	0.57
03114 2400	PVC Chamfer Strips, 1" Radius	CLF	0.625	ACARB	0.000	138.60	0.00	51.80	0.05
03170 4500	Straight Edge Metal Formwork, 4" High (10 cm), with Stakes, 10' Length, 100 Uses	LF	65	ACARI	0.003	1.60	0.00	0.58	0.01
03227 0700	Welded Wire Fabric in Slabs, 4 x 4 x #4 85 Lb/CSF, (W 4.0 x W 4.0), < 5	SF	312.5	SIWRB	1.200	0.31	0.00	0.31	0.74
03334 0300	Concrete Curing, Sprayed Membrane Curing Compound	CSF	11.88	ULABB	0.010	5.71	0.00	4.05	0.10
03372 4600	Pour Slab on Grade, >= 6", Direct Chute >= (15 cm) Place 3,000 PSI Concrete	CY	21.88	ALABE	0.012	7.70	0.29	84.22	1.14
03396 0150	Concrete Floor Finishes, Broom	SF	84.38	ACMAC	1.000	0.43	0.00	0.00	0.43
07192 0900	6 Mil Polyethylene Vapor Barrier	CSF	4.625	ACARA	0.011	9.36	0.00	3.30	0.14
18 02 0331	**6" Mesh Reinforced Slab on Grade**	**SF**				**1.75**	**0.18**	**2.31**	**4.23**
02227 4000	Trim & Shape for Slab on Grade by Hand Finish Grade	SF	75.38	ULABA	1.100	0.45	0.00	0.00	0.49
02611 2201	Graded Crushed Aggregate Base	CY	25	XSABA	0.012	9.51	13.80	22.77	0.57
03114 2400	PVC Chamfer Strips, 1" Radius	CLF	0.625	ACARB	0.000	138.60	0.00	51.80	0.05
03170 4560	Straight Edge Metal Formwork, 6" High (10 cm), with Stakes, 10' Length, 100 Uses	LF	65	ACARI	0.003	1.60	0.00	0.67	0.01
03227 0700	Welded Wire Fabric in Slabs, 4 x 4 x #4 85 Lb/CSF, (W 4.0 x W 4.0), < 5	SF	312.5	SIWRB	1.200	0.31	0.00	0.31	0.74
03334 0300	Concrete Curing, Sprayed Membrane Curing Compound	CSF	11.88	ULABB	0.010	5.71	0.00	4.05	0.10
03372 4600	Pour Slab on Grade, >= 6", Direct Chute >= (15 cm) Place 3,000 PSI Concrete	CY	21.88	ALABE	0.019	7.70	0.29	84.22	1.71
03396 0150	Concrete Floor Finishes, Broom	SF	84.38	ACMAC	1.000	0.43	0.00	0.00	0.43
07192 0900	6 Mil Polyethylene Vapor Barrier	CSF	4.625	ACARA	0.011	9.36	0.00	3.30	0.14
18 02 0332	**8" Mesh Reinforced Slab on Grade**	**SF**				**1.95**	**0.39**	**3.18**	**5.52**
02227 4000	Trim & Shape for Slab on Grade by Hand Finish Grade	SF	75.38	ULABA	1.100	0.45	0.00	0.00	0.49
02611 2201	Graded Crushed Aggregate Base	CY	25	XSABA	0.028	9.51	13.80	22.77	1.27
03114 2400	PVC Chamfer Strips, 1" Radius	CLF	0.625	ACARB	0.000	138.60	0.00	51.80	0.05
03170 4570	Straight Edge Metal Formwork, 8" High (10 cm), with Stakes, 10' Length, 100 Uses	LF	60	ACARI	0.003	1.74	0.00	0.82	0.01
03227 0700	Welded Wire Fabric in Slabs, 4 x 4 x #4 85 Lb/CSF, (W 4.0 x W 4.0), < 5	SF	312.5	SIWRB	1.200	0.31	0.00	0.31	0.74
03334 0300	Concrete Curing, Sprayed Membrane Curing Compound	CSF	11.88	ULABB	0.010	5.71	0.00	4.05	0.10
03372 4600	Pour Slab on Grade, >= 6", Direct Chute >= (15 cm) Place 3,000 PSI Concrete	CY	21.88	ALABE	0.025	7.70	0.29	84.22	2.29
03396 0150	Concrete Floor Finishes, Broom	SF	84.38	ACMAC	1.000	0.43	0.00	0.00	0.43
07192 0900	6 Mil Polyethylene Vapor Barrier	CSF	4.625	ACARA	0.011	9.36	0.00	3.30	0.14

18 02 03	Paved Surfaces	Unit	Hourly Output	Crew	Factor	Labor	Equip.	Matl.	Total
18 02 0333	**10" Mesh Reinforced Slab on Grade**	**SF**				**1.99**	**0.39**	**3.69**	**6.07**
02227 4000	Trim & Shape for Slab on Grade by Hand Finish Grade	SF	75.38	ULABA	1.100	0.45	0.00	0.00	0.49
02611 2201	Graded Crushed Aggregate Base	CY	25	XSABA	0.028	9.51	13.80	22.77	1.27
03114 2400	PVC Chamfer Strips, 1" Radius	CLF	0.625	ACARB	0.000	138.60	0.00	51.80	0.05
03170 4580	Straight Edge Metal Formwork, 12" High (10 cm), with Stakes, 10' Length, 100 Uses	LF	52.5	ACARI	0.003	1.99	0.00	1.04	0.01
03227 0700	Welded Wire Fabric in Slabs, 4 x 4 x #4 85 Lb/CSF, (W 4.0 x W 4.0), < 5	SF	312.5	SIWRB	1.200	0.31	0.00	0.31	0.74
03334 0300	Concrete Curing, Sprayed Membrane Curing Compound	CSF	11.88	ULABB	0.010	5.71	0.00	4.05	0.10
03372 4600	Pour Slab on Grade, >= 6", Direct Chute >= (15 cm) Place 3,000 PSI Concrete	CY	21.88	ALABE	0.031	7.70	0.29	84.22	2.85
03396 0150	Concrete Floor Finishes, Broom	SF	84.38	ACMAC	1.000	0.43	0.00	0.00	0.43
07192 0900	6 Mil Polyethylene Vapor Barrier	CSF	4.625	ACARA	0.011	9.36	0.00	3.30	0.14
18 02 0334	**12" Mesh Reinforced Slab on Grade**	**SF**				**2.06**	**0.42**	**4.26**	**6.74**
02227 4000	Trim & Shape for Slab on Grade by Hand Finish Grade	SF	75.38	ULABA	1.100	0.45	0.00	0.00	0.49
02611 2201	Graded Crushed Aggregate Base	CY	25	XSABA	0.030	9.51	13.80	22.77	1.37
03114 2400	PVC Chamfer Strips, 1" Radius	CLF	0.625	ACARB	0.000	138.60	0.00	51.80	0.05
03170 4580	Straight Edge Metal Formwork, 12" High (10 cm), with Stakes, 10' Length, 100 Uses	LF	52.5	ACARI	0.003	1.99	0.00	1.04	0.01
03227 0700	Welded Wire Fabric in Slabs, 4 x 4 x #4 85 Lb/CSF, (W 4.0 x W 4.0), < 5	SF	312.5	SIWRB	1.200	0.31	0.00	0.31	0.74
03334 0300	Concrete Curing, Sprayed Membrane Curing Compound	CSF	11.88	ULABB	0.010	5.71	0.00	4.05	0.10
03372 4600	Pour Slab on Grade, >= 6", Direct Chute >= (15 cm) Place 3,000 PSI Concrete	CY	21.88	ALABE	0.037	7.70	0.29	84.22	3.42
03396 0150	Concrete Floor Finishes, Broom	SF	84.38	ACMAC	1.000	0.43	0.00	0.00	0.43
07192 0900	6 Mil Polyethylene Vapor Barrier	CSF	4.625	ACARA	0.011	9.36	0.00	3.30	0.14
18 02 0340	**4" Unreinforced Slab on Grade**	**SF**				**1.34**	**0.17**	**1.41**	**2.92**
02227 4000	Trim & Shape for Slab on Grade by Hand Finish Grade	SF	75.38	ULABA	1.100	0.45	0.00	0.00	0.49
02611 2201	Graded Crushed Aggregate Base	CY	25	XSABA	0.012	9.51	13.80	22.77	0.57
03114 2400	PVC Chamfer Strips, 1" Radius	CLF	0.625	ACARB	0.000	138.60	0.00	51.80	0.05
03170 4500	Straight Edge Metal Formwork, 4" High (10 cm), with Stakes, 10' Length, 100 Uses	LF	65	ACARI	0.003	1.60	0.00	0.58	0.01
03334 0300	Concrete Curing, Sprayed Membrane Curing Compound	CSF	11.88	ULABB	0.010	5.71	0.00	4.05	0.10
03372 4600	Pour Slab on Grade, >= 6", Direct Chute >= (15 cm) Place 3,000 PSI Concrete	CY	21.88	ALABE	0.012	7.70	0.29	84.22	1.14
03396 0150	Concrete Floor Finishes, Broom	SF	84.38	ACMAC	1.000	0.43	0.00	0.00	0.43
07192 0900	6 Mil Polyethylene Vapor Barrier	CSF	4.625	ACARA	0.011	9.36	0.00	3.30	0.14

18 02 03	Paved Surfaces	Unit	Hourly Output	Crew	Factor	Labor	Equip.	Matl.	Total
18 02 0341	**6" Unreinforced Slab on Grade**	**SF**				**1.38**	**0.18**	**1.93**	**3.49**
02227 4000	Trim & Shape for Slab on Grade by Hand Finish Grade	SF	75.38	ULABA	1.100	0.45	0.00	0.00	0.49
02611 2201	Graded Crushed Aggregate Base	CY	25	XSABA	0.012	9.51	13.80	22.77	0.57
03114 2400	PVC Chamfer Strips, 1" Radius	CLF	0.625	ACARB	0.000	138.60	0.00	51.80	0.05
03170 4560	Straight Edge Metal Formwork, 6" High (10 cm), with Stakes, 10' Length, 100 Uses	LF	65	ACARI	0.003	1.60	0.00	0.67	0.01
03334 0300	Concrete Curing, Sprayed Membrane Curing Compound	CSF	11.88	ULABB	0.010	5.71	0.00	4.05	0.10
03372 4600	Pour Slab on Grade, >= 6", Direct Chute >= (15 cm) Place 3,000 PSI Concrete	CY	21.88	ALABE	0.019	7.70	0.29	84.22	1.71
03396 0150	Concrete Floor Finishes, Broom	SF	84.38	ACMAC	1.000	0.43	0.00	0.00	0.43
07192 0900	6 Mil Polyethylene Vapor Barrier	CSF	4.625	ACARA	0.011	9.36	0.00	3.30	0.14
18 02 0342	**8" Unreinforced Slab on Grade**	**SF**				**1.58**	**0.39**	**2.81**	**4.77**
02227 4000	Trim & Shape for Slab on Grade by Hand Finish Grade	SF	75.38	ULABA	1.100	0.45	0.00	0.00	0.49
02611 2201	Graded Crushed Aggregate Base	CY	25	XSABA	0.028	9.51	13.80	22.77	1.27
03114 2400	PVC Chamfer Strips, 1" Radius	CLF	0.625	ACARB	0.000	138.60	0.00	51.80	0.05
03170 4570	Straight Edge Metal Formwork, 8" High (10 cm), with Stakes, 10' Length, 100 Uses	LF	60	ACARI	0.003	1.74	0.00	0.82	0.01
03334 0300	Concrete Curing, Sprayed Membrane Curing Compound	CSF	11.88	ULABB	0.010	5.71	0.00	4.05	0.10
03372 4600	Pour Slab on Grade, >= 6", Direct Chute >= (15 cm) Place 3,000 PSI Concrete	CY	21.88	ALABE	0.025	7.70	0.29	84.22	2.29
03396 0150	Concrete Floor Finishes, Broom	SF	84.38	ACMAC	1.000	0.43	0.00	0.00	0.43
07192 0900	6 Mil Polyethylene Vapor Barrier	CSF	4.625	ACARA	0.011	9.36	0.00	3.30	0.14
18 02 0343	**10" Unreinforced Slab on Grade**	**SF**				**1.62**	**0.39**	**3.32**	**5.33**
02227 4000	Trim & Shape for Slab on Grade by Hand Finish Grade	SF	75.38	ULABA	1.100	0.45	0.00	0.00	0.49
02611 2201	Graded Crushed Aggregate Base	CY	25	XSABA	0.028	9.51	13.80	22.77	1.27
03114 2400	PVC Chamfer Strips, 1" Radius	CLF	0.625	ACARB	0.000	138.60	0.00	51.80	0.05
03170 4580	Straight Edge Metal Formwork, 12" High (10 cm), with Stakes, 10' Length, 100 Uses	LF	52.5	ACARI	0.003	1.99	0.00	1.04	0.01
03334 0300	Concrete Curing, Sprayed Membrane Curing Compound	CSF	11.88	ULABB	0.010	5.71	0.00	4.05	0.10
03372 4600	Pour Slab on Grade, >= 6", Direct Chute >= (15 cm) Place 3,000 PSI Concrete	CY	21.88	ALABE	0.031	7.70	0.29	84.22	2.85
03396 0150	Concrete Floor Finishes, Broom	SF	84.38	ACMAC	1.000	0.43	0.00	0.00	0.43
07192 0900	6 Mil Polyethylene Vapor Barrier	CSF	4.625	ACARA	0.011	9.36	0.00	3.30	0.14

18 02 03	Paved Surfaces	Unit	Hourly Output	Crew	Factor	Labor	Equip.	Matl.	Total
18 02 0344	**12" Unreinforced Slab on Grade**	**SF**				**1.69**	**0.42**	**3.89**	**6.00**
02227 4000	Trim & Shape for Slab on Grade by Hand Finish Grade	SF	75.38	ULABA	1.100	0.45	0.00	0.00	0.49
02611 2201	Graded Crushed Aggregate Base	CY	25	XSABA	0.030	9.51	13.80	22.77	1.37
03114 2400	PVC Chamfer Strips, 1" Radius	CLF	0.625	ACARB	0.000	138.60	0.00	51.80	0.05
03170 4580	Straight Edge Metal Formwork, 12" High (10 cm), with Stakes, 10' Length, 100 Uses	LF	52.5	ACARI	0.003	1.99	0.00	1.04	0.01
03334 0300	Concrete Curing, Sprayed Membrane Curing Compound	CSF	11.88	ULABB	0.010	5.71	0.00	4.05	0.10
03372 4600	Pour Slab on Grade, >= 6", Direct Chute >= (15 cm) Place 3,000 PSI Concrete	CY	21.88	ALABE	0.037	7.70	0.29	84.22	3.42
03396 0150	Concrete Floor Finishes, Broom	SF	84.38	ACMAC	1.000	0.43	0.00	0.00	0.43
07192 0900	6 Mil Polyethylene Vapor Barrier	CSF	4.625	ACARA	0.011	9.36	0.00	3.30	0.14

18 02 04	Markings and Signage	Unit	Hourly Output	Crew	Factor	Labor	Equip.	Matl.	Total
18 02 0401	**Parking Space Striping**	**EA**				**6.33**	**0.00**	**3.18**	**9.51**
02580 0800	4" (10 cm) Parking Stall Marking 4" (10 cm) Wide Stripe	EA	12.25	APTRB	1.000	6.33	0.00	3.18	9.51
18 02 0402	**Handicap Symbol, Painted**	**EA**				**18.24**	**0.00**	**12.12**	**30.36**
02580 1200	Pavement Marking Handicap Symbol	EA	4.25	APTRB	1.000	18.24	0.00	12.12	30.36
18 02 0403	**Handicap Parking Sign, with Post**	**EA**				**33.88**	**0.00**	**19.79**	**53.66**
02842 0580	Handicap Parking Sign, 12"x 18" with Post Reflectorized, OSHA Standard	EA	2	ULABB	1.000	33.88	0.00	19.79	53.66

18 02 05	Guardrails and Barriers	Unit	Hourly Output	Crew	Factor	Labor	Equip.	Matl.	Total
18 02 0501	**Precast Parking Stops**	**EA**				**8.21**	**4.37**	**17.16**	**29.73**
03414 5101	Precast Bumper Curbs, 6"x10"x 7' (.15 m x .25 m x 2.1 m)	EA	10	ULABL	1.000	8.21	4.37	17.16	29.73

18 03 03	Paved Surfaces	Unit	Hourly Output	Crew	Factor	Labor	Equip.	Matl.	Total
18 03 0301	**Standard 4" Sidewalk with Mesh, Formed**	**SF**				**1.60**	**0.00**	**1.37**	**2.97**
02514 0310	4" (10 cm) CIP Concrete Walk with Mesh, No Base	SF	82.5	ACMAE	1.000	1.14	0.00	1.35	2.49
03170 4600	Radius Metal Formwork, 4" High (10 cm), with Stakes,10' Length,100 Uses	LF	52.5	ACARI	0.020	1.99	0.00	0.78	0.06
03396 0150	Concrete Floor Finishes, Broom	SF	84.38	ACMAC	1.000	0.43	0.00	0.00	0.43
18 03 0302	**2" Thick Bituminous Sidewalk**	**SF**				**0.23**	**0.02**	**0.41**	**0.66**
02514 0020	2" (51 cm) Bituminous Sidewalk, No Base Included	SY	80.63	COFCB	0.111	2.09	0.17	3.65	0.66
18 03 0303	**Brick Sidewalk with Sand Joints (4.5 Bricks/SF)**	**SF**				**5.20**	**0.00**	**2.49**	**7.69**
02517 1500	Brick Walk on Sand Bed, Flat, 4.5 Bricks/SF, Sand Joint	SF	13.75	AMABD	1.000	5.20	0.00	2.49	7.69

18 03 03 Paved Surfaces		Unit	Hourly Output	Crew	Factor	Labor	Equip.	Matl.	Total
18 03 0304	**Standard 5" Sidewalk with Mesh, Formed**	**SF**				**1.64**	**0.00**	**1.82**	**3.46**
02514 0350	5" (13 cm) CIP Concrete Walk with Mesh, No Base	SF	80	ACMAE	1.000	1.17	0.00	1.80	2.97
03170 4660	Radius Metal Formwork, 6" High (10 cm), with Stakes, 10' Length, 100 Uses	LF	52.5	ACARI	0.020	1.99	0.00	0.90	0.06
03396 0150	Concrete Floor Finishes, Broom	SF	84.38	ACMAC	1.000	0.43	0.00	0.00	0.43
18 03 0305	**Standard 6" Sidewalk with Mesh, Formed**	**SF**				**1.66**	**0.00**	**2.12**	**3.78**
02514 0400	6" (15 cm) CIP Concrete Walk with Mesh, No Base	SF	78.75	ACMAE	1.000	1.19	0.00	2.10	3.29
03170 4660	Radius Metal Formwork, 6" High (10 cm), with Stakes, 10' Length, 100 Uses	LF	52.5	ACARI	0.020	1.99	0.00	0.90	0.06
03396 0150	Concrete Floor Finishes, Broom	SF	84.38	ACMAC	1.000	0.43	0.00	0.00	0.43
18 03 0306	**2 1/2" Thick Bituminous Sidewalk**	**SF**				**0.29**	**0.02**	**0.51**	**0.83**
02514 0100	2-1/2" (63 cm) Bituminous Sidewalk, No Base Included	SY	65	COFCB	0.111	2.59	0.21	4.63	0.83
18 03 0307	**Brick Sidewalk with Sand Joints (7.2 Bricks/SF)**	**SF**				**8.17**	**0.00**	**2.36**	**10.53**
02517 2000	Brick Walk on Sand Bed, on Edge, 7.2 Bricks/SF, Sand Joint	SF	8.75	AMABD	1.000	8.17	0.00	2.36	10.53
18 03 0308	**2" Prec Conc Patio Block, Nat**	**SF**				**2.16**	**0.00**	**1.26**	**3.42**
02518 0700	2" Prec Conc Patio Block, Nat	SF	33.13	AMABD	1.000	2.16	0.00	1.26	3.42
18 03 0310	**21" x 21" x 2 1/2" White Granite Paver**	**SF**				**1.80**	**0.00**	**13.12**	**14.92**
02630 5515	21" x 21" x 2 1/2" White Granite Paver	SF	109.6	AMABH	1.000	1.80	0.00	13.12	14.92
18 03 0311	**21" x 21" x 3" White Granite Paver**	**SF**				**2.40**	**0.00**	**15.12**	**17.52**
02630 5516	21" x 21" x 3" White Granite Paver	SF	82.1	AMABH	1.000	2.40	0.00	15.12	17.52
18 03 0312	**21" x 21" x 2 1/2" Exposed Aggregate Paver**	**SF**				**1.35**	**0.00**	**4.24**	**5.59**
02630 5613	21" x 21" x 2 1/2" Exposed Aggregate Paver	SF	146.1	AMABH	1.000	1.35	0.00	4.24	5.59
18 03 0313	**21" x 21" x 3" Exposed Aggregate Paver**	**SF**				**2.25**	**0.00**	**4.49**	**6.74**
02630 5614	21" x 21" x 3" Exposed Aggregate Paver	SF	87.6	AMABH	1.000	2.25	0.00	4.49	6.74

18 04 01 Fencing and Gates		Unit	Hourly Output	Crew	Factor	Labor	Equip.	Matl.	Total
18 04 0101	**Security Fence, 10' Galvanized with 3 Strands Barbed Wire**	**LF**				**9.63**	**11.69**	**15.91**	**37.22**
02834 4650	10' (3 m) High Galvanized Fencing	LF	11.96	ULABN	1.000	9.63	11.69	15.91	37.22
18 04 0102	**Security Fence, with 1' x 1' Grade Beam, 10' Galvanized with 3 Strands Barbed Wire**	**LF**				**9.96**	**11.70**	**19.03**	**40.69**
02834 4650	10' (3 m) High Galvanized Fencing	LF	11.96	ULABN	1.000	9.63	11.69	15.91	37.22
03372 3200	Pour Grade Beam, Regular, Direct Chute, Place 3,000 PSI Concrete Foundations	CY	18.75	ALABE	0.037	8.98	0.34	84.22	3.46

18 04 01 Fencing and Gates		Unit	Hourly Output	Crew	Factor	Labor	Equip.	Matl.	Total
18 04 0103	**Privacy Fence, 6' High, Wood**	**LF**				**5.63**	**0.00**	**9.12**	**14.75**
02836 0300	Board Fence, 3-2"x 4" Rails, 6' High 1"x 4" Boards, Preservative Treated	LF	15.63	XLABC	1.000	5.63	0.00	9.12	14.75
18 04 0104	**Privacy Fence, 5' High, PVC**	**LF**				**7.72**	**1.23**	**19.02**	**27.96**
02714 1001	PVC Picket Fences, 5' high, Top & Bottom Rails, 3" Slats on 3"	LF	16.25	XLABC	1.000	5.41	0.00	12.79	18.20
02714 1002	PVC Fence Post, 5" x 5"	EA	9	ULABL	0.150	9.12	4.85	21.00	5.25
02714 1003	PVC Picket Fence, Single Gate, 4' x 5'	EA	1.25	ULABL	0.014	65.68	34.92	215.67	4.52
18 04 0105	**Boundary Fence, 5' Galvanized**	**LF**				**2.52**	**1.92**	**1.70**	**6.14**
02712 2102	5' (1.5 m) High, 12 Gauge Galvanized Steel Fence, 2" x 4" Mesh, Posts at 5'	LF	37.5	ULABF	1.000	2.52	1.92	1.70	6.14
18 04 0106	**5' Galvanized Chain-link Fence**	**LF**				**1.45**	**0.00**	**22.21**	**23.66**
02834 4610	5' (1.5 m) High Fabric Security Fence Standard FE-5	LF	65	ULABC	1.000	1.45	0.00	22.21	23.66
18 04 0107	**6' Galvanized Chain-link Fence**	**LF**				**1.51**	**0.00**	**23.59**	**25.10**
02834 4620	6' (1.8 m) High Fabric Security Fence Standard FE-6	LF	62.5	ULABC	1.000	1.51	0.00	23.59	25.10
18 04 0108	**7' Galvanized Chain-link Fence**	**LF**				**1.57**	**0.00**	**32.34**	**33.91**
02834 4630	7' (2.1 m) High Fabric Security Fence Standard FE-7	LF	60	ULABC	1.000	1.57	0.00	32.34	33.91
18 04 0110	**Barbed-wire Fencing, 3-Strand**	**LF**				**0.22**	**0.12**	**2.33**	**2.67**
02834 7060	Barbed Wire, Standard 3-Strand Fence	LF	375	ULABL	1.000	0.22	0.12	2.33	2.67
18 04 0111	**Galvanized Barbed Wire, 3-Strand**	**LF**				**0.66**	**0.35**	**0.27**	**1.27**
02834 7040	Barbed Wire, Galvanized Per Strand	LF	375	ULABL	3.000	0.22	0.12	0.09	1.27
18 04 0115	**Swing Gates, Complete**	**EA**				**131.60**	**159.80**	**823.52**	**1,115**
02831 7385	10' (3 m) Wide x 8' (2.4 m) High Double Gate Galvanized Steel without Barbed Wire	EA	1.75	ULABN	2.000	65.80	79.90	411.76	1,115
18 04 0116	**5' Swing Gate, 12' Double**	**EA**				**101.05**	**53.73**	**686.46**	**841.23**
02831 7190	12' (3.7 m) Wide x 5' (1.5 m) High Double Gate Galvanized Steel without Barbed Wire	EA	1.625	ULABL	2.000	50.52	26.86	343.23	841.23
18 04 0117	**6' Swing Gate, 12' Double**	**EA**				**109.47**	**58.20**	**670.70**	**838.37**
02831 7260	12' (3.7 m) Wide x 6' (1.8 m) High Double Gate Galvanized Steel without Barbed Wire	EA	1.5	ULABL	2.000	54.73	29.10	335.35	838.37
18 04 0118	**7' Swing Gate, 12' Double**	**EA**				**79.42**	**96.44**	**428.69**	**604.55**
02831 7325	12' (3.7 m) Wide x 7' (2.1 m) High Double Gate Galvanized Steel without Barbed Wire	EA	1.45	ULABN	1.000	79.41	96.43	428.65	604.55

18 04 02 Retaining Walls		Unit	Hourly Output	Crew	Factor	Labor	Equip.	Matl.	Total

18 04 02 Retaining Walls		Unit	Hourly Output	Crew	Factor	Labor	Equip.	Matl.	Total
18 04 0201	**Continuous Footing, Edge Form, 4 Uses**	**SF**				**2.17**	**0.00**	**0.82**	**2.99**
03158 0150	Continuous Wall Footing Forms, 4 Uses Plywood Forms, Form & Strip with Accessories	SF	60.63	ACARJ	1.000	2.17	0.00	0.82	2.99
18 04 0202	**Footing, Rebar**	**LB**				**0.31**	**0.12**	**0.32**	**0.76**
03210 1007	Rebar #3-#7	TON	0.288	SIWRF	0.001	629.08	250.00	647.24	0.76
18 04 0203	**Pour & Cure Concrete, Continuous Footing**	**CY**				**23.75**	**7.55**	**87.58**	**118.88**
03372 2450	Pour Spread Footings, < 5 CY, Concrete Pump < (5 m3) Place 3,000 PSI Concrete Foundations	CY	12.5	ALABI	1.000	18.38	7.55	84.22	110.14
07174 0100	Silicone Retarder, 1 Coat on Concrete, Sprayed	CSF	1.875	ARFCD	0.270	19.90	0.00	12.47	8.74
18 04 0205	**CIP Walls Form & Strip (4 Uses)**	**SF**				**3.96**	**0.00**	**0.62**	**4.58**
03182 4750	Retaining Wall Forms - up to 8' 4 Uses (Below Grade) Plywood Forms Form and Strip	SF	40.63	ACARL	1.000	3.96	0.00	0.62	4.58
18 04 0206	**Reinforced Steel, Retaining Wall**	**LB**				**0.24**	**0.10**	**0.32**	**0.66**
03210 1011	Wall Rebar, #3-#7	TON	0.375	SIWRF	0.001	482.29	191.67	647.24	0.66
18 04 0207	**Pour & Cure Concrete, Retaining Wall**	**CY**				**33.95**	**10.51**	**90.95**	**135.41**
03372 5200	Pour Concrete Wall, 12" Thick, Crane & Bucket (31 cm) 3,000 PSI Concrete	CY	11.25	ALABG	1.000	23.20	10.51	84.22	117.94
07174 0100	Silicone Retarder, 1 Coat on Concrete, Sprayed	CSF	1.875	ARFCD	0.540	19.90	0.00	12.47	17.48
18 04 0208	**Bush Hammer Finish**	**SF**				**1.96**	**0.00**	**0.06**	**2.01**
03398 0300	Wall Finish, Bush Hammer Green Concrete	SF	21.25	ACMAA	1.000	1.96	0.00	0.06	2.01
18 04 0210	**Keyway**	**LF**				**0.65**	**0.00**	**0.24**	**0.90**
03158 1500	2x4 Keyway Forms, 4 Uses	LF	66.25	ACARA	1.000	0.65	0.00	0.24	0.90

18 04 05 Signage		Unit	Hourly Output	Crew	Factor	Labor	Equip.	Matl.	Total
18 04 0501	**Hazardous Waste Signing**	**EA**				**33.88**	**0.00**	**19.79**	**53.66**
02842 0590	Directional Sign, 12"x 18" with Post Reflectorized, OSHA Standard	EA	2	ULABB	1.000	33.88	0.00	19.79	53.66

18 04 07 Playing Fields		Unit	Hourly Output	Crew	Factor	Labor	Equip.	Matl.	Total

18 04 07	Playing Fields	Unit	Hourly Output	Crew	Factor	Labor	Equip.	Matl.	Total
18 04 0701	**100' x 100' Playground, Complete**	**EA**				**4,192**	**2,100**	**14,671**	**20,964**
02215 2460	Sand Bedding w/Sm FEnd Loader Without Compaction	CY	47.5	CODLB	50.000	1.01	0.97	17.75	986.55
02512 2520	Fin Grade Struc & Slabs, by HandHand Clean f/Concrete Placement	SF	215.3	ULABB	1352.000	0.31	0.00	0.00	425.54
02850 1098	Railroad Work, 8'-6" Wood Cross Tie	EA	11.75	ULABF	52.000	8.04	6.12	34.97	2,554
02863 0160	3' & 5',2 Deck Mdr Plygnd Pltfm Steel, Parks and School Yards	EA	0.1	XLABC	1.000	879.03	0.00	895.00	1,774
02864 0250	4-Way Climber	EA	0.625	XLABC	1.000	140.65	0.00	1,275	1,416
02864 0400	14'(4.3M)L Horiz Monkey Ladder	EA	0.75	ULABF	1.000	125.93	95.83	1,470	1,692
02864 1400	4 Unit Steel See-Saw	EA	0.275	ULABF	1.000	343.45	261.36	1,150	1,755
02864 1900	12'L x 6'H Slide w/Stain Stl Bed(3.7M) Long x (1.8M) High	EA	0.2	ULABF	1.000	472.25	359.38	2,275	3,107
02864 2300	8'(2.4M) High Swing with 8 Seats	EA	0.163	ULABF	1.000	581.23	442.31	1,875	2,899
02864 2900	10'(3M) Diameter Whirler	EA	0.125	ULABF	1.000	755.60	575.00	3,025	4,356
18 04 0702	**10'(3M)L Permanent Bike Rack**	**EA**				**160.77**	**122.34**	**470.00**	**753.11**
02864 0200	10'(3M)L Permanent Bike Rack	EA	0.588	ULABF	1.000	160.77	122.34	470.00	753.11
18 04 0703	**Sport Courts (Tennis & Basketball)**	**EA**				**27,945**	**18,645**	**28,507**	**75,097**
02215 2360	Spread Dumped Fill/Gravel, 6" Layers, without Compaction	SY	448.5	CODTG	1600.000	0.10	0.27	0.00	592.73
02224 7040	Rough Grade Small Area w/Dozer 75 HP	CY	115.1	CODTC	533.333	0.38	0.37	0.00	400.75
02512 2520	Fin Grade Struc & Slabs, by HandHand Clean f/Concrete Placement	SF	215.3	ULABB	14400.000	0.31	0.00	0.00	4,532
02532 0010	Tennis Ct Bit Pvmt, 2-1/2" Thick	SY	23.75	COKCC	1600.000	12.00	9.64	14.45	57,739
02532 1000	Tennis Court Colored Sealer 3 Coats, Acrylic Emulsion	SY	24.38	ULABF	800.000	3.87	2.95	3.72	8,435
02532 1500	Tennis Ct Net Post w/Pulley,Reel	PR	0.425	ULABF	1.000	222.24	169.12	239.00	630.35
02532 1700	42' Nylon Thread Tennis Ct Net Includes Binder	EA	1.25	ULABF	1.000	75.56	57.50	222.00	355.06
02861 0300	Sgl Goal Stl Basketball BackstopIncludes Pole	EA	0.375	ACARB	2.000	231.00	0.00	975.00	2,412
18 04 0704	**Paint Markings on Asph Tennis Ct2 Coats**	**EA**				**344.44**	**0.00**	**64.50**	**408.94**
02532 2000	Paint Markings on Asph Tennis Ct2 Coats	EA	0.113	APTRA	1.000	344.44	0.00	64.50	408.94

18 04 99	Miscellaneous Structures	Unit	Hourly Output	Crew	Factor	Labor	Equip.	Matl.	Total

18 04 Site Development

18 04 99 Miscellaneous Structures

18 04 99 Miscellaneous Structures		Unit	Hourly Output	Crew	Factor	Labor	Equip.	Matl.	Total
18 04 9901	**Covered Bus Stop, 50 SF**	**EA**				**1,033**	**36.45**	**1,060**	**2,129**
02280 3000	Soil Poisoning	SF	562.5	ULABA	50.000	0.06	0.00	0.06	6.04
02611 2201	Graded Crushed Aggregate Base	CY	25	XSABA	0.926	9.51	13.80	22.77	42.68
02871 1820	6' Port Colored Aluminum Bench	EA	1.25	XLABB	1.000	49.45	0.00	330.00	379.45
03170 3000	Slab on Gr Edge Forms, Up to 6"H(15 cm) High, Based on 4 Uses	LF	75	ACARJ	30.000	1.75	0.00	0.38	63.98
03227 0300	Welded Wire Fabric in Slabs, 6 x 6 x #6 42 Lb/CSF, (W 2.9 x W 2.9), < 5	SF	362.5	SIWRB	50.000	0.26	0.00	0.17	21.86
03334 0300	Concrete Curing, Sprayed Membrane Curing Compound	CSF	11.88	ULABB	0.500	5.71	0.00	4.05	4.88
03372 4300	Pour Slab on Grade, < 6", Direct Chute < (15 cm) Place 3,000 PSI Concrete	CY	16.25	ALABE	0.635	10.36	0.39	84.09	60.27
03396 0200	Conc Floor Finishes, Stl Trowel	SF	68.75	ACMAC	50.000	0.53	0.00	0.00	26.28
05008 2200	Bolts, Washers & Nuts	LB	37.5	ACARA	4.000	1.16	0.00	0.76	7.66
05020 1350	1/2"x4-1/4"(10.8cm)L Conc AnchorIncl Drilling,Matl,Installation	EA	5	USKCD	16.000	17.63	0.00	1.14	300.23
05178 0500	Alum Extrusions,Tubular Shapes 1"-10"Rolled/Plate Strl Members	LB	60	SIWSC	169.000	2.48	0.14	2.05	789.28
07411 0900	.032"Tk Corrugated Alum Roofing on Steel Frame, Natural Finish	SF	125	MSHMG	70.000	1.13	0.00	1.43	179.04
07420 1113	.024" Corr Alum Siding Pnl,PaintPainted Finish, Incl Fasteners	SF	200	MSHMG	120.000	0.71	0.00	1.36	247.40

18 05 01 Fine Grading and Soil Preparation

18 05 01 Fine Grading and Soil Preparation		Unit	Hourly Output	Crew	Factor	Labor	Equip.	Matl.	Total
18 05 0101	**Area Preparation, 67% Level & 33% Slope**	**ACRE**				**26.48**	**46.19**	**0.00**	**72.67**
02820 2004	Spread Soil with Machine	MSF	57	CODTR	43.560	0.61	1.06	0.00	72.67

18 05 02 Erosion Control Measures

18 05 02 Erosion Control Measures		Unit	Hourly Output	Crew	Factor	Labor	Equip.	Matl.	Total
18 05 0201	**Security Fence, Temporary**	**LF**				**2.74**	**2.02**	**1.45**	**6.21**
02215 0010	Backfill Trenches by Hand without Compaction	CY	1.625	ULABA	0.004	20.85	0.00	0.00	0.09
02230 0810	Trench, Chain Trencher, Light Soil 1' Wide x 4' Deep	CY	43	CODBB	0.004	0.96	0.14	0.00	0.00
02266 0010	Erosion Control Netting, 4' Wide Jute Mesh, 100 SY (83.6 m2)	SY	312.5	ULABF	0.444	0.30	0.23	0.69	0.54
02712 2101	3' (.9 m) High 12 Gauge Galvanized Steel Fence 2" x 4" Mesh, Posts at 5'	LF	37.5	ULABF	1.000	2.52	1.92	1.15	5.58
18 05 0202	**Rock Cover, Riprap, Light (10 to 100 Lb Pieces)**	**CY**				**3.61**	**2.44**	**18.62**	**24.67**
02261 1008	Rip Rap, 10 Lb to 100 Lb Pieces Random, Dumped from Truck	CY	32	COETF	1.000	3.61	2.44	18.62	24.67
18 05 0203	**Rock Cover, Riprap, Medium (10 to 200 Lb Pieces)**	**CY**				**3.61**	**2.44**	**19.05**	**25.10**
02261 1009	Rip Rap, 10 Lb to 200 Lb Pieces Random, Dumped from Truck	CY	32	COETF	1.000	3.61	2.44	19.05	25.10

18 05 02 Erosion Control Measures

18 05 02 Erosion Control Measures		Unit	Hourly Output	Crew	Factor	Labor	Equip.	Matl.	Total
18 05 0204	**Rock Cover, Riprap, Heavy (25 to 500 Lb Pieces)**	**CY**				**3.61**	**2.44**	**19.50**	**25.55**
02261 1010	Rip Rap, 25 Lb to 500 Lb Pieces Random, Dumped from Truck	CY	32	COETF	1.000	3.61	2.44	19.50	25.55
18 05 0205	**Rock Cover, Bank-run Gravel**	**CY**				**6.28**	**8.57**	**17.82**	**32.68**
02215 2360	Spread Dumped Fill/Gravel, 6" Layers, without Compaction	SY	448.5	CODTG	6.000	0.10	0.27	0.00	2.22
02220 5900	Compact Backfill, by Machine, Sheepsfoot Roller, 6" (15 cm) Lift	CY	117.1	COFCD	1.000	0.41	0.26	0.00	0.67
02234 1115	Haul, 16.5 CY (12.6 m3) Truck, 6 Miles (10 km), 40 MPH (60 km/Hour), 2.1 Cycles/Hour	CY	34.25	COETH	1.000	0.80	3.33	0.00	4.13
02244 1530	Bank Run Gravel Loaded at Pit	CY	12.38	CODEG	1.000	3.88	2.59	17.80	24.27
02976 4930	Watering by Truck	MSF	87.12	COFWI	1.000	0.61	0.76	0.02	1.38
18 05 0206	**Filter Barrier**	**LF**				**1.45**	**0.00**	**0.73**	**2.18**
02266 1120	Silt Fences, Vinyl, 3' High with 7.5' Posts	LF	46.88	ULABB	1.000	1.45	0.00	0.73	2.18

18 05 03 Topsoil and Planting Beds

18 05 03 Topsoil and Planting Beds		Unit	Hourly Output	Crew	Factor	Labor	Equip.	Matl.	Total
18 05 0301	**Topsoil, 6" Lifts, Off-Site**	**CY**				**4.17**	**3.05**	**21.00**	**28.22**
02241 0805	Furnish & Place Imported Topsoil, 6" Deep	CY	11.5	CODLA	1.000	4.17	3.05	21.00	28.22
18 05 0302	**Topsoil, 6" Lifts, On-Site**	**CY**				**1.81**	**6.11**	**0.01**	**7.92**
01418 4736	Soil Density Test, Nuclear Method ASTM D2922-71	EA	2	USKCB	0.000	44.06	0.00	40.00	0.03
02215 2360	Spread Dumped Fill/Gravel, 6" Layers, without Compaction	SY	448.5	CODTG	6.000	0.10	0.27	0.00	2.22
02232 0320	Excavate & Load, 2-3/4 CY Track Loader, Light Material 120 CY/Hour (92 m3)	CY	120	CODLD	1.000	0.40	1.14	0.00	1.54
02234 1115	Haul, 16.5 CY (12.6 m3) Truck, 6 Miles (10 km), 40 MPH (60 km/Hour), 2.1 Cycles/Hour	CY	34.25	COETH	1.000	0.80	3.33	0.00	4.13

18 05 04 Seeding and Sodding

18 05 04 Seeding and Sodding		Unit	Hourly Output	Crew	Factor	Labor	Equip.	Matl.	Total
18 05 0401	**Seeding, 67% Level & 33% Slope, Hydroseeding**	**ACRE**				**77.12**	**96.74**	**402.53**	**576.39**
02810 2002	Hydroseeding, 50 Lb/MSF	LB	164.1	COFWI	240.000	0.32	0.40	1.68	576.39
18 05 0402	**Seeding, Vegetative Cover**	**ACRE**				**70.01**	**55.78**	**3,665**	**3,790**
02830 2005	Large Power Mulcher, Oat Straw, 1" Deep	ACRE	1.521	COELD	1.000	35.18	22.70	1,728	1,786
02932 0300	Mechanical Seeding, 50 Lb/MSY	CSY	193.8	ULABE	48.400	0.17	0.00	40.00	1,944
02976 4930	Watering by Truck	MSF	87.12	COFWI	43.560	0.61	0.76	0.02	60.32
18 05 0404	**Sodding, Average CONUS (Continental U.S.)**	**ACRE**				**6,588**	**1,241**	**18,211**	**26,040**
02933 1200	Sod in Place Level on Prepared Subgrade, by Hand	MSF	1.125	CODEK	43.560	151.24	28.49	418.06	26,040

18 05 04 Seeding and Sodding		Unit	Hourly Output	Crew	Factor	Labor	Equip.	Matl.	Total
18 05 0405	**Sodding, Vegetative Cover**	**ACRE**				**6,615**	**1,274**	**18,212**	**26,100**
02933 1200	Sod in Place Level on Prepared Subgrade, by Hand	MSF	1.125	CODEK	43.560	151.24	28.49	418.06	26,040
02976 4930	Watering by Truck	MSF	87.12	COFWI	43.560	0.61	0.76	0.02	60.32
18 05 0406	**Landscaping, Trees**	**ACRE**				**4,298**	**3,239**	**8,132**	**15,669**
02911 0200	Guying Trees, 2" to 4" Diameter	EA	5.625	ULABF	24.000	16.79	12.78	17.85	1,138
02911 1400	Guying Trees, 6" to 8" Diameter	EA	1	ULABF	12.000	94.45	71.88	54.48	2,650
02947 0230	Trees, 7' to 8' High	EA	1.25	ULABF	36.000	75.56	57.50	194.00	11,774
02951 1900	Mulch, Wood Chips	SF	242.3	ULABA	300.000	0.14	0.00	0.22	107.41
18 05 0407	**Landscaping, Shrubs**	**ACRE**				**694.88**	**528.79**	**1,137**	**2,360**
02947 0440	Regular Shrubs, 15" to 18" Spread, 45 per acre	EA	10	ULABF	45.000	9.44	7.19	14.15	1,385
02947 0450	Regular Shrubs, 24" to 30" Spread, 20 per acre	EA	7	ULABF	20.000	13.49	10.27	25.00	975.21
18 05 0408	**Fertilizer, Hydro Spread**	**ACRE**				**30.40**	**32.22**	**94.09**	**156.71**
02932 7010	35 Lb/1,000 SF	MSF	75	COFWK	43.560	0.70	0.74	2.16	156.71
18 05 0409	**Fertilize, 800 Lbs/Acre, Push Rotary**	**ACRE**				**30.57**	**19.73**	**43.76**	**94.06**
02810 4003	Fertilize 800 Lbs/Acre	ACRE	1.75	COELD	1.000	30.57	19.73	43.76	94.06
18 05 0410	**Fertilize, 800 Lbs/Acre, Spray from Truck**	**ACRE**				**15.20**	**16.11**	**134.37**	**165.68**
02932 7020	18.36 Lb/1,000 SF Sprayed from Truck	MSF	75	COFWK	21.780	0.70	0.74	6.17	165.68
18 05 0411	**Crushed Limestone, pH Adjustment, 800 Lbs/Acre**	**ACRE**				**30.57**	**19.73**	**64.90**	**115.20**
02810 1004	pH Adjustment, 800 Lbs/Acre, Lime	ACRE	1.75	COELD	1.000	30.57	19.73	64.90	115.20
18 05 0412	**Purchase & Spread Dry Granular Limestone for pH Control**	**SY**				**0.01**	**0.00**	**0.03**	**0.04**
02932 7022	Limestone	ACRE	1.75	ULABE	0.001	19.36	0.00	61.00	0.04
18 05 0413	**Watering with 3,000-Gallon Tank Truck, per Pass**	**ACRE**				**26.37**	**33.08**	**4.23**	**63.67**
02555 6102	Process Water, Supplied by Water Line	KGAL	0	N/A	1.000	0.00	0.00	3.35	3.35
02976 4930	Watering by Truck	MSF	87.12	COFWI	43.560	0.61	0.76	0.02	60.32
18 05 0415	**Mowing**	**ACRE**				**20.54**	**11.04**	**0.00**	**31.57**
02973 4050	Mowing	MSF	147	COELC	43.560	0.47	0.25	0.00	31.57

18 05 07 Irrigation Systems		Unit	Hourly Output	Crew	Factor	Labor	Equip.	Matl.	Total
18 05 0701	**Full Circle Sprinkler Head, 30' Diameter**	**EA**				**37.98**	**0.00**	**20.41**	**58.39**
02737 1003	Rotary Impact Sprinkler Head with PVC Riser	EA	2.5	MPLUB	1.000	37.98	0.00	20.41	58.39
18 05 0702	**Semicircle Sprinkler Head, 20' Diameter**	**EA**				**18.99**	**0.00**	**4.85**	**23.85**
02737 1002	Impact Sprinkler Head with PVC Riser	EA	5	MPLUB	1.000	18.99	0.00	4.85	23.85

18 05 07	Irrigation Systems	Unit	Hourly Output	Crew	Factor	Labor	Equip.	Matl.	Total
18 05 0703	**Shrub/Tree Sprinkler Head**	**EA**				**18.99**	**0.00**	**1.35**	**20.34**
02737 1004	Shrub Sprinkler Head with PVC Riser	EA	5	MPLUB	1.000	18.99	0.00	1.35	20.34
18 05 0704	**Full Circle Sprinkler Head, 80' Diameter**	**EA**				**32.39**	**0.00**	**308.60**	**340.98**
02441 1064	Iron Case Rubber Cover	EA	2.747	MPLUE	1.000	32.39	0.00	308.60	340.98
18 05 0705	**Control Box**	**EA**				**288.80**	**0.00**	**1,311**	**1,600**
02737 1101	Control Box	EA	0.375	MPLUU	1.000	288.80	0.00	1,311	1,600
18 05 0706	**2 1/2", Cast-iron Body, Gate Valve**	**EA**				**36.48**	**0.00**	**458.33**	**494.81**
15194 3508	2-1/2" Cast Iron Gate Valve	EA	2.75	XPLUC	1.000	36.48	0.00	458.33	494.81
18 05 0707	**4" Reducer**	**EA**				**33.44**	**0.00**	**11.67**	**45.11**
16111 3991	PVC Conduit Schedule 40 4" x 3" Reducer	EA	3	XPLUC	1.000	33.44	0.00	11.67	45.11
18 05 0710	**Testing & Inspection of Sprinkler System**	**LS**				**412.50**	**0.00**	**0.00**	**412.50**
15048 1319	Test & Inspect, Labor Cost	EA	0.125	MPLUA	1.000	412.50	0.00	0.00	412.50
18 05 0711	**Automated Pop-Up Sprinkler System**	**ACRE**				**3,670**	**159.26**	**960.22**	**4,789**
02228 0306	Trench, 1/2 CY Hydraulic Excavator, Loose Rock 39 CY/Hour (29 m3)/Hour	CY	38.38	CODEA	77.000	1.63	1.82	0.00	265.51
02737 1003	Rotary Impact Sprinkler Head with PVC Riser	EA	2.5	MPLUB	14.000	37.98	0.00	20.41	817.52
15155 1910	2" Schedule 40 PVC Pressure Pipe Socket Weld	LF	33.38	MPLUE	1040.000	2.67	0.00	0.63	3,431
15157 2810	2" 90 Degree Elbow, PVC Schedule 40 Socket Weld	EA	4.625	MPLUE	3.000	19.24	0.00	1.54	62.34
15157 3230	2" Tee, PVC Schedule 40, Socket Weld	EA	3.125	MPLUH	6.000	30.39	3.23	1.90	213.07

18 06 01	Bridges	Unit	Hourly Output	Crew	Factor	Labor	Equip.	Matl.	Total
18 06 0101	**Form, Straight Beam Bottoms**	**SF**				**5.36**	**0.00**	**1.31**	**6.67**
03138 3650	Beam Forms, Bottom to 30" W, 4 Uses	SF	30	ACARL	1.000	5.36	0.00	1.31	6.67
18 06 0102	**Form, Straight Beam Sides**	**SF**				**3.43**	**0.00**	**1.31**	**4.74**
03138 4150	Beam Forms, Vertical Sides, 4 Uses	SF	46.88	ACARL	1.000	3.43	0.00	1.31	4.74
18 06 0103	**Rebar, CIP Beams**	**LB**				**0.31**	**0.12**	**0.32**	**0.76**
03210 1007	Rebar #3-#7	TON	0.288	SIWRF	0.001	629.08	250.00	647.24	0.76
18 06 0104	**Pour & Cure, CIP Beams**	**CY**				**30.93**	**9.44**	**89.20**	**129.57**
03372 0050	Pour Concrete Beam, Small, Concrete Pump	CY	10	ALABI	1.000	22.97	9.44	84.22	116.63
07174 0100	Silicone Retarder, 1 Coat on Concrete, Sprayed	CSF	1.875	ARFCD	0.400	19.90	0.00	12.47	12.95
18 06 0105	**Form, Beam Bracing Diaphragms**	**SF**				**3.41**	**0.00**	**0.87**	**4.29**
03138 2150	12" Beam/Gird Form, Interior Beam, 4 Uses	SF	47.13	ACARL	1.000	3.41	0.00	0.87	4.29
18 06 0106	**Rebar, Bracing Diaphragms**	**LB**				**0.31**	**0.12**	**0.32**	**0.76**
03210 1007	Rebar #3-#7	TON	0.288	SIWRF	0.001	629.08	250.00	647.24	0.76

18 06 01 Bridges		Unit	Hourly Output	Crew	Factor	Labor	Equip.	Matl.	Total
18 06 0107	**Pour & Cure, Bracing Diaphragms**	**CY**				**45.06**	**14.56**	**92.32**	**151.94**
03372 0250	Pour Concrete Beam, Large, Crane & Bucket	CY	8.125	ALABG	1.000	32.13	14.56	84.22	130.90
07174 0100	Silicone Retarder, 1 Coat on Concrete, Sprayed	CSF	1.875	ARFCD	0.650	19.90	0.00	12.47	21.04
18 06 0108	**Form Deck, 3 Uses**	**SF**				**3.22**	**0.00**	**0.44**	**3.65**
03150 7101	Elevated Slab Edge 7"-12", 4 Uses	SF	50	ACARL	1.000	3.22	0.00	0.44	3.65
18 06 0109	**Rebar, Bridge Deck**	**LB**				**0.26**	**0.10**	**0.31**	**0.67**
03210 1008	Rebar #8-#14	TON	0.35	SIWRF	0.001	516.74	205.36	616.05	0.67
18 06 0110	**Pour & Cure, Deck**	**CY**				**41.16**	**7.28**	**89.20**	**137.64**
03372 1550	Pour Elevated Slab, >= 6", Crane & Bucket >= (15 cm) Place 3,000 PSI Concrete	CY	16.25	ALABG	1.000	16.06	7.28	84.22	107.56
03396 0150	Concrete Floor Finishes, Broom	SF	84.38	ACMAC	40.000	0.43	0.00	0.00	17.13
07174 0100	Silicone Retarder, 1 Coat on Concrete, Sprayed	CSF	1.875	ARFCD	0.400	19.90	0.00	12.47	12.95
18 06 0111	**Bush Hammer Finish**	**SF**				**1.96**	**0.00**	**0.06**	**2.01**
03398 0300	Wall Finish, Bush Hammer Green Concrete	SF	21.25	ACMAA	1.000	1.96	0.00	0.06	2.01
18 06 0112	**Armor Joints**	**LF**				**1.94**	**0.15**	**143.85**	**145.94**
03413 1955	Armor Joint	LF	56.25	SIWWC	1.000	1.94	0.15	143.85	145.94
18 06 0113	**Beam Bearing Pads (Elastomeric)**	**SF**				**43.36**	**0.00**	**21.76**	**65.12**
05175 0600	Beam Bearing Pad Elastomeric	SF	0.781	ULABA	1.000	43.36	0.00	21.76	65.12
18 06 0120	**Precast I Beam, Type I, 1' 4" x 2' 4" Deep**	**LF**				**16.84**	**13.26**	**115.70**	**145.81**
03413 1951	P/C P/S I Beam AASH to 28" Deep	LF	21.25	SIWST	1.000	16.84	13.26	115.70	145.81
18 06 0125	**Precast Box Beam, 3' x 2' 9" Deep**	**LF**				**65.62**	**51.67**	**518.55**	**635.84**
03413 1952	P/C P/S Box Girder 36" x 33"	SF	45	SIWST	8.250	7.95	6.26	62.85	635.84
18 06 0126	**Precast Box Beam, 4' x 2' 9" Deep**	**LF**				**75.00**	**59.05**	**658.99**	**793.04**
03413 1953	P/C P/S Box Girder 48" x 33"	SF	52.5	SIWST	11.000	6.82	5.37	59.91	793.04
18 06 0127	**Pour & Cure Concrete Deck Topping**	**CY**				**58.99**	**9.96**	**91.57**	**160.53**
03372 1450	Pour Elevated Slab, < 6", Crane & Bucket < (15 cm) Place 3,000 PSI Concrete	CY	11.88	ALABG	1.000	21.98	9.96	84.22	116.16
03396 0150	Concrete Floor Finishes, Broom	SF	84.38	ACMAC	59.000	0.43	0.00	0.00	25.27
07174 0100	Silicone Retarder, 1 Coat on Concrete, Sprayed	CSF	1.875	ARFCD	0.590	19.90	0.00	12.47	19.10
18 06 0128	**Shear Key, High Strength, Nonshrink Grout**	**CY**				**23.20**	**10.51**	**80.60**	**114.32**
03354 3001	High Strength Concrete 5,000 PSI	CY	11.25	ALABG	1.000	23.20	10.51	80.60	114.32
18 06 0129	**1" Bituminous Fiber Expansion Joint**	**LF**				**1.16**	**0.00**	**1.82**	**2.97**
03132 2050	Premolded Asphalt Expansion Joint, 1"x 12" Premolded in Slabs	LF	37.5	ACARA	1.000	1.16	0.00	1.82	2.97

18 06 01	Bridges	Unit	Hourly Output	Crew	Factor	Labor	Equip.	Matl.	Total
18 06 0130	**1/2" x 1" Poured Rubber Joint**	**LF**				**0.60**	**0.00**	**0.73**	**1.33**
03132 0900	Rubberized Asphalt Expansion Joint, 1/2"x 1" Poured, Hot or Cold Applied	LF	56.25	ULABA	1.000	0.60	0.00	0.73	1.33
18 06 0131	**Structural Steel Beams, Rolled Shapes, A36, 2 - 6" Studs @ 3"**	**TON**				**1,035**	**225.45**	**1,420**	**2,680**
05060 0010	Studs 3/4" X 3"	EA	6.25	ACARB	54.000	13.86	0.00	0.35	767.09
05120 1007	A-36 Steel, Rolled Shapes	TON	1.25	SIWST	1.000	286.35	225.45	1,402	1,913
18 06 0132	**Structural Steel Beam Bracing**	**TON**				**354.71**	**140.96**	**1,174**	**1,670**
05120 3300	4"-6"Strl Tube,Light Sq SectionsA-36 Misc Steel Items	TON	0.794	SIWSM	1.000	354.71	140.96	1,174	1,670
18 06 0133	**Rocker/Fixed Bearing Assembly for Steel Beams, Complete**	**LB**				**0.14**	**0.08**	**5.00**	**5.22**
05120 2805	Rocker/Fixed Bearing Assembly	LB	930	SIWSE	1.000	0.14	0.08	5.00	5.22
18 06 0135	**Treated Timber Piles/Abutments, 12" Butt, 8" Tip**	**LF**				**4.28**	**4.86**	**9.56**	**18.70**
02318 1007	Treated Timber Piles, 12" Butt - 8" Point	LF	60	CPIDC	1.000	4.28	4.86	9.56	18.70
18 06 0136	**Treated Timber Piles (Bentcaps), 12" Butt, 8" Tip**	**LF**				**4.28**	**4.86**	**9.56**	**18.70**
02318 1007	Treated Timber Piles, 12" Butt - 8" Point	LF	60	CPIDC	1.000	4.28	4.86	9.56	18.70
18 06 0137	**Heavy Construction Timber, Treated with Misc Blocking**	**BF**				**0.69**	**0.00**	**2.59**	**3.28**
06103 2620	Blocking to Wood, 2x4 - 2x8 Southern Yellow Pine	BF	25	ACARC	0.100	1.73	0.00	0.68	0.24
06132 3002	Heavy Beam, 8x12 - 8x16 Douglas Fir Mill Frame Structure, Heavy Timber Construction	BF	150	ACARD	1.000	0.52	0.00	2.34	2.85
06311 1001	Creosote Treated, Dosage to 8 Lb/CF Preservative Treated Lumber	MBF	0	N/A	0.001	0.00	0.00	165.39	0.18
18 06 0138	**Bolts, Washers, & Nuts, Installed**	**LB**				**1.27**	**0.00**	**0.84**	**2.11**
05008 2200	Bolts, Washers & Nuts	LB	37.5	ACARA	1.100	1.16	0.00	0.76	2.11
18 06 0139	**Spikes & Nails, Installed**	**LB**				**7.97**	**0.00**	**6.95**	**14.92**
05120 2807	Spikes & Nails Installed	LB	6.25	ACARA	1.150	6.93	0.00	6.05	14.92
18 06 0140	**Shear Connectors, Dowels & Shear Plates**	**LB**				**3.64**	**0.00**	**6.23**	**9.87**
05120 2806	Shear Connector Dowels & Shear Plates, 5% Waste	LB	12.5	ACARA	1.050	3.46	0.00	5.93	9.87
18 06 0141	**Heavy Construction Timber (Decks), Treated with Misc Blocking**	**BF**				**1.51**	**0.57**	**1.41**	**3.49**
06103 2620	Blocking to Wood, 2x4 - 2x8 Southern Yellow Pine	BF	25	ACARC	0.100	1.73	0.00	0.68	0.24
06132 1003	Heavy Timber Beams	BF	125	ACARS	1.000	1.33	0.57	1.16	3.07
06311 1001	Creosote Treated, Dosage to 8 Lb/CF Preservative Treated Lumber	MBF	0	N/A	0.001	0.00	0.00	165.39	0.18

18 06 01	Bridges	Unit	Hourly Output	Crew	Factor	Labor	Equip.	Matl.	Total
18 06 0142	**Heavy Construction Timber (Beams), Treated with Misc Blocking**	**BF**				**0.69**	**0.00**	**2.59**	**3.28**
06103 2620	Blocking to Wood, 2x4 - 2x8 Southern Yellow Pine	BF	25	ACARC	0.100	1.73	0.00	0.68	0.24
06132 3002	Heavy Beam, 8x12 - 8x16 Douglas Fir Mill Frame Structure, Heavy Timber Construction	BF	150	ACARD	1.000	0.52	0.00	2.34	2.85
06311 1001	Creosote Treated, Dosage to 8 Lb/CF Preservative Treated Lumber	MBF	0	N/A	0.001	0.00	0.00	165.39	0.18
18 06 0145	**Large Spread Footing, Edge Form & Strip, 4 Uses**	**SF**				**2.45**	**0.00**	**0.64**	**3.09**
03158 5150	Column Spread Footing Forms, 4 Uses Plywood Forms, Form & Strip with Accessories	SF	53.75	ACARJ	1.000	2.45	0.00	0.64	3.09
18 06 0146	**Footing, Rebar**	**LB**				**0.31**	**0.12**	**0.32**	**0.76**
03210 1007	Rebar #3-#7	TON	0.288	SIWRF	0.001	629.08	250.00	647.24	0.76
18 06 0147	**Large Spread Footing, Pour & Cure Concrete**	**CY**				**23.35**	**7.55**	**87.33**	**118.24**
03372 2450	Pour Spread Footings, < 5 CY, Concrete Pump < (5 m3) Place 3,000 PSI Concrete Foundations	CY	12.5	ALABI	1.000	18.38	7.55	84.22	110.14
07174 0100	Silicone Retarder, 1 Coat on Concrete, Sprayed	CSF	1.875	ARFCD	0.250	19.90	0.00	12.47	8.09
18 06 0148	**Column Forms, Fiberglass, 24" Round**	**LF**				**6.42**	**0.00**	**10.76**	**17.18**
03142 1800	24" Diameter Round Fiber Forms, 1 Use	LF	16.25	ACARI	1.000	6.42	0.00	10.76	17.18
18 06 0149	**Columns, Rebar, Spiral**	**LB**				**0.33**	**0.13**	**0.52**	**0.98**
03217 0330	Spiral Reinforced 24"-36" Diameter	TON	0.275	SIWRF	0.001	657.67	261.36	1,046	0.98
18 06 0150	**Columns, Pour & Finish Concrete**	**CY**				**117.42**	**3.15**	**91.82**	**212.38**
03372 1000	Pour Concrete Columns, 36", Concrete Pump (91 cm), Square or Round, 3,000 PSI	CY	30	ALABI	1.000	7.66	3.15	84.22	95.02
03398 0150	Wall Finish, Carborundum Wet Rub	SF	21.88	ACMAA	54.000	1.90	0.00	0.06	105.71
07174 0100	Silicone Retarder, 1 Coat on Concrete, Sprayed	CSF	1.875	ARFCD	0.360	19.90	0.00	12.47	11.65
18 06 0151	**Form, Bentcap Bottom**	**SF**				**5.36**	**0.00**	**1.31**	**6.67**
03138 3650	Beam Forms, Bottom to 30" W, 4 Uses	SF	30	ACARL	1.000	5.36	0.00	1.31	6.67
18 06 0152	**Form, Bentcap Sides**	**SF**				**3.03**	**0.00**	**1.31**	**4.34**
03138 4650	Beam Forms, Sloped Sides, 4 Uses Up to 36" High	SF	53.13	ACARL	1.000	3.03	0.00	1.31	4.34
18 06 0153	**Bentcap, Rebar**	**LB**				**0.26**	**0.10**	**0.31**	**0.67**
03210 1008	Rebar #8-#14	TON	0.35	SIWRF	0.001	516.74	205.36	616.05	0.67
18 06 0154	**Bentcap, Pour & Cure**	**CY**				**36.51**	**14.56**	**86.96**	**138.03**
03372 0250	Pour Concrete Beam, Large, Crane & Bucket	CY	8.125	ALABG	1.000	32.13	14.56	84.22	130.90
07174 0100	Silicone Retarder, 1 Coat on Concrete, Sprayed	CSF	1.875	ARFCD	0.220	19.90	0.00	12.47	7.12
18 06 0155	**Edgeforms, 10" Approach Slab, 2 Uses**	**LF**				**4.21**	**0.00**	**1.00**	**5.21**
03150 7080	Elevated Slab Edge Form, 7-12", 2 Uses	LF	24.75	ACARI	1.000	4.21	0.00	1.00	5.21

18 06 01	Bridges	Unit	Hourly Output	Crew	Factor	Labor	Equip.	Matl.	Total
18 06 0156	**Welded Wire Mesh, Approach Slab, 6 x 6 x 4/4, 58 Lb/SQ**	**SQ**				**29.83**	**0.00**	**26.60**	**56.42**
03227 0400	Welded Wire Fabric in Slabs, 6 x 6 x #4 58 Lb/CSF, (W 4.0 x W 4.0), < 5	SF	337.5	SIWRB	105.000	0.28	0.00	0.25	56.42
18 06 0157	**Pour & Cure, Approach Slab**	**CY**				**34.48**	**6.53**	**88.21**	**129.21**
03372 4700	Pour Slab on Grade, >= 6", Crane & Bucket >= (15 cm) Place 3,000 PSI Concrete	CY	18.13	ALABG	1.000	14.40	6.53	84.22	105.15
03396 0150	Concrete Floor Finishes, Broom	SF	84.38	ACMAC	32.000	0.43	0.00	0.00	13.70
07174 0100	Silicone Retarder, 1 Coat on Concrete, Sprayed	CSF	1.875	ARFCD	0.320	19.90	0.00	12.47	10.36
18 06 0158	**Parapet, Form**	**SF**				**2.55**	**0.00**	**0.89**	**3.43**
03182 2150	Wall Forms, Up to 8' High Wall, 4 Uses (Below Grade) Plywood Forms, Form & Strip with Accessories	SF	63.13	ACARL	1.000	2.55	0.00	0.89	3.43
18 06 0159	**Parapet, Rebar**	**LB**				**0.25**	**0.00**	**0.30**	**0.55**
03217 0700	Grade 60 Rebar, Beam, Column, Wall, #3-#6	TON	0.313	SIWRC	0.001	492.16	0.00	609.86	0.55
18 06 0160	**Parapet, Pour & Cure**	**CY**				**42.22**	**11.83**	**94.32**	**148.37**
03372 5000	Pour Concrete Walls, 8" Thick, Crane & Bucket (21 cm) 3,000 PSI Concrete	CY	10	ALABG	1.000	26.11	11.83	84.22	122.15
07174 0100	Silicone Retarder, 1 Coat on Concrete, Sprayed	CSF	1.875	ARFCD	0.810	19.90	0.00	12.47	26.22

18 06 02	Railroad Spur	Unit	Hourly Output	Crew	Factor	Labor	Equip.	Matl.	Total
18 06 0201	**Ballast**	**CY**				**21.83**	**16.61**	**16.30**	**54.74**
02422 2160	Railroad Work, Ballast	TN	5.625	ULABF	1.300	16.79	12.78	12.54	54.74
18 06 0202	**Gravel (90%) & Sand Base (10%); with Calcium Chloride 3/4 - 1 Lb/CY**	**CY**				**2.03**	**1.78**	**23.39**	**27.20**
02244 1505	Washed & Graded Sand, Loaded at Pit	CY	12.38	CODEG	0.100	3.88	2.59	26.50	3.30
02611 2205	Graded Crushed Gravel Base	CY	90.88	XSABD	0.900	1.83	1.69	22.77	23.66
03318 0420	Admixture Calcium Chloride 100 Lb Bag	BAG	0	N/A	0.010	0.00	0.00	24.66	0.25
18 06 0210	**110 Lb #8 Turnout to New Track**	**EA**				**4,343**	**547.14**	**23,600**	**28,490**
02850 1155	Railroad Work, 100 Lb No. 8 Turnout In-place, Complete, ARA-A and ARAE	EA	0.035	UOEHD	1.000	4,343	547.14	23,600	28,490
18 06 0211	**115 Lb #8 Turnout to New Track**	**EA**				**4,343**	**547.14**	**28,057**	**32,947**
02850 1156	Railroad Work, 115 Lb No. 8 Turnout In-place, Complete, ARA-A and ARAE	EA	0.035	UOEHD	1.000	4,343	547.14	28,057	32,947
18 06 0212	**132 Lb #8 Turnout to New Track**	**EA**				**4,343**	**547.14**	**45,888**	**50,778**
02850 1157	Railroad Work, 132 Lb No. 8 Turnout In-place, Complete, ARA-A and ARAE	EA	0.035	UOEHD	1.000	4,343	547.14	45,888	50,778
18 06 0215	**110 Lb #8 Turnout to Existing Track**	**EA**				**4,343**	**547.14**	**22,408**	**27,298**
02850 1158	Railroad Work, 110 Lb No. 8 Turnout Relay In-place, Complete, ARA-A and ARAE	EA	0.035	UOEHD	1.000	4,343	547.14	22,408	27,298

18 06 02 Railroad Spur		Unit	Hourly Output	Crew	Factor	Labor	Equip.	Matl.	Total
18 06 0216	**115 Lb #8 Turnout to Existing Track**	**EA**				**4,343**	**547.14**	**27,342**	**32,232**
02850 1159	Railroad Work, 115 Lb No. 8 Turnout Relay In-place, Complete, ARA-A and ARAE	EA	0.035	UOEHD	1.000	4,343	547.14	27,342	32,232
18 06 0217	**132 Lb #8 Turnout to Existing Track**	**EA**				**4,343**	**547.14**	**30,601**	**35,491**
02850 1166	Railroad Work, 132 Lb No. 8 Turnout Relay In-place, Complete, ARA-A and ARAE	EA	0.035	UOEHD	1.000	4,343	547.14	30,601	35,491
18 06 0220	**Railroad Car Wheel Stops**	**EA**				**37.11**	**19.03**	**282.50**	**338.63**
02422 2400	Railroad Car Wheel Stops	PR	1.75	CODEF	0.500	74.21	38.05	565.00	338.63
18 06 0221	**Railroad Car Bumpers, Standard**	**EA**				**519.50**	**266.38**	**2,525**	**3,311**
02422 2190	Railroad Bumpers, Standard	EA	0.25	CODEF	1.000	519.50	266.38	2,525	3,311
18 06 0222	**Railroad Car Bumpers, Heavy Duty**	**EA**				**519.50**	**266.38**	**5,450**	**6,236**
02422 2195	Railroad Bumpers, Heavy Duty	EA	0.25	CODEF	1.000	519.50	266.38	5,450	6,236
18 06 0223	**Hand-throw Derailer with Standard Timbers, Open Stand & Target**	**EA**				**188.91**	**96.86**	**865.00**	**1,151**
02422 0300	Hand Throw Derailer with Standard Timbers	EA	0.688	CODEF	1.000	188.91	96.86	865.00	1,151
18 06 0224	**Railroad Crossing Precast Concrete Inserts, 12' Wide**	**EA**				**199.66**	**67.65**	**2,075**	**2,342**
02422 2700	Precast Concrete Inserts, Railroad Crossing, 12' Road, 6" Thick, Standard Gauge	EA	1.063	SIWSH	1.000	199.66	67.65	2,075	2,342
18 06 0225	**Railroad Crossing Molded Rubber with Headers, 12' Wide**	**EA**				**199.66**	**67.65**	**3,800**	**4,067**
02422 2800	Molded Rubber Inserts, Railroad Crossing, 12' Road 3' x 8'-6" Sections with Headers	EA	1.063	SIWSH	1.000	199.66	67.65	3,800	4,067
18 06 0230	**New 110 Lb Track**	**LF**				**2.15**	**0.27**	**13.59**	**16.01**
02422 3030	Railroad Work, 110 Lb Prime Rail ARA-A and ARAE	LF	141.3	UOEHD	2.000	1.08	0.14	6.79	16.01
18 06 0231	**New 115 Lb Track**	**LF**				**2.15**	**0.27**	**27.18**	**29.60**
02422 3040	Railroad Work, 115 Lb Prime Rail ARA-A and ARAE	LF	141.3	UOEHD	2.000	1.08	0.14	13.59	29.60
18 06 0232	**New 132 Lb Track**	**LF**				**2.15**	**0.27**	**30.63**	**33.05**
02422 3050	Railroad Work, 132 Lb Prime Rail ARA-A and ARAE	LF	141.3	UOEHD	2.000	1.08	0.14	15.31	33.05
18 06 0240	**110 Lb Angle Bar with Bolts and Washers**	**PAIR**				**9.68**	**7.36**	**33.91**	**50.95**
02422 2110	Railroad Work, 1" Track Bolts & Nuts	EA	109.6	ULABF	6.000	0.86	0.66	1.98	20.98
02422 2120	Railroad Work, 1" Track Lockwashers	EA	207.6	ULABF	6.000	0.45	0.35	0.22	6.13
02422 3130	Railroad Work, 110 Lb Angle Bar ARA-A and ARAE	PR	53.13	ULABF	1.000	1.78	1.35	20.71	23.84

18 06 02	Railroad Spur	Unit	Hourly Output	Crew	Factor	Labor	Equip.	Matl.	Total
18 06 0241	**115 Lb Angle Bar with Bolts and Washers**	**PAIR**				**9.68**	**7.36**	**106.20**	**123.24**
02422 2110	Railroad Work, 1" Track Bolts & Nuts	EA	109.6	ULABF	6.000	0.86	0.66	1.98	20.98
02422 2120	Railroad Work, 1" Track Lockwashers	EA	207.6	ULABF	6.000	0.45	0.35	0.22	6.13
02422 3140	Railroad Work, 115 Lb Angle Bar ARA-A and ARAE	PR	53.13	ULABF	1.000	1.78	1.35	93.00	96.13
18 06 0242	**132 Lb Angle Bar with Bolts and Washers**	**PAIR**				**9.68**	**7.36**	**114.20**	**131.24**
02422 2110	Railroad Work, 1" Track Bolts & Nuts	EA	109.6	ULABF	6.000	0.86	0.66	1.98	20.98
02422 2120	Railroad Work, 1" Track Lockwashers	EA	207.6	ULABF	6.000	0.45	0.35	0.22	6.13
02422 3150	Railroad Work, 132 Lb Angle Bar ARA-A and ARAE	PR	53.13	ULABF	1.000	1.78	1.35	101.00	104.13
18 06 0250	**Crossties with 110 Lb Tie Plates and Spikes**	**EA**				**24.87**	**18.92**	**44.93**	**88.72**
02422 2130	Railroad Work, 6" Track Spikes	EA	26.25	ULABF	4.000	3.60	2.74	0.46	27.18
02422 3230	Railroad Work, 110 Lb Tie Plate ARA-A and ARAE	EA	77.5	ULABF	2.000	1.22	0.93	4.06	12.41
02850 1098	Railroad Work, 8'-6" Wood Cross Tie	EA	11.75	ULABF	1.000	8.04	6.12	34.97	49.12
18 06 0251	**Crossties with 115 Lb Tie Plates and Spikes**	**EA**				**24.87**	**18.92**	**60.81**	**104.60**
02422 2130	Railroad Work, 6" Track Spikes	EA	26.25	ULABF	4.000	3.60	2.74	0.46	27.18
02422 3240	Railroad Work, 115 Lb Tie Plate ARA-A and ARAE	EA	77.5	ULABF	2.000	1.22	0.93	12.00	28.29
02850 1098	Railroad Work, 8'-6" Wood Cross Tie	EA	11.75	ULABF	1.000	8.04	6.12	34.97	49.12
18 06 0252	**Crossties with 132 Lb Tie Plates and Spikes**	**EA**				**24.87**	**18.92**	**45.90**	**89.69**
02422 2130	Railroad Work, 6" Track Spikes	EA	26.25	ULABF	4.000	3.60	2.74	0.46	27.18
02422 3250	Railroad Work, 132 Lb Tie Plate ARA-A and ARAE	EA	77.5	ULABF	2.000	1.22	0.93	4.55	13.38
02850 1098	Railroad Work, 8'-6" Wood Cross Tie	EA	11.75	ULABF	1.000	8.04	6.12	34.97	49.12

19 01 02	Potable Water Distribution	Unit	Hourly Output	Crew	Factor	Labor	Equip.	Matl.	Total
19 01 0201	**3/4", Class 200, PVC Piping**	**LF**				**3.23**	**0.16**	**0.13**	**3.52**
02667 4090	3/4" Class 200 PVC	LF	37.5	CODBC	1.000	3.23	0.16	0.13	3.52
19 01 0202	**1", Class 200, PVC Piping**	**LF**				**3.23**	**0.16**	**0.15**	**3.53**
02667 4100	1" Class 200 PVC	LF	37.5	CODBC	1.000	3.23	0.16	0.15	3.53
19 01 0203	**1 1/2", Class 200, PVC Piping**	**LF**				**3.23**	**0.16**	**0.30**	**3.69**
02667 4110	1-1/2" Class 200 PVC	LF	37.5	CODBC	1.000	3.23	0.16	0.30	3.69
19 01 0204	**2", Class 200, PVC Piping**	**LF**				**3.72**	**0.18**	**0.46**	**4.37**
02667 4120	2" Class 200 PVC	LF	32.5	CODBC	1.000	3.72	0.18	0.46	4.37
19 01 0205	**2 1/2", Class 200, PVC Piping**	**LF**				**3.72**	**0.18**	**0.67**	**4.57**
02667 4130	2-1/2" Class 200 PVC	LF	32.5	CODBC	1.000	3.72	0.18	0.67	4.57
19 01 0206	**3", Class 200, PVC Piping**	**LF**				**4.84**	**0.24**	**0.97**	**6.05**
02667 4140	3" Class 200 PVC	LF	25	CODBC	1.000	4.84	0.24	0.97	6.05
19 01 0207	**4", Class 200, PVC Piping**	**LF**				**4.84**	**0.24**	**1.65**	**6.73**
02667 4150	4" Class 200 PVC	LF	25	CODBC	1.000	4.84	0.24	1.65	6.73
19 01 0208	**6", Class 200, PVC Piping**	**LF**				**5.38**	**0.27**	**3.35**	**8.99**
02667 4160	6" Class 200 PVC	LF	22.5	CODBC	1.000	5.38	0.27	3.35	8.99
19 01 0210	**4", Class 150, PVC Piping**	**LF**				**4.84**	**0.24**	**1.82**	**6.90**
02464 2206	4" Class 150 PVC	LF	25	CODBC	1.000	4.84	0.24	1.82	6.90
19 01 0211	**6", Class 150, PVC Piping**	**LF**				**5.38**	**0.27**	**3.38**	**9.02**
02464 2207	6" Class 150 PVC	LF	22.5	CODBC	1.000	5.38	0.27	3.38	9.02
19 01 0212	**8", Class 150, PVC Piping**	**LF**				**6.05**	**0.30**	**5.99**	**12.34**
02464 2208	8" Class 150 PVC	LF	20	CODBC	1.000	6.05	0.30	5.99	12.34
19 01 0213	**10", Class 150, PVC Piping**	**LF**				**6.91**	**0.34**	**9.36**	**16.62**
02464 2209	10" Class 150 PVC	LF	17.5	CODBC	1.000	6.91	0.34	9.36	16.62
19 01 0214	**12", Class 150, PVC Piping**	**LF**				**8.06**	**0.40**	**13.58**	**22.05**
02464 2211	12" Class 150 PVC	LF	15	CODBC	1.000	8.06	0.40	13.58	22.05
19 01 0220	**4", Class 250, Mechanical Joint, Ductile Iron Pipe**	**LF**				**3.54**	**0.67**	**8.92**	**13.13**
02455 1014	4" Mechanical Joint, Ductile Iron Pipe	LF	48	CODEK	1.000	3.54	0.67	8.92	13.13
19 01 0221	**6", Class 250, Mechanical Joint, Ductile Iron Pipe**	**LF**				**3.87**	**0.73**	**10.31**	**14.90**
02455 1015	6" Mechanical Joint, Ductile Iron Pipe	LF	44	CODEK	1.000	3.87	0.73	10.31	14.90

19 01 02	Potable Water Distribution	Unit	Hourly Output	Crew	Factor	Labor	Equip.	Matl.	Total
19 01 0222	**8", Class 250, Mechanical Joint, Ductile Iron Pipe**	**LF**				**3.87**	**0.73**	**13.03**	**17.63**
02455 1016	8" Mechanical Joint, Ductile Iron Pipe	LF	44	CODEK	1.000	3.87	0.73	13.03	17.63
19 01 0223	**10", Class 250, Mechanical Joint, Ductile Iron Pipe**	**LF**				**4.73**	**0.89**	**19.60**	**25.22**
02455 1017	10" Mechanical Joint, Ductile Iron Pipe	LF	36	CODEK	1.000	4.73	0.89	19.60	25.22
19 01 0224	**12", Class 250, Mechanical Joint, Ductile Iron Pipe**	**LF**				**4.73**	**0.89**	**22.05**	**27.67**
02455 1018	12" Mechanical Joint, Ductile Iron Pipe	LF	36	CODEK	1.000	4.73	0.89	22.05	27.67
19 01 0225	**14", Class 250, Mechanical Joint, Ductile Iron Pipe**	**LF**				**5.00**	**0.94**	**30.81**	**36.76**
02455 1019	14" Mechanical Joint, Ductile Iron Pipe	LF	34	CODEK	1.000	5.00	0.94	30.81	36.76
19 01 0226	**16", Class 250, Mechanical Joint, Ductile Iron Pipe**	**LF**				**5.00**	**0.94**	**35.40**	**41.35**
02455 1021	16" Mechanical Joint, Ductile Iron Pipe	LF	34	CODEK	1.000	5.00	0.94	35.40	41.35
19 01 0227	**18", Class 250, Mechanical Joint, Ductile Iron Pipe**	**LF**				**6.08**	**1.14**	**36.82**	**44.04**
02455 1022	18" Mechanical Joint, Ductile Iron Pipe	LF	28	CODEK	1.000	6.08	1.14	36.82	44.04
19 01 0228	**20", Class 250, Mechanical Joint, Ductile Iron Pipe**	**LF**				**7.73**	**1.46**	**40.89**	**50.08**
02455 1023	20" Mechanical Joint, Ductile Iron Pipe	LF	22	CODEK	1.000	7.73	1.46	40.89	50.08
19 01 0229	**24", Class 250, Mechanical Joint, Ductile Iron Pipe**	**LF**				**23.40**	**7.19**	**70.00**	**100.58**
02664 2180	24" Diameter Mechanical Joint, Ductile Iron Pipe (61 cm) Diameter, No Fittings	LF	10	XXPLB	1.000	23.40	7.19	70.00	100.58
19 01 0235	**4", Class 250, Push-on Joint, Ductile Iron Pipe**	**LF**				**3.40**	**0.64**	**8.13**	**12.17**
02455 2017	4" Push-on Joint, Ductile Iron Pipe	LF	50	CODEK	1.000	3.40	0.64	8.13	12.17
19 01 0236	**6", Class 250, Push-on Joint, Ductile Iron Pipe**	**LF**				**3.54**	**0.67**	**9.32**	**13.53**
02455 2018	6" Push-on Joint, Ductile Iron Pipe	LF	48	CODEK	1.000	3.54	0.67	9.32	13.53
19 01 0237	**8", Class 250, Push-on Joint, Ductile Iron Pipe**	**LF**				**3.54**	**0.67**	**12.76**	**16.97**
02455 2019	8" Push-on Joint, Ductile Iron Pipe	LF	48	CODEK	1.000	3.54	0.67	12.76	16.97
19 01 0238	**10", Class 250, Push-on Joint, Ductile Iron Pipe**	**LF**				**4.25**	**0.80**	**16.19**	**21.25**
02455 2021	10" Push-on Joint, Ductile Iron Pipe	LF	40	CODEK	1.000	4.25	0.80	16.19	21.25
19 01 0239	**12", Class 250, Push-on Joint, Ductile Iron Pipe**	**LF**				**4.25**	**0.80**	**21.10**	**26.15**
02455 2022	12" Push-on Joint, Ductile Iron Pipe	LF	40	CODEK	1.000	4.25	0.80	21.10	26.15

19 01 02	Potable Water Distribution	Unit	Hourly Output	Crew	Factor	Labor	Equip.	Matl.	Total
19 01 0240	**14", Class 250, Push-on Joint, Ductile Iron Pipe**	**LF**				**4.48**	**0.84**	**25.63**	**30.95**
02455 2023	14" Push-on Joint, Ductile Iron Pipe	LF	38	CODEK	1.000	4.48	0.84	25.63	30.95
19 01 0241	**16", Class 250, Push-on Joint, Ductile Iron Pipe**	**LF**				**4.48**	**0.84**	**29.50**	**34.82**
02455 2024	16" Push-on Joint, Ductile Iron Pipe	LF	38	CODEK	1.000	4.48	0.84	29.50	34.82
19 01 0242	**18", Class 250, Push-on Joint, Ductile Iron Pipe**	**LF**				**5.32**	**1.00**	**36.30**	**42.62**
02455 2025	18" Push-on Joint, Ductile Iron Pipe	LF	32	CODEK	1.000	5.32	1.00	36.30	42.62
19 01 0243	**20", Class 250, Push-on Joint, Ductile Iron Pipe**	**LF**				**5.85**	**1.80**	**41.37**	**49.01**
02455 2009	20" (51 cm) Diameter Mechanical Ductile Iron Push-on Joint, No Fittings	LF	40	XXPLB	1.000	5.85	1.80	41.37	49.01
19 01 0244	**24", Class 250, Push-on Joint, Ductile Iron Pipe**	**LF**				**6.68**	**2.05**	**47.00**	**55.74**
02664 2250	24" (51 cm) Diameter Mechanical Ductile Iron Push-on Joint, No Fittings	LF	35	XXPLB	1.000	6.68	2.05	47.00	55.74
19 01 0250	**4", Schedule 80 Steel Pipe, Plain End, Tar Coated & Wrapped**	**LF**				**2.29**	**0.16**	**8.30**	**10.75**
02456 6129	10" or Less	LF	0	N/A	1.000	0.00	0.00	2.56	2.56
15061 2115	4" Black Steel	LF	62.5	MPLUO	1.000	2.29	0.16	5.74	8.19
19 01 0251	**6", Schedule 80 Steel Pipe, Plain End, Tar Coated & Wrapped**	**LF**				**2.62**	**0.65**	**12.34**	**15.60**
02456 6129	10" or Less	LF	0	N/A	1.000	0.00	0.00	2.56	2.56
15061 2117	6" Black Steel	LF	60	MPLUQ	1.000	2.62	0.65	9.78	13.04
19 01 0252	**8", Schedule 80 Steel Pipe, Plain End, Tar Coated & Wrapped**	**LF**				**2.62**	**0.65**	**17.25**	**20.51**
02456 6129	10" or Less	LF	0	N/A	1.000	0.00	0.00	2.56	2.56
15061 2118	8" Black Steel	LF	60	MPLUQ	1.000	2.62	0.65	14.69	17.96
19 01 0253	**12", Schedule 80 Steel Pipe, Plain End, Tar Coated & Wrapped**	**LF**				**3.92**	**0.97**	**24.00**	**28.89**
02661 1610	12" (31 cm) Outside Diameter Pipe Coating Coal Tar Epoxy, Kraft Paper Wrap	LF	0	N/A	1.000	0.00	0.00	3.32	3.32
15061 2119	12" Black Steel	LF	40	MPLUQ	1.000	3.92	0.97	20.68	25.57
19 01 0254	**16", Schedule 80 Steel Pipe, Plain End, Tar Coated & Wrapped**	**LF**				**5.73**	**0.85**	**32.08**	**38.65**
02456 6103	16" (41 cm) Outside Diameter Pipe Coating Coal Tar Epoxy, Kraft Paper Wrap	LF	0	N/A	1.000	0.00	0.00	4.31	4.31
15061 2121	16" Black Steel	LF	30	MSPFU	1.000	5.73	0.85	27.76	34.34

19 01 02	Potable Water Distribution	Unit	Hourly Output	Crew	Factor	Labor	Equip.	Matl.	Total
19 01 0255	**20", Schedule 80 Steel Pipe, Plain End, Tar Coated & Wrapped**	**LF**				**8.59**	**1.27**	**47.16**	**57.03**
02456 6105	20" (51 cm) Outside Diameter Pipe Coating Coal Tar Epoxy, Kraft Paper Wrap	LF	0	N/A	1.000	0.00	0.00	5.49	5.49
15061 2122	20" Black Steel	LF	20	MSPFU	1.000	8.59	1.27	41.67	51.54
19 01 0256	**24", Schedule 80 Steel Pipe, Plain End, Tar Coated & Wrapped**	**LF**				**10.74**	**1.59**	**65.72**	**78.05**
02661 1620	24" (61 cm) Outside Diameter Pipe Coating Coal Tar Epoxy, Kraft Paper Wrap	LF	0	N/A	1.000	0.00	0.00	7.35	7.35
15061 2123	24" Black Steel	LF	16	MSPFU	1.000	10.74	1.59	58.37	70.70
19 01 0260	**1", Schedule 40 Steel Pipe, Plain End, Tar Coated & Wrapped**	**LF**				**1.43**	**0.10**	**5.71**	**7.24**
02456 6129	10" or Less	LF	0	N/A	1.000	0.00	0.00	2.56	2.56
15061 2112	1" Black Steel	LF	100	MPLUO	1.000	1.43	0.10	3.15	4.68
19 01 0261	**2", Schedule 40 Steel Pipe, Plain End, Tar Coated & Wrapped**	**LF**				**1.79**	**0.13**	**7.13**	**9.04**
02456 6129	10" or Less	LF	0	N/A	1.000	0.00	0.00	2.56	2.56
15061 2113	2" Black Steel	LF	80	MPLUO	1.000	1.79	0.13	4.57	6.49
19 01 0262	**3", Schedule 40 Steel Pipe, Plain End, Tar Coated & Wrapped**	**LF**				**2.04**	**0.14**	**6.67**	**8.86**
02456 6129	10" or Less	LF	0	N/A	1.000	0.00	0.00	2.56	2.56
15061 2114	3" Black Steel	LF	70	MPLUO	1.000	2.04	0.14	4.11	6.30
19 01 0263	**4", Schedule 40 Steel Pipe, Plain End, Tar Coated & Wrapped**	**LF**				**2.29**	**0.16**	**8.30**	**10.75**
02456 6129	10" or Less	LF	0	N/A	1.000	0.00	0.00	2.56	2.56
15061 2115	4" Black Steel	LF	62.5	MPLUO	1.000	2.29	0.16	5.74	8.19
19 01 0264	**5", Schedule 40 Steel Pipe, Plain End, Tar Coated & Wrapped**	**LF**				**2.49**	**0.18**	**12.97**	**15.63**
02456 6129	10" or Less	LF	0	N/A	1.000	0.00	0.00	2.56	2.56
15061 2116	5" Black Steel	LF	57.5	MPLUO	1.000	2.49	0.18	10.41	13.08
19 01 0265	**6", Schedule 40 Steel Pipe, Plain End, Tar Coated & Wrapped**	**LF**				**2.62**	**0.65**	**12.34**	**15.60**
02456 6129	10" or Less	LF	0	N/A	1.000	0.00	0.00	2.56	2.56
15061 2117	6" Black Steel	LF	60	MPLUQ	1.000	2.62	0.65	9.78	13.04
19 01 0266	**8", Schedule 40 Steel Pipe, Plain End, Tar Coated & Wrapped**	**LF**				**2.62**	**0.65**	**17.25**	**20.51**
02456 6129	10" or Less	LF	0	N/A	1.000	0.00	0.00	2.56	2.56
15061 2118	8" Black Steel	LF	60	MPLUQ	1.000	2.62	0.65	14.69	17.96
19 01 0267	**10", Schedule 40 Steel Pipe, Plain End, Tar Coated & Wrapped**	**LF**				**3.92**	**0.97**	**23.92**	**28.81**
02456 6129	10" or Less	LF	0	N/A	1.000	0.00	0.00	2.56	2.56
15061 2124	10" Black Steel	LF	40	MPLUQ	1.000	3.92	0.97	21.36	26.26

19 01 Water Supply and Distribution Systems

19 01 02 Potable Water Distribution

19 01 02 Potable Water Distribution		Unit	Hourly Output	Crew	Factor	Labor	Equip.	Matl.	Total
19 01 0268	**12", Schedule 40 Steel Pipe, Plain End, Tar Coated & Wrapped**	**LF**				**3.92**	**0.97**	**24.00**	**28.89**
02661 1610	12" (31 cm) Outside Diameter Pipe Coating Coal Tar Epoxy, Kraft Paper Wrap	LF	0	N/A	1.000	0.00	0.00	3.32	3.32
15061 2119	12" Black Steel	LF	40	MPLUQ	1.000	3.92	0.97	20.68	25.57
19 01 0269	**14", Schedule 40 Steel Pipe, Plain End, Tar Coated & Wrapped**	**LF**				**5.23**	**1.29**	**27.41**	**33.94**
02456 6102	14" (36 cm) Outside Diameter Pipe Coating Coal Tar Epoxy, Kraft Paper Wrap	LF	0	N/A	1.000	0.00	0.00	3.76	3.76
15061 2125	14" Black Steel	LF	30	MPLUQ	1.000	5.23	1.29	23.65	30.17
19 01 0270	**16", Schedule 40 Steel Pipe, Plain End, Tar Coated & Wrapped**	**LF**				**5.73**	**0.85**	**32.08**	**38.65**
02456 6103	16" (41 cm) Outside Diameter Pipe Coating Coal Tar Epoxy, Kraft Paper Wrap	LF	0	N/A	1.000	0.00	0.00	4.31	4.31
15061 2121	16" Black Steel	LF	30	MSPFU	1.000	5.73	0.85	27.76	34.34
19 01 0271	**18", Schedule 40 Steel Pipe, Plain End, Tar Coated & Wrapped**	**LF**				**7.16**	**1.06**	**38.63**	**46.85**
02661 1615	18" (46 cm) Outside Diameter Pipe Coating Coal Tar Epoxy, Kraft Paper Wrap	LF	0	N/A	1.000	0.00	0.00	5.00	5.00
15061 2126	18" Black Steel	LF	24	MSPFU	1.000	7.16	1.06	33.63	41.85
19 01 0272	**20", Schedule 40 Steel Pipe, Plain End, Tar Coated & Wrapped**	**LF**				**8.59**	**1.27**	**47.16**	**57.03**
02456 6105	20" (51 cm) Outside Diameter Pipe Coating Coal Tar Epoxy, Kraft Paper Wrap	LF	0	N/A	1.000	0.00	0.00	5.49	5.49
15061 2122	20" Black Steel	LF	20	MSPFU	1.000	8.59	1.27	41.67	51.54
19 01 0273	**24", Schedule 40 Steel Pipe, Plain End, Tar Coated & Wrapped**	**LF**				**10.74**	**1.59**	**65.72**	**78.05**
02661 1620	24" (61 cm) Outside Diameter Pipe Coating Coal Tar Epoxy, Kraft Paper Wrap	LF	0	N/A	1.000	0.00	0.00	7.35	7.35
15061 2123	24" Black Steel	LF	16	MSPFU	1.000	10.74	1.59	58.37	70.70
19 01 0275	**4" Threaded Steel Pipe**	**LF**				**2.72**	**0.51**	**11.71**	**14.94**
15061 1616	4" Threaded Steel Pipe	LF	62.5	CODEK	1.000	2.72	0.51	11.71	14.94
19 01 0276	**8" Threaded Steel Pipe**	**LF**				**2.84**	**0.53**	**37.92**	**41.29**
15061 1617	8" Threaded Steel Pipe	LF	60	CODEK	1.000	2.84	0.53	37.92	41.29
19 01 0277	**12" Threaded Steel Pipe**	**LF**				**4.25**	**0.80**	**76.19**	**81.25**
15061 1618	12" Threaded Steel Pipe	LF	40	CODEK	1.000	4.25	0.80	76.19	81.25

19 01 03 Potable Water Storage

19 01 03 Potable Water Storage		Unit	Hourly Output	Crew	Factor	Labor	Equip.	Matl.	Total
19 01 0301	**50,000 Gallon Water Tank, Prestressed Concrete @ Grade**	**EA**				**19,026**	**4,985**	**18,715**	**42,727**
15240 1011	50,000 Gallon	EA	0.019	SIWSK	1.000	19,026	4,985	18,715	42,727

19 01 03	Potable Water Storage	Unit	Hourly Output	Crew	Factor	Labor	Equip.	Matl.	Total
19 01 0302	**100,000 Gallon Water Tank, Prestressed Concrete @ Grade**	**EA**				**32,345**	**8,475**	**129,000**	**169,820**
13205 0900	100, 000 Gallon Steel Water Tank, on Ground AWWA, with Low Pitch Cone Roof	EA	0.011	SIWSK	1.000	32,345	8,475	129,000	169,820
19 01 0303	**500,000 Gallon Water Tank, Prestressed Concrete @ Grade**	**EA**				**72,610**	**19,026**	**238,787**	**330,422**
13205 1200	500,000 Gallon Steel Water Tank, on Ground AWWA, with Low Pitch Cone Roof	EA	0.005	SIWSK	1.000	72,610	19,026	238,787	330,422
19 01 0304	**1,000,000 Gallon Water Tank, Prestressed Concrete @ Grade**	**EA**				**136,842**	**35,856**	**383,887**	**556,585**
13205 1300	1,000,000 Gallon Steel Water Tank, on Ground AWWA, with Low Pitch Cone Roof	EA	0.003	SIWSK	1.000	136,842	35,856	383,887	556,585
19 01 0310	**5,000 Gallon Water Tank, Steel, Horizontal @ Grade**	**EA**				**5,484**	**0.00**	**7,025**	**12,509**
13205 8010	5,000 Gallon	EA	0.025	SIWSV	1.000	5,484	0.00	7,025	12,509
19 01 0311	**10,000 Gallon Water Tank, Steel, Horizontal @ Grade**	**EA**				**5,484**	**0.00**	**12,000**	**17,484**
13205 8020	10,000 Gallon	EA	0.025	SIWSV	1.000	5,484	0.00	12,000	17,484
19 01 0312	**20,000 Gallon Water Tank, Steel, Horizontal @ Grade**	**EA**				**8,463**	**0.00**	**14,400**	**22,863**
13205 8030	20,000 Gallon	EA	0.016	SIWSV	1.000	8,463	0.00	14,400	22,863
19 01 0313	**35,000 Gallon Water Tank, Steel, Horizontal @ Grade**	**EA**				**10,968**	**0.00**	**38,600**	**49,568**
13205 8040	35,000 Gallon	EA	0.013	SIWSV	1.000	10,968	0.00	38,600	49,568
19 01 0320	**250,000 Gallon Water Tank, Steel Stand Pipe**	**EA**				**15,759**	**0.00**	**112,500**	**128,259**
13205 9050	250,000 Gallon	EA	0.009	SIWSV	1.000	15,759	0.00	112,500	128,259
19 01 0321	**500,000 Gallon Water Tank, Steel Stand Pipe**	**EA**				**13,710**	**0.00**	**129,500**	**143,210**
13205 9060	500,000 Gallon	EA	0.01	SIWSV	1.000	13,710	0.00	129,500	143,210
19 01 0325	**5,000 Gallon Water Tank, Elevated Steel on Towers**	**EA**				**26,500**	**12,761**	**8,275**	**47,536**
13205 9010	5,000 Gallon	EA	0	N/A	1.000	0.00	0.00	4,300	4,300
13210 4000	Tower for 5,000 Gallon	EA	0.009	SIWSW	1.000	26,500	12,761	3,975	43,236
19 01 0326	**10,000 Gallon Water Tank, Elevated Steel on Towers**	**EA**				**26,500**	**12,761**	**11,100**	**50,361**
13205 9020	10,000 Gallon	EA	0	N/A	1.000	0.00	0.00	7,000	7,000
13210 4100	Tower for 10,000 Gallon	EA	0.009	SIWSW	1.000	26,500	12,761	4,100	43,361
19 01 0327	**20,000 Gallon Water Tank, Elevated Steel on Towers**	**EA**				**26,500**	**12,761**	**17,071**	**56,333**
13205 9030	20,000 Gallon	EA	0	N/A	1.000	0.00	0.00	11,996	11,996
13210 4200	Tower for 20,000 Gallon	EA	0.009	SIWSW	1.000	26,500	12,761	5,075	44,336

19 01 03	Potable Water Storage	Unit	Hourly Output	Crew	Factor	Labor	Equip.	Matl.	Total
19 01 0328	**35,000 Gallon Water Tank, Elevated Steel on Towers**	**EA**				**30,740**	**14,803**	**29,000**	**74,543**
13205 9040	35,000 Gallon	EA	0	N/A	1.000	0.00	0.00	22,800	22,800
13210 4300	Tower for 35,000 Gallon	EA	0.008	SIWSW	1.000	30,740	14,803	6,200	51,743
19 01 0330	**50,000 Gallon Water Tanks, Elevated (100' +)**	**EA**				**95,363**	**31,354**	**189,658**	**316,375**
13210 3010	50,000 Gallon Elevated Water Tank, 100' (30 m) to Bottom Capacity Line	EA	0.004	SIWSO	1.000	95,363	31,354	189,658	316,375
19 01 0331	**100,000 Gallon Water Tanks, Elevated (100' +)**	**EA**				**118,482**	**38,955**	**261,637**	**419,074**
13210 3300	100,000 Gallon Elevated Water Tank, 100' (30 m) to Bottom Capacity Line	EA	0.003	SIWSO	1.000	118,482	38,955	261,637	419,074
19 01 0332	**250,000 Gallon Water Tanks, Elevated (100' +)**	**EA**				**169,996**	**55,891**	**361,036**	**586,923**
13210 3400	250,000 Gallon Elevated Water Tank, 100' (30 m) to Bottom Capacity Line	EA	0.002	SIWSO	1.000	169,996	55,891	361,036	586,923
19 01 0333	**500,000 Gallon Water Tanks, Elevated (100' +)**	**EA**				**260,660**	**85,700**	**582,685**	**929,045**
13210 3600	500,000 Gallon Elevated Water Tank, 100' (30 m) to Bottom Capacity Line	EA	0.002	SIWSO	1.000	260,660	85,700	582,685	929,045
19 01 0334	**750,000 Gallon Water Tanks, Elevated (100' +)**	**EA**				**390,990**	**128,550**	**803,192**	**1,322,732**
13210 3700	750,000 Gallon Elevated Water Tank, 100' (30 m) to Bottom Capacity Line	EA	0.001	SIWSO	1.000	390,990	128,550	803,192	1,322,732
19 01 0335	**1,000,000 Gallon Water Tanks, Elevated (100' +)**	**EA**				**488,737**	**160,688**	**927,726**	**1,577,151**
13210 3900	1,000,000 Gallon Elevated Water Tank, 100' (30 m) to Bottom Capacity Line	EA	0.0008	SIWSO	1.000	488,737	160,688	927,726	1,577,151

19 01 04	Fire Protection Water Distribution	Unit	Hourly Output	Crew	Factor	Labor	Equip.	Matl.	Total
19 01 0401	**Fire Hydrant**	**EA**				**1,354**	**354.80**	**2,463**	**4,172**
02215 0010	Backfill Trenches by Hand Without Compaction	CY	1.625	ULABA	3.330	20.85	0.00	0.00	69.42
02215 0500	Compaction, 6" Layers with Air Tamper by Hand, (15 cm) Layers	CY	50	CLACA	3.330	2.71	0.17	0.00	9.60
02215 2460	Sand Bedding w/Sm FEnd Loader Without Compaction	CY	47.5	CODLB	0.500	1.01	0.97	17.75	9.87
02227 4010	Fine Grd Gravel Bedding by Hand Finish Grade	SF	150	ULABA	18.000	0.23	0.00	0.00	4.06
02227 4020	Trim & Shape Machine Excavated Area by Hand - Finish Grade	SF	150	ULABA	18.000	0.23	0.00	0.00	4.06
02228 0322	Trench, 1 CY Hydraulic Excavator, Medium Soil, 4' to 6' Deep, 128 CY/Hour	CY	128	CODEO	3.330	0.49	0.70	0.00	3.97
02644 3814	6"(15cm) Cast Iron Gate Valve Includes Box	EA	0.375	XXPLA	1.000	467.20	191.67	214.00	872.87
02645 1020	6" Safetop Model Fire Hydrant Breakable w/5'(1.5M)Burial Secti	EA	0.438	MPLUJ	1.000	278.07	0.00	1,425	1,703
02664 5100	12" x 6" Mechanical Joint DI Tee	EA	0.45	XXPLB	1.000	519.94	159.72	815.00	1,495

19 01 08	Packaged Water Treatment Plants	Unit	Hourly Output	Crew	Factor	Labor	Equip.	Matl.	Total

19 01 08	Packaged Water Treatment Plants	Unit	Hourly Output	Crew	Factor	Labor	Equip.	Matl.	Total
19 01 0801	**360 GPD, Packaged Water Treatment Plant**	**EA**				**0.00**	**0.00**	**5,825**	**5,825**
02675 0100	360 GPD	EA	0	N/A	1.000	0.00	0.00	5,825	5,825
19 01 0802	**720 GPD, Packaged Water Treatment Plant, Skid Mounted**	**EA**				**0.00**	**0.00**	**7,275**	**7,275**
02675 0200	720 GPD	EA	0	N/A	1.000	0.00	0.00	7,275	7,275
19 01 0803	**2,880 GPD, Packaged Water Treatment Plant**	**EA**				**0.00**	**0.00**	**10,723**	**10,723**
02675 0300	2,880 GPD	EA	0	N/A	1.000	0.00	0.00	10,723	10,723
19 01 0804	**6,480 GPD, Packaged Water Treatment Plant**	**EA**				**0.00**	**0.00**	**15,318**	**15,318**
02675 0400	6,480 GPD	EA	0	N/A	1.000	0.00	0.00	15,318	15,318
19 01 0805	**12,240 GPD, Packaged Water Treatment Plant**	**EA**				**0.00**	**0.00**	**19,700**	**19,700**
02675 0500	12,240 GPD	EA	0	N/A	1.000	0.00	0.00	19,700	19,700
19 01 0806	**24,480 GPD, Packaged Water Treatment Plant**	**EA**				**0.00**	**0.00**	**24,900**	**24,900**
02675 0600	24,480 GPD	EA	0	N/A	1.000	0.00	0.00	24,900	24,900
19 01 0807	**36,000 GPD, Packaged Water Treatment Plant**	**EA**				**0.00**	**0.00**	**27,700**	**27,700**
02675 0700	36,000 GPD	EA	0	N/A	1.000	0.00	0.00	27,700	27,700

19 02 01	Sanitary Sewer Piping	Unit	Hourly Output	Crew	Factor	Labor	Equip.	Matl.	Total
19 02 0101	**4", Class 50, Bell & Spigot Sanitary Sewer, Cast-iron Pipe**	**LF**				**2.12**	**0.00**	**5.00**	**7.12**
15062 1011	4" Bell & Spigot Cast Iron	LF	42	MPLUE	1.000	2.12	0.00	5.00	7.12
19 02 0102	**5", Class 50, Bell & Spigot Sanitary Sewer, Cast-iron Pipe**	**LF**				**2.90**	**0.00**	**7.00**	**9.89**
15062 1012	5" Bell & Spigot Cast Iron	LF	42	MPLUJ	1.000	2.90	0.00	7.00	9.89
19 02 0103	**6", Class 50, Bell & Spigot Sanitary Sewer, Cast-iron Pipe**	**LF**				**3.04**	**0.00**	**8.40**	**11.44**
15062 1013	6" Bell & Spigot Cast Iron	LF	40	MPLUJ	1.000	3.04	0.00	8.40	11.44
19 02 0104	**8", Class 50, Bell & Spigot Sanitary Sewer, Cast-iron Pipe**	**LF**				**3.58**	**0.00**	**14.01**	**17.59**
15062 1014	8" Bell & Spigot Cast Iron	LF	40	MPLUM	1.000	3.58	0.00	14.01	17.59
19 02 0105	**10", Class 50, Bell & Spigot Sanitary Sewer, Cast-iron Pipe**	**LF**				**4.47**	**0.00**	**22.31**	**26.79**
15062 1015	10" Bell & Spigot Cast Iron	LF	32	MPLUM	1.000	4.47	0.00	22.31	26.79
19 02 0106	**12", Class 50, Bell & Spigot Sanitary Sewer, Cast-iron Pipe**	**LF**				**4.47**	**0.00**	**32.95**	**37.42**
15062 1016	12" Bell & Spigot Cast Iron	LF	32	MPLUM	1.000	4.47	0.00	32.95	37.42
19 02 0107	**15", Class 50, Bell & Spigot Sanitary Sewer, Cast-iron Pipe**	**LF**				**5.72**	**0.00**	**66.89**	**72.62**
15132 2270	15" Bell & Spigot Cast Iron	LF	25	MPLUM	1.000	5.72	0.00	66.89	72.62

19 02 01	Sanitary Sewer Piping	Unit	Hourly Output	Crew	Factor	Labor	Equip.	Matl.	Total
19 02 0110	**4" Extra-strength Vitrified Clay Pipe, Class 200, Premium Joints**	**LF**				**5.67**	**1.74**	**2.64**	**10.05**
02463 1013	4" Extra Strength Vitrified Clay Pipe	LF	41.25	XXPLB	1.000	5.67	1.74	2.64	10.05
19 02 0111	**5" Extra-strength Vitrified Clay Pipe, Class 200, Premium Joints**	**LF**				**6.04**	**1.85**	**3.33**	**11.22**
02463 1014	5" Extra Strength Vitrified Clay Pipe	LF	38.75	XXPLB	1.000	6.04	1.85	3.33	11.22
19 02 0112	**6" Extra-strength Vitrified Clay Pipe, Class 200, Premium Joints**	**LF**				**6.24**	**1.92**	**4.01**	**12.17**
02463 1015	6" Extra Strength Vitrified Clay Pipe	LF	37.5	XXPLB	1.000	6.24	1.92	4.01	12.17
19 02 0113	**8" Extra-strength Vitrified Clay Pipe, Class 200, Premium Joints**	**LF**				**6.45**	**1.98**	**5.99**	**14.43**
02463 1016	8" Extra Strength Vitrified Clay Pipe	LF	36.25	XXPLB	1.000	6.45	1.98	5.99	14.43
19 02 0114	**10" Extra-strength Vitrified Clay Pipe, Class 200, Premium Joints**	**LF**				**7.20**	**2.21**	**9.70**	**19.11**
02463 1017	10" Extra Strength Vitrified Clay Pipe	LF	32.5	XXPLB	1.000	7.20	2.21	9.70	19.11
19 02 0115	**12" Extra-strength Vitrified Clay Pipe, Class 200, Premium Joints**	**LF**				**7.49**	**2.30**	**10.29**	**20.08**
02463 1018	12" Extra Strength Vitrified Clay Pipe	LF	31.25	XXPLB	1.000	7.49	2.30	10.29	20.08
19 02 0116	**15" Extra-strength Vitrified Clay Pipe, Class 200, Premium Joints**	**LF**				**8.51**	**2.61**	**21.11**	**32.23**
02463 1019	15" Extra Strength Vitrified Clay Pipe	LF	27.5	XXPLB	1.000	8.51	2.61	21.11	32.23
19 02 0117	**18" Extra-strength Vitrified Clay Pipe, Class 200, Premium Joints**	**LF**				**9.85**	**3.03**	**31.47**	**44.34**
02463 1021	18" Extra Strength Vitrified Clay Pipe	LF	23.75	XXPLB	1.000	9.85	3.03	31.47	44.34
19 02 0118	**24" Extra-strength Vitrified Clay Pipe, Class 200, Premium Joints**	**LF**				**13.37**	**4.11**	**64.90**	**82.38**
02463 1022	24" Extra Strength Vitrified Clay Pipe	LF	17.5	XXPLB	1.000	13.37	4.11	64.90	82.38
19 02 0119	**30" Extra-strength Vitrified Clay Pipe, Class 200, Premium Joints**	**LF**				**18.72**	**5.75**	**68.18**	**92.64**
02463 1023	30" Extra Strength Vitrified Clay Pipe	LF	12.5	XXPLB	1.000	18.72	5.75	68.18	92.64
19 02 0120	**36" Extra-strength Vitrified Clay Pipe, Class 200, Premium Joints**	**LF**				**23.40**	**7.19**	**104.89**	**135.47**
02463 1024	36" Extra Strength Vitrified Clay Pipe	LF	10	XXPLB	1.000	23.40	7.19	104.89	135.47
19 02 0125	**4" PVC Pipe Sanitary**	**LF**				**3.74**	**1.53**	**1.53**	**6.80**
02766 9800	4" (10 cm) ABS Plastic Drainpipe Truss Type	LF	46.88	XXPLA	1.000	3.74	1.53	1.53	6.80
19 02 0126	**6" PVC Pipe Sanitary**	**LF**				**4.00**	**1.64**	**3.21**	**8.86**
02766 9810	6" (15 cm) ABS Plastic Drainpipe Truss Type	LF	43.75	XXPLA	1.000	4.00	1.64	3.21	8.86

19 02 Sanitary Sewer Systems

19 02 01 Sanitary Sewer Piping		Unit	Hourly Output	Crew	Factor	Labor	Equip.	Matl.	Total
19 02 0127	**8" PVC Pipe Sanitary**	**LF**				**4.18**	**1.72**	**6.15**	**12.05**
02766 9820	8" (21 cm) ABS Plastic Drainpipe Truss Type	LF	41.88	XXPLA	1.000	4.18	1.72	6.15	12.05
19 02 0128	**10" PVC Pipe Sanitary**	**LF**				**5.67**	**1.74**	**9.45**	**16.86**
02766 9830	10" (25 cm) ABS Plastic Drainpipe Truss Type	LF	41.25	XXPLB	1.000	5.67	1.74	9.45	16.86
19 02 0129	**12" PVC Pipe Sanitary**	**LF**				**5.85**	**1.80**	**13.85**	**21.50**
02766 9840	12" (31 cm) ABS Plastic Drainpipe Truss Type	LF	40	XXPLB	1.000	5.85	1.80	13.85	21.50
19 02 0130	**15" PVC Pipe Sanitary**	**LF**				**9.85**	**3.03**	**20.50**	**33.38**
02766 9850	15" (38 cm) ABS Plastic Drainpipe Truss Type	LF	23.75	XXPLB	1.000	9.85	3.03	20.50	33.38

19 02 02 Sanitary Sewer Manholes and Cleanouts		Unit	Hourly Output	Crew	Factor	Labor	Equip.	Matl.	Total
19 02 0201	**Precast, CIP Base, 4' Diameter, 6' Deep, Manhole**	**EA**				**312.98**	**88.60**	**568.89**	**970.46**
02752 1120	4' Diameter x 6' Deep Precast Manhole 8" (20 cm) Thick, (1.2 m Diameter x 1.8 m Deep)	EA	0.375	CODEJ	1.000	253.93	85.47	512.37	851.77
03354 3005	Concrete Redimix 2,500 PSI	CY	1.238	CLACC	0.890	66.34	3.52	63.50	118.69
19 02 0202	**Precast, CIP Base, 4' Diameter, 8' Deep, Manhole**	**EA**				**439.95**	**131.33**	**627.00**	**1,198**
02752 1130	4' Diameter x 8' Deep Precast Manhole 8" (20 cm) Thick, (1.2 m Diameter x 2.4 m Deep)	EA	0.25	CODEJ	1.000	380.90	128.20	570.48	1,080
03354 3005	Concrete Redimix 2,500 PSI	CY	1.238	CLACC	0.890	66.34	3.52	63.50	118.69
19 02 0203	**Precast, CIP Base, 4' Diameter, 12' Deep, Manhole**	**EA**				**630.40**	**195.43**	**1,037**	**1,863**
02752 1130	4' Diameter x 8' Deep Precast Manhole 8" (20 cm) Thick, (1.2 m Diameter x 2.4 m Deep)	EA	0.25	CODEJ	1.000	380.90	128.20	570.48	1,080
02752 1140	4' Diameter x Depth, Precast Manhole 8" (20 cm) Thick, (1.2 m Diameter x VLF)	VLF	2	CODEJ	4.000	47.61	16.02	102.46	664.38
03354 3005	Concrete Redimix 2,500 PSI	CY	1.238	CLACC	0.890	66.34	3.52	63.50	118.69
19 02 0204	**Precast, CIP Base, 2' Dia, 4' Deep, Manhole**	**EA**				**0.00**	**0.00**	**528.48**	**528.48**
01001 0008	Precast, CIP Base, 2' Dia, 4' Deep, Manhole	EA	0	N/A	1.000	0.00	0.00	528.48	528.48
19 02 0205	**Precast, CIP Base, 2.5' Dia, 4' Deep, Manhole**	**EA**				**0.00**	**0.00**	**566.35**	**566.35**
01001 0009	Precast, CIP Base, 2.5' Dia, 4' Deep, Manhole	EA	0	N/A	1.000	0.00	0.00	566.35	566.35
19 02 0206	**Precast, CIP Base, 3' Dia, 6' Deep, Manhole**	**EA**				**0.00**	**0.00**	**604.22**	**604.22**
01001 0010	Precast, CIP Base, 3' Dia, 6' Deep, Manhole	EA	0	N/A	1.000	0.00	0.00	604.22	604.22
19 02 0210	**Precast Manhole, Per Foot, 4' Diameter x Depth**	**LF**				**47.61**	**16.02**	**102.46**	**166.10**
02752 1140	4' Diameter x Depth, Precast Manhole 8" (20 cm) Thick, (1.2 m Diameter x VLF)	VLF	2	CODEJ	1.000	47.61	16.02	102.46	166.10

19 02 03 Lift Stations	Unit	Hourly Output	Crew	Factor	Labor	Equip.	Matl.	Total

19 02 03	Lift Stations	Unit	Hourly Output	Crew	Factor	Labor	Equip.	Matl.	Total
19 02 0301	**10,000 GPD (7 GPM) Lift Station**	**EA**				**355.90**	**0.00**	**841.76**	**1,198**
15294 0700	10 GPM Submersible Sewage Ejector with 230V 1/2 HP Motor, Simplex	EA	0.25	MPLUE	1.000	355.90	0.00	841.76	1,198
19 02 0302	**25,000 GPD (18 GPM) Lift Station**	**EA**				**474.53**	**0.00**	**1,126**	**1,600**
15294 0730	45 GPM Submersible Sewage Ejector with 230V 3/4 HP Motor, Simplex	EA	0.188	MPLUE	1.000	474.53	0.00	1,126	1,600
19 02 0303	**50,000 GPD (35 GPM) Lift Station**	**EA**				**711.80**	**0.00**	**1,178**	**1,890**
15294 0760	80 GPM Submersible Sewage Ejector with 230V 1 HP Motor, Simplex	EA	0.125	MPLUE	1.000	711.80	0.00	1,178	1,890
19 02 0304	**100,000 GPD (70 GPM) Lift Station**	**EA**				**695.17**	**0.00**	**7,851**	**8,546**
15294 4020	120 GPM Submersible Sewage Ejector, 1-1/2 HP, Alternator & Duplex	EA	0.175	MPLUJ	1.000	695.17	0.00	7,851	8,546
19 02 0305	**200,000 GPD Packaged Lift Station**	**EA**				**18,379**	**2,571**	**119,000**	**139,950**
02776 2500	200,000 GPD	EA	0.025	SIWSP	1.000	18,379	2,571	119,000	139,950
19 02 0306	**500,000 GPD Packaged Lift Station**	**EA**				**24,440**	**3,419**	**133,500**	**161,359**
02776 2510	500,000 GPD	EA	0.019	SIWSP	1.000	24,440	3,419	133,500	161,359
19 02 0307	**800,000 GPD Packaged Lift Station**	**EA**				**29,080**	**4,068**	**160,500**	**193,648**
02776 2520	800,000 GPD	EA	0.016	SIWSP	1.000	29,080	4,068	160,500	193,648

19 02 03	Lift Stations	Unit	Hourly Output	Crew	Factor	Labor	Equip.	Matl.	Total
19 02 0310	**24' x 36" Diameter Reinforced Concrete Pipe Wet Well**	**EA**				**3,938**	**4,122**	**8,852**	**16,912**
02215 1215	Backfill Trench with 200HP Dozer without Compaction	CY	284	CODTG	1280.700	0.15	0.43	0.00	749.25
02220 8800	Compact Backfill, by Hand with 1-Ton Roller, 6" (15 cm) Lift	CY	49.13	CLACD	1034.770	0.94	1.51	0.00	2,535
02221 5011	Sand Backfill, 950 Dumped	CY	83.25	CODLJ	242.930	0.40	0.46	0.00	207.69
02227 4000	Trim & Shape for Slab on Grade by Hand Finish Grade	SF	75.38	ULABA	116.500	0.45	0.00	0.00	52.36
02232 0155	Excavate & Load, 3 CY Hydraulic Excavator, Light Material, 218 CY/Hour (167 m3)	CY	218	XXQHK	1034.770	0.57	0.88	0.00	1,505
02244 1505	Washed & Graded Sand, Loaded at Pit	CY	12.38	CODEG	242.930	3.88	2.59	26.50	8,009
02452 1006	36" (91 cm) Diameter Class III Concrete Pipe Reinforced without Gaskets	LF	8.738	UOEHC	24.000	19.47	8.23	47.20	1,798
02611 2201	Graded Crushed Aggregate Base	CY	25	XSABA	3.139	9.51	13.80	22.77	144.65
02664 2040	6" Diameter Mechanical Joint, Ductile Iron Pipe (15 cm), No Fittings	LF	30	XXPLA	48.000	5.84	2.40	14.50	1,091
03114 2400	PVC Chamfer Strips, 1" Radius	CLF	0.625	ACARB	0.029	138.60	0.00	51.80	5.45
03170 3050	Slab on Grade Edge Forms, 7" to 12" High (18 cm to 31 cm), Based on 4 Uses	LF	54.38	ACARJ	36.000	2.42	0.00	0.66	110.65
03217 0600	Grade 60 Rebar, Footings & Slabs, #3-#6	TON	0.313	SIWRC	0.247	492.16	0.00	609.86	271.94
03334 0300	Concrete Curing, Sprayed Membrane Curing Compound	CSF	11.88	ULABB	1.059	5.71	0.00	4.05	10.33
03372 4600	Pour Slab on Grade, >= 6", Direct Chute >= (15 cm) Place 3,000 PSI Concrete	CY	21.88	ALABE	3.923	7.70	0.29	84.22	361.71
03396 0150	Concrete Floor Finishes, Broom	SF	84.38	ACMAC	106.000	0.43	0.00	0.00	45.40
07192 0900	6 Mil Polyethylene Vapor Barrier	CSF	4.625	ACARA	1.165	9.36	0.00	3.30	14.76

19 02 03	Lift Stations	Unit	Hourly Output	Crew	Factor	Labor	Equip.	Matl.	Total
19 02 0311	**12' x 36" Diameter Reinforced Concrete Pipe Wet Well for Lift Station**	**EA**				**1,817**	**1,658**	**4,577**	**8,052**
02215 1215	Backfill Trench with 200HP Dozer without Compaction	CY	284	CODTG	470.000	0.15	0.43	0.00	274.97
02220 8800	Compact Backfill, by Hand with 1-Ton Roller, 6" (15 cm) Lift	CY	49.13	CLACD	376.000	0.94	1.51	0.00	921.15
02221 5011	Sand Backfill, 950 Dumped	CY	83.25	CODLJ	117.000	0.40	0.46	0.00	100.03
02227 4000	Trim & Shape for Slab on Grade by Hand Finish Grade	SF	75.38	ULABA	116.500	0.45	0.00	0.00	52.36
02232 0155	Excavate & Load, 3 CY Hydraulic Excavator, Light Material, 218 CY/Hour (167 m3)	CY	218	XXQHK	376.000	0.57	0.88	0.00	546.88
02244 1505	Washed & Graded Sand, Loaded at Pit	CY	12.38	CODEG	117.000	3.88	2.59	26.50	3,857
02452 1006	36" (91 cm) Diameter Class III Concrete Pipe Reinforced without Gaskets	LF	8.738	UOEHC	12.000	19.47	8.23	47.20	898.79
02611 2201	Graded Crushed Aggregate Base	CY	25	XSABA	3.138	9.51	13.80	22.77	144.63
02664 2040	6" Diameter Mechanical Joint, Ductile Iron Pipe (15 cm), No Fittings	LF	30	XXPLA	24.000	5.84	2.40	14.50	545.66
03114 2400	PVC Chamfer Strips, 1" Radius	CLF	0.625	ACARB	0.029	138.60	0.00	51.80	5.45
03170 4580	Straight Edge Metal Formwork, 12" High (10 cm), with Stakes, 10' Length, 100 Uses	LF	52.5	ACARI	0.282	1.99	0.00	1.04	0.85
03217 0600	Grade 60 Rebar, Footings & Slabs, #3-#6	TON	0.313	SIWRC	0.247	492.16	0.00	609.86	271.94
03334 0300	Concrete Curing, Sprayed Membrane Curing Compound	CSF	11.88	ULABB	1.059	5.71	0.00	4.05	10.33
03372 4600	Pour Slab on Grade, >= 6", Direct Chute >= (15 cm) Place 3,000 PSI Concrete	CY	21.88	ALABE	3.923	7.70	0.29	84.22	361.71
03396 0150	Concrete Floor Finishes, Broom	SF	84.38	ACMAC	106.000	0.43	0.00	0.00	45.40
07192 0900	6 Mil Polyethylene Vapor Barrier	CSF	4.625	ACARA	1.165	9.36	0.00	3.30	14.76

19 02 03	Lift Stations	Unit	Hourly Output	Crew	Factor	Labor	Equip.	Matl.	Total
19 02 0312	**24' x 60" Diameter Reinforced Concrete Pipe Wet Well for Lift Station**	**EA**				**7,331**	**8,383**	**16,608**	**32,323**
02215 1215	Backfill Trench with 200HP Dozer without Compaction	CY	284	CODTG	2432.000	0.15	0.43	0.00	1,423
02220 8800	Compact Backfill, by Hand with 1-Ton Roller, 6" (15 cm) Lift	CY	49.13	CLACD	2432.000	0.94	1.51	0.00	5,958
02221 5011	Sand Backfill, 950 Dumped	CY	83.25	CODLJ	462.500	0.40	0.46	0.00	395.42
02227 4000	Trim & Shape for Slab on Grade by Hand Finish Grade	SF	75.38	ULABA	187.000	0.45	0.00	0.00	84.04
02232 0155	Excavate & Load, 3 CY Hydraulic Excavator, Light Material, 218 CY/Hour (167 m3)	CY	218	XXQHK	1969.000	0.57	0.88	0.00	2,864
02244 1505	Washed & Graded Sand, Loaded at Pit	CY	12.38	CODEG	462.500	3.88	2.59	26.50	15,248
02452 1009	60" (152 cm) Diameter Class III Concrete Pipe Reinforced without Gaskets	LF	5.125	UOEHC	24.000	33.20	14.02	114.72	3,887
02611 2201	Graded Crushed Aggregate Base	CY	25	XSABA	5.037	9.51	13.80	22.77	232.15
02664 2040	6" Diameter Mechanical Joint, Ductile Iron Pipe (15 cm), No Fittings	LF	30	XXPLA	48.000	5.84	2.40	14.50	1,091
03114 2400	PVC Chamfer Strips, 1" Radius	CLF	0.625	ACARB	0.046	138.60	0.00	51.80	8.74
03170 4580	Straight Edge Metal Formwork, 12" High (10 cm), with Stakes, 10' Length, 100 Uses	LF	52.5	ACARI	0.453	1.99	0.00	1.04	1.37
03217 0600	Grade 60 Rebar, Footings & Slabs, #3-#6	TON	0.313	SIWRC	0.396	492.16	0.00	609.86	436.51
03334 0300	Concrete Curing, Sprayed Membrane Curing Compound	CSF	11.88	ULABB	1.700	5.71	0.00	4.05	16.59
03372 4600	Pour Slab on Grade, >= 6", Direct Chute >= (15 cm) Place 3,000 PSI Concrete	CY	21.88	ALABE	6.297	7.70	0.29	84.22	580.59
03396 0150	Concrete Floor Finishes, Broom	SF	84.38	ACMAC	170.000	0.43	0.00	0.00	72.81
07192 0900	6 Mil Polyethylene Vapor Barrier	CSF	4.625	ACARA	1.870	9.36	0.00	3.30	23.69

19 02 03	Lift Stations	Unit	Hourly Output	Crew	Factor	Labor	Equip.	Matl.	Total
19 02 0313	**5' x 5' x 5' Reinforced Concrete Sump**	**EA**				**1,349**	**107.37**	**1,582**	**3,039**
02215 1215	Backfill Trench with 200HP Dozer without Compaction	CY	284	CODTG	23.140	0.15	0.43	0.00	13.54
02220 8800	Compact Backfill, by Hand with 1-Ton Roller, 6" (15 cm) Lift	CY	49.13	CLACD	23.140	0.94	1.51	0.00	56.69
02227 4000	Trim & Shape for Slab on Grade by Hand Finish Grade	SF	75.38	ULABA	27.500	0.45	0.00	0.00	12.36
02232 0155	Excavate & Load, 3 CY Hydraulic Excavator, Light Material, 218 CY/Hour (167 m3)	CY	218	XXQHK	23.140	0.57	0.88	0.00	33.66
02611 2201	Graded Crushed Aggregate Base	CY	25	XSABA	1.020	9.51	13.80	22.77	47.01
02752 4500	24" Diameter Cast Iron Catch Basin Frame & Cover NEENAH R2504C	EA	1.25	CODEJ	1.000	76.18	25.64	252.00	353.82
03114 2400	PVC Chamfer Strips, 1" Radius	CLF	0.625	ACARB	0.046	138.60	0.00	51.80	8.74
03150 2050	Elevated Slab Form with Drop Panels, 2 Use, Including Adjustable Shores, Form & Strip, Clean & Move	SF	51.88	ACARL	28.410	3.10	0.00	1.95	143.36
03170 3050	Slab on Grade Edge Forms, 7" to 12" High (18 cm to 31 cm), Based on 4 Uses	LF	54.38	ACARJ	25.320	2.42	0.00	0.66	77.83
03182 2000	Wall Forms, Up to 8' High Wall, 1 Use (Below Grade) Plywood Forms, Form & Strip with Accessories	SF	46.25	ACARL	250.000	3.48	0.00	2.69	1,543
03217 0600	Grade 60 Rebar, Footings & Slabs, #3-#6	TON	0.313	SIWRC	0.197	492.16	0.00	609.86	217.43
03334 0300	Concrete Curing, Sprayed Membrane Curing Compound	CSF	11.88	ULABB	0.400	5.71	0.00	4.05	3.90
03372 4600	Pour Slab on Grade, >= 6", Direct Chute >= (15 cm) Place 3,000 PSI Concrete	CY	21.88	ALABE	2.040	7.70	0.29	84.22	188.10
03372 4900	Pour Concrete Walls, 8" Thick, Direct Chute (21 cm) 3,000 PSI Concrete	CY	11.25	ALABE	3.140	14.96	0.57	84.22	313.21
03396 0150	Concrete Floor Finishes, Broom	SF	84.38	ACMAC	50.000	0.43	0.00	0.00	21.41
07192 0900	6 Mil Polyethylene Vapor Barrier	CSF	4.625	ACARA	0.400	9.36	0.00	3.30	5.07

19 02 04	Packaged Sanitary Sewer Treatment Plants	Unit	Hourly Output	Crew	Factor	Labor	Equip.	Matl.	Total
19 02 0401	**200,000 GPD, Packaged Sewage Treatment Plant**	**EA**				**0.00**	**0.00**	**396,500**	**396,500**
02776 2400	200,000 GPD (Steel)	EA	0	N/A	1.000	0.00	0.00	396,500	396,500
19 02 0402	**500,000 GPD, Packaged Sewage Treatment Plant**	**EA**				**0.00**	**0.00**	**955,000**	**955,000**
02776 2410	500,000 GPD (Steel)	EA	0	N/A	1.000	0.00	0.00	955,000	955,000
19 02 0403	**800,000 GPD, Packaged Sewage Treatment Plant**	**EA**				**0.00**	**0.00**	**1,505,000**	**1,505,000**
02776 2420	800,000 GPD (Steel)	EA	0	N/A	1.000	0.00	0.00	1,505,000	1,505,000

19 02 06	Drain Fields	Unit	Hourly Output	Crew	Factor	Labor	Equip.	Matl.	Total
19 02 0601	**12" x 12" Underground French Drain**	**LF**				**0.95**	**0.32**	**2.33**	**3.61**
02520 5014	12" x 12" Underground Trench Drain	LF	100	CODEJ	1.000	0.95	0.32	2.33	3.61

19 02 06	Drain Fields	Unit	Hourly Output	Crew	Factor	Labor	Equip.	Matl.	Total
19 02 0602	**18" x 18" Underground French Drain**	**LF**				**1.06**	**0.36**	**3.82**	**5.23**
02520 5015	18" x 18" Underground Trench Drain	LF	90	CODEJ	1.000	1.06	0.36	3.82	5.23
19 02 0603	**24" x 24" Underground French Drain**	**LF**				**1.19**	**0.40**	**5.28**	**6.87**
02520 5016	24" x 24" Underground Trench Drain	LF	80	CODEJ	1.000	1.19	0.40	5.28	6.87
19 02 0604	**12" x 12" CIP Concrete In-Ground Trench Drain with Metal Grate**	**LF**				**43.37**	**0.53**	**41.93**	**85.84**
02220 7260	Compact Backfill with Vibrating Plate Around Structures and Trenches	CY	14.75	CLACC	0.056	5.57	0.29	0.00	0.33
02228 0302	Trench, 1/2 CY Hydraulic Excavator, Medium Soil, 75 CY/Hour (58 m3)/Hour	CY	63.5	CODEA	0.301	0.99	1.10	0.00	0.63
03158 1500	2x4 Keyway Forms, 4 Uses	LF	66.25	ACARA	1.000	0.65	0.00	0.24	0.90
03170 3050	Slab on Grade Edge Forms, 7" to 12" High (18 cm to 31 cm), Based on 4 Uses	LF	54.38	ACARJ	2.120	2.42	0.00	0.66	6.52
03182 2050	Wall Forms, Up to 8' High Wall, 2 Uses (Below Grade) Plywood Forms, Form & Strip with Accessories	SF	54.38	ACARL	4.000	2.96	0.00	1.51	17.86
03217 0600	Grade 60 Rebar, Footings & Slabs, #3-#6	TON	0.313	SIWRC	0.010	492.16	0.00	609.86	11.12
03372 4600	Pour Slab on Grade, >= 6", Direct Chute >= (15 cm) Place 3,000 PSI Concrete	CY	21.88	ALABE	0.043	7.70	0.29	84.22	3.98
03372 4900	Pour Concrete Walls, 8" Thick, Direct Chute (21 cm) 3,000 PSI Concrete	CY	11.25	ALABE	0.049	14.96	0.57	84.22	4.93
03396 0010	Concrete Floor Finishes, Screed	SF	112.5	ACMAA	0.167	0.37	0.00	0.00	0.06
03398 0150	Wall Finish, Carborundum Wet Rub	SF	21.88	ACMAA	2.000	1.90	0.00	0.06	3.92
05120 2401	Miscellaneous Structural Shapes, Steel Angle A-36 Structural Steel	TON	0.05	SIWSB	0.007	1,928	0.00	1,630	23.26
05542 0696	1-3/4" x 3/16" Galvanized Steel Grating Welded Steel, 12.5 Lb/SF (2 Oz /SF)	SF	56.25	SIWSC	1.000	2.65	0.15	9.54	12.34

19 03 01	Storm Sewer Piping	Unit	Hourly Output	Crew	Factor	Labor	Equip.	Matl.	Total
19 03 0101	**8" Corrugated Metal Pipe, Bituminous Coated & Paved**	**LF**				**1.48**	**0.28**	**7.87**	**9.62**
02458 1017	8" Corrugated Metal Pipe Bituminous Coated & Paved	LF	115	CODEK	1.000	1.48	0.28	7.87	9.62
19 03 0102	**10" Corrugated Metal Pipe, Bituminous Coated & Paved**	**LF**				**1.48**	**0.28**	**9.83**	**11.59**
02458 1018	10" Corrugated Metal Pipe Bituminous Coated & Paved	LF	115	CODEK	1.000	1.48	0.28	9.83	11.59
19 03 0103	**12" Corrugated Metal Pipe, Bituminous Coated & Paved**	**LF**				**1.58**	**0.30**	**11.93**	**13.81**
02458 1019	12" Corrugated Metal Pipe Bituminous Coated & Paved	LF	107.5	CODEK	1.000	1.58	0.30	11.93	13.81
19 03 0104	**15" Corrugated Metal Pipe, Bituminous Coated & Paved**	**LF**				**1.77**	**0.33**	**14.95**	**17.05**
02458 1021	15" Corrugated Metal Pipe Bituminous Coated & Paved	LF	96.25	CODEK	1.000	1.77	0.33	14.95	17.05

19 03 01	Storm Sewer Piping	Unit	Hourly Output	Crew	Factor	Labor	Equip.	Matl.	Total
19 03 0105	**18" Corrugated Metal Pipe, Bituminous Coated & Paved**	**LF**				**1.77**	**0.33**	**17.31**	**19.41**
02458 1022	18" Corrugated Metal Pipe Bituminous Coated & Paved	LF	96.25	CODEK	1.000	1.77	0.33	17.31	19.41
19 03 0106	**24" Corrugated Metal Pipe, Bituminous Coated & Paved**	**LF**				**2.72**	**0.51**	**32.78**	**36.01**
02458 1023	24" Corrugated Metal Pipe Bituminous Coated & Paved	LF	62.5	CODEK	1.000	2.72	0.51	32.78	36.01
19 03 0107	**30" Corrugated Metal Pipe, Bituminous Coated & Paved**	**LF**				**3.58**	**0.67**	**39.33**	**43.59**
02458 1024	30" Corrugated Metal Pipe Bituminous Coated & Paved	LF	47.5	CODEK	1.000	3.58	0.67	39.33	43.59
19 03 0108	**36" Corrugated Metal Pipe, Bituminous Coated & Paved**	**LF**				**3.78**	**1.60**	**51.13**	**56.51**
02458 1025	36" Corrugated Metal Pipe Bituminous Coated & Paved	LF	45	UOEHC	1.000	3.78	1.60	51.13	56.51
19 03 0109	**48" Corrugated Metal Pipe, Bituminous Coated & Paved**	**LF**				**4.39**	**1.85**	**66.21**	**72.46**
02458 1026	48" Corrugated Metal Pipe Bituminous Coated & Paved	LF	38.75	UOEHC	1.000	4.39	1.85	66.21	72.46
19 03 0110	**60" Corrugated Metal Pipe, Bituminous Coated & Paved**	**LF**				**5.24**	**2.21**	**100.30**	**107.75**
02458 1027	60" Corrugated Metal Pipe Bituminous Coated & Paved	LF	32.5	UOEHC	1.000	5.24	2.21	100.30	107.75
19 03 0111	**72" Corrugated Metal Pipe, Bituminous Coated & Paved**	**LF**				**7.56**	**3.19**	**145.53**	**156.29**
02458 1028	72" Corrugated Metal Pipe Bituminous Coated & Paved	LF	22.5	UOEHC	1.000	7.56	3.19	145.53	156.29
19 03 0112	**8" Corrugated Metal Pipe, Plain**	**LF**				**1.43**	**0.27**	**6.69**	**8.39**
02458 2018	8" Corrugated Metal Pipe, Plain	LF	118.8	CODEK	1.000	1.43	0.27	6.69	8.39
19 03 0113	**10" Corrugated Metal Pipe, Plain**	**LF**				**1.43**	**0.27**	**8.39**	**10.09**
02458 2019	10" Corrugated Metal Pipe, Plain	LF	118.8	CODEK	1.000	1.43	0.27	8.39	10.09
19 03 0114	**12" Corrugated Metal Pipe, Plain**	**LF**				**1.55**	**0.29**	**9.64**	**11.47**
02458 2021	12" Corrugated Metal Pipe, Plain	LF	110	CODEK	1.000	1.55	0.29	9.64	11.47
19 03 0115	**15" Corrugated Metal Pipe, Plain**	**LF**				**1.72**	**0.32**	**11.87**	**13.91**
02458 2022	15" Corrugated Metal Pipe, Plain	LF	98.75	CODEK	1.000	1.72	0.32	11.87	13.91
19 03 0116	**18" Corrugated Metal Pipe, Plain**	**LF**				**2.62**	**0.49**	**13.37**	**16.48**
02458 2023	18" Corrugated Metal Pipe, Plain	LF	65	CODEK	1.000	2.62	0.49	13.37	16.48
19 03 0117	**24" Corrugated Metal Pipe, Plain**	**LF**				**3.49**	**0.66**	**20.72**	**24.86**
02458 2024	24" Corrugated Metal Pipe, Plain	LF	48.75	CODEK	1.000	3.49	0.66	20.72	24.86

19 03 01	Storm Sewer Piping	Unit	Hourly Output	Crew	Factor	Labor	Equip.	Matl.	Total
19 03 0118	**30" Corrugated Metal Pipe, Plain**	**LF**				**3.49**	**0.66**	**28.19**	**32.34**
02458 2025	30" Corrugated Metal Pipe, Plain	LF	48.75	CODEK	1.000	3.49	0.66	28.19	32.34
19 03 0119	**36" Corrugated Metal Pipe, Plain**	**LF**				**3.68**	**1.55**	**45.23**	**50.47**
02458 2026	36" Corrugated Metal Pipe, Plain	LF	46.25	UOEHC	1.000	3.68	1.55	45.23	50.47
19 03 0120	**48" Corrugated Metal Pipe, Plain**	**LF**				**4.25**	**1.80**	**60.97**	**67.02**
02458 2027	48" Corrugated Metal Pipe, Plain	LF	40	UOEHC	1.000	4.25	1.80	60.97	67.02
19 03 0121	**60" Corrugated Metal Pipe, Plain**	**LF**				**5.14**	**2.17**	**95.05**	**102.36**
02458 2028	60" Corrugated Metal Pipe, Plain	LF	33.13	UOEHC	1.000	5.14	2.17	95.05	102.36
19 03 0122	**18" x 11" (15" Equivalent) Corrugated Steel Pipe Arch, Coated & Paved**	**LF**				**1.77**	**0.33**	**15.67**	**17.77**
02458 3008	18" x 11" (15" Equivalent)	LF	96.25	CODEK	1.000	1.77	0.33	15.67	17.77
19 03 0123	**22" x 13" (18" Equivalent) Corrugated Steel Pipe Arch, Coated & Paved**	**LF**				**1.77**	**0.33**	**27.53**	**29.63**
02458 3009	22" x 13" (18" Equivalent)	LF	96.25	CODEK	1.000	1.77	0.33	27.53	29.63
19 03 0124	**29" x 18" (24" Equivalent) Corrugated Steel Pipe Arch, Coated & Paved**	**LF**				**2.72**	**0.51**	**31.47**	**34.70**
02458 3011	29" x 18" (24" Equivalent)	LF	62.5	CODEK	1.000	2.72	0.51	31.47	34.70
19 03 0125	**36" x 22" (30" Equivalent) Corrugated Steel Pipe Arch, Coated & Paved**	**LF**				**3.58**	**0.67**	**49.17**	**53.42**
02458 3012	36" x 22" (30" Equivalent)	LF	47.5	CODEK	1.000	3.58	0.67	49.17	53.42
19 03 0126	**43" x 27" (36" Equivalent) Corrugated Steel Pipe Arch, Coated & Paved**	**LF**				**3.78**	**1.60**	**51.79**	**57.17**
02458 3013	43" x 27" (36" Equivalent)	LF	45	UOEHC	1.000	3.78	1.60	51.79	57.17
19 03 0127	**50" x 31" (42" Equivalent) Corrugated Steel Pipe Arch, Coated & Paved**	**LF**				**4.25**	**1.80**	**59.65**	**65.71**
02458 3014	50" x 31" (42" Equivalent)	LF	40	UOEHC	1.000	4.25	1.80	59.65	65.71
19 03 0128	**58" x 36" (48" Equivalent) Corrugated Steel Pipe Arch, Coated & Paved**	**LF**				**4.25**	**1.80**	**68.18**	**74.23**
02458 3015	58" x 36" (48" Equivalent)	LF	40	UOEHC	1.000	4.25	1.80	68.18	74.23
19 03 0129	**72" x 44" (58" Equivalent) Corrugated Steel Pipe Arch, Coated & Paved**	**LF**				**4.25**	**1.80**	**82.60**	**88.65**
02458 3016	72" x 44" (58" Equivalent)	LF	40	UOEHC	1.000	4.25	1.80	82.60	88.65
19 03 0130	**72", 3 Gauge, Sectional Plate Pipe**	**LF**				**89.26**	**63.42**	**310.00**	**462.67**
02661 1250	72" (183 cm) Diameter Uncoated Steel Pipe, 7/8" Wall Thickness, Polyethylene, Welded	LF	5	XXPLD	1.000	89.26	63.42	310.00	462.67
19 03 0131	**78", 3 Gauge, Sectional Plate Pipe**	**LF**				**89.26**	**63.42**	**405.72**	**558.39**
02456 1402	78" (198 cm) Diameter Uncoated Steel Pipe, 7/8" Wall Thickness, Polyethylene, Welded	LF	5	XXPLD	1.000	89.26	63.42	405.72	558.39

19 03 01	Storm Sewer Piping	Unit	Hourly Output	Crew	Factor	Labor	Equip.	Matl.	Total
19 03 0132	**84", 3 Gauge, Sectional Plate Pipe**	**LF**				**97.43**	**65.43**	**415.00**	**577.86**
02661 1300	84" (213 cm) Diameter Uncoated Steel Pipe, 1" Wall Thickness, Polyethylene, Welded	LF	5	XXPLE	1.000	97.43	65.43	415.00	577.86
19 03 0133	**90", 3 Gauge, Sectional Plate Pipe**	**LF**				**97.43**	**65.43**	**484.45**	**647.31**
02456 1406	90" (229 cm) Diameter Uncoated Steel Pipe, 1" Wall Thickness, Polyethylene, Welded	LF	5	XXPLE	1.000	97.43	65.43	484.45	647.31
19 03 0134	**96", 3 Gauge, Sectional Plate Pipe**	**LF**				**105.60**	**66.89**	**475.00**	**647.49**
02661 1310	96" (244 cm) Diameter Uncoated Steel Pipe, 1" Wall Thickness, Polyethylene, Welded	LF	5	XXPLF	1.000	105.60	66.89	475.00	647.49
19 03 0135	**108", 3 Gauge, Sectional Plate Pipe**	**LF**				**194.85**	**130.87**	**620.00**	**945.72**
02661 1340	108" (274 cm) Diameter Uncoated Steel Pipe, 1-1/8" Wall Thickness, Polyethylene, Welded	LF	2.5	XXPLE	1.000	194.85	130.87	620.00	945.72
19 03 0136	**120", 3 Gauge, Sectional Plate Pipe**	**LF**				**194.85**	**130.87**	**690.00**	**1,016**
02661 1350	120" (305 cm) Diameter Uncoated Steel Pipe, 1-1/8" Wall Thickness, Polyethylene, Welded	LF	2.5	XXPLE	1.000	194.85	130.87	690.00	1,016
19 03 0137	**132", 3 Gauge, Sectional Plate Pipe**	**LF**				**389.70**	**261.73**	**890.00**	**1,541**
02661 1360	132" (335 cm) Diameter Uncoated Steel Pipe, 1-1/4" Wall Thickness, Polyethylene, Welded	LF	1.25	XXPLE	1.000	389.70	261.73	890.00	1,541
19 03 0138	**144", 3 Gauge, Sectional Plate Pipe**	**LF**				**389.70**	**261.73**	**890.00**	**1,541**
02661 1370	144" (366 cm) Diameter Uncoated Steel Pipe, 1-1/4" Wall Thickness, Polyethylene, Welded	LF	1.25	XXPLE	1.000	389.70	261.73	890.00	1,541
19 03 0139	**72", 5 Gauge, Sectional Plate Pipe**	**LF**				**89.26**	**63.42**	**266.00**	**418.67**
02661 1200	72" (183 cm) Diameter Uncoated Steel Pipe, 3/4" Wall Thickness, Polyethylene, Welded	LF	5	XXPLD	1.000	89.26	63.42	266.00	418.67
19 03 0140	**78", 5 Gauge, Sectional Plate Pipe**	**LF**				**89.26**	**63.42**	**351.37**	**504.05**
02456 1312	78" (198 cm) Diameter Uncoated Steel Pipe, 3/4" Wall Thickness, Polyethylene, Welded	LF	5	XXPLD	1.000	89.26	63.42	351.37	504.05
19 03 0141	**84", 5 Gauge, Sectional Plate Pipe**	**LF**				**89.26**	**63.42**	**310.00**	**462.67**
02661 1210	84" (213 cm) Diameter Uncoated Steel Pipe, 3/4" Wall Thickness, Polyethylene, Welded	LF	5	XXPLD	1.000	89.26	63.42	310.00	462.67
19 03 0142	**90", 5 Gauge, Sectional Plate Pipe**	**LF**				**97.43**	**65.43**	**386.25**	**549.11**
02456 1316	90" (229 cm) Diameter Uncoated Steel Pipe, 3/4" Wall Thickness, Polyethylene, Welded	LF	5	XXPLE	1.000	97.43	65.43	386.25	549.11
19 03 0143	**96", 5 Gauge, Sectional Plate Pipe**	**LF**				**97.43**	**65.43**	**355.00**	**517.86**
02661 1220	96" (244 cm) Diameter Uncoated Steel Pipe, 3/4" Wall Thickness, Polyethylene, Welded	LF	5	XXPLE	1.000	97.43	65.43	355.00	517.86
19 03 0144	**108", 5 Gauge, Sectional Plate Pipe**	**LF**				**105.60**	**66.89**	**614.29**	**786.77**
02456 1414	108" (274 cm) Diameter Uncoated Steel Pipe, 1" Wall Thickness, Polyethylene, Welded	LF	5	XXPLF	1.000	105.60	66.89	614.29	786.77

19 03 01 Storm Sewer Piping		Unit	Hourly Output	Crew	Factor	Labor	Equip.	Matl.	Total
19 03 0145	**120", 5 Gauge, Sectional Plate Pipe**	**LF**				**178.51**	**126.83**	**679.80**	**985.15**
02456 1421	120" (305 cm) Diameter Uncoated Steel Pipe, 1" Wall Thickness, Polyethylene, Welded	LF	2.5	XXPLD	1.000	178.51	126.83	679.80	985.15
19 03 0146	**132", 5 Gauge, Sectional Plate Pipe**	**LF**				**194.85**	**130.87**	**930.31**	**1,256**
02456 1428	132" (335 cm) Diameter Uncoated Steel Pipe, 1-1/8" Wall Thickness, Polyethylene, Welded	LF	2.5	XXPLE	1.000	194.85	130.87	930.31	1,256
19 03 0147	**144", 5 Gauge, Sectional Plate Pipe**	**LF**				**194.85**	**130.87**	**914.49**	**1,240**
02456 1437	144" (366 cm) Diameter Uncoated Steel Pipe, 1-1/8" Wall Thickness, Polyethylene, Welded	LF	2.5	XXPLE	1.000	194.85	130.87	914.49	1,240
19 03 0148	**72", 8 Gauge, Sectional Plate Pipe**	**LF**				**89.26**	**63.42**	**176.00**	**328.67**
02661 1130	72" (183 cm) Diameter Uncoated Steel Pipe, 1/2" Wall Thickness, Polyethylene, Welded	LF	5	XXPLD	1.000	89.26	63.42	176.00	328.67
19 03 0149	**78", 8 Gauge, Sectional Plate Pipe**	**LF**				**89.26**	**63.42**	**250.42**	**403.09**
02456 1229	78" (198 cm) Diameter Uncoated Steel Pipe, 1/2" Wall Thickness, Polyethylene, Welded	LF	5	XXPLD	1.000	89.26	63.42	250.42	403.09
19 03 0150	**84", 8 Gauge, Sectional Plate Pipe**	**LF**				**89.26**	**63.42**	**259.00**	**411.67**
02661 1170	84" (213 cm) Diameter Uncoated Steel Pipe, 5/8" Wall Thickness, Polyethylene, Welded	LF	5	XXPLD	1.000	89.26	63.42	259.00	411.67
19 03 0151	**90", 8 Gauge, Sectional Plate Pipe**	**LF**				**89.26**	**63.42**	**327.12**	**479.79**
02456 1315	90" (229 cm) Diameter Uncoated Steel Pipe, 5/8" Wall Thickness, Polyethylene, Welded	LF	5	XXPLD	1.000	89.26	63.42	327.12	479.79
19 03 0152	**96", 8 Gauge, Sectional Plate Pipe**	**LF**				**89.26**	**63.42**	**297.00**	**449.67**
02661 1180	96" (244 cm) Diameter Uncoated Steel Pipe, 5/8" Wall Thickness, Polyethylene, Welded	LF	5	XXPLD	1.000	89.26	63.42	297.00	449.67
19 03 0153	**108", 8 Gauge, Sectional Plate Pipe**	**LF**				**97.43**	**65.43**	**490.00**	**652.86**
02661 1280	108" (274 cm) Diameter Uncoated Steel Pipe, 7/8" Wall Thickness, Polyethylene, Welded	LF	5	XXPLE	1.000	97.43	65.43	490.00	652.86
19 03 0154	**120", 8 Gauge, Sectional Plate Pipe**	**LF**				**178.51**	**126.83**	**520.00**	**825.35**
02661 1290	120" (305 cm) Diameter Uncoated Steel Pipe, 7/8" Wall Thickness, Polyethylene, Welded	LF	2.5	XXPLD	1.000	178.51	126.83	520.00	825.35
19 03 0155	**132", 8 Gauge, Sectional Plate Pipe**	**LF**				**178.51**	**126.83**	**675.00**	**980.35**
02661 1320	132" (335 cm) Diameter Uncoated Steel Pipe, 1" Wall Thickness, Polyethylene, Welded	LF	2.5	XXPLD	1.000	178.51	126.83	675.00	980.35
19 03 0156	**144", 8 Gauge, Sectional Plate Pipe**	**LF**				**178.51**	**126.83**	**735.00**	**1,040**
02661 1330	144" (366 cm) Diameter Uncoated Steel Pipe, 1" Wall Thickness, Polyethylene, Welded	LF	2.5	XXPLD	1.000	178.51	126.83	735.00	1,040
19 03 0157	**6" Extra Strength, Nonreinforced Concrete Pipe**	**LF**				**3.40**	**0.64**	**2.95**	**6.99**
02451 1012	6" Extra Strength	LF	50	CODEK	1.000	3.40	0.64	2.95	6.99

19 03 01 Storm Sewer Piping		Unit	Hourly Output	Crew	Factor	Labor	Equip.	Matl.	Total
19 03 0158	**8" Extra Strength, Nonreinforced Concrete Pipe**	**LF**				**3.78**	**0.71**	**4.43**	**8.92**
02451 1013	8" Extra Strength	LF	45	CODEK	1.000	3.78	0.71	4.43	8.92
19 03 0159	**10" Extra Strength, Nonreinforced Concrete Pipe**	**LF**				**4.00**	**0.75**	**8.85**	**13.61**
02451 1014	10" Extra Strength	LF	42.5	CODEK	1.000	4.00	0.75	8.85	13.61
19 03 0160	**12" Extra Strength, Nonreinforced Concrete Pipe**	**LF**				**4.25**	**0.80**	**9.57**	**14.63**
02451 1015	12" Extra Strength	LF	40	CODEK	1.000	4.25	0.80	9.57	14.63
19 03 0161	**15" Extra Strength, Nonreinforced Concrete Pipe**	**LF**				**4.73**	**0.89**	**10.62**	**16.24**
02451 1016	15" Extra Strength	LF	36	CODEK	1.000	4.73	0.89	10.62	16.24
19 03 0162	**18" Extra Strength, Nonreinforced Concrete Pipe**	**LF**				**5.24**	**0.99**	**15.47**	**21.69**
02451 1017	18" Extra Strength	LF	32.5	CODEK	1.000	5.24	0.99	15.47	21.69
19 03 0163	**21" Extra Strength, Nonreinforced Concrete Pipe**	**LF**				**5.24**	**0.99**	**22.54**	**28.76**
02451 1018	21" Extra Strength	LF	32.5	CODEK	1.000	5.24	0.99	22.54	28.76
19 03 0164	**24" Extra Strength, Nonreinforced Concrete Pipe**	**LF**				**5.84**	**1.10**	**32.83**	**39.77**
02451 1019	24" Extra Strength	LF	29.13	CODEK	1.000	5.84	1.10	32.83	39.77
19 03 0165	**12" Reinforced Concrete Pipe, Class 3, with Gaskets**	**LF**				**2.13**	**0.90**	**10.82**	**13.84**
02452 1025	12" Reinforced Concrete Pipe	LF	80	UOEHC	1.000	2.13	0.90	10.42	13.45
02762 6200	12" Diameter Joint Gasket	EA	0	N/A	0.125	0.00	0.00	3.16	0.40
19 03 0166	**15" Reinforced Concrete Pipe, Class 3, with Gaskets**	**LF**				**2.64**	**1.11**	**11.51**	**15.27**
02452 1026	15" Reinforced Concrete Pipe	LF	64.5	UOEHC	1.000	2.64	1.11	11.21	14.96
02452 4002	15" Diameter Joint Gasket	EA	0	N/A	0.125	0.00	0.00	2.44	0.30
19 03 0167	**18" Reinforced Concrete Pipe, Class 3, with Gaskets**	**LF**				**4.25**	**1.80**	**14.86**	**20.91**
02452 1027	18" Reinforced Concrete Pipe	LF	40	UOEHC	1.000	4.25	1.80	14.42	20.47
02762 6210	18" Diameter Joint Gasket	EA	0	N/A	0.125	0.00	0.00	3.49	0.44
19 03 0168	**24" Reinforced Concrete Pipe, Class 3, with Gaskets**	**LF**				**5.44**	**2.30**	**24.78**	**32.53**
02452 1029	24" Reinforced Concrete Pipe	LF	31.25	UOEHC	1.000	5.44	2.30	24.06	31.80
02762 6220	24" Diameter Joint Gasket	EA	0	N/A	0.125	0.00	0.00	5.80	0.72

19 03 01 Storm Sewer Piping		Unit	Hourly Output	Crew	Factor	Labor	Equip.	Matl.	Total
19 03 0169	**30" Reinforced Concrete Pipe, Class 3, with Gaskets**	**LF**				**6.81**	**2.88**	**32.98**	**42.66**
02452 1032	30" Reinforced Concrete Pipe	LF	25	UOEHC	1.000	6.81	2.88	32.12	41.80
02762 6230	30" Diameter Joint Gasket	EA	0	N/A	0.125	0.00	0.00	6.85	0.86
19 03 0170	**36" Reinforced Concrete Pipe, Class 3, with Gaskets**	**LF**				**8.51**	**3.59**	**48.25**	**60.35**
02452 1033	36" Reinforced Concrete Pipe	LF	20	UOEHC	1.000	8.51	3.59	47.20	59.30
02762 6240	36" Diameter Joint Gasket	EA	0	N/A	0.125	0.00	0.00	8.40	1.05
19 03 0171	**48" Reinforced Concrete Pipe, Class 3, with Gaskets**	**LF**				**10.08**	**4.26**	**78.36**	**92.70**
02452 1034	48" Reinforced Concrete Pipe	LF	16.88	UOEHC	1.000	10.08	4.26	76.70	91.04
02762 6260	48" Diameter Joint Gasket	EA	0	N/A	0.125	0.00	0.00	13.25	1.66
19 03 0172	**60" Reinforced Concrete Pipe, Class 3, with Gaskets**	**LF**				**13.09**	**5.53**	**117.08**	**135.70**
02452 1035	60" Reinforced Concrete Pipe	LF	13	UOEHC	1.000	13.09	5.53	114.72	133.34
02762 6270	60" Diameter Joint Gasket	EA	0	N/A	0.125	0.00	0.00	18.90	2.36
19 03 0173	**72" Reinforced Concrete Pipe, Class 3, with Gaskets**	**LF**				**13.61**	**5.75**	**161.76**	**181.12**
02452 1036	72" Reinforced Concrete Pipe	LF	12.5	UOEHC	1.000	13.61	5.75	159.95	179.32
02762 6280	72" Diameter Joint Gasket	EA	0	N/A	0.125	0.00	0.00	14.42	1.80
19 03 0174	**12" Reinforced Concrete Pipe, Class 4, with Gaskets**	**LF**				**2.13**	**0.90**	**10.82**	**13.84**
02452 1025	12" Reinforced Concrete Pipe	LF	80	UOEHC	1.000	2.13	0.90	10.42	13.45
02762 6200	12" Diameter Joint Gasket	EA	0	N/A	0.125	0.00	0.00	3.16	0.40
19 03 0175	**15" Reinforced Concrete Pipe, Class 4, with Gaskets**	**LF**				**2.64**	**1.11**	**11.51**	**15.27**
02452 1026	15" Reinforced Concrete Pipe	LF	64.5	UOEHC	1.000	2.64	1.11	11.21	14.96
02452 4002	15" Diameter Joint Gasket	EA	0	N/A	0.125	0.00	0.00	2.44	0.30
19 03 0176	**18" Reinforced Concrete Pipe, Class 4, with Gaskets**	**LF**				**4.25**	**1.80**	**14.86**	**20.91**
02452 1027	18" Reinforced Concrete Pipe	LF	40	UOEHC	1.000	4.25	1.80	14.42	20.47
02762 6210	18" Diameter Joint Gasket	EA	0	N/A	0.125	0.00	0.00	3.49	0.44
19 03 0177	**21" Reinforced Concrete Pipe, Class 4, with Gaskets**	**LF**				**4.39**	**1.85**	**18.98**	**25.23**
02452 1028	21" Reinforced Concrete Pipe	LF	38.75	UOEHC	1.000	4.39	1.85	18.26	24.50
02762 6220	24" Diameter Joint Gasket	EA	0	N/A	0.125	0.00	0.00	5.80	0.72
19 03 0178	**24" Reinforced Concrete Pipe, Class 4, with Gaskets**	**LF**				**5.44**	**2.30**	**24.78**	**32.53**
02452 1029	24" Reinforced Concrete Pipe	LF	31.25	UOEHC	1.000	5.44	2.30	24.06	31.80
02762 6220	24" Diameter Joint Gasket	EA	0	N/A	0.125	0.00	0.00	5.80	0.72

19 03 01 Storm Sewer Piping		Unit	Hourly Output	Crew	Factor	Labor	Equip.	Matl.	Total
19 03 0179	**27" Reinforced Concrete Pipe, Class 4, with Gaskets**	**LF**				**6.05**	**2.56**	**29.04**	**37.65**
02452 1031	27" Reinforced Concrete Pipe	LF	28.13	UOEHC	1.000	6.05	2.56	28.19	36.79
02762 6230	30" Diameter Joint Gasket	EA	0	N/A	0.125	0.00	0.00	6.85	0.86
19 03 0180	**30" Reinforced Concrete Pipe, Class 4, with Gaskets**	**LF**				**6.81**	**2.88**	**32.98**	**42.66**
02452 1032	30" Reinforced Concrete Pipe	LF	25	UOEHC	1.000	6.81	2.88	32.12	41.80
02762 6230	30" Diameter Joint Gasket	EA	0	N/A	0.125	0.00	0.00	6.85	0.86
19 03 0181	**36" Reinforced Concrete Pipe, Class 4, with Gaskets**	**LF**				**8.51**	**3.59**	**48.25**	**60.35**
02452 1033	36" Reinforced Concrete Pipe	LF	20	UOEHC	1.000	8.51	3.59	47.20	59.30
02762 6240	36" Diameter Joint Gasket	EA	0	N/A	0.125	0.00	0.00	8.40	1.05
19 03 0182	**12" Reinforced Concrete Pipe, Class 5, with Gaskets**	**LF**				**2.13**	**0.90**	**10.82**	**13.84**
02452 1025	12" Reinforced Concrete Pipe	LF	80	UOEHC	1.000	2.13	0.90	10.42	13.45
02762 6200	12" Diameter Joint Gasket	EA	0	N/A	0.125	0.00	0.00	3.16	0.40
19 03 0183	**15" Reinforced Concrete Pipe, Class 5, with Gaskets**	**LF**				**2.64**	**1.11**	**11.51**	**15.27**
02452 1026	15" Reinforced Concrete Pipe	LF	64.5	UOEHC	1.000	2.64	1.11	11.21	14.96
02452 4002	15" Diameter Joint Gasket	EA	0	N/A	0.125	0.00	0.00	2.44	0.30
19 03 0184	**18" Reinforced Concrete Pipe, Class 5, with Gaskets**	**LF**				**4.25**	**1.80**	**14.86**	**20.91**
02452 1027	18" Reinforced Concrete Pipe	LF	40	UOEHC	1.000	4.25	1.80	14.42	20.47
02762 6210	18" Diameter Joint Gasket	EA	0	N/A	0.125	0.00	0.00	3.49	0.44
19 03 0185	**21" Reinforced Concrete Pipe, Class 5, with Gaskets**	**LF**				**4.39**	**1.85**	**18.98**	**25.23**
02452 1028	21" Reinforced Concrete Pipe	LF	38.75	UOEHC	1.000	4.39	1.85	18.26	24.50
02762 6220	24" Diameter Joint Gasket	EA	0	N/A	0.125	0.00	0.00	5.80	0.72
19 03 0186	**24" Reinforced Concrete Pipe, Class 5, with Gaskets**	**LF**				**5.44**	**2.30**	**24.78**	**32.53**
02452 1029	24" Reinforced Concrete Pipe	LF	31.25	UOEHC	1.000	5.44	2.30	24.06	31.80
02762 6220	24" Diameter Joint Gasket	EA	0	N/A	0.125	0.00	0.00	5.80	0.72
19 03 0187	**27" Reinforced Concrete Pipe, Class 5, with Gaskets**	**LF**				**6.05**	**2.56**	**29.04**	**37.65**
02452 1031	27" Reinforced Concrete Pipe	LF	28.13	UOEHC	1.000	6.05	2.56	28.19	36.79
02762 6230	30" Diameter Joint Gasket	EA	0	N/A	0.125	0.00	0.00	6.85	0.86
19 03 0188	**30" Reinforced Concrete Pipe, Class 5, with Gaskets**	**LF**				**6.81**	**2.88**	**32.98**	**42.66**
02452 1032	30" Reinforced Concrete Pipe	LF	25	UOEHC	1.000	6.81	2.88	32.12	41.80
02762 6230	30" Diameter Joint Gasket	EA	0	N/A	0.125	0.00	0.00	6.85	0.86

19 03 Storm Sewer Systems

19 03 01 Storm Sewer Piping		Unit	Hourly Output	Crew	Factor	Labor	Equip.	Matl.	Total
19 03 0189	**36" Reinforced Concrete Pipe, Class 5, with Gaskets**	**LF**				**8.51**	**3.59**	**48.25**	**60.35**
02452 1033	36" Reinforced Concrete Pipe	LF	20	UOEHC	1.000	8.51	3.59	47.20	59.30
02762 6240	36" Diameter Joint Gasket	EA	0	N/A	0.125	0.00	0.00	8.40	1.05
19 03 0190	**42" Reinforced Concrete Pipe, Class 5, with Gaskets**	**LF**				**8.51**	**3.59**	**62.22**	**74.32**
02452 1037	42" Reinforced Concrete Pipe	LF	20	UOEHC	1.000	8.51	3.59	60.97	73.07
02762 6250	42" Diameter Joint Gasket	EA	0	N/A	0.125	0.00	0.00	10.00	1.25
19 03 0191	**48" Reinforced Concrete Pipe, Class 5, with Gaskets**	**LF**				**10.08**	**4.26**	**78.36**	**92.70**
02452 1034	48" Reinforced Concrete Pipe	LF	16.88	UOEHC	1.000	10.08	4.26	76.70	91.04
02762 6260	48" Diameter Joint Gasket	EA	0	N/A	0.125	0.00	0.00	13.25	1.66
19 03 0192	**60" Reinforced Concrete Pipe, Class 5, with Gaskets**	**LF**				**13.09**	**5.53**	**117.08**	**135.70**
02452 1035	60" Reinforced Concrete Pipe	LF	13	UOEHC	1.000	13.09	5.53	114.72	133.34
02762 6270	60" Diameter Joint Gasket	EA	0	N/A	0.125	0.00	0.00	18.90	2.36
19 03 0193	**72" Reinforced Concrete Pipe, Class 5, with Gaskets**	**LF**				**13.61**	**5.75**	**161.76**	**181.12**
02452 1036	72" Reinforced Concrete Pipe	LF	12.5	UOEHC	1.000	13.61	5.75	159.95	179.32
02762 6280	72" Diameter Joint Gasket	EA	0	N/A	0.125	0.00	0.00	14.42	1.80

19 03 04 Culverts		Unit	Hourly Output	Crew	Factor	Labor	Equip.	Matl.	Total
19 03 0401	**18' Complete, 24" Corrugated Metal Pipe Culvert with Headwalls**	**EA**				**1,355**	**840.73**	**2,814**	**5,010**
02270 0110	Grouted Rip Rap, Random 3/8 CY to 1/4 CY Pieces	SY	6.25	COETF	65.000	18.50	12.49	36.50	4,387
02764 2140	24" (61 cm) 16 Gauge Corrugated Metal Pipe Galvanized or Aluminum, Bituminous Coated	LF	20	CODEK	18.000	8.51	1.60	24.50	622.98
19 03 0402	**34' Complete, 24" Corrugated Metal Pipe Culvert with Headwalls**	**EA**				**1,491**	**866.37**	**3,206**	**5,563**
02270 0110	Grouted Rip Rap, Random 3/8 CY to 1/4 CY Pieces	SY	6.25	COETF	65.000	18.50	12.49	36.50	4,387
02764 2140	24" (61 cm) 16 Gauge Corrugated Metal Pipe Galvanized or Aluminum, Bituminous Coated	LF	20	CODEK	34.000	8.51	1.60	24.50	1,177
19 03 0403	**52' Complete, 24" Corrugated Metal Pipe Culvert with Headwalls**	**EA**				**1,645**	**895.21**	**3,647**	**6,186**
02270 0110	Grouted Rip Rap, Random 3/8 CY to 1/4 CY Pieces	SY	6.25	COETF	65.000	18.50	12.49	36.50	4,387
02764 2140	24" (61 cm) 16 Gauge Corrugated Metal Pipe Galvanized or Aluminum, Bituminous Coated	LF	20	CODEK	52.000	8.51	1.60	24.50	1,800

19 03 04 Culverts		Unit	Hourly Output	Crew	Factor	Labor	Equip.	Matl.	Total
19 03 0404	**76' Complete, 24" Corrugated Metal Pipe Culvert with Headwalls**	**EA**				**1,849**	**933.67**	**4,235**	**7,017**
02270 0110	Grouted Rip Rap, Random 3/8 CY to 1/4 CY Pieces	SY	6.25	COETF	65.000	18.50	12.49	36.50	4,387
02764 2140	24" (61 cm) 16 Gauge Corrugated Metal Pipe Galvanized or Aluminum, Bituminous Coated	LF	20	CODEK	76.000	8.51	1.60	24.50	2,630
19 03 0410	**Continuous Footing, Edge Form, 4 Uses**	**SF**				**2.17**	**0.00**	**0.82**	**2.99**
03158 0150	Continuous Wall Footing Forms, 4 Uses Plywood Forms, Form & Strip with Accessories	SF	60.63	ACARJ	1.000	2.17	0.00	0.82	2.99
19 03 0411	**Continuous Footing, Pour Concrete**	**CY**				**11.22**	**0.42**	**84.22**	**95.86**
03372 1900	Pour Continuous Footings, Shallow, Direct Chute Place 3,000 PSI Concrete Foundations	CY	15	ALABE	1.000	11.22	0.42	84.22	95.86
19 03 0415	**CIP Wall Rebar**	**LB**				**0.24**	**0.10**	**0.32**	**0.66**
03210 1011	Wall Rebar, #3-#7	TON	0.375	SIWRF	0.001	482.29	191.67	647.24	0.66
19 03 0416	**Pour & Finish Concrete (1-Side), Retaining Wall**	**CY**				**33.95**	**10.51**	**90.95**	**135.41**
03372 5200	Pour Concrete Wall, 12" Thick, Crane & Bucket (31 cm) 3,000 PSI Concrete	CY	11.25	ALABG	1.000	23.20	10.51	84.22	117.94
07174 0100	Silicone Retarder, 1 Coat on Concrete, Sprayed	CSF	1.875	ARFCD	0.540	19.90	0.00	12.47	17.48
19 03 0417	**Pour & Cure CIP Walls**	**CY**				**42.22**	**11.83**	**94.32**	**148.37**
03372 5000	Pour Concrete Walls, 8" Thick, Crane & Bucket (21 cm) 3,000 PSI Concrete	CY	10	ALABG	1.000	26.11	11.83	84.22	122.15
07174 0100	Silicone Retarder, 1 Coat on Concrete, Sprayed	CSF	1.875	ARFCD	0.810	19.90	0.00	12.47	26.22
19 03 0419	**Slab on Grade, Edge Form, 4 Uses**	**LF**				**2.42**	**0.00**	**0.66**	**3.07**
03170 3050	Slab on Grade Edge Forms, 7" to 12" High (18 cm to 31 cm), Based on 4 Uses	LF	54.38	ACARJ	1.000	2.42	0.00	0.66	3.07
19 03 0420	**Slab on Grade, Rebar**	**LB**				**0.31**	**0.12**	**0.32**	**0.76**
03210 1007	Rebar #3-#7	TON	0.288	SIWRF	0.001	629.08	250.00	647.24	0.76
19 03 0421	**Pour & Cure Concrete, Slab on Grade**	**CY**				**39.44**	**2.29**	**90.95**	**132.68**
03372 4650	Pour Slab on Grade, >= 6", Concrete Pump >= (15 cm) Place 3,000 PSI Concrete	CY	41.25	ALABI	1.000	5.57	2.29	84.22	92.07
03396 0150	Concrete Floor Finishes, Broom	SF	84.38	ACMAC	54.000	0.43	0.00	0.00	23.13
07174 0100	Silicone Retarder, 1 Coat on Concrete, Sprayed	CSF	1.875	ARFCD	0.540	19.90	0.00	12.47	17.48
19 03 0424	**Top Slab Cover, Form & Strip, 4 Uses**	**SF**				**3.71**	**0.00**	**0.86**	**4.57**
03150 6650	Elevated Slab Curb Form, 6-12",4 Uses	SF	28.13	ACARI	1.000	3.71	0.00	0.86	4.57
19 03 0425	**Elevated Slab Rebar**	**LB**				**0.26**	**0.10**	**0.31**	**0.67**
03210 1008	Rebar #8-#14	TON	0.35	SIWRF	0.001	516.74	205.36	616.05	0.67

19 03 04	Culverts	Unit	Hourly Output	Crew	Factor	Labor	Equip.	Matl.	Total
19 03 0426	**Pour & Cure Top Slab Cover**	**CY**				**71.05**	**8.60**	**94.56**	**174.22**
03372 4400	Pour Slab on Grade, < 6", Crane & Bucket < (15 cm) Place 3,000 PSI Concrete	CY	13.75	ALABG	1.000	18.99	8.60	84.22	111.81
03396 0150	Concrete Floor Finishes, Broom	SF	84.38	ACMAC	83.000	0.43	0.00	0.00	35.55
07174 0100	Silicone Retarder, 1 Coat on Concrete, Sprayed	CSF	1.875	ARFCD	0.830	19.90	0.00	12.47	26.86
19 03 0429	**Bush Hammer Finish**	**SF**				**1.96**	**0.00**	**0.06**	**2.01**
03398 0300	Wall Finish, Bush Hammer Green Concrete	SF	21.25	ACMAA	1.000	1.96	0.00	0.06	2.01

19 03 05	Headwalls and Catch Basins	Unit	Hourly Output	Crew	Factor	Labor	Equip.	Matl.	Total
19 03 0501	**Lean Concrete Riprap/Headwall/Splash Slab, No Reinforcement**	**SF**				**2.06**	**1.39**	**4.06**	**7.50**
02270 0110	Grouted Rip Rap, Random 3/8 CY to 1/4 CY Pieces	SY	6.25	COETF	0.111	18.50	12.49	36.50	7.50
19 03 0505	**Headwall for 12" CIP Pipe, with Skewed Wingwall**	**EA**				**465.00**	**0.00**	**144.00**	**609.00**
02753 0100	CIP Concrete Headwall, 12" Pipe 30 Degree Skewed Wingwall	EA	0.4	ALABM	1.000	465.00	0.00	144.00	609.00
19 03 0506	**Headwall for 24" CIP Pipe, with Skewed Wingwall**	**EA**				**930.00**	**0.00**	**330.00**	**1,260**
02753 0120	CIP Concrete Headwall, 24" Pipe 30 Degree Skewed Wingwall	EA	0.2	ALABM	1.000	930.00	0.00	330.00	1,260
19 03 0507	**Headwall for 36" CIP Pipe, with Skewed Wingwall**	**EA**				**1,634**	**0.00**	**510.00**	**2,144**
02753 0140	CIP Concrete Headwall, 36" Pipe 30 Degree Skewed Wingwall	EA	0.114	ALABM	1.000	1,634	0.00	510.00	2,144
19 03 0508	**Headwall for 48" CIP Pipe, with Skewed Wingwall**	**EA**				**2,657**	**0.00**	**835.00**	**3,492**
02753 0150	CIP Concrete Headwall, 48" Pipe 30 Degree Skewed Wingwall	EA	0.07	ALABM	1.000	2,657	0.00	835.00	3,492
19 03 0509	**Headwall for 60" CIP Pipe, with Skewed Wingwall**	**EA**				**3,916**	**0.00**	**1,875**	**5,791**
02753 0160	CIP Concrete Headwall, 60" Pipe 30 Degree Skewed Wingwall	EA	0.048	ALABM	1.000	3,916	0.00	1,875	5,791
19 03 0515	**Headwall for 12" Precast Pipe, without Wingwall**	**EA**				**113.43**	**21.37**	**365.00**	**499.80**
02753 0200	Precast Concrete Headwall, 12" Pipe	EA	1.5	CODEK	1.000	113.43	21.37	365.00	499.80
19 03 0516	**Headwall for 24" Precast Pipe, without Wingwall**	**EA**				**136.12**	**25.64**	**418.22**	**579.98**
02753 0220	Precast Concrete Headwall, 24" Pipe	EA	1.25	CODEK	1.000	136.12	25.64	418.22	579.98
19 03 0517	**Headwall for 36" Precast Pipe, without Wingwall**	**EA**				**170.15**	**32.05**	**709.51**	**911.71**
02753 0240	Precast Concrete Headwall, 36" Pipe	EA	1	CODEK	1.000	170.15	32.05	709.51	911.71

19 03 05 Headwalls and Catch Basins		Unit	Hourly Output	Crew	Factor	Labor	Equip.	Matl.	Total
19 03 0518	**Headwall for 48" Precast Pipe, without Wingwall**	**EA**				**194.46**	**82.14**	**1,142**	**1,419**
02753 0250	Precast Concrete Headwall, 48" Pipe	EA	0.875	UOEHC	1.000	194.46	82.14	1,142	1,419
19 03 0519	**Headwall for 60" Precast Pipe, without Wingwall**	**EA**				**226.87**	**95.83**	**1,158**	**1,481**
02753 0260	Precast Concrete Headwall, 60" Pipe	EA	0.75	UOEHC	1.000	226.87	95.83	1,158	1,481

19 03 99 Other Storm Sewer		Unit	Hourly Output	Crew	Factor	Labor	Equip.	Matl.	Total
19 03 9901	**8" Corrugated Metal Pipe, Bituminous Coated, End Section**	**EA**				**90.75**	**17.09**	**2,025**	**2,133**
02764 2294	8" Diameter	EA	1.875	CODEK	1.000	90.75	17.09	2,025	2,133
19 03 9902	**10" Corrugated Metal Pipe, Bituminous Coated, End Section**	**EA**				**113.43**	**21.37**	**2,425**	**2,560**
02764 2296	10" Diameter	EA	1.5	CODEK	1.000	113.43	21.37	2,425	2,560
19 03 9903	**12" Corrugated Metal Pipe, Bituminous Coated, End Section**	**EA**				**136.12**	**25.64**	**52.50**	**214.26**
02764 2250	12" Diameter 16 Gauge	EA	1.25	CODEK	1.000	136.12	25.64	52.50	214.26
19 03 9904	**15" Corrugated Metal Pipe, Bituminous Coated, End Section**	**EA**				**151.24**	**28.49**	**59.50**	**239.23**
02764 2255	15" Diameter 16 Gauge	EA	1.125	CODEK	1.000	151.24	28.49	59.50	239.23
19 03 9905	**18" Corrugated Metal Pipe, Bituminous Coated, End Section**	**EA**				**170.15**	**32.05**	**46.46**	**248.66**
02764 2260	18" Diameter 16 Gauge	EA	1	CODEK	1.000	170.15	32.05	46.46	248.66
19 03 9906	**24" Corrugated Metal Pipe, Bituminous Coated, End Section**	**EA**				**194.46**	**36.63**	**54.94**	**286.03**
02764 2265	24" Diameter 16 Gauge	EA	0.875	CODEK	1.000	194.46	36.63	54.94	286.03
19 03 9907	**30" Corrugated Metal Pipe, Bituminous Coated, End Section**	**EA**				**209.42**	**39.45**	**115.00**	**363.86**
02764 2270	30" Diameter 16 Gauge	EA	0.813	CODEK	1.000	209.42	39.45	115.00	363.86
19 03 9908	**36" Corrugated Metal Pipe, Bituminous Coated, End Section**	**EA**				**226.87**	**42.73**	**164.00**	**433.60**
02764 2275	36" Diameter 14 Gauge	EA	0.75	CODEK	1.000	226.87	42.73	164.00	433.60
19 03 9909	**48" Corrugated Metal Pipe, Bituminous Coated, End Section**	**EA**				**272.24**	**51.28**	**1,175**	**1,499**
02764 2292	48" Diameter	EA	0.625	CODEK	1.000	272.24	51.28	1,175	1,499
19 03 9910	**60" Corrugated Metal Pipe, Bituminous Coated, End Section**	**EA**				**340.30**	**64.10**	**310.00**	**714.40**
02764 2280	60" Diameter	EA	0.5	CODEK	1.000	340.30	64.10	310.00	714.40

19 03 99	Other Storm Sewer	Unit	Hourly Output	Crew	Factor	Labor	Equip.	Matl.	Total
19 03 9911	**72" Corrugated Metal Pipe, Bituminous Coated, End Section**	**EA**				**388.91**	**73.26**	**410.00**	**872.17**
02764 2285	72" Diameter	EA	0.438	CODEK	1.000	388.91	73.26	410.00	872.17
19 03 9915	**8" Corrugated Metal Pipe, End Section**	**EA**				**36.79**	**6.93**	**71.50**	**115.22**
02764 2780	8" (21 cm) Diameter	EA	4.625	CODEK	1.000	36.79	6.93	71.50	115.22
19 03 9916	**10" Corrugated Metal Pipe, End Section**	**EA**				**36.79**	**6.93**	**73.00**	**116.72**
02764 2785	10" (25 cm) Diameter	EA	4.625	CODEK	1.000	36.79	6.93	73.00	116.72
19 03 9917	**12" Corrugated Metal Pipe, End Section**	**EA**				**38.89**	**7.33**	**82.50**	**128.72**
02764 2790	12" (31 cm) Corrugated Metal Pipe Ends	EA	4.375	CODEK	1.000	38.89	7.33	82.50	128.72
19 03 9918	**15" Corrugated Metal Pipe, End Section**	**EA**				**38.89**	**7.33**	**87.84**	**134.06**
02458 2602	15" (38 cm) Corrugated Metal Pipe Ends	EA	4.375	CODEK	1.000	38.89	7.33	87.84	134.06
19 03 9919	**18" Corrugated Metal Pipe, End Section**	**EA**				**45.37**	**8.55**	**98.00**	**151.92**
02764 2800	18" (46 cm) Corrugated Metal Pipe Ends	EA	3.75	CODEK	1.000	45.37	8.55	98.00	151.92
19 03 9920	**24" Corrugated Metal Pipe, End Section**	**EA**				**54.45**	**10.26**	**143.00**	**207.70**
02764 2810	24" (61 cm) Corrugated Metal Pipe Ends	EA	3.125	CODEK	1.000	54.45	10.26	143.00	207.70
19 03 9921	**30" Corrugated Metal Pipe, End Section**	**EA**				**54.45**	**10.26**	**272.00**	**336.70**
02764 2820	30" (76 cm) Corrugated Metal Pipe Ends	EA	3.125	CODEK	1.000	54.45	10.26	272.00	336.70
19 03 9922	**36" Corrugated Metal Pipe, End Section**	**EA**				**68.06**	**12.82**	**360.00**	**440.88**
02764 2825	36" (91 cm) Corrugated Metal Pipe Ends	EA	2.5	CODEK	1.000	68.06	12.82	360.00	440.88
19 03 9923	**48" Corrugated Metal Pipe, End Section**	**EA**				**90.75**	**17.09**	**745.00**	**852.84**
02764 2830	48" (122 cm) Corrugated Metal Pipe Ends	EA	1.875	CODEK	1.000	90.75	17.09	745.00	852.84
19 03 9924	**60" Corrugated Metal Pipe, End Section**	**EA**				**136.12**	**25.64**	**970.00**	**1,132**
02764 2835	60" (152 cm) Corrugated Metal Pipe Ends	EA	1.25	CODEK	1.000	136.12	25.64	970.00	1,132
19 03 9930	**12" Flared-end Reinforced Concrete Pipe, Class 3**	**EA**				**50.41**	**9.50**	**860.00**	**919.91**
02762 4800	12" Diameter Pipe	EA	3.375	CODEK	1.000	50.41	9.50	860.00	919.91
19 03 9931	**15" Flared-end Reinforced Concrete Pipe, Class 3**	**EA**				**54.45**	**10.26**	**211.96**	**276.67**
02452 5001	Precast End Section, 15" Diameter Pipe	EA	3.125	CODEK	1.000	54.45	10.26	211.96	276.67
19 03 9932	**18" Flared-end Reinforced Concrete Pipe, Class 3**	**EA**				**54.45**	**10.26**	**291.00**	**355.70**
02762 2900	Precast End Section, 18" Diameter Pipe	EA	3.125	CODEK	1.000	54.45	10.26	291.00	355.70
19 03 9933	**24" Flared-end Reinforced Concrete Pipe, Class 3**	**EA**				**68.06**	**12.82**	**340.00**	**420.88**
02762 2910	Precast End Section, 24" Diameter Pipe	EA	2.5	CODEK	1.000	68.06	12.82	340.00	420.88

19 03 99	Other Storm Sewer	Unit	Hourly Output	Crew	Factor	Labor	Equip.	Matl.	Total
19 03 9934	**30" Flared-end Reinforced Concrete Pipe, Class 3**	**EA**				**75.62**	**14.24**	**495.00**	**584.87**
02762 2960	Precast End Section, 30" Diameter Pipe	EA	2.25	CODEK	1.000	75.62	14.24	495.00	584.87
19 03 9935	**36" Flared-end Reinforced Concrete Pipe, Class 3**	**EA**				**90.75**	**17.09**	**660.00**	**767.84**
02762 4300	Precast End Section, 36" Diameter Pipe	EA	1.875	CODEK	1.000	90.75	17.09	660.00	767.84
19 03 9936	**48" Flared-end Reinforced Concrete Pipe, Class 3**	**EA**				**170.15**	**32.05**	**243.00**	**445.20**
02762 2850	48" Diameter Pipe	EA	1	CODEK	1.000	170.15	32.05	243.00	445.20
19 03 9937	**60" Flared-end Reinforced Concrete Pipe, Class 3**	**EA**				**272.24**	**51.28**	**425.00**	**748.52**
02762 2905	60" Diameter Pipe	EA	0.625	CODEK	1.000	272.24	51.28	425.00	748.52
19 03 9938	**72" Flared-end Reinforced Concrete Pipe, Class 3**	**EA**				**453.73**	**85.47**	**495.00**	**1,034**
02762 2908	72" Diameter Pipe	EA	0.375	CODEK	1.000	453.73	85.47	495.00	1,034
19 03 9940	**12" Flared-end Reinforced Concrete Pipe, Class 4**	**EA**				**50.41**	**9.50**	**860.00**	**919.91**
02762 4800	12" Diameter Pipe	EA	3.375	CODEK	1.000	50.41	9.50	860.00	919.91
19 03 9941	**15" Flared-end Reinforced Concrete Pipe, Class 4**	**EA**				**54.45**	**10.26**	**211.96**	**276.67**
02452 5001	Precast End Section, 15" Diameter Pipe	EA	3.125	CODEK	1.000	54.45	10.26	211.96	276.67
19 03 9942	**18" Flared-end Reinforced Concrete Pipe, Class 4**	**EA**				**54.45**	**10.26**	**291.00**	**355.70**
02762 2900	Precast End Section, 18" Diameter Pipe	EA	3.125	CODEK	1.000	54.45	10.26	291.00	355.70
19 03 9943	**21" Flared-end Reinforced Concrete Pipe, Class 4**	**EA**				**61.87**	**11.65**	**309.89**	**383.41**
02762 4900	21" Diameter Pipe	EA	2.75	CODEK	1.000	61.87	11.65	309.89	383.41
19 03 9944	**24" Flared-end Reinforced Concrete Pipe, Class 4**	**EA**				**68.06**	**12.82**	**340.00**	**420.88**
02762 2910	Precast End Section, 24" Diameter Pipe	EA	2.5	CODEK	1.000	68.06	12.82	340.00	420.88
19 03 9945	**27" Flared-end Reinforced Concrete Pipe, Class 4**	**EA**				**71.64**	**13.49**	**365.10**	**450.23**
02762 5000	27" Diameter Pipe	EA	2.375	CODEK	1.000	71.64	13.49	365.10	450.23
19 03 9946	**30" Flared-end Reinforced Concrete Pipe, Class 4**	**EA**				**75.62**	**14.24**	**495.00**	**584.87**
02762 2960	Precast End Section, 30" Diameter Pipe	EA	2.25	CODEK	1.000	75.62	14.24	495.00	584.87
19 03 9947	**36" Flared-end Reinforced Concrete Pipe, Class 4**	**EA**				**90.75**	**17.09**	**660.00**	**767.84**
02762 4300	Precast End Section, 36" Diameter Pipe	EA	1.875	CODEK	1.000	90.75	17.09	660.00	767.84

19 03 99	Other Storm Sewer	Unit	Hourly Output	Crew	Factor	Labor	Equip.	Matl.	Total
19 03 9950	**12" Flared-end Reinforced Concrete Pipe, Class 5**	**EA**				**50.41**	**9.50**	**860.00**	**919.91**
02762 4800	12" Diameter Pipe	EA	3.375	CODEK	1.000	50.41	9.50	860.00	919.91
19 03 9951	**15" Flared-end Reinforced Concrete Pipe, Class 5**	**EA**				**54.45**	**10.26**	**211.96**	**276.67**
02452 5001	Precast End Section, 15" Diameter Pipe	EA	3.125	CODEK	1.000	54.45	10.26	211.96	276.67
19 03 9952	**18" Flared-end Reinforced Concrete Pipe, Class 5**	**EA**				**54.45**	**10.26**	**291.00**	**355.70**
02762 2900	Precast End Section, 18" Diameter Pipe	EA	3.125	CODEK	1.000	54.45	10.26	291.00	355.70
19 03 9953	**21" Flared-end Reinforced Concrete Pipe, Class 5**	**EA**				**61.87**	**11.65**	**309.89**	**383.41**
02762 4900	21" Diameter Pipe	EA	2.75	CODEK	1.000	61.87	11.65	309.89	383.41
19 03 9954	**24" Flared-end Reinforced Concrete Pipe, Class 5**	**EA**				**68.06**	**12.82**	**340.00**	**420.88**
02762 2910	Precast End Section, 24" Diameter Pipe	EA	2.5	CODEK	1.000	68.06	12.82	340.00	420.88
19 03 9955	**27" Flared-end Reinforced Concrete Pipe, Class 5**	**EA**				**71.64**	**13.49**	**365.10**	**450.23**
02762 5000	27" Diameter Pipe	EA	2.375	CODEK	1.000	71.64	13.49	365.10	450.23
19 03 9956	**30" Flared-end Reinforced Concrete Pipe, Class 5**	**EA**				**75.62**	**14.24**	**495.00**	**584.87**
02762 2960	Precast End Section, 30" Diameter Pipe	EA	2.25	CODEK	1.000	75.62	14.24	495.00	584.87
19 03 9957	**36" Flared-end Reinforced Concrete Pipe, Class 5**	**EA**				**90.75**	**17.09**	**660.00**	**767.84**
02762 4300	Precast End Section, 36" Diameter Pipe	EA	1.875	CODEK	1.000	90.75	17.09	660.00	767.84
19 03 9958	**42" Flared-end Reinforced Concrete Pipe, Class 5**	**EA**				**136.12**	**25.64**	**830.00**	**991.76**
02762 4400	Precast End Section, 42" Diameter Pipe	EA	1.25	CODEK	1.000	136.12	25.64	830.00	991.76
19 03 9959	**48" Flared-end Reinforced Concrete Pipe, Class 5**	**EA**				**170.15**	**32.05**	**243.00**	**445.20**
02762 2850	48" Diameter Pipe	EA	1	CODEK	1.000	170.15	32.05	243.00	445.20
19 03 9960	**60" Flared-end Reinforced Concrete Pipe, Class 5**	**EA**				**272.24**	**51.28**	**425.00**	**748.52**
02762 2905	60" Diameter Pipe	EA	0.625	CODEK	1.000	272.24	51.28	425.00	748.52
19 03 9961	**72" Flared-end Reinforced Concrete Pipe, Class 5**	**EA**				**453.73**	**85.47**	**495.00**	**1,034**
02762 2908	72" Diameter Pipe	EA	0.375	CODEK	1.000	453.73	85.47	495.00	1,034

19 04 04	Holding Tanks and Separators	Unit	Hourly Output	Crew	Factor	Labor	Equip.	Matl.	Total

19 04 04	Holding Tanks and Separators	Unit	Hourly Output	Crew	Factor	Labor	Equip.	Matl.	Total
19 04 0401	**550 Gallon, Stainless Steel Aboveground Wastewater Holding Tank, Rental**	**MO**				**0.00**	**0.00**	**305.00**	**305.00**
01954 9111	550 Gallon, Stainless Steel, DOT Approved, Monthly Rental	EA	0	N/A	1.000	0.00	0.00	305.00	305.00
19 04 0402	**630 Gallon, Polyethylene Aboveground Wastewater Holding Tank, Rental**	**MO**				**0.00**	**0.00**	**325.00**	**325.00**
01954 9112	630 Gallon, Polyethylene, DOT Approved, Monthly Rental	EA	0	N/A	1.000	0.00	0.00	325.00	325.00
19 04 0403	**4,000 Gallon Polyethylene Wastewater Tank, Rental**	**MO**				**0.00**	**0.00**	**580.00**	**580.00**
01954 9114	4,000 Gallon, Polyethylene, Stackable, Monthly Rental	EA	0	N/A	1.000	0.00	0.00	580.00	580.00
19 04 0404	**4,000 Gallon, Polyethylene Trailer-mounted Wastewater Holding Tank, Rental**	**MO**				**0.00**	**0.00**	**1,125**	**1,125**
01954 9115	4,000 Gallon, Polyethylene, Trailer Mounted, Monthly Rental	EA	0	N/A	1.000	0.00	0.00	1,125	1,125
19 04 0405	**6,000 Gallon, Polyethylene Aboveground Wastewater Holding Tank, Rental**	**MO**				**0.00**	**0.00**	**645.00**	**645.00**
01954 9116	6,000 Gallon, Polyethylene, Total Drain, Monthly Rental	EA	0	N/A	1.000	0.00	0.00	645.00	645.00
19 04 0406	**21,000 Gallon Steel Wastewater Holding Tank, Rental**	**MO**				**0.00**	**0.00**	**1,225**	**1,225**
01954 9118	21,000 Gallon (500 Barrel), Steel, Closed, Stationary, Monthly Rental	EA	0	N/A	1.000	0.00	0.00	1,225	1,225
19 04 0407	**21,000 Gallon, Steel Closed Stationary Aboveground Wastewater Holding Tank, Rental**	**MO**				**0.00**	**0.00**	**1,225**	**1,225**
01954 9118	21,000 Gallon (500 Barrel), Steel, Closed, Stationary, Monthly Rental	EA	0	N/A	1.000	0.00	0.00	1,225	1,225
19 04 0408	**21,000 Gallon Steel, Open Top, Tank Rental**	**MO**				**0.00**	**0.00**	**1,175**	**1,175**
01954 9117	21,000 Gallon (500 Barrel), Steel, Open, Stationary, Monthly Rental	EA	0	N/A	1.000	0.00	0.00	1,175	1,175
19 04 0411	**Packaged Coalescing 20 GPM Oil/Water Separator**	**EA**				**1,400**	**139.75**	**9,174**	**10,714**
07121 1001	Fluid Neoprene, 1/8" Fiberglass Reinforced	CSF	0.625	ARFCB	0.700	119.40	0.00	178.30	208.39
15428 1101	20 GPM Coalescing Oil/Water Separator with 4" Flange Connection	EA	0.096	MPLUH	1.000	986.03	104.69	7,252	8,343
15428 1111	20 GPM Oil Stop Valve, Oil/Water Separator	EA	0.288	MPLUH	1.000	330.28	35.07	1,798	2,163
19 04 0412	**Packaged Coalescing 50 GPM Oil/Water Separator**	**EA**				**1,443**	**139.75**	**12,980**	**14,563**
07121 1001	Fluid Neoprene, 1/8" Fiberglass Reinforced	CSF	0.625	ARFCB	1.060	119.40	0.00	178.30	315.56
15428 1102	50 GPM Coalescing Oil/Water Separator with 4" Flange Connection	EA	0.096	MPLUH	1.000	986.03	104.69	10,994	12,085
15428 1112	50 GPM Oil Stop Valve, Oil/Water Separator	EA	0.288	MPLUH	1.000	330.28	35.07	1,798	2,163

19 04 04	Holding Tanks and Separators	Unit	Hourly Output	Crew	Factor	Labor	Equip.	Matl.	Total
19 04 0413	**Packaged Coalescing 100 GPM Oil/Water Separator**	**EA**				**1,528**	**139.75**	**15,051**	**16,719**
07121 1001	Fluid Neoprene, 1/8" Fiberglass Reinforced	CSF	0.625	ARFCB	1.770	119.40	0.00	178.30	526.92
15428 1103	100 GPM Coalescing Oil/Water Separator with 4" Flange Connection	EA	0.096	MPLUH	1.000	986.03	104.69	12,477	13,568
15428 1113	100 GPM Oil Stop Valve, Oil/Water Separator	EA	0.288	MPLUH	1.000	330.28	35.07	2,259	2,624
19 04 0414	**Packaged Coalescing 200 GPM Oil/Water Separator**	**EA**				**2,195**	**189.46**	**20,054**	**22,439**
07121 1001	Fluid Neoprene, 1/8" Fiberglass Reinforced	CSF	0.625	ARFCB	3.440	119.40	0.00	178.30	1,024
15428 1104	200 GPM Coalescing Oil/Water Separator with 4" Flange Connection	EA	0.073	MPLUH	1.000	1,310	139.05	17,643	19,092
15428 1114	200 GPM Oil Stop Valve, Oil/Water Separator	EA	0.2	MPLUH	1.000	474.77	50.41	1,798	2,323
19 04 0415	**Packaged Coalescing 300 GPM Oil/Water Separator**	**EA**				**2,371**	**189.46**	**23,238**	**25,798**
07121 1001	Fluid Neoprene, 1/8" Fiberglass Reinforced	CSF	0.625	ARFCB	4.910	119.40	0.00	178.30	1,462
15428 1105	300 GPM Coalescing Oil/Water Separator with 4" Flange Connection	EA	0.073	MPLUH	1.000	1,310	139.05	20,104	21,553
15428 1115	300 GPM Oil Stop Valve, Oil/Water Separator	EA	0.2	MPLUH	1.000	474.77	50.41	2,259	2,784
19 04 0416	**Packaged Coalescing 400 GPM Oil/Water Separator**	**EA**				**3,395**	**279.45**	**27,092**	**30,767**
07121 1001	Fluid Neoprene, 1/8" Fiberglass Reinforced	CSF	0.625	ARFCB	6.390	119.40	0.00	178.30	1,902
15428 1106	400 GPM Coalescing Oil/Water Separator with 4" Flange Connection	EA	0.048	MPLUH	1.000	1,999	212.24	23,129	25,340
15428 1116	400 GPM Oil Stop Valve, Oil/Water Separator	EA	0.15	MPLUH	1.000	633.03	67.21	2,824	3,524
19 04 0417	**Packaged Coalescing 500 GPM Oil/Water Separator**	**EA**				**3,208**	**238.08**	**40,043**	**43,489**
07121 1001	Fluid Neoprene, 1/8" Fiberglass Reinforced	CSF	0.625	ARFCB	8.090	119.40	0.00	178.30	2,408
15428 1107	8" Flanged Connection x 500 GPM	EA	0.059	MPLUH	1.000	1,609	170.87	36,027	37,807
15428 1117	500 GPM Oil Stop Valve	EA	0.15	MPLUH	1.000	633.03	67.21	2,573	3,274
19 04 0418	**55 Gallon Nalgene Horizontal XLPE Tank without legs**	**EA**				**0.00**	**0.00**	**342.59**	**342.59**
01954 9119	55 Gallon Nalgene Horizontal XLPE Tank without Legs	EA	0	N/A	1.000	0.00	0.00	342.59	342.59
19 04 0419	**110 Gallon Nalgene Horizontal XLPE Tank without legs**	**EA**				**0.00**	**0.00**	**476.14**	**476.14**
01954 9120	110 Gallon Nalgene Horizontal XLPE Tank without Legs	EA	0	N/A	1.000	0.00	0.00	476.14	476.14
19 04 0420	**200 Gallon Nalgene Horizontal XLPE Tank without legs**	**EA**				**0.00**	**0.00**	**592.27**	**592.27**
01954 9121	200 Gallon Nalgene Horizontal XLPE Tank without Legs	EA	0	N/A	1.000	0.00	0.00	592.27	592.27

19 04 04	Holding Tanks and Separators	Unit	Hourly Output	Crew	Factor	Labor	Equip.	Matl.	Total
19 04 0421	**300 Gallon Nalgene Horizontal XLPE Tank without legs**	**EA**				**0.00**	**0.00**	**644.53**	**644.53**
01954 9122	300 Gallon Nalgene Horizontal XLPE Tank without Legs	EA	0	N/A	1.000	0.00	0.00	644.53	644.53
19 04 0422	**500 Gallon Nalgene Horizontal XLPE Tank without legs**	**EA**				**0.00**	**0.00**	**807.12**	**807.12**
01954 9123	500 Gallon Nalgene Horizontal XLPE Tank without Legs	EA	0	N/A	1.000	0.00	0.00	807.12	807.12
19 04 0423	**1,000 Gallon Nalgene Horizontal XLPE Tank without legs**	**EA**				**0.00**	**0.00**	**1,736**	**1,736**
01954 9124	1,000 Gallon Nalgene Horizontal XLPE Tank without Legs	EA	0	N/A	1.000	0.00	0.00	1,736	1,736
19 04 0424	**1,650 Gallon Nalgene Horizontal XLPE Tank without legs**	**EA**				**0.00**	**0.00**	**2,317**	**2,317**
01954 9125	1,650 Gallon Nalgene Horizontal XLPE Tank without Legs	EA	0	N/A	1.000	0.00	0.00	2,317	2,317
19 04 0425	**2,500 Gallon Nalgene Horizontal XLPE Tank without legs**	**EA**				**0.00**	**0.00**	**3,774**	**3,774**
01954 9126	2,500 Gallon Nalgene Horizontal XLPE Tank without Legs	EA	0	N/A	1.000	0.00	0.00	3,774	3,774
19 04 0426	**55 Gallon Tank Fiberglass Saddle**	**EA**				**0.00**	**0.00**	**342.59**	**342.59**
01954 9127	55 Gallon Tank Fiberglass Saddle	EA	0	N/A	1.000	0.00	0.00	342.59	342.59
19 04 0427	**110 Gallon Tank Fiberglass Saddle**	**EA**				**0.00**	**0.00**	**476.14**	**476.14**
01954 9128	110 Gallon Tank Fiberglass Saddle	EA	0	N/A	1.000	0.00	0.00	476.14	476.14
19 04 0428	**200 Gallon Tank Fiberglass Saddle**	**EA**				**0.00**	**0.00**	**679.37**	**679.37**
01954 9129	200 Gallon Tank Fiberglass Saddle	EA	0	N/A	1.000	0.00	0.00	679.37	679.37
19 04 0429	**300 Gallon Tank Fiberglass Saddle**	**EA**				**0.00**	**0.00**	**1,156**	**1,156**
01954 9130	300 Gallon Tank Fiberglass Saddle	EA	0	N/A	1.000	0.00	0.00	1,156	1,156
19 04 0430	**500 Gallon Tank Fiberglass Saddle**	**EA**				**0.00**	**0.00**	**2,317**	**2,317**
01954 9131	500 Gallon Tank Fiberglass Saddle	EA	0	N/A	1.000	0.00	0.00	2,317	2,317
19 04 0431	**60 Gallon Nalgene Horizontal XLPE Tank with legs**	**EA**				**87.71**	**7.19**	**389.04**	**483.94**
01954 9132	60 Gallon Nalgene Horizontal XLPE Tank with Legs	EA	1	MPLUG	1.000	87.71	7.19	389.04	483.94
19 04 0432	**125 Gallon Nalgene Horizontal XLPE Tank with legs**	**EA**				**87.71**	**7.19**	**510.98**	**605.87**
01954 9133	125 Gallon Nalgene Horizontal XLPE Tank with Legs	EA	1	MPLUG	1.000	87.71	7.19	510.98	605.87

19 04 04	Holding Tanks and Separators	Unit	Hourly Output	Crew	Factor	Labor	Equip.	Matl.	Total
19 04 0433	**225 Gallon Nalgene Horizontal XLPE Tank with legs**	**EA**				**87.71**	**7.19**	**690.98**	**785.88**
01954 9134	225 Gallon Nalgene Horizontal XLPE Tank with Legs	EA	1	MPLUG	1.000	87.71	7.19	690.98	785.88
19 04 0434	**300 Gallon Nalgene Horizontal XLPE Tank with legs**	**EA**				**87.71**	**7.19**	**714.21**	**809.11**
01954 9135	300 Gallon Nalgene Horizontal XLPE Tank with Legs	EA	1	MPLUG	1.000	87.71	7.19	714.21	809.11
19 04 0435	**500 Gallon Nalgene Horizontal XLPE Tank with legs**	**EA**				**87.71**	**7.19**	**836.15**	**931.04**
01954 9136	500 Gallon Nalgene Horizontal XLPE Tank with Legs	EA	1	MPLUG	1.000	87.71	7.19	836.15	931.04
19 04 0436	**1,575 Gallon Conical Bottom Vertical XLPE Tank**	**EA**				**109.63**	**8.98**	**2,549**	**2,668**
01954 9137	1,575 Gallon Conical Bottom Vertical XLPE Tank	EA	0.8	MPLUG	1.000	109.63	8.98	2,549	2,668
19 04 0437	**2,200 Gallon Conical Bottom Vertical XLPE Tank**	**EA**				**109.63**	**8.98**	**3,066**	**3,184**
01954 9138	2,200 Gallon Conical Bottom Vertical XLPE Tank	EA	0.8	MPLUG	1.000	109.63	8.98	3,066	3,184
19 04 0438	**2,600 Gallon Conical Bottom Vertical XLPE Tank**	**EA**				**109.63**	**8.98**	**3,774**	**3,893**
01954 9139	2,600 Gallon Conical Bottom Vertical XLPE Tank	EA	0.8	MPLUG	1.000	109.63	8.98	3,774	3,893
19 04 0439	**3,000 Gallon Conical Bottom Vertical XLPE Tank**	**EA**				**131.55**	**10.78**	**4,239**	**4,381**
01954 9140	3,000 Gallon Conical Bottom Vertical XLPE Tank	EA	0.667	MPLUG	1.000	131.55	10.78	4,239	4,381
19 04 0440	**4,200 Gallon Conical Bottom Vertical XLPE Tank**	**EA**				**131.55**	**10.78**	**6,155**	**6,297**
01954 9141	4,200 Gallon Conical Bottom Vertical XLPE Tank	EA	0.667	MPLUG	1.000	131.55	10.78	6,155	6,297
19 04 0441	**6,000 Gallon Conical Bottom Vertical XLPE Tank**	**EA**				**131.55**	**10.78**	**8,710**	**8,852**
01954 9142	6,000 Gallon Conical Bottom Vertical XLPE Tank	EA	0.667	MPLUG	1.000	131.55	10.78	8,710	8,852
19 04 0442	**8,000 Gallon Conical Bottom Vertical XLPE Tank**	**EA**				**131.55**	**10.78**	**11,845**	**11,988**
01954 9143	8,000 Gallon Conical Bottom Vertical XLPE Tank	EA	0.667	MPLUG	1.000	131.55	10.78	11,845	11,988
19 04 0443	**1,575 Gallon Conical Tank Stand**	**EA**				**0.00**	**0.00**	**1,736**	**1,736**
01954 9144	1,575 Gallon Conical Tank Stand	EA	0	N/A	1.000	0.00	0.00	1,736	1,736
19 04 0444	**2,200 Gallon Conical Tank Stand**	**EA**				**0.00**	**0.00**	**1,968**	**1,968**
01954 9145	2,200 Gallon Conical Tank Stand	EA	0	N/A	1.000	0.00	0.00	1,968	1,968
19 04 0445	**2,600 Gallon Conical Tank Stand**	**EA**				**0.00**	**0.00**	**2,085**	**2,085**
01954 9146	2,600 Gallon Conical Tank Stand	EA	0	N/A	1.000	0.00	0.00	2,085	2,085

19 04 04 Holding Tanks and Separators		Unit	Hourly Output	Crew	Factor	Labor	Equip.	Matl.	Total
19 04 0446	**3,000 Gallon Conical Tank Stand**	**EA**				**0.00**	**0.00**	**2,085**	**2,085**
01954 9147	3,000 Gallon Conical Tank Stand	EA	0	N/A	1.000	0.00	0.00	2,085	2,085

19 04 06 Sumps		Unit	Hourly Output	Crew	Factor	Labor	Equip.	Matl.	Total
19 04 0601	**275 Gallon Steel Sump, Aboveground with Supports & Fittings**	**EA**				**359.57**	**0.00**	**305.00**	**664.57**
13205 3040	275 Gallon Steel Storage Tank with 3/16" Shell Above Ground with Supports, Coating & Fittings	EA	0.288	MSPFC	1.000	359.57	0.00	305.00	664.57
19 04 0602	**550 Gallon Steel Sump, Aboveground with Supports & Fittings**	**EA**				**435.26**	**0.00**	**931.59**	**1,367**
13205 3060	550 Gallon Steel Storage Tank with 3/16" Shell Above Ground with Supports, Coating & Fittings	EA	0.238	MSPFC	1.000	435.26	0.00	931.59	1,367
19 04 0603	**1,000 Gallon Steel Sump, Aboveground with Supports & Fittings**	**EA**				**573.44**	**0.00**	**1,166**	**1,740**
13205 3080	1,000 Gallon Steel Storage Tank with 3/16" Shell Above Ground with Supports, Coating & Fittings	EA	0.25	MSPFL	1.000	573.44	0.00	1,166	1,740
19 04 0604	**1,500 Gallon Steel Sump, Aboveground with Supports & Fittings**	**EA**				**716.80**	**0.00**	**1,725**	**2,442**
13205 3100	1,500 Gallon Steel Storage Tank with 3/16" Shell Above Ground with Supports, Coating & Fittings	EA	0.2	MSPFL	1.000	716.80	0.00	1,725	2,442
19 04 0605	**2,000 Gallon Steel Sump, Aboveground with Supports & Fittings**	**EA**				**878.60**	**123.21**	**2,239**	**3,241**
13205 3120	2,000 Gallon Steel Storage Tank with 3/16" Shell Above Ground with Supports, Coating & Fittings	EA	0.175	MSPFP	1.000	878.60	123.21	2,239	3,241
19 04 0606	**5,000 Gallon Steel Sump, Aboveground with Supports & Fittings**	**EA**				**1,118**	**156.82**	**5,175**	**6,450**
13205 3140	5,000 Gallon Steel Storage Tank with 3/16" Shell Above Ground with Supports, Coating & Fittings	EA	0.138	MSPFP	1.000	1,118	156.82	5,175	6,450
19 04 0610	**75 GPM Cast-iron Sump Pump, 2.0" Discharge, Single Stage**	**EA**				**400.78**	**0.00**	**2,754**	**3,155**
15292 2020	75 GPM Cast Iron Sump Pump, 2" Discharge Single Stage, 40' Head, 1-1/2 HP	EA	0.238	MSPFB	1.000	400.78	0.00	2,754	3,155
19 04 0611	**100 GPM Cast-iron Sump Pump, 2.5" Discharge, Single Stage**	**EA**				**475.93**	**0.00**	**2,849**	**3,325**
15292 2030	100 GPM Cast Iron Sump Pump, 2-1/2" Discharge Single Stage, 40' Head, 2 HP	EA	0.2	MSPFB	1.000	475.93	0.00	2,849	3,325
19 04 0621	**550 Gallon Horizontal Plastic Sump with 4" NPT Connection**	**EA**				**216.39**	**0.00**	**1,525**	**1,741**
13205 7710	550 Gallon Horizontal Cylindrical Plastic Tank with 4" NPT Connection	EA	0.663	MSPFL	1.000	216.39	0.00	1,525	1,741
19 04 0622	**1,000 Gallon Horizontal Plastic Sump with 4" NPT Connection**	**EA**				**286.72**	**0.00**	**1,925**	**2,212**
13205 7720	1,000 Gallon Horizontal Cylindrical Plastic Tank with 4" NPT Connection	EA	0.5	MSPFL	1.000	286.72	0.00	1,925	2,212

19 04 06	Sumps	Unit	Hourly Output	Crew	Factor	Labor	Equip.	Matl.	Total
19 04 0623	**2,000 Gallon Horizontal Plastic Sump with 6" NPT Connection**	**EA**				**358.40**	**0.00**	**2,400**	**2,758**
13205 7730	2,000 Gallon Horizontal Cylindrical Plastic Tank with 6" NPT Connection	EA	0.4	MSPFL	1.000	358.40	0.00	2,400	2,758
19 04 0624	**4,000 Gallon Horizontal Plastic Sump with 6" NPT Connection**	**EA**				**424.77**	**0.00**	**3,950**	**4,375**
13205 7740	4,000 Gallon Horizontal Cylindrical Plastic Tank with 6" NPT Connection	EA	0.338	MSPFL	1.000	424.77	0.00	3,950	4,375
19 04 0625	**6,000 Gallon Horizontal Plastic Sump with 6" NPT Connection**	**EA**				**573.44**	**0.00**	**4,925**	**5,498**
13205 7750	6,000 Gallon Horizontal Cylindrical Plastic Tank with 6" NPT Connection	EA	0.25	MSPFL	1.000	573.44	0.00	4,925	5,498
19 04 0626	**8,000 Gallon Horizontal Plastic Sump with 6" NPT Connection**	**EA**				**716.80**	**0.00**	**4,753**	**5,470**
13205 7760	8,000 Gallon Horizontal Cylindrical Plastic Tank with 6" NPT Connection	EA	0.2	MSPFL	1.000	716.80	0.00	4,753	5,470
19 04 0627	**10,000 Gallon Horizontal Plastic Sump with 6" NPT Connection**	**EA**				**878.60**	**123.21**	**6,075**	**7,077**
13205 7770	10,000 Gallon Horizontal Cylindrical Plastic Tank with 6" NPT Connection	EA	0.175	MSPFP	1.000	878.60	123.21	6,075	7,077
19 04 0628	**12,000 Gallon Horizontal Plastic Sump with 6" NPT Connection**	**EA**				**1,118**	**156.82**	**6,900**	**8,175**
13205 7780	12,000 Gallon Horizontal Cylindrical Plastic Tank with 6" NPT Connection	EA	0.138	MSPFP	1.000	1,118	156.82	6,900	8,175
19 04 0629	**15,000 Gallon Horizontal Plastic Sump with 6" NPT Connection**	**EA**				**1,230**	**172.50**	**12,363**	**13,766**
13205 7790	15,000 Gallon Horizontal Cylindrical Plastic Tank with 6" NPT Connection	EA	0.125	MSPFP	1.000	1,230	172.50	12,363	13,766
19 04 0631	**20,000 Gallon Horizontal Plastic Sump with 6" NPT Connection**	**EA**				**1,367**	**191.67**	**13,714**	**15,272**
13205 7800	20,000 Gallon Horizontal Cylindrical Plastic Tank with 6" NPT Connection	EA	0.113	MSPFP	1.000	1,367	191.67	13,714	15,272
19 04 0632	**25,000 Gallon Horizontal Plastic Sump with 6" NPT Connection**	**EA**				**1,757**	**246.43**	**21,400**	**23,404**
13205 7810	25,000 Gallon Horizontal Cylindrical Plastic Tank with 6" NPT Connection	EA	0.088	MSPFP	1.000	1,757	246.43	21,400	23,404
19 04 0633	**30,000 Gallon Horizontal Plastic Sump with 6" NPT Connection**	**EA**				**2,050**	**287.50**	**33,038**	**35,376**
13205 7820	30,000 Gallon Horizontal Cylindrical Plastic Tank with 6" NPT Connection	EA	0.075	MSPFP	1.000	2,050	287.50	33,038	35,376
19 04 0634	**40,000 Gallon Horizontal Plastic Sump with 6" NPT Connection**	**EA**				**3,075**	**431.25**	**36,674**	**40,181**
13205 7830	40,000 Gallon Horizontal Cylindrical Plastic Tank with 6" NPT Connection	EA	0.05	MSPFP	1.000	3,075	431.25	36,674	40,181

19 04 06	Sumps	Unit	Hourly Output	Crew	Factor	Labor	Equip.	Matl.	Total
19 04 0635	**48,000 Gallon Horizontal Plastic Sump with 6" NPT Connection**	**EA**				**4,100**	**575.00**	**61,000**	**65,675**
13205 7840	48,000 Gallon Horizontal Cylindrical Plastic Tank with 6" NPT Connection	EA	0.038	MSPFP	1.000	4,100	575.00	61,000	65,675

19 05 01	Overhead Hot Water Systems	Unit	Hourly Output	Crew	Factor	Labor	Equip.	Matl.	Total
19 05 0101	**Pre-Insulated Steel Pipe, 1 1/4" Supply/Return HTHW W/H**	**LF**				**6.25**	**2.89**	**25.07**	**34.21**
02663 1010	1-1/4" (32 mm) Diameter	LF	103.1	MSPFY	2.000	1.68	1.40	12.35	30.87
05130 0660	C2 x 1.78 - 2" Wide (Channels)	LF	14.25	MSPFZ	0.200	5.50	0.44	0.74	1.34
15190 1450	1-1/4" Adjustable Band Hanger Type 7	EA	10.63	MPLUB	0.200	8.94	0.00	1.12	2.01
19 05 0102	**Pre-Insulated Steel Pipe, 1 1/2" Supply/Return HTHW W/H**	**LF**				**6.43**	**2.84**	**27.90**	**37.17**
02663 1020	1-1/2" Diameter Insulated Standard Black Steel Pipe (3.7 cm) Align & Tackweld on Sleepers	LF	105	MSPFY	2.000	1.65	1.38	13.75	33.56
05130 0660	C2 x 1.78 - 2" Wide (Channels)	LF	14.25	MSPFZ	0.200	5.50	0.44	0.74	1.34
15190 1460	1-1/2" Adjustable Band Hanger Type 7	EA	9.375	MPLUB	0.200	10.13	0.00	1.25	2.28
19 05 0103	**Pre-Insulated Steel Pipe, 2" Supply/Return HTHW W/H**	**LF**				**7.33**	**3.48**	**29.49**	**40.30**
02663 1030	2" (50 mm) Diameter	LF	85.38	MSPFY	2.000	2.03	1.70	14.50	36.45
05130 0660	C2 x 1.78 - 2" Wide (Channels)	LF	14.25	MSPFZ	0.200	5.50	0.44	0.74	1.34
15190 1470	2" Adjustable Band Hanger Type 7	EA	8.75	MPLUB	0.200	10.85	0.00	1.73	2.52
19 05 0104	**Pre-Insulated Steel Pipe, 2 1/2" Supply/Return HTHW W/H**	**LF**				**8.03**	**3.92**	**31.71**	**43.66**
02663 1040	2-1/2" (63 mm) Diameter	LF	75.5	MSPFY	2.000	2.30	1.92	15.60	39.63
05130 0660	C2 x 1.78 - 2" Wide (Channels)	LF	14.25	MSPFZ	0.200	5.50	0.44	0.74	1.34
15190 1480	2-1/2" Adjustable Band Hanger Type 7	EA	8.125	MPLUB	0.200	11.69	0.00	1.82	2.70
19 05 0105	**Pre-Insulated Steel Pipe, 3" Supply/Return HTHW W/H**	**LF**				**9.38**	**4.54**	**32.65**	**46.56**
02663 1050	3" Diameter Insulated Standard Black Steel Pipe (7.6 cm) Align & Tackweld on Sleepers	LF	65.63	MSPFY	2.000	2.64	2.21	15.95	41.59
05130 0662	C3 x 4.1 - 3" Wide (Channels)	LF	10	MSPFZ	0.200	7.84	0.62	1.70	2.03
15190 1490	3" Adjustable Band Hanger Type 7	EA	7.5	MPLUB	0.200	12.66	0.00	2.03	2.94
19 05 0106	**Pre-Insulated Steel Pipe, 4" Supply/Return HTHW W/H**	**LF**				**11.87**	**6.19**	**35.97**	**54.02**
02663 1060	4" Diameter Insulated Standard Black Steel Pipe (10 cm) Align & Tackweld on Sleepers	LF	47.75	MSPFY	2.000	3.63	3.03	17.50	48.32
05130 0662	C3 x 4.1 - 3" Wide (Channels)	LF	10	MSPFZ	0.200	7.84	0.62	1.70	2.03
15190 1510	4" Adjustable Band Hanger Type 7	EA	6.25	MPLUB	0.200	15.19	0.00	3.14	3.67

19 05 01	Overhead Hot Water Systems	Unit	Hourly Output	Crew	Factor	Labor	Equip.	Matl.	Total
19 05 0107	**Pre-Insulated Steel Pipe, 6" Supply/Return HTHW W/H**	**LF**				**14.12**	**7.84**	**61.96**	**83.92**
02663 1080	6" Diameter Insulated Standard Black Steel Pipe (15 cm) Align & Tackweld on Sleepers	LF	37.5	MSPFY	2.000	4.62	3.86	28.00	72.96
05130 0662	C3 x 4.1 - 3" Wide (Channels)	LF	10	MSPFZ	0.200	7.84	0.62	1.70	2.03
15190 2970	6" Clevis Roller Hanger Type 43	EA	3.125	MSPFA	0.200	16.54	0.00	28.09	8.93
19 05 0108	**Pre-Insulated Steel Pipe, 8" Supply/Return HTHW W/H**	**LF**				**16.56**	**8.95**	**83.83**	**109.35**
02663 1090	8" Diameter Insulated Standard Black Steel Pipe (21 cm) Align & Tackweld on Sleepers	LF	32.88	MSPFY	2.000	5.27	4.40	37.50	94.35
05130 0664	C4 x 5.4 - 4" Wide (Channels)	LF	8.325	MSPFZ	0.200	9.42	0.75	2.23	2.48
15190 2980	8" Clevis Roller Hanger Type 43	EA	2.5	MSPFA	0.200	20.68	0.00	41.92	12.52
19 05 0109	**Pre-Insulated Steel Pipe, 10" Supply/Return HTHW W/H**	**LF**				**18.94**	**9.55**	**90.46**	**118.95**
02559 1116	10" Diameter Insulated Standard Black Steel Pipe (25 cm) Align & Tackweld on Sleepers	LF	30.88	MSPFY	2.000	5.61	4.69	39.96	100.53
05130 0666	C5 x 6.7 - 5" Wide (Channels)	LF	7.138	MSPFZ	0.200	10.98	0.87	2.77	2.93
15190 2990	10" Clevis Roller Hanger Type 43	EA	1.875	MSPFA	0.200	27.57	0.00	49.89	15.49

19 05 02	Overhead Steam Systems	Unit	Hourly Output	Crew	Factor	Labor	Equip.	Matl.	Total
19 05 0201	**Field Insulated Steel Pipe, 1 1/4" Supply, 3/4" Return Steam W/H**	**LF**				**13.50**	**0.09**	**13.35**	**26.94**
05130 0660	C2 x 1.78 - 2" Wide (Channels)	LF	14.25	MSPFZ	0.200	5.50	0.44	0.74	1.34
15172 0570	3/4" (20 mm) A-53 Pipe, Schedule 40, Not Including Hangers or Fittings	LF	38.75	MPLUE	1.000	2.30	0.00	0.75	3.04
15172 0590	1-1/4" (32 mm) A-53 Pipe, Schedule 40, Not Including Hangers or Fittings	LF	33.75	MPLUE	1.000	2.64	0.00	1.28	3.91
15185 1501	3/4" Pipe	LF	32.5	AASBC	1.000	2.88	0.00	5.12	8.00
15185 1503	1-1/4" Pipe	LF	32.5	AASBC	1.000	2.88	0.00	5.86	8.74
15190 1430	3/4" Adjustable Band Hanger Type 7	EA	11.88	MPLUB	0.100	8.00	0.00	0.94	0.89
15190 1450	1-1/4" Adjustable Band Hanger Type 7	EA	10.63	MPLUB	0.100	8.94	0.00	1.12	1.01
19 05 0202	**Field Insulated Steel Pipe, 1 1/2" Supply, 1" Return Steam W/H**	**LF**				**14.66**	**0.09**	**14.44**	**29.18**
05130 0660	C2 x 1.78 - 2" Wide (Channels)	LF	14.25	MSPFZ	0.200	5.50	0.44	0.74	1.34
15172 0580	1" (25 mm) A-53 Pipe, Schedule 40, Not Including Hangers or Fittings	LF	35.75	MPLUE	1.000	2.49	0.00	0.94	3.43
15172 0600	1-1/2" (40 mm) A-53 Pipe, Schedule 40, Not Including Hangers or Fittings	LF	28.75	MPLUE	1.000	3.09	0.00	1.52	4.61
15185 1502	1" Pipe	LF	32.5	AASBC	1.000	2.88	0.00	5.50	8.38
15185 1504	1-1/2" Pipe	LF	29	AASBC	1.000	3.23	0.00	6.10	9.33
15190 1440	1" Adjustable Band Hanger Type 7	EA	11.25	MPLUB	0.100	8.44	0.00	1.07	0.95
15190 1460	1-1/2" Adjustable Band Hanger Type 7	EA	9.375	MPLUB	0.100	10.13	0.00	1.25	1.14

19 05 02 Overhead Steam Systems		Unit	Hourly Output	Crew	Factor	Labor	Equip.	Matl.	Total
19 05 0203	**Field Insulated Steel Pipe, 2" Supply, 1" Return Steam W/H**	**LF**				**15.56**	**0.09**	**15.33**	**30.98**
05130 0660	C2 x 1.78 - 2" Wide (Channels)	LF	14.25	MSPFZ	0.200	5.50	0.44	0.74	1.34
15172 0580	1" (25 mm) A-53 Pipe, Schedule 40, Not Including Hangers or Fittings	LF	35.75	MPLUE	1.000	2.49	0.00	0.94	3.43
15172 0610	2" (50 mm) A-53 Pipe, Schedule 40, Not Including Hangers or Fittings	LF	23.75	MPLUE	1.000	3.75	0.00	2.07	5.82
15185 1502	1" Pipe	LF	32.5	AASBC	1.000	2.88	0.00	5.50	8.38
15185 1505	2" Pipe	LF	27.5	AASBC	1.000	3.41	0.00	6.39	9.80
15190 1440	1" Adjustable Band Hanger Type 7	EA	11.25	MPLUB	0.100	8.44	0.00	1.07	0.95
15190 1470	2" Adjustable Band Hanger Type 7	EA	8.75	MPLUB	0.100	10.85	0.00	1.73	1.26
19 05 0204	**Field Insulated Steel Pipe, 2 1/2" Supply, 1 1/2"Return Steam W/H**	**LF**				**17.73**	**0.09**	**18.24**	**36.06**
05130 0660	C2 x 1.78 - 2" Wide (Channels)	LF	14.25	MSPFZ	0.200	5.50	0.44	0.74	1.34
15172 0600	1-1/2" (40 mm) A-53 Pipe, Schedule 40, Not Including Hangers or Fittings	LF	28.75	MPLUE	1.000	3.09	0.00	1.52	4.61
15172 0620	2-1/2" (60 mm) A-53 Pipe, Schedule 40, Not Including Hangers or Fittings	LF	20	MPLUE	1.000	4.45	0.00	3.27	7.72
15185 1504	1-1/2" Pipe	LF	29	AASBC	1.000	3.23	0.00	6.10	9.33
15185 1506	2-1/2" Pipe	LF	25.5	AASBC	1.000	3.68	0.00	6.90	10.57
15190 1460	1-1/2" Adjustable Band Hanger Type 7	EA	9.375	MPLUB	0.100	10.13	0.00	1.25	1.14
15190 1480	2-1/2" Adjustable Band Hanger Type 7	EA	8.125	MPLUB	0.100	11.69	0.00	1.82	1.35
19 05 0205	**Field Insulated Steel Pipe, 3" Supply, 2" Return Steam W/H**	**LF**				**21.00**	**0.12**	**20.80**	**41.92**
05130 0662	C3 x 4.1 - 3" Wide (Channels)	LF	10	MSPFZ	0.200	7.84	0.62	1.70	2.03
15172 0610	2" (50 mm) A-53 Pipe, Schedule 40, Not Including Hangers or Fittings	LF	23.75	MPLUE	1.000	3.75	0.00	2.07	5.82
15172 0630	3" (80 mm) A-53 Pipe, Schedule 40, Not Including Hangers or Fittings	LF	15	MPLUE	1.000	5.93	0.00	4.28	10.21
15185 1505	2" Pipe	LF	27.5	AASBC	1.000	3.41	0.00	6.39	9.80
15185 1507	3" Pipe	LF	23.5	AASBC	1.000	3.99	0.00	7.34	11.33
15190 1470	2" Adjustable Band Hanger Type 7	EA	8.75	MPLUB	0.100	10.85	0.00	1.73	1.26
15190 1490	3" Adjustable Band Hanger Type 7	EA	7.5	MPLUB	0.100	12.66	0.00	2.03	1.47
19 05 0206	**Field Insulated Steel Pipe, 4" Supply, 2" Return Steam W/H**	**LF**				**22.80**	**0.12**	**24.37**	**47.29**
05130 0662	C3 x 4.1 - 3" Wide (Channels)	LF	10	MSPFZ	0.200	7.84	0.62	1.70	2.03
15172 0610	2" (50 mm) A-53 Pipe, Schedule 40, Not Including Hangers or Fittings	LF	23.75	MPLUE	1.000	3.75	0.00	2.07	5.82
15172 0650	4" (10 cm) A-53 Pipe, Schedule 40, Not Including Hangers or Fittings	LF	17.5	MPLUJ	1.000	6.95	0.00	6.12	13.08
15185 1505	2" Pipe	LF	27.5	AASBC	1.000	3.41	0.00	6.39	9.80
15185 1508	4" Pipe	LF	20.75	AASBC	1.000	4.52	0.00	8.95	13.47
15190 1470	2" Adjustable Band Hanger Type 7	EA	8.75	MPLUB	0.100	10.85	0.00	1.73	1.26
15190 1510	4" Adjustable Band Hanger Type 7	EA	6.25	MPLUB	0.100	15.19	0.00	3.14	1.83

19 05 02 Overhead Steam Systems		Unit	Hourly Output	Crew	Factor	Labor	Equip.	Matl.	Total
19 05 0207	**Field Insulated Steel Pipe, 6" Supply, 2 1/2" Return Steam W/H**	**LF**				**29.07**	**0.15**	**36.13**	**65.35**
05130 0664	C4 x 5.4 - 4" Wide (Channels)	LF	8.325	MSPFZ	0.200	9.42	0.75	2.23	2.48
15172 0620	2-1/2" (60 mm) A-53 Pipe, Schedule 40, Not Including Hangers or Fittings	LF	20	MPLUE	1.000	4.45	0.00	3.27	7.72
15172 0670	6" (15 cm) A-53 Pipe, Schedule 40, Not Including Hangers or Fittings	LF	11.63	MPLUJ	1.000	10.46	0.00	12.66	23.13
15185 1506	2-1/2" Pipe	LF	25.5	AASBC	1.000	3.68	0.00	6.90	10.57
15185 1509	6" Pipe	LF	16.25	AASBC	1.000	5.77	0.00	9.87	15.64
15190 1480	2-1/2" Adjustable Band Hanger Type 7	EA	8.125	MPLUB	0.100	11.69	0.00	1.82	1.35
15190 2970	6" Clevis Roller Hanger Type 43	EA	3.125	MSPFA	0.100	16.54	0.00	28.09	4.46
19 05 0208	**Field Insulated Steel Pipe, 8" Supply, 3" Return Steam W/H**	**LF**				**32.84**	**0.15**	**43.12**	**76.12**
05130 0664	C4 x 5.4 - 4" Wide (Channels)	LF	8.325	MSPFZ	0.200	9.42	0.75	2.23	2.48
15172 0630	3" (80 mm) A-53 Pipe, Schedule 40, Not Including Hangers or Fittings	LF	15	MPLUE	1.000	5.93	0.00	4.28	10.21
15172 0680	8", A-53 Pipe, Shedule 40, Not Including Hangers or Fittings	LF	9.625	MPLUJ	1.000	12.64	0.00	18.83	31.47
15185 1012	8" Diameter Pipe, 2" Thick Fiberglass Pipe Cover with Fire-retardant Jackets	LF	18.5	AASBC	1.000	5.07	0.00	7.82	12.89
15185 1507	3" Pipe	LF	23.5	AASBC	1.000	3.99	0.00	7.34	11.33
15190 1490	3" Adjustable Band Hanger Type 7	EA	7.5	MPLUB	0.100	12.66	0.00	2.03	1.47
15190 2980	8" Clevis Roller Hanger Type 43	EA	2.5	MSPFA	0.100	20.68	0.00	41.92	6.26
19 05 0209	**Field Insulated Steel Pipe, 10" Supply, 4" Return Steam W/H**	**LF**				**39.16**	**0.17**	**57.40**	**96.74**
05130 0666	C5 x 6.7 - 5" Wide (Channels)	LF	7.138	MSPFZ	0.200	10.98	0.87	2.77	2.93
15172 0650	4" (10 cm) A-53 Pipe, Schedule 40, Not Including Hangers or Fittings	LF	17.5	MPLUJ	1.000	6.95	0.00	6.12	13.08
15172 0690	10", A-53 Pipe, Shedule 40, Not Including Hangers or Fittings	LF	7.875	MPLUJ	1.000	15.45	0.00	25.32	40.77
15185 1508	4" Pipe	LF	20.75	AASBC	1.000	4.52	0.00	8.95	13.47
15190 1510	4" Adjustable Band Hanger Type 7	EA	6.25	MPLUB	0.100	15.19	0.00	3.14	1.83
15190 2990	10" Clevis Roller Hanger Type 43	EA	1.875	MSPFA	0.100	27.57	0.00	49.89	7.75
15567 7480	Fiberglass Cover for 10" Diameter Pipe, 2" Thick with Fire-retardant Jacket	LF	16.25	AASBC	1.000	5.77	0.00	11.14	16.91
19 05 03 Underground Hot Water Systems		**Unit**	**Hourly Output**	**Crew**	**Factor**	**Labor**	**Equip.**	**Matl.**	**Total**
19 05 0301	**Pre-Insulated Steel Pipe, 1 1/4" Supply/Return HTHW Direct Burial**	**LF**				**3.36**	**2.81**	**24.70**	**30.87**
02663 1010	1-1/4" (32 mm) Diameter	LF	103.1	MSPFY	2.000	1.68	1.40	12.35	30.87
19 05 0302	**Pre-Insulated Steel Pipe, 1 1/2" Supply/Return HTHW Direct Burial**	**LF**				**3.30**	**2.76**	**27.50**	**33.56**
02663 1020	1-1/2" Diameter Insulated Standard Black Steel Pipe (3.7 cm) Align & Tackweld on Sleepers	LF	105	MSPFY	2.000	1.65	1.38	13.75	33.56

19 05 03	Underground Hot Water Systems	Unit	Hourly Output	Crew	Factor	Labor	Equip.	Matl.	Total
19 05 0303	**Pre-Insulated Steel Pipe, 2" Supply/Return HTHW Direct Burial**	**LF**				**4.06**	**3.39**	**29.00**	**36.45**
02663 1030	2" (50 mm) Diameter	LF	85.38	MSPFY	2.000	2.03	1.70	14.50	36.45
19 05 0304	**Pre-Insulated Steel Pipe, 2 1/2" Supply/Return HTHW Direct Burial**	**LF**				**4.59**	**3.83**	**31.20**	**39.63**
02663 1040	2-1/2" (63 mm) Diameter	LF	75.5	MSPFY	2.000	2.30	1.92	15.60	39.63
19 05 0305	**Pre-Insulated Steel Pipe, 3" Supply/Return HTHW Direct Burial**	**LF**				**5.28**	**4.41**	**31.90**	**41.59**
02663 1050	3" Diameter Insulated Standard Black Steel Pipe (7.6 cm) Align & Tackweld on Sleepers	LF	65.63	MSPFY	2.000	2.64	2.21	15.95	41.59
19 05 0306	**Pre-Insulated Steel Pipe, 4" Supply/Return HTHW Direct Burial**	**LF**				**7.26**	**6.06**	**35.00**	**48.32**
02663 1060	4" Diameter Insulated Standard Black Steel Pipe (10 cm) Align & Tackweld on Sleepers	LF	47.75	MSPFY	2.000	3.63	3.03	17.50	48.32
19 05 0307	**Pre-Insulated Steel Pipe, 6" Supply/Return HTHW Direct Burial**	**LF**				**9.25**	**7.72**	**56.00**	**72.96**
02663 1080	6" Diameter Insulated Standard Black Steel Pipe (15 cm) Align & Tackweld on Sleepers	LF	37.5	MSPFY	2.000	4.62	3.86	28.00	72.96
19 05 0308	**Pre-Insulated Steel Pipe, 8" Supply/Return HTHW Direct Burial**	**LF**				**10.55**	**8.80**	**75.00**	**94.35**
02663 1090	8" Diameter Insulated Standard Black Steel Pipe (21 cm) Align & Tackweld on Sleepers	LF	32.88	MSPFY	2.000	5.27	4.40	37.50	94.35
19 05 0309	**Pre-Insulated Steel Pipe, 10" Supply/Return HTHW Direct Burial**	**LF**				**11.23**	**9.37**	**79.92**	**100.53**
02559 1116	10" Diameter Insulated Standard Black Steel Pipe (25 cm) Align & Tackweld on Sleepers	LF	30.88	MSPFY	2.000	5.61	4.69	39.96	100.53

19 05 04	Underground Steam Systems	Unit	Hourly Output	Crew	Factor	Labor	Equip.	Matl.	Total
19 05 0401	**Field Insulated Steel Pipe, 1 1/4" Supply, 3/4" Return Steam Direct Burial**	**LF**				**10.70**	**0.00**	**13.00**	**23.70**
15172 0570	3/4" (20 mm) A-53 Pipe, Schedule 40, Not Including Hangers or Fittings	LF	38.75	MPLUE	1.000	2.30	0.00	0.75	3.04
15172 0590	1-1/4" (32 mm) A-53 Pipe, Schedule 40, Not Including Hangers or Fittings	LF	33.75	MPLUE	1.000	2.64	0.00	1.28	3.91
15185 1501	3/4" Pipe	LF	32.5	AASBC	1.000	2.88	0.00	5.12	8.00
15185 1503	1-1/4" Pipe	LF	32.5	AASBC	1.000	2.88	0.00	5.86	8.74
19 05 0402	**Field Insulated Steel Pipe, 1 1/2" Supply, 1" Return Steam Direct Burial**	**LF**				**11.70**	**0.00**	**14.06**	**25.76**
15172 0580	1" (25 mm) A-53 Pipe, Schedule 40, Not Including Hangers or Fittings	LF	35.75	MPLUE	1.000	2.49	0.00	0.94	3.43
15172 0600	1-1/2" (40 mm) A-53 Pipe, Schedule 40, Not Including Hangers or Fittings	LF	28.75	MPLUE	1.000	3.09	0.00	1.52	4.61
15185 1502	1" Pipe	LF	32.5	AASBC	1.000	2.88	0.00	5.50	8.38
15185 1504	1-1/2" Pipe	LF	29	AASBC	1.000	3.23	0.00	6.10	9.33

19 05 04	Underground Steam Systems	Unit	Hourly Output	Crew	Factor	Labor	Equip.	Matl.	Total
19 05 0403	**Field Insulated Steel Pipe, 2" Supply, 1" Return Steam Direct Burial**	**LF**				**12.53**	**0.00**	**14.90**	**27.43**
15172 0580	1" (25 mm) A-53 Pipe, Schedule 40, Not Including Hangers or Fittings	LF	35.75	MPLUE	1.000	2.49	0.00	0.94	3.43
15172 0610	2" (50 mm) A-53 Pipe, Schedule 40, Not Including Hangers or Fittings	LF	23.75	MPLUE	1.000	3.75	0.00	2.07	5.82
15185 1502	1" Pipe	LF	32.5	AASBC	1.000	2.88	0.00	5.50	8.38
15185 1505	2" Pipe	LF	27.5	AASBC	1.000	3.41	0.00	6.39	9.80
19 05 0404	**Field Insulated Steel Pipe, 2 1/2" Supply, 1 1/2"Return Steam Direct Burial**	**LF**				**14.45**	**0.00**	**17.78**	**32.23**
15172 0600	1-1/2" (40 mm) A-53 Pipe, Schedule 40, Not Including Hangers or Fittings	LF	28.75	MPLUE	1.000	3.09	0.00	1.52	4.61
15172 0620	2-1/2" (60 mm) A-53 Pipe, Schedule 40, Not Including Hangers or Fittings	LF	20	MPLUE	1.000	4.45	0.00	3.27	7.72
15185 1504	1-1/2" Pipe	LF	29	AASBC	1.000	3.23	0.00	6.10	9.33
15185 1506	2-1/2" Pipe	LF	25.5	AASBC	1.000	3.68	0.00	6.90	10.57
19 05 0405	**Field Insulated Steel Pipe, 3" Supply, 2" Return Steam Direct Burial**	**LF**				**17.08**	**0.00**	**20.08**	**37.16**
15172 0610	2" (50 mm) A-53 Pipe, Schedule 40, Not Including Hangers or Fittings	LF	23.75	MPLUE	1.000	3.75	0.00	2.07	5.82
15172 0630	3" (80 mm) A-53 Pipe, Schedule 40, Not Including Hangers or Fittings	LF	15	MPLUE	1.000	5.93	0.00	4.28	10.21
15185 1505	2" Pipe	LF	27.5	AASBC	1.000	3.41	0.00	6.39	9.80
15185 1507	3" Pipe	LF	23.5	AASBC	1.000	3.99	0.00	7.34	11.33
19 05 0406	**Field Insulated Steel Pipe, 4" Supply, 2" Return Steam Direct Burial**	**LF**				**18.63**	**0.00**	**23.54**	**42.16**
15172 0610	2" (50 mm) A-53 Pipe, Schedule 40, Not Including Hangers or Fittings	LF	23.75	MPLUE	1.000	3.75	0.00	2.07	5.82
15172 0650	4" (10 cm) A-53 Pipe, Schedule 40, Not Including Hangers or Fittings	LF	17.5	MPLUJ	1.000	6.95	0.00	6.12	13.08
15185 1505	2" Pipe	LF	27.5	AASBC	1.000	3.41	0.00	6.39	9.80
15185 1508	4" Pipe	LF	20.75	AASBC	1.000	4.52	0.00	8.95	13.47
19 05 0407	**Field Insulated Steel Pipe, 6" Supply, 2 1/2" Return Steam Direct Burial**	**LF**				**24.36**	**0.00**	**32.70**	**57.06**
15172 0620	2-1/2" (60 mm) A-53 Pipe, Schedule 40, Not Including Hangers or Fittings	LF	20	MPLUE	1.000	4.45	0.00	3.27	7.72
15172 0670	6" (15 cm) A-53 Pipe, Schedule 40, Not Including Hangers or Fittings	LF	11.63	MPLUJ	1.000	10.46	0.00	12.66	23.13
15185 1506	2-1/2" Pipe	LF	25.5	AASBC	1.000	3.68	0.00	6.90	10.57
15185 1509	6" Pipe	LF	16.25	AASBC	1.000	5.77	0.00	9.87	15.64

19 05 04 Underground Steam Systems		Unit	Hourly Output	Crew	Factor	Labor	Equip.	Matl.	Total
19 05 0408	**Field Insulated Steel Pipe, 8" Supply, 3" Return Steam Direct Burial**	**LF**				**27.63**	**0.00**	**38.28**	**65.91**
15172 0630	3" (80 mm) A-53 Pipe, Schedule 40, Not Including Hangers or Fittings	LF	15	MPLUE	1.000	5.93	0.00	4.28	10.21
15172 0680	8", A-53 Pipe, Shedule 40, Not Including Hangers or Fittings	LF	9.625	MPLUJ	1.000	12.64	0.00	18.83	31.47
15185 1012	8" Diameter Pipe, 2" Thick Fiberglass Pipe Cover with Fire-retardant Jackets	LF	18.5	AASBC	1.000	5.07	0.00	7.82	12.89
15185 1507	3" Pipe	LF	23.5	AASBC	1.000	3.99	0.00	7.34	11.33
19 05 0409	**Field Insulated Steel Pipe, 10" Supply, 4" Return Steam Direct Burial**	**LF**				**32.69**	**0.00**	**51.55**	**84.23**
15172 0650	4" (10 cm) A-53 Pipe, Schedule 40, Not Including Hangers or Fittings	LF	17.5	MPLUJ	1.000	6.95	0.00	6.12	13.08
15172 0690	10", A-53 Pipe, Shedule 40, Not Including Hangers or Fittings	LF	7.875	MPLUJ	1.000	15.45	0.00	25.32	40.77
15185 1508	4" Pipe	LF	20.75	AASBC	1.000	4.52	0.00	8.95	13.47
15567 7480	Fiberglass Cover for 10" Diameter Pipe, 2" Thick with Fire-retardant Jacket	LF	16.25	AASBC	1.000	5.77	0.00	11.14	16.91

19 05 05 Trench Boxes		Unit	Hourly Output	Crew	Factor	Labor	Equip.	Matl.	Total
19 05 0501	**2' x 2' (Inside), Precast Trench Box**	**LF**				**13.61**	**5.75**	**108.00**	**127.36**
02752 5005	Box, 2' x 2' Inside	LF	12.5	UOEHC	1.000	13.61	5.75	108.00	127.36
19 05 0502	**2' 6" x 2' (Inside), Precast Trench Box**	**LF**				**14.33**	**6.05**	**116.00**	**136.38**
02752 5010	Box, 2'-6" x 2' Inside	LF	11.88	UOEHC	1.000	14.33	6.05	116.00	136.38
19 05 0503	**3' x 2' (Inside), Precast Trench Box**	**LF**				**15.12**	**6.39**	**112.00**	**133.51**
02752 5020	Box, 3' x 2' Inside	LF	11.25	UOEHC	1.000	15.12	6.39	112.00	133.51
19 05 0504	**3' 6" x 2' (Inside), Precast Trench Box**	**LF**				**16.01**	**6.76**	**122.00**	**144.78**
02752 5030	Box, 3'-6" x 2' Inside	LF	10.63	UOEHC	1.000	16.01	6.76	122.00	144.78
19 05 0505	**4' x 2' (Inside), Precast Trench Box**	**LF**				**17.01**	**7.19**	**129.00**	**153.20**
02752 5040	Box, 4' x 2' Inside	LF	10	UOEHC	1.000	17.01	7.19	129.00	153.20
19 05 0506	**5' x 2' (Inside), Precast Trench Box**	**LF**				**18.15**	**7.67**	**154.00**	**179.82**
02752 5050	Box, 5' x 2' Inside	LF	9.375	UOEHC	1.000	18.15	7.67	154.00	179.82
19 05 0507	**6' x 2' (Inside), Precast Trench Box**	**LF**				**19.45**	**8.21**	**183.00**	**210.66**
02752 5060	Box, 6' x 2' Inside	LF	8.75	UOEHC	1.000	19.45	8.21	183.00	210.66
19 05 0508	**4' x 2' 6" (Inside), Precast Trench Box**	**LF**				**18.15**	**7.67**	**181.00**	**206.82**
02752 5070	Box, 4' x 2'-6" Inside	LF	9.375	UOEHC	1.000	18.15	7.67	181.00	206.82
19 05 0509	**5' x 2' 6" (Inside), Precast Trench Box**	**LF**				**19.45**	**8.21**	**196.00**	**223.66**
02752 5080	Box, 5' x 2'-6" Inside	LF	8.75	UOEHC	1.000	19.45	8.21	196.00	223.66

19 05 05 Trench Boxes		Unit	Hourly Output	Crew	Factor	Labor	Equip.	Matl.	Total
19 05 0510	**6' x 2' 6" (Inside), Precast Trench Box**	**LF**				**20.94**	**8.85**	**182.00**	**211.79**
02752 5090	Box, 6' x 2'-6" Inside	LF	8.125	UOEHC	1.000	20.94	8.85	182.00	211.79

19 05 99 Other Heating		Unit	Hourly Output	Crew	Factor	Labor	Equip.	Matl.	Total
19 05 9901	**2 - 20' Wood Supports for Overhead Heat Distribution - Complete**	**EA**				**872.97**	**124.36**	**681.37**	**1,679**
02215 0010	Backfill Trenches by Hand without Compaction	CY	1.625	ULABA	1.800	20.85	0.00	0.00	37.52
02227 0390	Pier/Foundation Excavator, Heavy Clay, Hand Piled 2'-6' Deep	CY	0.438	ULABA	3.000	77.43	0.00	0.00	232.29
03158 5100	Column Spread Footing Forms, 3 Uses Plywood Forms, Form & Strip with Accessories	SF	51.25	ACARJ	2.000	2.56	0.00	0.78	6.70
03334 0300	Concrete Curing, Sprayed Membrane Curing Compound	CSF	11.88	ULABB	2.000	5.71	0.00	4.05	19.51
03372 2500	Pour Spread Footings, < 5 CY, Crane & Bucket < (5 m3) Place 3,000 PSI Concrete Foundations	CY	5.625	ALABG	2.000	46.41	21.03	84.22	303.31
16413 1001	Wood Pole, Class 3, Pine, Treated, 25' (7.6 m), Machine Dig & Set	EA	1.065	EELEJ	2.000	177.89	29.65	176.23	767.54
16921 0200	3-3/4" x 4-3/4" x 10'-0", Single Arm Wood Crossarm, with Hardware & Braces	EA	2.745	EELEJ	2.000	69.02	11.50	75.40	311.84
19 05 9902	**2 - 30' Wood Supports for Overhead Heat Distribution - Complete**	**EA**				**911.89**	**130.84**	**721.93**	**1,765**
02215 0010	Backfill Trenches by Hand without Compaction	CY	1.625	ULABA	1.800	20.85	0.00	0.00	37.52
02227 0390	Pier/Foundation Excavator, Heavy Clay, Hand Piled 2'-6' Deep	CY	0.438	ULABA	3.000	77.43	0.00	0.00	232.29
03158 5100	Column Spread Footing Forms, 3 Uses Plywood Forms, Form & Strip with Accessories	SF	51.25	ACARJ	2.000	2.56	0.00	0.78	6.70
03334 0300	Concrete Curing, Sprayed Membrane Curing Compound	CSF	11.88	ULABB	2.000	5.71	0.00	4.05	19.51
03372 2500	Pour Spread Footings, < 5 CY, Crane & Bucket < (5 m3) Place 3,000 PSI Concrete Foundations	CY	5.625	ALABG	2.000	46.41	21.03	84.22	303.31
16413 1002	Wood Pole, Class 3, Pine, Treated, 30' (9.1 m), Machine Dig & Set	EA	0.96	EELEJ	2.000	197.34	32.89	196.52	853.50
16921 0200	3-3/4" x 4-3/4" x 10'-0", Single Arm Wood Crossarm, with Hardware & Braces	EA	2.745	EELEJ	2.000	69.02	11.50	75.40	311.84
19 05 9905	**2 - 20' Galvanized Steel Supports for Overhead Heat Distribution - Complete**	**EA**				**993.50**	**125.31**	**1,911**	**3,030**
02215 0010	Backfill Trenches by Hand without Compaction	CY	1.625	ULABA	1.800	20.85	0.00	0.00	37.52
02227 0390	Pier/Foundation Excavator, Heavy Clay, Hand Piled 2'-6' Deep	CY	0.438	ULABA	3.000	77.43	0.00	0.00	232.29
03158 5100	Column Spread Footing Forms, 3 Uses Plywood Forms, Form & Strip with Accessories	SF	51.25	ACARJ	2.000	2.56	0.00	0.78	6.70
03334 0300	Concrete Curing, Sprayed Membrane Curing Compound	CSF	11.88	ULABB	2.000	5.71	0.00	4.05	19.51
03372 2500	Pour Spread Footings, < 5 CY, Crane & Bucket < (5 m3) Place 3,000 PSI Concrete Foundations	CY	5.625	ALABG	2.000	46.41	21.03	84.22	303.31
05130 0476	3" x 3" x 3/8" Angle Iron	LF	7.138	MSPFZ	20.000	10.98	0.87	2.98	296.75
16617 1000	Round Tapered Steel Roadway Pole, 1 - 6' Arm, 20' (6 m)	EA	0.96	EELEJ	2.000	197.34	32.89	836.73	2,134

19 05 99	Other Heating	Unit	Hourly Output	Crew	Factor	Labor	Equip.	Matl.	Total
19 05 9906	**2 - 30' Galvanized Steel Supports for Overhead Heat Distribution - Complete**	**EA**				**1,153**	**151.85**	**2,181**	**3,486**
02215 0010	Backfill Trenches by Hand without Compaction	CY	1.625	ULABA	1.800	20.85	0.00	0.00	37.52
02227 0390	Pier/Foundation Excavator, Heavy Clay, Hand Piled 2'-6' Deep	CY	0.438	ULABA	3.000	77.43	0.00	0.00	232.29
03158 5100	Column Spread Footing Forms, 3 Uses Plywood Forms, Form & Strip with Accessories	SF	51.25	ACARJ	2.000	2.56	0.00	0.78	6.70
03334 0300	Concrete Curing, Sprayed Membrane Curing Compound	CSF	11.88	ULABB	2.000	5.71	0.00	4.05	19.51
03372 2500	Pour Spread Footings, < 5 CY, Crane & Bucket < (5 m3) Place 3,000 PSI Concrete Foundations	CY	5.625	ALABG	2.000	46.41	21.03	84.22	303.31
05130 0476	3" x 3" x 3/8" Angle Iron	LF	7.138	MSPFZ	20.000	10.98	0.87	2.98	296.75
16617 1040	Round Tapered Steel Roadway Pole, 1 - 6' Arm, 30' (9.1 m)	EA	0.684	EELEJ	2.000	276.97	46.16	971.84	2,590
19 05 9910	**2 - 20' Aluminum Supports for Overhead Heat Distribution - Complete**	**EA**				**930.23**	**113.40**	**2,053**	**3,097**
02215 0010	Backfill Trenches by Hand without Compaction	CY	1.625	ULABA	1.800	20.85	0.00	0.00	37.52
02227 0390	Pier/Foundation Excavator, Heavy Clay, Hand Piled 2'-6' Deep	CY	0.438	ULABA	3.000	77.43	0.00	0.00	232.29
03158 5100	Column Spread Footing Forms, 3 Uses Plywood Forms, Form & Strip with Accessories	SF	51.25	ACARJ	2.000	2.56	0.00	0.78	6.70
03334 0300	Concrete Curing, Sprayed Membrane Curing Compound	CSF	11.88	ULABB	2.000	5.71	0.00	4.05	19.51
03372 2500	Pour Spread Footings, < 5 CY, Crane & Bucket < (5 m3) Place 3,000 PSI Concrete Foundations	CY	5.625	ALABG	2.000	46.41	21.03	84.22	303.31
05130 0668	C6 x 8.2 - 6" Wide (Channels)	LF	6.663	MSPFZ	20.000	11.77	0.94	3.28	319.64
16617 0100	Round Tapered Aluminum Roadway Pole, 1 - 6' Arm, 20' High (6 m)	EA	1.2	EELEJ	2.000	157.88	26.31	904.79	2,178
19 05 9911	**2 - 30' Aluminum Supports for Overhead Heat Distribution - Complete**	**EA**				**1,089**	**139.91**	**2,631**	**3,860**
02215 0010	Backfill Trenches by Hand without Compaction	CY	1.625	ULABA	1.800	20.85	0.00	0.00	37.52
02227 0390	Pier/Foundation Excavator, Heavy Clay, Hand Piled 2'-6' Deep	CY	0.438	ULABA	3.000	77.43	0.00	0.00	232.29
03158 5100	Column Spread Footing Forms, 3 Uses Plywood Forms, Form & Strip with Accessories	SF	51.25	ACARJ	2.000	2.56	0.00	0.78	6.70
03334 0300	Concrete Curing, Sprayed Membrane Curing Compound	CSF	11.88	ULABB	2.000	5.71	0.00	4.05	19.51
03372 2500	Pour Spread Footings, < 5 CY, Crane & Bucket < (5 m3) Place 3,000 PSI Concrete Foundations	CY	5.625	ALABG	2.000	46.41	21.03	84.22	303.31
05130 0668	C6 x 8.2 - 6" Wide (Channels)	LF	6.663	MSPFZ	20.000	11.77	0.94	3.28	319.64
16617 0140	Round Tapered Aluminum Roadway Pole, 1 - 6' Arm, 30' High (9.1 m)	EA	0.798	EELEJ	2.000	237.41	39.57	1,194	2,941

19 06 01	Overhead Cooling Systems	Unit	Hourly Output	Crew	Factor	Labor	Equip.	Matl.	Total
19 06 0101	**Pre-Insulated Steel Pipe, 1 1/4" Supply/Return Chilled Water W/H**	**LF**				**6.25**	**2.89**	**25.07**	**34.21**
02663 1010	1-1/4" (32 mm) Diameter	LF	103.1	MSPFY	2.000	1.68	1.40	12.35	30.87
05130 0660	C2 x 1.78 - 2" Wide (Channels)	LF	14.25	MSPFZ	0.200	5.50	0.44	0.74	1.34
15190 1450	1-1/4" Adjustable Band Hanger Type 7	EA	10.63	MPLUB	0.200	8.94	0.00	1.12	2.01

19 06 01	Overhead Cooling Systems	Unit	Hourly Output	Crew	Factor	Labor	Equip.	Matl.	Total
19 06 0102	**Pre-Insulated Steel Pipe, 1 1/2" Supply/Return Chilled Water W/H**	**LF**				**6.43**	**2.84**	**27.90**	**37.17**
02663 1020	1-1/2" Diameter Insulated Standard Black Steel Pipe (3.7 cm) Align & Tackweld on Sleepers	LF	105	MSPFY	2.000	1.65	1.38	13.75	33.56
05130 0660	C2 x 1.78 - 2" Wide (Channels)	LF	14.25	MSPFZ	0.200	5.50	0.44	0.74	1.34
15190 1460	1-1/2" Adjustable Band Hanger Type 7	EA	9.375	MPLUB	0.200	10.13	0.00	1.25	2.28
19 06 0103	**Pre-Insulated Steel Pipe, 2" Supply/Return Chilled Water W/H**	**LF**				**7.33**	**3.48**	**29.49**	**40.30**
02663 1030	2" (50 mm) Diameter	LF	85.38	MSPFY	2.000	2.03	1.70	14.50	36.45
05130 0660	C2 x 1.78 - 2" Wide (Channels)	LF	14.25	MSPFZ	0.200	5.50	0.44	0.74	1.34
15190 1470	2" Adjustable Band Hanger Type 7	EA	8.75	MPLUB	0.200	10.85	0.00	1.73	2.52
19 06 0104	**Pre-Insulated Steel Pipe, 2 1/2" Supply/Return Chilled Water W/H**	**LF**				**8.03**	**3.92**	**31.71**	**43.66**
02663 1040	2-1/2" (63 mm) Diameter	LF	75.5	MSPFY	2.000	2.30	1.92	15.60	39.63
05130 0660	C2 x 1.78 - 2" Wide (Channels)	LF	14.25	MSPFZ	0.200	5.50	0.44	0.74	1.34
15190 1480	2-1/2" Adjustable Band Hanger Type 7	EA	8.125	MPLUB	0.200	11.69	0.00	1.82	2.70
19 06 0105	**Pre-Insulated Steel Pipe, 3" Supply/Return Chilled Water W/H**	**LF**				**9.38**	**4.54**	**32.65**	**46.56**
02663 1050	3" Diameter Insulated Standard Black Steel Pipe (7.6 cm) Align & Tackweld on Sleepers	LF	65.63	MSPFY	2.000	2.64	2.21	15.95	41.59
05130 0662	C3 x 4.1 - 3" Wide (Channels)	LF	10	MSPFZ	0.200	7.84	0.62	1.70	2.03
15190 1490	3" Adjustable Band Hanger Type 7	EA	7.5	MPLUB	0.200	12.66	0.00	2.03	2.94
19 06 0106	**Pre-Insulated Steel Pipe, 4" Supply/Return Chilled Water W/H**	**LF**				**11.87**	**6.19**	**35.97**	**54.02**
02663 1060	4" Diameter Insulated Standard Black Steel Pipe (10 cm) Align & Tackweld on Sleepers	LF	47.75	MSPFY	2.000	3.63	3.03	17.50	48.32
05130 0662	C3 x 4.1 - 3" Wide (Channels)	LF	10	MSPFZ	0.200	7.84	0.62	1.70	2.03
15190 1510	4" Adjustable Band Hanger Type 7	EA	6.25	MPLUB	0.200	15.19	0.00	3.14	3.67
19 06 0107	**Pre-Insulated Steel Pipe, 6" Supply/Return Chilled Water W/H**	**LF**				**14.12**	**7.84**	**61.96**	**83.92**
02663 1080	6" Diameter Insulated Standard Black Steel Pipe (15 cm) Align & Tackweld on Sleepers	LF	37.5	MSPFY	2.000	4.62	3.86	28.00	72.96
05130 0662	C3 x 4.1 - 3" Wide (Channels)	LF	10	MSPFZ	0.200	7.84	0.62	1.70	2.03
15190 2970	6" Clevis Roller Hanger Type 43	EA	3.125	MSPFA	0.200	16.54	0.00	28.09	8.93
19 06 0108	**Pre-Insulated Steel Pipe, 8" Supply/Return Chilled Water W/H**	**LF**				**16.56**	**8.95**	**83.83**	**109.35**
02663 1090	8" Diameter Insulated Standard Black Steel Pipe (21 cm) Align & Tackweld on Sleepers	LF	32.88	MSPFY	2.000	5.27	4.40	37.50	94.35
05130 0664	C4 x 5.4 - 4" Wide (Channels)	LF	8.325	MSPFZ	0.200	9.42	0.75	2.23	2.48
15190 2980	8" Clevis Roller Hanger Type 43	EA	2.5	MSPFA	0.200	20.68	0.00	41.92	12.52

19 06 Cooling Distribution Systems

19 06 01 Overhead Cooling Systems		Unit	Hourly Output	Crew	Factor	Labor	Equip.	Matl.	Total
19 06 0109	**Pre-Insulated Steel Pipe, 10" Supply/Return Chilled Water W/H**	**LF**				**18.94**	**9.55**	**90.46**	**118.95**
02559 1116	10" Diameter Insulated Standard Black Steel Pipe (25 cm) Align & Tackweld on Sleepers	LF	30.88	MSPFY	2.000	5.61	4.69	39.96	100.53
05130 0666	C5 x 6.7 - 5" Wide (Channels)	LF	7.138	MSPFZ	0.200	10.98	0.87	2.77	2.93
15190 2990	10" Clevis Roller Hanger Type 43	EA	1.875	MSPFA	0.200	27.57	0.00	49.89	15.49

19 06 02 Underground Cooling Systems		Unit	Hourly Output	Crew	Factor	Labor	Equip.	Matl.	Total
19 06 0201	**Pre-Insulated Steel Pipe, 1 1/4" Supply/Return Chilled Water Direct Burial**	**LF**				**3.36**	**2.81**	**24.70**	**30.87**
02663 1010	1-1/4" (32 mm) Diameter	LF	103.1	MSPFY	2.000	1.68	1.40	12.35	30.87
19 06 0202	**Pre-Insulated Steel Pipe, 1 1/2" Supply/Return Chilled Water Direct Burial**	**LF**				**3.30**	**2.76**	**27.50**	**33.56**
02663 1020	1-1/2" Diameter Insulated Standard Black Steel Pipe (3.7 cm) Align & Tackweld on Sleepers	LF	105	MSPFY	2.000	1.65	1.38	13.75	33.56
19 06 0203	**Pre-Insulated Steel Pipe, 2" Supply/Return Chilled Water Direct Burial**	**LF**				**4.06**	**3.39**	**29.00**	**36.45**
02663 1030	2" (50 mm) Diameter	LF	85.38	MSPFY	2.000	2.03	1.70	14.50	36.45
19 06 0204	**Pre-Insulated Steel Pipe, 2 1/2" Supply/Return Chilled Water Direct Burial**	**LF**				**4.59**	**3.83**	**31.20**	**39.63**
02663 1040	2-1/2" (63 mm) Diameter	LF	75.5	MSPFY	2.000	2.30	1.92	15.60	39.63
19 06 0205	**Pre-Insulated Steel Pipe, 3" Supply/Return Chilled Water Direct Burial**	**LF**				**5.28**	**4.41**	**31.90**	**41.59**
02663 1050	3" Diameter Insulated Standard Black Steel Pipe (7.6 cm) Align & Tackweld on Sleepers	LF	65.63	MSPFY	2.000	2.64	2.21	15.95	41.59
19 06 0206	**Pre-Insulated Steel Pipe, 4" Supply/Return Chilled Water Direct Burial**	**LF**				**7.26**	**6.06**	**35.00**	**48.32**
02663 1060	4" Diameter Insulated Standard Black Steel Pipe (10 cm) Align & Tackweld on Sleepers	LF	47.75	MSPFY	2.000	3.63	3.03	17.50	48.32
19 06 0207	**Pre-Insulated Steel Pipe, 6" Supply/Return Chilled Water Direct Burial**	**LF**				**9.25**	**7.72**	**56.00**	**72.96**
02663 1080	6" Diameter Insulated Standard Black Steel Pipe (15 cm) Align & Tackweld on Sleepers	LF	37.5	MSPFY	2.000	4.62	3.86	28.00	72.96
19 06 0208	**Pre-Insulated Steel Pipe, 8" Supply/Return Chilled Water Direct Burial**	**LF**				**10.55**	**8.80**	**75.00**	**94.35**
02663 1090	8" Diameter Insulated Standard Black Steel Pipe (21 cm) Align & Tackweld on Sleepers	LF	32.88	MSPFY	2.000	5.27	4.40	37.50	94.35
19 06 0209	**Pre-Insulated Steel Pipe, 10" Supply/Return Chilled Water Direct Burial**	**LF**				**11.23**	**9.37**	**79.92**	**100.53**
02559 1116	10" Diameter Insulated Standard Black Steel Pipe (25 cm) Align & Tackweld on Sleepers	LF	30.88	MSPFY	2.000	5.61	4.69	39.96	100.53

19 06 03 Trench Boxes	Unit	Hourly Output	Crew	Factor	Labor	Equip.	Matl.	Total

19 06 Cooling Distribution Systems

19 06 03 Trench Boxes		Unit	Hourly Output	Crew	Factor	Labor	Equip.	Matl.	Total
19 06 0301	**2' x 2' (Inside), Precast Trench Box**	**LF**				**13.61**	**5.75**	**108.00**	**127.36**
02752 5005	Box, 2' x 2' Inside	LF	12.5	UOEHC	1.000	13.61	5.75	108.00	127.36
19 06 0302	**2' 6" x 2' (Inside), Precast Trench Box**	**LF**				**14.33**	**6.05**	**116.00**	**136.38**
02752 5010	Box, 2'-6" x 2' Inside	LF	11.88	UOEHC	1.000	14.33	6.05	116.00	136.38
19 06 0303	**3' x 2' (Inside), Precast Trench Box**	**LF**				**15.12**	**6.39**	**112.00**	**133.51**
02752 5020	Box, 3' x 2' Inside	LF	11.25	UOEHC	1.000	15.12	6.39	112.00	133.51
19 06 0304	**3' 6" x 2' (Inside), Precast Trench Box**	**LF**				**16.01**	**6.76**	**122.00**	**144.78**
02752 5030	Box, 3'-6" x 2' Inside	LF	10.63	UOEHC	1.000	16.01	6.76	122.00	144.78
19 06 0305	**4' x 2' (Inside), Precast Trench Box**	**LF**				**17.01**	**7.19**	**129.00**	**153.20**
02752 5040	Box, 4' x 2' Inside	LF	10	UOEHC	1.000	17.01	7.19	129.00	153.20
19 06 0306	**5' x 2' (Inside), Precast Trench Box**	**LF**				**18.15**	**7.67**	**154.00**	**179.82**
02752 5050	Box, 5' x 2' Inside	LF	9.375	UOEHC	1.000	18.15	7.67	154.00	179.82
19 06 0307	**6' x 2' (Inside), Precast Trench Box**	**LF**				**19.45**	**8.21**	**183.00**	**210.66**
02752 5060	Box, 6' x 2' Inside	LF	8.75	UOEHC	1.000	19.45	8.21	183.00	210.66
19 06 0308	**4' x 2' 6" (Inside), Precast Trench Box**	**LF**				**18.15**	**7.67**	**181.00**	**206.82**
02752 5070	Box, 4' x 2'-6" Inside	LF	9.375	UOEHC	1.000	18.15	7.67	181.00	206.82
19 06 0309	**5' x 2' 6" (Inside), Precast Trench Box**	**LF**				**19.45**	**8.21**	**196.00**	**223.66**
02752 5080	Box, 5' x 2'-6" Inside	LF	8.75	UOEHC	1.000	19.45	8.21	196.00	223.66
19 06 0310	**6' x 2' 6" (Inside), Precast Trench Box**	**LF**				**20.94**	**8.85**	**182.00**	**211.79**
02752 5090	Box, 6' x 2'-6" Inside	LF	8.125	UOEHC	1.000	20.94	8.85	182.00	211.79

19 07 01 Gas Distribution Piping		Unit	Hourly Output	Crew	Factor	Labor	Equip.	Matl.	Total
19 07 0101	**1" Black Steel Pipe, Welded T & C Schedule 40**	**LF**				**1.43**	**0.10**	**3.15**	**4.68**
15061 2112	1" Black Steel	LF	100	MPLUO	1.000	1.43	0.10	3.15	4.68
19 07 0102	**2" Black Steel Pipe, Welded T & C Schedule 40**	**LF**				**1.79**	**0.13**	**4.57**	**6.49**
15061 2113	2" Black Steel	LF	80	MPLUO	1.000	1.79	0.13	4.57	6.49
19 07 0103	**3" Black Steel Pipe, Welded T & C Schedule 40**	**LF**				**2.04**	**0.14**	**4.11**	**6.30**
15061 2114	3" Black Steel	LF	70	MPLUO	1.000	2.04	0.14	4.11	6.30
19 07 0104	**4" Black Steel Pipe, Welded T & C Schedule 40**	**LF**				**2.29**	**0.16**	**5.74**	**8.19**
15061 2115	4" Black Steel	LF	62.5	MPLUO	1.000	2.29	0.16	5.74	8.19

19 07 01 Gas Distribution Piping		Unit	Hourly Output	Crew	Factor	Labor	Equip.	Matl.	Total
19 07 0105	**5" Black Steel Pipe, Welded T & C Schedule 40**	**LF**				**2.49**	**0.18**	**10.41**	**13.08**
15061 2116	5" Black Steel	LF	57.5	MPLUO	1.000	2.49	0.18	10.41	13.08
19 07 0106	**6" Black Steel Pipe, Welded T & C Schedule 40**	**LF**				**2.62**	**0.65**	**9.78**	**13.04**
15061 2117	6" Black Steel	LF	60	MPLUQ	1.000	2.62	0.65	9.78	13.04
19 07 0107	**8" Black Steel Pipe, Welded T & C Schedule 40**	**LF**				**2.62**	**0.65**	**14.69**	**17.96**
15061 2118	8" Black Steel	LF	60	MPLUQ	1.000	2.62	0.65	14.69	17.96
19 07 0110	**4" Black Steel Pipe, Plain End**	**LF**				**2.29**	**0.16**	**5.74**	**8.19**
15061 2115	4" Black Steel	LF	62.5	MPLUO	1.000	2.29	0.16	5.74	8.19
19 07 0111	**6" Black Steel Pipe, Plain End**	**LF**				**2.62**	**0.65**	**9.78**	**13.04**
15061 2117	6" Black Steel	LF	60	MPLUQ	1.000	2.62	0.65	9.78	13.04
19 07 0112	**8" Black Steel Pipe, Plain End**	**LF**				**2.62**	**0.65**	**14.69**	**17.96**
15061 2118	8" Black Steel	LF	60	MPLUQ	1.000	2.62	0.65	14.69	17.96
19 07 0115	**4" Black Steel Pipe, Threaded**	**LF**				**2.72**	**0.51**	**11.71**	**14.94**
15061 1616	4" Threaded Steel Pipe	LF	62.5	CODEK	1.000	2.72	0.51	11.71	14.94
19 07 0116	**8" Black Steel Pipe, Threaded**	**LF**				**2.84**	**0.53**	**37.92**	**41.29**
15061 1617	8" Threaded Steel Pipe	LF	60	CODEK	1.000	2.84	0.53	37.92	41.29
19 07 0120	**1 1/4", 60 PSI, Polyethylene Pipe**	**LF**				**2.14**	**0.00**	**0.87**	**3.01**
02684 1040	1-1/4" (3.2 cm) Gas Distribution Line, Polyethylene, 60 PSI, Coils	LF	50	XPLUD	1.000	2.14	0.00	0.87	3.01
19 07 0121	**1 1/2", 60 PSI, Polyethylene Pipe**	**LF**				**2.14**	**0.00**	**0.79**	**2.92**
02551 1002	1-1/2" (3.7 cm) Gas Distribution Line, Polyethylene, 60 PSI, Coils	LF	50	XPLUD	1.000	2.14	0.00	0.79	2.92
19 07 0122	**2", 60 PSI, Polyethylene Pipe**	**LF**				**2.38**	**0.00**	**1.08**	**3.46**
02684 1100	2" (5.1 cm) Gas Distribution Line, Polyethylene, 60 PSI, Coils	LF	45	XPLUD	1.000	2.38	0.00	1.08	3.46
19 07 0123	**3" Polyethylene 40' Joints, 60 PSI**	**LF**				**2.85**	**0.00**	**2.25**	**5.10**
02684 1500	3" (7.6 cm) Gas Distribution Line, 60 PSI, Polyethylene, 38' Joint with Couplings	LF	37.5	XPLUD	1.000	2.85	0.00	2.25	5.10
19 07 0124	**4" Polyethylene 40' Joints, 60 PSI**	**LF**				**5.39**	**2.21**	**5.15**	**12.75**
02684 1540	4" (10 cm) Gas Distribution Line, 60 PSI, Polyethylene	LF	32.5	XXPLA	1.000	5.39	2.21	5.15	12.75
19 07 0125	**6" Polyethylene 40' Joints, 60 PSI**	**LF**				**5.84**	**2.40**	**16.10**	**24.34**
02684 1600	6" (15 cm) Gas Distribution Line, 60 PSI Polyethylene	LF	30	XXPLA	1.000	5.84	2.40	16.10	24.34

19 07 01	Gas Distribution Piping	Unit	Hourly Output	Crew	Factor	Labor	Equip.	Matl.	Total
19 07 0126	**8" Polyethylene 40' Joints, 60 PSI**	**LF**				**7.01**	**2.88**	**22.00**	**31.88**
02684 1640	8" (20 cm) Gas Distribution Line, 60 PSI Polyethylene	LF	25	XXPLA	1.000	7.01	2.88	22.00	31.88

20 02 Exterior Electrical Distribution

20 02 01	Transformers	Unit	Hourly Output	Crew	Factor	Labor	Equip.	Matl.	Total
20 02 0101	**Pole-mounted Transformer, 15 KV - 480/277 3 Phase**	**EA**				**1,789**	**76.33**	**8,934**	**10,799**
16416 1540	75 KVA, Primary 4160 Grounded Y/2400-34500 Grounded Y/19900, Secondary 240/120,480/240,277	EA	0.283	EELEE	3.000	596.34	25.44	2,978	10,799
20 02 0103	**Pole-mounted Capacitors, 6 KW**	**EA**				**152.78**	**25.46**	**574.99**	**753.24**
16925 2000	Capacitors-Pole Mounted 6 KVAR	EA	1.24	EELEJ	1.000	152.78	25.46	574.99	753.24
20 02 0105	**12.47 KV - 120/240V, 1, 50 KV, Oil, Pad-mounted Transformer**	**EA**				**534.81**	**22.82**	**2,639**	**3,197**
16416 1530	50 KVA, Primary 4160 Grounded Y/2400-34500 Grounded Y/19900, Secondary 240/120,480/240,277	EA	0.315	EELEE	1.000	534.81	22.82	2,639	3,197

20 02 03	Overhead Electrical Conductors	Unit	Hourly Output	Crew	Factor	Labor	Equip.	Matl.	Total
20 02 0301	**1/0 ACSR Conductor**	**LF**				**0.51**	**0.06**	**0.22**	**0.79**
16123 0180	#1/0 AWG ACRS Cable Raven 6/1, Installed on Poles, Aluminum Cable	MLF	0.56	EELEK	0.001	509.83	56.43	219.57	0.79
20 02 0302	**4/0 ACSR Conductor**	**LF**				**0.71**	**0.08**	**0.43**	**1.22**
16123 0240	#4/0 AWG ACRS Cable Penquin 6/1, Installed on Poles, Aluminum Cable	MLF	0.401	EELEK	0.001	712.23	78.84	432.65	1.22
20 02 0305	**477.0 ACSR Conductor**	**LF**				**4.60**	**0.77**	**1.87**	**7.24**
16120 6201	477.0 ACSR Conductor Aluminum	MLF	0.041	EELEJ	0.001	4,598	766.38	1,872	7.24
20 02 0306	**795.0 ACSR Conductor**	**LF**				**7.58**	**1.26**	**3.50**	**12.34**
16120 6202	795.0 ACSR Conductor Aluminum	MLF	0.025	EELEJ	0.001	7,578	1,263	3,500	12.34
20 02 0310	**1/C #2 Aluminum, Bare, Wire**	**LF**				**0.49**	**0.05**	**0.17**	**0.71**
16123 0160	#2 AWG ACRS Cable Separate 7/1, Installed on Poles, Aluminum Cable	MLF	0.579	EELEK	0.001	492.66	54.53	165.51	0.71

20 02 04	Towers, Poles, Crossarms, and Insulators	Unit	Hourly Output	Crew	Factor	Labor	Equip.	Matl.	Total
20 02 0401	**30' Class 3 Treated Power Pole**	**EA**				**197.34**	**32.89**	**196.52**	**426.75**
16413 1002	Wood Pole, Class 3, Pine, Treated, 30' (9.1 m), Machine Dig & Set	EA	0.96	EELEJ	1.000	197.34	32.89	196.52	426.75
20 02 0402	**35' Class 3 Treated Power Pole**	**EA**				**260.95**	**43.49**	**301.00**	**605.44**
16413 1003	Wood Pole, Class 3, Pine, Treated, 35' (10.7 m), Machine Dig & Set	EA	0.726	EELEJ	1.000	260.95	43.49	301.00	605.44
20 02 0403	**40' Class 3 Treated Power Pole**	**EA**				**287.05**	**47.84**	**325.96**	**660.85**
16413 1004	Wood Pole, Class 3, Pine, Treated, 40' (12.2 m), Machine Dig & Set	EA	0.66	EELEJ	1.000	287.05	47.84	325.96	660.85

20 02 04	Towers, Poles, Crossarms, and Insulators	Unit	Hourly Output	Crew	Factor	Labor	Equip.	Matl.	Total
20 02 0404	**45' Class 3 Treated Power Pole**	**EA**				**319.75**	**53.29**	**355.59**	**728.63**
16413 1005	Wood Pole, Class 3, Pine, Treated, 45' (13.7 m), Machine Dig & Set	EA	0.593	EELEJ	1.000	319.75	53.29	355.59	728.63
20 02 0405	**50' Class 3 Treated Power Pole**	**EA**				**358.81**	**59.80**	**388.34**	**806.95**
16413 1006	Wood Pole, Class 3, Pine, Treated, 50' (15.2 m), Machine Dig & Set	EA	0.528	EELEJ	1.000	358.81	59.80	388.34	806.95
20 02 0406	**55' Class 3 Treated Power Pole**	**EA**				**394.69**	**65.78**	**469.44**	**929.91**
16413 1007	Wood Pole, Class 3, Pine, Treated, 55' (16.8 m), Machine Dig & Set	EA	0.48	EELEJ	1.000	394.69	65.78	469.44	929.91
20 02 0407	**60' Class 3 Treated Power Pole**	**EA**				**434.02**	**72.34**	**617.60**	**1,124**
16413 1008	Wood Pole, Class 3, Pine, Treated, 60' (18.3 m), Machine Dig & Set	EA	0.437	EELEJ	1.000	434.02	72.34	617.60	1,124
20 02 0408	**65' Class 3 Treated Power Pole**	**EA**				**473.03**	**78.84**	**743.93**	**1,296**
16413 1009	Wood Pole, Class 3, Pine, Treated, 65' (19.8 m), Machine Dig & Set	EA	0.401	EELEJ	1.000	473.03	78.84	743.93	1,296
20 02 0409	**70' Class 3 Treated Power Pole**	**EA**				**503.19**	**83.86**	**952.92**	**1,540**
16413 1011	Wood Pole, Class 3, Pine, Treated, 70' (21.3 m), Machine Dig & Set	EA	0.377	EELEJ	1.000	503.19	83.86	952.92	1,540
20 02 0410	**75' Class 3 Treated Power Pole**	**EA**				**537.45**	**89.57**	**1,139**	**1,766**
16413 1012	Wood Pole, Class 3, Pine, Treated, 75' (22.9 m), Machine Dig & Set	EA	0.353	EELEJ	1.000	537.45	89.57	1,139	1,766
20 02 0420	**Straight-line Structure, 5 KV Pole Top**	**EA**				**258.67**	**43.11**	**111.36**	**413.14**
16921 0150	3-1/2" x 4-1/2" x 8'0", Single Arm Wood Crossarm, with Hardware & Braces	EA	2.91	EELEJ	1.000	65.10	10.85	68.26	144.22
16922 2050	5 KV Class 55-3, Pin Insulated with Pin	EA	3.915	EELEJ	4.000	48.39	8.07	10.77	268.92
20 02 0421	**Straight-line Structure, 15 KV Pole Top**	**EA**				**258.67**	**43.11**	**137.23**	**439.00**
16921 0150	3-1/2" x 4-1/2" x 8'0", Single Arm Wood Crossarm, with Hardware & Braces	EA	2.91	EELEJ	1.000	65.10	10.85	68.26	144.22
16922 2050	5 KV Class 55-3, Pin Insulated with Pin	EA	3.915	EELEJ	1.000	48.39	8.07	10.77	67.23
16922 2100	15 KV Class 55-5, Pin Insulated with Pin	EA	3.915	EELEJ	3.000	48.39	8.07	19.40	227.56
20 02 0422	**Straight-line Structure, 35 KV Pole Top**	**EA**				**426.31**	**71.05**	**343.00**	**840.37**
16921 0150	3-1/2" x 4-1/2" x 8'0", Single Arm Wood Crossarm, with Hardware & Braces	EA	2.91	EELEJ	2.000	65.10	10.85	68.26	288.43
16922 2500	5 KV Class 52-2, Suspension Insulated, with Clevis & Strain Clamp	EA	3.195	EELEJ	1.000	59.30	9.88	27.59	96.77
16922 2600	25 KV Class 2 x 52-4, Suspension Insulated, with Clevis & Strain Clamp	EA	2.4	EELEJ	3.000	78.94	13.16	59.63	455.17

20 02 04 Towers, Poles, Crossarms, and Insulators		Unit	Hourly Output	Crew	Factor	Labor	Equip.	Matl.	Total
20 02 0423	**Straight-line Structure, 70 KV Pole Top**	**EA**				**550.87**	**91.81**	**529.02**	**1,172**
16921 0150	3-1/2" x 4-1/2" x 8'0", Single Arm Wood Crossarm, with Hardware & Braces	EA	2.91	EELEJ	1.000	65.10	10.85	68.26	144.22
16921 0200	3-3/4" x 4-3/4" x 10'-0", Single Arm Wood Crossarm, with Hardware & Braces	EA	2.745	EELEJ	1.000	69.02	11.50	75.40	155.92
16922 2500	5 KV Class 52-2, Suspension Insulated, with Clevis & Strain Clamp	EA	3.195	EELEJ	1.000	59.30	9.88	27.59	96.77
16922 2650	50 KV Class 4 x 52-4, Suspension Insulated, with Clevis & Strain Clamp	EA	1.59	EELEJ	3.000	119.15	19.86	119.26	774.80
20 02 0430	**Terminal Structure, 5 KV Pole Top**	**EA**				**981.52**	**163.59**	**1,259**	**2,404**
16921 1050	3-1/2" x 4-1/2" x 8'0", Double Arm Wood Crossarm, with Hardware & Braces	EA	1.278	EELEJ	1.000	148.24	24.71	137.68	310.62
16922 0150	Down Guy 45' to 60' Pole	EA	0.906	EELEJ	2.000	209.11	34.85	463.91	1,416
16922 2500	5 KV Class 52-2, Suspension Insulated, with Clevis & Strain Clamp	EA	3.195	EELEJ	7.000	59.30	9.88	27.59	677.40
20 02 0431	**Terminal Structure, 15 KV Pole Top**	**EA**				**981.52**	**163.59**	**1,272**	**2,417**
16921 1050	3-1/2" x 4-1/2" x 8'0", Double Arm Wood Crossarm, with Hardware & Braces	EA	1.278	EELEJ	1.000	148.24	24.71	137.68	310.62
16922 0150	Down Guy 45' to 60' Pole	EA	0.906	EELEJ	2.000	209.11	34.85	463.91	1,416
16922 2500	5 KV Class 52-2, Suspension Insulated, with Clevis & Strain Clamp	EA	3.195	EELEJ	1.000	59.30	9.88	27.59	96.77
16922 2550	15 KV Class 52-4, Suspension Insulated with Clevis & Strain Clamp	EA	3.195	EELEJ	6.000	59.30	9.88	29.81	593.96
20 02 0432	**Terminal Structure, 35 KV Pole Top**	**EA**				**1,200**	**199.93**	**1,928**	**3,327**
16921 1100	3-3/4" x 4-3/4" x 10'0", Double Arm Wood Crossarm, with Hardware & Braces	EA	1.218	EELEJ	1.000	155.54	25.92	150.80	332.26
16922 0150	Down Guy 45' to 60' Pole	EA	0.906	EELEJ	3.000	209.11	34.85	463.91	2,124
16922 2500	5 KV Class 52-2, Suspension Insulated, with Clevis & Strain Clamp	EA	3.195	EELEJ	1.000	59.30	9.88	27.59	96.77
16922 2650	50 KV Class 4 x 52-4, Suspension Insulated, with Clevis & Strain Clamp	EA	1.59	EELEJ	3.000	119.15	19.86	119.26	774.80
20 02 0433	**Terminal Structure, 70 KV Pole Top**	**EA**				**1,557**	**259.49**	**2,258**	**4,074**
16921 0200	3-3/4" x 4-3/4" x 10'-0", Single Arm Wood Crossarm, with Hardware & Braces	EA	2.745	EELEJ	4.000	69.02	11.50	75.40	623.67
16922 0150	Down Guy 45' to 60' Pole	EA	0.906	EELEJ	3.000	209.11	34.85	463.91	2,124
16922 2500	5 KV Class 52-2, Suspension Insulated, with Clevis & Strain Clamp	EA	3.195	EELEJ	1.000	59.30	9.88	27.59	96.77
16922 2600	25 KV Class 2 x 52-4, Suspension Insulated, with Clevis & Strain Clamp	EA	2.4	EELEJ	3.000	78.94	13.16	59.63	455.17
16922 2650	50 KV Class 4 x 52-4, Suspension Insulated, with Clevis & Strain Clamp	EA	1.59	EELEJ	3.000	119.15	19.86	119.26	774.80

20 02 05 Underground Electric Conductors	Unit	Hourly Output	Crew	Factor	Labor	Equip.	Matl.	Total

20 02 05	Underground Electric Conductors	Unit	Hourly Output	Crew	Factor	Labor	Equip.	Matl.	Total
20 02 0501	**1/C #4 Copper Grounded 600V Direct Burial, Wire**	**LF**				**0.49**	**0.00**	**0.49**	**0.98**
16120 0130	#4 AWG Cable-Type USE Underground Feeder & Branch Circuits	MLF	0.39	EELEF	0.001	490.29	0.00	488.08	0.98
20 02 0502	**1/C #6 Copper Grounded 600V Direct Burial, Wire**	**LF**				**0.41**	**0.00**	**0.37**	**0.77**
16120 0120	#6 AWG Cable-Type USE Underground Feeder & Branch Circuits	MLF	0.47	EELEF	0.001	406.84	0.00	366.06	0.77
20 02 0503	**3/C #6, W/#6 Grounded 5 KV Direct Burial, Wire**	**LF**				**2.53**	**0.42**	**3.91**	**6.86**
16115 6010	#6 AWG Cable, XLP, 5 KV, Direct Burial Shielded, Single Conductor, 133% Insulation	MLF	0.3	EELEJ	0.004	631.50	105.25	978.46	6.86
20 02 0504	**3/c Underground, 15 KV Direct Burial, #1 with Bare Grounded, Wire**	**LF**				**3.03**	**0.50**	**8.47**	**12.00**
16115 6400	#1 AWG Cable, XLP, 15 KV, Direct Burial Shielded, Single Conductor, 133% Insulation	MLF	0.251	EELEJ	0.004	756.29	126.05	2,118	12.00
20 02 0505	**2 - #2 Underground, 600V Direct Burial, Wire**	**LF**				**1.26**	**0.21**	**1.19**	**2.66**
16114 2100	#2 AWG Cable-Type UF-NMC Underground Feeder & Branch Circuits	MLF	0.3	EELEJ	0.002	631.50	105.25	592.83	2.66
20 02 0506	**3/C #2 Underground, 600V Direct Burial, Wire**	**LF**				**3.11**	**0.00**	**2.83**	**5.94**
16109 4300	Romex Cable 600V Aluminum with GR 3/C #2	LF	61.5	EELEF	1.000	3.11	0.00	2.83	5.94
20 02 0507	**3/C Underground, 15 KV Direct Burial, #6 with Bare Grounded Wire**	**LF**				**1.02**	**0.00**	**7.02**	**8.04**
16115 6600	15 KV 3/C URD Cable Copper Direct Burial #6	LF	187.5	EELEF	1.000	1.02	0.00	7.02	8.04
20 02 0508	**#10 Insulated Strand Wire (THW)**	**LF**				**0.33**	**0.00**	**0.18**	**0.51**
16120 0100	#10 AWG Cable-Type USE Underground Feeder & Branch Circuits	MLF	0.58	EELEF	0.001	329.68	0.00	179.58	0.51
20 02 0510	**5 KV, 1/0, Shielded Cable, Copper**	**LF**				**1.01**	**0.17**	**1.90**	**3.08**
16115 0400	#1/0 AWG Cable, Cross Link Polyethylene, 5 KV, Placed in Conduit Shielded, Single Conductor, 12' Height	MLF	0.188	EELEJ	0.001	1,010	168.40	1,899	3.08
20 02 0511	**5 KV, 3/0, Shielded Cable, Copper**	**LF**				**1.26**	**0.21**	**2.73**	**4.21**
16115 0700	#3/0 AWG Cable, Cross Link Polyethylene, 5 KV, Placed in Conduit Shielded, Single Conductor, 12' Height	MLF	0.15	EELEJ	0.001	1,263	210.50	2,734	4.21
20 02 0512	**5 KV, 4/0, Shielded Cable, Copper**	**LF**				**1.42**	**0.24**	**3.11**	**4.76**
16115 0800	#4/0 AWG Cable, Cross Link Polyethylene, 5 KV, Placed in Conduit Shielded, Single Conductor, 12' Height	MLF	0.134	EELEJ	0.001	1,419	236.52	3,108	4.76

20 02 05	Underground Electric Conductors	Unit	Hourly Output	Crew	Factor	Labor	Equip.	Matl.	Total
20 02 0513	**5 KV, 350 MCM, Shielded Cable, Copper**	**LF**				**1.89**	**0.31**	**4.83**	**7.03**
16115 1200	350 MCM Cable, Cross Link Polyethylene, 5 KV, Placed in Conduit Shielded, Single Conductor, 12' Height	MLF	0.101	EELEJ	0.001	1,885	314.18	4,835	7.03
20 02 0514	**5 KV, 500 MCM, Shielded Cable, Copper**	**LF**				**2.22**	**0.37**	**5.99**	**8.57**
16115 1400	500 MCM Cable, Cross Link Polyethylene, 5 KV, Placed in Conduit Shielded, Single Conductor, 12' Height	MLF	0.086	EELEJ	0.001	2,216	369.30	5,986	8.57
20 02 0515	**5 KV, 750 MCM, Shielded Cable, Copper**	**LF**				**1.89**	**0.32**	**6.73**	**8.94**
16115 1450	Shielded Cable Copper, 5 KV, 750 MCM	MLF	0.1	EELEJ	0.001	1,894	315.75	6,734	8.94
20 02 0516	**5 KV, 1000 MCM, Shielded Cable, Copper**	**LF**				**3.03**	**0.51**	**9.78**	**13.32**
16115 1500	Shielded Cable Copper, 5 KV, 1,000 MCM	MLF	0.063	EELEJ	0.001	3,031	505.20	9,785	13.32
20 02 0520	**15 KV, 1/0, Shielded Cable, Copper**	**LF**				**1.33**	**0.22**	**2.53**	**4.08**
16115 6310	#1/0 AWG Cable, Cross Link Polyethylene, 15 KV, Placed in Duct Shielded, Single Conductor, 133% Insulation	MLF	0.143	EELEJ	0.001	1,329	221.58	2,532	4.08
20 02 0521	**15 KV, 3/0, Shielded Cable, Copper**	**LF**				**1.39**	**0.23**	**2.99**	**4.61**
16115 6330	#3/0 AWG Cable, Cross Link Polyethylene, 15 KV, Placed in Duct Shielded, Single Conductor, 133% Insulation	MLF	0.137	EELEJ	0.001	1,388	231.32	2,993	4.61
20 02 0522	**15 KV, 350 MCM, Shielded Cable, Copper**	**LF**				**1.49**	**0.25**	**5.30**	**7.03**
16115 6360	350 MCM Cable, Cross Link Polyethylene, 15 KV, Placed in Duct Shielded, Single Conductor, 133% Insulation	MLF	0.128	EELEJ	0.001	1,486	247.65	5,295	7.03
20 02 0525	**35 KV, 1/0, Shielded Cable, Copper**	**LF**				**2.43**	**0.27**	**3.68**	**6.38**
16115 7000	Copper 1/0	MLF	0.118	EELEK	0.001	2,428	268.72	3,684	6.38
20 02 0526	**35 KV, 3/0, Shielded Cable, Copper**	**LF**				**2.51**	**0.28**	**4.37**	**7.16**
16115 7020	Copper 3/0	MLF	0.114	EELEK	0.001	2,509	277.70	4,374	7.16
20 02 0528	**70 KV, 350 MCM, Shielded Cable, Copper**	**LF**				**2.85**	**0.32**	**4.98**	**8.15**
16115 7100	Copper, 350 MCM	MLF	0.1	EELEK	0.001	2,853	315.75	4,984	8.15
20 02 0530	**5 KV, 1/0, Nonshielded Cable, Copper**	**LF**				**1.33**	**0.22**	**1.09**	**2.64**
16116 0100	Copper, 1/0	MLF	0.143	EELEJ	0.001	1,329	221.58	1,094	2.64
20 02 0531	**5 KV, 3/0, Nonshielded Cable, Copper**	**LF**				**1.37**	**0.23**	**1.66**	**3.25**
16116 0110	Copper 3/0	MLF	0.139	EELEJ	0.001	1,366	227.65	1,658	3.25
20 02 0532	**5 KV, 4/0, Nonshielded Cable, Copper**	**LF**				**1.39**	**0.23**	**1.86**	**3.48**
16116 0120	Copper, 4/0	MLF	0.136	EELEJ	0.001	1,391	231.83	1,859	3.48
20 02 0533	**5 KV, 350 MCM, Nonshielded Cable, Copper**	**LF**				**1.44**	**0.24**	**3.76**	**5.45**
16116 0140	Copper 350 MCM	MLF	0.131	EELEJ	0.001	1,444	240.66	3,764	5.45

20 02 05	Underground Electric Conductors	Unit	Hourly Output	Crew	Factor	Labor	Equip.	Matl.	Total
20 02 0534	**5 KV, 500 MCM, Nonshielded Cable, Copper**	**LF**				**1.52**	**0.25**	**4.72**	**6.49**
16116 0160	Copper, 500 MCM	MLF	0.125	EELEJ	0.001	1,516	252.60	4,720	6.49
20 02 0535	**5 KV, 750 MCM, Nonshielded Cable, Copper**	**LF**				**1.89**	**0.32**	**5.72**	**7.93**
16116 0180	Copper, 750 MCM	MLF	0.1	EELEJ	0.001	1,894	315.75	5,721	7.93
20 02 0536	**5 KV, 1000 MCM, Nonshielded Cable, Copper**	**LF**				**2.53**	**0.42**	**8.32**	**11.26**
16116 0200	Copper, 1,000 MCM	MLF	0.075	EELEJ	0.001	2,526	421.00	8,317	11.26
20 02 0540	**15 KV, 1/0, Nonshielded Cable, Copper**	**LF**				**1.33**	**0.22**	**3.67**	**5.22**
16116 1000	Copper, 1/0	MLF	0.143	EELEJ	0.001	1,329	221.58	3,672	5.22
20 02 0541	**15 KV, 3/0, Nonshielded Cable, Copper**	**LF**				**1.39**	**0.23**	**6.78**	**8.40**
16116 1020	Copper, 3/0	MLF	0.136	EELEJ	0.001	1,391	231.83	6,780	8.40
20 02 0542	**15 KV, 350 MCM, Nonshielded Cable, Copper**	**LF**				**1.49**	**0.25**	**8.85**	**10.59**
16116 1040	Copper 350 MCM	MLF	0.128	EELEJ	0.001	1,486	247.65	8,852	10.59
20 02 0545	**5 KV, 1/0 to 4/0 Conductor, Terminations & Splicing**	**EA**				**195.95**	**0.00**	**485.81**	**681.76**
16121 2401	#6-#4/0 Connect to Pothead (incl) Medium Voltage to 5 KV	EA	0.525	EELEB	1.000	195.95	0.00	485.81	681.76
20 02 0546	**15 KV, 1/0 to 4/0 Conductor, Terminations & Splicing**	**EA**				**381.02**	**0.00**	**229.65**	**610.67**
16153 2000	#2-#4/0 Cable Splice, Grounded, Shielded 15 KV	EA	0.27	EELEB	1.000	381.02	0.00	229.65	610.67
20 02 0547	**35 KV, 1/0 to 4/0 Conductor, Terminations & Splicing**	**EA**				**633.08**	**0.00**	**244.04**	**877.12**
16153 3000	Cable Splicing, #1 Stranded to 4/0 Stranded, 35 KV	EA	0.163	EELEB	1.000	633.08	0.00	244.04	877.12
20 02 0548	**70 KV, 350 MCM Conductor, Terminations & Splicing**	**EA**				**857.29**	**0.00**	**5,986**	**6,843**
16153 4100	Cable Splicing, 350 MCM 70 KV	EA	0.12	EELEB	1.000	857.29	0.00	5,986	6,843
20 02 0549	**70 KV, #2 to 4/0 Conductor, Terminations & Splicing**	**EA**				**633.08**	**0.00**	**5,986**	**6,619**
16153 4000	Cable Splicing, #2 to 4/0 70 KV	EA	0.163	EELEB	1.000	633.08	0.00	5,986	6,619
20 02 0550	**5 KV, 3/0 to 500 MCM Conductor, Terminations & Splicing**	**EA**				**342.92**	**0.00**	**288.25**	**631.17**
16121 2502	250-500 MCM 2/c 5 KV Pothead (incl) Medium Voltage to 5 KV	EA	0.3	EELEB	1.000	342.92	0.00	288.25	631.17
20 02 0551	**5 KV, 400 to 700 MCM Conductor, Terminations & Splicing**	**EA**				**391.90**	**0.00**	**604.53**	**996.44**
16121 2503	400 MCM thru 700 MCM	EA	0.263	EELEB	1.000	391.90	0.00	604.53	996.44

20 02 05	Underground Electric Conductors	Unit	Hourly Output	Crew	Factor	Labor	Equip.	Matl.	Total
20 02 0552	**5 KV, 750 MCM Conductor, Terminations & Splicing**	**EA**				**457.22**	**0.00**	**1,209**	**1,666**
16121 2504	750 MCM	EA	0.225	EELEB	1.000	457.22	0.00	1,209	1,666
20 02 0553	**5 KV, 1,000 MCM Conductor, Terminations & Splicing**	**EA**				**685.83**	**0.00**	**2,418**	**3,104**
16121 2505	1,000 MCM	EA	0.15	EELEB	1.000	685.83	0.00	2,418	3,104
20 02 0555	**15 KV, 3/0 to 4/0 Conductor, Terminations & Splicing**	**EA**				**457.22**	**0.00**	**291.62**	**748.84**
16153 2010	250-500 MCM Cable Splice, Grounded, Shielded 15 KV	EA	0.225	EELEB	1.000	457.22	0.00	291.62	748.84

20 02 06	Ductbanks, Manholes, Handholes, Raceways	Unit	Hourly Output	Crew	Factor	Labor	Equip.	Matl.	Total
20 02 0601	**2" Steel Conduit**	**LF**				**2.78**	**0.00**	**9.45**	**12.23**
16712 2050	Rigid Galvanized Steel PVC Coated 2" Conduit with Coupling Direct Burial, 20 Mil Coating	LF	37.75	EELEC	1.000	2.78	0.00	9.45	12.23
20 02 0602	**3" Steel Conduit**	**LF**				**4.63**	**0.00**	**19.06**	**23.70**
16712 2060	Rigid Galvanized Steel PVC Coated 3" Conduit with Coupling Direct Burial, 20 Mil Coating	LF	22.63	EELEC	1.000	4.63	0.00	19.06	23.70
20 02 0603	**4" Steel Conduit**	**LF**				**6.50**	**0.00**	**27.66**	**34.16**
16712 2070	Rigid Galvanized Steel PVC Coated 4" Conduit with Coupling Direct Burial, 20 Mil Coating	LF	16.13	EELEC	1.000	6.50	0.00	27.66	34.16
20 02 0604	**6" Steel Conduit**	**LF**				**19.92**	**0.00**	**43.22**	**63.14**
16018 0720	6" Rigid Galvanized Steel Conduit with Couplings Service & Feeder Conduit, 10' Height	LF	9.6	EELEF	1.000	19.92	0.00	43.22	63.14
20 02 0610	**2" PVC Conduit**	**LF**				**2.08**	**0.00**	**1.12**	**3.20**
16712 1050	2" PVC Schedule 40 Conduit with Couplings Direct Burial, Glued Coupling	LF	50.38	EELEC	1.000	2.08	0.00	1.12	3.20
20 02 0611	**3" PVC Conduit**	**LF**				**3.48**	**0.00**	**2.22**	**5.70**
16712 1060	3" PVC Schedule 40 Conduit with Couplings Direct Burial, Glued Coupling	LF	30.13	EELEC	1.000	3.48	0.00	2.22	5.70
20 02 0612	**4" PVC Conduit**	**LF**				**4.88**	**0.00**	**3.04**	**7.92**
16712 1070	4" PVC Schedule 40 Conduit with Couplings Direct Burial, Glued Coupling	LF	21.5	EELEC	1.000	4.88	0.00	3.04	7.92
20 02 0613	**6" PVC Conduit**	**LF**				**8.39**	**0.00**	**5.95**	**14.34**
16712 1090	6" PVC Schedule 40 Conduit with Couplings Direct Burial, Glued Coupling	LF	12.5	EELEC	1.000	8.39	0.00	5.95	14.34
20 02 0615	**Concrete Encasement for Ductbank**	**CY**				**75.32**	**3.85**	**71.45**	**150.62**
03311 6008	Grade Beam Chute	CY	18.75	ALABE	1.000	8.98	0.34	0.00	9.32
03318 0600	Concrete Integral Colors	CY	0	N/A	1.000	0.00	0.00	7.94	7.94
03354 3005	Concrete Redimix 2,500 PSI	CY	1.238	CLACC	1.000	66.34	3.52	63.50	133.36

20 02 06	Ductbanks, Manholes, Handholes, Raceways	Unit	Hourly Output	Crew	Factor	Labor	Equip.	Matl.	Total
20 02 0616	**Electrical & Communication Manhole, 10.5' Square x 10.0' Deep, with Cable Tray**	**EA**				**9,160**	**4,273**	**6,101**	**19,534**
16711 1900	Single Manhole, 10.5' Square x 10' Deep Including Cable Racks, Supports, Pulling Irons	EA	0.03	ALABN	1.000	9,160	4,273	6,101	19,534

20 02 07	Grounding Systems	Unit	Hourly Output	Crew	Factor	Labor	Equip.	Matl.	Total
20 02 0701	**4/0 Bare Copper Ground Wire**	**LF**				**1.15**	**0.13**	**1.93**	**3.21**
16123 1110	#4/0 AWG 1/c Bare Strand Copper Cable Installed on Poles	MLF	0.248	EELEK	0.001	1,153	127.58	1,934	3.21

20 03 01	Transformers	Unit	Hourly Output	Crew	Factor	Labor	Equip.	Matl.	Total
20 03 0101	**4160V/480V, 1, 37.5 KVA Transformer, Oil, Pad Mounted**	**EA**				**452.26**	**19.30**	**1,531**	**2,003**
16416 1030	37.5 KVA, Primary 4160 Grounded Y/2400 - 24940 Grounded Y/14400, Secondary 240/120, 480/240, 277	EA	0.373	EELEE	1.000	452.26	19.30	1,531	2,003
20 03 0102	**5/15 KV - 277/480V, 75 KVA Transformer, Oil, Pad Mounted**	**EA**				**935.92**	**39.93**	**9,336**	**10,312**
16416 4010	75 KVA, Primary 4160 Grounded Y/-13800V, Secondary 480Y/277, 480V, 3 Phase, Oil Transformer, Pad	EA	0.18	EELEE	1.000	935.92	39.93	9,336	10,312

20 03 03	Towers, Poles, Crossarms, and Insulators	Unit	Hourly Output	Crew	Factor	Labor	Equip.	Matl.	Total
20 03 0301	**30' Area Lighting Pole**	**EA**				**237.41**	**39.57**	**1,194**	**1,471**
16617 0140	Round Tapered Aluminum Roadway Pole, 1 - 6' Arm, 30' High (9.1 m)	EA	0.798	EELEJ	1.000	237.41	39.57	1,194	1,471
20 03 0302	**30' Galvanized Steel Pole (Single Arm) with Base**	**EA**				**445.93**	**53.19**	**1,203**	**1,702**
03142 1900	36" Diameter Round Fiber Forms, 1 Use	LF	14.38	ACARI	5.000	7.25	0.00	20.80	140.27
03354 3005	Concrete Redimix 2,500 PSI	CY	1.238	CLACC	2.000	66.34	3.52	63.50	266.72
16617 1040	Round Tapered Steel Roadway Pole, 1 - 6' Arm, 30' (9.1 m)	EA	0.684	EELEJ	1.000	276.97	46.16	971.84	1,295
20 03 0303	**20' Brushed Aluminum Pole (Single Arm) with Base**	**EA**				**322.64**	**33.34**	**1,086**	**1,442**
03142 1800	24" Diameter Round Fiber Forms, 1 Use	LF	16.25	ACARI	5.000	6.42	0.00	10.76	85.90
03354 3005	Concrete Redimix 2,500 PSI	CY	1.238	CLACC	2.000	66.34	3.52	63.50	266.72
16617 0100	Round Tapered Aluminum Roadway Pole, 1 - 6' Arm, 20' High (6 m)	EA	1.2	EELEJ	1.000	157.88	26.31	904.79	1,089

20 03 06	Exterior Lighting, Fixtures and Controls	Unit	Hourly Output	Crew	Factor	Labor	Equip.	Matl.	Total
20 03 0601	**150W High Pressure Sodium Fixture**	**EA**				**144.31**	**0.00**	**274.83**	**419.15**
16622 2010	Surface Square High Intensity Discharge Fixture, 2x2, Type 304B 150 Watt High Pressure Sodium	EA	1.325	EELEF	1.000	144.31	0.00	274.83	419.15

20 03 Exterior Lighting

20 03 06 Exterior Lighting, Fixtures and Controls		Unit	Hourly Output	Crew	Factor	Labor	Equip.	Matl.	Total
20 03 0602	**400W High Intensity Sodium Fixture**	**EA**				**115.59**	**0.00**	**343.04**	**458.63**
16622 2060	Type 306 400 with High Pressure Sodium	EA	0.89	EELEB	1.000	115.59	0.00	343.04	458.63

20 03 99 Other Area Lighting		Unit	Hourly Output	Crew	Factor	Labor	Equip.	Matl.	Total
20 03 9901	**1" Rigid Steel Conduit**	**LF**				**2.77**	**0.00**	**1.49**	**4.26**
16111 1163	1" Rigid Steel Conduit with Coupling & Supports	LF	68.95	EELEF	1.000	2.77	0.00	1.49	4.26
20 03 9902	**4" Rigid Steel Conduit**	**LF**				**7.90**	**0.00**	**9.63**	**17.53**
16111 1170	4" Rigid Steel Conduit with Coupling & Supports	LF	24.21	EELEF	1.000	7.90	0.00	9.63	17.53
20 03 9903	**Concrete Backfill Around Conduit**	**CY**				**12.09**	**3.02**	**90.00**	**105.12**
02244 1505	Washed & Graded Sand, Loaded at Pit	CY	12.38	CODEG	1.000	3.88	2.59	26.50	32.97
03212 1200	Bar Chair 1-3/4" High	EA	0	N/A	0.002	0.00	0.00	0.05	0.00
03354 3013	2,500 PSI Concrete with Tamp Crew	CY	10	CLACC	1.000	8.21	0.44	63.50	72.15

20 04 01 Telephone Systems		Unit	Hourly Output	Crew	Factor	Labor	Equip.	Matl.	Total
20 04 0101	**27 Pair No. 22 AWG Wire, Communication Cable**	**LF**				**1.02**	**0.00**	**1.98**	**3.00**
16122 1200	27 Pair No. 22 AWG Phone, Direct Burial	MLF	0.188	EELEF	0.001	1,020	0.00	1,983	3.00
20 04 0102	**51 Pair No. 22 AWG Wire, Communication Cable**	**LF**				**1.34**	**0.00**	**4.59**	**5.93**
16122 1240	51 Pair No. 22 AWG Phone, Direct Burial	MLF	0.143	EELEF	0.001	1,342	0.00	4,588	5.93
20 04 0103	**100 Pair No. 22 AWG Wire, Communication Cable**	**LF**				**2.90**	**0.00**	**7.85**	**10.75**
16122 1280	100 Pair No. 22 AWG Phone, Direct Burial	MLF	0.066	EELEF	0.001	2,897	0.00	7,851	10.75
20 04 0105	**Electrical & Communications Manhole 10.5' Square x 10' Deep, Cable Tray**	**EA**				**9,160**	**4,273**	**6,101**	**19,534**
16711 1900	Single Manhole, 10.5' Square x 10' Deep Including Cable Racks, Supports, Pulling Irons	EA	0.03	ALABN	1.000	9,160	4,273	6,101	19,534
20 04 0106	**4 Fiber Optic 62.5 Microns Cable**	**LF**				**1.52**	**0.25**	**0.85**	**2.62**
16111 1000	Fiber Optic Direct Burial Cable, 4 Fiber 62.5 Microns Cable	LF	125	EELEJ	1.000	1.52	0.25	0.85	2.62
20 04 0107	**8 Fiber Optic 50 Microns Cable**	**LF**				**0.38**	**0.00**	**1.78**	**2.17**
16111 1060	Fiber Optic Direct Burial Cable, 8 Fiber 50 Microns Cable	LF	500	EELEF	1.000	0.38	0.00	1.78	2.17
20 04 0108	**12 Fiber Optic 50 Microns Cable**	**LF**				**0.38**	**0.00**	**2.42**	**2.80**
16111 1080	Fiber Optic Direct Burial Cable, 12 Fiber 50 Microns Cable	LF	500	EELEF	1.000	0.38	0.00	2.42	2.80

20 06 01 Sacrificial Anode System		Unit	Hourly Output	Crew	Factor	Labor	Equip.	Matl.	Total

20 06 Cathodic Protection

20 06 01 Sacrificial Anode System

20 06 01 Sacrificial Anode System		Unit	Hourly Output	Crew	Factor	Labor	Equip.	Matl.	Total
20 06 0101	**3 - 9 Lb Magnesium Anodes, Cathodic Protection Point**	**EA**				**340.92**	**48.58**	**223.74**	**613.24**
16181 6380	#8-1 AWG Cadweld Structural Steel Ground Conductor Bonding	EA	6.24	EELEB	3.000	16.49	0.00	19.90	109.17
16821 1000	9 Lb (4.08 kg) Magnesium Type Anodes	EA	1.95	EELEJ	3.000	97.15	16.19	54.68	504.07
20 06 0102	**3 - 17 Lb Magnesium Anodes, Cathodic Protection Point**	**EA**				**470.46**	**70.17**	**361.88**	**902.51**
16181 6380	#8-1 AWG Cadweld Structural Steel Ground Conductor Bonding	EA	6.24	EELEB	3.000	16.49	0.00	19.90	109.17
16821 1010	17 Lb (7.7 kg) Magnesium Type Anodes	EA	1.35	EELEJ	3.000	140.33	23.39	100.72	793.34
20 06 0103	**3 - 32 Lb Magnesium Anodes, Cathodic Protection Point**	**EA**				**590.74**	**90.21**	**615.70**	**1,297**
16181 6380	#8-1 AWG Cadweld Structural Steel Ground Conductor Bonding	EA	6.24	EELEB	3.000	16.49	0.00	19.90	109.17
16821 1020	32 Lb (14.5 kg) Magnesium Type Anode	EA	1.05	EELEJ	3.000	180.43	30.07	185.33	1,187
20 06 0104	**3 - 48 Lb Magnesium Anodes, Cathodic Protection Point**	**EA**				**807.26**	**126.30**	**898.88**	**1,832**
16181 6380	#8-1 AWG Cadweld Structural Steel Ground Conductor Bonding	EA	6.24	EELEB	3.000	16.49	0.00	19.90	109.17
16821 1030	48 Lb (21.8 kg) Magnesium Type Anode	EA	0.75	EELEJ	3.000	252.60	42.10	279.72	1,723

20 99 99 Diesel Generator

20 99 99 Diesel Generator		Unit	Hourly Output	Crew	Factor	Labor	Equip.	Matl.	Total
20 99 9901	**200 KW Diesel Generator, 3 Phase**	**EA**				**4,713**	**276.61**	**44,549**	**49,538**
16513 2700	200 KW, Diesel Generator Set, 3 Phase without Piping, Wiring, Concrete	EA	0.036	EELEH	1.000	4,713	276.61	44,549	49,538

25 01 01 Asbestos Abatement - Equipment		Unit	Hourly Output	Crew	Factor	Labor	Equip.	Matl.	Total
25 01 0101	**Air Filtration Device, 2000 C.F.M.**	**EA**				**0.00**	**0.00**	**3,009**	**3,009**
25010 1000	Air Filtration Device, 2000 C.F.M.	EA	0	N/A	1.000	0.00	0.00	3,009	3,009
25 01 0102	**Large Volume Air Sampling Pump, Minimum**	**EA**				**0.00**	**0.00**	**542.76**	**542.76**
25010 1001	Large Volume Air Sampling Pump, Minimum	EA	0	N/A	1.000	0.00	0.00	542.76	542.76
25 01 0103	**Large Volume Air Sampling Pump, Maximum**	**EA**				**0.00**	**0.00**	**1,180**	**1,180**
25010 1002	Large Volume Air Sampling Pump, Maximum	EA	0	N/A	1.000	0.00	0.00	1,180	1,180
25 01 0104	**Airless Sprayer unit, 2 gun**	**EA**				**0.00**	**0.00**	**2,448**	**2,448**
25010 1003	Airless Sprayer unit, 2 gun	EA	0	N/A	1.000	0.00	0.00	2,448	2,448
25 01 0105	**Light Stand, 500 watt**	**EA**				**0.00**	**0.00**	**265.48**	**265.48**
25010 1004	Light Stand, 500 watt	EA	0	N/A	1.000	0.00	0.00	265.48	265.48
25 01 0106	**Personal Respirators, Negative pressure, 1/2 face, dual operation, minimum**	**EA**				**0.00**	**0.00**	**21.83**	**21.83**
25010 1005	Personal Respirators, Negative pressure, 1/2 face, dual operation, minimum	EA	0	N/A	1.000	0.00	0.00	21.83	21.83
25 01 0107	**Personal Respirators, Negative pressure, 1/2 face, dual operation, maximum**	**EA**				**0.00**	**0.00**	**26.55**	**26.55**
25010 1006	Personal Respirators, Negative pressure, 1/2 face, dual operation, maximum	EA	0	N/A	1.000	0.00	0.00	26.55	26.55
25 01 0108	**P.A.P.R., Full Face, minimum**	**EA**				**0.00**	**0.00**	**530.96**	**530.96**
25010 1007	P.A.P.R., Full Face, minimum	EA	0	N/A	1.000	0.00	0.00	530.96	530.96
25 01 0109	**P.A.P.R., Full Face, maximum**	**EA**				**0.00**	**0.00**	**825.94**	**825.94**
25010 1008	P.A.P.R., Full Face, maximum	EA	0	N/A	1.000	0.00	0.00	825.94	825.94
25 01 0110	**Personal Respirator, supplied air, full face, incl. Air line, minimum**	**EA**				**0.00**	**0.00**	**448.37**	**448.37**
25010 1009	Personal Respirator, supplied air, full face, incl. Air line, minimum	EA	0	N/A	1.000	0.00	0.00	448.37	448.37
25 01 0111	**Personal Respirator, supplied air, full face, incl. Air line, maximum**	**EA**				**0.00**	**0.00**	**648.95**	**648.95**
25010 1010	Personal Respirator, supplied air, full face, incl. Air line, maximum	EA	0	N/A	1.000	0.00	0.00	648.95	648.95
25 01 0112	**Personal Sampling Pump, minimum**	**EA**				**0.00**	**0.00**	**519.16**	**519.16**
25010 1011	Personal Sampling Pump, minimum	EA	0	N/A	1.000	0.00	0.00	519.16	519.16
25 01 0113	**Personal Sampling Pump, maximum**	**EA**				**0.00**	**0.00**	**825.94**	**825.94**
25010 1012	Personal Sampling Pump, maximum	EA	0	N/A	1.000	0.00	0.00	825.94	825.94
25 01 0114	**Power Panel, 20 unit, incl. G.F.I.**	**EA**				**0.00**	**0.00**	**2,065**	**2,065**
25010 1013	Power Panel, 20 unit, incl. G.F.I.	EA	0	N/A	1.000	0.00	0.00	2,065	2,065

25 01 01	Asbestos Abatement - Equipment	Unit	Hourly Output	Crew	Factor	Labor	Equip.	Matl.	Total
25 01 0115	**Shower Unit, including pump and filters**	**EA**				**0.00**	**0.00**	**1,327**	**1,327**
25010 1014	Shower Unit, including pump and filters	EA	0	N/A	1.000	0.00	0.00	1,327	1,327
25 01 0116	**Supplied air system (type C)**	**EA**				**0.00**	**0.00**	**11,799**	**11,799**
25010 1015	Supplied air system (type C)	EA	0	N/A	1.000	0.00	0.00	11,799	11,799
25 01 0117	**Vacuum Cleaner, HEPA, 16 gal., stainless steel, wet/dry**	**EA**				**0.00**	**0.00**	**1,062**	**1,062**
25010 1016	Vacuum Cleaner, HEPA, 16 gal., stainless steel, wet/dry	EA	0	N/A	1.000	0.00	0.00	1,062	1,062
25 01 0118	**Vacuum Cleaner, HEPA, 55 gal., stainless steel, wet/dry**	**EA**				**0.00**	**0.00**	**2,655**	**2,655**
25010 1017	Vacuum Cleaner, HEPA, 55 gal., stainless steel, wet/dry	EA	0	N/A	1.000	0.00	0.00	2,655	2,655
25 01 0119	**Vacuum Loader, 9-18 ton/hr.**	**EA**				**0.00**	**0.00**	**100,292**	**100,292**
25010 1018	Vacuum Loader, 9-18 ton/hr.	EA	0	N/A	1.000	0.00	0.00	100,292	100,292
25 01 0120	**Water atomizer unit, including 55 gal. Drum**	**EA**				**0.00**	**0.00**	**265.48**	**265.48**
25010 1019	Water atomizer unit, including 55 gal. Drum	EA	0	N/A	1.000	0.00	0.00	265.48	265.48
25 01 0121	**Worker Protection, whole body, foot, head cover & gloves, plastic**	**EA**				**0.00**	**0.00**	**35.40**	**35.40**
25010 1020	Worker Protection, whole body, foot, head cover & gloves, plastic	EA	0	N/A	1.000	0.00	0.00	35.40	35.40
25 01 0122	**Respirator, single use**	**EA**				**0.00**	**0.00**	**10.50**	**10.50**
25010 1021	Respirator, single use	EA	0	N/A	1.000	0.00	0.00	10.50	10.50
25 01 0123	**Cartridge for respirator**	**EA**				**0.00**	**0.00**	**43.66**	**43.66**
25010 1022	Cartridge for respirator	EA	0	N/A	1.000	0.00	0.00	43.66	43.66
25 01 0124	**Glove bag, 7 mil, 50" x 64"**	**EA**				**0.00**	**0.00**	**10.32**	**10.32**
25010 1023	Glove bag, 7 mil, 50" x 64"	EA	0	N/A	1.000	0.00	0.00	10.32	10.32
25 01 0125	**Glove bag, 10 mil, 44" x 60"**	**EA**				**0.00**	**0.00**	**10.03**	**10.03**
25010 1024	Glove bag, 10 mil, 44" x 60"	EA	0	N/A	1.000	0.00	0.00	10.03	10.03
25 01 0126	**HEPA Vacuum for work area, minimum**	**EA**				**0.00**	**0.00**	**1,298**	**1,298**
25010 1025	HEPA Vacuum for work area, minimum	EA	0	N/A	1.000	0.00	0.00	1,298	1,298
25 01 0127	**HEPA Vacuum for work area, maximum**	**EA**				**0.00**	**0.00**	**5,192**	**5,192**
25010 1026	HEPA Vacuum for work area, maximum	EA	0	N/A	1.000	0.00	0.00	5,192	5,192
25 01 0128	**Disposable polyethylene bags, 6 mil, 3 C.F.**	**EA**				**0.00**	**0.00**	**1.30**	**1.30**
25010 1027	Disposable polyethylene bags, 6 mil, 3 C.F.	EA	0	N/A	1.000	0.00	0.00	1.30	1.30
25 01 0129	**Disposable fiber drums, 3 C.F.**	**EA**				**0.00**	**0.00**	**7.43**	**7.43**
25010 1028	Disposable fiber drums, 3 C.F.	EA	0	N/A	1.000	0.00	0.00	7.43	7.43

25 01 Asbestos Removal

25 01 01 Asbestos Abatement - Equipment		Unit	Hourly Output	Crew	Factor	Labor	Equip.	Matl.	Total
25 01 0130	**Pressure sensitive caution labels, 3" x 5"**	**EA**				**0.00**	**0.00**	**0.12**	**0.12**
25010 1029	Pressure sensitive caution labels, 3" x 5"	EA	0	N/A	1.000	0.00	0.00	0.12	0.12
25 01 0131	**Pressure sensitive caution labels, 11" x 17"**	**EA**				**0.00**	**0.00**	**0.24**	**0.24**
25010 1030	Pressure sensitive caution labels, 11" x 17"	EA	0	N/A	1.000	0.00	0.00	0.24	0.24
25 01 0132	**Negative Air Machine, 1800 C.F.M.**	**EA**				**0.00**	**0.00**	**914.43**	**914.43**
25010 1031	Negative Air Machine, 1800 C.F.M.	EA	0	N/A	1.000	0.00	0.00	914.43	914.43

25 01 02 Asbestos Abatement - Work Area		Unit	Hourly Output	Crew	Factor	Labor	Equip.	Matl.	Total
25 01 0201	**Pre-cleaning, HEPA vacuum and wet wipe, flat surfaces**	**SF**				**0.10**	**0.05**	**0.00**	**0.15**
25010 1032	Pre-cleaning, HEPA vacuum and wet wipe, flat surfaces	SF	1500	AASBH	1.000	0.10	0.05	0.00	0.15
25 01 0202	**Protect carpeted area, 2 layers 6 mil poly on 3/4" plywood**	**SF**				**1.19**	**0.61**	**1.65**	**3.46**
25010 1033	Protect carpeted area, 2 layers 6 mil poly on 3/4" plywood	SF	125	AASBH	1.000	1.19	0.61	1.65	3.46
25 01 0203	**Separation barrier, 2" x 4" @ 16", 1/2" plywood ea. side, 8' high**	**SF**				**2.09**	**0.00**	**1.42**	**3.50**
25010 1034	Separation barrier, 2" x 4" @ 16", 1/2" plywood ea. side, 8' high	SF	50	ACARI	1.000	2.09	0.00	1.42	3.50
25 01 0204	**Separation barrier, 2" x 4" @ 16", 1/2" plywood ea. side, 12' high**	**SF**				**2.61**	**0.00**	**1.59**	**4.20**
25010 1130	Separation barrier, 2" x 4" @ 16", 1/2" plywood ea. side, 12' high	SF	40	ACARI	1.000	2.61	0.00	1.59	4.20
25 01 0205	**Separation barrier, 2" x 4" @ 16", 1/2" plywood ea. side, 16' high**	**SF**				**4.17**	**0.00**	**1.89**	**6.06**
25010 1131	Separation barrier, 2" x 4" @ 16", 1/2" plywood ea. side, 16' high	SF	25	ACARI	1.000	4.17	0.00	1.89	6.06
25 01 0206	**Personnel Decontam. Chamber 2" x 4" @16", 3/4" ply ea. Side**	**SF**				**2.98**	**0.00**	**2.95**	**5.93**
25010 1132	Personnel Decontam. Chamber 2" x 4" @16", 3/4" ply ea. Side	SF	35	ACARI	1.000	2.98	0.00	2.95	5.93
25 01 0207	**Waste decontam. Chamber 2" x 4" studs @ 16", 3/4" ply. Ea side**	**SF**				**2.32**	**0.00**	**3.72**	**6.03**
25010 1133	Waste decontam. Chamber 2" x 4" studs @ 16", 3/4" ply. Ea side	SF	45	ACARI	1.000	2.32	0.00	3.72	6.03
25 01 0208	**Cover surfaces w/polyethylene sheeting floors 6 mil**	**SF**				**0.13**	**0.00**	**0.12**	**0.24**
25010 1134	Cover surfaces w/polyethylene sheeting floors 6 mil	SF	750	AASBC	1.000	0.13	0.00	0.12	0.24

25 01 02 Asbestos Abatement - Work Area		Unit	Hourly Output	Crew	Factor	Labor	Equip.	Matl.	Total
25 01 0209	**Cover surfaces w/polyethylene sheeting floors 4 mil**	**SF**				**0.11**	**0.00**	**0.06**	**0.17**
25010 1135	Cover surfaces w/polyethylene sheeting floors 4 mil	SF	875	AASBC	1.000	0.11	0.00	0.06	0.17
25 01 0210	**Cover surfaces w/polyethylene sheeting walls 6 mil**	**SF**				**0.19**	**0.00**	**0.11**	**0.29**
25010 1136	Cover surfaces w/polyethylene sheeting walls 6 mil	SF	500	AASBC	1.000	0.19	0.00	0.11	0.29
25 01 0211	**Cover surfaces w/polyethylene sheeting walls 4 mil**	**SF**				**0.15**	**0.00**	**0.07**	**0.22**
25010 1137	Cover surfaces w/polyethylene sheeting walls 4 mil	SF	625	AASBC	1.000	0.15	0.00	0.07	0.22
25 01 0212	**Seal floor penetrations w/ foam firestop to 36 Sq. In**	**EA**				**4.17**	**0.00**	**7.32**	**11.49**
25010 1138	Seal floor penetrations w/ foam firestop to 36 Sq. In	EA	25	ACARI	1.000	4.17	0.00	7.32	11.49
25 01 0213	**Seal floor penetrations w/ foam firestop 36 to 72 Sq. In**	**EA**				**6.67**	**0.00**	**14.63**	**21.30**
25010 1139	Seal floor penetrations w/ foam firestop 36 to 72 Sq. In	EA	15.63	ACARI	1.000	6.67	0.00	14.63	21.30
25 01 0214	**Seal floor penetrations w/ foam firestop 72 to 144 Sq. In**	**EA**				**10.43**	**0.00**	**29.50**	**39.92**
25010 1140	Seal floor penetrations w/ foam firestop 72 to 144 Sq. In	EA	10	ACARI	1.000	10.43	0.00	29.50	39.92
25 01 0215	**Seal wall penetrations w/ foam firestop to 36 Sq. In**	**EA**				**4.63**	**0.00**	**7.32**	**11.95**
25010 1141	Seal wall penetrations w/ foam firestop to 36 Sq. In	EA	22.5	ACARI	1.000	4.63	0.00	7.32	11.95
25 01 0216	**Seal wall penetrations w/ foam firestop 36 to 72 Sq. In**	**EA**				**8.34**	**0.00**	**14.63**	**22.97**
25010 1142	Seal wall penetrations w/ foam firestop 36 to 72 Sq. In	EA	12.5	ACARI	1.000	8.34	0.00	14.63	22.97
25 01 0217	**Seal wall penetrations w/ foam firestop 72 to 144 Sq. In**	**EA**				**13.90**	**0.00**	**29.50**	**43.40**
25010 1143	Seal wall penetrations w/ foam firestop 72 to 144 Sq. In	EA	7.5	ACARI	1.000	13.90	0.00	29.50	43.40
25 01 0218	**Caulk seams with Latex**	**LF**				**1.51**	**0.00**	**0.19**	**1.70**
25010 1144	Caulk seams with Latex	LF	28.75	ACARC	1.000	1.51	0.00	0.19	1.70
25 01 0219	**Setup neg. air machine, 12k C.F.M./25 MCF Volume**	**EA**				**87.21**	**0.00**	**0.00**	**87.21**
25010 1145	Setup neg. air machine, 12k C.F.M./25 MCF Volume	EA	0.538	AASBA	1.000	87.21	0.00	0.00	87.21

25 01 03 Asbestos Contaminated Area - Demolition		Unit	Hourly Output	Crew	Factor	Labor	Equip.	Matl.	Total
25 01 0301	**Ceiling including suspension system**	**SF**				**0.57**	**0.29**	**0.00**	**0.86**
25010 1146	Ceiling including suspension system	SF	262.5	AASBH	1.000	0.57	0.29	0.00	0.86
25 01 0302	**Ceiling including suspension system, Finished plaster, leaving wire lath`**	**SF**				**2.04**	**1.05**	**0.00**	**3.09**
25010 1147	Ceiling including suspension system, Finished plaster, leaving wire lath`	SF	73.13	AASBH	1.000	2.04	1.05	0.00	3.09
25 01 0303	**Ceiling including suspension system, suspended acoustic tile**	**SF**				**0.34**	**0.18**	**0.00**	**0.52**
25010 1148	Ceiling including suspension system, suspended acoustic tile	SF	437.5	AASBH	1.000	0.34	0.18	0.00	0.52
25 01 0304	**Ceiling including suspension system, concealed tile grid**	**SF**				**0.40**	**0.20**	**0.00**	**0.60**
25010 1149	Ceiling including suspension system, concealed tile grid	SF	375	AASBH	1.000	0.40	0.20	0.00	0.60
25 01 0305	**Ceiling including suspension system, metal pan grid**	**SF**				**0.80**	**0.41**	**0.00**	**1.20**
25010 1150	Ceiling including suspension system, metal pan grid	SF	187.5	AASBH	1.000	0.80	0.41	0.00	1.20
25 01 0306	**Ceiling including suspension system, gypsum board**	**SF**				**0.48**	**0.25**	**0.00**	**0.72**
25010 1151	Ceiling including suspension system, gypsum board	SF	312.5	AASBH	1.000	0.48	0.25	0.00	0.72
25 01 0307	**Ceiling including suspension system, lighting fixtures up to 2' x 4'**	**EA**				**16.58**	**8.51**	**0.00**	**25.09**
25010 1152	Ceiling including suspension system, lighting fixtures up to 2' x 4'	EA	9	AASBH	1.000	16.58	8.51	0.00	25.09
25 01 0308	**Partitions non load bearing, plaster lath and studs**	**SF**				**1.73**	**0.89**	**0.65**	**3.27**
25010 1153	Partitions non load bearing, plaster lath and studs	SF	86.25	AASBH	1.000	1.73	0.89	0.65	3.27
25 01 0309	**Partitions non load bearing, gypsum board and studs**	**SF**				**0.86**	**0.44**	**0.00**	**1.30**
25010 1154	Partitions non load bearing, gypsum board and studs	SF	173.8	AASBH	1.000	0.86	0.44	0.00	1.30

25 01 04 Asbestos - Bulk Removal		Unit	Hourly Output	Crew	Factor	Labor	Equip.	Matl.	Total
25 01 0401	**Bulk Asbestos Removal, beams W 10 x 19**	**LF**				**5.08**	**2.61**	**0.00**	**7.69**
25010 1155	Bulk Asbestos Removal, beams W 10 x 19	LF	29.38	AASBH	1.000	5.08	2.61	0.00	7.69
25 01 0402	**Bulk Asbestos Removal, beams W 12 x 22**	**LF**				**5.68**	**2.92**	**0.00**	**8.60**
25010 1156	Bulk Asbestos Removal, beams W 12 x 22	LF	26.25	AASBH	1.000	5.68	2.92	0.00	8.60
25 01 0403	**Bulk Asbestos Removal, beams W 14 x 26**	**LF**				**6.63**	**3.40**	**0.00**	**10.04**
25010 1157	Bulk Asbestos Removal, beams W 14 x 26	LF	22.5	AASBH	1.000	6.63	3.40	0.00	10.04

25 01 04 Asbestos - Bulk Removal		Unit	Hourly Output	Crew	Factor	Labor	Equip.	Matl.	Total
25 01 0404	**Bulk Asbestos Removal, beams W 16 x 31**	**LF**				**7.46**	**3.83**	**0.00**	**11.29**
25010 1158	Bulk Asbestos Removal, beams W 16 x 31	LF	20	AASBH	1.000	7.46	3.83	0.00	11.29
25 01 0405	**Bulk Asbestos Removal, beams W 18 x 40**	**LF**				**8.53**	**4.38**	**0.00**	**12.90**
25010 1159	Bulk Asbestos Removal, beams W 18 x 40	LF	17.5	AASBH	1.000	8.53	4.38	0.00	12.90
25 01 0406	**Bulk Asbestos Removal, beams W 24 x 55**	**LF**				**10.85**	**5.57**	**0.00**	**16.42**
25010 1160	Bulk Asbestos Removal, beams W 24 x 55	LF	13.75	AASBH	1.000	10.85	5.57	0.00	16.42
25 01 0407	**Bulk Asbestos Removal, beams W 30 x 108**	**LF**				**14.04**	**7.21**	**0.00**	**21.25**
25010 1161	Bulk Asbestos Removal, beams W 30 x 108	LF	10.63	AASBH	1.000	14.04	7.21	0.00	21.25
25 01 0408	**Bulk Asbestos Removal, beams W 36 x 150**	**LF**				**16.58**	**8.51**	**0.00**	**25.09**
25010 1162	Bulk Asbestos Removal, beams W 36 x 150	LF	9	AASBH	1.000	16.58	8.51	0.00	25.09
25 01 0409	**Boiler Insulation**	**SF**				**2.49**	**1.28**	**0.07**	**3.83**
25010 1163	Boiler Insulation	SF	60	AASBH	1.000	2.49	1.28	0.07	3.83
25 01 0410	**Boiler breeching or flue insulation**	**SF**				**2.30**	**1.18**	**0.00**	**3.47**
25010 1164	Boiler breeching or flue insulation	SF	65	AASBH	1.000	2.30	1.18	0.00	3.47
25 01 0411	**Duct or AHU insulation**	**SF**				**1.70**	**0.00**	**0.00**	**1.70**
25010 1165	Duct or AHU insulation	SF	55	AASBB	1.000	1.70	0.00	0.00	1.70
25 01 0412	**Duct vibration isolation joints, up to 24 Sq. In. duct**	**EA**				**21.31**	**10.94**	**0.00**	**32.26**
25010 1166	Duct vibration isolation joints, up to 24 Sq. In. duct	EA	7	AASBH	1.000	21.31	10.94	0.00	32.26
25 01 0413	**Duct vibration isolation joints, 25 to 48 Sq. In. duct**	**EA**				**25.95**	**13.32**	**0.00**	**39.27**
25010 1168	Duct vibration isolation joints, 25 to 48 Sq. In. duct	EA	5.75	AASBH	1.000	25.95	13.32	0.00	39.27
25 01 0414	**Pipe insulation, air cell type, up to 4" diameter pipe**	**LF**				**1.33**	**0.68**	**0.00**	**2.01**
25010 1169	Pipe insulation, air cell type, up to 4" diameter pipe	LF	112.5	AASBH	1.000	1.33	0.68	0.00	2.01
25 01 0415	**Pipe insulation, air cell type, 6" to 8" diameter pipe**	**LF**				**1.49**	**0.77**	**0.00**	**2.26**
25010 1170	Pipe insulation, air cell type, 6" to 8" diameter pipe	LF	100	AASBH	1.000	1.49	0.77	0.00	2.26
25 01 0416	**Pipe insulation, air cell type, 10" to 12" diameter pipe**	**LF**				**1.71**	**0.88**	**0.00**	**2.58**
25010 1171	Pipe insulation, air cell type, 10" to 12" diameter pipe	LF	87.5	AASBH	1.000	1.71	0.88	0.00	2.58

25 01 04 Asbestos - Bulk Removal		Unit	Hourly Output	Crew	Factor	Labor	Equip.	Matl.	Total
25 01 0417	**Pipe insulation, air cell type, 14" to 16" diameter pipe**	**LF**				**2.17**	**1.11**	**0.00**	**3.28**
25010 1172	Pipe insulation, air cell type, 14" to 16" diameter pipe	LF	68.75	AASBH	1.000	2.17	1.11	0.00	3.28
25 01 0418	**Pipe insulation, air cell type, over 16" diameter pipe**	**LF**				**1.84**	**0.94**	**0.00**	**2.78**
25010 1173	Pipe insulation, air cell type, over 16" diameter pipe	LF	81.25	AASBH	1.000	1.84	0.94	0.00	2.78
25 01 0419	**Pipe insulation, air cell type, with glove bag, up to 8" diameter pipe**	**LF**				**11.94**	**6.13**	**0.00**	**18.06**
25010 1174	Pipe insulation, air cell type, with glove bag, up to 8" diameter pipe	LF	12.5	AASBH	1.000	11.94	6.13	0.00	18.06
25 01 0420	**Scrape Foam fireproofing from flat surface**	**SF**				**0.50**	**0.26**	**0.00**	**0.75**
25010 1175	Scrape Foam fireproofing from flat surface	SF	300	AASBH	1.000	0.50	0.26	0.00	0.75
25 01 0421	**Scrape Foam fireproofing from irregular surface**	**SF**				**0.99**	**0.51**	**0.00**	**1.51**
25010 1176	Scrape Foam fireproofing from irregular surface	SF	150	AASBH	1.000	0.99	0.51	0.00	1.51
25 01 0422	**Remove cementitious material from flat surface**	**SF**				**1.49**	**0.77**	**0.00**	**2.26**
25010 1177	Remove cementitious material from flat surface	SF	100	AASBH	1.000	1.49	0.77	0.00	2.26
25 01 0423	**Remove cementitious material from irregular surface**	**SF**				**2.98**	**1.53**	**0.00**	**4.52**
25010 1178	Remove cementitious material from irregular surface	SF	50	AASBH	1.000	2.98	1.53	0.00	4.52
25 01 0424	**Scrape acoustical coating/fireproofing from ceiling**	**SF**				**0.37**	**0.19**	**0.00**	**0.56**
25010 1179	Scrape acoustical coating/fireproofing from ceiling	SF	400	AASBH	1.000	0.37	0.19	0.00	0.56
25 01 0425	**Remove VAT from floor by hand**	**SF**				**0.50**	**0.26**	**0.00**	**0.75**
25010 1180	Remove VAT from floor by hand	SF	300	AASBH	1.000	0.50	0.26	0.00	0.75
25 01 0426	**Remove VAT from floor by machine**	**SF**				**0.25**	**0.13**	**0.00**	**0.38**
25010 1181	Remove VAT from floor by machine	SF	600	AASBH	1.000	0.25	0.13	0.00	0.38
25 01 0427	**Remove contaminated soil from crawlspace by hand**	**CF**				**2.98**	**1.53**	**0.00**	**4.52**
25010 1182	Remove contaminated soil from crawlspace by hand	CF	50	AASBH	1.000	2.98	1.53	0.00	4.52
25 01 0428	**Remove contaminated soil from crawlspace w/ large prod. Vac loader**	**CF**				**1.71**	**0.88**	**0.00**	**2.58**
25010 1183	Remove contaminated soil from crawlspace w/ large prod. Vac loader	CF	87.5	AASBH	1.000	1.71	0.88	0.00	2.58

25 01 04 Asbestos - Bulk Removal		Unit	Hourly Output	Crew	Factor	Labor	Equip.	Matl.	Total
25 01 0429	**Radiator backing, not including radiator removal**	**SF**				**0.99**	**0.51**	**0.00**	**1.51**
25010 1184	Radiator backing, not including radiator removal	SF	150	AASBH	1.000	0.99	0.51	0.00	1.51
25 01 0430	**Cement asbestos transite baord**	**SF**				**0.19**	**0.00**	**0.09**	**0.28**
25010 1185	Cement asbestos transite baord	SF	250	AASBA	1.000	0.19	0.00	0.09	0.28
25 01 0431	**Transite shingle siding**	**SF**				**1.59**	**0.82**	**0.09**	**2.50**
25010 1186	Transite shingle siding	SF	93.75	AASBH	1.000	1.59	0.82	0.09	2.50
25 01 0432	**Shingle roofing**	**SF**				**0.19**	**0.00**	**0.08**	**0.27**
25010 1187	Shingle roofing	SF	250	AASBA	1.000	0.19	0.00	0.08	0.27
25 01 0433	**Shingle roofing, built up, no gravel, non-friable**	**SF**				**0.27**	**0.00**	**0.08**	**0.35**
25010 1188	Shingle roofing, built up, no gravel, non-friable	SF	175	AASBA	1.000	0.27	0.00	0.08	0.35
25 01 0434	**Asbestos millboard**	**SF**				**0.38**	**0.00**	**0.09**	**0.47**
25010 1189	Asbestos millboard	SF	125	AASBA	1.000	0.38	0.00	0.09	0.47
25 01 0435	**Duct vibration isolation joints, 49 to 76 Sq. In. duct**	**EA**				**29.84**	**15.32**	**0.00**	**45.16**
25010 1167	Duct vibration isolation joints, 49 to 76 Sq. In. duct	EA	5	AASBH	1.000	29.84	15.32	0.00	45.16

25 01 05 Abestos Waste Packaging, Handling & Disposal		Unit	Hourly Output	Crew	Factor	Labor	Equip.	Matl.	Total
25 01 0501	**Collect and bag bulk material, 3 C.F. bags, by hand**	**EA**				**2.98**	**1.53**	**1.30**	**5.81**
25010 1035	Collect and bag bulk material, 3 C.F. bags, by hand	EA	50	AASBH	1.000	2.98	1.53	1.30	5.81
25 01 0502	**Collect and bag bulk material, large production vacuum loader**	**EA**				**1.36**	**0.70**	**0.88**	**2.94**
25010 1036	Collect and bag bulk material, large production vacuum loader	EA	110	AASBH	1.000	1.36	0.70	0.88	2.94
25 01 0503	**Double bag and decontaminate**	**EA**				**1.24**	**0.64**	**20.06**	**21.94**
25010 1037	Double bag and decontaminate	EA	120	AASBH	1.000	1.24	0.64	20.06	21.94
25 01 0504	**Containerize bagged material in drums, per 3 C.F. drum**	**EA**				**1.49**	**0.77**	**7.43**	**9.69**
25010 1038	Containerize bagged material in drums, per 3 C.F. drum	EA	100	AASBH	1.000	1.49	0.77	7.43	9.69
25 01 0505	**Cart bags 50' to dumpster**	**EA**				**0.94**	**0.00**	**0.00**	**0.94**
25010 1039	Cart bags 50' to dumpster	EA	50	AASBA	1.000	0.94	0.00	0.00	0.94
25 01 0506	**Disposal charges, not including haul, minimum**	**CY**				**0.00**	**0.00**	**49.56**	**49.56**
25010 1040	Disposal charges, not including haul, minimum	CY	0	N/A	1.000	0.00	0.00	49.56	49.56

25 01 05 Abestos Waste Packaging, Handling & Disposal		Unit	Hourly Output	Crew	Factor	Labor	Equip.	Matl.	Total
25 01 0507	**Disposal charges, not including haul, maximum**	**CY**				**0.00**	**0.00**	**188.79**	**188.79**
25010 1041	Disposal charges, not including haul, maximum	CY	0	N/A	1.000	0.00	0.00	188.79	188.79
25 01 0508	**Remove refrigerant from system**	**LB**				**10.31**	**0.00**	**0.00**	**10.31**
25010 1042	Remove refrigerant from system	LB	5	MPLUA	1.000	10.31	0.00	0.00	10.31

25 01 06 Asbestos Decontamination/Containment - Demolition		Unit	Hourly Output	Crew	Factor	Labor	Equip.	Matl.	Total
25 01 0601	**Spray exposed substrate with surfactant (bridging) flat surface**	**SF**				**0.20**	**0.10**	**0.35**	**0.66**
25010 1043	Spray exposed substrate with surfactant (bridging) flat surface	SF	750	AASBH	1.000	0.20	0.10	0.35	0.66
25 01 0602	**Spray exposed substrate with surfactant (bridging) irregular surface**	**SF**				**0.30**	**0.15**	**0.41**	**0.86**
25010 1044	Spray exposed substrate with surfactant (bridging) irregular surface	SF	500	AASBH	1.000	0.30	0.15	0.41	0.86
25 01 0603	**Spray exposed substrate with surfactant (bridging) pipes and beams**	**LF**				**0.60**	**0.31**	**0.71**	**1.61**
25010 1045	Spray exposed substrate with surfactant (bridging) pipes and beams	LF	250	AASBH	1.000	0.60	0.31	0.71	1.61
25 01 0604	**Spray encapsulate polyethylene sheeting**	**SF**				**0.15**	**0.08**	**0.29**	**0.52**
25010 1047	Spray encapsulate polyethylene sheeting	SF	1000	AASBH	1.000	0.15	0.08	0.29	0.52
25 01 0605	**Roll down polyethylene sheeting**	**SF**				**0.15**	**0.08**	**0.00**	**0.23**
25010 1046	Roll down polyethylene sheeting	SF	1000	AASBH	1.000	0.15	0.08	0.00	0.23
25 01 0606	**Bag polyethylene sheeting**	**EA**				**2.98**	**1.53**	**0.83**	**5.34**
25010 1048	Bag polyethylene sheeting	EA	50	AASBH	1.000	2.98	1.53	0.83	5.34
25 01 0607	**Fine clean exposed substrate, with nylon brush**	**SF**				**0.50**	**0.26**	**0.00**	**0.75**
25010 1049	Fine clean exposed substrate, with nylon brush	SF	300	AASBH	1.000	0.50	0.26	0.00	0.75
25 01 0608	**Fine clean exposed substrate, wet wipe substrate**	**SF**				**0.25**	**0.13**	**0.00**	**0.38**
25010 1050	Fine clean exposed substrate, wet wipe substrate	SF	600	AASBH	1.000	0.25	0.13	0.00	0.38
25 01 0609	**Fine clean exposed substrate, vacuum surfaces, fine brush**	**SF**				**0.19**	**0.10**	**0.00**	**0.28**
25010 1051	Fine clean exposed substrate, vacuum surfaces, fine brush	SF	800	AASBH	1.000	0.19	0.10	0.00	0.28
25 01 0610	**Structural demolition, wood stud walls**	**SF**				**0.43**	**0.22**	**0.00**	**0.65**
25010 1052	Structural demolition, wood stud walls	SF	350	AASBH	1.000	0.43	0.22	0.00	0.65

25 01 Asbestos Removal

25 01 06	Asbestos Decontamination/Containment - Demolition	Unit	Hourly Output	Crew	Factor	Labor	Equip.	Matl.	Total
25 01 0611	**Structural demolition, window manifolds, not incl. window replacement**	**SF**				**0.28**	**0.15**	**0.00**	**0.43**
25010 1053	Structural demolition, window manifolds, not incl. window replacement	SF	525	AASBH	1.000	0.28	0.15	0.00	0.43
25 01 0612	**Structural demolition, plywood carpet protection**	**SF**				**0.30**	**0.15**	**0.00**	**0.45**
25010 1054	Structural demolition, plywood carpet protection	SF	500	AASBH	1.000	0.30	0.15	0.00	0.45
25 01 0613	**Remove custom decontamination facility**	**EA**				**93.75**	**0.00**	**12.98**	**106.73**
25010 1055	Remove custom decontamination facility	EA	1	AASBB	1.000	93.75	0.00	12.98	106.73
25 01 0614	**Remove portable decontamination facility**	**EA**				**62.50**	**0.00**	**13.57**	**76.07**
25010 1056	Remove portable decontamination facility	EA	1.5	AASBB	1.000	62.50	0.00	13.57	76.07
25 01 0615	**HEPA Vacuum, shampoo carpeting**	**SF**				**0.25**	**0.13**	**0.06**	**0.44**
25010 1057	HEPA Vacuum, shampoo carpeting	SF	600	AASBH	1.000	0.25	0.13	0.06	0.44
25 01 0616	**Final cleaning of protected surfaces**	**SF**				**0.09**	**0.00**	**0.00**	**0.09**
25010 1058	Final cleaning of protected surfaces	SF	1000	AASBB	1.000	0.09	0.00	0.00	0.09

25 01 07	Asbestos Encapsulation with Sealants	Unit	Hourly Output	Crew	Factor	Labor	Equip.	Matl.	Total
25 01 0701	**Encapsulation with sealants, ceilings and walls, minimum**	**SF**				**0.06**	**0.03**	**0.29**	**0.38**
25010 1059	Encapsulation with sealants, ceilings and walls, minimum	SF	2625	AASBH	1.000	0.06	0.03	0.29	0.38
25 01 0702	**Encapsulation with sealants, ceilings and walls, maximum**	**SF**				**0.11**	**0.06**	**0.41**	**0.58**
25010 1060	Encapsulation with sealants, ceilings and walls, maximum	SF	1325	AASBH	1.000	0.11	0.06	0.41	0.58
25 01 0703	**Encapsulation with sealants, columns and beams, minimum**	**SF**				**0.09**	**0.05**	**0.29**	**0.43**
25010 1061	Encapsulation with sealants, columns and beams, minimum	SF	1663	AASBH	1.000	0.09	0.05	0.29	0.43
25 01 0704	**Encapsulation with sealants, columns and beams, maximum**	**SF**				**0.22**	**0.12**	**0.41**	**0.75**
25010 1062	Encapsulation with sealants, columns and beams, maximum	SF	665.6	AASBH	1.000	0.22	0.12	0.41	0.75
25 01 0705	**Encapsulation with sealants, pipes to 12" diameter incl. minor repairs, minimum**	**LF**				**1.49**	**0.77**	**0.41**	**2.67**
25010 1063	Encapsulation with sealants, pipes to 12" diameter incl. minor repairs, minimum	LF	100	AASBH	1.000	1.49	0.77	0.41	2.67

25 01 07	Asbestos Encapsulation with Sealants	Unit	Hourly Output	Crew	Factor	Labor	Equip.	Matl.	Total
25 01 0706	**Encapsulation with sealants, pipes to 12" diameter incl. minor repairs, maximum**	**LF**				**2.98**	**1.53**	**1.18**	**5.70**
25010 1064	Encapsulation with sealants, pipes to 12" diameter incl. minor repairs, maximum	LF	50	AASBH	1.000	2.98	1.53	1.18	5.70

25 02 01	Lead Paint Removal - by Chemicals	Unit	Hourly Output	Crew	Factor	Labor	Equip.	Matl.	Total
25 02 0101	**Removal existing lead paint, by chemicals, baseboard to 6" wide**	**LF**				**4.84**	**0.00**	**1.56**	**6.40**
25010 1065	Removal existing lead paint, by chemicals, baseboard to 6" wide	LF	8	APTRA	1.000	4.84	0.00	1.56	6.40
25 02 0102	**Removal existing lead paint, by chemicals, baseboard to 12" wide**	**LF**				**9.69**	**0.00**	**3.07**	**12.76**
25010 1066	Removal existing lead paint, by chemicals, baseboard to 12" wide	LF	4	APTRA	1.000	9.69	0.00	3.07	12.76
25 02 0103	**Removal existing lead paint, by chemicals, balustrades, one side**	**SF**				**11.07**	**0.00**	**3.48**	**14.55**
25010 1067	Removal existing lead paint, by chemicals, balustrades, one side	SF	3.5	APTRA	1.000	11.07	0.00	3.48	14.55
25 02 0104	**Removal existing lead paint, by chemicals, cabinets, simple design**	**SF**				**9.69**	**0.00**	**3.04**	**12.73**
25010 1068	Removal existing lead paint, by chemicals, cabinets, simple design	SF	4	APTRA	1.000	9.69	0.00	3.04	12.73
25 02 0105	**Removal existing lead paint, by chemicals, cabinets, ornate design**	**SF**				**12.40**	**0.00**	**3.91**	**16.31**
25010 1069	Removal existing lead paint, by chemicals, cabinets, ornate design	SF	3.125	APTRA	1.000	12.40	0.00	3.91	16.31
25 02 0106	**Removal existing lead paint, by chemicals, cornice, simple design**	**SF**				**5.17**	**0.00**	**1.63**	**6.79**
25010 1070	Removal existing lead paint, by chemicals, cornice, simple design	SF	7.5	APTRA	1.000	5.17	0.00	1.63	6.79
25 02 0107	**Removal existing lead paint, by chemicals, cornice, ornate design**	**SF**				**15.50**	**0.00**	**4.81**	**20.31**
25010 1071	Removal existing lead paint, by chemicals, cornice, ornate design	SF	2.5	APTRA	1.000	15.50	0.00	4.81	20.31
25 02 0108	**Removal existing lead paint, by chemicals, doors, one side, flush**	**SF**				**3.69**	**0.00**	**1.18**	**4.87**
25010 1072	Removal existing lead paint, by chemicals, doors, one side, flush	SF	10.5	APTRA	1.000	3.69	0.00	1.18	4.87
25 02 0109	**Removal existing lead paint, by chemicals, doors, two panel**	**SF**				**3.88**	**0.00**	**1.23**	**5.10**
25010 1073	Removal existing lead paint, by chemicals, doors, two panel	SF	10	APTRA	1.000	3.88	0.00	1.23	5.10

25 02 01 Lead Paint Removal - by Chemicals		Unit	Hourly Output	Crew	Factor	Labor	Equip.	Matl.	Total
25 02 0110	**Removal existing lead paint, by chemicals, doors, four panel**	**SF**				**6.89**	**0.00**	**2.16**	**9.05**
25010 1074	Removal existing lead paint, by chemicals, doors, four panel	SF	5.625	APTRA	1.000	6.89	0.00	2.16	9.05
25 02 0111	**Removal existing lead paint, by chemicals, doors, for trim one side add**	**LF**				**4.84**	**0.00**	**1.56**	**6.40**
25010 1075	Removal existing lead paint, by chemicals, doors, for trim one side add	LF	8	APTRA	1.000	4.84	0.00	1.56	6.40
25 02 0112	**Removal existing lead paint, by chemicals, fence picket, one side**	**SF**				**10.33**	**0.00**	**3.27**	**13.60**
25010 1076	Removal existing lead paint, by chemicals, fence picket, one side	SF	3.75	APTRA	1.000	10.33	0.00	3.27	13.60
25 02 0113	**Removal existing lead paint, by chemicals, grills, one side, simple design**	**SF**				**10.33**	**0.00**	**3.27**	**13.60**
25010 1077	Removal existing lead paint, by chemicals, grills, one side, simple design	SF	3.75	APTRA	1.000	10.33	0.00	3.27	13.60
25 02 0114	**Removal existing lead paint, by chemicals, grills, one side, ornate design**	**SF**				**12.40**	**0.00**	**3.91**	**16.31**
25010 1078	Removal existing lead paint, by chemicals, grills, one side, ornate design	SF	3.125	APTRA	1.000	12.40	0.00	3.91	16.31
25 02 0115	**Removal existing lead paint, by chemicals, pipes, to 4" diameter**	**LF**				**3.44**	**0.00**	**1.11**	**4.55**
25010 1079	Removal existing lead paint, by chemicals, pipes, to 4" diameter	LF	11.25	APTRA	1.000	3.44	0.00	1.11	4.55
25 02 0116	**Removal existing lead paint, by chemicals, pipes, to 8" diameter**	**LF**				**6.20**	**0.00**	**1.95**	**8.15**
25010 1080	Removal existing lead paint, by chemicals, pipes, to 8" diameter	LF	6.25	APTRA	1.000	6.20	0.00	1.95	8.15
25 02 0117	**Removal existing lead paint, by chemicals, pipes, to 12" diameter**	**LF**				**8.61**	**0.00**	**2.70**	**11.31**
25010 1081	Removal existing lead paint, by chemicals, pipes, to 12" diameter	LF	4.5	APTRA	1.000	8.61	0.00	2.70	11.31
25 02 0118	**Removal existing lead paint, by chemicals, pipes, to 16" diameter**	**LF**				**15.50**	**0.00**	**4.85**	**20.35**
25010 1082	Removal existing lead paint, by chemicals, pipes, to 16" diameter	LF	2.5	APTRA	1.000	15.50	0.00	4.85	20.35
25 02 0119	**Removal existing lead paint, by chemicals, pipes, for hangers Add**	**EA**				**7.75**	**0.00**	**2.70**	**10.45**
25010 1083	Removal existing lead paint, by chemicals, pipes, for hangers Add	EA	5	APTRA	1.000	7.75	0.00	2.70	10.45
25 02 0120	**Removal existing lead paint, by chemicals, siding**	**SF**				**3.44**	**0.00**	**1.11**	**4.55**
25010 1084	Removal existing lead paint, by chemicals, siding	SF	11.25	APTRA	1.000	3.44	0.00	1.11	4.55

25 02 Lead Paint Removal

25 02 01 Lead Paint Removal - by Chemicals		Unit	Hourly Output	Crew	Factor	Labor	Equip.	Matl.	Total
25 02 0121	**Removal existing lead paint, by chemicals, trusses, open**	**SF**				**5.64**	**0.00**	**1.77**	**7.41**
25010 1085	Removal existing lead paint, by chemicals, trusses, open	SF	6.875	APTRA	1.000	5.64	0.00	1.77	7.41
25 02 0122	**Removal existing lead paint, by chemicals, windows, one side, double hung, 1/1 light, 24" x 48" high**	**EA**				**77.50**	**0.00**	**24.78**	**102.28**
25010 1086	Removal existing lead paint, by chemicals, windows, one side, double hung, 1/1 light, 24" x 48" high	EA	0.5	APTRA	1.000	77.50	0.00	24.78	102.28
25 02 0123	**Removal existing lead paint, by chemicals, windows, one side, double hung, 1/1 light, 30" x 60" high**	**EA**				**103.33**	**0.00**	**32.45**	**135.78**
25010 1087	Removal existing lead paint, by chemicals, windows, one side, double hung, 1/1 light, 30" x 60" high	EA	0.375	APTRA	1.000	103.33	0.00	32.45	135.78
25 02 0124	**Removal existing lead paint, by chemicals, windows, one side, double hung, 1/1 light, 36" x 72" high**	**EA**				**124.00**	**0.00**	**38.94**	**162.94**
25010 1088	Removal existing lead paint, by chemicals, windows, one side, double hung, 1/1 light, 36" x 72" high	EA	0.313	APTRA	1.000	124.00	0.00	38.94	162.94
25 02 0125	**Removal existing lead paint, by chemicals, windows, one side, double hung, 1/1 light, 40" x 80" high**	**EA**				**155.00**	**0.00**	**48.97**	**203.97**
25010 1089	Removal existing lead paint, by chemicals, windows, one side, double hung, 1/1 light, 40" x 80" high	EA	0.25	APTRA	1.000	155.00	0.00	48.97	203.97
25 02 0126	**Removal existing lead paint, by chemicals, colonial window, 6/6 light, 24" x 48" high**	**EA**				**155.00**	**0.00**	**48.97**	**203.97**
25010 1090	Removal existing lead paint, by chemicals, colonial window, 6/6 light, 24" x 48" high	EA	0.25	APTRA	1.000	155.00	0.00	48.97	203.97
25 02 0127	**Removal existing lead paint, by chemicals, colonial window, 6/6 light, 30" x 60" high**	**EA**				**206.67**	**0.00**	**65.48**	**272.15**
25010 1091	Removal existing lead paint, by chemicals, colonial window, 6/6 light, 30" x 60" high	EA	0.188	APTRA	1.000	206.67	0.00	65.48	272.15
25 02 0128	**Removal existing lead paint, by chemicals, colonial window, 6/6 light, 36" x 72" high**	**EA**				**310.00**	**0.00**	**97.93**	**407.93**
25010 1092	Removal existing lead paint, by chemicals, colonial window, 6/6 light, 36" x 72" high	EA	0.125	APTRA	1.000	310.00	0.00	97.93	407.93
25 02 0129	**Removal existing lead paint, by chemicals, colonial window, 6/6 light, 40" x 80" high**	**EA**				**310.00**	**0.00**	**97.93**	**407.93**
25010 1093	Removal existing lead paint, by chemicals, colonial window, 6/6 light, 40" x 80" high	EA	0.125	APTRA	1.000	310.00	0.00	97.93	407.93

25 02 Lead Paint Removal

25 02 01	Lead Paint Removal - by Chemicals	Unit	Hourly Output	Crew	Factor	Labor	Equip.	Matl.	Total
25 02 0130	**Removal existing lead paint, by chemicals, colonial window, 8/8 light, 24" x 48" high**	**EA**				**155.00**	**0.00**	**48.97**	**203.97**
25010 1094	Removal existing lead paint, by chemicals, colonial window, 8/8 light, 24" x 48" high	EA	0.25	APTRA	1.000	155.00	0.00	48.97	203.97
25 02 0131	**Removal existing lead paint, by chemicals, colonial window, 8/8 light, 40" x 80" high**	**EA**				**310.00**	**0.00**	**97.93**	**407.93**
25010 1095	Removal existing lead paint, by chemicals, colonial window, 8/8 light, 40" x 80" high	EA	0.125	APTRA	1.000	310.00	0.00	97.93	407.93
25 02 0132	**Removal existing lead paint, by chemicals, colonial window, 12/12 light, 24" x 48" high**	**EA**				**310.00**	**0.00**	**97.93**	**407.93**
25010 1096	Removal existing lead paint, by chemicals, colonial window, 12/12 light, 24" x 48" high	EA	0.125	APTRA	1.000	310.00	0.00	97.93	407.93
25 02 0133	**Removal existing lead paint, by chemicals, colonial window, 8/8 light, 40" x 80" high**	**EA**				**413.33**	**0.00**	**130.97**	**544.30**
25010 1097	Removal existing lead paint, by chemicals, colonial window, 12/12 light, 40" x 80" high	EA	0.094	APTRA	1.000	413.33	0.00	130.97	544.30
25 02 0134	**Removal existing lead paint, by chemicals, minimum labor/equipment charge**	**EA**				**103.33**	**0.00**	**0.00**	**103.33**
25010 1098	Removal existing lead paint, by chemicals, minimum labor/equipment charge	EA	0.375	APTRA	1.000	103.33	0.00	0.00	103.33

25 02 02	Lead Paint Removal - Lead Paint Encapsulation	Unit	Hourly Output	Crew	Factor	Labor	Equip.	Matl.	Total
25 02 0201	**Lead paint encapsulation, water based polymer, 14 mil DFT, interior brushwork, trim, under 6"**	**LF**				**1.29**	**0.00**	**2.29**	**3.58**
25010 1099	Lead paint encapsulation, water based polymer, 14 mil DFT, interior brushwork, trim, under 6"	LF	30	APTRA	1.000	1.29	0.00	2.29	3.58
25 02 0202	**Lead paint encapsulation, water based polymer, 14 mil DFT, interior brushwork, trim, 6" to 12"**	**LF**				**1.72**	**0.00**	**3.04**	**4.77**
25010 1100	Lead paint encapsulation, water based polymer, 14 mil DFT, interior brushwork, trim, 6" to 12"	LF	22.5	APTRA	1.000	1.72	0.00	3.04	4.77
25 02 0203	**Lead paint encapsulation, water based polymer, 14 mil DFT, interior brushwork, balustrades**	**LF**				**1.03**	**0.00**	**1.83**	**2.86**
25010 1101	Lead paint encapsulation, water based polymer, 14 mil DFT, interior brushwork, balustrades	LF	37.5	APTRA	1.000	1.03	0.00	1.83	2.86
25 02 0204	**Lead paint encapsulation, water based polymer, 14 mil DFT, interior brushwork, pipes to 4" diameter**	**LF**				**0.62**	**0.00**	**1.10**	**1.72**
25010 1102	Lead paint encapsulation, water based polymer, 14 mil DFT, interior brushwork, pipes to 4" diameter	LF	62.5	APTRA	1.000	0.62	0.00	1.10	1.72

25 02 02	Lead Paint Removal - Lead Paint Encapsulation	Unit	Hourly Output	Crew	Factor	Labor	Equip.	Matl.	Total
25 02 0205	**Lead paint encapsulation, water based polymer, 14 mil DFT, interior brushwork, pipes to 8" diameter**	**LF**				**0.83**	**0.00**	**1.46**	**2.29**
25010 1103	Lead paint encapsulation, water based polymer, 14 mil DFT, interior brushwork, pipes to 8" diameter	LF	46.88	APTRA	1.000	0.83	0.00	1.46	2.29
25 02 0206	**Lead paint encapsulation, water based polymer, 14 mil DFT, interior brushwork, pipes to 12" diameter**	**LF**				**1.24**	**0.00**	**2.18**	**3.42**
25010 1104	Lead paint encapsulation, water based polymer, 14 mil DFT, interior brushwork, pipes to 12" diameter	LF	31.25	APTRA	1.000	1.24	0.00	2.18	3.42
25 02 0207	**Lead paint encapsulation, water based polymer, 14 mil DFT, interior brushwork, pipes to 16" diameter**	**LF**				**1.82**	**0.00**	**3.22**	**5.04**
25010 1105	Lead paint encapsulation, water based polymer, 14 mil DFT, interior brushwork, pipes to 16" diameter	LF	21.25	APTRA	1.000	1.82	0.00	3.22	5.04
25 02 0208	**Lead paint encapsulation, water based polymer, 14 mil DFT, interior brushwork, cabinets, ornate**	**SF**				**0.78**	**0.00**	**2.74**	**3.51**
25010 1106	Lead paint encapsulation, water based polymer, 14 mil DFT, interior brushwork, cabinets, ornate	SF	50	APTRA	1.000	0.78	0.00	2.74	3.51
25 02 0209	**Lead paint encapsulation, water based polymer, 14 mil DFT, interior brushwork, cabinets, simple**	**SF**				**1.24**	**0.00**	**2.18**	**3.42**
25010 1107	Lead paint encapsulation, water based polymer, 14 mil DFT, interior brushwork, cabinets, simple	SF	31.25	APTRA	1.000	1.24	0.00	2.18	3.42
25 02 0210	**Lead paint encapsulation, water based polymer, 14 mil DFT, interior brushwork, doors, 3' x 7' flush**	**EA**				**51.67**	**0.00**	**28.32**	**79.98**
25010 1108	Lead paint encapsulation, water based polymer, 14 mil DFT, interior brushwork, doors, 3' x 7' flush	EA	0.75	APTRA	1.000	51.67	0.00	28.32	79.98
25 02 0211	**Lead paint encapsulation, water based polymer, 14 mil DFT, interior brushwork, doors, 3' x 7' french**	**EA**				**103.33**	**0.00**	**28.32**	**131.65**
25010 1109	Lead paint encapsulation, water based polymer, 14 mil DFT, interior brushwork, doors, 3' x 7' french	EA	0.375	APTRA	1.000	103.33	0.00	28.32	131.65
25 02 0212	**Lead paint encapsulation, water based polymer, 14 mil DFT, interior brushwork, doors, 3' x 7' panel**	**EA**				**77.50**	**0.00**	**5.62**	**83.12**
25010 1110	Lead paint encapsulation, water based polymer, 14 mil DFT, interior brushwork, doors, 3' x 7' panel	EA	0.5	APTRA	1.000	77.50	0.00	5.62	83.12

25 02 02	Lead Paint Removal - Lead Paint Encapsulation	Unit	Hourly Output	Crew	Factor	Labor	Equip.	Matl.	Total
25 02 0213	**Lead paint encapsulation, water based polymer, 14 mil DFT, interior brushwork, doors, 3' x 7' lovere**	**EA**				**112.73**	**0.00**	**30.68**	**143.40**
25010 1111	Lead paint encapsulation, water based polymer, 14 mil DFT, interior brushwork, doors, 3' x 7' lovere	EA	0.344	APTRA	1.000	112.73	0.00	30.68	143.40
25 02 0214	**Lead paint encapsulation, polymer, 14 mil DFT, interior brushwork, windows, side, 1 to 6 lite**	**EA**				**22.14**	**0.00**	**19.41**	**41.55**
25010 1112	Lead paint encapsulation, polymer, 14 mil DFT, interior brushwork, windows, side, 1 to 6 lite	EA	1.75	APTRA	1.000	22.14	0.00	19.41	41.55
25 02 0215	**Lead paint encapsulation, polymer, 14 mil DFT, interior brushwork, windows, side, 7 to 10 lite**	**EA**				**41.33**	**0.00**	**21.42**	**62.75**
25010 1113	Lead paint encapsulation, polymer, 14 mil DFT, interior brushwork, windows, side, 7 to 10 lite	EA	0.938	APTRA	1.000	41.33	0.00	21.42	62.75
25 02 0216	**Lead paint encapsulation, polymer, 14 mil DFT, interior brushwork, windows, side, 12 lite**	**EA**				**53.91**	**0.00**	**28.91**	**82.82**
25010 1114	Lead paint encapsulation, polymer, 14 mil DFT, interior brushwork, windows, side, 12 lite	EA	0.719	APTRA	1.000	53.91	0.00	28.91	82.82
25 02 0217	**Lead paint encapsulation, polymer, 14 mil DFT, interior brushwork, radiators**	**EA**				**38.75**	**0.00**	**68.43**	**107.18**
25010 1115	Lead paint encapsulation, polymer, 14 mil DFT, interior brushwork, radiators	EA	1	APTRA	1.000	38.75	0.00	68.43	107.18
25 02 0218	**Lead paint encapsulation, polymer, 14 mil DFT, interior brushwork, grilles, vents**	**SF**				**1.13**	**0.00**	**1.99**	**3.12**
25010 1116	Lead paint encapsulation, polymer, 14 mil DFT, interior brushwork, grilles, vents	SF	34.38	APTRA	1.000	1.13	0.00	1.99	3.12
25 02 0219	**Lead paint encapsulation, polymer, 14 mil DFT, interior brushwork, walls, roller, drywall, plaster**	**SF**				**0.31**	**0.00**	**0.54**	**0.85**
25010 1117	Lead paint encapsulation, polymer, 14 mil DFT, interior brushwork, walls, roller, drywall, plaster	SF	125	APTRA	1.000	0.31	0.00	0.54	0.85
25 02 0220	**Lead paint encapsulation, polymer, 14 mil DFT, interior brushwork, walls spunbond fabric**	**SF**				**0.43**	**0.00**	**0.61**	**1.04**
25010 1118	Lead paint encapsulation, polymer, 14 mil DFT, interior brushwork, walls spunbond fabric	SF	90	APTRA	1.000	0.43	0.00	0.61	1.04
25 02 0221	**Lead paint encapsulation, polymer, 14 mil DFT, interior brushwork, walls wood**	**SF**				**0.39**	**0.00**	**0.68**	**1.07**
25010 1119	Lead paint encapsulation, polymer, 14 mil DFT, interior brushwork, walls wood	SF	100	APTRA	1.000	0.39	0.00	0.68	1.07

25 02 02	Lead Paint Removal - Lead Paint Encapsulation	Unit	Hourly Output	Crew	Factor	Labor	Equip.	Matl.	Total
25 02 0222	**Lead paint encapsulation, polymer, 14 mil DFT, interior brushwork, ceilings**	**SF**				**0.34**	**0.00**	**0.61**	**0.96**
25010 1120	Lead paint encapsulation, polymer, 14 mil DFT, interior brushwork, ceilings	SF	112.5	APTRA	1.000	0.34	0.00	0.61	0.96
25 02 0223	**Lead paint encapsulation, polymer, 14 mil DFT, interior brushwork, ceilings wood**	**SF**				**0.44**	**0.00**	**0.78**	**1.22**
25010 1121	Lead paint encapsulation, polymer, 14 mil DFT, interior brushwork, ceilings wood	SF	87.5	APTRA	1.000	0.44	0.00	0.78	1.22
25 02 0224	**Lead paint encapsulation, polymer, 14 mil DFT, exterior brushwork, gutters, downspouts**	**LF**				**1.03**	**0.00**	**1.83**	**2.86**
25010 1122	Lead paint encapsulation, polymer, 14 mil DFT, exterior brushwork, gutters, downspouts	LF	37.5	APTRA	1.000	1.03	0.00	1.83	2.86
25 02 0225	**Lead paint encapsulation, polymer, 14 mil DFT, exterior brushwork, columns**	**SF**				**0.78**	**0.00**	**1.37**	**2.14**
25010 1123	Lead paint encapsulation, polymer, 14 mil DFT, exterior brushwork, columns	SF	50	APTRA	1.000	0.78	0.00	1.37	2.14
25 02 0226	**Lead paint encapsulation, polymer, 14 mil DFT, exterior brushwork, spray siding**	**SF**				**0.52**	**0.00**	**0.91**	**1.43**
25010 1124	Lead paint encapsulation, polymer, 14 mil DFT, exterior brushwork, spray siding	SF	75	APTRA	1.000	0.52	0.00	0.91	1.43
25 02 0227	**Lead paint encapsulation, polymer, 14 mil DFT, electrical conduit brushwork, 2" diameter**	**LF**				**0.62**	**0.00**	**1.10**	**1.72**
25010 1125	Lead paint encapsulation, polymer, 14 mil DFT, electrical conduit brushwork, 2" diameter	LF	62.5	APTRA	1.000	0.62	0.00	1.10	1.72
25 02 0228	**Lead paint encapsulation, polymer, 14 mil DFT, brick, block, or concrete spray**	**SF**				**0.62**	**0.00**	**1.10**	**1.72**
25010 1126	Lead paint encapsulation, polymer, 14 mil DFT, brick, block, or concrete spray	SF	62.5	APTRA	1.000	0.62	0.00	1.10	1.72
25 02 0229	**Lead paint encapsulation, polymer, 14 mil DFT, steel flat surface and tanks to 12"**	**SF**				**0.62**	**0.00**	**1.10**	**1.72**
25010 1127	Lead paint encapsulation, polymer, 14 mil DFT, steel flat surface and tanks to 12"	SF	62.5	APTRA	1.000	0.62	0.00	1.10	1.72
25 02 0230	**Lead paint encapsulation, polymer, 14 mil DFT, beams**	**SF**				**0.78**	**0.00**	**1.37**	**2.14**
25010 1128	Lead paint encapsulation, polymer, 14 mil DFT, beams	SF	50	APTRA	1.000	0.78	0.00	1.37	2.14
25 02 0231	**Lead paint encapsulation, polymer, 14 mil DFT, trusses**	**SF**				**0.78**	**0.00**	**1.37**	**2.14**
25010 1129	Lead paint encapsulation, polymer, 14 mil DFT, trusses	SF	50	APTRA	1.000	0.78	0.00	1.37	2.14

33 01 Mobilization and Preparatory Work

33 01 01 Mobilization of Construction Equipment and Facilities		Unit	Hourly Output	Crew	Factor	Labor	Equip.	Matl.	Total
33 01 0101	**Mobilize/DeMobilize Drilling Rig & Crew**	**LS**				**720.40**	**3,179**	**0.00**	**3,899**
02674 8570	Standby Well	HR	1	ULADB	8.000	90.05	397.38	0.00	3,899
33 01 0102	**Van or Pickup Rental**	**DAY**				**0.00**	**0.00**	**37.00**	**37.00**
01954 6121	Van or Pickup Rental	DAY	0	N/A	1.000	0.00	0.00	37.00	37.00
33 01 0103	**Flatbed Truck Rental**	**DAY**				**0.00**	**0.00**	**70.50**	**70.50**
01954 6122	Flat Bed Truck Rental	DAY	0	N/A	1.000	0.00	0.00	70.50	70.50
33 01 0104	**Car or Van Mileage Charge**	**MI**				**0.00**	**0.00**	**0.36**	**0.36**
01954 6131	Car or Van Mileage Charge	MI	0	N/A	1.000	0.00	0.00	0.36	0.36
33 01 0105	**Pickup Truck Mileage Charge**	**MI**				**0.00**	**0.00**	**0.27**	**0.27**
01954 6132	Pickup Truck Mileage Charge	MI	0	N/A	1.000	0.00	0.00	0.27	0.27
33 01 0106	**Flatbed Truck Mileage Charge**	**MI**				**0.00**	**0.00**	**0.39**	**0.39**
01954 6133	Flatbed Truck Mileage Charge	MI	0	N/A	1.000	0.00	0.00	0.39	0.39
33 01 0107	**Van or Pickup Rental**	**DAY**				**0.00**	**0.00**	**37.00**	**37.00**
01954 6121	Van or Pickup Rental	DAY	0	N/A	1.000	0.00	0.00	37.00	37.00
33 01 0108	**Sedan, Automobile, Rental**	**DAY**				**0.00**	**0.00**	**47.47**	**47.47**
01954 6125	Sedan, Automobile, Rental	DAY	0	N/A	1.000	0.00	0.00	47.47	47.47
33 01 0109	**Truck, 2 Axle, Highway, 21,700 GVW, 4 x 2, 2 Axle**	**DAY**				**0.00**	**0.00**	**127.75**	**127.75**
01954 6126	Truck, 2 Axle, Highway, 21,700 GVW, 4 x 2 2 Axle	DAY	0	N/A	1.000	0.00	0.00	127.75	127.75
33 01 0111	**Truck, 2 Axle, Highway, 33,000 GVW, 6 x 2, 2 Axle**	**DAY**				**0.00**	**0.00**	**226.46**	**226.46**
01954 6127	Truck, 2 Axle, Highway, 33,000 GVW, 6 x 2 2 Axle	DAY	0	N/A	1.000	0.00	0.00	226.46	226.46

33 01 02 Mobilization of Personnel		Unit	Hourly Output	Crew	Factor	Labor	Equip.	Matl.	Total
33 01 0201	**Mobilize Crew, >= 500 Miles, per Person**	**EA**				**0.00**	**0.00**	**475.21**	**475.21**
01954 6142	Mobilize/Demobilize 2-man Sampling Crew, 500 Miles	EA	0	N/A	0.500	0.00	0.00	950.43	475.21
33 01 0202	**Per Diem (per person)**	**DAY**				**0.00**	**0.00**	**90.02**	**90.02**
01954 6141	Personnel Per Diem	DAY	0	N/A	1.000	0.00	0.00	90.02	90.02
33 01 0203	**Mobilize Crew, 250 Miles, per Person**	**EA**				**0.00**	**0.00**	**237.61**	**237.61**
01954 6142	Mobilize/Demobilize 2-man Sampling Crew, 500 Miles	EA	0	N/A	0.250	0.00	0.00	950.43	237.61
33 01 0204	**Mobilize Crew, 100 Miles, per Person**	**EA**				**0.00**	**0.00**	**95.04**	**95.04**
01954 6142	Mobilize/Demobilize 2-man Sampling Crew, 500 Miles	EA	0	N/A	0.100	0.00	0.00	950.43	95.04

33 01 Mobilization and Preparatory Work

33 01 02 Mobilization of Personnel		Unit	Hourly Output	Crew	Factor	Labor	Equip.	Matl.	Total
33 01 0205	**Mobilize Crew, 50 Miles, per Person**	**EA**				**0.00**	**0.00**	**71.28**	**71.28**
01954 6142	Mobilize/Demobilize 2-man Sampling Crew, 500 Miles	EA	0	N/A	0.075	0.00	0.00	950.43	71.28
33 01 0206	**Mobilize Crew, Local, per Person**	**EA**				**0.00**	**0.00**	**23.76**	**23.76**
01954 6142	Mobilize/Demobilize 2-man Sampling Crew, 500 Miles	EA	0	N/A	0.025	0.00	0.00	950.43	23.76
33 01 0207	**Mobilize Crew, >=500 Miles, per Person**	**EA**				**0.00**	**0.00**	**475.21**	**475.21**
01954 6142	Mobilize/Demobilize 2-man Sampling Crew, 500 Miles	EA	0	N/A	0.500	0.00	0.00	950.43	475.21
33 01 0208	**Mobilize Crew, 250 Miles, per Person**	**EA**				**0.00**	**0.00**	**237.61**	**237.61**
01954 6142	Mobilize/Demobilize 2-man Sampling Crew, 500 Miles	EA	0	N/A	0.250	0.00	0.00	950.43	237.61
33 01 0209	**Mobilize Crew, 100 Miles, per Person**	**EA**				**0.00**	**0.00**	**95.04**	**95.04**
01954 6142	Mobilize/Demobilize 2-man Sampling Crew, 500 Miles	EA	0	N/A	0.100	0.00	0.00	950.43	95.04
33 01 0210	**Mobilize Crew, 50 Miles, per Person**	**EA**				**0.00**	**0.00**	**71.28**	**71.28**
01954 6142	Mobilize/Demobilize 2-man Sampling Crew, 500 Miles	EA	0	N/A	0.075	0.00	0.00	950.43	71.28
33 01 0211	**Mobilize Crew, Local, per Person**	**EA**				**0.00**	**0.00**	**23.76**	**23.76**
01954 6142	Mobilize/Demobilize 2-man Sampling Crew, 500 Miles	EA	0	N/A	0.025	0.00	0.00	950.43	23.76

33 01 04 Temporary Facilities/Personal Protective Equipment		Unit	Hourly Output	Crew	Factor	Labor	Equip.	Matl.	Total
33 01 0401	**Process Trailers (3)**	**MO**				**0.00**	**0.00**	**28,875**	**28,875**
02083 6132	10.25' x 5.5' x 8.33' - 10 Drum Capacity, 175 Gallon Sump, Two Doors	EA	0	N/A	3.000	0.00	0.00	9,625	28,875
33 01 0402	**Supplied Air**	**EA**				**0.00**	**0.00**	**1,405**	**1,405**
01951 5712	10 Minute Unit	EA	0	N/A	1.000	0.00	0.00	490.00	490.00
01951 5732	Supplied Air Respirator, Full Facepiece with Hose Assembly	EA	0	N/A	1.000	0.00	0.00	915.00	915.00
33 01 0403	**Breathing Apparatus (SCBA)**	**EA**				**0.00**	**0.00**	**3,125**	**3,125**
01951 5731	Self-contained Breathing Apparatus - 60 Minute Unit	EA	0	N/A	1.000	0.00	0.00	3,125	3,125
33 01 0404	**Level "C" Breathing Apparatus (Full Face Respirator)**	**EA**				**0.00**	**0.00**	**202.50**	**202.50**
01951 5723	Cartridges, Respirator, Organic Vapor	EA	0	N/A	2.000	0.00	0.00	7.25	14.50
01951 5725	Full Facepiece Air-purifying Respirators, Chemical Cartridge	EA	0	N/A	1.000	0.00	0.00	188.00	188.00

33 01 04	Temporary Facilities/Personal Protective Equipment	Unit	Hourly Output	Crew	Factor	Labor	Equip.	Matl.	Total
33 01 0405	**Level "C" Breathing Apparatus (Half-face Respirator)**	**EA**				**0.00**	**0.00**	**35.25**	**35.25**
01951 5723	Cartridges, Respirator, Organic Vapor	EA	0	N/A	1.000	0.00	0.00	7.25	7.25
01951 5726	Half-mask Air-purifying Respirators	EA	0	N/A	1.000	0.00	0.00	28.00	28.00
33 01 0406	**Level "C" Respirator Cartridges**	**EA**				**0.00**	**0.00**	**14.50**	**14.50**
01951 5723	Cartridges, Respirator, Organic Vapor	EA	0	N/A	2.000	0.00	0.00	7.25	14.50
33 01 0407	**Basic Level "A" Encapsulation Suit**	**EA**				**0.00**	**0.00**	**348.73**	**348.73**
01951 5302	Basic Level A Suit (Large)	EA	0	N/A	1.000	0.00	0.00	348.73	348.73
33 01 0408	**Butyl Level "A" Suit**	**EA**				**0.00**	**0.00**	**890.00**	**890.00**
01951 5304	Butyl, Level A (Large)	EA	0	N/A	1.000	0.00	0.00	890.00	890.00
33 01 0409	**PVC Level "A" Suit**	**EA**				**0.00**	**0.00**	**700.00**	**700.00**
01951 5305	PVC, Level A (Large)	EA	0	N/A	1.000	0.00	0.00	700.00	700.00
33 01 0410	**Teflon/Nomex Level "A" Suit**	**EA**				**0.00**	**0.00**	**625.00**	**625.00**
01951 5306	Teflon/Nomex, Level A (Large)	EA	0	N/A	1.000	0.00	0.00	625.00	625.00
33 01 0411	**Viton Level "A" Suit**	**EA**				**0.00**	**0.00**	**710.00**	**710.00**
01951 5307	Viton, Level A (Large)	EA	0	N/A	1.000	0.00	0.00	710.00	710.00
33 01 0412	**Basic Level "B" Suit**	**EA**				**0.00**	**0.00**	**210.00**	**210.00**
01951 5303	Basic Level B Suit (Large)	EA	0	N/A	1.000	0.00	0.00	210.00	210.00
33 01 0413	**Reusable PVC Boots**	**PAIR**				**0.00**	**0.00**	**12.50**	**12.50**
01951 5103	PVC Overboots (Reusable)	PR	0	N/A	1.000	0.00	0.00	12.50	12.50
33 01 0414	**Reusable Neoprene Chemical-resistant Boots**	**PAIR**				**0.00**	**0.00**	**51.50**	**51.50**
01951 5102	Neoprene Overboots (Reusable)	PR	0	N/A	1.000	0.00	0.00	51.50	51.50
33 01 0415	**Hard Hat**	**EA**				**0.00**	**0.00**	**9.60**	**9.60**
01951 5801	Hard Hats	EA	0	N/A	1.000	0.00	0.00	9.60	9.60
33 01 0416	**Reusable Gloves (Butyl)**	**PAIR**				**0.00**	**0.00**	**24.00**	**24.00**
01951 5501	Butyl, Medium Weight Gloves (Reusable)	PR	0	N/A	1.000	0.00	0.00	24.00	24.00
33 01 0417	**Reusable Gloves (Viton) for Chlorin/Aromatic Solvents**	**PAIR**				**0.00**	**0.00**	**0.45**	**0.45**
01951 5504	PVC/Nitrile Blend Gloves (Package of 12 Pair)	EA	0	N/A	0.083	0.00	0.00	5.40	0.45
33 01 0418	**Reusable Gloves (PVC/Nitrile)**	**PAIR**				**0.00**	**0.00**	**0.45**	**0.45**
01951 5504	PVC/Nitrile Blend Gloves (Package of 12 Pair)	EA	0	N/A	0.083	0.00	0.00	5.40	0.45
33 01 0419	**Reusable Gloves (Neoprene)**	**PAIR**				**0.00**	**0.00**	**0.74**	**0.74**
01951 5502	Neoprene Coated Gloves (Reusable) (Package of 12 Pair)	EA	0	N/A	0.083	0.00	0.00	8.90	0.74

33 01 04	Temporary Facilities/Personal Protective Equipment	Unit	Hourly Output	Crew	Factor	Labor	Equip.	Matl.	Total
33 01 0420	**Two-way Radio (Reusable)**	**EA**				**0.00**	**0.00**	**67.31**	**67.31**
01951 5811	Ear-Mike Two-way Radio	EA	0	N/A	1.000	0.00	0.00	67.31	67.31
33 01 0421	**Disposable Boot Covers (Tyvek)**	**PAIR**				**0.00**	**0.00**	**1.13**	**1.13**
01951 5202	Boot Covers, Tyvek (Bag of 10 Pair)	EA	0	N/A	0.100	0.00	0.00	11.30	1.13
33 01 0422	**Briefs (Disposable)**	**EA**				**0.00**	**0.00**	**0.07**	**0.07**
01951 5203	Briefs (Case of 100)	EA	0	N/A	0.010	0.00	0.00	7.25	0.07
33 01 0423	**Disposable Gloves (Latex)**	**PAIR**				**0.00**	**0.00**	**0.21**	**0.21**
01951 5206	Gloves, Latex (Box of 100) (Single Hand)	EA	0	N/A	0.020	0.00	0.00	10.25	0.21
33 01 0424	**Disposable Coveralls (Tyvek/Polycoated)**	**EA**				**0.00**	**0.00**	**6.45**	**6.45**
01951 5216	Coveralls, Tyvek, Polycoated (each)	EA	0	N/A	1.000	0.00	0.00	6.45	6.45
33 01 0425	**Disposable Coveralls (Tyvek)**	**EA**				**0.00**	**0.00**	**4.44**	**4.44**
01951 5215	Coveralls, Tyvek (each)	EA	0	N/A	1.000	0.00	0.00	4.44	4.44
33 01 0426	**Face Shield (Reusable)**	**EA**				**0.00**	**0.00**	**16.55**	**16.55**
01951 5401	Face Shield	EA	0	N/A	1.000	0.00	0.00	16.55	16.55
33 01 0427	**Safety Goggles**	**EA**				**0.00**	**0.00**	**6.50**	**6.50**
01951 5403	Safety Goggles	EA	0	N/A	1.000	0.00	0.00	6.50	6.50
33 01 0428	**Safety Glasses**	**EA**				**0.00**	**0.00**	**6.90**	**6.90**
01951 5402	Safety Glasses	EA	0	N/A	1.000	0.00	0.00	6.90	6.90
33 01 0429	**Disposable Ear Plugs**	**PAIR**				**0.00**	**0.00**	**0.12**	**0.12**
01951 5602	Ear Plugs, Disposable (Box of 200)	EA	0	N/A	0.005	0.00	0.00	24.00	0.12
33 01 0430	**Aluminized Outer Cover Suit for Level "A"**	**EA**				**0.00**	**0.00**	**1,850**	**1,850**
01951 5301	Aluminized Outercover	EA	0	N/A	1.000	0.00	0.00	1,850	1,850
33 01 0431	**Emergency Escape Breathing Apparatus (5 Minute)**	**EA**				**0.00**	**0.00**	**430.00**	**430.00**
01951 5711	5 Minute Unit	EA	0	N/A	1.000	0.00	0.00	430.00	430.00
33 01 0432	**Supplied Air**	**EA**				**0.00**	**0.00**	**1,405**	**1,405**
01951 5712	10 Minute Unit	EA	0	N/A	1.000	0.00	0.00	490.00	490.00
01951 5732	Supplied Air Respirator, Full Facepiece with Hose Assembly	EA	0	N/A	1.000	0.00	0.00	915.00	915.00
33 01 0433	**Breathing Apparatus (SCBA)**	**EA**				**0.00**	**0.00**	**3,125**	**3,125**
01951 5731	Self-contained Breathing Apparatus - 60 Minute Unit	EA	0	N/A	1.000	0.00	0.00	3,125	3,125

33 01 04	Temporary Facilities/Personal Protective Equipment	Unit	Hourly Output	Crew	Factor	Labor	Equip.	Matl.	Total
33 01 0434	**Level "C" Breathing Apparatus (Full Face Respirator)**	**EA**				**0.00**	**0.00**	**202.50**	**202.50**
01951 5723	Cartridges, Respirator, Organic Vapor	EA	0	N/A	2.000	0.00	0.00	7.25	14.50
01951 5725	Full Facepiece Air-purifying Respirators, Chemical Cartridge	EA	0	N/A	1.000	0.00	0.00	188.00	188.00
33 01 0435	**Level "C" Breathing Apparatus (Half Face Respirator)**	**EA**				**0.00**	**0.00**	**42.50**	**42.50**
01951 5723	Cartridges, Respirator, Organic Vapor	EA	0	N/A	2.000	0.00	0.00	7.25	14.50
01951 5726	Half-mask Air-purifying Respirators	EA	0	N/A	1.000	0.00	0.00	28.00	28.00
33 01 0436	**Level "C" Respirator Cartridges**	**EA**				**0.00**	**0.00**	**7.25**	**7.25**
01951 5723	Cartridges, Respirator, Organic Vapor	EA	0	N/A	1.000	0.00	0.00	7.25	7.25
33 01 0437	**Basic Level "A" Encapsulation Suit**	**EA**				**0.00**	**0.00**	**348.73**	**348.73**
01951 5302	Basic Level A Suit (Large)	EA	0	N/A	1.000	0.00	0.00	348.73	348.73
33 01 0438	**Butyl Level "A" Suit**	**EA**				**0.00**	**0.00**	**890.00**	**890.00**
01951 5304	Butyl, Level A (Large)	EA	0	N/A	1.000	0.00	0.00	890.00	890.00
33 01 0439	**PVC Level "A" Suit**	**EA**				**0.00**	**0.00**	**700.00**	**700.00**
01951 5305	PVC, Level A (Large)	EA	0	N/A	1.000	0.00	0.00	700.00	700.00
33 01 0440	**Teflon/Nomex Level "A" Suit**	**EA**				**0.00**	**0.00**	**625.00**	**625.00**
01951 5306	Teflon/Nomex, Level A (Large)	EA	0	N/A	1.000	0.00	0.00	625.00	625.00
33 01 0441	**Viton Level "A" Suit**	**EA**				**0.00**	**0.00**	**710.00**	**710.00**
01951 5307	Viton, Level A (Large)	EA	0	N/A	1.000	0.00	0.00	710.00	710.00
33 01 0442	**Basic Level "B" Suit**	**EA**				**0.00**	**0.00**	**210.00**	**210.00**
01951 5303	Basic Level B Suit (Large)	EA	0	N/A	1.000	0.00	0.00	210.00	210.00
33 01 0443	**Reusable PVC Boots**	**PAIR**				**0.00**	**0.00**	**12.50**	**12.50**
01951 5103	PVC Overboots (Reusable)	PR	0	N/A	1.000	0.00	0.00	12.50	12.50
33 01 0444	**Reusable Neoprene Chemical Resistant Boots**	**PAIR**				**0.00**	**0.00**	**51.50**	**51.50**
01951 5102	Neoprene Overboots (Reusable)	PR	0	N/A	1.000	0.00	0.00	51.50	51.50
33 01 0445	**Hard Hat**	**EA**				**0.00**	**0.00**	**9.60**	**9.60**
01951 5801	Hard Hats	EA	0	N/A	1.000	0.00	0.00	9.60	9.60
33 01 0446	**Reusable Gloves (Butyl)**	**PAIR**				**0.00**	**0.00**	**24.00**	**24.00**
01951 5501	Butyl, Medium Weight Gloves (Reusable)	PR	0	N/A	1.000	0.00	0.00	24.00	24.00
33 01 0447	**Reusable Gloves (Viton) for Chlorin./Aromatic Solvents**	**PAIR**				**0.00**	**0.00**	**5.37**	**5.37**
01951 5505	Safety, PPE, gloves (reusable),viton, pkg of 12	PR	0	N/A	0.083	0.00	0.00	64.50	5.37

33 01 04 Temporary Facilities/Personal Protective Equipment		Unit	Hourly Output	Crew	Factor	Labor	Equip.	Matl.	Total
33 01 0449	**Reusable Gloves (Neoprene)**	**PAIR**				**0.00**	**0.00**	**0.74**	**0.74**
01951 5502	Neoprene Coated Gloves (Reusable) (Package of 12 Pair)	EA	0	N/A	0.083	0.00	0.00	8.90	0.74
33 01 0450	**Two-Way Radio (Reusable)**	**EA**				**0.00**	**0.00**	**67.31**	**67.31**
01951 5811	Ear-Mike Two-way Radio	EA	0	N/A	1.000	0.00	0.00	67.31	67.31
33 01 0451	**Disposable Boot Covers (Tyvek)**	**PAIR**				**0.00**	**0.00**	**1.13**	**1.13**
01951 5202	Boot Covers, Tyvek (Bag of 10 Pair)	EA	0	N/A	0.100	0.00	0.00	11.30	1.13
33 01 0452	**Briefs (Disposable)**	**EA**				**0.00**	**0.00**	**0.07**	**0.07**
01951 5203	Briefs (Case of 100)	EA	0	N/A	0.010	0.00	0.00	7.25	0.07
33 01 0453	**Disposable Gloves (Latex)**	**PAIR**				**0.00**	**0.00**	**0.21**	**0.21**
01951 5206	Gloves, Latex (Box of 100) (Single Hand)	EA	0	N/A	0.020	0.00	0.00	10.25	0.21
33 01 0454	**Disposable Coveralls (Tyvek/Polycoated)**	**EA**				**0.00**	**0.00**	**6.45**	**6.45**
01951 5216	Coveralls, Tyvek, Polycoated (each)	EA	0	N/A	1.000	0.00	0.00	6.45	6.45
33 01 0455	**Disposable Coveralls (Tyvek)**	**EA**				**0.00**	**0.00**	**4.44**	**4.44**
01951 5215	Coveralls, Tyvek (each)	EA	0	N/A	1.000	0.00	0.00	4.44	4.44
33 01 0456	**Face Shield (Reusable)**	**EA**				**0.00**	**0.00**	**16.55**	**16.55**
01951 5401	Face Shield	EA	0	N/A	1.000	0.00	0.00	16.55	16.55
33 01 0457	**Safety Goggles**	**EA**				**0.00**	**0.00**	**6.50**	**6.50**
01951 5403	Safety Goggles	EA	0	N/A	1.000	0.00	0.00	6.50	6.50
33 01 0458	**Safety Glasses**	**EA**				**0.00**	**0.00**	**6.90**	**6.90**
01951 5402	Safety Glasses	EA	0	N/A	1.000	0.00	0.00	6.90	6.90
33 01 0459	**Disposable Ear Plugs**	**PAIR**				**0.00**	**0.00**	**0.12**	**0.12**
01951 5602	Ear Plugs, Disposable (Box of 200)	EA	0	N/A	0.005	0.00	0.00	24.00	0.12
33 01 0460	**Aluminized Outer Cover Suit For Level "A"**	**EA**				**0.00**	**0.00**	**1,850**	**1,850**
01951 5301	Aluminized Outercover	EA	0	N/A	1.000	0.00	0.00	1,850	1,850
33 01 0461	**Emergency Escape Breathing Apparatus (5 Minute)**	**EA**				**0.00**	**0.00**	**430.00**	**430.00**
01951 5711	5 Minute Unit	EA	0	N/A	1.000	0.00	0.00	430.00	430.00
33 01 0462	**Truck Scale Rental**	**MO**				**0.00**	**0.00**	**3,100**	**3,100**
13276 1275	Truck Scale Rental	MO	0	N/A	1.000	0.00	0.00	3,100	3,100
33 01 0463	**EOD Team Work Clothing Ensemble**	**EA**				**0.00**	**0.00**	**108.92**	**108.92**
01951 5104	Leather Work Gloves	EA	0	N/A	1.000	0.00	0.00	7.34	7.34
01951 5105	Boots, Leather, Safety Toe	PR	0	N/A	1.000	0.00	0.00	66.74	66.74
01951 5214	Coveralls, Cotton	EA	0	N/A	1.000	0.00	0.00	34.84	34.84

33 01 04 Temporary Facilities/Personal Protective Equipment		Unit	Hourly Output	Crew	Factor	Labor	Equip.	Matl.	Total
33 01 0464	**EOD Personal Safety Equipment**	**EA**				**0.00**	**0.00**	**26.15**	**26.15**
01951 5401	Face Shield	EA	0	N/A	1.000	0.00	0.00	16.55	16.55
01951 5801	Hard Hats	EA	0	N/A	1.000	0.00	0.00	9.60	9.60
33 01 0465	**Supervisor's Hazardous Training Course, 8-hours**	**EA**				**0.00**	**0.00**	**181.00**	**181.00**
01951 6102	8-Hour Supervisor Course	EA	0	N/A	1.000	0.00	0.00	181.00	181.00
33 01 0466	**Hazardous Training, per Worker, 40-hours**	**EA**				**0.00**	**0.00**	**445.00**	**445.00**
01951 6104	40-Hour Hazardous Worker Training	EA	0	N/A	1.000	0.00	0.00	445.00	445.00
33 01 0467	**Emergency Medical Kit**	**EA**				**0.00**	**0.00**	**84.50**	**84.50**
01951 1504	36 Ingredients	EA	0	N/A	1.000	0.00	0.00	84.50	84.50

33 01 05 Construct Temporary Utilities		Unit	Hourly Output	Crew	Factor	Labor	Equip.	Matl.	Total
33 01 0502	**4.0 KW, 120/240 VAC, Generator, Daily Rental**	**DAY**				**0.00**	**0.00**	**59.86**	**59.86**
16513 6020	Portable Generator Set, 4 KW, 120/241 VAC Gas Powered, Hourly Rental	HR	0	N/A	10.000	0.00	0.00	5.99	59.86
33 01 0503	**5.0 KW, 120/240 VAC, Generator, Daily Rental**	**DAY**				**0.00**	**0.00**	**63.31**	**63.31**
16513 6040	Portable Generator Set, 5 KW, 120/242 VAC Gas Powered, Hourly Rental	HR	0	N/A	10.000	0.00	0.00	6.33	63.31
33 01 0504	**7.5 KW, 120/240 VAC, Generator, Daily Rental**	**DAY**				**0.00**	**0.00**	**88.64**	**88.64**
16513 6060	Portable Generator Set, 7.5 KW, 120/243 VAC Gas Powered, Hourly Rental	HR	0	N/A	10.000	0.00	0.00	8.86	88.64
33 01 0505	**10 KW, 120/240 VAC, Generator, Daily Rental**	**DAY**				**0.00**	**0.00**	**90.94**	**90.94**
16513 6080	Portable Generator Set, 10 KW, 120/244 VAC, Gas Powered, Hourly Rental	HR	0	N/A	10.000	0.00	0.00	9.09	90.94
33 01 0507	**Portable Water Pump, 2" 6,000 GPH, Water/Fertilizer Transfer**	**EA**				**0.00**	**0.00**	**1,026**	**1,026**
15113 1134	2" Inside Diameter, 20' Rubber Discharge Hose	EA	0	N/A	1.000	0.00	0.00	26.25	26.25
15113 1138	2" Inside Diameter, 20' Rubber Intake Hose	EA	0	N/A	1.000	0.00	0.00	80.58	80.58
15290 0100	Portable Water Pump, 2", 6,000 GPH Water/Fertilizer Transfer	EA	0	N/A	1.000	0.00	0.00	904.79	904.79
15428 2101	2" Strainer, Pump Suction Side	EA	0	N/A	1.000	0.00	0.00	14.19	14.19
33 01 0508	**Water Pump, 2" 8,000 GPH, Portable, Gasoline Powered**	**EA**				**0.00**	**0.00**	**1,026**	**1,026**
15113 1134	2" Inside Diameter, 20' Rubber Discharge Hose	EA	0	N/A	1.000	0.00	0.00	26.25	26.25
15113 1138	2" Inside Diameter, 20' Rubber Intake Hose	EA	0	N/A	1.000	0.00	0.00	80.58	80.58
15290 0110	Portable Water Pump, 2", 10,000 GPH, Gas Powered, with Wheels	EA	0	N/A	1.000	0.00	0.00	904.79	904.79
15428 2101	2" Strainer, Pump Suction Side	EA	0	N/A	1.000	0.00	0.00	14.19	14.19

33 01 05	Construct Temporary Utilities	Unit	Hourly Output	Crew	Factor	Labor	Equip.	Matl.	Total
33 01 0509	**Portable Water Pump, 2", 10,000 GPH, Gas Powered, with Wheels**	**EA**				**0.00**	**0.00**	**1,026**	**1,026**
15113 1134	2" Inside Diameter, 20' Rubber Discharge Hose	EA	0	N/A	1.000	0.00	0.00	26.25	26.25
15113 1138	2" Inside Diameter, 20' Rubber Intake Hose	EA	0	N/A	1.000	0.00	0.00	80.58	80.58
15290 0110	Portable Water Pump, 2", 10,000 GPH, Gas Powered, with Wheels	EA	0	N/A	1.000	0.00	0.00	904.79	904.79
15428 2101	2" Strainer, Pump Suction Side	EA	0	N/A	1.000	0.00	0.00	14.19	14.19

33 02 02	Radiation Sample Preparation	Unit	Hourly Output	Crew	Factor	Labor	Equip.	Matl.	Total
33 02 0206	**Prefiltering Liquids**	**EA**				**0.00**	**0.00**	**16.77**	**16.77**
01954 8321	Prefiltering Liquids	EA	0	N/A	1.000	0.00	0.00	16.77	16.77
33 02 0207	**Acid Digestion**	**EA**				**0.00**	**0.00**	**26.50**	**26.50**
01954 8322	Acid Digestion for Air Filters	EA	0	N/A	1.000	0.00	0.00	26.50	26.50
33 02 0208	**Drying and Grinding**	**EA**				**0.00**	**0.00**	**32.00**	**32.00**
01954 8323	Drying & Grinding for Vegetation/Sediment/Soil	EA	0	N/A	1.000	0.00	0.00	32.00	32.00
33 02 0209	**Dissolution by Microwave**	**EA**				**0.00**	**0.00**	**24.00**	**24.00**
01954 8324	Dissolution by Microwave for Vegetation/Sediment/Soil	EA	0	N/A	1.000	0.00	0.00	24.00	24.00
33 02 0210	**Urine Preparation**	**EA**				**0.00**	**0.00**	**23.00**	**23.00**
01954 8325	Urine Preparation	EA	0	N/A	1.000	0.00	0.00	23.00	23.00
33 02 0211	**Feces Preparation**	**EA**				**0.00**	**0.00**	**32.00**	**32.00**
01954 8326	Feces Preparation	EA	0	N/A	1.000	0.00	0.00	32.00	32.00
33 02 0212	**Blood Preparation**	**EA**				**0.00**	**0.00**	**16.77**	**16.77**
01954 8327	Blood Preparation	EA	0	N/A	1.000	0.00	0.00	16.77	16.77
33 02 0213	**Animal Bone/Tissue Preparation**	**EA**				**0.00**	**0.00**	**32.00**	**32.00**
01954 8328	Animal Bone/Tissue Preparation	EA	0	N/A	1.000	0.00	0.00	32.00	32.00
33 02 0214	**Gas Flow Proportional Counting**	**EA**				**0.00**	**0.00**	**72.00**	**72.00**
01954 8331	Gas Flow Proportional Counting	EA	0	N/A	1.000	0.00	0.00	72.00	72.00
33 02 0215	**Liquid Scintillation**	**EA**				**0.00**	**0.00**	**90.50**	**90.50**
01954 8332	Liquid Scintillation	EA	0	N/A	1.000	0.00	0.00	90.50	90.50
33 02 0216	**Alpha Spectroscopy**	**EA**				**0.00**	**0.00**	**32.00**	**32.00**
01954 8333	Alpha Spectroscopy	EA	0	N/A	1.000	0.00	0.00	32.00	32.00
33 02 0217	**Gamma Spectroscopy**	**EA**				**0.00**	**0.00**	**98.50**	**98.50**
01954 8334	Gamma Spectroscopy	EA	0	N/A	1.000	0.00	0.00	98.50	98.50
33 02 0218	**Fluorometric or Laser**	**EA**				**0.00**	**0.00**	**55.50**	**55.50**
01954 8335	Fluorometric or Laser Phosphorometer	EA	0	N/A	1.000	0.00	0.00	55.50	55.50

33 02 Monitoring, Sampling, Testing, and Analysis

33 02 02	Radiation Sample Preparation	Unit	Hourly Output	Crew	Factor	Labor	Equip.	Matl.	Total
33 02 0219	**Alpha Scintillation Counting**	**EA**				**0.00**	**0.00**	**72.00**	**72.00**
01954 8336	Alpha Scintillation Counting	EA	0	N/A	1.000	0.00	0.00	72.00	72.00
33 02 0220	**Radon De-emanation**	**EA**				**0.00**	**0.00**	**72.00**	**72.00**
01954 8337	Radon De-emanation	EA	0	N/A	1.000	0.00	0.00	72.00	72.00
33 02 0221	**Beta/Gamma Coincidence**	**EA**				**0.00**	**0.00**	**55.50**	**55.50**
01954 8338	Beta/Gamma Coincidence	EA	0	N/A	1.000	0.00	0.00	55.50	55.50
33 02 0222	**Radioactivity Screening - Tritium Not Present**	**EA**				**0.00**	**0.00**	**16.05**	**16.05**
01954 8341	Radioactivity Screening (Tritium Not Present)	EA	0	N/A	1.000	0.00	0.00	16.05	16.05
33 02 0223	**Radioactivity Screening - Tritium Present**	**EA**				**0.00**	**0.00**	**16.05**	**16.05**
01954 8342	Radioactivity Screening (Tritium Present)	EA	0	N/A	1.000	0.00	0.00	16.05	16.05
33 02 0224	**Repackage Sample**	**EA**				**0.00**	**0.00**	**85.50**	**85.50**
01954 8343	Repackage Sample	EA	0	N/A	1.000	0.00	0.00	85.50	85.50
33 02 0225	**Repackage and Ship Sample**	**EA**				**0.00**	**0.00**	**24.00**	**24.00**
01954 8344	Repackage and Ship Sample	EA	0	N/A	1.000	0.00	0.00	24.00	24.00

33 02 03	Air Monitoring and Sampling	Unit	Hourly Output	Crew	Factor	Labor	Equip.	Matl.	Total
33 02 0301	**Air Monitoring Station**	**EA**				**0.00**	**0.00**	**855.00**	**855.00**
01954 4146	Remote Toxic Air Sampler, Manual	EA	0	N/A	1.000	0.00	0.00	855.00	855.00
33 02 0302	**Organic Vapor Analyzer Rental, per Month**	**MO**				**0.00**	**0.00**	**930.00**	**930.00**
01954 4138	Portable Organic Vapor Analyzers (Monthly Rental)	EA	0	N/A	1.000	0.00	0.00	930.00	930.00
33 02 0303	**Organic Vapor Analyzer Rental, per Day**	**DAY**				**0.00**	**0.00**	**112.20**	**112.20**
01001 1080	Portable Organic Vapor Analyzers (Daily Rental)	DAY	0	N/A	1.000	0.00	0.00	112.20	112.20
33 02 0306	**Monitoring Gas Vents**	**EA**				**9.56**	**0.00**	**0.00**	**9.56**
01050 0102	Field Engineer (Minimum Cost)	WK	0.025	XYBBC	0.013	765.00	0.00	0.00	9.56
33 02 0307	**Soil Gas Investigation & Analysis**	**DAY**				**0.00**	**0.00**	**78.50**	**78.50**
01954 5191	Soil Gas Investigation and Analysis	DAY	0	N/A	1.000	0.00	0.00	78.50	78.50
33 02 0308	**Soil Gas Probes**	**EA**				**0.00**	**0.00**	**2,497**	**2,497**
01954 5192	Soil Gas Probes	EA	0	N/A	1.000	0.00	0.00	2,497	2,497
33 02 0309	**Bacterial Air Sampler, Monthly Rental**	**MO**				**0.00**	**0.00**	**420.00**	**420.00**
01954 4111	Bacteriological Air Sampler (Monthly Rental)	EA	0	N/A	1.000	0.00	0.00	420.00	420.00
33 02 0310	**Microbial Air Sampler, Monthly Rental**	**MO**				**0.00**	**0.00**	**420.00**	**420.00**
01954 4112	Microbial Air Sampler (Monthly Rental)	EA	0	N/A	1.000	0.00	0.00	420.00	420.00

33 02 03	Air Monitoring and Sampling	Unit	Hourly Output	Crew	Factor	Labor	Equip.	Matl.	Total
33 02 0311	**Cotton Dust Sampler, Monthly Rental**	**MO**				**0.00**	**0.00**	**435.00**	**435.00**
01954 4113	Cotton Dust Sampler (Monthly Rental)	EA	0	N/A	1.000	0.00	0.00	435.00	435.00
33 02 0312	**Digital Dust Sampler, Monthly Rental**	**MO**				**0.00**	**0.00**	**1,100**	**1,100**
01954 4114	Digital Dust Sampler (Monthly Rental)	EA	0	N/A	1.000	0.00	0.00	1,100	1,100
33 02 0313	**Flow Calibrator, Monthly Rental**	**MO**				**0.00**	**0.00**	**410.00**	**410.00**
01954 4115	Flow Calibrator (Monthly Rental)	EA	0	N/A	1.000	0.00	0.00	410.00	410.00
33 02 0314	**Personal Low Flow Sampling Pump, Monthly Rental**	**MO**				**0.00**	**0.00**	**201.00**	**201.00**
01954 4116	Personal Low Flow Sampling Pump (Monthly Rental)	EA	0	N/A	1.000	0.00	0.00	201.00	201.00
33 02 0315	**Ambient Air Monitor, Monthly Rental**	**MO**				**0.00**	**0.00**	**725.00**	**725.00**
01954 4121	Ambient Air Monitor (Monthly Rental)	EA	0	N/A	1.000	0.00	0.00	725.00	725.00
33 02 0316	**Ambient CO Analyzer, Monthly Rental**	**MO**				**0.00**	**0.00**	**427.99**	**427.99**
01954 4122	Ambient CO Analyzer (Monthly Rental)	EA	0	N/A	1.000	0.00	0.00	427.99	427.99
33 02 0317	**Ambient Ozone Monitor, Monthly Rental**	**MO**				**0.00**	**0.00**	**1,925**	**1,925**
01954 4123	Ambient Ozone Monitor (Monthly Rental)	EA	0	N/A	1.000	0.00	0.00	1,925	1,925
33 02 0318	**Ambient Sulfur Dioxide Analyzer, Monthly Rental**	**MO**				**0.00**	**0.00**	**2,850**	**2,850**
01954 4124	Ambient Sulfur Dioxide Analyzer (Monthly Rental)	EA	0	N/A	1.000	0.00	0.00	2,850	2,850
33 02 0319	**Calibration Gas Monitor**	**EA**				**0.00**	**0.00**	**430.00**	**430.00**
01954 4125	Calibration Gas (Purchased)	EA	0	N/A	1.000	0.00	0.00	430.00	430.00
33 02 0320	**Chloride Analyzer, Monthly Rental**	**MO**				**0.00**	**0.00**	**669.68**	**669.68**
01954 4126	Chloride (HCl) Analyzer (Monthly Rental)	EA	0	N/A	1.000	0.00	0.00	669.68	669.68
33 02 0321	**Gas Monitors, Monthly Rental**	**MO**				**0.00**	**0.00**	**590.00**	**590.00**
01954 4127	Gas Monitors (Monthly Rental)	EA	0	N/A	1.000	0.00	0.00	590.00	590.00
33 02 0322	**Hydrogen Sulfide Analyzer, Monthly Rental**	**MO**				**0.00**	**0.00**	**555.00**	**555.00**
01954 4128	Hydrogen Sulfide Analyzer (Monthly Rental)	EA	0	N/A	1.000	0.00	0.00	555.00	555.00
33 02 0323	**Manual Air Sampling Kit, Monthly Rental**	**MO**				**0.00**	**0.00**	**83.00**	**83.00**
01954 4129	Manual Air Sampling Kit (Monthly Rental)	EA	0	N/A	1.000	0.00	0.00	83.00	83.00
33 02 0324	**Manual Air Sampling Kit, Detection Tubes**	**EA**				**0.00**	**0.00**	**72.50**	**72.50**
01954 4131	Manual Air Sampling Kit Detection Tubes (Purchase)	EA	0	N/A	1.000	0.00	0.00	72.50	72.50
33 02 0325	**Mercury Vapor Monitor, Monthly Rental**	**MO**				**0.00**	**0.00**	**1,350**	**1,350**
01954 4132	Mercury Vapor Monitor (Monthly Rental)	EA	0	N/A	1.000	0.00	0.00	1,350	1,350

33 02 03	Air Monitoring and Sampling	Unit	Hourly Output	Crew	Factor	Labor	Equip.	Matl.	Total
33 02 0326	**Nitrogen Dioxide Analyzer, Monthly Rental**	**MO**				**0.00**	**0.00**	**2,550**	**2,550**
01954 4133	Nitrogen Dioxide Analyzer (Monthly Rental)	EA	0	N/A	1.000	0.00	0.00	2,550	2,550
33 02 0327	**Personal Alarms, Monthly Rental**	**MO**				**0.00**	**0.00**	**425.00**	**425.00**
01954 4134	Personal Alarms (Monthly Rental)	EA	0	N/A	1.000	0.00	0.00	425.00	425.00
33 02 0328	**Portable Ambient Air Analyzer, Monthly Rental**	**MO**				**0.00**	**0.00**	**1,600**	**1,600**
01954 4135	Portable Ambient Air Analyzer (Monthly Rental)	EA	0	N/A	1.000	0.00	0.00	1,600	1,600
33 02 0329	**Portable CO2 Monitor, Monthly Rental**	**MO**				**0.00**	**0.00**	**385.00**	**385.00**
01954 4136	Portable CO2 Monitor (Monthly Rental)	EA	0	N/A	1.000	0.00	0.00	385.00	385.00
33 02 0330	**Portable Combustible Gas/Oxygen Indicator, Monthly Rental**	**MO**				**0.00**	**0.00**	**380.00**	**380.00**
01954 4137	Portable Combustible Gas/Oxygen Indicator (Monthly Rental)	EA	0	N/A	1.000	0.00	0.00	380.00	380.00
33 02 0331	**Portable Organic Vapor Analyzer, Monthly Rental**	**MO**				**0.00**	**0.00**	**930.00**	**930.00**
01954 4138	Portable Organic Vapor Analyzers (Monthly Rental)	EA	0	N/A	1.000	0.00	0.00	930.00	930.00
33 02 0332	**Portable Oxygen Deficiency Monitor, Monthly Rental**	**MO**				**0.00**	**0.00**	**355.00**	**355.00**
01954 4139	Portable Oxygen Deficiency Monitor (Monthly Rental)	EA	0	N/A	1.000	0.00	0.00	355.00	355.00
33 02 0333	**Portable Toxic Gas Analyzer, Monthly Rental**	**MO**				**0.00**	**0.00**	**330.00**	**330.00**
01954 4141	Portable Toxic Gas Analyzer (Monthly Rental)	EA	0	N/A	1.000	0.00	0.00	330.00	330.00
33 02 0334	**Solvent Vapor Indicator, Monthly Rental**	**MO**				**0.00**	**0.00**	**855.00**	**855.00**
01954 4142	Solvent Vapor Indicator (Monthly Rental)	EA	0	N/A	1.000	0.00	0.00	855.00	855.00
33 02 0335	**Specific Vapor Analyzer, Monthly Rental**	**MO**				**0.00**	**0.00**	**1,000**	**1,000**
01954 4143	Specific Vapor Analyzer (Monthly Rental)	EA	0	N/A	1.000	0.00	0.00	1,000	1,000
33 02 0336	**Stack Sampling Probe, Monthly Rental**	**MO**				**0.00**	**0.00**	**300.00**	**300.00**
01954 4144	Stack Sampling Probe (Monthly Rental)	EA	0	N/A	1.000	0.00	0.00	300.00	300.00
33 02 0337	**Trace-Gas Analyzer, Monthly Rental**	**MO**				**0.00**	**0.00**	**1,225**	**1,225**
01954 4145	Trace-Gas Analyzer (Monthly Rental)	EA	0	N/A	1.000	0.00	0.00	1,225	1,225
33 02 0338	**Instrument Shelter, 30.0" x 20.0" x 32.0", Pine with Latex Paint**	**EA**				**33.88**	**0.00**	**860.00**	**893.88**
01954 4221	30" x 20" x 32", Pine, 3 Coats Latex Paint	EA	1	ULABA	1.000	33.88	0.00	860.00	893.88
33 02 0339	**Instrument Shelter, 18.0" x 13.0" x 20.0", Wood Painted White**	**EA**				**33.88**	**0.00**	**410.00**	**443.88**
01954 4222	18" x 13" x 20", Wood, Painted White	EA	1	ULABA	1.000	33.88	0.00	410.00	443.88

33 02 03 Air Monitoring and Sampling

33 02 03 Air Monitoring and Sampling		Unit	Hourly Output	Crew	Factor	Labor	Equip.	Matl.	Total
33 02 0340	**Instrument Shelter, 81.0" x 5.5.0" x 16.5", Wood Painted White**	**EA**				**33.88**	**0.00**	**199.00**	**232.88**
01954 4223	8" x 5.5" x 16.5", Wood, Painted White	EA	1	ULABA	1.000	33.88	0.00	199.00	232.88
33 02 0341	**Alpha, Beta, Gamma Detector Equipment**	**EA**				**0.00**	**0.00**	**225.00**	**225.00**
01954 4522	Pancake G-M Detector	EA	0	N/A	1.000	0.00	0.00	225.00	225.00
33 02 0342	**Photo-Ionization Detector, HnU, Daily Rental**	**DAY**				**0.00**	**0.00**	**95.37**	**95.37**
01001 0011	Photo-Ionization Detector, HnU, Daily Rental	DAY	0	N/A	1.000	0.00	0.00	95.37	95.37
33 02 0343	**Photo-Ionization Detector, HnU, Weekly Rental**	**WK**				**0.00**	**0.00**	**336.61**	**336.61**
01001 0012	Photo-Ionization Detector, HnU, Weekly Rental	WK	0	N/A	1.000	0.00	0.00	336.61	336.61
33 02 0344	**Photo-Ionization Detector, HnU, Monthly Rental**	**MO**				**0.00**	**0.00**	**1,010**	**1,010**
01001 0013	Photo-Ionization Detector, HnU, Monthly Rental	MO	0	N/A	1.000	0.00	0.00	1,010	1,010
33 02 0345	**Portable Air Sampler, Continuous, Daily Rental**	**DAY**				**0.00**	**0.00**	**24.12**	**24.12**
01001 0014	Portable Air Sampler, Continuous, Daily Rental	DAY	0	N/A	1.000	0.00	0.00	24.12	24.12
33 02 0346	**Portable Air Sampler, Continuous, Weekly Rental**	**WK**				**0.00**	**0.00**	**84.15**	**84.15**
01001 0015	Portable Air Sampler, Continuous, Weekly Rental	WK	0	N/A	1.000	0.00	0.00	84.15	84.15

33 02 04 Monitoring Wells

33 02 04 Monitoring Wells		Unit	Hourly Output	Crew	Factor	Labor	Equip.	Matl.	Total
33 02 0401	**Disposable Materials per Sample**	**EA**				**0.00**	**0.00**	**8.36**	**8.36**
01951 5206	Gloves, Latex (Box of 100) (Single Hand)	EA	0	N/A	0.040	0.00	0.00	10.25	0.41
01954 6411	1 Liter (32 Oz) Clear Wide Mouth Jar (Case of 12)	EA	0	N/A	0.083	0.00	0.00	49.50	4.12
01954 6429	Packing Tape (per Roll)	EA	0	N/A	0.200	0.00	0.00	3.07	0.61
01954 6431	Plastic Sheeting (per Roll)	EA	0	N/A	0.050	0.00	0.00	46.00	2.30
01954 6432	Isopropanol	GAL	0	N/A	0.010	0.00	0.00	60.50	0.60
01954 6477	Bag Ice	EA	0	N/A	0.250	0.00	0.00	1.24	0.31
33 02 0402	**Decontamination Materials per Sample**	**EA**				**0.00**	**0.00**	**7.53**	**7.53**
01954 6433	Deionized Water	GAL	0	N/A	0.250	0.00	0.00	13.15	3.29
01954 6434	Alconox Soap	EA	0	N/A	0.250	0.00	0.00	16.95	4.24
33 02 0404	**12' Aluminum Pole with Polyhead Sampler**	**EA**				**0.00**	**0.00**	**390.00**	**390.00**
01954 2112	12' Aluminum with Polyhead with Narrow Mouth Borosilicate Bottle	EA	0	N/A	1.000	0.00	0.00	390.00	390.00
33 02 0405	**Monitoring Well Slug Testing Equipment Rental**	**WK**				**0.00**	**0.00**	**740.00**	**740.00**
01954 3273	Pressure Transducer Rental	WK	0	N/A	1.000	0.00	0.00	175.00	175.00
01954 3274	Data Logger Rental	WK	0	N/A	1.000	0.00	0.00	565.00	565.00

33 02 04 Monitoring Wells	Unit	Hourly Output	Crew	Factor	Labor	Equip.	Matl.	Total
33 02 0411 Pumping Test	**EA**				**0.00**	**0.00**	**7,607**	**7,607**
01954 3275 Pumping Test, per well	EA	0	N/A	1.000	0.00	0.00	7,607	7,607
33 02 0412 Tracer Test	**HR**				**0.00**	**0.00**	**370.79**	**370.79**
01954 3276 Tracer Test	HR	0	N/A	1.000	0.00	0.00	370.79	370.79
33 02 0413 Pneumatic Pumping Test, per well	**EA**				**0.00**	**0.00**	**3,703**	**3,703**
01001 0016 Pneumatic Pumping Test, per well	EA	0	N/A	1.000	0.00	0.00	3,703	3,703
33 02 0414 Hydraulic Pumping Test, per well	**EA**				**0.00**	**0.00**	**1,863**	**1,863**
01001 0017 Hydraulic Pumping Test, per well	EA	0	N/A	1.000	0.00	0.00	1,863	1,863
33 02 0415 Dye Tracer Test	**EA**				**0.00**	**0.00**	**7,013**	**7,013**
01001 0018 Dye Tracer Test	EA	0	N/A	1.000	0.00	0.00	7,013	7,013
33 02 0416 Slug Test, per well	**EA**				**0.00**	**0.00**	**544.19**	**544.19**
01001 0019 Slug Test, per well	EA	0	N/A	1.000	0.00	0.00	544.19	544.19
33 02 0417 Groundwater Fate and Transport Modeling, One Dimensional, UST Site Assessment	**LS**				**0.00**	**0.00**	**1,346**	**1,346**
01001 0020 Groundwater Fate and Transport Modeling, One Dimensional, UST Site Assessme	LS	0	N/A	1.000	0.00	0.00	1,346	1,346
33 02 0418 Groundwater Fate and Transport Modeling, Two Dimensional, Analytical, UST Site Assessment	**LS**				**0.00**	**0.00**	**2,020**	**2,020**
01001 0021 Groundwater Fate and Transport Modeling, Two Dimensional, Analytical, UST S	LS	0	N/A	1.000	0.00	0.00	2,020	2,020
33 02 0419 Groundwater Fate and Transport Modeling, Two Dimensional, Numerical, UST Site Assessment	**LS**				**0.00**	**0.00**	**4,376**	**4,376**
01001 0022 Groundwater Fate and Transport Modeling, Two Dimensional, Numerical, UST Si	LS	0	N/A	1.000	0.00	0.00	4,376	4,376
33 02 0420 Groundwater Fate and Transport Modeling, Three Dimensional, UST Site Assessment	**LS**				**0.00**	**0.00**	**4,713**	**4,713**
01001 0023 Groundwater Fate and Transport Modeling, Three Dimensional, UST Site Assess	LS	0	N/A	1.000	0.00	0.00	4,713	4,713
33 02 0421 Slug Test Equipment Rental, Day	**DAY**				**0.00**	**0.00**	**176.95**	**176.95**
01001 0024 Slug Test Equipment Rental, Day	DAY	0	N/A	1.000	0.00	0.00	176.95	176.95

33 02 05 Sampling Surface Water/Groundwater/Liquid Waste	Unit	Hourly Output	Crew	Factor	Labor	Equip.	Matl.	Total

33 02 05	Sampling Surface Water/Groundwater/Liquid Waste	Unit	Hourly Output	Crew	Factor	Labor	Equip.	Matl.	Total
33 02 0504	**Clarifier Indicator Analyses**	**EA**				**0.00**	**0.00**	**158.30**	**158.30**
01954 7224	Oil and Grease (EPA 413.1)	EA	0	N/A	1.000	0.00	0.00	53.00	53.00
01954 7225	pH (EPA 150.1)	EA	0	N/A	1.000	0.00	0.00	5.45	5.45
01954 7229	Settleable Matter (EPA 160.5)	EA	0	N/A	1.000	0.00	0.00	13.20	13.20
01954 7231	Dissolved Solids (EPA 160.1)	EA	0	N/A	1.000	0.00	0.00	11.80	11.80
01954 7232	Suspended Solids (EPA 160.2)	EA	0	N/A	1.000	0.00	0.00	10.40	10.40
01954 7233	Total Solids (EPA 160.3)	EA	0	N/A	1.000	0.00	0.00	11.85	11.85
01954 7234	Volatile Solids (EPA 160.4)	EA	0	N/A	1.000	0.00	0.00	16.70	16.70
01954 7237	Total Organic Carbons (EPA 415.1,9060)	EA	0	N/A	1.000	0.00	0.00	24.00	24.00
01954 7239	Turbidity (EPA 180.1)	EA	0	N/A	1.000	0.00	0.00	11.90	11.90
33 02 0505	**Groundwater Contamination Analysis, per Sample**	**EA**				**0.00**	**0.00**	**142.90**	**142.90**
01954 7112	Conductivity (EPA 120.1)	EA	0	N/A	1.000	0.00	0.00	14.45	14.45
01954 7225	pH (EPA 150.1)	EA	0	N/A	1.000	0.00	0.00	5.45	5.45
01954 7237	Total Organic Carbons (EPA 415.1,9060)	EA	0	N/A	1.000	0.00	0.00	24.00	24.00
01954 7238	Total Organic Halogens (Tox) (EPA 9022)	EA	0	N/A	1.000	0.00	0.00	99.00	99.00
33 02 0506	**Tank Decontamination Analysis, per Sample**	**EA**				**0.00**	**0.00**	**236.30**	**236.30**
01954 7108	Chloride (EPA 300.0/325.3)	EA	0	N/A	1.000	0.00	0.00	16.35	16.35
01954 7226	Phenols (EPA 420.1)	EA	0	N/A	1.000	0.00	0.00	35.73	35.73
01954 7235	Sulfur: Sulfate, Sulfide, Sulfite (EPA 300.0/376.2/377.1)	EA	0	N/A	2.000	0.00	0.00	28.00	56.00
01954 7241	Metals (EPA 200.7)	EA	0	N/A	2.000	0.00	0.00	64.11	128.21
33 02 0507	**Primary Drinking Water Parameters Analysis, per Sample**	**EA**				**0.00**	**0.00**	**792.82**	**792.82**
01954 7121	Coliform, Total, Method (SM 9222B)	EA	0	N/A	1.000	0.00	0.00	20.07	20.07
01954 7239	Turbidity (EPA 180.1)	EA	0	N/A	1.000	0.00	0.00	11.90	11.90
01954 7241	Metals (EPA 200.7)	EA	0	N/A	8.000	0.00	0.00	64.11	512.86
01954 7413	Pesticides/Herbicides (8081, 8151A)	EA	0	N/A	1.000	0.00	0.00	248.00	248.00
33 02 0508	**Detection/Compliance Monitoring Analysis, per Well**	**EA**				**0.00**	**0.00**	**754.31**	**754.31**
01954 7208	Pesticides and PCBs (608, 8081,8082)	EA	0	N/A	1.000	0.00	0.00	168.31	168.31
01954 7423	Metals (6010)	EA	0	N/A	1.000	0.00	0.00	95.50	95.50
01954 7425	Semivolatile Organics (8270C)	EA	0	N/A	1.000	0.00	0.00	325.00	325.00
01954 7426	Volatile Organics (8260B)	EA	0	N/A	1.000	0.00	0.00	139.00	139.00
01954 7613	Mercury (EPA 7470/7471)	EA	0	N/A	1.000	0.00	0.00	26.50	26.50

33 02 05	Sampling Surface Water/Groundwater/Liquid Waste	Unit	Hourly Output	Crew	Factor	Labor	Equip.	Matl.	Total
33 02 0509	**Compliance Monitoring 40 CFR 261, Appendix IX Constituents**	**EA**				**0.00**	**0.00**	**1,407**	**1,407**
01954 7113	Cyanide (EPA 335.1)	EA	0	N/A	1.000	0.00	0.00	33.50	33.50
01954 7203	Chlorinated Herbicides (615, 8151A)	EA	0	N/A	1.000	0.00	0.00	232.82	232.82
01954 7206	Organophosphorus Pesticides (614/8141A)	EA	0	N/A	1.000	0.00	0.00	168.49	168.49
01954 7208	Pesticides and PCBs (608, 8081,8082)	EA	0	N/A	1.000	0.00	0.00	168.31	168.31
01954 7215	Purgeable Organics (624, 8260B)	EA	0	N/A	1.000	0.00	0.00	139.00	139.00
01954 7217	Semivolatile Organics (625, 8270C)	EA	0	N/A	1.000	0.00	0.00	250.00	250.00
01954 7244	2,3,7,8-Tetrachloro-Dibzo-P-Dioxin (1613A)	EA	0	N/A	1.000	0.00	0.00	199.78	199.78
01954 7623	Sulfates (EPA 300.0)	EA	0	N/A	1.000	0.00	0.00	17.40	17.40
01954 7630	TAL Metals (6010/7000s)	EA	0	N/A	1.000	0.00	0.00	198.00	198.00
33 02 0510	**Analysis of Wastewater for POTW Permit**	**EA**				**0.00**	**0.00**	**427.99**	**427.99**
01954 7104	Biochemical Oxygen Demand (EPA 405.1)	EA	0	N/A	1.000	0.00	0.00	27.00	27.00
01954 7107	Chemical Oxygen Demand (COD) (EPA 410.4)	EA	0	N/A	1.000	0.00	0.00	19.30	19.30
01954 7113	Cyanide (EPA 335.1)	EA	0	N/A	1.000	0.00	0.00	33.50	33.50
01954 7225	pH (EPA 150.1)	EA	0	N/A	1.000	0.00	0.00	5.45	5.45
01954 7231	Dissolved Solids (EPA 160.1)	EA	0	N/A	1.000	0.00	0.00	11.80	11.80
01954 7232	Suspended Solids (EPA 160.2)	EA	0	N/A	1.000	0.00	0.00	10.40	10.40
01954 7241	Metals (EPA 200.7)	EA	0	N/A	5.000	0.00	0.00	64.11	320.54
33 02 0511	**Analysis of Lysimeters for Soil-Pore Liquid Monitoring**	**EA**				**0.00**	**0.00**	**455.01**	**455.01**
01954 7112	Conductivity (EPA 120.1)	EA	0	N/A	1.000	0.00	0.00	14.45	14.45
01954 7225	pH (EPA 150.1)	EA	0	N/A	1.000	0.00	0.00	5.45	5.45
01954 7237	Total Organic Carbons (EPA 415.1,9060)	EA	0	N/A	1.000	0.00	0.00	24.00	24.00
01954 7238	Total Organic Halogens (Tox) (EPA 9022)	EA	0	N/A	1.000	0.00	0.00	99.00	99.00
01954 7241	Metals (EPA 200.7)	EA	0	N/A	1.000	0.00	0.00	64.11	64.11
01954 7413	Pesticides/Herbicides (8081, 8151A)	EA	0	N/A	1.000	0.00	0.00	248.00	248.00
33 02 0512	**Rinsate Analysis**	**EA**				**0.00**	**0.00**	**236.30**	**236.30**
01954 7108	Chloride (EPA 300.0/325.3)	EA	0	N/A	1.000	0.00	0.00	16.35	16.35
01954 7226	Phenols (EPA 420.1)	EA	0	N/A	1.000	0.00	0.00	35.73	35.73
01954 7235	Sulfur: Sulfate, Sulfide, Sulfite (EPA 300.0/376.2/377.1)	EA	0	N/A	2.000	0.00	0.00	28.00	56.00
01954 7241	Metals (EPA 200.7)	EA	0	N/A	2.000	0.00	0.00	64.11	128.21
33 02 0520	**Hip Waders**	**EA**				**0.00**	**0.00**	**101.55**	**101.55**
01001 0025	Hip Waders	EA	0	N/A	1.000	0.00	0.00	101.55	101.55
33 02 0521	**Hip Waders, Weekly Rental**	**WK**				**0.00**	**0.00**	**81.50**	**81.50**
01951 5213	Hip Waders Rental	WK	0	N/A	1.000	0.00	0.00	81.50	81.50
33 02 0522	**Boat with Motor, Daily Rental**	**DAY**				**0.00**	**0.00**	**142.00**	**142.00**
01954 6124	Boat Rental (with Motor)	DAY	0	N/A	1.000	0.00	0.00	142.00	142.00

33 02 05	Sampling Surface Water/Groundwater/Liquid Waste	Unit	Hourly Output	Crew	Factor	Labor	Equip.	Matl.	Total
33 02 0523	**Boat without Motor, Daily Rental**	**DAY**				**0.00**	**0.00**	**29.00**	**29.00**
01954 6123	Boat Rental (without Motor)	DAY	0	N/A	1.000	0.00	0.00	29.00	29.00
33 02 0524	**Glass Coliwasas, Disposable, 7/8" x 42", 200 ml, Case of 12**	**EA**				**0.00**	**0.00**	**105.00**	**105.00**
01954 2121	Glass, Disposable, 7/8" x 42", 200 ml (Case of 12)	EA	0	N/A	1.000	0.00	0.00	105.00	105.00
33 02 0525	**Glass Coliwasas, Reusable, 7/8" x 42", 200 ml, Case of 4**	**EA**				**0.00**	**0.00**	**71.00**	**71.00**
01954 2122	Glass, Reusable, 7/8" x 42", 200 ml (Case of 4)	EA	0	N/A	1.000	0.00	0.00	71.00	71.00
33 02 0526	**PVC Coliwasas, 1.9" Outside Diameter x 49", 760 ml**	**EA**				**0.00**	**0.00**	**169.00**	**169.00**
01954 2123	PVC, 1.9" Outside Diameter x 49", 760 ml	EA	0	N/A	1.000	0.00	0.00	169.00	169.00
33 02 0527	**Bomb Sampler, 2.5" Diameter x 10", 500 ml**	**EA**				**0.00**	**0.00**	**590.00**	**590.00**
01954 2131	2.5" Diameter x 10", Holds 500 ml	EA	0	N/A	1.000	0.00	0.00	590.00	590.00
33 02 0528	**Drum Thief Tubes, 3/4" x 42", 150 ml, Disposable, Case of 25**	**EA**				**0.00**	**0.00**	**67.50**	**67.50**
01954 2141	3/4" x 42", 150 ml, Disposable (Case of 25)	EA	0	N/A	1.000	0.00	0.00	67.50	67.50
33 02 0529	**Drum Thief Tubes, 1/2" x 42", 75 ml, Disposable, Case of 25**	**EA**				**0.00**	**0.00**	**50.00**	**50.00**
01954 2142	1/2" x 42", 75 ml, Disposable (Case of 25)	EA	0	N/A	1.000	0.00	0.00	50.00	50.00
33 02 0530	**Bottom Sampler, 3 Lb Brass, 7.5" x 4.5", No Cable Required**	**EA**				**0.00**	**0.00**	**171.00**	**171.00**
01954 2161	3 Lb Brass (7.5" x 4.5") Includes Calibrated Cable	EA	0	N/A	1.000	0.00	0.00	171.00	171.00
33 02 0531	**Bottom Sampler, 17 Lb Stainless Steel, 6" x 6" x 6", with 100' Cable**	**EA**				**0.00**	**0.00**	**555.00**	**555.00**
01954 2162	17 Lb Stainless Steel (6" x 6" x 6") with 100' Line	EA	0	N/A	1.000	0.00	0.00	555.00	555.00
33 02 0532	**Bottom Sampler, 60 Lb Stainless Steel, 9" x 9"**	**EA**				**0.00**	**0.00**	**990.00**	**990.00**
01954 2163	60 Lb Stainless Steel (9" x 9")	EA	0	N/A	1.000	0.00	0.00	990.00	990.00
33 02 0533	**Water Level Indicator, 100' Tape, Electric, Light & Horn**	**EA**				**0.00**	**0.00**	**910.00**	**910.00**
01954 2241	Electronic, with Light and Horn, 100' Tape	EA	0	N/A	1.000	0.00	0.00	910.00	910.00
33 02 0534	**Water Level Indicator, 300' Tape, Electric, Light & Horn**	**EA**				**0.00**	**0.00**	**940.00**	**940.00**
01954 2242	Electronic, with Light and Horn, 300' Tape	EA	0	N/A	1.000	0.00	0.00	940.00	940.00
33 02 0535	**Water Level Indicator, 500' Tape, Electric, Light & Horn**	**EA**				**0.00**	**0.00**	**1,475**	**1,475**
01954 2243	Electronic, with Light and Horn, 500' Tape	EA	0	N/A	1.000	0.00	0.00	1,475	1,475

33 02 05	Sampling Surface Water/Groundwater/Liquid Waste	Unit	Hourly Output	Crew	Factor	Labor	Equip.	Matl.	Total
33 02 0536	**Water Level Indicator, Manual Steel Chalk Tape, with Weight, 100' Cable**	**EA**				**0.00**	**0.00**	**190.00**	**190.00**
01954 2244	Manual Steel Chalk Tape with Brass Weight - 100' Length	EA	0	N/A	1.000	0.00	0.00	190.00	190.00
33 02 0537	**Water Level Chart Recorder for Monitoring Wells on Casings**	**EA**				**0.00**	**0.00**	**920.00**	**920.00**
01954 2245	Water Level Chart Recorder, Battery Operated	EA	0	N/A	1.000	0.00	0.00	920.00	920.00
33 02 0538	**DO Meter, Analog, Portable, 6' Cable, Polarographic Probe**	**EA**				**0.00**	**0.00**	**400.00**	**400.00**
01954 2261	Analog, Portable, 6' Cable, Polarographic Probe (Purchase)	EA	0	N/A	1.000	0.00	0.00	400.00	400.00
33 02 0539	**DO Meter, Digital, Portable, 10' Cable, Small Diameter Probe**	**EA**				**0.00**	**0.00**	**565.00**	**565.00**
01954 2262	Digital, Portable, 10' Cable, Small Diameter Probe (Purchase)	EA	0	N/A	1.000	0.00	0.00	565.00	565.00
33 02 0540	**DO Meter, Portable, Probe, 10' Cable, Quick Readings**	**EA**				**0.00**	**0.00**	**775.00**	**775.00**
01954 2263	Highly Portable with Probe/10' Cable, Quick Readings	EA	0	N/A	1.000	0.00	0.00	775.00	775.00
33 02 0541	**Conductivity Meter with Dip-type Sensor, Digital, 9V Battery, 7 x 3.2 x 2"**	**EA**				**0.00**	**0.00**	**415.00**	**415.00**
01954 2271	Handheld with Dip-type Sensor Digital, 9V Battery, 7 x 3.2 x 2"	EA	0	N/A	1.000	0.00	0.00	415.00	415.00
33 02 0542	**Conductivity Meter, 1" Diameter Probe, 150' Cable, Also Temperature/Level**	**EA**				**0.00**	**0.00**	**1,125**	**1,125**
01954 2272	Digital, 1" Diameter Probe with 150' Cable (Also Measures Temperature/Water Level)	EA	0	N/A	1.000	0.00	0.00	1,125	1,125
33 02 0543	**Conductivity Meter, 10' Cable, Probe, Also Temperature/ Salinity**	**EA**				**0.00**	**0.00**	**805.00**	**805.00**
01954 2273	10' Cable with Probe, Compact (Also Measures Temperature/Salinity)	EA	0	N/A	1.000	0.00	0.00	805.00	805.00
33 02 0544	**Thermometer, Stainless Steel, Bi-Metal, 40 - 160 F, 12" Stem**	**EA**				**0.00**	**0.00**	**62.00**	**62.00**
01954 2281	Stainless Steel Bimetal Thermometer, -40 to 160 Degrees F, 12" Stem	EA	0	N/A	1.000	0.00	0.00	62.00	62.00
33 02 0545	**Thermometer, Thermocouple**	**EA**				**0.00**	**0.00**	**167.00**	**167.00**
01954 2282	Thermocouple Thermometer	EA	0	N/A	1.000	0.00	0.00	167.00	167.00
33 02 0546	**Thermometer, Digital**	**EA**				**0.00**	**0.00**	**72.00**	**72.00**
01954 2283	Digital Thermometer	EA	0	N/A	1.000	0.00	0.00	72.00	72.00

33 02 05	Sampling Surface Water/Groundwater/Liquid Waste	Unit	Hourly Output	Crew	Factor	Labor	Equip.	Matl.	Total
33 02 0547	**pH Meter, Test Paper Kits, 200 Strips, Charts**	**EA**				**0.00**	**0.00**	**25.00**	**25.00**
01954 2291	pH Test Paper Kits - 200 Paper Strips with Indicator Charts	EA	0	N/A	1.000	0.00	0.00	25.00	25.00
33 02 0548	**pH Meter, Digital, LCD, 0 - 14 pH Range**	**EA**				**0.00**	**0.00**	**775.00**	**775.00**
01954 2292	Digital pH Meter Kit - LCD pH Meter with 0-14 pH Range	EA	0	N/A	1.000	0.00	0.00	775.00	775.00
33 02 0549	**Lysimeter, 1.90" Outside Diameter, PVC Tube Type - PVC Body/Teflon Filter**	**EA**				**60.03**	**264.92**	**178.00**	**502.95**
01954 3251	1.90" Outside Diameter PVC Tube Type - PVC Body/Teflon Filter	EA	1.5	ULADB	1.000	60.03	264.92	178.00	502.95
33 02 0550	**Lysimeter, 1.90" Outside Diameter, Teflon Tube Type - Teflon Body/Filter**	**EA**				**60.03**	**264.92**	**238.00**	**562.95**
01954 3252	1.90" Outside Diameter Teflon Tube Type - Teflon Body/Filter	EA	1.5	ULADB	1.000	60.03	264.92	238.00	562.95
33 02 0551	**Lysimeter, 2.38" Outside Diameter, PVC Tube Type - PVC Body/Teflon Filter**	**EA**				**60.03**	**264.92**	**168.00**	**492.95**
01954 3253	2.38" Outside Diameter PVC Tube Type - PVC Body/Teflon Filter	EA	1.5	ULADB	1.000	60.03	264.92	168.00	492.95
33 02 0552	**Lysimeter, 2.38" Outside Diameter, Teflon Tube Type - Teflon Body/Filter**	**EA**				**60.03**	**264.92**	**266.00**	**590.95**
01954 3254	2.38" Outside Diameter Teflon Tube Type - Teflon Body/Filter	EA	1.5	ULADB	1.000	60.03	264.92	266.00	590.95
33 02 0553	**Lysimeter, 1.90" Outside Diameter, PVC Cup Type - PVC Body/Teflon Filter**	**EA**				**60.03**	**264.92**	**174.00**	**498.95**
01954 3255	1.90" Outside Diameter PVC Cup Type - PVC Body/Teflon Filter	EA	1.5	ULADB	1.000	60.03	264.92	174.00	498.95
33 02 0554	**Lysimeter, 1.90" Outside Diameter, Teflon Cup Type - Teflon Body/Filter**	**EA**				**60.03**	**264.92**	**239.00**	**563.95**
01954 3256	1.90" Outside Diameter Teflon Cup Type - Teflon Body/Filter	EA	1.5	ULADB	1.000	60.03	264.92	239.00	563.95
33 02 0555	**Lysimeter, 1.90" Outside Diameter, PVC Deep Sample Type - Teflon Body/Filter**	**EA**				**60.03**	**264.92**	**282.00**	**606.95**
01954 3257	1.90" Outside Diameter PVC Deep Sampling Type - Teflon Body/Filter	EA	1.5	ULADB	1.000	60.03	264.92	282.00	606.95
33 02 0556	**Lysimeter, 1.90" Outside Diameter, Teflon Deep Sample Type - Teflon Body/Filter**	**EA**				**60.03**	**264.92**	**350.00**	**674.95**
01954 3258	1.90" Outside Diameter Teflon Deep Sampling Type - Teflon Body/Filter	EA	1.5	ULADB	1.000	60.03	264.92	350.00	674.95
33 02 0557	**Lysimeter Head Assembly, All PVC Fittings**	**EA**				**60.03**	**264.92**	**268.00**	**592.95**
01954 3261	Lysimeter Head Assembly - All PVC Fittings	EA	1.5	ULADB	1.000	60.03	264.92	268.00	592.95

33 02 05 Sampling Surface Water/Groundwater/Liquid Waste		Unit	Hourly Output	Crew	Factor	Labor	Equip.	Matl.	Total
33 02 0558	**Lysimeter Head Assembly, Teflon & PVC Fittings**	**EA**				**60.03**	**264.92**	**292.00**	**616.95**
01954 3262	Lysimeter Head Assembly - Teflon and PVC Fittings	EA	1.5	ULADB	1.000	60.03	264.92	292.00	616.95
33 02 0559	**Silica Flour, 50 Lb Pail for Media Continuum**	**EA**				**18.01**	**79.47**	**59.00**	**156.48**
01954 3263	Silica Flour - 50 Lb Pail to Maintain Media Continuum	EA	5	ULADB	1.000	18.01	79.47	59.00	156.48
33 02 0560	**Vacuum Pressure Hand Pump**	**EA**				**0.00**	**0.00**	**55.00**	**55.00**
01954 3264	Vacuum Pressure Hand Pump	EA	0	N/A	1.000	0.00	0.00	55.00	55.00
33 02 0561	**Nylon Tubing, 1/4" Outside Diameter**	**LF**				**0.00**	**0.00**	**0.51**	**0.51**
01954 3265	1/4" Outside Diameter Nylon Tubing	LF	0	N/A	1.000	0.00	0.00	0.51	0.51
33 02 0562	**Teflon Tubing, 1/4" Outside Diameter**	**LF**				**0.00**	**0.00**	**2.06**	**2.06**
01954 3266	1/4" Outside Diameter Teflon Tubing	LF	0	N/A	1.000	0.00	0.00	2.06	2.06
33 02 0563	**DO Meter, Analog, 3' Cable, Daily Rental**	**DAY**				**0.00**	**0.00**	**33.66**	**33.66**
01001 0026	DO Meter, Analog, 3' Cable, Daily Rental	DAY	0	N/A	1.000	0.00	0.00	33.66	33.66
33 02 0564	**DO Meter, Analog, 3' Cable, Weekly Rental**	**WK**				**0.00**	**0.00**	**78.54**	**78.54**
01001 0027	DO Meter, Analog, 3' Cable, Weekly Rental	WK	0	N/A	1.000	0.00	0.00	78.54	78.54
33 02 0565	**DO Meter, Digital, 10' Cable, Daily Rental**	**DAY**				**0.00**	**0.00**	**56.10**	**56.10**
01001 0028	DO Meter, Digital, 10' Cable, Daily Rental	DAY	0	N/A	1.000	0.00	0.00	56.10	56.10
33 02 0566	**DO Meter, Digital, 10' Cable, Weekly Rental**	**WK**				**0.00**	**0.00**	**168.31**	**168.31**
01001 0029	DO Meter, Digital, 10' Cable, Weekly Rental	WK	0	N/A	1.000	0.00	0.00	168.31	168.31
33 02 0567	**pH-Conductivity-Temperature Meter, Daily Rental**	**DAY**				**0.00**	**0.00**	**30.86**	**30.86**
01001 0030	pH-Conductivity-Temperature Meter, Daily Rental	DAY	0	N/A	1.000	0.00	0.00	30.86	30.86
33 02 0568	**pH-Conductivity-Temperature Meter, Weekly Rental**	**WK**				**0.00**	**0.00**	**75.74**	**75.74**
01001 0031	pH-Conductivity-Temperature Meter, Weekly Rental	WK	0	N/A	1.000	0.00	0.00	75.74	75.74
33 02 0569	**pH-Conductivity-Temperature Meter, Monthly Rental**	**MO**				**0.00**	**0.00**	**218.80**	**218.80**
01001 0032	pH-Conductivity-Temperature Meter, Monthly Rental	MO	0	N/A	1.000	0.00	0.00	218.80	218.80
33 02 0570	**Downhole pH, Conductivity, Temperature and DO Meter, 50' Cable, Daily Rental**	**DAY**				**0.00**	**0.00**	**56.10**	**56.10**
01001 0033	Downhole pH, Conductivity, Temperature and DO Meter, 50' Cable, Daily Renta	DAY	0	N/A	1.000	0.00	0.00	56.10	56.10

33 02 05	Sampling Surface Water/Groundwater/Liquid Waste	Unit	Hourly Output	Crew	Factor	Labor	Equip.	Matl.	Total
33 02 0571	**Downhole pH, Conductivity, Temperature and DO Meter, 50' Cable, Weekly Rental**	**WK**				**0.00**	**0.00**	**75.74**	**75.74**
01001 0034	Downhole pH, Conductivity, Temperature and DO Meter, 50' Cable, Weekly Rent	WK	0	N/A	1.000	0.00	0.00	75.74	75.74
33 02 0572	**Water Level Indicator, Manual, Polyethylene Tape, 100' Cable, Daily Rental**	**DAY**				**0.00**	**0.00**	**25.25**	**25.25**
01001 0035	Water Level Indicator, Manual, Polyethylene Tape, 100' Cable, Daily Rental	DAY	0	N/A	1.000	0.00	0.00	25.25	25.25
33 02 0573	**Water Level Indicator, Manual, Polyethylene Tape, 100' Cable, Weekly Rental**	**WK**				**0.00**	**0.00**	**64.52**	**64.52**
01001 0036	Water Level Indicator, Manual, Polyethylene Tape, 100' Cable, Weekly Rental	WK	0	N/A	1.000	0.00	0.00	64.52	64.52
33 02 0574	**Water Level Indicator, Manual, Polyethylene Tape, 500' Cable, Daily Rental**	**DAY**				**0.00**	**0.00**	**39.27**	**39.27**
01001 0037	Water Level Indicator, Manual, Polyethylene Tape, 500' Cable, Daily Rental	DAY	0	N/A	1.000	0.00	0.00	39.27	39.27
33 02 0575	**Water Level Indicator, Manual, Polyethylene Tape, 500' Cable, Weekly Rental**	**WK**				**0.00**	**0.00**	**81.35**	**81.35**
01001 0038	Water Level Indicator, Manual, Polyethylene Tape, 500' Cable, Weekly Rental	WK	0	N/A	1.000	0.00	0.00	81.35	81.35
33 02 0576	**Flow Through Monitor, Weekly Rental**	**WK**				**0.00**	**0.00**	**218.80**	**218.80**
01001 0039	Flow Through Monitor, Weekly Rental	WK	0	N/A	1.000	0.00	0.00	218.80	218.80
33 02 0577	**pH/ORP/Temperature Meter - Daily Rental**	**DAY**				**0.00**	**0.00**	**58.50**	**58.50**
01954 4164	pH/ORP/Temperature Meter - Daily Rental	DAY	0	N/A	1.000	0.00	0.00	58.50	58.50
33 02 0578	**pH/ORP/Temperature Meter - Weekly Rental**	**WK**				**0.00**	**0.00**	**63.77**	**63.77**
01001 2164	pH/ORP/Temperature Meter - Weekly Rental	WK	0	N/A	1.000	0.00	0.00	63.77	63.77
33 02 0579	**pH/ORP/Temperature Meter - Monthly Rental**	**MO**				**0.00**	**0.00**	**244.44**	**244.44**
01001 2165	pH/ORP/Temperature Meter - Monthly Rental	MO	0	N/A	1.000	0.00	0.00	244.44	244.44
33 02 0580	**pH/ORP/Temperature Meter - Purchase**	**EA**				**0.00**	**0.00**	**607.92**	**607.92**
01001 2166	pH/ORP/Temperature Meter - Purchase	EA	0	N/A	1.000	0.00	0.00	607.92	607.92

33 02 06	Sampling Soil and Sediment	Unit	Hourly Output	Crew	Factor	Labor	Equip.	Matl.	Total
33 02 0601	**Drilling 2.5" Diameter Soil Borings, No Sampling**	**LF**				**3.57**	**5.61**	**8.40**	**17.58**
02010 0600	2-1/2" (6.4 cm) Diameter Augured Boring in Earth, No Samples	LF	23	CLADI	1.000	3.57	5.61	8.40	17.58
33 02 0602	**Stainless Steel Bottom Sampler**	**EA**				**0.00**	**0.00**	**525.00**	**525.00**
01954 5161	3.25" x 12", Stainless Steel Thread On	EA	0	N/A	1.000	0.00	0.00	525.00	525.00

33 02 06 Sampling Soil and Sediment	Unit	Hourly Output	Crew	Factor	Labor	Equip.	Matl.	Total
33 02 0603 Surface Soil Sampling Equipment	**EA**				**0.00**	**0.00**	**377.00**	**377.00**
01954 5131 High Carbon Steel, 1.5" x 6"	EA	0	N/A	1.000	0.00	0.00	111.00	111.00
01954 5143 Regular, 2" Diameter, Carbon/Stainless	EA	0	N/A	2.000	0.00	0.00	133.00	266.00
33 02 0605 Hand Auger Rental	**DAY**				**0.00**	**0.00**	**13.98**	**13.98**
01954 5133 Hand Auger Rental	DAY	0	N/A	1.000	0.00	0.00	13.98	13.98
33 02 0606 Power Auger Rental	**DAY**				**0.00**	**0.00**	**18.54**	**18.54**
01954 5134 Power Auger with Attachments, Rental	DAY	0	N/A	1.000	0.00	0.00	18.54	18.54
33 02 0612 Soil Analysis	**EA**				**0.00**	**0.00**	**1,036**	**1,036**
01954 7233 Total Solids (EPA 160.3)	EA	0	N/A	1.000	0.00	0.00	11.85	11.85
01954 7237 Total Organic Carbons (EPA 415.1,9060)	EA	0	N/A	1.000	0.00	0.00	24.00	24.00
01954 7241 Metals (EPA 200.7)	EA	0	N/A	8.000	0.00	0.00	64.11	512.86
01954 7413 Pesticides/Herbicides (8081, 8151A)	EA	0	N/A	1.000	0.00	0.00	248.00	248.00
01954 7527 Sulfur (Parr Bomb): ASTM D129	EA	0	N/A	1.000	0.00	0.00	121.00	121.00
01954 7607 Cyanide, Total, (9010B) Spectrophotometric	EA	0	N/A	1.000	0.00	0.00	59.63	59.63
01954 7613 Mercury (EPA 7470/7471)	EA	0	N/A	1.000	0.00	0.00	26.50	26.50
01954 7615 Nitrogen, Nitrate/Nitrite (EPA 300.0 / SM 4110B)	EA	0	N/A	1.000	0.00	0.00	25.50	25.50
01954 7617 pH, Electrometric (9045)	EA	0	N/A	1.000	0.00	0.00	6.75	6.75
33 02 0613 Soil Recovery Probe, Unslotted, 7/8" x 1'	**EA**				**0.00**	**0.00**	**158.00**	**158.00**
01954 5111 Unslotted, 7/8" x 1'	EA	0	N/A	1.000	0.00	0.00	158.00	158.00
33 02 0614 Soil Recovery Probe, Unslotted, 7/8" x 2' or 9/8" x 1'	**EA**				**0.00**	**0.00**	**171.00**	**171.00**
01954 5112 Unslotted, 7/8" x 2' or 9/8" x 1'	EA	0	N/A	1.000	0.00	0.00	171.00	171.00
33 02 0615 Soil Recovery Probe, Unslotted, 9/8" x 2'	**EA**				**0.00**	**0.00**	**183.00**	**183.00**
01954 5113 Unslotted, 9/8" x 2'	EA	0	N/A	1.000	0.00	0.00	183.00	183.00
33 02 0616 Soil Recovery Probe, Liners, 7/8" per LF, Butyrate Plast	**LF**				**0.00**	**0.00**	**1.12**	**1.12**
01954 5114 Liners, Butyrate Plastic, 7/8"	LF	0	N/A	1.000	0.00	0.00	1.12	1.12
33 02 0617 Soil Recovery Probe, Liners, 9/8" per LF, Butyrate Plast	**LF**				**0.00**	**0.00**	**1.96**	**1.96**
01954 5115 Liners, Butyrate Plastic, 9/8"	LF	0	N/A	1.000	0.00	0.00	1.96	1.96
33 02 0618 Soil Recovery Probe, Liners, Stainless Steel, 7/8"	**EA**				**0.00**	**0.00**	**7.50**	**7.50**
01954 5116 Liners, Stainless Steel, 7/8"	EA	0	N/A	1.000	0.00	0.00	7.50	7.50
33 02 0619 Soil Recovery Probe, Liners, Stainless Steel, 9/8"	**EA**				**0.00**	**0.00**	**9.20**	**9.20**
01954 5117 Liners, Stainless Steel, 9/8"	EA	0	N/A	1.000	0.00	0.00	9.20	9.20

33 02 06 Sampling Soil and Sediment		Unit	Hourly Output	Crew	Factor	Labor	Equip.	Matl.	Total
33 02 0620	**Soil Gas Vapor Probe, Stainless Steel, Manual, Nonremovable Tip**	**EA**				**0.00**	**0.00**	**262.00**	**262.00**
01954 5121	Stainless Steel, Nonremovable Tip, Ground Penetration by Hand	EA	0	N/A	1.000	0.00	0.00	262.00	262.00
33 02 0621	**Soil Gas Vapor Probe, Stainless Steel, Electric, Rotary, Up/Down Hammer**	**EA**				**0.00**	**0.00**	**331.54**	**331.54**
01954 5122	Stainless Steel, Removable Tip, Electric Rotary & Up/Down Hammer	EA	0	N/A	1.000	0.00	0.00	331.54	331.54
33 02 0622	**Screw Auger, High-carbon Steel, 1.5" x 6"**	**EA**				**0.00**	**0.00**	**111.00**	**111.00**
01954 5131	High Carbon Steel, 1.5" x 6"	EA	0	N/A	1.000	0.00	0.00	111.00	111.00
33 02 0623	**Screw Auger, High-carbon Steel, 1.75" x 6"**	**EA**				**0.00**	**0.00**	**138.00**	**138.00**
01954 5132	High Carbon Steel, 1.75" x 6"	EA	0	N/A	1.000	0.00	0.00	138.00	138.00
33 02 0624	**Soil Auger Buckets, Dutch, 3" Diameter, Carbon/Stainless Steel**	**EA**				**0.00**	**0.00**	**163.00**	**163.00**
01954 5141	Dutch, 3" Diameter, Carbon/Stainless	EA	0	N/A	1.000	0.00	0.00	163.00	163.00
33 02 0625	**Soil Auger Buckets, Mud, 2" Diameter, Carbon/Stainless Steel**	**EA**				**0.00**	**0.00**	**150.00**	**150.00**
01954 5142	Mud, 2" Diameter, Carbon/Stainless	EA	0	N/A	1.000	0.00	0.00	150.00	150.00
33 02 0626	**Soil Auger Buckets, Reg, 2" Diameter, Carbon/Stainless Steel**	**EA**				**0.00**	**0.00**	**133.00**	**133.00**
01954 5143	Regular, 2" Diameter, Carbon/Stainless	EA	0	N/A	1.000	0.00	0.00	133.00	133.00
33 02 0627	**Soil Auger Buckets, Sand, 2" Diameter, Carbon/Stainless Steel**	**EA**				**0.00**	**0.00**	**144.00**	**144.00**
01954 5144	Sand, 2" Diameter, Carbon/Stainless	EA	0	N/A	1.000	0.00	0.00	144.00	144.00
33 02 0628	**Soil Recovery Auger Buckets, Stainless Steel, 2.25" x 8"**	**EA**				**0.00**	**0.00**	**385.00**	**385.00**
01954 5151	Stainless Steel, 2.25" x 8"	EA	0	N/A	1.000	0.00	0.00	385.00	385.00
33 02 0629	**Soil Recovery Auger Buckets, Stainless Steel, 3.25" x 10"**	**EA**				**0.00**	**0.00**	**405.00**	**405.00**
01954 5152	Stainless Steel, 3.25" x 10"	EA	0	N/A	1.000	0.00	0.00	405.00	405.00
33 02 0630	**Soil Recovery Auger Buckets, Stainless Steel with Carbon Bits, 2.25" x 8"**	**EA**				**0.00**	**0.00**	**345.00**	**345.00**
01954 5153	Stainless Steel with Carbon Bits 2.25" x 8"	EA	0	N/A	1.000	0.00	0.00	345.00	345.00
33 02 0631	**Soil Recovery Auger Buckets, Stainless Steel with Carbon Bits, 3.25" x 10"**	**EA**				**0.00**	**0.00**	**360.00**	**360.00**
01954 5154	Stainless Steel with Carbon Bits 3.25" x 10"	EA	0	N/A	1.000	0.00	0.00	360.00	360.00
33 02 0632	**Soil Recovery Auger Buckets, Liners, Butyrate Plast, 3" x 10" or 2" x 8"**	**EA**				**0.00**	**0.00**	**5.35**	**5.35**
01954 5155	Liners, Butyrate Plastic, 3" x 10" or 2" x 8"	EA	0	N/A	1.000	0.00	0.00	5.35	5.35

33 02 06	Sampling Soil and Sediment	Unit	Hourly Output	Crew	Factor	Labor	Equip.	Matl.	Total
33 02 0633	**Soil Recovery Auger Buckets, Liners, Stainless Steel, 3" x 10" or 2" x 8"**	**EA**				**0.00**	**0.00**	**15.95**	**15.95**
01954 5156	Liners, Stainless Steel, 3" x 10" or 2" x 8"	EA	0	N/A	1.000	0.00	0.00	15.95	15.95
33 02 0634	**Sludge Sampler, Stainless Steel, 3.25" x 12", Thread On**	**EA**				**0.00**	**0.00**	**525.00**	**525.00**
01954 5161	3.25" x 12", Stainless Steel Thread On	EA	0	N/A	1.000	0.00	0.00	525.00	525.00
33 02 0635	**Tensiometer Soil Moist Probe, Standard, 7/8" Outside Diameter x 18", Fixed**	**EA**				**0.00**	**0.00**	**150.00**	**150.00**
01954 5171	Standard, 7/8" Outside Diameter x 18", Fixed	EA	0	N/A	1.000	0.00	0.00	150.00	150.00
33 02 0636	**Tensiometer Soil Moist Probe, Jet Fill, 7/8" Outside Diameter x 18", Fixed**	**EA**				**0.00**	**0.00**	**85.50**	**85.50**
01954 5172	Jet Fill, 7/8" x 18", Fixed, Recovery Time in Minutes	EA	0	N/A	1.000	0.00	0.00	85.50	85.50
33 02 0637	**Tensiometer Soil Moist Probe, Quick Draw, 18", Quick Read**	**EA**				**0.00**	**0.00**	**450.00**	**450.00**
01954 5173	Quick Draw, 18", Quick Readings Highly Mobile	EA	0	N/A	1.000	0.00	0.00	450.00	450.00
33 02 0638	**Soil Test Kit, 50 Tests**	**EA**				**0.00**	**0.00**	**345.00**	**345.00**
01954 5181	Detailed Soil Evaluation Kit (50 Tests)	EA	0	N/A	1.000	0.00	0.00	345.00	345.00
33 02 0639	**CPT Rig, Includes Labor, Sampling, Punching, Decontamination**	**DAY**				**0.00**	**0.00**	**3,350**	**3,350**
01954 6301	CPT Rig, Includes Labor, Sampling, Punching, Decontamination	DAY	0	N/A	1.000	0.00	0.00	3,350	3,350
33 02 0640	**Mobilize/Demobilize CPT Rig and Crew**	**EA**				**0.00**	**0.00**	**2,425**	**2,425**
01954 6302	Mobilize/Demobilize CPT Rig and Crew	EA	0	N/A	1.000	0.00	0.00	2,425	2,425
33 02 0641	**Standby Time for CPT Rig and Crew**	**HR**				**0.00**	**0.00**	**215.00**	**215.00**
01954 6303	Standby Time for CPT Rig and Crew	HR	0	N/A	1.000	0.00	0.00	215.00	215.00
33 02 0642	**Setup Cost per Each Hole, CPT Rig**	**EA**				**0.00**	**0.00**	**265.00**	**265.00**
01954 6304	Setup Cost per Each Hole, CPT Rig	EA	0	N/A	1.000	0.00	0.00	265.00	265.00
33 02 0643	**Decontamination Trailer Rental for CPT Rig**	**DAY**				**0.00**	**0.00**	**95.04**	**95.04**
01954 6305	Decontamination Trailer Rental	DAY	0	N/A	1.000	0.00	0.00	95.04	95.04
33 02 0644	**Decontaminate Cone on CPT Rig**	**DAY**				**0.00**	**0.00**	**278.00**	**278.00**
01954 6306	Decontaminate Cone on CPT Rig	DAY	0	N/A	1.000	0.00	0.00	278.00	278.00
33 02 0645	**Level "D" PPE Rental per 2-Man CPT Crew**	**DAY**				**0.00**	**0.00**	**2,850**	**2,850**
01954 6307	Level "D" PPE Rental per 2-Man CPT Crew (Rig & Equip)	DAY	0	N/A	1.000	0.00	0.00	2,850	2,850
33 02 0646	**Level "C" PPE Rental per 2-Man CPT Crew**	**DAY**				**0.00**	**0.00**	**3,075**	**3,075**
01954 6308	Level "C" PPE Rental per 2-Man CPT Crew (Rig & Equip)	DAY	0	N/A	1.000	0.00	0.00	3,075	3,075

33 02 06	Sampling Soil and Sediment	Unit	Hourly Output	Crew	Factor	Labor	Equip.	Matl.	Total
33 02 0647	**In Situ Batch Sampling for Groundwater, CPT Rig (per Sample)**	**EA**				**0.00**	**0.00**	**90.50**	**90.50**
01954 6309	In Situ Batch Sampling for Groundwater, CPT Rig	EA	0	N/A	1.000	0.00	0.00	90.50	90.50
33 02 0648	**In Situ Soil Sampling, CPT Rig (per Sample)**	**EA**				**0.00**	**0.00**	**58.50**	**58.50**
01954 6310	In Situ Soil Sampling, CPT Rig (per Sample)	EA	0	N/A	1.000	0.00	0.00	58.50	58.50
33 02 0649	**Hydropunch with CPT Rig, Hard Soil**	**FT**				**0.00**	**0.00**	**10.35**	**10.35**
01954 6311	Hydropunch with CPT Rig, Hard Soil	FT	0	N/A	1.000	0.00	0.00	10.35	10.35
33 02 0650	**Hydropunch with CPT Rig, Medium Soil**	**FT**				**0.00**	**0.00**	**10.35**	**10.35**
01954 6312	Hydropunch with CPT Rig, Medium Soil	FT	0	N/A	1.000	0.00	0.00	10.35	10.35
33 02 0651	**Hydropunch with CPT Rig, Soft Soil**	**FT**				**0.00**	**0.00**	**10.35**	**10.35**
01954 6313	Hydropunch with CPT Rig, Soft Soil	FT	0	N/A	1.000	0.00	0.00	10.35	10.35
33 02 0652	**Grout Hole After Hydropunching**	**FT**				**0.00**	**0.00**	**1.68**	**1.68**
01954 6314	Grout Hole After Hydropunching	FT	0	N/A	1.000	0.00	0.00	1.68	1.68
33 02 0653	**CPT Rig with ROST Analytical Equipment for Obtaining VOC in Soil**	**DAY**				**0.00**	**0.00**	**8,245**	**8,245**
01954 6315	CPT Rig with ROST Analytical Equipment, includes labor	DAY	0	N/A	1.000	0.00	0.00	8,245	8,245
33 02 0661	**CPT Rig with Resistivity Cone**	**DAY**				**0.00**	**0.00**	**4,181**	**4,181**
01954 6316	CPT Rig with Resistivity Cone	DAY	0	N/A	1.000	0.00	0.00	4,181	4,181
33 02 0662	**Chem-Probe (Geo-Probe) per point**	**EA**				**0.00**	**0.00**	**444.94**	**444.94**
01954 6317	Chem-Probe (Geo-Probe) per Point	EA	0	N/A	1.000	0.00	0.00	444.94	444.94
33 02 0663	**Drilling 4" Diameter Soil Borings, No Sampling, 0-10 Feet Deep**	**LF**				**0.00**	**0.00**	**27.15**	**27.15**
01001 0040	Drilling 4" Diameter Soil Borings, No Sampling, 0-10 Feet Deep	LF	0	N/A	1.000	0.00	0.00	27.15	27.15
33 02 0664	**Drilling 4" Diameter Soil Borings, No Sampling,11-20 Feet Deep**	**LF**				**0.00**	**0.00**	**17.39**	**17.39**
01001 0041	Drilling 4" Diameter Soil Borings, No Sampling,11-20 Feet Deep	LF	0	N/A	1.000	0.00	0.00	17.39	17.39
33 02 0665	**Drilling 4" Diameter Soil Borings, No Sampling, 21-30 Feet Deep**	**LF**				**0.00**	**0.00**	**14.81**	**14.81**
01001 0042	Drilling 4" Diameter Soil Borings, No Sampling, 21-30 Feet Deep	LF	0	N/A	1.000	0.00	0.00	14.81	14.81
33 02 0666	**Drilling 4" Diameter Soil Borings, No Sampling, > 30 Feet Deep**	**LF**				**0.00**	**0.00**	**13.29**	**13.29**
01001 0043	Drilling 4" Diameter Soil Borings, No Sampling, > 30 Feet Deep	LF	0	N/A	1.000	0.00	0.00	13.29	13.29

33 02 Monitoring, Sampling, Testing, and Analysis

33 02 06	Sampling Soil and Sediment	Unit	Hourly Output	Crew	Factor	Labor	Equip.	Matl.	Total
33 02 0667	**Direct Push Rig, Truck Mounted, Non Hydraulic, Includes Labor, Sampling, Decontamination**	**DAY**				**0.00**	**0.00**	**168.31**	**168.31**
01001 0044	Direct Push Rig, Truck Mounted, Non Hydraulic, Incl. Labor, Sampling, Decon	DAY	0	N/A	1.000	0.00	0.00	168.31	168.31
33 02 0668	**Mobilize Direct Push Rig and Crew**	**DAY**				**0.00**	**0.00**	**561.02**	**561.02**
01001 0045	Mobilize Direct Push Rig	DAY	0	N/A	1.000	0.00	0.00	561.02	561.02
33 02 0669	**Demobilize Direct Push Rig and Crew**	**DAY**				**0.00**	**0.00**	**561.02**	**561.02**
01001 0046	Demobilize Direct Push Rig	DAY	0	N/A	1.000	0.00	0.00	561.02	561.02

33 02 11	Geotechnical Testing	Unit	Hourly Output	Crew	Factor	Labor	Equip.	Matl.	Total
33 02 1101	**Geotechnical Characteristics Analysis**	**EA**				**81.84**	**0.00**	**93.80**	**175.64**
01418 4605	Grain Size (No. 200 Sieve) ASTM D421	EA	0.64	USKCA	1.000	68.85	0.00	59.00	127.85
01418 4755	Moisture	EA	3.393	USKCA	1.000	12.99	0.00	10.00	22.99
01954 7533	Bulk Density	EA	0	N/A	1.000	0.00	0.00	18.05	18.05
01954 7617	pH, Electrometric (9045)	EA	0	N/A	1.000	0.00	0.00	6.75	6.75
33 02 1102	**Soil Moisture Content ASTM D2216**	**EA**				**0.00**	**0.00**	**23.50**	**23.50**
01954 7538	Moisture Content ASTM D2216	EA	0	N/A	1.000	0.00	0.00	23.50	23.50

33 02 15	Monitoring Instrumentation	Unit	Hourly Output	Crew	Factor	Labor	Equip.	Matl.	Total
33 02 1501	**Saturation Indicator**	**EA**				**0.00**	**0.00**	**47.50**	**47.50**
13274 5135	Saturation Indicator	EA	0	N/A	1.000	0.00	0.00	47.50	47.50
33 02 1502	**Thermostat & Humidity Control Devices**	**EA**				**106.64**	**0.00**	**107.70**	**214.34**
15742 5210	Proportional Room Thermostat, 1 Potentiometer	EA	0.838	EELEA	1.000	61.42	0.00	78.92	140.34
16830 3600	Room Thermostats, Line Voltage Electric Resistance Heating	EA	1.138	EELEA	1.000	45.22	0.00	28.78	74.00
33 02 1506	**Monitoring Port with Gas Monitor**	**EA**				**13.34**	**0.00**	**1.09**	**14.43**
15174 8570	Plug, 1/4" Black Steel	EA	3.875	MSPFA	1.000	13.34	0.00	1.09	14.43
33 02 1507	**Continuous Monitoring and Recording of Air Flow**	**EA**				**0.00**	**0.00**	**5,699**	**5,699**
13273 4308	Continuous Monitoring and Recording (Air Flow)	EA	0	N/A	1.000	0.00	0.00	5,699	5,699
33 02 1509	**Water Quality Parameter Testing Device**	**WK**				**0.00**	**0.00**	**236.00**	**236.00**
01954 3272	Water Quality Testing Parameter Device Rental	WK	0	N/A	1.000	0.00	0.00	236.00	236.00
33 02 1510	**Water Quality Parameter Testing Device, DO, Temp., pH, Conductivity, Salinity, Turbidity, Daily Rent**	**DAY**				**0.00**	**0.00**	**67.32**	**67.32**
01001 0047	Water Quality Parameter Testing Device, DO, Temp., pH, Conductivity, Salini	DAY	0	N/A	1.000	0.00	0.00	67.32	67.32

33 02 15 Monitoring Instrumentation		Unit	Hourly Output	Crew	Factor	Labor	Equip.	Matl.	Total
33 02 1511	**Water Flowmeter**	**EA**				**0.00**	**0.00**	**79.00**	**79.00**
13274 2955	Recycle Flow Meter, 3/4 HP Unit	EA	0	N/A	1.000	0.00	0.00	79.00	79.00
33 02 1512	**pH Chemical Feed Controller without Input Card**	**EA**				**0.00**	**0.00**	**842.39**	**842.39**
13274 1151	pH Chemical Feed Controller without Input Card	EA	0	N/A	1.000	0.00	0.00	842.39	842.39
33 02 1513	**pH Input Card - Standard**	**EA**				**25.72**	**0.00**	**199.99**	**225.71**
13274 1152	pH Input Card - Standard	EA	4	EELEB	1.000	25.72	0.00	199.99	225.71
33 02 1514	**pH/ORP Output Card Dual Propor Alarm**	**EA**				**25.72**	**0.00**	**98.18**	**123.90**
13274 1153	pH/ORP Output Card Dual Propor Alarm	EA	4	EELEB	1.000	25.72	0.00	98.18	123.90
33 02 1515	**pH/ORP Input Card Isolated 0/4-20 mA**	**EA**				**25.72**	**0.00**	**155.15**	**180.86**
13274 1154	pH/ORP Input Card Isolated 0/4-20 mA	EA	4	EELEB	1.000	25.72	0.00	155.15	180.86
33 02 1516	**pH/ORP Input Card Isolated 0-5/10 VDC**	**EA**				**25.72**	**0.00**	**155.15**	**180.86**
13274 1155	pH/ORP Input Card Isolated 0-5/10 VDC	EA	4	EELEB	1.000	25.72	0.00	155.15	180.86
33 02 1517	**pH Controller with Non-isolated 4-20 mA Output**	**EA**				**25.72**	**0.00**	**1,194**	**1,220**
13274 1156	pH Controller with Non-isolated 4-20 mA Output	EA	4	EELEB	1.000	25.72	0.00	1,194	1,220
33 02 1518	**pH Controller with Isolated 4-20 mA Output**	**EA**				**25.72**	**0.00**	**1,303**	**1,329**
13274 1157	pH Controller with Isolated 4-20 mA Output	EA	4	EELEB	1.000	25.72	0.00	1,303	1,329
33 02 1519	**pH Pump Pulser Board for Controller**	**EA**				**25.72**	**0.00**	**261.81**	**287.53**
13274 1158	pH Pump Pulser Board for Controller	EA	4	EELEB	1.000	25.72	0.00	261.81	287.53
33 02 1520	**Analog pH Analyzer with 2 - 12 pH Dials**	**EA**				**12.86**	**0.00**	**700.58**	**713.44**
13274 1181	Analog pH Analyzer with 2 - 12 pH Dials	EA	8	EELEB	1.000	12.86	0.00	700.58	713.44
33 02 1521	**pH Coax Cable (4')**	**EA**				**12.86**	**0.00**	**69.33**	**82.19**
13274 1191	4' pH Coax Cable	EA	8	EELEB	1.000	12.86	0.00	69.33	82.19
33 02 1522	**pH Coax Cable (8')**	**EA**				**12.86**	**0.00**	**88.68**	**101.54**
13274 1192	8' pH Coax Cable	EA	8	EELEB	1.000	12.86	0.00	88.68	101.54
33 02 1523	**pH Coax Cable (12')**	**EA**				**12.86**	**0.00**	**108.03**	**120.89**
13274 1193	12' pH Coax Cable	EA	8	EELEB	1.000	12.86	0.00	108.03	120.89
33 02 1524	**Submersion Assembly**	**EA**				**12.86**	**0.00**	**236.35**	**249.21**
13274 1194	Submersion Assembly	EA	8	EELEB	1.000	12.86	0.00	236.35	249.21
33 02 1525	**Radiation Detector, Gamma/Alpha/Beta, Daily Rental**	**DAY**				**0.00**	**0.00**	**28.05**	**28.05**
01001 0048	Radiation Detector, Gamma/Alpha/Beta, Daily Rental	DAY	0	N/A	1.000	0.00	0.00	28.05	28.05

33 02 15	Monitoring Instrumentation	Unit	Hourly Output	Crew	Factor	Labor	Equip.	Matl.	Total
33 02 1526	**Radiation Detector, Gamma/Alpha/Beta,Weekly Rental**	**WK**				**0.00**	**0.00**	**67.32**	**67.32**
01001 0049	Radiation Detector, Gamma/Alpha/Beta,Weekly Rental	WK	0	N/A	1.000	0.00	0.00	67.32	67.32

33 02 16	Water/Liquid Laboratory Chemical Analysis	Unit	Hourly Output	Crew	Factor	Labor	Equip.	Matl.	Total
33 02 1601	**Specific Conductance (EPA 120.1), Water Analysis**	**EA**				**0.00**	**0.00**	**11.31**	**11.31**
01954 7122	Specific Conductance (EPA 120.1)	EA	0	N/A	1.000	0.00	0.00	11.31	11.31
33 02 1602	**pH (EPA 150.1), Water Analysis**	**EA**				**0.00**	**0.00**	**6.75**	**6.75**
01954 7617	pH, Electrometric (9045)	EA	0	N/A	1.000	0.00	0.00	6.75	6.75
33 02 1603	**Total Dissolved Solids (EPA 160.1), Water Analysis**	**EA**				**0.00**	**0.00**	**11.80**	**11.80**
01954 7231	Dissolved Solids (EPA 160.1)	EA	0	N/A	1.000	0.00	0.00	11.80	11.80
33 02 1604	**Total Suspended Solids (EPA 160.2), Water Analysis**	**EA**				**0.00**	**0.00**	**10.40**	**10.40**
01954 7232	Suspended Solids (EPA 160.2)	EA	0	N/A	1.000	0.00	0.00	10.40	10.40
33 02 1605	**Total Solids (EPA 160.3), Water Analysis**	**EA**				**0.00**	**0.00**	**11.85**	**11.85**
01954 7233	Total Solids (EPA 160.3)	EA	0	N/A	1.000	0.00	0.00	11.85	11.85
33 02 1606	**Total Volatile Solids (EPA 160.4), Water Analysis**	**EA**				**0.00**	**0.00**	**16.70**	**16.70**
01954 7234	Volatile Solids (EPA 160.4)	EA	0	N/A	1.000	0.00	0.00	16.70	16.70
33 02 1607	**Temperature (EPA 170.1), Water Analysis**	**EA**				**0.00**	**0.00**	**10.95**	**10.95**
01954 7123	Temperature (EPA 170.1)	EA	0	N/A	1.000	0.00	0.00	10.95	10.95
33 02 1608	**Nitrogen/Nitrite/Nitrate (EPA 300.0/SM 4110B, Water Analysis**	**EA**				**0.00**	**0.00**	**22.50**	**22.50**
01954 7221	Nitrogen, Nitrate/Nitrite (EPA 300.0 / SM 4110B)	EA	0	N/A	1.000	0.00	0.00	22.50	22.50
33 02 1609	**Acidity/Alkalinity (EPA 305.1/310.1), Water Analysis**	**EA**				**0.00**	**0.00**	**23.00**	**23.00**
01954 7101	Acidity/Alkalinity (EPA 305.1/310.1)	EA	0	N/A	1.000	0.00	0.00	23.00	23.00
33 02 1610	**Cyanide (EPA 335.2), Water Analysis**	**EA**				**0.00**	**0.00**	**33.50**	**33.50**
01954 7113	Cyanide (EPA 335.1)	EA	0	N/A	1.000	0.00	0.00	33.50	33.50
33 02 1611	**Ammonia Nitrogen (EPA 350.2), Water Analysis**	**EA**				**0.00**	**0.00**	**31.00**	**31.00**
01954 7601	Ammonia Nitrogen (350.2M)	EA	0	N/A	1.000	0.00	0.00	31.00	31.00
33 02 1612	**Phosphorus (EPA 365.3), Water Analysis**	**EA**				**0.00**	**0.00**	**32.00**	**32.00**
01954 7619	Phosphorus, All Forms (EPA 365.3M)	EA	0	N/A	1.000	0.00	0.00	32.00	32.00

33 02 16 Water/Liquid Laboratory Chemical Analysis		Unit	Hourly Output	Crew	Factor	Labor	Equip.	Matl.	Total
33 02 1613	**Oil And Grease (EPA 413.2), Water Analysis**	**EA**				**0.00**	**0.00**	**60.21**	**60.21**
01954 7128	Oil and Grease (EPA 413.2)	EA	0	N/A	1.000	0.00	0.00	60.21	60.21
33 02 1614	**Total Petroleum Hydrocarbons (EPA 418.1), Water Analysis**	**EA**				**0.00**	**0.00**	**54.50**	**54.50**
01954 7625	Total Petroleum Hydrocarbons (TPH)	EA	0	N/A	1.000	0.00	0.00	54.50	54.50
33 02 1615	**Phenols (Method 420.1), Water Analysis**	**EA**				**0.00**	**0.00**	**35.73**	**35.73**
01954 7226	Phenols (EPA 420.1)	EA	0	N/A	1.000	0.00	0.00	35.73	35.73
33 02 1617	**Pesticides/PCBs (EPA 608), Water Analysis**	**EA**				**0.00**	**0.00**	**168.31**	**168.31**
01954 7208	Pesticides and PCBs (608, 8081,8082)	EA	0	N/A	1.000	0.00	0.00	168.31	168.31
33 02 1618	**Volatile Organic Analysis (EPA 624), Water Analysis**	**EA**				**0.00**	**0.00**	**139.00**	**139.00**
01954 7215	Purgeable Organics (624, 8260B)	EA	0	N/A	1.000	0.00	0.00	139.00	139.00
33 02 1619	**Base Neutral & Acid Extractable Organics (EPA 625), Water Analysis**	**EA**				**0.00**	**0.00**	**250.00**	**250.00**
01954 7217	Semivolatile Organics (625, 8270C)	EA	0	N/A	1.000	0.00	0.00	250.00	250.00
33 02 1620	**TAL Metals (EPA 6010/7000s), Water, Water Analysis**	**EA**				**0.00**	**0.00**	**317.10**	**317.10**
01954 7120	TAL Metals (6010/7000s)	EA	0	N/A	1.000	0.00	0.00	317.10	317.10
33 02 1621	**Purgeable Halocarbons (EPA 601), Water Analysis**	**EA**				**0.00**	**0.00**	**108.70**	**108.70**
01954 7214	Purgeable Halocarbons (601, 8021B Halocarbons)	EA	0	N/A	1.000	0.00	0.00	108.70	108.70
33 02 1622	**Purgeable Aromatics (EPA 602), Water Analysis**	**EA**				**0.00**	**0.00**	**81.35**	**81.35**
01954 7213	Purgeable Aromatics (8021B Aromatics)	EA	0	N/A	1.000	0.00	0.00	81.35	81.35
33 02 1623	**Acrolein & Acrylonitrile (EPA 603)**	**EA**				**0.00**	**0.00**	**293.00**	**293.00**
01954 7201	Acrolein/Acrylonitrile (603, 8031)	EA	0	N/A	1.000	0.00	0.00	293.00	293.00
33 02 1624	**Phenols (EPA 604), Water Analysis**	**EA**				**0.00**	**0.00**	**156.00**	**156.00**
01954 7209	Phenols (EPA 604)	EA	0	N/A	1.000	0.00	0.00	156.00	156.00
33 02 1625	**Benzidines (EPA 605)**	**EA**				**0.00**	**0.00**	**365.00**	**365.00**
01954 7202	Benzidines (EPA 605)	EA	0	N/A	1.000	0.00	0.00	365.00	365.00
33 02 1626	**Phthalate Esters (EPA 606)**	**EA**				**0.00**	**0.00**	**172.00**	**172.00**
01954 7211	Phthalate Esters (EPA 606)	EA	0	N/A	1.000	0.00	0.00	172.00	172.00
33 02 1627	**Nitrosamines (EPA 607)**	**EA**				**0.00**	**0.00**	**187.00**	**187.00**
01954 7207	Nitrosoamines (607)	EA	0	N/A	1.000	0.00	0.00	187.00	187.00

33 02 16	Water/Liquid Laboratory Chemical Analysis	Unit	Hourly Output	Crew	Factor	Labor	Equip.	Matl.	Total
33 02 1628	**Nitroaromatics & Isophorone (EPA 609)**	**EA**				**0.00**	**0.00**	**195.00**	**195.00**
01954 7216	Nitroaromatics/Isophorone (609, 8091)	EA	0	N/A	1.000	0.00	0.00	195.00	195.00
33 02 1629	**Polynuclear Aromatic Hydrocarbons, PAH (EPA 610)**	**EA**				**0.00**	**0.00**	**117.81**	**117.81**
01954 7212	Polynuclear Aromatic Hydrocarbons (610, 8100,8310, HPLC)	EA	0	N/A	1.000	0.00	0.00	117.81	117.81
33 02 1630	**Haloethers (EPA 611)**	**EA**				**0.00**	**0.00**	**156.00**	**156.00**
01954 7205	Haloethers (EPA 611)	EA	0	N/A	1.000	0.00	0.00	156.00	156.00
33 02 1631	**Chlorinated Hydrocarbons (EPA 612), Water Analysis**	**EA**				**0.00**	**0.00**	**203.00**	**203.00**
01954 7204	Chlorinated Hydrocarbons (612, 8121)	EA	0	N/A	1.000	0.00	0.00	203.00	203.00
33 02 1632	**2,3,7,8-Tetrachloro-Dibzo-P-Dioxin (EPA 1613A), Water Analysis**	**EA**				**0.00**	**0.00**	**199.78**	**199.78**
01954 7244	2,3,7,8-Tetrachloro-Dibzo-P-Dioxin (1613A)	EA	0	N/A	1.000	0.00	0.00	199.78	199.78
33 02 1633	**Organophosphorus Pesticides (EPA 614)**	**EA**				**0.00**	**0.00**	**168.49**	**168.49**
01954 7206	Organophosphorus Pesticides (614/8141A)	EA	0	N/A	1.000	0.00	0.00	168.49	168.49
33 02 1634	**Chlorinated Phenoxy Acid Herbicide (EPA 615), Water Analysis**	**EA**				**0.00**	**0.00**	**232.82**	**232.82**
01954 7203	Chlorinated Herbicides (615, 8151A)	EA	0	N/A	1.000	0.00	0.00	232.82	232.82
33 02 1635	**Organochlorine Pesticides & PCBs (EPA 617), Water Analysis**	**EA**				**0.00**	**0.00**	**170.77**	**170.77**
01954 7243	Organochlorine Pesticides & PCBs (EPA 617)	EA	0	N/A	1.000	0.00	0.00	170.77	170.77
33 02 1636	**Triazine Pesticides (EPA 619)**	**EA**				**0.00**	**0.00**	**200.69**	**200.69**
01954 7245	Triazine Pesticides (EPA 619)	EA	0	N/A	1.000	0.00	0.00	200.69	200.69
33 02 1637	**Carbamate & Urea Pesticides (EPA 632)**	**EA**				**0.00**	**0.00**	**191.57**	**191.57**
01954 7246	Carbamate & Urea Pesticides (EPA 632)	EA	0	N/A	1.000	0.00	0.00	191.57	191.57
33 02 1638	**Color (EPA 110.2/110.3), Water Analysis**	**EA**				**0.00**	**0.00**	**14.75**	**14.75**
01954 7111	Color (EPA 110.2/110.3)	EA	0	N/A	1.000	0.00	0.00	14.75	14.75
33 02 1639	**Hardness, Total (EPA 130.2), Water Analysis**	**EA**				**0.00**	**0.00**	**17.10**	**17.10**
01954 7118	Hardness, Total (EPA 130.2)	EA	0	N/A	1.000	0.00	0.00	17.10	17.10
33 02 1640	**Odor (EPA 140.1), Water Analysis**	**EA**				**0.00**	**0.00**	**55.91**	**55.91**
01954 7223	Odor (EPA 140.1)	EA	0	N/A	1.000	0.00	0.00	55.91	55.91
33 02 1641	**Settable Solids (EPA 160.5), Water Analysis**	**EA**				**0.00**	**0.00**	**13.20**	**13.20**
01954 7229	Settleable Matter (EPA 160.5)	EA	0	N/A	1.000	0.00	0.00	13.20	13.20

33 02 16	Water/Liquid Laboratory Chemical Analysis	Unit	Hourly Output	Crew	Factor	Labor	Equip.	Matl.	Total
33 02 1642	**Total Solids (EPA 166.3)**	**EA**				**0.00**	**0.00**	**11.85**	**11.85**
01954 7233	Total Solids (EPA 160.3)	EA	0	N/A	1.000	0.00	0.00	11.85	11.85
33 02 1643	**Turbidity (EPA 180.1), Water Analysis**	**EA**				**0.00**	**0.00**	**11.90**	**11.90**
01954 7239	Turbidity (EPA 180.1)	EA	0	N/A	1.000	0.00	0.00	11.90	11.90
33 02 1644	**Metals Screen, ICP Flame, Per Each (EPA 200.7), Water Analysis**	**EA**				**0.00**	**0.00**	**64.11**	**64.11**
01954 7241	Metals (EPA 200.7)	EA	0	N/A	1.000	0.00	0.00	64.11	64.11
33 02 1645	**Metals Screen, ICP Furnace, Per Each (EPA 200.7), Water Analysis**	**EA**				**0.00**	**0.00**	**26.50**	**26.50**
01954 7613	Mercury (EPA 7470/7471)	EA	0	N/A	1.000	0.00	0.00	26.50	26.50
33 02 1646	**Arsenic, Hydride (EPA 206.3), Water Analysis**	**EA**				**0.00**	**0.00**	**32.84**	**32.84**
01954 7134	Arsenic, Hydride (EPA 206.3)	EA	0	N/A	1.000	0.00	0.00	32.84	32.84
33 02 1647	**ICP, Boron (EPA 200.7), Water Analysis**	**EA**				**0.00**	**0.00**	**16.50**	**16.50**
01954 7105	Boron (EPA 200.7)	EA	0	N/A	1.000	0.00	0.00	16.50	16.50
33 02 1648	**Chromium, Hexavalent (SW7196A), Water Analysis**	**EA**				**0.00**	**0.00**	**23.72**	**23.72**
01954 7135	Chromium, Hexavalent (EPA 218.5)	EA	0	N/A	1.000	0.00	0.00	23.72	23.72
33 02 1649	**Mercury, Cold Vapor (EPA 245.1), Water Analysis**	**EA**				**0.00**	**0.00**	**26.50**	**26.50**
01954 7613	Mercury (EPA 7470/7471)	EA	0	N/A	1.000	0.00	0.00	26.50	26.50
33 02 1650	**Ion Chromotography (EPA 300)**	**EA**				**0.00**	**0.00**	**20.07**	**20.07**
01954 7136	Ion Chromotography (EPA 300.0)	EA	0	N/A	1.000	0.00	0.00	20.07	20.07
33 02 1651	**Alkalinity (EPA 310.1/310.2), Water Analysis**	**EA**				**0.00**	**0.00**	**23.00**	**23.00**
01954 7101	Acidity/Alkalinity (EPA 305.1/310.1)	EA	0	N/A	1.000	0.00	0.00	23.00	23.00
33 02 1652	**Bromide, Titrimetric (EPA 300.0), Water Analysis**	**EA**				**0.00**	**0.00**	**25.00**	**25.00**
01954 7603	Bromide, Titrimetric (EPA 300.0/320.1M)	EA	0	N/A	1.000	0.00	0.00	25.00	25.00
33 02 1653	**Chloride (EPA 300), Water Analysis**	**EA**				**0.00**	**0.00**	**16.35**	**16.35**
01954 7108	Chloride (EPA 300.0/325.3)	EA	0	N/A	1.000	0.00	0.00	16.35	16.35
33 02 1654	**Chlorine, Total Residual (EPA 330.2/330.5), Water Analysis**	**EA**				**0.00**	**0.00**	**17.60**	**17.60**
01954 7109	Chlorine, Residual (EPA 330.2/330.5)	EA	0	N/A	1.000	0.00	0.00	17.60	17.60
33 02 1655	**Fluoride SPADNS (EPA 340.1)**	**EA**				**0.00**	**0.00**	**53.00**	**53.00**
01954 7117	Fluoride (with Distillation) (EPA 340.1)	EA	0	N/A	1.000	0.00	0.00	53.00	53.00

33 02 16 Water/Liquid Laboratory Chemical Analysis		Unit	Hourly Output	Crew	Factor	Labor	Equip.	Matl.	Total
33 02 1656	**Fluoride Electrode (EPA 300), Water Analysis**	**EA**				**0.00**	**0.00**	**23.50**	**23.50**
01954 7116	Fluoride (No Distillation) (EPA 300.0)	EA	0	N/A	1.000	0.00	0.00	23.50	23.50
33 02 1657	**Iodide (EPA 345.1), Water Analysis**	**EA**				**0.00**	**0.00**	**33.50**	**33.50**
01954 7119	Iodine (EPA 345.1)	EA	0	N/A	1.000	0.00	0.00	33.50	33.50
33 02 1658	**Nitrogen, Nitrate-Nitrite (EPA 300.0,353.2), Water Analysis**	**EA**				**0.00**	**0.00**	**22.50**	**22.50**
01954 7221	Nitrogen, Nitrate/Nitrite (EPA 300.0 / SM 4110B)	EA	0	N/A	1.000	0.00	0.00	22.50	22.50
33 02 1659	**Nitrogen, Kjeldahl, Total (EPA 351.2/351.4), Water Analysis**	**EA**				**0.00**	**0.00**	**28.50**	**28.50**
01954 7222	Nitrogen, Kjeldahl (EPA 351.2/351.4)	EA	0	N/A	1.000	0.00	0.00	28.50	28.50
33 02 1660	**Nitrogen, Organic (EPA 350.2), Water Analysis**	**EA**				**0.00**	**0.00**	**51.00**	**51.00**
01954 7221	Nitrogen, Nitrate/Nitrite (EPA 300.0 / SM 4110B)	EA	0	N/A	1.000	0.00	0.00	22.50	22.50
01954 7222	Nitrogen, Kjeldahl (EPA 351.2/351.4)	EA	0	N/A	1.000	0.00	0.00	28.50	28.50
33 02 1661	**Nitrogen, Nitrate (EPA 353.2), Water Analysis**	**EA**				**0.00**	**0.00**	**22.50**	**22.50**
01954 7221	Nitrogen, Nitrate/Nitrite (EPA 300.0 / SM 4110B)	EA	0	N/A	1.000	0.00	0.00	22.50	22.50
33 02 1662	**Nitrogen, Nitrate & Nitrite (EPA 353.2), Water Analysis**	**EA**				**0.00**	**0.00**	**51.00**	**51.00**
01954 7221	Nitrogen, Nitrate/Nitrite (EPA 300.0 / SM 4110B)	EA	0	N/A	1.000	0.00	0.00	22.50	22.50
01954 7222	Nitrogen, Kjeldahl (EPA 351.2/351.4)	EA	0	N/A	1.000	0.00	0.00	28.50	28.50
33 02 1663	**Dissolved Oxygen (EPA 360.1), Water Analysis**	**EA**				**0.00**	**0.00**	**13.15**	**13.15**
01954 7114	Dissolved Oxygen (EPA 360.1)	EA	0	N/A	1.000	0.00	0.00	13.15	13.15
33 02 1664	**Phosphorus, Ortho (EPA 365.1), Water Analysis**	**EA**				**0.00**	**0.00**	**26.50**	**26.50**
01954 7227	Phosphorus, Ortho/Hydrolizable (EPA 365.1)	EA	0	N/A	1.000	0.00	0.00	26.50	26.50
33 02 1665	**Phosphorus, Total (EPA 365.2/365.4), Water Analysis**	**EA**				**0.00**	**0.00**	**28.50**	**28.50**
01954 7228	Phosphorus Total (EPA 365.2/365.4)	EA	0	N/A	1.000	0.00	0.00	28.50	28.50
33 02 1666	**Silica, Dissolved (EPA 370.1), Water Analysis**	**EA**				**0.00**	**0.00**	**32.84**	**32.84**
01954 7137	Silica Dissolved (EPA 370.1)	EA	0	N/A	1.000	0.00	0.00	32.84	32.84
33 02 1667	**Sulfate (EPA 300.0), Water Analysis**	**EA**				**0.00**	**0.00**	**17.40**	**17.40**
01954 7623	Sulfates (EPA 300.0)	EA	0	N/A	1.000	0.00	0.00	17.40	17.40
33 02 1668	**Sulfide (EPA 376.1), Water Analysis**	**EA**				**0.00**	**0.00**	**28.00**	**28.00**
01954 7235	Sulfur: Sulfate, Sulfide, Sulfite (EPA 300.0/376.2/377.1)	EA	0	N/A	1.000	0.00	0.00	28.00	28.00

33 02 16	Water/Liquid Laboratory Chemical Analysis	Unit	Hourly Output	Crew	Factor	Labor	Equip.	Matl.	Total
33 02 1669	**Sulfite (EPA 377.1), Water Analysis**	**EA**				**0.00**	**0.00**	**28.00**	**28.00**
01954 7235	Sulfur: Sulfate, Sulfide, Sulfite (EPA 300.0/376.2/377.1)	EA	0	N/A	1.000	0.00	0.00	28.00	28.00
33 02 1670	**Metals Screen, 25 Metals Listed In Method EPA 200.7, Water Analysis**	**EA**				**0.00**	**0.00**	**1,603**	**1,603**
01954 7241	Metals (EPA 200.7)	EA	0	N/A	25.000	0.00	0.00	64.11	1,603
33 02 1671	**Biochemical Oxygen Demand, BOD (EPA 405.1), Water Analysis**	**EA**				**0.00**	**0.00**	**27.00**	**27.00**
01954 7104	Biochemical Oxygen Demand (EPA 405.1)	EA	0	N/A	1.000	0.00	0.00	27.00	27.00
33 02 1672	**Chem Oxy Demand, COD (EPA 410.1/410.2/410.3/410.4), Water Analysis**	**EA**				**0.00**	**0.00**	**19.30**	**19.30**
01954 7107	Chemical Oxygen Demand (COD) (EPA 410.4)	EA	0	N/A	1.000	0.00	0.00	19.30	19.30
33 02 1673	**Total Organic Carbon, TOC (EPA 415.1/415.2), Water Analysis**	**EA**				**0.00**	**0.00**	**24.00**	**24.00**
01954 7237	Total Organic Carbons (EPA 415.1,9060)	EA	0	N/A	1.000	0.00	0.00	24.00	24.00
33 02 1674	**MBAS, Surfactants (EPA 425.1), Water Analysis**	**EA**				**0.00**	**0.00**	**37.50**	**37.50**
01954 7236	Surfactant (MBAS) (EPA 425.1)	EA	0	N/A	1.000	0.00	0.00	37.50	37.50
33 02 1675	**NTA (EPA 430.2)**	**EA**				**0.00**	**0.00**	**27.37**	**27.37**
01954 7138	NTA (EPA 430.2)	EA	0	N/A	1.000	0.00	0.00	27.37	27.37
33 02 1676	**Total Organic Halogens, TOX (EPA 1650), Water Analysis**	**EA**				**0.00**	**0.00**	**99.00**	**99.00**
01954 7238	Total Organic Halogens (Tox) (EPA 9022)	EA	0	N/A	1.000	0.00	0.00	99.00	99.00
33 02 1677	**Trihalomethanes (EPA 501.3), Water Analysis**	**EA**				**0.00**	**0.00**	**114.94**	**114.94**
01954 7248	Trihalomethanes (EPA 501.3)	EA	0	N/A	1.000	0.00	0.00	114.94	114.94
33 02 1678	**Ferrous Iron (Standard Methods 3500 Fe-D)**	**EA**				**0.00**	**0.00**	**85.02**	**85.02**
01001 2167	Ferrous Iron (Standard Methods 3500 Fe-D)	EA	0	N/A	1.000	0.00	0.00	85.02	85.02
33 02 1688	**1,2-Dibromo-3-Chloropropane (EPA 504)**	**EA**				**0.00**	**0.00**	**114.94**	**114.94**
01954 7249	EDB & 1,2-Dibromo-3-Chloropropane (EPA 504)	EA	0	N/A	1.000	0.00	0.00	114.94	114.94
33 02 1689	**Organohalide Pesticides & Aroclors (EPA 505), Water Analysis**	**EA**				**0.00**	**0.00**	**170.77**	**170.77**
01954 7243	Organochlorine Pesticides & PCBs (EPA 617)	EA	0	N/A	1.000	0.00	0.00	170.77	170.77
33 02 1693	**Chlorinated Herbicides (EPA 515.2), Water Analysis**	**EA**				**0.00**	**0.00**	**232.82**	**232.82**
01954 7203	Chlorinated Herbicides (615, 8151A)	EA	0	N/A	1.000	0.00	0.00	232.82	232.82

33 02 16	Water/Liquid Laboratory Chemical Analysis	Unit	Hourly Output	Crew	Factor	Labor	Equip.	Matl.	Total
33 02 1694	**Total Petroleum Hydrocarbons (SW8015B), Water Analysis**	**EA**				**0.00**	**0.00**	**74.81**	**74.81**
01001 0050	Total Petroleum Hydrocarbons (SW8015B), Water Analysis	EA	0	N/A	1.000	0.00	0.00	74.81	74.81

33 02 17	Soil/Sediment Laboratory Chemical Analysis	Unit	Hourly Output	Crew	Factor	Labor	Equip.	Matl.	Total
33 02 1701	**EP Toxicity, Metals (EPA 1310A, TCLP), Soil Analysis**	**EA**				**0.00**	**0.00**	**190.00**	**190.00**
01954 7414	EP Toxicity, Metals (EPA 3010A)	EA	0	N/A	1.000	0.00	0.00	190.00	190.00
33 02 1702	**TCLP (RCRA) (EPA 1311), Soil Analysis**	**EA**				**0.00**	**0.00**	**617.97**	**617.97**
01954 7427	TCLP (RCRA) (EPA 1311)	EA	0	N/A	1.000	0.00	0.00	617.97	617.97
33 02 1703	**Chromium (SW 7191), with prep, Soil Analysis**	**EA**				**0.00**	**0.00**	**31.01**	**31.01**
01954 7132	Chromium (EPA 3005A/7191)	EA	0	N/A	1.000	0.00	0.00	31.01	31.01
33 02 1705	**Targeted TCLP (Metals, Volatiles, Semi-Volatiles only), Soil Analysis**	**EA**				**0.00**	**0.00**	**585.50**	**585.50**
01954 7422	Extraction: Volatiles (EPA 1311)	EA	0	N/A	1.000	0.00	0.00	165.00	165.00
01954 7423	Metals (6010)	EA	0	N/A	1.000	0.00	0.00	95.50	95.50
01954 7425	Semivolatile Organics (8270C)	EA	0	N/A	1.000	0.00	0.00	325.00	325.00
33 02 1706	**Chromium (EPA 6010), Soil Analysis**	**EA**				**0.00**	**0.00**	**14.60**	**14.60**
01954 7634	Chromium (EPA 3050B/6010)	EA	0	N/A	1.000	0.00	0.00	14.60	14.60
33 02 1707	**Target Analyte List (TAL) Metals, Soil Analysis**	**EA**				**0.00**	**0.00**	**342.22**	**342.22**
01001 0051	Target Analyte List (TAL) Metals, Soil Analysis	EA	0	N/A	1.000	0.00	0.00	342.22	342.22
33 02 1708	**Purgeable Aromatics (SW 5035/SW 8021B), Soil Analysis**	**EA**				**0.00**	**0.00**	**81.35**	**81.35**
01954 7213	Purgeable Aromatics (8021B Aromatics)	EA	0	N/A	1.000	0.00	0.00	81.35	81.35
33 02 1709	**TAL Metals (EPA 6010/7000s), Soil Analysis**	**EA**				**0.00**	**0.00**	**317.10**	**317.10**
01954 7120	TAL Metals (6010/7000s)	EA	0	N/A	1.000	0.00	0.00	317.10	317.10
33 02 1710	**Metals (EPA 6010), Per Each Metal, Soil Analysis**	**EA**				**0.00**	**0.00**	**11.45**	**11.45**
01954 7614	Metals (1Cp) (6010)	EA	0	N/A	1.000	0.00	0.00	11.45	11.45
33 02 1711	**Arsenic (SW 7060A), with prep, Soil Analysis**	**EA**				**0.00**	**0.00**	**22.63**	**22.63**
01954 7124	Arsenic (EPA 7060A)	EA	0	N/A	1.000	0.00	0.00	22.63	22.63
33 02 1712	**Mercury (SW 7471), with prep, Soil Analysis**	**EA**				**0.00**	**0.00**	**32.48**	**32.48**
01954 7126	Mercury (EPA 7470/7471)	EA	0	N/A	1.000	0.00	0.00	32.48	32.48
33 02 1713	**Selenium (SW 7740), with prep, Soil Analysis**	**EA**				**0.00**	**0.00**	**25.54**	**25.54**
01954 7127	Selenium (EPA 7740)	EA	0	N/A	1.000	0.00	0.00	25.54	25.54

33 02 17	Soil/Sediment Laboratory Chemical Analysis	Unit	Hourly Output	Crew	Factor	Labor	Equip.	Matl.	Total
33 02 1714	**BTEX Purgeable Aromatics (SW 5035/SW 8021B Aromatics), Soil Analysis**	**EA**				**0.00**	**0.00**	**73.00**	**73.00**
01954 7602	BTEX: Benzene, Toulene, Ethylbenzene, Xylene (8061A,8021B)	EA	0	N/A	1.000	0.00	0.00	73.00	73.00
33 02 1715	**BTEX/Gasoline Hydrocarbons (Mod. 8021B)(PID/FID), w/prep, Soil Analysis**	**EA**				**0.00**	**0.00**	**69.00**	**69.00**
01954 7636	BTEX/Gasoline Hydrocarbons EPA 8021B, (PID/FID)	EA	0	N/A	1.000	0.00	0.00	69.00	69.00
33 02 1716	**Phenols (SW 3550B/SW 8041), Soil Analysis**	**EA**				**0.00**	**0.00**	**82.46**	**82.46**
01954 7131	Phenols (EPA 8040)	EA	0	N/A	1.000	0.00	0.00	82.46	82.46
33 02 1717	**Pesticides/PCBs (SW 3550B/SW 8081/8082), Soil Analysis**	**EA**				**0.00**	**0.00**	**168.31**	**168.31**
01954 7208	Pesticides and PCBs (608, 8081,8082)	EA	0	N/A	1.000	0.00	0.00	168.31	168.31
33 02 1718	**Chlordane (SW 3550B/SW 8081), Soil Analysis**	**EA**				**0.00**	**0.00**	**95.50**	**95.50**
01954 7632	Chlordane (EPA 8081)	EA	0	N/A	1.000	0.00	0.00	95.50	95.50
33 02 1719	**Chlorinated Phenoxy Acid Herbicides (SW 3550B/SW 8151A), Soil Analysis**	**EA**				**0.00**	**0.00**	**223.00**	**223.00**
01954 7633	Chlorinated Phenoxy Acid Herbicides (EPA 8151A)	EA	0	N/A	1.000	0.00	0.00	223.00	223.00
33 02 1720	**Volatile Organic Analysis (SW 5035/SW 8260B), Soil Analysis**	**EA**				**0.00**	**0.00**	**139.00**	**139.00**
01954 7215	Purgeable Organics (624, 8260B)	EA	0	N/A	1.000	0.00	0.00	139.00	139.00
33 02 1721	**Base/Neutral & Acid Extractable Organics(SW3550B/SW8270C), Soil Analysis**	**EA**				**0.00**	**0.00**	**250.00**	**250.00**
01954 7217	Semivolatile Organics (625, 8270C)	EA	0	N/A	1.000	0.00	0.00	250.00	250.00
33 02 1722	**Polynuclear Aromatic Hydrocarbons(PAH) (SW 8310),w/prep, Soil Analysis**	**EA**				**0.00**	**0.00**	**117.81**	**117.81**
01954 7212	Polynuclear Aromatic Hydrocarbons (610, 8100,8310, HPLC)	EA	0	N/A	1.000	0.00	0.00	117.81	117.81
33 02 1723	**Cyanide (EPA 9010B), Soil Analysis**	**EA**				**0.00**	**0.00**	**59.63**	**59.63**
01954 7607	Cyanide, Total, (9010B) Spectrophotometric	EA	0	N/A	1.000	0.00	0.00	59.63	59.63
33 02 1724	**Phenols (EPA 9065), Soil Analysis**	**EA**				**0.00**	**0.00**	**34.00**	**34.00**
01954 7618	Phenols (9066) Total Recoverable	EA	0	N/A	1.000	0.00	0.00	34.00	34.00
33 02 1725	**CEC (EPA 9081), Soil Analysis**	**EA**				**0.00**	**0.00**	**69.50**	**69.50**
01954 7604	Cation Exchange Capacity (CEC) (9080, 9081)	EA	0	N/A	1.000	0.00	0.00	69.50	69.50
33 02 1726	**Diesel Hydrocarbon, Mod 8015B, GC/FID, with prep, Soil Analysis**	**EA**				**0.00**	**0.00**	**129.00**	**129.00**
01954 7637	Diesel Hydrocarbon (EPA 8015B) GC/FID	EA	0	N/A	1.000	0.00	0.00	129.00	129.00

33 02 17	Soil/Sediment Laboratory Chemical Analysis	Unit	Hourly Output	Crew	Factor	Labor	Equip.	Matl.	Total
33 02 1728	**Gasoline Hydrocarbon, Mod 8015B, GC/FID, with prep, Soil Analysis**	**EA**				**0.00**	**0.00**	**64.00**	**64.00**
01954 7638	Gasoline Hydrocarbon (EPA 8015B) GC/FID	EA	0	N/A	1.000	0.00	0.00	64.00	64.00
33 02 1729	**Hydrocarbon Screen GC/FID, Mod 8015B, with prep, Soil Analysis**	**EA**				**0.00**	**0.00**	**223.63**	**223.63**
01954 7635	Hydrocarbon Screen (EPA 8015B) GC/FID	EA	0	N/A	1.000	0.00	0.00	223.63	223.63
33 02 1730	**Sludge Constituents And Characteristics Analysis Tests, Soil Analysis**	**EA**				**0.00**	**0.00**	**134.15**	**134.15**
01954 7221	Nitrogen, Nitrate/Nitrite (EPA 300.0 / SM 4110B)	EA	0	N/A	1.000	0.00	0.00	22.50	22.50
01954 7222	Nitrogen, Kjeldahl (EPA 351.2/351.4)	EA	0	N/A	1.000	0.00	0.00	28.50	28.50
01954 7225	pH (EPA 150.1)	EA	0	N/A	1.000	0.00	0.00	5.45	5.45
01954 7234	Volatile Solids (EPA 160.4)	EA	0	N/A	1.000	0.00	0.00	16.70	16.70
01954 7619	Phosphorus, All Forms (EPA 365.3M)	EA	0	N/A	1.000	0.00	0.00	32.00	32.00
01954 7631	Ammonia (EPA 350.2/350.3)	EA	0	N/A	1.000	0.00	0.00	29.00	29.00
33 02 1731	**Purgeable Halocarbons (SW 5035/SW 8021B Halocarbons), Soil Analysis**	**EA**				**0.00**	**0.00**	**108.70**	**108.70**
01954 7214	Purgeable Halocarbons (601, 8021B Halocarbons)	EA	0	N/A	1.000	0.00	0.00	108.70	108.70
33 02 1732	**Total Petroleum Hydrocarbons (SW 5030/SW 8015B), Soil Analysis**	**EA**				**0.00**	**0.00**	**54.50**	**54.50**
01954 7625	Total Petroleum Hydrocarbons (TPH)	EA	0	N/A	1.000	0.00	0.00	54.50	54.50
33 02 1733	**Acrolein, Acrylonitrile, Acetonitrile (SW 5030/SW 8030A), Soil Analysis**	**EA**				**0.00**	**0.00**	**293.00**	**293.00**
01954 7201	Acrolein/Acrylonitrile (603, 8031)	EA	0	N/A	1.000	0.00	0.00	293.00	293.00
33 02 1734	**Phthalate Esters (SW 3550/SW 8060/SW 8061A), Soil Analysis**	**EA**				**0.00**	**0.00**	**172.00**	**172.00**
01954 7211	Phthalate Esters (EPA 606)	EA	0	N/A	1.000	0.00	0.00	172.00	172.00
33 02 1735	**Nitroaromatics & Cyclic Ketones(SW 3550/SW 8090/SW8091), Soil Analysis**	**EA**				**0.00**	**0.00**	**195.00**	**195.00**
01954 7216	Nitroaromatics/Isophorone (609, 8091)	EA	0	N/A	1.000	0.00	0.00	195.00	195.00
33 02 1736	**Polynuclear Aromatic Hydrocarbons (SW 3550/SW 8100), Soil Analysis**	**EA**				**0.00**	**0.00**	**117.81**	**117.81**
01954 7212	Polynuclear Aromatic Hydrocarbons (610, 8100,8310, HPLC)	EA	0	N/A	1.000	0.00	0.00	117.81	117.81
33 02 1737	**Chlorinated Hydrocarbons (SW 3550/SW 8121), Soil Analysis**	**EA**				**0.00**	**0.00**	**203.00**	**203.00**
01954 7204	Chlorinated Hydrocarbons (612, 8121)	EA	0	N/A	1.000	0.00	0.00	203.00	203.00
33 02 1738	**Organophosphorus Pesticides (SW 3550/SW 8141), Soil Analysis**	**EA**				**0.00**	**0.00**	**168.49**	**168.49**
01954 7206	Organophosphorus Pesticides (614/8141A)	EA	0	N/A	1.000	0.00	0.00	168.49	168.49

33 02 17	Soil/Sediment Laboratory Chemical Analysis	Unit	Hourly Output	Crew	Factor	Labor	Equip.	Matl.	Total
33 02 1739	**Semi-Volatile Organics, GC/MS (SW 8270C), with prep, Soil Analysis**	**EA**				**0.00**	**0.00**	**295.56**	**295.56**
01954 7247	Semivolatile Organics, GC/MS (8270C)	EA	0	N/A	1.000	0.00	0.00	295.56	295.56
33 02 1740	**Dioxins & Dibenzofurans (SW 3550/SW 8280B), Soil Analysis**	**EA**				**0.00**	**0.00**	**207.99**	**207.99**
01954 7242	Dioxins & Dibenzofurans (8280B)	EA	0	N/A	1.000	0.00	0.00	207.99	207.99
33 02 1741	**Total Organic Halides, TOX (EPA 9020B/9021), Soil Analysis**	**EA**				**0.00**	**0.00**	**99.00**	**99.00**
01954 7238	Total Organic Halogens (Tox) (EPA 9022)	EA	0	N/A	1.000	0.00	0.00	99.00	99.00
33 02 1742	**Total Dissolved Sulfide (EPA 9030B), Soil Analysis**	**EA**				**0.00**	**0.00**	**17.40**	**17.40**
01954 7623	Sulfates (EPA 300.0)	EA	0	N/A	1.000	0.00	0.00	17.40	17.40
33 02 1743	**Sulfate (EPA 9035/9036/9038), Soil Analysis**	**EA**				**0.00**	**0.00**	**28.00**	**28.00**
01954 7235	Sulfur: Sulfate, Sulfide, Sulfite (EPA 300.0/376.2/377.1)	EA	0	N/A	1.000	0.00	0.00	28.00	28.00
33 02 1744	**pH (EPA 9040/9041/9045), Soil Analysis**	**EA**				**0.00**	**0.00**	**6.75**	**6.75**
01954 7617	pH, Electrometric (9045)	EA	0	N/A	1.000	0.00	0.00	6.75	6.75
33 02 1745	**Specific Conductance, Conductivity (EPA 9050A), Soil Analysis**	**EA**				**0.00**	**0.00**	**11.15**	**11.15**
01954 7606	Conductivity (9050A) Specific Conductance	EA	0	N/A	1.000	0.00	0.00	11.15	11.15
33 02 1746	**Total Organic Carbon, TOC (EPA 9060), Soil Analysis**	**EA**				**0.00**	**0.00**	**25.50**	**25.50**
01954 7624	Total Organic Carbon (EPA 415.1)	EA	0	N/A	1.000	0.00	0.00	25.50	25.50
33 02 1747	**Total Recoverable Oil & Grease (EPA 9070), Soil Analysis**	**EA**				**0.00**	**0.00**	**53.00**	**53.00**
01954 7224	Oil and Grease (EPA 413.1)	EA	0	N/A	1.000	0.00	0.00	53.00	53.00
33 02 1748	**Oil & Grease Extraction Method (EPA 9071A), Soil Analysis**	**EA**				**0.00**	**0.00**	**44.00**	**44.00**
01954 7616	Oil and Grease (9071A) Total Recoverable	EA	0	N/A	1.000	0.00	0.00	44.00	44.00
33 02 1749	**Waste/Membrane Liner Compatibility Test (EPA 9090A), Soil Analysis**	**EA**				**0.00**	**0.00**	**114.00**	**114.00**
01954 7534	Compatibility Test for Wastes/Membrane Liners (9090A)	EA	0	N/A	1.000	0.00	0.00	114.00	114.00
33 02 1750	**Paint Filter Test (EPA 9095), Soil Analysis**	**EA**				**0.00**	**0.00**	**23.50**	**23.50**
01954 7535	Paint Filter Liquids Test (9095A)	EA	0	N/A	1.000	0.00	0.00	23.50	23.50
33 02 1751	**Sat Hydr Conduct/Leachate/Intrinsic Perm (EPA 9100), Soil Analysis**	**EA**				**0.00**	**0.00**	**355.00**	**355.00**
01954 7536	Saturated Hydraulic Conductivity, Leachate, Permeability (9100)	EA	0	N/A	1.000	0.00	0.00	355.00	355.00

33 02 17	Soil/Sediment Laboratory Chemical Analysis	Unit	Hourly Output	Crew	Factor	Labor	Equip.	Matl.	Total
33 02 1752	**Total Coliform (EPA 9131/9132), Soil Analysis**	**EA**				**0.00**	**0.00**	**27.50**	**27.50**
01954 7115	Fecal Coliform (SM9222D)	EA	0	N/A	1.000	0.00	0.00	27.50	27.50
33 02 1753	**Nitrate (EPA 9210), Soil Analysis**	**EA**				**0.00**	**0.00**	**25.50**	**25.50**
01954 7615	Nitrogen, Nitrate/Nitrite (EPA 300.0 / SM 4110B)	EA	0	N/A	1.000	0.00	0.00	25.50	25.50
33 02 1754	**Chloride (EPA 9250/9251/9253), Soil Analysis**	**EA**				**0.00**	**0.00**	**20.00**	**20.00**
01954 7605	Chloride, Titrimetric (9253)	EA	0	N/A	1.000	0.00	0.00	20.00	20.00
33 02 1756	**Ignitability (EPA 1010/1020), Soil Analysis**	**EA**				**0.00**	**0.00**	**27.00**	**27.00**
01954 7433	Ignitability (EPA 1010)	EA	0	N/A	1.000	0.00	0.00	27.00	27.00
33 02 1757	**Corrosivity (EPA 1110), Soil Analysis**	**EA**				**0.00**	**0.00**	**51.81**	**51.81**
01954 7432	Corrosivity (1110, NACE)	EA	0	N/A	1.000	0.00	0.00	51.81	51.81
33 02 1758	**Reactivity, SW846 Ch7 P7.4, Soil Analysis**	**EA**				**0.00**	**0.00**	**76.00**	**76.00**
01954 7434	Reactivity (Cyanide/Sulfide)	EA	0	N/A	1.000	0.00	0.00	76.00	76.00
33 02 1759	**Sorbent Cartridges From VOC Sampling (EPA 5041A), Soil Analysis**	**EA**				**0.00**	**0.00**	**158.00**	**158.00**
01954 7537	Sorbent Cartridges from VOC Sampling (5041A)	EA	0	N/A	1.000	0.00	0.00	158.00	158.00
33 02 1760	**Metals, Flame, 7000s, Per Each, Soil Analysis**	**EA**				**0.00**	**0.00**	**11.45**	**11.45**
01954 7614	Metals (1Cp) (6010)	EA	0	N/A	1.000	0.00	0.00	11.45	11.45
33 02 1761	**Metals, Furnace, 7000s, Per Each, Soil Analysis**	**EA**				**0.00**	**0.00**	**26.50**	**26.50**
01954 7613	Mercury (EPA 7470/7471)	EA	0	N/A	1.000	0.00	0.00	26.50	26.50
33 02 1762	**Chromium, Hexavalent (SW 7195), with prep, Soil Analysis**	**EA**				**0.00**	**0.00**	**27.50**	**27.50**
01954 7639	Chromium Hexavalent (EPA 7195/7196A/7197/7198)	EA	0	N/A	1.000	0.00	0.00	27.50	27.50
33 02 1764	**Selenium, Hydride (SW 7741A), with prep, Soil Analysis**	**EA**				**0.00**	**0.00**	**27.50**	**27.50**
01954 7641	Selenium, Hydride (EPA 7741A)	EA	0	N/A	1.000	0.00	0.00	27.50	27.50
33 02 1765	**Arsenic, Hydride (SW 7061A), with prep, Soil Analysis**	**EA**				**0.00**	**0.00**	**23.00**	**23.00**
01954 7642	Arsenic, Hydride (EPA 7061A)	EA	0	N/A	1.000	0.00	0.00	23.00	23.00
33 02 1766	**Total Petroleum Hydrocarbons (EPA 9074), Soil Analysis**	**EA**				**0.00**	**0.00**	**54.50**	**54.50**
01954 7625	Total Petroleum Hydrocarbons (TPH)	EA	0	N/A	1.000	0.00	0.00	54.50	54.50
33 02 1767	**Chromium, Total (SW 7190) with prep, Soil Analysis**	**EA**				**0.00**	**0.00**	**19.50**	**19.50**
01954 7650	Chromium, Total (SW 7190) with Prep	EA	0	N/A	1.000	0.00	0.00	19.50	19.50

33 02 17	Soil/Sediment Laboratory Chemical Analysis	Unit	Hourly Output	Crew	Factor	Labor	Equip.	Matl.	Total
33 02 1768	**Antimony (SW 7040) with prep, Soil Analysis**	**EA**				**0.00**	**0.00**	**19.55**	**19.55**
01954 7651	Antimony (SW 7040) with Prep	EA	0	N/A	1.000	0.00	0.00	19.55	19.55
33 02 1769	**Antimony (SW 7041) with prep, Soil Analysis**	**EA**				**0.00**	**0.00**	**20.50**	**20.50**
01954 7652	Antimony (SW 7041) with Prep	EA	0	N/A	1.000	0.00	0.00	20.50	20.50
33 02 1770	**Thallium (SW 7840) with prep, Soil Analysis**	**EA**				**0.00**	**0.00**	**20.50**	**20.50**
01954 7653	Thallium (SW 7840) with Prep	EA	0	N/A	1.000	0.00	0.00	20.50	20.50
33 02 1771	**Thallium (SW 7841) with prep, Soil Analysis**	**EA**				**0.00**	**0.00**	**35.50**	**35.50**
01954 7654	Thallium (SW 7841) with Prep	EA	0	N/A	1.000	0.00	0.00	35.50	35.50
33 02 1772	**Gasoline Group (8021B, Lead, EDB), Soil Analysis**	**EA**				**0.00**	**0.00**	**226.43**	**226.43**
01954 7656	Gasoline Group (EPA 8021B, Lead, EDB)	EA	0	N/A	1.000	0.00	0.00	226.43	226.43
33 02 1773	**Kerosene Group (8021B, 8100, TPH, Lead, EDB), Soil Analysis**	**EA**				**0.00**	**0.00**	**365.00**	**365.00**
01954 7657	Kerosene Group (EPA 8021B, 8100, TPH, Lead, EDB)	EA	0	N/A	1.000	0.00	0.00	365.00	365.00
33 02 1774	**BTEX/MTBE (8021B), Soil Analysis**	**EA**				**0.00**	**0.00**	**65.00**	**65.00**
01954 7658	BTEX/MTBE (EPA 8021B)	EA	0	N/A	1.000	0.00	0.00	65.00	65.00
33 02 1775	**Purgeable Organics (SW 3550/SW 8260B), Soil Analysis**	**EA**				**0.00**	**0.00**	**139.00**	**139.00**
01954 7215	Purgeable Organics (624, 8260B)	EA	0	N/A	1.000	0.00	0.00	139.00	139.00
33 02 1776	**BTEX/MTBE/TVPH (EPA 8021B/8015B), Soil Analysis**	**EA**				**0.00**	**0.00**	**74.81**	**74.81**
01001 0052	BTEX/MTBE/TVPH (EPA 8021B/8015B), Soil Analysis	EA	0	N/A	1.000	0.00	0.00	74.81	74.81
33 02 1777	**Total Petroleum Hydrocarbons (EPA 418.1), Soil Analysis**	**EA**				**0.00**	**0.00**	**117.81**	**117.81**
01001 0053	Total Petroleum Hydrocarbons (EPA 418.1), Soil Analysis	EA	0	N/A	1.000	0.00	0.00	117.81	117.81
33 02 1778	**Screening Test Method for PCB in Soil (9078)**	**EA**				**0.00**	**0.00**	**10.63**	**10.63**
01001 2170	Screening Test Method for PCB in Soil (9078)	EA	0	N/A	1.000	0.00	0.00	10.63	10.63

33 02 18	Air Laboratory Chemical Analysis	Unit	Hourly Output	Crew	Factor	Labor	Equip.	Matl.	Total
33 02 1801	**VOST Analysis, Air (30/5041/8260B), Air Analysis**	**EA**				**0.00**	**0.00**	**294.00**	**294.00**
01954 8401	VOST Analysis 30/5041/8260B	EA	0	N/A	1.000	0.00	0.00	294.00	294.00
33 02 1802	**Prncpl Organ Haz Constituents, Air (30/5041/8260B)**	**EA**				**0.00**	**0.00**	**259.00**	**259.00**
01954 8402	Principal Organic Hazardous Constituents 30/5041/8260B	EA	0	N/A	1.000	0.00	0.00	259.00	259.00

33 02 18	Air Laboratory Chemical Analysis	Unit	Hourly Output	Crew	Factor	Labor	Equip.	Matl.	Total
33 02 1803	**Tentative ID Compunds, GC/MS, Air (30/5041/8260B - TO-14), Air Analysis**	**EA**				**0.00**	**0.00**	**113.00**	**113.00**
01954 8403	Tentative ID of Compounds GC/MS 30/5041/8260B	EA	0	N/A	1.000	0.00	0.00	113.00	113.00
33 02 1804	**Tedlar Bag (Single Component), Air (30/5041/8260B), Air Analysis**	**EA**				**0.00**	**0.00**	**89.00**	**89.00**
01954 8404	Tedlar Bag (Single Component) 30/5041/8260B	EA	0	N/A	1.000	0.00	0.00	89.00	89.00
33 02 1805	**Sorbent Tube (Single Component), Air (5041/8260B), Air Analysis**	**EA**				**0.00**	**0.00**	**105.00**	**105.00**
01954 8405	Sorbent Tube (Single Component) 5041/8021B	EA	0	N/A	1.000	0.00	0.00	105.00	105.00
33 02 1806	**BTEX, Air (TO-14), Air Analysis**	**EA**				**0.00**	**0.00**	**60.00**	**60.00**
01954 8406	BTEX 5041/8021B	EA	0	N/A	1.000	0.00	0.00	60.00	60.00
33 02 1807	**BTEX/Total Volat Petro Hydrocarbs, Air (5041/8021B/8015B)**	**EA**				**0.00**	**0.00**	**84.50**	**84.50**
01954 8407	BTEX/Total Volatile Petroleum Hydrocarbons 5041/8021B/8015B	EA	0	N/A	1.000	0.00	0.00	84.50	84.50
33 02 1808	**Semivolatiles, Air (TO-13), Air Analysis**	**EA**				**0.00**	**0.00**	**440.00**	**440.00**
01954 8408	Semivolatiles 10/3550/8270C	EA	0	N/A	1.000	0.00	0.00	440.00	440.00
33 02 1809	**Multi-Metal Train, Air, Various Methods**	**EA**				**0.00**	**0.00**	**249.00**	**249.00**
01954 8409	Multimetal Train, Various Methods	EA	0	N/A	1.000	0.00	0.00	249.00	249.00
33 02 1810	**Pesticides/PCBs, GC, Air (TO-4), Air Analysis**	**EA**				**0.00**	**0.00**	**241.00**	**241.00**
01954 8411	Pesticides/PCBs, Various Methods	EA	0	N/A	1.000	0.00	0.00	241.00	241.00
33 02 1811	**Hydrocarbons, BTEX, Air (NIOSH 1500), Air Analysis**	**EA**				**0.00**	**0.00**	**90.00**	**90.00**
01954 8412	Hydrocarbons (BTEX) NIOSH 1500	EA	0	N/A	1.000	0.00	0.00	90.00	90.00
33 02 1812	**Polyaromatic Hydrocarbons, Air (NIOSH 5515), Air Analysis**	**EA**				**0.00**	**0.00**	**330.00**	**330.00**
01954 8413	Polyaromatic Hydrocarbons NIOSH 5515	EA	0	N/A	1.000	0.00	0.00	330.00	330.00
33 02 1813	**Metals by ICP, Air (NIOSH 7300)**	**EA**				**0.00**	**0.00**	**142.01**	**142.01**
01954 8414	Metals (by ICP) NIOSH 7300	EA	0	N/A	1.000	0.00	0.00	142.01	142.01
33 02 1814	**Metals by Graphite Furnace, Air (NIOSH 7300)**	**EA**				**0.00**	**0.00**	**279.17**	**279.17**
01954 8415	Metals (by Graphite Furnace) NIOSH 7300	EA	0	N/A	1.000	0.00	0.00	279.17	279.17
33 02 1815	**Mercury, Air (245.2)**	**EA**				**0.00**	**0.00**	**30.50**	**30.50**
01954 8416	Mercury (245.2)	EA	0	N/A	1.000	0.00	0.00	30.50	30.50
33 02 1816	**Cyanide, Air (335.2)**	**EA**				**0.00**	**0.00**	**37.00**	**37.00**
01954 8417	Cyanide (335.2)	EA	0	N/A	1.000	0.00	0.00	37.00	37.00

33 02 18 Air Laboratory Chemical Analysis	Unit	Hourly Output	Crew	Factor	Labor	Equip.	Matl.	Total
33 02 1817 Ion Chromatography, Air	**EA**				**0.00**	**0.00**	**17.75**	**17.75**
01954 8418 Ion Chromatography, per Each	EA	0	N/A	1.000	0.00	0.00	17.75	17.75
33 02 1818 Volatile Aromatic Halocarbons, Air, TO-1	**EA**				**0.00**	**0.00**	**116.00**	**116.00**
01954 8419 Volatile Aromatic Halocarbons TO-1	EA	0	N/A	1.000	0.00	0.00	116.00	116.00
33 02 1819 Highly Volatile Non-Polar Organics, TO-2	**EA**				**0.00**	**0.00**	**345.00**	**345.00**
01954 8421 Highly Volatile Nonpolar Organics TO-2	EA	0	N/A	1.000	0.00	0.00	345.00	345.00
33 02 1820 Volatile Non-Polar Organics, TO-3	**EA**				**0.00**	**0.00**	**122.00**	**122.00**
01954 8422 Volatile Nonpolar Organics TO-3	EA	0	N/A	1.000	0.00	0.00	122.00	122.00
33 02 1821 Organochlorine Pesticides/PCBs, Air, TO-4	**EA**				**0.00**	**0.00**	**253.00**	**253.00**
01954 8423 Organochlorine Pesticides/PCBs TO-4	EA	0	N/A	1.000	0.00	0.00	253.00	253.00
33 02 1822 Aldehydes & Ketones, Air, TO-5	**EA**				**0.00**	**0.00**	**44.73**	**44.73**
01954 8424 Aldehydes & Ketones TO-5	EA	0	N/A	1.000	0.00	0.00	44.73	44.73
33 02 1823 Phosgene in Ambient Air, TO-6	**EA**				**0.00**	**0.00**	**224.00**	**224.00**
01954 8425 Phosgene in Ambient Air TO-6	EA	0	N/A	1.000	0.00	0.00	224.00	224.00
33 02 1824 N-Nitrosodimethylamine, Air, TO-7	**EA**				**0.00**	**0.00**	**194.00**	**194.00**
01954 8426 N-Nitrosodimethylamine TO-7	EA	0	N/A	1.000	0.00	0.00	194.00	194.00
33 02 1825 Phenol & Methyl Phenols, Air, TO-8	**EA**				**0.00**	**0.00**	**71.00**	**71.00**
01954 8427 Phenol and Methyl Phenols TO-8	EA	0	N/A	1.000	0.00	0.00	71.00	71.00
33 02 1826 Dioxins, Air, TO-9	**EA**				**0.00**	**0.00**	**555.00**	**555.00**
01954 8428 Dioxins TO-9	EA	0	N/A	1.000	0.00	0.00	555.00	555.00
33 02 1827 Total Non-Methane Organics, Air, TO-12	**EA**				**0.00**	**0.00**	**40.50**	**40.50**
01954 8429 Total Non-methane Organics TO-12	EA	0	N/A	1.000	0.00	0.00	40.50	40.50
33 02 1828 Polyaromatic Hydrocarbons, Air, TO-13	**EA**				**0.00**	**0.00**	**478.94**	**478.94**
01954 8431 Polyaromatic Hydrocarbons TO-13	EA	0	N/A	1.000	0.00	0.00	478.94	478.94
33 02 1829 Canister Samples by GC or GC/MS, TO-14	**EA**				**0.00**	**0.00**	**450.00**	**450.00**
01954 8432 Summa Canisters by GC/MS TO-14	EA	0	N/A	1.000	0.00	0.00	450.00	450.00
33 02 1830 Hydrocarbon Speciation, C1-C8, GC/FID, Air (TO-12/14)	**EA**				**0.00**	**0.00**	**288.00**	**288.00**
01954 8433 Hydrocarbon Speciation C1-C8, TO-12/14	EA	0	N/A	1.000	0.00	0.00	288.00	288.00
33 02 1831 Hydrocarbon Speciation, C8-C22, GC/FID, Air (TO-12/14)	**EA**				**0.00**	**0.00**	**288.00**	**288.00**
01954 8434 Hydrocarbon Speciation C8-C22, TO-12/14	EA	0	N/A	1.000	0.00	0.00	288.00	288.00

33 02 18	Air Laboratory Chemical Analysis	Unit	Hourly Output	Crew	Factor	Labor	Equip.	Matl.	Total
33 02 1832	**Hydrocarbon Speciation, C1-C22, GC/FID, Air (TO-12/14)**	**EA**				**0.00**	**0.00**	**288.00**	**288.00**
01954 8435	Hydrocarbon Speciation C1-C22, TO-12/14	EA	0	N/A	1.000	0.00	0.00	288.00	288.00
33 02 1833	**Hydrocarbon Speciation, C4-C22, GC/FID, Air (TO-12/14)**	**EA**				**0.00**	**0.00**	**288.00**	**288.00**
01954 8436	Hydrocarbon Speciation C4-C22, TO-12/14	EA	0	N/A	1.000	0.00	0.00	288.00	288.00
33 02 1834	**Volatile Organic Compounds (TO-14)**	**DAY**				**0.00**	**0.00**	**106.84**	**106.84**
01954 8437	Volatile Organic Compounds (TO-14)	EA	0	N/A	1.000	0.00	0.00	106.84	106.84

33 02 19	Microbiological Laboratory Chemical Analysis	Unit	Hourly Output	Crew	Factor	Labor	Equip.	Matl.	Total
33 02 1901	**Freshwater Discharge, LC50, Bioassay Analysis**	**EA**				**0.00**	**0.00**	**637.00**	**637.00**
01954 7301	Daphnia Species 48-Hour, LC50, Definitive	EA	0	N/A	1.000	0.00	0.00	267.00	267.00
01954 7302	Pimephales Promelas 48-Hour, LC50, Definitive	EA	0	N/A	1.000	0.00	0.00	370.00	370.00
33 02 1902	**Saltwater Discharge, Lc50, Bioassay Analysis**	**EA**				**0.00**	**0.00**	**740.00**	**740.00**
01954 7303	Mysidopsis Species 48-Hour, LC50, Definitive	EA	0	N/A	1.000	0.00	0.00	370.00	370.00
01954 7304	Cyprindon Species 48-Hour, LC50, Definitive	EA	0	N/A	1.000	0.00	0.00	370.00	370.00
33 02 1903	**Drilling Mud, Lc50, Bioassay Analysis**	**EA**				**0.00**	**0.00**	**880.00**	**880.00**
01954 7305	Drilling Mud Bioassay, LC50 Mysidopsis Bahia	EA	0	N/A	1.000	0.00	0.00	880.00	880.00
33 02 1904	**Freshwater Chronic Toxicity Bioassay Analysis**	**EA**				**0.00**	**0.00**	**1,530**	**1,530**
01954 7306	Chronic Toxicity Bioassay Ceriodaphnia Dubia	EA	0	N/A	1.000	0.00	0.00	790.00	790.00
01954 7307	Chronic Toxicity Bioassay Pimephales Promelas Larval	EA	0	N/A	1.000	0.00	0.00	740.00	740.00
33 02 1905	**Saltwater Chronic Toxicity Bioassay Analysis**	**EA**				**0.00**	**0.00**	**1,920**	**1,920**
01954 7308	Chronic Toxicity Bioassay Mysidopsis Species	EA	0	N/A	1.000	0.00	0.00	1,025	1,025
01954 7309	Chronic Toxicity Bioassay Cyprinodon Species	EA	0	N/A	1.000	0.00	0.00	895.00	895.00
33 02 1906	**Ames (Full) Microbiological Analysis**	**EA**				**0.00**	**0.00**	**1,900**	**1,900**
01954 7312	Ames (Full)	EA	0	N/A	1.000	0.00	0.00	1,900	1,900
33 02 1907	**Ames (Screen) Microbiological Analysis**	**EA**				**0.00**	**0.00**	**124.00**	**124.00**
01954 7313	Ames (Screen)	EA	0	N/A	1.000	0.00	0.00	124.00	124.00
33 02 1908	**Daphnia IQ Toxicity Analysis**	**EA**				**0.00**	**0.00**	**335.00**	**335.00**
01954 7311	Daphnia IQ	EA	0	N/A	1.000	0.00	0.00	335.00	335.00
33 02 1909	**Microtox Microbiological Analysis**	**EA**				**0.00**	**0.00**	**310.00**	**310.00**
01954 7314	Microtox	EA	0	N/A	1.000	0.00	0.00	310.00	310.00
33 02 1910	**Total Heterotrophic Plate Count**	**EA**				**0.00**	**0.00**	**117.00**	**117.00**
01954 7315	Total Heterotrophic Plate Count	EA	0	N/A	1.000	0.00	0.00	117.00	117.00

33 02 19 Microbiological Laboratory Chemical Analysis		Unit	Hourly Output	Crew	Factor	Labor	Equip.	Matl.	Total
33 02 1911	**Hydrocarbon Degraders**	**EA**				**0.00**	**0.00**	**33.54**	**33.54**
01954 7316	Hydrocarbon Degraders	EA	0	N/A	1.000	0.00	0.00	33.54	33.54
33 02 1912	**Fluorescent Pseudomonas**	**EA**				**0.00**	**0.00**	**42.50**	**42.50**
01954 7317	Fluorescent Pseudomonas	EA	0	N/A	1.000	0.00	0.00	42.50	42.50
33 02 1913	**Laboratory Bench-scale Studies**	**EA**				**0.00**	**0.00**	**167.72**	**167.72**
01954 7318	Laboratory Bench-scale Studies	EA	0	N/A	1.000	0.00	0.00	167.72	167.72

33 02 20 Sample Preparation/Cleanup		Unit	Hourly Output	Crew	Factor	Labor	Equip.	Matl.	Total
33 02 2001	**Sample Preparation (3005A/3010A/3020A/3040A/3050B)**	**EA**				**0.00**	**0.00**	**14.35**	**14.35**
01954 7581	Sample Preparation (3005A/3010A/3020A/3040A/3050B)	EA	0	N/A	1.000	0.00	0.00	14.35	14.35
33 02 2002	**Sample Preparation (3510C/3520C/3540C/3550B/5030B)**	**EA**				**0.00**	**0.00**	**26.50**	**26.50**
01954 7582	Sample Preparation (3510C/3520C/3540C/3550B/5030B)	EA	0	N/A	1.000	0.00	0.00	26.50	26.50
33 02 2003	**Sample Preparation, Waste Dilution (3580A)**	**EA**				**0.00**	**0.00**	**53.11**	**53.11**
01954 7583	Sample Preparation, Waste Dilution (3580A)	EA	0	N/A	1.000	0.00	0.00	53.11	53.11
33 02 2004	**Cleanup (3610B/3611B/3620B/3630C/3650B/3660B)**	**EA**				**0.00**	**0.00**	**35.00**	**35.00**
01954 7584	Cleanup (3610B/3611B/3620B/3630C/3650B/3660B)	EA	0	N/A	1.000	0.00	0.00	35.00	35.00
33 02 2005	**Cleanup, Gel-Permeation (3640)**	**EA**				**0.00**	**0.00**	**100.00**	**100.00**
01954 7585	Cleanup, Gel-Permeation (3640)	EA	0	N/A	1.000	0.00	0.00	100.00	100.00
33 02 2019	**1 Gallon, 128 Oz, Glass Sample Container, Case of 4**	**EA**				**0.00**	**0.00**	**34.00**	**34.00**
01954 6435	1 Gallon, 128 Oz, Glass Sample Container, 4 per Case	EA	0	N/A	1.000	0.00	0.00	34.00	34.00
33 02 2020	**1 Liter, 32 Oz, Clear Wide Mouth Jar, Case of 12**	**EA**				**0.00**	**0.00**	**49.50**	**49.50**
01954 6411	1 Liter (32 Oz) Clear Wide Mouth Jar (Case of 12)	EA	0	N/A	1.000	0.00	0.00	49.50	49.50
33 02 2021	**500 ml, 16 Oz, Clear Wide Mouth Jar, Case of 12**	**EA**				**0.00**	**0.00**	**45.00**	**45.00**
01954 6412	500 ml (16 Oz) Clear Wide Mouth Jar (Case of 12)	EA	0	N/A	1.000	0.00	0.00	45.00	45.00
33 02 2022	**250 ml, 8 Oz, Clear Wide Mouth Jar, Case of 24**	**EA**				**0.00**	**0.00**	**53.50**	**53.50**
01954 6413	250 ml (8 Oz) Clear Wide Mouth Jar (Case of 24)	EA	0	N/A	1.000	0.00	0.00	53.50	53.50

33 02 20 Sample Preparation/Cleanup		Unit	Hourly Output	Crew	Factor	Labor	Equip.	Matl.	Total
33 02 2023	**120 ml, 4 Oz, Clear Wide Mouth Jar, Case of 24**	**EA**				**0.00**	**0.00**	**43.00**	**43.00**
01954 6414	120 ml (4 Oz) Clear Wide Mouth Jar (Case of 24)	EA	0	N/A	1.000	0.00	0.00	43.00	43.00
33 02 2024	**1 Liter, 32 Oz, Boston Round Bottle, Case of 12**	**EA**				**0.00**	**0.00**	**48.50**	**48.50**
01954 6415	1 Liter (32 Oz) Boston Round Bottle (Case of 12)	EA	0	N/A	1.000	0.00	0.00	48.50	48.50
33 02 2025	**500 ml, 16 Oz, Boston Round Bottle, Case of 12**	**EA**				**0.00**	**0.00**	**32.50**	**32.50**
01954 6416	500 ml (16 Oz) Boston Round Bottle (Case of 12)	EA	0	N/A	1.000	0.00	0.00	32.50	32.50
33 02 2026	**40 ml, Clear Vial, Case of 72**	**EA**				**0.00**	**0.00**	**64.00**	**64.00**
01954 6417	40 ml Clear Vials (Case of 72)	EA	0	N/A	1.000	0.00	0.00	64.00	64.00
33 02 2027	**20 ml, Clear Vial, Case of 72**	**EA**				**0.00**	**0.00**	**68.00**	**68.00**
01954 6418	20 ml Clear Vials (Case of 72)	EA	0	N/A	1.000	0.00	0.00	68.00	68.00
33 02 2028	**250 ml, 8 oz, Clear Wide Mouth Vial with Septa, Case of 24**	**EA**				**0.00**	**0.00**	**63.00**	**63.00**
01954 6419	250 ml (8 Oz) Clear Wide Mouth Vial with Septa (Case of 24)	EA	0	N/A	1.000	0.00	0.00	63.00	63.00
33 02 2029	**125 ml, 4 oz, Clear Wide Mouth Vial with Septa, Case of 24**	**EA**				**0.00**	**0.00**	**61.00**	**61.00**
01954 6421	125 ml (4 Oz) Clear Wide Mouth Vial with Septa (Case of 24)	EA	0	N/A	1.000	0.00	0.00	61.00	61.00
33 02 2030	**1 Liter, 32 oz, High-density Polyethylene Bottle, Case of 12**	**EA**				**0.00**	**0.00**	**35.00**	**35.00**
01954 6422	1 Liter (32 0z) High-density Polyethylene Bottle (Case of 12)	EA	0	N/A	1.000	0.00	0.00	35.00	35.00
33 02 2031	**500 ml, 16 oz, High-density Polyethylene Bottle, Case of 24**	**EA**				**0.00**	**0.00**	**43.50**	**43.50**
01954 6423	500 ml (16 Oz) High-density Polyethylene Bottle (Case of 24)	EA	0	N/A	1.000	0.00	0.00	43.50	43.50
33 02 2032	**250 ml, 8 oz, High-density Polyethylene Bottle, Case of 24**	**EA**				**0.00**	**0.00**	**19.60**	**19.60**
01954 6424	250 ml (8 Oz) High-density Polyethylene Bottle (Case of 24)	EA	0	N/A	1.000	0.00	0.00	19.60	19.60
33 02 2033	**125 ml, 4 oz, High-density Polyethylene Bottle, Case of 24**	**EA**				**0.00**	**0.00**	**14.75**	**14.75**
01954 6425	125 ml (4 Oz) High-density Polyethylene Bottle (Case of 24)	EA	0	N/A	1.000	0.00	0.00	14.75	14.75
33 02 2034	**Custody Seals, Package of 10**	**EA**				**0.00**	**0.00**	**1.76**	**1.76**
01954 6426	Custody Seals (Package of 10)	EA	0	N/A	1.000	0.00	0.00	1.76	1.76

33 02 20	Sample Preparation/Cleanup	Unit	Hourly Output	Crew	Factor	Labor	Equip.	Matl.	Total
33 02 2035	**Safe Transport Can Filled with Vermiculite, 1 Gallon, Case of 4**	**EA**				**0.00**	**0.00**	**23.00**	**23.00**
01954 6427	Safe Transport Can Filled with Vermiculite, 1 Gallon, Case of 4	EA	0	N/A	1.000	0.00	0.00	23.00	23.00
33 02 2036	**Documentation Package for QA Verification, Data & Benchwork**	**EA**				**0.00**	**0.00**	**67.00**	**67.00**
01954 6428	Documentation Package for QA Verification (Data & Benchwork)	EA	0	N/A	1.000	0.00	0.00	67.00	67.00
33 02 2037	**Overnight Delivery, 8 oz Letter**	**EA**				**0.00**	**0.00**	**14.54**	**14.54**
01954 6451	8 Oz Letter	EA	0	N/A	1.000	0.00	0.00	14.54	14.54
33 02 2038	**Overnight Delivery, 1 Lb Package**	**LB**				**0.00**	**0.00**	**15.35**	**15.35**
01954 6452	1 Lb Package (Price per Lb)	LB	0	N/A	1.000	0.00	0.00	15.35	15.35
33 02 2039	**Overnight Delivery, 2 - 5 Lb Package**	**LB**				**0.00**	**0.00**	**5.65**	**5.65**
01954 6453	2 Lb to 5 Lb Packages (Price per Lb)	LB	0	N/A	1.000	0.00	0.00	5.65	5.65
33 02 2040	**Overnight Delivery, 6 - 10 Lb Package**	**LB**				**0.00**	**0.00**	**3.42**	**3.42**
01954 6454	6 Lb to 10 Lb Packages (Price per Lb)	LB	0	N/A	1.000	0.00	0.00	3.42	3.42
33 02 2041	**Overnight Delivery, 11 - 20 Lb Package**	**LB**				**0.00**	**0.00**	**3.54**	**3.54**
01954 6455	11 Lb to 20 Lb Packages (Price per Lb)	LB	0	N/A	1.000	0.00	0.00	3.54	3.54
33 02 2042	**Overnight Delivery, 21 - 50 Lb Package**	**LB**				**0.00**	**0.00**	**1.82**	**1.82**
01954 6456	21 Lb to 50 Lb Packages (Price per Lb)	LB	0	N/A	1.000	0.00	0.00	1.82	1.82
33 02 2043	**Overnight Delivery, 51 - 70 Lb Package**	**LB**				**0.00**	**0.00**	**1.60**	**1.60**
01954 6457	51 Lb to 70 Lb Packages (Price per Lb)	LB	0	N/A	1.000	0.00	0.00	1.60	1.60
33 02 2044	**Overnight Delivery, 71 Lbs or Greater**	**LB**				**0.00**	**0.00**	**1.48**	**1.48**
01954 6458	71 Lb or Greater Packages (Price per Lb)	LB	0	N/A	1.000	0.00	0.00	1.48	1.48
33 02 2045	**48 Quart Ice Chest**	**EA**				**0.00**	**0.00**	**34.50**	**34.50**
01954 6471	48 Quart Ice Chest	EA	0	N/A	1.000	0.00	0.00	34.50	34.50
33 02 2046	**60 Quart Ice Chest**	**EA**				**0.00**	**0.00**	**62.00**	**62.00**
01954 6472	60 Quart Ice Chest	EA	0	N/A	1.000	0.00	0.00	62.00	62.00
33 02 2047	**80 Quart Ice Chest**	**EA**				**0.00**	**0.00**	**91.50**	**91.50**
01954 6473	80 Quart Ice Chest	EA	0	N/A	1.000	0.00	0.00	91.50	91.50
33 02 2048	**102 Quart Ice Chest**	**EA**				**0.00**	**0.00**	**120.00**	**120.00**
01954 6474	102 Quart Ice Chest	EA	0	N/A	1.000	0.00	0.00	120.00	120.00
33 02 2049	**120 Quart Ice Chest**	**EA**				**0.00**	**0.00**	**170.00**	**170.00**
01954 6475	120 Quart Ice Chest	EA	0	N/A	1.000	0.00	0.00	170.00	170.00

33 02 20 Sample Preparation/Cleanup		Unit	Hourly Output	Crew	Factor	Labor	Equip.	Matl.	Total
33 02 2050	**Blue Ice Soft Packs (Equivalent to 7 Lbs Ice)**	**EA**				**0.00**	**0.00**	**1.72**	**1.72**
01954 6476	Blue Ice Soft Packs (Equivalent to 4 Lbs Ice)	EA	0	N/A	1.000	0.00	0.00	1.72	1.72
33 02 2060	**Overnight Delivery, 2 - 5 Lb Package (Price per Package)**	**EA**				**0.00**	**0.00**	**18.79**	**18.79**
01001 2100	Overnight Delivery, 2 - 5 Lb Package (Price per Package)	EA	0	N/A	1.000	0.00	0.00	18.79	18.79
33 02 2061	**Overnight Delivery, 6 - 10 Lb Package (Price per Package)**	**EA**				**0.00**	**0.00**	**22.16**	**22.16**
01001 2101	Overnight Delivery, 6 - 10 Lb Package (Price per Package)	EA	0	N/A	1.000	0.00	0.00	22.16	22.16
33 02 2062	**Overnight Delivery, 11 - 20 Lb Package (Price per Package)**	**EA**				**0.00**	**0.00**	**28.89**	**28.89**
01001 2102	Overnight Delivery, 11 - 20 Lb Package (Price per Package)	EA	0	N/A	1.000	0.00	0.00	28.89	28.89
33 02 2063	**Overnight Delivery, 21 - 50 Lb Package (Price per Package)**	**EA**				**0.00**	**0.00**	**43.48**	**43.48**
01001 2103	Overnight Delivery, 21 - 50 Lb Package (Price per Package)	EA	0	N/A	1.000	0.00	0.00	43.48	43.48
33 02 2064	**Overnight Delivery, 51 - 75 Lb Package (Price per Package)**	**EA**				**0.00**	**0.00**	**69.57**	**69.57**
01001 2104	Overnight Delivery, 51 - 75 Lb Package (Price per Package)	EA	0	N/A	1.000	0.00	0.00	69.57	69.57
33 02 2065	**Overnight Delivery, 76 - 100 Lb Package (Price per Package)**	**EA**				**0.00**	**0.00**	**104.63**	**104.63**
01001 2105	Overnight Delivery, 76 - 100 Lb Package (Price per Package)	EA	0	N/A	1.000	0.00	0.00	104.63	104.63
33 02 2066	**Overnight Delivery, 101 - 125 Lb Package (Price per Package)**	**EA**				**0.00**	**0.00**	**135.67**	**135.67**
01001 2106	Overnight Delivery, 101 - 125 Lb Package (Price per Package)	EA	0	N/A	1.000	0.00	0.00	135.67	135.67
33 02 2067	**Overnight Delivery, 126 - 150 Lb Package (Price per Package)**	**EA**				**0.00**	**0.00**	**165.68**	**165.68**
01001 2107	Overnight Delivery, 126 - 150 Lb Package (Price per Package)	EA	0	N/A	1.000	0.00	0.00	165.68	165.68
33 02 2068	**Overnight Delivery, 151 Lbs or greater (Price per Lb)**	**LB**				**0.00**	**0.00**	**1.25**	**1.25**
01001 2108	Overnight Delivery, 151 Lbs or greater (Price per Lb)	LB	0	N/A	1.000	0.00	0.00	1.25	1.25

33 02 21 Water/Liquid Laboratory Chemical Analysis (continued)		Unit	Hourly Output	Crew	Factor	Labor	Equip.	Matl.	Total
33 02 2101	**Color (EPA 110.2/110.3)**	**EA**				**0.00**	**0.00**	**14.75**	**14.75**
01954 7111	Color (EPA 110.2/110.3)	EA	0	N/A	1.000	0.00	0.00	14.75	14.75

33 02 21	Water/Liquid Laboratory Chemical Analysis (continued)	Unit	Hourly Output	Crew	Factor	Labor	Equip.	Matl.	Total
33 02 2102	**Total Solids (EPA 160.3)**	**EA**				**0.00**	**0.00**	**11.85**	**11.85**
01954 7233	Total Solids (EPA 160.3)	EA	0	N/A	1.000	0.00	0.00	11.85	11.85
33 02 2103	**Total Dissolved Solids (EPA 160.1)**	**EA**				**0.00**	**0.00**	**11.80**	**11.80**
01954 7231	Dissolved Solids (EPA 160.1)	EA	0	N/A	1.000	0.00	0.00	11.80	11.80
33 02 2104	**Total Suspended Solids (EPA 160.2)**	**EA**				**0.00**	**0.00**	**10.40**	**10.40**
01954 7232	Suspended Solids (EPA 160.2)	EA	0	N/A	1.000	0.00	0.00	10.40	10.40
33 02 2105	**Total Volatile Solids (EPA 160.4)**	**EA**				**0.00**	**0.00**	**16.70**	**16.70**
01954 7234	Volatile Solids (EPA 160.4)	EA	0	N/A	1.000	0.00	0.00	16.70	16.70
33 02 2106	**Turbidity (EPA 180.1)**	**EA**				**0.00**	**0.00**	**11.90**	**11.90**
01954 7239	Turbidity (EPA 180.1)	EA	0	N/A	1.000	0.00	0.00	11.90	11.90
33 02 2107	**Hardness, Total (EPA 130.2)**	**EA**				**0.00**	**0.00**	**17.10**	**17.10**
01954 7118	Hardness, Total (EPA 130.2)	EA	0	N/A	1.000	0.00	0.00	17.10	17.10
33 02 2108	**Acidity (EPA 305.1)**	**EA**				**0.00**	**0.00**	**23.00**	**23.00**
01954 7101	Acidity/Alkalinity (EPA 305.1/310.1)	EA	0	N/A	1.000	0.00	0.00	23.00	23.00
33 02 2109	**Alkalinity (EPA 310.2)**	**EA**				**0.00**	**0.00**	**23.00**	**23.00**
01954 7101	Acidity/Alkalinity (EPA 305.1/310.1)	EA	0	N/A	1.000	0.00	0.00	23.00	23.00
33 02 2110	**Bromide (EPA 300.0)**	**EA**				**0.00**	**0.00**	**25.00**	**25.00**
01954 7603	Bromide, Titrimetric (EPA 300.0/320.1M)	EA	0	N/A	1.000	0.00	0.00	25.00	25.00
33 02 2111	**Chlorine, Residual (EPA 330.5)**	**EA**				**0.00**	**0.00**	**17.60**	**17.60**
01954 7109	Chlorine, Residual (EPA 330.2/330.5)	EA	0	N/A	1.000	0.00	0.00	17.60	17.60
33 02 2112	**Cyanide, Total (9010B)**	**EA**				**0.00**	**0.00**	**59.63**	**59.63**
01954 7607	Cyanide, Total, (9010B) Spectrophotometric	EA	0	N/A	1.000	0.00	0.00	59.63	59.63
33 02 2113	**Fluoride, Electrode (EPA 335.2)**	**EA**				**0.00**	**0.00**	**23.50**	**23.50**
01954 7116	Fluoride (No Distillation) (EPA 300.0)	EA	0	N/A	1.000	0.00	0.00	23.50	23.50
33 02 2114	**Fluoride, SPADNS (EPA 300.0)**	**EA**				**0.00**	**0.00**	**53.00**	**53.00**
01954 7117	Fluoride (with Distillation) (EPA 340.1)	EA	0	N/A	1.000	0.00	0.00	53.00	53.00
33 02 2115	**Nitrogen, Total (EPA 351.4)**	**EA**				**0.00**	**0.00**	**28.50**	**28.50**
01954 7222	Nitrogen, Kjeldahl (EPA 351.2/351.4)	EA	0	N/A	1.000	0.00	0.00	28.50	28.50
33 02 2116	**Sulfite (EPA 300.0)**	**EA**				**0.00**	**0.00**	**28.00**	**28.00**
01954 7235	Sulfur: Sulfate, Sulfide, Sulfite (EPA 300.0/376.2/377.1)	EA	0	N/A	1.000	0.00	0.00	28.00	28.00
33 02 2117	**Oil & Grease (1664)**	**EA**				**0.00**	**0.00**	**60.21**	**60.21**
01954 7128	Oil and Grease (EPA 413.2)	EA	0	N/A	1.000	0.00	0.00	60.21	60.21

33 02 21	Water/Liquid Laboratory Chemical Analysis (continued)	Unit	Hourly Output	Crew	Factor	Labor	Equip.	Matl.	Total
33 02 2118	**Total Organic Carbon, TOC (EPA 415.1)**	**EA**				**0.00**	**0.00**	**24.00**	**24.00**
01954 7237	Total Organic Carbons (EPA 415.1,9060)	EA	0	N/A	1.000	0.00	0.00	24.00	24.00
33 02 2119	**Biochemical Oxygen Demand, BOD (EPA 405.1)**	**EA**				**0.00**	**0.00**	**27.00**	**27.00**
01954 7104	Biochemical Oxygen Demand (EPA 405.1)	EA	0	N/A	1.000	0.00	0.00	27.00	27.00
33 02 2120	**Chemical Oxygen Demand, COD (EPA 410.4)**	**EA**				**0.00**	**0.00**	**19.30**	**19.30**
01954 7107	Chemical Oxygen Demand (COD) (EPA 410.4)	EA	0	N/A	1.000	0.00	0.00	19.30	19.30
33 02 2121	**Surfactants (EPA 425.1)**	**EA**				**0.00**	**0.00**	**37.50**	**37.50**
01954 7236	Surfactant (MBAS) (EPA 425.1)	EA	0	N/A	1.000	0.00	0.00	37.50	37.50
33 02 2122	**Fecal Coliform (9222B), Water Analysis**	**EA**				**0.00**	**0.00**	**27.50**	**27.50**
01954 7115	Fecal Coliform (SM9222D)	EA	0	N/A	1.000	0.00	0.00	27.50	27.50
33 02 2124	**Metal Analysis, Priority 17 Metals**	**EA**				**0.00**	**0.00**	**68.95**	**68.95**
01954 7412	Metals (6010, 7470)	EA	0	N/A	1.000	0.00	0.00	68.95	68.95
33 02 2125	**Arsenic, Hydride (SW 7061A) with prep, Water Analysis**	**EA**				**0.00**	**0.00**	**198.50**	**198.50**
01954 7090	Arsenic, Hydride (SW 7061A) with Prep	EA	0	N/A	1.000	0.00	0.00	198.50	198.50
33 02 2126	**Arsenic, (SW 7060A) with prep, Water Analysis**	**EA**				**0.00**	**0.00**	**31.75**	**31.75**
01954 7270	Arsenic, (SW 7060A) with Prep	EA	0	N/A	1.000	0.00	0.00	31.75	31.75
33 02 2127	**Metals Screen, 25 Listed in SW 3005A/SW 6010B**	**EA**				**0.00**	**0.00**	**310.16**	**310.16**
01954 7271	Metals Screen, 25 Listed in SW 3005/SW 6010	EA	0	N/A	1.000	0.00	0.00	310.16	310.16
33 02 2128	**Chromium, Hexavalent (SW 7195) with Prep**	**EA**				**0.00**	**0.00**	**23.72**	**23.72**
01954 7272	Chromium, Hexavalent (SW 7195) with Prep	EA	0	N/A	1.000	0.00	0.00	23.72	23.72
33 02 2129	**Mercury, Cold Vapor (SW 7470) with Prep**	**EA**				**0.00**	**0.00**	**45.62**	**45.62**
01954 7273	Mercury, Cold Vapor (SW 7470) with Prep	EA	0	N/A	1.000	0.00	0.00	45.62	45.62
33 02 2130	**Cyanide (SW 9010B) with Prep**	**EA**				**0.00**	**0.00**	**47.43**	**47.43**
01954 7274	Cyanide (SW 9010B) with Prep	EA	0	N/A	1.000	0.00	0.00	47.43	47.43
33 02 2131	**Purgeable Halocarbons (SW 5030B/SW 8021B Halocarbons), Water Analysis**	**EA**				**0.00**	**0.00**	**133.19**	**133.19**
01954 7275	Purgeable Halocarbons (SW 5030B/8021B Halocarbons)	EA	0	N/A	1.000	0.00	0.00	133.19	133.19
33 02 2132	**Purgeable Aromatics (SW 5030B/SW 8021B Aromatics), Water Analysis**	**EA**				**0.00**	**0.00**	**105.82**	**105.82**
01954 7276	Purgeable Aromatics (SW 5030B/8021B Aromatics)	EA	0	N/A	1.000	0.00	0.00	105.82	105.82

33 02 21	Water/Liquid Laboratory Chemical Analysis (continued)	Unit	Hourly Output	Crew	Factor	Labor	Equip.	Matl.	Total
33 02 2133	**Pesticides/PCB's (SW 3510C/SW 8081/8082), Water Analysis**	**EA**				**0.00**	**0.00**	**226.24**	**226.24**
01954 7277	Pesticides/PCBs (SW 3510C/SW 8081/8082)	EA	0	N/A	1.000	0.00	0.00	226.24	226.24
33 02 2134	**Polynuclear Aromatic Hydrocarbons,PAH (SW 3510C/SW 8310), Water Analysis**	**EA**				**0.00**	**0.00**	**206.16**	**206.16**
01954 7278	Polynuclear Aromatic Hydrocarbons, PAH (SW 3510C/SW 8310)	EA	0	N/A	1.000	0.00	0.00	206.16	206.16
33 02 2135	**Base Neutral & Acid Extractable Organics(SW3510C/SW8270C), Water Analysis**	**EA**				**0.00**	**0.00**	**475.82**	**475.82**
01954 7279	Base Neutral & Acid Extractable Organics (SW 3510C/SW 8270C)	EA	0	N/A	1.000	0.00	0.00	475.82	475.82
33 02 2136	**Carbamate & Urea Pesticides (SW 3510C/SW 8318)**	**EA**				**0.00**	**0.00**	**224.41**	**224.41**
01954 7280	Carbamate & Urea Pesticides (SW 3510C/SW 8318)	EA	0	N/A	1.000	0.00	0.00	224.41	224.41
33 02 2137	**Gasoline Group (Mod EPA 601/602, Lead, EDB), Water Analysis**	**EA**				**0.00**	**0.00**	**204.34**	**204.34**
01954 7290	Gasoline Group (Mod EPA 601/602, Lead, EDB)	EA	0	N/A	1.000	0.00	0.00	204.34	204.34
33 02 2138	**Kerosene Group (Mod EPA 601/602, 610, TPH, Lead, EDB), Water Analysis**	**EA**				**0.00**	**0.00**	**204.34**	**204.34**
01954 7291	Kerosene Group (Mod EPA 601/602, 610, TPH, Lead, EDB)	EA	0	N/A	1.000	0.00	0.00	204.34	204.34
33 02 2139	**BTEX/MTBE (Mod EPA 602), Water Analysis**	**EA**				**0.00**	**0.00**	**87.57**	**87.57**
01954 7292	BTEX/MTBE (Mod EPA 602)	EA	0	N/A	1.000	0.00	0.00	87.57	87.57
33 02 2140	**Ethylene Dibromide (EDB) (EPA 504.1)**	**EA**				**0.00**	**0.00**	**85.75**	**85.75**
01954 7293	Ethylene Dibromide (EDB) (EPA 504.1)	EA	0	N/A	1.000	0.00	0.00	85.75	85.75
33 02 2141	**Chromium, Total (EPA 218.1), Water Analysis**	**EA**				**0.00**	**0.00**	**20.80**	**20.80**
01954 7294	Chromium, Total (EPA 218.1)	EA	0	N/A	1.000	0.00	0.00	20.80	20.80
33 02 2142	**Titanium (EPA 283.2), Water Analysis**	**EA**				**0.00**	**0.00**	**19.70**	**19.70**
01954 7295	Titanium (EPA 283.2)	EA	0	N/A	1.000	0.00	0.00	19.70	19.70
33 02 2143	**Lead (SW 3005A/SW 7421), Water Analysis**	**EA**				**0.00**	**0.00**	**45.62**	**45.62**
01954 7281	Lead (SW 3005/SW 7421)	EA	0	N/A	1.000	0.00	0.00	45.62	45.62
33 02 2144	**Ethylene Dibromide (SW 8011) with prep, Water Analysis**	**EA**				**0.00**	**0.00**	**142.31**	**142.31**
01954 7282	Ethylene Dibromide (SW 8011) with Prep	EA	0	N/A	1.000	0.00	0.00	142.31	142.31
33 02 2145	**Dioxins (1613A) with prep, Water Analysis**	**EA**				**0.00**	**0.00**	**806.78**	**806.78**
01954 7283	Dioxins (1613A) with Prep	EA	0	N/A	1.000	0.00	0.00	806.78	806.78

33 02 21	Water/Liquid Laboratory Chemical Analysis (continued)	Unit	Hourly Output	Crew	Factor	Labor	Equip.	Matl.	Total
33 02 2146	**Selenium (SW 7740) with prep, Water Analysis**	**EA**				**0.00**	**0.00**	**40.14**	**40.14**
01954 7284	Selenium (SW 7740) with Prep	EA	0	N/A	1.000	0.00	0.00	40.14	40.14
33 02 2147	**Non Halogenated Volatile Organics (SW 5030B/SW 8015B), Water Analysis**	**EA**				**0.00**	**0.00**	**120.41**	**120.41**
01954 7285	Nonhalogenated Volatile Organics (SW 5030B/8015B)	EA	0	N/A	1.000	0.00	0.00	120.41	120.41
33 02 2148	**Organophosphorous Pesticides (SW 3510C/SW 8141A), Water Analysis**	**EA**				**0.00**	**0.00**	**235.36**	**235.36**
01954 7286	Organophosphorous Pesticides (SW 3510C/SW 8141A)	EA	0	N/A	1.000	0.00	0.00	235.36	235.36
33 02 2149	**Chlorinated Phenoxy Herbicides (SW 3510C/SW 8151A), Water Analysis**	**EA**				**0.00**	**0.00**	**255.42**	**255.42**
01954 7287	Chlorinated Phenoxy Herbicides (SW 3510C/SW 8151A)	EA	0	N/A	1.000	0.00	0.00	255.42	255.42
33 02 2150	**BTEX/MTBE/TVPH (EPA 8021B/8015B), Water Analysis**	**EA**				**0.00**	**0.00**	**117.81**	**117.81**
01001 0054	BTEX/MTBE/TVPH (EPA 8021B/8015B), Water Analysis	EA	0	N/A	1.000	0.00	0.00	117.81	117.81

33 02 22	Radiological Laboratory Chemical Analysis	Unit	Hourly Output	Crew	Factor	Labor	Equip.	Matl.	Total
33 02 2201	**Tissue/Bone, Americium Isotopic, Alpha Spectroscopy**	**EA**				**0.00**	**0.00**	**145.00**	**145.00**
01954 7711	Americium Isotopic	EA	0	N/A	1.000	0.00	0.00	145.00	145.00
33 02 2202	**Tissue/Bone, Curium Isotopic, Alpha Spectroscopy**	**EA**				**0.00**	**0.00**	**148.00**	**148.00**
01954 7712	Curium Isotopic	EA	0	N/A	1.000	0.00	0.00	148.00	148.00
33 02 2203	**Tissue/Bone, Neptunium Isotopic, Alpha Spectroscopy**	**EA**				**0.00**	**0.00**	**150.00**	**150.00**
01954 7713	Neptunium - 237	EA	0	N/A	1.000	0.00	0.00	150.00	150.00
33 02 2204	**Tissue/Bone, Plutonium Isotopic, Alpha Spectroscopy**	**EA**				**0.00**	**0.00**	**140.00**	**140.00**
01954 7714	Plutonium Isotopic	EA	0	N/A	1.000	0.00	0.00	140.00	140.00
33 02 2205	**Tissue/Bone, Thorium Isotopic, Alpha Spectroscopy**	**EA**				**0.00**	**0.00**	**128.00**	**128.00**
01954 7715	Thorium Isotopic	EA	0	N/A	1.000	0.00	0.00	128.00	128.00
33 02 2206	**Tissue/Bone, Uranium Isotopic, Alpha Spectroscopy**	**EA**				**0.00**	**0.00**	**128.00**	**128.00**
01954 7716	Uranium Isotopic	EA	0	N/A	1.000	0.00	0.00	128.00	128.00

33 02 22	Radiological Laboratory Chemical Analysis	Unit	Hourly Output	Crew	Factor	Labor	Equip.	Matl.	Total
33 02 2207	**Tissue/Bone, Strontium - 89, 90, Gas Flow Proportional Counting**	**EA**				**0.00**	**0.00**	**162.00**	**162.00**
01954 7721	Strontium - 89, 90	EA	0	N/A	1.000	0.00	0.00	162.00	162.00
33 02 2208	**Tissue/Bone, Strontium - 90, Gas Flow Proportional Counting**	**EA**				**0.00**	**0.00**	**143.00**	**143.00**
01954 7722	Strontium - 90	EA	0	N/A	1.000	0.00	0.00	143.00	143.00
33 02 2209	**Tissue/Bone, Technetium - 99, Gas Flow Proportional Counting**	**EA**				**0.00**	**0.00**	**132.00**	**132.00**
01954 7723	Technetium - 99	EA	0	N/A	1.000	0.00	0.00	132.00	132.00
33 02 2210	**Tissue/Bone, Americium Isotopic, Gamma Spectroscopy**	**EA**				**0.00**	**0.00**	**142.00**	**142.00**
01954 7731	Americium Isotopic	EA	0	N/A	1.000	0.00	0.00	142.00	142.00
33 02 2211	**Tissue/Bone, Gamma Isotopic, Gamma Spectroscopy**	**EA**				**0.00**	**0.00**	**112.00**	**112.00**
01954 7732	Gamma Isotopic	EA	0	N/A	1.000	0.00	0.00	112.00	112.00
33 02 2212	**Tissue/Bone, Iodine - 25 (Filter or Charcoal), Gamma Spectroscopy**	**EA**				**0.00**	**0.00**	**112.00**	**112.00**
01954 7733	Iodine - 25 (Filter or Charcoal)	EA	0	N/A	1.000	0.00	0.00	112.00	112.00
33 02 2213	**Tissue/Bone, Iodine - 129 (Filter or Charcoal), Gamma Spectroscopy**	**EA**				**0.00**	**0.00**	**125.00**	**125.00**
01954 7734	Iodine-129 (Filter or Charcoal)	EA	0	N/A	1.000	0.00	0.00	125.00	125.00
33 02 2214	**Tissue/Bone, Iodine - 131 (Filter or Charcoal), Gamma Spectroscopy**	**EA**				**0.00**	**0.00**	**135.00**	**135.00**
01954 7735	Iodine-131 (Filter or Charcoal)	EA	0	N/A	1.000	0.00	0.00	135.00	135.00
33 02 2215	**Tissue/Bone, Krypton - 85, Gamma Spectroscopy**	**EA**				**0.00**	**0.00**	**92.00**	**92.00**
01954 7736	Krypton - 85	EA	0	N/A	1.000	0.00	0.00	92.00	92.00
33 02 2216	**Tissue/Bone, Lead - 210, Gamma Spectroscopy**	**EA**				**0.00**	**0.00**	**129.00**	**129.00**
01954 7737	Lead - 210	EA	0	N/A	1.000	0.00	0.00	129.00	129.00
33 02 2217	**Tissue/Bone, Radium - 226, 228, Gamma Spectroscopy**	**EA**				**0.00**	**0.00**	**126.00**	**126.00**
01954 7738	Radium - 226, 228	EA	0	N/A	1.000	0.00	0.00	126.00	126.00
33 02 2218	**Tissue/Bone, Uranium - Total, Gamma Spectroscopy**	**EA**				**0.00**	**0.00**	**85.50**	**85.50**
01954 7739	Uranium - Total	EA	0	N/A	1.000	0.00	0.00	85.50	85.50
33 02 2219	**Tissue/Bone, Carbon - 14, Liquid Scintillation**	**EA**				**0.00**	**0.00**	**158.00**	**158.00**
01954 7751	Carbon - 14	EA	0	N/A	1.000	0.00	0.00	158.00	158.00

33 02 22	Radiological Laboratory Chemical Analysis	Unit	Hourly Output	Crew	Factor	Labor	Equip.	Matl.	Total
33 02 2220	**Tissue/Bone, Uranium Isotopic, Fluorometric**	**EA**				**0.00**	**0.00**	**80.50**	**80.50**
01954 7761	Uranium Isotopic Fluorometric	EA	0	N/A	1.000	0.00	0.00	80.50	80.50
33 02 2221	**Tissue/Bone, Uranium Isotopic, Laser Phosphorometer**	**EA**				**0.00**	**0.00**	**74.00**	**74.00**
01954 7762	Uranium Isotopic Laser Phosphorometer	EA	0	N/A	1.000	0.00	0.00	74.00	74.00
33 02 2222	**Air, Americium Isotopic, Alpha Spectroscopy**	**EA**				**0.00**	**0.00**	**142.00**	**142.00**
01954 7811	Americium Isotopic	EA	0	N/A	1.000	0.00	0.00	142.00	142.00
33 02 2223	**Air, Curium Isotopic, Alpha Spectroscopy**	**EA**				**0.00**	**0.00**	**158.00**	**158.00**
01954 7812	Curium Isotopic	EA	0	N/A	1.000	0.00	0.00	158.00	158.00
33 02 2224	**Air, Neptunium - 237, Alpha Spectroscopy**	**EA**				**0.00**	**0.00**	**158.00**	**158.00**
01954 7813	Neptunium - 237	EA	0	N/A	1.000	0.00	0.00	158.00	158.00
33 02 2225	**Air, Polonium - 210, Alpha Spectroscopy**	**EA**				**0.00**	**0.00**	**118.00**	**118.00**
01954 7814	Polonium - 210	EA	0	N/A	1.000	0.00	0.00	118.00	118.00
33 02 2226	**Air, Plutonium Isotopic, Alpha Spectroscopy**	**EA**				**0.00**	**0.00**	**140.00**	**140.00**
01954 7815	Plutonium Isotopic	EA	0	N/A	1.000	0.00	0.00	140.00	140.00
33 02 2227	**Air, Radium - 226, Alpha Spectroscopy**	**EA**				**0.00**	**0.00**	**112.00**	**112.00**
01954 7816	Radium - 226	EA	0	N/A	1.000	0.00	0.00	112.00	112.00
33 02 2228	**Air, Thorium Isotopic, Alpha Spectroscopy**	**EA**				**0.00**	**0.00**	**128.00**	**128.00**
01954 7817	Thorium Isotopic	EA	0	N/A	1.000	0.00	0.00	128.00	128.00
33 02 2229	**Air, Uranium Isotopic, Alpha Spectroscopy**	**EA**				**0.00**	**0.00**	**128.00**	**128.00**
01954 7818	Uranium Isotopic	EA	0	N/A	1.000	0.00	0.00	128.00	128.00
33 02 2230	**Air, Gross Beta - Total, Gas Flow Proportional Counting**	**EA**				**0.00**	**0.00**	**121.13**	**121.13**
01954 7821	Gross Beta - Total	EA	0	N/A	1.000	0.00	0.00	121.13	121.13
33 02 2231	**Air, Lead - 210, Gas Flow Proportional Counting**	**EA**				**0.00**	**0.00**	**125.00**	**125.00**
01954 7822	Lead - 210	EA	0	N/A	1.000	0.00	0.00	125.00	125.00
33 02 2232	**Air, Strontium - 89, 90, Gas Flow Proportional Counting**	**EA**				**0.00**	**0.00**	**131.38**	**131.38**
01954 7823	Strontium - 89,90	EA	0	N/A	1.000	0.00	0.00	131.38	131.38
33 02 2233	**Air, Strontium - 90, Gas Flow Proportional Counting**	**EA**				**0.00**	**0.00**	**110.00**	**110.00**
01954 7824	Strontium - 90	EA	0	N/A	1.000	0.00	0.00	110.00	110.00

33 02 22	Radiological Laboratory Chemical Analysis	Unit	Hourly Output	Crew	Factor	Labor	Equip.	Matl.	Total
33 02 2234	**Air, Technetium - 99, Gas Flow Proportional Counting**	**EA**				**0.00**	**0.00**	**126.00**	**126.00**
01954 7825	Technetium - 99	EA	0	N/A	1.000	0.00	0.00	126.00	126.00
33 02 2235	**Air, Gamma Isotopic, Gamma Spectroscopy**	**EA**				**0.00**	**0.00**	**101.00**	**101.00**
01954 7831	Gamma Isotopic	EA	0	N/A	1.000	0.00	0.00	101.00	101.00
33 02 2236	**Air, Radium - 226, 228, Gamma Spectroscopy**	**EA**				**0.00**	**0.00**	**130.00**	**130.00**
01954 7832	Radium - 226, 228	EA	0	N/A	1.000	0.00	0.00	130.00	130.00
33 02 2237	**Air, Carbon - 14, Liquid Scintillation**	**EA**				**0.00**	**0.00**	**187.00**	**187.00**
01954 7841	Carbon - 14	EA	0	N/A	1.000	0.00	0.00	187.00	187.00
33 02 2238	**Air, Iron - 55, Liquid Scintillation**	**EA**				**0.00**	**0.00**	**118.00**	**118.00**
01954 7842	Iron - 55	EA	0	N/A	1.000	0.00	0.00	118.00	118.00
33 02 2239	**Air, Gross Alpha - Total, Gas Flow Proportional Counting or Alpha Scintillation**	**EA**				**0.00**	**0.00**	**69.50**	**69.50**
01954 7851	Gross Alpha - Total Gas Flow Proportional Counting	EA	0	N/A	1.000	0.00	0.00	69.50	69.50
33 02 2240	**Air, Gross Alpha & Gross Beta - Total, Gas Flow**	**EA**				**0.00**	**0.00**	**46.50**	**46.50**
01954 7852	Gross Alpha & Gross Beta - Total Gas Flow Proportional Counting	EA	0	N/A	1.000	0.00	0.00	46.50	46.50
33 02 2241	**Air, Plutonium Isotopic, Plutonium - 241, Alpha Spectroscopy**	**EA**				**0.00**	**0.00**	**174.00**	**174.00**
01954 7853	Plutonium Isotopic (Including Plutonium - 241)	EA	0	N/A	1.000	0.00	0.00	174.00	174.00
33 02 2242	**Air, Radium - 226, Radon De-emanation**	**EA**				**0.00**	**0.00**	**113.00**	**113.00**
01954 7854	Radium - 226 Radon De-emanation	EA	0	N/A	1.000	0.00	0.00	113.00	113.00
33 02 2243	**Air, Uranium - Total, Laser Phosphorometer**	**EA**				**0.00**	**0.00**	**81.07**	**81.07**
01954 7855	Uranium - Total Laser Phosphorometer	EA	0	N/A	1.000	0.00	0.00	81.07	81.07
33 02 2244	**Air, Uranium - Total, Fluorometric**	**EA**				**0.00**	**0.00**	**53.50**	**53.50**
01954 7856	Uranium - Total Fluorometric	EA	0	N/A	1.000	0.00	0.00	53.50	53.50
33 02 2245	**Liquid, Americium Isotopic, Alpha Spectroscopy**	**EA**				**0.00**	**0.00**	**142.00**	**142.00**
01954 7911	Americium	EA	0	N/A	1.000	0.00	0.00	142.00	142.00
33 02 2246	**Liquid, Curium Isotopic, Alpha Spectroscopy**	**EA**				**0.00**	**0.00**	**142.00**	**142.00**
01954 7912	Curium Isotopic	EA	0	N/A	1.000	0.00	0.00	142.00	142.00
33 02 2247	**Liquid, Neptunium - 237, Alpha Spectroscopy**	**EA**				**0.00**	**0.00**	**142.00**	**142.00**
01954 7913	Neptunium - 237	EA	0	N/A	1.000	0.00	0.00	142.00	142.00

33 02 22	Radiological Laboratory Chemical Analysis	Unit	Hourly Output	Crew	Factor	Labor	Equip.	Matl.	Total
33 02 2248	**Liquid, Polonium - 210, Alpha Spectroscopy**	**EA**				**0.00**	**0.00**	**123.00**	**123.00**
01954 7914	Polonium - 210	EA	0	N/A	1.000	0.00	0.00	123.00	123.00
33 02 2249	**Liquid, Plutonium Isotopic, Alpha Spectroscopy**	**EA**				**0.00**	**0.00**	**140.00**	**140.00**
01954 7915	Plutonium - Isotopic	EA	0	N/A	1.000	0.00	0.00	140.00	140.00
33 02 2250	**Liquid, Radium - 226, Alpha Spectroscopy**	**EA**				**0.00**	**0.00**	**88.00**	**88.00**
01954 7916	Radium - 226	EA	0	N/A	1.000	0.00	0.00	88.00	88.00
33 02 2251	**Liquid, Radium - 226, 228, Alpha Spectroscopy**	**EA**				**0.00**	**0.00**	**129.00**	**129.00**
01954 7917	Radium - 226, 228	EA	0	N/A	1.000	0.00	0.00	129.00	129.00
33 02 2252	**Liquid, Thorium Isotopic, Alpha Spectroscopy**	**EA**				**0.00**	**0.00**	**128.00**	**128.00**
01954 7918	Thorium Isotopic	EA	0	N/A	1.000	0.00	0.00	128.00	128.00
33 02 2253	**Liquid, Uranium Isotopic, Alpha Spectroscopy**	**EA**				**0.00**	**0.00**	**128.00**	**128.00**
01954 7919	Uranium Isotopic	EA	0	N/A	1.000	0.00	0.00	128.00	128.00
33 02 2254	**Liquid, Iodine - 131, Beta/Gamma Coincidence**	**EA**				**0.00**	**0.00**	**106.00**	**106.00**
01954 7921	Iodine - 131	EA	0	N/A	1.000	0.00	0.00	106.00	106.00
33 02 2255	**Liquid, Radium - 226, 228, Beta/Gamma Coincidence**	**EA**				**0.00**	**0.00**	**107.00**	**107.00**
01954 7922	Radium - 226, 228	EA	0	N/A	1.000	0.00	0.00	107.00	107.00
33 02 2256	**Liquid, Radium - 228, Beta/Gamma Coincidence**	**EA**				**0.00**	**0.00**	**101.00**	**101.00**
01954 7923	Radium - 228	EA	0	N/A	1.000	0.00	0.00	101.00	101.00
33 02 2257	**Liquid, Gross Beta - Total, Gas Flow Proportional Count**	**EA**				**0.00**	**0.00**	**53.50**	**53.50**
01954 7931	Gross Beta - Total	EA	0	N/A	1.000	0.00	0.00	53.50	53.50
33 02 2258	**Liquid, Calcium - 45, Gas Flow Proportional Counting**	**EA**				**0.00**	**0.00**	**95.00**	**95.00**
01954 7932	Calcium - 45	EA	0	N/A	1.000	0.00	0.00	95.00	95.00
33 02 2259	**Liquid, Cerium - 144, Gas Flow Proportional Counting**	**EA**				**0.00**	**0.00**	**132.00**	**132.00**
01954 7933	Cerium - 144	EA	0	N/A	1.000	0.00	0.00	132.00	132.00
33 02 2260	**Liquid, Chlorine - 36, Gas Flow Proportional Counting**	**EA**				**0.00**	**0.00**	**83.00**	**83.00**
01954 7934	Chlorine - 36	EA	0	N/A	1.000	0.00	0.00	83.00	83.00
33 02 2261	**Liquid, Iodine - 131, Gas Flow Proportional Counting**	**EA**				**0.00**	**0.00**	**98.50**	**98.50**
01954 7935	Iodine - 131	EA	0	N/A	1.000	0.00	0.00	98.50	98.50

33 02 22	Radiological Laboratory Chemical Analysis	Unit	Hourly Output	Crew	Factor	Labor	Equip.	Matl.	Total
33 02 2262	**Liquid, Lead - 210, Gas Flow Proportional Counting**	**EA**				**0.00**	**0.00**	**121.00**	**121.00**
01954 7936	Lead - 210	EA	0	N/A	1.000	0.00	0.00	121.00	121.00
33 02 2263	**Liquid, Radium - 226, 228, Gas Flow Proportional Count**	**EA**				**0.00**	**0.00**	**101.00**	**101.00**
01954 7938	Radium - 226, 228	EA	0	N/A	1.000	0.00	0.00	101.00	101.00
33 02 2264	**Liquid, Radium - 228, Gas Flow Proportional Counting**	**EA**				**0.00**	**0.00**	**117.00**	**117.00**
01954 7939	Radium - 228	EA	0	N/A	1.000	0.00	0.00	117.00	117.00
33 02 2265	**Liquid, Ruthenium - 106, Gas Flow Proportional Counting**	**EA**				**0.00**	**0.00**	**105.00**	**105.00**
01954 7941	Ruthenium - 106	EA	0	N/A	1.000	0.00	0.00	105.00	105.00
33 02 2266	**Liquid, Strontium - 89, 90, Gas Flow Proportional Count**	**EA**				**0.00**	**0.00**	**192.00**	**192.00**
01954 7942	Strontium - 89,90	EA	0	N/A	1.000	0.00	0.00	192.00	192.00
33 02 2267	**Liquid, Strontium - 90, Gas Flow Proportional Counting**	**EA**				**0.00**	**0.00**	**122.00**	**122.00**
01954 7943	Strontium - 90	EA	0	N/A	1.000	0.00	0.00	122.00	122.00
33 02 2268	**Liquid, Sulfur - 35, Gas Flow Proportional Counting**	**EA**				**0.00**	**0.00**	**174.00**	**174.00**
01954 7944	Sulfur - 35	EA	0	N/A	1.000	0.00	0.00	174.00	174.00
33 02 2269	**Liquid, Technetium - 99, Gas Flow Proportional Counting**	**EA**				**0.00**	**0.00**	**126.00**	**126.00**
01954 7945	Technetium - 99	EA	0	N/A	1.000	0.00	0.00	126.00	126.00
33 02 2270	**Liquid, Cerium - 144, Gamma Spectroscopy**	**EA**				**0.00**	**0.00**	**118.00**	**118.00**
01954 7951	Cerium - 144	EA	0	N/A	1.000	0.00	0.00	118.00	118.00
33 02 2271	**Liquid, Gamma Isotopic, Gamma Spectroscopy**	**EA**				**0.00**	**0.00**	**117.00**	**117.00**
01954 7952	Gamma Isotopic (Cesium - 137 20 pci/i)	EA	0	N/A	1.000	0.00	0.00	117.00	117.00
33 02 2272	**Liquid, Iodine - 125, Gamma Spectroscopy**	**EA**				**0.00**	**0.00**	**117.00**	**117.00**
01954 7953	Iodine - 125	EA	0	N/A	1.000	0.00	0.00	117.00	117.00
33 02 2273	**Liquid Iodine - 129, Gamma Spectroscopy**	**EA**				**0.00**	**0.00**	**116.00**	**116.00**
01954 7954	Iodine - 129	EA	0	N/A	1.000	0.00	0.00	116.00	116.00
33 02 2274	**Liquid, Ruthenium - 106, Gamma Spectroscopy**	**EA**				**0.00**	**0.00**	**110.00**	**110.00**
01954 7955	Ruthenium - 106	EA	0	N/A	1.000	0.00	0.00	110.00	110.00

33 02 22	Radiological Laboratory Chemical Analysis	Unit	Hourly Output	Crew	Factor	Labor	Equip.	Matl.	Total
33 02 2275	**Liquid, Sulfur - 35, Gamma Spectroscopy**	**EA**				**0.00**	**0.00**	**119.00**	**119.00**
01954 7956	Sulfur - 35	EA	0	N/A	1.000	0.00	0.00	119.00	119.00
33 02 2276	**Liquid, Chlorine - 36, Liquid Scintillation**	**EA**				**0.00**	**0.00**	**117.00**	**117.00**
01954 7961	Chlorine - 36	EA	0	N/A	1.000	0.00	0.00	117.00	117.00
33 02 2277	**Liquid, Carbon - 14, Liquid Scintillation**	**EA**				**0.00**	**0.00**	**169.00**	**169.00**
01954 7962	Carbon - 14	EA	0	N/A	1.000	0.00	0.00	169.00	169.00
33 02 2278	**Liquid, Iron - 55, Liquid Scintillation**	**EA**				**0.00**	**0.00**	**150.00**	**150.00**
01954 7963	Iron - 55	EA	0	N/A	1.000	0.00	0.00	150.00	150.00
33 02 2279	**Liquid, Iron - 125, Liquid Scintillation**	**EA**				**0.00**	**0.00**	**105.00**	**105.00**
01954 7964	Iron - 125	EA	0	N/A	1.000	0.00	0.00	105.00	105.00
33 02 2280	**Liquid, Iron - 129, Liquid Scintillation**	**EA**				**0.00**	**0.00**	**105.00**	**105.00**
01954 7965	Iron - 129	EA	0	N/A	1.000	0.00	0.00	105.00	105.00
33 02 2281	**Liquid, Nickel - 63, Liquid Scintillation**	**EA**				**0.00**	**0.00**	**163.00**	**163.00**
01954 7966	Nickel - 63	EA	0	N/A	1.000	0.00	0.00	163.00	163.00
33 02 2282	**Liquid, Promethium - 147, Liquid Scintillation**	**EA**				**0.00**	**0.00**	**132.00**	**132.00**
01954 7967	Promethium - 147	EA	0	N/A	1.000	0.00	0.00	132.00	132.00
33 02 2283	**Liquid, Sulfur - 35, Liquid Scintillation**	**EA**				**0.00**	**0.00**	**200.00**	**200.00**
01954 7968	Sulfur - 35	EA	0	N/A	1.000	0.00	0.00	200.00	200.00
33 02 2284	**Liquid, Tritium, (Direct Counting), Liquid Scintillation**	**EA**				**0.00**	**0.00**	**78.00**	**78.00**
01954 7969	Tritium (Direct Counting)	EA	0	N/A	1.000	0.00	0.00	78.00	78.00
33 02 2285	**Liquid, Tritium, (Distillation), Liquid Scintillation**	**EA**				**0.00**	**0.00**	**71.50**	**71.50**
01954 7971	Tritium (Distillation)	EA	0	N/A	1.000	0.00	0.00	71.50	71.50
33 02 2286	**Liquid, Tritium, (Electrolytic Enrichment), Liquid Scintillation**	**EA**				**0.00**	**0.00**	**69.50**	**69.50**
01954 7972	Tritium (Electrolytic Enrichment)	EA	0	N/A	1.000	0.00	0.00	69.50	69.50
33 02 2287	**Liquid, Gross Alpha - Total, Gas Flow &/or Alpha Spectroscopy**	**EA**				**0.00**	**0.00**	**52.50**	**52.50**
01954 7981	Gross Alpha - Total Gas Flow Proportional Counting	EA	0	N/A	1.000	0.00	0.00	52.50	52.50
33 02 2288	**Liquid, Suspended or Dissolved, Gross Alpha/Gross Beta**	**EA**				**0.00**	**0.00**	**45.50**	**45.50**
01954 7982	Gross Alpha & Gross Beta Suspended or Dissolved, Gas Flow	EA	0	N/A	1.000	0.00	0.00	45.50	45.50

33 02 22	Radiological Laboratory Chemical Analysis	Unit	Hourly Output	Crew	Factor	Labor	Equip.	Matl.	Total
33 02 2289	**Liquid, Suspended & Dissolved, Gross Alpha/Gross Beta**	**EA**				**0.00**	**0.00**	**45.50**	**45.50**
01954 7982	Gross Alpha & Gross Beta Suspended or Dissolved, Gas Flow	EA	0	N/A	1.000	0.00	0.00	45.50	45.50
33 02 2290	**Liquid, Gross Alpha & Gross Beta - Total**	**EA**				**0.00**	**0.00**	**45.50**	**45.50**
01954 7984	Gross Alpha & Gross Beta - Total Gas Flow Proportional Counting	EA	0	N/A	1.000	0.00	0.00	45.50	45.50
33 02 2291	**Liquid, Plutonium Isotopic/Plutonium - 241, Alpha Spectroscopy**	**EA**				**0.00**	**0.00**	**239.00**	**239.00**
01954 7985	Plutonium Isotopic (Including Plutonium - 241)	EA	0	N/A	1.000	0.00	0.00	239.00	239.00
33 02 2292	**Liquid, Total Radium, Alpha Scintillation**	**EA**				**0.00**	**0.00**	**83.50**	**83.50**
01954 7986	Total Radium Alpha Scintillation Counting	EA	0	N/A	1.000	0.00	0.00	83.50	83.50
33 02 2293	**Liquid, Radium - 226, Radon De-emanation**	**EA**				**0.00**	**0.00**	**113.00**	**113.00**
01954 7987	Radium - 226 Radon De-emanation	EA	0	N/A	1.000	0.00	0.00	113.00	113.00
33 02 2294	**Liquid, Radium - 226, 228, Radon De-emanation**	**EA**				**0.00**	**0.00**	**140.00**	**140.00**
01954 7988	Radium - 226, 228 Radon De-emanation	EA	0	N/A	1.000	0.00	0.00	140.00	140.00
33 02 2295	**Liquid, Radon 222**	**EA**				**0.00**	**0.00**	**64.00**	**64.00**
01954 7989	Radon - 222	EA	0	N/A	1.000	0.00	0.00	64.00	64.00
33 02 2296	**Liquid, Uranium - Total, Laser Phosphorometer**	**EA**				**0.00**	**0.00**	**58.50**	**58.50**
01954 7991	Uranium - Total Laser Phosphorometer	EA	0	N/A	1.000	0.00	0.00	58.50	58.50
33 02 2297	**Liquid, Uranium - Total, Fluorometric**	**EA**				**0.00**	**0.00**	**53.50**	**53.50**
01954 7992	Uranium - Total (Fluorometric 1 mg/l)	EA	0	N/A	1.000	0.00	0.00	53.50	53.50

33 02 23	Radiological Laboratory Chemical Analysis (continued)	Unit	Hourly Output	Crew	Factor	Labor	Equip.	Matl.	Total
33 02 2301	**Urine & Feces, Americium Isotopic, Alpha Spectroscopy**	**EA**				**0.00**	**0.00**	**165.00**	**165.00**
01954 8111	Americium Isotopic	EA	0	N/A	1.000	0.00	0.00	165.00	165.00
33 02 2302	**Urine & Feces, Curium Isotopic, Alpha Spectroscopy**	**EA**				**0.00**	**0.00**	**165.00**	**165.00**
01954 8112	Curium Isotopic	EA	0	N/A	1.000	0.00	0.00	165.00	165.00
33 02 2303	**Urine & Feces, Neptunium - 237, Alpha Spectroscopy**	**EA**				**0.00**	**0.00**	**153.00**	**153.00**
01954 8113	Neptunium - 237	EA	0	N/A	1.000	0.00	0.00	153.00	153.00
33 02 2304	**Urine & Feces, Polonium - 210, Alpha Spectroscopy**	**EA**				**0.00**	**0.00**	**118.00**	**118.00**
01954 8114	Polonium - 210	EA	0	N/A	1.000	0.00	0.00	118.00	118.00

33 02 23	Radiological Laboratory Chemical Analysis (continued)	Unit	Hourly Output	Crew	Factor	Labor	Equip.	Matl.	Total
33 02 2305	**Urine & Feces, Plutonium Isotopic, Alpha Spectroscopy**	**EA**				**0.00**	**0.00**	**152.00**	**152.00**
01954 8115	Plutonium Isotopic	EA	0	N/A	1.000	0.00	0.00	152.00	152.00
33 02 2306	**Urine & Feces, Thorium Isotopic, Alpha Spectroscopy**	**EA**				**0.00**	**0.00**	**140.00**	**140.00**
01954 8116	Thorium Isotopic	EA	0	N/A	1.000	0.00	0.00	140.00	140.00
33 02 2307	**Urine & Feces, Uranium Isotopic, Alpha Spectroscopy**	**EA**				**0.00**	**0.00**	**140.00**	**140.00**
01954 8117	Uranium Isotopic	EA	0	N/A	1.000	0.00	0.00	140.00	140.00
33 02 2308	**Urine & Feces, Gross Beta - Total, Gas Flow Proportional Counting**	**EA**				**0.00**	**0.00**	**46.50**	**46.50**
01954 8121	Gross Beta - Total	EA	0	N/A	1.000	0.00	0.00	46.50	46.50
33 02 2309	**Urine & Feces, Ruthenium - 106, Gas Flow Proportional Counting**	**EA**				**0.00**	**0.00**	**104.00**	**104.00**
01954 8122	Ruthenium - 106	EA	0	N/A	1.000	0.00	0.00	104.00	104.00
33 02 2310	**Urine & Feces, Strontium - 89, 90, Gas Flow Proportional Counting**	**EA**				**0.00**	**0.00**	**172.00**	**172.00**
01954 8123	Strontium - 89, 90	EA	0	N/A	1.000	0.00	0.00	172.00	172.00
33 02 2311	**Urine & Feces, Strontium - 90, Gas Flow Proportional Counting**	**EA**				**0.00**	**0.00**	**138.00**	**138.00**
01954 8124	Strontium - 90	EA	0	N/A	1.000	0.00	0.00	138.00	138.00
33 02 2312	**Urine & Feces, Cerium - 144, Gamma Spectroscopy**	**EA**				**0.00**	**0.00**	**104.00**	**104.00**
01954 8131	Cerium - 144	EA	0	N/A	1.000	0.00	0.00	104.00	104.00
33 02 2313	**Urine & Feces, Gamma Isotopic, Gamma Spectroscopy**	**EA**				**0.00**	**0.00**	**109.00**	**109.00**
01954 8132	Gamma Isotopic	EA	0	N/A	1.000	0.00	0.00	109.00	109.00
33 02 2314	**Urine & Feces, Iodine - 131, Gamma Spectroscopy**	**EA**				**0.00**	**0.00**	**115.00**	**115.00**
01954 8133	Iodine - 131	EA	0	N/A	1.000	0.00	0.00	115.00	115.00
33 02 2315	**Urine & Feces, Radium - 228, Gamma Spectroscopy**	**EA**				**0.00**	**0.00**	**117.00**	**117.00**
01954 8134	Radium - 228	EA	0	N/A	1.000	0.00	0.00	117.00	117.00
33 02 2316	**Urine & Feces, Ruthenium - 106, Gamma Spectroscopy**	**EA**				**0.00**	**0.00**	**96.50**	**96.50**
01954 8135	Ruthenium - 106	EA	0	N/A	1.000	0.00	0.00	96.50	96.50
33 02 2317	**Urine & Feces, Carbon - 14, Liquid Scintillation**	**EA**				**0.00**	**0.00**	**161.00**	**161.00**
01954 8141	Carbon - 14	EA	0	N/A	1.000	0.00	0.00	161.00	161.00

33 02 23	Radiological Laboratory Chemical Analysis (continued)	Unit	Hourly Output	Crew	Factor	Labor	Equip.	Matl.	Total
33 02 2318	**Urine & Feces, Promethium - 147, Liquid Scintillation**	**EA**				**0.00**	**0.00**	**108.00**	**108.00**
01954 8142	Promethium - 147	EA	0	N/A	1.000	0.00	0.00	108.00	108.00
33 02 2319	**Urine & Feces, Phosphorus - 32, Liquid Scintillation**	**EA**				**0.00**	**0.00**	**273.95**	**273.95**
01954 8143	Phosphorus - 32	EA	0	N/A	1.000	0.00	0.00	273.95	273.95
33 02 2320	**Urine & Feces, Tritium (Direct Counting), Liquid Scintillation**	**EA**				**0.00**	**0.00**	**108.00**	**108.00**
01954 8144	Tritium (Direct Counting)	EA	0	N/A	1.000	0.00	0.00	108.00	108.00
33 02 2321	**Urine & Feces, Tritium (Distillation), Liquid Scintillation**	**EA**				**0.00**	**0.00**	**108.00**	**108.00**
01954 8145	Tritium (Distillation)	EA	0	N/A	1.000	0.00	0.00	108.00	108.00
33 02 2322	**Urine & Feces, Mixed Fission Products, Gas Flow**	**EA**				**0.00**	**0.00**	**113.00**	**113.00**
01954 8151	Mixed Fission Products Gamma Spectroscopy & Gas Flow	EA	0	N/A	1.000	0.00	0.00	113.00	113.00
33 02 2323	**Urine & Feces, Plutonium Isotopic, Plutonium - 241, Alpha Spectroscopy**	**EA**				**0.00**	**0.00**	**186.00**	**186.00**
01954 8152	Plutonium Isotopic (Including Plutonium - 241)	EA	0	N/A	1.000	0.00	0.00	186.00	186.00
33 02 2324	**Urine & Feces, Radium - 226, Radon De-emanation**	**EA**				**0.00**	**0.00**	**127.00**	**127.00**
01954 8153	Radium - 226 Radon De-emanation	EA	0	N/A	1.000	0.00	0.00	127.00	127.00
33 02 2325	**Urine & Feces, Radium - 226, 228, Radon De-emanation**	**EA**				**0.00**	**0.00**	**113.00**	**113.00**
01954 8154	Radium - 226, 228 Radon De-emanation, Gas Flow	EA	0	N/A	1.000	0.00	0.00	113.00	113.00
33 02 2326	**Urine & Feces, Uranium - Total, Laser Phosphorometer**	**EA**				**0.00**	**0.00**	**106.00**	**106.00**
01954 8155	Uranium - Total (5 mg/l) Fluorometric	EA	0	N/A	1.000	0.00	0.00	106.00	106.00
33 02 2327	**Urine & Feces, Uranium - Total, Fluorometric**	**EA**				**0.00**	**0.00**	**66.00**	**66.00**
01954 8156	Uranium - Total (5 mg/l) Laser Phosphorometer	EA	0	N/A	1.000	0.00	0.00	66.00	66.00
33 02 2328	**Urine & Feces, Uranium - Total, Radiometric, Alpha Scintillation**	**EA**				**0.00**	**0.00**	**55.91**	**55.91**
01954 8157	Uranium - Total (Radiometric) Alpha Scintillation Counting	EA	0	N/A	1.000	0.00	0.00	55.91	55.91
33 02 2329	**Urine & Feces, Uranium - Total, Laser Phosphorometer**	**EA**				**0.00**	**0.00**	**71.00**	**71.00**
01954 8158	Uranium - Total (.02 mg/l) Laser Phosphorometer	EA	0	N/A	1.000	0.00	0.00	71.00	71.00

33 02 23	Radiological Laboratory Chemical Analysis (continued)	Unit	Hourly Output	Crew	Factor	Labor	Equip.	Matl.	Total
33 02 2330	**Vegetation/Soil/Sediment, Americium Isotopic, Alpha Spectroscopy**	**EA**				**0.00**	**0.00**	**148.00**	**148.00**
01954 8211	Americium Isotopic	EA	0	N/A	1.000	0.00	0.00	148.00	148.00
33 02 2331	**Vegetation/Soil/Sediment, Curium Isotopic, Alpha Spectroscopy**	**EA**				**0.00**	**0.00**	**148.00**	**148.00**
01954 8212	Curium Isotopic	EA	0	N/A	1.000	0.00	0.00	148.00	148.00
33 02 2332	**Vegetation/Soil/Sediment, Neptunium - 237, Alpha Spectroscopy**	**EA**				**0.00**	**0.00**	**148.00**	**148.00**
01954 8213	Neptunium	EA	0	N/A	1.000	0.00	0.00	148.00	148.00
33 02 2333	**Vegetation/Soil/Sediment, Plutonium Isotopic, Alpha Spectroscopy**	**EA**				**0.00**	**0.00**	**145.00**	**145.00**
01954 8214	Plutonium Isotopic	EA	0	N/A	1.000	0.00	0.00	145.00	145.00
33 02 2334	**Vegetation/Soil/Sediment, Thorium Isotopic, Alpha Spectroscopy**	**EA**				**0.00**	**0.00**	**125.00**	**125.00**
01954 8215	Thorium Isotopic	EA	0	N/A	1.000	0.00	0.00	125.00	125.00
33 02 2335	**Vegetation/Soil/Sediment, Uranium Isotopic, Alpha Spectroscopy**	**EA**				**0.00**	**0.00**	**128.00**	**128.00**
01954 8216	Uranium Isotopic	EA	0	N/A	1.000	0.00	0.00	128.00	128.00
33 02 2336	**Vegetation/Soil/Sediment, Gross Beta - Total, Gas Flow Proportional Counting**	**EA**				**0.00**	**0.00**	**53.50**	**53.50**
01954 8221	Gross Beta - Total	EA	0	N/A	1.000	0.00	0.00	53.50	53.50
33 02 2337	**Vegetation/Soil/Sediment, Ruthenium - 106, Gas Flow Proportional Counting**	**EA**				**0.00**	**0.00**	**105.00**	**105.00**
01954 8222	Ruthenium - 106	EA	0	N/A	1.000	0.00	0.00	105.00	105.00
33 02 2338	**Vegetation/Soil/Sediment, Strontium - 89, 90, Gas Flow Proportional Counting**	**EA**				**0.00**	**0.00**	**163.00**	**163.00**
01954 8223	Strontium - 89, 90	EA	0	N/A	1.000	0.00	0.00	163.00	163.00
33 02 2339	**Vegetation/Soil/Sediment, Strontium - 90, Gas Flow Proportional Counting**	**EA**				**0.00**	**0.00**	**122.00**	**122.00**
01954 8224	Strontium - 90	EA	0	N/A	1.000	0.00	0.00	122.00	122.00
33 02 2340	**Vegetation/Soil/Sediment, Technetium - 99, Gas Flow Proportional Counting**	**EA**				**0.00**	**0.00**	**126.00**	**126.00**
01954 8225	Technetium - 99	EA	0	N/A	1.000	0.00	0.00	126.00	126.00
33 02 2341	**Vegetation/Soil/Sediment, Americium Isotopic, Gamma Spectroscopy**	**EA**				**0.00**	**0.00**	**138.00**	**138.00**
01954 8231	Americium Isotopic	EA	0	N/A	1.000	0.00	0.00	138.00	138.00
33 02 2342	**Vegetation/Soil/Sediment, Gamma Isotopic, Gamma Spectroscopy**	**EA**				**0.00**	**0.00**	**113.00**	**113.00**
01954 8232	Gamma Isotopic	EA	0	N/A	1.000	0.00	0.00	113.00	113.00

33 02 23	Radiological Laboratory Chemical Analysis (continued)	Unit	Hourly Output	Crew	Factor	Labor	Equip.	Matl.	Total
33 02 2343	**Vegetation/Soil/Sediment, Iodine - 129, Gamma Spectroscopy**	**EA**				**0.00**	**0.00**	**115.00**	**115.00**
01954 8233	Iodine - 129	EA	0	N/A	1.000	0.00	0.00	115.00	115.00
33 02 2344	**Vegetation/Soil/Sediment, Lead - 210, Gamma Spectroscopy**	**EA**				**0.00**	**0.00**	**122.00**	**122.00**
01954 8234	Lead - 210	EA	0	N/A	1.000	0.00	0.00	122.00	122.00
33 02 2345	**Vegetation/Soil/Sediment, Polonium - 210, Gamma Spectroscopy**	**EA**				**0.00**	**0.00**	**113.00**	**113.00**
01954 8235	Polonium - 210	EA	0	N/A	1.000	0.00	0.00	113.00	113.00
33 02 2346	**Vegetation/Soil/Sediment, Radium - 226, 228, Gamma Spectroscopy**	**EA**				**0.00**	**0.00**	**122.00**	**122.00**
01954 8236	Radium - 226, 228	EA	0	N/A	1.000	0.00	0.00	122.00	122.00
33 02 2347	**Vegetation/Soil/Sediment, Ruthenium - 106, Gamma Spectroscopy**	**EA**				**0.00**	**0.00**	**110.00**	**110.00**
01954 8237	Ruthenium - 106	EA	0	N/A	1.000	0.00	0.00	110.00	110.00
33 02 2348	**Vegetation/Soil/Sediment, Uranium - Total, Gamma Spectroscopy**	**EA**				**0.00**	**0.00**	**61.50**	**61.50**
01954 8238	Uranium - Total	EA	0	N/A	1.000	0.00	0.00	61.50	61.50
33 02 2349	**Vegetation/Soil/Sediment, Carbon - 14, Liquid Scintillation**	**EA**				**0.00**	**0.00**	**206.00**	**206.00**
01954 8241	Carbon - 14	EA	0	N/A	1.000	0.00	0.00	206.00	206.00
33 02 2350	**Vegetation/Soil/Sediment, Nickel - 63, Liquid Scintillation**	**EA**				**0.00**	**0.00**	**161.00**	**161.00**
01954 8242	Nickel - 63	EA	0	N/A	1.000	0.00	0.00	161.00	161.00
33 02 2351	**Vegetation/Soil/Sediment, Tritium, Liquid Scintillation**	**EA**				**0.00**	**0.00**	**89.50**	**89.50**
01954 8243	Tritium	EA	0	N/A	1.000	0.00	0.00	89.50	89.50
33 02 2352	**Vegetation/Soil/Sediment, Gross Alpha - Total, Gas Flow Proportional Counting**	**EA**				**0.00**	**0.00**	**80.50**	**80.50**
01954 8251	Gross Alpha - Total Gas Flow Proportional Counting	EA	0	N/A	1.000	0.00	0.00	80.50	80.50
33 02 2353	**Vegetation/Soil/Sediment, Gross Alpha/Gross Beta - Total, Gas Flow**	**EA**				**0.00**	**0.00**	**47.00**	**47.00**
01954 8252	Gross Alpha & Gross Beta - Total Gas Flow Proportional Counting	EA	0	N/A	1.000	0.00	0.00	47.00	47.00
33 02 2354	**Vegetation/Soil/Sediment, Plutonium Isotopic/Plutonium 241, Alpha Spectroscopy**	**EA**				**0.00**	**0.00**	**239.00**	**239.00**
01954 8253	Plutonium Isotopic (Including Plutonium - 241)	EA	0	N/A	1.000	0.00	0.00	239.00	239.00

33 02 23	Radiological Laboratory Chemical Analysis (continued)	Unit	Hourly Output	Crew	Factor	Labor	Equip.	Matl.	Total
33 02 2355	**Vegetation/Soil/Sediment, Uranium - Total, Laser Phosphorometer**	**EA**				**0.00**	**0.00**	**58.50**	**58.50**
01954 8254	Uranium - Total Laser Phosphorometer	EA	0	N/A	1.000	0.00	0.00	58.50	58.50
33 02 2356	**Vegetation/Soil/Sediment, Uranium - Total, Fluorometric**	**EA**				**0.00**	**0.00**	**53.50**	**53.50**
01954 8255	Uranium - Total Fluorometric	EA	0	N/A	1.000	0.00	0.00	53.50	53.50
33 02 2357	**Characterization of Any Unknown (Low Level) Radioactive Sample**	**EA**				**0.00**	**0.00**	**158.00**	**158.00**
01954 8351	Characterization of Any Unknown Radioactive Sample (Low Level)	EA	0	N/A	1.000	0.00	0.00	158.00	158.00

33 02 24	Explosive Laboratory Chemical Analysis	Unit	Hourly Output	Crew	Factor	Labor	Equip.	Matl.	Total
33 02 2401	**EPA Method 8330A (11 Compounds) Nitroaromatics/Nitramines**	**EA**				**0.00**	**0.00**	**333.71**	**333.71**
01954 7541	Method 8330A, Nitroaromatics & Nitramines (11 Compounds)	EA	0	N/A	1.000	0.00	0.00	333.71	333.71
33 02 2402	**Nitroglycerine**	**EA**				**0.00**	**0.00**	**274.38**	**274.38**
01954 7542	Nitroglycerin	EA	0	N/A	1.000	0.00	0.00	274.38	274.38
33 02 2403	**Nitrocellulose**	**EA**				**0.00**	**0.00**	**222.47**	**222.47**
01954 7543	Nitrocellulose	EA	0	N/A	1.000	0.00	0.00	222.47	222.47
33 02 2404	**Nitroquanadine**	**EA**				**0.00**	**0.00**	**222.47**	**222.47**
01954 7544	Nitroquanadine	EA	0	N/A	1.000	0.00	0.00	222.47	222.47
33 02 2405	**Ammonium Perchlorate**	**EA**				**0.00**	**0.00**	**126.07**	**126.07**
01954 7545	Ammonium Perchlorate	EA	0	N/A	1.000	0.00	0.00	126.07	126.07
33 02 2406	**Field Method for RDX in Soil (8510)**	**EA**				**0.00**	**0.00**	**26.57**	**26.57**
01001 2168	Field Method for RDX in Soil (8510)	EA	0	N/A	1.000	0.00	0.00	26.57	26.57
33 02 2407	**TNT in Soil by Colorimetric Screening (8515)**	**EA**				**0.00**	**0.00**	**24.44**	**24.44**
01001 2156	TNT in Soil by Colorimetric Screening (8515)	EA	0	N/A	1.000	0.00	0.00	24.44	24.44

33 02 25	Immunoassay Field Test	Unit	Hourly Output	Crew	Factor	Labor	Equip.	Matl.	Total
33 02 2501	**EPA 4010A PCP Immunoassay Field Test**	**EA**				**0.00**	**0.00**	**31.88**	**31.88**
01954 8601	EPA 4010A PCP Immunoassay Field Test	EA	0	N/A	1.000	0.00	0.00	31.88	31.88
33 02 2502	**EPA 4015 2,4-D Immunoassay Field Test**	**EA**				**0.00**	**0.00**	**24.44**	**24.44**
01954 8602	EPA 4015 2,4-D Immunoassay Field Test	EA	0	N/A	1.000	0.00	0.00	24.44	24.44
33 02 2503	**EPA 4020 PCBs Immunoassay Field Test**	**EA**				**0.00**	**0.00**	**37.20**	**37.20**
01954 8603	EPA 4020 PCBs Immunoassay Field Test	EA	0	N/A	1.000	0.00	0.00	37.20	37.20

33 02 25	Immunoassay Field Test	Unit	Hourly Output	Crew	Factor	Labor	Equip.	Matl.	Total
33 02 2504	**EPA 4030 Petroleum Hydrocarbons Immunoassay Field Test**	**EA**				**0.00**	**0.00**	**27.63**	**27.63**
01954 8604	EPA 4030 Petroleum Hydrocarbons Immunoassay Field Test	EA	0	N/A	1.000	0.00	0.00	27.63	27.63
33 02 2505	**EPA 4030/4031 BTEX Immunoassay Field Test**	**EA**				**0.00**	**0.00**	**51.91**	**51.91**
01954 8605	EPA 4030/4031 BTEX Immunoassay Field Test	EA	0	N/A	1.000	0.00	0.00	51.91	51.91
33 02 2506	**EPA 4035 PAH Immunoassay Field Test**	**EA**				**0.00**	**0.00**	**42.51**	**42.51**
01954 8606	EPA 4035 PAH Immunoassay Field Test	EA	0	N/A	1.000	0.00	0.00	42.51	42.51
33 02 2507	**EPA 4040 Toxaphene Immunoassay Field Test**	**EA**				**0.00**	**0.00**	**39.32**	**39.32**
01954 8607	EPA 4040 Toxaphene Immunoassay Field Test	EA	0	N/A	1.000	0.00	0.00	39.32	39.32
33 02 2508	**EPA 4041 Chlordane Immunoassay Field Test**	**EA**				**0.00**	**0.00**	**39.32**	**39.32**
01954 8608	EPA 4041 Chlordane Immunoassay Field Test	EA	0	N/A	1.000	0.00	0.00	39.32	39.32
33 02 2509	**EPA 4042 DDT Immunoassay Field Test**	**EA**				**0.00**	**0.00**	**39.32**	**39.32**
01954 8609	EPA 4042 DDT Immunoassay Field Test	EA	0	N/A	1.000	0.00	0.00	39.32	39.32
33 02 2510	**TNT Explosives in Water and Soil by Immunoassay (4050)**	**EA**				**0.00**	**0.00**	**41.45**	**41.45**
01001 2154	TNT Explosives in Water and Soil by Immunoassay (4050)	EA	0	N/A	1.000	0.00	0.00	41.45	41.45
33 02 2511	**RDX Explosives in Water and Soil by Immunoassay (4500)**	**EA**				**0.00**	**0.00**	**41.45**	**41.45**
01001 2155	RDX Explosives in Water and Soil by Immunoassay (4500)	EA	0	N/A	1.000	0.00	0.00	41.45	41.45
33 02 2512	**Mercury in Soil by Immunoassay (4670)**	**EA**				**0.00**	**0.00**	**24.64**	**24.64**
01001 2169	Mercury in Soil by Immunoassay (4670)	EA	0	N/A	1.000	0.00	0.00	24.64	24.64
33 02 2513	**Triazine Herbicides as Atrazine by Immunoassay (4032)**	**EA**				**0.00**	**0.00**	**11.69**	**11.69**
01001 2157	Triazine Herbicides as Atrazine by Immunoassay (4032)	EA	0	N/A	1.000	0.00	0.00	11.69	11.69
33 02 2514	**TCE in Soil by Immunoassay (4060)**	**EA**				**0.00**	**0.00**	**29.76**	**29.76**
01001 2171	TCE in Soil by Immunoassay (4060)	EA	0	N/A	1.000	0.00	0.00	29.76	29.76
33 02 2515	**Dioxin in Soil by Immunoassay (4025)**	**EA**				**0.00**	**0.00**	**63.77**	**63.77**
01001 2159	Dioxin in Soil by Immunoassay (4025)	EA	0	N/A	1.000	0.00	0.00	63.77	63.77
33 02 2516	**Immunoassay Test Apparatus - Daily Rental**	**DAY**				**0.00**	**0.00**	**185.99**	**185.99**
01001 2160	Immunoassay Test Apparatus - Daily Rental	DAY	0	N/A	1.000	0.00	0.00	185.99	185.99
33 02 2517	**Immunoassay Test Apparatus - Weekly Rental**	**WK**				**0.00**	**0.00**	**478.26**	**478.26**
01001 2161	Immunoassay Test Apparatus - Weekly Rental	WK	0	N/A	1.000	0.00	0.00	478.26	478.26

33 02 25	Immunoassay Field Test	Unit	Hourly Output	Crew	Factor	Labor	Equip.	Matl.	Total
33 02 2518	**Immunoassay Test Apparatus - Monthly Rental**	**MO**				**0.00**	**0.00**	**850.24**	**850.24**
01001 2162	Immunoassay Test Apparatus - Monthly Rental	MO	0	N/A	1.000	0.00	0.00	850.24	850.24
33 02 2519	**Immunoassay Test Apparatus - Purchase**	**EA**				**0.00**	**0.00**	**1,594**	**1,594**
01001 2163	Immunoassay Test Apparatus - Purchase	EA	0	N/A	1.000	0.00	0.00	1,594	1,594
33 02 2551	**EPA Draft 8515 TNT Field Test**	**EA**				**0.00**	**0.00**	**215.06**	**215.06**
01954 8651	EPA Draft 8515 TNT Field Test	EA	0	N/A	1.000	0.00	0.00	215.06	215.06
33 02 2552	**Dioxin Field Test**	**EA**				**0.00**	**0.00**	**35.59**	**35.59**
01954 8652	Dioxin Field Test	EA	0	N/A	1.000	0.00	0.00	35.59	35.59
33 02 2561	**Subcontracted Field Analysis with Portable Gas Chromotograph**	**DAY**				**0.00**	**0.00**	**1,137**	**1,137**
01954 6112	Daily Rate (Air, Water, Soil, Groundwater)	EA	0	N/A	1.000	0.00	0.00	605.00	605.00
01954 8502	Gas Chromatograph, HP5890A, Rental	MTH	0	N/A	0.160	0.00	0.00	3,325	532.00

33 02 99	Other	Unit	Hourly Output	Crew	Factor	Labor	Equip.	Matl.	Total
33 02 9901	**Magnetometer, Daily Rental**	**DAY**				**0.00**	**0.00**	**26.90**	**26.90**
01954 6211	Magnetometer, Daily Rental	DAY	0	N/A	1.000	0.00	0.00	26.90	26.90
33 02 9902	**Electromagnetics, Daily Rental**	**DAY**				**0.00**	**0.00**	**29.04**	**29.04**
01954 6212	Electromagnetics, Daily Rental	DAY	0	N/A	1.000	0.00	0.00	29.04	29.04
33 02 9903	**Ground Penetrating Radar, Daily Rental**	**DAY**				**0.00**	**0.00**	**29.04**	**29.04**
01954 6213	Ground Penetrating Radar, Daily Rental	DAY	0	N/A	1.000	0.00	0.00	29.04	29.04
33 02 9904	**Seismic Refraction, Daily Rental**	**DAY**				**0.00**	**0.00**	**2,450**	**2,450**
01954 6214	Seismic Refraction, Daily Rental	DAY	0	N/A	1.000	0.00	0.00	2,450	2,450
33 02 9905	**Resistivity, Daily Rental**	**DAY**				**0.00**	**0.00**	**2,075**	**2,075**
01954 6215	Resistivity, Daily Rental	DAY	0	N/A	1.000	0.00	0.00	2,075	2,075
33 02 9906	**Subcontracted Sampling per Hour, One-man Crew**	**HR**				**0.00**	**0.00**	**74.50**	**74.50**
01954 6111	Hourly Rate (Air, Water, Soil, Ground Water)	EA	0	N/A	1.000	0.00	0.00	74.50	74.50
33 02 9907	**Subcontracted Sampling per Day, One-man Crew**	**DAY**				**0.00**	**0.00**	**605.00**	**605.00**
01954 6112	Daily Rate (Air, Water, Soil, Groundwater)	EA	0	N/A	1.000	0.00	0.00	605.00	605.00
33 02 9908	**Double-ring Infiltrometer Test (DRIT Test)**	**EA**				**0.00**	**0.00**	**570.00**	**570.00**
01418 4785	Double Ring Infiltrometer Test	EA	0	N/A	1.000	0.00	0.00	570.00	570.00
33 02 9909	**Falling Head Rigid Wall Consolidometer Tech CAE 1110-2-1906**	**EA**				**0.00**	**0.00**	**286.00**	**286.00**
01418 5400	Falling Head Rigid Wall Consolidometer Technique CAE	EA	0	N/A	1.000	0.00	0.00	286.00	286.00

33 02 99 Other		Unit	Hourly Output	Crew	Factor	Labor	Equip.	Matl.	Total
33 02 9911	**Falling Head Permeability Test EPA-9100**	**EA**				**0.00**	**0.00**	**233.00**	**233.00**
01418 4820	Falling Head Permeability Test EPA-9100	EA	0	N/A	1.000	0.00	0.00	233.00	233.00
33 02 9912	**Constant Head Permeability Test ASTM D2432**	**EA**				**0.00**	**0.00**	**173.00**	**173.00**
01418 4830	Constant Head Permeability Test ASTM D2432	EA	0	N/A	1.000	0.00	0.00	173.00	173.00
33 02 9913	**Mobile Laboratory Trailer, 8' W x 30' L, Rental**	**MO**				**0.00**	**0.00**	**3,500**	**3,500**
01954 8501	Mobile Laboratory Trailer Rental, 8' W x 24' L x 7' H	MTH	0	N/A	1.000	0.00	0.00	3,500	3,500
33 02 9914	**Gas Chromatograph, HP5890A, Rental**	**MO**				**0.00**	**0.00**	**3,325**	**3,325**
01954 8502	Gas Chromatograph, HP5890A, Rental	MTH	0	N/A	1.000	0.00	0.00	3,325	3,325
33 02 9915	**Mass Spectrometer, HP5970B MD, Rental**	**MO**				**0.00**	**0.00**	**6,125**	**6,125**
01954 8503	Mass Spectrometer, HP5970B MD, Rental	MO	0	N/A	1.000	0.00	0.00	6,125	6,125
33 02 9916	**Unconfined Compressive Strength (Pocket Penetrometer)**	**EA**				**0.00**	**0.00**	**5.59**	**5.59**
01418 4745	Unconfined Compressive Strength (Pocket Penetrometer) Test	EA	0	N/A	1.000	0.00	0.00	5.59	5.59
33 02 9917	**Multi-purpose Groundwater Well Sampling Vehicle, Rental**	**DAY**				**0.00**	**0.00**	**261.30**	**261.30**
01954 6129	Multipurpose Groundwater Well Sampling Vehicle	DAY	0	N/A	1.000	0.00	0.00	261.30	261.30
33 02 9918	**Spectrophotometer, Monthly Rental**	**MO**				**0.00**	**0.00**	**319.36**	**319.36**
01954 8504	Spectrophotometer, Monthly Rental	MO	0	N/A	1.000	0.00	0.00	319.36	319.36
33 02 9919	**Spectrophotometer, Purchase**	**EA**				**0.00**	**0.00**	**2,207**	**2,207**
01954 8505	Spectrophotometer, Purchase	EA	0	N/A	1.000	0.00	0.00	2,207	2,207
33 02 9921	**Gas Chromatograph Weekly Rental**	**WK**				**0.00**	**0.00**	**1,057**	**1,057**
01954 8516	Gas Chromatograph, Weekly Rental	WK	0	N/A	1.000	0.00	0.00	1,057	1,057
33 02 9922	**Mobilize/Demobilize Laboratory Trailer**	**EA**				**0.00**	**0.00**	**4,065**	**4,065**
01954 8506	Mobilize/Demobilize Laboratory Trailer	EA	0	N/A	1.000	0.00	0.00	4,065	4,065
33 02 9923	**Radiological Scintillation Counter**	**MO**				**0.00**	**0.00**	**3,310**	**3,310**
01954 8507	Radiological Scintillation Counter	MO	0	N/A	1.000	0.00	0.00	3,310	3,310
33 02 9924	**pH/Temperature/Conductivity Tester Rental**	**MO**				**0.00**	**0.00**	**82.45**	**82.45**
01954 8508	pH/Temperature/Conductivity Tester	MO	0	N/A	1.000	0.00	0.00	82.45	82.45
33 02 9925	**Infrared Spectrometer Rental**	**MO**				**0.00**	**0.00**	**3,077**	**3,077**
01954 8509	Infrared Spectrometer	MO	0	N/A	1.000	0.00	0.00	3,077	3,077
33 02 9926	**Computer Rental**	**MO**				**0.00**	**0.00**	**232.26**	**232.26**
01954 8510	Computer Rental	MO	0	N/A	1.000	0.00	0.00	232.26	232.26

33 02 Monitoring, Sampling, Testing, and Analysis

33 02 99 Other

33 02 99 Other		Unit	Hourly Output	Crew	Factor	Labor	Equip.	Matl.	Total
33 02 9927	**Recalibration/Maintenance per Instrument per Month**	**MO**				**0.00**	**0.00**	**272.91**	**272.91**
01954 8511	Recalibration/Maintenance per Instrument	MO	0	N/A	1.000	0.00	0.00	272.91	272.91
33 02 9928	**Utilities Hook-up Fee**	**EA**				**0.00**	**0.00**	**2,622**	**2,622**
02560 8012	Utilities Hook-up Fee	EA	0	N/A	1.000	0.00	0.00	2,622	2,622
33 02 9929	**Laboratory Utilities Allowance**	**MO**				**0.00**	**0.00**	**232.26**	**232.26**
01954 8513	Laboratory Utilities Allowance	MO	0	N/A	1.000	0.00	0.00	232.26	232.26
33 02 9930	**Laboratory Expendables Allowance**	**EA**				**0.00**	**0.00**	**11.61**	**11.61**
01954 8514	Laboratory Expendables Allowance per Test	EA	0	N/A	1.000	0.00	0.00	11.61	11.61
33 02 9931	**Electric Laboratory Scales, Purchase**	**EA**				**0.00**	**0.00**	**226.46**	**226.46**
01954 8515	Electric Laboratory Scales	EA	0	N/A	1.000	0.00	0.00	226.46	226.46
33 02 9932	**Fire Extinguisher**	**EA**				**0.00**	**0.00**	**128.00**	**128.00**
01951 1201	CO2 Extinguisher (5 Lb)	EA	0	N/A	1.000	0.00	0.00	128.00	128.00
33 02 9938	**Screening Test Method for PCB in Transformer Oil (9079)**	**EA**				**0.00**	**0.00**	**21.26**	**21.26**
01001 2158	Screening Test Method for PCB in Transformer Oil (9079)	EA	0	N/A	1.000	0.00	0.00	21.26	21.26

33 04 01 Ordnance & Explosive Waste Remediation (see also 33 04 02)

33 04 01 Ordnance & Explosive Waste Remediation (see also 33 04 02)		Unit	Hourly Output	Crew	Factor	Labor	Equip.	Matl.	Total
33 04 0101	**UXO Disposal Technician**	**HR**				**33.04**	**0.00**	**0.00**	**33.04**
01951 4901	Explosive Ordnance Disposal Technician	HR	1	HORDA	1.000	33.04	0.00	0.00	33.04
33 04 0102	**UXO Supervisor**	**HR**				**38.68**	**0.00**	**0.00**	**38.68**
01951 4902	Master EOD Technician	HR	1	HORDB	0.930	41.59	0.00	0.00	38.68
33 04 0103	**UXO Site Setup**	**HR**				**74.64**	**0.00**	**0.00**	**74.64**
01951 4901	Explosive Ordnance Disposal Technician	HR	1	HORDA	1.000	33.04	0.00	0.00	33.04
01951 4902	Master EOD Technician	HR	1	HORDB	1.000	41.59	0.00	0.00	41.59
33 04 0104	**UXO Laborer**	**HR**				**26.09**	**0.00**	**0.00**	**26.09**
01951 4903	Laborer	HR	1	XLABA	1.000	26.09	0.00	0.00	26.09
33 04 0105	**Senior UXO Supervisor**	**HR**				**41.59**	**0.00**	**0.00**	**41.59**
01001 2146	Senior UXO Supervisor	HR	1	HORDB	1.000	41.59	0.00	0.00	41.59
33 04 0111	**Surface Towed Ordnance Locator, Minimum Charge**	**EA**				**6,225**	**74,526**	**464.53**	**81,215**
01953 3101	Surface Towed Ordnance Locator, System (Minimum Charge)	ACRE	0.364	HORDD	20.000	311.23	3,726	23.23	81,215

33 04 01 Ordnance & Explosive Waste Remediation (see also 33 04 02)		Unit	Hourly Output	Crew	Factor	Labor	Equip.	Matl.	Total
33 04 0112	**Surface Towed Ordnance Locator, Per Each Acre**	**ACRE**				**259.31**	**3,105**	**23.23**	**3,387**
01953 3102	Surface Towed Ordnance Locator, System	ACRE	0.437	HORDD	1.000	259.31	3,105	23.23	3,387
33 04 0113	**Ordnance Locator, Foerster Ferex K (M26), Weekly Rental**	**WK**				**0.00**	**0.00**	**162.58**	**162.58**
01953 3103	Ordnance Locator, Foerster Ferex K (M26)	WK	0	N/A	1.000	0.00	0.00	162.58	162.58
33 04 0114	**Ordnance Locator, Foerster Metex**	**EA**				**0.00**	**0.00**	**7,897**	**7,897**
01953 3104	Ordnance Locator, Foerster Metex	EA	0	N/A	1.000	0.00	0.00	7,897	7,897
33 04 0115	**Ordnance Locator, Foerster Minex 2 FD**	**EA**				**0.00**	**0.00**	**10,568**	**10,568**
01953 3105	Ordnance Locator, Foerster Minex 2D	EA	0	N/A	1.000	0.00	0.00	10,568	10,568
33 04 0116	**Underwater Ordnance Locator, Foerster Model W**	**EA**				**0.00**	**0.00**	**19,046**	**19,046**
01953 3106	Underwater Ordnance Locator, Foerster Model W	EA	0	N/A	1.000	0.00	0.00	19,046	19,046
33 04 0117	**Ordnance Locator, Schoenstedt, Model GA-72CV**	**EA**				**0.00**	**0.00**	**1,045**	**1,045**
01953 3107	Ordnance Locator, Schoenstedt, Model GA-72CV	EA	0	N/A	1.000	0.00	0.00	1,045	1,045
33 04 0118	**Ordnance Locator, Schoenstedt, Model GA-52CV**	**EA**				**0.00**	**0.00**	**1,045**	**1,045**
01953 3108	Ordnance Locator, Schoenstedt, Model GA-52CV	EA	0	N/A	1.000	0.00	0.00	1,045	1,045
33 04 0119	**Ordnance Locator, Schoenstedt, MG-220 w/Downhole Attach**	**EA**				**0.00**	**0.00**	**8,361**	**8,361**
01953 3109	Ordnance Locator, Schoenstedt, MG-220 with Downhole Attachment	EA	0	N/A	1.000	0.00	0.00	8,361	8,361
33 04 0120	**Portable Cesium Magnetometer, Geometrics, Model G-858, Weekly Rental**	**WK**				**0.00**	**0.00**	**127.75**	**127.75**
01953 3111	Portable Cesium Magnetometer, Geometrics, Model G-858	WK	0	N/A	1.000	0.00	0.00	127.75	127.75
33 04 0121	**D-Cell Batteries, Long Life Disposable**	**EA**				**0.00**	**0.00**	**0.54**	**0.54**
01953 3122	D-Cell Batteries, Long Life, Disposable	FT	0	N/A	1.000	0.00	0.00	0.54	0.54
33 04 0131	**ICAD, Miniature Chemical Agent Detector**	**EA**				**0.00**	**0.00**	**4,761**	**4,761**
01953 3201	ICAD, Miniature Chemical Agent Detector	EA	0	N/A	1.000	0.00	0.00	4,761	4,761
33 04 0132	**(MINICAM) Miniature Chemical Agent Monitor**	**EA**				**0.00**	**0.00**	**42,272**	**42,272**
01953 3202	(MINICAM) Miniature Chemical Agent Monitor	EA	0	N/A	1.000	0.00	0.00	42,272	42,272
33 04 0140	**4' X 4' X 4' Transportable Explosives Locker**	**EA**				**0.00**	**0.00**	**3,252**	**3,252**
01957 5102	4' x 4' x 4' Transportable Explosives Locker	EA	0	N/A	1.000	0.00	0.00	3,252	3,252

33 04 01	Ordnance & Explosive Waste Remediation (see also 33 04 02)	Unit	Hourly Output	Crew	Factor	Labor	Equip.	Matl.	Total
33 04 0141	**Explosives Storage Locker/Shelter, 16 x 9 x 8 ft.**	**EA**				**0.00**	**0.00**	**23,226**	**23,226**
01957 5101	Explosives Storage Locker/Shelter, 16 x 9 x 8'	EA	0	N/A	1.000	0.00	0.00	23,226	23,226
33 04 0142	**Rope, 5/8", Polyester, Kevlar Center**	**FT**				**0.00**	**0.00**	**2.85**	**2.85**
01953 3121	Rope, 5/8", Polyester, Kevlar Center	FT	0	N/A	1.000	0.00	0.00	2.85	2.85
33 04 0143	**Wood Box, Ordnance Transport Container**	**EA**				**163.27**	**0.00**	**33.34**	**196.61**
06113 3002	Subfloor, 3/4" Thick Plywood, Standard Grade	SF	156.1	ACARD	42.000	0.50	0.00	0.79	54.21
06120 7001	Framing, roofs, rafters, to 4 in12 pitch, 2" x 6"	BF	125	ACARF	26.000	1.06	0.00	0.00	27.49
06128 6141	Framing, walls, studs, 8' highwall, 2" x 4"	BF	115	ACARF	100.000	1.15	0.00	0.00	114.91
33 04 0144	**UXO Field Support Tools and Supplies, Common**	**EA**				**0.00**	**0.00**	**259.56**	**259.56**
01953 3304	EOD Field Support Tools and Supplies, Common	EA	0	N/A	1.000	0.00	0.00	259.56	259.56
33 04 0145	**Crane, 22 Ton Self Contained, Hydraulic, Rough Terrain, 4WD, for Ordinance Removal**	**HR**				**88.05**	**71.88**	**0.00**	**159.93**
01954 6128	Crane, 22 Ton Self-contained, Hydraulic, Rough Terrain, 4 WD, Ordnance & Explosive Waste Removal	HR	1	HORDC	1.000	88.05	71.88	0.00	159.93
33 04 0146	**Hoist, Heavy Duty, 5000 lb. Lift**	**MO**				**0.00**	**0.00**	**205.97**	**205.97**
01953 3305	2.5 Ton Hoist, Rental	MO	1	N/A	1.000	0.00	0.00	205.97	205.97
33 04 0147	**UXO Tool Kit**	**EA**				**0.00**	**0.00**	**370.79**	**370.79**
01953 3307	EOD Tool Kit	EA	0	N/A	1.000	0.00	0.00	370.79	370.79
33 04 0148	**Alloy Steel Chain, 3/8", Cut Length 100 ft.**	**EA**				**0.00**	**0.00**	**366.33**	**366.33**
01953 3308	3/8" Alloy Steel Chain, Cut Length Self-colored	CLF	0	N/A	1.000	0.00	0.00	366.33	366.33
33 04 0149	**Nonsparking UXO Shovels**	**EA**				**0.00**	**0.00**	**96.40**	**96.40**
01953 3306	Nonsparking EOD Shovels	EA	0	N/A	1.000	0.00	0.00	96.40	96.40
33 04 0150	**Standby Ambulance with EMT and Paramedic**	**HR**				**67.96**	**44.89**	**91.85**	**204.70**
01953 3309	Standby, Ambulance with EMT and Paramedic	HR	1	HORDE	1.000	67.96	44.89	91.85	204.70
33 04 0151	**Blast Shelter**	**EA**				**0.00**	**0.00**	**2,846**	**2,846**
01001 2112	Blast Shelter	EA	0	N/A	1.000	0.00	0.00	2,846	2,846
33 04 0152	**Firing Wire**	**LF**				**0.00**	**0.00**	**0.05**	**0.05**
02223 3721	Blasting only, detonator wire	LF	0	N/A	1.000	0.00	0.00	0.05	0.05
33 04 0153	**Reel for Firing Wire**	**EA**				**0.00**	**0.00**	**41.18**	**41.18**
02310 4314	Reel for Firing Wire	EA	0	N/A	1.000	0.00	0.00	41.18	41.18

33 04 01	Ordnance & Explosive Waste Remediation (see also 33 04 02)	Unit	Hourly Output	Crew	Factor	Labor	Equip.	Matl.	Total
33 04 0154	**Elec Blstng Set: Blasting Mach, Galvanometer, Test Set**	**EA**				**0.00**	**0.00**	**939.37**	**939.37**
02310 4315	Blasting Machine	EA	0	N/A	1.000	0.00	0.00	474.51	474.51
02310 4316	Galvanometer	EA	0	N/A	1.000	0.00	0.00	176.51	176.51
02310 4317	Blasting Cap Test Set	EA	0	N/A	1.000	0.00	0.00	288.35	288.35
33 04 0155	**Galvanometer battery**	**EA**				**0.00**	**0.00**	**59.06**	**59.06**
02310 4318	Galvanometer battery	EA	0	N/A	1.000	0.00	0.00	59.06	59.06
33 04 0156	**Electrical Blasting Caps**	**EA**				**0.00**	**0.00**	**17.22**	**17.22**
02323 0820	Dr & blstg, access, electriccaps, w/16' wire	EA	0	N/A	1.000	0.00	0.00	17.22	17.22
33 04 0157	**Non-electrical Blasting Caps**	**EA**				**0.00**	**0.00**	**0.67**	**0.67**
02323 0910	Dr & blstg, access, non-electriccaps, w/8' fuse line	EA	0	N/A	1.000	0.00	0.00	0.67	0.67
33 04 0158	**Igniter**	**EA**				**0.00**	**0.00**	**4.20**	**4.20**
02310 4319	Igniter	EA	0	N/A	1.000	0.00	0.00	4.20	4.20
33 04 0159	**Timed fuse**	**LF**				**0.00**	**0.00**	**0.18**	**0.18**
02310 4321	Timed fuse	LF	0	N/A	1.000	0.00	0.00	0.18	0.18
33 04 0160	**C4 Explosive**	**LB**				**0.00**	**0.00**	**16.88**	**16.88**
02310 4322	C4 Explosive	LB	0	N/A	1.000	0.00	0.00	16.88	16.88
33 04 0161	**Detonation Cord**	**LF**				**0.00**	**0.00**	**0.18**	**0.18**
02310 4323	Detonation Cord	LF	0	N/A	1.000	0.00	0.00	0.18	0.18
33 04 0162	**Lightning Detector**	**EA**				**0.00**	**0.00**	**1,535**	**1,535**
02310 4324	Lightning Detector	EA	0	N/A	1.000	0.00	0.00	1,535	1,535
33 04 0163	**Storm Detector**	**EA**				**0.00**	**0.00**	**191.80**	**191.80**
02310 4325	Storm Detector	EA	0	N/A	1.000	0.00	0.00	191.80	191.80
33 04 0164	**Polypropylene Shovel**	**EA**				**0.00**	**0.00**	**49.29**	**49.29**
01953 3311	Polypropylene Shovel	EA	0	N/A	1.000	0.00	0.00	49.29	49.29
33 04 0165	**22 Ton 4WD Rough Terrain Hydraulic Crane, No Operator**	**HR**				**0.00**	**71.88**	**0.00**	**71.88**
01001 2113	22 Ton 4WD Rough Terrain Hydraulic Crane, No Operator	HR	1	YAABD	1.000	0.00	71.88	0.00	71.88
33 04 0166	**4 x 2 All Terrain Utility Vehicle w/Trailer**	**EA**				**0.00**	**0.00**	**65,352**	**65,352**
01001 2114	4 x 2 All Terrain Utility Vehicle w/Trailer	EA	0	N/A	1.000	0.00	0.00	65,352	65,352
33 04 0167	**Explosives Day Locker**	**EA**				**0.00**	**0.00**	**2,892**	**2,892**
01001 2115	Explosives Day Locker	EA	0	N/A	1.000	0.00	0.00	2,892	2,892

33 04 01 Ordnance & Explosive Waste Remediation (see also 33 04 02)	Unit	Hourly Output	Crew	Factor	Labor	Equip.	Matl.	Total
33 04 0168 Commercial Explosive - 2 Part	**EA**				**0.00**	**0.00**	**22.22**	**22.22**
01001 2116 Commercial Explosive - 2 Part	EA	0	N/A	1.000	0.00	0.00	22.22	22.22

33 04 02 Ordnance & Explosive Waste Remediation (see also 33 04 01)	Unit	Hourly Output	Crew	Factor	Labor	Equip.	Matl.	Total
33 04 0201 Geometrics G858 Cesium magnetometer	**EA**				**0.00**	**0.00**	**7,115**	**7,115**
01001 2117 Geometrics G858 Cesium magnetometer	EA	0	N/A	1.000	0.00	0.00	7,115	7,115
33 04 0202 Geometrics G858 Cesium magnetometer, daily rental	**DAY**				**0.00**	**0.00**	**64.59**	**64.59**
01001 2118 Geometrics G858 Cesium magnetometer, daily rental	DAY	0	N/A	1.000	0.00	0.00	64.59	64.59
33 04 0203 Geometrics G858 Cesium magnetometer, monthly rental	**MO**				**0.00**	**0.00**	**1,938**	**1,938**
01001 2119 Geometrics G858 Cesium magnetometer, monthly rental	MO	0	N/A	1.000	0.00	0.00	1,938	1,938
33 04 0204 Geometrics G858G Cesium magnetometer w/ gradiometer	**EA**				**0.00**	**0.00**	**7,389**	**7,389**
01001 2120 Geometrics G858G Cesium magnetometer w/ gradiometer	EA	0	N/A	1.000	0.00	0.00	7,389	7,389
33 04 0205 Geometrics G858G Cesium magnetometer w/ gradiometer, daily rental	**DAY**				**0.00**	**0.00**	**97.43**	**97.43**
01001 2121 Geometrics G858G Cesium magnetometer w/ gradiometer, daily rental	DAY	0	N/A	1.000	0.00	0.00	97.43	97.43
33 04 0206 Geometrics G858G Cesium magnetometer w/ gradiometer, weekly rental	**WK**				**0.00**	**0.00**	**681.99**	**681.99**
01001 2122 Geometrics G858G Cesium magnetometer w/ gradiometer, weekly rental	WK	0	N/A	1.000	0.00	0.00	681.99	681.99
33 04 0207 Geometrics G858G Cesium magnetometer w/ gradiometer, monthly rental	**MO**				**0.00**	**0.00**	**2,923**	**2,923**
01001 2123 Geometrics G858G Cesium magnetometer w/ gradiometer, monthly rental	MO	0	N/A	1.000	0.00	0.00	2,923	2,923
33 04 0208 Geonics EM-61 Metal Locator, Towed	**EA**				**0.00**	**0.00**	**20,793**	**20,793**
01001 2124 Geonics EM-61 Metal Locator, Towed	EA	0	N/A	1.000	0.00	0.00	20,793	20,793
33 04 0209 Geonics EM-61 Metal Locator, Towed (Daily Rental)	**DAY**				**0.00**	**0.00**	**56.92**	**56.92**
01001 2125 Geonics EM-61 Metal Locator, Towed (Daily Rental)	DAY	0	N/A	1.000	0.00	0.00	56.92	56.92
33 04 0210 Geonics EM-61 Metal Locator, Towed (Weekly Rental)	**WK**				**0.00**	**0.00**	**398.46**	**398.46**
01001 2126 Geonics EM-61 Metal Locator, Towed (Weekly Rental)	WK	0	N/A	1.000	0.00	0.00	398.46	398.46

33 04 02	Ordnance & Explosive Waste Remediation (see also 33 04 01)	Unit	Hourly Output	Crew	Factor	Labor	Equip.	Matl.	Total
33 04 0211	**Geonics EM-61 Metal Locator, Towed (Monthly Rental)**	**MO**				**0.00**	**0.00**	**1,708**	**1,708**
01001 2127	Geonics EM-61 Metal Locator, Towed (Monthly Rental)	MO	0	N/A	1.000	0.00	0.00	1,708	1,708
33 04 0212	**Ordnance Locator, Foerster Metex, daily rental**	**DAY**				**0.00**	**0.00**	**18.61**	**18.61**
01001 2128	Ordnance Locator, Foerster Metex, daily rental	DAY	0	N/A	1.000	0.00	0.00	18.61	18.61
33 04 0213	**Ordnance Locator, Foerster Metex, weekly rental**	**WK**				**0.00**	**0.00**	**128.08**	**128.08**
01001 2129	Ordnance Locator, Foerster Metex, weekly rental	WK	0	N/A	1.000	0.00	0.00	128.08	128.08
33 04 0214	**Ordnance Locator, Foerster Metex, monthly rental**	**MO**				**0.00**	**0.00**	**512.31**	**512.31**
01001 2130	Ordnance Locator, Foerster Metex, monthly rental	MO	0	N/A	1.000	0.00	0.00	512.31	512.31
33 04 0215	**Ordnance Locator, Foerster Minex 2 FD, daily rental**	**DAY**				**0.00**	**0.00**	**18.61**	**18.61**
01001 2131	Ordnance Locator, Foerster Minex 2 FD, daily rental	DAY	0	N/A	1.000	0.00	0.00	18.61	18.61
33 04 0216	**Ordnance Locator, Foerster Minex 2 FD, weekly rental**	**WK**				**0.00**	**0.00**	**128.08**	**128.08**
01001 2132	Ordnance Locator, Foerster Minex 2 FD, weekly rental	WK	0	N/A	1.000	0.00	0.00	128.08	128.08
33 04 0217	**Ordnance Locator, Foerster Minex 2 FD, monthly rental**	**MO**				**0.00**	**0.00**	**512.31**	**512.31**
01001 2133	Ordnance Locator, Foerster Minex 2 FD, monthly rental	MO	0	N/A	1.000	0.00	0.00	512.31	512.31
33 04 0218	**Ordnance Locator, Schoenstedt, MG-230 magnetometer w/ Downhole Attachment**	**EA**				**0.00**	**0.00**	**7,115**	**7,115**
01001 2134	Ordnance Locator, Schoenstedt, MG-230 magnetometer w/ Downhole Attachment	EA	0	N/A	1.000	0.00	0.00	7,115	7,115
33 04 0219	**Ordnance Locator, Schoenstedt, MG-230 magnetometer w/ Downhole Attachment, daily rental**	**DAY**				**0.00**	**0.00**	**27.37**	**27.37**
01001 2135	Ordnance Locator, Schoenstedt, MG-230 magnetometer w/ Downhole Attachment, daily rental	DAY	0	N/A	1.000	0.00	0.00	27.37	27.37
33 04 0220	**Ordnance Locator, Schoenstedt, MG-230 magnetometer w/ Downhole Attachment, weekly rental**	**WK**				**0.00**	**0.00**	**191.57**	**191.57**
01001 2136	Ordnance Locator, Schoenstedt, MG-230 magnetometer w/ Downhole Attachment, weekly rental	WK	0	N/A	1.000	0.00	0.00	191.57	191.57

33 04 02 Ordnance & Explosive Waste Remediation (see also 33 04 01)		Unit	Hourly Output	Crew	Factor	Labor	Equip.	Matl.	Total
33 04 0221	**Ordnance Locator, Schoenstedt, MG-230 magnetometer w/ Downhole Attachment, monthly rental**	**MO**				**0.00**	**0.00**	**821.01**	**821.01**
01001 2137	Ordnance Locator, Schoenstedt, MG-230 magnetometer w/ Downhole Attachment, monthly rental	MO	0	N/A	1.000	0.00	0.00	821.01	821.01
33 04 0222	**Ordnance Locator, Schoenstedt, Model GA-72CD, daily rental**	**DAY**				**0.00**	**0.00**	**19.16**	**19.16**
01001 2138	Ordnance Locator, Schoenstedt, Model GA-72CD, daily rental	DAY	0	N/A	1.000	0.00	0.00	19.16	19.16
33 04 0223	**Ordnance Locator, Schoenstedt, Model GA-72CD, weekly rental**	**WK**				**0.00**	**0.00**	**45.98**	**45.98**
01001 2139	Ordnance Locator, Schoenstedt, Model GA-72CD, weekly rental	WK	0	N/A	1.000	0.00	0.00	45.98	45.98
33 04 0224	**Ordnance Locator, Schoenstedt, Model GA-72CD, monthly rental**	**MO**				**0.00**	**0.00**	**79.18**	**79.18**
01001 2140	Ordnance Locator, Schoenstedt, Model GA-72CD, monthly rental	MO	0	N/A	1.000	0.00	0.00	79.18	79.18
33 04 0225	**Underwater Ordnance Locator, Foerster Model W, daily rental**	**DAY**				**0.00**	**0.00**	**40.66**	**40.66**
01001 2141	Underwater Ordnance Locator, Foerster Model W, daily rental	DAY	0	N/A	1.000	0.00	0.00	40.66	40.66
33 04 0226	**Underwater Ordnance Locator, Foerster Model W, weekly rental**	**WK**				**0.00**	**0.00**	**284.62**	**284.62**
01001 2142	Underwater Ordnance Locator, Foerster Model W, weekly rental	WK	0	N/A	1.000	0.00	0.00	284.62	284.62
33 04 0227	**Underwater Ordnance Locator, Foerster Model W, monthly rental**	**MO**				**0.00**	**0.00**	**1,138**	**1,138**
01001 2143	Underwater Ordnance Locator, Foerster Model W, monthly rental	MO	0	N/A	1.000	0.00	0.00	1,138	1,138
33 04 0228	**Geonics EM-61 Metal Locator, Hand Held**	**EA**				**0.00**	**0.00**	**18,604**	**18,604**
01001 2150	Geonics EM-61 Metal Locator, Hand Held	EA	0	N/A	1.000	0.00	0.00	18,604	18,604
33 04 0229	**Geonics EM-61 Metal Locator, Hand Held (Daily Rental)**	**DAY**				**0.00**	**0.00**	**48.17**	**48.17**
01001 2151	Geonics EM-61 Metal Locator, Hand Held (Daily Rental)	DAY	0	N/A	1.000	0.00	0.00	48.17	48.17
33 04 0230	**Geonics EM-61 Metal Locator, Hand Held (Weekly Rental)**	**WK**				**0.00**	**0.00**	**337.16**	**337.16**
01001 2152	Geonics EM-61 Metal Locator, Hand Held (Weekly Rental)	WK	0	N/A	1.000	0.00	0.00	337.16	337.16
33 04 0231	**Geonics EM-61 Metal Locator, Hand Held (Monthly Rental)**	**MO**				**0.00**	**0.00**	**1,445**	**1,445**
01001 2153	Geonics EM-61 Metal Locator, Hand Held (Monthly Rental)	MO	0	N/A	1.000	0.00	0.00	1,445	1,445

33 05 08	Storm Drainage	Unit	Hourly Output	Crew	Factor	Labor	Equip.	Matl.	Total
33 05 0801	**10' Wide Grass Drainage Swale**	**LF**				**0.92**	**1.10**	**3.50**	**5.53**
02224 7040	Rough Grade Small Area w/Dozer 75 HP	CY	115.1	CODTC	1.000	0.38	0.37	0.00	0.75
02241 0020	Remove Topsoil, 6" (15 cm) Deep, Stockpile on Site	CY	49.88	CODTG	0.185	0.88	2.45	5.99	1.73
02241 0400	Spread Topsoil by Loader from Stockpile	CY	24.5	CODLA	0.185	1.96	1.43	8.38	2.18
02830 2005	Large Power Mulcher, Oat Straw, 1" Deep	ACRE	1.521	COELD	0.000	35.18	22.70	1,728	0.41
02932 0300	Mechanical Seeding, 50 Lb/MSY	CSY	193.8	ULABE	0.011	0.17	0.00	40.00	0.45
02976 4930	Watering by Truck	MSF	87.12	COFWI	0.010	0.61	0.76	0.02	0.01
33 05 0802	**Grass Ditching, 3' Bottom, 3' Deep, 2:1 Side Slopes**	**LF**				**3.08**	**4.99**	**5.81**	**13.88**
02220 7000	Compact Backfill, by Hand with Vibrating Plate, 6" (15 cm) Lift	CY	24.5	CLACC	0.304	3.35	0.18	0.00	1.07
02220 9000	Compaction Water, 3,000 Gallon Truck, 3-Mile Haul for Embankment Compaction, 5 km	CY	116.8	COFWI	0.304	0.45	0.57	0.20	0.37
02228 0322	Trench, 1 CY Hydraulic Excavator, Medium Soil, 4' to 6' Deep, 128 CY/Hour	CY	128	CODEO	1.000	0.49	0.70	0.00	1.19
02239 0210	Shape Embankment/Slope with Machine Greater than 1 on 4 Slope	SY	150	CODTK	1.820	0.30	1.57	0.00	3.41
02241 0020	Remove Topsoil, 6" (15 cm) Deep, Stockpile on Site	CY	49.88	CODTG	0.304	0.88	2.45	5.99	2.83
02241 0400	Spread Topsoil by Loader from Stockpile	CY	24.5	CODLA	0.304	1.96	1.43	8.38	3.58
02830 2005	Large Power Mulcher, Oat Straw, 1" Deep	ACRE	1.521	COELD	0.000	35.18	22.70	1,728	0.68
02932 0300	Mechanical Seeding, 50 Lb/MSY	CSY	193.8	ULABE	0.018	0.17	0.00	40.00	0.73
02976 4930	Watering by Truck	MSF	87.12	COFWI	0.016	0.61	0.76	0.02	0.02
33 05 0803	**Grass Ditching, 5' Bottom, 5' Deep, 4:1 Side Slopes**	**LF**				**9.57**	**15.34**	**15.53**	**40.44**
02220 7000	Compact Backfill, by Hand with Vibrating Plate, 6" (15 cm) Lift	CY	24.5	CLACC	0.856	3.35	0.18	0.00	3.02
02220 9000	Compaction Water, 3,000 Gallon Truck, 3-Mile Haul for Embankment Compaction, 5 km	CY	116.8	COFWI	0.856	0.45	0.57	0.20	1.04
02228 0322	Trench, 1 CY Hydraulic Excavator, Medium Soil, 4' to 6' Deep, 128 CY/Hour	CY	128	CODEO	4.630	0.49	0.70	0.00	5.52
02239 0210	Shape Embankment/Slope with Machine Greater than 1 on 4 Slope	SY	150	CODTK	5.137	0.30	1.57	0.00	9.62
02241 0020	Remove Topsoil, 6" (15 cm) Deep, Stockpile on Site	CY	49.88	CODTG	0.856	0.88	2.45	5.99	7.98
02241 0400	Spread Topsoil by Loader from Stockpile	CY	24.5	CODLA	0.856	1.96	1.43	8.38	10.08
02830 2005	Large Power Mulcher, Oat Straw, 1" Deep	ACRE	1.521	COELD	0.001	35.18	22.70	1,728	1.89
02932 0300	Mechanical Seeding, 50 Lb/MSY	CSY	193.8	ULABE	0.030	0.17	0.00	40.00	1.22
02976 4930	Watering by Truck	MSF	87.12	COFWI	0.046	0.61	0.76	0.02	0.06
33 05 0804	**Riprap Ditching, 3' Bottom, 3' Deep, 2:1 Side Slopes**	**LF**				**3.24**	**5.04**	**12.16**	**20.44**
02228 0322	Trench, 1 CY Hydraulic Excavator, Medium Soil, 4' to 6' Deep, 128 CY/Hour	CY	128	CODEO	1.000	0.49	0.70	0.00	1.19
02239 0210	Shape Embankment/Slope with Machine Greater than 1 on 4 Slope	SY	150	CODTK	1.820	0.30	1.57	0.00	3.41
02270 1100	Rip Rap, 10 Lb to 100 Lb Pieces Random, Dumped from Truck	CY	32	COETF	0.608	3.61	2.44	20.00	15.84

33 05 Surface Water Collection and Control

33 05 08 Storm Drainage	Unit	Hourly Output	Crew	Factor	Labor	Equip.	Matl.	Total
33 05 0805 Concrete Ditching, 3' Bottom, 3' Deep, 2:1 Side Slopes	**LF**				**19.67**	**2.11**	**27.64**	**49.42**
02083 7112 20 CY and 30 CY Roll-off Liner 6 Mil, Disposable	EA	0	N/A	0.051	0.00	0.00	25.50	1.31
02220 7000 Compact Backfill, by Hand with Vibrating Plate, 6" (15 cm) Lift	CY	24.5	CLACC	0.304	3.35	0.18	0.00	1.07
02220 9000 Compaction Water, 3,000 Gallon Truck, 3-Mile Haul for Embankment Compaction, 5 km	CY	116.8	COFWI	0.304	0.45	0.57	0.20	0.37
02228 0322 Trench, 1 CY Hydraulic Excavator, Medium Soil, 4' to 6' Deep, 128 CY/Hour	CY	128	CODEO	1.000	0.49	0.70	0.00	1.19
02239 0230 Shape Embankment/Slope by Hand Greater than 1 on 4 Slope	SY	12.5	ULABA	1.820	2.71	0.00	0.00	4.93
02611 2201 Graded Crushed Aggregate Base	CY	25	XSABA	0.080	9.51	13.80	22.77	3.69
03132 3750 Polyurethane Foam Expansion Joint, 1/2" x 6" Regular, Premolded in Slabs	LF	46.88	ACARA	0.340	0.92	0.00	1.28	0.75
03170 3060 Slab on Grade Edge Forms, Over 12" H (Over 31 cm), Based on 4 Uses	SF	41.88	ACARJ	1.000	3.14	0.00	0.96	4.09
03227 0700 Welded Wire Fabric in Slabs, 4 x 4 x #4 85 Lb/CSF, (W 4.0 x W 4.0), < 5	SF	312.5	SIWRB	16.400	0.31	0.00	0.31	10.13
03334 0200 Concrete Curing, Waterproof Paper, 2 Ply Reinforced	CSF	8.75	ULABB	0.170	7.74	0.00	5.63	2.27
03372 4300 Pour Slab on Grade, < 6", Direct Chute < (15 cm) Place 3,000 PSI Concrete	CY	16.25	ALABE	0.200	10.36	0.39	84.09	19.01
15155 1940 4" Schedule 40 PVC Pressure Pipe Socket Weld	LF	15.5	MPLUE	0.080	5.74	0.00	1.80	0.60

33 05 15 Passive Water Treatment	Unit	Hourly Output	Crew	Factor	Labor	Equip.	Matl.	Total
33 05 1501 Install Filter Compost in Confined Space	**CY**				**41.05**	**21.83**	**196.66**	**259.54**
02086 2001 Install Filter Compost in Confined Space	CY	2	ULABL	1.000	41.05	21.83	196.66	259.54
33 05 1502 Install Filter Compost in Open Space	**CY**				**47.22**	**0.00**	**177.00**	**224.22**
02086 2002 Install Filter Compost in Open Space	CY	2	ULABC	1.000	47.22	0.00	177.00	224.22
33 05 1503 Confined Space Filter Compost Removal	**CY**				**46.91**	**24.94**	**0.00**	**71.86**
02086 2003 Confined Space Filter Compost Removal	CY	1.75	ULABL	1.000	46.91	24.94	0.00	71.86
33 05 1504 Remove Filter Compost from Open-Top Unit	**CY**				**53.97**	**0.00**	**0.00**	**53.97**
02086 2005 Open Space Filter Compost Removal	CY	1.75	ULABC	1.000	53.97	0.00	0.00	53.97
33 05 1505 Tilling of Storm Water Filter Compost	**SF**				**1.98**	**0.00**	**0.00**	**1.98**
02086 2006 Till Filter Compost	SF	41.5	ULABG	1.000	1.98	0.00	0.00	1.98
33 05 1511 2' Wide Iron Sewer/Energy Dissipation Grating	**LF**				**18.07**	**0.00**	**73.51**	**91.58**
02086 2011 2' Wide Iron Sewer Grating	LF	5.333	SIWSB	1.000	18.07	0.00	73.51	91.58
33 05 1512 Galv. Debris Baffle	**SF**				**8.03**	**0.00**	**33.17**	**41.21**
02086 2012 Galv.Debris Baffle	SF	12	SIWSB	1.000	8.03	0.00	33.17	41.21
33 05 1513 4' High Concrete Flow Spreader Pre-Cast	**LF**				**14.18**	**5.99**	**16.88**	**37.05**
02086 2013 4' High Concrete Flow Spreader Pre-Cast	LF	12	UOEHC	1.000	14.18	5.99	16.88	37.05

33 05 15 Passive Water Treatment		Unit	Hourly Output	Crew	Factor	Labor	Equip.	Matl.	Total
33 05 1514	**8' High Weir Posts**	**EA**				**72.28**	**0.00**	**250.11**	**322.40**
02086 2014	8' High Weir Posts	EA	1.333	SIWSB	1.000	72.28	0.00	250.11	322.40
33 05 1515	**Galv. Overflow Weir**	**SF**				**11.92**	**0.67**	**34.61**	**47.19**
02086 2015	Galv. Overflow Weir	SF	12.5	SIWSC	1.000	11.92	0.67	34.61	47.19
33 05 1516	**6' x 12' x 6' I.D. Pre-Cast Vault**	**EA**				**2,722**	**1,150**	**7,408**	**11,280**
02086 2016	6' x 12' x 6' I.D. Pre-Cast Vault	EA	0.063	UOEHC	1.000	2,722	1,150	7,408	11,280
33 05 1517	**8' x 18' x 7' I.D. Pre-Cast Vault**	**EA**				**4,080**	**1,724**	**11,013**	**16,817**
02086 2017	8' x 18' x 7' I.D. Pre-Cast Vault	EA	0.042	UOEHC	1.000	4,080	1,724	11,013	16,817
33 05 1518	**Vault Access Doors**	**EA**				**96.38**	**0.00**	**458.88**	**555.26**
02086 2018	Vault Access Doors	EA	1	SIWSB	1.000	96.38	0.00	458.88	555.26
33 05 1519	**6' High Reinforced CIP Retaining Wall**	**LF**				**88.11**	**10.22**	**62.84**	**161.17**
02215 0500	Compaction, 6" Layers with Air Tamper by Hand, (15 cm) Layers	CY	50	CLACA	0.260	2.71	0.17	0.00	0.75
02220 8800	Compact Backfill, by Hand with 1-Ton Roller, 6" (15 cm) Lift	CY	49.13	CLACD	0.260	0.94	1.51	0.00	0.64
02221 5007	Trench Backfill, 3 CY, 950	CY	70.88	CODLJ	0.520	0.47	0.54	0.00	0.52
02227 4020	Trim & Shape Machine Excavated Area by Hand - Finish Grade	SF	150	ULABA	12.060	0.23	0.00	0.00	2.72
02228 0322	Trench, 1 CY Hydraulic Excavator, Medium Soil, 4' to 6' Deep, 128 CY/Hour	CY	128	CODEO	0.840	0.49	0.70	0.00	1.00
02234 4000	Load & Haul Soil, 12 CY, 5 Miles	CY	94.5	COEIE	0.410	1.52	2.37	0.00	1.59
02245 2208	Fine Grade, 120G, Double Pass	SY	450	COFGC	0.340	0.08	0.11	0.00	0.06
03158 0150	Continuous Wall Footing Forms, 4 Uses Plywood Forms, Form & Strip with Accessories	SF	60.63	ACARJ	2.000	2.17	0.00	0.82	5.97
03158 1500	2x4 Keyway Forms, 4 Uses	LF	66.25	ACARA	1.000	0.65	0.00	0.24	0.90
03182 4750	Retaining Wall Forms - up to 8' 4 Uses (Below Grade) Plywood Forms Form and Strip	SF	40.63	ACARL	12.000	3.96	0.00	0.62	54.97
03210 1007	Rebar #3-#7	TON	0.288	SIWRF	0.005	629.08	250.00	647.24	7.28
03210 1011	Wall Rebar, #3-#7	TON	0.375	SIWRF	0.013	482.29	191.67	647.24	16.70
03372 2450	Pour Spread Footings, < 5 CY, Concrete Pump < (5 m3) Place 3,000 PSI Concrete Foundations	CY	12.5	ALABI	0.220	18.38	7.55	84.22	24.23
03372 5200	Pour Concrete Wall, 12" Thick, Crane & Bucket (31 cm) 3,000 PSI Concrete	CY	11.25	ALABG	0.250	23.20	10.51	84.22	29.48
03398 0300	Wall Finish, Bush Hammer Green Concrete	SF	21.25	ACMAA	4.000	1.96	0.00	0.06	8.05
07174 0100	Silicone Retarder, 1 Coat on Concrete, Sprayed	CSF	1.875	ARFCD	0.194	19.90	0.00	12.47	6.29
33 05 1521	**Tipping Fee for Excavation Spoil**	**CY**				**0.00**	**0.00**	**6.82**	**6.82**
02086 2041	Spoil Tipping Fee	CY	0	N/A	1.000	0.00	0.00	6.82	6.82
33 05 1522	**Tipping Fee for Brush**	**CY**				**0.00**	**0.00**	**7.87**	**7.87**
02086 2042	Brush Tipping Fee	CY	0	N/A	1.000	0.00	0.00	7.87	7.87
33 05 1523	**Tipping Fee for Spent Filter Compost**	**CY**				**0.00**	**0.00**	**6.29**	**6.29**
02086 2043	Spent Compost Fee	CY	0	N/A	1.000	0.00	0.00	6.29	6.29

33 05 15 Passive Water Treatment		Unit	Hourly Output	Crew	Factor	Labor	Equip.	Matl.	Total
33 05 1531	**No. 3/4 Limestone - Bulk**	**TON**				**2.93**	**0.99**	**50.15**	**54.07**
02086 2031	No. 3/4 Limestone - Bulk	TN	32.5	CODEJ	1.000	2.93	0.99	50.15	54.07
33 05 1532	**Mushroom Compost - Bulk**	**CY**				**2.71**	**0.70**	**16.72**	**20.13**
02086 2032	Mushroom Compost - Bulk	CY	54.25	CODL2	1.000	2.71	0.70	16.72	20.13
33 05 1533	**Wetland Vegetation Planting**	**CSF**				**10.35**	**0.00**	**19.67**	**30.01**
02086 2033	Wetland Vegetation Planting	CSF	6.547	ULABB	1.000	10.35	0.00	19.67	30.01
33 05 1534	**Planting Mix - Hand Spread**	**CY**				**12.59**	**0.00**	**28.09**	**40.68**
02086 2034	Planting Mix - Hand Spread	CY	7.5	ULABC	1.000	12.59	0.00	28.09	40.68
33 05 1535	**Seed Wetland w/Bionutrients**	**LB**				**0.11**	**0.00**	**1.68**	**1.78**
02086 2035	Seed Wetland w/Bionutrients	LB	320	ULABE	1.000	0.11	0.00	1.68	1.78
33 05 1540	**6 Mil Polyethylene Vapor Barrier**	**CSF**				**9.36**	**0.00**	**3.30**	**12.67**
07192 0900	6 Mil Polyethylene Vapor Barrier	CSF	4.625	ACARA	1.000	9.36	0.00	3.30	12.67
33 05 1551	**36 inch Diameter Riser Outlet**	**EA**				**2,043**	**384.75**	**4,195**	**6,623**
02086 2051	36 inch Diameter Riser Outlet	EA	0.083	CODEK	1.000	2,043	384.75	4,195	6,623
33 05 1552	**48 inch Diameter Outlet Riser**	**EA**				**2,722**	**512.80**	**4,654**	**7,890**
02086 2052	48 inch Diameter Outlet Riser Outlet	EA	0.063	CODEK	1.000	2,722	512.80	4,654	7,890
33 05 1553	**60" Dia. Low Flow Outlet Riser**	**EA**				**34.03**	**6.41**	**4,983**	**5,023**
02086 2053	60" Dia. Outlet Riser Outlet	EA	5	CODEK	1.000	34.03	6.41	4,982	5,023

33 06 03 Slurry Walls		Unit	Hourly Output	Crew	Factor	Labor	Equip.	Matl.	Total
33 06 0301	**Level and Compact Working Surface**	**CY**				**1.78**	**4.14**	**0.00**	**5.92**
02081 3001	Level and Compact Working Surface	CY	106.4	COFCV	1.000	1.78	4.14	0.00	5.92
33 06 0302	**Construct Dike for Mixing Basin**	**CY**				**1.78**	**4.14**	**0.00**	**5.92**
02081 3002	Construct Dike for Mixing Basin	CY	106.4	COFCV	1.000	1.78	4.14	0.00	5.92
33 06 0303	**Normal Soil, to 25', Slurry Wall Excavation**	**CY**				**0.82**	**1.71**	**0.00**	**2.53**
02228 0388	Trench, 2-1/2 CY Hydraulic Excavator, 113 CY/Hour (86 m3)/Hour)	CY	112.9	CODEU	1.000	0.82	1.71	0.00	2.53
33 06 0304	**Clay/Sand with Boulders, to 25', Slurry Wall Excavation**	**CY**				**0.82**	**1.71**	**0.00**	**2.53**
02228 0388	Trench, 2-1/2 CY Hydraulic Excavator, 113 CY/Hour (86 m3)/Hour)	CY	112.9	CODEU	1.000	0.82	1.71	0.00	2.53
33 06 0305	**Normal Soil, 26' - 75', Slurry Wall Excavation**	**CY**				**1.78**	**3.71**	**0.00**	**5.49**
02081 3003	Normal Soil - 26' - 75' Slurry Wall Excavation	CY	52	CODEU	1.000	1.78	3.71	0.00	5.49

33 06 03	Slurry Walls	Unit	Hourly Output	Crew	Factor	Labor	Equip.	Matl.	Total
33 06 0306	**Clay/Sand with Boulders, 26' - 75', Slurry Wall Excavation**	**CY**				**2.65**	**5.51**	**0.00**	**8.16**
02081 3004	Clay/Sand w/Boulders - 26' - 75' Slurry Wall Excavation	CY	35	CODEU	1.000	2.65	5.51	0.00	8.16
33 06 0307	**Normal Soil, 76' - 120', Slurry Wall Excavation**	**CY**				**3.84**	**7.86**	**0.00**	**11.70**
02081 3005	Normal Soil - 76' - 120' Slurry Wall Excavation	CY	24.16	UOEDW	1.000	3.84	7.86	0.00	11.70
33 06 0308	**Clay/Sand with Boulders, 76' - 120', Slurry Wall Excavation**	**CY**				**5.73**	**11.73**	**0.00**	**17.45**
02081 3006	Clay/Sand w/Boulders - 76' - 120' Slurry Wall Excavation	CY	16.19	UOEDW	1.000	5.73	11.73	0.00	17.45
33 06 0309	**Bentonite, Material Purchase Price per Ton**	**TON**				**0.00**	**0.00**	**58.00**	**58.00**
02081 2225	90 Barrel Yield Bentonite No Contamination Resistance	TON	0	N/A	1.000	0.00	0.00	58.00	58.00
33 06 0310	**Slurry Mixing, Hydration, and Placement, per Gallon**	**GAL**				**0.05**	**0.01**	**0.00**	**0.05**
02081 3007	Slurry Mixing, Hydration, and Placement, per Gallon	GAL	2188	ULABR	1.000	0.05	0.01	0.00	0.05
33 06 0311	**Soil-Bentonite Backfill Mixing, per Cubic Yard**	**CY**				**0.80**	**2.06**	**0.00**	**2.87**
02081 3008	Soil, Bentonite Backfill Mixing, per Cubic Yard	CY	145.7	CODTU	1.000	0.80	2.06	0.00	2.87
33 06 0312	**Backfill Slurry Wall Trench, 1,000' Average Haul Distance**	**CY**				**0.59**	**1.49**	**0.00**	**2.08**
02081 3009	Backfill Slurry Wall Trench, 1,000' Average Haul Distance	CY	58.67	CODLK	1.000	0.59	1.49	0.00	2.08
33 06 0313	**Demolish Mixing Basins and Regrade Working Surface**	**SF**				**0.03**	**0.04**	**0.00**	**0.07**
02081 3011	Cleanup and Regrade Working Surface	SF	3000	CODTT	1.000	0.03	0.04	0.00	0.07

33 06 10	Permeable Barriers	Unit	Hourly Output	Crew	Factor	Labor	Equip.	Matl.	Total
33 06 1011	**Temporary Medium Wall Sheet Piling**	**SF**				**4.30**	**5.38**	**9.70**	**19.39**
02086 1011	Temporary Medium Wall Sheet Piling	SF	59.74	CPIDV	1.000	4.30	5.38	9.70	19.39
33 06 1012	**Temporary Heavy Wall Sheet Piling**	**SF**				**6.08**	**7.61**	**12.26**	**25.95**
02086 1012	Temporary Heavy Wall Sheet Piling	SF	42.26	CPIDV	1.000	6.08	7.61	12.26	25.95
33 06 1013	**Permanent Medium Wall Sheet Piling**	**SF**				**4.30**	**5.38**	**3.15**	**12.83**
02086 1013	Permanent Medium Wall Sheet Piling	SF	59.74	CPIDV	1.000	4.30	5.38	3.15	12.83
33 06 1014	**Permanent Heavy Wall Sheet Piling**	**SF**				**6.08**	**7.61**	**12.26**	**25.95**
02086 1014	Permanent Heavy Wall Sheet Piling	SF	42.26	CPIDV	1.000	6.08	7.61	12.26	25.95
33 06 1021	**Trench - 2-1/2 CY Hydraulic Excavator -113 CY/Hr (86M3)/Hr**	**CY**				**0.82**	**1.71**	**0.00**	**2.53**
02228 0388	Trench, 2-1/2 CY Hydraulic Excavator, 113 CY/Hour (86 m3)/Hour)	CY	112.9	CODEU	1.000	0.82	1.71	0.00	2.53

33 06 10 Permeable Barriers		Unit	Hourly Output	Crew	Factor	Labor	Equip.	Matl.	Total
33 06 1022	**Excavate Trench in Cobble to 25' Deep**	**CY**				**0.82**	**1.71**	**0.00**	**2.53**
02228 0388	Trench, 2-1/2 CY Hydraulic Excavator, 113 CY/Hour (86 m3)/Hour)	CY	112.9	CODEU	1.000	0.82	1.71	0.00	2.53
33 06 1023	**Normal Soil - 26' - 75' Slurry Wall Excavation**	**CY**				**1.78**	**3.71**	**0.00**	**5.49**
02081 3003	Normal Soil - 26' - 75' Slurry Wall Excavation	CY	52	CODEU	1.000	1.78	3.71	0.00	5.49
33 06 1024	**Clay/Sand w/Boulders - 26' - 75' Slurry Wall Excavation**	**CY**				**2.65**	**5.51**	**0.00**	**8.16**
02081 3004	Clay/Sand w/Boulders - 26' - 75' Slurry Wall Excavation	CY	35	CODEU	1.000	2.65	5.51	0.00	8.16
33 06 1025	**Normal Soil - 76' - 120' Slurry Wall Excavation**	**CY**				**3.84**	**7.86**	**0.00**	**11.70**
02081 3005	Normal Soil - 76' - 120' Slurry Wall Excavation	CY	24.16	UOEDW	1.000	3.84	7.86	0.00	11.70
33 06 1026	**Clay/Sand w/Boulders - 76' - 120' Slurry Wall Excavation**	**CY**				**5.73**	**11.73**	**0.00**	**17.45**
02081 3006	Clay/Sand w/Boulders - 76' - 120' Slurry Wall Excavation	CY	16.19	UOEDW	1.000	5.73	11.73	0.00	17.45
33 06 1027	**Key-in Treatment Wall**	**CY**				**40.26**	**23.37**	**32.94**	**96.57**
02221 0800	Rock Removal Drill for Pre-Splitting	LF	39.38	CLADH	2.679	1.90	3.40	0.16	14.65
02221 1605	Quarry Production Drill, 75 LF/Hour, (23m/Hour)	LF	75.5	CLADH	2.679	0.99	1.77	0.00	7.41
02223 3000	Blasting for Pre-Splitting, Load and Detonate	LF	59	CLADF	2.679	2.37	0.78	5.96	24.41
02223 3700	Ammonium Nitrite Explosives	LB	0	N/A	6.200	0.00	0.00	1.99	12.32
02223 3720	Detonator Wire	LF	0	N/A	5.000	0.00	0.00	0.03	0.15
02223 3740	Primacord (50 Grain)	MLF	0	N/A	0.003	0.00	0.00	390.00	1.21
02223 3762	Fuel Oil for Blasting	GAL	0	N/A	2.765	0.00	0.00	1.03	2.85
02232 0150	Excavate & Load, 2 CY Hydraulic Excavator, Blast Rock, 65 CY/Hour (49 m3)	CY	65	CODET	1.000	0.96	1.75	0.00	2.71
02238 0600	Break Rock with Air Hammer, Dense Rock	CY	2	CLADA	1.000	25.20	5.66	0.00	30.86
33 06 1028	**Level and Compact Working Surface**	**CY**				**1.78**	**4.14**	**0.00**	**5.92**
02081 3001	Level and Compact Working Surface	CY	106.4	COFCV	1.000	1.78	4.14	0.00	5.92
33 06 1031	**Iron Filings**	**CY**				**23.99**	**29.13**	**438.73**	**491.85**
02086 1031	Iron Filings	CY	4.8	ULABN	1.000	23.99	29.13	438.73	491.85
33 06 1032	**Activated Carbon - Coal Derived**	**CY**				**23.99**	**29.13**	**1,036**	**1,089**
02086 1032	Activated Carbon - Coal Derived	CY	4.8	ULABN	1.000	23.99	29.13	1,036	1,089
33 06 1033	**Crushed Limestone**	**CY**				**4.09**	**2.76**	**64.33**	**71.19**
02086 1033	Crushed Limestone	CY	28.25	COETF	1.000	4.09	2.76	64.33	71.19
33 06 1034	**Clinoptilolite**	**CY**				**23.99**	**29.13**	**284.38**	**337.50**
02086 1034	Clinoptilolite	CY	4.8	ULABN	1.000	23.99	29.13	284.38	337.50
33 06 1035	**Montmorillonite Clay**	**CY**				**23.99**	**29.13**	**291.48**	**344.60**
02086 1035	Montmorillonite Clay	CY	4.8	ULABN	1.000	23.99	29.13	291.48	344.60

33 06 10	Permeable Barriers	Unit	Hourly Output	Crew	Factor	Labor	Equip.	Matl.	Total
33 06 1036	**Glauconitic Greensands**	**CY**				**23.99**	**29.13**	**1,324**	**1,377**
02086 1036	Glauconitic Greensands	CY	4.8	ULABN	1.000	23.99	29.13	1,324	1,377
33 06 1037	**Fuller's (Diatomaceous) Earth**	**CY**				**23.99**	**29.13**	**406.27**	**459.38**
02086 1037	Fuller's (Diatomaceous) Earth	CY	4.8	ULABN	1.000	23.99	29.13	406.27	459.38
33 06 1038	**Peat Moss**	**CY**				**4.09**	**2.76**	**89.38**	**96.24**
02086 1038	Peat Moss	CY	28.25	COETF	1.000	4.09	2.76	89.38	96.24
33 06 1039	**Proprietary Metal Oxidizing Powder**	**CY**				**23.99**	**29.13**	**5,638**	**5,691**
02086 1039	Proprietary Metal Oxidizing Powder	CY	4.8	ULABN	1.000	23.99	29.13	5,638	5,691
33 06 1040	**Proprietary Iron-Foam Aggregate**	**CY**				**23.99**	**29.13**	**2,491**	**2,544**
02086 1040	Proprietary Iron-Foam Aggregate	CY	4.8	ULABN	1.000	23.99	29.13	2,491	2,544
33 06 1041	**Proprietary Humic-Acid Adsorbent**	**CY**				**23.99**	**29.13**	**753.88**	**807.00**
02086 1041	Proprietary Humic-Acid Adsorbent	CY	4.8	ULABN	1.000	23.99	29.13	753.88	807.00
33 06 1042	**Pea Gravel**	**CY**				**4.09**	**2.76**	**24.91**	**31.77**
02086 1042	Pea Gravel	CY	28.25	COETF	1.000	4.09	2.76	24.91	31.77

33 07 02	Gas/Vapor Collection Trench System	Unit	Hourly Output	Crew	Factor	Labor	Equip.	Matl.	Total
33 07 0201	**6" Inside Diameter (Vertical Pipe Spaced @ 200 LF), Gas Vent Piping System**	**LF**				**4.95**	**6.36**	**12.50**	**23.81**
02082 1313	6" Diameter, SDR 21, 2 to 6 Rows of Slots	LF	37.5	ULABD	1.000	2.82	0.00	9.45	12.27
02082 1418	6" Diameter - SDR 21	LF	40	XPLUD	0.050	2.67	0.00	2.65	0.27
02082 1458	6" Diameter - SDR 21	EA	3	XPLUD	0.010	35.64	0.00	67.00	1.03
02082 1496	6" Diameter - SDR 21/SDR 26	EA	2.75	XPLUD	0.005	38.88	0.00	93.00	0.66
02557 5483	6" Diameter Filter Sock	LF	125	ULADB	1.000	0.72	3.18	0.98	4.88
02557 6122	3/8" Bentonite Tablets, 10.62 Lbs/Gallon	CF	2.5	ULADB	0.020	36.02	158.95	36.58	4.63
03372 4600	Pour Slab on Grade, >= 6", Direct Chute >= (15 cm) Place 3,000 PSI Concrete	CY	21.88	ALABE	0.001	7.70	0.29	84.22	0.08

33 07 04	Thermal Catalytic Oxidation	Unit	Hourly Output	Crew	Factor	Labor	Equip.	Matl.	Total
33 07 0401	**150 SCFM Simple Thermal Oxidizer**	**EA**				**1,403**	**115.00**	**38,907**	**40,426**
13277 1161	150 SCFM Simple Thermal Oxidizer	EA	0.063	MPLUG	1.000	1,403	115.00	38,907	40,426
33 07 0402	**250 SCFM Simple Thermal Oxidizer**	**EA**				**1,403**	**115.00**	**41,635**	**43,153**
13277 1162	250 SCFM Simple Thermal Oxidizer	EA	0.063	MPLUG	1.000	1,403	115.00	41,635	43,153
33 07 0403	**500 SCFM Simple Thermal Oxidizer**	**EA**				**1,403**	**115.00**	**56,664**	**58,183**
13277 1163	500 SCFM Simple Thermal Oxidizer	EA	0.063	MPLUG	1.000	1,403	115.00	56,664	58,183
33 07 0404	**1,000 SCFM Simple Thermal Oxidizer**	**EA**				**2,103**	**172.36**	**78,179**	**80,454**
13277 1164	1,000 SCFM Simple Thermal Oxidizer	EA	0.042	MPLUG	1.000	2,103	172.36	78,179	80,454

33 07 04 Thermal Catalytic Oxidation		Unit	Hourly Output	Crew	Factor	Labor	Equip.	Matl.	Total
33 07 0405	**5,000 SCFM Simple Thermal Oxidizer**	**EA**				**2,103**	**172.36**	**309,684**	**311,960**
13277 1169	5,000 SCFM Simple Thermal Oxidizer	EA	0.042	MPLUG	1.000	2,103	172.36	309,684	311,960
33 07 0406	**150 SCFM Recuperative Thermal Oxidation Unit**	**EA**				**1,403**	**115.00**	**54,786**	**56,304**
13277 1165	150 SCFM Recuperative Thermal Oxidation Unit	EA	0.063	MPLUG	1.000	1,403	115.00	54,786	56,304
33 07 0407	**250 SCFM Recuperative Thermal Oxidation Unit**	**EA**				**1,403**	**115.00**	**57,876**	**59,395**
13277 1166	250 SCFM Recuperative Thermal Oxidation Unit	EA	0.063	MPLUG	1.000	1,403	115.00	57,876	59,395
33 07 0408	**500 SCFM Recuperative Thermal Oxidation Unit**	**EA**				**1,403**	**115.00**	**73,633**	**75,152**
13277 1167	500 SCFM Recuperative Thermal Oxidation Unit	EA	0.063	MPLUG	1.000	1,403	115.00	73,633	75,152
33 07 0409	**1,000 SCFM Regenerative Thermal Oxidation Unit**	**EA**				**2,103**	**172.36**	**98,178**	**100,453**
13277 1168	1,000 SCFM Recuperative Thermal Oxidation Unit	EA	0.042	MPLUG	1.000	2,103	172.36	98,178	100,453
33 07 0410	**5,000 SCFM Regenerative Thermal Oxidation Unit**	**EA**				**2,103**	**172.36**	**464,344**	**466,620**
13277 1170	5,000 SCFM Recuperative Thermal Oxidation Unit	EA	0.042	MPLUG	1.000	2,103	172.36	464,344	466,620
33 07 0414	**2,500 SCFM Regenerative Thermal Oxidation Unit**	**EA**				**2,103**	**172.36**	**170,296**	**172,572**
13277 1155	2,500 SCFM Regenerative Thermal Oxidation Unit	EA	0.042	MPLUG	1.000	2,103	172.36	170,296	172,572
33 07 0415	**5,000 SCFM Regenerative Thermal Oxidation Unit**	**EA**				**2,103**	**172.36**	**202,500**	**204,776**
13277 1151	5,000 SCFM - Paint Booth or Similar Exhaust Stream Emissions	EA	0	N/A	1.000	0.00	0.00	202,500	202,500
13277 1159	Oxidizer Installation	EA	0.042	MPLUG	1.000	2,103	172.36	0.00	2,276
33 07 0416	**5 SCFM Flameless Thermal Oxidation Unit**	**EA**				**1,403**	**115.00**	**61,937**	**63,455**
13277 1171	5 SCFM Flameless Thermal Oxidation Unit	EA	0.063	MPLUG	1.000	1,403	115.00	61,937	63,455
33 07 0417	**100 SCFM Flameless Thermal Oxidation Unit**	**EA**				**1,403**	**115.00**	**123,874**	**125,392**
13277 1172	100 SCFM Flameless Thermal Oxidation Unit	EA	0.063	MPLUG	1.000	1,403	115.00	123,874	125,392
33 07 0418	**1,000 SCFM Flameless Thermal Oxidation Unit**	**EA**				**2,103**	**172.36**	**178,174**	**180,450**
13277 1173	1,000 SCFM Flameless Thermal Oxidation Unit	EA	0.042	MPLUG	1.000	2,103	172.36	178,174	180,450
33 07 0419	**5,000 SCFM Flameless Thermal Oxidation Unit**	**EA**				**2,103**	**172.36**	**343,622**	**345,898**
13277 1174	5,000 SCFM Flameless Thermal Oxidation Unit	EA	0.042	MPLUG	1.000	2,103	172.36	343,622	345,898
33 07 0420	**1,000 SCFM Regenerative Flameless Thermal Unit**	**EA**				**2,103**	**172.36**	**204,234**	**206,510**
13277 1175	1,000 SCFM Recuperative Flameless Thermal Oxidation Unit	EA	0.042	MPLUG	1.000	2,103	172.36	204,234	206,510

33 07 04	Thermal Catalytic Oxidation	Unit	Hourly Output	Crew	Factor	Labor	Equip.	Matl.	Total
33 07 0421	**5,000 SCFM Regenerative Flameless Thermal Unit**	**EA**				**2,103**	**172.36**	**376,106**	**378,381**
13277 1176	5,000 SCFM Recuperative Flameless Thermal Oxidation Unit	EA	0.042	MPLUG	1.000	2,103	172.36	376,106	378,381
33 07 0441	**100 SCFM Fixed Bed Catalytic Unit**	**EA**				**1,403**	**115.00**	**46,422**	**47,941**
13277 1181	100 SCFM Fixed Bed Catalytic Oxidation Unit	EA	0.063	MPLUG	1.000	1,403	115.00	46,422	47,941
33 07 0442	**150 SCFM Fixed Bed Catalytic Unit**	**EA**				**1,403**	**115.00**	**46,786**	**48,304**
13277 1182	150 SCFM Fixed Bed Catalytic Oxidation Unit	EA	0.063	MPLUG	1.000	1,403	115.00	46,786	48,304
33 07 0443	**250 SCFM Fixed Bed Catalytic Unit**	**EA**				**1,403**	**115.00**	**50,907**	**52,425**
13277 1183	250 SCFM Fixed Bed Catalytic Oxidation Unit	EA	0.063	MPLUG	1.000	1,403	115.00	50,907	52,425
33 07 0444	**500 SCFM Fixed Bed Catalytic Unit**	**EA**				**2,103**	**172.36**	**72,906**	**75,182**
13277 1184	500 SCFM Fixed Bed Catalytic Oxidation Unit	EA	0.042	MPLUG	1.000	2,103	172.36	72,906	75,182
33 07 0445	**1,000 SCFM Fixed Bed Catalytic Unit**	**EA**				**2,103**	**172.36**	**109,086**	**111,362**
13277 1185	1,000 SCFM Fixed Bed Catalytic Oxidation Unit	EA	0.042	MPLUG	1.000	2,103	172.36	109,086	111,362
33 07 0446	**5,000 SCFM Fixed Bed Catalytic Unit**	**EA**				**2,103**	**172.36**	**132,116**	**134,391**
13277 1186	5,000 SCFM Fixed Bed Catalytic Oxidation Unit	EA	0.042	MPLUG	1.000	2,103	172.36	132,116	134,391
33 07 0447	**Precious Metal Catalyst**	**CF**				**0.00**	**0.00**	**4,848**	**4,848**
13277 1187	Precious Metal Catalyst	SCF	0	N/A	1.000	0.00	0.00	4,848	4,848
33 07 0451	**100 SCFM Recuperative Fixed Bed Catalytic**	**EA**				**1,403**	**115.00**	**61,816**	**63,334**
13277 1145	100 SCFM Recuperative Fixed Bed Catalytic Unit	EA	0.063	MPLUG	1.000	1,403	115.00	61,816	63,334
33 07 0452	**150 SCFM Recuperative Fixed Bed Catalytic**	**EA**				**1,403**	**115.00**	**63,028**	**64,546**
13277 1146	150 SCFM Recuperative Fixed Bed Catalytic Unit	EA	0.063	MPLUG	1.000	1,403	115.00	63,028	64,546
33 07 0453	**250 SCFM Recuperative Fixed Bed Catalytic**	**EA**				**1,403**	**115.00**	**67,270**	**68,788**
13277 1147	250 SCFM Recuperative Fixed Bed Catalytic Unit	EA	0.063	MPLUG	1.000	1,403	115.00	67,270	68,788
33 07 0454	**500 SCFM Recuperative Fixed Bed Catalytic**	**EA**				**1,403**	**115.00**	**89,936**	**91,454**
13277 1148	500 SCFM Recuperative Fixed Bed Catalytic Unit	EA	0.063	MPLUG	1.000	1,403	115.00	89,936	91,454
33 07 0455	**1,000 SCFM Regenerative Fixed Bed Catalytic**	**EA**				**3,262**	**459.27**	**76,500**	**80,222**
13277 1131	1,000 SCFM Recuperative Fixed Bed Catalytic Unit	EA	0.031	MSPFF	1.000	3,262	459.27	76,500	80,222
33 07 0456	**2,000 SCFM Regenerative Fixed Bed Catalytic**	**EA**				**3,262**	**459.27**	**109,000**	**112,722**
13277 1132	2,000 SCFM Recuperative Fixed Bed Catalytic Unit	EA	0.031	MSPFF	1.000	3,262	459.27	109,000	112,722
33 07 0457	**3,000 SCFM Regenerative Fixed Bed Catalytic**	**EA**				**4,085**	**575.00**	**131,000**	**135,660**
13277 1133	3,000 SCFM Recuperative Fixed Bed Catalytic Unit	EA	0.025	MSPFF	1.000	4,085	575.00	131,000	135,660

33 07 Air Pollution/Gas Collection and Control

33 07 04 Thermal Catalytic Oxidation	Unit	Hourly Output	Crew	Factor	Labor	Equip.	Matl.	Total
33 07 0458 4,000 SCFM Regenerative Fixed Bed Catalytic	**EA**				**4,085**	**575.00**	**154,000**	**158,660**
13277 1134 4,000 SCFM Recuperative Fixed Bed Catalytic Unit	EA	0.025	MSPFF	1.000	4,085	575.00	154,000	158,660
33 07 0459 5,000 SCFM Regenerative Fixed Bed Catalytic	**EA**				**4,085**	**575.00**	**171,000**	**175,660**
13277 1135 5,000 SCFM Recuperative Fixed Bed Catalytic Unit	EA	0.025	MSPFF	1.000	4,085	575.00	171,000	175,660
33 07 0461 Remote Monitoring Unit	**EA**				**0.00**	**0.00**	**4,666**	**4,666**
13277 1196 Remote Monitoring Unit	EA	0	N/A	1.000	0.00	0.00	4,666	4,666

33 07 99 Other	Unit	Hourly Output	Crew	Factor	Labor	Equip.	Matl.	Total
33 07 9901 Lease, Portable 40' W x 150' L x 18' High Emissions Containment Building	**YR**				**0.00**	**0.00**	**8,450**	**8,450**
02083 6171 Lease, Portable 40' W x 150' L x 18' H Emissions Containment Building	YR	0	N/A	1.000	0.00	0.00	8,450	8,450
33 07 9902 Buy, Portable 40' W x 150' L x 18' High Emissions Containment Building	**EA**				**0.00**	**0.00**	**23,700**	**23,700**
02083 6172 Purchase, Portable 40' W x 150' L x 18' H Emissions Containment Building	EA	0	N/A	1.000	0.00	0.00	23,700	23,700
33 07 9903 Assemble Emissions Containment Building, 5 Days	**EA**				**5,340**	**790.00**	**0.00**	**6,130**
02083 6173 Assemble Emissions Containment Building	EA	0.025	HBLDA	1.000	5,340	790.00	0.00	6,130
33 07 9904 Disassemble Emissions Containment Building, 3 Days	**EA**				**3,201**	**473.62**	**0.00**	**3,675**
02083 6174 Disassemble Emissions Containment Building	EA	0.042	HBLDA	1.000	3,201	473.62	0.00	3,675
33 07 9905 Purchase & Install Portable Building, 18' Ceiling	**SF**				**0.91**	**0.13**	**4.03**	**5.07**
02083 6172 Purchase, Portable 40' W x 150' L x 18' H Emissions Containment Building	EA	0	N/A	0.000	0.00	0.00	23,700	4.03
02083 6173 Assemble Emissions Containment Building	EA	0.025	HBLDA	0.000	5,340	790.00	0.00	1.04
33 07 9907 Disassemble Portable Building, 18' Ceiling, without Slab	**SF**				**0.54**	**0.08**	**0.00**	**0.62**
02083 6174 Disassemble Emissions Containment Building	EA	0.042	HBLDA	0.000	3,201	473.62	0.00	0.62

33 08 05 Capping of Contaminated Area/Waste Pile	Unit	Hourly Output	Crew	Factor	Labor	Equip.	Matl.	Total
33 08 0501 Site Inspections/Reporting, Quarterly	**EA**				**479.75**	**0.00**	**0.00**	**479.75**
01050 0101 Field Office Clerk	WK	0.025	XYBBA	0.100	320.00	0.00	0.00	32.00
01050 0103 Field Engineer (Average Cost)	WK	0.025	XYBBD	0.450	995.00	0.00	0.00	447.75
33 08 0502 Post-Closure Administrative	**YR**				**29,062**	**0.00**	**0.00**	**29,062**
01050 0101 Field Office Clerk	WK	0.025	XYBBA	26.000	320.00	0.00	0.00	8,320
01050 0107 Project Manager - Average Cost	WK	0.025	XYBBH	13.000	1,596	0.00	0.00	20,742

33 08 05	Capping of Contaminated Area/Waste Pile	Unit	Hourly Output	Crew	Factor	Labor	Equip.	Matl.	Total
33 08 0503	**Polymeric Liner Anchor Trench, 3' x 1.5'**	**LF**				**0.54**	**0.21**	**0.06**	**0.81**
02215 0500	Compaction, 6" Layers with Air Tamper by Hand, (15 cm) Layers	CY	50	CLACA	0.167	2.71	0.17	0.00	0.48
02215 1215	Backfill Trench with 200HP Dozer without Compaction	CY	284	CODTG	0.167	0.15	0.43	0.00	0.10
02223 1001	Compaction Water Price $.005/Gallon ($0.454/100 Liters)	CY	174.3	COFWK	0.167	0.30	0.32	0.34	0.16
02230 0870	Trench, FW Trencher, Medium Soil 24" Wide x 9' Deep	CY	498	CODBG	0.167	0.08	0.37	0.00	0.07
33 08 0504	**Herbicide Application**	**ACRE**				**105.47**	**132.31**	**64.86**	**302.64**
02280 3100	Herbicide Application from Truck	SY	2420	COFWI	4840.000	0.02	0.03	0.01	302.64
33 08 0505	**Soil/Bentonite Liner - 6" Lifts**	**SF**				**0.03**	**0.00**	**1.37**	**1.40**
02081 2228	Liner/Cap Installed - Cost per 6 Lift-4 Lb/SF Bentonite @ $243/Ton	SF	1250	CODTF	1.000	0.03	0.00	1.37	1.40
33 08 0506	**Clay 10E-7, 6" Lifts, On-Site**	**CY**				**6.01**	**10.52**	**0.98**	**17.51**
01418 4405	Atterberg Limits	EA	0.429	USKCA	0.001	102.76	0.00	60.00	0.14
01418 4605	Grain Size (No. 200 Sieve) ASTM D421	EA	0.64	USKCA	0.001	68.85	0.00	59.00	0.13
01418 4620	Grain Size (2 Micron)	EA	0.64	USKCA	0.001	68.85	0.00	93.50	0.11
01418 4736	Soil Density Test, Nuclear Method ASTM D2922-71	EA	2	USKCB	0.004	44.06	0.00	40.00	0.34
01418 4751	Moisture Content	EA	0.471	USKCA	0.004	93.65	0.00	5.49	0.40
01418 4755	Moisture	EA	3.393	USKCA	0.001	12.99	0.00	10.00	0.02
01418 4760	Moisture Content Undisturbed Sample	EA	0.848	USKCA	0.001	51.99	0.00	10.00	0.04
01418 4770	Dry Density (Undisturbed Sample)	EA	0.369	USKCA	0.001	119.48	0.00	33.00	0.10
01418 4810	Undisturbed Permeability	EA	0.089	USKCA	0.001	496.20	0.00	246.00	0.50
01418 4855	Laboratory Permeability (Remolded Sample)	EA	0.081	USKCA	0.000	541.97	0.00	243.00	0.08
01418 4970	Moisture Density, Relative Moisture Content of Soils ASTM D1557-70	EA	0.5	USKCA	0.000	88.13	0.00	144.00	0.09
02212 3106	By 654 HP Sheepsfoot Roller, 6 Passes	CY	37.5	COFCQ	1.000	2.45	3.75	0.00	6.20
02215 2360	Spread Dumped Fill/Gravel, 6" Layers, without Compaction	SY	448.5	CODTG	6.000	0.10	0.27	0.00	2.22
02223 1001	Compaction Water Price $.005/Gallon ($0.454/100 Liters)	CY	174.3	COFWK	1.000	0.30	0.32	0.34	0.96
02232 0325	Excavate & Load, 2-3/4 CY Track Loader, Medium Material, 100 CY/Hour (76 m3)	CY	100	CODLD	1.000	0.48	1.37	0.00	1.85
02234 1115	Haul, 16.5 CY (12.6 m3) Truck, 6 Miles (10 km), 40 MPH (60 km/Hour), 2.1 Cycles/Hour	CY	34.25	COETH	1.000	0.80	3.33	0.00	4.13
13273 9112	Till Soil with D4 Dozer & Disc Attachment	HR	1	CODTW	0.003	34.65	47.24	0.00	0.20

33 08 05	Capping of Contaminated Area/Waste Pile	Unit	Hourly Output	Crew	Factor	Labor	Equip.	Matl.	Total
33 08 0507	**Clay 10E-7, 6" Lifts, Off-Site**	**CY**				**5.36**	**9.15**	**7.55**	**22.06**
01418 4405	Atterberg Limits	EA	0.429	USKCA	0.001	102.76	0.00	60.00	0.11
01418 4620	Grain Size (2 Micron)	EA	0.64	USKCA	0.001	68.85	0.00	93.50	0.11
01418 4736	Soil Density Test, Nuclear Method ASTM D2922-71	EA	2	USKCB	0.004	44.06	0.00	40.00	0.34
01418 4751	Moisture Content	EA	0.471	USKCA	0.004	93.65	0.00	5.49	0.40
01418 4760	Moisture Content Undisturbed Sample	EA	0.848	USKCA	0.001	51.99	0.00	10.00	0.04
01418 4770	Dry Density (Undisturbed Sample)	EA	0.369	USKCA	0.001	119.48	0.00	33.00	0.10
01418 4810	Undisturbed Permeability	EA	0.089	USKCA	0.001	496.20	0.00	246.00	0.50
01418 4970	Moisture Density, Relative Moisture Content of Soils ASTM D1557-70	EA	0.5	USKCA	0.000	88.13	0.00	144.00	0.05
02081 2241	Clay Backfill	CY	0	N/A	1.000	0.00	0.00	6.70	6.70
02212 3106	By 654 HP Sheepsfoot Roller, 6 Passes	CY	37.5	COFCQ	1.000	2.45	3.75	0.00	6.20
02215 2360	Spread Dumped Fill/Gravel, 6" Layers, without Compaction	SY	448.5	CODTG	6.000	0.10	0.27	0.00	2.22
02223 1001	Compaction Water Price $.005/Gallon ($0.454/100 Liters)	CY	174.3	COFWK	1.000	0.30	0.32	0.34	0.96
02234 1115	Haul, 16.5 CY (12.6 m3) Truck, 6 Miles (10 km), 40 MPH (60 km/Hour), 2.1 Cycles/Hour	CY	34.25	COETH	1.000	0.80	3.33	0.00	4.13
13273 9112	Till Soil with D4 Dozer & Disc Attachment	HR	1	CODTW	0.003	34.65	47.24	0.00	0.20
33 08 0508	**Fabric Membrane Liner/Bentonite Liner**	**SF**				**0.09**	**0.07**	**0.73**	**0.90**
02081 2311	1/4" Thick Geocomposite Installed on Soil Or Aggregate Base	SF	1250	CODEZ	1.000	0.09	0.07	0.73	0.90
33 08 0511	**Drainage Netting, 1/4" Thick High-density Polyethylene**	**SF**				**0.04**	**0.01**	**0.20**	**0.24**
02082 2111	Drainage Net - 1/4" Thick High-density Polyethylene	SF	6250	USKCF	1.000	0.04	0.01	0.20	0.24
33 08 0512	**Drainage Netting, Geotextile Fabric Heat-bonded 1 Side**	**SF**				**0.04**	**0.01**	**0.43**	**0.48**
02082 2112	Drainage Net with Geotextile Fabric Heat-bonded to One Side	SF	5625	USKCF	1.000	0.04	0.01	0.43	0.48
33 08 0513	**Drainage Netting, Geotextile Fabric Heat-bonded 2 Sides**	**SF**				**0.05**	**0.01**	**0.48**	**0.54**
02082 2113	Drainage Net with Geotextile Fabric Heat-bonded to Both Sides	SF	5000	USKCF	1.000	0.05	0.01	0.48	0.54
33 08 0520	**Geosynthetic Clay Liner**	**SF**				**0.00**	**0.00**	**0.38**	**0.38**
01001 0081	Geosynthetic Clay Liner	SF	0	N/A	1.000	0.00	0.00	0.38	0.38
33 08 0521	**Geogrid, Biaxial 9.8' x 164' Roll, TM = 14,000 PSF**	**SF**				**0.02**	**0.00**	**0.19**	**0.22**
02081 2411	Biaxial, 9.8' x 164' Rolls Tensile Modulus = 14,000 Lb/SF	SF	9375	USKCF	1.000	0.02	0.00	0.19	0.22

33 08 05	Capping of Contaminated Area/Waste Pile	Unit	Hourly Output	Crew	Factor	Labor	Equip.	Matl.	Total
33 08 0522	**Geogrid, Biaxial 9.8' x 164' Roll, TM = 18,500 PSF**	**SF**				**0.03**	**0.01**	**0.33**	**0.37**
02081 2413	Biaxial, 9.8' x 164' Rolls Tensile Modulus = 18,500 Lb/SF	SF	6875	USKCF	1.000	0.03	0.01	0.33	0.37
33 08 0523	**Geogrid, Uniaxial 4.3' x 98' Roll, TM = 50,000 PSF**	**SF**				**0.04**	**0.01**	**0.58**	**0.62**
02081 2415	Uniaxial, 4.3' x 98' Rolls Tensile Modulus = 50,000 Lb/SF	SF	6250	USKCF	1.000	0.04	0.01	0.58	0.62
33 08 0524	**Geogrid, Uniaxial 4.3' x 98' Roll, TM = 130,000 PSF**	**SF**				**0.05**	**0.01**	**1.19**	**1.25**
02081 2418	Uniaxial, 4.3' x 98' Rolls Tensile Modulus = 130,000 Lb/SF	SF	4375	USKCF	1.000	0.05	0.01	1.19	1.25
33 08 0525	**Geogrid, Nylon Geomatrix/PVC-coated Polyester**	**SF**				**0.04**	**0.01**	**1.07**	**1.12**
02081 2419	Composite Grip Layer Material - Nylon Geometric/PVC Coated Polyester	SF	5625	USKCF	1.000	0.04	0.01	1.07	1.12
33 08 0531	**6 oz/sy Geotextile, Non-Woven (60 Mil)**	**SY**				**0.44**	**0.02**	**0.74**	**1.20**
02250 2130	Geotextile Fabric, 60 Mil Thick Non-Woven Polypropylene	SY	187.5	ULABJ	1.000	0.44	0.02	0.74	1.20
33 08 0532	**8 oz/sy Erosion Control/Drainage Filter Fabric (80 Mil)**	**SY**				**0.44**	**0.02**	**0.79**	**1.25**
02535 1014	Geotextile Fabric, 80 Mil Thick Non-Woven Polypropylene	SY	187.5	ULABJ	1.000	0.44	0.02	0.79	1.25
33 08 0533	**105 Mil Geotextile, Nonwoven**	**SY**				**0.55**	**0.03**	**0.93**	**1.50**
02535 1016	Geotextile Fabric, 105 Mil Thick Non-Woven Polypropylene	SY	150	ULABJ	1.000	0.55	0.03	0.93	1.50
33 08 0534	**12 oz/sy Geotextile/Drainage Fabric (130 Mil)**	**SY**				**0.73**	**0.03**	**1.25**	**2.01**
02535 1018	Geotextile Fabric, 130 Mil Thick Non-Woven Polypropylene	SY	112.5	ULABJ	1.000	0.73	0.03	1.25	2.01
33 08 0535	**16 oz/sy Geotextile/Drainage Fabric (170 Mil)**	**SY**				**0.73**	**0.03**	**1.77**	**2.53**
02535 1021	Geotextile Fabric, 170 Mil Thick Non-Woven Polypropylene	SY	112.5	ULABJ	1.000	0.73	0.03	1.77	2.53
33 08 0541	**20 Mil Polymeric Liner, Very Low Density Polyethylene**	**SF**				**0.46**	**0.09**	**0.28**	**0.83**
02081 2111	20 Mil Very Low Density Polyethylene	SF	501	USKCF	1.000	0.46	0.09	0.28	0.83
33 08 0542	**30 Mil Polymeric Liner, Very Low Density Polyethylene**	**SF**				**0.77**	**0.15**	**0.35**	**1.27**
02081 2112	30 Mil Very Low Density Polyethylene	SF	300.6	USKCF	1.000	0.77	0.15	0.35	1.27
33 08 0543	**40 Mil Polymeric Liner, Very Low Density Polyethylene**	**SF**				**0.93**	**0.17**	**0.28**	**1.38**
02081 2113	40 Mil Very Low Density Polyethylene	SF	250.5	USKCF	1.000	0.93	0.17	0.28	1.38

33 08 05	Capping of Contaminated Area/Waste Pile	Unit	Hourly Output	Crew	Factor	Labor	Equip.	Matl.	Total
33 08 0544	**60 Mil Polymeric Liner, Very Low Density Polyethylene**	**SF**				**1.39**	**0.26**	**0.45**	**2.10**
02081 2115	60 Mil Very Low Density Polyethylene	SF	167	USKCF	1.000	1.39	0.26	0.45	2.10
33 08 0545	**80 Mil Polymeric Liner, Very Low Density Polyethylene**	**SF**				**1.70**	**0.32**	**0.60**	**2.62**
02081 2117	80 Mil Very Low Density Polyethylene	SF	136.6	USKCF	1.000	1.70	0.32	0.60	2.62
33 08 0546	**100 Mil Polymeric Liner, Very Low Density Polyethylene**	**SF**				**2.16**	**0.41**	**0.74**	**3.31**
02081 2119	100 Mil Very Low Density Polyethylene	SF	107.4	USKCF	1.000	2.16	0.41	0.74	3.31
33 08 0551	**36 Mil Polymeric Liner, Chlorosulfanated Polyethylene**	**SF**				**0.77**	**0.15**	**0.97**	**1.89**
02081 2133	36 Mil Hypalon	SF	300.6	USKCF	1.000	0.77	0.15	0.97	1.89
33 08 0552	**45 Mil Polymeric Liner, Chlorosulfanated Polyethylene**	**SF**				**0.93**	**0.17**	**1.04**	**2.14**
02081 2135	45 Mil Hypalon	SF	250.5	USKCF	1.000	0.93	0.17	1.04	2.14
33 08 0561	**20 Mil Polymeric Liner, PVC**	**SF**				**0.46**	**0.09**	**0.23**	**0.78**
02081 2141	20 Mil PVC	SF	501	USKCF	1.000	0.46	0.09	0.23	0.78
33 08 0562	**30 Mil Polymeric Liner, PVC**	**SF**				**0.77**	**0.15**	**0.31**	**1.23**
02081 2142	30 Mil PVC	SF	300.6	USKCF	1.000	0.77	0.15	0.31	1.23
33 08 0563	**40 Mil Polymeric Liner, PVC**	**SF**				**0.93**	**0.17**	**0.35**	**1.45**
02081 2143	40 Mil PVC	SF	250.5	USKCF	1.000	0.93	0.17	0.35	1.45
33 08 0564	**60 Mil Polymeric Liner, PVC**	**SF**				**1.39**	**0.26**	**0.61**	**2.26**
02081 2145	60 Mil PVC	SF	167	USKCF	1.000	1.39	0.26	0.61	2.26
33 08 0565	**80 Mil Polymeric Liner, PVC**	**SF**				**1.70**	**0.32**	**1.15**	**3.17**
02081 2147	80 Mil PVC	SF	136.6	USKCF	1.000	1.70	0.32	1.15	3.17
33 08 0566	**100 Mil Polymeric Liner, PVC**	**SF**				**2.16**	**0.41**	**1.72**	**4.29**
02081 2149	100 Mil PVC	SF	107.4	USKCF	1.000	2.16	0.41	1.72	4.29
33 08 0571	**40 Mil Polymeric Liner, High-density Polyethylene**	**SF**				**0.93**	**0.17**	**0.32**	**1.42**
02081 2151	40 Mil High-density Polyethylene	SF	250	USKCF	1.000	0.93	0.17	0.32	1.42
33 08 0572	**60 Mil Polymeric Liner, High-density Polyethylene**	**SF**				**1.37**	**0.26**	**0.46**	**2.09**
02081 2152	60 Mil High-density Polyethylene	SF	168.8	USKCF	1.000	1.37	0.26	0.46	2.09
33 08 0573	**80 Mil Polymeric Liner, High-density Polyethylene**	**SF**				**1.69**	**0.32**	**0.62**	**2.62**
02081 2153	80 Mil High-density Polyethylene	SF	137.5	USKCF	1.000	1.69	0.32	0.62	2.62

33 08 05 Capping of Contaminated Area/Waste Pile		Unit	Hourly Output	Crew	Factor	Labor	Equip.	Matl.	Total
33 08 0574	**Treesap-based, Overall Dust Suppressant**	**SY**				**0.86**	**0.79**	**0.34**	**1.99**
02082 7101	Overall Dust Suppression and Erosion Control, Treesap Based	SY	125	COKBD	1.000	0.86	0.79	0.34	1.99
33 08 0575	**Chloride-based, Overall Dust Suppressant**	**SY**				**0.86**	**0.79**	**0.32**	**1.97**
02082 7105	Overall Dust Suppression and Erosion Control, Chloride Based	SY	125	COKBD	1.000	0.86	0.79	0.32	1.97
33 08 0576	**Oil Emulsion-based, Overall Dust Suppressant**	**SY**				**0.86**	**0.79**	**0.23**	**1.88**
02082 7109	Overall Dust Suppression and Erosion Control, Oil Emulsion	SY	125	COKBD	1.000	0.86	0.79	0.23	1.88
33 08 0577	**Asphalt Emulsion-based, Overall Dust Suppressant**	**SY**				**0.96**	**0.88**	**0.32**	**2.16**
02082 7114	Overall Dust Suppression and Erosion Control, Asphalt	SY	112.5	COKBD	1.000	0.96	0.88	0.32	2.16
33 08 0578	**Lignosulphonate-based, Overall Dust Suppressant**	**SY**				**0.86**	**0.79**	**0.26**	**1.91**
02082 7118	Overall Dust Suppression and Erosion Control, Lignosulphonate	SY	125	COKBD	1.000	0.86	0.79	0.26	1.91
33 08 0579	**Treesap-based, Stockpile Dust Suppressant**	**SY**				**0.86**	**0.79**	**0.39**	**2.04**
02082 7103	Stockpile Dust Suppression Treesap-Based Dust Suppressant	SY	125	COKBD	1.000	0.86	0.79	0.39	2.04
33 08 0580	**Chloride-based, Stockpile Dust Suppressant**	**SY**				**0.86**	**0.79**	**0.37**	**2.02**
02082 7107	Stockpile Dust Suppression, Chloride-Based Sprayed Dust	SY	125	COKBD	1.000	0.86	0.79	0.37	2.02
33 08 0581	**Oil Emulsion-based, Stockpile Dust Suppressant**	**SY**				**0.86**	**0.79**	**0.31**	**1.96**
02082 7112	Stockpile Dust Suppression, Oil Emulsion Sprayed Dust	SY	125	COKBD	1.000	0.86	0.79	0.31	1.96
33 08 0582	**Asphalt Emulsion-based, Stockpile Dust Suppressant**	**SY**				**0.96**	**0.88**	**0.37**	**2.21**
02082 7116	Stockpile Dust Suppression, Asphalt Emulsion Sprayed Dust	SY	112.5	COKBD	1.000	0.96	0.88	0.37	2.21
33 08 0583	**Lignosulphonate-based, Stockpile Dust Suppressant**	**SY**				**0.86**	**0.79**	**0.32**	**1.97**
02082 7121	Stockpile Dust Suppression, Lignosulphonate Sprayed Dust	SY	125	COKBD	1.000	0.86	0.79	0.32	1.97
33 08 0584	**Plastic Laminate Waste Pile Cover**	**SF**				**0.02**	**0.00**	**0.12**	**0.14**
02082 7211	Plastic Laminate Waste Pile Cover, 130 Lb Tear Strength	SF	5000	ULABC	1.000	0.02	0.00	0.12	0.14
33 08 0585	**Sprayed Water Dust Suppressant**	**SY**				**0.01**	**0.01**	**0.00**	**0.01**
02976 4930	Watering by Truck	MSF	87.12	COFWI	0.009	0.61	0.76	0.02	0.01

33 08 Solids Collection and Containment

33 08 05	Capping of Contaminated Area/Waste Pile	Unit	Hourly Output	Crew	Factor	Labor	Equip.	Matl.	Total
33 08 0590	**Waste Pile Cover, 135 Lb Tear, 2 - 2.5 Year Life**	**SY**				**0.19**	**0.00**	**1.62**	**1.81**
02082 7212	Plastic Laminate Waste Pile Cover, 135 Lb Tear Strength	SF	4375	ULABC	9.000	0.02	0.00	0.18	1.81
33 08 0591	**Waste Pile Cover, 185 Lb Tear, 3 - 4 Year Life**	**SY**				**0.19**	**0.00**	**2.52**	**2.71**
02082 7213	Plastic Laminate Waste Pile Cover, 185 Lb Tear Strength	SF	4375	ULABC	9.000	0.02	0.00	0.28	2.71
33 08 0592	**Waste Pile Cover, 250 Lb Tear, 4 - 5 Year Life**	**SY**				**0.23**	**0.00**	**3.51**	**3.74**
02082 7214	Plastic Laminate Waste Pile Cover, 250 Lb Tear Strength	SF	3750	ULABC	9.000	0.03	0.00	0.39	3.74

33 09 07	Lagoons/Basins/Tanks/Dikes	Unit	Hourly Output	Crew	Factor	Labor	Equip.	Matl.	Total
33 09 0701	**CIP Reinforced Concrete Containment Wall**	**CY**				**60.48**	**21.06**	**126.55**	**208.08**
03210 1011	Wall Rebar, #3-#7	TON	0.375	SIWRF	0.055	482.29	191.67	647.24	72.67
03372 5200	Pour Concrete Wall, 12" Thick, Crane & Bucket (31 cm) 3,000 PSI Concrete	CY	11.25	ALABG	1.000	23.20	10.51	84.22	117.94
07174 0100	Silicone Retarder, 1 Coat on Concrete, Sprayed	CSF	1.875	ARFCD	0.540	19.90	0.00	12.47	17.48
33 09 0702	**CIP, Reinforced Concrete Containment Wall, Including Forms**	**CY**				**363.23**	**0.30**	**165.93**	**529.45**
03170 2150	Slab Curb Forms, 6" to 12" High, 4 Uses	SF	34.38	ACARI	108.000	3.03	0.00	0.53	384.76
03217 0600	Grade 60 Rebar, Footings & Slabs, #3-#6	TON	0.313	SIWRC	0.036	492.16	0.00	609.86	39.75
03372 4600	Pour Slab on Grade, >= 6", Direct Chute >= (15 cm) Place 3,000 PSI Concrete	CY	21.88	ALABE	1.030	7.70	0.29	84.22	94.97
03396 0010	Concrete Floor Finishes, Screed	SF	112.5	ACMAA	27.000	0.37	0.00	0.00	9.98

33 10 01	Drum Removal	Unit	Hourly Output	Crew	Factor	Labor	Equip.	Matl.	Total
33 10 0101	**Transfer Drums to Staging Area with Grappler**	**EA**				**14.00**	**17.15**	**0.00**	**31.15**
02083 6153	Drum Staging	EA	5	HLABN	1.000	14.00	17.15	0.00	31.15
33 10 0102	**Consolidate Drums into Tanker Truck**	**GAL**				**0.31**	**0.11**	**0.00**	**0.42**
02083 6154	Drummed Liquid Waste	GAL	220	HLACA	1.000	0.31	0.11	0.00	0.42
33 10 0103	**Consolidate Drums into Dump Truck**	**CY**				**49.64**	**18.97**	**0.00**	**68.62**
02083 6155	Drummed Solid Waste Consolidation, per Cubic Yard	CY	1.362	HLACB	1.000	49.64	18.97	0.00	68.62
33 10 0104	**Drum Crushing, per Individual Drum**	**EA**				**1.02**	**0.48**	**0.00**	**1.50**
02083 6162	Drum Crushing, per Individual Drum	EA	60	HLABK	1.000	1.02	0.48	0.00	1.50
33 10 0105	**Polyethylene Spill Pallet, 4 55-Gallon Drum Capacity, 85 Gallon Sump**	**EA**				**0.00**	**0.00**	**355.00**	**355.00**
02083 6111	Polyethylene Spill Pallet, 4 55-Gallon Drum Capacity, 85 Gallon Sump, Liftable	EA	0	N/A	1.000	0.00	0.00	355.00	355.00

33 10 01 Drum Removal		Unit	Hourly Output	Crew	Factor	Labor	Equip.	Matl.	Total
33 10 0106	**Polyethylene Drum Storage System, 2 Drum Capacity, 135 Gallon Sump, Lockable**	**EA**				**0.00**	**0.00**	**670.00**	**670.00**
02083 6112	Polyethylene Drum Storage System, 2 Drum Capacity, 135 Gallon Sump, Lockable	EA	0	N/A	1.000	0.00	0.00	670.00	670.00
33 10 0107	**7 Gauge Steel Drum Storage, 4 55-Gallon Drum Capacity, 85 Gallon Sump**	**EA**				**0.00**	**0.00**	**269.00**	**269.00**
02083 6113	7 Gauge Steel, 4 55-Gallon Drum Capacity, 85 Gallon Sump	EA	0	N/A	1.000	0.00	0.00	269.00	269.00
33 10 0108	**Polyethylene Drum Storage System, 1 Drum Capacity, 85 Gallon Sump**	**EA**				**0.00**	**0.00**	**226.00**	**226.00**
02083 6114	Polyethylene Drum Storage, 1 Drum Capacity, 85 Gallon Sump	EA	0	N/A	1.000	0.00	0.00	226.00	226.00
33 10 0109	**250 Gallon Bulk Container, Polyethylene with Metal Case**	**EA**				**0.00**	**0.00**	**475.00**	**475.00**
02083 6121	250 Gallon Bulk Container, 2" Ball Valve, Polyethylene with Metal Case	EA	0	N/A	1.000	0.00	0.00	475.00	475.00
33 10 0110	**350 Gallon Bulk Container, Polyethylene**	**EA**				**0.00**	**0.00**	**335.00**	**335.00**
02083 6122	350 Gallon Bulk Container, Polyethylene	EA	0	N/A	1.000	0.00	0.00	335.00	335.00
33 10 0111	**450 Gallon Bulk Container, Polyethylene**	**EA**				**0.00**	**0.00**	**430.00**	**430.00**
02083 6123	450 Gallon Bulk Container, Polyethylene, Non-DOT	EA	0	N/A	1.000	0.00	0.00	430.00	430.00
33 10 0112	**550 Gallon Bulk Container, Polyethylene**	**EA**				**0.00**	**0.00**	**460.00**	**460.00**
02083 6124	550 Gallon Bulk Container, Polyethylene, Non-DOT	EA	0	N/A	1.000	0.00	0.00	460.00	460.00
33 10 0113	**35 CF Dry Flow Hopper, Solids/Granular**	**EA**				**0.00**	**0.00**	**1,000**	**1,000**
02083 6125	Dry Flow Hopper, Solids/Granular Polyethylene, Non-DOT, 35 CF	EA	0	N/A	1.000	0.00	0.00	1,000	1,000
33 10 0114	**70 CF Dry Flow Hopper, Solids/Granular**	**EA**				**0.00**	**0.00**	**1,475**	**1,475**
02083 6126	Dry Flow Hopper, Solids/Granular Polyethylene, Non-DOT, 70 CF	EA	0	N/A	1.000	0.00	0.00	1,475	1,475
33 10 0115	**5.58' x 5.58' x 8.33' Hazardous Material Storage Building, 4 Drum Capacity**	**EA**				**0.00**	**0.00**	**3,725**	**3,725**
02083 6131	5.58' x 5.5' x 8.33' - 4 Drum Capacity, 88 Gallon Sump, One Door	EA	0	N/A	1.000	0.00	0.00	3,725	3,725
33 10 0116	**10.25' x 10.25' x 8.33' Hazardous Material Storage Building, 10 Drum Capacity**	**EA**				**0.00**	**0.00**	**9,625**	**9,625**
02083 6132	10.25' x 5.5' x 8.33' - 10 Drum Capacity, 175 Gallon Sump, Two Doors	EA	0	N/A	1.000	0.00	0.00	9,625	9,625
33 10 0117	**10.25' x 10.25' x 8.33' Hazardous Material Storage Building, 4-Hour Rated**	**EA**				**0.00**	**0.00**	**11,300**	**11,300**
02083 6133	10.25' x 5.5' x 8.33' - 10 Drum Capacity, 175 Gallon Sump, 4-Hour Rated	EA	0	N/A	1.000	0.00	0.00	11,300	11,300

33 10 Drums/Tanks/Structures/Misc. Demolition/Removal

33 10 01	Drum Removal	Unit	Hourly Output	Crew	Factor	Labor	Equip.	Matl.	Total
33 10 0118	**R60 Rough Terrain Forklift, 6,000 Lb @ 24" LC**	**HR**				**33.05**	**29.13**	**27.19**	**89.36**
02083 6156	R60 Rough Terrain Forklift, 6,000 Lb @ 24" Load Center	HR	1	XXQEW	1.000	33.05	29.13	27.19	89.36
33 10 0120	**20 CY Open Top Roll-Off Container**	**EA**				**0.00**	**0.00**	**5,450**	**5,450**
02083 7136	20' Long x 4' High Container, 20 CY	EA	0	N/A	1.000	0.00	0.00	5,450	5,450
33 10 0121	**25 CY Open Top Roll-Off Container**	**EA**				**0.00**	**0.00**	**5,675**	**5,675**
02083 7138	20' Long x 5' High Container, 25 CY	EA	0	N/A	1.000	0.00	0.00	5,675	5,675
33 10 0122	**30 CY Open Top Roll-Off Container**	**EA**				**0.00**	**0.00**	**5,900**	**5,900**
02083 7141	20' Long x 6' High Container, 30 CY	EA	0	N/A	1.000	0.00	0.00	5,900	5,900
33 10 0123	**20 CY Closed Top Roll-Off Container**	**EA**				**0.00**	**0.00**	**5,450**	**5,450**
02083 7136	20' Long x 4' High Container, 20 CY	EA	0	N/A	1.000	0.00	0.00	5,450	5,450
33 10 0124	**19 CY Chassis-mounted Ejector Body**	**EA**				**0.00**	**0.00**	**15,900**	**15,900**
02083 7192	19 CY Chassis Mounted Ejector Body	EA	0	N/A	1.000	0.00	0.00	15,900	15,900
33 10 0125	**24 CY Chassis-mounted Ejector Body**	**EA**				**0.00**	**0.00**	**19,200**	**19,200**
02083 7194	24 CY Chassis Mounted Ejector Body	EA	0	N/A	1.000	0.00	0.00	19,200	19,200
33 10 0126	**25 CY Horizontal Discharge Trailer**	**EA**				**0.00**	**0.00**	**38,300**	**38,300**
02083 7204	25 CY Horizontal Discharge Trailer, 30' Long x 44" High	EA	0	N/A	1.000	0.00	0.00	38,300	38,300
33 10 0127	**34 CY Horizontal Discharge Trailer**	**EA**				**0.00**	**0.00**	**39,100**	**39,100**
02083 7205	34 CY Horizontal Discharge Trailer, 30' Long x 60" High	EA	0	N/A	1.000	0.00	0.00	39,100	39,100
33 10 0128	**42 CY Horizontal Discharge Trailer**	**EA**				**0.00**	**0.00**	**41,800**	**41,800**
02083 7208	42 CY Horizontal Discharge Trailer, 37' Long x 60" High	EA	0	N/A	1.000	0.00	0.00	41,800	41,800
33 10 0130	**4,000 Gallon Tanker Trailer**	**EA**				**0.00**	**0.00**	**53,755**	**53,755**
02083 6185	4,000 Gallon Tanker Trailer	EA	0	N/A	1.000	0.00	0.00	53,755	53,755
33 10 0131	**5,000 Gallon Tanker Trailer**	**EA**				**0.00**	**0.00**	**60,310**	**60,310**
02083 6186	5,000 Gallon Tanker Trailer	EA	0	N/A	1.000	0.00	0.00	60,310	60,310
33 10 0132	**6,000 Gallon Tanker Trailer**	**EA**				**0.00**	**0.00**	**64,244**	**64,244**
02083 6187	6,000 Gallon Tanker Trailer	EA	0	N/A	1.000	0.00	0.00	64,244	64,244

33 10 02	Tank Removal	Unit	Hourly Output	Crew	Factor	Labor	Equip.	Matl.	Total
33 10 0202	**Tank Purging with Dry Ice**	**KGAL**				**26.27**	**1.25**	**0.02**	**27.54**
01957 3115	Purge Tank with Dry Ice	KGAL	3.125	ULABJ	1.000	26.27	1.25	0.02	27.54
33 10 0203	**Pump <=220 Gallon Liquid from Tank**	**GAL**				**0.73**	**0.09**	**0.00**	**0.82**
02083 7411	Loading Tank Contents, <= 220 Gal	GAL	112.5	ULABI	1.000	0.73	0.09	0.00	0.82

33 10 02	Tank Removal	Unit	Hourly Output	Crew	Factor	Labor	Equip.	Matl.	Total
33 10 0204	**Pump 221 - 500 Gallon Liquid from Tank**	**GAL**				**0.24**	**0.39**	**0.00**	**0.63**
02083 7412	Loading Tank Contents, 221 - 500 Gal	GAL	112.5	XTRLC	1.000	0.24	0.39	0.00	0.63
33 10 0205	**Pump 501 - 3,000 Gallon Liquid from Tank**	**GAL**				**0.15**	**0.29**	**0.00**	**0.44**
02083 7413	Loading Tank Contents, 501 - 3,000 Gal	GAL	187.5	XTRHC	1.000	0.15	0.29	0.00	0.44
33 10 0206	**Pump >3000 Gallon Liquid from Tank**	**GAL**				**0.09**	**0.19**	**0.00**	**0.28**
02083 7414	Loading Tank Contents, > 3,000 Gal	GAL	312.5	COETI	1.000	0.09	0.19	0.00	0.28
33 10 0207	**Drain/Flush Liquids in Pipes, 1 Pipe**	**EA**				**25.62**	**0.00**	**176.07**	**201.69**
15292 2020	75 GPM Cast Iron Sump Pump, 2" Discharge Single Stage, 40' Head, 1-1/2 HP	EA	0.238	MSPFB	0.064	400.78	0.00	2,754	201.69
33 10 0208	**Disposable Materials UST Removal**	**EA**				**0.00**	**0.00**	**111.22**	**111.22**
01951 5206	Gloves, Latex (Box of 100) (Single Hand)	EA	0	N/A	0.040	0.00	0.00	10.25	0.41
01954 6429	Packing Tape (per Roll)	EA	0	N/A	1.000	0.00	0.00	3.07	3.07
01954 6431	Plastic Sheeting (per Roll)	EA	0	N/A	1.000	0.00	0.00	46.00	46.00
01954 6432	Isopropanol	GAL	0	N/A	1.000	0.00	0.00	60.50	60.50
01954 6477	Bag Ice	EA	0	N/A	1.000	0.00	0.00	1.24	1.24

33 10 05	Ordnance Disposal	Unit	Hourly Output	Crew	Factor	Labor	Equip.	Matl.	Total
33 10 0501	**Explosive Ordnance Disposal Technician**	**HR**				**33.04**	**0.00**	**0.00**	**33.04**
01951 4901	Explosive Ordnance Disposal Technician	HR	1	HORDA	1.000	33.04	0.00	0.00	33.04
33 10 0502	**Master EOD Technician**	**HR**				**41.59**	**0.00**	**0.00**	**41.59**
01951 4902	Master EOD Technician	HR	1	HORDB	1.000	41.59	0.00	0.00	41.59
33 10 0503	**EOD Site Setup**	**HR**				**74.64**	**0.00**	**0.00**	**74.64**
01951 4901	Explosive Ordnance Disposal Technician	HR	1	HORDA	1.000	33.04	0.00	0.00	33.04
01951 4902	Master EOD Technician	HR	1	HORDB	1.000	41.59	0.00	0.00	41.59
33 10 0504	**Laborer**	**HR**				**26.09**	**0.00**	**0.00**	**26.09**
01951 4903	Laborer	HR	1	XLABA	1.000	26.09	0.00	0.00	26.09
33 10 0511	**Surface Towed Ordnance Locator, Minimum Charge**	**EA**				**6,225**	**74,526**	**464.53**	**81,215**
01953 3101	Surface Towed Ordnance Locator, System (Minimum Charge)	ACRE	0.364	HORDD	20.000	311.23	3,726	23.23	81,215
33 10 0512	**Surface Towed Ordnance Locator, per Each Acre**	**ACRE**				**259.31**	**3,105**	**23.23**	**3,387**
01953 3102	Surface Towed Ordnance Locator, System	ACRE	0.437	HORDD	1.000	259.31	3,105	23.23	3,387
33 10 0513	**Ordnance Locator, Foerster Ferex K (M26)**	**WK**				**0.00**	**0.00**	**162.58**	**162.58**
01953 3103	Ordnance Locator, Foerster Ferex K (M26)	WK	0	N/A	1.000	0.00	0.00	162.58	162.58
33 10 0514	**Ordnance Locator, Foerster Metex**	**EA**				**0.00**	**0.00**	**7,897**	**7,897**
01953 3104	Ordnance Locator, Foerster Metex	EA	0	N/A	1.000	0.00	0.00	7,897	7,897

33 10 05	Ordnance Disposal	Unit	Hourly Output	Crew	Factor	Labor	Equip.	Matl.	Total
33 10 0515	**Ordnance Locator, Foerster Minex 2 FD**	**EA**				**0.00**	**0.00**	**10,568**	**10,568**
01953 3105	Ordnance Locator, Foerster Minex 2D	EA	0	N/A	1.000	0.00	0.00	10,568	10,568
33 10 0516	**Underwater Ordnance Locator, Foerster Model W**	**EA**				**0.00**	**0.00**	**19,046**	**19,046**
01953 3106	Underwater Ordnance Locator, Foerster Model W	EA	0	N/A	1.000	0.00	0.00	19,046	19,046
33 10 0517	**Ordnance Locator, Schoenstedt, Model GA-72CV**	**EA**				**0.00**	**0.00**	**1,045**	**1,045**
01953 3107	Ordnance Locator, Schoenstedt, Model GA-72CV	EA	0	N/A	1.000	0.00	0.00	1,045	1,045
33 10 0518	**Ordnance Locator, Schoenstedt, Model GA-52CV**	**EA**				**0.00**	**0.00**	**1,045**	**1,045**
01953 3108	Ordnance Locator, Schoenstedt, Model GA-52CV	EA	0	N/A	1.000	0.00	0.00	1,045	1,045
33 10 0519	**Ordnance Locator, Schoenstedt, MG-220 with Downhole Attach**	**EA**				**0.00**	**0.00**	**8,361**	**8,361**
01953 3109	Ordnance Locator, Schoenstedt, MG-220 with Downhole Attachment	EA	0	N/A	1.000	0.00	0.00	8,361	8,361
33 10 0520	**Portable Cesium Magnetometer, Geometrics, Model G-858**	**WK**				**0.00**	**0.00**	**127.75**	**127.75**
01953 3111	Portable Cesium Magnetometer, Geometrics, Model G-858	WK	0	N/A	1.000	0.00	0.00	127.75	127.75
33 10 0521	**D-Cell Batteries, Long Life Disposable**	**EA**				**0.00**	**0.00**	**0.54**	**0.54**
01953 3122	D-Cell Batteries, Long Life, Disposable	EA	0	N/A	1.000	0.00	0.00	0.54	0.54
33 10 0531	**ICAD, Miniature Chemical Agent Detector**	**EA**				**0.00**	**0.00**	**4,761**	**4,761**
01953 3201	ICAD, Miniature Chemical Agent Detector	EA	0	N/A	1.000	0.00	0.00	4,761	4,761
33 10 0532	**(MINICAM) Miniature Chemical Agent Monitor**	**EA**				**0.00**	**0.00**	**42,272**	**42,272**
01953 3202	(MINICAM) Miniature Chemical Agent Monitor	EA	0	N/A	1.000	0.00	0.00	42,272	42,272
33 10 0540	**4' x 4' x 4' Transportable Explosives Locker**	**EA**				**0.00**	**0.00**	**3,252**	**3,252**
01957 5102	4' x 4' x 4' Transportable Explosives Locker	EA	0	N/A	1.000	0.00	0.00	3,252	3,252
33 10 0541	**Explosives Storage Locker/Shelter, 16 x 9 x 8'**	**EA**				**0.00**	**0.00**	**23,226**	**23,226**
01957 5101	Explosives Storage Locker/Shelter, 16 x 9 x 8'	EA	0	N/A	1.000	0.00	0.00	23,226	23,226
33 10 0542	**Rope, 5/8", Polyester, Kevlar Center**	**FT**				**0.00**	**0.00**	**2.85**	**2.85**
01953 3121	Rope, 5/8", Polyester, Kevlar Center	FT	0	N/A	1.000	0.00	0.00	2.85	2.85
33 10 0543	**Wood Box, Ordnance Transport Container**	**EA**				**138.91**	**0.00**	**118.91**	**257.83**
06113 3002	Subfloor, 3/4" Thick Plywood, Standard Grade	SF	156.1	ACARD	42.000	0.50	0.00	0.79	54.21
06120 7000	Rafters, 2 x 6 to 2 x 8 Southern Yellow Pine Random Lengths for Light Structural Framing	BF	68.88	ACARD	26.000	1.13	0.00	0.68	46.94
06128 6140	Stud Framing, 2x3 - 2x4 Southern Yellow Pine for Partition Walls	BF	87.38	ACARD	100.000	0.89	0.00	0.68	156.68

33 10 Drums/Tanks/Structures/Misc. Demolition/Removal

33 10 05	Ordnance Disposal	Unit	Hourly Output	Crew	Factor	Labor	Equip.	Matl.	Total
33 10 0544	**EOD Field Support Tools and Supplies, Common**	**EA**				**0.00**	**0.00**	**259.56**	**259.56**
01953 3304	EOD Field Support Tools and Supplies, Common	EA	0	N/A	1.000	0.00	0.00	259.56	259.56
33 10 0545	**Crane, 22 Ton Self-contained, Hydraulic, Rough Terrain, 4 WD, for Ordnance Removal**	**HR**				**88.05**	**71.88**	**0.00**	**159.93**
01954 6128	Crane, 22 Ton Self-contained, Hydraulic, Rough Terrain, 4 WD, Ordnance & Explosive Waste Removal	HR	1	HORDC	1.000	88.05	71.88	0.00	159.93
33 10 0546	**Hoist, Heavy Duty, 5,000 Lb Lift**	**MO**				**0.00**	**0.00**	**205.97**	**205.97**
01953 3305	2.5 Ton Hoist, Rental	MO	1	N/A	1.000	0.00	0.00	205.97	205.97
33 10 0547	**EOD Tool Kit**	**EA**				**0.00**	**0.00**	**370.79**	**370.79**
01953 3307	EOD Tool Kit	EA	0	N/A	1.000	0.00	0.00	370.79	370.79
33 10 0548	**Alloy Steel Chain, 3/8", Cut Length 100'**	**EA**				**0.00**	**0.00**	**366.33**	**366.33**
01953 3308	3/8" Alloy Steel Chain, Cut Length Self-colored	CLF	0	N/A	1.000	0.00	0.00	366.33	366.33
33 10 0549	**Nonsparking EOD Shovels**	**EA**				**0.00**	**0.00**	**96.40**	**96.40**
01953 3306	Nonsparking EOD Shovels	EA	0	N/A	1.000	0.00	0.00	96.40	96.40
33 10 0550	**Standby Ambulance with EMT and Paramedic**	**HR**				**67.96**	**44.89**	**91.85**	**204.70**
01953 3309	Standby, Ambulance with EMT and Paramedic	HR	1	HORDE	1.000	67.96	44.89	91.85	204.70

33 10 95	Underground Storage Tank Closure	Unit	Hourly Output	Crew	Factor	Labor	Equip.	Matl.	Total
33 10 9501	**Fill Tank/Vault with Sand**	**GAL**				**0.09**	**0.12**	**0.04**	**0.25**
02081 2242	Select Granular Fill (Sand)	CY	0	N/A	0.005	0.00	0.00	7.90	0.04
02084 6101	Fill Tank/Vault with Sand, Not Including Material	GAL	1936	COETL	1.000	0.09	0.12	0.00	0.21
33 10 9502	**Remove Steel/Fiberglass UST, up to 2,000 gallons**	**EA**				**349.28**	**1,540**	**0.00**	**1,889**
02084 6201	Remove Steel/Fiberglass Underground Storage Tank	EA	0.265	CODEV	1.000	349.28	1,540	0.00	1,889
33 10 9503	**Concrete Tank/Vault Demolition, to 6" Thick Mesh Reinforced/Power**	**CY**				**29.69**	**15.22**	**0.00**	**44.91**
02046 1900	Demolish 6" (15 cm) Thick Concrete Pavement with Wire Mesh	SY	26.25	CODEF	6.000	4.95	2.54	0.00	44.91
33 10 9504	**Concrete Tank/Vault Demolition, to 6" Thick Rod Reinforced/Power**	**CY**				**32.81**	**16.82**	**0.00**	**49.63**
02046 2000	Demolish 6" (15 cm) Thick Concrete Pavement with Reinforcing Rod	SY	23.75	CODEF	6.000	5.47	2.80	0.00	49.63
33 10 9505	**Remove Steel/Fiberglass UST, 2,001 - 5,000 gallon**	**EA**				**741.60**	**3,269**	**0.00**	**4,010**
01001 0082	Remove Steel/Fiberglass UST, 2,001 - 5,000 gallon	EA	0.125	CODEV	1.000	741.60	3,269	0.00	4,010

33 10 95 Underground Storage Tank Closure		Unit	Hourly Output	Crew	Factor	Labor	Equip.	Matl.	Total
33 10 9506	**Remove Steel/Fiberglass UST, 5,001 - 15,000 gallon**	**EA**				**1,117**	**4,923**	**0.00**	**6,040**
01001 0083	Remove Steel/Fiberglass UST, 5,001 - 15,000 gallon	EA	0.083	CODEV	1.000	1,117	4,923	0.00	6,040
33 10 9507	**Remove Steel/Fiberglass UST, 15,001 - 20,000 gallon**	**EA**				**1,306**	**5,755**	**0.00**	**7,061**
01001 0084	Remove Steel/Fiberglass UST, 15,001 - 20,000 gallon	EA	0.071	CODEV	1.000	1,306	5,755	0.00	7,061
33 10 9508	**Remove Steel/Fiberglass UST, 20,001 - 30,000 gallon**	**EA**				**1,655**	**7,297**	**0.00**	**8,952**
01001 0085	Remove Steel/Fiberglass UST, 20,001 - 30,000 gallon	EA	0.056	CODEV	1.000	1,655	7,297	0.00	8,952

33 10 96 Tank Installation		Unit	Hourly Output	Crew	Factor	Labor	Equip.	Matl.	Total
33 10 9601	**550 Gallon Single-wall Steel UST**	**EA**				**494.34**	**0.00**	**1,065**	**1,559**
13214 5705	550 Gallon Single Wall Steel Storage Tank, Underground, with Coating, Fittings & Straps	EA	0.29	MSPFL	1.000	494.34	0.00	1,065	1,559
33 10 9602	**1,000 Gallon Single-wall Steel UST**	**EA**				**530.96**	**0.00**	**1,650**	**2,181**
13214 5520	1,000 Gallon Steel Storage Tank, with 7-Gauge Shell, Underground, Including Setting & Hold-down Bars	EA	0.27	MSPFL	1.000	530.96	0.00	1,650	2,181
33 10 9603	**2,000 Gallon Single-wall Steel UST**	**EA**				**615.02**	**86.25**	**2,450**	**3,151**
13214 5710	2,000 Gallon Single Wall Steel Storage Tank, Underground, with Coating, Fittings & Straps	EA	0.25	MSPFP	1.000	615.02	86.25	2,450	3,151
33 10 9604	**3,000 Gallon Single-wall Steel UST**	**EA**				**698.89**	**98.01**	**3,525**	**4,322**
13214 5720	3,000 Gallon Single Wall Steel Storage Tank, Underground, with Coating, Fittings & Straps	EA	0.22	MSPFP	1.000	698.89	98.01	3,525	4,322
33 10 9605	**5,000 Gallon Single-wall Steel UST**	**EA**				**1,230**	**172.50**	**6,850**	**8,253**
13214 5540	5,000 Gallon Steel Storage Tank, with 1/4" Shell, Underground, Including Setting & Hold-down Bars	EA	0.125	MSPFP	1.000	1,230	172.50	6,850	8,253
33 10 9606	**8,000 Gallon Single-wall Steel UST**	**EA**				**1,398**	**196.02**	**5,870**	**7,464**
13214 5730	8,000 Gallon Single Wall Steel Storage Tank, Underground, with Coating, Fittings & Straps	EA	0.11	MSPFP	1.000	1,398	196.02	5,870	7,464
33 10 9607	**10,000 Gallon Single-wall Steel UST**	**EA**				**1,538**	**215.63**	**9,841**	**11,594**
15176 2013	10,000 Gallon Single Wall Steel Storage Tank, Underground, with Coating, Fittings & Straps	EA	0.1	MSPFP	1.000	1,538	215.63	9,841	11,594
33 10 9608	**12,000 Gallon Single-wall Steel UST**	**EA**				**1,708**	**239.58**	**9,475**	**11,423**
13214 5740	12,000 Gallon Single Wall Steel Storage Tank, Underground, with Coating, Fittings & Straps	EA	0.09	MSPFP	1.000	1,708	239.58	9,475	11,423
33 10 9609	**15,000 Gallon Single-wall Steel UST**	**EA**				**1,922**	**269.53**	**10,597**	**12,789**
13214 5750	15,000 Gallon Single Wall Steel Storage Tank, Underground, with Coating, Fittings & Straps	EA	0.08	MSPFP	1.000	1,922	269.53	10,597	12,789

33 10 96	Tank Installation	Unit	Hourly Output	Crew	Factor	Labor	Equip.	Matl.	Total
33 10 9610	**20,000 Gallon Single-wall Steel UST**	**EA**				**2,460**	**345.00**	**17,800**	**20,605**
13214 5600	20,000 Gallon Steel Storage Tank, with 5/6" Shell, Underground, Including Setting & Hold-down Bars	EA	0.063	MSPFQ	1.000	2,460	345.00	17,800	20,605
33 10 9611	**30,000 Gallon Single-wall Steel UST**	**EA**				**3,075**	**431.25**	**28,600**	**32,106**
13214 5620	30,000 Gallon Steel Storage Tank, with 3/8" Shell, Underground, Including Setting & Hold Down Bars	EA	0.05	MSPFQ	1.000	3,075	431.25	28,600	32,106
33 10 9612	**550 Gallon Double-wall Steel UST**	**EA**				**307.51**	**43.13**	**3,091**	**3,441**
13214 9100	550 Gallon Underground Double Wall Steel Tank Coated, in Place with Hold-down Bars	EA	0.5	MSPFP	1.000	307.51	43.13	3,091	3,441
33 10 9613	**1,000 Gallon Double-wall Steel UST**	**EA**				**1,922**	**269.53**	**4,775**	**6,966**
13214 9120	1,000 Gallon Underground Double Wall Steel Tank Coated, in Place with Hold-down Bars	EA	0.08	MSPFP	1.000	1,922	269.53	4,775	6,966
33 10 9614	**2,000 Gallon Double-wall Steel UST**	**EA**				**473.09**	**66.35**	**6,150**	**6,689**
13214 9140	2,000 Gallon Underground Double Wall Steel Tank Coated, in Place with Hold-down Bars	EA	0.325	MSPFP	1.000	473.09	66.35	6,150	6,689
33 10 9615	**3,000 Gallon Double-wall Steel UST**	**EA**				**530.19**	**74.35**	**6,850**	**7,455**
13214 9180	3,000 Gallon Underground Double Wall Steel Tank Coated, in Place with Hold-down Bars	EA	0.29	MSPFP	1.000	530.19	74.35	6,850	7,455
33 10 9616	**5,000 Gallon Double-wall Steel UST**	**EA**				**1,538**	**215.63**	**11,100**	**12,853**
13214 9220	5,000 Gallon Underground Double Wall Steel Tank Coated, in Place with Hold-down Bars	EA	0.1	MSPFP	1.000	1,538	215.63	11,100	12,853
33 10 9617	**8,000 Gallon Double-wall Steel UST**	**EA**				**1,757**	**246.43**	**14,400**	**16,404**
13214 9260	8,000 Gallon Underground Double Wall Steel Tank Coated, in Place with Hold-down Bars	EA	0.088	MSPFP	1.000	1,757	246.43	14,400	16,404
33 10 9618	**10,000 Gallon Double-wall Steel UST**	**EA**				**1,922**	**269.53**	**16,100**	**18,291**
13214 9280	10,000 Gallon Underground Double Wall Steel Tank Coated, in Place with Hold-down Bars	EA	0.08	MSPFP	1.000	1,922	269.53	16,100	18,291
33 10 9619	**12,000 Gallon Double-wall Steel UST**	**EA**				**2,135**	**299.48**	**15,896**	**18,331**
13214 9300	12,000 Gallon Underground Double Wall Steel Tank Coated, in Place with Hold-down Bars	EA	0.072	MSPFP	1.000	2,135	299.48	15,896	18,331
33 10 9620	**15,000 Gallon Double-wall Steel UST**	**EA**				**2,410**	**337.97**	**21,300**	**24,048**
13214 9320	15,000 Gallon Underground Double Wall Steel Tank Coated, in Place with Hold-down Bars	EA	0.064	MSPFQ	1.000	2,410	337.97	21,300	24,048
33 10 9621	**20,000 Gallon Double-wall Steel UST**	**EA**				**3,075**	**431.25**	**25,600**	**29,106**
13214 9340	20,000 Gallon Underground Double Wall Steel Tank Coated, in Place with Hold-down Bars	EA	0.05	MSPFQ	1.000	3,075	431.25	25,600	29,106

33 10 96	Tank Installation	Unit	Hourly Output	Crew	Factor	Labor	Equip.	Matl.	Total
33 10 9622	**30,000 Gallon Double-wall Steel UST**	**EA**				**3,844**	**539.06**	**50,000**	**54,383**
13214 9380	30,000 Gallon Underground Double Wall Steel Tank Coated, in Place with Hold-down Bars	EA	0.04	MSPFQ	1.000	3,844	539.06	50,000	54,383
33 10 9623	**550 Gallon Single-wall Fiberglass UST**	**EA**				**651.64**	**0.00**	**1,725**	**2,377**
13214 2225	550 Gallon Single Wall Fiberglass Tank Including Fittings & Anchor Straps	EA	0.22	MSPFL	1.000	651.64	0.00	1,725	2,377
33 10 9624	**1,000 Gallon Single-wall Fiberglass UST**	**EA**				**716.80**	**0.00**	**2,125**	**2,842**
13214 2230	1,000 Gallon Single Wall Fiberglass Tank Including Fittings & Anchor Straps	EA	0.2	MSPFL	1.000	716.80	0.00	2,125	2,842
33 10 9625	**2,000 Gallon Single-wall Fiberglass UST**	**EA**				**960.97**	**134.77**	**2,750**	**3,846**
13214 2240	2,000 Gallon Single Wall Fiberglass Tank Including Fittings & Anchor Straps	EA	0.16	MSPFP	1.000	960.97	134.77	2,750	3,846
33 10 9626	**3,000 Gallon Single-wall Fiberglass UST**	**EA**				**960.97**	**134.77**	**4,400**	**5,496**
13214 2245	3,000 Gallon Single Wall Fiberglass Tank Including Fittings & Anchor Straps	EA	0.16	MSPFP	1.000	960.97	134.77	4,400	5,496
33 10 9627	**5,000 Gallon Single-wall Fiberglass UST**	**EA**				**1,183**	**165.87**	**5,400**	**6,749**
13214 2255	5,000 Gallon Single Wall Fiberglass Tank Including Fittings & Anchor Straps	EA	0.13	MSPFP	1.000	1,183	165.87	5,400	6,749
33 10 9628	**8,000 Gallon Single-wall Fiberglass UST**	**EA**				**1,398**	**196.02**	**5,325**	**6,918**
13214 2270	8,000 Gallon Single Wall Fiberglass Tank Including Fittings & Anchor Straps	EA	0.11	MSPFP	1.000	1,398	196.02	5,325	6,918
33 10 9629	**10,000 Gallon Single-wall Fiberglass UST**	**EA**				**1,538**	**215.63**	**6,025**	**7,778**
13214 2280	10,000 Gallon Single Wall Fiberglass Tank Including Fittings & Anchor Straps	EA	0.1	MSPFP	1.000	1,538	215.63	6,025	7,778
33 10 9630	**12,000 Gallon Single-wall Fiberglass UST**	**EA**				**1,708**	**239.58**	**7,714**	**9,662**
13214 2282	12,000 Gallon Single Wall Fiberglass Tank Including Fittings & Anchor Straps	EA	0.09	MSPFP	1.000	1,708	239.58	7,714	9,662
33 10 9631	**15,000 Gallon Single-wall Fiberglass UST**	**EA**				**1,922**	**269.53**	**8,208**	**10,399**
13214 2284	15,000 Gallon Single Wall Fiberglass Tank Including Fittings & Anchor Straps	EA	0.08	MSPFP	1.000	1,922	269.53	8,208	10,399
33 10 9632	**20,000 Gallon Single-wall Fiberglass UST**	**EA**				**2,563**	**359.38**	**11,900**	**14,822**
13214 2290	20,000 Gallon Single Wall Fiberglass Tank Including Fittings & Anchor Straps	EA	0.06	MSPFP	1.000	2,563	359.38	11,900	14,822
33 10 9633	**30,000 Gallon Single-wall Fiberglass UST**	**EA**				**3,075**	**431.25**	**23,584**	**27,090**
13214 2320	30,000 Gallon Single Wall Fiberglass Tank Including Fittings & Anchor Straps	EA	0.05	MSPFP	1.000	3,075	431.25	23,584	27,090
33 10 9634	**550 Gallon Double-wall Fiberglass UST**	**EA**				**307.51**	**43.13**	**3,850**	**4,201**
13214 3035	550 Gallon Double Wall Fiberglass Tank Including Fittings & Anchor Straps	EA	0.5	MSPFP	1.000	307.51	43.13	3,850	4,201

33 10 96 Tank Installation		Unit	Hourly Output	Crew	Factor	Labor	Equip.	Matl.	Total
33 10 9635	**1,000 Gallon Double-wall Fiberglass UST**	**EA**				**384.39**	**53.91**	**4,875**	**5,313**
13214 3050	1,000 Gallon Double Wall Fiberglass Tank Including Fittings & Anchor Straps	EA	0.4	MSPFP	1.000	384.39	53.91	4,875	5,313
33 10 9636	**2,000 Gallon Double-wall Fiberglass UST**	**EA**				**465.92**	**65.34**	**6,375**	**6,906**
13214 3055	2,000 Gallon Double Wall Fiberglass Tank Including Fittings & Anchor Straps	EA	0.33	MSPFP	1.000	465.92	65.34	6,375	6,906
33 10 9637	**3,000 Gallon Double-wall Fiberglass UST**	**EA**				**465.92**	**65.34**	**7,575**	**8,106**
13214 3070	3,000 Gallon Double Wall Fiberglass Tank Including Fittings & Anchor Straps	EA	0.33	MSPFP	1.000	465.92	65.34	7,575	8,106
33 10 9638	**5,000 Gallon Double-wall Fiberglass UST**	**EA**				**1,538**	**215.63**	**9,875**	**11,628**
13214 3085	5,000 Gallon Double Wall Fiberglass Tank Including Fittings & Anchor Straps	EA	0.1	MSPFP	1.000	1,538	215.63	9,875	11,628
33 10 9639	**8,000 Gallon Double-wall Fiberglass UST**	**EA**				**1,708**	**239.58**	**11,300**	**13,248**
13214 3100	8,000 Gallon Double Wall Fiberglass Tank Including Fittings & Anchor Straps	EA	0.09	MSPFP	1.000	1,708	239.58	11,300	13,248
33 10 9640	**10,000 Gallon Double-wall Fiberglass UST**	**EA**				**1,922**	**269.53**	**12,500**	**14,691**
13214 3110	10000 Gallon Double Wall Fiberglass Tank Including Fittings & Anchor Straps	EA	0.08	MSPFP	1.000	1,922	269.53	12,500	14,691
33 10 9641	**12,000 Gallon Double-wall Fiberglass UST**	**EA**				**2,196**	**308.04**	**15,300**	**17,805**
13214 3120	12,000 Gallon Double Wall Fiberglass Tank Including Fittings & Anchor Straps	EA	0.07	MSPFP	1.000	2,196	308.04	15,300	17,805
33 10 9642	**15,000 Gallon Double-wall Fiberglass UST**	**EA**				**2,563**	**359.38**	**19,636**	**22,558**
13214 3122	15,000 Gallon Double Wall Fiberglass Tank Including Fittings & Anchor Straps	EA	0.06	MSPFQ	1.000	2,563	359.38	19,636	22,558
33 10 9643	**20,000 Gallon Double-wall Fiberglass UST**	**EA**				**3,075**	**431.25**	**25,400**	**28,906**
13214 3124	20,000 Gallon Double Wall Fiberglass Tank Including Fittings & Anchor Straps	EA	0.05	MSPFQ	1.000	3,075	431.25	25,400	28,906
33 10 9644	**30,000 Gallon Double-wall Fiberglass UST**	**EA**				**3,844**	**539.06**	**37,817**	**42,200**
13214 3128	30,000 Gallon Double Wall Fiberglass Tank Including Fittings & Anchor Straps	EA	0.04	MSPFQ	1.000	3,844	539.06	37,817	42,200
33 10 9645	**Tank Fittings**	**EA**				**452.78**	**48.07**	**295.21**	**796.06**
15525 0164	1" Double Poppet Fort Valve	EA	1.75	MPLUH	1.000	54.26	5.76	95.83	155.85
15525 0640	Fill Adapter, 4"	EA	0.75	MPLUH	1.000	126.61	13.44	36.17	176.22
15525 0680	Fill Cap, 4"	EA	2	MPLUH	1.000	47.48	5.04	29.12	81.64
15525 0740	Vapor Hose Adapter	EA	1.5	MPLUH	1.000	63.30	6.72	80.12	150.14
15525 5000	Float Vent Valve	EA	1	MPLUH	1.000	94.95	10.08	28.78	133.81
15525 5540	Vent Cap, 2"	EA	3	MPLUH	1.000	31.65	3.36	14.73	49.75
15525 5600	Dust Cap	EA	2.75	MPLUH	1.000	34.53	3.67	10.45	48.65
33 10 9646	**Overfill Spill Container**	**EA**				**189.91**	**20.16**	**545.17**	**755.25**
15525 0620	Manhole Spill Containment	EA	0.5	MPLUH	1.000	189.91	20.16	545.18	755.25

33 10 96	Tank Installation	Unit	Hourly Output	Crew	Factor	Labor	Equip.	Matl.	Total
33 10 9647	**30" Diameter Manhole at Grade**	**EA**				**108.83**	**36.63**	**305.00**	**450.46**
02752 4620	30" Diameter x 620 Lb Reg Manhole Cover (76 cm Diameter x 281 kg) with Frame, City Type	EA	0.875	CODEJ	1.000	108.83	36.63	305.00	450.46
33 10 9648	**Turbine Enclosure with Secondary Containment Collar**	**EA**				**0.00**	**0.00**	**2,417**	**2,417**
15605 1501	Secondary Containment Collar	EA	0	N/A	1.000	0.00	0.00	779.80	779.80
15605 1502	Turbine Enclosure with Lid, 4' L x 42" Diameter	EA	0	N/A	1.000	0.00	0.00	779.80	779.80
15605 1503	Fitting Kit for Turbine Enclosure	EA	0	N/A	1.000	0.00	0.00	155.96	155.96
15605 1504	Grommet Kits for Plumbing & Electrical	EA	0	N/A	1.000	0.00	0.00	311.93	311.93
15605 1505	Turbine Electrical Coupling Kit	EA	0	N/A	1.000	0.00	0.00	389.91	389.91
33 10 9649	**Pump, Cast-iron Close Coupling, 2 HP, 50 GPM**	**EA**				**237.96**	**0.00**	**1,220**	**1,458**
15286 4090	Pump Cast Iron Close Coupling 2 HP 50 GPM	EA	0.4	MSPFB	1.000	237.96	0.00	1,220	1,458
33 10 9650	**Control Panel, 1 - 8 Single-wall Tanks, Monitoring System**	**EA**				**0.00**	**0.00**	**2,300**	**2,300**
13219 0181	Control Panel, 1 - 8 Tanks, Single Wall Tank Monitoring System	EA	0	N/A	1.000	0.00	0.00	2,300	2,300
33 10 9651	**Control Panel, 1 - 4 Double-wall Tanks, Monitoring System**	**EA**				**0.00**	**0.00**	**2,300**	**2,300**
13219 0180	Control Panel, 1 - 4 Tanks, Double Wall Tank Monitoring System	EA	0	N/A	1.000	0.00	0.00	2,300	2,300
33 10 9652	**Control Panel, 1 - 8 Double-wall Tanks, Monitoring System**	**EA**				**0.00**	**0.00**	**2,625**	**2,625**
13219 0190	Control Panel, 1 - 8 Tanks, Double Wall Tank Monitoring System	EA	0	N/A	1.000	0.00	0.00	2,625	2,625
33 10 9653	**Tank Leak Detection Probe**	**EA**				**0.00**	**0.00**	**600.00**	**600.00**
13219 0240	Tank Leak Detection Probe	EA	0	N/A	1.000	0.00	0.00	600.00	600.00
33 10 9654	**Secondary Containment Collar Probe**	**EA**				**0.00**	**0.00**	**325.00**	**325.00**
13219 0310	Secondary Containment Collar Probe	EA	0	N/A	1.000	0.00	0.00	325.00	325.00
33 10 9655	**Interstitial Leak Detection Probe**	**EA**				**0.00**	**0.00**	**295.00**	**295.00**
13219 0330	Interstitial Leak Detection Probe	EA	0	N/A	1.000	0.00	0.00	295.00	295.00
33 10 9656	**550 Gallon Single-wall Steel Aboveground Tank**	**EA**				**435.26**	**0.00**	**931.59**	**1,367**
13205 3060	550 Gallon Steel Storage Tank with 3/16" Shell Above Ground with Supports, Coating & Fittings	EA	0.238	MSPFC	1.000	435.26	0.00	931.59	1,367
33 10 9657	**1,000 Gallon Single-wall Steel Aboveground Tank**	**EA**				**573.44**	**0.00**	**1,166**	**1,740**
13205 3080	1,000 Gallon Steel Storage Tank with 3/16" Shell Above Ground with Supports, Coating & Fittings	EA	0.25	MSPFL	1.000	573.44	0.00	1,166	1,740

33 10 96	Tank Installation	Unit	Hourly Output	Crew	Factor	Labor	Equip.	Matl.	Total
33 10 9658	**2,000 Gallon Single-wall Steel Aboveground Tank**	**EA**				**878.60**	**123.21**	**2,239**	**3,241**
13205 3120	2,000 Gallon Steel Storage Tank with 3/16" Shell Above Ground with Supports, Coating & Fittings	EA	0.175	MSPFP	1.000	878.60	123.21	2,239	3,241
33 10 9659	**3,000 Gallon Single-wall Steel Aboveground Tank**	**EA**				**904.44**	**126.84**	**3,250**	**4,281**
13205 3130	3,000 Gallon Single Wall Steel Storage Tank, Above Ground with Supports, Coating & Fittings	EA	0.17	MSPFP	1.000	904.44	126.84	3,250	4,281
33 10 9660	**5,000 Gallon Single-wall Steel Aboveground Tank**	**EA**				**1,118**	**156.82**	**5,175**	**6,450**
13205 3140	5,000 Gallon Steel Storage Tank with 3/16" Shell Above Ground with Supports, Coating & Fittings	EA	0.138	MSPFP	1.000	1,118	156.82	5,175	6,450
33 10 9661	**8,000 Gallon Single-wall Steel Aboveground Tank**	**EA**				**1,281**	**179.69**	**7,406**	**8,867**
13205 3145	8,000 Gallon Single Wall Steel Storage Tank, Above Ground with Supports, Coating & Fittings	EA	0.12	MSPFP	1.000	1,281	179.69	7,406	8,867
33 10 9662	**10,000 Gallon Single-wall Steel Aboveground Tank**	**EA**				**1,398**	**196.02**	**7,864**	**9,458**
13205 3150	10,000 Gallon Single Wall Steel Storage Tank, Above Ground with Supports, Coating & Fittings	EA	0.11	MSPFP	1.000	1,398	196.02	7,864	9,458
33 10 9663	**12,000 Gallon Single-wall Steel Aboveground Tank**	**EA**				**1,538**	**215.63**	**9,149**	**10,902**
13205 3155	12,000 Gallon Single Wall Steel Storage Tank, Above Ground with Supports, Coating & Fittings	EA	0.1	MSPFP	1.000	1,538	215.63	9,149	10,902
33 10 9664	**15,000 Gallon Single-wall Steel Aboveground Tank**	**EA**				**1,708**	**239.58**	**11,355**	**13,303**
13205 3160	15,000 Gallon Single Wall Steel Storage Tank, Above Ground with Supports, Coating & Fittings	EA	0.09	MSPFP	1.000	1,708	239.58	11,355	13,303
33 10 9665	**20,000 Gallon Single-wall Steel Aboveground Tank**	**EA**				**2,196**	**308.04**	**14,225**	**16,730**
13205 3170	20,000 Gallon Single Wall Steel Storage Tank, Above Ground with Supports, Coating & Fittings	EA	0.07	MSPFQ	1.000	2,196	308.04	14,225	16,730
33 10 9666	**30,000 Gallon Single-wall Steel Aboveground Tank**	**EA**				**2,563**	**359.38**	**21,255**	**24,177**
13205 3190	30,000 Gallon Single Wall Steel Storage Tank, Above Ground with Supports, Coating & Fittings	EA	0.06	MSPFQ	1.000	2,563	359.38	21,255	24,177
33 10 9667	**Steel Dike for 550 Gallon Tank, Uncoated**	**EA**				**0.00**	**0.00**	**1,225**	**1,225**
13216 0020	Steel Dike for 550 Gallon Tank, Uncoated	EA	0	N/A	1.000	0.00	0.00	1,225	1,225
33 10 9668	**Steel Dike for 1,000 Gallon Tank, Uncoated**	**EA**				**0.00**	**0.00**	**1,922**	**1,922**
13216 0040	Steel Dike for 1,000 Gallon Tank, Uncoated	EA	0	N/A	1.000	0.00	0.00	1,922	1,922
33 10 9669	**Steel Dike for 2,000 Gallon Tank, Uncoated**	**EA**				**0.00**	**0.00**	**3,800**	**3,800**
13216 0060	Steel Dike for 2,000 Gallon Tank, Uncoated	EA	0	N/A	1.000	0.00	0.00	3,800	3,800

33 10 96 Tank Installation		Unit	Hourly Output	Crew	Factor	Labor	Equip.	Matl.	Total
33 10 9670	**Steel Dike for 3,000 Gallon Tank, Uncoated**	**EA**				**0.00**	**0.00**	**4,961**	**4,961**
13216 0080	Steel Dike for 3,000 Gallon Tank, Uncoated	EA	0	N/A	1.000	0.00	0.00	4,961	4,961
33 10 9671	**Steel Dike for 5,000 Gallon Tank, Uncoated**	**EA**				**0.00**	**0.00**	**6,000**	**6,000**
13216 0100	Steel Dike for 5,000 Gallon Tank, Uncoated	EA	0	N/A	1.000	0.00	0.00	6,000	6,000
33 10 9672	**Steel Dike for 8,000 Gallon Tank, Uncoated**	**EA**				**0.00**	**0.00**	**11,215**	**11,215**
13216 0140	Steel Dike for 8,000 Gallon Tank, Uncoated	EA	0	N/A	1.000	0.00	0.00	11,215	11,215
33 10 9673	**Steel Dike for 10,000 Gallon Tank, Uncoated**	**EA**				**0.00**	**0.00**	**12,600**	**12,600**
13216 0160	Steel Dike for 10,000 Gallon Tank, Uncoated	EA	0	N/A	1.000	0.00	0.00	12,600	12,600
33 10 9674	**Steel Dike for 12,000 Gallon Tank, Uncoated**	**EA**				**0.00**	**0.00**	**14,000**	**14,000**
13216 0180	Steel Dike for 12,000 Gallon Tank, Uncoated	EA	0	N/A	1.000	0.00	0.00	14,000	14,000
33 10 9675	**Steel Dike for 15,000 Gallon Tank, Uncoated**	**EA**				**0.00**	**0.00**	**16,901**	**16,901**
13216 0200	Steel Dike for 15,000 Gallon Tank, Uncoated	EA	0	N/A	1.000	0.00	0.00	16,901	16,901
33 10 9676	**Steel Dike for 20,000 Gallon Tank, Uncoated**	**EA**				**0.00**	**0.00**	**19,800**	**19,800**
13216 0220	Steel Dike for 20,000 Gallon Tank, Uncoated	EA	0	N/A	1.000	0.00	0.00	19,800	19,800
33 10 9677	**500 Barrel Fuel Oil Tank, 15' Diameter x 16' H, Cone Roof, Including Foundation**	**EA**				**19,747**	**6,492**	**99,500**	**125,739**
13205 4050	500 Barrel Fuel Oil Tank, Steel, 15' Diameter x 16' High, Cone Roof, Including Foundation, Fabricate, & Erection	EA	0.02	SIWSO	1.000	19,747	6,492	99,500	125,739
33 10 9678	**2,000 Barrel Fuel Oil Tank, 25' Diameter x 24' H, Cone Roof, Including Foundation**	**EA**				**39,897**	**13,117**	**127,000**	**180,014**
13205 4100	2,000 Barrel Fuel Oil Tank, Steel, 25' Diameter x 24' High, Cone Roof, Including Foundation, Fabricate, & Erection	EA	0.01	SIWSO	1.000	39,897	13,117	127,000	180,014
33 10 9679	**5,000 Barrel Fuel Oil Tank, 40' Diameter x 24' H, Cone Roof, Including Foundation**	**EA**				**52,836**	**17,372**	**173,457**	**243,665**
13205 4200	5,000 Barrel Fuel Oil Tank, Steel, 40' Diameter x 24' High, Cone Roof, Including Foundation, Fabricate, & Erection	EA	0.007	SIWSO	1.000	52,836	17,372	173,457	243,665
33 10 9680	**12,000 Barrel Fuel Oil Tank, 60' Diameter x 24' H, Cone Roof, Including Foundation**	**EA**				**86,887**	**28,567**	**240,000**	**355,453**
13205 4210	12,000 Barrel Fuel Oil Tank, Steel, 60' Diameter x 24' High, Cone Roof, Including Foundation, Fabricate, & Erection	EA	0.005	SIWSO	1.000	86,887	28,567	240,000	355,453
33 10 9681	**20,000 Barrel Fuel Oil Tank, 60' Diameter x 40' H, Cone Roof, Including Foundation**	**EA**				**118,482**	**38,955**	**309,000**	**466,436**
13205 4220	20,000 Barrel Fuel Oil Tank, Steel, 60' Diameter x 40' High, Cone Roof, Including Foundation, Fabricate, & Erection	EA	0.003	SIWSO	1.000	118,482	38,955	309,000	466,436

33 10 96	Tank Installation	Unit	Hourly Output	Crew	Factor	Labor	Equip.	Matl.	Total
33 10 9682	**24,000 Barrel Fuel Oil Tank, 70' Diameter x 32' H, Cone Roof, Including Foundation**	**EA**				**126,126**	**41,468**	**296,500**	**464,094**
13205 4230	24,000 Barrel Fuel Oil Tank, Steel, 70' Diameter x 32' High, Cone Roof, Including Foundation, Fabricate, & Erection	EA	0.003	SIWSO	1.000	126,126	41,468	296,500	464,094
33 10 9683	**36,000 Barrel Fuel Oil Tank, 80' Diameter x 40' H, Cone Roof, Including Foundation**	**EA**				**169,996**	**55,891**	**344,927**	**570,814**
13205 4350	36,000 Barrel Fuel Oil Tank, Steel, 80' Diameter x 40' High, Cone Roof, Including Foundation, Fabricate, & Erection	EA	0.002	SIWSO	1.000	169,996	55,891	344,927	570,814
33 10 9684	**45,000 Barrel Fuel Oil Tank, 90' Diameter x 40' H, Cone Roof, Including Foundation**	**EA**				**205,784**	**67,658**	**482,500**	**755,942**
13205 4400	45,000 Barrel Fuel Oil Tank, Steel, 90' Diameter x 40' High, Cone Roof, Including Foundation, Fabricate, & Erection	EA	0.002	SIWSO	1.000	205,784	67,658	482,500	755,942
33 10 9685	**56,000 Barrel Fuel Oil Tank, 100' Diameter x 40' H, Cone Roof, with Foundation**	**EA**				**244,369**	**80,344**	**488,820**	**813,533**
13205 4510	56,000 Barrel Fuel Oil Tank, Steel, 100' Diameter x 40' High, Cone Roof, Including Foundation, Fabricate, & Erection	EA	0.002	SIWSO	1.000	244,369	80,344	488,820	813,533
33 10 9686	**64,000 Barrel Fuel Oil Tank, 120' Diameter x 32' H, Cone Roof, with Foundation**	**EA**				**279,279**	**91,821**	**515,313**	**886,413**
13205 4520	64,000 Barrel Fuel Oil Tank, Steel, 120' Diameter x 32' High, Cone Roof, Including Foundation, Fabricate, & Erection	EA	0.001	SIWSO	1.000	279,279	91,821	515,313	886,413
33 10 9687	**80,000 Barrel Fuel Oil Tank, 120' Diameter x 40' H, Cone Roof, with Foundation**	**EA**				**355,445**	**116,864**	**691,500**	**1,163,809**
13205 4530	80,000 Barrel Fuel Oil Tank, Steel, 120' Diameter x 40' High, Cone Roof, Including Foundation, Fabricate, & Erection	EA	0.001	SIWSO	1.000	355,445	116,864	691,500	1,163,809
33 10 9688	**88,000 Barrel Fuel Oil Tank, 140' Diameter x 32' H, Cone Roof, with Foundation**	**EA**				**355,445**	**116,864**	**630,635**	**1,102,944**
13205 4540	88,000 Barrel Fuel Oil Tank, Steel, 140' Diameter x 32' High, Cone Roof, Including Foundation, Fabricate, & Erection	EA	0.001	SIWSO	1.000	355,445	116,864	630,635	1,102,944
33 10 9689	**110,000 Barrel Fuel Oil Tank, 140' Diameter x 40' H, Cone Roof, with Foundation**	**EA**				**355,445**	**116,864**	**855,000**	**1,327,309**
13205 4650	110,000 Barrel Fuel Oil Tank, Steel, 140' Diameter x 40' High, Cone Roof, Including Foundation, Fabricate, & Erection	EA	0.001	SIWSO	1.000	355,445	116,864	855,000	1,327,309
33 10 9690	**143,000 Barrel Fuel Oil Tank, 160' Diameter x 40' H, Cone Roof, with Foundation**	**EA**				**390,990**	**128,550**	**1,096,500**	**1,616,040**
13205 4700	143,000 Barrel Fuel Oil Tank, Steel, 160' Diameter x 40' High, Cone Roof, Including Foundation, Fabricate, & Erection	EA	0.001	SIWSO	1.000	390,990	128,550	1,096,500	1,616,040

33 10 Drums/Tanks/Structures/Misc. Demolition/Removal

33 10 96	Tank Installation	Unit	Hourly Output	Crew	Factor	Labor	Equip.	Matl.	Total
33 10 9691	**180,000 Barrel Fuel Oil Tank, 180' Diameter x 40' H, Cone Roof, with Foundation**	**EA**				**488,737**	**160,688**	**1,188,025**	**1,837,450**
13205 4810	180,000 Barrel Fuel Oil Tank, Steel, 180' Diameter x 40' High, Cone Roof, Including Foundation, Fabricate, & Erection	EA	0.0008	SIWSO	1.000	488,737	160,688	1,188,025	1,837,450
33 10 9692	**224,000 Barrel Fuel Oil Tank, 200' Diameter x 40' H, Cone Roof, with Foundation**	**EA**				**651,650**	**214,250**	**1,645,500**	**2,511,400**
13205 4820	224,000 Barrel Fuel Oil Tank, Steel, 200' Diameter x 40' High, Cone Roof, Including Foundation, Fabricate, & Erection	EA	0.0006	SIWSO	1.000	651,650	214,250	1,645,500	2,511,400
33 10 9693	**Tank Tightness Testing, <= 12,000 Gallon Tank**	**EA**				**0.00**	**0.00**	**325.00**	**325.00**
01418 7520	Tank Volumetric Tightness Testing <= 12,000 Gallons	EA	0	N/A	1.000	0.00	0.00	325.00	325.00
33 10 9694	**Tank Tightness Testing, 12,001 - 29,999 Gallon Tank**	**EA**				**0.00**	**0.00**	**460.00**	**460.00**
01418 7530	Tank Volumetric Tightness Testing 12,001 - 29,999 Gallons	EA	0	N/A	1.000	0.00	0.00	460.00	460.00
33 10 9695	**Tank Tightness Testing, >= 30,000 Gallon Tank**	**EA**				**0.00**	**0.00**	**560.00**	**560.00**
01418 7540	Tank Volumetric Tightness Testing >= 30,000 Gallons	EA	0	N/A	1.000	0.00	0.00	560.00	560.00

33 10 97	Tank Installation (see also 33 10 96)	Unit	Hourly Output	Crew	Factor	Labor	Equip.	Matl.	Total
33 10 9701	**Control Panel, 1-4 Single-wall ASTs, Monitoring System (Including Sensor)**	**EA**				**0.00**	**0.00**	**2,062**	**2,062**
01001 0087	Control Panel, 1-4 Single-wall ASTs, Monitoring System (Including Sensor)	EA	0	N/A	1.000	0.00	0.00	2,062	2,062
33 10 9702	**Control Panel, 1-8 Single-wall ASTs, Monitoring System (Including Sensor)**	**EA**				**0.00**	**0.00**	**2,445**	**2,445**
01001 0088	Control Panel, 1-8 Single-wall ASTs, Monitoring System (Including Sensor)	EA	0	N/A	1.000	0.00	0.00	2,445	2,445
33 10 9703	**Control Panel, 1-4 Double-wall ASTs, Monitoring System (Including Sensor)**	**EA**				**0.00**	**0.00**	**2,664**	**2,664**
01001 0089	Control Panel, 1-4 Double-wall ASTs, Monitoring System (Including Sensor)	EA	0	N/A	1.000	0.00	0.00	2,664	2,664
33 10 9704	**Control Panel, 1-8 Double-wall ASTs, Monitoring System (Including Sensor)**	**EA**				**0.00**	**0.00**	**3,399**	**3,399**
01001 0090	Control Panel, 1-8 Double-wall ASTs, Monitoring System (Including Sensor)	EA	0	N/A	1.000	0.00	0.00	3,399	3,399
33 10 9705	**550 Gallon Single-walled Storage Tank, Leak Detection Sensor**	**EA**				**0.00**	**0.00**	**673.23**	**673.23**
01001 0091	550 Gallon Single-walled Storage Tank, Leak Detection Sensor	EA	0	N/A	1.000	0.00	0.00	673.23	673.23

33 10 97	Tank Installation (see also 33 10 96)	Unit	Hourly Output	Crew	Factor	Labor	Equip.	Matl.	Total
33 10 9706	**1,000 Gallon Single-walled Storage Tank, Leak Detection Sensor**	**EA**				**0.00**	**0.00**	**907.58**	**907.58**
01001 0092	1,000 Gallon Single-walled Storage Tank, Leak Detection Sensor	EA	0	N/A	1.000	0.00	0.00	907.58	907.58
33 10 9707	**2,000 Gallon Single-walled Storage Tank, Leak Detection Sensor**	**EA**				**0.00**	**0.00**	**1,142**	**1,142**
01001 0093	2,000 Gallon Single-walled Storage Tank, Leak Detection Sensor	EA	0	N/A	1.000	0.00	0.00	1,142	1,142
33 10 9708	**3,000 Gallon Single-walled Storage Tank, Leak Detection Sensor**	**EA**				**0.00**	**0.00**	**1,376**	**1,376**
01001 0094	3,000 Gallon Single-walled Storage Tank, Leak Detection Sensor	EA	0	N/A	1.000	0.00	0.00	1,376	1,376
33 10 9709	**5,000 Gallon Single-walled Storage Tank, Leak Detection Sensor**	**EA**				**0.00**	**0.00**	**1,611**	**1,611**
01001 0095	5,000 Gallon Single-walled Storage Tank, Leak Detection Sensor	EA	0	N/A	1.000	0.00	0.00	1,611	1,611
33 10 9710	**10,000 Gallon Single-walled Storage Tank, Leak Detection Sensor**	**EA**				**0.00**	**0.00**	**1,845**	**1,845**
01001 0096	10,000 Gallon Single-walled Storage Tank, Leak Detection Sensor	EA	0	N/A	1.000	0.00	0.00	1,845	1,845
33 10 9711	**12,000 Gallon Single-walled Storage Tank, Leak Detection Sensor**	**EA**				**0.00**	**0.00**	**2,079**	**2,079**
01001 0097	12,000 Gallon Single-walled Storage Tank, Leak Detection Sensor	EA	0	N/A	1.000	0.00	0.00	2,079	2,079
33 10 9712	**15,000 Gallon Single-walled Storage Tank, Leak Detection Sensor**	**EA**				**0.00**	**0.00**	**2,314**	**2,314**
01001 0098	15,000 Gallon Single-walled Storage Tank, Leak Detection Sensor	EA	0	N/A	1.000	0.00	0.00	2,314	2,314
33 10 9713	**20,000 Gallon Single-walled Storage Tank, Leak Detection Sensor**	**EA**				**0.00**	**0.00**	**2,810**	**2,810**
01001 0099	20,000 Gallon Single-walled Storage Tank, Leak Detection Sensor	EA	0	N/A	1.000	0.00	0.00	2,810	2,810
33 10 9714	**30,000 Gallon Single-walled Storage Tank, Leak Detection Sensor**	**EA**				**0.00**	**0.00**	**4,439**	**4,439**
01001 0100	30,000 Gallon Single-walled Storage Tank, Leak Detection Sensor	EA	0	N/A	1.000	0.00	0.00	4,439	4,439
33 10 9715	**550 Gallon Double-walled Storage Tank, Leak Detection Sensor**	**EA**				**0.00**	**0.00**	**673.23**	**673.23**
01001 0101	550 Gallon Double-walled Storage Tank, Leak Detection Sensor	EA	0	N/A	1.000	0.00	0.00	673.23	673.23
33 10 9716	**1,000 Gallon Double-walled Storage Tank, Leak Detection Sensor**	**EA**				**0.00**	**0.00**	**907.58**	**907.58**
01001 0102	1,000 Gallon Double-walled Storage Tank, Leak Detection Sensor	EA	0	N/A	1.000	0.00	0.00	907.58	907.58

33 10 97 Tank Installation (see also 33 10 96)		Unit	Hourly Output	Crew	Factor	Labor	Equip.	Matl.	Total
33 10 9717	**2,000 Gallon Double-walled Storage Tank, Leak Detection Sensor**	**EA**				**0.00**	**0.00**	**1,142**	**1,142**
01001 0103	2,000 Gallon Double-walled Storage Tank, Leak Detection Sensor	EA	0	N/A	1.000	0.00	0.00	1,142	1,142
33 10 9718	**3,000 Gallon Double-walled Storage Tank, Leak Detection Sensor**	**EA**				**0.00**	**0.00**	**1,376**	**1,376**
01001 0104	3,000 Gallon Double-walled Storage Tank, Leak Detection Sensor	EA	0	N/A	1.000	0.00	0.00	1,376	1,376
33 10 9719	**5,000 Gallon Double-walled Storage Tank, Leak Detection Sensor**	**EA**				**0.00**	**0.00**	**1,611**	**1,611**
01001 0105	5,000 Gallon Double-walled Storage Tank, Leak Detection Sensor	EA	0	N/A	1.000	0.00	0.00	1,611	1,611
33 10 9720	**10,000 Gallon Double-walled Storage Tank, Leak Detection Sensor**	**EA**				**0.00**	**0.00**	**1,845**	**1,845**
01001 0106	10,000 Gallon Double-walled Storage Tank, Leak Detection Sensor	EA	0	N/A	1.000	0.00	0.00	1,845	1,845
33 10 9721	**12,000 Gallon Double-walled Storage Tank, Leak Detection Sensor**	**EA**				**0.00**	**0.00**	**2,079**	**2,079**
01001 0107	12,000 Gallon Double-walled Storage Tank, Leak Detection Sensor	EA	0	N/A	1.000	0.00	0.00	2,079	2,079
33 10 9722	**15,000 Gallon Double-walled Storage Tank, Leak Detection Sensor**	**EA**				**0.00**	**0.00**	**2,314**	**2,314**
01001 0108	15,000 Gallon Double-walled Storage Tank, Leak Detection Sensor	EA	0	N/A	1.000	0.00	0.00	2,314	2,314
33 10 9723	**20,000 Gallon Double-walled Storage Tank, Leak Detection Sensor**	**EA**				**0.00**	**0.00**	**2,810**	**2,810**
01001 0109	20,000 Gallon Double-walled Storage Tank, Leak Detection Sensor	EA	0	N/A	1.000	0.00	0.00	2,810	2,810
33 10 9724	**30,000 Gallon Double-walled Storage Tank, Leak Detection Sensor**	**EA**				**0.00**	**0.00**	**4,439**	**4,439**
01001 0110	30,000 Gallon Double-walled Storage Tank, Leak Detection Sensor	EA	0	N/A	1.000	0.00	0.00	4,439	4,439
33 10 9730	**550 Gallon Double-walled Steel Aboveground Tank**	**EA**				**0.00**	**0.00**	**1,675**	**1,675**
01001 0111	550 Gallon Double-walled Steel Aboveground Tank	EA	0	N/A	1.000	0.00	0.00	1,675	1,675
33 10 9731	**1,000 Gallon Double-walled Steel Aboveground Tank**	**EA**				**0.00**	**0.00**	**2,275**	**2,275**
01001 0112	1,000 Gallon Double-walled Steel Aboveground Tank	EA	0	N/A	1.000	0.00	0.00	2,275	2,275
33 10 9732	**2,000 Gallon Double-walled Steel Aboveground Tank**	**EA**				**0.00**	**0.00**	**3,508**	**3,508**
01001 0113	2,000 Gallon Double-walled Steel Aboveground Tank	EA	0	N/A	1.000	0.00	0.00	3,508	3,508

33 10 97 Tank Installation (see also 33 10 96)		Unit	Hourly Output	Crew	Factor	Labor	Equip.	Matl.	Total
33 10 9733	**3,000 Gallon Double-walled Steel Aboveground Tank**	**EA**				**0.00**	**0.00**	**4,725**	**4,725**
01001 0114	3,000 Gallon Double-walled Steel Aboveground Tank	EA	0	N/A	1.000	0.00	0.00	4,725	4,725
33 10 9734	**5,000 Gallon Double-walled Steel Aboveground Tank**	**EA**				**0.00**	**0.00**	**9,397**	**9,397**
01001 0115	5,000 Gallon Double-walled Steel Aboveground Tank	EA	0	N/A	1.000	0.00	0.00	9,397	9,397
33 10 9735	**10,000 Gallon Double-walled Steel Aboveground Tank**	**EA**				**0.00**	**0.00**	**12,762**	**12,762**
01001 0116	10,000 Gallon Double-walled Steel Aboveground Tank	EA	0	N/A	1.000	0.00	0.00	12,762	12,762
33 10 9736	**12,000 Gallon Double-walled Steel Aboveground Tank**	**EA**				**0.00**	**0.00**	**15,069**	**15,069**
01001 0117	12,000 Gallon Double-walled Steel Aboveground Tank	EA	0	N/A	1.000	0.00	0.00	15,069	15,069
33 10 9737	**15,000 Gallon Double-walled Steel Aboveground Tank**	**EA**				**0.00**	**0.00**	**16,601**	**16,601**
01001 0118	15,000 Gallon Double-walled Steel Aboveground Tank	EA	0	N/A	1.000	0.00	0.00	16,601	16,601
33 10 9738	**20,000 Gallon Double-walled Steel Aboveground Tank**	**EA**				**0.00**	**0.00**	**20,224**	**20,224**
01001 0119	20,000 Gallon Double-walled Steel Aboveground Tank	EA	0	N/A	1.000	0.00	0.00	20,224	20,224
33 10 9740	**550 Gallon Single-walled Fiberglass Aboveground Tank**	**EA**				**0.00**	**0.00**	**4,728**	**4,728**
01001 0120	550 Gallon Single-walled Fiberglass Aboveground Tank	EA	0	N/A	1.000	0.00	0.00	4,728	4,728
33 10 9741	**1,000 Gallon Single-walled Fiberglass Aboveground Tank**	**EA**				**0.00**	**0.00**	**6,336**	**6,336**
01001 0121	1,000 Gallon Single-walled Fiberglass Aboveground Tank	EA	0	N/A	1.000	0.00	0.00	6,336	6,336
33 10 9742	**2,000 Gallon Single-walled Fiberglass Aboveground Tank**	**EA**				**0.00**	**0.00**	**11,023**	**11,023**
01001 0122	2,000 Gallon Single-walled Fiberglass Aboveground Tank	EA	0	N/A	1.000	0.00	0.00	11,023	11,023
33 10 9743	**3,000 Gallon Single-walled Fiberglass Aboveground Tank**	**EA**				**0.00**	**0.00**	**14,396**	**14,396**
01001 0123	3,000 Gallon Single-walled Fiberglass Aboveground Tank	EA	0	N/A	1.000	0.00	0.00	14,396	14,396
33 10 9744	**5,000 Gallon Single-walled Fiberglass Aboveground Tank**	**EA**				**0.00**	**0.00**	**18,611**	**18,611**
01001 0124	5,000 Gallon Single-walled Fiberglass Aboveground Tank	EA	0	N/A	1.000	0.00	0.00	18,611	18,611

33 10 97 Tank Installation (see also 33 10 96)		Unit	Hourly Output	Crew	Factor	Labor	Equip.	Matl.	Total
33 10 9745	**10,000 Gallon Single-walled Fiberglass Aboveground Tank**	**EA**				**0.00**	**0.00**	**20,144**	**20,144**
01001 0125	10,000 Gallon Single-walled Fiberglass Aboveground Tank	EA	0	N/A	1.000	0.00	0.00	20,144	20,144
33 10 9746	**12,000 Gallon Single-walled Fiberglass Aboveground Tank**	**EA**				**0.00**	**0.00**	**20,144**	**20,144**
01001 0126	12,000 Gallon Single-walled Fiberglass Aboveground Tank	EA	0	N/A	1.000	0.00	0.00	20,144	20,144
33 10 9747	**15,000 Gallon Single-walled Fiberglass Aboveground Tank**	**EA**				**0.00**	**0.00**	**21,553**	**21,553**
01001 0127	15,000 Gallon Single-walled Fiberglass Aboveground Tank	EA	0	N/A	1.000	0.00	0.00	21,553	21,553
33 10 9748	**20,000 Gallon Single-walled Fiberglass Aboveground Tank**	**EA**				**0.00**	**0.00**	**23,063**	**23,063**
01001 0128	20,000 Gallon Single-walled Fiberglass Aboveground Tank	EA	0	N/A	1.000	0.00	0.00	23,063	23,063
33 10 9750	**550 Gallon Double-walled Fiberglass Aboveground Tank**	**EA**				**0.00**	**0.00**	**9,458**	**9,458**
01001 0129	550 Gallon Double-walled Fiberglass Aboveground Tank	EA	0	N/A	1.000	0.00	0.00	9,458	9,458
33 10 9751	**1,000 Gallon Double-walled Fiberglass Aboveground Tank**	**EA**				**0.00**	**0.00**	**12,672**	**12,672**
01001 0130	1,000 Gallon Double-walled Fiberglass Aboveground Tank	EA	0	N/A	1.000	0.00	0.00	12,672	12,672
33 10 9752	**2,000 Gallon Double-walled Fiberglass Aboveground Tank**	**EA**				**0.00**	**0.00**	**22,046**	**22,046**
01001 0131	2,000 Gallon Double-walled Fiberglass Aboveground Tank	EA	0	N/A	1.000	0.00	0.00	22,046	22,046
33 10 9753	**3,000 Gallon Double-walled Fiberglass Aboveground Tank**	**EA**				**0.00**	**0.00**	**28,793**	**28,793**
01001 0132	3,000 Gallon Double-walled Fiberglass Aboveground Tank	EA	0	N/A	1.000	0.00	0.00	28,793	28,793
33 10 9754	**5,000 Gallon Double-walled Fiberglass Aboveground Tank**	**EA**				**0.00**	**0.00**	**37,224**	**37,224**
01001 0133	5,000 Gallon Double-walled Fiberglass Aboveground Tank	EA	0	N/A	1.000	0.00	0.00	37,224	37,224
33 10 9755	**10,000 Gallon Double-walled Fiberglass Aboveground Tank**	**EA**				**0.00**	**0.00**	**40,287**	**40,287**
01001 0134	10,000 Gallon Double-walled Fiberglass Aboveground Tank	EA	0	N/A	1.000	0.00	0.00	40,287	40,287
33 10 9756	**12,000 Gallon Double-walled Fiberglass Aboveground Tank**	**EA**				**0.00**	**0.00**	**40,287**	**40,287**
01001 0135	12,000 Gallon Double-walled Fiberglass Aboveground Tank	EA	0	N/A	1.000	0.00	0.00	40,287	40,287

33 10 97	Tank Installation (see also 33 10 96)	Unit	Hourly Output	Crew	Factor	Labor	Equip.	Matl.	Total
33 10 9757	**15,000 Gallon Double-walled Fiberglass Aboveground Tank**	**EA**				**0.00**	**0.00**	**43,107**	**43,107**
01001 0136	15,000 Gallon Double-walled Fiberglass Aboveground Tank	EA	0	N/A	1.000	0.00	0.00	43,107	43,107
33 10 9758	**20,000 Gallon Double-walled Fiberglass Aboveground Tank**	**EA**				**0.00**	**0.00**	**46,124**	**46,124**
01001 0137	20,000 Gallon Double-walled Fiberglass Aboveground Tank	EA	0	N/A	1.000	0.00	0.00	46,124	46,124

33 11 03	Land Treatment/Farming (Solid Phase Bio)	Unit	Hourly Output	Crew	Factor	Labor	Equip.	Matl.	Total
33 11 0301	**Soil Tilling, D3 Dozer with Tiller Attachment**	**HR**				**34.65**	**46.71**	**0.00**	**81.36**
13273 9111	Till Soil with D3 Dozer & Disc Attachment	HR	1	CODTV	1.000	34.65	46.71	0.00	81.36
33 11 0302	**Soil Tilling, D4 Dozer with Tiller Attachment**	**HR**				**34.65**	**47.24**	**0.00**	**81.89**
13273 9112	Till Soil with D4 Dozer & Disc Attachment	HR	1	CODTW	1.000	34.65	47.24	0.00	81.89
33 11 0303	**Soil Tilling, D7 Dozer with Tiller Attachment**	**HR**				**34.65**	**126.57**	**0.00**	**161.22**
13273 9113	Till Soil with D7 Dozer & Disc Attachment	HR	1	CODTX	1.000	34.65	126.57	0.00	161.22
33 11 0304	**Soil Tilling, D9 Dozer with Tiller Attachment**	**HR**				**34.65**	**207.61**	**0.00**	**242.26**
13273 9114	Till Soil with D9 Dozer & Disc Attachment	HR	1	CODTY	1.000	34.65	207.61	0.00	242.26
33 11 0305	**Soil Tilling, D3 Dozer with Plow**	**HR**				**34.65**	**46.71**	**0.00**	**81.36**
13273 9111	Till Soil with D3 Dozer & Disc Attachment	HR	1	CODTV	1.000	34.65	46.71	0.00	81.36
33 11 0306	**Soil Tilling, D4 Dozer with Plow**	**HR**				**34.65**	**47.24**	**0.00**	**81.89**
13273 9112	Till Soil with D4 Dozer & Disc Attachment	HR	1	CODTW	1.000	34.65	47.24	0.00	81.89
33 11 0307	**Soil Tilling, D6 Dozer with Plow**	**HR**				**34.65**	**81.13**	**0.00**	**115.78**
13273 9115	Till Soil with D6 Dozer & Disc Attachment	HR	1	CODTZ	1.000	34.65	81.13	0.00	115.78
33 11 0308	**Soil Tilling, D7 Dozer with Plow**	**HR**				**34.65**	**126.57**	**0.00**	**161.22**
13273 9113	Till Soil with D7 Dozer & Disc Attachment	HR	1	CODTX	1.000	34.65	126.57	0.00	161.22
33 11 0309	**Soil Tilling, D9 Dozer with Plow**	**HR**				**34.65**	**207.61**	**0.00**	**242.26**
13273 9114	Till Soil with D9 Dozer & Disc Attachment	HR	1	CODTY	1.000	34.65	207.61	0.00	242.26

33 11 93	Bioremediation Water (Ex Situ)	Unit	Hourly Output	Crew	Factor	Labor	Equip.	Matl.	Total
33 11 9301	**200 MBH Gas-fired Water Boiler**	**EA**				**1,025**	**143.75**	**4,641**	**5,809**
15624 3015	200 MBH Gas Fired Hot Water Packaged System with Insulated Jacket & Controls	EA	0.15	MSPFP	1.000	1,025	143.75	4,641	5,809
33 11 9302	**275 MBH Gas-fired Water Boiler**	**EA**				**1,757**	**246.43**	**4,913**	**6,917**
15624 3016	275 MBH Gas Fired Hot Water Packaged System with Insulated Jacket & Controls	EA	0.088	MSPFP	1.000	1,757	246.43	4,913	6,917

33 11 93	Bioremediation Water (Ex Situ)	Unit	Hourly Output	Crew	Factor	Labor	Equip.	Matl.	Total
33 11 9303	**360 MBH Gas-fired Water Boiler**	**EA**				**1,757**	**246.43**	**5,008**	**7,012**
15624 3017	360 MBH Gas Fired Hot Water Packaged System with Insulated Jacket & Controls	EA	0.088	MSPFP	1.000	1,757	246.43	5,008	7,012
33 11 9304	**520 MBH Gas-fired Water Boiler**	**EA**				**2,050**	**287.50**	**5,751**	**8,088**
15624 3018	520 MBH Gas Fired Hot Water Packaged System with Insulated Jacket & Controls	EA	0.075	MSPFP	1.000	2,050	287.50	5,751	8,088
33 11 9305	**600 MBH Gas-fired Water Boiler**	**EA**				**2,050**	**287.50**	**6,493**	**8,831**
15624 3019	600 MBH Gas Fired Hot Water Packaged System with Insulated Jacket & Controls	EA	0.075	MSPFP	1.000	2,050	287.50	6,493	8,831
33 11 9306	**720 MBH Gas-fired Water Boiler**	**EA**				**2,460**	**345.00**	**7,237**	**10,042**
15624 3021	720 MBH Gas Fired Hot Water Packaged System with Insulated Jacket & Controls	EA	0.063	MSPFP	1.000	2,460	345.00	7,237	10,042
33 11 9307	**960 MBH Gas-fired Water Boiler**	**EA**				**3,075**	**431.25**	**7,979**	**11,486**
15624 3022	960 MBH Gas Fired Hot Water Packaged System with Insulated Jacket & Controls	EA	0.05	MSPFP	1.000	3,075	431.25	7,979	11,486
33 11 9308	**1,220 MBH Gas-fired Water Boiler**	**EA**				**3,075**	**431.25**	**8,723**	**12,229**
15624 3023	1,220 MBH Gas Fired Hot Water Packaged System with Insulated Jacket & Controls	EA	0.05	MSPFP	1.000	3,075	431.25	8,723	12,229
33 11 9309	**1,440 MBH Gas-fired Water Boiler**	**EA**				**3,510**	**492.29**	**9,751**	**13,754**
15513 2236	1,440 MBH Gas Fired Water Tube Boiler, Steel Shell with Insulated Jacket & Controls	EA	0.044	MSPFP	1.000	3,510	492.29	9,751	13,754
33 11 9310	**1,680 MBH Gas-fired Water Boiler**	**EA**				**4,100**	**575.00**	**10,878**	**15,553**
15513 2240	1,680 MBH Gas Fired Water Tube Boiler, Steel Shell with Insulated Jacket & Controls	EA	0.038	MSPFP	1.000	4,100	575.00	10,878	15,553
33 11 9311	**1,920 MBH Gas-fired Water Boiler**	**EA**				**4,731**	**663.46**	**11,518**	**16,913**
15513 2244	1,920 MBH Gas Fired Water Tube Boiler, Steel Shell with Insulated Jacket & Controls	EA	0.033	MSPFP	1.000	4,731	663.46	11,518	16,913
33 11 9312	**2,160 MBH Gas-fired Water Boiler**	**EA**				**6,150**	**862.50**	**12,417**	**19,430**
15513 2248	2,160 MBH Gas Fired Water Tube Boiler, Steel Shell with Insulated Jacket & Controls	EA	0.025	MSPFP	1.000	6,150	862.50	12,417	19,430
33 11 9313	**2,400 MBH Gas-fired Water Boiler**	**EA**				**6,150**	**862.50**	**13,940**	**20,953**
15513 2252	2,400 MBH Gas Fired Water Tube Boiler, Steel Shell with Insulated Jacket & Controls	EA	0.025	MSPFP	1.000	6,150	862.50	13,940	20,953
33 11 9314	**7 GPM Water Heat Exchanger**	**EA**				**137.83**	**0.00**	**1,237**	**1,375**
15561 1020	Heat Exchanger by 200 Degree Water, 7 GPM 40 - 140 Degree, 4 Pass 3/4" Outside Diameter Copper Tube	EA	0.75	MSPFC	1.000	137.83	0.00	1,237	1,375

33 11 93	Bioremediation Water (Ex Situ)	Unit	Hourly Output	Crew	Factor	Labor	Equip.	Matl.	Total
33 11 9315	**16 GPM Water Heat Exchanger**	**EA**				**165.40**	**0.00**	**1,757**	**1,922**
15561 1040	Heat Exchanger by 200 Degree Water, 16 GPM 40 - 140 Degree, 4 Pass 3/4" Outside Diameter Copper Tube	EA	0.625	MSPFC	1.000	165.40	0.00	1,757	1,922
33 11 9316	**34 GPM Water Heat Exchanger**	**EA**				**206.75**	**0.00**	**2,663**	**2,870**
15561 1060	Heat Exchanger by 200 Degree Water, 34 GPM 40 - 140 Degree, 4 Pass 3/4" Outside Diameter Copper Tube	EA	0.5	MSPFC	1.000	206.75	0.00	2,663	2,870
33 11 9317	**55 GPM Water Heat Exchanger**	**EA**				**275.67**	**0.00**	**3,848**	**4,123**
15561 1080	Heat Exchanger by 200 Degree Water, 55 GPM 40 - 140 Degree, 4 Pass 3/4" Outside Diameter Copper Tube	EA	0.375	MSPFC	1.000	275.67	0.00	3,848	4,123
33 11 9318	**74 GPM Water Heat Exchanger**	**EA**				**359.57**	**0.00**	**4,762**	**5,122**
15561 1100	Heat Exchanger by 200 Degree Water, 74 GPM 40 - 140 Degree, 4 Pass 3/4" Outside Diameter Copper Tube	EA	0.288	MSPFC	1.000	359.57	0.00	4,762	5,122
33 11 9319	**86 GPM Water Heat Exchanger**	**EA**				**435.26**	**0.00**	**6,378**	**6,813**
15561 1120	Heat Exchanger by 200 Degree Water, 86 GPM 40 - 140 Degree, 4 Pass 3/4" Outside Diameter Copper Tube	EA	0.238	MSPFC	1.000	435.26	0.00	6,378	6,813
33 11 9320	**112 GPM Water Heat Exchanger**	**EA**				**480.54**	**67.65**	**7,941**	**8,489**
15561 1140	Heat Exchanger by 200 Degree Water, 112 GPM 40 - 140 Degree, 4 Pass 3/4" Outside Diameter Copper Tube	EA	0.213	MSPFF	1.000	480.54	67.65	7,941	8,489
33 11 9321	**3,000 Gallon Fixed Film Bioreactor**	**EA**				**0.00**	**0.00**	**64,500**	**64,500**
13273 2305	1 - 10 GPM Fixed Film Bioreactor (3,000 Gallon)	EA	1	N/A	1.000	0.00	0.00	64,500	64,500
33 11 9322	**5,000 Gallon Fixed Film Bioreactor**	**EA**				**0.00**	**0.00**	**81,500**	**81,500**
13273 2306	11 - 23 GPM Fixed Film Bioreactor (5,000 Gallon)	EA	1	N/A	1.000	0.00	0.00	81,500	81,500
33 11 9323	**9,000 Gallon Fixed Film Bioreactor**	**EA**				**0.00**	**0.00**	**133,000**	**133,000**
13273 2307	24 - 33 GPM Fixed Film Bioreactor (9,000 Gallon)	EA	1	N/A	1.000	0.00	0.00	133,000	133,000
33 11 9324	**11,500 Gallon Fixed Film Bioreactor**	**EA**				**0.00**	**0.00**	**140,500**	**140,500**
13273 2308	34 - 45 GPM Fixed Film Bioreactor (11,500 Gallon)	EA	1	N/A	1.000	0.00	0.00	140,500	140,500
33 11 9325	**17,500 Gallon Fixed Film Bioreactor, Short Retention**	**EA**				**0.00**	**0.00**	**180,500**	**180,500**
13273 2309	46 - 60 GPM Fixed Film Bioreactor (17,500 Gallon)	EA	1	N/A	1.000	0.00	0.00	180,500	180,500
33 11 9326	**17,500 Gallon Fixed Film Bioreactor, Long Retention**	**EA**				**0.00**	**0.00**	**194,000**	**194,000**
13273 2310	61 - 82 GPM Fixed Film Bioreactor(17,500 Gallon)	EA	1	N/A	1.000	0.00	0.00	194,000	194,000

33 11 93	Bioremediation Water (Ex Situ)	Unit	Hourly Output	Crew	Factor	Labor	Equip.	Matl.	Total
33 11 9327	**18,500 Gallon Fixed Film Bioreactor**	**EA**				**0.00**	**0.00**	**198,500**	**198,500**
13273 2311	83 - 100 GPM Fixed Film Bioreactor (18,500 Gallon)	EA	1	N/A	1.000	0.00	0.00	198,500	198,500
33 11 9328	**24,000 Gallon Fixed Film Bioreactor**	**EA**				**0.00**	**0.00**	**218,500**	**218,500**
13273 2312	> 100 GPM Fixed Film Bioreactor (24,000 Gallon)	EA	1	N/A	1.000	0.00	0.00	218,500	218,500
33 11 9329	**27,000 Gallon Fixed Film Bioreactor**	**EA**				**0.00**	**0.00**	**243,000**	**243,000**
13273 2313	> 100 GPM Fixed Film Bioreactor (27,000 Gallon)	EA	1	N/A	1.000	0.00	0.00	243,000	243,000
33 11 9330	**29,500 Gallon Fixed Film Bioreactor**	**EA**				**0.00**	**0.00**	**261,500**	**261,500**
13273 2314	> 100 GPM Fixed Film Bioreactor (29,500 Gallon)	EA	1	N/A	1.000	0.00	0.00	261,500	261,500
33 11 9331	**Expansion Tank, Pipe, & Fittings, 1 - 1,220 MBH Boiler**	**EA**				**1,629**	**0.00**	**1,309**	**2,938**
15172 0610	2" (50 mm) A-53 Pipe, Schedule 40, Not Including Hangers or Fittings	LF	23.75	MPLUE	100.000	3.75	0.00	2.07	581.83
15174 4720	2" Nipple, Black Steel Schedule 40, 4"	EA	11.63	MPLUE	4.000	7.65	0.00	3.80	45.81
15174 4818	1-1/2" Nipple, Galvanized Steel Schedule 40, 4" (10 cm) Long	EA	15	MPLUE	2.000	5.93	0.00	2.83	17.53
15174 5130	2" 90 Degree Elbow, 150 Lb Malleable Iron Black	EA	4.625	MPLUE	10.000	19.24	0.00	9.55	287.92
15174 5740	2" Coupling, Straight 150 Lb Malleable Iron Black	EA	4.625	MPLUE	6.000	19.24	0.00	7.66	161.36
15174 6128	2" x 1-1/2" 90 Degree Reducing Elbow 150 Lb Malleable Iron Galvanized	EA	4.625	MPLUE	2.000	19.24	0.00	13.33	65.14
15174 7080	2" Union, 150 Lb Malleable Iron Black	EA	3.125	MPLUE	2.000	28.47	0.00	16.70	90.35
15185 1301	0.006" Aluminum Insulated Jacket with Bands	SF	17.5	AASBC	65.000	5.36	0.00	0.24	363.91
15185 1505	2" Pipe	LF	27.5	AASBC	100.000	3.41	0.00	6.39	979.75
15635 0220	2" Diameter x 17" Flexible Hose Connection, for Vibration Isolation Bronze Hose with Hex Threaded Nipple Ends	EA	1.175	MSPFA	2.000	43.99	0.00	83.34	254.66
15698 0180	2" Strainer (Bronze Body) Y-Type, 150 Lb (68 kg) Screwed	EA	1.625	MSPFA	1.000	31.81	0.00	58.28	90.09

33 11 93	Bioremediation Water (Ex Situ)	Unit	Hourly Output	Crew	Factor	Labor	Equip.	Matl.	Total
33 11 9332	**Expansion Tank, Pipe, & Fittings, 1,440 - 2,400 MBH Boiler**	**EA**				**5,817**	**39.80**	**10,209**	**16,065**
11261 0320	10 Gallon Bypass Shot Chemical Feeder Floor Mount, 150 Lb ASME Code	EA	0.146	MSPFD	1.000	650.62	0.00	1,499	2,149
15103 1120	1" Screwed Backflow Preventer with 2 Gate & 2 Check Valve & Automatic Dif Relief Valve	EA	1.038	MPLUA	1.000	49.70	0.00	711.40	761.10
15172 0580	1" (25 mm) A-53 Pipe, Schedule 40, Not Including Hangers or Fittings	LF	35.75	MPLUE	30.000	2.49	0.00	0.94	102.98
15172 0650	4" (10 cm) A-53 Pipe, Schedule 40, Not Including Hangers or Fittings	LF	17.5	MPLUJ	100.000	6.95	0.00	6.12	1,308
15174 5100	1" 90 Degree Elbow, 150 Lb Malleable Iron Black	EA	7.5	MPLUE	4.000	11.86	0.00	2.57	57.72
15174 5170	4" 90 Degree Elbow, 150 Lb Malleable Iron Black	EA	3	MPLUJ	10.000	40.55	0.00	66.82	1,074
15174 5510	1" Tee, Straight 150 Lb Malleable Iron Black	EA	5.5	MPLUE	4.000	16.18	0.00	4.01	80.73
15174 5580	4" Tee, Straight 150 Lb Malleable Iron Black	EA	2.375	MPLUJ	2.000	51.22	0.00	105.67	313.79
15174 5914	4" Coupling, Red 150 Lb Malleable Iron Black	EA	3	MPLUJ	6.000	40.55	0.00	62.16	616.28
15174 7050	1" Union, 150 Lb Malleable Iron Black	EA	5.5	MPLUE	2.000	16.18	0.00	8.08	48.52
15178 0670	4" Bolt, Nut, & Gasket Set, A-307 MB	EA	3.038	MPLUJ	6.000	40.05	0.00	11.28	307.99
15178 2390	4" Screwed Flange, Black 125 Lb Cast Iron	EA	2.188	MPLUJ	6.000	55.61	0.00	37.50	558.71
15178 4666	4" x 2" Reducer, Eccentric	EA	0.713	MSPFE	2.000	133.59	14.15	25.05	345.58
15185 1009	4" Diameter Pipe, 1" Thick Fiberglass Pipe Cover with Fire Retardant Jackets	LF	30.63	AASBC	100.000	3.06	0.00	2.49	555.04
15185 1301	0.006" Aluminum Insulated Jacket with Bands	SF	17.5	AASBC	110.000	5.36	0.00	0.24	615.84
15193 2950	1" Bronze 125 Lb Gate Valve Threaded, Brazed or Soldered	EA	2.375	MPLUA	2.000	21.71	0.00	26.52	96.47
15193 4970	1" Globe Valve, Bronze, Threaded 125 Lb	EA	2.375	MPLUA	1.000	21.71	0.00	72.52	94.23
15193 6450	1" x 1" Bronze Pressure Relief Valve, Threaded	EA	1.875	MPLUA	1.000	27.50	0.00	104.70	132.20
15193 7790	1" Water Pressure Reduction Valve -Threaded Iron Construction	EA	2.125	MPLUA	1.000	24.26	0.00	184.18	208.45
15194 2280	4" Iron Body Gate Valve, NRS	EA	0.588	MPLUE	2.000	151.45	0.00	501.43	1,306
15194 6060	4" Iron Body Swing Check Valve, Bronze, Flanged & 125 Lb	EA	0.588	MPLUE	1.000	151.45	0.00	329.04	480.49
15568 2140	144 Gallon 24-1/4" x 83-1/2" Compression Tank 125 PSI	EA	0.588	MSPFC	1.000	175.96	0.00	889.25	1,065
15605 0120	Air Separator, Rolairtrol, 300 GPM with 4" Boiler Connection, Flanged	EA	0.375	MSPFB	1.000	253.83	0.00	1,458	1,712
15605 2420	Compression Tank Fittings for 100 Gallon & Larger	EA	2.5	MSPFA	1.000	20.68	0.00	107.91	128.58
15622 4510	Brass Gauge Cock, 1/4" FMPT	EA	4	MSPFA	3.000	12.92	0.00	6.70	58.86
15635 3470	4" Diameter x 9" Metal Wire Braid Connection Over Corrugated Stainless Steel 150 PSI for Vibration Isolation	EA	0.586	MSPFC	2.000	176.32	0.00	120.87	594.37
15680 0240	4" Carbon Steel Flow Venturi BW	EA	1.25	MSPFN	1.000	120.23	11.50	433.98	565.71
15699 1060	4" Strainer-Iron Body Y-Type, 125 Lb (58 kg) Flanged Ends	EA	1	MSPFC	1.000	103.38	0.00	157.41	260.78
15742 3014	4" Diameter Dial Pressure Gauge Steel Case 0-300 PSI	EA	2	MSPFA	1.000	25.84	0.00	39.18	65.03
15742 4694	Thermometer, 12" Scale & Angle with Socket, Industrial with Hg, Multi-range with Socket	EA	2	MSPFA	2.000	25.84	0.00	175.05	401.79

33 11 99	Other	Unit	Hourly Output	Crew	Factor	Labor	Equip.	Matl.	Total

33 11 Biological Treatment

33 11 99	Other	Unit	Hourly Output	Crew	Factor	Labor	Equip.	Matl.	Total
33 11 9901	**Application of Bioculture to Contaminated Soil**	**ACRE**				**26.37**	**33.08**	**0.87**	**60.32**
02976 4930	Watering by Truck	MSF	87.12	COFWI	43.560	0.61	0.76	0.02	60.32
33 11 9902	**Light Petroleum Hydrocarbon Degraders, Microorganisms**	**LB**				**0.00**	**0.00**	**24.00**	**24.00**
13273 9101	L-103 Light Petroleum Biocultures, per Pale	LB	0	N/A	1.000	0.00	0.00	24.00	24.00
33 11 9903	**Light Petroleum Hydrocarbon Degraders, 100 Lb Bag, Microorganisms**	**EA**				**0.00**	**0.00**	**2,475**	**2,475**
13273 9102	L-103 Light Petroleum Biocultures, per 100 Lb Bag	EA	0	N/A	1.000	0.00	0.00	2,475	2,475
33 11 9904	**Light Petroleum Hydrocarbon Degraders, 1,000 Lb Bag, Microorganisms**	**EA**				**0.00**	**0.00**	**23,700**	**23,700**
13273 9103	L-103 Light Petroleum Biocultures, per 1,000 Lb Bag	EA	0	N/A	1.000	0.00	0.00	23,700	23,700
33 11 9905	**Heavy Petroleum Hydrocarbon/Creosote Degraders, Microorganisms**	**LB**				**0.00**	**0.00**	**17.95**	**17.95**
13273 9104	L-104 Heavy Petroleum/Creosote Biocultures, per Pale	LB	0	N/A	1.000	0.00	0.00	17.95	17.95
33 11 9906	**Heavy Petroleum Hydrocarbon/Creosote Degraders, 100 Lb Bag, Microorganisms**	**EA**				**0.00**	**0.00**	**1,050**	**1,050**
13273 9105	L-104 Heavy Petroleum/Creosote Biocultures, per 100 Lb Bag	EA	0	N/A	1.000	0.00	0.00	1,050	1,050
33 11 9907	**Heavy Petroleum Hydrocarbon/Creosote Degraders, 1,000 Lb Bag, Microorganisms**	**EA**				**0.00**	**0.00**	**8,775**	**8,775**
13273 9106	L-104 Heavy Petroleum/Creosote Biocultures, per 1,000 Lb Bag	EA	0	N/A	1.000	0.00	0.00	8,775	8,775
33 11 9951	**Bionutrients, 50 Lb Bag**	**EA**				**0.00**	**0.00**	**68.00**	**68.00**
13273 9151	Bionutrients, 50 Lb Bag	EA	0	N/A	1.000	0.00	0.00	68.00	68.00

33 12 02	Solvent Extraction	Unit	Hourly Output	Crew	Factor	Labor	Equip.	Matl.	Total
33 12 0201	**1,500 CY/Month Unit, Site Preparation Charge**	**EA**				**0.00**	**0.00**	**163,500**	**163,500**
13276 3621	1,500 CY/Month Unit - Site Preparation Charge	EA	0	N/A	1.000	0.00	0.00	163,500	163,500
33 12 0202	**6,000 CY/Month Unit - Site Preparation Charge**	**EA**				**0.00**	**0.00**	**329,500**	**329,500**
13276 3622	6,000 CY/Month Unit - Site Preparation Charge	EA	0	N/A	1.000	0.00	0.00	329,500	329,500
33 12 0203	**18,000 CY/Month Unit - Site Preparation Charge**	**EA**				**0.00**	**0.00**	**540,500**	**540,500**
13276 3623	18,000 CY/Month Unit - Site Preparation Charge	EA	0	N/A	1.000	0.00	0.00	540,500	540,500
33 12 0204	**1,500 CY/Month Unit - Mobilize and Assemble**	**EA**				**0.00**	**0.00**	**217,500**	**217,500**
13276 3624	1,500 CY/Month Unit - Mobilization and Assembly	EA	0	N/A	1.000	0.00	0.00	217,500	217,500

33 12 02 Solvent Extraction

33 12 02 Solvent Extraction		Unit	Hourly Output	Crew	Factor	Labor	Equip.	Matl.	Total
33 12 0205	**6,000 CY/Month Unit - Mobilize and Assemble**	**EA**				**0.00**	**0.00**	**517,000**	**517,000**
13276 3625	6,000 CY/Month Unit - Mobilization and Assembly	EA	0	N/A	1.000	0.00	0.00	517,000	517,000
33 12 0206	**18,000 CY/Month Unit - Mobilize and Assemble**	**EA**				**0.00**	**0.00**	**127,500**	**127,500**
13276 3626	18,000 CY/Month Unit - Mobilization and Assembly	EA	0	N/A	1.000	0.00	0.00	127,500	127,500
33 12 0207	**1,500 CY/Month Unit - Start-up Charge**	**EA**				**0.00**	**0.00**	**44,800**	**44,800**
13276 3627	1,500 CY/Month Unit - Start-up Charge	EA	0	N/A	1.000	0.00	0.00	44,800	44,800
33 12 0208	**6,000 CY/Month Unit - Start-up Charge**	**EA**				**0.00**	**0.00**	**86,500**	**86,500**
13276 3628	6,000 CY/Month Unit - Start-up Charge	EA	0	N/A	1.000	0.00	0.00	86,500	86,500
33 12 0209	**18,000 CY/Month Unit - Start-up Charge**	**EA**				**0.00**	**0.00**	**171,500**	**171,500**
13276 3629	18,000 CY/Month Unit - Start-up Charge	EA	0	N/A	1.000	0.00	0.00	171,500	171,500
33 12 0210	**1,500 CY/Month Unit - Decontaminate/DeMobilize**	**EA**				**0.00**	**0.00**	**99,000**	**99,000**
13276 3631	1,500 CY/Month Unit - Decontamination/Demobilization	EA	0	N/A	1.000	0.00	0.00	99,000	99,000
33 12 0211	**6,000 CY/Month Unit - Decontaminate/DeMobilize**	**EA**				**0.00**	**0.00**	**198,000**	**198,000**
13276 3632	6,000 CY/Month Unit - Decontamination/Demobilization	EA	0	N/A	1.000	0.00	0.00	198,000	198,000
33 12 0212	**18,000 CY/Month Unit - Decontaminate/DeMobilize**	**EA**				**0.00**	**0.00**	**329,500**	**329,500**
13276 3633	18,000 CY/Month Unit - Decontamination/Demobilization	EA	0	N/A	1.000	0.00	0.00	329,500	329,500
33 12 0213	**1,500 CY/Month Unit - Process Equipment Rental**	**MO**				**0.00**	**181,457**	**278.78**	**181,736**
13276 3634	1,500 CY/Month Unit - Process Equipment Rental	MO	0.001	XYSEA	1.000	0.00	181,457	278.78	181,736
33 12 0214	**6,000 CY/Month Unit - Process Equipment Rental**	**MO**				**0.00**	**400,537**	**599.98**	**401,137**
13276 3635	6,000 CY/Month Unit - Process Equipment Rental	MO	0.001	XYSEB	1.000	0.00	400,537	599.98	401,137
33 12 0215	**18,000 CY/Month Unit - Process Equipment Rental**	**MO**				**0.00**	**825,338**	**1,200**	**826,538**
13276 3636	18,000 CY/Month Unit - Process Equipment Rental	MO	0.001	XYSEC	1.000	0.00	825,338	1,200	826,538
33 12 0216	**1,500 CY/Month Unit - Process Labor**	**MO**				**70,555**	**0.00**	**90.37**	**70,645**
13276 3637	1,500 CY/Month Unit - Process Labor	MO	0.001	HLABS	1.000	70,555	0.00	90.37	70,645
33 12 0217	**6,000 CY/Month Unit - Process Labor**	**MO**				**132,314**	**0.00**	**165.41**	**132,479**
13276 3638	6,000 CY/Month Unit - Process Labor	MO	0.001	HLABT	1.000	132,314	0.00	165.41	132,479

33 12 Chemical Treatment

33 12 02 Solvent Extraction		Unit	Hourly Output	Crew	Factor	Labor	Equip.	Matl.	Total
33 12 0218	**18,000 CY/Month Unit - Process Labor**	**MO**				**132,314**	**0.00**	**165.41**	**132,479**
13276 3639	18,000 CY/Month Unit - Process Labor	MO	0.001	HLABT	1.000	132,314	0.00	165.41	132,479
33 12 0219	**1,500 CY/Month Unit - Consumables**	**CY**				**0.00**	**0.00**	**44.00**	**44.00**
13276 3641	1,500 CY/Month Unit - Consumables	CY	0	N/A	1.000	0.00	0.00	44.00	44.00
33 12 0220	**6,000 CY/Month Unit - Consumables**	**CY**				**0.00**	**0.00**	**28.50**	**28.50**
13276 3642	6,000 CY/Month Unit - Consumables	CY	0	N/A	1.000	0.00	0.00	28.50	28.50
33 12 0221	**18,000 CY/Month Unit - Consumables**	**CY**				**0.00**	**0.00**	**25.00**	**25.00**
13276 3643	18,000 CY/Month Unit - Consumables	CY	0	N/A	1.000	0.00	0.00	25.00	25.00
33 12 0222	**Pretreatment Unit - Electric Hammermill (25 hp, 5 CY/HR, 1,500 CY/Mo)**	**EA**				**0.00**	**0.00**	**72,000**	**72,000**
13276 3644	1,500 CY/Month Unit - Pretreatment Unit	EA	0	N/A	1.000	0.00	0.00	72,000	72,000
33 12 0223	**Pretreatment Unit - Electric Hammermill (75 hp, 20 CY/HR, 6,000 CY/Mo)**	**EA**				**0.00**	**0.00**	**91,000**	**91,000**
13276 3645	6,000 CY/Month Unit - Pretreatment Unit	EA	0	N/A	1.000	0.00	0.00	91,000	91,000
33 12 0224	**Pretreatment Unit - Electric Hammermill (200 hp, 60 CY/HR, 18,000 CY/Mo)**	**EA**				**0.00**	**0.00**	**127,500**	**127,500**
13276 3646	18,000 CY/Month Unit - Pretreatment Unit	EA	0	N/A	1.000	0.00	0.00	127,500	127,500

33 12 06 Neutralization		Unit	Hourly Output	Crew	Factor	Labor	Equip.	Matl.	Total
33 12 0601	**10 Gallon Batch Neutralizer**	**EA**				**492.20**	**0.00**	**18,700**	**19,192**
13274 1111	10 Gallon Batch Neutralizer	EA	0.25	UEMCD	1.000	492.20	0.00	18,700	19,192
33 12 0602	**25 Gallon Batch Neutralizer**	**EA**				**492.20**	**0.00**	**23,100**	**23,592**
13274 1112	25 Gallon Batch Neutralizer	EA	0.25	UEMCD	1.000	492.20	0.00	23,100	23,592
33 12 0603	**50 Gallon Batch Neutralizer**	**EA**				**492.20**	**0.00**	**29,400**	**29,892**
13274 1113	50 Gallon Batch Neutralizer	EA	0.25	UEMCD	1.000	492.20	0.00	29,400	29,892
33 12 0604	**100 Gallon Batch Neutralizer**	**EA**				**492.20**	**0.00**	**34,100**	**34,592**
13274 1114	100 Gallon Batch Neutralizer	EA	0.25	UEMCD	1.000	492.20	0.00	34,100	34,592
33 12 0605	**200 Gallon Batch Neutralizer**	**EA**				**492.20**	**0.00**	**41,800**	**42,292**
13274 1115	200 Gallon Batch Neutralizer	EA	0.25	UEMCD	1.000	492.20	0.00	41,800	42,292
33 12 0606	**Pumping System**	**EA**				**0.00**	**0.00**	**2,025**	**2,025**
13274 1116	Pumping System Added to System	EA	0	N/A	1.000	0.00	0.00	2,025	2,025
33 12 0607	**5 GPM Continuous Neutralization System**	**EA**				**0.00**	**0.00**	**28,612**	**28,612**
01001 0138	5 GPM Continuous Neutralization System	EA	0	N/A	1.000	0.00	0.00	28,612	28,612
33 12 0608	**10 GPM Continuous Neutralization System**	**EA**				**0.00**	**0.00**	**31,530**	**31,530**
01001 0139	10 GPM Continuous Neutralization System	EA	0	N/A	1.000	0.00	0.00	31,530	31,530

33 12 Chemical Treatment

33 12 06 Neutralization		Unit	Hourly Output	Crew	Factor	Labor	Equip.	Matl.	Total
33 12 0609	**30 GPM Continuous Neutralization System**	**EA**				**0.00**	**0.00**	**48,921**	**48,921**
01001 0140	30 GPM Continuous Neutralization System	EA	0	N/A	1.000	0.00	0.00	48,921	48,921
33 12 0610	**70 GPM Continuous Neutralization System**	**EA**				**0.00**	**0.00**	**81,685**	**81,685**
01001 0141	70 GPM Continuous Neutralization System	EA	0	N/A	1.000	0.00	0.00	81,685	81,685
33 12 0611	**100 GPM Continuous Neutralization System**	**EA**				**0.00**	**0.00**	**107,155**	**107,155**
01001 0142	100 GPM Continuous Neutralization System	EA	0	N/A	1.000	0.00	0.00	107,155	107,155

33 12 08 Ultraviolet Oxidation		Unit	Hourly Output	Crew	Factor	Labor	Equip.	Matl.	Total
33 12 0801	**Cavitation System Treatability Study**	**EA**				**0.00**	**0.00**	**3,024**	**3,024**
13272 6101	Cavitation System Treatability Study	EA	0	N/A	1.000	0.00	0.00	3,024	3,024
33 12 0802	**Peroxide Treatability/Pilot Plant Study**	**EA**				**0.00**	**0.00**	**7,394**	**7,394**
13272 6102	Peroxide Treatability/Pilot Plant Study	EA	0	N/A	1.000	0.00	0.00	7,394	7,394
33 12 0803	**Peroxide System Mob/Assembly/Shakedown**	**EA**				**0.00**	**0.00**	**10,084**	**10,084**
13272 6103	Peroxide System Mob/Assembly/Shakedown	EA	0	N/A	1.000	0.00	0.00	10,084	10,084
33 12 0804	**Ozone System On-site Operator Training**	**EA**				**3,630**	**0.00**	**0.00**	**3,630**
13272 6104	Ozone System On-site Operator Training	EA	0.025	UOEHB	1.000	3,630	0.00	0.00	3,630
33 12 0805	**Operator Health and Safety Course**	**EA**				**1,088**	**0.00**	**424.22**	**1,512**
13272 6105	Operator Health and Safety Course	EA	0.042	UOEHA	1.000	1,088	0.00	424.22	1,512
33 12 0806	**Supervisor Health and Safety Course**	**EA**				**363.00**	**0.00**	**72.72**	**435.72**
13272 6106	Supervisor Health and Safety Course	EA	0.125	UOEHA	1.000	363.00	0.00	72.72	435.72
33 12 0807	**Cavitation Permit & Regulatory Fee**	**EA**				**0.00**	**0.00**	**24,241**	**24,241**
13272 6107	Cavitation Permit & Regulatory Fee	EA	0	N/A	1.000	0.00	0.00	24,241	24,241
33 12 0808	**Cavitation System Demobilization & Disposal**	**EA**				**21,290**	**24,710**	**3,636**	**49,636**
13272 6108	Cavitation System Demobilization & Disposal	EA	0.003	UOEHE	1.000	21,290	24,710	3,636	49,636
33 12 0809	**5 KW High Intensity Ultraviolet, H2O2 Capital Equipment**	**EA**				**3,929**	**4,560**	**54,543**	**63,031**
13272 6110	5 KW High Intensity Ultraviolet, H2O2 Capital Equipment	EA	0.017	UOEHE	1.000	3,929	4,560	54,543	63,031
33 12 0810	**10 KW High Intensity Ultraviolet, H2O2 Capital Equipment**	**EA**				**4,400**	**5,107**	**63,634**	**73,140**
13272 6111	10 KW High Intensity Ultraviolet, H2O2 Capital Equipment	EA	0.015	UOEHE	1.000	4,400	5,107	63,634	73,140
33 12 0811	**15 KW High Intensity Ultraviolet, H2O2 Capital Equipment**	**EA**				**4,853**	**5,632**	**70,118**	**80,604**
13272 6112	15 KW High Intensity Ultraviolet, H2O2 Capital Equipment	EA	0.014	UOEHE	1.000	4,853	5,632	70,118	80,604

33 12 08 Ultraviolet Oxidation		Unit	Hourly Output	Crew	Factor	Labor	Equip.	Matl.	Total
33 12 0812	**20 KW High Intensity Ultraviolet, H2O2 Capital Equipment**	**EA**				**5,323**	**6,177**	**76,724**	**88,224**
13272 6113	20 KW High Intensity Ultraviolet, H2O2 Capital Equipment	EA	0.012	UOEHE	1.000	5,323	6,177	76,724	88,224
33 12 0813	**25 KW High Intensity Ultraviolet, H2O2 Capital Equipment**	**EA**				**5,789**	**6,719**	**83,512**	**96,020**
13272 6114	25 KW High Intensity Ultraviolet, H2O2 Capital Equipment	EA	0.011	UOEHE	1.000	5,789	6,719	83,512	96,020
33 12 0814	**30 KW High Intensity Ultraviolet, H2O2 Capital Equipment**	**EA**				**6,226**	**7,226**	**90,057**	**103,510**
13272 6115	30 KW High Intensity Ultraviolet, H2O2 Capital Equipment	EA	0.011	UOEHE	1.000	6,226	7,226	90,057	103,510
33 12 0815	**35 KW High Intensity Ultraviolet, H2O2 Capital Equipment**	**EA**				**6,667**	**7,737**	**96,723**	**111,127**
13272 6116	35 KW High Intensity Ultraviolet, H2O2 Capital Equipment	EA	0.01	UOEHE	1.000	6,667	7,737	96,723	111,127
33 12 0816	**40 KW High Intensity Ultraviolet, H2O2 Capital Equipment**	**EA**				**7,174**	**8,326**	**103,390**	**118,890**
13272 6117	40 KW High Intensity Ultraviolet, H2O2 Capital Equipment	EA	0.009	UOEHE	1.000	7,174	8,326	103,390	118,890
33 12 0817	**45 KW High Intensity Ultraviolet, H2O2 Capital Equipment**	**EA**				**7,586**	**8,805**	**110,056**	**126,447**
13272 6118	45 KW High Intensity Ultraviolet, H2O2 Capital Equipment	EA	0.009	UOEHE	1.000	7,586	8,805	110,056	126,447
33 12 0818	**50 KW High Intensity Ultraviolet, H2O2 Capital Equipment**	**EA**				**8,049**	**9,341**	**116,722**	**134,113**
13272 6120	50 KW High Intensity Ultraviolet, H2O2 Capital Equipment	EA	0.008	UOEHE	1.000	8,049	9,341	116,722	134,113
33 12 0819	**60 KW High Intensity Ultraviolet, H2O2 Capital Equipment**	**EA**				**8,919**	**10,351**	**129,934**	**149,204**
13272 6121	60 KW High Intensity Ultraviolet, H2O2 Capital Equipment	EA	0.007	UOEHE	1.000	8,919	10,351	129,934	149,204
33 12 0820	**70 KW High Intensity Ultraviolet, H2O2 Capital Equipment**	**EA**				**9,851**	**11,433**	**143,267**	**164,550**
13272 6122	70 KW High Intensity Ultraviolet, H2O2 Capital Equipment	EA	0.007	UOEHE	1.000	9,851	11,433	143,267	164,550
33 12 0821	**80 KW High Intensity Ultraviolet, H2O2 Capital Equipment**	**EA**				**10,820**	**12,557**	**156,478**	**179,855**
13272 6123	80 KW High Intensity Ultraviolet, H2O2 Capital Equipment	EA	0.006	UOEHE	1.000	10,820	12,557	156,478	179,855
33 12 0822	**90 KW High Intensity Ultraviolet, H2O2 Capital Equipment**	**EA**				**11,786**	**13,679**	**169,811**	**195,275**
13272 6124	90 KW High Intensity Ultraviolet, H2O2 Capital Equipment	EA	0.006	UOEHE	1.000	11,786	13,679	169,811	195,275

33 12 08	Ultraviolet Oxidation	Unit	Hourly Output	Crew	Factor	Labor	Equip.	Matl.	Total
33 12 0823	**100 KW High Intensity Ultraviolet, H2O2 Capital Equipment**	**EA**				**12,692**	**14,731**	**183,144**	**210,567**
13272 6125	100 KW High Intensity Ultraviolet, H2O2 Capital Equipment	EA	0.005	UOEHE	1.000	12,692	14,731	183,144	210,567
33 12 0824	**125 KW High Intensity Ultraviolet, H2O2 Capital Equipment**	**EA**				**15,000**	**17,409**	**216,355**	**248,764**
13272 6126	125 KW High Intensity Ultraviolet, H2O2 Capital Equipment	EA	0.004	UOEHE	1.000	15,000	17,409	216,355	248,764
33 12 0825	**150 KW High Intensity Ultraviolet, H2O2 Capital Equipment**	**EA**				**17,368**	**20,158**	**249,687**	**287,213**
13272 6127	150 KW High Intensity Ultraviolet, H2O2 Capital Equipment	EA	0.004	UOEHE	1.000	17,368	20,158	249,687	287,213
33 12 0826	**175 KW High Intensity Ultraviolet, H2O2 Capital Equipment**	**EA**				**17,368**	**20,158**	**282,776**	**320,303**
13272 6128	175 KW High Intensity Ultraviolet, H2O2 Capital Equipment	EA	0.004	UOEHE	1.000	17,368	20,158	282,776	320,303
33 12 0827	**200 KW High Intensity Ultraviolet, H2O2 Capital Equipment**	**EA**				**22,000**	**25,533**	**316,108**	**363,642**
13272 6129	200 KW High Intensity Ultraviolet, H2O2 Capital Equipment	EA	0.003	UOEHE	1.000	22,000	25,533	316,108	363,642
33 12 0828	**225 KW High Intensity Ultraviolet, H2O2 Capital Equipment**	**EA**				**24,444**	**28,370**	**349,077**	**401,891**
13272 6130	225 KW High Intensity Ultraviolet, H2O2 Capital Equipment	EA	0.003	UOEHE	1.000	24,444	28,370	349,077	401,891
33 12 0829	**250 KW High Intensity Ultraviolet, H2O2 Capital Equipment**	**EA**				**26,400**	**30,640**	**382,530**	**439,570**
13272 6131	250 KW High Intensity Ultraviolet, H2O2 Capital Equipment	EA	0.003	UOEHE	1.000	26,400	30,640	382,530	439,570
33 12 0830	**10 GPM Low Energy Cavitation System Capital Equipment**	**EA**				**5,000**	**5,803**	**42,907**	**53,710**
13272 6132	10 GPM Low Energy Cavitation System Capital Equipment	EA	0.013	UOEHE	1.000	5,000	5,803	42,907	53,710
33 12 0831	**25 GPM Low Energy Cavitation System Capital Equipment**	**EA**				**6,667**	**7,737**	**58,301**	**72,705**
13272 6133	25 GPM Low Energy Cavitation System Capital Equipment	EA	0.01	UOEHE	1.000	6,667	7,737	58,301	72,705
33 12 0832	**10 GPM High Energy Cavitation System Capital Equipment**	**EA**				**5,000**	**5,803**	**43,877**	**54,680**
13272 6134	10 GPM High Energy Cavitation System Capital Equipment	EA	0.013	UOEHE	1.000	5,000	5,803	43,877	54,680
33 12 0833	**25 GPM High Energy Cavitation System Capital Equipment**	**EA**				**6,667**	**7,737**	**58,785**	**73,189**
13272 6135	25 GPM High Energy Cavitation System Capital Equipment	EA	0.01	UOEHE	1.000	6,667	7,737	58,785	73,189

33 12 08	Ultraviolet Oxidation	Unit	Hourly Output	Crew	Factor	Labor	Equip.	Matl.	Total
33 12 0834	**20 GPM Ultraviolet Reactor, Ozone and Peroxide Capital Equipment**	**EA**				**4,853**	**5,632**	**66,664**	**77,149**
13272 6136	20 GPM Ultraviolet Reactor, Ozone and Peroxide Capital Equipment	EA	0.014	UOEHE	1.000	4,853	5,632	66,664	77,149
33 12 0835	**100 GPM Ultraviolet Reactor, Ozone and Peroxide Capital Equipment**	**EA**				**10,476**	**12,159**	**150,660**	**173,295**
13272 6137	100 GPM Ultraviolet Reactor, Ozone and Peroxide Capital Equipment	EA	0.006	UOEHE	1.000	10,476	12,159	150,660	173,295
33 12 0836	**250 GPM Ultraviolet Reactor, Ozone and Peroxide Capital Equipment**	**EA**				**17,368**	**20,158**	**251,783**	**289,309**
13272 6138	250 GPM Ultraviolet Reactor, Ozone and Peroxide Capital Equipment	EA	0.004	UOEHE	1.000	17,368	20,158	251,783	289,309
33 12 0837	**Cavitation System Peroxide Consumption**	**LB**				**0.00**	**0.00**	**0.52**	**0.52**
13272 6139	Cavitation System Peroxide Consumption	LB	0	N/A	1.000	0.00	0.00	0.52	0.52
33 12 0838	**20 GPM Ozone System Consumables**	**WK**				**0.00**	**0.00**	**303.02**	**303.02**
13272 6140	20 GPM Ozone System Consumables	WK	0	N/A	1.000	0.00	0.00	303.02	303.02
33 12 0839	**100 GPM Ozone System Consumables**	**WK**				**0.00**	**0.00**	**757.55**	**757.55**
13272 6141	100 GPM Ozone System Consumables	WK	0	N/A	1.000	0.00	0.00	757.55	757.55
33 12 0840	**250 GPM Ozone System Consumables**	**WK**				**0.00**	**0.00**	**1,212**	**1,212**
13272 6142	250 GPM Ozone System Consumables	WK	0	N/A	1.000	0.00	0.00	1,212	1,212
33 12 0845	**Ozone System Demobilization**	**EA**				**3,708**	**4,303**	**0.00**	**8,011**
13272 6147	Ozone System Demobilization	EA	0.018	UOEHE	1.000	3,708	4,303	0.00	8,011
33 12 0847	**Peroxide System Sodium Hydroxide**	**KGAL**				**0.00**	**0.00**	**0.59**	**0.59**
13272 6149	Peroxide System Sodium Hydroxide	KGAL	0	N/A	1.000	0.00	0.00	0.59	0.59
33 12 0848	**Fugitive Emission Control System**	**EA**				**1,618**	**1,877**	**24,241**	**27,737**
13272 6150	Fugitive Emission Control System	EA	0.041	UOEHE	1.000	1,618	1,877	24,241	27,737
33 12 0849	**Fugitive Emission System Maintenance**	**MO**				**1,448**	**0.00**	**0.00**	**1,448**
13272 6151	Fugitive Emission System Maintenance	MO	0.023	ULABA	1.000	1,448	0.00	0.00	1,448
33 12 0850	**60 watt Ultraviolet Source Low Intensity Lamp**	**EA**				**0.00**	**0.00**	**115.15**	**115.15**
13272 6152	60 W Ultraviolet Source Low Intensity Lamp	EA	0	N/A	1.000	0.00	0.00	115.15	115.15
33 12 0851	**7.5 KW Ultraviolet Source High Intensity Lamp**	**EA**				**0.00**	**0.00**	**181.81**	**181.81**
13272 6153	7.5 KW Ultraviolet Source High Intensity Lamp	EA	0	N/A	1.000	0.00	0.00	181.81	181.81
33 12 0852	**20 KW Ultraviolet Source High Intensity Lamp**	**EA**				**0.00**	**0.00**	**515.13**	**515.13**
13272 6154	20 KW Ultraviolet Source High Intensity Lamp	EA	0	N/A	1.000	0.00	0.00	515.13	515.13
33 12 0853	**5 KW Ultraviolet Source High Intensity Lamp**	**EA**				**0.00**	**0.00**	**151.51**	**151.51**
13272 6155	5 KW Ultraviolet Source High Intensity Lamp	EA	0	N/A	1.000	0.00	0.00	151.51	151.51

33 12 Chemical Treatment

33 12 08 Ultraviolet Oxidation

33 12 08 Ultraviolet Oxidation		Unit	Hourly Output	Crew	Factor	Labor	Equip.	Matl.	Total
33 12 0854	**15 KW Ultraviolet Source High Intensity Lamp**	**EA**				**0.00**	**0.00**	**363.62**	**363.62**
13272 6156	15 KW Ultraviolet Source High Intensity Lamp	EA	0	N/A	1.000	0.00	0.00	363.62	363.62
33 12 0855	**10 GPM Cavitation System Maintenance**	**EA**				**0.00**	**0.00**	**1,333**	**1,333**
13272 6157	10 GPM Cavitation System Maintenance	EA	0	N/A	1.000	0.00	0.00	1,333	1,333
33 12 0856	**25 GPM Cavitation System Maintenance**	**EA**				**0.00**	**0.00**	**1,818**	**1,818**
13272 6158	25 GPM Cavitation System Maintenance	EA	0	N/A	1.000	0.00	0.00	1,818	1,818
33 12 0860	**Ozone System Operational/Sampling Labor**	**HR**				**39.43**	**0.00**	**0.00**	**39.43**
13272 6162	Ozone System Operational/Sampling Labor	HR	1	COEMA	0.900	43.81	0.00	0.00	39.43
33 12 0861	**Ozone System Lamp Replacement (Installed)**	**EA**				**0.00**	**0.00**	**92.88**	**92.88**
13272 6163	Ozone System Lamp Replacement (Installed)	EA	0	N/A	1.000	0.00	0.00	92.88	92.88
33 12 0862	**20 GPM Ozone System Maintenance & Repair**	**EA**				**5,553**	**0.00**	**454.53**	**6,008**
13272 6164	20 GPM Ozone System Maintenance & Repair	EA	0.006	ULABA	1.000	5,553	0.00	454.53	6,008
33 12 0863	**100 GPM Ozone System Maintenance & Repair**	**EA**				**28,229**	**0.00**	**2,182**	**30,411**
13272 6165	100 GPM Ozone System Maintenance & Repair	EA	0.001	ULABA	1.000	28,229	0.00	2,182	30,411
33 12 0864	**250 GPM Ozone System Maintenance & Repair**	**EA**				**56,458**	**0.00**	**4,485**	**60,943**
13272 6166	250 GPM Ozone System Maintenance & Repair	EA	0.0006	ULABA	1.000	56,458	0.00	4,485	60,943
33 12 0865	**Peroxide System Sulfuric Acid Consumption**	**KGAL**				**0.00**	**0.00**	**0.18**	**0.18**
13272 6167	Peroxide System Sulfuric Acid Consumption	KGAL	0	N/A	1.000	0.00	0.00	0.18	0.18
33 12 0866	**System Operational/Sampling Labor**	**HR**				**26.29**	**0.00**	**0.00**	**26.29**
13272 6168	System Operational/Sampling Labor	HR	1	COEMA	0.600	43.81	0.00	0.00	26.29
33 12 0867	**Peroxide System Demobilizaton**	**EA**				**0.00**	**0.00**	**42,422**	**42,422**
13272 6169	Peroxide System Demobilization	EA	0	N/A	1.000	0.00	0.00	42,422	42,422

33 12 10 Environmental Treatment and Process Chemicals

33 12 10 Environmental Treatment and Process Chemicals		Unit	Hourly Output	Crew	Factor	Labor	Equip.	Matl.	Total
33 12 1001	**Sodium Hypochlorite (NaOCl), Lab Usage, 500 mL**	**EA**				**0.00**	**0.00**	**31.08**	**31.08**
01001 2147	Sodium Hypochlorite (NaOCl), 500 mL	EA	0	N/A	1.000	0.00	0.00	31.08	31.08
33 12 1002	**Sodium Hypochlorite (NaOCl), Lab Usage, 4 L**	**EA**				**0.00**	**0.00**	**168.69**	**168.69**
01001 2148	Sodium Hypochlorite (NaOCl), 4 L	EA	0	N/A	1.000	0.00	0.00	168.69	168.69
33 12 1003	**Sodium Hypochlorite (NaOCl), Field Usage, 55-gal drum**	**EA**				**0.00**	**0.00**	**72.14**	**72.14**
01001 2149	Sodium Hypochlorite (NaOCl), 55-gal drum	EA	0	N/A	1.000	0.00	0.00	72.14	72.14

33 12 99 Chemical Shot Feeder

33 12 99 Chemical Shot Feeder	Unit	Hourly Output	Crew	Factor	Labor	Equip.	Matl.	Total

33 12 99	Chemical Shot Feeder	Unit	Hourly Output	Crew	Factor	Labor	Equip.	Matl.	Total
33 12 9901	**1.7 Gallon Bypass Chemical Shot Feeder, In-line Mount, 125 PSIG**	**EA**				**326.76**	**0.00**	**247.49**	**574.25**
11261 0200	1.7 Gallon Bypass Shot Chemical Feeder In-Line Mount, 125 PSIG	EA	0.291	MSPFD	1.000	326.76	0.00	247.49	574.25
33 12 9902	**5 Gallon Bypass Chemical Shot Feeder, Floor Mount, 175 PSIG**	**EA**				**491.15**	**0.00**	**328.07**	**819.22**
11261 0220	5 Gallon Bypass Shot Chemical Feeder Floor Mount, 175 PSIG	EA	0.194	MSPFD	1.000	491.15	0.00	328.07	819.22
33 12 9903	**12 Gallon Bypass Chemical Shot Feeder, Floor Mount, 175 PSIG**	**EA**				**650.62**	**0.00**	**526.07**	**1,177**
11261 0230	12 Gallon Bypass Shot Chemical Feeder Floor Mount, 175 PSIG	EA	0.146	MSPFD	1.000	650.62	0.00	526.07	1,177
33 12 9904	**5 Gallon Bypass Chemical Shot Feeder, Floor Mount, 150 Lb ASME**	**EA**				**491.15**	**0.00**	**1,376**	**1,867**
11261 0300	5 Gallon Bypass Shot Chemical Feeder Floor Mount, 150 Lb ASME Code	EA	0.194	MSPFD	1.000	491.15	0.00	1,376	1,867
33 12 9905	**10 Gallon Bypass Chemical Shot Feeder, Floor Mount, 150 Lb ASME**	**EA**				**650.62**	**0.00**	**1,499**	**2,149**
11261 0320	10 Gallon Bypass Shot Chemical Feeder Floor Mount, 150 Lb ASME Code	EA	0.146	MSPFD	1.000	650.62	0.00	1,499	2,149
33 12 9906	**5 Gallon Bypass Chemical Shot Feeder, Floor Mount, 300 Lb ASME**	**EA**				**491.15**	**0.00**	**3,701**	**4,192**
11261 0400	5 Gallon Bypass Shot Chemical Feeder Floor Mount, 300 Lb ASME Code	EA	0.194	MSPFD	1.000	491.15	0.00	3,701	4,192
33 12 9907	**10 Gallon Bypass Chemical Shot Feeder, Floor Mount, 300 Lb ASME**	**EA**				**650.62**	**0.00**	**4,617**	**5,268**
11261 0420	10 Gallon Bypass Shot Chemical Feeder Floor Mount, 300 Lb ASME Code	EA	0.146	MSPFD	1.000	650.62	0.00	4,617	5,268

33 13 01	Filtration	Unit	Hourly Output	Crew	Factor	Labor	Equip.	Matl.	Total
33 13 0101	**3' Diameter Electric Automatic Pressure Filter Unit**	**EA**				**2,320**	**916.77**	**3,762**	**6,998**
11225 0200	3' Diameter, 36 GPM Electric Operation Pressure Filter for Pretreatment of Raw Water	EA	0.059	MPLUT	1.000	2,320	916.77	3,762	6,998
33 13 0102	**4' Diameter Electric Automatic Pressure Filter Unit**	**EA**				**2,598**	**1,027**	**8,452**	**12,077**
11225 0220	4' Diameter, 65 GPM Electric Operation Pressure Filter for Pretreatment of Raw Water	EA	0.053	MPLUT	1.000	2,598	1,027	8,452	12,077
33 13 0103	**5' Diameter Electric Automatic Pressure Filter Unit**	**EA**				**2,598**	**1,027**	**10,363**	**13,987**
11225 0230	5' Diameter, 100 GPM Electric Operation Pressure Filter for Pretreatment of Raw Water	EA	0.053	MPLUT	1.000	2,598	1,027	10,363	13,987

33 13 01 Filtration		Unit	Hourly Output	Crew	Factor	Labor	Equip.	Matl.	Total
33 13 0104	**6' Diameter Electric Automatic Pressure Filter Unit**	**EA**				**2,871**	**1,135**	**13,443**	**17,449**
11225 0240	6' Diameter, 140 GPM Electric Operation Pressure Filter for Pretreatment of Raw Water	EA	0.048	MPLUT	1.000	2,871	1,135	13,443	17,449
33 13 0105	**7' Diameter Electric Automatic Pressure Filter Unit**	**EA**				**2,871**	**1,135**	**19,590**	**23,597**
11225 0250	7' Diameter, 200 GPM Electric Operation Pressure Filter for Pretreatment of Raw Water	EA	0.048	MPLUT	1.000	2,871	1,135	19,590	23,597
33 13 0106	**8' Diameter Electric Automatic Pressure Filter Unit**	**EA**				**3,515**	**1,389**	**21,367**	**26,271**
11225 0260	8' Diameter, 250 GPM Electric Operation Pressure Filter for Pretreatment of Raw Water	EA	0.039	MPLUT	1.000	3,515	1,389	21,367	26,271
33 13 0107	**5 GPM Polypropylene (PP) Filter Chamber**	**EA**				**3.22**	**0.00**	**66.84**	**70.07**
15251 1301	5 GPM Polypropylene Filter Chamber	EA	16	MPLUA	1.000	3.22	0.00	66.84	70.07
33 13 0108	**5 GPM Polypropylene/Styrene Filter Chamber**	**EA**				**3.22**	**0.00**	**66.84**	**70.07**
15251 1302	5 GPM Polypropylene/Styrene Filter Chamber	EA	16	MPLUA	1.000	3.22	0.00	66.84	70.07
33 13 0109	**5 GPM Nylon Filter Chamber**	**EA**				**3.22**	**0.00**	**85.55**	**88.77**
15251 1303	5 GPM Nylon Filter Chamber	EA	16	MPLUA	1.000	3.22	0.00	85.55	88.77
33 13 0110	**15 GPM CPVC Filter Chamber**	**EA**				**11.87**	**0.00**	**1,185**	**1,196**
15251 1304	15 GPM CPVC Filter Chamber	EA	8	MPLUB	1.000	11.87	0.00	1,185	1,196
33 13 0111	**30 GPM CPVC Filter Chamber**	**EA**				**11.87**	**0.00**	**1,293**	**1,305**
15251 1305	30 GPM CPVC Filter Chamber	EA	8	MPLUB	1.000	11.87	0.00	1,293	1,305
33 13 0112	**45 GPM CPVC Filter Chamber**	**EA**				**11.87**	**0.00**	**1,431**	**1,443**
15251 1306	45 GPM CPVC Filter Chamber	EA	8	MPLUB	1.000	11.87	0.00	1,431	1,443
33 13 0113	**25 GPM Polypropylene Filter Chamber**	**EA**				**11.87**	**0.00**	**1,491**	**1,502**
15251 1307	25 GPM Polypropylene Filter Chamber	EA	8	MPLUB	1.000	11.87	0.00	1,491	1,502
33 13 0114	**42 GPM Polypropylene Filter Chamber**	**EA**				**11.87**	**0.00**	**1,579**	**1,591**
15251 1308	42 GPM Polypropylene Filter Chamber	EA	8	MPLUB	1.000	11.87	0.00	1,579	1,591
33 13 0115	**60 GPM Polypropylene Filter Chamber**	**EA**				**11.87**	**0.00**	**1,629**	**1,641**
15251 1309	60 GPM Polypropylene Filter Chamber	EA	8	MPLUB	1.000	11.87	0.00	1,629	1,641
33 13 0116	**0 - 50 GPM Cartridge Filter Equipment**	**EA**				**47.48**	**0.00**	**2,339**	**2,387**
15251 1310	0 - 50 GPM Cartridge Filter Equipment	EA	2	MPLUB	1.000	47.48	0.00	2,339	2,387
33 13 0117	**50 - 100 GPM Cartridge Filter Equipment**	**EA**				**47.48**	**0.00**	**4,679**	**4,726**
15251 1311	50 - 100 GPM Cartridge Filter Equipment	EA	2	MPLUB	1.000	47.48	0.00	4,679	4,726
33 13 0118	**Replacement Cartridge Filter up to 100 GPM**	**EA**				**3.22**	**0.00**	**233.94**	**237.17**
15251 1312	Replacement Cartridge Filter up to 100 GPM	EA	16	MPLUA	1.000	3.22	0.00	233.94	237.17

33 13 04	Coagulation/Flocculation	Unit	Hourly Output	Crew	Factor	Labor	Equip.	Matl.	Total
33 13 0401	**9' Diameter, 25 GPM, Waste Flow, Contact Clarifier**	**EA**				**19,549**	**6,428**	**51,000**	**76,977**
13272 1411	9' Diameter, 25 GPM Waste Flow, Purchase and Install Only	EA	0.02	SIWSO	1.000	19,549	6,428	51,000	76,977
33 13 0402	**12' Diameter, 45 GPM, Waste Flow, Contact Clarifier**	**EA**				**21,722**	**7,142**	**69,000**	**97,863**
13272 1412	12' Diameter, 45 GPM Waste Flow, Purchase and Install Only	EA	0.018	SIWSO	1.000	21,722	7,142	69,000	97,863
33 13 0403	**15' Diameter, 70 GPM, Waste Flow, Contact Clarifier**	**EA**				**26,066**	**8,570**	**78,500**	**113,136**
13272 1413	15' Diameter, 70 GPM Waste Flow, Purchase and Install Only	EA	0.015	SIWSO	1.000	26,066	8,570	78,500	113,136
33 13 0404	**20' Diameter, 130 GPM, Waste Flow, Contact Clarifier**	**EA**				**27,928**	**9,182**	**67,500**	**104,610**
13272 1414	20' Diameter, 130 GPM Waste Flow, Purchase and Install Only	EA	0.014	SIWSO	1.000	27,928	9,182	67,500	104,610
33 13 0405	**25' Diameter, 208 GPM, Waste Flow, Contact Clarifier**	**EA**				**32,582**	**10,713**	**71,500**	**114,795**
13272 1415	25' Diameter, 208 GPM Waste Flow, Purchase and Install Only	EA	0.012	SIWSO	1.000	32,582	10,713	71,500	114,795
33 13 0406	**30' Diameter, 305 GPM, Waste Flow, Contact Clarifier**	**EA**				**35,545**	**11,686**	**82,500**	**129,731**
13272 1416	30' Diameter, 305 GPM Waste Flow, Purchase and Install Only	EA	0.011	SIWSO	1.000	35,545	11,686	82,500	129,731
33 13 0407	**40' Diameter, 546 GPM, Waste Flow, Contact Clarifier**	**EA**				**48,874**	**16,069**	**176,500**	**241,443**
13272 1417	40' Diameter, 546 GPM Waste Flow, Purchase and Install Only	EA	0.008	SIWSO	1.000	48,874	16,069	176,500	241,443
33 13 0408	**50' Diameter, 850 GPM, Waste Flow, Contact Clarifier**	**EA**				**65,165**	**21,425**	**222,500**	**309,090**
13272 1418	50' Diameter, 850 GPM Waste Flow, Purchase and Install Only	EA	0.006	SIWSO	1.000	65,165	21,425	222,500	309,090
33 13 0409	**60' Diameter, 1,194 GPM, Waste Flow, Contact Clarifier**	**EA**				**78,198**	**25,710**	**277,500**	**381,408**
13272 1419	60' Diameter, 1,194 GPM Waste Flow, Purchase and Install Only	EA	0.005	SIWSO	1.000	78,198	25,710	277,500	381,408
33 13 0410	**70' Diameter, 1,645 GPM, Waste Flow, Contact Clarifier**	**EA**				**97,747**	**32,138**	**327,000**	**456,885**
13272 1421	70' Diameter, 1,645 GPM Waste Flow, Purchase and Install Only	EA	0.004	SIWSO	1.000	97,747	32,138	327,000	456,885
33 13 0411	**2 GPM Vertical Plate Clarifier with Mix Tank**	**EA**				**43.85**	**3.59**	**8,969**	**9,017**
13272 1441	2 GPM Vertical Plate Clarifier with Mix Tank	EA	2	MPLUG	1.000	43.85	3.59	8,969	9,017

33 13 04 Coagulation/Flocculation		Unit	Hourly Output	Crew	Factor	Labor	Equip.	Matl.	Total
33 13 0412	**10 GPM Vertical Plate Clarifier with Mix Tank**	**EA**				**43.85**	**3.59**	**12,969**	**13,017**
13272 1442	10 GPM Vertical Plate Clarifier with Mix Tank	EA	2	MPLUG	1.000	43.85	3.59	12,969	13,017
33 13 0413	**15 GPM Vertical Plate Clarifier with Mix Tank**	**EA**				**43.85**	**3.59**	**17,090**	**17,138**
13272 1443	15 GPM Vertical Plate Clarifier with Mix Tank	EA	2	MPLUG	1.000	43.85	3.59	17,090	17,138
33 13 0414	**30 GPM Vertical Plate Clarifier with Mix Tank**	**EA**				**43.85**	**3.59**	**23,151**	**23,198**
13272 1444	30 GPM Vertical Plate Clarifier with Mix Tank	EA	2	MPLUG	1.000	43.85	3.59	23,151	23,198
33 13 0415	**1/4 HP, Single Propeller 3-1/2" Diameter Mixer**	**EA**				**51.56**	**0.00**	**1,394**	**1,445**
13272 1281	1/4 HP Single Propeller 3-1/2" Diameter Mixer	EA	1	MPLUA	1.000	51.56	0.00	1,394	1,445
33 13 0416	**1/3 HP, Single Propeller 4" Diameter Mixer**	**EA**				**51.56**	**0.00**	**1,576**	**1,627**
13272 1282	1/3 HP Single Propeller 4" Diameter Mixer	EA	1	MPLUA	1.000	51.56	0.00	1,576	1,627
33 13 0417	**1/2 HP, Single Propeller 4-1/2" Diameter Mixer**	**EA**				**51.56**	**0.00**	**1,788**	**1,839**
13272 1283	1/2 HP Single Propeller 4-1/2" Diameter Mixer	EA	1	MPLUA	1.000	51.56	0.00	1,788	1,839
33 13 0418	**3/4 HP, Single Propeller 5" Diameter Mixer**	**EA**				**51.56**	**0.00**	**1,988**	**2,039**
13272 1284	3/4 HP Single Propeller 5" Diameter Mixer	EA	1	MPLUA	1.000	51.56	0.00	1,988	2,039
33 13 0419	**1 HP, Single Propeller 5-1/2" Diameter Mixer**	**EA**				**51.56**	**0.00**	**2,303**	**2,354**
13272 1285	1 HP Single Propeller 5-1/2" Diameter Mixer	EA	1	MPLUA	1.000	51.56	0.00	2,303	2,354
33 13 0420	**1-1/2 HP, Single Propeller 6" Diameter Mixer**	**EA**				**51.56**	**0.00**	**2,618**	**2,670**
13272 1286	1-1/2 HP Single Propeller 6" Diameter Mixer	EA	1	MPLUA	1.000	51.56	0.00	2,618	2,670
33 13 0421	**2 HP, Single Propeller 6-1/2" Diameter Mixer**	**EA**				**51.56**	**0.00**	**2,939**	**2,991**
13272 1287	2 HP Single Propeller 6-1/2" Diameter Mixer	EA	1	MPLUA	1.000	51.56	0.00	2,939	2,991
33 13 0422	**1/4 HP, Double Propeller 3" Diameter Mixer**	**EA**				**51.56**	**0.00**	**1,576**	**1,627**
13272 1291	1/4 HP Double Propeller 3" Diameter Mixer	EA	1	MPLUA	1.000	51.56	0.00	1,576	1,627
33 13 0423	**1/3 HP, Double Propeller 3-1/2" Diameter Mixer**	**EA**				**51.56**	**0.00**	**1,667**	**1,718**
13272 1292	1/3 HP Double Propeller 3-1/2" Diameter Mixer	EA	1	MPLUA	1.000	51.56	0.00	1,667	1,718
33 13 0424	**1/2 HP, Double Propeller 4" Diameter Mixer**	**EA**				**51.56**	**0.00**	**1,879**	**1,930**
13272 1293	1/2 HP Double Propeller 4" Diameter Mixer	EA	1	MPLUA	1.000	51.56	0.00	1,879	1,930
33 13 0425	**3/4 HP, Double Propeller 4-1/2" Diameter Mixer**	**EA**				**51.56**	**0.00**	**2,061**	**2,112**
13272 1294	3/4 HP Double Propeller 4-1/2" Diameter Mixer	EA	1	MPLUA	1.000	51.56	0.00	2,061	2,112
33 13 0426	**1 HP, Double Propeller 5" Diameter Mixer**	**EA**				**51.56**	**0.00**	**2,412**	**2,464**
13272 1295	1 HP Double Propeller 5" Diameter Mixer	EA	1	MPLUA	1.000	51.56	0.00	2,412	2,464

33 13 04	Coagulation/Flocculation	Unit	Hourly Output	Crew	Factor	Labor	Equip.	Matl.	Total
33 13 0427	**1-1/2 HP, Double Propeller 5-1/2" Diameter Mixer**	**EA**				**51.56**	**0.00**	**2,727**	**2,779**
13272 1296	1-1/2 HP Double Propeller 5-1/2" Diameter Mixer	EA	1	MPLUA	1.000	51.56	0.00	2,727	2,779
33 13 0428	**2 HP, Double Propeller 6" Diameter Mixer**	**EA**				**51.56**	**0.00**	**3,091**	**3,142**
13272 1297	2 HP Double Propeller 6" Diameter Mixer	EA	1	MPLUA	1.000	51.56	0.00	3,091	3,142

33 13 07	Air Stripping	Unit	Hourly Output	Crew	Factor	Labor	Equip.	Matl.	Total
33 13 0701	**Packing Reconditioning**	**EA**				**1,056**	**1,226**	**0.00**	**2,282**
13273 4196	Packing Reconditioning	EA	0.063	UOEHE	1.000	1,056	1,226	0.00	2,282
33 13 0702	**Daily Inspection of Air Stripper**	**HR**				**22.25**	**0.00**	**0.00**	**22.25**
01050 0112	Timekeeper - Average Cost	WK	0.025	XYBBM	0.025	890.00	0.00	0.00	22.25
33 13 0703	**General Maintenance of Air Stripper**	**EA**				**72.99**	**0.00**	**0.00**	**72.99**
01050 0117	Master Mechanic	HR	1	XYBBP	2.000	36.50	0.00	0.00	72.99
33 13 0704	**Install Air Stripper Tower, 1' - 3' Diameter, <= 12' High**	**EA**				**3,237**	**453.95**	**0.00**	**3,691**
13273 4137	Install Air Stripper Tower, 1' - 3' Diameter, <= 12' High	EA	0.048	MSPFP	1.000	3,237	453.95	0.00	3,691
33 13 0705	**Install Air Stripper Tower, 1' - 3' Diameter, 13' - 20' High**	**EA**				**3,479**	**487.84**	**0.00**	**3,966**
13273 4138	Install Air Stripper Tower, 1' - 3' Diameter, 13'-20' High	EA	0.044	MSPFP	1.000	3,479	487.84	0.00	3,966
33 13 0706	**Install Air Stripper Tower, 1' - 3' Diameter, 21' - 30' High**	**EA**				**3,723**	**522.09**	**0.00**	**4,245**
13273 4139	Install Air Stripper Tower, 1' - 3' Diameter, 21' - 30' High	EA	0.041	MSPFP	1.000	3,723	522.09	0.00	4,245
33 13 0707	**Install Air Stripper Tower, 1' - 3' Diameter, > 30' High**	**EA**				**4,004**	**561.52**	**0.00**	**4,566**
13273 4156	Install Air Stripper Tower, 1' - 3' Diameter, > 30' High	EA	0.038	MSPFP	1.000	4,004	561.52	0.00	4,566
33 13 0711	**1.0' Diameter, 14.5' Packing Height, with Blower, Fiberglass Reinforced Plastic Shell**	**EA**				**3,497**	**0.00**	**7,100**	**10,597**
13273 4111	1' Diameter, 14.5' Packing Height, 30 GPM, 850 CFM, Fiberglass Reinforced Plastic Shell	EA	0.041	MSPFL	1.000	3,497	0.00	7,100	10,597
33 13 0712	**1.0' Diameter, 19.0' Packing Height, with Blower, Fiberglass Reinforced Plastic Shell**	**EA**				**4,595**	**0.00**	**7,775**	**12,370**
13273 4112	1' Diameter, 19.0' Packing Height, 30 GPM, 800 CFM, Fiberglass Reinforced Plastic Shell	EA	0.031	MSPFL	1.000	4,595	0.00	7,775	12,370
33 13 0713	**1.0' Diameter, 23.5' Packing Height, with Blower, Fiberglass Reinforced Plastic Shell**	**EA**				**4,625**	**0.00**	**8,900**	**13,525**
13273 4113	1' Diameter, 23.5' Packing Height, 30 GPM, 750 CFM, Fiberglass Reinforced Plastic Shell	EA	0.031	MSPFL	1.000	4,625	0.00	8,900	13,525

33 13 07	Air Stripping	Unit	Hourly Output	Crew	Factor	Labor	Equip.	Matl.	Total
33 13 0714	**1.5' Diameter, 14.5' Packing Height, with Blower, Fiberglass Reinforced Plastic Shell**	**EA**				**4,595**	**0.00**	**10,000**	**14,595**
13273 4114	1.5' Diameter, 14.5' Packing Height, 50 GPM, 1,050 CFM, Fiberglass Reinforced Plastic Shell	EA	0.031	MSPFL	1.000	4,595	0.00	10,000	14,595
33 13 0715	**1.5' Diameter, 19.0' Packing Height, with Blower, Fiberglass Reinforced Plastic Shell**	**EA**				**4,595**	**0.00**	**10,900**	**15,495**
13273 4115	1.5' Diameter, 19.0' Packing Height, 50 GPM, 1,000 CFM, Fiberglass Reinforced Plastic Shell	EA	0.031	MSPFL	1.000	4,595	0.00	10,900	15,495
33 13 0716	**1.5' Diameter, 23.5' Packing Height, with Blower, Fiberglass Reinforced Plastic Shell**	**EA**				**4,595**	**0.00**	**12,100**	**16,695**
13273 4116	1.5' Diameter, 23.5' Packing Height, 50 GPM, 950 CFM, Fiberglass Reinforced Plastic Shell	EA	0.031	MSPFL	1.000	4,595	0.00	12,100	16,695
33 13 0717	**2.0' Diameter, 14.5' Packing Height, with Blower, Fiberglass Reinforced Plastic Shell**	**EA**				**4,595**	**0.00**	**15,300**	**19,895**
13273 4117	2' Diameter, 14.5' Packing Height, 90 GPM, 1,600 CFM, Fiberglass Reinforced Plastic Shell	EA	0.031	MSPFL	1.000	4,595	0.00	15,300	19,895
33 13 0718	**2.0' Diameter, 19.0' Packing Height, with Blower, Fiberglass Reinforced Plastic Shell**	**EA**				**4,595**	**0.00**	**16,800**	**21,395**
13273 4118	2' Diameter, 19.0' Packing Height, 90 GPM, 1,550 CFM, Fiberglass Reinforced Plastic Shell	EA	0.031	MSPFL	1.000	4,595	0.00	16,800	21,395
33 13 0719	**2.0' Diameter, 23.5' Packing Height, with Blower, Fiberglass Reinforced Plastic Shell**	**EA**				**5,734**	**0.00**	**18,900**	**24,634**
13273 4119	2' Diameter, 23.5' Packing Height, 90 GPM, 1,550 CFM, Fiberglass Reinforced Plastic Shell	EA	0.025	MSPFL	1.000	5,734	0.00	18,900	24,634
33 13 0720	**3.0' Diameter, 5.5' Packing Height, with Blower, Polyethylene Shell**	**EA**				**3,471**	**0.00**	**7,175**	**10,646**
13273 4121	3' Diameter, 5.5' Packing Height, 90 GPM, 4,800 CFM, Polyethylene Shell	EA	0.041	MSPFL	1.000	3,471	0.00	7,175	10,646
33 13 0721	**3.0' Diameter, 5.5' Packing Height, with Blower, Polyethylene Shell, Truss**	**EA**				**3,723**	**522.09**	**10,200**	**14,445**
13273 4122	3' Diameter, 5.5' Packing Height, 90 GPM, 4,800 CFM, Polyethylene Shell, Truss	EA	0.041	MSPFP	1.000	3,723	522.09	10,200	14,445
33 13 0722	**3.0' Diameter, 9.5' Packing Height, with Blower, Polyethylene Shell**	**EA**				**5,734**	**0.00**	**10,300**	**16,034**
13273 4123	3' Diameter, 9.5' Packing Height, 90 GPM, 4,600 CFM, Polyethylene Shell	EA	0.025	MSPFL	1.000	5,734	0.00	10,300	16,034
33 13 0723	**3.0' Diameter, 9.5' Packing Height, with Blower, Polyethylene Shell, Truss**	**EA**				**6,150**	**862.50**	**13,300**	**20,313**
13273 4124	3' Diameter, 9.5' Packing Height, 90 GPM, 4,600 CFM, Polyethylene Shell, Truss	EA	0.025	MSPFP	1.000	6,150	862.50	13,300	20,313
33 13 0724	**3.0' Diameter, 13.5' Packing Height, with Blower, Polyethylene Shell, Truss**	**EA**				**6,150**	**862.50**	**16,400**	**23,413**
13273 4125	3' Diameter, 13.5' Packing Height, 90 GPM, 5,000 CFM, Polyethylene Shell, Truss	EA	0.025	MSPFP	1.000	6,150	862.50	16,400	23,413

33 13 07 Air Stripping		Unit	Hourly Output	Crew	Factor	Labor	Equip.	Matl.	Total
33 13 0725	**3.0' Diameter, 17.5' Packing Height, with Blower, Polyethylene Shell, Truss**	**EA**				**6,150**	**862.50**	**19,600**	**26,613**
13273 4126	3' Diameter, 17.5' Packing Height, 90 GPM, 4,800 CFM, Polyethylene Shell, Truss	EA	0.025	MSPFP	1.000	6,150	862.50	19,600	26,613
33 13 0726	**4.0' Diameter, 6.5' Packing Height, with Blower, Polyethylene Shell**	**EA**				**3,723**	**522.09**	**10,900**	**15,145**
13273 4127	4' Diameter, 6.5' Packing Height, 150 GPM, 7,500 CFM, Polyethylene Shell	EA	0.041	MSPFP	1.000	3,723	522.09	10,900	15,145
33 13 0727	**4.0' Diameter, 6.5' Packing Height, with Blower, Polyethylene Shell, Truss**	**EA**				**3,723**	**522.09**	**14,600**	**18,845**
13273 4128	4' Diameter, 6.5' Packing Height, 150 GPM, 7,500 CFM, Polyethylene Shell, Truss	EA	0.041	MSPFP	1.000	3,723	522.09	14,600	18,845
33 13 0728	**4.0' Diameter, 9.5' Packing Height, with Blower, Polyethylene Shell, Truss**	**EA**				**6,150**	**862.50**	**16,200**	**23,213**
13273 4129	4' Diameter, 9.5' Packing Height, 150 GPM, 7,300 CFM, Polyethylene Shell, Truss	EA	0.025	MSPFP	1.000	6,150	862.50	16,200	23,213
33 13 0729	**4.0' Diameter, 12.5' Packing Height, with Blower, Polyethylene Shell, Truss**	**EA**				**6,150**	**862.50**	**20,000**	**27,013**
13273 4131	4' Diameter, 12.5' Packing Height, 150 GPM, 8,400 CFM, Polyethylene Shell, Truss	EA	0.025	MSPFP	1.000	6,150	862.50	20,000	27,013
33 13 0730	**4.0' Diameter, 15.5' Packing Height, with Blower, Polyethylene Shell, Truss**	**EA**				**6,150**	**862.50**	**21,800**	**28,813**
13273 4132	4' Diameter, 15.5' Packing Height, 150 GPM, 8,200 CFM, Polyethylene Shell, Truss	EA	0.025	MSPFP	1.000	6,150	862.50	21,800	28,813
33 13 0731	**4.0' Diameter, 18.5' Packing Height, with Blower, Polyethylene Shell, Truss**	**EA**				**6,150**	**862.50**	**25,500**	**32,513**
13273 4133	4' Diameter, 18.5' Packing Height, 150 GPM, 8,000 CFM, Polyethylene Shell, Truss	EA	0.025	MSPFP	1.000	6,150	862.50	25,500	32,513
33 13 0736	**Internal Parts for Air Stripper, < 20' High, per Foot of Tower Diameter**	**FT**				**0.00**	**0.00**	**3,438**	**3,438**
13273 4135	Internal Parts for Air Stripper, < 20' High	FT	0	N/A	1.000	0.00	0.00	3,438	3,438
33 13 0737	**Internal Parts for Air Stripper, >= 20' High, per Foot of Tower Diameter**	**FT**				**0.00**	**0.00**	**5,155**	**5,155**
13273 4136	Internal Parts for Air Stripper, >= 20' High	FT	0	N/A	1.000	0.00	0.00	5,155	5,155
33 13 0738	**1" - 3.5" Packing for Air Stripper Tower**	**CF**				**0.00**	**0.00**	**16.50**	**16.50**
13273 4134	1" - 3.5" Packing for Air Stripper Tower	CF	0	N/A	1.000	0.00	0.00	16.50	16.50

33 13 Physical Treatment

33 13 07	Air Stripping	Unit	Hourly Output	Crew	Factor	Labor	Equip.	Matl.	Total
33 13 0741	**Electrical Controls for Air Stripper**	**EA**				**1,380**	**86.25**	**4,341**	**5,807**
13273 4161	Air Flow Switch to Indicate Loss of Air Flow Due to Motor Failure	EA	0	N/A	1.000	0.00	0.00	520.00	520.00
13273 4162	High Sump Level Switch for Avoiding Overflow	EA	0	N/A	1.000	0.00	0.00	211.00	211.00
13273 4163	Liquid Level Control Package for Control of Effluent Water Levels	EA	0.25	MSPFP	1.000	615.02	86.25	1,350	2,051
13273 4164	Differential Pressure Gauge to Indicate Packing Pressure Loss	EA	0.5	MSPFL	1.000	286.72	0.00	635.00	921.72
13273 4165	Vibration Switch for Shutoff During Excessive Vibration	EA	0	N/A	1.000	0.00	0.00	465.00	465.00
13273 4166	Drain Valve Actuator to Prevent Freezing During System Shutdown	EA	0.75	MSPFL	1.000	191.15	0.00	635.00	826.15
13273 4167	Monitoring Well Liquid Level Control to Avoid Pump Exposure in Drawdown	EA	0.5	MSPFL	1.000	286.72	0.00	525.00	811.72
33 13 0742	**Install Air Stripper Tower, 4' - 8' Diameter, <= 12' High**	**EA**				**3,543**	**496.83**	**0.00**	**4,040**
13273 4157	Install Air Stripper Tower, 4' - 8' Diameter, <= 12' High	EA	0.043	MSPFP	1.000	3,543	496.83	0.00	4,040
33 13 0743	**Install Air Stripper Tower, 4' - 8' Diameter, 13' - 20' High**	**EA**				**3,723**	**522.09**	**0.00**	**4,245**
13273 4158	Install Air Stripper Tower, 4' - 8' Diameter, 13' - 20' High	EA	0.041	MSPFP	1.000	3,723	522.09	0.00	4,245
33 13 0744	**Install Air Stripper Tower, 4' - 8' Diameter, 21' - 30' High**	**EA**				**4,897**	**686.70**	**0.00**	**5,583**
13273 4159	Install Air Stripper Tower, 4' - 8' Diameter, 21' - 30' High	EA	0.031	MSPFP	1.000	4,897	686.70	0.00	5,583
33 13 0745	**Install Air Stripper Tower, 4' - 8' Diameter, > 30' High**	**EA**				**6,406**	**898.44**	**0.00**	**7,305**
13273 4174	Install Air Stripper Tower, 4' - 8' Diameter, > 30' High	EA	0.024	MSPFP	1.000	6,406	898.44	0.00	7,305
33 13 0746	**Install Air Stripper Tower, 9' - 12' Diameter, <= 12' High**	**EA**				**3,723**	**522.09**	**0.00**	**4,245**
13273 4175	Install Air Stripper Tower, 9' - 12' Diameter, <= 12' High	EA	0.041	MSPFP	1.000	3,723	522.09	0.00	4,245
33 13 0747	**Install Air Stripper Tower, 9' - 12' Diameter, 13' - 20' High**	**EA**				**6,006**	**842.29**	**0.00**	**6,848**
13273 4176	Install Air Stripper Tower, 9' - 12' Diameter, 13' - 20' High	EA	0.026	MSPFP	1.000	6,006	842.29	0.00	6,848
33 13 0748	**Install Air Stripper Tower, 9' - 12' Diameter, 21' - 30' High**	**EA**				**10,604**	**1,487**	**0.00**	**12,091**
13273 4177	Install Air Stripper Tower, 9' - 12' Diameter, 21' - 30' High	EA	0.015	MSPFP	1.000	10,604	1,487	0.00	12,091
33 13 0749	**Install Air Stripper Tower, 9' - 12' Diameter, > 30' High**	**EA**				**14,928**	**2,093**	**0.00**	**17,021**
13273 4178	Install Air Stripper Tower, 9' - 12' Diameter, > 30' High	EA	0.010	MSPFP	1.000	14,928	2,093	0.00	17,021

33 13 Physical Treatment

33 13 07	Air Stripping	Unit	Hourly Output	Crew	Factor	Labor	Equip.	Matl.	Total
33 13 0751	**1.0' Diameter Tower, Skid Mount**	**EA**				**0.00**	**0.00**	**705.00**	**705.00**
13273 4151	Skid Mount for 1.0' Diameter Tower	EA	0	N/A	1.000	0.00	0.00	705.00	705.00
33 13 0752	**1.5' Diameter Tower, Skid Mount**	**EA**				**0.00**	**0.00**	**830.00**	**830.00**
13273 4152	Skid Mount for 1.5' Diameter Tower	EA	0	N/A	1.000	0.00	0.00	830.00	830.00
33 13 0753	**2.0' Diameter Tower, Skid Mount**	**EA**				**0.00**	**0.00**	**1,025**	**1,025**
13273 4153	Skid Mount for 2.0' Diameter Tower	EA	0	N/A	1.000	0.00	0.00	1,025	1,025
33 13 0754	**3.0' Diameter Tower, Skid Mount**	**EA**				**0.00**	**0.00**	**1,225**	**1,225**
13273 4154	Skid Mount for 3.0' Diameter Tower	EA	0	N/A	1.000	0.00	0.00	1,225	1,225
33 13 0755	**4.0' Diameter Tower, Skid Mount**	**EA**				**0.00**	**0.00**	**1,925**	**1,925**
13273 4155	Skid Mount for 4.0' Diameter Tower	EA	0	N/A	1.000	0.00	0.00	1,925	1,925
33 13 0761	**3' Diameter Tower, 90 - 150 GPM Range, High Water Flow Pump**	**EA**				**0.00**	**0.00**	**1,350**	**1,350**
13273 4168	High Water Flow Pump - 3' Diameter Tower - 90 to 150 GPM Range	EA	0	N/A	1.000	0.00	0.00	1,350	1,350
33 13 0762	**4' Diameter Tower, 150 - 350 GPM Range, High Water Flow Pump**	**EA**				**0.00**	**0.00**	**2,400**	**2,400**
13273 4169	High Water Flow Pump - 4' Diameter Tower - 150 to 350 GPM Range	EA	0	N/A	1.000	0.00	0.00	2,400	2,400
33 13 0771	**12' Truss-mounted Towers, Aluminum Safety Ladder & Cage**	**EA**				**615.02**	**86.25**	**2,950**	**3,651**
13273 4171	Aluminum Safety Ladder and Cage 12' Truss-mounted Towers	EA	0.25	MSPFP	1.000	615.02	86.25	2,950	3,651
33 13 0772	**15' - 19' Truss-mounted Towers, Aluminum Safety Ladder & Cage**	**EA**				**820.03**	**115.00**	**3,500**	**4,435**
13273 4172	Aluminum Safety Ladder and Cage 15 to 19' Truss-mounted Towers	EA	0.188	MSPFP	1.000	820.03	115.00	3,500	4,435
33 13 0781	**19' - 25' Truss-mounted Towers, Aluminum Safety Ladder & Cage**	**EA**				**1,230**	**172.50**	**4,200**	**5,603**
13273 4173	Aluminum Safety Ladder and Cage 19 to 25' Truss-mounted Towers	EA	0.125	MSPFP	1.000	1,230	172.50	4,200	5,603
33 13 0782	**3' - 4' Diameter Tower Blowers, Blower Weather Enclosure**	**EA**				**114.69**	**0.00**	**485.00**	**599.69**
13273 4181	Blower Weather Enclosure, 3' to 4' Diameter Tower Blowers	EA	1.25	MSPFL	1.000	114.69	0.00	485.00	599.69
33 13 0783	**1' - 3' Diameter Tower Blowers, Mastic Coating of Blower**	**EA**				**0.00**	**0.00**	**690.00**	**690.00**
13273 4182	Mastic Coating of Blower, 1' to 3' Diameter Tower Blowers	EA	0	N/A	1.000	0.00	0.00	690.00	690.00

33 13 07 Air Stripping		Unit	Hourly Output	Crew	Factor	Labor	Equip.	Matl.	Total
33 13 0784	**4' Diameter Tower Blowers, Mastic Coating of Blower**	**EA**				**0.00**	**0.00**	**945.00**	**945.00**
13273 4183	Mastic Coating of Blower, 4' Diameter Tower Blowers	EA	0	N/A	1.000	0.00	0.00	945.00	945.00
33 13 0785	**1' - 2' Diameter Tower Blowers, PVC Coating of Blower**	**EA**				**0.00**	**0.00**	**930.00**	**930.00**
13273 4184	PVC Coating of Blower, 1' to 2' Diameter Tower Blowers	EA	0	N/A	1.000	0.00	0.00	930.00	930.00
33 13 0786	**3' Diameter Tower Blowers, PVC Coating of Blower**	**EA**				**0.00**	**0.00**	**1,150**	**1,150**
13273 4185	PVC Coating of Blower, 3' Diameter Tower Blowers	EA	0	N/A	1.000	0.00	0.00	1,150	1,150
33 13 0787	**4' Diameter Tower Blowers, PVC Coating of Blower**	**EA**				**0.00**	**0.00**	**1,350**	**1,350**
13273 4186	PVC Coating of Blower, 4' Diameter Tower Blowers	EA	0	N/A	1.000	0.00	0.00	1,350	1,350
33 13 0788	**1.0' Diameter x Height, Prefabricated, Fiberglass Reinforced Plastic, Air Stripper Column/Shell Only**	**FT**				**0.00**	**0.00**	**345.00**	**345.00**
13273 4401	1' Diameter x Height, Prefabricated, Fiberglass Reinforced Plastic, Air Stripper Column/Shell Only	FT	0	N/A	1.000	0.00	0.00	345.00	345.00
33 13 0789	**1.5' Diameter x Height, Prefabricated, Fiberglass Reinforced Plastic Air Stripper Column/Shell Only**	**FT**				**0.00**	**0.00**	**370.00**	**370.00**
13273 4402	1.5' Diameter x Height, Prefabricated, Fiberglass Reinforced Plastic, Air Stripper Column/Shell Only	FT	0	N/A	1.000	0.00	0.00	370.00	370.00
33 13 0790	**2.0' Diameter x Height, Prefabricated, Fiberglass Reinforced Plastic, Air Stripper Column/Shell Only**	**FT**				**0.00**	**0.00**	**345.00**	**345.00**
13273 4403	2' Diameter x Height, Prefabricated, Fiberglass Reinforced Plastic, Air Stripper Column/Shell Only	FT	0	N/A	1.000	0.00	0.00	345.00	345.00
33 13 0791	**3.0' Diameter x Height, Prefabricated, Fiberglass Reinforced Plastic, Air Stripper Column/Shell Only**	**FT**				**0.00**	**0.00**	**635.00**	**635.00**
13273 4404	3' Diameter x Height, Prefabricated, Fiberglass Reinforced Plastic, Air Stripper Column/Shell Only	FT	0	N/A	1.000	0.00	0.00	635.00	635.00
33 13 0792	**4.0' Diameter x Height, Prefabricated, Fiberglass Reinforced Plastic, Air Stripper Column/Shell Only**	**FT**				**0.00**	**0.00**	**645.00**	**645.00**
13273 4405	4' Diameter x Height, Prefabricated, Fiberglass Reinforced Plastic, Air Stripper Column/Shell Only	FT	0	N/A	1.000	0.00	0.00	645.00	645.00

33 13 07	Air Stripping	Unit	Hourly Output	Crew	Factor	Labor	Equip.	Matl.	Total
33 13 0793	**5.0' Diameter x Height, Prefabricated, Fiberglass Reinforced Plastic, Air Stripper Column/Shell Only**	**FT**				**0.00**	**0.00**	**770.00**	**770.00**
13273 4406	5' Diameter x Height, Prefabricated, Fiberglass Reinforced Plastic, Air Stripper Column/Shell Only	FT	0	N/A	1.000	0.00	0.00	770.00	770.00
33 13 0794	**6.0' Diameter x Height, Prefabricated, Fiberglass Reinforced Plastic, Air Stripper Column/Shell Only**	**FT**				**0.00**	**0.00**	**895.00**	**895.00**
13273 4407	6' Diameter x Height, Prefabricated, Fiberglass Reinforced Plastic, Air Stripper Column/Shell Only	FT	0	N/A	1.000	0.00	0.00	895.00	895.00
33 13 0795	**8.0' Diameter x Height, Prefabricated, Fiberglass Reinforced Plastic, Air Stripper Column/Shell Only**	**FT**				**0.00**	**0.00**	**1,050**	**1,050**
13273 4408	8' Diameter x Height, Prefabricated, Fiberglass Reinforced Plastic, Air Stripper Column/Shell Only	FT	0	N/A	1.000	0.00	0.00	1,050	1,050
33 13 0796	**10' Diameter x Height, Prefabricated, Fiberglass Reinforced Plastic, Air Stripper Column/Shell Only**	**FT**				**0.00**	**0.00**	**1,375**	**1,375**
13273 4409	10' Diameter x Height, Prefabricated, Fiberglass Reinforced Plastic, Air Stripper Column/Shell Only	FT	0	N/A	1.000	0.00	0.00	1,375	1,375
33 13 0797	**12' Diameter x Height, Prefabricated, Fiberglass Reinforced Plastic, Air Stripper Column/Shell Only**	**FT**				**0.00**	**0.00**	**1,600**	**1,600**
13273 4411	12' Diameter x Height, Prefabricated, Fiberglass Reinforced Plastic, Air Stripper Column/Shell Only	FT	0	N/A	1.000	0.00	0.00	1,600	1,600

33 13 09	Soil Washing	Unit	Hourly Output	Crew	Factor	Labor	Equip.	Matl.	Total
33 13 0901	**Treat 0 - 9,999 Tons of Soil, Including Residual Water**	**TON**				**2.40**	**1.60**	**125.00**	**129.00**
13273 5101	Treatment of 5,000 Tons of Soil, Including Residual Water	TON	20	CODEG	1.000	2.40	1.60	125.00	129.00
33 13 0902	**Treat 10,000 - 14,999 Tons of Soil, Including Residual Water**	**TON**				**2.40**	**1.60**	**120.00**	**124.00**
13273 5102	Treatment of 10,000 Tons of Soil, Including Residual Water	TON	20	CODEG	1.000	2.40	1.60	120.00	124.00
33 13 0903	**Treat 15,000 - 19,999 Tons of Soil, Including Residual Water**	**TON**				**2.40**	**1.60**	**116.00**	**120.00**
13273 5103	Treatment of 15,000 Tons of Soil, Including Residual Water	TON	20	CODEG	1.000	2.40	1.60	116.00	120.00
33 13 0904	**Treat 20,000 - 24,999 Tons of Soil, Including Residual Water**	**TON**				**2.40**	**1.60**	**108.00**	**112.00**
13273 5104	Treatment of 20,000 Tons of Soil, Including Residual Water	TON	20	CODEG	1.000	2.40	1.60	108.00	112.00

33 13 09	Soil Washing	Unit	Hourly Output	Crew	Factor	Labor	Equip.	Matl.	Total
33 13 0905	**Treat 25,000 - 29,999 Tons of Soil, Including Residual Water**	**TON**				**2.40**	**1.60**	**107.00**	**111.00**
13273 5105	Treatment of 25,000 Tons of Soil, Including Residual Water	TON	20	CODEG	1.000	2.40	1.60	107.00	111.00
33 13 0906	**Treat 30,000 - 34,999 Tons of Soil, Including Residual Water**	**TON**				**2.40**	**1.60**	**98.00**	**102.00**
13273 5106	Treatment of 30,000 Tons of Soil, Including Residual Water	TON	20	CODEG	1.000	2.40	1.60	98.00	102.00
33 13 0907	**Treat 35,000 - 39,999 Tons of Soil, Including Residual Water**	**TON**				**2.40**	**1.60**	**97.00**	**101.00**
13273 5107	Treatment of 35,000 Tons of Soil, Including Residual Water	TON	20	CODEG	1.000	2.40	1.60	97.00	101.00
33 13 0908	**Treat 40,000 - 44,999 Tons of Soil, Including Residual Water**	**TON**				**2.40**	**1.60**	**88.50**	**92.50**
13273 5108	Treatment of 40,000 Tons of Soil, Including Residual Water	TON	20	CODEG	1.000	2.40	1.60	88.50	92.50
33 13 0909	**Treat 45,000 - 49,999 Tons of Soil, Including Residual Water**	**TON**				**2.40**	**1.60**	**87.50**	**91.50**
13273 5109	Treatment of 45,000 Tons of Soil, Including Residual Water	TON	20	CODEG	1.000	2.40	1.60	87.50	91.50
33 13 0910	**Treat 50,000 - 54,999 Tons of Soil, Including Residual Water**	**TON**				**2.40**	**1.60**	**86.00**	**90.00**
13273 5111	Treatment of 50,000 Tons of Soil, Including Residual Water	TON	20	CODEG	1.000	2.40	1.60	86.00	90.00
33 13 0911	**Treat 55,000 - 59,999 Tons of Soil, Including Residual Water**	**TON**				**2.40**	**1.60**	**80.50**	**84.50**
13273 5112	Treatment of 55,000 Tons of Soil, Including Residual Water	TON	20	CODEG	1.000	2.40	1.60	80.50	84.50
33 13 0912	**Treat >= 60,000 Tons of Soil, Including Residual Water**	**TON**				**2.40**	**1.60**	**76.50**	**80.50**
13273 5113	Treatment of 60,000 Tons of Soil, Including Residual Water	TON	20	CODEG	1.000	2.40	1.60	76.50	80.50
33 13 0915	**Mobilize/Demobilize Soil Washing System**	**MI**				**0.00**	**0.00**	**2.56**	**2.56**
13273 5114	Mobilize/Demobilize Soil Washing System	MI	0	N/A	1.000	0.00	0.00	2.56	2.56
33 13 0916	**Assemble/Disassemble Soil Washing System**	**EA**				**20,418**	**8,625**	**0.00**	**29,043**
13273 5115	Assemble/Disassemble Soil Washing System	EA	1	UOEHC	120.000	170.15	71.88	0.00	29,043
33 13 0917	**Startup/Shakedown - Soil Washing System**	**EA**				**17,525**	**0.00**	**0.00**	**17,525**
13273 5116	Startup/Shakedown - Soil Washing System	EA	0.025	COEMA	10.000	1,753	0.00	0.00	17,525
33 13 0918	**Decontaminate Soil Washing System**	**EA**				**246.74**	**0.00**	**0.00**	**246.74**
01957 3112	Decontaminate Heavy Equipment	EA	0.338	XXBBW	1.000	246.74	0.00	0.00	246.74

33 13 Physical Treatment

33 13 09	Soil Washing	Unit	Hourly Output	Crew	Factor	Labor	Equip.	Matl.	Total
33 13 0920	**25 Tons per Hour Soil Washing System, Monthly Rental**	**MO**				**0.00**	**0.00**	**98,178**	**98,178**
13273 5118	25 Tons per Hour Soil Washing System, Monthly Rental	MO	0	N/A	1.000	0.00	0.00	98,178	98,178
33 13 0921	**Operational Labor, 10 - 25 Tons per Hour System**	**HR**				**262.88**	**0.00**	**0.00**	**262.88**
13273 5122	Operational Labor, 10 - 25 Tons per Hour System	HR	1	COEMA	6.000	43.81	0.00	0.00	262.88
33 13 0922	**50 Tons per Hour Soil Washing System, Monthly Rental**	**MO**				**0.00**	**0.00**	**115,147**	**115,147**
13273 5119	50 Tons per Hour Soil Washing System, Monthly Rental	MO	0	N/A	1.000	0.00	0.00	115,147	115,147
33 13 0923	**Operational Labor, 50 Tons per Hour System**	**HR**				**350.50**	**0.00**	**0.00**	**350.50**
13273 5123	Operational Labor, 50 Tons per Hour System	HR	1	COEMA	8.000	43.81	0.00	0.00	350.50
33 13 0924	**100 Tons per Hour Soil Washing System, Monthly Rental**	**MO**				**0.00**	**0.00**	**145,449**	**145,449**
13273 5121	100 Tons per Hour Soil Washing System, Monthly Rental	MO	0	N/A	1.000	0.00	0.00	145,449	145,449
33 13 0925	**Operational Labor, 100 Tons per Hour System**	**HR**				**438.13**	**0.00**	**0.00**	**438.13**
13273 5124	Operational Labor, 100 Tons per Hour System	HR	1	COEMA	10.000	43.81	0.00	0.00	438.13
33 13 0927	**Maintenance/Spare Parts, 25 Tons per Hour System**	**TON**				**0.00**	**0.00**	**2.71**	**2.71**
13273 5125	Maintenance/Spare Parts, 25 Tons per Hour System	TON	0	N/A	1.000	0.00	0.00	2.71	2.71
33 13 0928	**Maintenance/Spare Parts, 50 Tons per Hour System**	**TON**				**0.00**	**0.00**	**1.59**	**1.59**
13273 5126	Maintenance/Spare Parts, 50 Tons per Hour System	TON	0	N/A	1.000	0.00	0.00	1.59	1.59
33 13 0929	**Maintenance/Spare Parts, 100 Tons per Hour System**	**TON**				**0.00**	**0.00**	**1.02**	**1.02**
13273 5127	Maintenance/Spare Parts, 100 Tons per Hour System	TON	0	N/A	1.000	0.00	0.00	1.02	1.02

33 13 11	Media Filtration	Unit	Hourly Output	Crew	Factor	Labor	Equip.	Matl.	Total
33 13 1101	**3.1 SF of Nominal Filtering Area, 1' Drum Diameter, 4 HP**	**EA**				**975.08**	**0.00**	**28,000**	**28,975**
13276 4101	3 HP, 3.1 SF of Nominal Filtering Area, 1' Drum Diameter	EA	0.125	MSPFI	1.000	975.08	0.00	28,000	28,975
33 13 1102	**6.2 SF of Nominal Filtering Area, 2' Drum Diameter, 6 HP**	**EA**				**975.08**	**0.00**	**33,400**	**34,375**
13276 4102	3 HP, 6.2 SF of Nominal Filtering Area, 2' Drum Diameter	EA	0.125	MSPFI	1.000	975.08	0.00	33,400	34,375

33 13 11	Media Filtration	Unit	Hourly Output	Crew	Factor	Labor	Equip.	Matl.	Total
33 13 1103	**12.6 SF of Nominal Filtering Area, 2' Drum Diameter, 9 HP**	**EA**				**1,299**	**0.00**	**35,800**	**37,099**
13276 4103	9 HP, 12.6 SF of Nominal Filtering Area, 2' Drum Diameter	EA	0.094	MSPFI	1.000	1,299	0.00	35,800	37,099
33 13 1104	**18.8 SF of Nominal Filtering Area, 2' Drum Diameter, 9 HP**	**EA**				**1,299**	**0.00**	**41,000**	**42,299**
13276 4104	9 HP, 18.8 SF of Nominal Filtering Area, 2' Drum Diameter	EA	0.094	MSPFI	1.000	1,299	0.00	41,000	42,299
33 13 1105	**25.1 SF of Nominal Filtering Area, 2' Drum Diameter, 11 HP**	**EA**				**1,625**	**0.00**	**46,200**	**47,825**
13276 4105	11 HP, 25.1 SF of Nominal Filtering Area, 2' Drum Diameter	EA	0.075	MSPFI	1.000	1,625	0.00	46,200	47,825
33 13 1106	**28.3 SF of Nominal Filtering Area, 3' Drum Diameter, 11 HP**	**EA**				**1,625**	**0.00**	**48,700**	**50,325**
13276 4106	11 HP, 28.3 SF of Nominal Filtering Area, 3' Drum Diameter	EA	0.075	MSPFI	1.000	1,625	0.00	48,700	50,325
33 13 1107	**37.7 SF of Nominal Filtering Area, 3' Drum Diameter, 13 HP**	**EA**				**1,625**	**0.00**	**56,000**	**57,625**
13276 4107	13 HP, 37.7 SF of Nominal Filtering Area, 3' Drum Diameter	EA	0.075	MSPFI	1.000	1,625	0.00	56,000	57,625
33 13 1108	**56.5 SF of Nominal Filtering Area, 3' Drum Diameter, 24 HP**	**EA**				**1,950**	**0.00**	**65,000**	**66,950**
13276 4108	24 HP, 56.5 SF of Nominal Filtering Area, 3' Drum Diameter	EA	0.063	MSPFI	1.000	1,950	0.00	65,000	66,950
33 13 1109	**75.4 SF of Nominal Filtering Area, 6' Drum Diameter, 33 HP**	**EA**				**1,950**	**0.00**	**72,500**	**74,450**
13276 4109	33 HP, 75.4 SF of Nominal Filtering Area, 6' Drum Diameter	EA	0.063	MSPFI	1.000	1,950	0.00	72,500	74,450
33 13 1110	**94.2 SF of Nominal Filtering Area, 6' Drum Diameter, 38 HP**	**EA**				**2,438**	**0.00**	**81,000**	**83,438**
13276 4111	38 HP, 94.2 SF of Nominal Filtering Area, 6' Drum Diameter	EA	0.05	MSPFI	1.000	2,438	0.00	81,000	83,438
33 13 1111	**113.0 SF of Nominal Filtering Area, 6' Drum Diameter, 49 HP**	**EA**				**2,438**	**0.00**	**89,500**	**91,938**
13276 4112	49 HP, 113 SF of Nominal Filtering Area, 6' Drum Diameter	EA	0.05	MSPFI	1.000	2,438	0.00	89,500	91,938
33 13 1112	**170.0 SF of Nominal Filtering Area, 6' Drum Diameter, 60 HP**	**EA**				**2,973**	**0.00**	**130,500**	**133,473**
13276 4113	60 HP, 170 SF of Nominal Filtering Area, 6' Drum Diameter	EA	0.041	MSPFI	1.000	2,973	0.00	130,500	133,473
33 13 1113	**226.0 SF of Nominal Filtering Area, 6' Drum Diameter, 79 HP**	**EA**				**2,973**	**0.00**	**158,500**	**161,473**
13276 4114	79 HP, 226 SF of Nominal Filtering Area, 6' Drum Diameter	EA	0.041	MSPFI	1.000	2,973	0.00	158,500	161,473

33 13 Physical Treatment

33 13 11	Media Filtration	Unit	Hourly Output	Crew	Factor	Labor	Equip.	Matl.	Total
33 13 1114	**302.0 SF of Nominal Filtering Area, 8' Drum Diameter, 99 HP**	**EA**				**3,932**	**0.00**	**182,000**	**185,932**
13276 4115	99 HP, 302 SF of Nominal Filtering Area, 8' Drum Diameter	EA	0.031	MSPFI	1.000	3,932	0.00	182,000	185,932
33 13 1115	**402.0 SF of Nominal Filtering Area, 8' Drum Diameter, 120 HP**	**EA**				**4,353**	**0.00**	**201,500**	**205,853**
13276 4116	120 HP, 402 SF of Nominal Filtering Area, 8' Drum Diameter	EA	0.028	MSPFI	1.000	4,353	0.00	201,500	205,853
33 13 1116	**503.0 SF of Nominal Filtering Area, 8' Drum Diameter, 140 HP**	**EA**				**4,875**	**0.00**	**232,500**	**237,375**
13276 4117	140 HP, 503 SF of Nominal Filtering Area, 8' Drum Diameter	EA	0.025	MSPFI	1.000	4,875	0.00	232,500	237,375
33 13 1117	**Rotary Drum Filter, 5 GPM, Minimum Rental Charge**	**EA**				**0.00**	**0.00**	**2,200**	**2,200**
13276 4118	Minimum Rental (1 Month), 9 HP, 12.6 SF Nominal Filtering Area, 2' Drum Diameter	EA	0	N/A	1.000	0.00	0.00	2,200	2,200
33 13 1118	**Rotary Drum Filter, 5 GPM, Additional Rental Charge**	**EA**				**0.00**	**0.00**	**490.00**	**490.00**
13276 4119	Additional Weekly Rental, 9 HP, 12.6 SF Nominal Filtering Area, 2' Drum Diameter	EA	0	N/A	1.000	0.00	0.00	490.00	490.00
33 13 1119	**Rotary Drum Filter, 5 GPM, Mobilize/DeMobilize**	**EA**				**0.00**	**0.00**	**4,950**	**4,950**
13276 4121	Mobilization/Demobilization 9 HP, 12.6 SF Nominal Filtering Area, 2' Drum Diameter	EA	0	N/A	1.000	0.00	0.00	4,950	4,950
33 13 1120	**Pretreatment System Operational Labor Cost**	**DAY**				**0.00**	**0.00**	**220.00**	**220.00**
13276 4122	Equipment Operator Costs, Based on 8 Hours/Day	EA	0	N/A	1.000	0.00	0.00	220.00	220.00
33 13 1131	**Operation and Maintenance of Vacuum Filtration Unit**	**HR**				**122.14**	**35.94**	**79.62**	**237.70**
13276 4151	2-Man Operations & Maintenance Crew	HR	1	USKCG	1.000	122.14	35.94	79.62	237.70

33 13 12	Oil/Water Separation	Unit	Hourly Output	Crew	Factor	Labor	Equip.	Matl.	Total

33 13 12 Oil/Water Separation		Unit	Hourly Output	Crew	Factor	Labor	Equip.	Matl.	Total
33 13 1201	**Install Underground 20 GPM Oil/Water Separator**	**EA**				**443.80**	**83.20**	**597.58**	**1,125**
02215 0500	Compaction, 6" Layers with Air Tamper by Hand, (15 cm) Layers	CY	50	CLACA	25.000	2.71	0.17	0.00	72.10
02221 5007	Trench Backfill, 3 CY, 950	CY	70.88	CODLJ	25.000	0.47	0.54	0.00	25.11
02221 5011	Sand Backfill, 950 Dumped	CY	83.25	CODLJ	2.000	0.40	0.46	0.00	1.71
02228 0320	Trench, 1 CY Hydraulic Excavator, Light Soil, 4' to 6' Deep, 155 CY/Hour	CY	156	CODEO	25.000	0.40	0.58	0.00	24.46
02244 1505	Washed & Graded Sand, Loaded at Pit	CY	12.38	CODEG	2.400	3.88	2.59	26.50	79.12
02560 3108	36" Diameter 12 Gauge Galvanized Corrugated Metal Pipe (91 cm)	LF	13.5	UOEHC	3.000	12.60	5.32	52.56	211.47
02667 4590	2" Diameter PVC Schedule 40 Pipe (3.7 cm) Diameter	LF	65	XPLUB	15.000	1.14	0.00	0.53	25.08
02667 4620	4" Diameter PVC Schedule 40 Pipe (10 cm) Diameter	LF	47.5	XPLUC	6.000	2.11	0.00	1.56	22.03
02667 4800	2" Diameter PVC Schedule 40 Tee (5.1 cm) Diameter	EA	3.125	XPLUB	1.000	23.75	0.00	2.07	25.82
02667 5330	2" Diameter PVC Schedule 80 Coupling (5.1 cm) Diameter	EA	4.625	XPLUB	1.000	16.05	0.00	3.48	19.53
02752 4600	24" Diameter x 200 Lb Reg Manhole Cover (61 cm Diameter x 90 kg) with Frame, City Type	EA	1.15	CODEJ	1.000	82.80	27.87	168.00	278.67
05108 0140	3/4" Diameter x 12" Length, J-type Anchor Bolt (31 cm) with Nut, Washer, Template	EA	5.625	ACARB	4.000	15.40	0.00	2.43	71.32
05180 2320	3/4" x 12" Jaw & Jaw Turnbuckles	EA	4.375	SIWSB	2.000	22.03	0.00	57.76	159.58
15190 2727	3/4" Diameter Threaded Rod, Bright Finish	LF	22.5	SIWSA	6.000	2.14	0.00	8.96	66.58
16018 1000	1/2" IMC Conduit with Coupling Mount Exposed on Flat Wall, 10' Height	LF	98	EELEF	6.000	1.95	0.00	0.93	17.30
16024 0011	1/2" IMC Bushing Set with Locknut Mount Exposed on Flat Wall, 10' Height	EA	16.05	EELEF	2.000	11.91	0.00	0.43	24.69

33 13 12 Oil/Water Separation		Unit	Hourly Output	Crew	Factor	Labor	Equip.	Matl.	Total
33 13 1202	**Install Underground 50 GPM Oil/Water Separator**	**EA**				**449.50**	**90.33**	**656.98**	**1,197**
02215 0500	Compaction, 6" Layers with Air Tamper by Hand, (15 cm) Layers	CY	50	CLACA	25.000	2.71	0.17	0.00	72.10
02221 5007	Trench Backfill, 3 CY, 950	CY	70.88	CODLJ	25.000	0.47	0.54	0.00	25.11
02221 5011	Sand Backfill, 950 Dumped	CY	83.25	CODLJ	4.000	0.40	0.46	0.00	3.42
02228 0320	Trench, 1 CY Hydraulic Excavator, Light Soil, 4' to 6' Deep, 155 CY/Hour	CY	156	CODEO	25.000	0.40	0.58	0.00	24.46
02244 1505	Washed & Graded Sand, Loaded at Pit	CY	12.38	CODEG	4.800	3.88	2.59	26.50	158.25
02560 3108	36" Diameter 12 Gauge Galvanized Corrugated Metal Pipe (91 cm)	LF	13.5	UOEHC	3.000	12.60	5.32	52.56	211.47
02667 4570	1" Diameter PVC Schedule 40 Pipe (2.5 cm)	LF	87.5	XPLUB	15.000	0.85	0.00	0.25	16.47
02667 4620	4" Diameter PVC Schedule 40 Pipe (10 cm) Diameter	LF	47.5	XPLUC	6.000	2.11	0.00	1.56	22.03
02667 4800	2" Diameter PVC Schedule 40 Tee (5.1 cm) Diameter	EA	3.125	XPLUB	1.000	23.75	0.00	2.07	25.82
02667 5330	2" Diameter PVC Schedule 80 Coupling (5.1 cm) Diameter	EA	4.625	XPLUB	1.000	16.05	0.00	3.48	19.53
02752 4600	24" Diameter x 200 Lb Reg Manhole Cover (61 cm Diameter x 90 kg) with Frame, City Type	EA	1.15	CODEJ	1.000	82.80	27.87	168.00	278.67
05108 0140	3/4" Diameter x 12" Length, J-type Anchor Bolt (31 cm) with Nut, Washer, Template	EA	5.625	ACARB	4.000	15.40	0.00	2.43	71.32
05180 2320	3/4" x 12" Jaw & Jaw Turnbuckles	EA	4.375	SIWSB	2.000	22.03	0.00	57.76	159.58
15190 2727	3/4" Diameter Threaded Rod, Bright Finish	LF	22.5	SIWSA	6.000	2.14	0.00	8.96	66.58
16018 1000	1/2" IMC Conduit with Coupling Mount Exposed on Flat Wall, 10' Height	LF	98	EELEF	6.000	1.95	0.00	0.93	17.30
16024 0011	1/2" IMC Bushing Set with Locknut Mount Exposed on Flat Wall, 10' Height	EA	16.05	EELEF	2.000	11.91	0.00	0.43	24.69

33 13 12	Oil/Water Separation	Unit	Hourly Output	Crew	Factor	Labor	Equip.	Matl.	Total
33 13 1203	**Install Underground 100 GPM Oil/Water Separator**	**EA**				**494.61**	**106.79**	**706.69**	**1,308**
02215 0500	Compaction, 6" Layers with Air Tamper by Hand, (15 cm) Layers	CY	50	CLACA	35.000	2.71	0.17	0.00	100.94
02221 5007	Trench Backfill, 3 CY, 950	CY	70.88	CODLJ	35.000	0.47	0.54	0.00	35.15
02221 5011	Sand Backfill, 950 Dumped	CY	83.25	CODLJ	5.000	0.40	0.46	0.00	4.27
02228 0320	Trench, 1 CY Hydraulic Excavator, Light Soil, 4' to 6' Deep, 155 CY/Hour	CY	156	CODEO	35.000	0.40	0.58	0.00	34.25
02244 1505	Washed & Graded Sand, Loaded at Pit	CY	12.38	CODEG	6.000	3.88	2.59	26.50	197.81
02560 3108	36" Diameter 12 Gauge Galvanized Corrugated Metal Pipe (91 cm)	LF	13.5	UOEHC	3.000	12.60	5.32	52.56	211.47
02667 4570	1" Diameter PVC Schedule 40 Pipe (2.5 cm)	LF	87.5	XPLUB	15.000	0.85	0.00	0.25	16.47
02667 4620	4" Diameter PVC Schedule 40 Pipe (10 cm) Diameter	LF	47.5	XPLUC	6.000	2.11	0.00	1.56	22.03
02667 4800	2" Diameter PVC Schedule 40 Tee (5.1 cm) Diameter	EA	3.125	XPLUB	1.000	23.75	0.00	2.07	25.82
02667 5330	2" Diameter PVC Schedule 80 Coupling (5.1 cm) Diameter	EA	4.625	XPLUB	1.000	16.05	0.00	3.48	19.53
02752 4600	24" Diameter x 200 Lb Reg Manhole Cover (61 cm Diameter x 90 kg) with Frame, City Type	EA	1.15	CODEJ	1.000	82.80	27.87	168.00	278.67
05108 0140	3/4" Diameter x 12" Length, J-type Anchor Bolt (31 cm) with Nut, Washer, Template	EA	5.625	ACARB	4.000	15.40	0.00	2.43	71.32
05180 2320	3/4" x 12" Jaw & Jaw Turnbuckles	EA	4.375	SIWSB	2.000	22.03	0.00	57.76	159.58
15190 2727	3/4" Diameter Threaded Rod, Bright Finish	LF	22.5	SIWSA	8.000	2.14	0.00	8.96	88.78
16018 1000	1/2" IMC Conduit with Coupling Mount Exposed on Flat Wall, 10' Height	LF	98	EELEF	6.000	1.95	0.00	0.93	17.30
16024 0011	1/2" IMC Bushing Set with Locknut Mount Exposed on Flat Wall, 10' Height	EA	16.05	EELEF	2.000	11.91	0.00	0.43	24.69

33 13 12 Oil/Water Separation		Unit	Hourly Output	Crew	Factor	Labor	Equip.	Matl.	Total
33 13 1204	**Install Underground 200 GPM Oil/Water Separator**	**EA**				**671.54**	**176.19**	**901.51**	**1,749**
02215 0500	Compaction, 6" Layers with Air Tamper by Hand, (15 cm) Layers	CY	50	CLACA	75.000	2.71	0.17	0.00	216.30
02221 5007	Trench Backfill, 3 CY, 950	CY	70.88	CODLJ	75.000	0.47	0.54	0.00	75.32
02221 5011	Sand Backfill, 950 Dumped	CY	83.25	CODLJ	10.000	0.40	0.46	0.00	8.55
02228 0320	Trench, 1 CY Hydraulic Excavator, Light Soil, 4' to 6' Deep, 155 CY/Hour	CY	156	CODEO	75.000	0.40	0.58	0.00	73.39
02244 1505	Washed & Graded Sand, Loaded at Pit	CY	12.38	CODEG	12.000	3.88	2.59	26.50	395.62
02560 3108	36" Diameter 12 Gauge Galvanized Corrugated Metal Pipe (91 cm)	LF	13.5	UOEHC	3.000	12.60	5.32	52.56	211.47
02667 4570	1" Diameter PVC Schedule 40 Pipe (2.5 cm)	LF	87.5	XPLUB	15.000	0.85	0.00	0.25	16.47
02667 4620	4" Diameter PVC Schedule 40 Pipe (10 cm) Diameter	LF	47.5	XPLUC	6.000	2.11	0.00	1.56	22.03
02667 4800	2" Diameter PVC Schedule 40 Tee (5.1 cm) Diameter	EA	3.125	XPLUB	1.000	23.75	0.00	2.07	25.82
02667 5330	2" Diameter PVC Schedule 80 Coupling (5.1 cm) Diameter	EA	4.625	XPLUB	1.000	16.05	0.00	3.48	19.53
02752 4600	24" Diameter x 200 Lb Reg Manhole Cover (61 cm Diameter x 90 kg) with Frame, City Type	EA	1.15	CODEJ	1.000	82.80	27.87	168.00	278.67
05108 0140	3/4" Diameter x 12" Length, J-type Anchor Bolt (31 cm) with Nut, Washer, Template	EA	5.625	ACARB	4.000	15.40	0.00	2.43	71.32
05180 2320	3/4" x 12" Jaw & Jaw Turnbuckles	EA	4.375	SIWSB	2.000	22.03	0.00	57.76	159.58
15190 2727	3/4" Diameter Threaded Rod, Bright Finish	LF	22.5	SIWSA	12.000	2.14	0.00	8.96	133.17
16018 1000	1/2" IMC Conduit with Coupling Mount Exposed on Flat Wall, 10' Height	LF	98	EELEF	6.000	1.95	0.00	0.93	17.30
16024 0011	1/2" IMC Bushing Set with Locknut Mount Exposed on Flat Wall, 10' Height	EA	16.05	EELEF	2.000	11.91	0.00	0.43	24.69

33 13 12	Oil/Water Separation	Unit	Hourly Output	Crew	Factor	Labor	Equip.	Matl.	Total
33 13 1205	**Install Underground 300 GPM Oil/Water Separator**	**EA**				**771.08**	**215.55**	**965.11**	**1,952**
02215 0500	Compaction, 6" Layers with Air Tamper by Hand, (15 cm) Layers	CY	50	CLACA	100.000	2.71	0.17	0.00	288.40
02221 5007	Trench Backfill, 3 CY, 950	CY	70.88	CODLJ	100.000	0.47	0.54	0.00	100.42
02221 5011	Sand Backfill, 950 Dumped	CY	83.25	CODLJ	12.000	0.40	0.46	0.00	10.26
02228 0320	Trench, 1 CY Hydraulic Excavator, Light Soil, 4' to 6' Deep, 155 CY/Hour	CY	156	CODEO	100.000	0.40	0.58	0.00	97.85
02244 1505	Washed & Graded Sand, Loaded at Pit	CY	12.38	CODEG	14.400	3.88	2.59	26.50	474.75
02560 3108	36" Diameter 12 Gauge Galvanized Corrugated Metal Pipe (91 cm)	LF	13.5	UOEHC	3.000	12.60	5.32	52.56	211.47
02667 4570	1" Diameter PVC Schedule 40 Pipe (2.5 cm)	LF	87.5	XPLUB	15.000	0.85	0.00	0.25	16.47
02667 4620	4" Diameter PVC Schedule 40 Pipe (10 cm) Diameter	LF	47.5	XPLUC	6.000	2.11	0.00	1.56	22.03
02667 4800	2" Diameter PVC Schedule 40 Tee (5.1 cm) Diameter	EA	3.125	XPLUB	1.000	23.75	0.00	2.07	25.82
02667 5330	2" Diameter PVC Schedule 80 Coupling (5.1 cm) Diameter	EA	4.625	XPLUB	1.000	16.05	0.00	3.48	19.53
02752 4600	24" Diameter x 200 Lb Reg Manhole Cover (61 cm Diameter x 90 kg) with Frame, City Type	EA	1.15	CODEJ	1.000	82.80	27.87	168.00	278.67
05108 0140	3/4" Diameter x 12" Length, J-type Anchor Bolt (31 cm) with Nut, Washer, Template	EA	5.625	ACARB	4.000	15.40	0.00	2.43	71.32
05180 2320	3/4" x 12" Jaw & Jaw Turnbuckles	EA	4.375	SIWSB	2.000	22.03	0.00	57.76	159.58
15190 2727	3/4" Diameter Threaded Rod, Bright Finish	LF	22.5	SIWSA	12.000	2.14	0.00	8.96	133.17
16018 1000	1/2" IMC Conduit with Coupling Mount Exposed on Flat Wall, 10' Height	LF	98	EELEF	6.000	1.95	0.00	0.93	17.30
16024 0011	1/2" IMC Bushing Set with Locknut Mount Exposed on Flat Wall, 10' Height	EA	16.05	EELEF	2.000	11.91	0.00	0.43	24.69

33 13 12	Oil/Water Separation	Unit	Hourly Output	Crew	Factor	Labor	Equip.	Matl.	Total
33 13 1206	**Install Underground 400 GPM Oil/Water Separator**	**EA**				**852.74**	**248.46**	**1,029**	**2,130**
02215 0500	Compaction, 6" Layers with Air Tamper by Hand, (15 cm) Layers	CY	50	CLACA	120.000	2.71	0.17	0.00	346.08
02221 5007	Trench Backfill, 3 CY, 950	CY	70.88	CODLJ	120.000	0.47	0.54	0.00	120.51
02221 5011	Sand Backfill, 950 Dumped	CY	83.25	CODLJ	14.000	0.40	0.46	0.00	11.97
02228 0320	Trench, 1 CY Hydraulic Excavator, Light Soil, 4' to 6' Deep, 155 CY/Hour	CY	156	CODEO	120.000	0.40	0.58	0.00	117.42
02244 1505	Washed & Graded Sand, Loaded at Pit	CY	12.38	CODEG	16.800	3.88	2.59	26.50	553.87
02560 3108	36" Diameter 12 Gauge Galvanized Corrugated Metal Pipe (91 cm)	LF	13.5	UOEHC	3.000	12.60	5.32	52.56	211.47
02667 4570	1" Diameter PVC Schedule 40 Pipe (2.5 cm)	LF	87.5	XPLUB	15.000	0.85	0.00	0.25	16.47
02667 4620	4" Diameter PVC Schedule 40 Pipe (10 cm) Diameter	LF	47.5	XPLUC	6.000	2.11	0.00	1.56	22.03
02667 4800	2" Diameter PVC Schedule 40 Tee (5.1 cm) Diameter	EA	3.125	XPLUB	1.000	23.75	0.00	2.07	25.82
02667 5330	2" Diameter PVC Schedule 80 Coupling (5.1 cm) Diameter	EA	4.625	XPLUB	1.000	16.05	0.00	3.48	19.53
02752 4600	24" Diameter x 200 Lb Reg Manhole Cover (61 cm Diameter x 90 kg) with Frame, City Type	EA	1.15	CODEJ	1.000	82.80	27.87	168.00	278.67
05108 0140	3/4" Diameter x 12" Length, J-type Anchor Bolt (31 cm) with Nut, Washer, Template	EA	5.625	ACARB	4.000	15.40	0.00	2.43	71.32
05180 2320	3/4" x 12" Jaw & Jaw Turnbuckles	EA	4.375	SIWSB	2.000	22.03	0.00	57.76	159.58
15190 2727	3/4" Diameter Threaded Rod, Bright Finish	LF	22.5	SIWSA	12.000	2.14	0.00	8.96	133.17
16018 1000	1/2" IMC Conduit with Coupling Mount Exposed on Flat Wall, 10' Height	LF	98	EELEF	6.000	1.95	0.00	0.93	17.30
16024 0011	1/2" IMC Bushing Set with Locknut Mount Exposed on Flat Wall, 10' Height	EA	16.05	EELEF	2.000	11.91	0.00	0.43	24.69

33 13 12 Oil/Water Separation		Unit	Hourly Output	Crew	Factor	Labor	Equip.	Matl.	Total
33 13 1207	**Install Underground 500 GPM Oil/Water Separator**	**EA**				**974.05**	**268.49**	**1,114**	**2,357**
02215 0500	Compaction, 6" Layers with Air Tamper by Hand, (15 cm) Layers	CY	50	CLACA	130.000	2.71	0.17	0.00	374.92
02221 5007	Trench Backfill, 3 CY, 950	CY	70.88	CODLJ	130.000	0.47	0.54	0.00	130.55
02221 5011	Sand Backfill, 950 Dumped	CY	83.25	CODLJ	16.000	0.40	0.46	0.00	13.68
02228 0320	Trench, 1 CY Hydraulic Excavator, Light Soil, 4' to 6' Deep, 155 CY/Hour	CY	156	CODEO	130.000	0.40	0.58	0.00	127.21
02244 1505	Washed & Graded Sand, Loaded at Pit	CY	12.38	CODEG	19.200	3.88	2.59	26.50	633.00
02560 3108	36" Diameter 12 Gauge Galvanized Corrugated Metal Pipe (91 cm)	LF	13.5	UOEHC	3.000	12.60	5.32	52.56	211.47
02667 4570	1" Diameter PVC Schedule 40 Pipe (2.5 cm)	LF	87.5	XPLUB	15.000	0.85	0.00	0.25	16.47
02667 4620	4" Diameter PVC Schedule 40 Pipe (10 cm) Diameter	LF	47.5	XPLUC	6.000	2.11	0.00	1.56	22.03
02667 4800	2" Diameter PVC Schedule 40 Tee (5.1 cm) Diameter	EA	3.125	XPLUB	1.000	23.75	0.00	2.07	25.82
02667 5330	2" Diameter PVC Schedule 80 Coupling (5.1 cm) Diameter	EA	4.625	XPLUB	1.000	16.05	0.00	3.48	19.53
02752 4600	24" Diameter x 200 Lb Reg Manhole Cover (61 cm Diameter x 90 kg) with Frame, City Type	EA	1.15	CODEJ	1.000	82.80	27.87	168.00	278.67
05180 2360	1" x 12" Jaw & Jaw Turnbuckles	EA	2	SIWSB	2.000	48.19	0.00	91.46	279.29
05501 1011	7/8" Diameter x 12" Length, J-type Anchor Bolt (31 cm) with Nut, Washer, Template	EA	5	ACARB	4.000	17.32	0.00	3.18	82.00
15190 2728	1" Diameter Threaded Rod, Bright Finish	LF	18.75	SIWSA	16.000	2.57	0.00	3.67	99.87
16018 1000	1/2" IMC Conduit with Coupling Mount Exposed on Flat Wall, 10' Height	LF	98	EELEF	6.000	1.95	0.00	0.93	17.30
16024 0011	1/2" IMC Bushing Set with Locknut Mount Exposed on Flat Wall, 10' Height	EA	16.05	EELEF	2.000	11.91	0.00	0.43	24.69
33 13 1211	**20 GPM Oil Pump-out Unit with Controls**	**EA**				**737.23**	**78.27**	**5,791**	**6,606**
15428 1121	20 GPM Pump-out Unit with Controls	EA	0.129	MPLUH	1.000	737.23	78.27	5,791	6,606
33 13 1212	**50 GPM Oil Pump-out Unit with Controls**	**EA**				**737.23**	**78.27**	**6,136**	**6,951**
15428 1122	50 GPM Pump-out Unit with Controls	EA	0.129	MPLUH	1.000	737.23	78.27	6,136	6,951
33 13 1213	**100 GPM Oil Pump-out Unit with Controls**	**EA**				**737.23**	**78.27**	**9,675**	**10,491**
15428 1123	100 GPM Pump-out Unit with Controls	EA	0.129	MPLUH	1.000	737.23	78.27	9,675	10,491
33 13 1214	**200 GPM Oil Pump-out Unit with Controls**	**EA**				**973.90**	**103.40**	**12,165**	**13,242**
15428 1124	200 GPM Pump-out Unit with Controls	EA	0.098	MPLUH	1.000	973.90	103.40	12,165	13,242
33 13 1215	**300 GPM Oil Pump-out Unit with Controls**	**EA**				**973.90**	**103.40**	**15,634**	**16,711**
15428 1125	300 GPM Pump-out Unit with Controls	EA	0.098	MPLUH	1.000	973.90	103.40	15,634	16,711
33 13 1216	**400 GPM Oil Pump-out Unit with Controls**	**EA**				**1,310**	**139.05**	**17,592**	**19,041**
15428 1126	400 GPM Pump-out Unit with Controls	EA	0.073	MPLUH	1.000	1,310	139.05	17,592	19,041
33 13 1217	**500 GPM Oil Pump-out Unit with Controls**	**EA**				**1,312**	**139.24**	**18,713**	**20,164**
15428 1127	500 GPM Pump-out Unit with Controls	EA	0.072	MPLUH	1.000	1,312	139.24	18,713	20,164

33 13 12 Oil/Water Separation		Unit	Hourly Output	Crew	Factor	Labor	Equip.	Matl.	Total
33 13 1218	**10 GPM Oil Water Separator**	**EA**				**4,012**	**0.00**	**3,080**	**7,092**
15428 1108	10 GPM Oil/Water Separator	EA	0.033	MPLUI	1.000	4,012	0.00	3,080	7,092
33 13 1219	**Emulsion Tank**	**EA**				**0.00**	**0.00**	**1,025**	**1,025**
13205 7721	Tanks, gnd stor, 1000 gal, 4" NPconn, plastic, horiz cyl	EA	0	N/A	1.000	0.00	0.00	1,025	1,025

33 13 13 Air Sparged Hydrocyclone		Unit	Hourly Output	Crew	Factor	Labor	Equip.	Matl.	Total
33 13 1301	**Rental of 10-20 GPM Turnkey Mobile ASH Units**	**MO**				**0.00**	**0.00**	**2,000**	**2,000**
13272 7101	10-20 GPM Mobile ASH Unit,	MO	0	N/A	1.000	0.00	0.00	2,000	2,000
33 13 1302	**Rental of 100-400 GPM Turnkey Mobile ASH Units**	**MO**				**0.00**	**0.00**	**10,304**	**10,304**
13272 7102	100-400 GPM Mobile ASH Unit,	MO	0	N/A	1.000	0.00	0.00	10,303	10,304
33 13 1305	**Purchase of 10-20 GPM Turnkey Mobile ASH Units**	**EA**				**27.54**	**1.41**	**24,244**	**24,273**
13272 7103	10-20 GPM Mobile ASH Unit	EA	2.5	HLABD	1.000	27.54	1.41	24,241	24,273
33 13 1306	**Purchase of 100-400 GPM Turnkey Mobile ASH Units**	**EA**				**55.08**	**2.81**	**193,951**	**194,009**
13272 7104	100-400 GPM Mobile ASH Unit	EA	1.25	HLABD	1.000	55.07	2.81	193,931	194,009
33 13 1307	**Install a 10-20 GPM Unit (equip. not included)**	**EA**				**27.54**	**1.41**	**0.00**	**28.94**
13272 7117	Install & Setup a 10-20 GPM Unit	EA	2.5	HLABD	1.000	27.54	1.41	0.00	28.94
33 13 1308	**Install a 100-400 GPM Unit (equip. not included)**	**EA**				**55.08**	**2.81**	**0.00**	**57.89**
13272 7118	Install & Setup a 100-400 GPM	EA	1.25	HLABD	1.000	55.07	2.81	0.00	57.89
33 13 1310	**Purchase of a 2 in., 10-20 GPM ASH device**	**EA**				**22.77**	**2.42**	**6,061**	**6,086**
13272 7105	2 in. 10-20 GPM ASH Device	EA	4.17	MPLUH	1.000	22.77	2.42	6,060	6,086
33 13 1311	**Purchase of a 6 in., 100-400 GPM ASH device**	**EA**				**30.34**	**3.22**	**18,183**	**18,216**
13272 7106	6 in. 100-400 GPM ASH Device	EA	3.13	MPLUH	1.000	30.34	3.22	18,181	18,216
33 13 1312	**Purchase of a 15 in., 1,200-1,600 GPM ASH device**	**EA**				**37.99**	**4.03**	**36,366**	**36,408**
13272 7107	15 in. 1,200 - 1,600 GPM ASH	EA	2.5	MPLUH	1.000	37.98	4.03	36,362	36,408
33 13 1320	**Transport of 10-20 GPM Turnkey Mobile ASH Units**	**MI**				**0.00**	**0.00**	**2.00**	**2.00**
13272 7108	Transport 10-20 GPM Mobile ASH	MI	0	N/A	1.000	0.00	0.00	2.00	2.00
33 13 1321	**Transport of 100-400 GPM Turnkey Mobile ASH Units**	**MI**				**0.00**	**0.00**	**2.48**	**2.48**
13272 7109	Transport 100-400 GPM Mobile ASH	MI	0	N/A	1.000	0.00	0.00	2.48	2.48

33 13 13 Air Sparged Hydrocyclone		Unit	Hourly Output	Crew	Factor	Labor	Equip.	Matl.	Total
33 13 1322	**Startup of a 10-20 GPM Unit**	**EA**				**1,753**	**0.00**	**0.00**	**1,753**
13272 7119	Startup of a 10-20 GPM Unit	EA	0.025	COEMA	1.000	1,753	0.00	0.00	1,753
33 13 1323	**Startup of a 100-400 GPM Unit Treating Solids**	**EA**				**7,011**	**0.00**	**0.00**	**7,011**
13272 7120	Startup of a 100-400 GPM Unit (Treating Solids)	EA	0.013	COEMB	1.000	7,010	0.00	0.00	7,011
33 13 1324	**Startup of a 100-400 GPM Unit Treating Liquids**	**EA**				**3,505**	**0.00**	**0.00**	**3,505**
13272 7121	Startup of a 100-400 GPM Unit (Treating Liquids)	EA	0.013	COEMA	1.000	3,505	0.00	0.00	3,505
33 13 1330	**Decon/Breakdown of 10-20 GPM Turnkey Mobile ASH Units**	**EA**				**27.54**	**1.41**	**0.00**	**28.94**
13272 7110	Decon/Breakdown 10-20 GPM Mobile	EA	2.5	HLABD	1.000	27.54	1.41	0.00	28.94
33 13 1331	**Decon/Breakdown of 100-400 GPM Turnkey Mobile ASH Units**	**EA**				**55.08**	**2.81**	**0.00**	**57.89**
13272 7111	Decon/Breakdown 100-400 GPM	EA	1.25	HLABD	1.000	55.07	2.81	0.00	57.89
33 13 1340	**Operational Labor for a 10-20 GPM ASH Unit**	**HR**				**0.44**	**0.00**	**0.00**	**0.44**
13272 7112	Operational Labor of 10-20 GPM	HR	100	COEMA	1.000	0.44	0.00	0.00	0.44
33 13 1341	**Oper. Labor for a 100-400 GPM ASH Unit Treating Solids**	**HR**				**0.88**	**0.00**	**0.00**	**0.88**
13272 7113	Operational Labor of 100-400 GPM	HR	100	COEMB	1.000	0.88	0.00	0.00	0.88
33 13 1342	**Oper. Labor for a 100 - 400 GPM Unit Treating Liquids**	**HR**				**0.44**	**0.00**	**0.00**	**0.44**
13272 7124	Oper. Labor for a 100 - 400 GPM	HR	100	COEMA	1.000	0.44	0.00	0.00	0.44
33 13 1345	**Maintenance for 10-20 GPM Turnkey Mobile ASH Units**	**HR**				**0.44**	**0.00**	**0.00**	**0.44**
13272 7114	Maintenance of 10-20 GPM	HR	100	COEMA	1.000	0.44	0.00	0.00	0.44
33 13 1346	**Maintenance for 100-400 GPM Turnkey Mobile ASH Units**	**HR**				**0.88**	**0.00**	**0.00**	**0.88**
13272 7115	Maintenance of 100-400 GPM	HR	100	COEMB	1.000	0.88	0.00	0.00	0.88
33 13 1347	**Monthly Spare Parts Charge for 10-20 GPM ASH Units**	**MO**				**0.00**	**0.00**	**103.04**	**103.04**
13272 7122	Monthly Spare Parts Charge for	MO	0	N/A	1.000	0.00	0.00	103.03	103.04
33 13 1348	**Monthly Spare Parts Charge for 100-400 GPM ASH Units**	**MO**				**0.00**	**0.00**	**787.93**	**787.93**
13272 7123	Monthly Spare Parts Charge for	MO	0	N/A	1.000	0.00	0.00	787.85	787.93
33 13 1360	**Reagent Frother Usage**	**LB**				**0.00**	**0.00**	**0.87**	**0.87**
13272 7116	Reagent Frother Usage	LB	1E+06	XTRHC	1.000	0.00	0.00	0.87	0.87

33 13 19 Carbon Adsorption - Gases		Unit	Hourly Output	Crew	Factor	Labor	Equip.	Matl.	Total

33 13 19 Carbon Adsorption - Gases		Unit	Hourly Output	Crew	Factor	Labor	Equip.	Matl.	Total
33 13 1901	**50 CFM, 110 Lb Fill, Closed Upflow, 7.0" Pressure Drop**	**EA**				**71.68**	**0.00**	**460.00**	**531.68**
13274 5111	50 CFM, 110 Lb Fill, Closed Upflow, 7.0" Pressure Drop	EA	2	MSPFL	1.000	71.68	0.00	460.00	531.68
33 13 1902	**50 CFM, 110 Lb Fill, Closed Upflow, 7.0" Pressure Drop High-density Polyethylene**	**EA**				**71.68**	**0.00**	**460.00**	**531.68**
13274 5112	50 CFM, 110 Lb Fill, Closed Upflow, High-density Polyethylene, 7.0" Pressure Drop	EA	2	MSPFL	1.000	71.68	0.00	460.00	531.68
33 13 1903	**100 CFM, 150 Lb Fill, Closed Upflow, 5" Pressure Drop**	**EA**				**95.57**	**0.00**	**575.00**	**670.57**
13274 5113	100 CFM, 150 Lb Fill, Closed Upflow, 5.0" Pressure Drop	EA	1.5	MSPFL	1.000	95.57	0.00	575.00	670.57
33 13 1904	**100 CFM, 200 Lb Fill, Closed Upflow, 6.8" Pressure Drop**	**EA**				**95.57**	**0.00**	**660.00**	**755.57**
13274 5114	100 CFM, 200 Lb Fill, Closed Upflow, 6.8" Pressure Drop	EA	1.5	MSPFL	1.000	95.57	0.00	660.00	755.57
33 13 1905	**100 CFM, 200 Lb Fill, Closed Upflow, 6.8" Pressure Drop High-density Polyethylene**	**EA**				**95.57**	**0.00**	**660.00**	**755.57**
13274 5115	100 CFM, 200 Lb Fill, Closed Upflow, High-density Polyethylene, 6.8" Pressure Drop	EA	1.5	MSPFL	1.000	95.57	0.00	660.00	755.57
33 13 1906	**150 CFM, 300 Lb Fill, Closed Upflow, 7.9" Pressure Drop**	**EA**				**191.15**	**0.00**	**1,025**	**1,216**
13274 5116	150 CFM, 300 Lb Fill, Closed Upflow, 7.9" Pressure Drop	EA	0.75	MSPFL	1.000	191.15	0.00	1,025	1,216
33 13 1907	**250 CFM, 400 Lb Fill, Closed Upflow, 11.3" Pressure Drop**	**EA**				**286.72**	**0.00**	**1,375**	**1,662**
13274 5117	250 CFM, 400 Lb Fill, Closed Upflow, 11.3" Pressure Drop	EA	0.5	MSPFL	1.000	286.72	0.00	1,375	1,662
33 13 1908	**250 CFM, 400 Lb Fill, Closed Upflow, 11.3" Pressure Drop High-density Polyethylene**	**EA**				**286.72**	**0.00**	**2,375**	**2,662**
13274 5118	250 CFM, 400 Lb Fill, Closed Upflow, High-density Polyethylene, 11.3" Pressure Drop	EA	0.5	MSPFL	1.000	286.72	0.00	2,375	2,662
33 13 1909	**500 CFM, 1,200 Lb Fill, Closed Upflow, 8.5" Pressure Drop High-density Polyethylene**	**EA**				**307.51**	**43.13**	**4,550**	**4,901**
13274 5119	500 CFM, 1,200 Lb Fill, Closed Upflow, High-density Polyethylene, 8.5" Pressure Drop	EA	0.5	MSPFP	1.000	307.51	43.13	4,550	4,901
33 13 1910	**500 CFM, 1,400 Lb Fill, Closed Upflow, 11.5" Pressure Drop**	**EA**				**615.02**	**86.25**	**6,650**	**7,351**
13274 5121	500 CFM, 1,400 Lb Fill, Closed Upflow, 11.5" Pressure Drop	EA	0.25	MSPFP	1.000	615.02	86.25	6,650	7,351
33 13 1911	**500 CFM, 200 Lb Fill, Radial Flow, 4.8" Pressure Drop**	**EA**				**191.15**	**0.00**	**1,375**	**1,566**
13274 5122	500 CFM, 200 Lb Fill, Radial Flow, 4.8" Pressure Drop	EA	0.75	MSPFL	1.000	191.15	0.00	1,375	1,566

33 13 19 Carbon Adsorption - Gases		Unit	Hourly Output	Crew	Factor	Labor	Equip.	Matl.	Total
33 13 1912	**750 CFM, 3,200 Lb Fill, Closed Upflow, 11.5" Pressure Drop**	**EA**				**615.02**	**86.25**	**14,800**	**15,501**
13274 5123	750 CFM, 3,200 Lb Fill, Closed Upflow, 11.5" Pressure Drop	EA	0.25	MSPFP	1.000	615.02	86.25	14,800	15,501
33 13 1913	**1,000 CFM, 400 Lb Fill, Radial Flow, 4.8" Pressure Drop**	**EA**				**286.72**	**0.00**	**2,475**	**2,762**
13274 5124	1,000 CFM, 400 Lb Fill, Radial Flow, 4.8" Pressure Drop	EA	0.5	MSPFL	1.000	286.72	0.00	2,475	2,762
33 13 1914	**1,500 CFM, 5,700 Lb Fill, Closed Upflow, 11.5" Pressure Drop**	**EA**				**1,230**	**172.50**	**15,200**	**16,603**
13274 5125	1,500 CFM, 5,700 Lb Fill, Closed Upflow, 11.5" Pressure Drop	EA	0.125	MSPFP	1.000	1,230	172.50	15,200	16,603
33 13 1915	**1,500 CFM, 300 Lb Fill, Radial Flow, 4.8" Pressure Drop**	**EA**				**191.15**	**0.00**	**2,350**	**2,541**
13274 5126	1,500 CFM, 300 Lb Fill, Radial Flow, 4.8" Pressure Drop	EA	0.75	MSPFL	1.000	191.15	0.00	2,350	2,541
33 13 1916	**3,000 CFM, 1,600 Lb Fill, Radial Flow, 4.8" Pressure Drop**	**EA**				**615.02**	**86.25**	**5,125**	**5,826**
13274 5127	3,000 CFM, 1,600 Lb Fill, Radial Flow, 4.8" Pressure Drop	EA	0.25	MSPFP	1.000	615.02	86.25	5,125	5,826
33 13 1917	**5,000 CFM, 4,700 Lb Fill, Radial Flow**	**EA**				**615.02**	**86.25**	**11,900**	**12,601**
13274 5128	5,000 CFM, 4,700 Lb Fill, Radial Flow	EA	0.25	MSPFP	1.000	615.02	86.25	11,900	12,601
33 13 1918	**8,000 CFM, 6,300 Lb Fill, Radial Flow**	**EA**				**1,230**	**172.50**	**23,100**	**24,503**
13274 5129	8,000 CFM, 6,300 Lb Fill, Radial Flow	EA	0.125	MSPFP	1.000	1,230	172.50	23,100	24,503
33 13 1919	**2,200 CFM, 3,000 Lb Fill, 6 x 6 Closed Traverse 8" Pressure Drop**	**EA**				**615.02**	**86.25**	**25,100**	**25,801**
13274 5131	2,200 CFM, 3,000 Lb Fill, 6 x 6 Closed Traverse, 8" Pressure Drop	EA	0.25	MSPFP	1.000	615.02	86.25	25,100	25,801
33 13 1920	**4,000 CFM, 5,300 Lb Fill, 8 x 8 Closed Traverse 8" Pressure Drop**	**EA**				**1,230**	**172.50**	**28,000**	**29,403**
13274 5132	4,000 CFM, 5,300 Lb Fill, 8 x 8 Closed Traverse, 8" Pressure Drop	EA	0.125	MSPFP	1.000	1,230	172.50	28,000	29,403
33 13 1921	**Dual Bed, 500 CFM Series/1,000 CFM Parallel, 2,000 Lb Fill**	**EA**				**615.02**	**86.25**	**7,821**	**8,522**
13274 5133	Dual Bed, 500 CFM Series/1,000 CFM Parallel, 2,000 Lb Fill, 316L Stainless Steel	EA	0.25	MSPFP	1.000	615.02	86.25	7,821	8,522
33 13 1922	**Dual Bed, 1,000 CFM Series/2,000 CFM Parallel, 2,000 Lb Fill**	**EA**				**1,230**	**172.50**	**5,865**	**7,268**
13274 5134	Dual Bed, 1,000 CFM Series/1,000 CFM Parallel, 2,000 Lb Fill, 316L Stainless Steel	EA	0.125	MSPFP	1.000	1,230	172.50	5,865	7,268
33 13 1931	**500 CFM Radial Unit Accumulator Cabinet**	**EA**				**307.51**	**43.13**	**895.00**	**1,246**
13274 5136	500 CFM Radial Unit Accumulator Cabinet	EA	0.5	MSPFP	1.000	307.51	43.13	895.00	1,246

33 13 19 Carbon Adsorption - Gases		Unit	Hourly Output	Crew	Factor	Labor	Equip.	Matl.	Total
33 13 1932	**1,000 CFM & 1,500 CFM Radial Unit Accumulator Cabinet**	**EA**				**307.51**	**43.13**	**1,375**	**1,726**
13274 5137	1,000 CFM & 1,500 CFM Radial Unit Accumulator Cabinet	EA	0.5	MSPFP	1.000	307.51	43.13	1,375	1,726
33 13 1933	**3,000 CFM Radial Unit Accumulator Cabinet**	**EA**				**615.02**	**86.25**	**3,800**	**4,501**
13274 5138	3,000 CFM Radial Unit Accumulator Cabinet	EA	0.25	MSPFP	1.000	615.02	86.25	3,800	4,501
33 13 1941	**Coal-based, 4-mm Pellet, for Solvent Recovery < 2,000 Lb**	**LB**				**0.00**	**0.00**	**1.48**	**1.48**
13274 5151	Coal Based, 4 mm Pellet, for Solvent Recovery, < 2,000 Lb	LB	0	N/A	1.000	0.00	0.00	1.48	1.48
33 13 1942	**Coal-based, 4 mm Pellet, for Solvent Recovery 2,000 - 10,000 Lb**	**LB**				**0.00**	**0.00**	**1.04**	**1.04**
13274 5152	Coal Based, 4 mm Pellet, for Solvent Recovery, 2,000 - 10,000 Lb	LB	0	N/A	1.000	0.00	0.00	1.04	1.04
33 13 1943	**Coal-based, 4 mm Pellet, for Solvent Recovery > 10,000 Lb**	**LB**				**0.00**	**0.00**	**0.90**	**0.90**
13274 5153	Coal Based, 4 mm Pellet, for Solvent Recovery, > 10,000 Lb	LB	0	N/A	1.000	0.00	0.00	0.90	0.90
33 13 1944	**KOH Impregnated for H2S, Acidic Gas or Mercaptans, < 2,000 Lb**	**LB**				**0.00**	**0.00**	**2.48**	**2.48**
13274 5154	KOH Impregnated for H2S, Acidic Gases or Mercaptans, < 2,000 Lb	LB	0	N/A	1.000	0.00	0.00	2.48	2.48
33 13 1945	**KOH Impregnated for H2S, Acidic Gas or Mercaptans, 2,000 - 10,000 Lb**	**LB**				**0.00**	**0.00**	**2.27**	**2.27**
13274 5155	KOH Impregnated for H2S, Acidic Gas or Mercaptans, 2,000 - 10,000 Lb	LB	0	N/A	1.000	0.00	0.00	2.27	2.27
33 13 1946	**KOH Impregnated for H2S, Acidic Gas or Mercaptans, >10,000 Lb**	**LB**				**0.00**	**0.00**	**1.90**	**1.90**
13274 5156	KOH Impregnated for H2S, Acidic Gases or Mercaptans, > 10,000 Lb	LB	0	N/A	1.000	0.00	0.00	1.90	1.90
33 13 1947	**Coconut-based, 4 x 8 Sieve, General Purpose, < 2,000 Lb**	**LB**				**0.00**	**0.00**	**1.48**	**1.48**
13274 5157	Coconut Base, 4 x 8 Sieve, General Purpose, < 2,000 Lb	LB	0	N/A	1.000	0.00	0.00	1.48	1.48
33 13 1948	**Coconut-based, 4 x 8 Sieve, General Purpose, 2,000 - 10,000 Lb**	**LB**				**0.00**	**0.00**	**1.04**	**1.04**
13274 5158	Coconut Base, 4 x 8 Sieve, General Purpose, 2,000 - 10,000 Lb	LB	0	N/A	1.000	0.00	0.00	1.04	1.04
33 13 1949	**Coconut-based, 4 x 8 Sieve, General Purpose, > 10,000 Lb**	**LB**				**0.00**	**0.00**	**0.90**	**0.90**
13274 5159	Coconut Base, 4x8 Sieve, General Purpose, > 10,000 Lb	LB	0	N/A	1.000	0.00	0.00	0.90	0.90

33 13 Physical Treatment

33 13 19 Carbon Adsorption - Gases		Unit	Hourly Output	Crew	Factor	Labor	Equip.	Matl.	Total
33 13 1950	**25' x 6" Flexible Stainless Steel High-pressure Hose**	**EA**				**99.98**	**5.76**	**242.03**	**347.77**
15414 1420	Hose Coupling with 2-1/2" Pin Lug	EA	2.25	MSPIA	2.000	22.86	0.00	34.25	114.21
15635 3390	25' x 6", High Pressure, Flexible Stainless Steel Hose	EA	1.75	MPLUH	1.000	54.26	5.76	173.53	233.55
33 13 1970	**500 watt Hazardous Air Heater**	**EA**				**0.00**	**0.00**	**1,249**	**1,249**
01001 0179	500 watt Hazardous Air Heater	EA	0	N/A	1.000	0.00	0.00	1,249	1,249
33 13 1971	**1 KW Hazardous Air Heater**	**EA**				**0.00**	**0.00**	**2,123**	**2,123**
01001 0180	1 KW Hazardous Air Heater	EA	0	N/A	1.000	0.00	0.00	2,123	2,123
33 13 1972	**2.5 KW Hazardous Air Heater**	**EA**				**0.00**	**0.00**	**3,389**	**3,389**
01001 0181	2.5 KW Hazardous Air Heater	EA	0	N/A	1.000	0.00	0.00	3,389	3,389
33 13 1973	**5 KW Hazardous Air Heater**	**EA**				**0.00**	**0.00**	**4,776**	**4,776**
01001 0182	5 KW Hazardous Air Heater	EA	0	N/A	1.000	0.00	0.00	4,776	4,776
33 13 1980	**Dual Bed,500 CFM Series/1000 CFM Parallel, 1000 Lb Fill each**	**EA**				**0.00**	**0.00**	**12,513**	**12,513**
01001 0183	Dual Bed,500 CFM Series/1000 CFM Parallel, 1000 Lb Fill each	EA	0	N/A	1.000	0.00	0.00	12,513	12,513
33 13 1981	**Dual Bed,1000 CFM Series/2000 CFM Parallel,2000 Lb Fill each**	**EA**				**0.00**	**0.00**	**21,039**	**21,039**
01001 0184	Dual Bed,1000 CFM Series/2000 CFM Parallel,2000 Lb Fill each	EA	0	N/A	1.000	0.00	0.00	21,039	21,039
33 13 1982	**Dual Bed,2500 CFM Series/5000 CFM Parallel,5000 Lb Fill each**	**EA**				**0.00**	**0.00**	**33,881**	**33,881**
01001 0185	Dual Bed,2500 CFM Series/5000 CFM Parallel,5000 Lb Fill each	EA	0	N/A	1.000	0.00	0.00	33,881	33,881
33 13 1983	**Dual Bed,5000 CFM Series/10000 CFM Parallel,10000 Lb Fill each**	**EA**				**0.00**	**0.00**	**46,150**	**46,150**
01001 0186	Dual Bed,5000 CFM Series/10000 CFM Parallel,10000 Lb Fill each	EA	0	N/A	1.000	0.00	0.00	46,150	46,150
33 13 1984	**Dual Bed,10000 CFM Series/20000 CFM Parallel,20000 Lb Fill each**	**EA**				**0.00**	**0.00**	**74,568**	**74,568**
01001 0187	Dual Bed,10000 CFM Series/20000 CFM Parallel,20000 Lb Fill each	EA	0	N/A	1.000	0.00	0.00	74,568	74,568

33 13 20 Carbon Adsorption - Liquids		Unit	Hourly Output	Crew	Factor	Labor	Equip.	Matl.	Total
33 13 2001	**5 GPM, 85 Lb Fill, DOT 5B Drum, Disposable**	**EA**				**71.54**	**0.00**	**495.00**	**566.54**
13274 1211	5 GPM, 85 Lb Fill, DOT 5B Drum, Disposable	EA	2	MPLUM	1.000	71.54	0.00	495.00	566.54
33 13 2002	**5 GPM, 85 Lb Fill, High-density Polyethylene DOT Spec 34, Disposable**	**EA**				**71.54**	**0.00**	**490.00**	**561.54**
13274 1212	5 GPM, 85 Lb Fill, High-density Polyethylene DOT Spec 34, Disposable	EA	2	MPLUM	1.000	71.54	0.00	490.00	561.54

33 13 20 Carbon Adsorption - Liquids		Unit	Hourly Output	Crew	Factor	Labor	Equip.	Matl.	Total
33 13 2003	**15 GPM, 165 Lb Fill, DOT 5B Drum, Disposable**	**EA**				**95.39**	**0.00**	**695.00**	**790.39**
13274 1213	15 GPM, 165 Lb Fill, DOT 5B Drum, Disposable	EA	1.5	MPLUM	1.000	95.39	0.00	695.00	790.39
33 13 2004	**15 GPM, 165 Lb Fill, High-density Polyethylene DOT Spec 34, Disposable**	**EA**				**95.39**	**0.00**	**695.00**	**790.39**
13274 1214	15 GPM, 165 Lb Fill, High-density Polyethylene DOT Spec 34, Disposable	EA	1.5	MPLUM	1.000	95.39	0.00	695.00	790.39
33 13 2005	**20 GPM, 250 Lb Fill, Disposable**	**EA**				**190.77**	**0.00**	**975.00**	**1,166**
13274 1215	20 GPM, 250 Lb Fill, Disposable	EA	0.75	MPLUM	1.000	190.77	0.00	975.00	1,166
33 13 2006	**25 GPM, 330 Lb Fill, Disposable**	**EA**				**286.16**	**0.00**	**1,600**	**1,886**
13274 1216	25 GPM, 330 Lb Fill, Disposable	EA	0.5	MPLUM	1.000	286.16	0.00	1,600	1,886
33 13 2007	**25 GPM, 330 Lb Fill, High-density Polyethylene, Disposable**	**EA**				**286.16**	**0.00**	**2,550**	**2,836**
13274 1217	25 GPM, 330 Lb Fill, High-density Polyethylene, Disposable	EA	0.5	MPLUM	1.000	286.16	0.00	2,550	2,836
33 13 2008	**25 GPM, 330 Lb Fill, High-density Polyethylene, Permanent**	**EA**				**286.16**	**0.00**	**2,550**	**2,836**
13274 1218	25 GPM, 330 Lb Fill, High-density Polyethylene, Permanent	EA	0.5	MPLUM	1.000	286.16	0.00	2,550	2,836
33 13 2009	**25 GPM, 330 Lb Fill, High-density Polyethylene-lined Steel, Permanent**	**EA**				**102.18**	**28.75**	**2,750**	**2,881**
13274 1219	25 GPM, 330 Lb Fill, High-density Polyethylene Lined Steel, Permanent	EA	1.25	MPLUS	1.000	102.18	28.75	2,750	2,881
33 13 2010	**25 GPM, 330 Lb Fill, 316L Stainless Steel, Permanent**	**EA**				**102.18**	**28.75**	**2,750**	**2,881**
13274 1221	25 GPM, 330 Lb Fill, 316L Stainless Steel, Permanent	EA	1.25	MPLUS	1.000	102.18	28.75	2,750	2,881
33 13 2011	**35 GPM, 1,050 Lb Fill, Disposable**	**EA**				**255.45**	**71.88**	**3,575**	**3,902**
13274 1222	35 GPM, 1,050 Lb Fill, Disposable	EA	0.5	MPLUS	1.000	255.45	71.88	3,575	3,902
33 13 2012	**35 GPM, 660 Lb Fill, High-density Polyethylene-lined Steel, Permanent**	**EA**				**255.45**	**71.88**	**4,625**	**4,952**
13274 1223	35 GPM, 660 Lb Fill, High-density Polyethylene Lined Steel, Permanent	EA	0.5	MPLUS	1.000	255.45	71.88	4,625	4,952
33 13 2013	**35 GPM, 660 Lb Fill, 316L Stainless Steel, Permanent**	**EA**				**255.45**	**71.88**	**4,850**	**5,177**
13274 1224	35 GPM, 660 Lb Fill, 316L Stainless Steel, Permanent	EA	0.5	MPLUS	1.000	255.45	71.88	4,850	5,177
33 13 2014	**50 GPM, 1,650 Lb Fill, Disposable**	**EA**				**340.60**	**95.83**	**4,600**	**5,036**
13274 1225	50 GPM, 1,650 Lb Fill, Disposable	EA	0.375	MPLUS	1.000	340.60	95.83	4,600	5,036

33 13 20 Carbon Adsorption - Liquids		Unit	Hourly Output	Crew	Factor	Labor	Equip.	Matl.	Total
33 13 2015	**50 GPM, 1,050 Lb Fill, High-density Polyethylene Disposable**	**EA**				**340.60**	**95.83**	**4,525**	**4,961**
13274 1226	50 GPM, 1,050 Lb Fill, High-density Polyethylene, Disposable	EA	0.375	MPLUS	1.000	340.60	95.83	4,525	4,961
33 13 2016	**50 GPM, 1,050 Lb Fill, High-density Polyethylene Permanent**	**EA**				**340.60**	**95.83**	**4,525**	**4,961**
13274 1227	50 GPM, 1,050 Lb Fill, High-density Polyethylene, Permanent	EA	0.375	MPLUS	1.000	340.60	95.83	4,525	4,961
33 13 2017	**50 GPM, 880 Lb Fill, High-density Polyethylene-lined Steel, Permanent**	**EA**				**340.60**	**95.83**	**6,075**	**6,511**
13274 1228	50 GPM, 880 Lb Fill, High-density Polyethylene Lined Steel, Permanent	EA	0.375	MPLUS	1.000	340.60	95.83	6,075	6,511
33 13 2018	**50 GPM, 880 Lb Fill, 316L Stainless Steel, Permanent**	**EA**				**340.60**	**95.83**	**6,475**	**6,911**
13274 1229	50 GPM, 880 Lb Fill, 316L Stainless Steel, Permanent	EA	0.375	MPLUS	1.000	340.60	95.83	6,475	6,911
33 13 2019	**75 GPM, 1,650 Lb Fill, High-density Polyethylene-lined Steel, Permanent**	**EA**				**510.90**	**143.75**	**8,525**	**9,180**
13274 1231	75 GPM, 1,650 Lb Fill, High-density Polyethylene Lined Steel, Permanent	EA	0.25	MPLUS	1.000	510.90	143.75	8,525	9,180
33 13 2020	**75 GPM, 1,650 Lb Fill, 316L Stainless Steel, Permanent**	**EA**				**510.90**	**143.75**	**9,700**	**10,355**
13274 1232	75 GPM, 1,650 Lb Fill, 316L Stainless Steel, Permanent	EA	0.25	MPLUS	1.000	510.90	143.75	9,700	10,355
33 13 2021	**100 GPM, 3,000 Lb Fill, Disposable**	**EA**				**510.90**	**143.75**	**8,525**	**9,180**
13274 1233	100 GPM, 3,000 Lb Fill, Disposable	EA	0.25	MPLUS	1.000	510.90	143.75	8,525	9,180
33 13 2022	**100 GPM, 3,000 Lb Fill, High-density Polyethylene-lined Steel, Permanent**	**EA**				**510.90**	**143.75**	**11,700**	**12,355**
13274 1234	100 GPM, 3,000 Lb Fill, High-density Polyethylene Lined Steel, Permanent	EA	0.25	MPLUS	1.000	510.90	143.75	11,700	12,355
33 13 2023	**100 GPM, 3,000 Lb Fill, 316L Stainless Steel, Permanent**	**EA**				**510.90**	**143.75**	**13,400**	**14,055**
13274 1235	100 GPM, 3,000 Lb Fill, 316L Stainless Steel, Permanent	EA	0.25	MPLUS	1.000	510.90	143.75	13,400	14,055
33 13 2024	**200 GPM, 6,000 Lb Fill, Disposable**	**EA**				**1,022**	**287.50**	**13,500**	**14,809**
13274 1236	200 GPM, 6,000 Lb Fill, Disposable	EA	0.125	MPLUS	1.000	1,022	287.50	13,500	14,809
33 13 2025	**200 GPM, 6,000 Lb Fill, High-density Polyethylene-lined Steel, Permanent**	**EA**				**1,022**	**287.50**	**16,900**	**18,209**
13274 1237	200 GPM, 6,000 Lb Fill, High-density Polyethylene Lined Steel, Permanent	EA	0.125	MPLUS	1.000	1,022	287.50	16,900	18,209

33 13 20 Carbon Adsorption - Liquids		Unit	Hourly Output	Crew	Factor	Labor	Equip.	Matl.	Total
33 13 2026	**200 GPM, 6,000 Lb Fill, 316L Stainless Steel, Permanent**	**EA**				**1,022**	**287.50**	**21,900**	**23,209**
13274 1238	200 GPM, 6,000 Lb Fill, 316L Stainless Steel, Permanent	EA	0.125	MPLUS	1.000	1,022	287.50	21,900	23,209
33 13 2027	**Dual Bed, 50 GPM Series, 100 GPM Parallel, 1,760 Lb Fill Each**	**EA**				**1,022**	**287.50**	**13,100**	**14,409**
13274 1239	Dual Bed, 50 GPM Series, 100 GPM Parallel, 1,760 Lb Fill, 316L Stainless Steel	EA	0.125	MPLUS	1.000	1,022	287.50	13,100	14,409
33 13 2028	**Dual Bed, 75 GPM Series, 150 GPM Parallel, 3,300 Lb Fill Each**	**EA**				**1,022**	**287.50**	**19,300**	**20,609**
13274 1241	Dual Bed, 75 GPM Series, 150 GPM Parallel, 3,300 Lb Fill, 316L Stainless Steel	EA	0.125	MPLUS	1.000	1,022	287.50	19,300	20,609
33 13 2029	**Dual Bed, 2 - 4' Diameter, 65 GPM Series, 130 GPM Parallel, 2,000 Lb Each**	**EA**				**7,836**	**2,205**	**30,900**	**40,941**
13274 1244	Dual Bed, 65 GPM Series, 130 Parallel, 4' Diameter, 2,000 Lb Carbon	EA	0.016	MPLUS	1.000	7,836	2,205	30,900	40,941
33 13 2030	**Dual Bed, 2 - 7.5' Diameter, 175 GPM Series, 350 GPM Parallel, 10,000 Lb Each**	**EA**				**11,496**	**2,205**	**88,500**	**102,201**
13274 1245	Dual Bed, 175 Series, 350 Parallel, 7.5' Diameter, 10,000 Lb	EA	0.016	MSPFW	1.000	11,496	2,205	88,500	102,201
33 13 2031	**Dual Bed, 2 - 10' Diameter, 350 GPM Series, 700 GPM Parallel, 20,000 Lb Each**	**EA**				**13,172**	**2,350**	**140,000**	**155,522**
13274 1246	Dual Bed, 350 Series, 700 Parallel, 10' Diameter, 20,000 Lb	EA	0.016	MSPFX	1.000	13,172	2,350	140,000	155,522
33 13 2032	**Dual Bed, 2 - 12' Diameter, 350 GPM Series, 700 GPM Parallel, 20,000 Lb Each**	**EA**				**13,172**	**2,350**	**150,000**	**165,522**
13274 1247	Dual Bed, 350 Series, 700 Parallel, 12' Diameter, 20,000 Lb	EA	0.016	MSPFX	1.000	13,172	2,350	150,000	165,522
33 13 2041	**Prefilter/Postfilter Housing & Cartridge to 20 GPM**	**EA**				**63.86**	**17.97**	**246.00**	**327.83**
13274 1242	Prefilter/Postfilter Housing and Cartridge to 20 GPM	EA	2	MPLUS	1.000	63.86	17.97	246.00	327.83
33 13 2042	**Replacement High-capacity Prefilter Cartridges**	**EA**				**0.00**	**0.00**	**42.00**	**42.00**
13274 1243	Replacement High-capacity Pre-filter Cartridges	EA	0	N/A	1.000	0.00	0.00	42.00	42.00
33 13 2051	**Coal-based General Purpose, 8 x 30 Sieve, 900 Iodine, < 2,000 Lb**	**LB**				**0.00**	**0.00**	**1.30**	**1.30**
13274 1251	Coal Based General Purpose, 8 x 30 Sieve, 900 Iodine, < 2,000 Lb	LB	0	N/A	1.000	0.00	0.00	1.30	1.30
33 13 2052	**Coal-based General Purpose, 8 x 30 Sieve, 900 Iodine, 2,000 - 10,000 Lb**	**LB**				**0.00**	**0.00**	**0.99**	**0.99**
13274 1252	Coal Based General Purpose, 8 x 30 Sieve, 900 Iodine, 2,000 - 10,000 Lb	LB	0	N/A	1.000	0.00	0.00	0.99	0.99

33 13 20 Carbon Adsorption - Liquids		Unit	Hourly Output	Crew	Factor	Labor	Equip.	Matl.	Total
33 13 2053	**Coal-based General Purpose, 8 x 30 Sieve, 900 Iodine, > 10,000 Lb**	**LB**				**0.00**	**0.00**	**0.79**	**0.79**
13274 1253	Coal Based General Purpose, 8 x 30 Sieve, 1,100 Iodine, > 10,000 Lb	LB	0	N/A	1.000	0.00	0.00	0.79	0.79
33 13 2054	**Coconut-based, High Capacity, 12 x 30 Sieve, 1100 Iodine, < 2,000 Lb**	**LB**				**0.00**	**0.00**	**1.30**	**1.30**
13274 1254	Coconut Based, High Capacity, 12 x 30 Sieve, 1,100 Iodine, < 2000 Lb	LB	0	N/A	1.000	0.00	0.00	1.30	1.30
33 13 2055	**Coconut-based, High Capacity, 12 x 30 Sieve, 1100 Iodine, 2,000 - 10,000 Lb**	**LB**				**0.00**	**0.00**	**0.99**	**0.99**
13274 1255	Coconut Based, High Capacity, 12 x 30 Sieve, 1,100 Iodine, 2,000 - 10,000 Lb	LB	0	N/A	1.000	0.00	0.00	0.99	0.99
33 13 2056	**Coconut-based, High Capacity, 12 x 30 Sieve, 1100 Iodine, > 10,000 Lb**	**LB**				**0.00**	**0.00**	**0.79**	**0.79**
13274 1256	Coconut Based, High Capacity, 12 x 30 Sieve, 1,100 Iodine, > 10,000 Lb	LB	0	N/A	1.000	0.00	0.00	0.79	0.79
33 13 2057	**Activated Aluminas for Highly Oxidizable Contaminants**	**LB**				**0.00**	**0.00**	**0.68**	**0.68**
13274 1257	Activated Aluminas for Highly Oxidizable Contaminants	LB	0	N/A	1.000	0.00	0.00	0.68	0.68
33 13 2058	**Reactivation or Thermal Regeneration of Carbon**	**LB**				**0.00**	**0.00**	**0.05**	**0.05**
13274 1258	Reactivation or Thermal Regeneration of Carbon	LB	0	N/A	1.000	0.00	0.00	0.05	0.05
33 13 2059	**Remove Carbon from Vessels, 10,000 - 20,000 Lb Minimum, Transport & Reactivate**	**LB**				**0.00**	**0.00**	**0.02**	**0.02**
13274 1259	Remove Carbon from Vessels, 10,000 -20,000 Lb Minimum, Transport & Reactivate	LB	0	N/A	1.000	0.00	0.00	0.02	0.02
33 13 2065	**Removal, Transport, Regeneration of Spent Carbon, < 2K lb**	**LB**				**0.00**	**0.00**	**0.62**	**0.62**
01001 0188	Removal, Transport, Regeneration of Spent Carbon, < 2K lb	LB	0	N/A	1.000	0.00	0.00	0.62	0.62
33 13 2066	**Removal, Transport, Regeneration of Spent Carbon, < 2K to 10 K lb**	**LB**				**0.00**	**0.00**	**0.28**	**0.28**
01001 0189	Removal, Transport, Regeneration of Spent Carbon, < 2K to 10 K lb	LB	0	N/A	1.000	0.00	0.00	0.28	0.28
33 13 2067	**Coal-based General Purpose, 8 x 30 Sieve, 900 Iodine, < 2,000 Lb**	**LB**				**0.00**	**0.00**	**1.30**	**1.30**
13274 1251	Coal Based General Purpose, 8 x 30 Sieve, 900 Iodine, < 2,000 Lb	LB	0	N/A	1.000	0.00	0.00	1.30	1.30
33 13 2068	**Coal-based General Purpose, 8 x 30 Sieve, 900 Iodine, 2,000 - 10,000 Lb**	**LB**				**0.00**	**0.00**	**0.99**	**0.99**
13274 1252	Coal Based General Purpose, 8 x 30 Sieve, 900 Iodine, 2,000 - 10,000 Lb	LB	0	N/A	1.000	0.00	0.00	0.99	0.99

33 13 Physical Treatment

33 13 20 Carbon Adsorption - Liquids	Unit	Hourly Output	Crew	Factor	Labor	Equip.	Matl.	Total
33 13 2069 Coal-based General Purpose, 8 x 30 Sieve, 900 Iodine, > 10,000 Lb	**LB**				**0.00**	**0.00**	**0.79**	**0.79**
13274 1253 Coal Based General Purpose, 8 x 30 Sieve, 1,100 Iodine, > 10,000 Lb	LB	0	N/A	1.000	0.00	0.00	0.79	0.79
33 13 2070 Coal Based, 4MM Pellet, For Solvent Recovery < 2000 Lb Disposable	**LB**				**0.00**	**0.00**	**1.91**	**1.91**
01001 0190 Coal Based, 4MM Pellet, For Solvent Recovery < 2000 Lb Disposable	LB	0	N/A	1.000	0.00	0.00	1.91	1.91
33 13 2071 Coal Based, 4MM Pellet, For Solvent Recovery 2K-10K Lb Disposable	**LB**				**0.00**	**0.00**	**1.57**	**1.57**
01001 0191 Coal Based, 4MM Pellet, For Solvent Recovery 2K-10K Lb Disposable	LB	0	N/A	1.000	0.00	0.00	1.57	1.57
33 13 2072 Coal Based, 4MM Pellet, For Solvent Recovery > 10000 Lb Disposable	**LB**				**0.00**	**0.00**	**1.46**	**1.46**
01001 0192 Coal Based, 4MM Pellet, For Solvent Recovery > 10000 Lb Disposable	LB	0	N/A	1.000	0.00	0.00	1.46	1.46
33 13 2073 Remove and Dispose of Spent Carbon from Water Treatment	**LB**				**0.00**	**0.00**	**0.28**	**0.28**
01001 0193 Remove and Dispose of Spent Carbon from Water Treatment	LB	0	N/A	1.000	0.00	0.00	0.28	0.28
33 13 2074 Remove and Dispose of Spent Carbon from Offgas Treatment	**LB**				**0.00**	**0.00**	**0.28**	**0.28**
01001 0194 Remove and Dispose of Spent Carbon from Offgas Treatment	LB	0	N/A	1.000	0.00	0.00	0.28	0.28

33 13 23 Vapor Extraction	Unit	Hourly Output	Crew	Factor	Labor	Equip.	Matl.	Total
33 13 2301 1 HP, 115V, 98 SCFM, Vapor Recovery System	**EA**				**375.20**	**0.00**	**3,550**	**3,925**
13273 4314 Vapor Recovery System, 1 HP, 115V, 98 SCFM	EA	0.25	ACMAE	1.000	375.20	0.00	3,550	3,925
33 13 2302 1 HP, 230V, 98 SCFM, Vapor Recovery System	**EA**				**375.20**	**0.00**	**3,600**	**3,975**
13273 4315 Vapor Recovery System, 1 HP, 230V, 98 SCFM	EA	0.25	ACMAE	1.000	375.20	0.00	3,600	3,975
33 13 2303 1 1/2 HP, 230V, 127 SCFM, Vapor Recovery System	**EA**				**375.20**	**0.00**	**4,475**	**4,850**
13273 4316 Vapor Recovery System, 1-1/2 HP, 230V, 127 SCFM	EA	0.25	ACMAE	1.000	375.20	0.00	4,475	4,850
33 13 2304 2 HP, 230V, 160 SCFM, Vapor Recovery System	**EA**				**375.20**	**0.00**	**5,350**	**5,725**
13273 4317 Vapor Recovery System, 2 HP, 230V, 160 SCFM	EA	0.25	ACMAE	1.000	375.20	0.00	5,350	5,725
33 13 2305 5 HP, 230V, 280 SCFM, Vapor Recovery System	**EA**				**375.20**	**0.00**	**6,825**	**7,200**
13273 4318 Vapor Recovery System, 5 HP, 230V, 280 SCFM	EA	0.25	ACMAE	1.000	375.20	0.00	6,825	7,200
33 13 2306 Propane Conversion for Soil Vapor Extractor	**EA**				**0.00**	**0.00**	**284.94**	**284.94**
13273 4306 Propane Conversion for Soil	EA	0	N/A	1.000	0.00	0.00	284.94	284.94

33 13 23	Vapor Extraction	Unit	Hourly Output	Crew	Factor	Labor	Equip.	Matl.	Total
33 13 2307	**Modulating Motor for Automatic Fuel Adjustment**	**EA**				**0.00**	**0.00**	**2,125**	**2,125**
13273 4307	Modulating Motor for Automatic Fuel Adjustment	EA	0	N/A	1.000	0.00	0.00	2,125	2,125
33 13 2308	**Trailer Mounting for Vapor Extractor Unit**	**EA**				**0.00**	**0.00**	**2,223**	**2,223**
13273 4309	Trailer Mounting for Vapor Extractor Unit	EA	0	N/A	1.000	0.00	0.00	2,223	2,223
33 13 2309	**Security Fence for Vapor Extractor**	**EA**				**0.00**	**0.00**	**609.20**	**609.20**
13273 4311	Security Fence for Vapor Extractor	EA	0	N/A	1.000	0.00	0.00	609.20	609.20
33 13 2310	**Operations Procedure Training for Mechanical System**	**LS**				**0.00**	**0.00**	**805.00**	**805.00**
13273 4312	Operations Procedure Training for Mechanical System	EA	0	N/A	1.000	0.00	0.00	805.00	805.00
33 13 2311	**Operational Labor Cost**	**DAY**				**761.48**	**80.65**	**0.00**	**842.13**
13273 4313	Operational Labor Cost for SVE	DAY	0.125	MSPFE	1.000	761.48	80.65	0.00	842.13
33 13 2315	**2" Diameter, PVC @ 5' Depth, Horizontal Vent Pipe Installed**	**LF**				**3.31**	**0.48**	**34.00**	**37.79**
13273 4391	Vent Pipe Installed, 2" Diameter PVC, 5' Depth Maximum	LF	12.5	CODBB	1.000	3.31	0.48	34.00	37.79
33 13 2316	**4" Diameter, PVC @ 5' Depth, Horizontal Vent Pipe Installed**	**LF**				**3.31**	**0.48**	**40.00**	**43.79**
13273 4392	Vent Pipe Installed, 4" Diameter PVC, 5' Depth Maximum	LF	12.5	CODBB	1.000	3.31	0.48	40.00	43.79
33 13 2317	**2" Diameter, Stainless Steel @ 5' Depth, Horizontal Vent Pipe Installed**	**LF**				**3.31**	**0.48**	**42.50**	**46.29**
13273 4393	Vent Pipe Installed, 2" Diameter Stainless Steel, 5' Depth Maximum	LF	12.5	CODBB	1.000	3.31	0.48	42.50	46.29
33 13 2318	**4" Diameter, Stainless Steel @ 5' Depth, Horizontal Vent Pipe Installed**	**LF**				**3.31**	**0.48**	**54.50**	**58.29**
13273 4394	Vent Pipe Installed, 4" Diameter Stainless Steel, 5' Depth Maximum	LF	12.5	CODBB	1.000	3.31	0.48	54.50	58.29
33 13 2320	**Installation Using Chain Trencher, Depth <= 4'**	**CY**				**0.96**	**0.14**	**0.00**	**1.10**
02230 0810	Trench, Chain Trencher, Light Soil 1' Wide x 4' Deep	CY	43	CODBB	1.000	0.96	0.14	0.00	1.10
33 13 2322	**1 HP, 230V, 98 SCFM, Weekly Rental, Vapor Recovery System**	**WK**				**0.00**	**0.00**	**286.00**	**286.00**
13273 4332	Vapor Recovery System, 1 HP, 230V, 98 SCFM, Rental, Per Week	WK	0.25	N/A	1.000	0.00	0.00	286.00	286.00
33 13 2332	**1 HP, 230V, 98 SCFM, Monthly Rental, Vapor Recovery System**	**MO**				**0.00**	**0.00**	**795.00**	**795.00**
13273 4342	Vapor Recovery System, 1 HP, 230V, 98 SCFM, Rental, Per Month	MO	0.25	N/A	1.000	0.00	0.00	795.00	795.00

33 13 23	Vapor Extraction	Unit	Hourly Output	Crew	Factor	Labor	Equip.	Matl.	Total
33 13 2333	**5 HP, 90 SCFM Vapor Extraction Blower, Monthly Rental**	**MO**				**0.00**	**0.00**	**895.00**	**895.00**
13273 4343	5 HP, 90 SCFM Vapor Extraction Blower, Monthly Rental	MO	0.25	N/A	1.000	0.00	0.00	895.00	895.00
33 13 2334	**7.5 HP, 140 SCFM Vapor Extraction Blower, Monthly Rental**	**MO**				**0.00**	**0.00**	**1,025**	**1,025**
13273 4344	7.5 HP 140 SCFM Vapor Extraction Blower, Monthly Rental	MO	0.25	N/A	1.000	0.00	0.00	1,025	1,025
33 13 2335	**10 HP, 190 SCFM Vapor Extraction Blower, Monthly Rental**	**MO**				**0.00**	**0.00**	**1,275**	**1,275**
13273 4345	10 HP, 190 SCFM Vapor Extraction Blower, Monthly Rental	MO	0.25	N/A	1.000	0.00	0.00	1,275	1,275
33 13 2336	**15 HP, 290 SCFM Vapor Extraction Blower, Monthly Rental**	**MO**				**0.00**	**0.00**	**1,525**	**1,525**
13273 4346	15 HP, 290 SCFM Vapor Extraction Blower, Monthly Rental	MO	0.25	N/A	1.000	0.00	0.00	1,525	1,525
33 13 2337	**30 HP, 580 SCFM Vapor Extraction Blower, Monthly Rental**	**MO**				**0.00**	**0.00**	**2,250**	**2,250**
13273 4347	30 HP, 580 SCFM Vapor Extraction Blower, Monthly Rental	MO	0.25	N/A	1.000	0.00	0.00	2,250	2,250
33 13 2338	**Purchase, 5.0 HP, 90 SCFM Vapor Extraction Blower**	**EA**				**0.00**	**0.00**	**2,450**	**2,450**
13273 4321	Purchase, 5 HP, 90 SCFM Vapor Extraction Blower	EA	0.25	N/A	1.000	0.00	0.00	2,450	2,450
33 13 2339	**Purchase, 7.5 HP, 140 SCFM Vapor Extraction Blower**	**EA**				**375.20**	**0.00**	**3,000**	**3,375**
13273 4322	Purchase, 7.5 HP, 140 SCFM Vapor Extraction Blower	EA	0.25	ACMAE	1.000	375.20	0.00	3,000	3,375
33 13 2340	**Purchase, 10.0 HP, 190 SCFM Vapor Extraction Blower**	**EA**				**375.20**	**0.00**	**3,025**	**3,400**
13273 4323	Purchase, 10 HP, 190 SCFM Vapor Extraction Blower	EA	0.25	ACMAE	1.000	375.20	0.00	3,025	3,400
33 13 2341	**Purchase, 15.0 HP, 280 SCFM Vapor Extraction Blower**	**EA**				**375.20**	**0.00**	**5,375**	**5,750**
13273 4324	Purchase, 15 HP, 280 SCFM Vapor Extraction Blower	EA	0.25	ACMAE	1.000	375.20	0.00	5,375	5,750
33 13 2342	**Purchase, 30.0 HP, 580 SCFM Vapor Extraction Blower**	**EA**				**375.20**	**0.00**	**10,100**	**10,475**
13273 4325	Purchase, 30 HP, 580 SCFM Vapor Extraction Blower	EA	0.25	ACMAE	1.000	375.20	0.00	10,100	10,475
33 13 2343	**Knockout Drum**	**EA**				**0.00**	**0.00**	**89.50**	**89.50**
02083 5211	55 Gallon, 17C, Open	EA	0	N/A	1.000	0.00	0.00	89.50	89.50

33 13 23 Vapor Extraction

33 13 23 Vapor Extraction	Unit	Hourly Output	Crew	Factor	Labor	Equip.	Matl.	Total
33 13 2348 Install/Assemble Rental Vacuum Extraction Blower	**EA**				**375.20**	**0.00**	**0.00**	**375.20**
13273 4348 Install/Assemble Rental Vacuum Extraction Blower	EA	0.25	ACMAE	1.000	375.20	0.00	0.00	375.20
33 13 2360 500 SCFM, Vapor Recovery System	**EA**				**0.00**	**0.00**	**19,102**	**19,102**
01001 0195 500 SCFM, Vapor Recovery System	EA	0	N/A	1.000	0.00	0.00	19,102	19,102
33 13 2361 1000 SCFM, Vapor Recovery System	**EA**				**0.00**	**0.00**	**24,576**	**24,576**
01001 0196 1000 SCFM, Vapor Recovery System	EA	0	N/A	1.000	0.00	0.00	24,576	24,576
33 13 2370 Trailer for Blower System, 8ft x 20 ft	**EA**				**0.00**	**0.00**	**5,879**	**5,879**
01001 0197 Trailer for Blower System, 8ft x 20 ft	EA	0	N/A	1.000	0.00	0.00	5,879	5,879
33 13 2371 Trailer for Blower System, 8ft x 32 ft	**EA**				**0.00**	**0.00**	**8,707**	**8,707**
01001 0198 Trailer for Blower System, 8ft x 32 ft	EA	0	N/A	1.000	0.00	0.00	8,707	8,707
33 13 2372 Trailer for Blower System, 8ft x 20 ft (Monthly Rental)	**MO**				**0.00**	**0.00**	**168.31**	**168.31**
01001 0199 Trailer for Blower System, 8ft x 20 ft (Monthly Rental)	MO	0	N/A	1.000	0.00	0.00	168.31	168.31
33 13 2373 Trailer for Blower System, 8ft x 32 ft (Monthly Rental)	**MO**				**0.00**	**0.00**	**210.94**	**210.94**
01001 0200 Trailer for Blower System, 8ft x 32 ft (Monthly Rental)	MO	0	N/A	1.000	0.00	0.00	210.94	210.94
33 13 2375 Equipment Enclosure, 6' x 8', Portable Building/Shed; lined, insulated, skid mounted, w/ exhaust fan	**EA**				**0.00**	**0.00**	**674.10**	**674.10**
01001 0201 Equipment Enclosure, 6' x 8', Portable Building/Shed (lined, insulated, ski	EA	0	N/A	1.000	0.00	0.00	674.10	674.10
33 13 2376 Equipment Enclosure, 6' x 8', Portable Building/Shed; Weekly Rental	**WK**				**0.00**	**0.00**	**224.41**	**224.41**
01001 0202 Equipment Enclosure, 6' x 8', Portable Building/Shed (lined, insulated, ski	WK	0	N/A	1.000	0.00	0.00	224.41	224.41
33 13 2377 Equipment Enclosure, 8' x 15', Portable Building/Shed; lined, insulated, skid mounted, w/exhaust fan	**EA**				**0.00**	**0.00**	**2,434**	**2,434**
01001 0203 Equipment Enclosure, 8' x 15', Portable Building/Shed (lined, insulated, sk	EA	0	N/A	1.000	0.00	0.00	2,434	2,434

33 13 24 Shredding

33 13 24 Shredding	Unit	Hourly Output	Crew	Factor	Labor	Equip.	Matl.	Total
33 13 2401 Weekly Rental, Portable Drum Shredder, 60 - 100 Drums/Hour	**WK**				**0.00**	**0.00**	**11,300**	**11,300**
13271 2411 Portable Shredder, One Week Rental, 60 to 100 Drums/Hour	EA	0	N/A	1.000	0.00	0.00	11,300	11,300

33 13 Physical Treatment

33 13 24 Shredding		Unit	Hourly Output	Crew	Factor	Labor	Equip.	Matl.	Total
33 13 2402	**Daily Rental, Portable Drum Shredder, 60 - 100 Drums/Hour**	**DAY**				**0.00**	**0.00**	**2,275**	**2,275**
13271 2412	Portable Shredder, Daily Rental After First Week, 60 to 100 Drums/Hours	EA	0	N/A	1.000	0.00	0.00	2,275	2,275
33 13 2403	**Monthly Rental, Portable Drum Shredder, 60 - 100 Drums/Hour**	**MO**				**0.00**	**0.00**	**45,300**	**45,300**
13271 2413	Portable Shredder, One Month Rental	EA	0	N/A	1.000	0.00	0.00	45,300	45,300
33 13 2404	**Shredder Mobilize/DeMobilize**	**EA**				**0.00**	**0.00**	**1,800**	**1,800**
13271 2424	Mobilization/Demobilization of a Shredder	EA	0	N/A	1.000	0.00	0.00	1,800	1,800
33 13 2405	**Electric Shredder, Minimum Rental Charge, 50 - 75 Tons/Hour**	**EA**				**0.00**	**0.00**	**43,300**	**43,300**
13271 2426	Electric Shredder, Minimum Rental, 1 Month, 50 - 75 Tons/Hour Output	EA	0	N/A	1.000	0.00	0.00	43,300	43,300
33 13 2406	**Electric Shredder, Additional Monthly Rental, 50 - 75 Tons/Hour**	**MO**				**0.00**	**0.00**	**19,900**	**19,900**
13271 2427	Electric Shredder, Monthly Rental After 1 Month, 50 - 75 Tons/Hour Output	EA	0	N/A	1.000	0.00	0.00	19,900	19,900
33 13 2407	**Electric Shredder, Purchase, 35 Tons/Hour Output**	**EA**				**0.00**	**0.00**	**281,500**	**281,500**
13271 2428	Electric Shredder, Purchase, 35 Tons/Hour Output	EA	0	N/A	1.000	0.00	0.00	281,500	281,500
33 13 2408	**Portable Drum Shredder, 60 - 100 Drums/Hour Output**	**EA**				**0.00**	**0.00**	**270,500**	**270,500**
13271 2429	Portable Drum Shredder, 60 - 100 Drums per Hour	EA	0	N/A	1.000	0.00	0.00	270,500	270,500

33 13 29 In Situ Steam Extraction		Unit	Hourly Output	Crew	Factor	Labor	Equip.	Matl.	Total
33 13 2901	**1,700 pph 100 HP Steam Boiler, Monthly Rental**	**MO**				**0.00**	**0.00**	**4,075**	**4,075**
13273 4384	1,700 PPH 100 HP Steam Boiler Monthly Rental	MO	0	N/A	1.000	0.00	0.00	4,075	4,075
33 13 2902	**3,450 pph, 100 HP Steam Boiler, Monthly Rental**	**MO**				**0.00**	**0.00**	**4,225**	**4,225**
13273 4385	3,450 PPH, 100 HP Steam Boiler Monthly Rental	MO	0	N/A	1.000	0.00	0.00	4,225	4,225
33 13 2903	**5,000 pph 150 HP Steam Boiler, Monthly Rental**	**MO**				**0.00**	**0.00**	**6,475**	**6,475**
13273 4386	5,000 PPH 150 HP Steam Boiler Monthly Rental	MO	0	N/A	1.000	0.00	0.00	6,475	6,475
33 13 2904	**12,000 pph 350 HP Steam Boiler, Monthly Rental**	**MO**				**0.00**	**0.00**	**11,400**	**11,400**
13273 4387	12,000 PPH 350 HP Steam Boiler Monthly Rental	MO	0	N/A	1.000	0.00	0.00	11,400	11,400
33 13 2905	**500 pph Vapor Condenser**	**EA**				**813.26**	**132.34**	**7,650**	**8,596**
13273 4388	500 PPH Vapor Condenser	EA	0.185	MSPFO	1.000	813.26	132.34	7,650	8,596

33 13 29 In Situ Steam Extraction		Unit	Hourly Output	Crew	Factor	Labor	Equip.	Matl.	Total
33 13 2906	**75 Ton Centrifugal Cooling Tower with Pump**	**EA**				**1,718**	**279.50**	**9,556**	**11,554**
15730 2520	75 Ton Centrifugal-type Cooling Tower with Pump, 20' Pipe & Motor	EA	0.088	MSPFO	1.000	1,718	279.50	9,556	11,554
33 13 2907	**1,500 pph Vapor Condenser**	**EA**				**862.25**	**140.31**	**11,000**	**12,003**
13273 4389	1,500 PPH Vapor Condenser	EA	0.174	MSPFO	1.000	862.25	140.31	11,000	12,003
33 13 2908	**200 Ton Centrifugal Cooling Tower with Pump**	**EA**				**2,452**	**398.96**	**17,263**	**20,114**
15730 2532	200 Ton Centrifugal-type Cooling Tower with Pump, 20' Pipe & Motor	EA	0.061	MSPFO	1.000	2,452	398.96	17,263	20,114
33 13 2909	**2,500 pph Vapor Condenser**	**EA**				**950.60**	**154.69**	**14,400**	**15,505**
13273 4395	2,500 PPH Vapor Condenser	EA	0.158	MSPFO	1.000	950.60	154.69	14,400	15,505
33 13 2910	**300 Ton Centrifugal Cooling Tower with Pump**	**EA**				**2,669**	**434.39**	**23,675**	**26,779**
15730 2540	300 Ton Centrifugal-type Cooling Tower with Pump, 20' Pipe & Motor	EA	0.056	MSPFO	1.000	2,669	434.39	23,675	26,779
33 13 2911	**6,000 pph Vapor Condenser**	**EA**				**1,256**	**204.31**	**26,400**	**27,860**
13273 4396	6,000 PPH Vapor Condenser	EA	0.12	MSPFO	1.000	1,256	204.31	26,400	27,860
33 13 2912	**800 Ton Centrifugal Cooling Tower with Pump**	**EA**				**3,972**	**852.09**	**61,161**	**65,986**
15730 2580	800 Ton Centrifugal-type Cooling Tower with Pump, 20' Pipe & Motor	EA	0.039	MSPFT	1.000	3,972	852.09	61,161	65,986
33 13 2913	**10,000 pph Vapor Condenser**	**EA**				**1,402**	**228.14**	**39,800**	**41,430**
13273 4397	10,000 PPH Vapor Condenser	EA	0.107	MSPFO	1.000	1,402	228.14	39,800	41,430
33 13 2914	**1,200 Ton Centrifugal Cooling Tower with Pump**	**EA**				**6,863**	**1,117**	**130,167**	**138,147**
15730 2564	600 Ton Centrifugal-type Cooling Tower with Pump, 20' Pipe & Motor	EA	0.044	MSPFO	2.000	3,431	558.36	65,084	138,147
33 13 2915	**Cooling Water Chemical Feeder**	**EA**				**169.36**	**11.91**	**302.17**	**483.44**
15730 4100	Cooling Water Chemical Feeder	EA	0.847	MSPFM	1.000	169.36	11.91	302.17	483.44
33 13 2916	**Natural Gas Usage, per 1,000 CF**	**MCF**				**0.00**	**0.00**	**6.80**	**6.80**
02688 5010	Natural Gas Usage, per 1,000 CF	MCF	0	N/A	1.000	0.00	0.00	6.80	6.80
33 13 2917	**Propane**	**GAL**				**0.00**	**0.00**	**1.68**	**1.68**
02554 2301	Propane	GAL	0	N/A	1.000	0.00	0.00	1.68	1.68

33 13 90 Air Sparging		Unit	Hourly Output	Crew	Factor	Labor	Equip.	Matl.	Total
33 13 9001	**Blower 98 SCFM, 3.2 HP, 5 PSI**	**EA**				**1,147**	**0.00**	**4,500**	**5,647**
13272 5101	Blower 98 SCFM, 3.2 HP, 5 PSI	EA	0.125	MSPFL	1.000	1,147	0.00	4,500	5,647
33 13 9002	**Blower 170 SCFM, 10.3 HP, 10 PSI**	**EA**				**1,147**	**0.00**	**4,550**	**5,697**
13272 5102	Blower 170 SCFM, 10.3 HP, 10 PSI	EA	0.125	MSPFL	1.000	1,147	0.00	4,550	5,697

33 13 90 Air Sparging		Unit	Hourly Output	Crew	Factor	Labor	Equip.	Matl.	Total
33 13 9003	**Blower 163 SCFM, 15.0 HP, 15 PSI**	**EA**				**1,147**	**0.00**	**4,825**	**5,972**
13272 5103	Blower 163 SCFM, 15 HP, 15 PSI	EA	0.125	MSPFL	1.000	1,147	0.00	4,825	5,972
33 13 9004	**Blower 426 SCFM, 84.0 HP, 30 PSI**	**EA**				**1,147**	**0.00**	**11,600**	**12,747**
13272 5104	Blower 426 SCFM, 84 HP, 30 PSI	EA	0.125	MSPFL	1.000	1,147	0.00	11,600	12,747
33 13 9014	**Trailer Mounting for Small Blower System**	**EA**				**0.00**	**0.00**	**2,223**	**2,223**
13273 4309	Trailer Mounting for Vapor Extractor Unit	EA	0	N/A	1.000	0.00	0.00	2,223	2,223

33 13 91 Gas Absorption (Packed Column)		Unit	Hourly Output	Crew	Factor	Labor	Equip.	Matl.	Total
33 13 9101	**1,500 CFM Fluidized Bed Gas Scrubber, Single Stage, Off-Gas**	**EA**				**0.00**	**0.00**	**33,000**	**33,000**
02082 6311	1,500 CFM Fluidized Bed Gas Scrubber - Single Stage, Purchase Only	EA	0	N/A	1.000	0.00	0.00	33,000	33,000
33 13 9102	**3,000 CFM Fluidized Bed Gas Scrubber, Single Stage, Off-Gas**	**EA**				**0.00**	**0.00**	**50,000**	**50,000**
02082 6312	3,000 CFM Fluidized Bed Gas Scrubber - Single Stage, Purchase Only	EA	0	N/A	1.000	0.00	0.00	50,000	50,000
33 13 9103	**4,400 CFM Fluidized Bed Gas Scrubber, Single Stage, Off-Gas**	**EA**				**0.00**	**0.00**	**58,500**	**58,500**
02082 6313	4,400 CFM Fluidized Bed Gas Scrubber - Single Stage, Purchase Only	EA	0	N/A	1.000	0.00	0.00	58,500	58,500
33 13 9104	**5,800 CFM Fluidized Bed Gas Scrubber, Single Stage, Off-Gas**	**EA**				**0.00**	**0.00**	**68,000**	**68,000**
02082 6314	5,800 CFM Fluidized Bed Gas Scrubber - Single Stage, Purchase Only	EA	0	N/A	1.000	0.00	0.00	68,000	68,000
33 13 9105	**7,700 CFM Fluidized Bed Gas Scrubber, Single Stage, Off-Gas**	**EA**				**0.00**	**0.00**	**85,000**	**85,000**
02082 6315	7,700 CFM Fluidized Bed Gas Scrubber - Single Stage, Purchase Only	EA	0	N/A	1.000	0.00	0.00	85,000	85,000
33 13 9106	**9,600 CFM Fluidized Bed Gas Scrubber, Single Stage, Off-Gas**	**EA**				**0.00**	**0.00**	**94,500**	**94,500**
02082 6316	9,600 CFM Fluidized Bed Gas Scrubber - Single Stage, Purchase Only	EA	0	N/A	1.000	0.00	0.00	94,500	94,500
33 13 9107	**12,000 CFM Fluidized Bed Gas Scrubber, Single Stage, Off-Gas**	**EA**				**0.00**	**0.00**	**106,500**	**106,500**
02082 6317	12,000 CFM Fluidized Bed Gas Scrubber - Single Stage, Purchase Only	EA	0	N/A	1.000	0.00	0.00	106,500	106,500
33 13 9108	**17,000 CFM Fluidized Bed Gas Scrubber, Single Stage, Off-Gas**	**EA**				**0.00**	**0.00**	**129,500**	**129,500**
02082 6318	17,000 CFM Fluidized Bed Gas Scrubber - Single Stage, Purchase Only	EA	0	N/A	1.000	0.00	0.00	129,500	129,500

33 13 Physical Treatment

33 13 91	Gas Absorption (Packed Column)	Unit	Hourly Output	Crew	Factor	Labor	Equip.	Matl.	Total
33 13 9109	**23,600 CFM Fluidized Bed Gas Scrubber, Single Stage, Off-Gas**	**EA**				**0.00**	**0.00**	**158,000**	**158,000**
02082 6319	23,600 CFM Fluidized Bed Gas Scrubber - Single Stage, Purchase Only	EA	0	N/A	1.000	0.00	0.00	158,000	158,000
33 13 9110	**30,200 CFM Fluidized Bed Gas Scrubber, Single Stage, Off-Gas**	**EA**				**0.00**	**0.00**	**184,000**	**184,000**
02082 6321	30,200 CFM Fluidized Bed Gas Scrubber - Single Stage, Purchase Only	EA	0	N/A	1.000	0.00	0.00	184,000	184,000
33 13 9111	**38,700 CFM Fluidized Bed Gas Scrubber, Single Stage, Off-Gas**	**EA**				**0.00**	**0.00**	**218,000**	**218,000**
02082 6322	38,700 CFM Fluidized Bed Gas Scrubber - Single Stage, Purchase Only	EA	0	N/A	1.000	0.00	0.00	218,000	218,000
33 13 9112	**47,200 CFM Fluidized Bed Gas Scrubber, Single Stage, Off-Gas**	**EA**				**0.00**	**0.00**	**238,000**	**238,000**
02082 6323	47,200 CFM Fluidized Bed Gas Scrubber - Single Stage, Purchase Only	EA	0	N/A	1.000	0.00	0.00	238,000	238,000
33 13 9113	**58,000 CFM Fluidized Bed Gas Scrubber, Single Stage, Off-Gas**	**EA**				**0.00**	**0.00**	**264,000**	**264,000**
02082 6324	58,000 CFM Fluidized Bed Gas Scrubber - Single Stage, Purchase Only	EA	0	N/A	1.000	0.00	0.00	264,000	264,000
33 13 9114	**79,700 CFM Fluidized Bed Gas Scrubber, Single Stage, Off-Gas**	**EA**				**0.00**	**0.00**	**323,500**	**323,500**
02082 6325	79,700 CFM Fluidized Bed Gas Scrubber - Single Stage, Purchase Only	EA	0	N/A	1.000	0.00	0.00	323,500	323,500
33 13 9115	**90,900 CFM Fluidized Bed Gas Scrubber, Single Stage, Off-Gas**	**EA**				**0.00**	**0.00**	**346,000**	**346,000**
02082 6326	90,900 CFM Fluidized Bed Gas Scrubber - Single Stage, Purchase Only	EA	0	N/A	1.000	0.00	0.00	346,000	346,000
33 13 9116	**113,300 CFM Fluidized Bed Gas Scrubber, Single Stage, Off-Gas**	**EA**				**0.00**	**0.00**	**461,500**	**461,500**
02082 6327	113,300 CFM Fluidized Bed Gas Scrubber - Single Stage, Purchase Only	EA	0	N/A	1.000	0.00	0.00	461,500	461,500

33 14 01	Incineration	Unit	Hourly Output	Crew	Factor	Labor	Equip.	Matl.	Total
33 14 0101	**Permitting Nonhazardous Waste Site**	**EA**				**0.00**	**0.00**	**815.00**	**815.00**
13277 2216	Permitting for Site	EA	0	N/A	1.000	0.00	0.00	815.00	815.00
33 14 0102	**Permitting/Trial Burn/Engineering for Site**	**EA**				**0.00**	**0.00**	**168,047**	**168,047**
13277 2228	Mobile Incineration Trial Burn Plan	EA	0	N/A	1.000	0.00	0.00	19,547	19,547
13277 2232	Trial Burn, TSCA or RCRA, 3 - 4 Hour Runs Minimum with Analysis	EA	0	N/A	1.000	0.00	0.00	148,500	148,500
33 14 0103	**Minimum Mobilization/Demobilization Charge**	**EA**				**0.00**	**0.00**	**457,000**	**457,000**
13277 2261	1,000 Mile Limit, Minimum Mobilization/ Demobilization Charge	EA	0	N/A	1.000	0.00	0.00	457,000	457,000

33 14 01	Incineration	Unit	Hourly Output	Crew	Factor	Labor	Equip.	Matl.	Total
33 14 0104	**Additional Mobilization/Demobilization Charge**	**MI**				**0.00**	**0.00**	**36.50**	**36.50**
13277 2262	Additional Mobilization/Demobilization Charge for Mileage Greater than 1,000	MI	0	N/A	1.000	0.00	0.00	36.50	36.50
33 14 0105	**Rotary Kiln Incinerator, Operations Cost**	**TON**				**0.00**	**0.00**	**330.00**	**330.00**
13277 2263	Rotary Kiln, without Mobilization/ Demobilization Pretreatment	TON	0	N/A	1.000	0.00	0.00	330.00	330.00
33 14 0106	**Rotary Kiln Incinerator, Fixed Cost**	**EA**				**0.00**	**0.00**	**1,263,000**	**1,263,000**
13277 2264	Rotary Kiln, Fixed Cost, with Unit Installation, Monitoring, Operation, Maintenance	EA	0	N/A	1.000	0.00	0.00	1,263,000	1,263,000
33 14 0107	**Infrared Furnace, Operations Cost**	**TON**				**0.00**	**0.00**	**345.00**	**345.00**
13277 2265	Infrared Furnace, without Mobilization/ Demobilization, Pretreatment	TON	0	N/A	1.000	0.00	0.00	345.00	345.00
33 14 0108	**Infrared Furnace, Fixed Cost**	**EA**				**0.00**	**0.00**	**385,000**	**385,000**
13277 2266	Infrared Furnace, Fixed Cost, with Unit Installation, Monitoring, Operation, Maintenance	EA	0	N/A	1.000	0.00	0.00	385,000	385,000
33 14 0109	**Vitrification, Operations Cost**	**TON**				**0.00**	**0.00**	**540.00**	**540.00**
13277 2267	Vitrification, Glass Melters, without Mobilization/Demobilization, Pretreatment	TON	0	N/A	1.000	0.00	0.00	540.00	540.00
33 14 0110	**Vitrification, Fixed Cost**	**EA**				**0.00**	**0.00**	**39,892**	**39,892**
13277 2268	Vitrification, Glass Melters, Fixed Cost with Unit Installation, Monitoring, Operation, Maintenance	EA	0	N/A	1.000	0.00	0.00	39,892	39,892
33 14 0111	**Liquid Injection Incinerator, Operations Cost**	**TON**				**0.00**	**0.00**	**470.00**	**470.00**
13277 2269	Liquid Injection, without Mobilization/Demobilization, Pretreatment	TON	0	N/A	1.000	0.00	0.00	470.00	470.00
33 14 0112	**Liquid Injection Incinerator, Fixed Cost**	**EA**				**0.00**	**0.00**	**1,054,500**	**1,054,500**
13277 2271	Liquid Injection, Fixed Cost, with Unit Installation, Monitoring, Operation, Maintenance	EA	0	N/A	1.000	0.00	0.00	1,054,500	1,054,500
33 14 0113	**Circulating Bed Combustor, Operations Cost**	**TON**				**0.00**	**0.00**	**570.00**	**570.00**
13277 2272	Circulating Bed Combustor, without Mobilization/Demobilization, Pretreatment	TON	0	N/A	1.000	0.00	0.00	570.00	570.00
33 14 0114	**Circulating Bed Combustor, Fixed Cost**	**EA**				**0.00**	**0.00**	**1,107,500**	**1,107,500**
13277 2273	Circulating Bed Combustor, Fixed Cost with Unit Installation, Monitoring, Operation, Maintenance	EA	0	N/A	1.000	0.00	0.00	1,107,500	1,107,500
33 14 0115	**Supercritical Water Oxidation, Operations Cost**	**TON**				**0.00**	**0.00**	**505.00**	**505.00**
13277 2274	Supercritical Water Oxidation, without Mobilization/Demobilization, Pretreatment	TON	0	N/A	1.000	0.00	0.00	505.00	505.00

33 14 Thermal Treatment

33 14 01 Incineration		Unit	Hourly Output	Crew	Factor	Labor	Equip.	Matl.	Total
33 14 0116	**Supercritical Water Oxidation, Fixed Cost**	**EA**				**0.00**	**0.00**	**1,318,500**	**1,318,500**
13277 2275	Supercritical Water Oxidation, Fixed Cost with Unit Installation, Monitoring, Operation, Maintenance	EA	0	N/A	1.000	0.00	0.00	1,318,500	1,318,500
33 14 0117	**Advanced Electric Reactor, Operations Cost**	**TON**				**0.00**	**0.00**	**700.00**	**700.00**
13277 2276	Advanced Electric Reactor, without Mobilization/Demobilization, Pretreatment	TON	0	N/A	1.000	0.00	0.00	700.00	700.00
33 14 0118	**Advanced Electric Reactor, Fixed Cost**	**EA**				**0.00**	**0.00**	**3,981,500**	**3,981,500**
13277 2277	Advanced Electric Reactor, Fixed Cost with Unit Installation, Monitoring, Operation, Maintenance	EA	0	N/A	1.000	0.00	0.00	3,981,500	3,981,500
33 14 0119	**Rotary Kiln Incinerator Weekly Operation Cost**	**WK**				**0.00**	**0.00**	**504,921**	**504,921**
01001 1100	Rotary Kiln Incinerator Weekly Operation Cost	WK	0	N/A	1.000	0.00	0.00	504,921	504,921

33 14 02 Low Temperature Thermal Desorption		Unit	Hourly Output	Crew	Factor	Labor	Equip.	Matl.	Total
33 14 0201	**Minimum Mobilize/DeMobilize Charge <=1,000 Mile, Mobile Process Unit**	**EA**				**0.00**	**0.00**	**126,500**	**126,500**
13277 3511	Minimum Mobilization/Demobilization Charge, <= 1,000 Miles, for Mobile Process Unit	EA	0	N/A	1.000	0.00	0.00	126,500	126,500
33 14 0202	**Additional Mobilize/DeMobilize Charge per Mile, Mobile Process Unit**	**MI**				**0.00**	**0.00**	**68.50**	**68.50**
13277 3512	Additional Mobilization/Demobilization Charge, > 1,000 Miles, for Mobile Process Unit	MI	0	N/A	1.000	0.00	0.00	68.50	68.50
33 14 0203	**Permitting/Engineering for Site**	**EA**				**0.00**	**0.00**	**42,422**	**42,422**
13277 3513	Permitting Site with Treatability Studies, Interfacing with Regulator	EA	0	N/A	1.000	0.00	0.00	42,422	42,422
33 14 0204	**Direct Firing, Rental and Operations Cost**	**TON**				**1.92**	**1.28**	**22.80**	**26.00**
13277 2211	Soil Volatilization in Southeast Region	TON	25	CODEG	1.000	1.92	1.28	22.80	26.00
33 14 0205	**Indirect Firing, Rental and Operations Cost**	**TON**				**1.92**	**1.28**	**141.00**	**144.20**
13277 2213	Soil Volatilization in Northeast Region	TON	25	CODEG	1.000	1.92	1.28	141.00	144.20
33 14 0206	**Min Mob/Demob Chrg for Sm Portable LTTD Units <= 1000 mi**	**EA**				**0.00**	**0.00**	**6,060**	**6,060**
13277 3516	Min. Mob/Demob Charge for SmallPortable LTTD Units <= 1000 mi	EA	0	N/A	1.000	0.00	0.00	6,060	6,060
33 14 0207	**Add. Mob/Demob Chrg for Sm. Port LTTD Units <= 1000 mi**	**MI**				**0.00**	**0.00**	**1.89**	**1.89**
13277 3517	Additional Mob/Demob Charge forPortable LTTD Units <= 1000 mi	MI	0	N/A	1.000	0.00	0.00	1.89	1.89
33 14 0210	**LTTD, bulk solid haz waste, off-site**	**CY**				**0.00**	**0.00**	**65.69**	**65.69**
01001 0204	LTTD, bulk solid haz waste, off-site	CY	0	N/A	1.000	0.00	0.00	65.69	65.69

33 14 02	Low Temperature Thermal Desorption	Unit	Hourly Output	Crew	Factor	Labor	Equip.	Matl.	Total
33 14 0211	**LTTD, drum solid haz waste, off-site**	**EA**				**0.00**	**0.00**	**92.57**	**92.57**
01001 0205	LTTD, drum solid haz waste, off-site	EA	0	N/A	1.000	0.00	0.00	92.57	92.57
33 14 0212	**LTTD, bulk liquid haz waste, off-site**	**GAL**				**0.00**	**0.00**	**0.50**	**0.50**
01001 0206	LTTD, bulk liquid haz waste, off-site	GAL	0	N/A	1.000	0.00	0.00	0.50	0.50
33 14 0213	**LTTD, drum liquid haz waste, off-site**	**EA**				**0.00**	**0.00**	**95.37**	**95.37**
01001 0207	LTTD, drum liquid haz waste, off-site	EA	0	N/A	1.000	0.00	0.00	95.37	95.37
33 14 0220	**LTTD, VOC/Fuel Contaminated Soil, Fixed Costs (Mob/Demob, Engineering), 500-5,000 Tons**	**EA**				**0.00**	**0.00**	**16,831**	**16,831**
01001 0573	LTTD, VOC/Fuel Contaminated Soil, Fixed Costs (Mobilization/Demobilization, Engineering), 500-5,000 Tons	EA	0	N/A	1.000	0.00	0.00	16,831	16,831
33 14 0221	**LTTD, VOC/Fuel Contaminated Soil, Fixed Costs (Mob/Demob, Engineering), 5,001-50,000 Tons**	**EA**				**0.00**	**0.00**	**112,205**	**112,205**
01001 0574	LTTD, VOC/Fuel Contaminated Soil, Fixed Costs (Mobilization/Demobilization, Engineering), 5,001-50,000 Tons	EA	0	N/A	1.000	0.00	0.00	112,205	112,205
33 14 0222	**LTTD, VOC/Fuel Contaminated Soil, Fixed Costs (Mob/Demob, Engineering), >50,000 Tons**	**EA**				**0.00**	**0.00**	**561,023**	**561,023**
01001 0575	LTTD, VOC/Fuel Contaminated Soil, Fixed Costs (Mobilization/Demobilization, Engineering), >50,000 Tons	EA	0	N/A	1.000	0.00	0.00	561,023	561,023
33 14 0223	**LTTD, SVOC Contaminated Soil, Fixed Costs (Mob/Demob, Engineering), 500-10,000 Tons**	**EA**				**0.00**	**0.00**	**112,205**	**112,205**
01001 0576	LTTD, SVOC Contaminated Soil, Fixed Costs (Mobilization/Demobilization, Engineering), 500-10,000 Tons	EA	0	N/A	1.000	0.00	0.00	112,205	112,205
33 14 0224	**LTTD, SVOC Contaminated Soil, Fixed Costs (Mob/Demob, Engineering), 10,001-50,000 Tons**	**EA**				**0.00**	**0.00**	**561,023**	**561,023**
01001 0577	LTTD, SVOC Contaminated Soil, Fixed Costs (Mobilization/Demobilization, Engineering), 10,001-50,000 Tons	EA	0	N/A	1.000	0.00	0.00	561,023	561,023
33 14 0225	**LTTD, SVOC Contaminated Soil, Fixed Costs (Mob/Demob, Engineering), >50,000 Tons**	**EA**				**0.00**	**0.00**	**1,683,070**	**1,683,070**
01001 0578	LTTD, SVOC Contaminated Soil, Fixed Costs (Mobilization/Demobilization, Engineering), >50,000 Tons	EA	0	N/A	1.000	0.00	0.00	1,683,070	1,683,070

33 14 02	Low Temperature Thermal Desorption	Unit	Hourly Output	Crew	Factor	Labor	Equip.	Matl.	Total
33 14 0230	**LTTD, VOC/Fuel Contaminated Soil, Low Moisture Content, Service Contract, 500-2,500 Tons**	**TON**				**0.00**	**0.00**	**84.15**	**84.15**
01001 0579	LTTD, VOC/Fuel Contaminated Soil, Low Moisture Content, Service Contract, 500-2,500 Tons	TON	0	N/A	1.000	0.00	0.00	84.15	84.15
33 14 0231	**LTTD, VOC/Fuel Contaminated Soil, Low Moisture Content, Service Contract 2,501-5,000 Tons**	**TON**				**0.00**	**0.00**	**67.32**	**67.32**
01001 0580	LTTD, VOC/Fuel Contaminated Soil, Low Moisture Content, Service Contract 2,501-5,000 Tons	TON	0	N/A	1.000	0.00	0.00	67.32	67.32
33 14 0232	**LTTD, VOC/Fuel Contaminated Soil, Low Moisture Content, Service Contract 5,001-10,000 Tons**	**TON**				**0.00**	**0.00**	**44.88**	**44.88**
01001 0581	LTTD, VOC/Fuel Contaminated Soil, Low Moisture Content, Service Contract 5,001-10,000 Tons	TON	0	N/A	1.000	0.00	0.00	44.88	44.88
33 14 0233	**LTTD, VOC/Fuel Contaminated Soil, Low Moisture Content, Service Contract 10,001-50,000 Tons**	**TON**				**0.00**	**0.00**	**33.66**	**33.66**
01001 0582	LTTD, VOC/Fuel Contaminated Soil, Low Moisture Content, Service Contract 10,001-50,000 Tons	TON	0	N/A	1.000	0.00	0.00	33.66	33.66
33 14 0234	**LTTD, VOC/Fuel Contaminated Soil, Low Moisture Content, Service Contract >50,000 Tons**	**TON**				**0.00**	**0.00**	**22.44**	**22.44**
01001 0583	LTTD, VOC/Fuel Contaminated Soil, Low Moisture Content, Service Contract >50,000 Tons	TON	0	N/A	1.000	0.00	0.00	22.44	22.44
33 14 0235	**LTTD, SVOC Contaminated Soil, Low Moisture Content, Service Contract 500-2,500 Tons**	**TON**				**0.00**	**0.00**	**168.31**	**168.31**
01001 0584	LTTD, SVOC Contaminated Soil, Low Moisture Content, Service Contract 500-2,500 Tons	TON	0	N/A	1.000	0.00	0.00	168.31	168.31
33 14 0236	**LTTD, SVOC Contaminated Soil, Low Moisture Content, Service Contract 2,501-10,000 Tons**	**TON**				**0.00**	**0.00**	**140.26**	**140.26**
01001 0585	LTTD, SVOC Contaminated Soil, Low Moisture Content, Service Contract 2,501-10,000 Tons	TON	0	N/A	1.000	0.00	0.00	140.26	140.26
33 14 0237	**LTTD, SVOC Contaminated Soil, Low Moisture Content, Service Contract 10,001-50,000 Tons**	**TON**				**0.00**	**0.00**	**84.15**	**84.15**
01001 0586	LTTD, SVOC Contaminated Soil, Low Moisture Content, Service Contract 10,001-50,000 Tons	TON	0	N/A	1.000	0.00	0.00	84.15	84.15

33 14 Thermal Treatment

33 14 02	Low Temperature Thermal Desorption	Unit	Hourly Output	Crew	Factor	Labor	Equip.	Matl.	Total
33 14 0238	**LTTD, SVOC Contaminated Soil, Low Moisture Content, Service Contract >50,000 Tons**	**TON**				**0.00**	**0.00**	**56.10**	**56.10**
01001 0587	LTTD, SVOC Contaminated Soil, Low Moisture Content, Service Contract >50,000 Tons	TON	0	N/A	1.000	0.00	0.00	56.10	56.10

33 14 93	Flaring	Unit	Hourly Output	Crew	Factor	Labor	Equip.	Matl.	Total
33 14 9301	**Flare, 2.25 Million BTU/Hour, Landfill Off-Gas**	**EA**				**583.51**	**82.14**	**43,200**	**43,866**
02082 6221	2.25 Million BTU/Hour Landfill Flare	EA	0.175	MSPFF	1.000	583.51	82.14	43,200	43,866
33 14 9302	**Flare, 7.5 Million BTU/Hour, Landfill Off-Gas**	**EA**				**680.77**	**95.83**	**56,500**	**57,277**
02082 6222	7.5 Million BTU/Hour Landfill Flare	EA	0.15	MSPFF	1.000	680.77	95.83	56,500	57,277
33 14 9303	**Flare, 10.5 Million BTU/Hour, Landfill Off-Gas**	**EA**				**1,249**	**239.58**	**66,000**	**67,489**
02082 6223	10.5 Million BTU/Hour Landfill Flare	EA	0.15	MSPFW	1.000	1,249	239.58	66,000	67,489
33 14 9304	**Flare, 22.5 Million BTU/Hour, Landfill Off-Gas**	**EA**				**1,326**	**254.33**	**91,000**	**92,580**
02082 6224	22.5 Million BTU/Hour Landfill Flare	EA	0.141	MSPFW	1.000	1,326	254.33	91,000	92,580
33 14 9305	**Flare, 30.0 Million BTU/Hour, Landfill Off-Gas**	**EA**				**1,363**	**261.36**	**28,200**	**29,824**
02082 6225	30 Million BTU/Hour Landfill Flare	EA	0.138	MSPFW	1.000	1,363	261.36	28,200	29,824
33 14 9306	**Flare, 36.0 Million BTU/Hour, Landfill Off-Gas**	**EA**				**2,015**	**386.42**	**116,000**	**118,401**
02082 6226	36 Million BTU/Hour Landfill Flare	EA	0.093	MSPFW	1.000	2,015	386.42	116,000	118,401
33 14 9307	**Flare, 60.0 Million BTU/Hour, Landfill Off-Gas**	**EA**				**2,059**	**394.92**	**159,000**	**161,454**
02082 6227	60 Million BTU/Hour Landfill Flare	EA	0.091	MSPFW	1.000	2,059	394.92	159,000	161,454
33 14 9308	**Flare, 120.0 Million BTU/Hour, Landfill Off-Gas**	**EA**				**2,974**	**570.44**	**303,000**	**306,545**
02082 6228	120 Million BTU/Hour Landfill Flare	EA	0.063	MSPFW	1.000	2,974	570.44	303,000	306,545

33 15 02	In Situ Vitrification	Unit	Hourly Output	Crew	Factor	Labor	Equip.	Matl.	Total
33 15 0202	**Setup/Move ISV Equipment Between Settings**	**EA**				**0.00**	**0.00**	**39,988**	**39,988**
13273 7021	Setup/Move ISV Equipment Between Settings	EA	0	N/A	1.000	0.00	0.00	39,988	39,988
33 15 0203	**Initial Setup of ISV Equipment (incl. Labor and Equipment costs)**	**LS**				**0.00**	**0.00**	**2,571**	**2,571**
13273 7024	Initial Setup of ISV Equipment	LS	0	N/A	1.000	0.00	0.00	2,571	2,571
33 15 0204	**Mobilize ISV Equipment**	**MI**				**0.00**	**0.00**	**199.94**	**199.94**
13273 7022	Mobilize ISV Equipment	MI	0	N/A	1.000	0.00	0.00	199.94	199.94
33 15 0205	**Demobilize ISV Equipment**	**MI**				**0.00**	**0.00**	**199.94**	**199.94**
13273 7023	Demobilize ISV Equipment	MI	0	N/A	1.000	0.00	0.00	199.94	199.94
33 15 0206	**Vitrify Soil**	**TON**				**0.00**	**0.00**	**428.44**	**428.44**
13273 7011	Vitrify Soil, 20% Moisture, $0.05/KWH, 94 PCF	TON	0	N/A	1.000	0.00	0.00	428.45	428.44

33 15 04	Pozzolan Process (Lime/Portland Cement)	Unit	Hourly Output	Crew	Factor	Labor	Equip.	Matl.	Total
33 15 0401	**Cement Kiln Dust (Bulk)**	**TON**				**0.00**	**0.00**	**6.87**	**6.87**
13275 2131	Bulk Quantities	TON	0	N/A	1.000	0.00	0.00	6.87	6.87
33 15 0402	**Fly Ash, Class C (Bulk)**	**TON**				**0.00**	**0.00**	**30.50**	**30.50**
13275 2111	Class C Fly Ash, Bulk Quantities	TON	0	N/A	1.000	0.00	0.00	30.50	30.50
33 15 0403	**Fly Ash, Class F (Bulk)**	**TON**				**0.00**	**0.00**	**9.50**	**9.50**
13275 2112	Class F Fly Ash, Bulk Quantities	TON	0	N/A	1.000	0.00	0.00	9.50	9.50
33 15 0404	**Bottom Ash (Bulk)**	**TON**				**0.00**	**0.00**	**2.28**	**2.28**
13275 2113	Bottom Ash, Bulk Quantities	TON	0	N/A	1.000	0.00	0.00	2.28	2.28
33 15 0405	**Portland Cement Type I (Bulk)**	**TON**				**0.00**	**0.00**	**82.00**	**82.00**
13275 1112	Portland Cement, Type I, Bulk, FOB Plant	TON	0	N/A	1.000	0.00	0.00	82.00	82.00
33 15 0406	**Portland Cement Type K (Bulk)**	**TON**				**0.00**	**0.00**	**111.00**	**111.00**
13276 6112	Portland Cement, Type K, Bulk Quantity	TON	0	N/A	1.000	0.00	0.00	111.00	111.00
33 15 0407	**Lime (Crushed Limestone) Bulk**	**TON**				**0.00**	**0.00**	**7.90**	**7.90**
13275 1137	Lime (Crushed Limestone) Bulk	TON	0	N/A	1.000	0.00	0.00	7.90	7.90
33 15 0408	**Urrichem Proprietary Additive (Bulk)**	**TON**				**0.00**	**0.00**	**1,188**	**1,188**
13276 6134	Urrichem by Soliditech	TON	0	N/A	1.000	0.00	0.00	1,188	1,188
33 15 0409	**Bitumen (Bulk)**	**TON**				**0.00**	**0.00**	**40.00**	**40.00**
13275 1141	Bitumen (Bulk)	TON	0	N/A	1.000	0.00	0.00	40.00	40.00
33 15 0410	**Chloranan Proprietary Additive (Bulk)**	**TON**				**0.00**	**0.00**	**635.00**	**635.00**
13276 6131	Chloranan by EM-Tech	TON	0	N/A	1.000	0.00	0.00	635.00	635.00
33 15 0411	**P4 Proprietary Reagent (Bulk)**	**TON**				**0.00**	**0.00**	**878.75**	**878.75**
13276 6135	P4 Reagent (Bulk)	TON	0	N/A	1.000	0.00	0.00	878.75	878.75
33 15 0412	**P27 Proprietary Reagent (Bulk)**	**TON**				**0.00**	**0.00**	**333.32**	**333.32**
13276 6136	P27 Reagent (Bulk)	TON	0	N/A	1.000	0.00	0.00	333.32	333.32
33 15 0413	**Activated Carbon, Granular or Powdered Form (Bulk)**	**TON**				**0.00**	**0.00**	**3,160**	**3,160**
13276 6133	Activated Carbon, Granule or Powder Form, Bulk Quantity	LB	0	N/A	2000.000	0.00	0.00	1.58	3,160
33 15 0414	**Sodium Silicate (Bulk)**	**TON**				**0.00**	**0.00**	**340.00**	**340.00**
13276 6212	Sodium Silicate, Bulk Quantity	LB	0	N/A	2000.000	0.00	0.00	0.17	340.00
33 15 0415	**Tank Truck Standby Time for Solidification/Stabilization Unit**	**HR**				**0.00**	**12.33**	**0.00**	**12.33**
13275 7149	Tank Truck Standby Time for Solidification/Stabilization	HR	1	HLACD	0.220	0.00	56.01	0.00	12.33

33 15 04 Pozzolan Process (Lime/Portland Cement)		Unit	Hourly Output	Crew	Factor	Labor	Equip.	Matl.	Total
33 15 0417	**7.5 HP Sludge Pump, 1" Maximum Particle Size, Rental**	**MO**				**0.00**	**0.00**	**1,151**	**1,151**
13276 8236	7.5 HP Sludge Pump, 1" Maximum Particle Size, Rental	MO	0	N/A	1.000	0.00	0.00	1,151	1,151
33 15 0418	**1 CY Plywood Boxes**	**EA**				**26.57**	**0.00**	**28.02**	**54.59**
06120 2003	Structural Plywood on Walls, 5/8" Thick	SF	131.4	ACARD	45.000	0.59	0.00	0.62	54.59
33 15 0419	**Process and Field Support Training**	**HR**				**33.37**	**0.00**	**0.00**	**33.37**
13275 7146	Process and Field Support Training	HR	1	HLABV	1.000	33.37	0.00	0.00	33.37
33 15 0420	**Operational Labor for Process Equipment**	**HR**				**33.37**	**0.00**	**0.00**	**33.37**
13275 7147	Operational Labor for Process Equipment	HR	1	HLABW	1.000	33.37	0.00	0.00	33.37
33 15 0421	**Bulk Chemical Transport (40,000 Lb Truckload)**	**EA**				**0.00**	**0.00**	**2,200**	**2,200**
13277 2651	Minimum Charges for Shipment	EA	0	N/A	1.000	0.00	0.00	2,200	2,200
33 15 0422	**2 CY Mixing System**	**MO**				**0.00**	**0.00**	**1,775**	**1,775**
13275 7124	2 CY Waste Mixer, 12-Month Rental, Trailer Mounted	EA	0	N/A	0.083	0.00	0.00	21,300	1,775
33 15 0423	**10 CY Mixing System**	**MO**				**0.00**	**0.00**	**5,291**	**5,291**
13275 7126	10 CY Waste Mixer, 12-Month Rental, Electric Powered	EA	0	N/A	0.083	0.00	0.00	63,500	5,291
33 15 0425	**High-pressure Water System for 2 CY Waste Mixer, 28 Gallon**	**EA**				**0.00**	**0.00**	**3,225**	**3,225**
13275 7131	High-pressure Water System for 2 CY Waste Mixer, 28 Gallon	EA	0	N/A	1.000	0.00	0.00	3,225	3,225
33 15 0426	**Nonpressurized Water System for 10 CY Waste Mixer**	**EA**				**0.00**	**0.00**	**2,200**	**2,200**
13275 7132	Non-Pressurized Water System for 10 CY Waste Mixer, 100 Gallon	EA	0	N/A	1.000	0.00	0.00	2,200	2,200
33 15 0427	**Barrel Loader for 2 CY Waste Mixer**	**EA**				**0.00**	**0.00**	**6,150**	**6,150**
13275 7133	Barrel Loader for 2 CY Mixer	EA	0	N/A	1.000	0.00	0.00	6,150	6,150
33 15 0428	**Belt Feeder for 10 CY Mixer, 13' Long**	**EA**				**0.00**	**0.00**	**11,100**	**11,100**
13275 7134	Belt Feeder for 10 CY Mixer, 13' Long	EA	0	N/A	1.000	0.00	0.00	11,100	11,100
33 15 0429	**50 CY/Hour Decumulative Batch Plant, 8 CY Capacity**	**EA**				**0.00**	**0.00**	**205,000**	**205,000**
13275 7135	50 CY/Hour Decumulative Batch Plant, 8 CY Capacity, 912 CF Dry	EA	0	N/A	1.000	0.00	0.00	205,000	205,000
33 15 0430	**Dust Collection with 2 HP Blower and Controls**	**EA**				**0.00**	**0.00**	**8,075**	**8,075**
13275 7137	Dust Collection With 2 HP Blower And Controls - 50 CY/Hour Batch	EA	0	N/A	1.000	0.00	0.00	8,075	8,075

33 15 Stabilization/Fixation/Encapsulation

33 15 04	Pozzolan Process (Lime/Portland Cement)	Unit	Hourly Output	Crew	Factor	Labor	Equip.	Matl.	Total
33 15 0431	**Water Pump, 3" Self-priming with 10 HP Motor**	**EA**				**0.00**	**0.00**	**8,075**	**8,075**
13275 7139	Water Pump, 2" Self-priming with 5 HP Motor - 50 CY/Hour Batch	EA	0	N/A	1.000	0.00	0.00	8,075	8,075
33 15 0432	**Radial Stacking Conveyor with 2 CY Hopper, 55' Long**	**EA**				**0.00**	**0.00**	**58,500**	**58,500**
13275 7144	Radial Stacking Conveyor with 2 CY Hopper, 55' Long - 50 CY/Hour	EA	0	N/A	1.000	0.00	0.00	58,500	58,500
33 15 0433	**5 CY Waste Mixer**	**MO**				**0.00**	**0.00**	**1,869**	**1,869**
13275 7152	5 CY Waste Mixer, Annual Rental Charge	YR	0	N/A	0.083	0.00	0.00	22,423	1,869
33 15 0434	**15 CY Waste Mixer**	**MO**				**0.00**	**0.00**	**5,656**	**5,656**
13275 7150	15 CY Waste Mixer, Annual Rental Charge	YR	0	N/A	0.083	0.00	0.00	67,876	5,656
33 15 0435	**Solidification/Stabilization Ancillary Equipment**	**EA**				**0.00**	**0.00**	**8,606**	**8,606**
13275 7151	Solidification/Stabilization Ancillary Equipment	EA	0	N/A	1.000	0.00	0.00	8,606	8,606
33 15 0436	**Mobilize/DeMobilize of Solidification/Stabilization Equipment**	**EA**				**1,084**	**0.00**	**0.00**	**1,084**
13275 7147	Operational Labor for Process Equipment	HR	1	HLABW	32.500	33.37	0.00	0.00	1,084
33 15 0437	**Maintenance of Solidification/Stabilization Unit**	**YR**				**5,205**	**0.00**	**0.00**	**5,205**
13275 7147	Operational Labor for Process Equipment	HR	1	HLABW	156.000	33.37	0.00	0.00	5,205
33 15 0438	**Equipment Cost - In Situ Solidification/Stabilization**	**MO**				**0.00**	**40,372**	**0.00**	**40,372**
13275 7161	Solidification/Stabilization Equipment Cost	MO	0.003	XYSEO	1.000	0.00	40,372	0.00	40,372
33 15 0439	**Operational Labor -In Situ Solidification/Stabilization**	**HR**				**372.77**	**0.00**	**0.00**	**372.77**
13275 7160	Solidification/Stabilization Operational Labor	HR	1	HLABX	1.000	372.77	0.00	0.00	372.77
33 15 0440	**Mobilize/DeMobilize of In Situ Solidification/Stabilization Equipment**	**EA**				**2,609**	**0.00**	**0.00**	**2,609**
13275 7160	Solidification/Stabilization Operational Labor	HR	1	HLABX	7.000	372.77	0.00	0.00	2,609

33 15 95	Ash Stabilization	Unit	Hourly Output	Crew	Factor	Labor	Equip.	Matl.	Total
33 15 9501	**Metal Stabilization of Ash**	**TON**				**0.00**	**0.00**	**69.00**	**69.00**
13275 7211	Stabilization of Metals in the Incineration Ash	TON	0	N/A	1.000	0.00	0.00	69.00	69.00

33 17 08	Decontamination	Unit	Hourly Output	Crew	Factor	Labor	Equip.	Matl.	Total
33 17 0801	**Decontaminate Light Equipment**	**EA**				**83.28**	**0.00**	**0.00**	**83.28**
01957 3109	Decontaminate Light Equipment	EA	1	XXBBW	1.000	83.28	0.00	0.00	83.28

33 17 08 Decontamination		Unit	Hourly Output	Crew	Factor	Labor	Equip.	Matl.	Total
33 17 0802	**Decontaminate Medium Equipment**	**EA**				**166.55**	**0.00**	**0.00**	**166.55**
01957 3111	Decontaminate Medium Equipment	EA	0.5	XXBBW	1.000	166.55	0.00	0.00	166.55
33 17 0803	**Decontaminate Heavy Equipment**	**EA**				**287.16**	**0.00**	**0.00**	**287.16**
01957 3112	Decontaminate Heavy Equipment	EA	0.29	XXBBW	1.000	287.16	0.00	0.00	287.16
33 17 0804	**Decontaminate Tank By Steam Cleaning, 50 SF/Hour**	**SF**				**1.07**	**0.00**	**0.00**	**1.07**
01957 3114	Tank Decontamination, Steam Clean	SF	50	XXBBZ	1.000	1.07	0.00	0.00	1.07
33 17 0805	**Decontaminate Tank By High-pressure Wash, 40 SF/Hour**	**SF**				**1.49**	**0.94**	**0.00**	**2.44**
01957 3113	Tank Decontamination, High Pressure Wash	SF	40	XXBBX	1.000	1.49	0.94	0.00	2.44
33 17 0806	**Decontaminate Tank By Sandblasting, 40 SF/Hour**	**SF**				**0.98**	**0.14**	**0.23**	**1.35**
03398 0700	Wall Finishes, Light Sand Blast	SF	137.5	ACMAF	1.000	0.98	0.14	0.23	1.35
33 17 0807	**Decontaminate Surface By Sandblasting, 137.5 SF/Hour**	**SF**				**0.98**	**0.14**	**0.23**	**1.35**
03398 0700	Wall Finishes, Light Sand Blast	SF	137.5	ACMAF	1.000	0.98	0.14	0.23	1.35
33 17 0808	**Decontaminate Rig, Augers, Screen (Rental Equipment)**	**DAY**				**0.00**	**0.00**	**111.95**	**111.95**
01954 6434	Alconox Soap	EA	0	N/A	1.000	0.00	0.00	16.95	16.95
01957 3916	Decontaminate Rig, Augers, Screen (Rental Equipment)	DAY	0	N/A	1.000	0.00	0.00	95.00	95.00
33 17 0809	**Triple-rinse 55 Gallon Drums**	**EA**				**6.09**	**0.00**	**0.00**	**6.09**
02084 2143	Triple-rinse 55 Gallon Drums	EA	12	HLABQ	1.000	6.09	0.00	0.00	6.09
33 17 0810	**Decontaminate Split Spoon, Bailers, Hand Auger (Rental Equipment)**	**EA**				**0.00**	**0.00**	**43.00**	**43.00**
01957 3917	Decontaminate Split Spoon, Bailers, Hand Auger (Rental Equipment)	EA	0	N/A	1.000	0.00	0.00	43.00	43.00
33 17 0811	**Decontaminate Trenching/Drilling Equipment**	**EA**				**122.46**	**0.00**	**0.00**	**122.46**
01001 1082	Decontaminate Trenching/Drilling Equipment	HR	0.68	XXBBW	1.000	122.46	0.00	0.00	122.46
33 17 0812	**Decontaminate Surface By Steam Cleaning, 137.5 SF/Hour**	**SF**				**0.39**	**0.00**	**0.00**	**0.39**
01957 3118	Decontaminate Surface by Steam Cleaning, 137.5 SF/Hour	SF	137.5	XXBBZ	1.000	0.39	0.00	0.00	0.39
33 17 0813	**Decontaminate Surface By Pressure Washing, 105 SF/Hour**	**SF**				**0.57**	**0.36**	**0.00**	**0.93**
01957 3119	Decontaminate Surface by Pressure Washing, 105 SF/Hour	SF	105	XXBBX	1.000	0.57	0.36	0.00	0.93
33 17 0814	**1,800 PSI Pressure Washer, 6 HP, 4.8 GPM**	**EA**				**0.00**	**0.00**	**1,675**	**1,675**
01957 3121	Electric, 1,800 PSI, 4.8 GPM, 6 HP, Pressure Washer, with 50' Hose	EA	0	N/A	1.000	0.00	0.00	1,675	1,675

33 17 Decontamination and Decommissioning (D&D)

33 17 08 Decontamination		Unit	Hourly Output	Crew	Factor	Labor	Equip.	Matl.	Total
33 17 0815	**1,800 PSI Steam Cleaner, 6 HP**	**EA**				**0.00**	**0.00**	**3,075**	**3,075**
01957 3122	Electric, 1,800 PSI, 6 HP, Steam Cleaner, with 50' Hose, Thermostat	EA	0	N/A	1.000	0.00	0.00	3,075	3,075
33 17 0816	**3,000 PSI Pressure Washer, 4.5 GPM**	**EA**				**0.00**	**0.00**	**6,875**	**6,875**
01957 3123	Diesel, 3,000 PSI, 4-1/2 GPM, Pressure Washer	EA	0	N/A	1.000	0.00	0.00	6,875	6,875
33 17 0817	**55 Gallon Drum of Liquid Soap**	**EA**				**0.00**	**0.00**	**1,100**	**1,100**
01957 3151	55 Gallon Container of Liquid Decontamination Soap	EA	0	N/A	1.000	0.00	0.00	1,100	1,100
33 17 0818	**1,800 PSI Pressure Washer Rental**	**MO**				**0.00**	**0.00**	**1,100**	**1,100**
01957 3102	Cold Water, Electric, 1,800 PSI, 5 GPM, 5 HP (Monthly Rental)	EA	0	N/A	1.000	0.00	0.00	1,100	1,100
33 17 0819	**1,800 PSI Steam Cleaner Rental**	**MO**				**0.00**	**0.00**	**1,525**	**1,525**
01957 3104	Hot Water, Electric, 1,800 PSI, 5 GPM, 5 HP (Monthly Rental)	EA	0	N/A	1.000	0.00	0.00	1,525	1,525
33 17 0821	**8' x 24' Decontamination Trailer with 4 Showers, HVAC, 2 Sinks**	**MO**				**0.00**	**0.00**	**2,475**	**2,475**
01957 4302	8' x 24', 4 Showers, HVAC, 2 Sinks (Monthly Rental)	EA	0	N/A	1.000	0.00	0.00	2,475	2,475
33 17 0822	**8' x 36' Decontamination Trailer with 2 Showers, Fans**	**MO**				**0.00**	**0.00**	**2,325**	**2,325**
01957 4301	8' x 36', 2 Showers, 2 Wall Fans (Monthly Rental)	EA	0	N/A	1.000	0.00	0.00	2,325	2,325
33 17 0823	**Operation of Pressure Washer, Including Water, Soap, Electricity, Labor**	**HR**				**33.88**	**0.00**	**6.56**	**40.44**
01957 3106	Cold Water, Electric, 1,800 PSI, 5 GPM, 5 HP (Daily Operation)	EA	0.125	ULABA	0.125	271.00	0.00	52.50	40.44
33 17 0824	**Operation of Steam Cleaner, Including Water, Soap, Electricity, Labor**	**HR**				**33.88**	**0.00**	**13.38**	**47.25**
01957 3108	Hot Water, Electric, 1,800 PSI, 5 GPM, 5 HP (Daily Operation)	EA	0.125	ULABA	0.125	271.00	0.00	107.00	47.25
33 17 0825	**8' - 6" Railroad Wood Crossties**	**EA**				**8.04**	**6.12**	**35.50**	**49.66**
02422 1600	Railroad Work 8' - 6" Wood Cross Tie	EA	11.75	ULABF	1.000	8.04	6.12	35.50	49.66

33 18 04 Sorting		Unit	Hourly Output	Crew	Factor	Labor	Equip.	Matl.	Total
33 18 0401	**Mobilize/Demobilize Radiological Screening System**	**MI**				**0.00**	**0.00**	**2.56**	**2.56**
13271 5201	Mobilize/Demobilize Radiological Screening System	MI	0	N/A	1.000	0.00	0.00	2.56	2.56
33 18 0402	**Assemble/Shakedown Radiological Screening System**	**EA**				**39,853**	**0.00**	**0.00**	**39,853**
13271 5202	Assemble/Shakedown Radiological Screening System	EA	0.003	ULABH	1.000	39,853	0.00	0.00	39,853

33 18 Disposal (Other than Commercial)

33 18 04 Sorting	Unit	Hourly Output	Crew	Factor	Labor	Equip.	Matl.	Total
33 18 0403 Disassemble/Decontaminate Radiological Screening System	**EA**				**89,967**	**0.00**	**0.00**	**89,967**
13271 5203 Disassemble/Decontaminate Radiological Screening System	EA	0.003	HRADB	1.000	89,967	0.00	0.00	89,967
33 18 0404 Rental - 13 Tons per Hour Radiological Screening System	**MO**				**0.00**	**0.00**	**96,966**	**96,966**
13271 5204 13 Tons per Hour Radiological Screening System - Monthly Rental	MO	0	N/A	1.000	0.00	0.00	96,966	96,966
33 18 0405 Operational Labor - 13 Tons per Hour Radiological Screening System	**HR**				**105.15**	**0.00**	**0.00**	**105.15**
13271 5205 Operational Labor - 13 Tons per Hour Radiological Screening System	HR	1	COEMA	2.400	43.81	0.00	0.00	105.15
33 18 0406 Soil Conditioning Charges	**TON**				**0.00**	**0.00**	**7.88**	**7.88**
13271 5206 Soil Conditioning Charges	TON	0	N/A	1.000	0.00	0.00	7.88	7.88
33 18 0407 Maintenance/Spare Parts - 13 Tons per Hour Radiological Screening System	**TON**				**0.00**	**0.00**	**6.06**	**6.06**
13271 5207 Maintenance/Spare Parts - 13 Tons per Hour Radiological Screening System	TON	0	N/A	1.000	0.00	0.00	6.06	6.06

33 18 09 Disposal Fees and Taxes	Unit	Hourly Output	Crew	Factor	Labor	Equip.	Matl.	Total
33 18 0901 Radioactive Waste Disposal Fee	**CY**				**0.00**	**0.00**	**262.22**	**262.22**
02083 8932 Radioactive Waste Disposal Fee	CY	0	N/A	1.000	0.00	0.00	262.22	262.22

33 18 84 Conveyors for Waste Handling	Unit	Hourly Output	Crew	Factor	Labor	Equip.	Matl.	Total
33 18 8401 41.5' Automatic Conveyor, 45 FPM, Horizontal 24" Belt, Center Drive	**EA**				**1,420**	**0.00**	**4,256**	**5,676**
14551 0450 41.5' Length, Horizontal 24" Belt, Center Drive and Takeup, 45 FPM	EA	0.05	ACAMC	1.000	1,420	0.00	4,256	5,676
33 18 8402 61.5' Automatic Conveyor, 45 FPM, Horizontal 24" Belt, Center Drive	**EA**				**1,893**	**0.00**	**5,513**	**7,406**
14551 0500 61.5' Length, Horizontal 24" Belt, Center Drive and Takeup, 45 FPM	EA	0.038	ACAMC	1.000	1,893	0.00	5,513	7,406
33 18 8403 34' Automatic Inclined Conveyor, 25 Degree, 24" Belt, Loader/End Idler	**EA**				**1,893**	**0.00**	**5,976**	**7,869**
14551 0710 34' Length, 25 Degree, Inclined 24" Belt, with Horizontal Loader & End Idler	EA	0.038	ACAMC	1.000	1,893	0.00	5,976	7,869

33 18 85 Hoppers	Unit	Hourly Output	Crew	Factor	Labor	Equip.	Matl.	Total
33 18 8501 Refuse Bottom Hopper, Aluminized Steel, 18" Diameter	**EA**				**181.33**	**0.00**	**414.98**	**596.32**
14561 1807 Refuse Bottom Hopper, Aluminized Steel, 18" Diameter	EA	0.375	APLHB	1.000	181.33	0.00	414.98	596.32

33 18 86 Vibrating Screens	Unit	Hourly Output	Crew	Factor	Labor	Equip.	Matl.	Total

33 18 86 Vibrating Screens		Unit	Hourly Output	Crew	Factor	Labor	Equip.	Matl.	Total
33 18 8601	**4' x 12' Single-tray Vibrating Screening Unit, with Motor & Accessories**	**EA**				**1,131**	**332.75**	**19,066**	**20,530**
13271 2114	4' x 12' Single Tray Unit	EA	0.108	USKCG	1.000	1,131	332.75	16,500	17,964
13271 2131	Motor Base	EA	0	N/A	1.000	0.00	0.00	980.00	980.00
13271 2132	Drive Guard	EA	0	N/A	1.000	0.00	0.00	640.00	640.00
13271 2133	4' Wide Feed Box	EA	0	N/A	1.000	0.00	0.00	820.00	820.00
13271 2137	Deck Liners	EA	0	N/A	1.000	0.00	0.00	14.25	14.25
13271 2138	Tension Plate Liner, per Deck	EA	0	N/A	1.000	0.00	0.00	41.50	41.50
13271 2139	Discharge Slip, per Deck	EA	0	N/A	1.000	0.00	0.00	67.50	67.50
13271 2141	Wire Cloth, per Deck	EA	0	N/A	1.000	0.00	0.00	2.82	2.82
33 18 8602	**5' x 16' Double-tray Vibrating Screening Unit, with Motor & Accessories**	**EA**				**1,299**	**382.31**	**29,692**	**31,374**
13271 2116	5' x 16' Double Tray Unit	EA	0.094	USKCG	1.000	1,299	382.31	27,000	28,682
13271 2131	Motor Base	EA	0	N/A	1.000	0.00	0.00	980.00	980.00
13271 2132	Drive Guard	EA	0	N/A	1.000	0.00	0.00	640.00	640.00
13271 2133	4' Wide Feed Box	EA	0	N/A	1.000	0.00	0.00	820.00	820.00
13271 2137	Deck Liners	EA	0	N/A	2.000	0.00	0.00	14.25	28.50
13271 2138	Tension Plate Liner, per Deck	EA	0	N/A	2.000	0.00	0.00	41.50	83.00
13271 2139	Discharge Slip, per Deck	EA	0	N/A	2.000	0.00	0.00	67.50	135.00
13271 2141	Wire Cloth, per Deck	EA	0	N/A	2.000	0.00	0.00	2.82	5.64
33 18 8603	**6' x 20' Triple-tray Vibrating Screening Unit, with Motor & Accessories**	**EA**				**2,262**	**665.51**	**25,666**	**28,593**
13271 2118	6' x 20' Triple Tray Unit	EA	0.054	USKCG	1.000	2,262	665.51	23,100	26,027
13271 2131	Motor Base	EA	0	N/A	1.000	0.00	0.00	980.00	980.00
13271 2132	Drive Guard	EA	0	N/A	1.000	0.00	0.00	640.00	640.00
13271 2133	4' Wide Feed Box	EA	0	N/A	1.000	0.00	0.00	820.00	820.00
13271 2137	Deck Liners	EA	0	N/A	1.000	0.00	0.00	14.25	14.25
13271 2138	Tension Plate Liner, per Deck	EA	0	N/A	1.000	0.00	0.00	41.50	41.50
13271 2139	Discharge Slip, per Deck	EA	0	N/A	1.000	0.00	0.00	67.50	67.50
13271 2141	Wire Cloth, per Deck	EA	0	N/A	1.000	0.00	0.00	2.82	2.82
33 18 8604	**7' x 24' Triple-tray Vibrating Screening Unit, with Motor & Accessories**	**EA**				**3,214**	**945.72**	**27,418**	**31,578**
13271 2119	7' x 24' Triple Tray Unit	EA	0.038	USKCG	1.000	3,214	945.72	24,600	28,760
13271 2131	Motor Base	EA	0	N/A	1.000	0.00	0.00	980.00	980.00
13271 2132	Drive Guard	EA	0	N/A	1.000	0.00	0.00	640.00	640.00
13271 2133	4' Wide Feed Box	EA	0	N/A	1.000	0.00	0.00	820.00	820.00
13271 2137	Deck Liners	EA	0	N/A	3.000	0.00	0.00	14.25	42.75
13271 2138	Tension Plate Liner, per Deck	EA	0	N/A	3.000	0.00	0.00	41.50	124.50
13271 2139	Discharge Slip, per Deck	EA	0	N/A	3.000	0.00	0.00	67.50	202.50
13271 2141	Wire Cloth, per Deck	EA	0	N/A	3.000	0.00	0.00	2.82	8.46

33 19 01 Container Handling	Unit	Hourly Output	Crew	Factor	Labor	Equip.	Matl.	Total

33 19 Disposal (Commercial)

33 19 01	Container Handling	Unit	Hourly Output	Crew	Factor	Labor	Equip.	Matl.	Total
33 19 0101	**Liquid Loading Into 5,000 Gallon Bulk Tank Truck**	**EA**				**243.09**	**111.94**	**0.00**	**355.04**
02083 6152	Load Liquid or Sludge into 5,000 Gallon Bulk Tank Truck	EA	0.333	COFWL	1.000	243.09	111.94	0.00	355.04
33 19 0102	**Bulk Solid Hazardous Waste Loading Into Truck**	**CY**				**0.66**	**1.21**	**0.00**	**1.87**
02225 2332	Excavate & Load, 1-1/2 CY Wheel Loader, Medium Material 75 CY/Hour (57 m3)	CY	72.5	CODLC	1.000	0.66	1.21	0.00	1.87
33 19 0103	**Load Drums on Disposal Vehicle**	**EA**				**1.67**	**1.38**	**0.00**	**3.05**
02083 6151	Load Drums on Disposal Truck	EA	40	HLACC	1.000	1.67	1.38	0.00	3.05
33 19 0104	**Load Bulk Liquid/Sludge Waste Into 55 Gallon Drums, Drums Separate**	**EA**				**8.47**	**0.00**	**0.00**	**8.47**
02083 6141	Bulk Liquid/Sludge into 55 Gallon Drums, Drums Separate	EA	4	ULABA	1.000	8.47	0.00	0.00	8.47
33 19 0105	**Load Bulk Solid Waste Into 55 Gallon Drums, Drums Separate**	**EA**				**30.11**	**0.00**	**0.00**	**30.11**
02083 6142	Bulk Solids into 55 Gallon Drums, Drums Separate	EA	1.125	ULABA	1.000	30.11	0.00	0.00	30.11
33 19 0106	**Recontainerize Drums, Not Including Salvage Drums**	**EA**				**23.34**	**28.58**	**0.00**	**51.92**
02083 6161	Recontainerize Leaking 55 Gallon Drums, Not Including Salvage Drums	EA	3	HLABN	1.000	23.34	28.58	0.00	51.92
33 19 0107	**Remove/Reinstall Carbon Adsorber Unit**	**EA**				**206.30**	**0.00**	**0.00**	**206.30**
01050 0103	Field Engineer (Average Cost)	WK	0.025	XYBBD	0.100	995.00	0.00	0.00	99.50
01050 0105	General Purpose Laborer	WK	0.025	XYBBF	0.100	1,068	0.00	0.00	106.80
33 19 0108	**Tanker Pumping Equipment to Load Liquid**	**HR**				**0.00**	**0.00**	**62.50**	**62.50**
02083 6164	Tanker Pumping Equipment to Load Liquid	HR	0	N/A	1.000	0.00	0.00	62.50	62.50
33 19 0109	**Dredging Sludge < 1,000' to Loading Area**	**DAY**				**619.92**	**942.25**	**0.00**	**1,562**
02083 2118	Rental & Operation of Dredging Unit < 1,000' to Loading Area	DAY	0.1	CDRGA	1.000	619.92	942.25	0.00	1,562
33 19 0110	**Dredging Sludge > 1,000' to Loading Area**	**DAY**				**619.92**	**1,363**	**0.00**	**1,983**
02083 2119	Rental & Operation of Dredging Unit > 1,000' to Loading Area	DAY	0.1	CDRGB	1.000	619.92	1,363	0.00	1,983
33 19 0111	**Vacuum Truck on Board Loader**	**HR**				**0.00**	**0.00**	**56.50**	**56.50**
02083 6163	Vacuum Truck on Board Loader	HR	0	N/A	1.000	0.00	0.00	56.50	56.50
33 19 0112	**Load Bulk Liquid/Sludge Waste Into 350 Gallon Container**	**EA**				**53.86**	**0.00**	**0.00**	**53.86**
02083 6141	Bulk Liquid/Sludge into 55 Gallon Drums, Drums Separate	EA	4	ULABA	6.360	8.47	0.00	0.00	53.86

33 19 01	Container Handling	Unit	Hourly Output	Crew	Factor	Labor	Equip.	Matl.	Total
33 19 0113	**Load Bulk Liquid/Sludge Waste Into 250 Gallon Container**	**EA**				**38.50**	**0.00**	**0.00**	**38.50**
02083 6141	Bulk Liquid/Sludge into 55 Gallon Drums, Drums Separate	EA	4	ULABA	4.546	8.47	0.00	0.00	38.50
33 19 0114	**Load Bulk Liquid/Sludge Waste Into 110 Gallon Drum**	**EA**				**16.94**	**0.00**	**0.00**	**16.94**
02083 6141	Bulk Liquid/Sludge into 55 Gallon Drums, Drums Separate	EA	4	ULABA	2.000	8.47	0.00	0.00	16.94
33 19 0115	**Load Bulk Liquid/Sludge Waste Into 95 Gallon Drum**	**EA**				**14.63**	**0.00**	**0.00**	**14.63**
02083 6141	Bulk Liquid/Sludge into 55 Gallon Drums, Drums Separate	EA	4	ULABA	1.727	8.47	0.00	0.00	14.63
33 19 0116	**Load Bulk Liquid/Sludge Waste Into 85 Gallon Drum**	**EA**				**13.09**	**0.00**	**0.00**	**13.09**
02083 6141	Bulk Liquid/Sludge into 55 Gallon Drums, Drums Separate	EA	4	ULABA	1.546	8.47	0.00	0.00	13.09
33 19 0117	**Load Bulk Liquid/Sludge Waste Into 83 Gallon Drum**	**EA**				**12.79**	**0.00**	**0.00**	**12.79**
02083 6141	Bulk Liquid/Sludge into 55 Gallon Drums, Drums Separate	EA	4	ULABA	1.510	8.47	0.00	0.00	12.79
33 19 0118	**Load Bulk Liquid/Sludge Waste Into 35 Gallon Drum**	**EA**				**5.39**	**0.00**	**0.00**	**5.39**
02083 6141	Bulk Liquid/Sludge into 55 Gallon Drums, Drums Separate	EA	4	ULABA	0.636	8.47	0.00	0.00	5.39
33 19 0119	**Load Bulk Liquid/Sludge Waste Into 30 Gallon Drum**	**EA**				**4.62**	**0.00**	**0.00**	**4.62**
02083 6141	Bulk Liquid/Sludge into 55 Gallon Drums, Drums Separate	EA	4	ULABA	0.546	8.47	0.00	0.00	4.62
33 19 0120	**Load Bulk Liquid/Sludge Waste Into 20 Gallon Drum**	**EA**				**3.08**	**0.00**	**0.00**	**3.08**
02083 6141	Bulk Liquid/Sludge into 55 Gallon Drums, Drums Separate	EA	4	ULABA	0.364	8.47	0.00	0.00	3.08
33 19 0121	**Load Bulk Liquid/Sludge Waste Into 15 Gallon Drum**	**EA**				**2.31**	**0.00**	**0.00**	**2.31**
02083 6141	Bulk Liquid/Sludge into 55 Gallon Drums, Drums Separate	EA	4	ULABA	0.273	8.47	0.00	0.00	2.31
33 19 0122	**Load Bulk Liquid/Sludge Waste Into 10 Gallon Drum**	**EA**				**1.54**	**0.00**	**0.00**	**1.54**
02083 6141	Bulk Liquid/Sludge into 55 Gallon Drums, Drums Separate	EA	4	ULABA	0.182	8.47	0.00	0.00	1.54
33 19 0123	**Load Bulk Solid Waste Into 110 Gallon Drum**	**EA**				**60.22**	**0.00**	**0.00**	**60.22**
02083 6142	Bulk Solids into 55 Gallon Drums, Drums Separate	EA	1.125	ULABA	2.000	30.11	0.00	0.00	60.22

33 19 Disposal (Commercial)

33 19 01	Container Handling	Unit	Hourly Output	Crew	Factor	Labor	Equip.	Matl.	Total
33 19 0124	**Load Bulk Solid Waste Into 95 Gallon Drum**	**EA**				**52.00**	**0.00**	**0.00**	**52.00**
02083 6142	Bulk Solids into 55 Gallon Drums, Drums Separate	EA	1.125	ULABA	1.727	30.11	0.00	0.00	52.00
33 19 0125	**Load Bulk Solid Waste Into 85 Gallon Drum**	**EA**				**46.55**	**0.00**	**0.00**	**46.55**
02083 6142	Bulk Solids into 55 Gallon Drums, Drums Separate	EA	1.125	ULABA	1.546	30.11	0.00	0.00	46.55
33 19 0126	**Load Bulk Solid Waste Into 83 Gallon Drum**	**EA**				**45.47**	**0.00**	**0.00**	**45.47**
02083 6142	Bulk Solids into 55 Gallon Drums, Drums Separate	EA	1.125	ULABA	1.510	30.11	0.00	0.00	45.47
33 19 0127	**Load Bulk Solid Waste Into 35 Gallon Drum**	**EA**				**19.15**	**0.00**	**0.00**	**19.15**
02083 6142	Bulk Solids into 55 Gallon Drums, Drums Separate	EA	1.125	ULABA	0.636	30.11	0.00	0.00	19.15
33 19 0128	**Load Bulk Solid Waste Into 30 Gallon Drum**	**EA**				**16.44**	**0.00**	**0.00**	**16.44**
02083 6142	Bulk Solids into 55 Gallon Drums, Drums Separate	EA	1.125	ULABA	0.546	30.11	0.00	0.00	16.44
33 19 0129	**Load Bulk Solid Waste Into 20 Gallon Drum**	**EA**				**10.96**	**0.00**	**0.00**	**10.96**
02083 6142	Bulk Solids into 55 Gallon Drums, Drums Separate	EA	1.125	ULABA	0.364	30.11	0.00	0.00	10.96
33 19 0130	**Load Bulk Solid Waste Into 15 Gallon Drum**	**EA**				**8.22**	**0.00**	**0.00**	**8.22**
02083 6142	Bulk Solids into 55 Gallon Drums, Drums Separate	EA	1.125	ULABA	0.273	30.11	0.00	0.00	8.22
33 19 0131	**Load Bulk Solid Waste Into 10 Gallon Drum**	**EA**				**5.48**	**0.00**	**0.00**	**5.48**
02083 6142	Bulk Solids into 55 Gallon Drums, Drums Separate	EA	1.125	ULABA	0.182	30.11	0.00	0.00	5.48
33 19 0132	**Load Bulk Liquid/Sludge into HazMax**	**EA**				**31.11**	**0.00**	**0.00**	**31.11**
02083 6141	Bulk Liquid/Sludge into 55 Gallon Drums, Drums Separate	EA	4	ULABA	3.673	8.47	0.00	0.00	31.11
33 19 0133	**Load Bulk Solid Waste into HazMax**	**EA**				**110.60**	**0.00**	**0.00**	**110.60**
02083 6142	Bulk Solids into 55 Gallon Drums, Drums Separate	EA	1.125	ULABA	3.673	30.11	0.00	0.00	110.60
33 19 0150	**Bulk Solid Hazardous Waste Loading Into Truck**	**CY**				**0.66**	**1.21**	**0.00**	**1.87**
02225 2332	Excavate & Load, 1-1/2 CY Wheel Loader, Medium Material 75 CY/Hour (57 m3)	CY	72.5	CODLC	1.000	0.66	1.21	0.00	1.87

33 19 02	Transportation to Storage/Disposal Facility	Unit	Hourly Output	Crew	Factor	Labor	Equip.	Matl.	Total
33 19 0201	**Drummed Waste, Minimum Charge for Shipment**	**EA**				**0.00**	**0.00**	**2,200**	**2,200**
13277 2632	Minimum Charges for Shipment	EA	0	N/A	1.000	0.00	0.00	2,200	2,200

33 19 02 Transportation to Storage/Disposal Facility		Unit	Hourly Output	Crew	Factor	Labor	Equip.	Matl.	Total
33 19 0202	**Bulk Hazardous Waste, Minimum Charge for Shipment**	**EA**				**0.00**	**0.00**	**2,200**	**2,200**
13277 2651	Minimum Charges for Shipment	EA	0	N/A	1.000	0.00	0.00	2,200	2,200
33 19 0203	**PCB Hazardous Waste, Minimum Charge for Shipment**	**EA**				**0.00**	**0.00**	**2,475**	**2,475**
13277 2744	Minimum Charge for Each Shipment	EA	0	N/A	1.000	0.00	0.00	2,475	2,475
33 19 0204	**Transport 55 Gallon Drums of Hazardous Waste, Max 80 drums (per Mile)**	**MI**				**0.00**	**0.00**	**1.71**	**1.71**
02083 7301	Transport 80 55-Gallon Drums of Solid Hazardous Waste	MI	0	N/A	1.000	0.00	0.00	1.71	1.71
33 19 0205	**Transport Bulk Solid Hazardous Waste, Maximum 20 CY (per Mile)**	**MI**				**0.00**	**0.00**	**1.65**	**1.65**
02083 7302	Transport 20 CY Bulk Solid Hazardous Waste	MI	0	N/A	1.000	0.00	0.00	1.65	1.65
33 19 0206	**Transport Bulk Solid Hazardous Waste, Maximum 18 Ton (per Mile)**	**MI**				**0.00**	**0.00**	**1.65**	**1.65**
02083 7303	Transport 18 Ton Bulk Solid Hazardous Waste	MI	0	N/A	1.000	0.00	0.00	1.65	1.65
33 19 0207	**Transport Bulk Liquid/Sludge Hazardous Waste, Maximum 5,000 Gallon (per Mile)**	**MI**				**0.00**	**0.00**	**1.71**	**1.71**
02083 7304	Transport 5,000 Gallon Bulk Sludge/Liquid Hazardous Waste	MI	0	N/A	1.000	0.00	0.00	1.71	1.71
33 19 0208	**Haul 6,000 Gallon Wastewater, Includes Loading & Unloading**	**MI**				**4.19**	**4.44**	**6.40**	**15.03**
13278 1261	Transport/Load/Unload 6,000 Gallon of Wastewater	MI	12.5	COFWK	1.000	4.19	4.44	6.40	15.03
33 19 0209	**Dump Truck Transportation Hazardous Waste Minimum Charge**	**EA**				**0.00**	**0.00**	**735.00**	**735.00**
13278 8111	Dump Truck Transportation Hazardous Waste Minimum Charge	EA	0	N/A	1.000	0.00	0.00	735.00	735.00
33 19 0210	**Dump Truck Transportation Hazardous Waste 200 - 299 Miles**	**MI**				**0.00**	**0.00**	**2.32**	**2.32**
13278 8112	Dump Truck Transportation Hazardous Waste 200 - 299 Miles	MI	0	N/A	1.000	0.00	0.00	2.32	2.32
33 19 0211	**Dump Truck Transportation Hazardous Waste 300 - 399 Miles**	**MI**				**0.00**	**0.00**	**2.35**	**2.35**
13278 8113	Dump Truck Transportation Hazardous Waste 300 - 399 Miles	MI	0	N/A	1.000	0.00	0.00	2.35	2.35
33 19 0212	**Dump Truck Transportation Hazardous Waste 400 - 499 Miles**	**MI**				**0.00**	**0.00**	**2.22**	**2.22**
13278 8114	Dump Truck Transportation Hazardous Waste 400 - 499 Miles	MI	0	N/A	1.000	0.00	0.00	2.22	2.22

33 19 02	Transportation to Storage/Disposal Facility	Unit	Hourly Output	Crew	Factor	Labor	Equip.	Matl.	Total
33 19 0213	**Dump Truck Transportation Hazardous Waste 500 - 599 Miles**	**MI**				**0.00**	**0.00**	**2.14**	**2.14**
13278 8115	Dump Truck Transportation Hazardous Waste 500 - 599 Miles	MI	0	N/A	1.000	0.00	0.00	2.14	2.14
33 19 0214	**Dump Truck Transportation Hazardous Waste 600 - 699 Miles**	**MI**				**0.00**	**0.00**	**2.12**	**2.12**
13278 8116	Dump Truck Transportation Hazardous Waste 600 - 699 Miles	MI	0	N/A	1.000	0.00	0.00	2.12	2.12
33 19 0215	**Dump Truck Transportation Hazardous Waste 700 - 799 Miles**	**MI**				**0.00**	**0.00**	**2.05**	**2.05**
13278 8117	Dump Truck Transportation Hazardous Waste 700 - 799 Miles	MI	0	N/A	1.000	0.00	0.00	2.05	2.05
33 19 0216	**Dump Truck Transportation Hazardous Waste 800 - 899 Miles**	**MI**				**0.00**	**0.00**	**2.02**	**2.02**
13278 8118	Dump Truck Transportation Hazardous Waste 800 - 899 Miles	MI	0	N/A	1.000	0.00	0.00	2.02	2.02
33 19 0217	**Dump Truck Transportation Hazardous Waste 900 - 999 Miles**	**MI**				**0.00**	**0.00**	**2.00**	**2.00**
13278 8119	Dump Truck Transportation Hazardous Waste 900 - 999 Miles	MI	0	N/A	1.000	0.00	0.00	2.00	2.00
33 19 0218	**Dump Truck Transportation Hazardous Waste 1000+ Miles**	**MI**				**0.00**	**0.00**	**1.97**	**1.97**
13278 8121	Dump Truck Transportation Hazardous Waste 1,000+ Miles	MI	0	N/A	1.000	0.00	0.00	1.97	1.97
33 19 0219	**Dump Truck Transportation Explosive Waste Minimum Charge**	**EA**				**0.00**	**0.00**	**760.00**	**760.00**
13278 8211	Dump Truck Transportation Explosive Waste Minimum Charge	EA	0	N/A	1.000	0.00	0.00	760.00	760.00
33 19 0220	**Dump Truck Transportation Explosive Waste 200 - 299 Miles**	**MI**				**0.00**	**0.00**	**2.75**	**2.75**
13278 8212	Dump Truck Transportation Explosive Waste 200 - 299 Miles	MI	0	N/A	1.000	0.00	0.00	2.75	2.75
33 19 0221	**Dump Truck Transportation Explosive Waste 300 - 399 Miles**	**MI**				**0.00**	**0.00**	**2.50**	**2.50**
13278 8213	Dump Truck Transportation Explosive Waste 300 - 399 Miles	MI	0	N/A	1.000	0.00	0.00	2.50	2.50
33 19 0222	**Dump Truck Transportation Explosive Waste 400 - 499 Miles**	**MI**				**0.00**	**0.00**	**2.28**	**2.28**
13278 8214	Dump Truck Transportation Explosive Waste 400 - 499 Miles	MI	0	N/A	1.000	0.00	0.00	2.28	2.28
33 19 0223	**Dump Truck Transportation Explosive Waste 500 - 599 Miles**	**MI**				**0.00**	**0.00**	**2.20**	**2.20**
13278 8215	Dump Truck Transportation Explosive Waste 500 - 599 Miles	MI	0	N/A	1.000	0.00	0.00	2.20	2.20

33 19 02 Transportation to Storage/Disposal Facility		Unit	Hourly Output	Crew	Factor	Labor	Equip.	Matl.	Total
33 19 0224	**Dump Truck Transportation Explosive Waste 600 - 699 Miles**	**MI**				**0.00**	**0.00**	**2.18**	**2.18**
13278 8216	Dump Truck Transportation Explosive Waste 600 - 699 Miles	MI	0	N/A	1.000	0.00	0.00	2.18	2.18
33 19 0225	**Dump Truck Transportation Explosive Waste 700 - 799 Miles**	**MI**				**0.00**	**0.00**	**2.11**	**2.11**
13278 8217	Dump Truck Transportation Explosive Waste 700 - 799 Miles	MI	0	N/A	1.000	0.00	0.00	2.11	2.11
33 19 0226	**Dump Truck Transportation Explosive Waste 800 - 899 Miles**	**MI**				**0.00**	**0.00**	**2.08**	**2.08**
13278 8218	Dump Truck Transportation Explosive Waste 800 - 899 Miles	MI	0	N/A	1.000	0.00	0.00	2.08	2.08
33 19 0227	**Dump Truck Transportation Explosive Waste 900 - 999 Miles**	**MI**				**0.00**	**0.00**	**2.06**	**2.06**
13278 8219	Dump Truck Transportation Explosive Waste 900 - 999 Miles	MI	0	N/A	1.000	0.00	0.00	2.06	2.06
33 19 0228	**Dump Truck Transportation Explosive Waste 1,000+ Miles**	**MI**				**0.00**	**0.00**	**2.03**	**2.03**
13278 8221	Dump Truck Transportation Explosive Waste 1,000+ Miles	MI	0	N/A	1.000	0.00	0.00	2.03	2.03
33 19 0229	**Van Trailer Transportation Hazardous Waste Minimum Charge**	**EA**				**0.00**	**0.00**	**745.00**	**745.00**
13278 8311	Van Trailer Transportation Hazardous Waste Minimum Charge	EA	0	N/A	1.000	0.00	0.00	745.00	745.00
33 19 0230	**Van Trailer Transportation Hazardous Waste 200 - 299 Miles**	**MI**				**0.00**	**0.00**	**2.69**	**2.69**
13278 8312	Van Trailer Transportation Hazardous Waste 200 - 299 Miles	MI	0	N/A	1.000	0.00	0.00	2.69	2.69
33 19 0231	**Van Trailer Transportation Hazardous Waste 300 - 399 Miles**	**MI**				**0.00**	**0.00**	**2.49**	**2.49**
13278 8313	Van Trailer Transportation Hazardous Waste 300 - 399 Miles	MI	0	N/A	1.000	0.00	0.00	2.49	2.49
33 19 0232	**Van Trailer Transportation Hazardous Waste 400 - 499 Miles**	**MI**				**0.00**	**0.00**	**2.25**	**2.25**
13278 8314	Van Trailer Transportation Hazardous Waste 400 - 499 Miles	MI	0	N/A	1.000	0.00	0.00	2.25	2.25
33 19 0233	**Van Trailer Transportation Hazardous Waste 500 - 599 Miles**	**MI**				**0.00**	**0.00**	**2.13**	**2.13**
13278 8315	Van Trailer Transportation Hazardous Waste 500 - 599 Miles	MI	0	N/A	1.000	0.00	0.00	2.13	2.13
33 19 0234	**Van Trailer Transportation Hazardous Waste 600 - 699 Miles**	**MI**				**0.00**	**0.00**	**2.09**	**2.09**
13278 8316	Van Trailer Transportation Hazardous Waste 600 - 699 Miles	MI	0	N/A	1.000	0.00	0.00	2.09	2.09

33 19 02 Transportation to Storage/Disposal Facility		Unit	Hourly Output	Crew	Factor	Labor	Equip.	Matl.	Total
33 19 0235	**Van Trailer Transportation Hazardous Waste 700 - 799 Miles**	**MI**				**0.00**	**0.00**	**2.01**	**2.01**
13278 8317	Van Trailer Transportation Hazardous Waste 700 - 799 Miles	MI	0	N/A	1.000	0.00	0.00	2.01	2.01
33 19 0236	**Van Trailer Transportation Hazardous Waste 800 - 899 Miles**	**MI**				**0.00**	**0.00**	**2.01**	**2.01**
13278 8318	Van Trailer Transportation Hazardous Waste 800 - 899 Miles	MI	0	N/A	1.000	0.00	0.00	2.01	2.01
33 19 0237	**Van Trailer Transportation Hazardous Waste 900 - 999 Miles**	**MI**				**0.00**	**0.00**	**2.19**	**2.19**
13278 8319	Van Trailer Transportation Hazardous Waste 900 - 999 Miles	MI	0	N/A	1.000	0.00	0.00	2.19	2.19
33 19 0238	**Van Trailer Transportation Hazardous Waste 1,000 - 1,099 Miles**	**MI**				**0.00**	**0.00**	**2.10**	**2.10**
13278 8321	Van Trailer Transportation Hazardous Waste 1,000 - 1,099 Miles	MI	0	N/A	1.000	0.00	0.00	2.10	2.10
33 19 0239	**Van Trailer Transportation Hazardous Waste 1,100 - 1,199 Miles**	**MI**				**0.00**	**0.00**	**2.10**	**2.10**
13278 8322	Van Trailer Transportation Hazardous Waste 1,100 - 1,199 Miles	MI	0	N/A	1.000	0.00	0.00	2.10	2.10
33 19 0240	**Van Trailer Transportation Hazardous Waste 1,200+ Miles**	**MI**				**0.00**	**0.00**	**2.10**	**2.10**
13278 8323	Van Trailer Transportation Hazardous Waste 1,200+ Miles	MI	0	N/A	1.000	0.00	0.00	2.10	2.10
33 19 0241	**Van Trailer Transportation Explosive Waste Minimum Charge**	**EA**				**0.00**	**0.00**	**1,275**	**1,275**
13278 8411	Van Trailer Transportation Explosive Waste Minimum Charge	EA	0	N/A	1.000	0.00	0.00	1,275	1,275
33 19 0242	**Van Trailer Transportation Explosive Waste 200 - 299 Miles**	**MI**				**0.00**	**0.00**	**2.32**	**2.32**
13278 8412	Van Trailer Transportation Explosive Waste 200 - 299 Miles	MI	0	N/A	1.000	0.00	0.00	2.32	2.32
33 19 0243	**Van Trailer Transportation Explosive Waste 300 - 399 Miles**	**MI**				**0.00**	**0.00**	**2.26**	**2.26**
13278 8413	Van Trailer Transportation Explosive Waste 300 - 399 Miles	MI	0	N/A	1.000	0.00	0.00	2.26	2.26
33 19 0244	**Van Trailer Transportation Explosive Waste 400 - 499 Miles**	**MI**				**0.00**	**0.00**	**2.32**	**2.32**
13278 8414	Van Trailer Transportation Explosive Waste 400 - 499 Miles	MI	0	N/A	1.000	0.00	0.00	2.32	2.32
33 19 0245	**Van Trailer Transportation Explosive Waste 500 - 599 Miles**	**MI**				**0.00**	**0.00**	**2.15**	**2.15**
13278 8415	Van Trailer Transportation Explosive Waste 500 - 599 Miles	MI	0	N/A	1.000	0.00	0.00	2.15	2.15

33 19 02 Transportation to Storage/Disposal Facility		Unit	Hourly Output	Crew	Factor	Labor	Equip.	Matl.	Total
33 19 0246	**Van Trailer Transportation Explosive Waste 600 - 699 Miles**	**MI**				**0.00**	**0.00**	**3.96**	**3.96**
13278 8416	Van Trailer Transportation Explosive Waste 600 - 699 Miles	MI	0	N/A	1.000	0.00	0.00	3.96	3.96
33 19 0247	**Van Trailer Transportation Explosive Waste 700 - 799 Miles**	**MI**				**0.00**	**0.00**	**3.96**	**3.96**
13278 8417	Van Trailer Transportation Explosive Waste 700 - 799 Miles	MI	0	N/A	1.000	0.00	0.00	3.96	3.96
33 19 0248	**Van Trailer Transportation Explosive Waste 800 - 899 Miles**	**MI**				**0.00**	**0.00**	**1.92**	**1.92**
13278 8418	Van Trailer Transportation Explosive Waste 800 - 899 Miles	MI	0	N/A	1.000	0.00	0.00	1.92	1.92
33 19 0249	**Van Trailer Transportation Explosive Waste 900 - 999 Miles**	**MI**				**0.00**	**0.00**	**3.96**	**3.96**
13278 8419	Van Trailer Transportation Explosive Waste 900 - 999 Miles	MI	0	N/A	1.000	0.00	0.00	3.96	3.96
33 19 0250	**Van Trailer Transportation Explosive Waste 1,000 - 1,099 Miles**	**MI**				**0.00**	**0.00**	**1.81**	**1.81**
13278 8421	Van Trailer Transportation Explosive Waste 1,000 - 1,099 Miles	MI	0	N/A	1.000	0.00	0.00	1.81	1.81
33 19 0251	**Van Trailer Transportation Explosive Waste 1,100 - 1,199 Miles**	**MI**				**0.00**	**0.00**	**1.76**	**1.76**
13278 8422	Van Trailer Transportation Explosive Waste 1,100 - 1,199 Miles	MI	0	N/A	1.000	0.00	0.00	1.76	1.76
33 19 0252	**Van Trailer Transportation Explosive Waste 1,200+ Miles**	**MI**				**0.00**	**0.00**	**1.71**	**1.71**
13278 8423	Van Trailer Transportation Explosive Waste 1,200+ Miles	MI	0	N/A	1.000	0.00	0.00	1.71	1.71
33 19 0253	**Tanker Trailer Transport Hazardous Waste Minimum Charge**	**EA**				**0.00**	**0.00**	**765.00**	**765.00**
13278 8511	Tanker Trailer Transport Hazardous Waste Minimum Charge	EA	0	N/A	1.000	0.00	0.00	765.00	765.00
33 19 0254	**Tanker Trailer Transport Hazardous Waste 200 - 299 Miles**	**MI**				**0.00**	**0.00**	**2.73**	**2.73**
13278 8512	Tanker Trailer Transport Hazardous Waste 200 - 299 Miles	MI	0	N/A	1.000	0.00	0.00	2.73	2.73
33 19 0255	**Tanker Trailer Transport Hazardous Waste 300 - 399 Miles**	**MI**				**0.00**	**0.00**	**2.36**	**2.36**
13278 8513	Tanker Trailer Transport Hazardous Waste 300 - 399 Miles	MI	0	N/A	1.000	0.00	0.00	2.36	2.36
33 19 0256	**Tanker Trailer Transport Hazardous Waste 400 - 499 Miles**	**MI**				**0.00**	**0.00**	**2.40**	**2.40**
13278 8514	Tanker Trailer Transport Hazardous Waste 400 - 499 Miles	MI	0	N/A	1.000	0.00	0.00	2.40	2.40

33 19 02	Transportation to Storage/Disposal Facility	Unit	Hourly Output	Crew	Factor	Labor	Equip.	Matl.	Total
33 19 0257	**Tanker Trailer Transport Hazardous Waste 500 - 599 Miles**	**MI**				**0.00**	**0.00**	**2.46**	**2.46**
13278 8515	Tanker Trailer Transport Hazardous Waste 500 - 599 Miles	MI	0	N/A	1.000	0.00	0.00	2.46	2.46
33 19 0258	**Tanker Trailer Transport Hazardous Waste 600 - 699 Miles**	**MI**				**0.00**	**0.00**	**2.32**	**2.32**
13278 8516	Tanker Trailer Transport Hazardous Waste 600 - 699 Miles	MI	0	N/A	1.000	0.00	0.00	2.32	2.32
33 19 0259	**Tanker Trailer Transport Hazardous Waste 700 - 799 Miles**	**MI**				**0.00**	**0.00**	**2.27**	**2.27**
13278 8517	Tanker Trailer Transport Hazardous Waste 700 - 799 Miles	MI	0	N/A	1.000	0.00	0.00	2.27	2.27
33 19 0260	**Tanker Trailer Transport Hazardous Waste 800 - 899 Miles**	**MI**				**0.00**	**0.00**	**2.24**	**2.24**
13278 8518	Tanker Trailer Transport Hazardous Waste 800 - 899 Miles	MI	0	N/A	1.000	0.00	0.00	2.24	2.24
33 19 0261	**Tanker Trailer Transport Hazardous Waste 900 - 999 Miles**	**MI**				**0.00**	**0.00**	**2.23**	**2.23**
13278 8519	Tanker Trailer Transport Hazardous Waste 900 - 999 Miles	MI	0	N/A	1.000	0.00	0.00	2.23	2.23
33 19 0262	**Tanker Trailer Transport Hazardous Waste 1,000+ Miles**	**MI**				**0.00**	**0.00**	**2.20**	**2.20**
13278 8521	Tanker Trailer Transport Hazardous Waste 1,000+ Miles	MI	0	N/A	1.000	0.00	0.00	2.20	2.20
33 19 0263	**Tanker Trailer Transport Explosive Waste Minimum Charge**	**EA**				**0.00**	**0.00**	**775.00**	**775.00**
13278 8611	Tanker Trailer Transport Explosive Waste Minimum Charge	EA	0	N/A	1.000	0.00	0.00	775.00	775.00
33 19 0264	**Tanker Trailer Transport Explosive Waste 200 - 299 Miles**	**MI**				**0.00**	**0.00**	**2.86**	**2.86**
13278 8612	Tanker Trailer Transport Explosive Waste 200 - 299 Miles	MI	0	N/A	1.000	0.00	0.00	2.86	2.86
33 19 0265	**Tanker Trailer Transport Explosive Waste 300 - 399 Miles**	**MI**				**0.00**	**0.00**	**2.75**	**2.75**
13278 8613	Tanker Trailer Transport Explosive Waste 300 - 399 Miles	MI	0	N/A	1.000	0.00	0.00	2.75	2.75
33 19 0266	**Tanker Trailer Transport Explosive Waste 400 - 499 Miles**	**MI**				**0.00**	**0.00**	**2.69**	**2.69**
13278 8614	Tanker Trailer Transport Explosive Waste 400 - 499 Miles	MI	0	N/A	1.000	0.00	0.00	2.69	2.69
33 19 0267	**Tanker Trailer Transport Explosive Waste 500 - 599 Miles**	**MI**				**0.00**	**0.00**	**2.65**	**2.65**
13278 8615	Tanker Trailer Transport Explosive Waste 500 - 599 Miles	MI	0	N/A	1.000	0.00	0.00	2.65	2.65

33 19 02	Transportation to Storage/Disposal Facility	Unit	Hourly Output	Crew	Factor	Labor	Equip.	Matl.	Total
33 19 0268	**Tanker Trailer Transport Explosive Waste 600 - 699 Miles**	**MI**				**0.00**	**0.00**	**2.64**	**2.64**
13278 8616	Tanker Trailer Transport Explosive Waste 600 - 699 Miles	MI	0	N/A	1.000	0.00	0.00	2.64	2.64
33 19 0269	**Tanker Trailer Transport Explosive Waste 700 - 799 Miles**	**MI**				**0.00**	**0.00**	**2.60**	**2.60**
13278 8617	Tanker Trailer Transport Explosive Waste 700 - 799 Miles	MI	0	N/A	1.000	0.00	0.00	2.60	2.60
33 19 0270	**Tanker Trailer Transport Explosive Waste 800 - 899 Miles**	**MI**				**0.00**	**0.00**	**2.28**	**2.28**
13278 8618	Tanker Trailer Transport Explosive Waste 800 - 899 Miles	MI	0	N/A	1.000	0.00	0.00	2.28	2.28
33 19 0271	**Tanker Trailer Transport Explosive Waste 900 - 999 Miles**	**MI**				**0.00**	**0.00**	**2.58**	**2.58**
13278 8619	Tanker Trailer Transport Explosive Waste 900 - 999 Miles	MI	0	N/A	1.000	0.00	0.00	2.58	2.58
33 19 0272	**Tanker Trailer Transport Explosive Waste 1,000+ Miles**	**MI**				**0.00**	**0.00**	**2.56**	**2.56**
13278 8621	Tanker Trailer Transport Explosive Waste 1,000+ Miles	MI	0	N/A	1.000	0.00	0.00	2.56	2.56
33 19 0273	**Vacuum Trailer Transport Hazardous Waste Minimum Charge**	**EA**				**0.00**	**0.00**	**830.00**	**830.00**
13278 8711	Vacuum Trailer Transport Hazardous Waste Minimum Charge	EA	0	N/A	1.000	0.00	0.00	830.00	830.00
33 19 0274	**Vacuum Trailer Transport Hazardous Waste 200 - 299 Miles**	**MI**				**0.00**	**0.00**	**3.19**	**3.19**
13278 8712	Vacuum Trailer Transport Hazardous Waste 200 - 299 Miles	MI	0	N/A	1.000	0.00	0.00	3.19	3.19
33 19 0275	**Vacuum Trailer Transport Hazardous Waste 300 - 399 Miles**	**MI**				**0.00**	**0.00**	**3.19**	**3.19**
13278 8713	Vacuum Trailer Transport Hazardous Waste 300 - 399 Miles	MI	0	N/A	1.000	0.00	0.00	3.19	3.19
33 19 0276	**Vacuum Trailer Transport Hazardous Waste 400 - 899 Miles**	**MI**				**0.00**	**0.00**	**3.19**	**3.19**
13278 8714	Vacuum Trailer Transport Hazardous Waste 400 - 899 Miles	MI	0	N/A	1.000	0.00	0.00	3.19	3.19
33 19 0277	**Vacuum Trailer Transport Hazardous Waste 900+ Miles**	**MI**				**0.00**	**0.00**	**2.94**	**2.94**
13278 8715	Vacuum Trailer Transport Hazardous Waste 900+ Miles	MI	0	N/A	1.000	0.00	0.00	2.94	2.94
33 19 0278	**Vacuum Trailer Transport Explosive Waste Minimum Charge**	**EA**				**0.00**	**0.00**	**820.00**	**820.00**
13278 8811	Vacuum Trailer Transport Explosive Waste Minimum Charge	EA	0	N/A	1.000	0.00	0.00	820.00	820.00

33 19 02	Transportation to Storage/Disposal Facility	Unit	Hourly Output	Crew	Factor	Labor	Equip.	Matl.	Total
33 19 0279	**Vacuum Trailer Transport Explosive Waste 200 - 299 Miles**	**MI**				**0.00**	**0.00**	**2.94**	**2.94**
13278 8812	Vacuum Trailer Transport Explosive Waste 200 - 299 Miles	MI	0	N/A	1.000	0.00	0.00	2.94	2.94
33 19 0280	**Vacuum Trailer Transport Explosive Waste 300 - 399 Miles**	**MI**				**0.00**	**0.00**	**2.94**	**2.94**
13278 8813	Vacuum Trailer Transport Explosive Waste 300 - 399 Miles	MI	0	N/A	1.000	0.00	0.00	2.94	2.94
33 19 0281	**Vacuum Trailer Transport Explosive Waste 400 - 899 Miles**	**MI**				**0.00**	**0.00**	**2.94**	**2.94**
13278 8814	Vacuum Trailer Transport Explosive Waste 400 - 899 Miles	MI	0	N/A	1.000	0.00	0.00	2.94	2.94
33 19 0282	**Vacuum Trailer Transport Explosive Waste 900+Miles**	**MI**				**0.00**	**0.00**	**2.94**	**2.94**
13278 8815	Vacuum Trailer Transport Explosive Waste 900+ Miles	MI	0	N/A	1.000	0.00	0.00	2.94	2.94
33 19 0283	**Radioactive Truck Haul Over 500 Miles**	**MI**				**0.00**	**0.00**	**3.87**	**3.87**
02083 8722	Loaded Container, 42,000 Lbs 500 Miles - Over	MI	0	N/A	1.000	0.00	0.00	3.87	3.87
33 19 0284	**Radioactive Truck Haul 251 - 500 Miles**	**MI**				**0.00**	**0.00**	**4.66**	**4.66**
02083 8732	Loaded Container, 42,000 Lbs 251 - 500 Miles	MI	0	N/A	1.000	0.00	0.00	4.66	4.66
33 19 0285	**Radioactive Truck Haul 101 - 250 Miles**	**MI**				**0.00**	**0.00**	**2.97**	**2.97**
02083 8742	Loaded Container, 42,000 Lbs 101 - 250 Miles	MI	0	N/A	1.000	0.00	0.00	2.97	2.97
33 19 0286	**Radioactive Truck Haul 0 - 100 Miles**	**MI**				**0.00**	**0.00**	**1,311**	**1,311**
02085 8701	Radioactive Truck Haul 0 - 100 Miles, Minimum Charge	EA	0	N/A	1.000	0.00	0.00	1,311	1,311
33 19 0287	**Rail Boxcar Transport**	**CWT**				**0.00**	**0.00**	**3.16**	**3.16**
13278 9101	Rail Box Car Transport	CWT	0	N/A	1.000	0.00	0.00	3.16	3.16
33 19 0288	**Rail Tanker Transport**	**CWT**				**0.00**	**0.00**	**4.22**	**4.22**
13278 9201	Rail Tanker Transport	CWT	0	N/A	1.000	0.00	0.00	4.22	4.22
33 19 0289	**Rail Gondola Transport**	**CWT**				**0.00**	**0.00**	**5.25**	**5.25**
13278 9301	Rail Gondola Transport	CWT	0	N/A	1.000	0.00	0.00	5.25	5.25
33 19 0290	**Rail Flatbed Transport**	**CWT**				**0.00**	**0.00**	**6.35**	**6.35**
13278 9401	Rail Flatbed Transport	CWT	0	N/A	1.000	0.00	0.00	6.35	6.35

33 19 03	Disposal Fees and Taxes	Unit	Hourly Output	Crew	Factor	Labor	Equip.	Matl.	Total
33 19 0301	**Extra Charges for 85 Gallon Overpack**	**EA**				**0.00**	**0.00**	**61.00**	**61.00**
13278 1177	Extra Charges for 85 Gallon Overpack	EA	0	N/A	1.000	0.00	0.00	61.00	61.00

33 19 03 Disposal Fees and Taxes		Unit	Hourly Output	Crew	Factor	Labor	Equip.	Matl.	Total
33 19 0302	**Extra Charges for Leaking Drums**	**EA**				**0.00**	**0.00**	**204.00**	**204.00**
13278 1178	Extra Charges for Leaking Drums	EA	0	N/A	1.000	0.00	0.00	204.00	204.00
33 19 0303	**Minimum Charges for Drummed Shipments**	**EA**				**0.00**	**0.00**	**256.45**	**256.45**
13278 1179	Minimum Charges for Drummed Shipments	EA	0	N/A	1.000	0.00	0.00	256.45	256.45
33 19 0304	**Extra Charges for Drums < 90% Full, < 30 Gallon Drum**	**EA**				**0.00**	**0.00**	**22.50**	**22.50**
13278 1181	Extra Charges for Drums Less than 90% Full, Drums Less than 55 Gallons	EA	0	N/A	1.000	0.00	0.00	22.50	22.50
33 19 0305	**Extra Charges for Drums < 90% Full, 55 Gallon Drum**	**EA**				**0.00**	**0.00**	**24.00**	**24.00**
13278 1182	Extra Charges for Drums Less than 90% Full, 55 Gallon Drums	EA	0	N/A	1.000	0.00	0.00	24.00	24.00
33 19 0306	**Extra Charges for Drums < 90% Full, > 55 Gallon Drum**	**EA**				**0.00**	**0.00**	**23.00**	**23.00**
13278 1183	Extra Charges for Drums Less than 90% Full, Drums Greater than 55 Gallons	EA	0	N/A	1.000	0.00	0.00	23.00	23.00
33 19 0307	**Minimum Charges for Bulk Shipments**	**EA**				**0.00**	**0.00**	**2,000**	**2,000**
13278 1229	Minimum Charges for Bulk Shipments	EA	0	N/A	1.000	0.00	0.00	2,000	2,000
33 19 0308	**Minimum Charges for Bulk Shipments Requiring Treatment or Stabilization**	**EA**				**0.00**	**0.00**	**2,275**	**2,275**
13278 1231	Minimum Charges for Bulk Shipments Requiring Treatment or Stabilization	EA	0	N/A	1.000	0.00	0.00	2,275	2,275
33 19 0309	**Minimum Charges on Nonfuel Bulk Liquid Shipments**	**EA**				**0.00**	**0.00**	**2,375**	**2,375**
13278 1232	Minimum Charges on Non-fuel Bulk Liquid Shipments	EA	0	N/A	1.000	0.00	0.00	2,375	2,375
33 19 0310	**Extra Charges for Leaking Loads, per Drum**	**EA**				**0.00**	**0.00**	**305.00**	**305.00**
13278 1233	Extra Charges for Leaking Loads, per Drum	EA	0	N/A	1.000	0.00	0.00	305.00	305.00
33 19 0311	**Truck Washout/Decontamination**	**EA**				**0.00**	**0.00**	**158.00**	**158.00**
13278 1234	Truck Washout	EA	0	N/A	1.000	0.00	0.00	158.00	158.00
33 19 0312	**Land Disposal Tax, In-State Waste, Nevada**	**TON**				**0.00**	**0.00**	**105.00**	**105.00**
13278 5111	Land Disposal Tax, In-State Wastes, State of Nevada	TON	0	N/A	1.000	0.00	0.00	105.00	105.00
33 19 0313	**Land Disposal Tax, Out-of-State Waste, Nevada**	**TON**				**0.00**	**0.00**	**127.00**	**127.00**
13278 5112	Land Disposal Tax, Out-of-State Wastes, Higher of Origin or Host	TON	0	N/A	1.000	0.00	0.00	127.00	127.00
33 19 0314	**Deep Well Injection Tax, In-State Waste, Ohio**	**TON**				**0.00**	**0.00**	**2.11**	**2.11**
13278 5115	Deep Well Injection Tax - In-State Waste, State of Ohio	TON	0	N/A	1.000	0.00	0.00	2.11	2.11

33 19 03 Disposal Fees and Taxes		Unit	Hourly Output	Crew	Factor	Labor	Equip.	Matl.	Total
33 19 0315	**Deep Well Injection Tax, Out-of-State Waste, Ohio**	**TON**				**0.00**	**0.00**	**4.75**	**4.75**
13278 5116	Deep Well Injection Tax - Out-of-State Wastes, State of Ohio	TON	0	N/A	1.000	0.00	0.00	4.75	4.75
33 19 0316	**Storage of Hazardous Wastes, South Carolina**	**TON**				**0.00**	**0.00**	**3.70**	**3.70**
13278 5117	Storage of Hazardous Wastes, State of South Carolina	TON	0	N/A	1.000	0.00	0.00	3.70	3.70
33 19 0317	**Waste Stream Evaluation Fee, Not Including 50% Rebate on 1st Shipment**	**EA**				**0.00**	**0.00**	**445.00**	**445.00**
13278 1251	Waste Stream Evaluation Fee Not Including 50% Rebate on First Shipment	EA	0	N/A	1.000	0.00	0.00	445.00	445.00
33 19 0318	**Manifest Discrepancies**	**EA**				**0.00**	**0.00**	**52.50**	**52.50**
13278 1252	Manifest Discrepancies	EA	0	N/A	1.000	0.00	0.00	52.50	52.50
33 19 0319	**Lab Pack Profile Evaluation Fee**	**EA**				**0.00**	**0.00**	**500.00**	**500.00**
13278 1253	Lab Pack Profile Evaluation Fee	EA	0	N/A	1.000	0.00	0.00	500.00	500.00
33 19 0320	**Priority Waste Stream Evaluation Fee**	**EA**				**0.00**	**0.00**	**770.00**	**770.00**
13278 1254	Priority Waste Stream Evaluation Fee	EA	0	N/A	1.000	0.00	0.00	770.00	770.00
33 19 0321	**Disposal of Empty Drum, Metal 55 Gallon RCRA Empty**	**EA**				**0.00**	**0.00**	**50.00**	**50.00**
13278 1255	Disposal of Empty Drums, Metal 55 Gallon RCRA Empty	EA	0	N/A	1.000	0.00	0.00	50.00	50.00
33 19 0322	**Disposal of Empty Drum, Metal 85 Gallon RCRA Empty**	**EA**				**0.00**	**0.00**	**56.50**	**56.50**
13278 1256	Disposal of Empty Drums, Metal 85 Gallon RCRA Empty	EA	0	N/A	1.000	0.00	0.00	56.50	56.50
33 19 0323	**Disposal of Empty Drum, Plastic 55 Gallon RCRA Empty**	**EA**				**0.00**	**0.00**	**58.00**	**58.00**
13278 1257	Disposal of Empty Drums, Plastic 55 Gallon RCRA Empty	EA	0	N/A	1.000	0.00	0.00	58.00	58.00
33 19 0324	**State HTW Disposal Tax/Fee (Bulk Solid)**	**CY**				**0.00**	**0.00**	**106.00**	**106.00**
13278 5101	State HTW Disposal Tax/Fee (Bulk Solid)	CY	0	N/A	1.000	0.00	0.00	106.00	106.00
33 19 0325	**State HTW Disposal Tax/Fee (Bulk Liquid)**	**GAL**				**0.00**	**0.00**	**1.02**	**1.02**
13278 5102	State HTW Disposal Tax/Fee (Bulk Liquid)	GAL	0	N/A	1.000	0.00	0.00	1.02	1.02
33 19 0326	**State HTW Disposal Tax/Fee (Drums/Overpacks)**	**EA**				**0.00**	**0.00**	**29.50**	**29.50**
13278 5103	State HTW Disposal Tax/Fee (Drums)	DRM	0	N/A	1.000	0.00	0.00	29.50	29.50
33 19 0327	**Demurrage Vacuum, Dump, and Tanker**	**HR**				**0.00**	**0.00**	**69.00**	**69.00**
13278 7111	Demurrage Vacuum, Dump, and Tanker (Hourly)	HR	0	N/A	1.000	0.00	0.00	69.00	69.00

33 19 Disposal (Commercial)

33 19 03 Disposal Fees and Taxes		Unit	Hourly Output	Crew	Factor	Labor	Equip.	Matl.	Total
33 19 0328	**Demurrage Van Trailer**	**HR**				**0.00**	**0.00**	**66.00**	**66.00**
13278 7112	Demurrage Van Trailer (Hourly)	HR	0	N/A	1.000	0.00	0.00	66.00	66.00
33 19 0329	**Overnight Demurrage**	**EA**				**0.00**	**0.00**	**525.00**	**525.00**
13278 7113	Overnight Demurrage	EA	0	N/A	1.000	0.00	0.00	525.00	525.00
33 19 0330	**Maximum 24 Hour Demurrage**	**EA**				**0.00**	**0.00**	**720.00**	**720.00**
13278 7114	Maximum 24 Hour Demurrage	EA	0	N/A	1.000	0.00	0.00	720.00	720.00
33 19 0331	**Additional Stops <= 3**	**EA**				**0.00**	**0.00**	**59.50**	**59.50**
13278 7121	Additional Demurrage Stops <= 3	EA	0	N/A	1.000	0.00	0.00	59.50	59.50
33 19 0332	**Additional Stops > 3**	**EA**				**0.00**	**0.00**	**52.50**	**52.50**
13278 7122	Additional Demurrage Stops > 3	EA	0	N/A	1.000	0.00	0.00	52.50	52.50
33 19 0333	**Tanker Trailer Drop (Daily)**	**DAY**				**0.00**	**0.00**	**133.00**	**133.00**
13278 7131	Tanker Trailer Drop (Daily)	DAY	0	N/A	1.000	0.00	0.00	133.00	133.00
33 19 0334	**Tanker Trailer Drop (Monthly)**	**MO**				**0.00**	**0.00**	**193.76**	**193.76**
13278 7132	Tanker Trailer Drop (Monthly)	MO	0	N/A	1.000	0.00	0.00	193.76	193.76
33 19 0335	**Dump Trailer Drop (Daily)**	**DAY**				**0.00**	**0.00**	**15.30**	**15.30**
13278 7133	Dump Trailer Drop (Daily)	DAY	0	N/A	1.000	0.00	0.00	15.30	15.30
33 19 0336	**Dump Trailer Drop (Monthly)**	**MO**				**0.00**	**0.00**	**1,140**	**1,140**
13278 7134	Dump Trailer Drop (Monthly)	MO	0	N/A	1.000	0.00	0.00	1,140	1,140
33 19 0337	**Van Trailer Drop (Daily)**	**DAY**				**0.00**	**0.00**	**53.57**	**53.57**
13278 7135	Van Trailer Drop (Daily)	DAY	0	N/A	1.000	0.00	0.00	53.57	53.57
33 19 0338	**Van Trailer Drop (Monthly)**	**MO**				**0.00**	**0.00**	**810.00**	**810.00**
13278 7136	Van Trailer Drop (Monthly)	MO	0	N/A	1.000	0.00	0.00	810.00	810.00
33 19 0340	**Non Haz Drummed Site Waste - Load, Transp, & Landfill Disp (55-Gal Drums)**	**EA**				**0.00**	**0.00**	**204.30**	**204.30**
01001 0212	Non Haz Drummed Site Waste - Load, Transp, & Landfill Disp (55-Gal Drums)	EA	0	N/A	1.000	0.00	0.00	204.30	204.30
33 19 0341	**Free Product Disposal**	**GAL**				**0.00**	**0.00**	**0.28**	**0.28**
01001 0213	Free Product Disposal	GAL	0	N/A	1.000	0.00	0.00	0.28	0.28

33 19 04 Steel Drums		Unit	Hourly Output	Crew	Factor	Labor	Equip.	Matl.	Total
33 19 0401	**55 Gallon 17C Closed Head Steel Drum**	**EA**				**0.00**	**0.00**	**81.00**	**81.00**
02083 5212	55 Gal, 17C, Closed	EA	0	N/A	1.000	0.00	0.00	81.00	81.00
33 19 0402	**55 Gallon 17H Open Head Steel Drum**	**EA**				**0.00**	**0.00**	**80.00**	**80.00**
02083 5213	55 Gal, 17H, Open Only	EA	0	N/A	1.000	0.00	0.00	80.00	80.00

33 19 04 Steel Drums		Unit	Hourly Output	Crew	Factor	Labor	Equip.	Matl.	Total
33 19 0403	**55 Gallon 17H Closed Head Steel Drum**	**EA**				**0.00**	**0.00**	**67.00**	**67.00**
02083 5214	55 Gal, 17H, Closed Only	EA	0	N/A	1.000	0.00	0.00	67.00	67.00
33 19 0404	**20 Gallon 17C Open Head Steel Drum**	**EA**				**0.00**	**0.00**	**61.50**	**61.50**
02083 5218	20 Gal, 17C, Open	EA	0	N/A	1.000	0.00	0.00	61.50	61.50
33 19 0405	**20 Gallon 17E Open Head Steel Drum**	**EA**				**0.00**	**0.00**	**67.50**	**67.50**
02083 5219	20 Gal, 17E, Open	EA	0	N/A	1.000	0.00	0.00	67.50	67.50
33 19 0406	**55 Gallon Stainless Steel DOT 5C Closed Head Seamless 16 Gauge (1 - 9)**	**EA**				**0.00**	**0.00**	**822.70**	**822.70**
02083 5252	55 Gallon Stainless Steel DOT 5C Closed Head Seamless 16-gauge (1 - 9)	EA	0	N/A	1.000	0.00	0.00	822.70	822.70
33 19 0407	**55 Gallon Stainless Steel DOT 5C Closed Head Seamless 16 Gauge (10 - 24)**	**EA**				**0.00**	**0.00**	**815.00**	**815.00**
02083 5222	55 Gallon Stainless Steel DOT 5C Closed Head Seamless Drum 16-Gauge (10 - 24)	EA	0	N/A	1.000	0.00	0.00	815.00	815.00
33 19 0408	**55 Gallon Stainless Steel DOT 5C Closed Head Seamless 16 Gauge (25 - 49)**	**EA**				**0.00**	**0.00**	**810.00**	**810.00**
02083 5223	55 Gallon Stainless Steel DOT 5C Closed Head Seamless Drum 16-Gauge (25 - 49)	EA	0	N/A	1.000	0.00	0.00	810.00	810.00
33 19 0409	**55 Gallon Stainless Steel DOT 5C Closed Head Seamless 16 Gauge (50+)**	**EA**				**0.00**	**0.00**	**805.00**	**805.00**
02083 5224	55 Gallon Stainless Steel DOT 5C Closed Head Seamless Drum 16-Gauge (50+)	EA	0	N/A	1.000	0.00	0.00	805.00	805.00
33 19 0410	**55 Gallon Stainless Steel DOT 5C Closed Head 16 Gauge (1 - 9)**	**EA**				**0.00**	**0.00**	**805.00**	**805.00**
02083 5225	55 Gallon Stainless Steel DOT 5C Closed Head 16-Gauge Drum (1 - 9)	EA	0	N/A	1.000	0.00	0.00	805.00	805.00
33 19 0411	**55 Gallon Stainless Steel DOT 5C Closed Head 16 Gauge (10 - 24)**	**EA**				**0.00**	**0.00**	**720.00**	**720.00**
02083 5226	55 Gallon Stainless Steel DOT 5C Closed Head 16-Gauge Drum (10 - 24)	EA	0	N/A	1.000	0.00	0.00	720.00	720.00
33 19 0412	**55 Gallon Stainless Steel DOT 5C Closed Head 16 Gauge (25 - 49)**	**EA**				**0.00**	**0.00**	**800.00**	**800.00**
02083 5227	55 Gallon Stainless Steel DOT 5C Closed Head 16-Gauge Drum (25 - 49)	EA	0	N/A	1.000	0.00	0.00	800.00	800.00
33 19 0413	**55 Gallon Stainless Steel DOT 5C Closed Head 16 Gauge (50+)**	**EA**				**0.00**	**0.00**	**800.00**	**800.00**
02083 5228	55 Gallon Stainless Steel DOT 5C Closed Head 16-Gauge Drum (50+)	EA	0	N/A	1.000	0.00	0.00	800.00	800.00
33 19 0414	**55 Gallon Stainless Steel DOT 17E Closed Head 18 Gauge (1 - 9)**	**EA**				**0.00**	**0.00**	**695.00**	**695.00**
02083 5229	55 Gallon Stainless Steel DOT 17E Closed Head 18-Gauge Drum (1 - 9)	EA	0	N/A	1.000	0.00	0.00	695.00	695.00

33 19 04	Steel Drums	Unit	Hourly Output	Crew	Factor	Labor	Equip.	Matl.	Total
33 19 0415	**55 Gallon Stainless Steel DOT 17E Closed Head 18 Gauge (10 - 24)**	**EA**				**0.00**	**0.00**	**620.00**	**620.00**
02083 5231	55 Gallon Stainless Steel DOT 17E Closed Head 18-Gauge Drum (10 - 24)	EA	0	N/A	1.000	0.00	0.00	620.00	620.00
33 19 0416	**55 Gallon Stainless Steel DOT 17E Closed Head 18 Gauge (25 - 49)**	**EA**				**0.00**	**0.00**	**610.00**	**610.00**
02083 5232	55 Gallon Stainless Steel DOT 17E Closed Head 18-Gauge Drum (25 - 49)	EA	0	N/A	1.000	0.00	0.00	610.00	610.00
33 19 0417	**55 Gallon Stainless Steel DOT 17E Closed Head 18 Gauge (50+)**	**EA**				**0.00**	**0.00**	**610.00**	**610.00**
02083 5233	55 Gallon Stainless Steel DOT 17E Closed Head 18-Gauge Drum (50+)	EA	0	N/A	1.000	0.00	0.00	610.00	610.00
33 19 0418	**30 Gallon Stainless Steel DOT 5C Closed Head 16 Gauge (1 - 9)**	**EA**				**0.00**	**0.00**	**690.00**	**690.00**
02083 5234	30 Gallon Stainless Steel DOT 5C Closed Head 16-Gauge Drum (1 - 9)	EA	0	N/A	1.000	0.00	0.00	690.00	690.00
33 19 0419	**30 Gallon Stainless Steel DOT 5C Closed Head 16 Gauge (10 - 24)**	**EA**				**0.00**	**0.00**	**645.00**	**645.00**
02083 5235	30 Gallon Stainless Steel DOT 5C Closed Head 16-Gauge Drum (10 - 24)	EA	0	N/A	1.000	0.00	0.00	645.00	645.00
33 19 0420	**30 Gallon Stainless Steel DOT 5C Closed Head 16 Gauge (25 - 49)**	**EA**				**0.00**	**0.00**	**640.00**	**640.00**
02083 5236	30 Gallon Stainless Steel DOT 5C Closed Head 16-Gauge Drum (25 - 49)	EA	0	N/A	1.000	0.00	0.00	640.00	640.00
33 19 0421	**30 Gallon Stainless Steel DOT 5C Closed Head 16 Gauge (50+)**	**EA**				**0.00**	**0.00**	**640.00**	**640.00**
02083 5237	30 Gallon Stainless Steel DOT 5C Closed Head 16-Gauge Drum (50+)	EA	0	N/A	1.000	0.00	0.00	640.00	640.00
33 19 0422	**55 Gallon Stainless Steel DOT 5C Seamless Open Head 16 Gauge (1 - 9)**	**EA**				**0.00**	**0.00**	**845.00**	**845.00**
02083 5238	55 Gallon Stainless Steel DOT 5C Seam-less Open Head Drum 16-Gauge (1 - 9)	EA	0	N/A	1.000	0.00	0.00	845.00	845.00
33 19 0423	**55 Gallon Stainless Steel DOT 5C Seamless Open Head 16 Gauge (10 - 24)**	**EA**				**0.00**	**0.00**	**810.00**	**810.00**
02083 5239	55 Gallon Stainless Steel DOT 5C Seam-less Open Head Drum 16-Gauge (10 - 24)	EA	0	N/A	1.000	0.00	0.00	810.00	810.00
33 19 0424	**55 Gallon Stainless Steel DOT 5C Seamless Open Head 16 Gauge (25 - 49)**	**EA**				**0.00**	**0.00**	**805.00**	**805.00**
02083 5241	55 Gallon Stainless Steel DOT 5C Seam-less Open Head Drum 16-Gauge (25 - 49)	EA	0	N/A	1.000	0.00	0.00	805.00	805.00
33 19 0425	**55 Gallon Stainless Steel DOT 5C Seamless Open Head 16 Gauge (50+)**	**EA**				**0.00**	**0.00**	**795.00**	**795.00**
02083 5242	55 Gallon Stainless Steel DOT 5C Seam-less Open Head Drum 16-Gauge (50+)	EA	0	N/A	1.000	0.00	0.00	795.00	795.00

33 19 04	Steel Drums	Unit	Hourly Output	Crew	Factor	Labor	Equip.	Matl.	Total
33 19 0426	**55 Gallon Stainless Steel DOT 5C Open Head 16 Gauge (1 - 9)**	**EA**				**0.00**	**0.00**	**735.00**	**735.00**
02083 5243	55 Gallon Stainless Steel DOT 5C Open Head 16-Gauge Drum (1 - 9)	EA	0	N/A	1.000	0.00	0.00	735.00	735.00
33 19 0427	**55 Gallon Stainless Steel DOT 5C Open Head 16 Gauge (10 - 24)**	**EA**				**0.00**	**0.00**	**685.00**	**685.00**
02083 5244	55 Gallon Stainless Steel DOT 5C Open Head 16-Gauge Drum (10 - 24)	EA	0	N/A	1.000	0.00	0.00	685.00	685.00
33 19 0428	**55 Gallon Stainless Steel DOT 5C Open Head 16 Gauge (25 - 49)**	**EA**				**0.00**	**0.00**	**685.00**	**685.00**
02083 5245	55 Gallon Stainless Steel DOT 5C Open Head 16-Gauge Drum (25 - 49)	EA	0	N/A	1.000	0.00	0.00	685.00	685.00
33 19 0429	**55 Gallon Stainless Steel DOT 5C Open Head 16 Gauge (50+)**	**EA**				**0.00**	**0.00**	**685.00**	**685.00**
02083 5246	55 Gallon Stainless Steel DOT 5C Open Head 16-Gauge Drum (50+)	EA	0	N/A	1.000	0.00	0.00	685.00	685.00
33 19 0430	**55 Gallon Reconditioned Steel 17H (1 - 9)**	**EA**				**0.00**	**0.00**	**79.00**	**79.00**
02083 5247	55 Gallon Reconditioned Steel 17H Drum (1 - 9)	EA	0	N/A	1.000	0.00	0.00	79.00	79.00
33 19 0431	**56 Gallon Reconditioned Steel 17H (10 - 24)**	**EA**				**0.00**	**0.00**	**71.00**	**71.00**
02083 5248	55 Gallon Reconditioned Steel 17H Drum (10 - 24)	EA	0	N/A	1.000	0.00	0.00	71.00	71.00
33 19 0432	**57 Gallon Reconditioned Steel 17H (25 - 49)**	**EA**				**0.00**	**0.00**	**71.00**	**71.00**
02083 5249	55 Gallon Reconditioned Steel 17H Drum (25 - 49)	EA	0	N/A	1.000	0.00	0.00	71.00	71.00
33 19 0433	**58 Gallon Reconditioned Steel 17H (50+)**	**EA**				**0.00**	**0.00**	**71.00**	**71.00**
02083 5251	55 Gallon Reconditioned Steel 17H Drum (50+)	EA	0	N/A	1.000	0.00	0.00	71.00	71.00
33 19 0434	**110 Gallon Steel Overpack 16 Gauge**	**EA**				**0.00**	**0.00**	**179.00**	**179.00**
02083 5111	110 Gallon, 30" Diameter x 41" High	EA	0	N/A	1.000	0.00	0.00	179.00	179.00
33 19 0435	**85 Gallon Steel Overpack 16 Gauge**	**EA**				**0.00**	**0.00**	**75.77**	**75.77**
02083 5112	85 Gallon, 26" Diameter x 38" High	EA	0	N/A	1.000	0.00	0.00	75.77	75.77
33 19 0436	**83 Gallon Steel Overpack 16 Gauge**	**EA**				**0.00**	**0.00**	**147.00**	**147.00**
02083 5113	83 Gallon, 26" Diameter x 36" High	EA	0	N/A	1.000	0.00	0.00	147.00	147.00

33 19 05	Polyethylene Drums	Unit	Hourly Output	Crew	Factor	Labor	Equip.	Matl.	Total
33 19 0501	**35 Gallon Polyethylene Closed Head**	**EA**				**0.00**	**0.00**	**46.50**	**46.50**
02083 5312	35 Gallon	EA	0	N/A	1.000	0.00	0.00	46.50	46.50
33 19 0502	**30 Gallon Polyethylene Closed Head**	**EA**				**0.00**	**0.00**	**46.00**	**46.00**
02083 5313	30 Gallon	EA	0	N/A	1.000	0.00	0.00	46.00	46.00

33 19 Disposal (Commercial)

33 19 05 Polyethylene Drums		Unit	Hourly Output	Crew	Factor	Labor	Equip.	Matl.	Total
33 19 0503	**20 Gallon Polyethylene Closed Head**	**EA**				**0.00**	**0.00**	**30.50**	**30.50**
02083 5314	20 Gallon	EA	0	N/A	1.000	0.00	0.00	30.50	30.50
33 19 0504	**15 Gallon Polyethylene Closed Head**	**EA**				**0.00**	**0.00**	**28.50**	**28.50**
02083 5315	15 Gallon	EA	0	N/A	1.000	0.00	0.00	28.50	28.50
33 19 0505	**10 Gallon DOT 21C-250 Closed Head PET/Poly Lined**	**EA**				**0.00**	**0.00**	**27.50**	**27.50**
02083 5411	10 Gallon, 20" Diameter, DOT 21C-250 Polyethylene/Polyester Lined	EA	0	N/A	1.000	0.00	0.00	27.50	27.50
33 19 0506	**20 Gallon 20" Diameter DOT 21C-250 Closed Head PET/Poly Lined**	**EA**				**0.00**	**0.00**	**19.15**	**19.15**
02083 5412	20 Gallon, 20" Diameter, DOT 21C-250 Polyethylene/Polyester Lined	EA	0	N/A	1.000	0.00	0.00	19.15	19.15
33 19 0507	**20 Gallon 23" Diameter DOT 21C-250 Closed Head PET/Poly Lined**	**EA**				**0.00**	**0.00**	**31.00**	**31.00**
02083 5413	20 Gallon, 23" Diameter, DOT 21C-250 Polyethylene/Polyester Lined	EA	0	N/A	1.000	0.00	0.00	31.00	31.00
33 19 0508	**35 Gallon Polyethylene Open Head**	**EA**				**0.00**	**0.00**	**56.50**	**56.50**
02083 5322	35 Gallon	EA	0	N/A	1.000	0.00	0.00	56.50	56.50
33 19 0509	**30 Gallon Polyethylene Open Head**	**EA**				**0.00**	**0.00**	**64.50**	**64.50**
02083 5323	30 Gallon	EA	0	N/A	1.000	0.00	0.00	64.50	64.50
33 19 0510	**20 Gallon Polyethylene Open Head**	**EA**				**0.00**	**0.00**	**40.00**	**40.00**
02083 5324	20 Gallon	EA	0	N/A	1.000	0.00	0.00	40.00	40.00
33 19 0511	**15 Gallon Polyethylene Open Head**	**EA**				**0.00**	**0.00**	**29.50**	**29.50**
02083 5325	15 Gallon	EA	0	N/A	1.000	0.00	0.00	29.50	29.50
33 19 0512	**95 Gallon Poly Overpack DOT E9618**	**EA**				**0.00**	**0.00**	**164.00**	**164.00**
02083 5151	95 Gallon, Dot E9618, Twist-on Lid 31" Diameter x 41" High	EA	0	N/A	1.000	0.00	0.00	164.00	164.00
33 19 0513	**85 Gallon Poly Overpack DOT E9775**	**EA**				**0.00**	**0.00**	**160.00**	**160.00**
02083 5153	85 Gallon, DOT E9775, Clamp-ring Lid, 27" Diameter x 41" High	EA	0	N/A	1.000	0.00	0.00	160.00	160.00

33 19 06 Fiberglass/HazMax Drums		Unit	Hourly Output	Crew	Factor	Labor	Equip.	Matl.	Total
33 19 0603	**55 Gallon Fiberglass DOT-21C (1 - 9)**	**EA**				**0.00**	**0.00**	**36.00**	**36.00**
02083 5414	55 Gallon Fiberglass Drum DOT-21C (1 - 9)	EA	0	N/A	1.000	0.00	0.00	36.00	36.00
33 19 0604	**55 Gallon Fiberglass DOT-21C (10 - 24)**	**EA**				**0.00**	**0.00**	**35.00**	**35.00**
02083 5416	55 Gallon Fiberglass Drum DOT-21C (10 - 24)	EA	0	N/A	1.000	0.00	0.00	35.00	35.00
33 19 0605	**55 Gallon Fiberglass DOT-21C (25 - 49)**	**EA**				**0.00**	**0.00**	**30.50**	**30.50**
02083 5417	55 Gallon Fiberglass Drum DOT-21C (25 - 49)	EA	0	N/A	1.000	0.00	0.00	30.50	30.50

33 19 Disposal (Commercial)

33 19 06 Fiberglass/HazMax Drums		Unit	Hourly Output	Crew	Factor	Labor	Equip.	Matl.	Total
33 19 0606	**55 Gallon Fiberglass DOT-21C (50 - 99)**	**EA**				**0.00**	**0.00**	**29.50**	**29.50**
02083 5418	55 Gallon Fiberglass Drum DOT-21C (50 - 99)	EA	0	N/A	1.000	0.00	0.00	29.50	29.50
33 19 0607	**55 Gallon Fiberglass DOT-21C (100+)**	**EA**				**0.00**	**0.00**	**27.50**	**27.50**
02083 5419	55 Gallon Fiberglass Drum DOT-21C (100+)	EA	0	N/A	1.000	0.00	0.00	27.50	27.50
33 19 0608	**30 Gallon Fiberglass DOT-21C (1 - 9)**	**EA**				**0.00**	**0.00**	**26.50**	**26.50**
02083 5421	30 Gallon Fiberglass Drum DOT-21C (1 - 9)	EA	0	N/A	1.000	0.00	0.00	26.50	26.50
33 19 0609	**30 Gallon Fiberglass DOT-21C (10 - 24)**	**EA**				**0.00**	**0.00**	**25.50**	**25.50**
02083 5422	30 Gallon Fiberglass Drum DOT-21C (10 - 24)	EA	0	N/A	1.000	0.00	0.00	25.50	25.50
33 19 0610	**30 Gallon Fiberglass DOT-21C (25 - 49)**	**EA**				**0.00**	**0.00**	**26.00**	**26.00**
02083 5423	30 Gallon Fiberglass Drum DOT-21C (25 - 49)	EA	0	N/A	1.000	0.00	0.00	26.00	26.00
33 19 0611	**30 Gallon Fiberglass DOT-21C (50+)**	**EA**				**0.00**	**0.00**	**25.50**	**25.50**
02083 5424	30 Gallon Fiberglass Drum DOT-21C (50+)	EA	0	N/A	1.000	0.00	0.00	25.50	25.50
33 19 0616	**HazMax Box 1 CY with Liner (1 - 9)**	**EA**				**0.00**	**0.00**	**102.00**	**102.00**
02083 6143	HazMax Box 1 CY with Liner (1 - 9)	EA	0	N/A	1.000	0.00	0.00	102.00	102.00
33 19 0617	**HazMax Box 1 CY with Liner (10 - 24)**	**EA**				**0.00**	**0.00**	**94.00**	**94.00**
02083 6144	HazMax Box 1 CY with Liner (10 - 24)	EA	0	N/A	1.000	0.00	0.00	94.00	94.00
33 19 0618	**HazMax Box 1 CY with Liner (25 - 49)**	**EA**				**0.00**	**0.00**	**87.00**	**87.00**
02083 6145	HazMax Box 1 CY with Liner (25 - 49)	EA	0	N/A	1.000	0.00	0.00	87.00	87.00
33 19 0619	**HazMax Box 1 CY with Liner (50+)**	**EA**				**0.00**	**0.00**	**83.50**	**83.50**
02083 6146	HazMax Box 1 CY with Liner (50+)	EA	0	N/A	1.000	0.00	0.00	83.50	83.50

33 19 07 Radiological Containers		Unit	Hourly Output	Crew	Factor	Labor	Equip.	Matl.	Total
33 19 0701	**Tall 55 Gallon Drum for 250 Lb UO2**	**EA**				**0.00**	**0.00**	**1,250**	**1,250**
02083 8311	Tall 55 Gallon Drum for 250 Lb UO2	EA	0	N/A	1.000	0.00	0.00	1,250	1,250
33 19 0702	**5,020 Lb, 30" Cylinder Enriched UF6**	**EA**				**0.00**	**0.00**	**3,625**	**3,625**
02083 8315	5,020 Lb Payload, 30" Cylinder to Ship Enriched UF6	EA	0	N/A	1.000	0.00	0.00	3,625	3,625
33 19 0703	**27,560 Lb, Cylinder UF6**	**EA**				**0.00**	**0.00**	**8,250**	**8,250**
02083 8317	27,560 Lb Payload, Cylinder to Ship UF6 as Solid	EA	0	N/A	1.000	0.00	0.00	8,250	8,250
33 19 0704	**55 Gallon Drum for 2 5-Gallon Pails UO2**	**EA**				**0.00**	**0.00**	**865.00**	**865.00**
02083 8321	55 Gallon Drum for 2 5-Gallon Pails UO2	EA	0	N/A	1.000	0.00	0.00	865.00	865.00
33 19 0705	**5" 55 Lb UF6 Cylinder Overpack**	**EA**				**0.00**	**0.00**	**2,225**	**2,225**
02083 8322	5" 55 Lb UF6 Cylinder Overpack	EA	0	N/A	1.000	0.00	0.00	2,225	2,225

33 19 07	Radiological Containers	Unit	Hourly Output	Crew	Factor	Labor	Equip.	Matl.	Total
33 19 0706	**8" 250 Lb UF6 Cylinder Overpack**	**EA**				**0.00**	**0.00**	**4,425**	**4,425**
02083 8323	8" 250 Lb UF6 Cylinder Overpack	EA	0	N/A	1.000	0.00	0.00	4,425	4,425
33 19 0707	**12" 430 Lb UF6 Cylinder Overpack**	**EA**				**0.00**	**0.00**	**4,975**	**4,975**
02083 8324	12" 430 Lb UF6 Cylinder Overpack	EA	0	N/A	1.000	0.00	0.00	4,975	4,975
33 19 0708	**30" 2.5 Ton UF6 Cylinder Overpack**	**EA**				**0.00**	**0.00**	**14,300**	**14,300**
02083 8325	30" 2.5 Ton UF6 Cylinder Overpack	EA	0	N/A	1.000	0.00	0.00	14,300	14,300
33 19 0709	**48" 10 Ton UF6 Cylinder Overpack**	**EA**				**0.00**	**0.00**	**66,500**	**66,500**
02083 8326	48" 10 Ton UF6 Cylinder Overpack	EA	0	N/A	1.000	0.00	0.00	66,500	66,500
33 19 0710	**48" 14 Ton UF6 Cylinder Overpack**	**EA**				**0.00**	**0.00**	**111,000**	**111,000**
02083 8327	48" 14 Ton UF6 Cylinder Overpack	EA	0	N/A	1.000	0.00	0.00	111,000	111,000
33 19 0711	**14 - 55 Gallon Drum Cask <= 20 R/Hour (Daily Rent)**	**DAY**				**0.00**	**0.00**	**289.00**	**289.00**
02083 8197	20,000 Lb Payload, 14 x 55 Gallon Drums or 215 CF Liner, 20R/Hour	DAY	0	N/A	1.000	0.00	0.00	289.00	289.00
33 19 0712	**14 - 55 Gallon Drum Cask <= 20 R/Hour (Monthly Rental)**	**MO**				**0.00**	**0.00**	**7,975**	**7,975**
02083 8198	20,000 Lb Payload, 14 x 55 Gallon Drums or 215 CF Liner, 20R/Hour	MO	0	N/A	1.000	0.00	0.00	7,975	7,975

33 19 08	Liners - Dump Truck/Van Transportation	Unit	Hourly Output	Crew	Factor	Labor	Equip.	Matl.	Total
33 19 0801	**55 Gallon, 15 Mil, Corrugated**	**EA**				**0.00**	**0.00**	**9.70**	**9.70**
02083 5931	55 Gallon, 15 Mil, Corrugated	EA	0	N/A	1.000	0.00	0.00	9.70	9.70
33 19 0802	**55 Gallon, 10 Mil, Corrugated**	**EA**				**0.00**	**0.00**	**8.05**	**8.05**
02083 5932	55 Gallon, 10 Mil, Corrugated	EA	0	N/A	1.000	0.00	0.00	8.05	8.05
33 19 0803	**55 Gallon, 6 Mil**	**EA**				**0.00**	**0.00**	**7.10**	**7.10**
02083 5933	55 Gallon, 6 Mil	EA	0	N/A	1.000	0.00	0.00	7.10	7.10
33 19 0804	**55 Gallon, 90 Mil High-density Polyethylene Radioactive Waste**	**EA**				**0.00**	**0.00**	**102.00**	**102.00**
02083 8131	90 Mil High Density Polyethylene Liner, 55 Gallon Drum	EA	0	N/A	1.000	0.00	0.00	102.00	102.00
33 19 0805	**85 Gallon, 8 Mil**	**EA**				**0.00**	**0.00**	**5.25**	**5.25**
02083 5934	85 Gallon, 8 Mil	EA	0	N/A	1.000	0.00	0.00	5.25	5.25
33 19 0806	**32 Ft. Dump Truck, 4 Mil Liner, disposable**	**EA**				**0.00**	**0.00**	**26.00**	**26.00**
02083 7113	32' Dump Truck Body Liner 4 Mil, Disposable	EA	0	N/A	1.000	0.00	0.00	26.00	26.00
33 19 0807	**32 Ft. Dump Truck, 6 Mil Liner, disposable**	**EA**				**0.00**	**0.00**	**37.50**	**37.50**
02083 7114	32' Dump Truck Body Liner 6 Mil, Disposable	EA	0	N/A	1.000	0.00	0.00	37.50	37.50

33 19 08 Liners - Dump Truck/Van Transportation		Unit	Hourly Output	Crew	Factor	Labor	Equip.	Matl.	Total
33 19 0808	**39 Ft. Dump Truck, 72" Deep, 4 Mil Liner, disposable**	**EA**				**0.00**	**0.00**	**28.00**	**28.00**
02083 7116	39' Dump Truck, 72" Deep, 4 Mil Liner	EA	0	N/A	1.000	0.00	0.00	28.00	28.00
33 19 0809	**39 Ft. Dump Truck, 96" Deep, 4 Mil Liner, disposable**	**EA**				**0.00**	**0.00**	**35.00**	**35.00**
02083 7117	39' Dump Truck, 96" Deep, 4 Mil Liner	EA	0	N/A	1.000	0.00	0.00	35.00	35.00
33 19 0810	**40 Ft. Dump Truck, 96" Deep, 8 Mil Liner, disposable**	**EA**				**0.00**	**0.00**	**38.00**	**38.00**
02083 7118	40' Dump Truck, 96" Deep, 8 Mil	EA	0	N/A	1.000	0.00	0.00	38.00	38.00
33 19 0811	**40 to 44 Van, 96" Deep, 8 Mil Liner, disposable**	**EA**				**0.00**	**0.00**	**90.00**	**90.00**
02083 7119	40 to 44' Van, 96" Deep, 8 Mil Liner	EA	0	N/A	1.000	0.00	0.00	90.00	90.00
33 19 0812	**50 to 54 Ft Rail Gondola, 10 Mil Liner, disposable**	**EA**				**0.00**	**0.00**	**159.00**	**159.00**
02083 7115	50 to 54' Rail Gondolas Liner 10 Mil, Disposable	EA	0	N/A	1.000	0.00	0.00	159.00	159.00

33 19 09 U.N. Code Drums		Unit	Hourly Output	Crew	Factor	Labor	Equip.	Matl.	Total
33 19 0901	**Composite Drum (Plastic/Steel), Open Head, 30 Gallon, UN Code 1A2/X150/S**	**EA**				**0.00**	**0.00**	**274.90**	**274.90**
01001 0215	Composite Drum (Plastic/Steel), Open Head, 30 Gallon, UN Code 1A2/X150/S	EA	0	N/A	1.000	0.00	0.00	274.90	274.90
33 19 0902	**Composite Drum (Plastic/Steel), Closed Head, 55 Gallon, UN Code 6HA1/Y1.2/100**	**EA**				**0.00**	**0.00**	**95.61**	**95.61**
01001 0216	Composite Drum (Plastic/Steel), Closed Head, 55 Gallon, UN Code 6HA1/Y1.2/1	EA	0	N/A	1.000	0.00	0.00	95.61	95.61
33 19 0903	**Composite Drum (Plastic/Steel), Open Head, 55 Gallon, UN Code 1A2/X400/S**	**EA**				**0.00**	**0.00**	**308.56**	**308.56**
01001 0217	Composite Drum (Plastic/Steel), Open Head, 55 Gallon, UN Code 1A2/X400/S	EA	0	N/A	1.000	0.00	0.00	308.56	308.56
33 19 0904	**Composite Drum (Plastic/Steel), Open Head, 55 Gallon, UN Code 6HA1/Y1.8/100**	**EA**				**0.00**	**0.00**	**102.41**	**102.41**
01001 0218	Composite Drum (Plastic/Steel), Open Head, 55 Gallon, UN Code 6HA1/Y1.8/100	EA	0	N/A	1.000	0.00	0.00	102.41	102.41
33 19 0905	**Composite Drum (Plastic/Steel), Open Head, 85 Gallon, UN Code 1A2/Y400/S**	**EA**				**0.00**	**0.00**	**342.22**	**342.22**
01001 0219	Composite Drum (Plastic/Steel), Open Head, 85 Gallon, UN Code 1A2/Y400/S	EA	0	N/A	1.000	0.00	0.00	342.22	342.22
33 19 0906	**Plastic Drum, Closed Head, 20 Gallon, UN Code 1H1/Y1.3/100**	**EA**				**0.00**	**0.00**	**34.22**	**34.22**
01001 0220	Plastic Drum, Closed Head, 20 Gallon, UN Code 1H1/Y1.3/100	EA	0	N/A	1.000	0.00	0.00	34.22	34.22

33 19 09 U.N. Code Drums	Unit	Hourly Output	Crew	Factor	Labor	Equip.	Matl.	Total
33 19 0907 Plastic Drum, Closed Head, 20 Gallon, UN Code 1H1/Y1.8/100	**EA**				**0.00**	**0.00**	**44.40**	**44.40**
01001 0221 Plastic Drum, Closed Head, 20 Gallon, UN Code 1H1/Y1.8/100	EA	0	N/A	1.000	0.00	0.00	44.40	44.40
33 19 0908 Plastic Drum, Closed Head, 20 Gallon, UN Code 1H1/Y2.0/100	**EA**				**0.00**	**0.00**	**34.78**	**34.78**
01001 0222 Plastic Drum, Closed Head, 20 Gallon, UN Code 1H1/Y2.0/100	EA	0	N/A	1.000	0.00	0.00	34.78	34.78
33 19 0909 Plastic Drum, Open Head, 20 Gallon, UN Code 1H2/X75/S	**EA**				**0.00**	**0.00**	**36.47**	**36.47**
01001 0223 Plastic Drum, Open Head, 20 Gallon, UN Code 1H2/X75/S	EA	0	N/A	1.000	0.00	0.00	36.47	36.47
33 19 0910 Plastic Drum, Open Head, 20 Gallon, UN Code 1H2/Y130/S	**EA**				**0.00**	**0.00**	**43.20**	**43.20**
01001 0224 Plastic Drum, Open Head, 20 Gallon, UN Code 1H2/Y130/S	EA	0	N/A	1.000	0.00	0.00	43.20	43.20
33 19 0911 Plastic Drum, Open Head, 20 Gallon, UN Code 1H2/Y80/F	**EA**				**0.00**	**0.00**	**45.98**	**45.98**
01001 0225 Plastic Drum, Open Head, 20 Gallon, UN Code 1H2/Y80/F	EA	0	N/A	1.000	0.00	0.00	45.98	45.98
33 19 0912 Plastic Drum, Closed Head, 30 Gallon, UN Code 1H1/Y1.3/100	**EA**				**0.00**	**0.00**	**49.23**	**49.23**
01001 0226 Plastic Drum, Closed Head, 30 Gallon, UN Code 1H1/Y1.3/100	EA	0	N/A	1.000	0.00	0.00	49.23	49.23
33 19 0913 Plastic Drum, Open Head, 30 Gallon, UN Code 1H2/Y1.2/80	**EA**				**0.00**	**0.00**	**84.79**	**84.79**
01001 0227 Plastic Drum, Open Head, 30 Gallon, UN Code 1H2/Y1.2/80	EA	0	N/A	1.000	0.00	0.00	84.79	84.79
33 19 0914 Plastic Drum, Open Head, 30 Gallon, UN Code 1H2/Y120/F	**EA**				**0.00**	**0.00**	**62.39**	**62.39**
01001 0228 Plastic Drum, Open Head, 30 Gallon, UN Code 1H2/Y120/F	EA	0	N/A	1.000	0.00	0.00	62.39	62.39
33 19 0915 Plastic Drum, Open Head, 30 Gallon, UN Code 1H2/Y150/S	**EA**				**0.00**	**0.00**	**38.71**	**38.71**
01001 0229 Plastic Drum, Open Head, 30 Gallon, UN Code 1H2/Y150/S	EA	0	N/A	1.000	0.00	0.00	38.71	38.71
33 19 0916 Plastic Drum, Closed Head, 35 Gallon, UN Code 1H1/Y1.3/100	**EA**				**0.00**	**0.00**	**54.17**	**54.17**
01001 0230 Plastic Drum, Closed Head, 35 Gallon, UN Code 1H1/Y1.3/100	EA	0	N/A	1.000	0.00	0.00	54.17	54.17
33 19 0917 Plastic Drum, Open Head, 35 Gallon, UN Code 1H2/Y165/F	**EA**				**0.00**	**0.00**	**65.64**	**65.64**
01001 0231 Plastic Drum, Open Head, 35 Gallon, UN Code 1H2/Y165/F	EA	0	N/A	1.000	0.00	0.00	65.64	65.64

33 19 Disposal (Commercial)

33 19 09 U.N. Code Drums		Unit	Hourly Output	Crew	Factor	Labor	Equip.	Matl.	Total
33 19 0918	**Plastic Drum, Open Head, 40 Gallon, UN Code 1H2/X100/S**	**EA**				**0.00**	**0.00**	**54.98**	**54.98**
01001 0232	Plastic Drum, Open Head, 40 Gallon, UN Code 1H2/X100/S	EA	0	N/A	1.000	0.00	0.00	54.98	54.98
33 19 0919	**Plastic Drum, Open Head, 47 Gallon, UN Code 1H2/X95/S**	**EA**				**0.00**	**0.00**	**62.83**	**62.83**
01001 0233	Plastic Drum, Open Head, 47 Gallon, UN Code 1H2/X95/S	EA	0	N/A	1.000	0.00	0.00	62.83	62.83
33 19 0920	**Plastic Drum, Open Head, 53 Gallon, UN Code 1H2/Y130/S**	**EA**				**0.00**	**0.00**	**55.26**	**55.26**
01001 0234	Plastic Drum, Open Head, 53 Gallon, UN Code 1H2/Y130/S	EA	0	N/A	1.000	0.00	0.00	55.26	55.26
33 19 0921	**Plastic Drum, Closed Head, 55 Gallon, UN Code 1H1/Y1.3/100**	**EA**				**0.00**	**0.00**	**70.63**	**70.63**
01001 0235	Plastic Drum, Closed Head, 55 Gallon, UN Code 1H1/Y1.3/100	EA	0	N/A	1.000	0.00	0.00	70.63	70.63
33 19 0922	**Plastic Drum, Closed Head, 55 Gallon, UN Code 1H1/Y1.8/100**	**EA**				**0.00**	**0.00**	**57.32**	**57.32**
01001 0236	Plastic Drum, Closed Head, 55 Gallon, UN Code 1H1/Y1.8/100	EA	0	N/A	1.000	0.00	0.00	57.32	57.32
33 19 0923	**Plastic Drum, Closed Head, 55 Gallon, UN Code 1H1/Y1.9/150**	**EA**				**0.00**	**0.00**	**42.64**	**42.64**
01001 0237	Plastic Drum, Closed Head, 55 Gallon, UN Code 1H1/Y1.9/150	EA	0	N/A	1.000	0.00	0.00	42.64	42.64
33 19 0924	**Plastic Drum, Open Head, 55 Gallon, UN Code 1H2/X250/S**	**EA**				**0.00**	**0.00**	**118.71**	**118.71**
01001 0238	Plastic Drum, Open Head, 55 Gallon, UN Code 1H2/X250/S	EA	0	N/A	1.000	0.00	0.00	118.71	118.71
33 19 0925	**Plastic Drum, Open Head, 55 Gallon, UN Code 1H2/Y140/S**	**EA**				**0.00**	**0.00**	**64.24**	**64.24**
01001 0239	Plastic Drum, Open Head, 55 Gallon, UN Code 1H2/Y140/S	EA	0	N/A	1.000	0.00	0.00	64.24	64.24
33 19 0926	**Plastic Drum, Open Head, 55 Gallon, UN Code 1H2/Y150/S**	**EA**				**0.00**	**0.00**	**60.96**	**60.96**
01001 0240	Plastic Drum, Open Head, 55 Gallon, UN Code 1H2/Y150/S	EA	0	N/A	1.000	0.00	0.00	60.96	60.96
33 19 0927	**Plastic Drum, Open Head, 55 Gallon, UN Code 1H2/Y195/F**	**EA**				**0.00**	**0.00**	**89.26**	**89.26**
01001 0241	Plastic Drum, Open Head, 55 Gallon, UN Code 1H2/Y195/F	EA	0	N/A	1.000	0.00	0.00	89.26	89.26
33 19 0928	**Steel Drum, Closed Head, 20 Gallon, UN Code 1A1/X1.2/300**	**EA**				**0.00**	**0.00**	**37.59**	**37.59**
01001 0242	Steel Drum, Closed Head, 20 Gallon, UN Code 1A1/X1.2/300	EA	0	N/A	1.000	0.00	0.00	37.59	37.59

33 19 09 U.N. Code Drums		Unit	Hourly Output	Crew	Factor	Labor	Equip.	Matl.	Total
33 19 0929	**Steel Drum, Closed Head, 20 Gallon, UN Code 1A1/Y1.8/300**	**EA**				**0.00**	**0.00**	**44.21**	**44.21**
01001 0243	Steel Drum, Closed Head, 20 Gallon, UN Code 1A1/Y1.8/300	EA	0	N/A	1.000	0.00	0.00	44.21	44.21
33 19 0930	**Steel Drum, Open Head, 20 Gallon, UN Code 1A2/X114/S**	**EA**				**0.00**	**0.00**	**47.52**	**47.52**
01001 0244	Steel Drum, Open Head, 20 Gallon, UN Code 1A2/X114/S	EA	0	N/A	1.000	0.00	0.00	47.52	47.52
33 19 0931	**Steel Drum, Open Head, 20 Gallon, UN Code 1A2/Y1.5/100**	**EA**				**0.00**	**0.00**	**43.76**	**43.76**
01001 0245	Steel Drum, Open Head, 20 Gallon, UN Code 1A2/Y1.5/100	EA	0	N/A	1.000	0.00	0.00	43.76	43.76
33 19 0932	**Steel Drum, Open Head, 20 Gallon, UN Code 1A2/Y1.5/150**	**EA**				**0.00**	**0.00**	**79.84**	**79.84**
01001 0246	Steel Drum, Open Head, 20 Gallon, UN Code 1A2/Y1.5/150	EA	0	N/A	1.000	0.00	0.00	79.84	79.84
33 19 0933	**Steel Drum, Closed Head, 30 Gallon, UN Code 1A1/X1.3/250**	**EA**				**0.00**	**0.00**	**47.69**	**47.69**
01001 0247	Steel Drum, Closed Head, 30 Gallon, UN Code 1A1/X1.3/250	EA	0	N/A	1.000	0.00	0.00	47.69	47.69
33 19 0934	**Steel Drum, Closed Head, 30 Gallon, UN Code 1A1/Y1.4/300**	**EA**				**0.00**	**0.00**	**62.92**	**62.92**
01001 0248	Steel Drum, Closed Head, 30 Gallon, UN Code 1A1/Y1.4/300	EA	0	N/A	1.000	0.00	0.00	62.92	62.92
33 19 0935	**Steel Drum, Closed Head, 30 Gallon, UN Code 1A1/Y1.5/300**	**EA**				**0.00**	**0.00**	**57.25**	**57.25**
01001 0249	Steel Drum, Closed Head, 30 Gallon, UN Code 1A1/Y1.5/300	EA	0	N/A	1.000	0.00	0.00	57.25	57.25
33 19 0936	**Steel Drum, Open Head, 30 Gallon, UN Code 1A2/X135/S**	**EA**				**0.00**	**0.00**	**48.25**	**48.25**
01001 0250	Steel Drum, Open Head, 30 Gallon, UN Code 1A2/X135/S	EA	0	N/A	1.000	0.00	0.00	48.25	48.25
33 19 0937	**Steel Drum, Open Head, 30 Gallon, UN Code 1A2/X160/S**	**EA**				**0.00**	**0.00**	**55.20**	**55.20**
01001 0251	Steel Drum, Open Head, 30 Gallon, UN Code 1A2/X160/S	EA	0	N/A	1.000	0.00	0.00	55.20	55.20
33 19 0938	**Steel Drum, Open Head, 30 Gallon, UN Code 1A2/Y1.5/150**	**EA**				**0.00**	**0.00**	**76.21**	**76.21**
01001 0252	Steel Drum, Open Head, 30 Gallon, UN Code 1A2/Y1.5/150	EA	0	N/A	1.000	0.00	0.00	76.21	76.21
33 19 0939	**Steel Drum, Closed Head, 55 Gallon, UN Code 1A1/Y1.8/300**	**EA**				**0.00**	**0.00**	**86.34**	**86.34**
01001 0253	Steel Drum, Closed Head, 55 Gallon, UN Code 1A1/Y1.8/300	EA	0	N/A	1.000	0.00	0.00	86.34	86.34

33 19 09 U.N. Code Drums		Unit	Hourly Output	Crew	Factor	Labor	Equip.	Matl.	Total
33 19 0940	**Steel Drum, Open Head, 55 Gallon, UN Code 1A2/X400/S**	**EA**				**0.00**	**0.00**	**72.34**	**72.34**
01001 0254	Steel Drum, Open Head, 55 Gallon, UN Code 1A2/X400/S	EA	0	N/A	1.000	0.00	0.00	72.34	72.34
33 19 0941	**Steel Drum, Open Head, 55 Gallon, UN Code 1A2/X430/S**	**EA**				**0.00**	**0.00**	**74.95**	**74.95**
01001 0255	Steel Drum, Open Head, 55 Gallon, UN Code 1A2/X430/S	EA	0	N/A	1.000	0.00	0.00	74.95	74.95
33 19 0942	**Steel Drum, Open Head, 55 Gallon, UN Code 1A2/Y1.6/150**	**EA**				**0.00**	**0.00**	**110.65**	**110.65**
01001 0256	Steel Drum, Open Head, 55 Gallon, UN Code 1A2/Y1.6/150	EA	0	N/A	1.000	0.00	0.00	110.65	110.65
33 19 0943	**Steel Drum, Open Head, 55 Gallon, UN Code 1A2/Y1.8/200**	**EA**				**0.00**	**0.00**	**63.40**	**63.40**
01001 0257	Steel Drum, Open Head, 55 Gallon, UN Code 1A2/Y1.8/200	EA	0	N/A	1.000	0.00	0.00	63.40	63.40
33 19 0944	**Steel Drum, Open Head, 55 Gallon, UN Code 1A2/Y228/S**	**EA**				**0.00**	**0.00**	**58.57**	**58.57**
01001 0258	Steel Drum, Open Head, 55 Gallon, UN Code 1A2/Y228/S	EA	0	N/A	1.000	0.00	0.00	58.57	58.57
33 19 0945	**SST Drum, Closed Head, 20 Gallon, UN Code 1A1/X1.2/300**	**EA**				**0.00**	**0.00**	**494.82**	**494.82**
01001 0259	SST Drum, Closed Head, 20 Gallon, UN Code 1A1/X1.2/300	EA	0	N/A	1.000	0.00	0.00	494.82	494.82
33 19 0946	**SST Drum, Open Head, 20 Gallon, UN Code 1A2/X106/S**	**EA**				**0.00**	**0.00**	**536.82**	**536.82**
01001 0260	SST Drum, Open Head, 20 Gallon, UN Code 1A2/X106/S	EA	0	N/A	1.000	0.00	0.00	536.82	536.82
33 19 0947	**SST Drum, Open Head, 20 Gallon, UN Code 1A2/X70/S**	**EA**				**0.00**	**0.00**	**200.81**	**200.81**
01001 0261	SST Drum, Open Head, 20 Gallon, UN Code 1A2/X70/S	EA	0	N/A	1.000	0.00	0.00	200.81	200.81
33 19 0948	**SST Drum, Closed Head, 30 Gallon, UN Code 1A1/X1.2/300**	**EA**				**0.00**	**0.00**	**582.86**	**582.86**
01001 0262	SST Drum, Closed Head, 30 Gallon, UN Code 1A1/X1.2/300	EA	0	N/A	1.000	0.00	0.00	582.86	582.86
33 19 0949	**SST Drum, Closed Head, 30 Gallon, UN Code 1A1/X1.8/550**	**EA**				**0.00**	**0.00**	**739.43**	**739.43**
01001 0263	SST Drum, Closed Head, 30 Gallon, UN Code 1A1/X1.8/550	EA	0	N/A	1.000	0.00	0.00	739.43	739.43
33 19 0950	**SST Drum, Open Head, 30 Gallon, UN Code 1A2/X160/S**	**EA**				**0.00**	**0.00**	**594.33**	**594.33**
01001 0264	SST Drum, Open Head, 30 Gallon, UN Code 1A2/X160/S	EA	0	N/A	1.000	0.00	0.00	594.33	594.33

33 19 09	U.N. Code Drums	Unit	Hourly Output	Crew	Factor	Labor	Equip.	Matl.	Total
33 19 0951	**SST Drum, Closed Head, 55 Gallon, UN Code 1A1/X1.2/300**	**EA**				**0.00**	**0.00**	**679.55**	**679.55**
01001 0265	SST Drum, Closed Head, 55 Gallon, UN Code 1A1/X1.2/300	EA	0	N/A	1.000	0.00	0.00	679.55	679.55
33 19 0952	**SST Drum, Closed Head, 55 Gallon, UN Code 1A1/X1.8/300**	**EA**				**0.00**	**0.00**	**700.31**	**700.31**
01001 0266	SST Drum, Closed Head, 55 Gallon, UN Code 1A1/X1.8/300	EA	0	N/A	1.000	0.00	0.00	700.31	700.31
33 19 0953	**SST Drum, Closed Head, 55 Gallon, UN Code 1A1/X1.8/550**	**EA**				**0.00**	**0.00**	**813.48**	**813.48**
01001 0267	SST Drum, Closed Head, 55 Gallon, UN Code 1A1/X1.8/550	EA	0	N/A	1.000	0.00	0.00	813.48	813.48
33 19 0954	**SST Drum, Closed Head, 55 Gallon, UN Code 1A1/X2.0/600**	**EA**				**0.00**	**0.00**	**1,167**	**1,167**
01001 0268	SST Drum, Closed Head, 55 Gallon, UN Code 1A1/X2.0/600	EA	0	N/A	1.000	0.00	0.00	1,167	1,167
33 19 0955	**SST Drum, Closed Head, 55 Gallon, UN Code 1A1/X2.0/600**	**EA**				**0.00**	**0.00**	**1,167**	**1,167**
01001 0269	SST Drum, Closed Head, 55 Gallon, UN Code 1A1/X2.0/600	EA	0	N/A	1.000	0.00	0.00	1,167	1,167
33 19 0956	**SST Drum, Open Head, 55 Gallon, UN Code 1A2/X430/S**	**EA**				**0.00**	**0.00**	**817.41**	**817.41**
01001 0270	SST Drum, Open Head, 55 Gallon, UN Code 1A2/X430/S	EA	0	N/A	1.000	0.00	0.00	817.41	817.41
33 19 0957	**SST Drum, Open Head, 55 Gallon, UN Code 1A2/X437/S**	**EA**				**0.00**	**0.00**	**807.87**	**807.87**
01001 0271	SST Drum, Open Head, 55 Gallon, UN Code 1A2/X437/S	EA	0	N/A	1.000	0.00	0.00	807.87	807.87
33 19 0958	**Composite Overpack (Plastic/Steel), 20 Gallon, UN Code 1H2/Y262/S**	**EA**				**0.00**	**0.00**	**35.91**	**35.91**
01001 0272	Composite Overpack (Plastic/Steel), 20 Gallon, UN Code 1H2/Y262/S	EA	0	N/A	1.000	0.00	0.00	35.91	35.91
33 19 0959	**Composite Overpack (Plastic/Steel), 30 Gallon, UN Code 1H2/X136/S**	**EA**				**0.00**	**0.00**	**51.61**	**51.61**
01001 0273	Composite Overpack (Plastic/Steel), 30 Gallon, UN Code 1H2/X136/S	EA	0	N/A	1.000	0.00	0.00	51.61	51.61
33 19 0960	**Composite Overpack (Plastic/Steel), 50 Gallon, UN Code 1H2/Y125/S**	**EA**				**0.00**	**0.00**	**80.79**	**80.79**
01001 0274	Composite Overpack (Plastic/Steel), 50 Gallon, UN Code 1H2/Y125/S	EA	0	N/A	1.000	0.00	0.00	80.79	80.79
33 19 0961	**Composite Overpack (Plastic/Steel), 55 Gallon, UN Code 1H2/X250/S**	**EA**				**0.00**	**0.00**	**75.74**	**75.74**
01001 0275	Composite Overpack (Plastic/Steel), 55 Gallon, UN Code 1H2/X250/S	EA	0	N/A	1.000	0.00	0.00	75.74	75.74

33 19 09 U.N. Code Drums		Unit	Hourly Output	Crew	Factor	Labor	Equip.	Matl.	Total
33 19 0962	**Composite Overpack (Plastic/Steel), 65 Gallon, UN Code 1H2/X200/S**	**EA**				**0.00**	**0.00**	**143.62**	**143.62**
01001 0276	Composite Overpack (Plastic/Steel), 65 Gallon, UN Code 1H2/X200/S	EA	0	N/A	1.000	0.00	0.00	143.62	143.62
33 19 0963	**Composite Overpack (Plastic/Steel), 85 Gallon, UN Code 1H2/Y262/S**	**EA**				**0.00**	**0.00**	**146.99**	**146.99**
01001 0277	Composite Overpack (Plastic/Steel), 85 Gallon, UN Code 1H2/Y262/S	EA	0	N/A	1.000	0.00	0.00	146.99	146.99
33 19 0964	**Composite Overpack (Plastic/Steel), 95 Gallon, UN Code 1H2/X295/S**	**EA**				**0.00**	**0.00**	**154.84**	**154.84**
01001 0278	Composite Overpack (Plastic/Steel), 95 Gallon, UN Code 1H2/X295/S	EA	0	N/A	1.000	0.00	0.00	154.84	154.84
33 19 0965	**Composite Overpack (Plastic/Steel), 95 Gallon, UN Code 1H2/X340/S**	**EA**				**0.00**	**0.00**	**176.72**	**176.72**
01001 0279	Composite Overpack (Plastic/Steel), 95 Gallon, UN Code 1H2/X340/S	EA	0	N/A	1.000	0.00	0.00	176.72	176.72
33 19 0966	**Composite Overpack (Plastic/Steel), 110 Gallon, UN Code 1H2/Y262/S**	**EA**				**0.00**	**0.00**	**228.34**	**228.34**
01001 0280	Composite Overpack (Plastic/Steel), 110 Gallon, UN Code 1H2/Y262/S	EA	0	N/A	1.000	0.00	0.00	228.34	228.34
33 19 0967	**Steel Overpack, 20 Gallon, UN Code 1A2/X106/S**	**EA**				**0.00**	**0.00**	**58.21**	**58.21**
01001 0281	Steel Overpack, 20 Gallon, UN Code 1A2/X106/S	EA	0	N/A	1.000	0.00	0.00	58.21	58.21
33 19 0968	**Steel Overpack, 20 Gallon, UN Code 1A2/X114/S**	**EA**				**0.00**	**0.00**	**66.85**	**66.85**
01001 0282	Steel Overpack, 20 Gallon, UN Code 1A2/X114/S	EA	0	N/A	1.000	0.00	0.00	66.85	66.85
33 19 0969	**Steel Overpack, 30 Gallon, UN Code 1A2/X150/S**	**EA**				**0.00**	**0.00**	**55.54**	**55.54**
01001 0283	Steel Overpack, 30 Gallon, UN Code 1A2/X150/S	EA	0	N/A	1.000	0.00	0.00	55.54	55.54
33 19 0970	**Steel Overpack, 30 Gallon, UN Code 1A2/X160/S**	**EA**				**0.00**	**0.00**	**98.56**	**98.56**
01001 0284	Steel Overpack, 30 Gallon, UN Code 1A2/X160/S	EA	0	N/A	1.000	0.00	0.00	98.56	98.56
33 19 0971	**Steel Overpack, 55 Gallon, UN Code 1A2/X430/S**	**EA**				**0.00**	**0.00**	**88.25**	**88.25**
01001 0285	Steel Overpack, 55 Gallon, UN Code 1A2/X430/S	EA	0	N/A	1.000	0.00	0.00	88.25	88.25
33 19 0972	**Steel Overpack, 85 Gallon, UN Code 1A2/X440/S**	**EA**				**0.00**	**0.00**	**134.78**	**134.78**
01001 0286	Steel Overpack, 85 Gallon, UN Code 1A2/X440/S	EA	0	N/A	1.000	0.00	0.00	134.78	134.78
33 19 0973	**Steel Overpack, 96 Gallon, UN Code 1A2/X440/S**	**EA**				**0.00**	**0.00**	**150.34**	**150.34**
01001 0287	Steel Overpack, 96 Gallon, UN Code 1A2/X440/S	EA	0	N/A	1.000	0.00	0.00	150.34	150.34

33 19 Disposal (Commercial)

33 19 09	U.N. Code Drums	Unit	Hourly Output	Crew	Factor	Labor	Equip.	Matl.	Total
33 19 0974	**Steel Overpack, 96 Gallon, UN Code 1A2/X440/S**	**EA**				**0.00**	**0.00**	**137.45**	**137.45**
01001 0288	Steel Overpack, 96 Gallon, UN Code 1A2/X440/S	EA	0	N/A	1.000	0.00	0.00	137.45	137.45
33 19 0975	**Steel Overpack, 110 Gallon, UN Code 1A2/Y409/S**	**EA**				**0.00**	**0.00**	**320.87**	**320.87**
01001 0289	Steel Overpack, 110 Gallon, UN Code 1A2/Y409/S	EA	0	N/A	1.000	0.00	0.00	320.87	320.87

33 19 71	Discharge to POTW	Unit	Hourly Output	Crew	Factor	Labor	Equip.	Matl.	Total
33 19 7101	**Sewer Connection Fee**	**EA**				**0.00**	**0.00**	**2,657**	**2,657**
02752 8000	Sewer Connection Charge	EA	0	N/A	1.000	0.00	0.00	2,657	2,657
33 19 7102	**Wastewater Disposal Fee**	**KGAL**				**0.00**	**0.00**	**1.90**	**1.90**
02778 1000	Treatment of Waste Water at POTW	KGAL	0	N/A	1.000	0.00	0.00	1.90	1.90
33 19 7103	**Process Water Hauling Fee, Subcontracted**	**GAL**				**0.00**	**0.00**	**0.31**	**0.31**
02560 8031	Process Water Hauling Fee, Subcontracted	GAL	0	N/A	1.000	0.00	0.00	0.31	0.31
33 19 7104	**Radioactive Process Water Treatment/Disposal, Subcontracted**	**GAL**				**0.00**	**0.00**	**2.63**	**2.63**
02560 8032	Radioactive Process Water Treatment/Disposal, Subcontracted	GAL	0	N/A	1.000	0.00	0.00	2.63	2.63
33 19 7105	**Non-Radioactive Process Water Treatment/Disposal, Subcontracted**	**GAL**				**0.00**	**0.00**	**1.32**	**1.32**
02560 8033	Non-Radioactive Process Water Treatment/Disposal, Subcontracted	GAL	0	N/A	1.000	0.00	0.00	1.32	1.32

33 19 72	Landfill Disposal	Unit	Hourly Output	Crew	Factor	Labor	Equip.	Matl.	Total
33 19 7201	**Landfill Hazardous Solid Waste, <= 30 Gallon Drum**	**EA**				**0.00**	**0.00**	**98.50**	**98.50**
13278 1111	Hazardous Solid Waste, Drums 30 Gallons or Less	EA	0	N/A	1.000	0.00	0.00	98.50	98.50
33 19 7202	**Landfill Hazardous Solid Waste, 55 Gallon Drum**	**EA**				**0.00**	**0.00**	**136.00**	**136.00**
13278 1112	Hazardous Solid Waste, 55 Gallon Drums	EA	0	N/A	1.000	0.00	0.00	136.00	136.00
33 19 7203	**Landfill Hazardous Solid Waste, > 55 Gallon Drum**	**EA**				**0.00**	**0.00**	**182.00**	**182.00**
13278 1113	Hazardous Solid Waste, Drums Greater than 55 Gallons	EA	0	N/A	1.000	0.00	0.00	182.00	182.00
33 19 7204	**Landfill Nonhazardous Solid Waste, <= 30 Gallon Drum**	**EA**				**0.00**	**0.00**	**55.65**	**55.65**
13278 1114	Non-Hazardous Solid Wastes, Drums 30 Gallons or Less	EA	0	N/A	1.000	0.00	0.00	55.65	55.65

33 19 72	Landfill Disposal	Unit	Hourly Output	Crew	Factor	Labor	Equip.	Matl.	Total
33 19 7205	**Landfill Nonhazardous Solid Waste, 55 Gallon Drum**	**EA**				**0.00**	**0.00**	**129.00**	**129.00**
13278 1115	Non-Hazardous Solid Wastes, 55 Gallon Drums	EA	0	N/A	1.000	0.00	0.00	129.00	129.00
33 19 7206	**Landfill Nonhazardous Solid Waste, > 55 Gallon Drum**	**EA**				**0.00**	**0.00**	**149.00**	**149.00**
13278 1116	Non-Hazardous Solid Wastes, Drums Greater than 55 Gallons	EA	0	N/A	1.000	0.00	0.00	149.00	149.00
33 19 7207	**Landfill Drummed Solid Waste Requiring Stabilization, <= 30 Gallon Drum**	**EA**				**0.00**	**0.00**	**234.00**	**234.00**
13278 1117	Drummed Solid Waste Requiring Stabilization, Drums 30 Gallons or Less	EA	0	N/A	1.000	0.00	0.00	234.00	234.00
33 19 7208	**Landfill Drummed Solid Waste Requiring Stabilization, 55 Gallon Drum**	**EA**				**0.00**	**0.00**	**235.00**	**235.00**
13278 1118	Drummed Solid Waste Requiring Stabilization, 55 Gallon Drums	EA	0	N/A	1.000	0.00	0.00	235.00	235.00
33 19 7209	**Landfill Drummed Solid Waste Requiring Stabilization, > 55 Gallon Drum**	**EA**				**0.00**	**0.00**	**340.00**	**340.00**
13278 1119	Drummed Solid Waste Requiring Stabilization, Drums Greater than 55 Gallons	EA	0	N/A	1.000	0.00	0.00	340.00	340.00
33 19 7210	**Landfill Drummed Solid/Sludge Requiring Stabilization, <= 30 Gallon Drum**	**EA**				**0.00**	**0.00**	**234.00**	**234.00**
13278 1121	Drummed Solid/Sludge Requiring Stabilization, Drums 30 Gallons or Less	EA	0	N/A	1.000	0.00	0.00	234.00	234.00
33 19 7211	**Landfill Drummed Solid/Sludge Requiring Stabilization, 55 Gallon Drum**	**EA**				**0.00**	**0.00**	**293.00**	**293.00**
13278 1122	Drummed Solid/Sludge Requiring Stabilization, 55 Gallon Drums	EA	0	N/A	1.000	0.00	0.00	293.00	293.00
33 19 7212	**Landfill Drummed Solid/Sludge Requiring Stabilization, > 55 Gallon Drum**	**EA**				**0.00**	**0.00**	**571.51**	**571.51**
13278 1123	Drummed Solid/Sludge Requiring Stabilization, Drums Greater than 55 Gallons	EA	0	N/A	1.000	0.00	0.00	571.51	571.51
33 19 7213	**Landfill Drummed Liquid Waste, <= 30 Gallon Drum**	**EA**				**0.00**	**0.00**	**250.00**	**250.00**
13278 1124	Drummed Liquid Waste, Drums 30 Gallons or Less	EA	0	N/A	1.000	0.00	0.00	250.00	250.00
33 19 7214	**Landfill Drummed Liquid Waste, 55 Gallon Drum**	**EA**				**0.00**	**0.00**	**145.00**	**145.00**
13278 1125	Drummed Liquid Waste, 55 Gallon Drums	EA	0	N/A	1.000	0.00	0.00	145.00	145.00
33 19 7215	**Landfill Drummed Liquid Waste, > 55 Gallon Drum**	**EA**				**0.00**	**0.00**	**282.00**	**282.00**
13278 1126	Drummed Liquid Waste, Drums Greater than 55 Gallons	EA	0	N/A	1.000	0.00	0.00	282.00	282.00

33 19 Disposal (Commercial)

33 19 72	Landfill Disposal	Unit	Hourly Output	Crew	Factor	Labor	Equip.	Matl.	Total
33 19 7216	**Landfill Drummed Liquid Waste Requiring Stabilization, <= 30 Gallon Drum**	**EA**				**0.00**	**0.00**	**295.00**	**295.00**
13278 1127	Drummed Liquid Waste Requiring Stabilization, Drums 30 Gallons or Less	EA	0	N/A	1.000	0.00	0.00	295.00	295.00
33 19 7217	**Landfill Drummed Liquid Waste Requiring Stabilization, 55 Gallon Drum**	**EA**				**0.00**	**0.00**	**310.00**	**310.00**
13278 1128	Drummed Liquid Waste Requiring Stabilization, 55 Gallon Drums	EA	0	N/A	1.000	0.00	0.00	310.00	310.00
33 19 7218	**Landfill Drummed Liquid Waste Requiring Stabilization, > 55 Gallon Drum**	**EA**				**0.00**	**0.00**	**455.00**	**455.00**
13278 1129	Drummed Liquid Waste Requiring Stabilization, Drums Greater than 55 Gallons	EA	0	N/A	1.000	0.00	0.00	455.00	455.00
33 19 7219	**Landfill Drummed Liquid Waste, Solar Evaporation, <= 30 Gallon Drum, pH>2.0**	**EA**				**0.00**	**0.00**	**155.00**	**155.00**
13278 1131	Drummed Liquid Waste Treated by Solar Evaporation, Drums 30 Gallons or Less	EA	0	N/A	1.000	0.00	0.00	155.00	155.00
33 19 7220	**Landfill Drummed Liquid Waste, Solar Evaporation, 55 Gallon Drum, pH > 2.0**	**EA**				**0.00**	**0.00**	**170.00**	**170.00**
13278 1132	Drummed Liquid Waste Treated by Solar Evaporation, 55 Gallon Drums	EA	0	N/A	1.000	0.00	0.00	170.00	170.00
33 19 7221	**Landfill Drummed Liquid Waste, Solar Evaporation, > 55 Gallon Drum, pH>2.0**	**EA**				**0.00**	**0.00**	**196.00**	**196.00**
13278 1133	Drummed Liquid Waste Treated by Solar Evaporation, Drums Greater than 55 Gallons	EA	0	N/A	1.000	0.00	0.00	196.00	196.00
33 19 7222	**Landfill Drummed Inorganic/Organic Nonhazardous Liquid/Sludge, <= 30 Gallon Drum**	**EA**				**0.00**	**0.00**	**133.00**	**133.00**
13278 1134	Drummed Inorganic or Organic Non-Hazardous Liquid/Sludge, Drums 30 Gallons or Less	EA	0	N/A	1.000	0.00	0.00	133.00	133.00
33 19 7223	**Landfill Drummed Inorganic/Organic Nonhazardous Liquid/Sludge, 55 Gallon Drum**	**EA**				**0.00**	**0.00**	**162.00**	**162.00**
13278 1135	Drummed Inorganic or Organic Non-Hazardous Liquid/Sludge, 55 Gallon Drums	EA	0	N/A	1.000	0.00	0.00	162.00	162.00
33 19 7224	**Landfill Drummed Inorganic/Organic Nonhazardous Liquid/Sludge, >55 Gallon Drum**	**EA**				**0.00**	**0.00**	**380.00**	**380.00**
13278 1136	Drummed Inorganic or Organic Non-Hazardous Liquid/Sludge, Drums Greater than 55 Gallons	EA	0	N/A	1.000	0.00	0.00	380.00	380.00
33 19 7225	**Landfill Drummed Organic Liquid/Sludge Requiring Stabilization, <= 30 Gallon Drum**	**EA**				**0.00**	**0.00**	**335.00**	**335.00**
13278 1137	Drummed Organic Liquid/Sludge Requiring Stabilization, Drums 30 Gallons or Less	EA	0	N/A	1.000	0.00	0.00	335.00	335.00

33 19 72	Landfill Disposal	Unit	Hourly Output	Crew	Factor	Labor	Equip.	Matl.	Total
33 19 7226	**Landfill Drummed Organic Liquid/Sludge Requiring Stabilization, 55 Gallon Drum**	**EA**				**0.00**	**0.00**	**300.00**	**300.00**
13278 1138	Drummed Organic Liquid/Sludge Requiring Stabilization, 55 Gallon Drums	EA	0	N/A	1.000	0.00	0.00	300.00	300.00
33 19 7227	**Landfill Drummed Organic Liquid/Sludge Requiring Stabilization, > 55 Gallon Drum**	**EA**				**0.00**	**0.00**	**455.00**	**455.00**
13278 1139	Drummed Organic Liquid/Sludge Requiring Stabilization, Drums Greater than 55 Gallons	EA	0	N/A	1.000	0.00	0.00	455.00	455.00
33 19 7228	**Landfill Drummed Liquid/Sludge, Solar Evaporation, <= 30 Gallon Drum, pH>2.0**	**EA**				**0.00**	**0.00**	**690.00**	**690.00**
13278 1141	Drummed Liquid/Sludge Treated By Solar Evaporation, Drums 30 Gallons or Less	EA	0	N/A	1.000	0.00	0.00	690.00	690.00
33 19 7229	**Landfill Drummed Liquid/Sludge, Solar Evaporation, 55 Gallon Drum, pH > 2.0**	**EA**				**0.00**	**0.00**	**212.00**	**212.00**
13278 1142	Drummed Liquid/Sludge Treated By Solar Evaporation, 55 Gallon Drums	EA	0	N/A	1.000	0.00	0.00	212.00	212.00
33 19 7230	**Landfill Drummed Liquid/Sludge, Solar Evaporation, > 55 Gallon Drum, pH > 2.0**	**EA**				**0.00**	**0.00**	**246.00**	**246.00**
13278 1143	Drummed Liquid/Sludge Treated By Solar Evaporation, Drums Greater than 55 Gallons	EA	0	N/A	1.000	0.00	0.00	246.00	246.00
33 19 7231	**Landfill Mineral Acids, 0% - 40% Concentration, <= 30 Gallon Drum**	**EA**				**0.00**	**0.00**	**395.00**	**395.00**
13278 1144	Mineral Acids with Concentration of 0% -40%, Drums 30 Gallons or Less	EA	0	N/A	1.000	0.00	0.00	395.00	395.00
33 19 7232	**Landfill Mineral Acids, 0% - 40% Concentration, 55 Gallon Drum**	**EA**				**0.00**	**0.00**	**525.00**	**525.00**
13278 1145	Mineral Acids with Concentration of 0% -40%, 55 Gallon Drums	EA	0	N/A	1.000	0.00	0.00	525.00	525.00
33 19 7233	**Landfill Mineral Acids, 0% - 40% Concentration, > 55 Gallon Drum**	**EA**				**0.00**	**0.00**	**800.00**	**800.00**
13278 1146	Mineral Acids with Concentration of 0% -40%, Drums Greater than 55 Gallons	EA	0	N/A	1.000	0.00	0.00	800.00	800.00
33 19 7234	**Landfill Mineral Acids, 41% - 60% Concentration, <= 30 Gallon Drum**	**EA**				**0.00**	**0.00**	**450.00**	**450.00**
13278 1147	Mineral Acids with Concentration of 41% -60%, Drums 30 Gallons or Less	EA	0	N/A	1.000	0.00	0.00	450.00	450.00
33 19 7235	**Landfill Mineral Acids, 41% - 60% Concentration, 55 Gallon Drum**	**EA**				**0.00**	**0.00**	**595.00**	**595.00**
13278 1148	Mineral Acids with Concentration of 41% -60%, 55 Gallon Drums	EA	0	N/A	1.000	0.00	0.00	595.00	595.00
33 19 7236	**Landfill Mineral Acids, 41% - 60% Concentration, > 55 Gallon Drum**	**EA**				**0.00**	**0.00**	**900.00**	**900.00**
13278 1149	Mineral Acids with Concentration of 41% -60%, Drums Greater than 55 Gallons	EA	0	N/A	1.000	0.00	0.00	900.00	900.00

33 19 72 Landfill Disposal		Unit	Hourly Output	Crew	Factor	Labor	Equip.	Matl.	Total
33 19 7237	**Landfill Mineral Acids, > 61% Concentration, <= 30 Gallon Drum**	**EA**				**0.00**	**0.00**	**535.00**	**535.00**
13278 1151	Mineral Acids with Concentration of Greater than 61%, Drums 30 Gallons or Less	EA	0	N/A	1.000	0.00	0.00	535.00	535.00
33 19 7238	**Landfill Mineral Acids, > 61% Concentration, 55 Gallon Drum**	**EA**				**0.00**	**0.00**	**715.00**	**715.00**
13278 1152	Mineral Acids with Concentration of Greater than 61%, 55 Gallon Drums	EA	0	N/A	1.000	0.00	0.00	715.00	715.00
33 19 7239	**Landfill Mineral Acids, > 61% Concentration, > 55 Gallon Drum**	**EA**				**0.00**	**0.00**	**1,150**	**1,150**
13278 1153	Mineral Acids with Concentration of Greater than 61%, Drums Greater than 55 Gallons	EA	0	N/A	1.000	0.00	0.00	1,150	1,150
33 19 7240	**Landfill Fuel Substitutes, 0% - 10% Solids, <= 30 Gallon Drum**	**EA**				**0.00**	**0.00**	**80.00**	**80.00**
13278 1154	Fuels Substitution Comprised of 0% - 10% Solids, Drums 30 Gallons or Less	EA	0	N/A	1.000	0.00	0.00	80.00	80.00
33 19 7241	**Landfill Fuel Substitutes, 0% - 10% Solids, 55 Gallon Drum**	**EA**				**0.00**	**0.00**	**125.00**	**125.00**
13278 1155	Fuels Substitution Comprised of 0% - 10% Solids, 55 Gallon Drums	EA	0	N/A	1.000	0.00	0.00	125.00	125.00
33 19 7242	**Landfill Fuel Substitutes, 0% - 10% Solids, > 55 Gallon Drum**	**EA**				**0.00**	**0.00**	**150.00**	**150.00**
13278 1156	Fuels Substitution Comprised of 0% - 10% Solids, Drums Greater than 55 Gallons	EA	0	N/A	1.000	0.00	0.00	150.00	150.00
33 19 7243	**Landfill Fuel Substitutes, 10.1% - 25% Solids, <= 30 Gallon Drum**	**EA**				**0.00**	**0.00**	**150.00**	**150.00**
13278 1157	Fuels Substitution Comprised of 10.1% - 25% Solids, Drums 30 Gallons or Less	EA	0	N/A	1.000	0.00	0.00	150.00	150.00
33 19 7244	**Landfill Fuel Substitutes, 10.1% - 25% Solids, 55 Gallon Drum**	**EA**				**0.00**	**0.00**	**150.00**	**150.00**
13278 1158	Fuels Substitution Comprised of 10.1% -25% Solids, 55 Gallon Drums	EA	0	N/A	1.000	0.00	0.00	150.00	150.00
33 19 7245	**Landfill Fuel Substitutes, 10.1% - 25% Solids, > 55 Gallon Drum**	**EA**				**0.00**	**0.00**	**225.00**	**225.00**
13278 1159	Fuels Substitution Comprised of 10.1% -25% Solids, Drums Greater than 55 Gallons	EA	0	N/A	1.000	0.00	0.00	225.00	225.00
33 19 7246	**Landfill Fuel Substitutes, 25.1% - 50% Solids, <= 30 Gallon Drum**	**EA**				**0.00**	**0.00**	**200.00**	**200.00**
13278 1161	Fuels Substitution Comprised of 25.1% -50% Solids, Drums 30 Gallons or Less	EA	0	N/A	1.000	0.00	0.00	200.00	200.00
33 19 7247	**Landfill Fuel Substitutes, 25.1% - 50% Solids, 55 Gallon Drum**	**EA**				**0.00**	**0.00**	**225.00**	**225.00**
13278 1162	Fuels Substitution Comprised of 25.1% -50% Solids, 55 Gallon Drums	EA	0	N/A	1.000	0.00	0.00	225.00	225.00

33 19 72	Landfill Disposal	Unit	Hourly Output	Crew	Factor	Labor	Equip.	Matl.	Total
33 19 7248	**Landfill Fuel Substitutes, 25.1% - 50% Solids, > 55 Gallon Drum**	**EA**				**0.00**	**0.00**	**471.24**	**471.24**
13278 1163	Fuels Substitution Comprised of 25.1% -50% Solids, Drums Greater than 55 Gallons	EA	0	N/A	1.000	0.00	0.00	471.24	471.24
33 19 7249	**Landfill Fuel Substitutes, 50.1% - 75% Solids, <= 30 Gallon Drum**	**EA**				**0.00**	**0.00**	**225.00**	**225.00**
13278 1164	Fuels Substitution Comprised of 50.1% -75% Solids, Drums 30 Gallons or Less	EA	0	N/A	1.000	0.00	0.00	225.00	225.00
33 19 7250	**Landfill Fuel Substitutes, 50.1% - 75% Solids, 55 Gallon Drum**	**EA**				**0.00**	**0.00**	**225.00**	**225.00**
13278 1165	Fuels Substitution Comprised of 50.1% -75% Solids, 55 Gallon Drums	EA	0	N/A	1.000	0.00	0.00	225.00	225.00
33 19 7251	**Landfill Fuel Substitutes, 50.1% - 75% Solids, > 55 Gallon Drum**	**EA**				**0.00**	**0.00**	**656.73**	**656.73**
13278 1166	Fuels Substitution Comprised of 50.1% -75% Solids, Drums Greater than 55 Gallons	EA	0	N/A	1.000	0.00	0.00	656.73	656.73
33 19 7252	**Landfill Fuel Substitutes, 75.1% - 100% Solids, <= 30 Gallon Drum**	**EA**				**0.00**	**0.00**	**466.23**	**466.23**
13278 1167	Fuels Substitution Comprised of 75.1% -100% Solids, Drums 30 Gallons or Less	EA	0	N/A	1.000	0.00	0.00	466.23	466.23
33 19 7253	**Landfill Fuel Substitutes, 75.1% - 100% Solids, 55 Gallon Drum**	**EA**				**0.00**	**0.00**	**621.64**	**621.64**
13278 1168	Fuels Substitution Comprised of 75.1% -100% Solids, 55 Gallons	EA	0	N/A	1.000	0.00	0.00	621.64	621.64
33 19 7254	**Landfill Fuel Substitutes, 75.1% - 100% Solids, > 55 Gallon Drum**	**EA**				**0.00**	**0.00**	**947.50**	**947.50**
13278 1169	Fuels Substitution Comprised of 75.1% -100% Solids, Drums Greater than 55 Gallons	EA	0	N/A	1.000	0.00	0.00	947.50	947.50
33 19 7255	**Landfill Packaged Lab Chemicals, 5 - 25 Gallon Drum**	**EA**				**0.00**	**0.00**	**215.00**	**215.00**
13278 1171	Packaged Lab Chemicals, Drums 5 to 25 Gallons	EA	0	N/A	1.000	0.00	0.00	215.00	215.00
33 19 7256	**Landfill Packaged Lab Chemicals, 30 - 40 Gallon Drum**	**EA**				**0.00**	**0.00**	**250.00**	**250.00**
13278 1172	Packaged Lab Chemicals, Drums 30 to 40 Gallons	EA	0	N/A	1.000	0.00	0.00	250.00	250.00
33 19 7257	**Landfill Packaged Lab Chemicals, > 45 Gallon Drum**	**EA**				**0.00**	**0.00**	**262.00**	**262.00**
13278 1173	Packaged Lab Chemicals, Drums Greater than 45 Gallons	EA	0	N/A	1.000	0.00	0.00	262.00	262.00
33 19 7258	**Landfill Packaged Lab Chemicals Requiring Stabilization, 5 - 25 Gallon Drum**	**EA**				**0.00**	**0.00**	**784.73**	**784.73**
13278 1174	Packaged Lab Chemicals Requiring Stabilization, Drums 5 to 25 Gallons	EA	0	N/A	1.000	0.00	0.00	784.73	784.73

33 19 72	Landfill Disposal	Unit	Hourly Output	Crew	Factor	Labor	Equip.	Matl.	Total
33 19 7259	**Landfill Packaged Lab Chemicals Requiring Stabilization, 30 - 40 Gallon Drum**	**EA**				**0.00**	**0.00**	**784.73**	**784.73**
13278 1175	Packaged Lab Chemicals Requiring Stabilization, Drums 30 to 40 Gallons	EA	0	N/A	1.000	0.00	0.00	784.73	784.73
33 19 7260	**Landfill Packaged Lab Chemicals Requiring Stabilization, > 45 Gallon Drum**	**EA**				**0.00**	**0.00**	**1,046**	**1,046**
13278 1176	Packaged Lab Chemicals Requiring Stabilization, Drums Greater than 45 Gallons	EA	0	N/A	1.000	0.00	0.00	1,046	1,046
33 19 7261	**Landfill Solid Palletized Waste**	**EA**				**0.00**	**0.00**	**440.00**	**440.00**
13278 1241	Solid Palletized Wastes	EA	0	N/A	1.000	0.00	0.00	440.00	440.00
33 19 7262	**Landfill Liquid or Sludge Palletized Waste**	**EA**				**0.00**	**0.00**	**480.00**	**480.00**
13278 1242	Liquid or Sludge Palletized Wastes	EA	0	N/A	1.000	0.00	0.00	480.00	480.00
33 19 7263	**Landfill Hazardous Solid Bulk Waste by Ton**	**TON**				**0.00**	**0.00**	**169.45**	**169.45**
13278 1211	Hazardous Solid Bulk Waste, Based on 2,000 Lb/CY	TON	0	N/A	1.000	0.00	0.00	169.45	169.45
33 19 7264	**Landfill Hazardous Solid Bulk Waste by CY**	**CY**				**0.00**	**0.00**	**148.00**	**148.00**
13278 1212	Hazardous Solid Bulk Waste, Less than 2,000 Lb/CY	CY	0	N/A	1.000	0.00	0.00	148.00	148.00
33 19 7265	**Landfill Hazardous Solid Bulk Waste Requiring Stabilization**	**TON**				**0.00**	**0.00**	**251.00**	**251.00**
13278 1213	Hazardous Solid Bulk Waste Requiring Stabilization, Based on 2,000 Lb/CY	TON	0	N/A	1.000	0.00	0.00	251.00	251.00
33 19 7266	**Landfill Hazardous Solid Bulk Waste, Odd-shaped Container, < 10 CF/Container**	**CF**				**0.00**	**0.00**	**10.45**	**10.45**
13278 1214	Hazardous Solid Bulk Waste in Odd-shaped Containers and Boxes	CF	0	N/A	1.000	0.00	0.00	10.45	10.45
33 19 7267	**Landfill Hazardous Solid Bulk Waste, Odd-shaped Container, >= 10 CF/Container**	**CF**				**0.00**	**0.00**	**10.90**	**10.90**
13278 1215	Hazardous Solid Bulk Waste in Odd-shaped Containers and Boxes	CF	0	N/A	1.000	0.00	0.00	10.90	10.90
33 19 7268	**Landfill Hazardous Solid Bulk Waste, Odd-shaped Container Requiring Stabilization**	**CF**				**0.00**	**0.00**	**101.44**	**101.44**
13278 1216	Hazardous Solid Bulk Waste in Odd-shaped Containers and Boxes	CF	0	N/A	1.000	0.00	0.00	101.44	101.44
33 19 7269	**Landfill Nonhazardous Solid Bulk Waste by Ton**	**TON**				**0.00**	**0.00**	**72.00**	**72.00**
13278 1217	Non-Hazardous Solid Bulk Waste, Based on 2,000 Lb/CY	TON	0	N/A	1.000	0.00	0.00	72.00	72.00
33 19 7270	**Landfill Nonhazardous Solid Bulk Waste by CY**	**CY**				**0.00**	**0.00**	**93.50**	**93.50**
13278 1218	Non-Hazardous Solid Bulk Waste, Less than 2,000 Lb/CY	CY	0	N/A	1.000	0.00	0.00	93.50	93.50

33 19 72 Landfill Disposal		Unit	Hourly Output	Crew	Factor	Labor	Equip.	Matl.	Total
33 19 7271	**Landfill Jumbo Bags, Direct**	**EA**				**0.00**	**0.00**	**275.00**	**275.00**
13278 1219	Jumbo Bags, Direct Landfill (1 CY/Bag)	EA	0	N/A	1.000	0.00	0.00	275.00	275.00
33 19 7272	**Landfill Jumbo Bags Requiring Stabilization**	**EA**				**0.00**	**0.00**	**279.00**	**279.00**
13278 1221	Jumbo Bags Requiring Stabilization (1 CY/Bag)	EA	0	N/A	1.000	0.00	0.00	279.00	279.00
33 19 7273	**Landfill Hazardous Liquid Bulk Waste**	**GAL**				**0.00**	**0.00**	**1.81**	**1.81**
13278 1222	Hazardous Liquid Bulk Waste	GAL	0	N/A	1.000	0.00	0.00	1.81	1.81
33 19 7274	**Landfill Nonhazardous Liquid Bulk Waste**	**GAL**				**0.00**	**0.00**	**2.50**	**2.50**
13278 1223	Non-Hazardous Liquid Bulk Waste	GAL	0	N/A	1.000	0.00	0.00	2.50	2.50
33 19 7275	**Landfill Hazard Liquid Bulk Waste Requiring Stabilization**	**GAL**				**0.00**	**0.00**	**3.16**	**3.16**
13278 1224	Hazardous Liquid Bulk Waste Requiring Stabilization	GAL	0	N/A	1.000	0.00	0.00	3.16	3.16
33 19 7276	**Landfill Hazard Liquid Bulk Waste, Solar Evaporation, pH > 2.0**	**GAL**				**0.00**	**0.00**	**2.58**	**2.58**
13278 1225	Hazardous Bulk Liquid Waste Treated by Solar Evaporation, pH	GAL	0	N/A	1.000	0.00	0.00	2.58	2.58
33 19 7277	**Landfill Hazardous Nonfuel Liquid/Sludge**	**GAL**				**0.00**	**0.00**	**1.33**	**1.33**
13278 1226	Hazardous Non-Fuel Liquid/Sludge	GAL	0	N/A	1.000	0.00	0.00	1.33	1.33
33 19 7278	**Landfill Nonhazardous Nonfuel Liquid/Sludge**	**GAL**				**0.00**	**0.00**	**2.58**	**2.58**
13278 1227	Non-Hazardous, Non-Fuel Liquid/Sludge	GAL	0	N/A	1.000	0.00	0.00	2.58	2.58
33 19 7279	**Landfill Fuel Substitution Liquids**	**GAL**				**0.00**	**0.00**	**1.11**	**1.11**
13278 1228	Fuel Substitution Liquids	GAL	0	N/A	1.000	0.00	0.00	1.11	1.11
33 19 7280	**30 Gallon, 17C, Open**	**EA**				**0.00**	**0.00**	**86.50**	**86.50**
02083 5215	30 Gallon, 17C, Open	EA	0	N/A	1.000	0.00	0.00	86.50	86.50
33 19 7281	**30 Gallon, 17H, Open**	**EA**				**0.00**	**0.00**	**86.00**	**86.00**
02083 5216	30 Gallon, 17H, Open	EA	0	N/A	1.000	0.00	0.00	86.00	86.00
33 19 7282	**30 Gallon, 17E, Open**	**EA**				**0.00**	**0.00**	**69.00**	**69.00**
02083 5217	30 Gallon, 17E, Open	EA	0	N/A	1.000	0.00	0.00	69.00	69.00

33 19 73 Commercial Disposal		Unit	Hourly Output	Crew	Factor	Labor	Equip.	Matl.	Total
33 19 7301	**Haul & Dispose Debris, 16.5 CY Truck, 10 Mile, Landfill**	**CY**				**0.80**	**3.33**	**92.50**	**96.63**
02234 1116	Haul, 16.5 CY (12.6 m3) Truck, 10 Miles (15 km), 40 MPH (60 km/Hour), 2.1 Cycles/Hour	CY	34.25	COETH	1.000	0.80	3.33	0.00	4.13
13278 1235	Non-Hazardous Bulk Solid Waste @ Sanitary Landfill	CY	0	N/A	1.000	0.00	0.00	92.50	92.50

33 19 Disposal (Commercial)

33 19 73 Commercial Disposal		Unit	Hourly Output	Crew	Factor	Labor	Equip.	Matl.	Total
33 19 7302	**Disposal of Nonhazardous Liquid Bulk Waste**	**GAL**				**0.00**	**0.00**	**2.50**	**2.50**
13278 1223	Non-Hazardous Liquid Bulk Waste	GAL	0	N/A	1.000	0.00	0.00	2.50	2.50
33 19 7303	**Disposal of Hazardous Liquid Bulk Waste**	**GAL**				**0.00**	**0.00**	**3.16**	**3.16**
13278 1224	Hazardous Liquid Bulk Waste Requiring Stabilization	GAL	0	N/A	1.000	0.00	0.00	3.16	3.16

33 19 95 Commercial Disposal (Incineration)		Unit	Hourly Output	Crew	Factor	Labor	Equip.	Matl.	Total
33 19 9501	**Energetic Drummed Solid Waste Incineration, 55 Gallon**	**EA**				**0.00**	**0.00**	**1,225**	**1,225**
13277 2611	Drummed Solid Waste, Energetic, Amenable to Bulking, 55 Gallon	EA	0	N/A	1.000	0.00	0.00	1,225	1,225
33 19 9502	**Nonenergetic Drummed Solid Waste Incineration, 55 Gallon**	**EA**				**0.00**	**0.00**	**940.31**	**940.31**
13277 2612	Drummed Solid Waste, Non-Energetic, Amenable to Bulking	EA	0	N/A	1.000	0.00	0.00	940.31	940.31
33 19 9503	**Energetic, Requiring Repack, Drummed Solid Waste Incineration, 55 Gallon**	**EA**				**0.00**	**0.00**	**1,175**	**1,175**
13277 2613	Drummed Solid Waste, Energetic, Requiring Repacking, 55 Gallon	EA	0	N/A	1.000	0.00	0.00	1,175	1,175
33 19 9504	**Nonenergetic, Requiring Repack, Drummed Solid Waste Incineration, 55 Gallon**	**EA**				**0.00**	**0.00**	**850.00**	**850.00**
13277 2614	Drummed Solid Waste, Non-Energetic, Requiring Repacking	EA	0	N/A	1.000	0.00	0.00	850.00	850.00
33 19 9505	**Reactive or Corrosive Drummed Solids Incineration**	**LB**				**0.00**	**0.00**	**2.79**	**2.79**
13277 2615	Reactive or Corrosive Drummed Solids, per Pound	LB	0	N/A	1.000	0.00	0.00	2.79	2.79
33 19 9506	**Amenable-to-Bulking Drummed Liquids Incineration, 55 Gallon**	**EA**				**0.00**	**0.00**	**795.00**	**795.00**
13277 2616	Drummed Liquids, Amenable to Bulking, 55 Gallon Drum	EA	0	N/A	1.000	0.00	0.00	795.00	795.00
33 19 9507	**Drummed Liquids Requiring Repack Incineration, 55 Gallon**	**EA**				**0.00**	**0.00**	**1,400**	**1,400**
13277 2617	Drummed Liquids Requiring Repacking, 55 Gallon Drum	EA	0	N/A	1.000	0.00	0.00	1,400	1,400
33 19 9508	**Drummed Reactive or Corrosive Liquids Incineration, 55 Gallon**	**EA**				**0.00**	**0.00**	**1,050**	**1,050**
13277 2618	Reactive or Corrosive Drummed Liquids, 55 Gallon Drum	EA	0	N/A	1.000	0.00	0.00	1,050	1,050
33 19 9509	**Energetic Drummed Sludge Incineration, 55 Gallon**	**EA**				**0.00**	**0.00**	**510.00**	**510.00**
13277 2619	Drummed Sludge, Energetic, Amenable to Bulking, 55 Gallon Drum	EA	0	N/A	1.000	0.00	0.00	510.00	510.00

33 19 95	Commercial Disposal (Incineration)	Unit	Hourly Output	Crew	Factor	Labor	Equip.	Matl.	Total
33 19 9510	**Nonenergetic Drummed Sludge Incineration, 55 Gallon**	**EA**				**0.00**	**0.00**	**450.00**	**450.00**
13277 2621	Drummed Sludge, Non-Energetic, Amenable to Bulking, 55 Gallon Drum	EA	0	N/A	1.000	0.00	0.00	450.00	450.00
33 19 9511	**Drummed Sludge Requiring Repack Incineration, 55 Gallon**	**EA**				**0.00**	**0.00**	**1,250**	**1,250**
13277 2622	Drummed Sludge Requiring Repacking, 55 Gallon Drum	EA	0	N/A	1.000	0.00	0.00	1,250	1,250
33 19 9512	**Lean Water Incineration for Non-PCB 55 Gallon Drummed Waste**	**EA**				**0.00**	**0.00**	**785.00**	**785.00**
13277 2623	Lean Water Incineration for Non-PCB Drummed Waste, 55 Gallon	EA	0	N/A	1.000	0.00	0.00	785.00	785.00
33 19 9513	**Lab Packs Containing Nonreactive Material Incineration**	**LB**				**0.00**	**0.00**	**2.27**	**2.27**
13277 2624	Lab Packs Containing Non-Reactive Material	LB	0	N/A	1.000	0.00	0.00	2.27	2.27
33 19 9514	**Fluorinated Aerosol Cans Incineration**	**LB**				**0.00**	**0.00**	**3.96**	**3.96**
13277 2625	Aerosol Cans, Fluorinated	LB	0	N/A	1.000	0.00	0.00	3.96	3.96
33 19 9515	**Nonfluorinated Aerosol Cans Incineration**	**LB**				**0.00**	**0.00**	**2.99**	**2.99**
13277 2626	Aerosol Cans, Non-Fluorinated	LB	0	N/A	1.000	0.00	0.00	2.99	2.99
33 19 9516	**Drummed Waste Containing Over 5% Halogen, Extra Charges**	**EA**				**0.00**	**0.00**	**8.00**	**8.00**
13277 2627	Extra Charges for Waste Containing Over 5% Halogen	EA	0	N/A	1.000	0.00	0.00	8.00	8.00
33 19 9517	**Drummed Waste Containing Over 10% Ash, Extra Charges**	**EA**				**0.00**	**0.00**	**4.70**	**4.70**
13277 2628	Extra Charges for Waste Containing Over 10% Ash	EA	0	N/A	1.000	0.00	0.00	4.70	4.70
33 19 9518	**Waste Packed in 85 Gallon Metal Drums, Extra Charges**	**EA**				**0.00**	**0.00**	**69.00**	**69.00**
13277 2629	Extra Charges for Waste Packed in Metal 85 Gallon Drums	EA	0	N/A	1.000	0.00	0.00	69.00	69.00
33 19 9519	**Waste Packed in 85 Gallon Plastic Drums, Extra Charges**	**EA**				**0.00**	**0.00**	**174.00**	**174.00**
13277 2631	Extra Charges for Waste Packed in Plastic 85 Gallon Drums	EA	0	N/A	1.000	0.00	0.00	174.00	174.00
33 19 9520	**Incineration of Bulk Solid Hazardous Waste (2,000 Lb/CY)**	**CY**				**0.00**	**0.00**	**616.00**	**616.00**
13277 2653	Incinerate Bulk Solid Waste	LB	0	N/A	700.000	0.00	0.00	0.88	616.00
33 19 9521	**Bulk Liquids, 2,000 to 12,000 BTU, Incineration**	**LB**				**0.00**	**0.00**	**1.39**	**1.39**
13277 2642	Bulk Liquids, 2,000 BTU to 12,000 BTU, per Pound	LB	0	N/A	1.000	0.00	0.00	1.39	1.39

33 19 95 Commercial Disposal (Incineration)		Unit	Hourly Output	Crew	Factor	Labor	Equip.	Matl.	Total
33 19 9522	**Reactive or Corrosive Bulk Liquids Incineration**	**GAL**				**0.00**	**0.00**	**14.25**	**14.25**
13277 2643	Reactive or Corrosive Bulk Liquids	GAL	0	N/A	1.000	0.00	0.00	14.25	14.25
33 19 9523	**Lean Water Incineration for Non-PCB Bulk Liquids**	**GAL**				**0.00**	**0.00**	**3.82**	**3.82**
13277 2644	Lean Water Incineration for Non-PCB Bulk Liquids	GAL	0	N/A	1.000	0.00	0.00	3.82	3.82
33 19 9524	**Aqueous Liquids, Approximately 90% Water Content, Incineration**	**LB**				**0.00**	**0.00**	**1.22**	**1.22**
13277 2645	Aqueous Liquids, Approximately 90% Water	LB	0	N/A	1.000	0.00	0.00	1.22	1.22
33 19 9525	**55 Gallon Drum with Chlorine Content 26 - 50%, Extra Charges**	**EA**				**0.00**	**0.00**	**168.00**	**168.00**
13277 2646	Extra Charges for Chlorine Content of 26% to 50%, 55 Gallon Drum	EA	0	N/A	1.000	0.00	0.00	168.00	168.00
33 19 9526	**55 Gallon Drum with Chlorine Content > 50%, Extra Charges**	**EA**				**0.00**	**0.00**	**213.00**	**213.00**
13277 2647	Extra Charges for Chlorine Content > 50%, 55 Gallon Drum	EA	0	N/A	1.000	0.00	0.00	213.00	213.00
33 19 9527	**Bulk Waste Containing Over 5% Halogen, Extra Charges**	**GAL**				**0.00**	**0.00**	**0.30**	**0.30**
13277 2648	Extra Charges for Waste Containing Over 5% Halogen	GAL	0	N/A	1.000	0.00	0.00	0.30	0.30
33 19 9528	**Bulk Waste Containing Over 10% Ash, Extra Charges**	**GAL**				**0.00**	**0.00**	**0.12**	**0.12**
13277 2649	Extra Charges for Waste Containing Over 10% Ash	GAL	0	N/A	1.000	0.00	0.00	0.12	0.12
33 19 9529	**Drummed or Boxed PCB Capacitors Incineration**	**LB**				**0.00**	**0.00**	**1.95**	**1.95**
13277 2711	Drummed or Boxed PCB Capacitors	LB	0	N/A	1.000	0.00	0.00	1.95	1.95
33 19 9530	**PCB Debris (Clothing, Rags, etc.) Incineration**	**LB**				**0.00**	**0.00**	**1.82**	**1.82**
13277 2712	Drummed PCB Debris - Clothing, Rags, Etc.	LB	0	N/A	1.000	0.00	0.00	1.82	1.82
33 19 9531	**PCB Contaminated Soil Incineration**	**LB**				**0.00**	**0.00**	**1.95**	**1.95**
13277 2713	Drummed PCB Contaminated Soil	LB	0	N/A	1.000	0.00	0.00	1.95	1.95
33 19 9532	**PCB Light Ballasts Incineration**	**LB**				**0.00**	**0.00**	**1.95**	**1.95**
13277 2714	Drummed PCB Light Ballasts	LB	0	N/A	1.000	0.00	0.00	1.95	1.95
33 19 9533	**Drummed PCB Transformer Incineration, 55 Gallon**	**EA**				**0.00**	**0.00**	**805.00**	**805.00**
13277 2715	Drummed PCB Transformer, 55 Gallon Drum	EA	0	N/A	1.000	0.00	0.00	805.00	805.00

33 19 Disposal (Commercial)

33 19 95	Commercial Disposal (Incineration)	Unit	Hourly Output	Crew	Factor	Labor	Equip.	Matl.	Total
33 19 9534	**Incineration of Drummed PCB Solid Waste, 55 Gallon**	**EA**				**0.00**	**0.00**	**74.50**	**74.50**
13277 2732	Incinerate Drummed PCB Solid Waste, 55 Gallon	EA	0	N/A	1.000	0.00	0.00	74.50	74.50
33 19 9535	**Drummed PCB Liquids < 100,000 PPM Incineration, 55 Gallon**	**EA**				**0.00**	**0.00**	**435.00**	**435.00**
13277 2719	Drummed PCB Liquids Less than 100,000 PPM, 55 Gallon Drum	EA	0	N/A	1.000	0.00	0.00	435.00	435.00
33 19 9536	**Drummed PCB Liquids > 100,000 PPM Incineration, 55 Gallon**	**EA**				**0.00**	**0.00**	**510.00**	**510.00**
13277 2721	Drummed PCB Liquids Greater than 100,000 PPM, 55 Gallon Drum	EA	0	N/A	1.000	0.00	0.00	510.00	510.00
33 19 9537	**Lean Water Incineration for 55 Gallon Drummed PCB Waste**	**EA**				**0.00**	**0.00**	**430.00**	**430.00**
13277 2733	Lean Water Incineration, Drummed PCB Waste, 55 Gallon	EA	0	N/A	1.000	0.00	0.00	430.00	430.00
33 19 9538	**Palletized PCB Solid Waste, Single-layer Stack, Incineration**	**EA**				**0.00**	**0.00**	**650.00**	**650.00**
13277 2723	Palletized PCB Solid Waste, Single Layer Stack	EA	0	N/A	1.000	0.00	0.00	650.00	650.00
33 19 9539	**Empty Uncrushed Drums Incineration**	**EA**				**0.00**	**0.00**	**80.00**	**80.00**
13277 2729	Empty Drums, Not Crushed, 55 Gallon Drum	EA	0	N/A	1.000	0.00	0.00	80.00	80.00
33 19 9540	**Incineration of Bulk PCB Solid Waste (2,000 Lb/CY)**	**CY**				**0.00**	**0.00**	**2,999**	**2,999**
13277 2746	Incinerate Bulk PCB Solid Waste	CY	0	N/A	1.000	0.00	0.00	2,999	2,999
33 19 9541	**Lean Water Incineration of Bulk PCB Waste**	**GAL**				**0.00**	**0.00**	**6.55**	**6.55**
13277 2745	Lean Water Incineration of Bulk PCB Waste	GAL	0	N/A	1.000	0.00	0.00	6.55	6.55
33 19 9542	**Incineration of Boxes & Nondrummed PCB Solid Wastes**	**CY**				**0.00**	**0.00**	**369.90**	**369.90**
13277 2743	Boxes and Non-Drummed PCB Solid Waste	CF	0	N/A	27.000	0.00	0.00	13.70	369.90
33 19 9543	**Initial Waste Stream Evaluation, Non-PCB**	**EA**				**0.00**	**0.00**	**2,350**	**2,350**
13277 2633	Initial Waste Stream Evaluation - Incineration	EA	0	N/A	1.000	0.00	0.00	2,350	2,350
33 19 9544	**Initial PCB Waste Stream Evaluation**	**EA**				**0.00**	**0.00**	**745.00**	**745.00**
13277 2731	Initial Waste Stream Evaluation Fee	EA	0	N/A	1.000	0.00	0.00	745.00	745.00

33 19 99	Other	Unit	Hourly Output	Crew	Factor	Labor	Equip.	Matl.	Total
33 19 9901	**750 Gallon, 10' x 10' x 17" Inflatable Containment Berm, Single Lined**	**EA**				**35.20**	**0.00**	**1,750**	**1,785**
02083 6211	10' x 10' x 17", 750 Gallon Capacity, 70 CF of Air, Single Lined	EA	3	ULABD	1.000	35.20	0.00	1,750	1,785

33 19 99 Other		Unit	Hourly Output	Crew	Factor	Labor	Equip.	Matl.	Total
33 19 9902	**1,800 Gallon, 13' x 13' x 17" Inflatable Containment Berm, Single Lined**	**EA**				**35.20**	**0.00**	**2,925**	**2,960**
02083 6212	13' x 13' x 17", 1,800 Gallon Capacity, 80 CF of Air, Single Lined	EA	3	ULABD	1.000	35.20	0.00	2,925	2,960
33 19 9903	**3,400 Gallon, 18' x 18' x 17" Inflatable Containment Berm, Single Lined**	**EA**				**70.41**	**0.00**	**4,025**	**4,095**
02083 6213	18' x 18' x 17", 3,400 Gallon Capacity, 130 CF of Air, Single Lined	EA	1.5	ULABD	1.000	70.41	0.00	4,025	4,095
33 19 9904	**6,100 Gallon, 24' x 24' x 17" Inflatable Containment Berm, Single Lined**	**EA**				**84.49**	**0.00**	**6,125**	**6,209**
02083 6214	24' x 24' x 17", 6,100 Gallon Capacity, 170 CF of Air, Single Lined	EA	1.25	ULABD	1.000	84.49	0.00	6,125	6,209
33 19 9905	**10,800 Gallon, 32' x 32' x 17" Inflatable Containment Berm, Single Lined**	**EA**				**105.61**	**0.00**	**6,500**	**6,606**
02083 6215	32' x 32' x 17", 10,800 Gallon Capacity, 230 CF of Air, Single Lined	EA	1	ULABD	1.000	105.61	0.00	6,500	6,606
33 19 9906	**21,600 Gallon, 32' x 32' x 34" Inflatable Containment Berm, Single Lined**	**EA**				**168.98**	**0.00**	**9,600**	**9,769**
02083 6216	32' x 32' x 34", 21,600 Gallon Capacity, 920 CF of Air, Single Lined	EA	0.625	ULABD	1.000	168.98	0.00	9,600	9,769
33 19 9907	**53,400 Gallon, 74' x 34' x 34" Inflatable Containment Berm, Single Lined**	**EA**				**281.63**	**0.00**	**16,900**	**17,182**
02083 6217	74' x 34' x 34", 53,400 Gallon Capacity, 1,500 CF of Air, Single Lined	EA	0.375	ULABD	1.000	281.63	0.00	16,900	17,182
33 19 9908	**7,600 Gallon, 45' x 15' x 17" Inflatable Containment Berm, Double Lined**	**EA**				**105.61**	**0.00**	**5,975**	**6,081**
02083 6218	45' x 15' x 17", 7,600 Gallon Capacity, 220 CF of Air, Double Lined	EA	1	ULABD	1.000	105.61	0.00	5,975	6,081
33 19 9909	**11,000 Gallon, 65' x 16' x 17" Inflatable Containment Berm, Double Lined**	**EA**				**105.61**	**0.00**	**7,225**	**7,331**
02083 6219	65' x 16' x 17", 11,000 Gallon Capacity, 290 CF of Air, Double Lined	EA	1	ULABD	1.000	105.61	0.00	7,225	7,331
33 19 9910	**11,600 Gallon, 50' x 22' x 17" Inflatable Containment Berm, Double Lined**	**EA**				**105.61**	**0.00**	**7,150**	**7,256**
02083 6221	50' x 22' x 17", 11,600 Gallon Capacity, 260 CF of Air, Double Lined	EA	1	ULABD	1.000	105.61	0.00	7,150	7,256
33 19 9911	**Air Blower to Inflate Portable Containment Berm**	**EA**				**0.00**	**0.00**	**190.00**	**190.00**
02083 6222	Air Blower to Inflate Portable Containment Berms	EA	0	N/A	1.000	0.00	0.00	190.00	190.00
33 19 9921	**DOT Steel Drum, 55 Gallon**	**EA**				**0.00**	**0.00**	**89.50**	**89.50**
02083 5211	55 Gallon, 17C, Open	EA	0	N/A	1.000	0.00	0.00	89.50	89.50
33 19 9922	**Polyethylene Closed Head Drum, 55 Gallon**	**EA**				**0.00**	**0.00**	**67.50**	**67.50**
02083 5311	55 Gallon	EA	0	N/A	1.000	0.00	0.00	67.50	67.50

33 19 99 Other		Unit	Hourly Output	Crew	Factor	Labor	Equip.	Matl.	Total
33 19 9923	**Polyethylene Open Head Drum, 55 Gallon**	**EA**				**0.00**	**0.00**	**62.36**	**62.36**
02083 5321	55 Gallon	EA	0	N/A	1.000	0.00	0.00	62.36	62.36
33 19 9924	**DOT Steel Salvage Drum, 85 Gallon, 16 Gauge**	**EA**				**0.00**	**0.00**	**75.77**	**75.77**
02083 5112	85 Gallon, 26" Diameter x 38" High	EA	0	N/A	1.000	0.00	0.00	75.77	75.77
33 19 9925	**DOT Steel Salvage Drum Composite Overpack, 85 Gallon, 16 Gauge**	**EA**				**0.00**	**0.00**	**108.00**	**108.00**
02083 5173	85 Gallon, 16 Gauge Salvage Drum Nested with 3-20 Gallon Fiberglass Drums	EA	0	N/A	1.000	0.00	0.00	108.00	108.00
33 19 9926	**20 Gallon, Fiberglass Drums, DOT 21C-250**	**EA**				**0.00**	**0.00**	**31.00**	**31.00**
02083 5413	20 Gallon, 23" Diameter, DOT 21C-250 Polyethylene/Polyester Lined	EA	0	N/A	1.000	0.00	0.00	31.00	31.00

33 23 01 Casing		Unit	Hourly Output	Crew	Factor	Labor	Equip.	Matl.	Total
33 23 0101	**2" PVC, Schedule 40, Well Casing**	**LF**				**2.40**	**10.60**	**1.25**	**14.24**
02557 5216	2" Diameter (50 mm) Casing, Standard Length 10'	LF	37.5	ULADB	1.000	2.40	10.60	1.25	14.24
33 23 0102	**4" PVC, Schedule 40, Well Casing**	**LF**				**3.60**	**15.90**	**2.92**	**22.42**
02557 5219	4" Diameter (100 mm) Casing, Standard Length 10'	LF	25	ULADB	1.000	3.60	15.90	2.92	22.42
33 23 0103	**6" PVC, Schedule 40, Well Casing**	**LF**				**3.46**	**15.26**	**4.34**	**23.06**
02557 5223	6" Diameter (150 mm) Casing, Standard Length 10'	LF	15.63	ULADB	0.600	5.76	25.43	7.24	23.06
33 23 0104	**8" PVC, Schedule 40, Well Casing**	**LF**				**4.11**	**18.12**	**8.25**	**30.48**
02557 5224	8" Diameter (200 mm) Casing, Standard Length 10'	LF	12.5	ULADB	0.570	7.20	31.79	14.47	30.48
33 23 0105	**10" PVC, Schedule 40, Well Casing**	**LF**				**4.11**	**18.12**	**13.20**	**35.42**
02557 5225	10" Diameter (250 mm) Casing, Standard Length 10'	LF	12.5	ULADB	0.570	7.20	31.79	23.15	35.42
33 23 0106	**12" PVC, Schedule 40, Well Casing**	**LF**				**4.11**	**18.12**	**21.12**	**43.35**
02557 5226	12" Diameter (300 mm) Casing, Standard Length 10'	LF	12.5	ULADB	0.570	7.20	31.79	37.05	43.35
33 23 0111	**2" PVC, Schedule 80, Well Casing**	**LF**				**2.40**	**10.60**	**1.68**	**14.68**
02557 5236	2" Diameter (50 mm) Casing, Standard Length 10'	LF	37.5	ULADB	1.000	2.40	10.60	1.68	14.68
33 23 0112	**4" PVC, Schedule 80, Well Casing**	**LF**				**3.60**	**15.90**	**4.98**	**24.48**
02557 5239	4" Diameter (100 mm) Casing, Standard Length 10'	LF	25	ULADB	1.000	3.60	15.90	4.98	24.48
33 23 0113	**6" PVC, Schedule 80, Well Casing**	**LF**				**5.76**	**25.43**	**10.40**	**41.59**
02557 5242	6" Diameter (150 mm) Casing, Standard Length 10'	LF	15.63	ULADB	1.000	5.76	25.43	10.40	41.59

33 23 Wells

33 23 01 Casing		Unit	Hourly Output	Crew	Factor	Labor	Equip.	Matl.	Total
33 23 0115	**8" PVC, Schedule 80, Well Casing**	**LF**				**7.20**	**31.79**	**16.39**	**55.38**
02557 5243	8" Diameter (200 mm) Casing, Standard Length 10'	LF	12.5	ULADB	1.000	7.20	31.79	16.39	55.38
33 23 0116	**10" PVC, Schedule 80, Well Casing**	**LF**				**7.20**	**31.79**	**20.66**	**59.66**
02557 5244	10" Diameter (250 mm) Casing, Standard Length 10'	LF	12.5	ULADB	1.000	7.20	31.79	20.66	59.66
33 23 0117	**12" PVC, Schedule 80, Well Casing**	**LF**				**7.20**	**31.79**	**25.05**	**64.05**
02557 5245	12" Diameter (300 mm) Casing, Standard Length 10'	LF	12.5	ULADB	1.000	7.20	31.79	25.05	64.05
33 23 0121	**2" Stainless Steel, Well Casing**	**LF**				**2.83**	**12.50**	**16.11**	**31.44**
02557 5271	2" Diameter (50 mm) Casing No. 304 Stainless Steel	LF	37.5	ULADB	1.180	2.40	10.60	13.65	31.44
33 23 0122	**4" Stainless Steel, Well Casing**	**LF**				**3.60**	**15.90**	**30.04**	**49.53**
02557 5272	4' Diameter (100 mm) Casing No. 304 Stainless Steel	LF	25	ULADB	1.000	3.60	15.90	30.04	49.53
33 23 0123	**6" Stainless Steel Well Casing, 5' Sections, Flush Threaded**	**LF**				**46.09**	**203.39**	**70.93**	**320.41**
02557 5301	6" Diameter, Stainless Steel Well Casing, 5' Sections, Flush Threaded	LF	1.954	ULADB	1.000	46.09	203.39	70.93	320.41
33 23 0124	**6" Stainless Steel Well Casing, 10' Sections, Flush Threaded**	**LF**				**46.09**	**203.39**	**41.82**	**291.30**
02557 5311	6" Diameter, Stainless Steel Well Casing, 10' Sections, Flush Threaded	LF	1.954	ULADB	1.000	46.09	203.39	41.82	291.30
33 23 0125	**8" Stainless Steel Well Casing, 5' Sections, Flush Threaded**	**LF**				**57.63**	**254.32**	**171.10**	**483.05**
02557 5302	8" Diameter, Stainless Steel Well Casing, 5' Sections, Flush Threaded	LF	1.563	ULADB	1.000	57.63	254.32	171.10	483.05
33 23 0126	**8" Stainless Steel Well Casing, 10' Sections, Flush Threaded**	**LF**				**57.63**	**254.32**	**107.64**	**419.59**
02557 5312	8" Diameter, Stainless Steel Well Casing, 10' Sections, Flush Threaded	LF	1.563	ULADB	1.000	57.63	254.32	107.64	419.59
33 23 0127	**10" Stainless Steel Well Casing, 5' Sections, Flush Threaded**	**LF**				**57.63**	**254.32**	**180.58**	**492.53**
02557 5303	10" Diameter, Stainless Steel Well Casing, 5' Sections, Flush Threaded	LF	1.563	ULADB	1.000	57.63	254.32	180.58	492.53
33 23 0128	**10" Stainless Steel Well Casing, 10' Sections, Flush Threaded**	**LF**				**57.63**	**254.32**	**153.27**	**465.22**
02557 5313	10" Diameter, Stainless Steel Well Casing, 10' Sections, Flush Threaded	LF	1.563	ULADB	1.000	57.63	254.32	153.27	465.22
33 23 0129	**12" Stainless Steel Well Casing, 5' Sections, Flush Threaded**	**LF**				**57.63**	**254.32**	**217.18**	**529.13**
02557 5304	12" Diameter, Stainless Steel Well Casing, 5' Sections, Flush Threaded	LF	1.563	ULADB	1.000	57.63	254.32	217.18	529.13

33 23 01 Casing

33 23 01 Casing		Unit	Hourly Output	Crew	Factor	Labor	Equip.	Matl.	Total
33 23 0130	**12" Stainless Steel Well Casing, 10' Sections, Flush Threaded**	**LF**				**57.63**	**254.32**	**180.80**	**492.75**
02557 5314	12" Diameter, Stainless Steel Well Casing, 10' Sections, Flush Threaded	LF	1.563	ULADB	1.000	57.63	254.32	180.80	492.75
33 23 0131	**10" Stainless Steel Well Casing, 5' Sections, Welded Ring**	**LF**				**57.63**	**254.32**	**112.87**	**424.82**
02557 5321	10" Diameter, Stainless Steel Well Casing, 5' Sections, Welded Ring	LF	1.563	ULADB	1.000	57.63	254.32	112.87	424.82
33 23 0132	**10" Stainless Steel Well Casing, 10' Sections, Welded Ring**	**LF**				**57.63**	**254.32**	**83.31**	**395.26**
02557 5331	10" Diameter, Stainless Steel Well Casing, 10' Sections, Welded Ring	LF	1.563	ULADB	1.000	57.63	254.32	83.31	395.26
33 23 0133	**12" Stainless Steel Well Casing, 5' Sections, Welded Ring**	**LF**				**57.63**	**254.32**	**162.52**	**474.48**
02557 5322	12" Diameter, Stainless Steel Well Casing, 5' Sections, Welded Ring	LF	1.563	ULADB	1.000	57.63	254.32	162.52	474.48
33 23 0134	**12" Stainless Steel Well Casing, 10' Sections, Welded Ring**	**LF**				**57.63**	**254.32**	**116.52**	**428.47**
02557 5332	12" Diameter, Stainless Steel Well Casing, 10' Sections, Welded Ring	LF	1.563	ULADB	1.000	57.63	254.32	116.52	428.47
33 23 0135	**2" PVC, Schedule 40, Horizontal Well Casing, Material Only**	**LF**				**0.00**	**0.00**	**1.25**	**1.25**
02557 7201	2" PVC, Schedule 40, Horizontal Well Casing, Material Only	LF	0	N/A	1.000	0.00	0.00	1.25	1.25
33 23 0136	**4" PVC, Schedule 40, Horizontal Well Casing, Material Only**	**LF**				**0.00**	**0.00**	**2.92**	**2.92**
02557 7202	4" PVC, Schedule 40, Horizontal Well Casing, Material Only	LF	0	N/A	1.000	0.00	0.00	2.92	2.92
33 23 0137	**6" PVC, Schedule 40, Horizontal Well Casing, Material Only**	**LF**				**0.00**	**0.00**	**7.24**	**7.24**
02557 7203	6" PVC, Schedule 40, Horizontal Well Casing, Material Only	LF	0	N/A	1.000	0.00	0.00	7.24	7.24
33 23 0138	**8" PVC, Schedule 40, Horizontal Well Casing, Material Only**	**LF**				**0.00**	**0.00**	**14.47**	**14.47**
02557 7204	8" PVC, Schedule 40, Horizontal Well Casing, Material Only	LF	0	N/A	1.000	0.00	0.00	14.47	14.47
33 23 0139	**2" Stainless Steel, Horizontal Well Casing, Material Only**	**LF**				**0.00**	**0.00**	**13.65**	**13.65**
02557 7211	2" Stainless Steel, Horizontal Well Casing, Material Only	LF	0	N/A	1.000	0.00	0.00	13.65	13.65
33 23 0140	**4" Stainless Steel, Horizontal Well Casing, Material Only**	**LF**				**0.00**	**0.00**	**30.04**	**30.04**
02557 7212	4" Stainless Steel Horizontal Well Casing, Material Only	LF	0	N/A	1.000	0.00	0.00	30.04	30.04

33 23 Wells

33 23 01 Casing		Unit	Hourly Output	Crew	Factor	Labor	Equip.	Matl.	Total
33 23 0141	**6" Diameter, Stainless Steel, Horizontal Well Casing, Material Only**	**LF**				**0.00**	**0.00**	**70.93**	**70.93**
02557 7213	6" Diameter, Stainless Steel Horizontal Well Casing, Material Only	LF	0	N/A	1.000	0.00	0.00	70.93	70.93
33 23 0142	**8" Diameter, Stainless Steel, Horizontal Well Casing, Material Only**	**LF**				**0.00**	**0.00**	**171.10**	**171.10**
02557 7214	8" Diameter, Stainless Steel Horizontal Well Casing, Material Only	LF	0	N/A	1.000	0.00	0.00	171.10	171.10
33 23 0143	**2" High-density Polyethylene Horizontal Well Casing, Material Only**	**LF**				**0.00**	**0.00**	**1.68**	**1.68**
02557 7221	2" High-density Polyethylene Horizontal Well Casing, Material Only	LF	0	N/A	1.000	0.00	0.00	1.68	1.68
33 23 0144	**4" High-density Polyethylene Horizontal Well Casing, Material Only**	**LF**				**0.00**	**0.00**	**6.69**	**6.69**
02557 7222	4" High-density Polyethylene Horizontal Well Casing, Material Only	LF	0	N/A	1.000	0.00	0.00	6.69	6.69
33 23 0145	**6" High-density Polyethylene Horizontal Well Casing, Material Only**	**LF**				**0.00**	**0.00**	**16.74**	**16.74**
02557 7223	6" High-density Polyethylene Horizontal Well Casing, Material Only	LF	0	N/A	1.000	0.00	0.00	16.74	16.74
33 23 0146	**8" High-density Polyethylene Horizontal Well Casing, Material Only**	**LF**				**0.00**	**0.00**	**26.79**	**26.79**
02557 7224	8" High-density Polyethylene Horizontal Well Casing, Material Only	LF	0	N/A	1.000	0.00	0.00	26.79	26.79
33 23 0147	**2" PVC, Schedule 40, Well Casing, Slant Well**	**LF**				**3.12**	**13.78**	**1.25**	**18.14**
02557 7231	2" PVC, Schedule 40, Well Casing, Slant Well	LF	28.84	ULADB	1.000	3.12	13.78	1.25	18.14
33 23 0148	**4" PVC, Schedule 40, Well Casing, Slant Well**	**LF**				**4.68**	**20.67**	**2.92**	**28.27**
02557 7232	4" PVC, Schedule 40 Well Casing, Slant Well	LF	19.23	ULADB	1.000	4.68	20.67	2.92	28.27
33 23 0150	**2" PVC, Schedule 80, Well Casing, Slant Well**	**LF**				**3.12**	**13.78**	**1.68**	**18.58**
02557 7241	2" PVC, Schedule 80, Well Casing, Slant Well	LF	28.84	ULADB	1.000	3.12	13.78	1.68	18.58
33 23 0151	**4" PVC, Schedule 80, Well Casing, Slant Well**	**LF**				**4.68**	**20.67**	**4.98**	**30.33**
02557 7242	4" PVC, Schedule 80 Well Casing, Slant Well	LF	19.23	ULADB	1.000	4.68	20.67	4.98	30.33
33 23 0153	**2" Stainless Steel Well Casing, Slant Well**	**LF**				**3.12**	**13.78**	**13.65**	**30.55**
02557 7251	2" Stainless Steel Well Casing, Slant Well	LF	28.84	ULADB	1.000	3.12	13.78	13.65	30.55
33 23 0154	**4" Stainless Steel Well Casing, Slant Well**	**LF**				**4.68**	**20.67**	**30.04**	**55.39**
02557 7252	4" Stainless Steel Well Casing, Slant Well	LF	19.23	ULADB	1.000	4.68	20.67	30.04	55.39

33 23 02 Screens	Unit	Hourly Output	Crew	Factor	Labor	Equip.	Matl.	Total

33 23 02	Screens	Unit	Hourly Output	Crew	Factor	Labor	Equip.	Matl.	Total
33 23 0201	**2" PVC, Schedule 40, Well Screen**	**LF**				**3.10**	**13.67**	**2.88**	**19.64**
02557 5416	2" Diameter (50 mm) M-Screens, Standard Length 10'	LF	37.5	ULADB	1.290	2.40	10.60	2.23	19.64
33 23 0202	**4" PVC, Schedule 40, Well Screen**	**LF**				**4.83**	**21.30**	**6.55**	**32.68**
02557 5419	4" Diameter (100 mm) M-Screens, Standard Length 10'	LF	25	ULADB	1.340	3.60	15.90	4.89	32.68
33 23 0203	**6" PVC, Schedule 40, Well Screen**	**LF**				**5.76**	**25.43**	**9.89**	**41.08**
02557 5423	6" Diameter (150 mm) M-Screens, Standard Length 10'	LF	15.63	ULADB	1.000	5.76	25.43	9.89	41.08
33 23 0204	**8" PVC, Schedule 40, Well Screen**	**LF**				**7.20**	**31.79**	**16.43**	**55.42**
02557 5424	8" Diameter (200 mm) M-Screens, Standard Length 10'	LF	12.5	ULADB	1.000	7.20	31.79	16.43	55.42
33 23 0205	**10" PVC, Schedule 40, Well Screen**	**LF**				**7.20**	**31.79**	**24.08**	**63.08**
02557 5425	10" Diameter (250 mm) M-Screens, Standard Length 10'	LF	12.5	ULADB	1.000	7.20	31.79	24.08	63.08
33 23 0206	**12" PVC, Schedule 40, Well Screen**	**LF**				**7.20**	**31.79**	**28.08**	**67.08**
02557 5426	12" Diameter (300 mm) M-Screens, Standard Length 10'	LF	12.5	ULADB	1.000	7.20	31.79	28.08	67.08
33 23 0211	**2" PVC, Schedule 80, Well Screen**	**LF**				**2.40**	**10.60**	**2.88**	**15.88**
02557 5436	2" Diameter (50 mm) M-Screens, Standard Length 10'	LF	37.5	ULADB	1.000	2.40	10.60	2.88	15.88
33 23 0212	**4" PVC, Schedule 80, Well Screen**	**LF**				**3.60**	**15.90**	**6.90**	**26.39**
02557 5439	4" Diameter (100 mm) M-Screens, Standard Length 10'	LF	25	ULADB	1.000	3.60	15.90	6.90	26.39
33 23 0213	**6" PVC, Schedule 80, Well Screen**	**LF**				**5.76**	**25.43**	**13.28**	**44.48**
02557 5442	6" Diameter (150 mm) M-Screens, Standard Length 10'	LF	15.63	ULADB	1.000	5.76	25.43	13.28	44.48
33 23 0214	**8" PVC, Schedule 80, Well Screen**	**LF**				**7.20**	**31.79**	**21.53**	**60.52**
02557 5443	8" Diameter (200 mm) M-Screens, Standard Length 10'	LF	12.5	ULADB	1.000	7.20	31.79	21.53	60.52
33 23 0215	**10" PVC, Schedule 80, Well Screen**	**LF**				**7.20**	**31.79**	**31.75**	**70.75**
02557 5444	10" Diameter (250 mm) M-Screens, Standard Length 10'	LF	12.5	ULADB	1.000	7.20	31.79	31.75	70.75
33 23 0216	**12" PVC, Schedule 80, Well Screen**	**LF**				**7.20**	**31.79**	**36.61**	**75.60**
02557 5445	12" Diameter (300 mm) M-Screens, Standard Length 10'	LF	12.5	ULADB	1.000	7.20	31.79	36.61	75.60
33 23 0221	**2" Stainless Steel, Well Screen**	**LF**				**2.40**	**10.60**	**13.65**	**26.65**
02557 5461	2" Diameter (50 mm) M-Screens, No. 304 Stainless Steel	LF	37.5	ULADB	1.000	2.40	10.60	13.65	26.65

33 23 02	Screens	Unit	Hourly Output	Crew	Factor	Labor	Equip.	Matl.	Total
33 23 0222	**4" Stainless Steel, Well Screen**	**LF**				**3.60**	**15.90**	**30.04**	**49.53**
02557 5462	4" Diameter (100 mm) M-Screens, No. 304 Stainless Steel	LF	25	ULADB	1.000	3.60	15.90	30.04	49.53
33 23 0223	**6" Stainless Steel Well Screen, 5' Sec, Flush Threaded**	**LF**				**46.09**	**203.39**	**64.79**	**314.27**
02557 5501	6" Diameter, Stainless Steel Well Screen, 5' Sections, Flush Threaded	LF	1.954	ULADB	1.000	46.09	203.39	64.79	314.27
33 23 0224	**6" Stainless Steel Well Screen, 10' Sec, Flush Threaded**	**LF**				**46.09**	**203.39**	**51.57**	**301.04**
02557 5511	6" Diameter, Stainless Steel Well Screen, 10' Sections, Flush Threaded	LF	1.954	ULADB	1.000	46.09	203.39	51.57	301.04
33 23 0225	**8" Stainless Steel Well Screen, 5' Sec, Flush Threaded**	**LF**				**57.63**	**254.32**	**86.93**	**398.88**
02557 5502	8" Diameter, Stainless Steel Well Screen, 5' Sections, Flush Threaded	LF	1.563	ULADB	1.000	57.63	254.32	86.93	398.88
33 23 0226	**8" Stainless Steel Well Screen, 10' Sec, Flush Threaded**	**LF**				**57.63**	**254.32**	**70.01**	**381.96**
02557 5512	8" Diameter, Stainless Steel Well Screen, 10' Sections, Flush Threaded	LF	1.563	ULADB	1.000	57.63	254.32	70.01	381.96
33 23 0227	**10" Stainless Steel Well Screen, 5' Sec, Flush Threaded**	**LF**				**57.63**	**254.32**	**131.35**	**443.30**
02557 5503	10" Diameter, Stainless Steel Well Screen, 5' Sections, Flush Threaded	LF	1.563	ULADB	1.000	57.63	254.32	131.35	443.30
33 23 0228	**10" Stainless Steel Well Screen, 10' Sec, Flush Threaded**	**LF**				**57.63**	**254.32**	**104.31**	**416.26**
02557 5513	10" Diameter, Stainless Steel Well Screen, 10' Sections, Flush Threaded	LF	1.563	ULADB	1.000	57.63	254.32	104.31	416.26
33 23 0229	**12" Stainless Steel Well Screen, 5' Sec, Flush Threaded**	**LF**				**57.63**	**254.32**	**161.58**	**473.53**
02557 5504	12" Diameter, Stainless Steel Well Screen, 5' Sections, Flush Threaded	LF	1.563	ULADB	1.000	57.63	254.32	161.58	473.53
33 23 0230	**12" Stainless Steel Well Screen, 10' Sec, Flush Threaded**	**LF**				**57.63**	**254.32**	**124.61**	**436.56**
02557 5514	12" Diameter, Stainless Steel Well Screen, 10' Sections, Flush Threaded	LF	1.563	ULADB	1.000	57.63	254.32	124.61	436.56
33 23 0231	**10" Dia, SS Well Screen, 5' Sec, Welded Ring**	**LF**				**57.63**	**274.59**	**86.00**	**418.22**
02086 5521	Monitor well, M-screens, 10"dia, sst, welded ring, 5'	LF	1.563	ULABW8	1.000	57.63	274.59	86.00	418.22
33 23 0232	**10" Stainless Steel Well Screen, 10' Sec, Welded Ring**	**LF**				**57.63**	**254.32**	**91.07**	**403.02**
02557 5531	10" Diameter, Stainless Steel Well Screen, 10' Sections, Welded Ring	LF	1.563	ULADB	1.000	57.63	254.32	91.07	403.02

33 23 02 Screens		Unit	Hourly Output	Crew	Factor	Labor	Equip.	Matl.	Total
33 23 0233	**12" Stainless Steel Well Screen, 5' Sec, Welded Ring**	**LF**				**57.63**	**254.32**	**145.39**	**457.34**
02557 5522	12" Diameter, Stainless Steel Well Screen, 5' Sections, Welded Ring	LF	1.563	ULADB	1.000	57.63	254.32	145.39	457.34
33 23 0234	**12" Stainless Steel Well Screen, 10' Sec, Welded Ring**	**LF**				**57.63**	**254.32**	**116.53**	**428.48**
02557 5532	12" Diameter, Stainless Steel Well Screen, 10' Sections, Welded Ring	LF	1.563	ULADB	1.000	57.63	254.32	116.53	428.48
33 23 0235	**4" SDR 17, High-density Polyethylene, Well Screen**	**LF**				**0.00**	**0.00**	**11.80**	**11.80**
15157 7724	4" SDR-17 Screened Pipe	LF	0	N/A	1.000	0.00	0.00	11.80	11.80
33 23 0236	**6" SDR 17, High-density Polyethylene, Well Screen**	**LF**				**0.00**	**0.00**	**14.16**	**14.16**
15157 7726	6" SDR-17 Screened Pipe	LF	0	N/A	1.000	0.00	0.00	14.16	14.16
33 23 0237	**2" PVC, Schedule 40, Horizontal Well Screen, Material Only**	**LF**				**0.00**	**0.00**	**2.23**	**2.23**
02557 7301	2" PVC, Schedule 40, Horizontal Well Screen, Material Only	LF	0	N/A	1.000	0.00	0.00	2.23	2.23
33 23 0238	**4" PVC, Schedule 40, Horizontal Well Screen, Material Only**	**LF**				**0.00**	**0.00**	**4.89**	**4.89**
02557 7302	4" PVC, Schedule 40, Horizontal Well Screen, Material Only	LF	0	N/A	1.000	0.00	0.00	4.89	4.89
33 23 0239	**6" PVC, Schedule 40, Horizontal Well Screen, Material Only**	**LF**				**0.00**	**0.00**	**9.89**	**9.89**
02557 7303	6" PVC, Schedule 40, Horizontal Well Screen, Material Only	LF	0	N/A	1.000	0.00	0.00	9.89	9.89
33 23 0240	**8" PVC, Schedule 40, Horizontal Well Screen, Material Only**	**LF**				**0.00**	**0.00**	**16.43**	**16.43**
02557 7304	8" PVC, Schedule 40, Horizontal Well Screen, Material Only	LF	0	N/A	1.000	0.00	0.00	16.43	16.43
33 23 0241	**2" Stainless Steel Horizontal Well Screen, Material Only**	**LF**				**0.00**	**0.00**	**13.65**	**13.65**
02557 7311	2" Stainless Steel Horizontal Well Screen, Material Only	LF	0	N/A	1.000	0.00	0.00	13.65	13.65
33 23 0242	**4" Stainless Steel Horizontal Well Screen, Material Only**	**LF**				**0.00**	**0.00**	**30.04**	**30.04**
02557 7312	4" Stainless Steel Horizontal Well Screen, Material Only	LF	0	N/A	1.000	0.00	0.00	30.04	30.04
33 23 0243	**6" Diameter, Stainless Steel Horizontal Well Screen, Material Only**	**LF**				**0.00**	**0.00**	**64.79**	**64.79**
02557 7313	6" Diameter, Stainless Steel Horizontal Well Screen, Material Only	LF	0	N/A	1.000	0.00	0.00	64.79	64.79

33 23 Wells

33 23 02 Screens		Unit	Hourly Output	Crew	Factor	Labor	Equip.	Matl.	Total
33 23 0244	**8" Diameter, Stainless Steel Horizontal Well Screen, Material Only**	**LF**				**0.00**	**0.00**	**86.93**	**86.93**
02557 7314	8" Diameter, Stainless Steel Horizontal Well Screen, Material Only	LF	0	N/A	1.000	0.00	0.00	86.93	86.93
33 23 0245	**2" High-density Polyethylene, Horizontal Well Screen, Material Only**	**LF**				**0.00**	**0.00**	**5.86**	**5.86**
02557 7321	2" High-density Polyethylene, Horizontal Well Screen, Material Only	LF	0	N/A	1.000	0.00	0.00	5.86	5.86
33 23 0246	**4" High-density Polyethylene, Horizontal Well Screen, Material Only**	**LF**				**0.00**	**0.00**	**13.89**	**13.89**
02557 7322	4" High-density Polyethylene, Horizontal Well Screen, Material Only	LF	0	N/A	1.000	0.00	0.00	13.89	13.89
33 23 0247	**6" High-density Polyethylene, Horizontal Well Screen, Material Only**	**LF**				**0.00**	**0.00**	**30.13**	**30.13**
02557 7323	6" High-density Polyethylene, Horizontal Well Screen, Material Only	LF	0	N/A	1.000	0.00	0.00	30.13	30.13
33 23 0248	**8" High-density Polyethylene Horizontal Well Screen, Material Only**	**LF**				**0.00**	**0.00**	**40.18**	**40.18**
02557 7324	8" High-density Polyethylene Horizontal Well Screen, Material Only	LF	0	N/A	1.000	0.00	0.00	40.18	40.18
33 23 0249	**2" PVC, Schedule 40, Horizontal Prepack, Material Only**	**LF**				**0.00**	**0.00**	**25.12**	**25.12**
02557 7331	2" PVC, Schedule 40, Horizontal Prepack, Material Only	LF	0	N/A	1.000	0.00	0.00	25.12	25.12
33 23 0250	**4" PVC, Schedule 40, Horizontal Prepack, Material Only**	**LF**				**0.00**	**0.00**	**46.88**	**46.88**
02557 7332	4" PVC, Schedule 40, Horizontal Prepack, Material Only	LF	0	N/A	1.000	0.00	0.00	46.88	46.88
33 23 0251	**6" PVC, Schedule 40, Horizontal Prepack, Material Only**	**LF**				**0.00**	**0.00**	**66.98**	**66.98**
02557 7333	6" PVC, Schedule 40, Horizontal Prepack, Material Only	LF	0	N/A	1.000	0.00	0.00	66.98	66.98
33 23 0253	**2" Stainless Steel, Horizontal Prepack, Material Only**	**LF**				**0.00**	**0.00**	**83.72**	**83.72**
02557 7341	2" Stainless Steel, Horizontal Prepack, Material Only	LF	0	N/A	1.000	0.00	0.00	83.72	83.72
33 23 0254	**4" Stainless Steel, Horizontal Prepack, Material Only**	**LF**				**0.00**	**0.00**	**142.33**	**142.33**
02557 7342	4" Stainless Steel, Horizontal Prepack, Material Only	LF	0	N/A	1.000	0.00	0.00	142.33	142.33
33 23 0255	**6" Stainless Steel, Horizontal Prepack, Material Only**	**LF**				**0.00**	**0.00**	**293.02**	**293.02**
02557 7343	6" Stainless Steel, Horizontal Prepack, Material Only	LF	0	N/A	1.000	0.00	0.00	293.02	293.02

33 23 02 Screens		Unit	Hourly Output	Crew	Factor	Labor	Equip.	Matl.	Total
33 23 0256	**2" PVC, Schedule 40, Well Screen, Slant Well**	**LF**				**4.03**	**17.77**	**2.23**	**24.03**
02557 7334	2" PVC, Schedule 40, Well Screen, Slant Well	LF	22.36	ULADB	1.000	4.03	17.77	2.23	24.03
33 23 0257	**2" High-density Polyethylene, Horizontal Prepack, Material Only**	**LF**				**0.00**	**0.00**	**25.12**	**25.12**
02557 7351	2" High-density Polyethylene, Horizontal Prepack, Material Only	LF	0	N/A	1.000	0.00	0.00	25.12	25.12
33 23 0258	**4" High-density Polyethylene, Horizontal Prepack, Material Only**	**LF**				**0.00**	**0.00**	**46.88**	**46.88**
02557 7352	4" High-density Polyethylene, Horizontal Prepack, Material Only	LF	0	N/A	1.000	0.00	0.00	46.88	46.88
33 23 0259	**6" High-density Polyethylene, Horizontal Prepack, Material Only**	**LF**				**0.00**	**0.00**	**75.35**	**75.35**
02557 7353	6" High-density Polyethylene, Horizontal Prepack, Material Only	LF	0	N/A	1.000	0.00	0.00	75.35	75.35
33 23 0260	**2" PVC, Schedule 80, Well Screen, Slant Well**	**LF**				**6.43**	**28.37**	**5.39**	**40.19**
02557 5253	2" Diameter (50 mm) Casing, Standard Length 5'	LF	37.5	ULADB	1.000	2.40	10.60	2.49	15.49
02557 7344	2" PVC, Schedule 80, Well Screen, Slant Well	LF	22.36	ULADB	1.000	4.03	17.77	2.90	24.70
33 23 0261	**4" PVC, Schedule 40, Well Screen, Slant Well**	**LF**				**9.07**	**40.04**	**10.74**	**59.85**
02557 5255	4" Diameter (100 mm) Casing, Standard Length 5'	LF	31.25	ULADB	1.000	2.88	12.72	5.85	21.45
02557 7335	4" PVC, Schedule 40, Well Screen, Slant Well	LF	14.54	ULADB	1.000	6.19	27.33	4.89	38.41
33 23 0262	**4" PVC, Schedule 80, Well Screen, Slant Well**	**LF**				**6.19**	**27.33**	**6.36**	**39.88**
02557 7345	4" PVC, Schedule 80, Well Screen, Slant Well	LF	14.54	ULADB	1.000	6.19	27.33	6.36	39.88
33 23 0264	**2" Stainless Steel Well Screen, Slant Well**	**LF**				**4.03**	**17.77**	**13.65**	**35.45**
02557 7354	2" Stainless Steel Well Screen, Slant Well	LF	22.36	ULADB	1.000	4.03	17.77	13.65	35.45
33 23 0265	**4" Stainless Steel Well Screen, Slant Well**	**LF**				**6.19**	**27.33**	**30.04**	**63.55**
02557 7355	4" Stainless Steel Well Screen, Slant Well	LF	14.54	ULADB	1.000	6.19	27.33	30.04	63.55
33 23 0270	**2" PVC Wire Wrapped Well Screen, 10' Sec**	**LF**				**24.01**	**105.97**	**4.49**	**134.47**
02557 7361	2" PVC, Wire Wrapped on PipeScreen, 10' Sec	LF	3.75	ULADB	1.000	24.01	105.97	4.49	134.47
33 23 0271	**4" PVC Wire Wrapped Well Screen, 10' Sec**	**LF**				**36.02**	**158.95**	**9.11**	**204.08**
02557 7362	4" PVC, Wire Wrapped on PipeScreen, 10' Sec	LF	2.5	ULADB	1.000	36.02	158.95	9.11	204.08
33 23 0272	**6" PVC Wire Wrapped Well Screen, 10' Sec**	**LF**				**57.61**	**254.24**	**18.60**	**330.45**
02557 7363	6" PVC, Wire Wrapped on PipeScreen, 10' Sec	LF	1.563	ULADB	1.000	57.61	254.24	18.60	330.45
33 23 0273	**8" PVC Wire Wrapped Well Screen, 10' Sec**	**LF**				**57.61**	**254.24**	**26.30**	**338.15**
02557 7364	8" PVC, Wire Wrapped on PipeScreen, 10' Sec	LF	1.563	ULADB	1.000	57.61	254.24	26.30	338.15

33 23 Wells

33 23 02 Screens		Unit	Hourly Output	Crew	Factor	Labor	Equip.	Matl.	Total
33 23 0274	**2" Stainless Steel Wire Wrapped Well Screen, 10' Sec**	**LF**				**24.01**	**105.97**	**29.45**	**159.43**
02557 7365	2" Stainless Steel, Wire Wrapped on Pipe Screen, 10' Sec	LF	3.75	ULADB	1.000	24.01	105.97	29.45	159.43
33 23 0275	**4" Stainless Steel Wire Wrapped Well Screen, 10' Sec**	**LF**				**36.02**	**158.95**	**48.81**	**243.78**
02557 7366	4" Stainless Steel, Wire Wrapped on Pipe Screen, 10' Sec	LF	2.5	ULADB	1.000	36.02	158.95	48.81	243.78
33 23 0276	**6" Stainless Steel Wire Wrapped Well Screen, 10' Sec**	**LF**				**57.61**	**254.24**	**74.65**	**386.51**
02557 7367	6" Stainless Steel, Wire Wrapped on Pipe Screen, 10' Sec	LF	1.563	ULADB	1.000	57.61	254.24	74.65	386.51
33 23 0277	**8" Stainless Steel Wire Wrapped Well Screen, 10' Sec**	**LF**				**57.61**	**254.24**	**108.67**	**420.53**
02557 7368	8" Stainless Steel, Wire Wrapped on Pipe Screen, 10' Sec	LF	1.563	ULADB	1.000	57.61	254.24	108.67	420.53
33 23 0280	**2" Teflon Screen, Flush Threaded, 5' Sec, Slotted**	**LF**				**2.40**	**11.44**	**3.00**	**16.84**
02086 5453	Monitor well, M-screens, 5' L,teflon, flush threaded, 2" dia	LF	37.5	ULABW8	1.000	2.40	11.44	3.00	16.84
33 23 0281	**4" Teflon Screen, Flush Threaded, 5' Sec, Slotted**	**LF**				**2.88**	**13.73**	**5.00**	**21.61**
02086 5455	Monitor well, M-screens, 5' L,teflon, flush threaded, 4" dia	LF	31.25	ULABW8	1.000	2.88	13.73	5.00	21.61

33 23 03 Plugs		Unit	Hourly Output	Crew	Factor	Labor	Equip.	Matl.	Total
33 23 0301	**2" PVC, Well Plug**	**EA**				**3.60**	**15.90**	**6.06**	**25.55**
02557 5622	2" Solid Plug - PVC	EA	12.5	ULADB	0.500	7.20	31.79	12.11	25.55
33 23 0302	**4" PVC, Well Plug**	**EA**				**5.28**	**23.31**	**25.38**	**53.98**
02557 5625	4" Solid Plug - PVC	EA	11.25	ULADB	0.660	8.00	35.32	38.45	53.98
33 23 0303	**6" PVC, Well Plug**	**EA**				**9.00**	**39.74**	**66.56**	**115.31**
02557 5628	6" Solid Plug - PVC	EA	10	ULADB	1.000	9.00	39.74	66.56	115.31
33 23 0304	**8" PVC, Well Plug**	**EA**				**10.29**	**45.41**	**81.29**	**136.99**
02557 5629	8" Solid Plug - PVC	EA	8.75	ULADB	1.000	10.29	45.41	81.29	136.99
33 23 0305	**10" PVC, Well Plug**	**EA**				**12.01**	**52.98**	**96.80**	**161.79**
02557 5631	10" Solid Plug - PVC	EA	7.5	ULADB	1.000	12.01	52.98	96.80	161.79
33 23 0306	**12" PVC, Well Plug**	**EA**				**14.41**	**63.58**	**129.65**	**207.64**
02557 5632	12" Solid Plug - PVC	EA	6.25	ULADB	1.000	14.41	63.58	129.65	207.64
33 23 0311	**2" Stainless Steel, Well Plug**	**EA**				**7.20**	**31.79**	**44.25**	**83.24**
02557 5633	2" Solid Plug - No. 304 Stainless Steel	EA	12.5	ULADB	1.000	7.20	31.79	44.25	83.24

33 23 03 Plugs	Unit	Hourly Output	Crew	Factor	Labor	Equip.	Matl.	Total
33 23 0312 4" Stainless Steel, Well Plug	**EA**				**8.00**	**35.32**	**75.06**	**118.39**
02557 5634 4" Solid Plug - No. 304 Stainless Steel	EA	11.25	ULADB	1.000	8.00	35.32	75.06	118.39
33 23 0313 6" Stainless Steel, Well Plug	**EA**				**18.01**	**79.47**	**102.44**	**199.92**
02557 5638 6" Stainless Steel Plug	EA	5	ULADB	1.000	18.01	79.47	102.44	199.92
33 23 0314 8" Stainless Steel, Well Plug	**EA**				**22.51**	**99.34**	**179.55**	**301.41**
02557 5639 8" Stainless Steel Plug	EA	4	ULADB	1.000	22.51	99.34	179.55	301.41
33 23 0315 10" Stainless Steel, Well Plug	**EA**				**36.02**	**158.95**	**303.26**	**498.23**
02557 5640 10" Stainless Steel Plug	EA	2.5	ULADB	1.000	36.02	158.95	303.26	498.23
33 23 0316 4" High-density Polyethylene, Well Plug	**EA**				**0.00**	**0.00**	**92.62**	**92.62**
02557 5641 4" High-density Polyethylene Plug	EA	0	N/A	1.000	0.00	0.00	92.62	92.62
33 23 0317 6" High-density Polyethylene, Well Plug	**EA**				**0.00**	**0.00**	**112.86**	**112.86**
02557 5642 6" High-density Polyethylene Plug	EA	0	N/A	1.000	0.00	0.00	112.86	112.86
33 23 0325 12" Stainless Steel, Well Plug	**EA**				**60.03**	**264.92**	**331.44**	**656.39**
02557 5643 12" Stainless Steel Plug	EA	1.5	ULADB	1.000	60.03	264.92	331.44	656.39
33 23 0326 2" PVC Plug for Horizontal Well, Material Only	**EA**				**0.00**	**0.00**	**12.11**	**12.11**
02557 7401 2" PVC Plug for Horizontal Well, Material Only	LF	0	N/A	1.000	0.00	0.00	12.11	12.11
33 23 0327 4" PVC Plug for Horizontal Well, Material Only	**EA**				**0.00**	**0.00**	**38.45**	**38.45**
02557 7402 4" PVC Plug for Horizontal Well, Material Only	LF	0	N/A	1.000	0.00	0.00	38.45	38.45
33 23 0328 6" PVC Plug for Horizontal Well, Material Only	**LF**				**0.00**	**0.00**	**66.56**	**66.56**
02557 7403 6" PVC Plug for Horizontal Well, Material Only	LF	0	N/A	1.000	0.00	0.00	66.56	66.56
33 23 0329 8" PVC Plug for Horizontal Well, Material Only	**EA**				**0.00**	**0.00**	**81.29**	**81.29**
02557 7404 8" PVC Plug for Horizontal Well, Material Only	LF	0	N/A	1.000	0.00	0.00	81.29	81.29
33 23 0330 2" Stainless Steel Plug for Horizontal, Material Only	**EA**				**0.00**	**0.00**	**35.17**	**35.17**
02557 7411 2" Stainless Steel Plug for Horizontal Well, Material Only	LF	0	N/A	1.000	0.00	0.00	35.17	35.17
33 23 0331 4" Stainless Steel Plug for Horizontal, Material Only	**EA**				**0.00**	**0.00**	**71.99**	**71.99**
02557 7412 4" Stainless Steel Plug for Horizontal Well, Material Only	LF	0	N/A	1.000	0.00	0.00	71.99	71.99
33 23 0332 6" Stainless Steel Plug for Horizontal, Material Only	**LF**				**0.00**	**0.00**	**133.95**	**133.95**
02557 7413 6" Stainless Steel Plug for Horizontal Well, Material Only	LF	0	N/A	1.000	0.00	0.00	133.95	133.95

33 23 03 Plugs

33 23 03 Plugs		Unit	Hourly Output	Crew	Factor	Labor	Equip.	Matl.	Total
33 23 0333	**8" Stainless Steel Plug for Horizontal, Material Only**	**EA**				**0.00**	**0.00**	**277.96**	**277.96**
02557 7414	8" Stainless Steel Plug for Horizontal Well, Material Only	LF	0	N/A	1.000	0.00	0.00	277.96	277.96
33 23 0334	**2" High-density Polyethylene Plug for Horizontal Well, Material Only**	**EA**				**0.00**	**0.00**	**8.37**	**8.37**
02557 7421	2" High-density Polyethylene Plug for Horizontal Well, Material Only	LF	0	N/A	1.000	0.00	0.00	8.37	8.37
33 23 0335	**4" High-density Polyethylene Plug for Horizontal Well, Material Only**	**EA**				**0.00**	**0.00**	**38.51**	**38.51**
02557 7422	4" High-density Polyethylene Plug for Horizontal Well, Material Only	LF	0	N/A	1.000	0.00	0.00	38.51	38.51
33 23 0336	**6" High-density Polyethylene Plug for Horizontal Well, Material Only**	**LF**				**0.00**	**0.00**	**92.09**	**92.09**
02557 7423	6" High-density Polyethylene Plug for Horizontal Well, Material Only	LF	0	N/A	1.000	0.00	0.00	92.09	92.09
33 23 0337	**8" High-density Polyethylene Plug for Horizontal Well, Material Only**	**EA**				**0.00**	**0.00**	**150.70**	**150.70**
02557 7424	8" High-density Polyethylene Plug for Horizontal Well, Material Only	LF	0	N/A	1.000	0.00	0.00	150.70	150.70
33 23 0338	**2" PVC, Schedule 40, Well Plug, Slant Well**	**EA**				**4.68**	**20.67**	**12.11**	**37.46**
02557 7431	2" PVC, Schedule 40, Well Plug, Slant Well	EA	19.23	ULADB	1.000	4.68	20.67	12.11	37.46
33 23 0339	**4" PVC, Schedule 40, Well Plug, Slant Well**	**EA**				**6.87**	**30.31**	**38.45**	**75.63**
02557 7432	4" PVC, Schedule 40, Well Plug, Slant Well	LF	13.11	ULADB	1.000	6.87	30.31	38.45	75.63
33 23 0341	**2" Stainless Steel Well Plug, Slant Well**	**EA**				**9.37**	**41.33**	**35.17**	**85.86**
02557 7441	2" Stainless Steel Well Plug, Slant Well	EA	9.614	ULADB	1.000	9.37	41.33	35.17	85.86
33 23 0342	**4" Stainless Steel Well Plug, Slant Well**	**EA**				**10.41**	**45.92**	**71.99**	**128.32**
02557 7442	4" Stainless Steel Well Plug, Slant Well	LF	8.653	ULADB	1.000	10.41	45.92	71.99	128.32

33 23 04 Filter Socks

33 23 04 Filter Socks		Unit	Hourly Output	Crew	Factor	Labor	Equip.	Matl.	Total
33 23 0401	**2" Filter Sock**	**LF**				**0.72**	**3.18**	**0.59**	**4.49**
02557 5481	2" Diameter Filter Sock	LF	125	ULADB	1.000	0.72	3.18	0.59	4.49
33 23 0402	**4" Filter Sock**	**LF**				**0.72**	**3.18**	**0.81**	**4.71**
02557 5482	4" Diameter Filter Sock	LF	125	ULADB	1.000	0.72	3.18	0.81	4.71
33 23 0403	**6" Filter Sock**	**LF**				**0.72**	**3.18**	**0.98**	**4.88**
02557 5483	6" Diameter Filter Sock	LF	125	ULADB	1.000	0.72	3.18	0.98	4.88
33 23 0404	**8" Filter Sock**	**LF**				**0.72**	**3.18**	**1.84**	**5.73**
02557 5484	8" Diameter Filter Sock	LF	125	ULADB	1.000	0.72	3.18	1.84	5.73

33 23 05	Submersible Pumps	Unit	Hourly Output	Crew	Factor	Labor	Equip.	Matl.	Total
33 23 0501	**4" Submersible Pump, 4 - 13 GPM, < 180' Head**	**EA**				**444.26**	**125.00**	**933.45**	**1,503**
15298 2020	250 - 770 GPH Submersible Pump, 4" Discharge for Wells, 180' (55 m) Deep	EA	0.288	MPLUS	1.000	444.26	125.00	933.45	1,503
33 23 0502	**4" Submersible Pump, 9 - 26 GPM, < 180' Head**	**EA**				**510.90**	**143.75**	**1,974**	**2,629**
15298 2030	550 - 1,600 GPH Submersible Pump, 4" Discharge for Wells, 180' (55 m) Deep	EA	0.25	MPLUS	1.000	510.90	143.75	1,974	2,629
33 23 0503	**4" Submersible Pump, 26 - 37 GPM, < 180' Head**	**EA**				**510.90**	**143.75**	**2,025**	**2,680**
15298 2040	800 - 2,200 GPH Submersible Pump, 4" Discharge for Wells, 180' (55 m) Deep	EA	0.25	MPLUS	1.000	510.90	143.75	2,025	2,680
33 23 0504	**6" Submersible Pump, 50 - 125 GPM, < 150' Head**	**EA**				**729.86**	**205.36**	**1,292**	**2,227**
15298 2050	50 - 125 GPM Submersible Pump, 6" Discharge for Wells, 50 - 150' Deep	EA	0.175	MPLUS	1.000	729.86	205.36	1,292	2,227
33 23 0505	**6" Submersible Pump, 15 - 135 GPM, 200' < Head < 500'**	**EA**				**1,022**	**287.50**	**2,061**	**3,370**
15298 2060	15 - 135 GPM Submersible Pump, 6" Discharge for Wells, 200 - 500' Deep	EA	0.125	MPLUS	1.000	1,022	287.50	2,061	3,370
33 23 0506	**2" Submersible Pump Rental, Day**	**DAY**				**0.00**	**0.00**	**77.98**	**77.98**
15146 1004	2" Submersible Pump Rental, Day	DAY	0	N/A	1.000	0.00	0.00	77.98	77.98
33 23 0507	**2" Submersible Pump Rental, Week**	**WK**				**0.00**	**0.00**	**233.94**	**233.94**
15146 1005	2" Submersible Pump Rental, Week	WK	0	N/A	1.000	0.00	0.00	233.94	233.94
33 23 0508	**2" Submersible Pump Rental, Month**	**MO**				**0.00**	**0.00**	**584.86**	**584.86**
15146 1006	2" Submersible Pump Rental, Month	MO	0	N/A	1.000	0.00	0.00	584.86	584.86
33 23 0509	**4" Submersible Pump Rental, Day**	**DAY**				**0.00**	**0.00**	**77.98**	**77.98**
15146 1007	4" Submersible Pump Rental, Day	DAY	0	N/A	1.000	0.00	0.00	77.98	77.98
33 23 0510	**4" Submersible Pump Rental, Week**	**WK**				**0.00**	**0.00**	**233.94**	**233.94**
15146 1008	4" Submersible Pump Rental, Week	WK	0	N/A	1.000	0.00	0.00	233.94	233.94
33 23 0511	**4" Submersible Pump Rental, Month**	**MO**				**0.00**	**0.00**	**584.86**	**584.86**
15146 1009	4" Submersible Pump Rental, Month	MO	0	N/A	1.000	0.00	0.00	584.86	584.86
33 23 0512	**1" Submersible Pump Rental, Month**	**MO**				**0.00**	**0.00**	**109.17**	**109.17**
15146 1011	1" Submersible Pump Rental, Month	MO	0	N/A	1.000	0.00	0.00	109.17	109.17
33 23 0513	**6" Submersible Pump Rental, Day**	**DAY**				**0.00**	**0.00**	**105.10**	**105.10**
01001 0318	6" Submersible Pump Rental, Day	DAY	0	N/A	1.000	0.00	0.00	105.10	105.10
33 23 0514	**6" Submersible Pump Rental, Week**	**WK**				**0.00**	**0.00**	**335.12**	**335.12**
01001 0319	6" Submersible Pump Rental, Week	WK	0	N/A	1.000	0.00	0.00	335.12	335.12

33 23 05 Submersible Pumps	Unit	Hourly Output	Crew	Factor	Labor	Equip.	Matl.	Total
33 23 0515 6" Submersible Pump Rental, Month	**MO**				**0.00**	**0.00**	**1,001**	**1,001**
01001 0320 6" Submersible Pump Rental, Month	MO	0	N/A	1.000	0.00	0.00	1,001	1,001
33 23 0516 8" Submersible Pump Rental, Day	**DAY**				**0.00**	**0.00**	**333.81**	**333.81**
01001 0321 8" Submersible Pump Rental, Day	DAY	0	N/A	1.000	0.00	0.00	333.81	333.81
33 23 0517 8" Submersible Pump Rental, Week	**WK**				**0.00**	**0.00**	**1,001**	**1,001**
01001 0322 8" Submersible Pump Rental, Week	WK	0	N/A	1.000	0.00	0.00	1,001	1,001
33 23 0518 8" Submersible Pump Rental, Month	**MO**				**0.00**	**0.00**	**3,004**	**3,004**
01001 0323 8" Submersible Pump Rental, Month	MO	0	N/A	1.000	0.00	0.00	3,004	3,004
33 23 0521 4" Submersible Pump, 0.3-7 GPM, Head <=140', 1/3 hp, w/ controls	**EA**				**0.00**	**0.00**	**1,841**	**1,841**
01001 0324 4" Submersible Pump, 0.3-7 GPM, Head <=140', 1/3 hp, w/ controls	EA	0	N/A	1.000	0.00	0.00	1,841	1,841
33 23 0522 4" Submersible Pump, 0.3-7 GPM, 141'< Head <=240', 1/2 hp, w/ controls	**EA**				**0.00**	**0.00**	**1,995**	**1,995**
01001 0325 4" Submersible Pump, 0.3-7 GPM, 141'< Head <=240', 1/2 hp, w/ controls	EA	0	N/A	1.000	0.00	0.00	1,995	1,995
33 23 0523 4" Submersible Pump, 0.3-7 GPM, 241'< Head <=340', 3/4 hp, w/ controls	**EA**				**0.00**	**0.00**	**2,318**	**2,318**
01001 0326 4" Submersible Pump, 0.3-7 GPM, 241'< Head <=340', 3/4 hp, w/ controls	EA	0	N/A	1.000	0.00	0.00	2,318	2,318
33 23 0524 4" Submersible Pump, 0.3-7 GPM, 341'< Head <=520', 1 hp, w/ controls	**EA**				**0.00**	**0.00**	**2,622**	**2,622**
01001 0327 4" Submersible Pump, 0.3-7 GPM, 341'< Head <=520', 1 hp, w/ controls	EA	0	N/A	1.000	0.00	0.00	2,622	2,622
33 23 0525 4" Submersible Pump, 0.3-7 GPM, 521'< Head <=800', 1 1/2 hp, w/ controls	**EA**				**0.00**	**0.00**	**3,433**	**3,433**
01001 0328 4" Submersible Pump, 0.3-7 GPM, 521'< Head <=800', 1 1/2 hp, w/ controls	EA	0	N/A	1.000	0.00	0.00	3,433	3,433
33 23 0526 4" Submersible Pump, 8-14 GPM, Head <=80', 1/3 hp, w/ controls	**EA**				**0.00**	**0.00**	**1,595**	**1,595**
01001 0329 4" Submersible Pump, 8-14 GPM, Head <=80', 1/3 hp, w/ controls	EA	0	N/A	1.000	0.00	0.00	1,595	1,595
33 23 0527 4" Submersible Pump, 8-14 GPM, 81'< Head <=140', 1/2 hp, w/ controls	**EA**				**0.00**	**0.00**	**1,697**	**1,697**
01001 0330 4" Submersible Pump, 8-14 GPM, 81'< Head <=140', 1/2 hp, w/ controls	EA	0	N/A	1.000	0.00	0.00	1,697	1,697
33 23 0528 4" Submersible Pump, 8-14 GPM, 141'< Head <=220', 3/4 hp, w/ controls	**EA**				**0.00**	**0.00**	**1,969**	**1,969**
01001 0331 4" Submersible Pump, 8-14 GPM, 141'< Head <=220', 3/4 hp, w/ controls	EA	0	N/A	1.000	0.00	0.00	1,969	1,969

33 23 05	Submersible Pumps	Unit	Hourly Output	Crew	Factor	Labor	Equip.	Matl.	Total
33 23 0529	**4" Submersible Pump, 8-14 GPM, 221< Head <=280', 1 hp, w/ controls**	**EA**				**0.00**	**0.00**	**2,195**	**2,195**
01001 0332	4" Submersible Pump, 8-14 GPM, 221< Head <=280', 1 hp, w/ controls	EA	0	N/A	1.000	0.00	0.00	2,195	2,195
33 23 0530	**4" Submersible Pump, 8-14 GPM, 281'< Head <=460', 1 1/2 hp, w/ controls**	**EA**				**0.00**	**0.00**	**2,631**	**2,631**
01001 0333	4" Submersible Pump, 8-14 GPM, 281'< Head <=460', 1 1/2 hp, w/ controls	EA	0	N/A	1.000	0.00	0.00	2,631	2,631
33 23 0531	**4" Submersible Pump, 8-14 GPM, 461'< Head <=600', 2 hp, w/ controls**	**EA**				**0.00**	**0.00**	**3,021**	**3,021**
01001 0334	4" Submersible Pump, 8-14 GPM, 461'< Head <=600', 2 hp, w/ controls	EA	0	N/A	1.000	0.00	0.00	3,021	3,021
33 23 0532	**4" Submersible Pump, 8-14 GPM, 601'< Head <=800', 3 hp, w/ controls**	**EA**				**0.00**	**0.00**	**4,242**	**4,242**
01001 0335	4" Submersible Pump, 8-14 GPM, 601'< Head <=800', 3 hp, w/ controls	EA	0	N/A	1.000	0.00	0.00	4,242	4,242
33 23 0533	**4" Submersible Pump,15-20 GPM, Head <=80', 1/2 hp, w/ controls**	**EA**				**0.00**	**0.00**	**1,605**	**1,605**
01001 0336	4" Submersible Pump,15-20 GPM, Head <=80', 1/2 hp, w/ controls	EA	0	N/A	1.000	0.00	0.00	1,605	1,605
33 23 0534	**4" Submersible Pump, 15-20 GPM, 81< Head <=160', 3/4 hp, w/ controls**	**EA**				**0.00**	**0.00**	**1,837**	**1,837**
01001 0337	4" Submersible Pump, 15-20 GPM, 81< Head <=160', 3/4 hp, w/ controls	EA	0	N/A	1.000	0.00	0.00	1,837	1,837
33 23 0535	**4" Submersible Pump, 15-20 GPM, 161'< Head <=240', 1 hp, w/ controls**	**EA**				**0.00**	**0.00**	**2,028**	**2,028**
01001 0338	4" Submersible Pump, 15-20 GPM, 161'< Head <=240', 1 hp, w/ controls	EA	0	N/A	1.000	0.00	0.00	2,028	2,028
33 23 0536	**4" Submersible Pump, 15-20 GPM, 241'< Head <=300', 1 1/2 hp, w/ controls**	**EA**				**0.00**	**0.00**	**2,294**	**2,294**
01001 0339	4" Submersible Pump, 15-20 GPM, 241'< Head <=300', 1 1/2 hp, w/ controls	EA	0	N/A	1.000	0.00	0.00	2,294	2,294
33 23 0537	**4" Submersible Pump, 15-20 GPM, 301'< Head <=400', 2 hp, w/ controls**	**EA**				**0.00**	**0.00**	**2,694**	**2,694**
01001 0340	4" Submersible Pump, 15-20 GPM, 301'< Head <=400', 2 hp, w/ controls	EA	0	N/A	1.000	0.00	0.00	2,694	2,694
33 23 0538	**4" Submersible Pump, 15-20 GPM, 401'< Head <=600', 3 hp, w/ controls**	**EA**				**0.00**	**0.00**	**3,591**	**3,591**
01001 0341	4" Submersible Pump, 15-20 GPM, 401'< Head <=600', 3 hp, w/ controls	EA	0	N/A	1.000	0.00	0.00	3,591	3,591
33 23 0539	**4" Submersible Pump, 15-20 GPM, 601'< Head <=1000', 5 hp, w/ controls**	**EA**				**0.00**	**0.00**	**4,984**	**4,984**
01001 0342	4" Submersible Pump, 15-20 GPM, 601'< Head <=1000', 5 hp, w/ controls	EA	0	N/A	1.000	0.00	0.00	4,984	4,984

33 23 Wells

33 23 05 Submersible Pumps		Unit	Hourly Output	Crew	Factor	Labor	Equip.	Matl.	Total
33 23 0540	**4" Submersible Pump, 21-32 GPM, Head <=60', 1/2 hp, w/ controls**	**EA**				**0.00**	**0.00**	**1,644**	**1,644**
01001 0343	4" Submersible Pump, 21-32 GPM, Head <=60', 1/2 hp, w/ controls	EA	0	N/A	1.000	0.00	0.00	1,644	1,644
33 23 0541	**4" Submersible Pump, 21-32 GPM, 61'< Head <=120', 3/4 hp, w/ controls**	**EA**				**0.00**	**0.00**	**1,815**	**1,815**
01001 0344	4" Submersible Pump, 21-32 GPM, 61'< Head <=120', 3/4 hp, w/ controls	EA	0	N/A	1.000	0.00	0.00	1,815	1,815
33 23 0542	**4" Submersible Pump, 21-32 GPM, 121'< Head <=160', 1 hp, w/ controls**	**EA**				**0.00**	**0.00**	**1,954**	**1,954**
01001 0345	4" Submersible Pump, 21-32 GPM, 121'< Head <=160', 1 hp, w/ controls	EA	0	N/A	1.000	0.00	0.00	1,954	1,954
33 23 0543	**4" Submersible Pump, 21-32 GPM, 161'< Head <=200', 1 1/2 hp, w/ controls**	**EA**				**0.00**	**0.00**	**2,173**	**2,173**
01001 0346	4" Submersible Pump, 21-32 GPM, 161'< Head <=200', 1 1/2 hp, w/ controls	EA	0	N/A	1.000	0.00	0.00	2,173	2,173
33 23 0544	**4" Submersible Pump, 21-32 GPM, 201'< Head <=280', 2 hp, w/ controls**	**EA**				**0.00**	**0.00**	**2,450**	**2,450**
01001 0347	4" Submersible Pump, 21-32 GPM, 201'< Head <=280', 2 hp, w/ controls	EA	0	N/A	1.000	0.00	0.00	2,450	2,450
33 23 0545	**4" Submersible Pump, 21-32 GPM, 281'< Head <=340', 3 hp, w/ controls**	**EA**				**0.00**	**0.00**	**3,268**	**3,268**
01001 0348	4" Submersible Pump, 21-32 GPM, 281'< Head <=340', 3 hp, w/ controls	EA	0	N/A	1.000	0.00	0.00	3,268	3,268
33 23 0546	**4" Submersible Pump, 21-32 GPM, 341'< Head <=600', 5 hp, w/ controls**	**EA**				**0.00**	**0.00**	**4,377**	**4,377**
01001 0349	4" Submersible Pump, 21-32 GPM, 341'< Head <=600', 5 hp, w/ controls	EA	0	N/A	1.000	0.00	0.00	4,377	4,377
33 23 0547	**4" Submersible Pump, 33-55 GPM, Head <=20', 1 hp, w/ controls**	**EA**				**0.00**	**0.00**	**2,045**	**2,045**
01001 0350	4" Submersible Pump, 33-55 GPM, Head <=20', 1 hp, w/ controls	EA	0	N/A	1.000	0.00	0.00	2,045	2,045
33 23 0548	**4" Submersible Pump, 33-55 GPM, 21'< Head <=100', 1 1/2 hp, w/ controls**	**EA**				**0.00**	**0.00**	**2,183**	**2,183**
01001 0351	4" Submersible Pump, 33-55 GPM, 21'< Head <=100', 1 1/2 hp, w/ controls	EA	0	N/A	1.000	0.00	0.00	2,183	2,183
33 23 0549	**4" Submersible Pump, 33-55 GPM, 101'< Head <=160', 2 hp, w/ controls**	**EA**				**0.00**	**0.00**	**2,322**	**2,322**
01001 0352	4" Submersible Pump, 33-55 GPM, 101'< Head <=160', 2 hp, w/ controls	EA	0	N/A	1.000	0.00	0.00	2,322	2,322
33 23 0550	**4" Submersible Pump, 33-55 GPM, 161'< Head <=220', 3 hp, w/ controls**	**EA**				**0.00**	**0.00**	**3,170**	**3,170**
01001 0353	4" Submersible Pump, 33-55 GPM, 161'< Head <=220', 3 hp, w/ controls	EA	0	N/A	1.000	0.00	0.00	3,170	3,170

33 23 05	Submersible Pumps	Unit	Hourly Output	Crew	Factor	Labor	Equip.	Matl.	Total
33 23 0551	**4" Submersible Pump, 33-55 GPM, 221'< Head <=340', 5 hp, w/ controls**	**EA**				**0.00**	**0.00**	**4,483**	**4,483**
01001 0354	4" Submersible Pump, 33-55 GPM, 221'< Head <=340', 5 hp, w/ controls	EA	0	N/A	1.000	0.00	0.00	4,483	4,483
33 23 0552	**6" Submersible Pump, 33-55 GPM, 341'< Head <=600', 7 1/2 hp, w/ controls**	**EA**				**0.00**	**0.00**	**5,951**	**5,951**
01001 0355	6" Submersible Pump, 33-55 GPM, 341'< Head <=600', 7 1/2 hp, w/ controls	EA	0	N/A	1.000	0.00	0.00	5,951	5,951
33 23 0553	**6" Submersible Pump, 33-55 GPM, 601'< Head <=800', 10 hp, w/ controls**	**EA**				**0.00**	**0.00**	**8,304**	**8,304**
01001 0356	6" Submersible Pump, 33-55 GPM, 601'< Head <=800', 10 hp, w/ controls	EA	0	N/A	1.000	0.00	0.00	8,304	8,304
33 23 0554	**4" Submersible Pump, 56-95 GPM, Head <=40', 2 hp, w/ controls**	**EA**				**0.00**	**0.00**	**2,185**	**2,185**
01001 0357	4" Submersible Pump, 56-95 GPM, Head <=40', 2 hp, w/ controls	EA	0	N/A	1.000	0.00	0.00	2,185	2,185
33 23 0555	**4" Submersible Pump, 56-95 GPM, 41'< Head <=100', 3 hp, w/ controls**	**EA**				**0.00**	**0.00**	**3,063**	**3,063**
01001 0358	4" Submersible Pump, 56-95 GPM, 41'< Head <=100', 3 hp, w/ controls	EA	0	N/A	1.000	0.00	0.00	3,063	3,063
33 23 0556	**4" Submersible Pump, 56-95 GPM, 101'< Head <=220', 5 hp, w/ controls**	**EA**				**0.00**	**0.00**	**4,142**	**4,142**
01001 0359	4" Submersible Pump, 56-95 GPM, 101'< Head <=220', 5 hp, w/ controls	EA	0	N/A	1.000	0.00	0.00	4,142	4,142
33 23 0557	**6" Submersible Pump, 56-95 GPM, 221'< Head <=300', 7 1/2 hp, w/ controls**	**EA**				**0.00**	**0.00**	**5,108**	**5,108**
01001 0360	6" Submersible Pump, 56-95 GPM, 221'< Head <=300', 7 1/2 hp, w/ controls	EA	0	N/A	1.000	0.00	0.00	5,108	5,108
33 23 0558	**6" Submersible Pump, 56-95 GPM, 301'< Head <=400', 10 hp, w/ controls**	**EA**				**0.00**	**0.00**	**6,842**	**6,842**
01001 0361	6" Submersible Pump, 56-95 GPM, 301'< Head <=400', 10 hp, w/ controls	EA	0	N/A	1.000	0.00	0.00	6,842	6,842
33 23 0559	**4" Submersible Pump, 96-200 GPM, Head <=50', 2 hp, w/ controls**	**EA**				**0.00**	**0.00**	**2,297**	**2,297**
01001 0362	4" Submersible Pump, 96-200 GPM, Head <=50', 2 hp, w/ controls	EA	0	N/A	1.000	0.00	0.00	2,297	2,297
33 23 0560	**4" Submersible Pump, 96-200 GPM, 51'< Head <=100', 5 hp, w/ controls**	**EA**				**0.00**	**0.00**	**3,806**	**3,806**
01001 0363	4" Submersible Pump, 96-200 GPM, 51'< Head <=100', 5 hp, w/ controls	EA	0	N/A	1.000	0.00	0.00	3,806	3,806
33 23 0561	**4" Submersible Pump, 96-200 GPM, 101'< Head <=150', 7 1/2 hp, w/ controls**	**EA**				**0.00**	**0.00**	**4,512**	**4,512**
01001 0364	4" Submersible Pump, 96-200 GPM, 101'< Head <=150', 7 1/2 hp, w/ controls	EA	0	N/A	1.000	0.00	0.00	4,512	4,512

33 23 05	Submersible Pumps	Unit	Hourly Output	Crew	Factor	Labor	Equip.	Matl.	Total
33 23 0562	**6" Submersible Pump, 96-200 GPM, 151'< Head <=200', 10 hp, w/ controls**	**EA**				**0.00**	**0.00**	**6,043**	**6,043**
01001 0365	6" Submersible Pump, 96-200 GPM, 151'< Head <=200', 10 hp, w/ controls	EA	0	N/A	1.000	0.00	0.00	6,043	6,043
33 23 0563	**6" Submersible Pump, 96-200 GPM, 201'< Head <=300', 15 hp, w/ controls**	**EA**				**0.00**	**0.00**	**7,452**	**7,452**
01001 0366	6" Submersible Pump, 96-200 GPM, 201'< Head <=300', 15 hp, w/ controls	EA	0	N/A	1.000	0.00	0.00	7,452	7,452
33 23 0564	**6" Submersible Pump, 96-200 GPM, 301'< Head <=400', 20 hp, w/ controls**	**EA**				**0.00**	**0.00**	**9,426**	**9,426**
01001 0367	6" Submersible Pump, 96-200 GPM, 301'< Head <=400', 20 hp, w/ controls	EA	0	N/A	1.000	0.00	0.00	9,426	9,426
33 23 0565	**6" Submersible Pump, 96-200 GPM, 401'< Head <=500', 25 hp, w/ controls**	**EA**				**0.00**	**0.00**	**10,579**	**10,579**
01001 0368	6" Submersible Pump, 96-200 GPM, 401'< Head <=500', 25 hp, w/ controls	EA	0	N/A	1.000	0.00	0.00	10,579	10,579
33 23 0566	**6" Submersible Pump, 96-200 GPM, 501'< Head <=600', 30 hp, w/ controls**	**EA**				**0.00**	**0.00**	**13,897**	**13,897**
01001 0369	6" Submersible Pump, 96-200 GPM, 501'< Head <=600', 30 hp, w/ controls	EA	0	N/A	1.000	0.00	0.00	13,897	13,897
33 23 0567	**6" Submersible Pump, 96-200 GPM, 601'< Head <=725', 40 hp, w/ controls**	**EA**				**0.00**	**0.00**	**16,876**	**16,876**
01001 0370	6" Submersible Pump, 96-200 GPM, 601'< Head <=725', 40 hp, w/ controls	EA	0	N/A	1.000	0.00	0.00	16,876	16,876
33 23 0568	**6" Submersible Pump, 96-200 GPM, 726'< Head <=950', 50 hp, w/ controls**	**EA**				**0.00**	**0.00**	**20,651**	**20,651**
01001 0371	6" Submersible Pump, 96-200 GPM, 726'< Head <=950', 50 hp, w/ controls	EA	0	N/A	1.000	0.00	0.00	20,651	20,651
33 23 0569	**6" Submersible Pump, 96-200 GPM, 951'< Head <=1100', 60 hp, w/ controls**	**EA**				**0.00**	**0.00**	**35,679**	**35,679**
01001 0372	6" Submersible Pump, 96-200 GPM, 951'< Head <=1100', 60 hp, w/ controls	EA	0	N/A	1.000	0.00	0.00	35,679	35,679
33 23 0570	**8" Submersible Pump, 96-200 GPM, 1101'< Head <=1400', 75 hp, w/ controls**	**EA**				**0.00**	**0.00**	**42,680**	**42,680**
01001 0373	8" Submersible Pump, 96-200 GPM, 1101'< Head <=1400', 75 hp, w/ controls	EA	0	N/A	1.000	0.00	0.00	42,680	42,680
33 23 0571	**6" Submersible Pump, 201-410 GPM, Head <=75', 7 1/2 hp, w/ controls**	**EA**				**0.00**	**0.00**	**4,096**	**4,096**
01001 0374	6" Submersible Pump, 201-410 GPM, Head <=75', 7 1/2 hp, w/ controls	EA	0	N/A	1.000	0.00	0.00	4,096	4,096
33 23 0572	**6" Submersible Pump, 201-410 GPM, 76'< Head <=150', 15 hp, w/ controls**	**EA**				**0.00**	**0.00**	**6,704**	**6,704**
01001 0375	6" Submersible Pump, 201-410 GPM, 76'< Head <=150', 15 hp, w/ controls	EA	0	N/A	1.000	0.00	0.00	6,704	6,704

33 23 05	Submersible Pumps	Unit	Hourly Output	Crew	Factor	Labor	Equip.	Matl.	Total
33 23 0573	**6" Submersible Pump, 201-410 GPM, 151'< Head <=175', 20 hp, w/ controls**	**EA**				**0.00**	**0.00**	**8,270**	**8,270**
01001 0376	6" Submersible Pump, 201-410 GPM, 151'< Head <=175', 20 hp, w/ controls	EA	0	N/A	1.000	0.00	0.00	8,270	8,270
33 23 0574	**6" Submersible Pump, 201-410 GPM, 176'< Head <=225', 25 hp, w/ controls**	**EA**				**0.00**	**0.00**	**9,140**	**9,140**
01001 0377	6" Submersible Pump, 201-410 GPM, 176'< Head <=225', 25 hp, w/ controls	EA	0	N/A	1.000	0.00	0.00	9,140	9,140
33 23 0575	**6" Submersible Pump, 201-410 GPM, 226'< Head <=300', 30 hp, w/ controls**	**EA**				**0.00**	**0.00**	**12,073**	**12,073**
01001 0378	6" Submersible Pump, 201-410 GPM, 226'< Head <=300', 30 hp, w/ controls	EA	0	N/A	1.000	0.00	0.00	12,073	12,073
33 23 0576	**6" Submersible Pump, 201-410 GPM, 301'< Head <=400', 40 hp, w/ controls**	**EA**				**0.00**	**0.00**	**13,543**	**13,543**
01001 0379	6" Submersible Pump, 201-410 GPM, 301'< Head <=400', 40 hp, w/ controls	EA	0	N/A	1.000	0.00	0.00	13,543	13,543
33 23 0577	**6" Submersible Pump, 201-410 GPM, 401'< Head <=500', 50 hp, w/ controls**	**EA**				**0.00**	**0.00**	**15,556**	**15,556**
01001 0380	6" Submersible Pump, 201-410 GPM, 401'< Head <=500', 50 hp, w/ controls	EA	0	N/A	1.000	0.00	0.00	15,556	15,556
33 23 0578	**6" Submersible Pump, 201-410 GPM, 501'< Head <=600', 60 hp, w/ controls**	**EA**				**0.00**	**0.00**	**20,933**	**20,933**
01001 0381	6" Submersible Pump, 201-410 GPM, 501'< Head <=600', 60 hp, w/ controls	EA	0	N/A	1.000	0.00	0.00	20,933	20,933
33 23 0579	**8" Submersible Pump, 201-410 GPM, 601'< Head <=750', 75 hp, w/ controls**	**EA**				**0.00**	**0.00**	**26,682**	**26,682**
01001 0382	8" Submersible Pump, 201-410 GPM, 601'< Head <=750', 75 hp, w/ controls	EA	0	N/A	1.000	0.00	0.00	26,682	26,682
33 23 0580	**6" Submersible Pump, 411-680 GPM, Head <=50', 10 hp, w/ controls**	**EA**				**0.00**	**0.00**	**6,573**	**6,573**
01001 0383	6" Submersible Pump, 411-680 GPM, Head <=50', 10 hp, w/ controls	EA	0	N/A	1.000	0.00	0.00	6,573	6,573
33 23 0581	**6" Submersible Pump, 411-680 GPM, 51'< Head <=125', 20 hp, w/ controls**	**EA**				**0.00**	**0.00**	**8,868**	**8,868**
01001 0384	6" Submersible Pump, 411-680 GPM, 51'< Head <=125', 20 hp, w/ controls	EA	0	N/A	1.000	0.00	0.00	8,868	8,868
33 23 0582	**6" Submersible Pump, 411-680 GPM, 126'< Head <=200', 30 hp, w/ controls**	**EA**				**0.00**	**0.00**	**12,095**	**12,095**
01001 0385	6" Submersible Pump, 411-680 GPM, 126'< Head <=200', 30 hp, w/ controls	EA	0	N/A	1.000	0.00	0.00	12,095	12,095
33 23 0583	**6" Submersible Pump, 411-680 GPM, 201'< Head <=275', 40 hp, w/ controls**	**EA**				**0.00**	**0.00**	**13,063**	**13,063**
01001 0386	6" Submersible Pump, 411-680 GPM, 201'< Head <=275', 40 hp, w/ controls	EA	0	N/A	1.000	0.00	0.00	13,063	13,063

33 23 Wells

33 23 05	Submersible Pumps	Unit	Hourly Output	Crew	Factor	Labor	Equip.	Matl.	Total
33 23 0584	**6" Submersible Pump, 411-680 GPM, 276'< Head <=350', 50 hp, w/ controls**	**EA**				**0.00**	**0.00**	**16,131**	**16,131**
01001 0387	6" Submersible Pump, 411-680 GPM, 276'< Head <=350', 50 hp, w/ controls	EA	0	N/A	1.000	0.00	0.00	16,131	16,131
33 23 0585	**6" Submersible Pump, 411-680 GPM, 351'< Head <=400', 60 hp, w/ controls**	**EA**				**0.00**	**0.00**	**18,435**	**18,435**
01001 0388	6" Submersible Pump, 411-680 GPM, 351'< Head <=400', 60 hp, w/ controls	EA	0	N/A	1.000	0.00	0.00	18,435	18,435
33 23 0586	**8" Submersible Pump, 411-680 GPM, 401'< Head <=500', 75 hp, w/ controls**	**EA**				**0.00**	**0.00**	**23,514**	**23,514**
01001 0389	8" Submersible Pump, 411-680 GPM, 401'< Head <=500', 75 hp, w/ controls	EA	0	N/A	1.000	0.00	0.00	23,514	23,514
33 23 0587	**8" Submersible Pump, 411-680 GPM, 501'< Head <=700', 100 hp, w/ controls**	**EA**				**0.00**	**0.00**	**30,719**	**30,719**
01001 0390	8" Submersible Pump, 411-680 GPM, 501'< Head <=700', 100 hp, w/ controls	EA	0	N/A	1.000	0.00	0.00	30,719	30,719
33 23 0588	**8" Submersible Pump, 411-680 GPM, 701'< Head <=900', 125 hp, w/ controls**	**EA**				**0.00**	**0.00**	**40,547**	**40,547**
01001 0391	8" Submersible Pump, 411-680 GPM, 701'< Head <=900', 125 hp, w/ controls	EA	0	N/A	1.000	0.00	0.00	40,547	40,547
33 23 0589	**6" Submersible Pump, 681-1400 GPM, Head <=100', 40 hp, w/ controls**	**EA**				**0.00**	**0.00**	**13,475**	**13,475**
01001 0392	6" Submersible Pump, 681-1400 GPM, Head <=100', 40 hp, w/ controls	EA	0	N/A	1.000	0.00	0.00	13,475	13,475
33 23 0590	**6" Submersible Pump, 681-1400 GPM, 101'< Head <=175', 60 hp, w/ controls**	**EA**				**0.00**	**0.00**	**19,603**	**19,603**
01001 0393	6" Submersible Pump, 681-1400 GPM, 101'< Head <=175', 60 hp, w/ controls	EA	0	N/A	1.000	0.00	0.00	19,603	19,603
33 23 0591	**8" Submersible Pump, 681-1400 GPM, 176'< Head <=200', 75 hp, w/ controls**	**EA**				**0.00**	**0.00**	**23,741**	**23,741**
01001 0394	8" Submersible Pump, 681-1400 GPM, 176'< Head <=200', 75 hp, w/ controls	EA	0	N/A	1.000	0.00	0.00	23,741	23,741
33 23 0592	**8" Submersible Pump, 681-1400 GPM, 201'< Head <=225', 100 hp, w/ controls**	**EA**				**0.00**	**0.00**	**30,988**	**30,988**
01001 0395	8" Submersible Pump, 681-1400 GPM, 201'< Head <=225', 100 hp, w/ controls	EA	0	N/A	1.000	0.00	0.00	30,988	30,988
33 23 0593	**8" Submersible Pump, 681-1400 GPM, 226'< Head <=350', 125 hp, w/ controls**	**EA**				**0.00**	**0.00**	**35,291**	**35,291**
01001 0396	8" Submersible Pump, 681-1400 GPM, 226'< Head <=350', 125 hp, w/ controls	EA	0	N/A	1.000	0.00	0.00	35,291	35,291
33 23 0594	**8" Submersible Pump, 681-1400 GPM, 351'< Head <=400', 150 hp, w/ controls**	**EA**				**0.00**	**0.00**	**40,501**	**40,501**
01001 0397	8" Submersible Pump, 681-1400 GPM, 351'< Head <=400', 150 hp, w/ controls	EA	0	N/A	1.000	0.00	0.00	40,501	40,501

33 23 Wells

33 23 05 Submersible Pumps		Unit	Hourly Output	Crew	Factor	Labor	Equip.	Matl.	Total
33 23 0595	**8" Submersible Pump, 681-1400 GPM, 401'< Head <=475', 175 hp, w/ controls**	**EA**				**0.00**	**0.00**	**52,317**	**52,317**
01001 0398	8" Submersible Pump, 681-1400 GPM, 401'< Head <=475', 175 hp, w/ controls	EA	0	N/A	1.000	0.00	0.00	52,317	52,317
33 23 0596	**8" Submersible Pump, 681-1400 GPM, 476'< Head <=600', 200 hp, w/ controls**	**EA**				**0.00**	**0.00**	**59,268**	**59,268**
01001 0399	8" Submersible Pump, 681-1400 GPM, 476'< Head <=600', 200 hp, w/ controls	EA	0	N/A	1.000	0.00	0.00	59,268	59,268
33 23 0597	**8" Submersible Pump, 681-1400 GPM, 601'< Head <=750', 250 hp, w/ controls**	**EA**				**0.00**	**0.00**	**66,745**	**66,745**
01001 0400	8" Submersible Pump, 681-1400 GPM, 601'< Head <=750', 250 hp, w/ controls	EA	0	N/A	1.000	0.00	0.00	66,745	66,745

33 23 06 Groundwater Pumps with Controls and Probes		Unit	Hourly Output	Crew	Factor	Labor	Equip.	Matl.	Total
33 23 0601	**Groundwater Pump, 1/3 HP, 230V, Controls, Probe**	**EA**				**422.45**	**0.00**	**2,050**	**2,472**
13273 6115	1/3 HP Groundwater Drawdown Unit with Controls and Probe, 230V	EA	0.25	ULABD	1.000	422.45	0.00	2,050	2,472
33 23 0602	**Groundwater Pump, 3/4 HP, 230V, Controls, Probe**	**EA**				**422.45**	**0.00**	**2,225**	**2,647**
13273 6116	3/4 HP Groundwater Drawdown Unit with Controls and Probe, 230V	EA	0.25	ULABD	1.000	422.45	0.00	2,225	2,647
33 23 0603	**Groundwater Pump, 1 HP, 230V, Controls, Probe**	**EA**				**422.45**	**0.00**	**2,350**	**2,772**
13273 6117	1 HP Groundwater Drawdown Unit with Controls and Probe, 230V	EA	0.25	ULABD	1.000	422.45	0.00	2,350	2,772
33 23 0604	**Groundwater Pump, 1 1/2 HP, 230V, Controls, Probe**	**EA**				**422.45**	**0.00**	**2,575**	**2,997**
13273 6118	1-1/2 HP Groundwater Drawdown Unit with Controls and Probe,	EA	0.25	ULABD	1.000	422.45	0.00	2,575	2,997
33 23 0605	**Groundwater Pump, 2 HP, 230V, Controls, Probe**	**EA**				**422.45**	**0.00**	**2,925**	**3,347**
13273 6119	2 HP Groundwater Drawdown Unit with Controls and Probe, 230V	EA	0.25	ULABD	1.000	422.45	0.00	2,925	3,347
33 23 0606	**Groundwater Pump, 5 HP, 230V, Controls, Probe**	**EA**				**422.45**	**0.00**	**3,700**	**4,122**
13273 6121	5 HP Groundwater Drawdown Unit with Controls and Probe, 230V	EA	0.25	ULABD	1.000	422.45	0.00	3,700	4,122
33 23 0611	**Groundwater Pump, 360 GPD, Pneumatic, with Controls & Compressor**	**EA**				**422.45**	**0.00**	**3,405**	**3,828**
13273 6135	Product Pump, 2" Oil Skimmer, 360 GPD, Controls, Probe, Pneumatic	EA	0.25	ULABD	1.000	422.45	0.00	3,405	3,828

33 23 06	Groundwater Pumps with Controls and Probes	Unit	Hourly Output	Crew	Factor	Labor	Equip.	Matl.	Total
33 23 0612	**Groundwater Pump, 700 GPD, Pneumatic, with Controls & Compressor**	**EA**				**422.45**	**0.00**	**3,715**	**4,138**
13273 6136	Product Pump, 2" Oil Skimmer, 700 GPD, Controls, Probe, Pneumatic	EA	0.25	ULABD	1.000	422.45	0.00	3,715	4,138
33 23 0613	**Groundwater Pump, 2160 GPD, Pneumatic, with Controls & Compressor**	**EA**				**422.45**	**0.00**	**4,025**	**4,447**
13273 6137	Product Pump, 4" Oil Skimmer, 2,160 GPD, Controls, Probe, Pneumatic	EA	0.25	ULABD	1.000	422.45	0.00	4,025	4,447
33 23 0614	**Peristaltic Pump, Weekly Rental**	**WK**				**0.00**	**0.00**	**112.20**	**112.20**
01001 0401	Peristaltic Pump, Weekly Rental	WK	0	N/A	1.000	0.00	0.00	112.20	112.20

33 23 07	Groundwater Pumps without Controls	Unit	Hourly Output	Crew	Factor	Labor	Equip.	Matl.	Total
33 23 0701	**Groundwater Pump, 1 1/2 HP, 230V, without Controls**	**EA**				**422.45**	**0.00**	**1,575**	**1,997**
13273 6133	1-1/2 HP Groundwater Pump without Controls, Includes 50' of Wire	EA	0.25	ULABD	1.000	422.45	0.00	1,575	1,997
33 23 0702	**Groundwater Pump, 5 HP, 230V, without Controls**	**EA**				**422.45**	**0.00**	**2,475**	**2,897**
13273 6134	5 HP Groundwater Pump without Controls, Includes 50' of Wire	EA	0.25	ULABD	1.000	422.45	0.00	2,475	2,897

33 23 08	Product Recovery Pumps with Controls, Probes	Unit	Hourly Output	Crew	Factor	Labor	Equip.	Matl.	Total
33 23 0801	**Product Recovery Pump, 1/3 HP, 230V, Controls, Probe (>3" dia.)**	**EA**				**422.45**	**0.00**	**6,675**	**7,097**
13273 6111	1/3 HP Product Recovery Pump with Controls and Probe, 230V	EA	0.25	ULABD	1.000	422.45	0.00	6,675	7,097
33 23 0802	**Product Recovery Pump, 1/3 HP, 115V, Controls, Probe (>3" dia.)**	**EA**				**422.45**	**0.00**	**8,475**	**8,897**
13273 6112	1/3 HP Product Recovery Pump with Controls and Probe, 115V	EA	0.25	ULABD	1.000	422.45	0.00	8,475	8,897
33 23 0803	**Product Recovery Pump, 1/3 HP, 230V, Controls, Probe, Highly Corrosive Liq. (>3" dia.)**	**EA**				**422.45**	**0.00**	**8,025**	**8,447**
13273 6113	1/3 HP Product Recovery Pump with Controls and Probe	EA	0.25	ULABD	1.000	422.45	0.00	8,025	8,447
33 23 0804	**Product Recovery Pump, 2" Oil Skimmer, 360 GPD, Controls & Compressor, Pneumatic**	**EA**				**422.45**	**0.00**	**3,405**	**3,828**
13273 6135	Product Pump, 2" Oil Skimmer, 360 GPD, Controls, Probe, Pneumatic	EA	0.25	ULABD	1.000	422.45	0.00	3,405	3,828

33 23 08	Product Recovery Pumps with Controls, Probes	Unit	Hourly Output	Crew	Factor	Labor	Equip.	Matl.	Total
33 23 0805	**Product Recovery Pump, 2" Oil Skimmer, 700 GPD, Controls & Compressor, Pneumatic**	**EA**				**422.45**	**0.00**	**3,715**	**4,138**
13273 6136	Product Pump, 2" Oil Skimmer, 700 GPD, Controls, Probe, Pneumatic	EA	0.25	ULABD	1.000	422.45	0.00	3,715	4,138
33 23 0806	**Product Recovery Pump, 4" Oil Skimmer, 2,160 GPD, Controls & Compressor, Pneumatic**	**EA**				**422.45**	**0.00**	**4,025**	**4,447**
13273 6137	Product Pump, 4" Oil Skimmer, 2,160 GPD, Controls, Probe, Pneumatic	EA	0.25	ULABD	1.000	422.45	0.00	4,025	4,447
33 23 0807	**Product Recovery Pump, Shallow Depths (<=20 ft), 6 gpm, Controls & Compressor, Pneumatic**	**EA**				**422.45**	**0.00**	**3,560**	**3,983**
13273 6139	Product Recovery Pump, Pneumatic, Automatic Pump AP-40, Shallow System, 6 GPM	EA	0.25	ULABD	1.000	422.45	0.00	3,560	3,983
33 23 0808	**Product Recovery Pump, Deep Depths (>20 ft), 6 gpm, Controls & Compressor, Pneumatic**	**EA**				**422.45**	**0.00**	**3,792**	**4,215**
13273 6138	Product Recovery Pump, Pneumatic, Automatic Pump AP-40, Deep System, 6 GPM	EA	0.25	ULABD	1.000	422.45	0.00	3,792	4,215
33 23 0809	**Product Recovery Pump, 1/2 HP, Controls, Probe**	**EA**				**422.45**	**0.00**	**2,225**	**2,647**
13273 6114	1/2 HP Narrow Well Product Recovery Pump with Controls	EA	0.25	ULABD	1.000	422.45	0.00	2,225	2,647
33 23 0811	**Explosionproof Electrical Receptacle for Product Recovery Pump**	**EA**				**62.35**	**0.00**	**230.23**	**292.57**
16232 2180	Simplex Receptacle, Explosion-proof, 120V, 20A, Exp Specialty Devices	EA	1.65	EELEB	1.000	62.35	0.00	230.23	292.57
33 23 0820	**Product Recovery Pump, Shallow Depths (<=20 ft), Controls, Electric**	**EA**				**0.00**	**0.00**	**1,122**	**1,122**
01001 0402	Product Recovery Pump, Shallow Depths (<=20 ft), Controls, Electric	EA	0	N/A	1.000	0.00	0.00	1,122	1,122
33 23 0821	**Product Recovery Pump, Deep Depths (>20 ft), Controls, Electric**	**EA**				**0.00**	**0.00**	**1,234**	**1,234**
01001 0403	Product Recovery Pump, Deep Depths (>20 ft), Controls, Electric	EA	0	N/A	1.000	0.00	0.00	1,234	1,234
33 23 0822	**Product Recovery Pump, 2" Oil Skimmer, Controls, Electric**	**EA**				**0.00**	**0.00**	**1,346**	**1,346**
01001 0404	Product Recovery Pump, 2" Oil Skimmer, Controls, Electric	EA	0	N/A	1.000	0.00	0.00	1,346	1,346
33 23 0823	**Product Recovery Pump, 4" Oil Skimmer, Controls, Electric**	**EA**				**0.00**	**0.00**	**1,571**	**1,571**
01001 0405	Product Recovery Pump, 4" Oil Skimmer, Controls, Electric	EA	0	N/A	1.000	0.00	0.00	1,571	1,571

33 23 08	Product Recovery Pumps with Controls, Probes	Unit	Hourly Output	Crew	Factor	Labor	Equip.	Matl.	Total
33 23 0824	**Product Recovery Pump, 2" Oil Skimmer, Controls, Pneumatic, Monthly Rental**	**MO**				**0.00**	**0.00**	**224.41**	**224.41**
01001 0406	Product Recovery Pump, 2" Oil Skimmer, Controls, Pneumatic, Monthly Rental	MO	0	N/A	1.000	0.00	0.00	224.41	224.41
33 23 0825	**Product Recovery Pump, 4" Oil Skimmer, Controls, Pneumatic, Monthly Rental**	**MO**				**0.00**	**0.00**	**230.02**	**230.02**
01001 0407	Product Recovery Pump, 4" Oil Skimmer, Controls, Pneumatic, Monthly Rental	MO	0	N/A	1.000	0.00	0.00	230.02	230.02
33 23 0826	**Product Recovery Pump, 2" Oil Skimmer, Controls, Electric, Mo. Rental**	**MO**				**0.00**	**0.00**	**224.41**	**224.41**
01001 0408	Product Recovery Pump, 2" Oil Skimmer, Controls, Electric, Mo. Rental	MO	0	N/A	1.000	0.00	0.00	224.41	224.41
33 23 0827	**Product Recovery Pump, 4" Oil Skimmer, Controls, Electric, Mo. Rental**	**MO**				**0.00**	**0.00**	**230.02**	**230.02**
01001 0409	Product Recovery Pump, 4" Oil Skimmer, Controls, Electric, Mo. Rental	MO	0	N/A	1.000	0.00	0.00	230.02	230.02
33 23 0828	**Product Recovery Pump, Shallow Depths (<=20 ft), 6 gpm, Controls, Pneumatic, Mo. Rental**	**MO**				**0.00**	**0.00**	**224.41**	**224.41**
01001 0410	Product Recovery Pump, Shallow Depths (<=20 ft), 6 gpm, Controls, Pneumatic	MO	0	N/A	1.000	0.00	0.00	224.41	224.41
33 23 0829	**Product Recovery Pump, Deep Depths (>20 ft), 6 gpm, Controls, Pneumatic, Mo. Rental**	**MO**				**0.00**	**0.00**	**230.02**	**230.02**
01001 0411	Product Recovery Pump, Deep Depths (>20 ft), 6 gpm, Controls, Pneumatic, Mo	MO	0	N/A	1.000	0.00	0.00	230.02	230.02
33 23 0830	**Product Recovery Pump, Shallow Depths (<=20 ft), Controls, Electric, Mo. Rental**	**MO**				**0.00**	**0.00**	**224.41**	**224.41**
01001 0412	Product Recovery Pump, Shallow Depths (<=20 ft), Controls, Electric, Mo. Re	MO	0	N/A	1.000	0.00	0.00	224.41	224.41
33 23 0831	**Product Recovery Pump, Deep Depths (>20 ft), Controls, Electric, Mo. Rental**	**MO**				**0.00**	**0.00**	**230.02**	**230.02**
01001 0413	Product Recovery Pump, Deep Depths (>20 ft), Controls, Electric, Mo. Rental	MO	0	N/A	1.000	0.00	0.00	230.02	230.02
33 23 0832	**Holding Tank Belt-Skimmer, 3 gph, 1 in. wide belt**	**EA**				**0.00**	**0.00**	**456.67**	**456.67**
01001 0414	Holding Tank Belt-Skimmer, 3 gph, 1 in. wide belt	EA	0	N/A	1.000	0.00	0.00	456.67	456.67
33 23 0840	**Product Recovery Probe/Sensor, with 100' Probe Wire**	**EA**				**0.00**	**0.00**	**448.82**	**448.82**
01001 0415	Product Recovery Probe/Sensor, with 100' Probe Wire	EA	0	N/A	1.000	0.00	0.00	448.82	448.82

33 23 08	Product Recovery Pumps with Controls, Probes	Unit	Hourly Output	Crew	Factor	Labor	Equip.	Matl.	Total
33 23 0841	**Free Product Recovery System, 2" Well (coellescing filter regulator,pump,contols,probe,piping)**	**EA**				**0.00**	**0.00**	**1,515**	**1,515**
01001 0416	Free Product Recovery System, 2" Well (incl. filter reg,pump,contrls,piping)	EA	0	N/A	1.000	0.00	0.00	1,515	1,515
33 23 0842	**Free Product Removal, Product Skimmer, 2" Well**	**EA**				**0.00**	**0.00**	**662.01**	**662.01**
01001 0417	Free Product Removal, Product Skimmer, 2" Well	EA	0	N/A	1.000	0.00	0.00	662.01	662.01
33 23 0843	**Skimmer, 1" Diam (2" Well or larger), 47" Length, 0.10 gal Capacity**	**EA**				**0.00**	**0.00**	**402.81**	**402.81**
01001 0418	Skimmer, 1" Diam (2" Well or larger), 47" Length, 0.10 gal Capacity	EA	0	N/A	1.000	0.00	0.00	402.81	402.81
33 23 0844	**Skimmer, 2" Diam (4" Well or larger), 46" Length, 0.30 gal Capacity**	**EA**				**0.00**	**0.00**	**403.94**	**403.94**
01001 0419	Skimmer, 2" Diam (4" Well or larger), 46" Length, 0.30 gal Capacity	EA	0	N/A	1.000	0.00	0.00	403.94	403.94
33 23 0845	**Skimmer, 3" Diam (4" Well or larger), 60" Length, 1.0 gal Capacity**	**EA**				**0.00**	**0.00**	**483.60**	**483.60**
01001 0420	Skimmer, 3" Diam (4" Well or larger), 60" Length, 1.0 gal Capacity	EA	0	N/A	1.000	0.00	0.00	483.60	483.60
33 23 0846	**Skimmer, Pneumatic Pump w/ Floating Skimmer, 1.75" Diam**	**EA**				**0.00**	**0.00**	**1,070**	**1,070**
01001 0421	Skimmer, Pneumatic Pump w/ Floating Skimmer, 1.75" Diam	EA	0	N/A	1.000	0.00	0.00	1,070	1,070

33 23 09	Product Recovery Pumps without Controls, Probes	Unit	Hourly Output	Crew	Factor	Labor	Equip.	Matl.	Total
33 23 0901	**Product Recovery Pump, 1/3 HP, without Controls, Includes Probe**	**EA**				**422.45**	**0.00**	**2,100**	**2,522**
13273 6131	1/3 HP Product Recovery Pump without Controls, Includes Probe and 50' of Wire	EA	0.25	ULABD	1.000	422.45	0.00	2,100	2,522
33 23 0902	**Product Recovery Pump, 3/4 HP, without Controls**	**EA**				**422.45**	**0.00**	**1,275**	**1,697**
13273 6132	3/4 HP Groundwater Pump without Controls, Includes 50' of Wire	EA	0.25	ULABD	1.000	422.45	0.00	1,275	1,697
33 23 0910	**Pneumatic Product Recovery Pump, 2" Well**	**EA**				**0.00**	**0.00**	**658.64**	**658.64**
01001 0422	Pneumatic Product Recovery Pump, 2" Well	EA	0	N/A	1.000	0.00	0.00	658.64	658.64
33 23 0911	**Pneumatic Product Recovery Pump, 4" Well**	**EA**				**0.00**	**0.00**	**778.70**	**778.70**
01001 0423	Pneumatic Product Recovery Pump, 4" Well	EA	0	N/A	1.000	0.00	0.00	778.70	778.70
33 23 0912	**Pneumatic Product Recovery Pump, Cycle Control**	**EA**				**0.00**	**0.00**	**1,257**	**1,257**
01001 0424	Pneumatic Product Recovery Pump, Cycle Control	EA	0	N/A	1.000	0.00	0.00	1,257	1,257

33 23 11	Drilling	Unit	Hourly Output	Crew	Factor	Labor	Equip.	Matl.	Total
33 23 1101	**Hollow Stem Auger, 8" Dia Borehole, Depth <= 100 ft**	**LF**				**6.59**	**29.06**	**0.00**	**35.64**
01001 0600	Hollow Stem Auger, 8" Dia Borehole, Depth <= 100 ft	LF	13.68	ULADB	1.000	6.59	29.06	0.00	35.64
33 23 1102	**Hollow Stem Auger, 8" Dia Borehole, Depth > 100ft**	**LF**				**9.05**	**39.96**	**0.00**	**49.01**
01001 0601	Hollow Stem Auger, 8" Dia Borehole, Depth > 100ft	LF	9.945	ULADB	1.000	9.05	39.96	0.00	49.01
33 23 1103	**Hollow Stem Auger, 11" Dia Borehole, Depth <= 100 ft**	**LF**				**10.29**	**45.40**	**0.00**	**55.69**
01001 0602	Hollow Stem Auger, 11" Dia Borehole, Depth <= 100 ft	LF	8.752	ULADB	1.000	10.29	45.40	0.00	55.69
33 23 1104	**Hollow Stem Auger, 11" Dia Borehole, Depth > 100 ft**	**LF**				**11.94**	**52.67**	**0.00**	**64.60**
01001 0425	Hollow Stem Auger, 11" Dia Borehole, Depth > 100 ft	LF	7.545	ULADB	1.000	11.94	52.67	0.00	64.60
33 23 1105	**Hollow Stem Auger, 13-3/4" Dia Borehole, Depth <= 100 ft**	**LF**				**10.29**	**45.40**	**0.00**	**55.69**
01001 0603	Hollow Stem Auger, 13-3/4" Dia Borehole, Depth <= 100 ft	LF	8.752	ULADB	1.000	10.29	45.40	0.00	55.69
33 23 1106	**Hollow Stem Auger, 13-3/4" Dia Borehole, Depth > 100 ft**	**LF**				**11.94**	**52.67**	**0.00**	**64.60**
01001 0604	Hollow Stem Auger, 13-3/4" Dia Borehole, Depth > 100 ft	LF	7.545	ULADB	1.000	11.94	52.67	0.00	64.60
33 23 1107	**Hollow Stem Auger, 16" Dia Borehole, Depth <= 100 ft**	**LF**				**14.82**	**65.38**	**0.00**	**80.20**
01001 0605	Hollow Stem Auger, 16" Dia Borehole, Depth <= 100 ft	LF	6.078	ULADB	1.000	14.82	65.38	0.00	80.20
33 23 1108	**Hollow Stem Auger, 16" Dia Borehole, Depth > 100 ft**	**LF**				**16.87**	**74.46**	**0.00**	**91.34**
01001 0606	Hollow Stem Auger, 16" Dia Borehole, Depth > 100 ft	LF	5.337	ULADB	1.000	16.87	74.46	0.00	91.34
33 23 1109	**Mud Drilling, 4" Dia Borehole, Depth <= 100 ft**	**LF**				**8.64**	**38.14**	**0.00**	**46.78**
01001 0607	Mud Drilling, 4" Dia Borehole, Depth <= 100 ft	LF	10.42	ULADB	1.000	8.64	38.14	0.00	46.78
33 23 1110	**Mud Drilling, 4" Dia Borehole, Depth > 100 ft**	**LF**				**9.47**	**41.77**	**0.00**	**51.24**
01001 0608	Mud Drilling, 4" Dia Borehole, Depth > 100 ft	LF	9.513	ULADB	1.000	9.47	41.77	0.00	51.24
33 23 1111	**Mud Drilling, 6" Dia Borehole, Depth <= 100 ft**	**LF**				**8.64**	**38.14**	**0.00**	**46.78**
01001 0609	Mud Drilling, 6" Dia Borehole, Depth <= 100 ft	LF	10.42	ULADB	1.000	8.64	38.14	0.00	46.78
33 23 1112	**Mud Drilling, 6" Dia Borehole, 100 ft < Depth <= 500 ft**	**LF**				**8.23**	**36.32**	**0.00**	**44.55**
01001 0426	Mud Drilling, 6" Dia Borehole, 100 ft < Depth <= 500 ft	LF	10.94	ULADB	1.000	8.23	36.32	0.00	44.55

33 23 11 Drilling		Unit	Hourly Output	Crew	Factor	Labor	Equip.	Matl.	Total
33 23 1113	**Mud Drilling, 6" Dia Borehole, Depth > 500 ft**	**LF**				**8.64**	**38.14**	**0.00**	**46.78**
01001 0427	Mud Drilling, 6" Dia Borehole, Depth > 500 ft	LF	10.42	ULADB	1.000	8.64	38.14	0.00	46.78
33 23 1114	**Mud Drilling, 8" Dia Borehole, Depth <= 100 ft**	**LF**				**12.96**	**57.21**	**0.00**	**70.17**
01001 0428	Mud Drilling, 8" Dia Borehole, Depth <= 100 ft	LF	6.946	ULADB	1.000	12.96	57.21	0.00	70.17
33 23 1115	**Mud Drilling, 8" Dia Borehole,100 ft < Depth <= 500 ft**	**LF**				**11.32**	**49.94**	**0.00**	**61.26**
01001 0429	Mud Drilling, 8" Dia Borehole,100 ft < Depth <= 500 ft	LF	7.956	ULADB	1.000	11.32	49.94	0.00	61.26
33 23 1116	**Mud Drilling, 8" Dia Borehole, Depth > 500 ft**	**LF**				**12.14**	**53.58**	**0.00**	**65.72**
01001 0430	Mud Drilling, 8" Dia Borehole, Depth > 500 ft	LF	7.417	ULADB	1.000	12.14	53.58	0.00	65.72
33 23 1117	**Mud Drilling, 10" Dia Borehole, Depth <= 100 ft**	**LF**				**17.70**	**78.09**	**0.00**	**95.79**
01001 0431	Mud Drilling, 10" Dia Borehole, Depth <= 100 ft	LF	5.088	ULADB	1.000	17.70	78.09	0.00	95.79
33 23 1118	**Mud Drilling, 10" Dia Borehole, 100 ft < Depth <= 500 ft**	**LF**				**16.26**	**71.74**	**0.00**	**87.99**
01001 0432	Mud Drilling, 10" Dia Borehole, 100 ft < Depth <= 500 ft	LF	5.539	ULADB	1.000	16.26	71.74	0.00	87.99
33 23 1119	**Mud Drilling, 10" Dia Borehole, Depth > 500 ft**	**LF**				**17.29**	**76.28**	**0.00**	**93.56**
01001 0433	Mud Drilling, 10" Dia Borehole, Depth > 500 ft	LF	5.21	ULADB	1.000	17.29	76.28	0.00	93.56
33 23 1120	**Mud Drilling, 12" Dia Borehole, Depth <= 100 ft**	**LF**				**20.58**	**90.81**	**0.00**	**111.39**
01001 0434	Mud Drilling, 12" Dia Borehole, Depth <= 100 ft	LF	4.376	ULADB	1.000	20.58	90.81	0.00	111.39
33 23 1121	**Mud Drilling, 12" Dia Borehole, 100 ft < Depth <= 500 ft**	**LF**				**16.05**	**70.83**	**0.00**	**86.88**
01001 0610	Mud Drilling, 12" Dia Borehole, 100 ft < Depth <= 500 ft	LF	5.610	ULADB	1.000	16.05	70.83	0.00	86.88
33 23 1122	**Mud Drilling, 12" Dia Borehole, Depth > 500 ft**	**LF**				**17.29**	**76.28**	**0.00**	**93.56**
01001 0611	Mud Drilling, 12" Dia Borehole, Depth > 500 ft	LF	5.21	ULADB	1.000	17.29	76.28	0.00	93.56
33 23 1123	**Mud Drilling, 15" Dia Borehole, Depth <= 100 ft**	**LF**				**27.99**	**123.50**	**0.00**	**151.48**
01001 0612	Mud Drilling, 15" Dia Borehole, Depth <= 100 ft	LF	3.218	ULADB	1.000	27.99	123.50	0.00	151.48
33 23 1124	**Mud Drilling, 15" Dia Borehole, 100 ft < Depth <= 500 ft**	**LF**				**27.16**	**119.87**	**0.00**	**147.03**
01001 0613	Mud Drilling, 15" Dia Borehole, 100 ft < Depth <= 500 ft	LF	3.315	ULADB	1.000	27.16	119.87	0.00	147.03
33 23 1125	**Mud Drilling, 15" Dia Borehole, Depth > 500 ft**	**LF**				**28.40**	**125.31**	**0.00**	**153.71**
01001 0614	Mud Drilling, 15" Dia Borehole, Depth > 500 ft	LF	3.171	ULADB	1.000	28.40	125.31	0.00	153.71

33 23 11	Drilling	Unit	Hourly Output	Crew	Factor	Labor	Equip.	Matl.	Total
33 23 1126	**Air Rotary, 6" Dia Borehole (Consolidated), Depth <= 100 ft**	**LF**				**9.88**	**43.59**	**0.00**	**53.47**
01001 0615	Air Rotary, 6" Dia Borehole (Consolidated), Depth <= 100 ft	LF	9.117	ULADB	1.000	9.88	43.59	0.00	53.47
33 23 1127	**Air Rotary, 6" Dia Borehole (Consolidated), Depth > 100 ft**	**LF**				**8.64**	**38.14**	**0.00**	**46.78**
01001 0616	Air Rotary, 6" Dia Borehole (Consolidated), Depth > 100 ft	LF	10.42	ULADB	1.000	8.64	38.14	0.00	46.78
33 23 1128	**Air Rotary, 8" Dia Borehole (Consolidated), Depth <= 100 ft**	**LF**				**12.76**	**56.30**	**0.00**	**69.06**
01001 0617	Air Rotary, 8" Dia Borehole (Consolidated), Depth <= 100 ft	LF	7.058	ULADB	1.000	12.76	56.30	0.00	69.06
33 23 1129	**Air Rotary, 8" Dia Borehole (Consolidated), 100 < Depth <= 500 ft**	**LF**				**11.11**	**49.04**	**0.00**	**60.15**
01001 0618	Air Rotary, 8" Dia Borehole (Consolidated), 100 < Depth <= 500 ft	LF	8.104	ULADB	1.000	11.11	49.04	0.00	60.15
33 23 1130	**Air Rotary, 8" Dia Borehole (Consolidated), Depth > 500 ft**	**LF**				**28.40**	**125.31**	**0.00**	**153.71**
01001 0435	Air Rotary, 8" Dia Borehole (Consolidated), Depth > 500 ft	LF	3.171	ULADB	1.000	28.40	125.31	0.00	153.71
33 23 1131	**Air Rotary, 10" Dia Borehole (Consolidated), Depth <= 100 ft**	**LF**				**17.29**	**76.28**	**0.00**	**93.56**
01001 0619	Air Rotary, 10" Dia Borehole (Consolidated), Depth <= 100 ft	LF	5.21	ULADB	1.000	17.29	76.28	0.00	93.56
33 23 1132	**Air Rotary, 10" Dia Borehole (Consolidated), 100 ft < Depth <= 500 ft**	**LF**				**16.05**	**70.83**	**0.00**	**86.88**
01001 0620	Air Rotary, 10" Dia Borehole (Consolidated), 100 ft < Depth <= 500 ft	LF	5.610	ULADB	1.000	16.05	70.83	0.00	86.88
33 23 1133	**Air Rotary, 10" Dia Borehole (Consolidated), Depth > 500 ft**	**LF**				**18.11**	**79.91**	**0.00**	**98.02**
01001 0621	Air Rotary, 10" Dia Borehole (Consolidated), Depth > 500 ft	LF	4.973	ULADB	1.000	18.11	79.91	0.00	98.02
33 23 1134	**Air Rotary, 12" Dia Borehole (Consolidated), Depth <= 100 ft**	**LF**				**20.17**	**88.99**	**0.00**	**109.16**
01001 0622	Air Rotary, 12" Dia Borehole (Consolidated), Depth <= 100 ft	LF	4.465	ULADB	1.000	20.17	88.99	0.00	109.16
33 23 1135	**Air Rotary, 12" Dia Borehole (Consolidated), 100 ft < Depth <= 500 ft**	**LF**				**18.52**	**81.73**	**0.00**	**100.25**
01001 0623	Air Rotary, 12" Dia Borehole (Consolidated), 100 ft < Depth <= 500 ft	LF	4.862	ULADB	1.000	18.52	81.73	0.00	100.25
33 23 1136	**Air Rotary, 12" Dia Borehole (Consolidated), Depth > 500 ft**	**LF**				**20.99**	**92.62**	**0.00**	**113.61**
01001 0624	Air Rotary, 12" Dia Borehole (Consolidated), Depth > 500 ft	LF	4.290	ULADB	1.000	20.99	92.62	0.00	113.61

33 23 11	Drilling	Unit	Hourly Output	Crew	Factor	Labor	Equip.	Matl.	Total
33 23 1137	**Air Rotary, 16" Dia Borehole (Consolidated), Depth <= 100 ft**	**LF**				**29.63**	**130.76**	**0.00**	**160.40**
01001 0436	Air Rotary, 16" Dia Borehole (Consolidated), Depth <= 100 ft	LF	3.039	ULADB	1.000	29.63	130.76	0.00	160.40
33 23 1138	**Air Rotary, 16" Dia Borehole (Consolidated), 100 ft < Depth <= 500 ft**	**LF**				**37.04**	**163.45**	**0.00**	**200.49**
01001 0437	Air Rotary, 16" Dia Borehole (Consolidated), 100 ft < Depth <= 500 ft	LF	2.431	ULADB	1.000	37.04	163.45	0.00	200.49
33 23 1139	**Air Rotary, 16" Dia Borehole (Consolidated), Depth > 500 ft**	**LF**				**30.87**	**136.21**	**0.00**	**167.08**
01001 0438	Air Rotary, 16" Dia Borehole (Consolidated), Depth > 500 ft	LF	2.917	ULADB	1.000	30.87	136.21	0.00	167.08
33 23 1140	**Air Rotary, 18" Dia Borehole (Consolidated), Depth <= 100 ft**	**LF**				**30.46**	**134.40**	**0.00**	**164.85**
01001 0439	Air Rotary, 18" Dia Borehole (Consolidated), Depth <= 100 ft	LF	2.957	ULADB	1.000	30.46	134.40	0.00	164.85
33 23 1141	**Air Rotary, 18" Dia Borehole (Consolidated), 100 ft < Depth <= 500 ft**	**LF**				**30.46**	**134.40**	**0.00**	**164.85**
01001 0625	Air Rotary, 18" Dia Borehole (Consolidated), 100 ft < Depth <= 500 ft	LF	2.957	ULADB	1.000	30.46	134.40	0.00	164.85
33 23 1142	**Air Rotary, 18" Dia Borehole (Consolidated), Depth > 500 ft**	**LF**				**33.75**	**148.92**	**0.00**	**182.67**
01001 0626	Air Rotary, 18" Dia Borehole (Consolidated), Depth > 500 ft	LF	2.668	ULADB	1.000	33.75	148.92	0.00	182.67
33 23 1143	**Air Rotary, 22" Dia Borehole (Consolidated), Depth <= 100 ft**	**LF**				**62.97**	**277.87**	**0.00**	**340.84**
01001 0627	Air Rotary, 22" Dia Borehole (Consolidated), Depth <= 100 ft	LF	1.430	ULADB	1.000	62.97	277.87	0.00	340.84
33 23 1144	**Air Rotary, 22" Dia Borehole (Consolidated), 100 ft < Depth <= 500 ft**	**LF**				**74.08**	**326.91**	**0.00**	**400.99**
01001 0628	Air Rotary, 22" Dia Borehole (Consolidated), 100 ft < Depth <= 500 ft	LF	1.216	ULADB	1.000	74.08	326.91	0.00	400.99
33 23 1145	**Air Rotary, 22" Dia Borehole (Consolidated), Depth > 500 ft**	**LF**				**70.38**	**310.56**	**0.00**	**380.94**
01001 0629	Air Rotary, 22" Dia Borehole (Consolidated), Depth > 500 ft	LF	1.28	ULADB	1.000	70.38	310.56	0.00	380.94
33 23 1146	**Air Rotary, 6" Dia Borehole (Unconsolidated), Depth <= 100 ft**	**LF**				**10.29**	**45.40**	**0.00**	**55.69**
01001 0630	Air Rotary, 6" Dia Borehole (Unconsolidated), Depth <= 100 ft	LF	8.752	ULADB	1.000	10.29	45.40	0.00	55.69
33 23 1147	**Air Rotary, 6" Dia Borehole (Unconsolidated), Depth > 100 ft**	**LF**				**12.76**	**56.30**	**0.00**	**69.06**
01001 0440	Air Rotary, 6" Dia Borehole (Unconsolidated), Depth > 100 ft	LF	7.058	ULADB	1.000	12.76	56.30	0.00	69.06

33 23 11 Drilling		Unit	Hourly Output	Crew	Factor	Labor	Equip.	Matl.	Total
33 23 1148	**Air Rotary, 8" Dia Borehole (Unconsolidated), Depth <= 100 ft**	**LF**				**11.11**	**49.04**	**0.00**	**60.15**
01001 0441	Air Rotary, 8" Dia Borehole (Unconsolidated), Depth <= 100 ft	LF	8.104	ULADB	1.000	11.11	49.04	0.00	60.15
33 23 1149	**Air Rotary, 8" Dia Borehole (Unconsolidated), 100 < Depth <= 500 ft**	**LF**				**17.29**	**76.28**	**0.00**	**93.56**
01001 0442	Air Rotary, 8" Dia Borehole (Unconsolidated), 100 < Depth <= 500 ft	LF	5.21	ULADB	1.000	17.29	76.28	0.00	93.56
33 23 1150	**Air Rotary, 8" Dia Borehole (Unconsolidated), Depth > 500 ft**	**LF**				**16.05**	**70.83**	**0.00**	**86.88**
01001 0443	Air Rotary, 8" Dia Borehole (Unconsolidated), Depth > 500 ft	LF	5.610	ULADB	1.000	16.05	70.83	0.00	86.88
33 23 1151	**Air Rotary, 10" Dia Borehole (Unconsolidated), Depth <= 100 ft**	**LF**				**18.11**	**79.91**	**0.00**	**98.02**
01001 0631	Air Rotary, 10" Dia Borehole (Unconsolidated), Depth <= 100 ft	LF	4.973	ULADB	1.000	18.11	79.91	0.00	98.02
33 23 1152	**Air Rotary, 10" Dia Borehole (Unconsolidated), 100 ft < Depth <= 500 ft**	**LF**				**16.46**	**72.65**	**0.00**	**89.11**
01001 0632	Air Rotary, 10" Dia Borehole (Unconsolidated), 100 ft < Depth <= 500 ft	LF	5.47	ULADB	1.000	16.46	72.65	0.00	89.11
33 23 1153	**Air Rotary, 10" Dia Borehole (Unconsolidated), Depth > 500 ft**	**LF**				**23.05**	**101.70**	**0.00**	**124.75**
01001 0633	Air Rotary, 10" Dia Borehole (Unconsolidated), Depth > 500 ft	LF	3.907	ULADB	1.000	23.05	101.70	0.00	124.75
33 23 1154	**Air Rotary, 12" Dia Borehole (Unconsolidated), Depth <= 100 ft**	**LF**				**20.99**	**92.62**	**0.00**	**113.61**
01001 0634	Air Rotary, 12" Dia Borehole (Unconsolidated), Depth <= 100 ft	LF	4.290	ULADB	1.000	20.99	92.62	0.00	113.61
33 23 1155	**Air Rotary, 12" Dia Borehole (Unconsolidated), 100 ft < Depth <= 500 ft**	**LF**				**20.58**	**90.81**	**0.00**	**111.39**
01001 0635	Air Rotary, 12" Dia Borehole (Unconsolidated), 100 ft < Depth <= 500 ft	LF	4.376	ULADB	1.000	20.58	90.81	0.00	111.39
33 23 1156	**Air Rotary, 12" Dia Borehole (Unconsolidated), Depth > 500 ft**	**LF**				**26.75**	**118.05**	**0.00**	**144.80**
01001 0636	Air Rotary, 12" Dia Borehole (Unconsolidated), Depth > 500 ft	LF	3.366	ULADB	1.000	26.75	118.05	0.00	144.80
33 23 1157	**Air Rotary, 16" Dia Borehole (Unconsolidated), Depth <= 100 ft**	**LF**				**28.40**	**125.31**	**0.00**	**153.71**
01001 0637	Air Rotary, 16" Dia Borehole (Unconsolidated), Depth <= 100 ft	LF	3.171	ULADB	1.000	28.40	125.31	0.00	153.71
33 23 1158	**Air Rotary, 16" Dia Borehole (Unconsolidated), 100 ft < Depth <= 500 ft**	**LF**				**31.69**	**139.84**	**0.00**	**171.53**
01001 0444	Air Rotary, 16" Dia Borehole (Unconsolidated), 100 ft < Depth <= 500 ft	LF	2.842	ULADB	1.000	31.69	139.84	0.00	171.53

33 23 11 Drilling		Unit	Hourly Output	Crew	Factor	Labor	Equip.	Matl.	Total
33 23 1159	**Air Rotary, 16" Dia Borehole (Unconsolidated), Depth > 500 ft**	**LF**				**32.93**	**145.29**	**0.00**	**178.22**
01001 0445	Air Rotary, 16" Dia Borehole (Unconsolidated), Depth > 500 ft	LF	2.735	ULADB	1.000	32.93	145.29	0.00	178.22
33 23 1160	**Air Rotary, 18" Dia Borehole (Unconsolidated), Depth <= 100 ft**	**LF**				**41.16**	**181.62**	**0.00**	**222.77**
01001 0446	Air Rotary, 18" Dia Borehole (Unconsolidated), Depth <= 100 ft	LF	2.188	ULADB	1.000	41.16	181.62	0.00	222.77
33 23 1161	**Air Rotary, 18" Dia Borehole (Unconsolidated), 100 ft < Depth <= 500 ft**	**LF**				**32.93**	**145.29**	**0.00**	**178.22**
01001 0638	Air Rotary, 18" Dia Borehole (Unconsolidated), 100 ft < Depth <= 500 ft	LF	2.735	ULADB	1.000	32.93	145.29	0.00	178.22
33 23 1162	**Air Rotary, 18" Dia Borehole (Unconsolidated), Depth > 500 ft**	**LF**				**41.16**	**181.62**	**0.00**	**222.77**
01001 0639	Air Rotary, 18" Dia Borehole (Unconsolidated), Depth > 500 ft	LF	2.188	ULADB	1.000	41.16	181.62	0.00	222.77
33 23 1163	**Air Rotary, 22" Dia Borehole (Unconsolidated), Depth <= 100 ft**	**LF**				**64.62**	**285.14**	**0.00**	**349.75**
01001 0640	Air Rotary, 22" Dia Borehole (Unconsolidated), Depth <= 100 ft	LF	1.394	ULADB	1.000	64.62	285.14	0.00	349.75
33 23 1164	**Air Rotary, 22" Dia Borehole (Unconsolidated), 100 ft < Depth <= 500 ft**	**LF**				**80.25**	**354.15**	**0.00**	**434.41**
01001 0641	Air Rotary, 22" Dia Borehole (Unconsolidated), 100 ft < Depth <= 500 ft	LF	1.122	ULADB	1.000	80.25	354.15	0.00	434.41
33 23 1165	**Air Rotary, 22" Dia Borehole (Unconsolidated), Depth > 500 ft**	**LF**				**87.25**	**385.03**	**0.00**	**472.28**
01001 0642	Air Rotary, 22" Dia Borehole (Unconsolidated), Depth > 500 ft	LF	1.032	ULADB	1.000	87.25	385.03	0.00	472.28
33 23 1166	**Slant Well, Hollow Stem Auger, 2" Outside Diameter**	**LF**				**6.95**	**30.66**	**0.00**	**37.60**
01001 0643	Slant Well, Hollow Stem Auger, 2" Outside Diameter	LF	12.96	ULADB	1.000	6.95	30.66	0.00	37.60
33 23 1167	**Slant Well, Hollow Stem Auger, 4" Outside Diameter**	**LF**				**10.53**	**46.46**	**0.00**	**56.99**
01001 0644	Slant Well, Hollow Stem Auger, 4" Outside Diameter	LF	8.554	ULADB	1.000	10.53	46.46	0.00	56.99
33 23 1168	**Slant Well, Air Rotary, 6" Outside Diameter**	**LF**				**18.10**	**79.89**	**0.00**	**98.00**
01001 0645	Slant Well, Air Rotary, 6" Outside Diameter	LF	4.974	ULADB	1.000	18.10	79.89	0.00	98.00
33 23 1169	**Slant Well, Air Rotary, 8" Outside Diameter**	**LF**				**21.52**	**94.95**	**0.00**	**116.47**
01001 0646	Slant Well, Air Rotary, 8" Outside Diameter	LF	4.185	ULADB	1.000	21.52	94.95	0.00	116.47
33 23 1170	**Cable Tool Drilling, 10" Dia Borehole**	**LF**				**85.19**	**375.94**	**0.00**	**461.14**
01001 0588	Cable Tool Drilling, 10" Dia Borehole	LF	1.057	ULADB	1.000	85.19	375.94	0.00	461.14

33 23 11	Drilling	Unit	Hourly Output	Crew	Factor	Labor	Equip.	Matl.	Total
33 23 1171	**Cable Tool Drilling, 14" Dia Borehole**	**LF**				**76.67**	**338.35**	**0.00**	**415.02**
01001 0589	Cable Tool Drilling, 14" Dia Borehole	LF	1.174	ULADB	1.000	76.67	338.35	0.00	415.02
33 23 1172	**Split Spoon Sample, 2" x 24", During Drilling**	**EA**				**0.00**	**0.00**	**45.21**	**45.21**
02557 7121	4" Borehole, Slant Well, Subcontract Price	LF	0	N/A	1.000	0.00	0.00	45.21	45.21
33 23 1173	**Split Spoon Sampling**	**LF**				**10.29**	**45.41**	**0.00**	**55.71**
02557 5111	2-1/4" Inside Diameter x 6" Outside Diameter for Split Spoon Sampling	LF	8.75	ULADB	1.000	10.29	45.41	0.00	55.71
33 23 1174	**Continuous Split Spoon Sample**	**LF**				**87.13**	**270.55**	**0.00**	**357.68**
02557 5119	Continuous Split Spoon Sample	LF	0.835	ULADD	1.000	87.13	270.55	0.00	357.68
33 23 1175	**Coring (Fluid or Air)**	**LF**				**16.67**	**26.19**	**14.20**	**57.06**
02010 1050	Drill in Rock, BX Core, with Casing, Includes Sampling	LF	4.925	CLADI	1.000	16.67	26.19	14.20	57.06
33 23 1176	**Standby for Drilling**	**EA**				**90.05**	**397.38**	**55.50**	**542.92**
02674 8560	Standby, Drill Rig/Trencher	HR	1	ULADB	1.000	90.05	397.38	55.50	542.92
33 23 1177	**Standby for Trenching**	**EA**				**90.05**	**397.38**	**55.50**	**542.92**
02674 8560	Standby, Drill Rig/Trencher	HR	1	ULADB	1.000	90.05	397.38	55.50	542.92
33 23 1178	**Move Rig/Equipment Around Site**	**EA**				**103.56**	**456.98**	**63.82**	**624.36**
02674 8560	Standby, Drill Rig/Trencher	HR	1	ULADB	1.150	90.05	397.38	55.50	624.36
33 23 1179	**Move Trencher/Equipment Around Site**	**EA**				**103.56**	**456.98**	**63.82**	**624.36**
02674 8560	Standby, Drill Rig/Trencher	HR	1	ULADB	1.150	90.05	397.38	55.50	624.36
33 23 1180	**Mobilization/Demobilization, Drill Equipment or Trencher, Crew**	**EA**				**450.25**	**1,987**	**277.50**	**2,715**
02674 8560	Standby, Drill Rig/Trencher	HR	1	ULADB	5.000	90.05	397.38	55.50	2,715
33 23 1181	**Concrete Coring (Minimum Charge)**	**DAY**				**360.20**	**1,590**	**222.00**	**2,172**
02674 8560	Standby, Drill Rig/Trencher	HR	1	ULADB	4.000	90.05	397.38	55.50	2,172
33 23 1182	**Furnish 55 Gallon Drum for Drill Cuttings & Development Water**	**EA**				**0.00**	**0.00**	**89.50**	**89.50**
02083 5211	55 Gallon, 17C, Open	EA	0	N/A	1.000	0.00	0.00	89.50	89.50
33 23 1183	**90 lb. Jackhammer w/Compressor, Bits, and Hose (rental, per day)**	**DAY**				**0.00**	**0.00**	**217.68**	**217.68**
02557 5152	Jack Hammer Rental, Per Day	DAY	0	N/A	1.000	0.00	0.00	217.68	217.68
33 23 1184	**Concrete Saw Rental + 14" Blade (per day)**	**DAY**				**0.00**	**0.00**	**133.73**	**133.73**
02557 5151	Concrete Saw Rental, Per Day	DAY	0	N/A	1.000	0.00	0.00	133.73	133.73
33 23 1185	**Bucket Auger, 18" Diameter Borehole**	**LF**				**0.00**	**0.00**	**59.89**	**59.89**
02557 5117	Bucket Auger, 18" Diameter Borehole, Subcontract Price	LF	0	N/A	1.000	0.00	0.00	59.89	59.89

33 23 Wells

33 23 11 Drilling		Unit	Hourly Output	Crew	Factor	Labor	Equip.	Matl.	Total
33 23 1186	**Well Development Equipment Rental (weekly)**	**WK**				**0.00**	**0.00**	**236.00**	**236.00**
01954 3272	Water Quality Testing Parameter Device Rental	WK	0	N/A	1.000	0.00	0.00	236.00	236.00
33 23 1187	**Load Supplies/Equipment**	**LS**				**270.15**	**1,192**	**166.50**	**1,629**
02674 8560	Standby, Drill Rig/Trencher	HR	1	ULADB	3.000	90.05	397.38	55.50	1,629
33 23 1188	**Security Pass/Protocol**	**LS**				**45.02**	**198.69**	**27.75**	**271.46**
02674 8560	Standby, Drill Rig/Trencher	HR	1	ULADB	0.500	90.05	397.38	55.50	271.46
33 23 1189	**Furnish 55 Gallon Drum for Development/Purge Water**	**EA**				**0.00**	**0.00**	**89.50**	**89.50**
02083 5211	55 Gallon, 17C, Open	EA	0	N/A	1.000	0.00	0.00	89.50	89.50
33 23 1190	**Trenching for Horizontal Well Installation**	**LF**				**0.00**	**0.00**	**41.86**	**41.86**
02221 1951	Trenching using HDSI Trencher	LF	0	N/A	1.000	0.00	0.00	41.86	41.86
33 23 1191	**Additional Trenching Cost, Horizontal Well Installation**	**LF**				**0.00**	**0.00**	**83.72**	**83.72**
02221 1954	Additional Trenching Cost, Depth > 20'	LF	0	N/A	1.000	0.00	0.00	83.72	83.72
33 23 1192	**Thin Wall Tube**	**EA**				**0.00**	**0.00**	**69.68**	**69.68**
01001 0590	Thin Wall Tube	EA	0	N/A	1.000	0.00	0.00	69.68	69.68
33 23 1193	**Well Development Equipment Rental (daily)**	**DAY**				**0.00**	**0.00**	**139.47**	**139.47**
01001 0447	Well Development Equipment Rental (daily)	DAY	0	N/A	1.000	0.00	0.00	139.47	139.47

33 23 12 Well Drilling		Unit	Hourly Output	Crew	Factor	Labor	Equip.	Matl.	Total
33 23 1201	**Mobilize/Demobilize Directional Drill Rig**	**EA**				**450.25**	**1,987**	**277.50**	**2,715**
02674 8560	Standby, Drill Rig/Trencher	HR	1	ULADB	5.000	90.05	397.38	55.50	2,715
33 23 1203	**Fluid Drilling to 400' Length, Consolidated, Continuous**	**LF**				**0.00**	**0.00**	**58.60**	**58.60**
02557 7102	Fluid Drilling to 400' Length, Consolidated, Continuous	LF	0	N/A	1.000	0.00	0.00	58.60	58.60
33 23 1204	**Mud Drilling, 400 - 1,200' Length, Consolidated, Continuous**	**LF**				**0.00**	**0.00**	**98.79**	**98.79**
02557 7103	Mud Drilling, 400 - 1,200' Length, Consolidated, Continuous	LF	0	N/A	1.000	0.00	0.00	98.79	98.79
33 23 1205	**Mud Drilling, > 1,200' Length, Consolidated, Continuous**	**LF**				**0.00**	**0.00**	**152.38**	**152.38**
02557 7104	Mud Drilling, > 1,200' Length, Consolidated, Continuous	LF	0	N/A	1.000	0.00	0.00	152.38	152.38
33 23 1206	**Fluid Drilling to 400' Length, Consolidated, Blind Well**	**LF**				**0.00**	**0.00**	**78.70**	**78.70**
02557 7105	Fluid Drilling to 400' Length, Consolidated, Blind Well	LF	0	N/A	1.000	0.00	0.00	78.70	78.70

33 23 12	Well Drilling	Unit	Hourly Output	Crew	Factor	Labor	Equip.	Matl.	Total
33 23 1207	**Mud Drilling, 400 - 1,200' Length, Consolidated, Blind Well**	**LF**				**0.00**	**0.00**	**133.95**	**133.95**
02557 7106	Mud Drilling, 400 - 1,200' Length, Consolidated, Blind Well	LF	0	N/A	1.000	0.00	0.00	133.95	133.95
33 23 1208	**Mud Drilling, > 1,200' Length, Consolidated, Blind Well**	**LF**				**0.00**	**0.00**	**204.27**	**204.27**
02557 7107	Mud Drilling, > 1,200' Length, Consolidated, Blind Well	LF	0	N/A	1.000	0.00	0.00	204.27	204.27
33 23 1209	**Fluid Drilling to 400' Length, Unconsolidated, Continuous**	**LF**				**0.00**	**0.00**	**75.35**	**75.35**
02557 7108	Fluid Drilling to 400' Length, Unconsolidated, Continuous	LF	0	N/A	1.000	0.00	0.00	75.35	75.35
33 23 1210	**Mud Drilling, 400 - 1,200' Length, Unconsolidated, Continuous**	**LF**				**0.00**	**0.00**	**118.89**	**118.89**
02557 7109	Mud Drilling, 400 - 1,200' Length, Unconsolidated, Continuous	LF	0	N/A	1.000	0.00	0.00	118.89	118.89
33 23 1211	**Mud Drilling, > 1,200' Length, Unconsolidated, Continuous**	**LF**				**0.00**	**0.00**	**184.19**	**184.19**
02557 7111	Mud Drilling, > 1,200' Length, Unconsolidated, Continuous	LF	0	N/A	1.000	0.00	0.00	184.19	184.19

33 23 13	Control Panels	Unit	Hourly Output	Crew	Factor	Labor	Equip.	Matl.	Total
33 23 1301	**4 Product Recovery/4 Groundwater Pump Control Panel**	**EA**				**422.45**	**0.00**	**3,700**	**4,122**
13273 6126	4 Pump Multiwell Control Panel (4 Product and 4 Groundwater)	EA	0.25	ULABD	1.000	422.45	0.00	3,700	4,122
33 23 1302	**3 Product Recovery/3 Groundwater Pump Control Panel**	**EA**				**422.45**	**0.00**	**4,275**	**4,697**
13273 6127	3 Pump Multiwell Control Panel (3 Product and 3 Groundwater)	EA	0.25	ULABD	1.000	422.45	0.00	4,275	4,697
33 23 1303	**1 Product Recovery/4 Groundwater Pump Control Panel**	**EA**				**422.45**	**0.00**	**4,175**	**4,597**
13273 6128	Multiwell Control Panel (1 Product and 4 Groundwater)	EA	0.25	ULABD	1.000	422.45	0.00	4,175	4,597
33 23 1304	**2 Product Recovery/2 Groundwater Pump Control Panel**	**EA**				**422.45**	**0.00**	**3,375**	**3,797**
13273 6129	Multiwell Control Panel (2 Product and 2 Groundwater)	EA	0.25	ULABD	1.000	422.45	0.00	3,375	3,797
33 23 1305	**Water Level Sensor, Float Switch, with 50' Cable**	**EA**				**211.23**	**0.00**	**494.99**	**706.21**
16820 3574	Water Level Sensor, Float Switch, with 50' Cable	EA	0.5	ULABD	1.000	211.23	0.00	494.99	706.21
33 23 1306	**High Sump Level Switch for Avoiding Overflow**	**EA**				**140.82**	**0.00**	**221.85**	**362.67**
13273 4170	High Sump Level Switch for Avoiding Overflow	EA	0.75	ULABD	1.000	140.82	0.00	221.85	362.67

33 23 Wells

33 23 14 Gravel Packs	Unit	Hourly Output	Crew	Factor	Labor	Equip.	Matl.	Total
33 23 1401 2" Screen, Filter Pack	**LF**				**2.04**	**9.01**	**3.23**	**14.28**
02557 6111 No. 2 Morie Silica Sand - Contaminant Free	TON	0.75	ULADB	0.017	120.07	529.83	190.11	14.28
33 23 1402 4" Screen, Filter Pack	**LF**				**3.60**	**15.90**	**5.70**	**25.20**
02557 6111 No. 2 Morie Silica Sand - Contaminant Free	TON	0.75	ULADB	0.030	120.07	529.83	190.11	25.20
33 23 1403 6" Screen, Filter Pack	**LF**				**5.22**	**23.05**	**8.27**	**36.54**
02557 6111 No. 2 Morie Silica Sand - Contaminant Free	TON	0.75	ULADB	0.043	120.07	529.83	190.11	36.54
33 23 1404 Gravel Pack, 8" Well	**LF**				**5.24**	**16.27**	**3.78**	**25.30**
02674 8405 Artificial Gravel Pack, 8" Casing (20 cm) Casing, 2" (5.1 cm) Screen	LF	13.88	ULADD	1.000	5.24	16.27	3.78	25.30
33 23 1405 Gravel Pack, 10" Well	**LF**				**7.81**	**24.25**	**4.75**	**36.81**
02674 8410 Artificial Gravel Pack, 10" Casing (25 cm) Casing, 2" (5.1 cm) Screen	LF	9.313	ULADD	1.000	7.81	24.25	4.75	36.81
33 23 1406 Gravel Pack, 12" Well	**LF**				**9.70**	**30.11**	**6.40**	**46.20**
02674 8415 Artificial Gravel Pack, 12" Casing (31 cm) Casing, 2" (5.1 cm) Screen	LF	7.5	ULADD	1.000	9.70	30.11	6.40	46.20
33 23 1407 Gravel Pack for Horizontal Well Installation	**CF**				**0.04**	**0.04**	**0.84**	**0.93**
02215 1220 Backfill Trench with Small Front-end Loader, Without Compaction	CY	70	CODLB	0.037	0.69	0.66	0.00	0.05
02611 2206 Crushed Stone	CY	72.38	XSABC	0.037	0.46	0.53	22.77	0.88
33 23 1408 2" Screen, Filter Pack, Slant Well	**LF**				**2.65**	**11.70**	**8.13**	**22.47**
02557 7521 2" Screen, Filter Pack, Slant Well	LF	33.97	ULADB	1.000	2.65	11.70	8.13	22.47
33 23 1409 4" Screen, Filter Pack, Slant Well	**LF**				**4.68**	**20.67**	**14.34**	**39.69**
02557 7522 4" Screen, Filter Pack, Slant Well	LF	19.23	ULADB	1.000	4.68	20.67	14.34	39.69

33 23 15 Surface Pads	Unit	Hourly Output	Crew	Factor	Labor	Equip.	Matl.	Total
33 23 1501 Surface Seal, Concrete Filled	**LS**				**720.40**	**3,179**	**540.00**	**4,439**
02674 8580 Surface Seal Well, Concrete Filled	LS	0.125	ULADB	1.000	720.40	3,179	540.00	4,439
33 23 1502 Surface Pad, Concrete, 4' x 4' x 4"	**EA**				**117.84**	**4.45**	**60.97**	**183.27**
03362 0010 Pour Spread Footing,< 5 CY, Direct Chute < (5 m3) Place 3,000 PSI Concrete Foundations	CY	1	ALABE	0.700	168.35	6.36	87.11	183.27
33 23 1503 3,000 PSI, 6" Thick, Concrete Pad	**CY**				**168.35**	**6.36**	**78.39**	**253.10**
03362 0100 Pour Slab on Grade, >= 6", Direct Chute >= (15 cm) Place 3,000 PSI Concrete	CY	1	ALABE	1.000	168.35	6.36	78.39	253.10
33 23 1504 Surface Pad, Concrete, 2' x 2' x 4"	**EA**				**75.76**	**2.86**	**39.20**	**117.82**
03362 0010 Pour Spread Footing,< 5 CY, Direct Chute < (5 m3) Place 3,000 PSI Concrete Foundations	CY	1	ALABE	0.450	168.35	6.36	87.11	117.82

33 23 16 Pump Probes	Unit	Hourly Output	Crew	Factor	Labor	Equip.	Matl.	Total

33 23 16 Pump Probes	Unit	Hourly Output	Crew	Factor	Labor	Equip.	Matl.	Total
33 23 1601 Groundwater or Product Recovery Probe, 100' Cable	**EA**				**0.00**	**0.00**	**345.00**	**345.00**
13273 6124 Groundwater or Product Recovery Probe with 100' Probe Wire	EA	0	N/A	1.000	0.00	0.00	345.00	345.00
33 23 1602 Dual Pump Probe, 100' Cable	**EA**				**0.00**	**0.00**	**275.00**	**275.00**
13273 6125 Multiwell Dual Pump Probe with 100' of Cable	EA	0	N/A	1.000	0.00	0.00	275.00	275.00

33 23 17 Well Caps	Unit	Hourly Output	Crew	Factor	Labor	Equip.	Matl.	Total
33 23 1701 2" Well, Locking Cap	**EA**				**22.51**	**99.34**	**12.81**	**134.67**
02557 5713 2" Aluminum Locking Cap, Slip-type, Not Water Tight	EA	4	ULADB	1.000	22.51	99.34	12.48	134.34
02557 5731 Warning Labels for Caps	EA	0	N/A	1.000	0.00	0.00	0.33	0.33
33 23 1702 4" Well, Locking Cap	**EA**				**25.73**	**113.54**	**15.76**	**155.02**
02557 5714 4" Aluminum Locking Cap, Slip-type, Not Water Tight	EA	3.5	ULADB	1.000	25.73	113.54	15.43	154.70
02557 5731 Warning Labels for Caps	EA	0	N/A	1.000	0.00	0.00	0.33	0.33
33 23 1703 6" Well, Locking Cap	**EA**				**30.02**	**132.46**	**26.64**	**189.12**
02557 5716 6" Aluminum Locking Cap, Slip-type, Not Water Tight	EA	3	ULADB	1.000	30.02	132.46	26.31	188.79
02557 5731 Warning Labels for Caps	EA	0	N/A	1.000	0.00	0.00	0.33	0.33
33 23 1711 2" PVC, Vented Cap	**EA**				**7.20**	**31.79**	**7.05**	**46.05**
02557 5666 2" Diameter Vented Cap - PVC	EA	12.5	ULADB	1.000	7.20	31.79	7.05	46.05
33 23 1712 4" PVC, Vented Cap	**EA**				**8.00**	**35.32**	**10.70**	**54.03**
02557 5669 4" Diameter Vented Cap - PVC	EA	11.25	ULADB	1.000	8.00	35.32	10.70	54.03
33 23 1713 6" PVC, Vented Cap	**EA**				**9.00**	**39.74**	**40.97**	**89.71**
02557 5673 6" Diameter Vented Cap - PVC	EA	10	ULADB	1.000	9.00	39.74	40.97	89.71

33 23 18 Annular Seals	Unit	Hourly Output	Crew	Factor	Labor	Equip.	Matl.	Total
33 23 1801 2" Well, Grout (Annular Seal)	**LF**				**11.89**	**52.45**	**3.05**	**67.39**
02557 6128 Bentonite Grout - 3.04 Lbs/Gallon 50 Lb Yields 2.2 CF	CF	2.5	ULADB	0.330	36.02	158.95	9.24	67.39
33 23 1802 4" Well, Grout (Annular Seal)	**LF**				**20.53**	**90.60**	**5.27**	**116.40**
02557 6128 Bentonite Grout - 3.04 Lbs/Gallon 50 Lb Yields 2.2 CF	CF	2.5	ULADB	0.570	36.02	158.95	9.24	116.40
33 23 1803 6" Well, Grout (Annular Seal)	**LF**				**30.26**	**133.52**	**7.76**	**171.54**
02557 6128 Bentonite Grout - 3.04 Lbs/Gallon 50 Lb Yields 2.2 CF	CF	2.5	ULADB	0.840	36.02	158.95	9.24	171.54
33 23 1804 8" Well, Grout (Annular Seal)	**LF**				**40.34**	**178.02**	**10.35**	**228.72**
02557 6128 Bentonite Grout - 3.04 Lbs/Gallon 50 Lb Yields 2.2 CF	CF	2.5	ULADB	1.120	36.02	158.95	9.24	228.72

33 23 18 Annular Seals		Unit	Hourly Output	Crew	Factor	Labor	Equip.	Matl.	Total
33 23 1805	**10" Well, Grout (Annular Seal)**	**LF**				**50.43**	**222.53**	**12.94**	**285.90**
02557 6128	Bentonite Grout - 3.04 Lbs/Gallon 50 Lb Yields 2.2 CF	CF	2.5	ULADB	1.400	36.02	158.95	9.24	285.90
33 23 1806	**12" Well, Grout (Annular Seal)**	**LF**				**60.51**	**267.04**	**15.53**	**343.08**
02557 6128	Bentonite Grout - 3.04 Lbs/Gallon 50 Lb Yields 2.2 CF	CF	2.5	ULADB	1.680	36.02	158.95	9.24	343.08
33 23 1811	**2" Well, Portland Cement Grout**	**LF**				**0.00**	**0.00**	**1.20**	**1.20**
02557 6142	Purchase Portland Cement Grout	CF	0	N/A	0.200	0.00	0.00	6.02	1.20
33 23 1812	**4" Well, Portland Cement Grout**	**LF**				**0.00**	**0.00**	**1.81**	**1.81**
02557 6142	Purchase Portland Cement Grout	CF	0	N/A	0.300	0.00	0.00	6.02	1.81
33 23 1813	**6" Well, Portland Cement Grout**	**LF**				**0.00**	**0.00**	**10.23**	**10.23**
02557 6142	Purchase Portland Cement Grout	CF	0	N/A	1.700	0.00	0.00	6.02	10.23
33 23 1814	**8" Well, Portland Cement Grout**	**LF**				**0.00**	**0.00**	**14.44**	**14.44**
02557 6142	Purchase Portland Cement Grout	CF	0	N/A	2.400	0.00	0.00	6.02	14.44
33 23 1815	**10" Well, Portland Cement Grout**	**LF**				**0.00**	**0.00**	**16.85**	**16.85**
02557 6142	Purchase Portland Cement Grout	CF	0	N/A	2.800	0.00	0.00	6.02	16.85
33 23 1816	**12" Well, Portland Cement Grout**	**LF**				**0.00**	**0.00**	**19.26**	**19.26**
02557 6142	Purchase Portland Cement Grout	CF	0	N/A	3.200	0.00	0.00	6.02	19.26
33 23 1820	**Grout Continuous Borehole**	**CF**				**0.00**	**0.00**	**6.02**	**6.02**
02557 6142	Purchase Portland Cement Grout	CF	0	N/A	1.000	0.00	0.00	6.02	6.02
33 23 1821	**Sealing Existing Well**	**LF**				**2.13**	**9.39**	**0.55**	**12.07**
02557 6128	Bentonite Grout - 3.04 Lbs/Gallon 50 Lb Yields 2.2 CF	CF	2.5	ULADB	0.059	36.02	158.95	9.24	12.07
33 23 1822	**Well Abandonment, 2" Well**	**LF**				**4.64**	**20.47**	**0.78**	**25.89**
02557 6141	Monitor well, portland cement,94# bag, portland cement grout	CF	2.5	ULADB	0.129	36.02	158.95	6.02	25.89
33 23 1823	**Well Abandonment, 4" Well**	**LF**				**8.00**	**35.32**	**1.20**	**44.53**
02557 6141	Monitor well, portland cement,94# bag, portland cement grout	CF	2.25	ULADB	0.200	40.02	176.61	6.02	44.53
33 23 1824	**Well Abandonment, 6" Well**	**LF**				**14.86**	**65.57**	**1.73**	**82.16**
02086 6141	Monitor well, portland cement,94# bag, portland cement grout	CF	2	ULADB	0.330	45.02	198.69	5.25	82.16
33 23 1825	**Well Abandonment, 8" Well**	**LF**				**23.16**	**102.18**	**2.36**	**127.70**
02086 6141	Monitor well, portland cement,94# bag, portland cement grout	CF	1.75	ULADB	0.450	51.46	227.07	5.25	127.70

33 23 21 Bentonite Seals		Unit	Hourly Output	Crew	Factor	Labor	Equip.	Matl.	Total

33 23 Wells

33 23 21 Bentonite Seals		Unit	Hourly Output	Crew	Factor	Labor	Equip.	Matl.	Total
33 23 2101	**2" Well, Bentonite Seal**	**EA**				**8.10**	**35.76**	**9.61**	**53.47**
02557 6123	1/4" Bentonite Tablets, 10.80 Lbs/Gallon	CF	2.5	ULADB	0.225	36.02	158.95	42.69	53.47
33 23 2102	**4" Well, Bentonite Seal**	**EA**				**20.26**	**89.43**	**24.02**	**133.71**
02557 6123	1/4" Bentonite Tablets, 10.80 Lbs/Gallon	CF	2.5	ULADB	0.563	36.02	158.95	42.69	133.71
33 23 2103	**6" Well, Bentonite Seal**	**EA**				**32.42**	**143.06**	**38.42**	**213.89**
02557 6123	1/4" Bentonite Tablets, 10.80 Lbs/Gallon	CF	2.5	ULADB	0.900	36.02	158.95	42.69	213.89
33 23 2104	**3/8", 1' Thick, Sodium Bentonite Pellets**	**CF**				**36.02**	**158.95**	**36.58**	**231.55**
02557 6122	3/8" Bentonite Tablets, 10.62 Lbs/Gallon	CF	2.5	ULADB	1.000	36.02	158.95	36.58	231.55
33 23 2105	**8" Well, Bentonite Seal**	**EA**				**44.59**	**196.75**	**52.84**	**294.17**
02557 6123	1/4" Bentonite Tablets, 10.80 Lbs/Gallon	CF	2.5	ULADB	1.238	36.02	158.95	42.69	294.17
33 23 2106	**10" Well, Bentonite Seal**	**EA**				**56.75**	**250.41**	**67.25**	**374.41**
02557 6123	1/4" Bentonite Tablets, 10.80 Lbs/Gallon	CF	2.5	ULADB	1.575	36.02	158.95	42.69	374.41
33 23 2107	**12" Well, Bentonite Seal**	**EA**				**68.91**	**304.07**	**81.66**	**454.64**
02557 6123	1/4" Bentonite Tablets, 10.80 Lbs/Gallon	CF	2.5	ULADB	1.913	36.02	158.95	42.69	454.64
33 23 2108	**2" Well, Bentonite Seal, Slant Well**	**EA**				**10.54**	**46.50**	**15.29**	**72.32**
02557 7501	2" Well, Bentonite Seal, Slant Well	EA	8.546	ULADB	1.000	10.54	46.50	15.29	72.32
33 23 2109	**4" Well, Bentonite Seal, Slant Well**	**EA**				**22.91**	**101.08**	**50.22**	**174.20**
02557 7502	4" Well, Bentonite Seal, Slant Well	EA	3.931	ULADB	1.000	22.91	101.08	50.22	174.20

33 23 22 Protective Enclosures		Unit	Hourly Output	Crew	Factor	Labor	Equip.	Matl.	Total
33 23 2201	**Protective Enclosure with Cover**	**EA**				**90.05**	**397.38**	**169.17**	**656.60**
02557 5731	Warning Labels for Caps	EA	0	N/A	1.000	0.00	0.00	0.33	0.33
02557 5741	8" x 7.5" Manhole Cover - Nonlocking	EA	2	ULADB	1.000	45.02	198.69	44.41	288.12
02557 5759	6" x 3' Protective Enclosure Schedule 40, Hinged Lid, Lockable	EA	2	ULADB	1.000	45.02	198.69	124.44	368.15
33 23 2202	**4" x 4" x 5' Stand-Up Cover**	**EA**				**72.04**	**317.90**	**65.61**	**455.55**
02557 5732	4" x 4" x 5' Standup Cover	EA	1.25	ULADB	1.000	72.04	317.90	65.61	455.55
33 23 2203	**8" x 7.5" Manhole Cover**	**EA**				**45.02**	**198.69**	**44.41**	**288.12**
02557 5741	8" x 7.5" Manhole Cover - Nonlocking	EA	2	ULADB	1.000	45.02	198.69	44.41	288.12

33 23 22 Protective Enclosures		Unit	Hourly Output	Crew	Factor	Labor	Equip.	Matl.	Total
33 23 2204	**Hazardous Area, Pedestrian Load, Well Protection**	**EA**				**857.57**	**3,213**	**992.31**	**5,063**
02227 4000	Trim & Shape for Slab on Grade by Hand Finish Grade	SF	75.38	ULABA	80.000	0.45	0.00	0.00	35.95
02232 0100	Excavate & Load, 1 CY Hydraulic Excavator, Medium Material, 40 CY/Hour (31 m3)	CY	40	CODEO	15.000	1.56	2.25	0.00	57.24
02557 5771	48" x 48" Steel Trough Frame Pedestrian-load Double Door	EA	0.125	ULADB	1.000	720.40	3,179	593.65	4,493
03372 4600	Pour Slab on Grade, >= 6", Direct Chute >= (15 cm) Place 3,000 PSI Concrete	CY	21.88	ALABE	2.000	7.70	0.29	84.22	184.41
16232 2180	Simplex Receptacle, Explosion-proof, 120V, 20A, Exp Specialty Devices	EA	1.65	EELEB	1.000	62.35	0.00	230.23	292.57
33 23 2205	**Hazardous Area, Traffic Load, Well Protection**	**EA**				**857.57**	**3,213**	**998.51**	**5,069**
02227 4000	Trim & Shape for Slab on Grade by Hand Finish Grade	SF	75.38	ULABA	80.000	0.45	0.00	0.00	35.95
02232 0100	Excavate & Load, 1 CY Hydraulic Excavator, Medium Material, 40 CY/Hour (31 m3)	CY	40	CODEO	15.000	1.56	2.25	0.00	57.24
02557 5772	48" x 48" Steel Angle Frame Heavy Duty Double Door Hatch	EA	0.125	ULADB	1.000	720.40	3,179	599.85	4,499
03372 4600	Pour Slab on Grade, >= 6", Direct Chute >= (15 cm) Place 3,000 PSI Concrete	CY	21.88	ALABE	2.000	7.70	0.29	84.22	184.41
16232 2180	Simplex Receptacle, Explosion-Proof, 120V, 20A, Exp Specialty Devices	EA	1.65	EELEB	1.000	62.35	0.00	230.23	292.57
33 23 2206	**Restricted Area, Well Protection (with 4 Posts & Explosionproof Receptacle)**	**EA**				**274.06**	**1.23**	**549.08**	**824.38**
03372 2400	Pour Spread Footing,< 5 CY, Direct Chute < (5 m3) Place 3,000 PSI Concrete Foundations	CY	6.875	ALABE	1.333	24.49	0.93	84.22	146.17
09910 2002	Paint Exterior Steel Pipes, 6"-8" Diameter, One Coat Primer, One Coat Paint	LF	40	APTRA	12.000	0.97	0.00	0.33	15.59
15172 0670	6" (15 cm) A-53 Pipe, Schedule 40, Not Including Hangers or Fittings	LF	11.63	MPLUJ	16.000	10.46	0.00	12.66	370.04
16232 2180	Simplex Receptacle, Explosion-Proof, 120V, 20A, Exp Specialty Devices	EA	1.65	EELEB	1.000	62.35	0.00	230.23	292.57
33 23 2207	**48" x 48" Steel Frame, Pedestrian Load, 2-door Hatch**	**EA**				**720.40**	**3,179**	**593.65**	**4,493**
02557 5771	48" x 48" Steel Trough Frame Pedestrian-load Double Door	EA	0.125	ULADB	1.000	720.40	3,179	593.65	4,493
33 23 2208	**48" x 48" Steel Frame, Heavy Duty, 2-door Hatch**	**EA**				**720.40**	**3,179**	**599.85**	**4,499**
02557 5772	48" x 48" Steel Angle Frame Heavy Duty Double Door Hatch	EA	0.125	ULADB	1.000	720.40	3,179	599.85	4,499
33 23 2209	**10" x 7.5" Manhole Cover**	**EA**				**51.46**	**227.07**	**52.37**	**330.89**
02557 5742	10" x 7.5" Manhole Cover - Nonlocking	EA	1.75	ULADB	1.000	51.46	227.07	52.37	330.89
33 23 2210	**12" x 7.5" Manhole Cover**	**EA**				**60.03**	**264.92**	**84.60**	**409.55**
02557 5743	12" x 7.5" Manhole Cover - Nonlocking	EA	1.5	ULADB	1.000	60.03	264.92	84.60	409.55

33 23 22	Protective Enclosures	Unit	Hourly Output	Crew	Factor	Labor	Equip.	Matl.	Total
33 23 2211	**2" Well Finish, Flush with Grade, 8" x 7.5" Manhole, Locking Cap**	**EA**				**67.54**	**298.03**	**61.20**	**426.77**
02557 5711	2" PVC Locking Cap - Waterproof with Grout Depth Control Ring	EA	4	ULADB	1.000	22.51	99.34	16.80	138.65
02557 5741	8" x 7.5" Manhole Cover - Nonlocking	EA	2	ULADB	1.000	45.02	198.69	44.41	288.12
33 23 2212	**4" Well Finish, Flush with Grade, 8" x 7.5" Manhole, Locking Cap**	**EA**				**70.75**	**312.22**	**66.97**	**449.95**
02557 5712	4" PVC Locking Cap - Waterproof with Grout Depth Control Ring	EA	3.5	ULADB	1.000	25.73	113.54	22.56	161.83
02557 5741	8" x 7.5" Manhole Cover - Nonlocking	EA	2	ULADB	1.000	45.02	198.69	44.41	288.12
33 23 2213	**8" x 7.5" Locking Manhole Cover, Watertight**	**EA**				**45.02**	**198.69**	**47.47**	**291.19**
02557 5744	8" x 7.5" Manhole Cover - Locking and Watertight	EA	2	ULADB	1.000	45.02	198.69	47.47	291.19
33 23 2214	**12" x 7.5" Locking Manhole Cover, Watertight**	**EA**				**60.03**	**264.92**	**61.82**	**386.77**
02557 5745	12" x 7.5" Manhole Cover - Locking and Watertight	EA	1.5	ULADB	1.000	60.03	264.92	61.82	386.77
33 23 2215	**4" x 4" x 5' Steel Protective Cover, Lockable**	**EA**				**72.04**	**317.90**	**59.05**	**448.99**
02557 5733	4" x 4" x 5' Steel Protective Cover, Lockable	EA	1.25	ULADB	1.000	72.04	317.90	59.05	448.99
33 23 2216	**6" x 6" x 5' Steel Protective Cover, Lockable**	**EA**				**72.04**	**317.90**	**86.57**	**476.51**
02557 5734	6" x 6" x 5' Steel Protective Cover, Lockable	EA	1.25	ULADB	1.000	72.04	317.90	86.57	476.51
33 23 2217	**8" x 8" x 5' Steel Protective Cover, Lockable**	**EA**				**72.04**	**317.90**	**164.54**	**554.48**
02557 5735	8" x 8" x 5' Steel Protective Cover, Lockable	EA	1.25	ULADB	1.000	72.04	317.90	164.54	554.48
33 23 2220	**6" Well Finish, Cover, Flush w/Grade, Manhole, Lock Cap**	**EA**				**0.00**	**0.00**	**58.91**	**58.91**
01001 0448	6" Well Finish, Cover, Flush w/Grade, Manhole, Lock Cap	EA	0	N/A	1.000	0.00	0.00	58.91	58.91
33 23 2221	**8" Well Finish, Cover, Flush w/Grade, Manhole, Lock Cap**	**EA**				**0.00**	**0.00**	**60.06**	**60.06**
01001 0449	8" Well Finish, Cover, Flush w/Grade, Manhole, Lock Cap	EA	0	N/A	1.000	0.00	0.00	60.06	60.06
33 23 2222	**10" Well Finish, Cover, Flush w/Grade, Manhole, Lock Cap**	**EA**				**0.00**	**0.00**	**83.33**	**83.33**
01001 0450	10" Well Finish, Cover, Flush w/Grade, Manhole, Lock Cap	EA	0	N/A	1.000	0.00	0.00	83.33	83.33
33 23 2223	**12" Well Finish, Cover, Flush w/Grade, Manhole, Lock Cap**	**EA**				**0.00**	**0.00**	**106.59**	**106.59**
01001 0451	12" Well Finish, Cover, Flush w/Grade, Manhole, Lock Cap	EA	0	N/A	1.000	0.00	0.00	106.59	106.59
33 23 2224	**24" Well Finish, Cover, Flush w/Grade, Manhole, Lock Cap**	**EA**				**0.00**	**0.00**	**246.19**	**246.19**
01001 0452	24" Well Finish, Cover, Flush w/Grade, Manhole, Lock Cap	EA	0	N/A	1.000	0.00	0.00	246.19	246.19

33 23 Wells

33 23 22 Protective Enclosures		Unit	Hourly Output	Crew	Factor	Labor	Equip.	Matl.	Total
33 23 2225	**24" X 7.5" Manhole Cover**	**EA**				**0.00**	**0.00**	**155.21**	**155.21**
01001 0453	24" X 7.5" Manhole Cover	EA	0	N/A	1.000	0.00	0.00	155.21	155.21
33 23 2226	**30" X 7.5" Manhole Cover**	**EA**				**0.00**	**0.00**	**196.62**	**196.62**
01001 0454	30" X 7.5" Manhole Cover	EA	0	N/A	1.000	0.00	0.00	196.62	196.62
33 23 2227	**36" X 7.5" Manhole Cover**	**EA**				**0.00**	**0.00**	**238.02**	**238.02**
01001 0455	36" X 7.5" Manhole Cover	EA	0	N/A	1.000	0.00	0.00	238.02	238.02
33 23 2250	**4" x 3' Protective Enclosure, Schedule 40, Hinged Lid, Lockable**	**EA**				**45.02**	**198.69**	**79.85**	**323.56**
02557 5751	4" x 3' Protective Enclosure Schedule 40, Hinged Lid, Lockable	EA	2	ULADB	1.000	45.02	198.69	79.85	323.56
33 23 2251	**4" x 4' Protective Enclosure, Schedule 40, Hinged Lid, Lockable**	**EA**				**60.03**	**264.92**	**78.00**	**402.95**
02557 5752	4" x 4' Protective Enclosure Schedule 40, Hinged Lid, Lockable	EA	1.5	ULADB	1.000	60.03	264.92	78.00	402.95
33 23 2252	**6" x 3' Protective Enclosure, Schedule 40, Hinged Lid, Lockable**	**EA**				**45.02**	**198.69**	**124.44**	**368.15**
02557 5759	6" x 3' Protective Enclosure Schedule 40, Hinged Lid, Lockable	EA	2	ULADB	1.000	45.02	198.69	124.44	368.15

33 23 23 Guard Posts		Unit	Hourly Output	Crew	Factor	Labor	Equip.	Matl.	Total
33 23 2301	**5' Guard Posts, Cast Iron, Concrete Fill**	**EA**				**32.41**	**0.03**	**24.92**	**57.36**
03372 2400	Pour Spread Footing,< 5 CY, Direct Chute < (5 m3) Place 3,000 PSI Concrete Foundations	CY	6.875	ALABE	0.032	24.49	0.93	84.22	3.54
09910 2001	Paint Exterior Steel Pipes, 4" Diameter, One Coat Primer, One Coat Paint	LF	55.63	APTRA	2.814	0.70	0.00	0.28	2.75
15172 0630	3" (80 mm) A-53 Pipe, Schedule 40, Not Including Hangers or Fittings	LF	15	MPLUE	5.000	5.93	0.00	4.28	51.07

33 23 24 Bailers		Unit	Hourly Output	Crew	Factor	Labor	Equip.	Matl.	Total
33 23 2401	**Teflon Bailer, 3/4" Outside Diameter x 1', 60 cc**	**EA**				**0.00**	**0.00**	**213.00**	**213.00**
01954 3214	3/4" Outside Diameter x 1', 60 cc	EA	0	N/A	1.000	0.00	0.00	213.00	213.00
33 23 2402	**Teflon Bailer, 3/4" Outside Diameter x 3', 180 cc**	**EA**				**0.00**	**0.00**	**294.00**	**294.00**
01954 3215	3/4" Outside Diameter x 3', 180 cc	EA	0	N/A	1.000	0.00	0.00	294.00	294.00
33 23 2403	**Teflon Bailer, 1" Outside Diameter x 1', 80 cc**	**EA**				**0.00**	**0.00**	**141.00**	**141.00**
01954 3216	1" Outside Diameter x 1', 80 cc	EA	0	N/A	1.000	0.00	0.00	141.00	141.00
33 23 2404	**Teflon Bailer, 1 7/8" Outside Diameter x 1', 350 cc**	**EA**				**0.00**	**0.00**	**219.00**	**219.00**
01954 3211	1-7/8" Outside Diameter x 1', 350 cc	EA	0	N/A	1.000	0.00	0.00	219.00	219.00

33 23 Wells

33 23 24 Bailers		Unit	Hourly Output	Crew	Factor	Labor	Equip.	Matl.	Total
33 23 2405	**Teflon Bailer, 1 7/8" Outside Diameter x 2', 700 cc**	**EA**				**0.00**	**0.00**	**216.00**	**216.00**
01954 3212	1-7/8" Outside Diameter x 2', 700 cc	EA	0	N/A	1.000	0.00	0.00	216.00	216.00
33 23 2406	**Teflon Bailer, 1 7/8" Outside Diameter x 3', 1,050 cc**	**EA**				**0.00**	**0.00**	**312.18**	**312.18**
01954 3213	1-7/8" Outside Diameter x 3', 1,050 cc	EA	0	N/A	1.000	0.00	0.00	312.18	312.18
33 23 2407	**Disposable Bailer, Polyethylene, 1.5" Outside Diameter x 36"**	**EA**				**0.00**	**0.00**	**6.05**	**6.05**
01954 3232	1.5" Outside Diameter x 36", Disposable Polyethylene Bailer	EA	0	N/A	1.000	0.00	0.00	6.05	6.05
33 23 2408	**Teflon Bailer, 1-2/3" OD X 34"**	**EA**				**0.00**	**0.00**	**35.04**	**35.04**
01001 0456	Teflon Bailer, 1-2/3" OD X 34"	EA	0	N/A	1.000	0.00	0.00	35.04	35.04
33 23 2411	**Stainless Steel Bailer, 1 3/4" Outside Diameter x 2'**	**EA**				**0.00**	**0.00**	**117.00**	**117.00**
01954 3221	1.75" Outside Diameter x 3', 1,000 ml	EA	0	N/A	1.000	0.00	0.00	117.00	117.00
33 23 2412	**Stainless Steel Bailer, 1 3/4" Outside Diameter x 3'**	**EA**				**0.00**	**0.00**	**30.00**	**30.00**
01954 3231	1.66" Outside Diameter x 34"	EA	0	N/A	1.000	0.00	0.00	30.00	30.00
33 23 2421	**Emptying Stand (1-7/8" Outside Diameter)**	**EA**				**0.00**	**0.00**	**360.00**	**360.00**
01954 3241	Emptying Stand (1-7/8" Outside Diameter)	EA	0	N/A	1.000	0.00	0.00	360.00	360.00
33 23 2422	**Suspension Cable, Teflon Coated**	**FT**				**0.00**	**0.00**	**1.49**	**1.49**
01954 3242	Suspension Cable, Teflon Coated 304 Stainless Steel	FT	0	N/A	1.000	0.00	0.00	1.49	1.49
33 23 2423	**Hand Reel**	**EA**				**0.00**	**0.00**	**10.20**	**10.20**
01954 3243	Hand Reel (Holds 300-500')	EA	0	N/A	1.000	0.00	0.00	10.20	10.20
33 23 2430	**Hand Reel/Winch, Monthly Rental**	**MO**				**0.00**	**0.00**	**336.61**	**336.61**
01001 0457	Bailer w/ winch, Free Product Removal, Monthly Rental	MO	0	N/A	1.000	0.00	0.00	336.61	336.61

33 23 25 Well Installation		Unit	Hourly Output	Crew	Factor	Labor	Equip.	Matl.	Total
33 23 2501	**Drill and Install 2" PVC Monitoring Well, 0-10 Feet Deep**	**LF**				**0.00**	**0.00**	**42.97**	**42.97**
01001 0459	Drill and Install 2" PVC Monitoring Well, 0-10 Feet Deep	LF	0	N/A	1.000	0.00	0.00	42.97	42.97
33 23 2502	**Drill and Install 2" PVC Monitoring Well, 11-20 Feet Deep**	**LF**				**0.00**	**0.00**	**29.00**	**29.00**
01001 0460	Drill and Install 2" PVC Monitoring Well,11-20 Feet Deep	LF	0	N/A	1.000	0.00	0.00	29.00	29.00

33 23 25	Well Installation	Unit	Hourly Output	Crew	Factor	Labor	Equip.	Matl.	Total
33 23 2503	**Drill and Install 2" PVC Monitoring Well, 21-30 Feet Deep**	**LF**				**0.00**	**0.00**	**29.11**	**29.11**
01001 0461	Drill and Install 2" PVC Monitoring Well, 21-30 Feet Deep	LF	0	N/A	1.000	0.00	0.00	29.11	29.11
33 23 2504	**Drill and Install 2" PVC Monitoring Well, 31-40 Feet Deep**	**LF**				**0.00**	**0.00**	**28.33**	**28.33**
01001 0462	Drill and Install 2" PVC Monitoring Well,31-40 Feet Deep	LF	0	N/A	1.000	0.00	0.00	28.33	28.33
33 23 2505	**Drill and Install 2" PVC Monitoring Well, >40 Feet Deep**	**LF**				**0.00**	**0.00**	**27.83**	**27.83**
01001 0463	Drill and Install 2" PVC Monitoring Well, >40 Feet Deep	LF	0	N/A	1.000	0.00	0.00	27.83	27.83

33 23 26	Free Product Removal (Eductors & Filters)	Unit	Hourly Output	Crew	Factor	Labor	Equip.	Matl.	Total
33 23 2601	**Free Product Removal, Eductor**	**EA**				**0.00**	**0.00**	**336.61**	**336.61**
01001 0465	Free Product Removal, Eductor	EA	0	N/A	1.000	0.00	0.00	336.61	336.61
33 23 2602	**Free Product Removal, 5 HP Eductor Drive Pump**	**EA**				**0.00**	**0.00**	**2,805**	**2,805**
01001 0466	Free Product Removal, 5 HP Eductor Drive Pump	EA	0	N/A	1.000	0.00	0.00	2,805	2,805
33 23 2603	**Portable Eductor Driven Free Product Removal System, Weekly Rental**	**WK**				**0.00**	**0.00**	**561.02**	**561.02**
01001 0467	Portable Eductor Driven Free Product Removal System, Weekly Rental	WK	0	N/A	1.000	0.00	0.00	561.02	561.02
33 23 2610	**Product Recovery Filter, 1.5" OD, 1" ID, 14" length**	**EA**				**0.00**	**0.00**	**302.95**	**302.95**
01001 0468	Product Recovery Filter, 1.5" OD, 1" ID, 14" length	EA	0	N/A	1.000	0.00	0.00	302.95	302.95
33 23 2611	**Product Recovery Filter, 3.0" OD, 2" ID, 15" length**	**EA**				**0.00**	**0.00**	**359.05**	**359.05**
01001 0469	Product Recovery Filter, 3.0" OD, 2" ID, 15" length	EA	0	N/A	1.000	0.00	0.00	359.05	359.05

33 26 01	Carbon Steel Pipe	Unit	Hourly Output	Crew	Factor	Labor	Equip.	Matl.	Total
33 26 0101	**1" Carbon Steel Piping**	**LF**				**3.11**	**0.23**	**0.94**	**4.29**
15172 0581	1" (2.5 cm) A-53 Pipe, Schedule 40, Not Including Hangers or Fittings	LF	35.75	MSPFH	1.000	3.11	0.23	0.94	4.29
33 26 0102	**2" Carbon Steel Piping**	**LF**				**4.68**	**0.35**	**2.07**	**7.10**
15172 0611	2" (5 cm) A-53 Pipe, Schedule 40, Not Including Hangers or Fittings	LF	23.75	MSPFH	1.000	4.68	0.35	2.07	7.10
33 26 0103	**3" Carbon Steel Piping**	**LF**				**6.35**	**0.47**	**4.28**	**11.11**
15172 0632	3" (8 cm) A-53 Pipe, Schedule 40, Not Including Hangers or Fittings	LF	17.5	MSPFH	1.000	6.35	0.47	4.28	11.11

33 26 01	Carbon Steel Pipe	Unit	Hourly Output	Crew	Factor	Labor	Equip.	Matl.	Total
33 26 0104	**4" Carbon Steel Piping**	**LF**				**7.41**	**0.55**	**6.12**	**14.09**
15172 0651	4" (10 cm) A-53 Pipe, Schedule 40, Not Including Hangers or Fittings	LF	15	MSPFH	1.000	7.41	0.55	6.12	14.09
33 26 0105	**6" Carbon Steel Piping**	**LF**				**10.84**	**0.81**	**12.86**	**24.51**
15172 2130	6" (15 cm) A-53 Pipe, Schedule 40, Not Including Hangers or Fittings	LF	10.25	MSPFH	1.000	10.84	0.81	12.86	24.51
33 26 0106	**8" Carbon Steel Piping**	**LF**				**13.68**	**1.02**	**19.09**	**33.79**
15172 2140	8" (20 cm), A-53 Pipe, Shedule 40, Not Including Hangers or Fittings	LF	8.125	MSPFH	1.000	13.68	1.02	19.09	33.79
33 26 0107	**10" Carbon Steel Piping**	**LF**				**17.10**	**1.28**	**27.10**	**45.48**
15172 2150	10" (25 cm), A-53 Pipe, Shedule 40, Not Including Hangers or Fittings	LF	6.5	MSPFH	1.000	17.10	1.28	27.10	45.48
33 26 0108	**12" Carbon Steel Piping**	**LF**				**20.68**	**1.55**	**38.16**	**60.39**
15172 2160	12" (30.5 cm), A-53 Pipe, Shedule 40, Not Including Hangers or Fittings	LF	5.375	MSPFH	1.000	20.68	1.55	38.16	60.39
33 26 0114	**4" Carbon Steel Schedule 80 Piping**	**LF**				**8.24**	**0.62**	**8.92**	**17.78**
15172 4900	4" Carbon Steel Schedule 80 Pipe, A53	LF	13.49	MSPFH	1.000	8.24	0.62	8.92	17.78
33 26 0115	**6" Carbon Steel Schedule 80 Piping**	**LF**				**11.74**	**0.88**	**18.31**	**30.94**
15172 4920	6" Carbon Steel Schedule 80 Pipe, A53	LF	9.465	MSPFH	1.000	11.74	0.88	18.31	30.94
33 26 0131	**1" Carbon Steel Piping Including Fittings & Hangers**	**LF**				**6.41**	**0.00**	**1.59**	**8.00**
15061 2103	1" (25 mm) A53 Pipe, T&C Schedule 40 Including Hanger & 125 Lb Cast Iron Fitting	LF	13.88	MPLUE	1.000	6.41	0.00	1.59	8.00
33 26 0132	**2" Carbon Steel Piping Including Fittings & Hangers**	**LF**				**11.12**	**0.00**	**3.33**	**14.45**
15061 2106	2" (50 mm) A53 Pipe, T&C Schedule 40 Including Hanger & 125 Lb Cast Iron Fitting	LF	8	MPLUE	1.000	11.12	0.00	3.33	14.45
33 26 0133	**3" Carbon Steel Piping Including Fittings & Hangers**	**LF**				**16.55**	**0.00**	**4.39**	**20.94**
15172 0631	3" (80 mm) A53 Pipe, T&C Schedule 40 Including Hanger & 125 Lb Cast Iron Fitting	LF	5.375	MPLUE	1.000	16.55	0.00	4.39	20.94
33 26 0134	**4" Carbon Steel Piping Including Fittings & Hangers**	**LF**				**18.36**	**0.00**	**10.54**	**28.90**
15061 2109	4" (10 cm) A53 Pipe, T&C Schedule 40 Including Hanger & 125 Lb Cast Iron Fitting	LF	6.625	MPLUJ	1.000	18.36	0.00	10.54	28.90

33 26 02	Stainless Steel Pipe	Unit	Hourly Output	Crew	Factor	Labor	Equip.	Matl.	Total
33 26 0201	**1" Stainless Steel Piping, Schedule 40, Threaded**	**LF**				**5.34**	**0.00**	**8.67**	**14.00**
15162 3014	1" Stainless Steel Pipe, Schedule 40 Type 304 with 150 Lb Threaded Fitting	LF	19.38	MSPFC	1.000	5.34	0.00	8.67	14.00

33 26 02	Stainless Steel Pipe	Unit	Hourly Output	Crew	Factor	Labor	Equip.	Matl.	Total
33 26 0202	**2" Stainless Steel Piping, Schedule 40, Threaded**	**LF**				**8.89**	**0.00**	**17.15**	**26.04**
15162 3018	2" Stainless Steel Pipe, Schedule 40 Type 304 with 150 Lb Threaded Fitting	LF	11.63	MSPFC	1.000	8.89	0.00	17.15	26.04
33 26 0203	**3" Stainless Steel Piping, Schedule 40, Threaded**	**LF**				**13.34**	**0.00**	**35.59**	**48.92**
15066 2118	3" Stainless Steel Pipe, Schedule 40 Type 304 with 150 Lb Threaded Fitting	LF	7.75	MSPFC	1.000	13.34	0.00	35.59	48.92
33 26 0204	**4" Stainless Steel Piping, Schedule 40, Threaded**	**LF**				**17.60**	**0.00**	**49.96**	**67.55**
15162 3020	4" Stainless Steel Pipe, Schedule 40 Type 304 with 150 Lb Threaded Fitting	LF	5.875	MSPFC	1.000	17.60	0.00	49.96	67.55
33 26 0205	**6" Stainless Steel Piping, Schedule 10, Type 316**	**LF**				**17.15**	**1.82**	**46.74**	**65.70**
15162 2910	6" Stainless Steel Pipe, Schedule 10 Type 316	LF	5.55	MSPFE	1.000	17.15	1.82	46.74	65.70
33 26 0206	**8" Stainless Steel Piping, Schedule 40, Welded**	**LF**				**27.33**	**2.61**	**155.61**	**185.55**
15162 5370	8" Stainless Steel Pipe, Schedule 40, Type 316L with Welded Butt Fitting	LF	5.5	MSPFN	1.000	27.33	2.61	155.61	185.55
33 26 0207	**8" Stainless Steel, Schedule 10, Type 316, Manifold Piping**	**LF**				**27.33**	**2.61**	**74.52**	**104.46**
15162 5320	8" Stainless Steel Pipe, Schedule 10 Type 316L with Welded Butt Fitting	LF	5.5	MSPFN	1.000	27.33	2.61	74.52	104.46
33 26 0208	**10" Stainless Steel Piping, Schedule 40, Welded**	**LF**				**30.06**	**2.88**	**244.64**	**277.57**
15162 5372	10" Stainless Steel Pipe, Schedule 40, Type 316L with Welded Butt Fitting	LF	5	MSPFN	1.000	30.06	2.88	244.64	277.57
33 26 0209	**12" Stainless Steel Piping, Schedule 40, Welded**	**LF**				**33.40**	**3.19**	**272.36**	**308.95**
15162 5374	12" Stainless Steel Pipe, Schedule 40, Type 316L with Welded Butt Fitting	LF	4.5	MSPFN	1.000	33.40	3.19	272.36	308.95
33 26 0210	**2" Stainless Steel, Schedule 40, Type 304, Connection Piping**	**LF**				**8.89**	**0.00**	**17.15**	**26.04**
15162 3018	2" Stainless Steel Pipe, Schedule 40 Type 304 with 150 Lb Threaded Fitting	LF	11.63	MSPFC	1.000	8.89	0.00	17.15	26.04
33 26 0211	**4" Stainless Steel, Schedule 40, Type 304, Connection Piping**	**LF**				**17.60**	**0.00**	**49.96**	**67.55**
15162 3020	4" Stainless Steel Pipe, Schedule 40 Type 304 with 150 Lb Threaded Fitting	LF	5.875	MSPFC	1.000	17.60	0.00	49.96	67.55
33 26 0212	**4" Stainless Steel, Schedule 40, Type 304, Manifold Piping**	**LF**				**17.60**	**0.00**	**49.96**	**67.55**
15162 3020	4" Stainless Steel Pipe, Schedule 40 Type 304 with 150 Lb Threaded Fitting	LF	5.875	MSPFC	1.000	17.60	0.00	49.96	67.55

33 26 Hazardous Waste Piping/Hoses

33 26 02 Stainless Steel Pipe		Unit	Hourly Output	Crew	Factor	Labor	Equip.	Matl.	Total
33 26 0213	**2" Stainless Steel, Schedule 40, Type 304, Manifold Piping**	**LF**				**8.89**	**0.00**	**17.15**	**26.04**
15162 3018	2" Stainless Steel Pipe, Schedule 40 Type 304 with 150 Lb Threaded Fitting	LF	11.63	MSPFC	1.000	8.89	0.00	17.15	26.04
33 26 0231	**1" Stainless Steel Piping Including Fittings & Hangers**	**LF**				**8.78**	**0.00**	**9.20**	**17.97**
15066 1003	1" (25 mm) Stainless Steel Pipe, Schedule 40 Type 304 with 150 Lb Screwed Fittings	LF	5.875	MPLUA	1.000	8.78	0.00	9.20	17.97
33 26 0232	**2" Stainless Steel Piping Including Fittings & Hangers**	**LF**				**13.43**	**0.00**	**18.20**	**31.63**
15066 1006	2" (50 mm) Stainless Steel Pipe, Schedule 40 Type 304 with 150 Lb Screwed Footings	LF	6.625	MPLUE	1.000	13.43	0.00	18.20	31.63
33 26 0233	**3" Stainless Steel Piping Including Fittings & Hangers**	**LF**				**19.24**	**0.00**	**27.90**	**47.14**
15066 1008	3" (80 mm) Stainless Steel Pipe, Schedule 40 Type 304 with 150 Lb Screwed Footings	LF	4.625	MPLUE	1.000	19.24	0.00	27.90	47.14
33 26 0234	**4" Stainless Steel Piping Including Fittings & Hangers**	**LF**				**34.22**	**0.46**	**64.07**	**98.75**
15094 3015	4" Hanger	EA	2.175	MPLUK	0.100	55.93	4.64	12.22	7.28
15162 3020	4" Stainless Steel Pipe, Schedule 40 Type 304 with 150 Lb Threaded Fitting	LF	5.875	MSPFC	1.000	17.60	0.00	49.96	67.55
15164 6370	4" Stainless Steel Flanges, Schedule 40 Type 304	EA	0.938	MSPFC	0.100	110.27	0.00	128.93	23.92

33 26 03 Polypropylene Pipe		Unit	Hourly Output	Crew	Factor	Labor	Equip.	Matl.	Total
33 26 0301	**1 1/2" Polypropylene Pipe Including Fittings**	**LF**				**3.05**	**0.00**	**1.01**	**4.07**
15145 5000	1-1/2" Polypropylene Acid-Resistant Schedule 40 Drain Waste Vent Pipe	LF	29.13	MPLUE	1.000	3.05	0.00	1.01	4.07
33 26 0302	**2" Polypropylene Pipe Including Fittings**	**LF**				**3.44**	**0.00**	**1.36**	**4.80**
15145 5100	2" Polypropylene Acid-Resistant Schedule 40 Pipe Drain Waste Vent Pipe	LF	25.88	MPLUE	1.000	3.44	0.00	1.36	4.80
33 26 0303	**3" Polypropylene Pipe Including Fittings**	**LF**				**4.98**	**0.00**	**2.85**	**7.83**
15145 5120	3" Polypropylene Acid-Resistant Schedule 40 Pipe Drain Waste Vent Pipe	LF	17.88	MPLUE	1.000	4.98	0.00	2.85	7.83
33 26 0304	**4" Polypropylene Pipe Including Fittings**	**LF**				**6.47**	**0.00**	**4.06**	**10.53**
15145 5140	4" Polypropylene Acid-Resistant Schedule 40 Pipe Drain Waste Vent Pipe	LF	13.75	MPLUE	1.000	6.47	0.00	4.06	10.53

33 26 04 PVC Pipe		Unit	Hourly Output	Crew	Factor	Labor	Equip.	Matl.	Total
33 26 0401	**1" PVC Piping Including Fittings & Hangers**	**LF**				**9.57**	**0.79**	**0.61**	**10.98**
01001 0900	1" Hanger	EA	1.269	MPLUH	0.100	74.83	7.94	2.87	8.56
01001 0901	1" Schedule 40 PVC Pressure Pipe Socket Weld	LF	42.55	MPLUE	1.000	2.09	0.00	0.33	2.42

33 26 Hazardous Waste Piping/Hoses

33 26 04	PVC Pipe	Unit	Hourly Output	Crew	Factor	Labor	Equip.	Matl.	Total
33 26 0402	**1-1/4" PVC Piping Including Fittings & Hangers**	**LF**				**8.32**	**0.65**	**0.86**	**9.83**
01001 0902	1-1/4" Hanger	EA	1.55	MPLUH	0.100	61.26	6.50	4.22	7.20
01001 0903	1-1/4" Schedule 40 PVC Pressure Pipe Socket Weld	LF	40.63	MPLUE	1.000	2.19	0.00	0.44	2.63
33 26 0403	**1-1/2" PVC Piping Including Fittings & Hangers**	**LF**				**7.26**	**0.54**	**0.99**	**8.79**
01001 0904	1-1/2" Hanger	EA	1.875	MPLUH	0.100	50.64	5.38	4.74	6.08
01001 0905	1-1/2" Schedule 40 PVC Pressure Pipe Socket Weld	LF	40.51	MPLUE	1.000	2.20	0.00	0.52	2.71
33 26 0404	**2" PVC Piping Including Fittings & Hangers**	**LF**				**7.73**	**0.54**	**1.25**	**9.52**
01001 0906	2" Hanger	EA	1.875	MPLUH	0.100	50.64	5.38	6.37	6.24
01001 0907	2" Schedule 40 PVC Pressure Pipe Socket Weld	LF	33.38	MPLUE	1.000	2.67	0.00	0.62	3.28
33 26 0405	**3" PVC Piping Including Fittings & Hangers**	**LF**				**7.28**	**0.37**	**2.07**	**9.72**
01001 0908	3" Hanger	EA	2.75	MPLUH	0.100	34.53	3.67	8.47	4.67
01001 0909	3" Schedule 40 PVC Pressure Pipe Socket Weld	LF	23.25	MPLUE	1.000	3.83	0.00	1.22	5.05
33 26 0406	**4" PVC Piping Including Fittings & Hangers**	**LF**				**10.11**	**0.46**	**2.98**	**13.55**
01001 0910	4" Hanger	EA	2.175	MPLUH	0.100	43.66	4.64	11.62	5.99
01001 0911	4" Schedule 40 PVC Pressure Pipe Socket Weld	LF	15.5	MPLUE	1.000	5.74	0.00	1.82	7.56
33 26 0410	**1" PVC, Schedule 40, Connection Piping**	**LF**				**1.93**	**0.00**	**0.25**	**2.17**
01001 0667	1" PVC, Sch 40, Connection Piping	LF	46.21	MPLUE	1.000	1.93	0.00	0.25	2.17
33 26 0411	**1-1/4" PVC, Schedule 40, Connection Piping**	**LF**				**2.20**	**0.00**	**0.30**	**2.50**
01001 0649	1-1/4" PVC, Sch 40, Connection Piping	LF	40.51	MPLUE	1.000	2.20	0.00	0.30	2.50
33 26 0412	**1-1/2" PVC, Schedule 40, Connection Piping**	**LF**				**2.41**	**0.00**	**0.36**	**2.77**
01001 0650	1-1/2" PVC, Sch 40, Connection Piping	LF	36.89	MPLUE	1.000	2.41	0.00	0.36	2.77
33 26 0413	**2" PVC, Schedule 40, Connection Piping**	**LF**				**2.67**	**0.00**	**0.62**	**3.28**
01001 0912	2" PVC, Sch 40, Connection Piping	LF	33.38	MPLUE	1.000	2.67	0.00	0.62	3.28
33 26 0414	**3" PVC, Schedule 40, Connection Piping**	**LF**				**3.83**	**0.00**	**1.02**	**4.85**
01001 0651	3" PVC, Sch 40, Connection Piping	LF	23.25	MPLUE	1.000	3.83	0.00	1.02	4.85
33 26 0415	**4" PVC, Schedule 40, Connection Piping**	**LF**				**5.74**	**0.00**	**1.82**	**7.56**
01001 0913	4" PVC, Sch 40, Connection Piping	LF	15.5	MPLUE	1.000	5.74	0.00	1.82	7.56
33 26 0416	**6" PVC, Schedule 40, Connection Piping**	**LF**				**7.65**	**0.00**	**3.19**	**10.84**
01001 0914	6" PVC, Sch 40, Connection Piping	LF	11.63	MPLUE	1.000	7.65	0.00	3.19	10.84
33 26 0417	**8" PVC, Schedule 40, Connection Piping**	**LF**				**9.13**	**0.00**	**4.82**	**13.95**
01001 0915	8" PVC, Sch 40, Connection Piping	LF	9.75	MPLUE	1.000	9.13	0.00	4.82	13.95

33 26 04	PVC Pipe	Unit	Hourly Output	Crew	Factor	Labor	Equip.	Matl.	Total
33 26 0418	**10" PVC, Schedule 40, Connection Piping**	**LF**				**15.66**	**0.00**	**7.44**	**23.10**
01001 0916	10" PVC, Sch 40, Connection Piping	LF	5.68	MPLUE	1.000	15.66	0.00	7.44	23.10
33 26 0419	**12" PVC, Schedule 40, Connection Piping**	**LF**				**11.14**	**0.00**	**7.85**	**18.99**
01001 0917	12" PVC, Sch 40, Connection Piping	LF	7.987	MPLUE	1.000	11.14	0.00	7.85	18.99
33 26 0425	**1" PVC, Schedule 80, Connection Piping**	**LF**				**1.93**	**0.00**	**0.36**	**2.28**
01001 0470	1" PVC, Sch 80, Connection Piping	LF	46.21	MPLUE	1.000	1.93	0.00	0.36	2.28
33 26 0426	**1-1/4" PVC, Schedule 80, Connection Piping**	**LF**				**2.20**	**0.00**	**0.49**	**2.68**
01001 0471	1-1/4" PVC, Sch 80, Connection Piping	LF	40.51	MPLUE	1.000	2.20	0.00	0.49	2.68
33 26 0427	**1-1/2" PVC, Schedule 80, Connection Piping**	**LF**				**2.41**	**0.00**	**0.59**	**3.01**
01001 0472	1-1/2" PVC, Sch 80, Connection Piping	LF	36.89	MPLUE	1.000	2.41	0.00	0.59	3.01
33 26 0428	**2" PVC, Schedule 80, Connection Piping**	**LF**				**2.67**	**0.00**	**0.82**	**3.48**
01001 0473	2" PVC, Sch 80, Connection Piping	LF	33.38	MPLUE	1.000	2.67	0.00	0.82	3.48
33 26 0429	**3" PVC, Schedule 80, Connection Piping**	**LF**				**3.83**	**0.00**	**1.69**	**5.52**
01001 0474	3" PVC, Sch 80, Connection Piping	LF	23.25	MPLUE	1.000	3.83	0.00	1.69	5.52
33 26 0430	**4" PVC, Schedule 80, Connection Piping**	**LF**				**5.74**	**0.00**	**2.44**	**8.18**
01001 0475	4" PVC, Sch 80, Connection Piping	LF	15.5	MPLUE	1.000	5.74	0.00	2.44	8.18
33 26 0431	**6" PVC, Schedule 80, Connection Piping**	**LF**				**7.65**	**0.00**	**4.60**	**12.25**
01001 0476	6" PVC, Sch 80, Connection Piping	LF	11.63	MPLUE	1.000	7.65	0.00	4.60	12.25
33 26 0432	**8" PVC, Schedule 80, Connection Piping**	**LF**				**9.13**	**0.00**	**6.98**	**16.10**
01001 0477	8" PVC, Sch 80, Connection Piping	LF	9.75	MPLUE	1.000	9.13	0.00	6.98	16.10
33 26 0433	**10" PVC, Schedule 80, Connection Piping**	**LF**				**12.34**	**0.00**	**10.38**	**22.72**
01001 0478	10" PVC, Sch 80, Connection Piping	LF	7.21	MPLUE	1.000	12.34	0.00	10.38	22.72
33 26 0434	**12" PVC, Schedule 80, Connection Piping**	**LF**				**15.66**	**0.00**	**14.30**	**29.96**
01001 0479	12" PVC, Sch 80, Connection Piping	LF	5.68	MPLUE	1.000	15.66	0.00	14.30	29.96
33 26 0440	**1" PVC, Schedule 40, Manifold Piping**	**LF**				**1.93**	**0.00**	**0.25**	**2.17**
01001 0799	1" PVC, Sch 40, Manifold Piping	LF	46.21	MPLUE	1.000	1.93	0.00	0.25	2.17
33 26 0441	**1-1/4" PVC,Schedule 40, Manifold Piping**	**LF**				**2.20**	**0.00**	**1.89**	**4.08**
01001 0480	1-1/4" PVC, Sch 40, Manifold Piping	LF	40.51	MPLUE	1.000	2.20	0.00	1.89	4.08
33 26 0442	**1-1/2" PVC, Schedule 40, Manifold Piping**	**LF**				**2.41**	**0.00**	**1.92**	**4.33**
01001 0481	1-1/2" PVC, Sch 40, Manifold Piping	LF	36.89	MPLUE	1.000	2.41	0.00	1.92	4.33
33 26 0443	**2" PVC, Schedule 40, Manifold Piping**	**LF**				**2.67**	**0.00**	**0.62**	**3.28**
01001 0800	2" PVC, Sch 40, Manifold Piping	LF	33.38	MPLUE	1.000	2.67	0.00	0.62	3.28

33 26 Hazardous Waste Piping/Hoses

33 26 04 PVC Pipe	Unit	Hourly Output	Crew	Factor	Labor	Equip.	Matl.	Total
33 26 0444 3" PVC, Schedule 40, Manifold Piping	**LF**				**3.83**	**0.00**	**3.30**	**7.13**
01001 0482 3" PVC, Sch 40, Manifold Piping	LF	23.25	MPLUE	1.000	3.83	0.00	3.30	7.13
33 26 0445 4" PVC, Schedule 40, Manifold Piping	**LF**				**5.74**	**0.00**	**1.82**	**7.56**
01001 0801 4" PVC, Sch 40, Manifold Piping	LF	15.5	MPLUE	1.000	5.74	0.00	1.82	7.56
33 26 0446 6" PVC, Schedule 40, Manifold Piping	**LF**				**7.65**	**0.00**	**3.19**	**10.84**
01001 0802 6" PVC, Sch 40, Manifold Piping	LF	11.63	MPLUE	1.000	7.65	0.00	3.19	10.84
33 26 0447 8" PVC, Schedule 40, Manifold Piping	**LF**				**9.13**	**0.00**	**4.82**	**13.95**
01001 0803 8" PVC, Sch 40, Manifold Piping	LF	9.75	MPLUE	1.000	9.13	0.00	4.82	13.95
33 26 0448 10" PVC, Schedule 40, Manifold Piping	**LF**				**12.34**	**0.00**	**7.44**	**19.78**
01001 0804 10" PVC, Sch 40, Manifold Piping	LF	7.21	MPLUE	1.000	12.34	0.00	7.44	19.78
33 26 0449 12" PVC, Schedule 40, Manifold Piping	**LF**				**15.66**	**0.00**	**7.85**	**23.52**
01001 0805 12" PVC, Sch 40, Manifold Piping	LF	5.68	MPLUE	1.000	15.66	0.00	7.85	23.52
33 26 0455 1" PVC, Schedule 80, Manifold Piping	**LF**				**1.93**	**0.00**	**0.36**	**2.28**
01001 0483 1" PVC, Sch 80, Manifold Piping	LF	46.21	MPLUE	1.000	1.93	0.00	0.36	2.28
33 26 0456 1-1/4" PVC, Schedule 80, Manifold Piping	**LF**				**2.20**	**0.00**	**0.49**	**2.68**
01001 0484 1-1/4" PVC, Sch 80, Manifold Piping	LF	40.51	MPLUE	1.000	2.20	0.00	0.49	2.68
33 26 0457 1-1/2" PVC, Schedule 80, Manifold Piping	**LF**				**2.41**	**0.00**	**0.59**	**3.01**
01001 0485 1-1/2" PVC, Sch 80, Manifold Piping	LF	36.89	MPLUE	1.000	2.41	0.00	0.59	3.01
33 26 0458 2" PVC, Schedule 80, Manifold Piping	**LF**				**2.67**	**0.00**	**0.82**	**3.48**
01001 0486 2" PVC, Sch 80, Manifold Piping	LF	33.38	MPLUE	1.000	2.67	0.00	0.82	3.48
33 26 0459 3" PVC, Schedule 80, Manifold Piping	**LF**				**3.83**	**0.00**	**1.69**	**5.52**
01001 0487 3" PVC, Sch 80, Manifold Piping	LF	23.25	MPLUE	1.000	3.83	0.00	1.69	5.52
33 26 0460 4" PVC, Schedule 80, Manifold Piping	**LF**				**5.74**	**0.00**	**2.44**	**8.18**
01001 0488 4" PVC, Sch 80, Manifold Piping	LF	15.5	MPLUE	1.000	5.74	0.00	2.44	8.18
33 26 0461 6" PVC, Schedule 80, Manifold Piping	**LF**				**7.65**	**0.00**	**4.60**	**12.25**
01001 0489 6" PVC, Sch 80, Manifold Piping	LF	11.63	MPLUE	1.000	7.65	0.00	4.60	12.25
33 26 0462 8" PVC, Schedule 80, Manifold Piping	**LF**				**9.13**	**0.00**	**6.98**	**16.10**
01001 0490 8" PVC, Sch 80, Manifold Piping	LF	9.75	MPLUE	1.000	9.13	0.00	6.98	16.10
33 26 0463 10" PVC, Schedule 80, Manifold Piping	**LF**				**12.34**	**0.00**	**10.38**	**22.72**
01001 0491 10" PVC, Sch 80, Manifold Piping	LF	7.21	MPLUE	1.000	12.34	0.00	10.38	22.72
33 26 0464 12" PVC, Schedule 80, Manifold Piping	**LF**				**15.66**	**0.00**	**14.30**	**29.96**
01001 0492 12" PVC, Sch 80, Manifold Piping	LF	5.68	MPLUE	1.000	15.66	0.00	14.30	29.96

33 26 05 Polyethylene Pipe	Unit	Hourly Output	Crew	Factor	Labor	Equip.	Matl.	Total

33 26 Hazardous Waste Piping/Hoses

33 26 05 Polyethylene Pipe		Unit	Hourly Output	Crew	Factor	Labor	Equip.	Matl.	Total
33 26 0501	**1" Polyethylene (SDR 21) Piping**	**LF**				**4.31**	**0.00**	**0.29**	**4.60**
15154 8512	1" Polyethylene SDR-21	LF	31	MPLUI	1.000	4.31	0.00	0.29	4.60
33 26 0502	**2" Polyethylene (SDR 21) Piping**	**LF**				**5.16**	**0.00**	**0.44**	**5.60**
15154 8514	2" Polyethylene SDR-21	LF	25.88	MPLUI	1.000	5.16	0.00	0.44	5.60
33 26 0503	**3" Polyethylene (SDR 21) Piping**	**LF**				**5.31**	**0.56**	**1.53**	**7.40**
15064 8111	3" Polyethylene Pipe, SDR-21	LF	17.88	MPLUH	1.000	5.31	0.56	1.53	7.40
33 26 0504	**4" Polyethylene (SDR 21) Piping**	**LF**				**6.91**	**0.73**	**1.06**	**8.70**
15154 8516	4" Polyethylene Pipe, SDR-21	LF	13.75	MPLUH	1.000	6.91	0.73	1.06	8.70
33 26 0505	**6" Polyethylene (SDR 21) Piping**	**LF**				**10.13**	**1.08**	**2.81**	**14.01**
15154 8517	6" Polyethylene Pipe, SDR-21	LF	9.375	MPLUH	1.000	10.13	1.08	2.81	14.01
33 26 0506	**8" Polyethylene (SDR 21) Piping**	**LF**				**11.73**	**2.90**	**7.56**	**22.20**
15064 8114	8" Polyethylene Pipe, SDR-21	LF	13.38	MPLUQ	1.000	11.73	2.90	7.56	22.20
33 26 0507	**10" Polyethylene (SDR 21) Piping**	**LF**				**14.98**	**3.71**	**6.06**	**24.75**
15154 8518	10" Polyethylene Pipe, SDR-21	LF	10.48	MPLUQ	1.000	14.98	3.71	6.06	24.75
33 26 0508	**12" Polyethylene (SDR 21) Piping**	**LF**				**17.83**	**4.41**	**8.52**	**30.77**
15154 8519	12" Polyethylene Pipe, SDR-21	LF	8.8	MPLUQ	1.000	17.83	4.41	8.52	30.77
33 26 0512	**4" High-density Polyethylene, Transfer Pipe**	**LF**				**2.25**	**0.00**	**1.84**	**4.09**
02082 1415	4" Diameter - SDR-21	LF	47.5	XPLUD	1.000	2.25	0.00	1.84	4.09
33 26 0513	**6" High-density Polyethylene, Transfer Pipe**	**LF**				**2.67**	**0.00**	**2.65**	**5.32**
02082 1418	6" Diameter - SDR-21	LF	40	XPLUD	1.000	2.67	0.00	2.65	5.32
33 26 0514	**8" High-density Polyethylene, Transfer Pipe**	**LF**				**2.85**	**0.00**	**4.48**	**7.33**
02082 1422	8" Diameter - SDR-21	LF	37.5	XPLUD	1.000	2.85	0.00	4.48	7.33
33 26 0516	**12" High-density Polyethylene, Transfer Pipe**	**LF**				**3.89**	**0.00**	**9.80**	**13.69**
02082 1428	12" Diameter - SDR-21	LF	27.5	XPLUD	1.000	3.89	0.00	9.80	13.69
33 26 0518	**16" High-density Polyethylene, Transfer Pipe**	**LF**				**6.11**	**0.00**	**15.45**	**21.56**
02082 1435	16" Diameter - SDR-21	LF	17.5	XPLUD	1.000	6.11	0.00	15.45	21.56
33 26 0520	**24" High-density Polyethylene, Transfer Pipe**	**LF**				**10.69**	**0.00**	**35.00**	**45.69**
02082 1442	24" Diameter - SDR-21	LF	10	XPLUD	1.000	10.69	0.00	35.00	45.69
33 26 0550	**2" Polyethylene, flexible piping, SDR15, 125 psi**	**LF**				**0.00**	**0.00**	**1.20**	**1.20**
01001 0493	2" Polyethylene, flexible piping, SDR15, 125 psi	LF	0	N/A	1.000	0.00	0.00	1.20	1.20

33 26 06 Double-wall Pipe	Unit	Hourly Output	Crew	Factor	Labor	Equip.	Matl.	Total

33 26 06 Double-wall Pipe		Unit	Hourly Output	Crew	Factor	Labor	Equip.	Matl.	Total
33 26 0601	**(1 1/2", 3") Stainless Steel Double-wall Piping, with Fittings**	**LF**				**36.70**	**0.00**	**71.77**	**108.46**
15066 2118	3" Stainless Steel Pipe, Schedule 40 Type 304 with 150 Lb Threaded Fitting	LF	7.75	MSPFC	1.000	13.34	0.00	35.59	48.92
15066 2128	3" Stainless Steel 90 Degree Elbow, Schedule 40 Type 304	EA	1.163	MSPFC	0.050	88.92	0.00	112.57	10.07
15066 2148	3" Stainless Steel Tee, Schedule 40 Type 304	EA	0.775	MSPFC	0.050	133.39	0.00	163.30	14.83
15162 3016	1-1/2" Stainless Steel Pipe, Schedule 40 Type 304 with 150 Lb Threaded Fitting	LF	14.63	MSPFC	1.000	7.07	0.00	12.94	20.01
15164 7120	1-1/2" Stainless Steel 90 Degree Elbow, Schedule 40 Type 304	EA	2.875	MSPFC	0.050	35.96	0.00	18.61	2.73
15164 7400	1-1/2" Stainless Steel Tee, Schedule 40 Type 304	EA	2	MSPFC	0.050	51.69	0.00	28.61	4.01
15193 3630	2" Bronze Flanged Gate Valve - 150 Lb	EA	1.125	MPLUE	0.010	79.09	0.00	708.64	7.88
33 26 0602	**(2", 4") Stainless Steel Double-wall Piping, with Fittings**	**LF**				**47.08**	**0.00**	**105.55**	**152.63**
15162 3018	2" Stainless Steel Pipe, Schedule 40 Type 304 with 150 Lb Threaded Fitting	LF	11.63	MSPFC	1.000	8.89	0.00	17.15	26.04
15162 3020	4" Stainless Steel Pipe, Schedule 40 Type 304 with 150 Lb Threaded Fitting	LF	5.875	MSPFC	1.000	17.60	0.00	49.96	67.55
15164 7130	2" Stainless Steel 90 Degree Elbow, Schedule 40 Type 304	EA	2.375	MSPFC	0.050	43.53	0.00	26.91	3.52
15164 7160	4" Stainless Steel 90 Degree Elbow, Schedule 40 Type 304	EA	0.938	MSPFC	0.050	110.27	0.00	185.33	14.78
15164 7410	2" Stainless Steel Tee, Schedule 40 Type 304	EA	1.5	MSPFC	0.050	68.92	0.00	33.84	5.14
15164 7440	4" Stainless Steel Tee, Schedule 40 Type 304	EA	0.625	MSPFC	0.050	165.40	0.00	273.97	21.97
15193 3640	3" Bronze Flanged Gate Valve -150 Lb	EA	0.75	MPLUE	0.010	118.63	0.00	1,243	13.62
33 26 0603	**(2 1/2", 4") Stainless Steel Double-wall Piping, with Fittings**	**LF**				**51.60**	**0.00**	**120.76**	**172.35**
15066 2117	2-1/2" Stainless Steel Pipe, Schedule 40 Type 304 with 150 Lb Threaded Fitting	LF	9.75	MSPFC	1.000	10.60	0.00	26.89	37.49
15066 2127	2-1/2" Stainless Steel 90 Degree Elbow, Schedule 40 Type 304	EA	1.5	MSPFC	0.050	68.92	0.00	70.56	6.97
15066 2147	2-1/2" Stainless Steel Tee, Schedule 40 Type 304	EA	1.038	MSPFC	0.050	99.64	0.00	99.60	9.96
15162 3020	4" Stainless Steel Pipe, Schedule 40 Type 304 with 150 Lb Threaded Fitting	LF	5.875	MSPFC	1.000	17.60	0.00	49.96	67.55
15164 7160	4" Stainless Steel 90 Degree Elbow, Schedule 40 Type 304	EA	0.938	MSPFC	0.050	110.27	0.00	185.33	14.78
15164 7440	4" Stainless Steel Tee, Schedule 40 Type 304	EA	0.625	MSPFC	0.050	165.40	0.00	273.97	21.97
15193 3640	3" Bronze Flanged Gate Valve -150 Lb	EA	0.75	MPLUE	0.010	118.63	0.00	1,243	13.62

33 26 06	Double-wall Pipe	Unit	Hourly Output	Crew	Factor	Labor	Equip.	Matl.	Total
33 26 0604	**(4", 6") Stainless Steel Double-wall Piping, with Fittings**	**LF**				**117.88**	**12.32**	**94.41**	**224.61**
15101 1406	5" Iron Body Gate Valve, NRS	EA	0.775	MPLUJ	0.010	156.97	0.00	758.60	9.16
15162 2140	4" Stainless Steel Pipe, Schedule 10 Type 304 with Butt Weld Fitting	LF	7.288	MSPFE	1.000	13.06	1.38	22.24	36.68
15162 2160	6" Stainless Steel Pipe, Schedule 10 Type 304 with Butt Weld Fitting	LF	5.55	MSPFE	1.000	17.15	1.82	35.45	54.42
15164 2130	4" Stainless Steel 90 Degree Elbow, Schedule 10 Type 304	EA	0.313	MSPFE	0.050	304.59	32.26	73.38	20.51
15164 2150	6" Stainless Steel 90 Degree Elbow, Schedule 10 Type 304	EA	0.248	MSPFE	0.050	384.59	40.73	154.71	29.00
15164 3120	4" Stainless Steel Tees, Schedule 10 Type 304	EA	0.209	MSPFE	0.050	455.87	48.28	88.64	29.64
15164 3140	6" Stainless Steel Tees, Schedule 10 Type 304	EA	0.165	MSPFE	0.050	576.88	61.10	265.91	45.19
33 26 0621	**(1 1/2", 3") PVC Double-wall Piping, with Fittings**	**LF**				**12.84**	**0.00**	**9.76**	**22.60**
15064 5118	3" Schedule 40 PVC Pressure Pipe Socket Weld	LF	23.25	MPLUE	1.000	3.83	0.00	1.28	5.11
15064 5138	3" 90 Degree Elbow, PVC Schedule 40 Socket Weld	EA	2.375	MPLUE	0.050	37.46	0.00	6.66	2.21
15064 5178	3" Tee, PVC Schedule 40, Socket Weld	EA	2	MPLUE	0.050	44.49	0.00	9.65	2.71
15155 1900	1-1/2" Schedule 40 PVC Pressure Pipe Socket Weld	LF	40	MPLUE	1.000	2.22	0.00	0.46	2.68
15157 2800	1-1/2" 90 Degree Elbow, PVC Schedule 40 Socket Weld	EA	5.875	MPLUE	0.050	15.14	0.00	0.98	0.81
15157 3220	1-1/2" Tee, PVC Schedule 40, Socket Weld	EA	3.875	MPLUE	0.050	22.96	0.00	1.30	1.21
15193 3630	2" Bronze Flanged Gate Valve - 150 Lb	EA	1.125	MPLUE	0.010	79.09	0.00	708.64	7.88
33 26 0622	**(2", 4") PVC Double-wall Piping, with Fittings**	**LF**				**17.26**	**0.16**	**16.28**	**33.70**
15155 1910	2" Schedule 40 PVC Pressure Pipe Socket Weld	LF	33.38	MPLUE	1.000	2.67	0.00	0.63	3.30
15155 1940	4" Schedule 40 PVC Pressure Pipe Socket Weld	LF	15.5	MPLUE	1.000	5.74	0.00	1.80	7.54
15157 2810	2" 90 Degree Elbow, PVC Schedule 40 Socket Weld	EA	4.625	MPLUE	0.050	19.24	0.00	1.54	1.04
15157 2840	4" 90 Degree Elbow, PVC Schedule 40 Socket Weld	EA	2	MPLUE	0.050	44.49	0.00	10.01	2.73
15157 3230	2" Tee, PVC Schedule 40, Socket Weld	EA	3.125	MPLUH	0.050	30.39	3.23	1.90	1.78
15157 3260	4" Tee, PVC Schedule 40, Socket Weld	EA	1.5	MPLUE	0.050	59.32	0.00	14.86	3.71
15193 3640	3" Bronze Flanged Gate Valve -150 Lb	EA	0.75	MPLUE	0.010	118.63	0.00	1,243	13.62
33 26 0623	**(2 1/2", 4") PVC Double-wall Piping, with Fittings**	**LF**				**18.58**	**0.00**	**16.99**	**35.57**
15064 5117	2-1/2" Schedule 40 PVC Pressure Pipe Socket Weld	LF	28.13	MPLUE	1.000	3.16	0.00	1.01	4.18
15064 5137	2-1/2" 90 Degree Elbow, PVC Schedule 40 Socket Weld	EA	3.125	MPLUE	0.050	28.47	0.00	4.21	1.63
15064 5177	2-1/2" Tee, PVC Schedule 40, Socket Weld	EA	2.375	MPLUE	0.050	37.46	0.00	5.91	2.17
15155 1940	4" Schedule 40 PVC Pressure Pipe Socket Weld	LF	15.5	MPLUE	1.000	5.74	0.00	1.80	7.54
15157 2840	4" 90 Degree Elbow, PVC Schedule 40 Socket Weld	EA	2	MPLUE	0.050	44.49	0.00	10.01	2.73
15157 3260	4" Tee, PVC Schedule 40, Socket Weld	EA	1.5	MPLUE	0.050	59.32	0.00	14.86	3.71
15193 3640	3" Bronze Flanged Gate Valve -150 Lb	EA	0.75	MPLUE	0.010	118.63	0.00	1,243	13.62

33 26 06	Double-wall Pipe	Unit	Hourly Output	Crew	Factor	Labor	Equip.	Matl.	Total
33 26 0624	**(4", 6") PVC Double-wall Piping, with Fittings**	**LF**				**26.52**	**0.00**	**17.88**	**44.41**
15101 1406	5" Iron Body Gate Valve, NRS	EA	0.775	MPLUJ	0.010	156.97	0.00	758.60	9.16
15155 1940	4" Schedule 40 PVC Pressure Pipe Socket Weld	LF	15.5	MPLUE	1.000	5.74	0.00	1.80	7.54
15155 1960	6" Schedule 40 PVC Pressure Pipe Socket Weld	LF	11.63	MPLUE	1.000	7.65	0.00	3.17	10.82
15157 2840	4" 90 Degree Elbow, PVC Schedule 40 Socket Weld	EA	2	MPLUE	0.050	44.49	0.00	10.01	2.73
15157 2860	6" 90 Degree Elbow, PVC Schedule 40 Socket Weld	EA	1.75	MPLUE	0.050	50.84	0.00	31.84	4.13
15157 3260	4" Tee, PVC Schedule 40, Socket Weld	EA	1.5	MPLUE	0.050	59.32	0.00	14.86	3.71
15157 3280	6" Tee, PVC Schedule 40, Socket Weld	EA	1.163	MPLUE	0.050	76.54	0.00	50.03	6.33
33 26 0641	**(1 1/2", 3") Carbon Steel Double-wall Piping, with Fittings**	**LF**				**15.82**	**0.00**	**17.30**	**33.12**
15172 0600	1-1/2" (40 mm) A-53 Pipe, Schedule 40, Not Including Hangers or Fittings	LF	28.75	MPLUE	1.000	3.09	0.00	1.52	4.61
15172 0630	3" (80 mm) A-53 Pipe, Schedule 40, Not Including Hangers or Fittings	LF	15	MPLUE	1.000	5.93	0.00	4.28	10.21
15174 5120	1-1/2" 90 Degree Elbow, 150 Lb Malleable Iron Black	EA	5.875	MPLUE	0.050	15.14	0.00	5.54	1.03
15174 5150	3" 90 Degree Elbow, 150 Lb Malleable Iron Black	EA	2.375	MPLUE	0.050	37.46	0.00	31.08	3.43
15174 5530	1-1/2" Tee, Straight 150 Lb Malleable Iron Black	EA	3.875	MPLUE	0.050	22.96	0.00	8.08	1.55
15174 5560	3" Tee, Straight 150 Lb Malleable Iron Black	EA	2	MPLUE	0.050	44.49	0.00	43.51	4.40
15193 3630	2" Bronze Flanged Gate Valve - 150 Lb	EA	1.125	MPLUE	0.010	79.09	0.00	708.64	7.88
33 26 0642	**(2", 4") Carbon Steel Double-wall Piping, with Fittings**	**LF**				**18.86**	**0.00**	**30.42**	**49.28**
15172 0610	2" (50 mm) A-53 Pipe, Schedule 40, Not Including Hangers or Fittings	LF	23.75	MPLUE	1.000	3.75	0.00	2.07	5.82
15172 0650	4" (10 cm) A-53 Pipe, Schedule 40, Not Including Hangers or Fittings	LF	17.5	MPLUJ	1.000	6.95	0.00	6.12	13.08
15174 5130	2" 90 Degree Elbow, 150 Lb Malleable Iron Black	EA	4.625	MPLUE	0.050	19.24	0.00	9.55	1.44
15174 5170	4" 90 Degree Elbow, 150 Lb Malleable Iron Black	EA	3	MPLUJ	0.050	40.55	0.00	66.82	5.37
15174 5540	2" Tee, Straight 150 Lb Malleable Iron Black	EA	3.125	MPLUE	0.050	28.47	0.00	13.76	2.11
15174 5580	4" Tee, Straight 150 Lb Malleable Iron Black	EA	2.375	MPLUJ	0.050	51.22	0.00	105.67	7.84
15193 3640	3" Bronze Flanged Gate Valve -150 Lb	EA	0.75	MPLUE	0.010	118.63	0.00	1,243	13.62
33 26 0643	**(2 1/2", 4") Carbon Steel Double-wall Piping, with Fittings**	**LF**				**20.47**	**0.00**	**32.99**	**53.47**
15172 0620	2-1/2" (60 mm) A-53 Pipe, Schedule 40, Not Including Hangers or Fittings	LF	20	MPLUE	1.000	4.45	0.00	3.27	7.72
15172 0650	4" (10 cm) A-53 Pipe, Schedule 40, Not Including Hangers or Fittings	LF	17.5	MPLUJ	1.000	6.95	0.00	6.12	13.08
15174 5140	2-1/2" 90 Degree Elbow, 150 Lb Malleable Iron Black	EA	3.125	MPLUE	0.050	28.47	0.00	21.37	2.49
15174 5170	4" 90 Degree Elbow, 150 Lb Malleable Iron Black	EA	3	MPLUJ	0.050	40.55	0.00	66.82	5.37
15174 5550	2-1/2" Tee, Straight 150 Lb Malleable Iron Black	EA	2.375	MPLUE	0.050	37.46	0.00	29.53	3.35
15174 5580	4" Tee, Straight 150 Lb Malleable Iron Black	EA	2.375	MPLUJ	0.050	51.22	0.00	105.67	7.84
15193 3640	3" Bronze Flanged Gate Valve -150 Lb	EA	0.75	MPLUE	0.010	118.63	0.00	1,243	13.62

33 26 06	Double-wall Pipe	Unit	Hourly Output	Crew	Factor	Labor	Equip.	Matl.	Total
33 26 0644	**(4", 6") Carbon Steel Double-wall Piping, with Fittings**	**LF**				**29.26**	**0.00**	**60.64**	**89.90**
15101 1406	5" Iron Body Gate Valve, NRS	EA	0.775	MPLUJ	0.010	156.97	0.00	758.60	9.16
15172 0650	4" (10 cm) A-53 Pipe, Schedule 40, Not Including Hangers or Fittings	LF	17.5	MPLUJ	1.000	6.95	0.00	6.12	13.08
15172 0670	6" (15 cm) A-53 Pipe, Schedule 40, Not Including Hangers or Fittings	LF	11.63	MPLUJ	1.000	10.46	0.00	12.66	23.13
15174 5170	4" 90 Degree Elbow, 150 Lb Malleable Iron Black	EA	3	MPLUJ	0.050	40.55	0.00	66.82	5.37
15174 5190	6" 90 Degree Elbow, 150 Lb Malleable Iron Black	EA	2.75	MPLUJ	0.050	44.24	0.00	205.91	12.51
15174 5580	4" Tee, Straight 150 Lb Malleable Iron Black	EA	2.375	MPLUJ	0.050	51.22	0.00	105.67	7.84
15174 5600	6" Tee, Straight 150 Lb Malleable Iron Black	EA	1.75	MPLUJ	0.050	69.52	0.00	306.93	18.82
33 26 0645	**(2", 4") Carbon Steel Double-wall Piping**	**EA**				**10.70**	**0.00**	**8.20**	**18.89**
15172 0610	2" (50 mm) A-53 Pipe, Schedule 40, Not Including Hangers or Fittings	LF	23.75	MPLUE	1.000	3.75	0.00	2.07	5.82
15172 0650	4" (10 cm) A-53 Pipe, Schedule 40, Not Including Hangers or Fittings	LF	17.5	MPLUJ	1.000	6.95	0.00	6.12	13.08

33 26 07	Chemical-resistant Hoses	Unit	Hourly Output	Crew	Factor	Labor	Equip.	Matl.	Total
33 26 0702	**50' x 6" Brown Gum Rubber, Chemical-resistant, Flexible Hose**	**EA**				**183.25**	**0.00**	**1,396**	**1,579**
15113 1114	50' x 6" Brown Gum Rubber, Chemical Resistant Flexible Hose	EA	0.51	MSHMD	1.000	183.25	0.00	1,396	1,579
33 26 0703	**50' X 6" Nitrile Rubber, Low pH Res, Flex Hose**	**EA**				**311.53**	**0.00**	**1,718**	**2,030**
15113 1124	Rubber hose, chem resist,6"x50', nitrile, low pH resist	EA	0.3	MSHMD	1.000	311.53	0.00	1,718	2,030
33 26 0704	**50' Flexible, Product Discharge Hose**	**EA**				**0.00**	**0.00**	**185.49**	**185.49**
13273 6123	50' Section of Gasoline-grade Discharge Hose Extension	EA	0	N/A	1.000	0.00	0.00	185.49	185.49

33 26 08	Slotted PVC Pipe	Unit	Hourly Output	Crew	Factor	Labor	Equip.	Matl.	Total
33 26 0801	**2" Slotted PVC Pipe**	**LF**				**1.69**	**0.00**	**3.69**	**5.38**
02082 1311	2" Diameter, Schedule 40, 2 to 3 Rows of Slots	LF	62.5	ULABD	1.000	1.69	0.00	3.69	5.38
33 26 0802	**4" Slotted PVC Pipe**	**LF**				**2.11**	**0.00**	**5.25**	**7.36**
02082 1312	4" Diameter, Schedule 40, 2 to 4 Rows of Slots	LF	50	ULABD	1.000	2.11	0.00	5.25	7.36
33 26 0803	**6" Slotted PVC Pipe**	**LF**				**2.82**	**0.00**	**9.45**	**12.27**
02082 1313	6" Diameter, SDR-21, 2 to 6 Rows of Slots	LF	37.5	ULABD	1.000	2.82	0.00	9.45	12.27
33 26 0804	**8" Slotted PVC Pipe**	**LF**				**4.22**	**0.00**	**15.05**	**19.27**
02082 1315	8" Diameter, SDR-21, 2 to 6 Rows of Slots	LF	25	ULABD	1.000	4.22	0.00	15.05	19.27
33 26 0805	**10" Slotted PVC Pipe**	**LF**				**5.63**	**0.00**	**30.50**	**36.13**
02082 1317	10" Diameter, SDR-21, 2 to 8 Rows of Slots	LF	18.75	ULABD	1.000	5.63	0.00	30.50	36.13

33 26 Hazardous Waste Piping/Hoses

33 26 08 Slotted PVC Pipe		Unit	Hourly Output	Crew	Factor	Labor	Equip.	Matl.	Total
33 26 0806	**12" Slotted PVC Pipe**	**LF**				**8.45**	**0.00**	**39.50**	**47.95**
02082 1319	12" Diameter, SDR-21, 2 to 10 Rows of Slots	LF	12.5	ULABD	1.000	8.45	0.00	39.50	47.95

33 26 09 Perforated Pipe		Unit	Hourly Output	Crew	Factor	Labor	Equip.	Matl.	Total
33 26 0901	**4" Diameter Perforated PVC Pipe**	**LF**				**3.69**	**2.29**	**1.75**	**7.73**
02709 2100	4" Diameter Perforated PVC Pipe, Underdrain (10 cm) Diameter	LF	39.25	CODEX	1.000	3.69	2.29	1.75	7.73
33 26 0902	**6" Diameter Perforated PVC Pipe**	**LF**				**3.69**	**2.29**	**3.22**	**9.20**
02709 2110	6" Diameter Perforated PVC Pipe, Underdrain (15 cm) Diameter	LF	39.25	CODEX	1.000	3.69	2.29	3.22	9.20
33 26 0903	**8" Diameter Perforated PVC Pipe**	**LF**				**3.73**	**2.32**	**3.66**	**9.72**
02709 2120	8" Diameter Perforated PVC Pipe, Underdrain (21 cm) Diameter	LF	38.75	CODEX	1.000	3.73	2.32	3.66	9.72
33 26 0904	**10" Diameter Perforated PVC Pipe**	**LF**				**4.06**	**2.53**	**5.67**	**12.26**
02709 2130	10" Diameter Perforated PVC Pipe, Underdrain (25 cm) Diameter	LF	35.63	CODEX	1.000	4.06	2.53	5.67	12.26
33 26 0905	**12" Diameter Perforated PVC Pipe**	**LF**				**4.29**	**2.67**	**7.85**	**14.81**
02709 2140	12" Diameter Perforated PVC Pipe, Underdrain (31 cm) Diameter	LF	33.75	CODEX	1.000	4.29	2.67	7.85	14.81
33 26 0951	**6-hole Distribution Box**	**EA**				**0.00**	**0.00**	**41.86**	**41.86**
02511 2206	6-Hole Distribution Box	EA	0	N/A	1.000	0.00	0.00	41.86	41.86

33 26 10 Galvanized Pipe		Unit	Hourly Output	Crew	Factor	Labor	Equip.	Matl.	Total
33 26 1004	**4" Galvanized Steel Pipe, Schedule 80**	**LF**				**7.06**	**0.75**	**30.77**	**38.58**
15061 1209	4" Galvanized Steel Pipe, Schedule 80	LF	13.49	MSPFE	1.000	7.06	0.75	30.77	38.58
33 26 1005	**6" Galvanized Steel Pipe, Schedule 80**	**LF**				**10.06**	**1.07**	**43.26**	**54.38**
15061 1211	6" Galvanized Steel Pipe, Schedule 80	LF	9.465	MSPFE	1.000	10.06	1.07	43.26	54.38

33 27 01 PVC Fittings		Unit	Hourly Output	Crew	Factor	Labor	Equip.	Matl.	Total
33 27 0101	**1" PVC, Schedule 40, Tee**	**EA**				**0.00**	**0.00**	**0.66**	**0.66**
01001 1001	1" PVC, Sch 40, Tee	EA	0	N/A	1.000	0.00	0.00	0.66	0.66
33 27 0102	**1-1/4" PVC, Schedule 40, Tee**	**EA**				**0.00**	**0.00**	**0.90**	**0.90**
01001 1002	1-1/4" PVC, Sch 40, Tee	EA	0	N/A	1.000	0.00	0.00	0.90	0.90
33 27 0103	**1-1/2" PVC, Schedule 40, Tee**	**EA**				**0.00**	**0.00**	**1.08**	**1.08**
01001 1003	1-1/2" PVC, Sch 40, Tee	EA	0	N/A	1.000	0.00	0.00	1.08	1.08
33 27 0104	**2" PVC, Schedule 40, Tee**	**EA**				**0.00**	**0.00**	**1.85**	**1.85**
01001 1004	2" PVC, Sch 40, Tee	EA	0	N/A	1.000	0.00	0.00	1.85	1.85

33 27 01 PVC Fittings	Unit	Hourly Output	Crew	Factor	Labor	Equip.	Matl.	Total
33 27 0105 3" PVC, Schedule 40, Tee	**EA**				**0.00**	**0.00**	**9.17**	**9.17**
01001 1005 3" PVC, Sch 40, Tee	EA	0	N/A	1.000	0.00	0.00	9.17	9.17
33 27 0106 4" PVC, Schedule 40, Tee	**EA**				**0.00**	**0.00**	**14.49**	**14.49**
01001 1006 4" PVC, Sch 40, Tee	EA	0	N/A	1.000	0.00	0.00	14.49	14.49
33 27 0107 6" PVC, Schedule 40, Tee	**EA**				**0.00**	**0.00**	**21.53**	**21.53**
01001 1007 6" PVC, Sch 40, Tee	EA	0	N/A	1.000	0.00	0.00	21.53	21.53
33 27 0108 8" PVC, Schedule 40, Tee	**EA**				**0.00**	**0.00**	**28.91**	**28.91**
01001 1008 8" PVC, Sch 40, Tee	EA	0	N/A	1.000	0.00	0.00	28.91	28.91
33 27 0109 10" PVC, Schedule 40, Tee	**EA**				**0.00**	**0.00**	**39.05**	**39.05**
01001 1009 10" PVC, Sch 40, Tee	EA	0	N/A	1.000	0.00	0.00	39.05	39.05
33 27 0110 12" PVC, Schedule 40, Tee	**EA**				**0.00**	**0.00**	**61.79**	**61.79**
01001 1010 12" PVC, Sch 40, Tee	EA	0	N/A	1.000	0.00	0.00	61.79	61.79
33 27 0111 1" PVC, Schedule 40, 90 Degree, Elbow	**EA**				**0.00**	**0.00**	**0.50**	**0.50**
01001 1011 1" PVC, Sch 40, 90 Degree, Elbow	EA	0	N/A	1.000	0.00	0.00	0.50	0.50
33 27 0112 1-1/4" PVC, Schedule 40, 90 Degree, Elbow	**EA**				**0.00**	**0.00**	**0.75**	**0.75**
01001 1012 1-1/4" PVC, Sch 40, 90 Degree, Elbow	EA	0	N/A	1.000	0.00	0.00	0.75	0.75
33 27 0113 1-1/2" PVC, Schedule 40, 90 Degree, Elbow	**EA**				**0.00**	**0.00**	**0.81**	**0.81**
01001 1013 1-1/2" PVC, Sch 40, 90 Degree, Elbow	EA	0	N/A	1.000	0.00	0.00	0.81	0.81
33 27 0114 2" PVC, Schedule 40, 90 Degree, Elbow	**EA**				**0.00**	**0.00**	**1.50**	**1.50**
01001 1014 2" PVC, Sch 40, 90 Degree, Elbow	EA	0	N/A	1.000	0.00	0.00	1.50	1.50
33 27 0115 3" PVC, Schedule 40, 90 Degree, Elbow	**EA**				**0.00**	**0.00**	**6.33**	**6.33**
01001 1015 3" PVC, Sch 40, 90 Degree, Elbow	EA	0	N/A	1.000	0.00	0.00	6.33	6.33
33 27 0116 4" PVC, Schedule 40, 90 Degree, Elbow	**EA**				**0.00**	**0.00**	**9.76**	**9.76**
01001 1016 4" PVC, Sch 40, 90 Degree, Elbow	EA	0	N/A	1.000	0.00	0.00	9.76	9.76
33 27 0117 6" PVC, Schedule 40, 90 Degree, Elbow	**EA**				**0.00**	**0.00**	**31.04**	**31.04**
01001 1017 6" PVC, Sch 40, 90 Degree, Elbow	EA	0	N/A	1.000	0.00	0.00	31.04	31.04
33 27 0118 8" PVC, Schedule 40, 90 Degree, Elbow	**EA**				**0.00**	**0.00**	**86.57**	**86.57**
01001 1018 8" PVC, Sch 40, 90 Degree, Elbow	EA	0	N/A	1.000	0.00	0.00	86.57	86.57
33 27 0119 10" PVC, Schedule 40, 90 Degree, Elbow	**EA**				**0.00**	**0.00**	**275.98**	**275.98**
01001 1019 10" PVC, Sch 40, 90 Degree, Elbow	EA	0	N/A	1.000	0.00	0.00	275.98	275.98
33 27 0120 12" PVC, Schedule 40, 90 Degree, Elbow	**EA**				**0.00**	**0.00**	**367.88**	**367.88**
01001 1020 12" PVC, Sch 40, 90 Degree, Elbow	EA	0	N/A	1.000	0.00	0.00	367.88	367.88

33 27 01 PVC Fittings		Unit	Hourly Output	Crew	Factor	Labor	Equip.	Matl.	Total
33 27 0121	**1" PVC, Schedule 80, Tee**	**EA**				**0.00**	**0.00**	**3.13**	**3.13**
01001 1021	1" PVC, Sch 80, Tee	EA	0	N/A	1.000	0.00	0.00	3.13	3.13
33 27 0122	**1-1/4" PVC, Schedule 80, Tee**	**EA**				**0.00**	**0.00**	**8.61**	**8.61**
01001 1022	1-1/4" PVC, Sch 80, Tee	EA	0	N/A	1.000	0.00	0.00	8.61	8.61
33 27 0123	**1-1/2" PVC, Scheduleh 80, Tee**	**EA**				**0.00**	**0.00**	**9.37**	**9.37**
01001 1023	1-1/2" PVC, Sch 80, Tee	EA	0	N/A	1.000	0.00	0.00	9.37	9.37
33 27 0124	**2" PVC, Schedule 80, Tee**	**EA**				**0.00**	**0.00**	**11.72**	**11.72**
01001 1024	2" PVC, Sch 80, Tee	EA	0	N/A	1.000	0.00	0.00	11.72	11.72
33 27 0125	**3" PVC, Schedule 80, Tee**	**EA**				**0.00**	**0.00**	**15.93**	**15.93**
01001 1025	3" PVC, Sch 80, Tee	EA	0	N/A	1.000	0.00	0.00	15.93	15.93
33 27 0126	**4" PVC, Schedule 80, Tee**	**EA**				**0.00**	**0.00**	**18.45**	**18.45**
01001 1026	4" PVC, Sch 80, Tee	EA	0	N/A	1.000	0.00	0.00	18.45	18.45
33 27 0127	**6" PVC, Schedule 80, Tee**	**EA**				**0.00**	**0.00**	**63.01**	**63.01**
01001 1027	6" PVC, Sch 80, Tee	EA	0	N/A	1.000	0.00	0.00	63.01	63.01
33 27 0128	**8" PVC, Schedule 80, Tee**	**EA**				**0.00**	**0.00**	**152.19**	**152.19**
01001 1028	8" PVC, Sch 80, Tee	EA	0	N/A	1.000	0.00	0.00	152.19	152.19
33 27 0129	**10" PVC, Schedule 80, Tee**	**EA**				**0.00**	**0.00**	**592.12**	**592.12**
01001 1029	10" PVC, Sch 80, Tee	EA	0	N/A	1.000	0.00	0.00	592.12	592.12
33 27 0130	**12" PVC, Schedule 80, Tee**	**EA**				**0.00**	**0.00**	**836.24**	**836.24**
01001 1030	12" PVC, Sch 80, Tee	EA	0	N/A	1.000	0.00	0.00	836.24	836.24
33 27 0131	**1" PVC, Schedule 80, 90 Degree, Elbow**	**EA**				**0.00**	**0.00**	**1.90**	**1.90**
01001 1031	1" PVC, Sch 80, 90 Degree, Elbow	EA	0	N/A	1.000	0.00	0.00	1.90	1.90
33 27 0132	**1-1/4" PVC, Schedule 80, 90 Degree, Elbow**	**EA**				**0.00**	**0.00**	**2.54**	**2.54**
01001 1032	1-1/4" PVC, Sch 80, 90 Degree, Elbow	EA	0	N/A	1.000	0.00	0.00	2.54	2.54
33 27 0133	**1-1/2" PVC, Sch 80, 90 Degree, Elbow**	**EA**				**0.00**	**0.00**	**2.73**	**2.73**
01001 1033	1-1/2" PVC, Sch 80, 90 Degree, Elbow	EA	0	N/A	1.000	0.00	0.00	2.73	2.73
33 27 0134	**2" PVC, Schedule 80, 90 Degree, Elbow**	**EA**				**0.00**	**0.00**	**3.19**	**3.19**
01001 1034	2" PVC, Sch 80, 90 Degree, Elbow	EA	0	N/A	1.000	0.00	0.00	3.19	3.19
33 27 0135	**3" PVC, Schedule 80, 90 Degree, Elbow**	**EA**				**0.00**	**0.00**	**8.66**	**8.66**
01001 1035	3" PVC, Sch 80, 90 Degree, Elbow	EA	0	N/A	1.000	0.00	0.00	8.66	8.66
33 27 0136	**4" PVC, Schedule, 90 Degree, Elbow**	**EA**				**0.00**	**0.00**	**13.17**	**13.17**
01001 1036	4" PVC, Sch 80, 90 Degree, Elbow	EA	0	N/A	1.000	0.00	0.00	13.17	13.17

33 27 Fittings/Valves

33 27 01 PVC Fittings		Unit	Hourly Output	Crew	Factor	Labor	Equip.	Matl.	Total
33 27 0137	**6" PVC, Schedule, 90 Degree, Elbow**	**EA**				**0.00**	**0.00**	**37.47**	**37.47**
01001 1037	6" PVC, Sch 80, 90 Degree, Elbow	EA	0	N/A	1.000	0.00	0.00	37.47	37.47
33 27 0138	**8" PVC, Schedule, 90 Degree, Elbow**	**EA**				**0.00**	**0.00**	**103.28**	**103.28**
01001 1038	8" PVC, Sch 80, 90 Degree, Elbow	EA	0	N/A	1.000	0.00	0.00	103.28	103.28
33 27 0139	**10" PVC, Schedule, 90 Degree, Elbow**	**EA**				**0.00**	**0.00**	**566.82**	**566.82**
01001 1039	10" PVC, Sch 80, 90 Degree, Elbow	EA	0	N/A	1.000	0.00	0.00	566.82	566.82
33 27 0140	**12" PVC, Schedule, 90 Degree, Elbow**	**EA**				**0.00**	**0.00**	**733.19**	**733.19**
01001 1040	12" PVC, Sch 80, 90 Degree, Elbow	EA	0	N/A	1.000	0.00	0.00	733.19	733.19
33 27 0141	**1" Flange Assemblies, PVC, Schedule 40**	**EA**				**0.00**	**0.00**	**4.71**	**4.71**
01001 1041	1" Flange Assemblies, PVC, Sch 40	EA	0	N/A	1.000	0.00	0.00	4.71	4.71
33 27 0142	**1-1/4" Flange Assemblies, PVC, Schedule 40**	**EA**				**0.00**	**0.00**	**4.86**	**4.86**
01001 1042	1-1/4" Flange Assemblies, PVC, Sch 40	EA	0	N/A	1.000	0.00	0.00	4.86	4.86
33 27 0143	**1-1/2" Flange Assemblies, PVC, Schedule 40**	**EA**				**0.00**	**0.00**	**4.96**	**4.96**
01001 1043	1-1/2" Flange Assemblies, PVC, Sch 40	EA	0	N/A	1.000	0.00	0.00	4.96	4.96
33 27 0144	**2" Flange Assemblies, PVC, Schedule 40**	**EA**				**0.00**	**0.00**	**6.60**	**6.60**
01001 1044	2" Flange Assemblies, PVC, Sch 40	EA	0	N/A	1.000	0.00	0.00	6.60	6.60
33 27 0145	**3" Flange Assemblies, PVC, Schedule 40**	**EA**				**0.00**	**0.00**	**10.94**	**10.94**
01001 1045	3" Flange Assemblies, PVC Sch 40	EA	0	N/A	1.000	0.00	0.00	10.94	10.94
33 27 0146	**4" Flange Assemblies, PVC, Schedule 40**	**EA**				**0.00**	**0.00**	**13.74**	**13.74**
01001 1046	4" Flange Assemblies, PVC Sch 40	EA	0	N/A	1.000	0.00	0.00	13.74	13.74
33 27 0147	**6" Flange Assemblies, PVC, Schedule 40**	**EA**				**0.00**	**0.00**	**19.05**	**19.05**
01001 1047	6" Flange Assemblies, PVC Sch 40	EA	0	N/A	1.000	0.00	0.00	19.05	19.05
33 27 0148	**8" Flange Assemblies, PVC, Schedule 40**	**EA**				**0.00**	**0.00**	**32.67**	**32.67**
01001 1048	8" Flange Assemblies, PVC Sch 40	EA	0	N/A	1.000	0.00	0.00	32.67	32.67
33 27 0149	**10" Flange Assemblies, PVC, Schedule 40**	**EA**				**0.00**	**0.00**	**49.53**	**49.53**
01001 1049	10" Flange Assemblies, PVC Sch 40	EA	0	N/A	1.000	0.00	0.00	49.53	49.53
33 27 0150	**12" Flange Assemblies, PVC, Sch 40**	**EA**				**0.00**	**0.00**	**74.50**	**74.50**
01001 1050	12" Flange Assemblies, PVC, Sch 40	EA	0	N/A	1.000	0.00	0.00	74.50	74.50
33 27 0151	**1" Flange Assemblies, PVC, Schedule 80**	**EA**				**0.00**	**0.00**	**6.60**	**6.60**
01001 1051	1" Flange Assemblies, PVC, Sch 80	EA	0	N/A	1.000	0.00	0.00	6.60	6.60
33 27 0152	**1-1/4" Flange Assemblies, PVC, Schedule 80**	**EA**				**0.00**	**0.00**	**6.80**	**6.80**
01001 1052	1-1/4" Flange Assemblies, PVC, Sch 80	EA	0	N/A	1.000	0.00	0.00	6.80	6.80

33 27 01	PVC Fittings	Unit	Hourly Output	Crew	Factor	Labor	Equip.	Matl.	Total
33 27 0153	**1-1/2" Flange Assemblies, PVC, Schedule 80**	**EA**				**0.00**	**0.00**	**6.94**	**6.94**
01001 1053	1-1/2" Flange Assemblies, PVC, Sch 80	EA	0	N/A	1.000	0.00	0.00	6.94	6.94
33 27 0154	**2" Flange Assemblies, PVC, Schedule 80**	**EA**				**0.00**	**0.00**	**10.29**	**10.29**
01001 1054	2" Flange Assemblies, PVC, Sch 80	EA	0	N/A	1.000	0.00	0.00	10.29	10.29
33 27 0155	**3" Flange Assemblies, PVC, Schedule 80**	**EA**				**0.00**	**0.00**	**17.56**	**17.56**
01001 1055	3" Flange Assemblies, PVC, Sch 80	EA	0	N/A	1.000	0.00	0.00	17.56	17.56
33 27 0156	**4" Flange Assemblies, PVC, Schedule 80**	**EA**				**0.00**	**0.00**	**22.22**	**22.22**
01001 1056	4" Flange Assemblies, PVC, Sch 80	EA	0	N/A	1.000	0.00	0.00	22.22	22.22
33 27 0157	**6" Flange Assemblies, PVC, Schedule 80**	**EA**				**0.00**	**0.00**	**28.59**	**28.59**
01001 1057	6" Flange Assemblies, PVC, Sch 80	EA	0	N/A	1.000	0.00	0.00	28.59	28.59
33 27 0158	**8" Flange Assemblies, PVC, Schedule 80**	**EA**				**0.00**	**0.00**	**46.54**	**46.54**
01001 1058	8" Flange Assemblies, PVC, Sch 80	EA	0	N/A	1.000	0.00	0.00	46.54	46.54
33 27 0159	**10" Flange Assemblies, PVC, Schedule 80**	**EA**				**0.00**	**0.00**	**68.14**	**68.14**
01001 1059	10" Flange Assemblies, PVC, Sch 80	EA	0	N/A	1.000	0.00	0.00	68.14	68.14
33 27 0160	**12" Flange Assemblies, PVC, Schedule 80**	**EA**				**0.00**	**0.00**	**73.45**	**73.45**
01001 1060	12" Flange Assemblies, PVC, Sch 80	EA	0	N/A	1.000	0.00	0.00	73.45	73.45
33 27 0161	**4" x 2" Reducer, PVC Schedule 40**	**EA**				**0.00**	**0.00**	**3.50**	**3.50**
01001 1061	4" x 2" Reducer, PVC Sch 40	EA	0	N/A	1.000	0.00	0.00	3.50	3.50
33 27 0162	**6" x 4" Reducer, PVC Schedule 40**	**EA**				**0.00**	**0.00**	**21.14**	**21.14**
01001 1062	6" x 4" Reducer, PVC Sch 40	EA	0	N/A	1.000	0.00	0.00	21.14	21.14
33 27 0163	**8" x 6" Reducer, PVC Schedule 40**	**EA**				**0.00**	**0.00**	**31.88**	**31.88**
01001 1063	8" x 6" Reducer, PVC Sch 40	EA	0	N/A	1.000	0.00	0.00	31.88	31.88
33 27 0164	**10" x 8" Reducer, PVC Schedule 40**	**EA**				**0.00**	**0.00**	**107.39**	**107.39**
01001 1064	10" x 8" Reducer, PVC Sch 40	EA	0	N/A	1.000	0.00	0.00	107.39	107.39
33 27 0165	**8" x 4" Reducer, PVC Schedule 40**	**EA**				**0.00**	**0.00**	**34.84**	**34.84**
01001 1065	8" x 4" Reducer, PVC Sch 40	EA	0	N/A	1.000	0.00	0.00	34.84	34.84
33 27 0166	**12" x 8" Reducer, PVC Schedule 40**	**EA**				**0.00**	**0.00**	**145.11**	**145.11**
01001 1066	12" x 8" Reducer, PVC Sch 40	EA	0	N/A	1.000	0.00	0.00	145.11	145.11
33 27 0167	**4" x 2" Reducer, PVC Schedule 80**	**EA**				**0.00**	**0.00**	**34.33**	**34.33**
01001 1067	4" x 2" Reducer, PVC Sch 80	EA	0	N/A	1.000	0.00	0.00	34.33	34.33
33 27 0168	**6" x 4" Reducer, PVC Schedule 80**	**EA**				**0.00**	**0.00**	**64.73**	**64.73**
01001 1068	6" x 4" Reducer, PVC Sch 80	EA	0	N/A	1.000	0.00	0.00	64.73	64.73

33 27 Fittings/Valves

33 27 01	PVC Fittings	Unit	Hourly Output	Crew	Factor	Labor	Equip.	Matl.	Total
33 27 0169	**8" x 6" Reducer, PVC Schedule 80**	**EA**				**0.00**	**0.00**	**83.84**	**83.84**
01001 1069	8" x 6" Reducer, PVC Sch 80	EA	0	N/A	1.000	0.00	0.00	83.84	83.84
33 27 0170	**10" x 8" Reducer, PVC Schedule 80**	**EA**				**0.00**	**0.00**	**130.66**	**130.66**
01001 1070	10" x 8" Reducer, PVC Sch 80	EA	0	N/A	1.000	0.00	0.00	130.66	130.66
33 27 0171	**8" x 4" Reducer, PVC Schedule 80**	**EA**				**0.00**	**0.00**	**78.50**	**78.50**
01001 1071	8" x 4" Reducer, PVC Sch 80	EA	0	N/A	1.000	0.00	0.00	78.50	78.50
33 27 0172	**12" x 8" Reducer, PVC Schedule 80**	**EA**				**0.00**	**0.00**	**160.56**	**160.56**
01001 1072	12" x 8" Reducer, PVC Sch 80	EA	0	N/A	1.000	0.00	0.00	160.56	160.56

33 27 02	Stainless Steel Fittings	Unit	Hourly Output	Crew	Factor	Labor	Equip.	Matl.	Total
33 27 0201	**2" Stainless Steel, Tee**	**EA**				**68.92**	**0.00**	**33.84**	**102.76**
15164 7410	2" Stainless Steel Tee, Schedule 40 Type 304	EA	1.5	MSPFC	1.000	68.92	0.00	33.84	102.76
33 27 0202	**4" Stainless Steel, Tee**	**EA**				**165.40**	**0.00**	**273.97**	**439.37**
15164 7440	4" Stainless Steel Tee, Schedule 40 Type 304	EA	0.625	MSPFC	1.000	165.40	0.00	273.97	439.37
33 27 0203	**8" Stainless Steel, Schedule 10, Type 316, Tee**	**EA**				**686.88**	**65.70**	**900.18**	**1,653**
15164 3243	8" Stainless Steel Tee, Type 316L	EA	0.219	MSPFN	1.000	686.88	65.70	900.18	1,653
33 27 0204	**6" Stainless Steel, Tee**	**EA**				**576.88**	**61.10**	**265.91**	**903.89**
15164 3140	6" Stainless Steel Tees, Schedule 10 Type 304	EA	0.165	MSPFE	1.000	576.88	61.10	265.91	903.89
33 27 0211	**2" Stainless Steel, 90 Degree, Elbow**	**EA**				**43.53**	**0.00**	**26.91**	**70.44**
15164 7130	2" Stainless Steel 90 Degree Elbow, Schedule 40 Type 304	EA	2.375	MSPFC	1.000	43.53	0.00	26.91	70.44
33 27 0212	**4" Stainless Steel, 90 Degree, Elbow**	**EA**				**110.27**	**0.00**	**185.33**	**295.60**
15164 7160	4" Stainless Steel 90 Degree Elbow, Schedule 40 Type 304	EA	0.938	MSPFC	1.000	110.27	0.00	185.33	295.60
33 27 0213	**6" Stainless Steel, 90 Degree, Elbow**	**EA**				**106.79**	**0.00**	**322.32**	**429.11**
15164 3370	6" Stainless Steel 90 Degree Elbow, Schedule 40, Type 304	EA	0.968	MSPFC	1.000	106.79	0.00	322.32	429.11
33 27 0214	**8" Stainless Steel, 90 Degree, Elbow**	**EA**				**103.38**	**0.00**	**539.88**	**643.25**
15164 3380	8" Stainless Steel 90 Degree Elbow, Schedule 40, Type 304	EA	1	MSPFC	1.000	103.38	0.00	539.88	643.25
33 27 0231	**4" x 2" Stainless Steel Reducer, Schedule 40, Type 304**	**EA**				**110.27**	**0.00**	**135.83**	**246.10**
15164 7624	4" x 2" Stainless Steel Reducer, Schedule 40, Type 304	EA	0.938	MSPFC	1.000	110.27	0.00	135.83	246.10
33 27 0241	**4" x 2" Stainless Steel Reducer, Schedule 10, Type 316**	**EA**				**110.27**	**0.00**	**48.35**	**158.61**
15164 3198	4" x 2" Stainless Steel Reducer, Schedule 10, Type 316	EA	0.938	MSPFC	1.000	110.27	0.00	48.35	158.61

33 27 02	Stainless Steel Fittings	Unit	Hourly Output	Crew	Factor	Labor	Equip.	Matl.	Total
33 27 0242	**8" x 4" Stainless Steel Reducer, Schedule 10, Type 316**	**EA**				**103.38**	**0.00**	**293.54**	**396.91**
15164 3202	8" x 4" Stainless Steel Reducer, Schedule 10, Type 316	EA	1	MSPFC	1.000	103.38	0.00	293.54	396.91

33 27 03	Polyethylene Fittings	Unit	Hourly Output	Crew	Factor	Labor	Equip.	Matl.	Total
33 27 0301	**4" High-density Polyethylene, Tee**	**LF**				**32.89**	**0.00**	**35.50**	**68.39**
02082 1494	4" Diameter - SDR-21/SDR-26	EA	3.25	XPLUD	1.000	32.89	0.00	35.50	68.39
33 27 0302	**6" High-density Polyethylene, Tee**	**LF**				**38.88**	**0.00**	**93.00**	**131.88**
02082 1496	6" Diameter - SDR-21/SDR-26	EA	2.75	XPLUD	1.000	38.88	0.00	93.00	131.88
33 27 0303	**8" High-density Polyethylene, Tee**	**LF**				**61.09**	**0.00**	**236.00**	**297.09**
02082 1498	8" Diameter - SDR-21/SDR-26	EA	1.75	XPLUD	1.000	61.09	0.00	236.00	297.09
33 27 0304	**12" High-density Polyethylene, Tee**	**LF**				**106.91**	**0.00**	**425.00**	**531.91**
02082 1501	12" Diameter - SDR-21/SDR-26	EA	1	XPLUD	1.000	106.91	0.00	425.00	531.91
33 27 0305	**18" High-density Polyethylene, Tee**	**LF**				**427.63**	**0.00**	**800.00**	**1,228**
02082 1503	18" Diameter - SDR-21/SDR-26	EA	0.25	XPLUD	1.000	427.63	0.00	800.00	1,228
33 27 0306	**24" High-density Polyethylene, Tee**	**LF**				**568.65**	**0.00**	**1,650**	**2,219**
02082 1505	24" Diameter - SDR-21/SDR-26	EA	0.188	XPLUD	1.000	568.65	0.00	1,650	2,219
33 27 0311	**4", 90 Degree, High-density Polyethylene, Elbow**	**EA**				**30.54**	**0.00**	**25.50**	**56.04**
02082 1455	4" Diameter - SDR-21	EA	3.5	XPLUD	1.000	30.54	0.00	25.50	56.04
33 27 0312	**6", 90 Degree, High-density Polyethylene, Elbow**	**EA**				**35.64**	**0.00**	**67.00**	**102.64**
02082 1458	6" Diameter - SDR-21	EA	3	XPLUD	1.000	35.64	0.00	67.00	102.64
33 27 0313	**8", 90 Degree, High-density Polyethylene, Elbow**	**EA**				**47.51**	**0.00**	**168.00**	**215.51**
02082 1462	8" Diameter - SDR-21	EA	2.25	XPLUD	1.000	47.51	0.00	168.00	215.51
33 27 0314	**10", 90 Degree, High-density Polyethylene, Elbow**	**EA**				**71.27**	**0.00**	**290.00**	**361.27**
02082 1465	10" Diameter - SDR-21	EA	1.5	XPLUD	1.000	71.27	0.00	290.00	361.27
33 27 0315	**12", 90 Degree, High-density Polyethylene, Elbow**	**EA**				**95.03**	**0.00**	**490.00**	**585.03**
02082 1468	12" Diameter - SDR-21	EA	1.125	XPLUD	1.000	95.03	0.00	490.00	585.03
33 27 0321	**10" Polyethylene SDR-21, Tee**	**EA**				**133.57**	**33.05**	**223.89**	**390.50**
15157 7647	10" Tees, Polyethylene SDR-21	EA	1.175	MPLUQ	1.000	133.57	33.05	223.89	390.50
33 27 0322	**12" Polyethylene SDR-21, Tee**	**EA**				**171.99**	**42.55**	**298.52**	**513.06**
15157 7648	12" Tees, Polyethylene SDR-21	EA	0.913	MPLUQ	1.000	171.99	42.55	298.52	513.06

33 27 Fittings/Valves

33 27 03 Polyethylene Fittings		Unit	Hourly Output	Crew	Factor	Labor	Equip.	Matl.	Total
33 27 0331	**10" Polyethylene SDR-21, 90-degree Elbow**	**EA**				**66.43**	**16.44**	**138.54**	**221.41**
15157 7617	10" 90 Degree Elbows, Polyethylene SDR-21	EA	2.363	MPLUQ	1.000	66.43	16.44	138.54	221.41
33 27 0332	**12" Polyethylene SDR-21, 90-degree Elbow**	**EA**				**86.59**	**21.42**	**328.56**	**436.57**
15157 7618	12" 90 Degree Elbows, Polyethylene SDR-21	EA	1.813	MPLUQ	1.000	86.59	21.42	328.56	436.57
33 27 0351	**3" SDR-17 Flange Assemblies**	**EA**				**209.25**	**51.77**	**28.54**	**289.57**
15157 7732	3" Flange Assemblies, Polyethylene SDR-17	EA	0.75	MPLUQ	1.000	209.25	51.77	28.54	289.57
33 27 0352	**4" SDR-17 Flange Assemblies**	**EA**				**176.34**	**43.63**	**35.14**	**255.11**
15157 7733	4" Flange Assemblies, Polyethylene SDR-17	EA	0.89	MPLUQ	1.000	176.34	43.63	35.14	255.11
33 27 0353	**6" SDR-17 Flange Assemblies**	**EA**				**155.39**	**38.45**	**40.81**	**234.65**
15157 7734	6" Flange Assemblies, Polyethylene SDR-17	EA	1.01	MPLUQ	1.000	155.39	38.45	40.81	234.65

33 27 04 Valves		Unit	Hourly Output	Crew	Factor	Labor	Equip.	Matl.	Total
33 27 0401	**1" Iron Body Check Valve**	**EA**				**43.88**	**0.00**	**406.12**	**450.00**
15194 5994	1" Iron Body Swing Check Valve, Bronze, Flanged - 125 Lb	EA	1.175	MPLUA	1.000	43.88	0.00	406.12	450.00
33 27 0402	**2" Iron Body Check Valve**	**EA**				**30.63**	**0.00**	**161.16**	**191.79**
15194 6640	2" Single Disc-type Check Valve Iron Body Wafer-type - 125 Lb	EA	1.688	MSPFA	1.000	30.63	0.00	161.16	191.79
33 27 0403	**3" Iron Body Check Valve**	**EA**				**36.09**	**0.00**	**189.71**	**225.80**
15194 6660	3" Single Disc-type Check Valve Iron Body Wafer-type - 125 Lb	EA	2.638	MSPFD	1.000	36.09	0.00	189.71	225.80
33 27 0404	**4" Iron Body Check Valve**	**EA**				**80.16**	**0.00**	**242.20**	**322.35**
15194 6670	4" Single Disc-type Check Valve Iron Body Wafer-type - 125 Lb	EA	1.188	MSPFD	1.000	80.16	0.00	242.20	322.35
33 27 0405	**6" Iron Body Check Valve**	**EA**				**121.93**	**17.16**	**423.62**	**562.71**
15194 6690	6" Single Disc-type Check Valve Iron Body Wafer-type - 125 Lb	EA	0.838	MSPFF	1.000	121.93	17.16	423.62	562.71
33 27 0406	**8" Iron Body Check Valve**	**EA**				**214.98**	**30.26**	**756.98**	**1,002**
15194 6700	8" Single Disc-type Check Valve Iron Body Wafer-type - 125 Lb	EA	0.475	MSPFF	1.000	214.98	30.26	756.98	1,002
33 27 0407	**10" Iron Body Check Valve**	**EA**				**302.56**	**42.59**	**1,312**	**1,657**
15194 6710	10" Single Disc-type Check Valve Iron Body Wafer-type - 125 Lb	EA	0.338	MSPFF	1.000	302.56	42.59	1,312	1,657
33 27 0408	**12" Iron Body Check Valve**	**EA**				**355.18**	**50.00**	**1,439**	**1,844**
15194 6762	12" Twin Disc-type Check Valve	EA	0.288	MSPFF	1.000	355.18	50.00	1,439	1,844
33 27 0411	**1/2" x 1/2" Bronze Pressure Relief Valve**	**EA**				**17.19**	**0.00**	**15.15**	**32.34**
15193 6000	1/2" x 1/2" Bronze Pressure Relief Valve, Threaded	EA	3	MPLUA	1.000	17.19	0.00	15.15	32.34

33 27 04 Valves	Unit	Hourly Output	Crew	Factor	Labor	Equip.	Matl.	Total
33 27 0412 3/4" x 3/4" Bronze Pressure Relief Valve	**EA**				**20.63**	**0.00**	**44.89**	**65.52**
15193 6040 3/4" x 3/4" Bronze Pressure Relief Valve, Threaded	EA	2.5	MPLUA	1.000	20.63	0.00	44.89	65.52
33 27 0413 1" x 1" Bronze Pressure Relief Valve	**EA**				**27.50**	**0.00**	**104.70**	**132.20**
15193 6450 1" x 1" Bronze Pressure Relief Valve, Threaded	EA	1.875	MPLUA	1.000	27.50	0.00	104.70	132.20
33 27 0414 1 1/4" x 1 1/4" Bronze Pressure Relief Valve	**EA**				**29.46**	**0.00**	**221.04**	**250.50**
15122 1104 1-1/4" x 1-1/4" Bronze Pressure Relief Valve, Threaded	EA	1.75	MPLUA	1.000	29.46	0.00	221.04	250.50
33 27 0415 1 1/2" x 1 1/2" Bronze Pressure Relief Valve	**EA**				**33.00**	**0.00**	**236.79**	**269.79**
15193 6470 1-1/2" x 1-1/2" Bronze Pressure Relief Valve, Threaded	EA	1.563	MPLUA	1.000	33.00	0.00	236.79	269.79
33 27 0416 2" x 2" Bronze Pressure Relief Valve	**EA**				**44.49**	**0.00**	**342.46**	**386.95**
15193 6480 2" x 2" Bronze Pressure Relief Valve, Threaded	EA	2	MPLUE	1.000	44.49	0.00	342.46	386.95
33 27 0417 2 1/2" x 2 1/2" Bronze Pressure Relief Valve	**EA**				**50.84**	**0.00**	**957.74**	**1,009**
15193 6490 2-1/2" x 2-1/2" Bronze Pressure Relief Valve, Threaded	EA	1.75	MPLUE	1.000	50.84	0.00	957.74	1,009
33 27 0418 1" NPT Brass Vacuum Breaker	**EA**				**25.78**	**0.00**	**43.70**	**69.48**
15118 1100 1" NPTM (Brass) Inlet Vacuum Breaker with Stainless Steel Ball & Spring	EA	2	MPLUA	1.000	25.78	0.00	43.70	69.48
33 27 0421 1" Ball Valve, Carbon Steel Trim	**EA**				**21.71**	**0.00**	**102.45**	**124.16**
15197 5380 1" Threaded Ball Valve, Carbon Steel Trim Regular Port	EA	2.375	MPLUA	1.000	21.71	0.00	102.45	124.16
33 27 0422 2" Ball Valve, Carbon Steel Trim	**EA**				**37.50**	**0.00**	**235.98**	**273.48**
15197 5410 2" Threaded Ball Valve, Carbon Steel Trim Regular Port	EA	1.375	MPLUA	1.000	37.50	0.00	235.98	273.48
33 27 0423 3" Ball Valve, Carbon Steel Trim	**EA**				**58.93**	**0.00**	**105.87**	**164.80**
15104 1105 3" Threaded Ball Valve, Carbon Steel Trim Regular Port	EA	0.875	MPLUA	1.000	58.93	0.00	105.87	164.80
33 27 0424 4" Ball Valve, Carbon Steel Trim	**EA**				**151.45**	**0.00**	**447.33**	**598.78**
15197 5560 4" Ball Valve, 150 Lb Flanged Carbon Steel Trim	EA	0.588	MPLUE	1.000	151.45	0.00	447.33	598.78
33 27 0425 2" Bronze Gate Valve	**EA**				**37.50**	**0.00**	**62.99**	**100.49**
15101 1121 Cost for each additional valve; 150 Lb Rated Valve	EA	0	N/A	0.000	0.00	0.00	77.05	0.00
15193 2980 2" Bronze 125 Lb Gate Valve Threaded, Brazed, or Soldered	EA	1.375	MPLUA	1.000	37.50	0.00	62.99	100.49
33 27 0440 2" PVC, Sch 80, Ball Valve	**LF**				**0.00**	**0.00**	**81.92**	**81.92**
01001 0527 2" PVC, Sch 80, Ball Valve	LF	0	N/A	1.000	0.00	0.00	81.92	81.92

33 27 Fittings/Valves

33 27 04 Valves		Unit	Hourly Output	Crew	Factor	Labor	Equip.	Matl.	Total
33 27 0441	**4" PVC, Sch 80, Ball Valve**	**EA**				**0.00**	**0.00**	**260.34**	**260.34**
01001 0528	4" PVC, Sch 80, Ball Valve	EA	0	N/A	1.000	0.00	0.00	260.34	260.34
33 27 0450	**2" Stainless Steel, Ball Valve**	**EA**				**0.00**	**0.00**	**163.62**	**163.62**
01001 0529	2" Stainless Steel, Ball Valve	EA	0	N/A	1.000	0.00	0.00	163.62	163.62
33 27 0451	**4" Stainless Steel, Ball Valve**	**EA**				**0.00**	**0.00**	**1,076**	**1,076**
01001 0530	4" Stainless Steel, Ball Valve	EA	0	N/A	1.000	0.00	0.00	1,076	1,076

33 27 05 Carbon Steel Fittings		Unit	Hourly Output	Crew	Factor	Labor	Equip.	Matl.	Total
33 27 0502	**2" Steel Tee**	**EA**				**28.47**	**0.00**	**13.76**	**42.23**
15174 5540	2" Tee, Straight 150 Lb Malleable Iron Black	EA	3.125	MPLUE	1.000	28.47	0.00	13.76	42.23
33 27 0504	**(2", 4") Tee, Carbon Steel**	**EA**				**79.70**	**0.00**	**119.43**	**199.12**
15174 5540	2" Tee, Straight 150 Lb Malleable Iron Black	EA	3.125	MPLUE	1.000	28.47	0.00	13.76	42.23
15174 5580	4" Tee, Straight 150 Lb Malleable Iron Black	EA	2.375	MPLUJ	1.000	51.22	0.00	105.67	156.90
33 27 0512	**2" Steel 90-degree Elbow**	**EA**				**19.24**	**0.00**	**9.55**	**28.79**
15174 5130	2" 90 Degree Elbow, 150 Lb Malleable Iron Black	EA	4.625	MPLUE	1.000	19.24	0.00	9.55	28.79
33 27 0514	**(2", 4") 90-degree Elbow, Carbon Steel**	**EA**				**59.79**	**0.00**	**76.38**	**136.17**
15174 5130	2" 90 Degree Elbow, 150 Lb Malleable Iron Black	EA	4.625	MPLUE	1.000	19.24	0.00	9.55	28.79
15174 5170	4" 90 Degree Elbow, 150 Lb Malleable Iron Black	EA	3	MPLUJ	1.000	40.55	0.00	66.82	107.37
33 27 0522	**2" Steel 45-degree Elbow**	**EA**				**19.24**	**0.00**	**10.37**	**29.61**
15174 5340	2" 45 Degree Elbow, 150 Lb Malleable Iron Black	EA	4.625	MPLUE	1.000	19.24	0.00	10.37	29.61
33 27 0524	**(2", 4") 45-degree Elbow, Carbon Steel**	**EA**				**59.79**	**0.00**	**86.91**	**146.70**
15174 5340	2" 45 Degree Elbow, 150 Lb Malleable Iron Black	EA	4.625	MPLUE	1.000	19.24	0.00	10.37	29.61
15174 5380	4" 45 Degree Elbow, 150 Lb Malleable Iron Black	EA	3	MPLUJ	1.000	40.55	0.00	76.54	117.09
33 27 0532	**2" Steel Coupling**	**EA**				**19.24**	**0.00**	**7.66**	**26.89**
15174 5740	2" Coupling, Straight 150 Lb Malleable Iron Black	EA	4.625	MPLUE	1.000	19.24	0.00	7.66	26.89
33 27 0534	**(2", 4") Coupling, Carbon Steel**	**EA**				**59.79**	**0.00**	**65.15**	**124.94**
15174 5740	2" Coupling, Straight 150 Lb Malleable Iron Black	EA	4.625	MPLUE	1.000	19.24	0.00	7.66	26.89
15174 5780	4" Coupling, Straight 150 Lb Malleable Iron Black	EA	3	MPLUJ	1.000	40.55	0.00	57.50	98.05
33 27 0544	**4" x 2" Eccentric Reducer, Steel**	**EA**				**133.59**	**14.15**	**25.05**	**172.79**
15178 4666	4" x 2" Reducer, Eccentric	EA	0.713	MSPFE	1.000	133.59	14.15	25.05	172.79
33 27 0552	**4" Carbon Steel 90-degree Elbow, Schedule 80**	**EA**				**75.72**	**8.02**	**29.10**	**112.84**
15178 3600	4" Carbon Steel 90 Degree Elbow, Schedule 80	EA	1.257	MSPFE	1.000	75.72	8.02	29.10	112.84
33 27 0554	**6" Carbon Steel 90-degree Elbow, Schedule 80**	**EA**				**104.46**	**11.06**	**69.99**	**185.51**
15178 3610	6" Carbon Steel 90 Degree Elbow, Schedule 80	EA	0.911	MSPFE	1.000	104.46	11.06	69.99	185.51

33 27 05 Carbon Steel Fittings	Unit	Hourly Output	Crew	Factor	Labor	Equip.	Matl.	Total
33 27 0562 4" Carbon Steel Tee, Schedule 80	**EA**				**113.59**	**12.03**	**61.88**	**187.50**
15178 3890 4" Carbon Steel Tee, Schedule 80	EA	0.838	MSPFE	1.000	113.59	12.03	61.88	187.50
33 27 0564 6" Carbon Steel Tee, Schedule 80	**EA**				**155.63**	**16.48**	**97.25**	**269.36**
15178 3900 6" Carbon Steel Tee, Schedule 80	EA	0.612	MSPFE	1.000	155.63	16.48	97.25	269.36

33 27 06 Fiberglass Reinforced Fittings	Unit	Hourly Output	Crew	Factor	Labor	Equip.	Matl.	Total
33 27 0602 2" Tee, Fiberglass Reinforced	**EA**				**45.92**	**0.00**	**39.50**	**85.42**
02681 0200 2" Tee, Fiberglass Reinforced	EA	1.938	MPLUE	1.000	45.92	0.00	39.50	85.42
33 27 0604 (2", 4") Tee, Fiberglass Reinforced	**EA**				**91.85**	**0.00**	**98.50**	**190.35**
02681 0200 2" Tee, Fiberglass Reinforced	EA	1.938	MPLUE	1.000	45.92	0.00	39.50	85.42
02681 1270 4" Tee, Fiberglass Reinforced	EA	1.938	MPLUE	1.000	45.92	0.00	59.00	104.92
33 27 0612 2" 90-degree Elbow, Fiberglass Reinforced	**EA**				**45.92**	**0.00**	**39.47**	**85.40**
02681 0111 2" 90 Degree Elbow, Fiberglass Reinforced	EA	1.938	MPLUE	1.000	45.92	0.00	39.47	85.40
33 27 0614 (2", 4") 90-degree Elbow, Fiberglass Reinforced	**EA**				**91.85**	**0.00**	**86.60**	**178.44**
02681 0111 2" 90 Degree Elbow, Fiberglass Reinforced	EA	1.938	MPLUE	1.000	45.92	0.00	39.47	85.40
02681 1231 4" 90 Degree Elbow, Fiberglass Reinforced	EA	1.938	MPLUE	1.000	45.92	0.00	47.12	93.05
33 27 0622 2" 45-degree Elbow, Fiberglass Reinforced	**EA**				**45.92**	**0.00**	**32.50**	**78.42**
02681 0110 2" 45 Degree Elbow, Fiberglass Reinforced	EA	1.938	MPLUE	1.000	45.92	0.00	32.50	78.42
33 27 0624 (2", 4") 45-degree Elbow, Fiberglass Reinforced	**EA**				**91.85**	**0.00**	**72.50**	**164.35**
02681 0110 2" 45 Degree Elbow, Fiberglass Reinforced	EA	1.938	MPLUE	1.000	45.92	0.00	32.50	78.42
02681 1230 4" 45 Degree Elbow, Fiberglass Reinforced	EA	1.938	MPLUE	1.000	45.92	0.00	40.00	85.92
33 27 0632 2" Coupling, Fiberglass Reinforced	**EA**				**45.92**	**0.00**	**8.35**	**54.27**
02681 0260 2" Coupling, Fiberglass Reinforced	EA	1.938	MPLUE	1.000	45.92	0.00	8.35	54.27
33 27 0634 (2", 4") Coupling, Fiberglass Reinforced	**EA**				**91.85**	**0.00**	**36.35**	**128.20**
02681 0260 2" Coupling, Fiberglass Reinforced	EA	1.938	MPLUE	1.000	45.92	0.00	8.35	54.27
02681 1310 4" Coupling, Fiberglass Reinforced	EA	1.938	MPLUE	1.000	45.92	0.00	28.00	73.92
33 27 0642 2" Termination Reducer, Fiberglass Reinforced	**EA**				**45.92**	**0.00**	**14.60**	**60.52**
02681 0320 2" Termination Concentric Reducer, Fiberglass Reinforced	EA	1.938	MPLUE	1.000	45.92	0.00	14.60	60.52
33 27 0644 (2", 4") Termination Reducer, Fiberglass Reinforced	**EA**				**91.85**	**0.00**	**36.10**	**127.95**
02681 0320 2" Termination Concentric Reducer, Fiberglass Reinforced	EA	1.938	MPLUE	1.000	45.92	0.00	14.60	60.52
02681 1400 4" Termination Concentric Reducer, Fiberglass Reinforced	EA	1.938	MPLUE	1.000	45.92	0.00	21.50	67.42

33 27 09	Fiberglass Reinforced Piping	Unit	Hourly Output	Crew	Factor	Labor	Equip.	Matl.	Total
33 27 0902	**2" Fiberglass Reinforced Piping**	**EA**				**4.59**	**0.00**	**3.23**	**7.82**
02681 0040	2" Fiberglass Reinforced Piping	LF	19.38	MPLUE	1.000	4.59	0.00	3.23	7.82
33 27 0904	**(2", 4") Fiberglass Reinforced Double-wall Piping**	**LF**				**9.18**	**0.00**	**8.78**	**17.96**
02681 0040	2" Fiberglass Reinforced Piping	LF	19.38	MPLUE	1.000	4.59	0.00	3.23	7.82
02681 1120	4" Fiberglass Reinforced Piping	LF	19.38	MPLUE	1.000	4.59	0.00	5.55	10.14

33 27 10	Galvanized Fittings	Unit	Hourly Output	Crew	Factor	Labor	Equip.	Matl.	Total
33 27 1002	**4" Galvanized Steel Tee, Schedule 80**	**EA**				**113.59**	**12.03**	**76.11**	**201.73**
15061 1279	4" Galvanized Steel Tee, Schedule 80	EA	0.838	MSPFE	1.000	113.59	12.03	76.11	201.73
33 27 1004	**6" Galvanized Steel Tee, Schedule 80**	**EA**				**155.63**	**16.48**	**147.36**	**319.48**
15061 1282	6" Galvanized Steel Tee, Schedule 80	EA	0.612	MSPFE	1.000	155.63	16.48	147.36	319.48
33 27 1012	**4" Galvanized Steel 90-degree Elbow, Schedule 80**	**EA**				**75.72**	**8.02**	**31.00**	**114.75**
15061 1239	4" Galvanized Steel 90 Degree Elbow, Schedule 80	EA	1.257	MSPFE	1.000	75.72	8.02	31.00	114.75
33 27 1013	**6" Galvanized Steel 90-degree Elbow, Schedule 80**	**EA**				**104.46**	**11.06**	**90.58**	**206.10**
15061 1242	6" Galvanized Steel 90 Degree Elbow, Schedule 80	EA	0.911	MSPFE	1.000	104.46	11.06	90.58	206.10

33 28 01	Pipe Racks	Unit	Hourly Output	Crew	Factor	Labor	Equip.	Matl.	Total
33 28 0101	**Concrete Pier Pipe Rack**	**EA**				**302.08**	**2.40**	**167.35**	**471.83**
03182 8730	Column, Pier & Pilaster Form, 3 Uses (Below Grade) Plywood Forms, Form & Strip with Accessories	SF	38.75	ACARL	37.819	4.15	0.00	1.16	200.96
03217 0700	Grade 60 Rebar, Beam, Column, Wall, #3-#6	TON	0.313	SIWRC	0.060	492.16	0.00	609.86	66.12
03372 2400	Pour Spread Footing,< 5 CY, Direct Chute < (5 m3) Place 3,000 PSI Concrete Foundations	CY	6.875	ALABE	0.559	24.49	0.93	84.22	61.24
03398 0010	Wall Finishes, Break Ties & Patch	SF	67.5	ACMAA	37.819	0.62	0.00	0.06	25.47
05130 0660	C2 x 1.78 - 2" Wide (Channels)	LF	14.25	MSPFZ	2.000	5.50	0.44	0.74	13.35
05501 1012	7/8" Diameter x 18" Length, J-type Anchor Bolt (46 cm Long) with Nut, Washer, Template	EA	4.375	ACARB	2.000	19.80	0.00	3.24	46.07
15190 2503	1/2" Washer, Flat	EA	34.38	MPLUA	4.000	1.50	0.00	0.08	6.32
15190 2690	1/2" Threaded Rod	LF	43.75	MPLUA	2.000	1.18	0.00	0.85	4.06
15190 3850	4" Pipe Saddle, Type 39 (1" Insulation)	EA	10	MSPFE	1.000	9.52	1.01	18.26	28.78
15190 4240	1/2" Eye Socket	EA	34.38	MPLUA	4.000	1.50	0.00	2.05	14.20
15190 4420	1/2" U-Bolt (Light Duty)Type 24	EA	12.5	MPLUA	1.000	4.13	0.00	1.14	5.26
33 28 0102	**Seismic Expansion Joint**	**EA**				**245.64**	**0.00**	**561.75**	**807.39**
15625 1160	4" Stainless Steel Expansion Joint, 150 Lb with Flanged Ends Bellows with Internal Sleeves & Exterior Covers	EA	0.388	MSPFB	1.000	245.64	0.00	561.75	807.39

33 28 01 Pipe Racks

		Unit	Hourly Output	Crew	Factor	Labor	Equip.	Matl.	Total
33 28 0103	**Heat Trace for Piping**	**LF**				**2.50**	**0.00**	**6.68**	**9.18**
16830 4030	Heat Tracing Tape - 120V	LF	37.5	AASBC	1.000	2.50	0.00	6.68	9.18

33 28 02 Hangers

		Unit	Hourly Output	Crew	Factor	Labor	Equip.	Matl.	Total
33 28 0201	**1" Hanger**	**EA**				**50.64**	**5.38**	**5.26**	**61.28**
15094 3012	1" Hanger	EA	1.875	MPLUH	1.000	50.64	5.38	5.26	61.28
33 28 0202	**2" Hanger**	**EA**				**50.64**	**5.38**	**6.71**	**62.73**
15094 3013	2" Hanger	EA	1.875	MPLUH	1.000	50.64	5.38	6.71	62.73
33 28 0203	**3" Hanger**	**EA**				**44.24**	**3.67**	**8.90**	**56.81**
15094 3014	3" Hanger	EA	2.75	MPLUK	1.000	44.24	3.67	8.90	56.81
33 28 0204	**4" Hanger**	**EA**				**55.93**	**4.64**	**12.22**	**72.79**
15094 3015	4" Hanger	EA	2.175	MPLUK	1.000	55.93	4.64	12.22	72.79
33 28 0205	**6" Hanger**	**EA**				**59.34**	**4.92**	**20.05**	**84.32**
15094 3016	6" Hanger	EA	2.05	MPLUK	1.000	59.34	4.92	20.05	84.32
33 28 0206	**8" Hanger**	**EA**				**68.06**	**5.64**	**32.35**	**106.05**
15094 3017	8" Hanger	EA	1.788	MPLUK	1.000	68.06	5.64	32.35	106.05
33 28 0207	**12" Hanger**	**EA**				**68.06**	**5.64**	**49.72**	**123.42**
15094 3019	12" Hanger	EA	1.788	MPLUK	1.000	68.06	5.64	49.72	123.42

33 29 01 Centrifugal Pumps

		Unit	Hourly Output	Crew	Factor	Labor	Equip.	Matl.	Total
33 29 0101	**10 GPM, 10' Head, 1/6 HP, Centrifugal Pump**	**EA**				**122.82**	**0.00**	**854.14**	**976.96**
15286 1140	10 GPM Bronze Centrifugal Pump, with 10' Head & 1/6 HP	EA	0.775	MSPFB	1.000	122.82	0.00	854.14	976.96
33 29 0102	**10 GPM, 1/2 HP, Centrifugal Pump**	**EA**				**230.75**	**0.00**	**709.62**	**940.37**
15141 5103	Centrifugal, Iron Circulating Pump 1/2 HP	EA	0.413	MSPFB	1.000	230.75	0.00	709.62	940.37
33 29 0103	**50 GPM, 100' Head, 3 HP, Centrifugal Pump**	**EA**				**331.08**	**0.00**	**570.96**	**902.04**
15288 5020	50 GPM Iron Centrifugal Pump, 1-1/2" Discharge Single Stage with 100' Head & 3 HP	EA	0.288	MSPFB	1.000	331.08	0.00	570.96	902.04
33 29 0104	**75 GPM, 100' Head, 5 HP, Centrifugal Pump**	**EA**				**400.78**	**0.00**	**964.65**	**1,365**
15288 5030	75 GPM Iron Centrifugal Pump, 1-1/2" Discharge Single Stage with 100' Head & 5 HP	EA	0.238	MSPFB	1.000	400.78	0.00	964.65	1,365
33 29 0105	**5 HP, 100 GPM, Centrifugal Pump**	**EA**				**362.61**	**0.00**	**2,468**	**2,831**
15288 4040	100 GPM Cast Iron Centrifugal Pump, 1-1/2" Discharge Single Stage, Vertical Split Case, 5 HP	EA	0.263	MSPFB	1.000	362.61	0.00	2,468	2,831
33 29 0106	**100 GPM, 150' Head, 7.5 HP, Centrifugal Pump**	**EA**				**475.93**	**0.00**	**1,039**	**1,515**
15288 5040	100 GPM Iron Centrifugal Pump, 1-1/2" Discharge Single Stage with 150' Head 7-1/2 HP	EA	0.2	MSPFB	1.000	475.93	0.00	1,039	1,515

33 29 Pumps Other than Submersible/Groundwater

33 29 01	Centrifugal Pumps	Unit	Hourly Output	Crew	Factor	Labor	Equip.	Matl.	Total
33 29 0107	**125 GPM, 150' Head, 10 HP, Centrifugal Pump**	**EA**				**634.57**	**0.00**	**1,290**	**1,925**
15288 5050	125 GPM Iron Centrifugal Pump, 1-1/2" Discharge Single Stage with 150' Head 10 HP	EA	0.15	MSPFB	1.000	634.57	0.00	1,290	1,925
33 29 0108	**10 HP, 200 GPM, Centrifugal Pump**	**EA**				**475.93**	**0.00**	**2,633**	**3,109**
15288 4050	200 GPM Cast Iron Centrifugal Pump, with 2" Discharge Single Stage, Vertical Split Case, 10 HP	EA	0.2	MSPFB	1.000	475.93	0.00	2,633	3,109
33 29 0109	**10 HP, 250 GPM, Centrifugal Pump**	**EA**				**475.93**	**0.00**	**2,745**	**3,221**
15288 4060	250 GPM Cast Iron Centrifugal Pump, with 3" Discharge Single Stage, Vertical Split Case, 10 HP	EA	0.2	MSPFB	1.000	475.93	0.00	2,745	3,221
33 29 0110	**15 HP, 300 GPM, Centrifugal Pump**	**EA**				**585.75**	**0.00**	**3,015**	**3,601**
15288 4070	300 GPM Cast Iron Centrifugal Pump, with 2" Discharge Single Stage, Vertical Split Case, 15 HP	EA	0.163	MSPFB	1.000	585.75	0.00	3,015	3,601
33 29 0111	**20 HP, 500 GPM, Centrifugal Pump**	**EA**				**634.57**	**0.00**	**4,017**	**4,651**
15288 4080	500 GPM Cast Iron Centrifugal Pump, with 4" Discharge Single Stage, Vertical Split Case, 20 HP	EA	0.15	MSPFB	1.000	634.57	0.00	4,017	4,651
33 29 0112	**30 HP, 750 GPM, Centrifugal Pump**	**EA**				**846.09**	**0.00**	**4,710**	**5,556**
15288 4090	750 GPM Cast Iron Centrifugal Pump, with 4" Discharge Single Stage, Vertical Split Case, 30 HP	EA	0.113	MSPFB	1.000	846.09	0.00	4,710	5,556
33 29 0113	**40 HP, 1050 GPM, Centrifugal Pump**	**EA**				**951.85**	**0.00**	**5,081**	**6,032**
15288 4100	1,050 GPM Cast Iron Centrifugal Pump, with 5" Discharge Single Stage, Vertical Split Case, 40 HP	EA	0.1	MSPFB	1.000	951.85	0.00	5,081	6,032
33 29 0114	**60 HP, 1500 GPM, Centrifugal Pump**	**EA**				**2,004**	**191.67**	**7,406**	**9,602**
15288 4110	1,500 GPM Cast Iron Centrifugal Pump, with 6" Discharge Single Stage, Vertical Split Case, 60 HP	EA	0.075	MSPFN	1.000	2,004	191.67	7,406	9,602
33 29 0115	**75 HP, 2000 GPM, Centrifugal Pump**	**EA**				**2,405**	**230.00**	**10,221**	**12,856**
15288 4120	2,000 GPM Cast Iron Centrifugal Pump, with 6" Discharge Single Stage, Vertical Split Case, 75 HP	EA	0.063	MSPFN	1.000	2,405	230.00	10,221	12,856
33 29 0116	**100 HP, 3000 GPM, Centrifugal Pump**	**EA**				**3,006**	**287.50**	**12,783**	**16,076**
15288 4130	3,000 GPM Cast Iron Centrifugal Pump, with 8" Discharge Single Stage, Vertical Split Case, 100 HP	EA	0.05	MSPFN	1.000	3,006	287.50	12,783	16,076

33 29 Pumps Other than Submersible/Groundwater

33 29 01	Centrifugal Pumps	Unit	Hourly Output	Crew	Factor	Labor	Equip.	Matl.	Total
33 29 0117	**15 GPM, 1/2 HP, Transfer Pump with Motor, Valves, Piping**	**EA**				**395.03**	**0.00**	**925.20**	**1,320**
15141 5103	Centrifugal, Iron Circulating Pump 1/2 HP	EA	0.413	MSPFB	1.000	230.75	0.00	709.62	940.37
15162 3012	3/4" Stainless Steel Pipe, Schedule 40 Type 304 with 150 Lb Threaded Fitting	LF	23.25	MSPFC	25.000	4.45	0.00	6.65	277.49
15164 7090	3/4" Stainless Steel 90 Degree Elbow, Schedule 40 Type 304	EA	5.875	MSPFC	2.000	17.60	0.00	8.22	51.63
15193 1860	3/4" Swing Check Valve Bronze 125 Lb for Threaded, Brazed, or Soldered Installation	EA	2.875	MPLUA	1.000	17.93	0.00	32.81	50.74
33 29 0118	**20 GPM, 1/2 HP, Transfer Pump with Motor, Valves, Piping**	**EA**				**452.01**	**0.00**	**1,004**	**1,456**
15141 5103	Centrifugal, Iron Circulating Pump 1/2 HP	EA	0.413	MSPFB	1.000	230.75	0.00	709.62	940.37
15162 3014	1" Stainless Steel Pipe, Schedule 40 Type 304 with 150 Lb Threaded Fitting	LF	19.38	MSPFC	25.000	5.34	0.00	8.67	350.09
15164 7380	1" Stainless Steel Tee, Schedule 40 Type 304	EA	3.125	MSPFC	2.000	33.08	0.00	14.02	94.20
15193 1870	1" Swing Check Valve Bronze 125 Lb for Threaded, Brazed, or Soldered Installation	EA	2.375	MPLUA	1.000	21.71	0.00	49.50	71.21
33 29 0119	**25 GPM, 1 HP, Transfer Pump with Motor, Valves, Piping**	**EA**				**526.24**	**0.00**	**1,816**	**2,342**
15066 2114	1-1/4" Stainless Steel Pipe, Schedule 40 Type 304 with 150 Lb Threaded Fitting	LF	16.63	MSPFC	25.000	6.22	0.00	11.47	442.13
15066 2124	1-1/4" Stainless Steel 90 Degree Elbow, Schedule 40 Type 304	EA	3.375	MSPFC	2.000	30.63	0.00	16.07	93.40
15111 1107	1-1/4" Swing Check Valve Bronze 125 Lb for Threaded, Brazed, or Soldered Installation	EA	1.875	MPLUA	1.000	27.50	0.00	61.06	88.56
15286 2380	Centrifugal, Iron Circulating Pump 1 HP	EA	0.338	MSPFB	1.000	282.03	0.00	1,436	1,718
33 29 0120	**35 GPM, 1 HP, Transfer Pump with Motor, Valves, Piping**	**EA**				**597.18**	**0.00**	**2,259**	**2,856**
15162 3016	1-1/2" Stainless Steel Pipe, Schedule 40 Type 304 with 150 Lb Threaded Fitting	LF	14.63	MSPFC	25.000	7.07	0.00	12.94	500.17
15164 7120	1-1/2" Stainless Steel 90 Degree Elbow, Schedule 40 Type 304	EA	2.875	MSPFC	2.000	35.96	0.00	18.61	109.14
15194 5998	1-1/2" Iron Body Swing Check Valve, Bronze, Flanged 125 Lb	EA	0.775	MPLUA	1.000	66.53	0.00	462.48	529.01
15286 2380	Centrifugal, Iron Circulating Pump 1 HP	EA	0.338	MSPFB	1.000	282.03	0.00	1,436	1,718
33 29 0121	**50 GPM, 1.5 HP, Transfer Pump with Motor, Valves, Piping**	**EA**				**867.19**	**0.00**	**2,704**	**3,571**
13273 4192	Transfer Pump, 50 GPM, 1.5 HP	EA	0.375	MSPFL	1.000	382.29	0.00	1,900	2,282
15162 3018	2" Stainless Steel Pipe, Schedule 40 Type 304 with 150 Lb Threaded Fitting	LF	11.63	MSPFC	25.000	8.89	0.00	17.15	651.11
15164 7130	2" Stainless Steel 90 Degree Elbow, Schedule 40 Type 304	EA	2.375	MSPFC	2.000	43.53	0.00	26.91	140.88
15194 6000	2" Iron Body Swing Check Valve, Bronze, Flanged 125 Lb	EA	0.588	MPLUA	2.000	87.77	0.00	160.79	497.11

33 29 01	Centrifugal Pumps	Unit	Hourly Output	Crew	Factor	Labor	Equip.	Matl.	Total
33 29 0122	**75 GPM, 3 HP, Transfer Pump with Motor, Valves, Piping**	**EA**				**1,166**	**0.00**	**3,793**	**4,959**
13273 4193	Transfer Pump, 90 GPM, 3 HP	EA	0.25	MSPFL	1.000	573.44	0.00	2,625	3,198
15066 2117	2-1/2" Stainless Steel Pipe, Schedule 40 Type 304 with 150 Lb Threaded Fitting	LF	9.75	MSPFC	25.000	10.60	0.00	26.89	937.34
15066 2127	2-1/2" Stainless Steel 90 Degree Elbow, Schedule 40 Type 304	EA	1.5	MSPFC	2.000	68.92	0.00	70.56	278.96
15111 1205	2-1/2" Iron Body Swing Check Valve, Bronze, Flanged 125 Lb	EA	0.938	MPLUE	2.000	94.91	0.00	177.07	543.96
33 29 0123	**100 GPM, 5 HP, Transfer Pump with Motor, Valves, Piping**	**EA**				**955.32**	**0.00**	**3,636**	**4,591**
15066 2117	2-1/2" Stainless Steel Pipe, Schedule 40 Type 304 with 150 Lb Threaded Fitting	LF	9.75	MSPFC	25.000	10.60	0.00	26.89	937.34
15066 2127	2-1/2" Stainless Steel 90 Degree Elbow, Schedule 40 Type 304	EA	1.5	MSPFC	2.000	68.92	0.00	70.56	278.96
15111 1205	2-1/2" Iron Body Swing Check Valve, Bronze, Flanged 125 Lb	EA	0.938	MPLUE	2.000	94.91	0.00	177.07	543.96
15288 4040	100 GPM Cast Iron Centrifugal Pump, 1-1/2" Discharge Single Stage, Vertical Split Case, 5 HP	EA	0.263	MSPFB	1.000	362.61	0.00	2,468	2,831
33 29 0124	**150 GPM, 5 HP, Transfer Pump with Motor, Valves, Piping**	**EA**				**1,888**	**0.00**	**4,346**	**6,234**
13273 4194	Transfer Pump, 150 GPM, 5 HP	EA	0.125	MSPFL	1.000	1,147	0.00	2,850	3,997
15066 2118	3" Stainless Steel Pipe, Schedule 40 Type 304 with 150 Lb Threaded Fitting	LF	7.75	MSPFC	25.000	13.34	0.00	35.59	1,223
15066 2128	3" Stainless Steel 90 Degree Elbow, Schedule 40 Type 304	EA	1.163	MSPFC	2.000	88.92	0.00	112.57	402.99
15111 1206	3" Iron Body Swing Check Valve, Bronze, Flanged 125 Lb	EA	0.775	MPLUE	2.000	114.81	0.00	190.51	610.63
33 29 0125	**200 GPM, 10 HP, Transfer Pump with Motor, Valves, Piping**	**EA**				**1,217**	**0.00**	**4,129**	**5,346**
15066 2118	3" Stainless Steel Pipe, Schedule 40 Type 304 with 150 Lb Threaded Fitting	LF	7.75	MSPFC	25.000	13.34	0.00	35.59	1,223
15066 2128	3" Stainless Steel 90 Degree Elbow, Schedule 40 Type 304	EA	1.163	MSPFC	2.000	88.92	0.00	112.57	402.99
15111 1206	3" Iron Body Swing Check Valve, Bronze, Flanged 125 Lb	EA	0.775	MPLUE	2.000	114.81	0.00	190.51	610.63
15288 4050	200 GPM Cast Iron Centrifugal Pump, with 2" Discharge Single Stage, Vertical Split Case, 10 HP	EA	0.2	MSPFB	1.000	475.93	0.00	2,633	3,109
33 29 0126	**350 GPM, 10 HP, Transfer Pump with Motor, Valves, Piping**	**EA**				**2,110**	**0.00**	**4,278**	**6,388**
13273 4195	Transfer Pump, 350 GPM, 10 HP	EA	0.125	MSPFL	1.000	1,147	0.00	2,000	3,147
15162 3020	4" Stainless Steel Pipe, Schedule 40 Type 304 with 150 Lb Threaded Fitting	LF	5.875	MSPFC	25.000	17.60	0.00	49.96	1,689
15164 7160	4" Stainless Steel 90 Degree Elbow, Schedule 40 Type 304	EA	0.938	MSPFC	2.000	110.27	0.00	185.33	591.20
15194 6060	4" Iron Body Swing Check Valve, Bronze, Flanged & 125 Lb	EA	0.588	MPLUE	2.000	151.45	0.00	329.04	960.97

33 29 01	Centrifugal Pumps	Unit	Hourly Output	Crew	Factor	Labor	Equip.	Matl.	Total
33 29 0127	**700 GPM, 30 HP, Transfer Pump with Motor, Valves, Piping**	**EA**				**2,392**	**126.88**	**7,066**	**9,584**
15162 2160	6" Stainless Steel Pipe, Schedule 10 Type 304 with Butt Weld Fitting	LF	5.55	MSPFE	25.000	17.15	1.82	35.45	1,361
15164 2150	6" Stainless Steel 90 Degree Elbow, Schedule 10 Type 304	EA	0.248	MSPFE	2.000	384.59	40.73	154.71	1,160
15194 6070	6" Iron Body Swing Check Valve, Bronze, Flanged 125 Lb	EA	0.7	MPLUJ	2.000	173.79	0.00	580.17	1,508
15288 4090	750 GPM Cast Iron Centrifugal Pump, with 4" Discharge Single Stage, Vertical Split Case, 30 HP	EA	0.113	MSPFB	1.000	846.09	0.00	4,710	5,556
33 29 0128	**100 GPM, Centrifugal Pump, 100' Head, 7.5 HP, Cast Iron**	**EA**				**362.61**	**0.00**	**3,412**	**3,775**
15288 4210	100 GPM Cast Iron Centrifugal Pump, 1-1/2" Discharge Single Stage, Horizontal Split Case, 7.5 HP	EA	0.263	MSPFB	1.000	362.61	0.00	3,412	3,775
33 29 0129	**250 GPM, Centrifugal Pump, 100' Head, 15 HP, Cast Iron**	**EA**				**432.66**	**0.00**	**4,477**	**4,910**
15288 4220	250 GPM Cast Iron Centrifugal Pump, 2-1/2" Discharge Single Stage, Horizontal Split Case, 15 HP	EA	0.22	MSPFB	1.000	432.66	0.00	4,477	4,910
33 29 0130	**500 GPM, Centrifugal Pump, 100' Head, 20 HP, Cast Iron**	**EA**				**475.93**	**0.00**	**4,899**	**5,375**
15288 4230	500 GPM Cast Iron Centrifugal Pump, with 4" Discharge Single Stage, Horizontal Split Case, 20 HP	EA	0.2	MSPFB	1.000	475.93	0.00	4,899	5,375
33 29 0131	**750 GPM, Centrifugal Pump, 100' Head, 25 HP, Cast Iron**	**EA**				**528.81**	**0.00**	**5,210**	**5,739**
15288 4240	750 GPM Cast Iron Centrifugal Pump, with 5" Discharge Single Stage, Horizontal Split Case, 25 HP	EA	0.18	MSPFB	1.000	528.81	0.00	5,210	5,739
33 29 0132	**1,000 GPM, Centrifugal Pump, 100' Head, 40 HP, Cast Iron**	**EA**				**576.88**	**0.00**	**6,314**	**6,891**
15288 4250	1,000 GPM Cast Iron Centrifugal Pump, 5" Discharge Single Stage, Horizontal Split Case ,40 HP	EA	0.165	MSPFB	1.000	576.88	0.00	6,314	6,891
33 29 0133	**1,500 GPM, Centrifugal Pump, 100' Head, 50 HP, Cast Iron**	**EA**				**626.22**	**0.00**	**11,583**	**12,209**
15288 4260	1,500 GPM Cast Iron Centrifugal Pump, 6" Discharge Single Stage, Horizontal Split Case, 50 HP	EA	0.152	MSPFB	1.000	626.22	0.00	11,583	12,209
33 29 0134	**2,000 GPM, Centrifugal Pump, 100' Head, 75 HP, Cast Iron**	**EA**				**684.78**	**0.00**	**13,678**	**14,363**
15288 4270	2,000 GPM Cast Iron Centrifugal Pump, 8" Discharge Single Stage, Horizontal Split Case, 75 HP	EA	0.139	MSPFB	1.000	684.78	0.00	13,678	14,363

33 29 01 Centrifugal Pumps		Unit	Hourly Output	Crew	Factor	Labor	Equip.	Matl.	Total
33 29 0135	**3,000 GPM, Centrifugal Pump, 100' Head, 100 HP, Cast Iron**	**EA**				**761.48**	**0.00**	**14,041**	**14,802**
15288 4280	3,000 GPM Cast Iron Centrifugal Pump, 10" Discharge Single Stage, Horizontal Split Case, 100 HP	EA	0.125	MSPFB	1.000	761.48	0.00	14,041	14,802
33 29 0136	**3,500 GPM, Centrifugal Pump, 125' Head, 150 HP, Cast Iron**	**EA**				**846.09**	**0.00**	**15,737**	**16,583**
15288 4290	3,500 GPM Cast Iron Centrifugal Pump, 125' Head 10" Discharge, Single Stage, Horizontal Split Case, 150 HP	EA	0.113	MSPFB	1.000	846.09	0.00	15,737	16,583
33 29 0137	**4,000 GPM, Centrifugal Pump, 150' Head, 200 HP, Cast Iron**	**EA**				**1,088**	**0.00**	**20,362**	**21,450**
15288 4300	4,000 GPM Cast Iron Centrifugal Pump, 150' Head 10" Discharge, Single Stage, Horizontal Split Case, 200 HP	EA	0.088	MSPFB	1.000	1,088	0.00	20,362	21,450
33 29 0138	**100 GPM, Centrifugal Pump, 500' Head, 40 HP, Cast Iron**	**EA**				**583.96**	**0.00**	**4,763**	**5,347**
15288 4340	100 GPM Cast Iron Centrifugal Pump, 1-1/2" Discharge, Two Stage, Horizontal Split Case, 40 HP	EA	0.163	MSPFB	1.000	583.96	0.00	4,763	5,347
33 29 0139	**200 GPM, Centrifugal Pump, 500' Head, 50 HP, Cast Iron**	**EA**				**634.57**	**0.00**	**5,307**	**5,942**
15288 4350	200 GPM Cast Iron Centrifugal Pump,1-1/2" Discharge, Two Stage, Horizontal Split Case, 50 HP	EA	0.15	MSPFB	1.000	634.57	0.00	5,307	5,942
33 29 0140	**300 GPM, Centrifugal Pump, 500' Head, 75 HP, Cast Iron**	**EA**				**689.75**	**0.00**	**5,640**	**6,330**
15288 4360	300 GPM Cast Iron Centrifugal Pump, 2" Discharge, Two Stage, Horizontal Split Case, 75 HP	EA	0.138	MSPFB	1.000	689.75	0.00	5,640	6,330
33 29 0141	**400 GPM, Centrifugal Pump, 500' Head, 100 HP, Cast Iron**	**EA**				**951.85**	**0.00**	**7,352**	**8,304**
15288 4370	400 GPM Cast Iron Centrifugal Pump, 3" Discharge, Two Stage, Horizontal Split Case, 100 HP	EA	0.1	MSPFB	1.000	951.85	0.00	7,352	8,304
33 29 0142	**800 GPM, Centrifugal Pump, 500' Head, 200 HP, Cast Iron**	**EA**				**1,269**	**0.00**	**17,233**	**18,503**
15288 4380	800 GPM Cast Iron Centrifugal Pump, 4" Discharge, Two Stage, Horizontal Split Case, 200 HP	EA	0.075	MSPFB	1.000	1,269	0.00	17,233	18,503
33 29 0143	**15 GPM, Centrifugal Pump, 6' Head, 1/8 HP, Bronze**	**EA**				**101.53**	**0.00**	**596.29**	**697.82**
15286 1060	15 GPM Bronze Centrifugal Pump, 3/4" Discharge Booster Pump with 6' Head and 1/8 HP	EA	0.938	MSPFB	1.000	101.53	0.00	596.29	697.82

33 29 01	Centrifugal Pumps	Unit	Hourly Output	Crew	Factor	Labor	Equip.	Matl.	Total
33 29 0144	**20 GPM, Centrifugal Pump, 5.5' Head, 1/8 HP, Bronze**	**EA**				**114.68**	**0.00**	**611.19**	**725.87**
15286 1061	20 GPM Bronze Centrifugal Pump, 1" Discharge Booster Pump with 5-1/2' Head and 1/8 HP	EA	0.83	MSPFB	1.000	114.68	0.00	611.19	725.87
33 29 0145	**25 GPM, Centrifugal Pump, 4.5' Head, 1/8 HP, Bronze**	**EA**				**122.82**	**0.00**	**623.85**	**746.67**
15286 1062	25 GPM Bronze Centrifugal Pump, 1-1/4" Discharge Booster Pump with 4-1/2' Head and 1/8 HP	EA	0.775	MSPFB	1.000	122.82	0.00	623.85	746.67
33 29 0146	**30 GPM, Centrifugal Pump, 2' Head, 1/8 HP, Bronze**	**EA**				**146.21**	**0.00**	**623.85**	**770.06**
15286 1064	30 GPM Bronze Centrifugal Pump, 1-1/2" Discharge Booster Pump with 2' Head and 1/8 HP	EA	0.651	MSPFB	1.000	146.21	0.00	623.85	770.06
33 29 0147	**30 GPM, Centrifugal Pump, High Velocity, 7' Head, 1/8 HP**	**EA**				**148.96**	**0.00**	**623.85**	**772.81**
15286 1063	30 GPM Bronze Centrifugal Pump with 1" Discharge Booster Pump with 7' Head and 1/8 HP	EA	0.639	MSPFB	1.000	148.96	0.00	623.85	772.81
33 29 0148	**35 GPM, Centrifugal Pump, High Velocity, 4' Head, 1/8 HP**	**EA**				**159.97**	**0.00**	**623.85**	**783.82**
15286 1066	35 GPM Bronze Centrifugal Pump 1-1/4" Discharge Booster Pump with 4' Head and 1/8 HP	EA	0.595	MSPFB	1.000	159.97	0.00	623.85	783.82
33 29 0149	**35 GPM, Centrifugal Pump, High Velocity, 5' Head, 1/8 HP**	**EA**				**163.83**	**0.00**	**623.85**	**787.68**
15286 1065	35 GPM Bronze Centrifugal Pump 1-1/2" Discharge Booster Pump with 5' Head and 1/8 HP	EA	0.581	MSPFB	1.000	163.83	0.00	623.85	787.68
33 29 0150	**50 GPM, Turbine Pump, 100' Head, 2 HP, Iron/Bronze**	**EA**				**842.35**	**0.00**	**1,209**	**2,051**
15290 1010	50 GPM Cast Iron Turbine Pump, 3" Discharge 100' Head, 10 Stage, 2 HP	EA	0.113	MSPFB	1.000	842.35	0.00	1,209	2,051
33 29 0151	**100 GPM, Turbine Pump, 100' Head, 3 HP, Iron/Bronze**	**EA**				**951.85**	**0.00**	**1,813**	**2,765**
15290 1020	100 GPM Cast Iron Turbine Pump, 4" Discharge 100' Head, 7 Stage, 3 HP	EA	0.1	MSPFB	1.000	951.85	0.00	1,813	2,765
33 29 0152	**250 GPM, Turbine Pump, 150' Head, 15 HP, Iron/Bronze**	**EA**				**1,082**	**0.00**	**3,626**	**4,708**
15290 1030	250 GPM Cast Iron Turbine Pump, 6" Discharge 150' Head, 5 Stage, 15 HP	EA	0.088	MSPFB	1.000	1,082	0.00	3,626	4,708
33 29 0153	**500 GPM, Turbine Pump, 150' Head, 25 HP, Iron/Bronze**	**EA**				**1,205**	**0.00**	**6,043**	**7,248**
15290 1040	500 GPM Cast Iron Turbine Pump, 6" Discharge 150' Head, 3 Stage, 25 HP	EA	0.079	MSPFB	1.000	1,205	0.00	6,043	7,248

33 29 01	Centrifugal Pumps	Unit	Hourly Output	Crew	Factor	Labor	Equip.	Matl.	Total
33 29 0154	**1,000 GPM, Turbine Pump, 150' Head, 50 HP, Iron/Bronze**	**EA**				**1,341**	**0.00**	**9,669**	**11,010**
15290 1050	1,000 GPM Cast Iron Turbine Pump, 8" Discharge 150' Head, 2 Stage, 50 HP	EA	0.071	MSPFB	1.000	1,341	0.00	9,669	11,010
33 29 0155	**2,000 GPM, Turbine Pump, 150' Head, 100 HP, Iron/Bronze**	**EA**				**1,421**	**0.00**	**10,878**	**12,299**
15290 1060	2,000 GPM Cast Iron Turbine Pump, 10" Discharge 150' Head, 3 Stage, 100 HP	EA	0.067	MSPFB	1.000	1,421	0.00	10,878	12,299
33 29 0156	**3,000 GPM, Turbine Pump, 150' Head, 150 HP, Iron/Bronze**	**EA**				**1,511**	**0.00**	**12,087**	**13,598**
15290 1070	3,000 GPM Cast Iron Turbine Pump, 10" Discharge 150' Head, 2 Stage, 150 HP	EA	0.063	MSPFB	1.000	1,511	0.00	12,087	13,598
33 29 0157	**4,000 GPM, Turbine Pump, 150' Head, 200 HP, Iron/Bronze**	**EA**				**1,740**	**0.00**	**18,130**	**19,870**
15290 1080	4,000 GPM Cast Iron Turbine Pump, 12" Discharge 150' Head, 2 Stage, 200 HP	EA	0.055	MSPFB	1.000	1,740	0.00	18,130	19,870
33 29 0158	**6,000 GPM, Turbine Pump, 150' Head, 300 HP, Iron/Bronze**	**EA**				**2,025**	**0.00**	**28,778**	**30,803**
15290 1090	6,000 GPM Cast Iron Turbine Pump, 14" Discharge 150' Head, 2 Stage, 300 HP	EA	0.047	MSPFB	1.000	2,025	0.00	28,778	30,803
33 29 0159	**10,000 GPM, Turbine Pump, 100' Head, 300 HP, Iron/Bronze**	**EA**				**2,615**	**0.00**	**40,290**	**42,905**
15290 1100	10,000 GPM Cast Iron Turbine Pump, 18" Discharge 100' Head, 2 Stage, 300 HP	EA	0.036	MSPFB	1.000	2,615	0.00	40,290	42,905
33 29 0170	**1,000 GPM, Centrifugal Pump, 100' Head, 40 HP, Cast Iron**	**EA**				**696.49**	**0.00**	**6,314**	**7,011**
15288 4251	1,000 GPM Cast Iron Centrifugal Pump, 5" Discharge Single Stage, Horizontal Split Case ,40 HP	EA	0.175	MSPFI	1.000	696.49	0.00	6,314	7,011
33 29 0171	**1,500 GPM, Centrifugal Pump, 100' Head, 50 HP, Cast Iron**	**EA**				**801.55**	**76.67**	**11,583**	**12,461**
15288 4261	1,500 GPM Cast Iron Centrifugal Pump, 6" Discharge Single Stage, Horizontal Split Case, 50 HP	EA	0.188	MSPFN	1.000	801.55	76.67	11,583	12,461
33 29 0172	**2,000 GPM, Centrifugal Pump, 100' Head, 75 HP, Cast Iron**	**EA**				**1,002**	**95.83**	**13,678**	**14,776**
15288 4271	2,000 GPM Cast Iron Centrifugal Pump, 8" Discharge Single Stage, Horizontal Split Case, 75 HP	EA	0.15	MSPFN	1.000	1,002	95.83	13,678	14,776
33 29 0173	**3,000 GPM, Centrifugal Pump, 100' Head, 100 HP, Cast Iron**	**EA**				**1,202**	**115.00**	**14,041**	**15,358**
15288 4281	3,000 GPM Cast Iron Centrifugal Pump, 10" Discharge Single Stage, Horizontal Split Case, 100 HP	EA	0.125	MSPFN	1.000	1,202	115.00	14,041	15,358

33 29 Pumps Other than Submersible/Groundwater

33 29 01	Centrifugal Pumps	Unit	Hourly Output	Crew	Factor	Labor	Equip.	Matl.	Total
33 29 0174	**3,500 GPM, Centrifugal Pump, 125' Head, 150 HP, Cast Iron**	**EA**				**1,336**	**127.78**	**15,737**	**17,201**
15288 4291	3,500 GPM Cast Iron Centrifugal Pump, 125' Head 10" Discharge, Single Stage, Horizontal Split Case, 150 HP	EA	0.113	MSPFN	1.000	1,336	127.78	15,737	17,201
33 29 0175	**4,000 GPM, Centrifugal Pump, 150' Head, 200 HP, Cast Iron**	**EA**				**1,718**	**164.29**	**20,362**	**22,244**
15288 4301	4,000 GPM Cast Iron Centrifugal Pump, 150' Head 10" Discharge, Single Stage, Horizontal Split Case, 200 HP	EA	0.088	MSPFN	1.000	1,718	164.29	20,362	22,244

33 29 02	Pneumatic Ejectors	Unit	Hourly Output	Crew	Factor	Labor	Equip.	Matl.	Total
33 29 0201	**30 GPM Pneumatic Ejector (Includes Air Compressor and Controls)**	**EA**				**510.90**	**143.75**	**15,916**	**16,570**
15295 0050	30 GPM Pneumatic Ejector	EA	0.25	MPLUS	1.000	510.90	143.75	15,916	16,570
33 29 0202	**50 GPM Pneumatic Ejector (Includes Air Compressor and Controls)**	**EA**				**567.67**	**159.72**	**23,617**	**24,344**
15295 0060	50 GPM Pneumatic Ejector	EA	0.225	MPLUS	1.000	567.67	159.72	23,617	24,344
33 29 0203	**100 GPM Pneumatic Ejector (Includes Air Compressor and Controls)**	**EA**				**681.20**	**191.67**	**28,557**	**29,430**
15295 0070	100 GPM Pneumatic Ejector	EA	0.188	MPLUS	1.000	681.20	191.67	28,557	29,430
33 29 0204	**150 GPM Pneumatic Ejector (Includes Air Compressor and Controls)**	**EA**				**786.00**	**221.15**	**55,070**	**56,077**
15295 0080	150 GPM Pneumatic Ejector	EA	0.163	MPLUS	1.000	786.00	221.15	55,070	56,077
33 29 0205	**200 GPM Pneumatic Ejector (Includes Air Compressor and Controls)**	**EA**				**1,022**	**287.50**	**70,090**	**71,399**
15295 0090	200 GPM Pneumatic Ejector	EA	0.125	MPLUS	1.000	1,022	287.50	70,090	71,399
33 29 0206	**250 GPM Pneumatic Ejector (Includes Air Compressor and Controls)**	**EA**				**1,277**	**359.38**	**90,116**	**91,753**
15295 0100	250 GPM Pneumatic Ejector	EA	0.1	MPLUS	1.000	1,277	359.38	90,116	91,753
33 29 0207	**300 GPM Pneumatic Ejector (Includes Air Compressor and Controls)**	**EA**				**2,044**	**575.00**	**110,141**	**112,760**
15295 0110	300 GPM Pneumatic Ejector	EA	0.063	MPLUS	1.000	2,044	575.00	110,141	112,760

33 29 03	Iron Rotary Pumps	Unit	Hourly Output	Crew	Factor	Labor	Equip.	Matl.	Total
33 29 0301	**10 GPM, 20 PSI, 1/2 HP, Iron Rotary Pump**	**EA**				**331.08**	**0.00**	**1,046**	**1,377**
15290 0610	10 GPM Cast Iron Rotary Pump, 20 PSI 1/2 HP with 1" Discharge	EA	0.288	MSPFB	1.000	331.08	0.00	1,046	1,377
33 29 0302	**10 GPM, 40 PSI, 3/4 HP, Iron Rotary Pump**	**EA**				**400.78**	**0.00**	**1,046**	**1,447**
15290 0614	10 GPM Cast Iron Rotary Pump, 40 PSI 3/4 HP with 1" Discharge	EA	0.238	MSPFB	1.000	400.78	0.00	1,046	1,447

33 29 Pumps Other than Submersible/Groundwater

33 29 03 Iron Rotary Pumps		Unit	Hourly Output	Crew	Factor	Labor	Equip.	Matl.	Total
33 29 0303	**10 GPM, 60 PSI, 1 HP, Iron Rotary Pump**	**EA**				**475.93**	**0.00**	**1,087**	**1,563**
15290 0618	10 GPM Cast Iron Rotary Pump, 60 PSI 1 HP with 1" Discharge	EA	0.2	MSPFB	1.000	475.93	0.00	1,087	1,563
33 29 0304	**25 GPM, 20 PSI, 1 HP, Iron Rotary Pump**	**EA**				**331.08**	**0.00**	**1,087**	**1,418**
15290 0620	25 GPM Cast Iron Rotary Pump, 20 PSI 1 HP with 1-1/4" Discharge	EA	0.288	MSPFB	1.000	331.08	0.00	1,087	1,418
33 29 0305	**25 GPM, 40 PSI, 1 1/2 HP, Iron Rotary Pump**	**EA**				**400.78**	**0.00**	**1,184**	**1,585**
15290 0624	25 GPM Cast Iron Rotary Pump, 40 PSI 1.5 HP with 1-1/4" Discharge	EA	0.238	MSPFB	1.000	400.78	0.00	1,184	1,585
33 29 0306	**25 GPM, 60 PSI, 2 HP, Iron Rotary Pump**	**EA**				**475.93**	**0.00**	**1,184**	**1,660**
15290 0628	25 GPM Cast Iron Rotary Pump, 60 PSI 2 HP with 1-1/4" Discharge	EA	0.2	MSPFB	1.000	475.93	0.00	1,184	1,660

33 29 04 Sump Pumps		Unit	Hourly Output	Crew	Factor	Labor	Equip.	Matl.	Total
33 29 0401	**25 GPM, 1 1/2" Discharge, Cast-iron Sump Pump**	**EA**				**331.08**	**0.00**	**2,205**	**2,536**
15292 2010	25 GPM Cast Iron Sump Pump, 1-1/2" Discharge Single Stage, 40' Head, 1 HP	EA	0.288	MSPFB	1.000	331.08	0.00	2,205	2,536
33 29 0402	**75 GPM, 2" Discharge, Cast-iron Sump Pump**	**EA**				**400.78**	**0.00**	**2,754**	**3,155**
15292 2020	75 GPM Cast Iron Sump Pump, 2" Discharge Single Stage, 40' Head, 1-1/2 HP	EA	0.238	MSPFB	1.000	400.78	0.00	2,754	3,155
33 29 0403	**100 GPM, 2 1/2" Discharge, Cast-iron Sump Pump**	**EA**				**475.93**	**0.00**	**2,849**	**3,325**
15292 2030	100 GPM Cast Iron Sump Pump, 2-1/2" Discharge Single Stage, 40' Head, 2 HP	EA	0.2	MSPFB	1.000	475.93	0.00	2,849	3,325
33 29 0404	**150 GPM, 3" Discharge, Cast-iron Sump Pump**	**EA**				**585.75**	**0.00**	**2,659**	**3,245**
15292 2040	150 GPM Cast Iron Sump Pump, 3" Discharge Single Stage, 40' Head, 3 HP	EA	0.163	MSPFB	1.000	585.75	0.00	2,659	3,245
33 29 0405	**200 GPM, 3" Discharge, Cast-iron Sump Pump**	**EA**				**634.57**	**0.00**	**2,687**	**3,322**
15292 2050	200 GPM Cast Iron Sump Pump, 3" Discharge Single Stage, 40' Head, 3 HP	EA	0.15	MSPFB	1.000	634.57	0.00	2,687	3,322
33 29 0406	**300 GPM, 4" Discharge, Cast-iron Sump Pump**	**EA**				**761.48**	**0.00**	**4,087**	**4,849**
15292 2060	300 GPM Cast Iron Sump Pump, 4" Discharge Single Stage, 70' Head, 10 HP	EA	0.125	MSPFB	1.000	761.48	0.00	4,087	4,849
33 29 0407	**500 GPM, 5" Discharge, Cast-iron Sump Pump**	**EA**				**750.06**	**0.00**	**4,309**	**5,059**
15292 2070	500 GPM Cast Iron Sump Pump, 5" Discharge Single Stage, 70' Head, 15 HP	EA	0.163	MSPFI	1.000	750.06	0.00	4,309	5,059
33 29 0408	**800 GPM, 6" Discharge, Cast-iron Sump Pump**	**EA**				**858.80**	**82.14**	**13,422**	**14,363**
15292 2080	800 GPM Cast Iron Sump Pump, 6" Discharge Single Stage, 70' Head, 20 HP	EA	0.175	MSPFN	1.000	858.80	82.14	13,422	14,363

33 29 Pumps Other than Submersible/Groundwater

33 29 04 Sump Pumps		Unit	Hourly Output	Crew	Factor	Labor	Equip.	Matl.	Total
33 29 0409	**1,000 GPM, 6" Discharge, Cast-iron Sump Pump**	**EA**				**1,002**	**95.83**	**16,383**	**17,480**
15292 2090	1,000 GPM Cast Iron Sump Pump, 6" Discharge Single Stage, 70' Head, 30 HP	EA	0.15	MSPFN	1.000	1,002	95.83	16,383	17,480
33 29 0410	**1,600 GPM, 8" Discharge, Cast-iron Sump Pump**	**EA**				**1,202**	**115.00**	**18,780**	**20,097**
15292 2100	1,600 GPM Cast Iron Sump Pump, 8" Discharge Single Stage, 70' Head, 50 HP	EA	0.125	MSPFN	1.000	1,202	115.00	18,780	20,097
33 29 0411	**2,000 GPM, 8" Discharge, Cast-iron Sump Pump**	**EA**				**1,336**	**127.78**	**20,763**	**22,226**
15292 2110	2,000 GPM Cast Iron Sump Pump, 8" Discharge Single Stage, 70' Head, 60 HP	EA	0.113	MSPFN	1.000	1,336	127.78	20,763	22,226
33 29 0412	**15 GPM Submersible Sump Pump**	**EA**				**68.75**	**0.00**	**823.06**	**891.81**
15296 8010	15 GPM Submersible Sump Pump, 15' Head, 1-1/2" Discharge	EA	0.75	MPLUA	1.000	68.75	0.00	823.06	891.81
33 29 0413	**42 GPM Submersible Sump Pump**	**EA**				**68.75**	**0.00**	**823.06**	**891.81**
15296 8020	42 GPM Submersible Sump Pump, 15' Head, 1-1/2" Discharge	EA	0.75	MPLUA	1.000	68.75	0.00	823.06	891.81
33 29 0414	**52 GPM Submersible Sump Pump**	**EA**				**82.50**	**0.00**	**823.06**	**905.56**
15296 8030	52 GPM Submersible Sump Pump, 15' Head, 1-1/2" Discharge	EA	0.625	MPLUA	1.000	82.50	0.00	823.06	905.56
33 29 05 200 PSI Pump with Motor		**Unit**	**Hourly Output**	**Crew**	**Factor**	**Labor**	**Equip.**	**Matl.**	**Total**
33 29 0501	**Adjustable Flow 0 - 100% (25 GPM), 200 PSI, Pump with Motor**	**EA**				**48.99**	**0.00**	**1,484**	**1,533**
15293 0120	Adjustable Flow 0 - 100% (25 GPM) 200 PSI Neoprene Packing	EA	0	N/A	1.000	0.00	0.00	1,468	1,468
16353 0030	> 1/2 HP AC Motor 1 Phase, 230 Less than 1/2 HP	EA	2.1	EELEB	1.000	48.99	0.00	16.38	65.37
33 29 06 250 PSI Pump with Motor		**Unit**	**Hourly Output**	**Crew**	**Factor**	**Labor**	**Equip.**	**Matl.**	**Total**
33 29 0601	**Adjustable Flow 0 - 100% (28 GPM), 250 PSI, Pump with Motor**	**EA**				**48.99**	**0.00**	**1,829**	**1,878**
15293 0130	Adjustable Flow 0 - 100% (28 GPM) 250 PSI	EA	0	N/A	1.000	0.00	0.00	1,813	1,813
16353 0030	> 1/2 HP AC Motor 1 Phase, 230 Less than 1/2 HP	EA	2.1	EELEB	1.000	48.99	0.00	16.38	65.37
33 31 01 Blowers		**Unit**	**Hourly Output**	**Crew**	**Factor**	**Labor**	**Equip.**	**Matl.**	**Total**
33 31 0101	**50 CFM, 7" Pressure, 3/4 HP, Blower System**	**EA**				**127.16**	**0.00**	**800.00**	**927.16**
13274 5216	50 CFM, 7" Pressure, 3/4 HP, 3,500 RPM, Blower	EA	0.735	MSHMD	1.000	127.16	0.00	800.00	927.16
33 31 0102	**100 CFM, 5" Pressure, 1/3 HP, Blower System**	**EA**				**135.94**	**0.00**	**848.74**	**984.68**
13274 5231	100 CFM, 5" Pressure, 1/3 HP, 3,450 RPM, Blower	EA	0.688	MSHMD	1.000	135.94	0.00	848.74	984.68

33 31 01	Blowers	Unit	Hourly Output	Crew	Factor	Labor	Equip.	Matl.	Total
33 31 0103	**150 CFM, 8" Pressure, 3/4 HP, Blower System**	**EA**				**149.54**	**0.00**	**895.00**	**1,045**
13274 5217	150 CFM, 8" Pressure, 3/4 HP, 3,750 RPM, Blower	EA	0.625	MSHMD	1.000	149.54	0.00	895.00	1,045
33 31 0104	**250 CFM, 12" Pressure, 1 1/2 HP, Blower System**	**EA**				**173.88**	**0.00**	**1,075**	**1,249**
13274 5211	250 CFM, 12" Pressure, 1-1/2 HP, 4,625 RPM, Blower	EA	0.538	MSHMD	1.000	173.88	0.00	1,075	1,249
33 31 0105	**500 CFM, 5" Pressure, 1 HP, Blower System**	**EA**				**173.88**	**0.00**	**1,023**	**1,197**
13274 5232	500 CFM, 5" Pressure, 1 HP, 3,450 RPM, Blower	EA	0.538	MSHMD	1.000	173.88	0.00	1,023	1,197
33 31 0106	**500 CFM, 9" Pressure, 2 HP, Blower System**	**EA**				**196.76**	**0.00**	**1,200**	**1,397**
13274 5212	500 CFM, 9" Pressure, 2 HP, 3,500 RPM, Blower	EA	0.475	MSHMD	1.000	196.76	0.00	1,200	1,397
33 31 0107	**500 CFM, 12" Pressure, 5 HP, Blower System**	**EA**				**373.84**	**0.00**	**2,350**	**2,724**
13274 5221	500 CFM, 12" Pressure, 5 HP, 3,870 RPM, Blower	EA	0.25	MSHMD	1.000	373.84	0.00	2,350	2,724
33 31 0108	**750 CFM, 12" Pressure, 5 HP, Blower System**	**EA**				**373.84**	**0.00**	**1,400**	**1,774**
13274 5213	750 CFM, 12" Pressure, 5 HP, 3,920 RPM, Blower	EA	0.25	MSHMD	1.000	373.84	0.00	1,400	1,774
33 31 0109	**1,000 CFM, 5" Pressure, 1 1/2 HP, Blower System**	**EA**				**173.88**	**0.00**	**1,125**	**1,299**
13274 5214	1,000 CFM, 5" Pressure, 1-1/2 HP, 2,670 RPM, Blower	EA	0.538	MSHMD	1.000	173.88	0.00	1,125	1,299
33 31 0110	**1,020 SCFM, 2 HP, Replacement Blower Motor**	**EA**				**72.99**	**0.00**	**400.00**	**472.99**
01050 0117	Master Mechanic	HR	1	XYBBP	2.000	36.50	0.00	0.00	72.99
13273 4142	Blower Motor, 1,020 SCFM, 2 HP	EA	0	N/A	1.000	0.00	0.00	400.00	400.00
33 31 0111	**1,050 SCFM, 1.5 HP, Replacement Blower Motor**	**EA**				**72.99**	**0.00**	**320.00**	**392.99**
01050 0117	Master Mechanic	HR	1	XYBBP	2.000	36.50	0.00	0.00	72.99
13273 4141	Blower Motor, 1,050 SCFM, 1.5 HP	EA	0	N/A	1.000	0.00	0.00	320.00	320.00
33 31 0112	**1,500 CFM, 5" Pressure, 3 HP, Blower System**	**EA**				**249.23**	**0.00**	**1,200**	**1,449**
13274 5215	1,500 CFM, 5" Pressure, 3 HP, 3,340 RPM, Blower	EA	0.375	MSHMD	1.000	249.23	0.00	1,200	1,449
33 31 0113	**1,500 CFM, 12" Pressure, 10 HP, Blower System**	**EA**				**575.14**	**0.00**	**2,825**	**3,400**
13274 5222	1,500 CFM, 12" Pressure, 10 HP, 3,180 RPM, Blower	EA	0.163	MSHMD	1.000	575.14	0.00	2,825	3,400
33 31 0114	**2,200 CFM, 8" Pressure, 7 1/2 HP, Blower System**	**EA**				**575.14**	**0.00**	**2,425**	**3,000**
13274 5223	2,200 CFM, 8" Pressure, 7-1/2 HP, 2,920 RPM, Blower	EA	0.163	MSHMD	1.000	575.14	0.00	2,425	3,000

33 31 01	Blowers	Unit	Hourly Output	Crew	Factor	Labor	Equip.	Matl.	Total
33 31 0115	**2,560 SCFM, 3 HP, Replacement Blower Motor**	**EA**				**72.99**	**0.00**	**420.00**	**492.99**
01050 0117	Master Mechanic	HR	1	XYBBP	2.000	36.50	0.00	0.00	72.99
13273 4143	Blower Motor, 2,560 SCFM, 3 HP	EA	0	N/A	1.000	0.00	0.00	420.00	420.00
33 31 0116	**3,000 CFM, 5" Pressure, 7 1/2 HP, Blower System**	**EA**				**575.14**	**0.00**	**2,750**	**3,325**
13274 5224	3,000 CFM, 5" Pressure, 7-1/2 HP, 2,125 RPM, Blower	EA	0.163	MSHMD	1.000	575.14	0.00	2,750	3,325
33 31 0117	**4,000 CFM, 8" Pressure, 15 HP, Blower System**	**EA**				**475.93**	**0.00**	**3,400**	**3,876**
13274 5225	4,000 CFM, 8" Pressure, 15 HP, 2,590 RPM, Blower	EA	0.2	MSPFB	1.000	475.93	0.00	3,400	3,876
33 31 0118	**4,800 SCFM, 2 HP, Replacement Blower Motor**	**EA**				**72.99**	**0.00**	**400.00**	**472.99**
01050 0117	Master Mechanic	HR	1	XYBBP	2.000	36.50	0.00	0.00	72.99
13273 4144	Blower Motor, 4,800 SCFM, 2 HP	EA	0	N/A	1.000	0.00	0.00	400.00	400.00
33 31 0119	**5,000 SCFM, 3 HP, Replacement Blower Motor**	**EA**				**72.99**	**0.00**	**420.00**	**492.99**
01050 0117	Master Mechanic	HR	1	XYBBP	2.000	36.50	0.00	0.00	72.99
13273 4145	Blower Motor, 5,000 SCFM, 3 HP	EA	0	N/A	1.000	0.00	0.00	420.00	420.00
33 31 0120	**5,000 CFM, 8" Pressure, 20 HP, Blower System**	**EA**				**585.75**	**0.00**	**4,900**	**5,486**
13274 5226	5,000 CFM, 8" Pressure, 20 HP, 2,090 RPM, Blower	EA	0.163	MSPFB	1.000	585.75	0.00	4,900	5,486
33 31 0121	**7,500 SCFM, 3 HP, Replacement Blower Motor**	**EA**				**72.99**	**0.00**	**435.00**	**507.99**
01050 0117	Master Mechanic	HR	1	XYBBP	2.000	36.50	0.00	0.00	72.99
13273 4146	Blower Motor, 7,500 SCFM, 3 HP	EA	0	N/A	1.000	0.00	0.00	435.00	435.00
33 31 0122	**8,000 CFM, 8" Pressure, 25 HP, Blower System**	**EA**				**423.04**	**0.00**	**5,750**	**6,173**
13274 5227	8,000 CFM, 8" Pressure, 25 HP, 1,860 RPM, Blower	EA	0.225	MSPFB	1.000	423.04	0.00	5,750	6,173
33 31 0123	**9,050 SCFM, 5 HP, Replacement Blower Motor**	**EA**				**72.99**	**0.00**	**475.00**	**547.99**
01050 0117	Master Mechanic	HR	1	XYBBP	2.000	36.50	0.00	0.00	72.99
13273 4147	Blower Motor, 9,050 SCFM, 5 HP	EA	0	N/A	1.000	0.00	0.00	475.00	475.00
33 31 0124	**10,000 SCFM, 7.5 HP, Replacement Blower Motor**	**EA**				**72.99**	**0.00**	**605.00**	**677.99**
01050 0117	Master Mechanic	HR	1	XYBBP	2.000	36.50	0.00	0.00	72.99
13273 4149	Blower Motor, 10,000 SCFM, 7.5 HP	EA	0	N/A	1.000	0.00	0.00	605.00	605.00
33 31 0125	**11,000 SCFM, 5 HP, Replacement Blower Motor**	**EA**				**72.99**	**0.00**	**550.00**	**622.99**
01050 0117	Master Mechanic	HR	1	XYBBP	2.000	36.50	0.00	0.00	72.99
13273 4148	Blower Motor, 11,000 SCFM, 5 HP	EA	0	N/A	1.000	0.00	0.00	550.00	550.00

33 31 01	Blowers	Unit	Hourly Output	Crew	Factor	Labor	Equip.	Matl.	Total
33 31 0131	**48 CFM, 4.9 HP Maximum, Blower, Positive Displacement, with Motor**	**EA**				**636.15**	**0.00**	**1,000**	**1,636**
02082 6451	Positive Displacement Blower, 4.9 Maximum HP, 48 CFM	EA	0.163	MSPFC	1.000	636.15	0.00	1,000	1,636
33 31 0132	**111 CFM, 5.6 HP Maximum, Blower, Positive Displacement, with Motor**	**EA**				**689.17**	**0.00**	**1,175**	**1,864**
02082 6452	Positive Displacement Blower, 5.6 Maximum HP, 111 CFM	EA	0.15	MSPFC	1.000	689.17	0.00	1,175	1,864
33 31 0133	**143 CFM, 12.2 HP Maximum, Blower, Positive Displacement, with Motor**	**EA**				**751.82**	**0.00**	**1,200**	**1,952**
02082 6453	Positive Displacement Blower, 12.2 Maximum HP, 143 CFM	EA	0.138	MSPFC	1.000	751.82	0.00	1,200	1,952
33 31 0134	**271 CFM, 11.8 HP Maximum, Blower, Positive Displacement, with Motor**	**EA**				**827.00**	**0.00**	**1,175**	**2,002**
02082 6454	Positive Displacement Blower, 11.8 Maximum HP, 271 CFM	EA	0.125	MSPFC	1.000	827.00	0.00	1,175	2,002
33 31 0135	**134 CFM, 15 HP Maximum, Blower, Positive Displacement, with Motor**	**EA**				**827.00**	**0.00**	**1,350**	**2,177**
02082 6455	Positive Displacement Blower, 15 Maximum HP, 134 CFM, Relief Valve not Included	EA	0.125	MSPFC	1.000	827.00	0.00	1,350	2,177
33 31 0136	**278 CFM, 19.7 HP Maximum, Blower, Positive Displacement, with Motor**	**EA**				**827.00**	**0.00**	**1,400**	**2,227**
02082 6456	Positive Displacement Blower, 19.7 Maximum HP, 278 CFM	EA	0.125	MSPFC	1.000	827.00	0.00	1,400	2,227
33 31 0137	**399 CFM, 18.3 HP Maximum, Blower, Positive Displacement, with Motor**	**EA**				**995.91**	**0.00**	**1,525**	**2,521**
02082 6457	Positive Displacement Blower, 18.3 Maximum HP, 399 CFM	EA	0.104	MSPFC	1.000	995.91	0.00	1,525	2,521
33 31 0138	**247 CFM, 25.4 HP Maximum, Blower, Positive Displacement, with Motor**	**EA**				**995.91**	**0.00**	**1,950**	**2,946**
02082 6458	Positive Displacement Blower, 254 Maximum HP, 247 CFM	EA	0.104	MSPFC	1.000	995.91	0.00	1,950	2,946
33 31 0139	**427 CFM, 28.3 HP Maximum, Blower, Positive Displacement, with Motor**	**EA**				**995.91**	**0.00**	**2,050**	**3,046**
02082 6459	Positive Displacement Blower, 28.3 Maximum HP, 427 CFM	EA	0.104	MSPFC	1.000	995.91	0.00	2,050	3,046
33 31 0140	**665 CFM, 29 HP Maximum, Blower, Positive Displacement, with Motor**	**EA**				**995.91**	**0.00**	**2,225**	**3,221**
02082 6461	Positive Displacement Blower, 29 Maximum HP, 665 CFM	EA	0.104	MSPFC	1.000	995.91	0.00	2,225	3,221
33 31 0141	**393 CFM, 38.8 HP Maximum, Blower, Positive Displacement, with Motor**	**EA**				**1,046**	**0.00**	**3,275**	**4,321**
02082 6462	Positive Displacement Blower, 38.8 Maximum HP, 393 CFM, Relief Valve not Included	EA	0.099	MSPFC	1.000	1,046	0.00	3,275	4,321

33 31 01 Blowers		Unit	Hourly Output	Crew	Factor	Labor	Equip.	Matl.	Total
33 31 0142	**615 CFM, 49.6 HP Maximum, Blower, Positive Displacement, with Motor**	**EA**				**1,046**	**0.00**	**3,475**	**4,521**
02082 6463	Positive Displacement Blower, 49.6 Maximum HP, 615 CFM	EA	0.099	MSPFC	1.000	1,046	0.00	3,475	4,521
33 31 0143	**1,248 CFM, 46.5 HP Maximum, Blower, Positive Displacement, with Motor**	**EA**				**1,334**	**0.00**	**4,150**	**5,484**
02082 6464	Positive Displacement Blower, 46.5 Maximum HP, 1,248 CFM	EA	0.078	MSPFC	1.000	1,334	0.00	4,150	5,484
33 31 0144	**574 CFM, 55 HP Maximum, Blower, Positive Displacement, with Motor**	**EA**				**1,334**	**0.00**	**4,900**	**6,234**
02082 6465	Positive Displacement Blower, 55 Maximum HP, 574 CFM, Relief Valve not Included	EA	0.078	MSPFC	1.000	1,334	0.00	4,900	6,234
33 31 0145	**1,071 CFM, 67.1 HP Maximum, Blower, Positive Displacement, with Motor**	**EA**				**1,401**	**0.00**	**5,100**	**6,501**
02082 6466	Positive Displacement Blower, 67.1 Maximum HP, 1,071 CFM	EA	0.074	MSPFC	1.000	1,401	0.00	5,100	6,501
33 31 0146	**1,878 CFM, 65.5 HP Maximum, Blower, Positive Displacement, with Motor**	**EA**				**1,452**	**0.00**	**5,925**	**7,377**
02082 6467	Positive Displacement Blower, 65.5 Maximum HP, 1,878 CFM	EA	0.071	MSPFC	1.000	1,452	0.00	5,925	7,377
33 31 0147	**100 CFM, 6" Pressure, 1/3 HP, Blower**	**EA**				**135.94**	**0.00**	**695.00**	**830.94**
13273 4501	100 CFM, 6" Pressure, 1/3 HP, Blower	EA	0.688	MSHMD	1.000	135.94	0.00	695.00	830.94
33 31 0148	**150 CFM, 6" Pressure, 1/2 HP, Blower**	**EA**				**149.54**	**0.00**	**695.00**	**844.54**
13273 4502	150 CFM, 6" Pressure, 1/2 HP, Blower	EA	0.625	MSHMD	1.000	149.54	0.00	695.00	844.54
33 31 0149	**250 CFM, 6" Pressure, 3/4 HP, Blower**	**EA**				**173.88**	**0.00**	**705.00**	**878.88**
13273 4503	250 CFM, 6" Pressure, 3/4 HP, Blower	EA	0.538	MSHMD	1.000	173.88	0.00	705.00	878.88
33 31 0150	**500 CFM, 6" Pressure, 1 HP, Blower**	**EA**				**196.76**	**0.00**	**705.00**	**901.76**
13273 4504	500 CFM, 6" Pressure, 1 HP, Blower	EA	0.475	MSHMD	1.000	196.76	0.00	705.00	901.76
33 31 0151	**750 CFM, 8" Pressure, 1.5 HP, Blower**	**EA**				**227.40**	**0.00**	**865.00**	**1,092**
13273 4505	750 CFM, 8" Pressure, 1.5 HP, Blower	EA	0.411	MSHMD	1.000	227.40	0.00	865.00	1,092
33 31 0152	**1,000 CFM, 8" Pressure, 2.5 HP, Blower**	**EA**				**245.62**	**0.00**	**1,200**	**1,446**
13273 4506	1,000 CFM, 8" Pressure, 2.5 HP, Blower	EA	0.381	MSHMD	1.000	245.62	0.00	1,200	1,446
33 31 0153	**1,500 CFM, 8" Pressure, 4 HP, Blower**	**EA**				**300.80**	**0.00**	**1,775**	**2,076**
13273 4507	1,500 CFM, 8" Pressure, 4 HP, Blower	EA	0.311	MSHMD	1.000	300.80	0.00	1,775	2,076
33 31 0154	**2,000 CFM, 7" Pressure, 6 HP, Blower**	**EA**				**575.14**	**0.00**	**2,100**	**2,675**
13273 4508	2,000 CFM, 7" Pressure, 6 HP, Blower	EA	0.163	MSHMD	1.000	575.14	0.00	2,100	2,675
33 31 0155	**3,000 CFM, 6" Pressure, 7.5 HP, Blower**	**EA**				**575.14**	**0.00**	**2,125**	**2,700**
13273 4509	3,000 CFM, 6" Pressure, 7.5 HP, Blower	EA	0.163	MSHMD	1.000	575.14	0.00	2,125	2,700

33 31 01 Blowers		Unit	Hourly Output	Crew	Factor	Labor	Equip.	Matl.	Total
33 31 0156	**4,000 CFM, 6" Pressure, 10 HP, Blower**	**EA**				**623.07**	**0.00**	**2,275**	**2,898**
13273 4511	4,000 CFM, 6" Pressure, 10 HP, Blower	EA	0.15	MSHMD	1.000	623.07	0.00	2,275	2,898
33 31 0157	**5,000 CFM, 6" Pressure, 10 HP, Blower**	**EA**				**623.07**	**0.00**	**2,400**	**3,023**
13273 4512	5,000 CFM, 6" Pressure, 10 HP, Blower	EA	0.15	MSHMD	1.000	623.07	0.00	2,400	3,023
33 31 0158	**8,000 CFM, 6" Pressure, 15 HP, Blower**	**EA**				**667.57**	**0.00**	**3,200**	**3,868**
13273 4513	8,000 CFM, 6" Pressure, 15 HP, Blower	EA	0.14	MSHMD	1.000	667.57	0.00	3,200	3,868
33 31 0159	**11,000 CFM, 6" Pressure, 20 HP, Blower**	**EA**				**667.57**	**0.00**	**3,625**	**4,293**
13273 4514	11,000 CFM, 6" Pressure, 20 HP, Blower	EA	0.14	MSHMD	1.000	667.57	0.00	3,625	4,293
33 31 0160	**15,000 CFM, 6" Pressure, 25 HP, Blower**	**EA**				**718.92**	**0.00**	**4,625**	**5,344**
13273 4515	15,000 CFM, 6" Pressure, 25 HP, Blower	EA	0.13	MSHMD	1.000	718.92	0.00	4,625	5,344
33 31 0161	**18,000 CFM, 6" Pressure, 25 HP, Blower**	**EA**				**718.92**	**0.00**	**4,000**	**4,719**
13273 4516	18,000 CFM, 6" Pressure, 25 HP, Blower	EA	0.13	MSHMD	1.000	718.92	0.00	4,000	4,719
33 31 0162	**20,000 CFM, 6" Pressure, 30 HP, Blower**	**EA**				**778.83**	**0.00**	**5,300**	**6,079**
13273 4517	20,000 CFM, 6" Pressure, 30 HP, Blower	EA	0.12	MSHMD	1.000	778.83	0.00	5,300	6,079

33 31 02 Air Compressors		Unit	Hourly Output	Crew	Factor	Labor	Equip.	Matl.	Total
33 31 0201	**6 HP, 80 Gallon, 200 PSI, 13 SCFM, Air Compressor**	**EA**				**750.06**	**0.00**	**4,978**	**5,728**
15726 6062	T-30 Air Compressor, 6 HP, 13 CFM 80 Gallon Tank, 200 PSIG	EA	0.163	MSPFI	1.000	750.06	0.00	4,978	5,728
33 31 0202	**6 HP, 80 Gallon, 200 PSI, 28 SCFM, Air Compressor**	**EA**				**812.57**	**0.00**	**4,919**	**5,731**
15726 6063	T-30 Air Compressor, 6 HP, 28.2 CFM 80 Gallon Tank, 200 PSIG	EA	0.15	MSPFI	1.000	812.57	0.00	4,919	5,731
33 31 0203	**10 HP, 80 Gallon, 200 PSI, 35 SCFM, Air Compressor**	**EA**				**886.44**	**0.00**	**5,149**	**6,035**
15726 5676	2-Stage Air Compressor, 10 HP - 80 Gallon 35 ACFM, 80 - 200 PSIG Discharge Pressure	EA	0.138	MSPFI	1.000	886.44	0.00	5,149	6,035
33 31 0204	**15 HP, 120 Gallon, 200 PSI, 50 SCFM, Air Compressor**	**EA**				**1,037**	**0.00**	**7,047**	**8,084**
15726 5684	2-Stage Air Compressor, 15 HP - 120 Gallon 50 ACFM, 80 - 200 PSIG Discharge Pressure	EA	0.118	MSPFI	1.000	1,037	0.00	7,047	8,084
33 31 0205	**20 HP, 120 Gallon, 200 PSI, 70 SCFM, Air Compressor**	**EA**				**1,174**	**0.00**	**9,044**	**10,219**
15726 5686	2-Stage Air Compressor, 20 HP - 120 Gallon 70 ACFM, 80-200 PSIG Discharge Pressure	EA	0.104	MSPFI	1.000	1,174	0.00	9,044	10,219
33 31 0206	**30 HP, 240 Gallon, 200 PSI, 143 SCFM, Air Compressor**	**EA**				**1,299**	**0.00**	**13,212**	**14,512**
15726 6130	T-30 Air Compressor, 30 HP, 143 CFM 240 Gallon Tank, 200 PSIG	EA	0.094	MSPFI	1.000	1,299	0.00	13,212	14,512

33 31 02 Air Compressors		Unit	Hourly Output	Crew	Factor	Labor	Equip.	Matl.	Total
33 31 0207	**2" Steel, 0 - 300 PSI Pressure Gauge**	**EA**				**25.84**	**0.00**	**23.31**	**49.15**
15742 3012	2" Diameter Dial Pressure Gauge Steel Case 0 - 300 PSI	EA	2	MSPFA	1.000	25.84	0.00	23.31	49.15
33 31 0208	**2" Brass, 0 - 300 PSI Pressure Gauge**	**EA**				**25.84**	**0.00**	**26.35**	**52.19**
15742 3032	2" Diameter Dial Pressure Gauge Brass Case 0 - 300 PSI	EA	2	MSPFA	1.000	25.84	0.00	26.35	52.19
33 31 0209	**Pressure Gauge**	**EA**				**38.77**	**0.00**	**68.00**	**106.76**
15622 4512	Brass Gauge Cock, 1/4" MPT	EA	4	MSPFA	1.000	12.92	0.00	7.74	20.66
15742 3034	4.5" Diameter Dial Pressure Gauge Brass Case 0 - 300 PSI	EA	2	MSPFA	1.000	25.84	0.00	60.26	86.11
33 31 0210	**Air Compressor, 5 CFM**	**EA**				**0.00**	**0.00**	**223.29**	**223.29**
01001 0531	Air Compressor, 5 CFM	EA	0	N/A	1.000	0.00	0.00	223.29	223.29
33 31 0211	**Air Compressor, 8 CFM**	**EA**				**0.00**	**0.00**	**334.37**	**334.37**
01001 0532	Air Compressor, 8 CFM	EA	0	N/A	1.000	0.00	0.00	334.37	334.37
33 31 0212	**Air Compressor, 10 CFM**	**EA**				**0.00**	**0.00**	**436.48**	**436.48**
01001 0533	Air Compressor, 10 CFM	EA	0	N/A	1.000	0.00	0.00	436.48	436.48
33 31 0213	**Air Compressor, 12 CFM**	**EA**				**0.00**	**0.00**	**447.70**	**447.70**
01001 0534	Air Compressor, 12 CFM	EA	0	N/A	1.000	0.00	0.00	447.70	447.70
33 31 0214	**Air Compressor, 20 CFM**	**EA**				**0.00**	**0.00**	**895.39**	**895.39**
01001 0535	Air Compressor, 20 CFM	EA	0	N/A	1.000	0.00	0.00	895.39	895.39
33 31 0215	**Air Compressor, 5 CFM (Daily Rental)**	**DAY**				**0.00**	**0.00**	**31.79**	**31.79**
01001 0536	Air Compressor, 5 CFM (Daily Rental)	DAY	0	N/A	1.000	0.00	0.00	31.79	31.79
33 31 0216	**Air Compressor, 8 CFM (Daily Rental)**	**DAY**				**0.00**	**0.00**	**33.66**	**33.66**
01001 0537	Air Compressor, 8 CFM (Daily Rental)	DAY	0	N/A	1.000	0.00	0.00	33.66	33.66
33 31 0217	**Air Compressor, 10 CFM (Daily Rental)**	**DAY**				**0.00**	**0.00**	**41.89**	**41.89**
01001 0538	Air Compressor, 10 CFM (Daily Rental)	DAY	0	N/A	1.000	0.00	0.00	41.89	41.89
33 31 0218	**Air Compressor, 12 CFM (Daily Rental)**	**DAY**				**0.00**	**0.00**	**41.89**	**41.89**
01001 0539	Air Compressor, 12 CFM (Daily Rental)	DAY	0	N/A	1.000	0.00	0.00	41.89	41.89
33 31 0219	**Air Compressor, 20 CFM (Daily Rental)**	**DAY**				**0.00**	**0.00**	**56.10**	**56.10**
01001 0540	Air Compressor, 20 CFM (Daily Rental)	DAY	0	N/A	1.000	0.00	0.00	56.10	56.10
33 31 0220	**Air Compressor, 100 CFM (Daily Rental)**	**DAY**				**0.00**	**0.00**	**89.76**	**89.76**
01001 0541	Air Compressor, 100 CFM (Daily Rental)	DAY	0	N/A	1.000	0.00	0.00	89.76	89.76
33 31 0221	**Air Compressor, 160 CFM (Daily Rental)**	**DAY**				**0.00**	**0.00**	**100.98**	**100.98**
01001 0542	Air Compressor, 160 CFM (Daily Rental)	DAY	0	N/A	1.000	0.00	0.00	100.98	100.98

33 31 Blowers/Pumps/Compressors/Heaters

33 31 02	Air Compressors	Unit	Hourly Output	Crew	Factor	Labor	Equip.	Matl.	Total
33 31 0222	**Air Compressor, 185 CFM (Daily Rental)**	**DAY**				**0.00**	**0.00**	**112.20**	**112.20**
01001 0543	Air Compressor, 185 CFM (Daily Rental)	DAY	0	N/A	1.000	0.00	0.00	112.20	112.20

33 31 03	Heaters	Unit	Hourly Output	Crew	Factor	Labor	Equip.	Matl.	Total
33 31 0301	**7.5 KW, 25,600 BTU, Hazardous Air Heater**	**EA**				**706.30**	**0.00**	**5,600**	**6,306**
13274 5301	7.5 KW, 25,600 BTU, 1/4 HP, Hazardous Air Heater	EA	0.2	MSHMG	1.000	706.30	0.00	5,600	6,306
33 31 0302	**10 KW, 34,150 BTU, Hazardous Air Heater**	**EA**				**706.30**	**0.00**	**5,950**	**6,656**
13274 5302	10 KW, 34,150 BTU, 1/4 HP, Hazardous Air Heater	EA	0.2	MSHMG	1.000	706.30	0.00	5,950	6,656
33 31 0303	**15 KW, 51,200 BTU, Hazardous Air Heater**	**EA**				**753.39**	**0.00**	**6,225**	**6,978**
13274 5303	15 KW, 51,200 BTU, 1/4 HP, Hazardous Air Heater	EA	0.188	MSHMG	1.000	753.39	0.00	6,225	6,978
33 31 0304	**20 KW, 68,300 BTU, Hazardous Air Heater**	**EA**				**753.39**	**0.00**	**6,450**	**7,203**
13274 5304	20 KW, 68,300 BTU, 1/4 HP, Hazardous Air Heater	EA	0.188	MSHMG	1.000	753.39	0.00	6,450	7,203

33 31 04	Heat Exchanger	Unit	Hourly Output	Crew	Factor	Labor	Equip.	Matl.	Total
33 31 0401	**Heat Exchanger, 8 GPM, Purchase and Installation**	**EA**				**137.83**	**0.00**	**11,998**	**12,136**
15561 0120	Heat Exchanger by Steam @ 10 PSI, 8 GPM 40-180 Degree, 4 Pass 3/4" Outside Diameter Copper Tube	EA	0.75	MSPFC	1.000	137.83	0.00	1,007	1,145
15732 1109	Heat Exchanger Mobilization/Demobilization, Installation, Modification for Sludge/Liquid	EA	0	N/A	1.000	0.00	0.00	10,991	10,991

33 32 01	Chemical Feed Systems	Unit	Hourly Output	Crew	Factor	Labor	Equip.	Matl.	Total
33 32 0101	**7.5 GPH, 860 PSI, 55 Gallon Chemical Feed System with Polyethylene Tank with Steel Overpack**	**EA**				**661.39**	**0.00**	**1,325**	**1,986**
13272 1211	Polyethylene Tank with Steel Overpack, 7.5 GPH, 860 PSI, 55 Gallon	EA	0.156	MSPFC	1.000	661.39	0.00	1,325	1,986
33 32 0102	**20.8 GPH, 150 PSI, 55 Gallon Chemical Feed System with Polyethylene Tank with Steel Overpack**	**EA**				**661.39**	**0.00**	**1,400**	**2,061**
13272 1212	Polyethylene Tank with Steel Overpack, 20.8 GPH, 150 PSI, 55 Gallon	EA	0.156	MSPFC	1.000	661.39	0.00	1,400	2,061
33 32 0103	**7.5 GPH, 860 PSI, 102 Gallon Chemical Feed System with Polyethylene Tank with Steel Overpack**	**EA**				**827.00**	**0.00**	**1,850**	**2,677**
13272 1213	Polyethylene Tank with Steel Overpack, 7.5 GPH, 860 PSI, 102 Gallon	EA	0.125	MSPFC	1.000	827.00	0.00	1,850	2,677

33 32 01	Chemical Feed Systems	Unit	Hourly Output	Crew	Factor	Labor	Equip.	Matl.	Total
33 32 0104	**20.8 GPH, 150 PSI, 102 Gallon Chemical Feed System with Polyethylene Tank with Steel Overpack**	**EA**				**827.00**	**0.00**	**1,900**	**2,727**
13272 1214	Polyethylene Tank with Steel Overpack, 20.8 GPH, 150 PSI, 102 Gallon	EA	0.125	MSPFC	1.000	827.00	0.00	1,900	2,727
33 32 0105	**7.5 GPH, 860 PSI, 167 Gallon Chemical Feed System with Polyethylene Tank with Steel Overpack**	**EA**				**827.00**	**0.00**	**2,075**	**2,902**
13272 1215	Polyethylene Tank with Steel Overpack, 7.5 GPH, 860 PSI, 167 Gallon	EA	0.125	MSPFC	1.000	827.00	0.00	2,075	2,902
33 32 0106	**20.8 GPH, 150 PSI, 167 Gallon Chemical Feed System with Polyethylene Tank with Steel Overpack**	**EA**				**827.00**	**0.00**	**2,100**	**2,927**
13272 1216	Polyethylene Tank with Steel Overpack, 20.8 GPH, 150 PSI, 167 Gallon	EA	0.125	MSPFC	1.000	827.00	0.00	2,100	2,927
33 32 0107	**7.5 GPH, 860 PSI, 260 Gallon Chemical Feed System with Polyethylene Tank with Steel Overpack**	**EA**				**827.00**	**0.00**	**2,325**	**3,152**
13272 1217	Polyethylene Tank with Steel Overpack, 7.5 GPH, 860 PSI, 260 Gallon	EA	0.125	MSPFC	1.000	827.00	0.00	2,325	3,152
33 32 0108	**20.8 GPH, 150 PSI, 260 Gallon Chemical Feed System with Polyethylene Tank with Steel Overpack**	**EA**				**827.00**	**0.00**	**2,350**	**3,177**
13272 1218	Polyethylene Tank with Steel Overpack, 20.8 GPH, 150 PSI, 260 Gallon	EA	0.125	MSPFC	1.000	827.00	0.00	2,350	3,177
33 32 0109	**7.5 GPH, 860 PSI, 50 Gallon Chemical Feed System with 304 Stainless Steel Tank**	**EA**				**661.39**	**0.00**	**2,275**	**2,936**
13272 1222	304 Stainless Steel Tank, 7.5 GPH, 860 PSI, 50 Gallon Tank	EA	0.156	MSPFC	1.000	661.39	0.00	2,275	2,936
33 32 0110	**20.8 GPH, 150 PSI, 50 Gallon Chemical Feed System with 304 Stainless Steel Tank**	**EA**				**661.39**	**0.00**	**2,325**	**2,986**
13272 1223	304 Stainless Steel Tank, 20.8 GPH, 150 PSI, 50 Gallon Tank	EA	0.156	MSPFC	1.000	661.39	0.00	2,325	2,986
33 32 0111	**7.5 GPH, 860 PSI, 100 Gallon Chemical Feed System with 304 Stainless Steel Tank**	**EA**				**827.00**	**0.00**	**2,700**	**3,527**
13272 1224	304 Stainless Steel Tank, 7.5 GPH, 860 PSI, 100 Gallon Tank	EA	0.125	MSPFC	1.000	827.00	0.00	2,700	3,527
33 32 0112	**20.8 GPH, 150 PSI, 100 Gallon Chemical Feed System with 304 Stainless Steel Tank**	**EA**				**827.00**	**0.00**	**2,725**	**3,552**
13272 1225	304 Stainless Steel Tank, 20.8 GPH, 150 PSI, 100 Gallon Tank	EA	0.125	MSPFC	1.000	827.00	0.00	2,725	3,552
33 32 0113	**7.5 GPH, 860 PSI, 150 Gallon Chemical Feed System with 304 Stainless Steel Tank**	**EA**				**827.00**	**0.00**	**3,125**	**3,952**
13272 1226	304 Stainless Steel Tank, 7.5 GPH, 860 PSI, 150 Gallon Tank	EA	0.125	MSPFC	1.000	827.00	0.00	3,125	3,952

33 32 01	Chemical Feed Systems	Unit	Hourly Output	Crew	Factor	Labor	Equip.	Matl.	Total
33 32 0114	**20.8 GPH, 150 PSI, 150 Gallon Chemical Feed System with 304 Stainless Steel Tank**	**EA**				**827.00**	**0.00**	**3,150**	**3,977**
13272 1227	304 Stainless Steel Tank, 20.8 GPH, 150 PSI, 150 Gallon Tank	EA	0.125	MSPFC	1.000	827.00	0.00	3,150	3,977
33 32 0115	**7.5 GPH, 860 PSI, 250 Gallon Chemical Feed System with 304 Stainless Steel Tank**	**EA**				**827.00**	**0.00**	**3,950**	**4,777**
13272 1228	304 Stainless Steel Tank, 7.5 GPH, 860 PSI, 250 Gallon Tank	EA	0.125	MSPFC	1.000	827.00	0.00	3,950	4,777
33 32 0116	**20.8 GPH, 150 PSI, 250 Gallon Chemical Feed System with 304 Stainless Steel Tank**	**EA**				**827.00**	**0.00**	**3,975**	**4,802**
13272 1229	304 Stainless Steel Tank, 20.8 GPH, 150 PSI, 250 Gallon Tank	EA	0.125	MSPFC	1.000	827.00	0.00	3,975	4,802
33 32 0117	**7.5 GPH, 860 PSI, 500 Gallon Chemical Feed System with 304 Stainless Steel Tank**	**EA**				**551.33**	**0.00**	**6,025**	**6,576**
13272 1231	304 Stainless Steel Tank, 7.5 GPH, 860 PSI, 500 Gallon Tank	EA	0.188	MSPFC	1.000	551.33	0.00	6,025	6,576
33 32 0118	**20.8 GPH, 150 PSI, 500 Gallon Chemical Feed System with 304 Stainless Steel Tank**	**EA**				**551.33**	**0.00**	**6,050**	**6,601**
13272 1232	304 Stainless Steel Tank, 20.8 GPH, 150 PSI, 500 Gallon Tank	EA	0.188	MSPFC	1.000	551.33	0.00	6,050	6,601
33 32 0119	**Chemical Feed Pump, 7.5 GPH, 860 PSI, Simplex, Packed Plunger**	**EA**				**551.33**	**0.00**	**3,050**	**3,601**
13272 1261	7.5 GPH, 860 PSI, Simplex Packed Plunger	EA	0.188	MSPFC	1.000	551.33	0.00	3,050	3,601
33 32 0120	**Chemical Feed Pump, 1.0 GPH, 2,000 PSI, Simplex, Packed Plunger**	**EA**				**551.33**	**0.00**	**3,225**	**3,776**
13272 1262	1.0 GPH, 2,000 PSI, Simplex Packed Plunger	EA	0.188	MSPFC	1.000	551.33	0.00	3,225	3,776
33 32 0121	**Chemical Feed Pump, 20.8 GPH, 150 PSI, Simplex, Diaphragm**	**EA**				**551.33**	**0.00**	**1,625**	**2,176**
13272 1263	20.8 GPH, 150 PSI, Simplex Diaphragm Type	EA	0.188	MSPFC	1.000	551.33	0.00	1,625	2,176
33 32 0122	**Chemical Feed Pump, 105 GPH, 50 PSI, Simplex, Diaphragm**	**EA**				**551.33**	**0.00**	**4,550**	**5,101**
13272 1265	105 GPH, 50 PSI, Simplex, Diaphragm Type	EA	0.188	MSPFC	1.000	551.33	0.00	4,550	5,101
33 32 0123	**Chemical Feed Pump, 0.86 GPH, 700 PSI, Simplex, 316 Stainless Steel, Liquid End**	**EA**				**551.33**	**0.00**	**1,850**	**2,401**
13272 1266	0.86 GPH, 700 PSI, Simplex, 316 Stainless Steel, Liquid End	EA	0.188	MSPFC	1.000	551.33	0.00	1,850	2,401
33 32 0124	**Chemical Feed Pump, 2.70 GPH, 1,100 PSI, Simplex, 316 Stainless Steel, Liquid End**	**EA**				**551.33**	**0.00**	**1,600**	**2,151**
13272 1268	2.70 GPH, 1,100 PSI, Simplex, 316 Stainless Steel, Liquid End	EA	0.188	MSPFC	1.000	551.33	0.00	1,600	2,151

33 32 Tanks

33 32 01	Chemical Feed Systems	Unit	Hourly Output	Crew	Factor	Labor	Equip.	Matl.	Total
33 32 0125	**Chemical Feed Pump, 7.0 GPH, 925 PSI, Simplex, 316 Stainless Steel, Liquid End**	**EA**				**551.33**	**0.00**	**3,100**	**3,651**
13272 1271	7.0 GPH, 925 PSI, Simplex, 316 Stainless Steel, Liquid End	EA	0.188	MSPFC	1.000	551.33	0.00	3,100	3,651
33 32 0126	**Chemical Feed Pump, 18.0 GPH, 350 PSI, Simplex, 316 Stainless Steel, Liquid End**	**EA**				**551.33**	**0.00**	**2,225**	**2,776**
13272 1273	18.0 GPH, 350 PSI, Simplex, 316 Stainless Steel, Liquid End	EA	0.188	MSPFC	1.000	551.33	0.00	2,225	2,776
33 32 0127	**Chemical Feed Pump, 41.6 GPH, 150 PSI, Duplex, Diaphragm**	**EA**				**551.33**	**0.00**	**2,850**	**3,401**
13272 1264	41.6 GPH, 150 PSI, Duplex Diaphragm-type	EA	0.188	MSPFC	1.000	551.33	0.00	2,850	3,401
33 32 0128	**Chemical Feed Pump, 0.86 GPH, 700 PSI, Duplex, 316 Stainless Steel, Liquid End**	**EA**				**551.33**	**0.00**	**2,525**	**3,076**
13272 1267	0.86 GPH, 700 PSI, Duplex, 316 Stainless Steel, Liquid End	EA	0.188	MSPFC	1.000	551.33	0.00	2,525	3,076
33 32 0129	**Chemical Feed Pump, 2.70 GPH, 1,100 PSI, Duplex, 316 Stainless Steel, Liquid End**	**EA**				**551.33**	**0.00**	**2,100**	**2,651**
13272 1269	2.70 GPH, 1,100 PSI, Duplex, 316 Stainless Steel, Liquid End	EA	0.188	MSPFC	1.000	551.33	0.00	2,100	2,651
33 32 0130	**Chemical Feed Pump, 7.0 GPH, 925 PSI, Duplex, 316 Stainless Steel, Liquid End**	**EA**				**551.33**	**0.00**	**3,725**	**4,276**
13272 1272	7.0 GPH, 925 PSI, Duplex, 316 Stainless Steel, Liquid End	EA	0.188	MSPFC	1.000	551.33	0.00	3,725	4,276
33 32 0131	**Chemical Feed Pump, 18 GPH, 350 PSI, Duplex, 316 Stainless Steel, Liquid End**	**EA**				**551.33**	**0.00**	**2,850**	**3,401**
13272 1274	18.0 GPH, 350 PSI, Duplex, 316 Stainless Steel, Liquid End	EA	0.188	MSPFC	1.000	551.33	0.00	2,850	3,401
33 32 0132	**Mixing Tank for Dispersant (3 Month Minimum Rental)**	**EA**				**0.00**	**0.00**	**29,991**	**29,991**
13275 7117	2 CY Waste Mixer, 3-month Rental, Trailer Mounted	EA	0	N/A	1.000	0.00	0.00	2,591	2,591
13275 7145	2 CY Waste Mixer, Mobilization/Demobilization	EA	0	N/A	1.000	0.00	0.00	27,400	27,400
33 32 0133	**Monthly Rental of Mixing Tank**	**EA**				**0.00**	**0.00**	**863.54**	**863.54**
13275 7117	2 CY Waste Mixer, 3-month Rental, Trailer Mounted	EA	0	N/A	0.333	0.00	0.00	2,591	863.54

33 33 01	Chemical Additives	Unit	Hourly Output	Crew	Factor	Labor	Equip.	Matl.	Total
33 33 0101	**Sodium Hydroxide Solution, 190 Lb Drummed Liquid**	**EA**				**6.49**	**3.51**	**49.40**	**59.40**
13272 1351	Containerized Staging	EA	10	CODLA	1.000	4.80	3.51	0.00	8.31
13272 1353	Containerized Feed System Loading	EA	20	ULABA	1.000	1.69	0.00	0.00	1.69
13274 1125	Sodium Hydroxide Solution, 190 Lb Drummed Liquid	LB	0	N/A	190.000	0.00	0.00	0.26	49.40

33 33 Chemical Additives

33 33 01	Chemical Additives	Unit	Hourly Output	Crew	Factor	Labor	Equip.	Matl.	Total
33 33 0102	**Sodium Hydroxide Solution, 700 Lb Drummed Liquid**	**EA**				**6.49**	**3.51**	**133.00**	**143.00**
13272 1351	Containerized Staging	EA	10	CODLA	1.000	4.80	3.51	0.00	8.31
13272 1353	Containerized Feed System Loading	EA	20	ULABA	1.000	1.69	0.00	0.00	1.69
13274 1126	Sodium Hydroxide Solution, 700 Lb Drummed Liquid	LB	0	N/A	700.000	0.00	0.00	0.19	133.00
33 33 0103	**Sodium Hydroxide, 500 Lb Container, Beads**	**EA**				**6.49**	**3.51**	**185.00**	**195.00**
13272 1351	Containerized Staging	EA	10	CODLA	1.000	4.80	3.51	0.00	8.31
13272 1353	Containerized Feed System Loading	EA	20	ULABA	1.000	1.69	0.00	0.00	1.69
13274 1127	Sodium Hydroxide, 500 Lb Container, Beads	LB	0	N/A	500.000	0.00	0.00	0.37	185.00
33 33 0104	**Sodium Hydroxide, 100 Lb Container, Flakes**	**EA**				**6.49**	**3.51**	**60.00**	**70.00**
13272 1351	Containerized Staging	EA	10	CODLA	1.000	4.80	3.51	0.00	8.31
13272 1353	Containerized Feed System Loading	EA	20	ULABA	1.000	1.69	0.00	0.00	1.69
13274 1128	Sodium Hydroxide, 100 Lb Container, Flakes	LB	0	N/A	100.000	0.00	0.00	0.60	60.00
33 33 0105	**Sodium Hydroxide, 400 Lb Container, Flakes**	**EA**				**0.00**	**0.00**	**156.00**	**156.00**
13274 1129	Sodium Hydroxide, 400 Lb Container, Flakes	LB	0	N/A	400.000	0.00	0.00	0.39	156.00
33 33 0106	**Magnesium Hydroxide Slurry, 55 Gallon Drums, < 40 Drums**	**EA**				**0.00**	**0.00**	**201.00**	**201.00**
13274 1141	Magnesium Hydroxide Slurry, 55 Gallon Drums, Less than 40 Drums	EA	0	N/A	1.000	0.00	0.00	201.00	201.00
33 33 0107	**Magnesium Hydroxide Slurry, 55 Gallon Drums, 40 - 71 Drums**	**EA**				**0.00**	**0.00**	**192.00**	**192.00**
13274 1142	Magnesium Hydroxide Slurry, 55 Gallon Drums, 40 to 71 Drum Order	EA	0	N/A	1.000	0.00	0.00	192.00	192.00
33 33 0108	**Magnesium Hydroxide Slurry, 55 Gallon Drums, 72 Drums**	**EA**				**0.00**	**0.00**	**180.00**	**180.00**
13274 1143	Magnesium Hydroxide Slurry, 55 Gallon Drums, Order of 72 Drums	EA	0	N/A	1.000	0.00	0.00	180.00	180.00
33 33 0109	**Magnesium Hydroxide, Bulk Quantity, 20 Tons**	**EA**				**0.00**	**0.00**	**6,400**	**6,400**
13274 1144	Magnesium Hydroxide Slurry, Bulk Quantity, Minimum Order of 20	TON	0	N/A	20.000	0.00	0.00	320.00	6,400
33 33 0110	**Sulfuric Acid Solution, 220 Lb Drummed Liquid**	**EA**				**6.49**	**3.51**	**28.60**	**38.60**
13272 1351	Containerized Staging	EA	10	CODLA	1.000	4.80	3.51	0.00	8.31
13272 1353	Containerized Feed System Loading	EA	20	ULABA	1.000	1.69	0.00	0.00	1.69
13274 1134	Sulfuric Acid Solution, 220 Lb Drummed Liquid	LB	0	N/A	220.000	0.00	0.00	0.13	28.60
33 33 0111	**Sulfuric Acid Solution, 750 Lb Drummed Liquid**	**EA**				**6.49**	**3.51**	**82.50**	**92.50**
13272 1351	Containerized Staging	EA	10	CODLA	1.000	4.80	3.51	0.00	8.31
13272 1353	Containerized Feed System Loading	EA	20	ULABA	1.000	1.69	0.00	0.00	1.69
13274 1135	Sulfuric Acid Solution, 750 Lb Drummed Liquid	LB	0	N/A	750.000	0.00	0.00	0.11	82.50

33 33 01	Chemical Additives	Unit	Hourly Output	Crew	Factor	Labor	Equip.	Matl.	Total
33 33 0112	**Soda Ash, 50 Lb Bag, Powdered**	**EA**				**6.49**	**3.51**	**5.71**	**15.72**
13272 1351	Containerized Staging	EA	10	CODLA	1.000	4.80	3.51	0.00	8.31
13272 1353	Containerized Feed System Loading	EA	20	ULABA	1.000	1.69	0.00	0.00	1.69
13274 1161	Soda Ash, 50 Lb Bag, Powdered	LB	0	N/A	50.000	0.00	0.00	0.11	5.71
33 33 0113	**Soda Ash, 100 Lb Bag, Powdered**	**EA**				**6.49**	**3.51**	**14.85**	**24.85**
13272 1351	Containerized Staging	EA	10	CODLA	1.000	4.80	3.51	0.00	8.31
13272 1353	Containerized Feed System Loading	EA	20	ULABA	1.000	1.69	0.00	0.00	1.69
13274 1162	Soda Ash, 100 Lb Bag, Powdered	LB	0	N/A	100.000	0.00	0.00	0.15	14.85
33 33 0114	**Quicklime, 1/4" Nominal Granules, Bulk Quantity**	**TON**				**0.01**	**0.01**	**99.02**	**99.04**
13272 1352	Bulk Chemical Staging	TON	8	CODLA	0.001	6.00	4.38	0.00	0.01
13272 1354	Bulk Feed System Loading	TON	4	CODLA	0.001	12.00	8.77	0.00	0.01
13274 1171	Quicklime, 1/4" Nominal Granules, Bulk Quantity	TON	0	N/A	1.000	0.00	0.00	99.02	99.02
33 33 0115	**Quicklime, 3/4" Nominal Granules, Bulk Quantity**	**TON**				**0.01**	**0.01**	**99.02**	**99.04**
13272 1352	Bulk Chemical Staging	TON	8	CODLA	0.001	6.00	4.38	0.00	0.01
13272 1354	Bulk Feed System Loading	TON	4	CODLA	0.001	12.00	8.77	0.00	0.01
13274 1172	Quicklime, 3/4" Nominal Granules, Bulk Quantity	TON	0	N/A	1.000	0.00	0.00	99.02	99.02
33 33 0116	**Quicklime, Combination 1/4" & 3/4" Granules, Bulk Quantity**	**TON**				**0.01**	**0.01**	**99.02**	**99.04**
13272 1352	Bulk Chemical Staging	TON	8	CODLA	0.001	6.00	4.38	0.00	0.01
13272 1354	Bulk Feed System Loading	TON	4	CODLA	0.001	12.00	8.77	0.00	0.01
13274 1173	Quicklime, Combination of 1/4" and 3/4" Granules, Bulk	TON	0	N/A	1.000	0.00	0.00	99.02	99.02
33 33 0117	**Hydrated Lime, Powdered, Bulk**	**TON**				**0.01**	**0.01**	**102.06**	**102.08**
13272 1352	Bulk Chemical Staging	TON	8	CODLA	0.001	6.00	4.38	0.00	0.01
13272 1354	Bulk Feed System Loading	TON	4	CODLA	0.001	12.00	8.77	0.00	0.01
13274 1174	Hydrated Lime, Powdered, Bulk	TON	0	N/A	1.000	0.00	0.00	102.06	102.06
33 33 0118	**Hydrated Lime, Powdered, 50 Lb Bag**	**EA**				**6.49**	**3.51**	**4.42**	**14.42**
13272 1351	Containerized Staging	EA	10	CODLA	1.000	4.80	3.51	0.00	8.31
13272 1353	Containerized Feed System Loading	EA	20	ULABA	1.000	1.69	0.00	0.00	1.69
13274 1176	Hydrated Lime, Powdered, 50 Lb Bag	EA	0	N/A	1.000	0.00	0.00	4.42	4.42
33 33 0119	**Aluminum Sulfate (Alum), Bulk**	**TON**				**0.01**	**0.01**	**252.00**	**252.02**
13272 1331	Aluminum Sulfate (Alum), Bulk	TON	0	N/A	1.000	0.00	0.00	252.00	252.00
13272 1352	Bulk Chemical Staging	TON	8	CODLA	0.001	6.00	4.38	0.00	0.01
13272 1354	Bulk Feed System Loading	TON	4	CODLA	0.001	12.00	8.77	0.00	0.01
33 33 0120	**Ferric Sulfate, Bulk**	**TON**				**0.01**	**0.01**	**200.00**	**200.02**
13272 1332	Ferric Sulfate, Bulk	TON	0	N/A	1.000	0.00	0.00	200.00	200.00
13272 1352	Bulk Chemical Staging	TON	8	CODLA	0.001	6.00	4.38	0.00	0.01
13272 1354	Bulk Feed System Loading	TON	4	CODLA	0.001	12.00	8.77	0.00	0.01

33 33 Chemical Additives

33 33 01	Chemical Additives	Unit	Hourly Output	Crew	Factor	Labor	Equip.	Matl.	Total
33 33 0121	**Ferric Chloride, Bulk**	**TON**				**0.01**	**0.01**	**274.00**	**274.02**
13272 1333	Ferric Chloride, Bulk	TON	0	N/A	1.000	0.00	0.00	274.00	274.00
13272 1352	Bulk Chemical Staging	TON	8	CODLA	0.001	6.00	4.38	0.00	0.01
13272 1354	Bulk Feed System Loading	TON	4	CODLA	0.001	12.00	8.77	0.00	0.01
33 33 0122	**Montmorillonite Clay Flocculant Aid, Bulk**	**LB**				**0.01**	**0.01**	**0.18**	**0.20**
13272 1314	Montmorillonite Clay Flocculant Aid, Powdered or Granular, Bulk	LB	0	N/A	1.000	0.00	0.00	0.18	0.18
13272 1352	Bulk Chemical Staging	TON	8	CODLA	0.001	6.00	4.38	0.00	0.01
13272 1354	Bulk Feed System Loading	TON	4	CODLA	0.001	12.00	8.77	0.00	0.01
33 33 0123	**Sodium Hectorite Clay Flocculant Aid, Bulk**	**LB**				**0.01**	**0.01**	**0.14**	**0.16**
13272 1318	Sodium Hectorite Clay Flocculant Aid, Finely Ground, Bulk	LB	0	N/A	1.000	0.00	0.00	0.14	0.14
13272 1352	Bulk Chemical Staging	TON	8	CODLA	0.001	6.00	4.38	0.00	0.01
13272 1354	Bulk Feed System Loading	TON	4	CODLA	0.001	12.00	8.77	0.00	0.01
33 33 0124	**Super Dispersible Sodium Flocculant Aid, Bulk**	**LB**				**0.01**	**0.01**	**0.26**	**0.28**
13272 1322	Super Dispersible Sodium Bentonite Flocculant Aid, Bulk	LB	0	N/A	1.000	0.00	0.00	0.26	0.26
13272 1352	Bulk Chemical Staging	TON	8	CODLA	0.001	6.00	4.38	0.00	0.01
13272 1354	Bulk Feed System Loading	TON	4	CODLA	0.001	12.00	8.77	0.00	0.01
33 33 0125	**Alkylbenzene Sulfonate Surfactant, Bulk**	**LB**				**0.00**	**0.00**	**1.12**	**1.12**
13273 8151	Surfactant (Alkylbenzene Sulfonate)	LB	0	N/A	1.000	0.00	0.00	1.12	1.12
33 33 0126	**Dispersant, Viscosity Reduction of Liquids/Sludges**	**LB**				**0.00**	**0.00**	**0.37**	**0.37**
13272 1319	Super Dispersible Sodium Bentonite Flocculant Aid	LB	0	N/A	1.000	0.00	0.00	0.37	0.37
33 33 0127	**Non-ionic Polymeric Flocculant**	**LB**				**0.00**	**0.00**	**3.65**	**3.65**
13272 1334	Non-ionic Polymeric Flocculant	LB	0	N/A	1.000	0.00	0.00	3.65	3.65
33 33 0151	**Oxygen, 192 Liter Cylinders**	**EA**				**0.00**	**0.00**	**89.50**	**89.50**
13276 2311	Oxygen, 192 Liter Oxygen Gas Dewar	EA	0	N/A	1.000	0.00	0.00	89.50	89.50
33 33 0171	**Hydrogen Peroxide, 50% Solution, 500 Lb Drums**	**EA**				**0.00**	**0.00**	**940.00**	**940.00**
13274 1319	Hydrogen Peroxide, 50% Solution, 500 Lb Drum	LB	0	N/A	500.000	0.00	0.00	1.88	940.00
33 33 0172	**Hydrogen Peroxide Feed Tank, Regulator, Injector & Panel Rental**	**EA**				**889.18**	**66.50**	**1,825**	**2,781**
13274 1712	Gas Feeder, 100 Lb/Day Feed Capacity, Vacuum Regulator, Injector, and Separate Meter Panel	EA	0.125	MSPFH	1.000	889.18	66.50	1,825	2,781

33 33 30	Filterpress Moisture Reduction	Unit	Hourly Output	Crew	Factor	Labor	Equip.	Matl.	Total

33 33 Chemical Additives

33 33 30 Filterpress Moisture Reduction		Unit	Hourly Output	Crew	Factor	Labor	Equip.	Matl.	Total
33 33 3001	**Filter Press, Moisture Reduction (Rental)**	**TON**				**0.00**	**0.00**	**55.85**	**55.85**
13276 5228	Retrofitted Filter Belt Press, Dewatering Solid Waste	TON	0	N/A	1.000	0.00	0.00	55.85	55.85
33 33 3010	**1 CF Filter Press system w/sludge tanks,pumps,mixers,sludge cart**	**EA**				**0.00**	**0.00**	**27,545**	**27,545**
01001 0544	1 CF Filter Press system w/sludge tanks,pumps,mixers,sludge cart	EA	0	N/A	1.000	0.00	0.00	27,545	27,545
33 33 3011	**2 CF Filter Press system w/sludge tanks,pumps,mixers,sludge cart**	**EA**				**0.00**	**0.00**	**33,076**	**33,076**
01001 0545	2 CF Filter Press system w/sludge tanks,pumps,mixers,sludge cart	EA	0	N/A	1.000	0.00	0.00	33,076	33,076
33 33 3012	**5 CF Filter Press system w/sludge tanks,pumps,mixers,sludge cart**	**EA**				**0.00**	**0.00**	**49,666**	**49,666**
01001 0546	5 CF Filter Press system w/sludge tanks,pumps,mixers,sludge cart	EA	0	N/A	1.000	0.00	0.00	49,666	49,666
33 33 3013	**10 CF Filter Press system w/sludge tanks,pumps,mixers,sludge cart**	**EA**				**0.00**	**0.00**	**77,318**	**77,318**
01001 0547	10 CF Filter Press system w/sludge tanks,pumps,mixers,sludge cart	EA	0	N/A	1.000	0.00	0.00	77,318	77,318
33 33 3014	**25 CF Filter Press system w/sludge tanks,pumps,mixers,sludge cart**	**EA**				**0.00**	**0.00**	**139,482**	**139,482**
01001 0548	25 CF Filter Press system w/sludge tanks,pumps,mixers,sludge cart	EA	0	N/A	1.000	0.00	0.00	139,482	139,482
33 33 3015	**50 CF Filter Press system w/sludge tanks,pumps,mixers,sludge cart**	**EA**				**0.00**	**0.00**	**298,532**	**298,532**
01001 0549	50 CF Filter Press system w/sludge tanks,pumps,mixers,sludge cart	EA	0	N/A	1.000	0.00	0.00	298,532	298,532

33 34 01 Dozer Rental		Unit	Hourly Output	Crew	Factor	Labor	Equip.	Matl.	Total
33 34 0101	**Standby D3 Bulldozer with Tiller**	**HR**				**0.00**	**9.64**	**0.00**	**9.64**
13273 9121	Standby D3 Bulldozer with Tiller	HR	1	XYCDL	0.218	0.00	44.24	0.00	9.64
33 34 0102	**Standby D4 Bulldozer with Tiller**	**HR**				**0.00**	**9.48**	**0.00**	**9.48**
13273 9122	Standby D4 Bulldozer with Tiller	HR	1	XYCDI	0.214	0.00	44.24	0.00	9.48
33 34 0103	**Standby D7 Bulldozer with Tiller**	**HR**				**0.00**	**26.82**	**0.00**	**26.82**
13273 9123	Standby D7 Bulldozer with Tiller	HR	1	XYCDJ	0.224	0.00	119.84	0.00	26.82
33 34 0104	**Standby D9 Bulldozer with Tiller**	**HR**				**0.00**	**48.42**	**0.00**	**48.42**
13273 9124	Standby D9 Bulldozer with Tiller	HR	1	XYCDK	0.248	0.00	195.39	0.00	48.42
33 34 0105	**Standby D6 Bulldozer with Tiller**	**HR**				**0.00**	**10.04**	**0.00**	**10.04**
13273 9125	Standby D6 Bulldozer with Tiller	HR	1	XYCDI	0.227	0.00	44.24	0.00	10.04

33 41 01 Pumps and Motors	Unit	Hourly Output	Crew	Factor	Labor	Equip.	Matl.	Total

33 41 Mechanical Maintenance

33 41 01 Pumps and Motors		Unit	Hourly Output	Crew	Factor	Labor	Equip.	Matl.	Total
33 41 0101	**Pump & Motor Maintenance/Repair**	**EA**				**371.20**	**53.60**	**0.00**	**424.80**
13273 4197	Rework Motor Gasket, Bearings, and Shaft	EA	0.125	COFWB	1.000	371.20	53.60	0.00	424.80

33 41 02 Blowers		Unit	Hourly Output	Crew	Factor	Labor	Equip.	Matl.	Total
33 41 0201	**Blower and Motor Maintenance and Repair**	**EA**				**371.20**	**53.60**	**0.00**	**424.80**
13273 4197	Rework Motor Gasket, Bearings, and Shaft	EA	0.125	COFWB	1.000	371.20	53.60	0.00	424.80

33 42 01 Electrical Usage Fees		Unit	Hourly Output	Crew	Factor	Labor	Equip.	Matl.	Total
33 42 0101	**Electrical Charge**	**KWH**				**0.00**	**0.00**	**0.07**	**0.07**
02688 5001	Electrical Charge Industrial Use	KWH	0	N/A	1.000	0.00	0.00	0.07	0.07
33 42 0102	**1.5 HP Motor, Electric Charge**	**MO**				**0.00**	**0.00**	**66.69**	**66.69**
02688 5001	Electrical Charge Industrial Use	KWH	0	N/A	900.000	0.00	0.00	0.07	66.69
33 42 0103	**2 HP Motor, Electric Charge**	**MO**				**0.00**	**0.00**	**96.33**	**96.33**
02688 5001	Electrical Charge Industrial Use	KWH	0	N/A	1300.000	0.00	0.00	0.07	96.33
33 42 0104	**3 HP Motor, Electric Charge**	**MO**				**0.00**	**0.00**	**140.79**	**140.79**
02688 5001	Electrical Charge Industrial Use	KWH	0	N/A	1900.000	0.00	0.00	0.07	140.79
33 42 0105	**5 HP Motor, Electric Charge**	**MO**				**0.00**	**0.00**	**214.89**	**214.89**
02688 5001	Electrical Charge Industrial Use	KWH	0	N/A	2900.000	0.00	0.00	0.07	214.89
33 42 0106	**Miscellaneous Electrical Site Usage**	**MO**				**0.00**	**0.00**	**296.40**	**296.40**
02688 5001	Electrical Charge Industrial Use	KWH	0	N/A	4000.000	0.00	0.00	0.07	296.40
33 42 0110	**Electrical Charge for Oil Pump-Out Pump Operation**	**KWH**				**0.00**	**0.00**	**0.07**	**0.07**
02688 5001	Electrical Charge Industrial Use	KWH	0	N/A	1.000	0.00	0.00	0.07	0.07
33 42 0120	**Generator Supplied Electricity**	**KWH**				**0.00**	**0.00**	**0.11**	**0.11**
02554 1112	Generator Supplied Electricity	KWH	0	N/A	1.000	0.00	0.00	0.11	0.11

33 42 02 Diesel Fuel		Unit	Hourly Output	Crew	Factor	Labor	Equip.	Matl.	Total
33 42 0201	**Diesel Fuel**	**GAL**				**0.00**	**0.00**	**1.24**	**1.24**
02688 5005	Diesel Fuel	GAL	0	N/A	1.000	0.00	0.00	1.24	1.24

33 42 03 Water		Unit	Hourly Output	Crew	Factor	Labor	Equip.	Matl.	Total
33 42 0301	**Process Water, Supplied by Tanker Truck**	**KGAL**				**0.00**	**0.00**	**11.00**	**11.00**
02555 6101	Water	KGAL	0	N/A	1.000	0.00	0.00	11.00	11.00
33 42 0302	**Process Water, Supplied by Water Line**	**KGAL**				**0.00**	**0.00**	**3.35**	**3.35**
02555 6102	Process Water, Supplied by Water Line	KGAL	0	N/A	1.000	0.00	0.00	3.35	3.35

33 43 Electrical Usage Fees

33 43 01	Equipment/Hazardous Material Storage Building	Unit	Hourly Output	Crew	Factor	Labor	Equip.	Matl.	Total
33 43 0101	**Equipment Building 10' Ceiling, Built-Up Roof, Concrete Block Exterior**	**SF**				**0.00**	**0.00**	**77.00**	**77.00**
02083 6134	Equipment Building 10' Ceiling, Built-up Roof, Concrete Block Exterior	SF	0	N/A	1.000	0.00	0.00	77.00	77.00
33 43 0102	**5.58' x 5.5' x 8.33' Hazardous Material Storage Building, 4 Drum Capacity**	**EA**				**0.00**	**0.00**	**3,725**	**3,725**
02083 6131	5.58' x 5.5' x 8.33' - 4 Drum Capacity, 88 Gallon Sump, One Door	EA	0	N/A	1.000	0.00	0.00	3,725	3,725
33 43 0103	**10.25' x 5.5' x 8.33' Hazardous Material Storage Building, 4 Drum Capacity**	**EA**				**0.00**	**0.00**	**9,625**	**9,625**
02083 6132	10.25' x 5.5' x 8.33' - 10 Drum Capacity, 175 Gallon Sump, Two Doors	EA	0	N/A	1.000	0.00	0.00	9,625	9,625
33 43 0104	**10.25' x 5.5' x 8.33' Hazardous Material Storage Building, 8 Drum Capacity**	**EA**				**0.00**	**0.00**	**11,300**	**11,300**
02083 6133	10.25' x 5.5' x 8.33' - 10 Drum Capacity, 175 Gallon Sump, 4-Hour Rated	EA	0	N/A	1.000	0.00	0.00	11,300	11,300
33 43 0106	**10.0' x 12.0' x 8.67' Hazardous Material Storage Building**	**EA**				**201.50**	**189.88**	**11,669**	**12,060**
02083 6135	10.0' x 12.0' x 8.67' Hazardous Material Storage Building Precast Concrete	EA	1	UOEHH	1.000	201.50	189.88	11,669	12,060
33 43 0107	**12.0' x 20.0' x 8.67' Hazardous Material Storage Building**	**EA**				**268.67**	**253.17**	**15,733**	**16,255**
02083 6136	12.0' x 20.0' x 8.67' Hazardous Material Storage Building Precast Concrete	EA	0.75	UOEHH	1.000	268.67	253.17	15,733	16,255
33 43 0108	**20.0' x 20.0' x 10.0' Hazardous Material Storage Building**	**EA**				**403.00**	**379.75**	**28,844**	**29,627**
02083 6137	20.0' x 20.0' x 10.0' Hazardous Material Storage Building Precast Concrete	EA	0.5	UOEHH	1.000	403.00	379.75	28,844	29,627
33 43 0109	**24.0' x 20.0' x 10.0' Hazardous Material Storage Building**	**EA**				**503.75**	**474.69**	**36,711**	**37,689**
02083 6138	24.0' x 20.0' x 10.0' Hazardous Material Storage Building Precast Concrete	EA	0.4	UOEHH	1.000	503.75	474.69	36,711	37,689
33 43 0110	**20.0' x 10.0' x 10.0' Hazardous Material Storage Building Expansion**	**EA**				**268.67**	**253.17**	**11,669**	**12,191**
02083 6181	20.0' x 10.0' x 10.0' Hazardous Material Storage Building Expansion Piece, Precast	EA	0.75	UOEHH	1.000	268.67	253.17	11,669	12,191
33 43 0111	**24.0' x 10.0' x 10.0' Hazardous Material Storage Building Expansion**	**EA**				**268.67**	**253.17**	**14,422**	**14,944**
02083 6182	24.0' x 10.0' x 10.0' Hazardous Material Storage Building Expansion Piece, Precast	EA	0.75	UOEHH	1.000	268.67	253.17	14,422	14,944
33 43 0115	**Non-Skid Concrete Ramp**	**EA**				**247.50**	**0.00**	**393.33**	**640.83**
02083 6183	Non-Skid Concrete Ramp	EA	0.35	ACARB	1.000	247.50	0.00	393.33	640.83

33 99 Other

33 99 01	Community Relations Support	Unit	Hourly Output	Crew	Factor	Labor	Equip.	Matl.	Total
33 99 0101	**Meeting Facility Rental, per Day**	**DAY**				**0.00**	**0.00**	**383.14**	**383.14**
01001 2144	Meeting Facility Rental, per Day	DAY	0	N/A	1.000	0.00	0.00	383.14	383.14
33 99 0105	**Letter/Brochure Printing and Distribution, per Page**	**EA**				**0.00**	**0.00**	**1.64**	**1.64**
01001 2145	Letter/Brochure Printing and Distribution, per Page	EA	0	N/A	1.000	0.00	0.00	1.64	1.64

99 01 Prime Contractor Field Office Staff

99 01 01 Site Project Manager			Unit	Hourly Output	Crew	Factor	Labor	Equip.	Matl.	Total
99 01	**0101**	**Site Project Manager - Minimum Cost**	**MWK**				**1,424**	**0.00**	**0.00**	**1,424**
01050	0106	Project Manager - Minimum Cost	WK	0.025	XYBBG	1.000	1,424	0.00	0.00	1,424
99 01	**0102**	**Site Project Manager - Average Cost**	**MWK**				**1,596**	**0.00**	**0.00**	**1,596**
01050	0107	Project Manager - Average Cost	WK	0.025	XYBBH	1.000	1,596	0.00	0.00	1,596
99 01	**0103**	**Site Project Manager - Maximum Cost**	**MWK**				**1,803**	**0.00**	**0.00**	**1,803**
01050	0108	Project Manager - Maximum Cost	WK	0.025	XYBBI	1.000	1,803	0.00	0.00	1,803

99 01 02 Superintendent			Unit	Hourly Output	Crew	Factor	Labor	Equip.	Matl.	Total
99 01	**0201**	**Superintendent - Minimum Cost**	**MWK**				**1,375**	**0.00**	**0.00**	**1,375**
01050	0109	Field Superintendent - Minimum Cost	WK	0.025	XYBBJ	1.000	1,375	0.00	0.00	1,375
99 01	**0202**	**Superintendent - Average Cost**	**MWK**				**1,500**	**0.00**	**0.00**	**1,500**
01050	0110	Field Superintendent - Average Cost	WK	0.025	XYBBK	1.000	1,500	0.00	0.00	1,500
99 01	**0203**	**Superintendent - Maximum Cost**	**MWK**				**1,725**	**0.00**	**0.00**	**1,725**
01050	0111	Field Superintendent - Maximum Cost	WK	0.025	XYBBL	1.000	1,725	0.00	0.00	1,725

99 01 03 Clerk			Unit	Hourly Output	Crew	Factor	Labor	Equip.	Matl.	Total
99 01	**0301**	**Clerk - Average Cost**	**MWK**				**320.00**	**0.00**	**0.00**	**320.00**
01050	0101	Field Office Clerk	WK	0.025	XYBBA	1.000	320.00	0.00	0.00	320.00

99 01 04 Field Engineer			Unit	Hourly Output	Crew	Factor	Labor	Equip.	Matl.	Total
99 01	**0401**	**Field Engineer - Minimum Cost**	**MWK**				**765.00**	**0.00**	**0.00**	**765.00**
01050	0102	Field Engineer (Minimum Cost)	WK	0.025	XYBBC	1.000	765.00	0.00	0.00	765.00
99 01	**0402**	**Field Engineer - Average Cost**	**MWK**				**995.00**	**0.00**	**0.00**	**995.00**
01050	0103	Field Engineer (Average Cost)	WK	0.025	XYBBD	1.000	995.00	0.00	0.00	995.00
99 01	**0403**	**Field Engineer - Maximum Cost**	**MWK**				**1,150**	**0.00**	**0.00**	**1,150**
01050	0104	Field Engineer - Maximum Cost	WK	0.025	XYBBE	1.000	1,150	0.00	0.00	1,150

99 01 05 Timekeeper			Unit	Hourly Output	Crew	Factor	Labor	Equip.	Matl.	Total
99 01	**0501**	**Timekeeper - Average Cost**	**MWK**				**890.00**	**0.00**	**0.00**	**890.00**
01050	0112	Timekeeper - Average Cost	WK	0.025	XYBBM	1.000	890.00	0.00	0.00	890.00

99 01 06 General Purpose Laborer			Unit	Hourly Output	Crew	Factor	Labor	Equip.	Matl.	Total
99 01	**0601**	**General-purpose Laborer**	**MWK**				**1,068**	**0.00**	**0.00**	**1,068**
01050	0105	General Purpose Laborer	WK	0.025	XYBBF	1.000	1,068	0.00	0.00	1,068

99 01 07 Safety Engineer			Unit	Hourly Output	Crew	Factor	Labor	Equip.	Matl.	Total
99 01	**0701**	**Safety Engineer - Minimum Cost**	**MWK**				**1,545**	**0.00**	**0.00**	**1,545**
01050	0121	Safety Engineer - Minimum Cost	WK	0.025	XYBBQ	1.000	1,545	0.00	0.00	1,545

99 01 Prime Contractor Field Office Staff

99 01 07 Safety Engineer		Unit	Hourly Output	Crew	Factor	Labor	Equip.	Matl.	Total
99 01 0702	**Safety Engineer - Average Cost**	**MWK**				**1,545**	**0.00**	**0.00**	**1,545**
01050 0122	Safety Engineer - Average Cost	WK	0.025	XYBBR	1.000	1,545	0.00	0.00	1,545
99 01 0703	**Safety Engineer - Maximum Cost**	**MWK**				**1,545**	**0.00**	**0.00**	**1,545**
01050 0123	Safety Engineer - Maximum Cost	WK	0.025	XYBBS	1.000	1,545	0.00	0.00	1,545

99 01 08 Quality Control		Unit	Hourly Output	Crew	Factor	Labor	Equip.	Matl.	Total
99 01 0801	**Quality Control - Minimum Cost**	**MWK**				**1,224**	**0.00**	**0.00**	**1,224**
01050 0124	Quality Control - Minimum Cost	WK	0.025	XYBBT	1.000	1,224	0.00	0.00	1,224
99 01 0802	**Quality Control - Average Cost**	**MWK**				**1,402**	**0.00**	**0.00**	**1,402**
01050 0125	Quality Control - Average Cost	WK	0.025	XYBBU	1.000	1,402	0.00	0.00	1,402
99 01 0803	**Quality Control - Maximum Cost**	**MWK**				**1,291**	**0.00**	**0.00**	**1,291**
01050 0126	Quality Control - Maximum Cost	WK	0.025	XYBBV	1.000	1,291	0.00	0.00	1,291

99 04 01 Temporary Offices		Unit	Hourly Output	Crew	Factor	Labor	Equip.	Matl.	Total
99 04 0101	**Temporary Office 20' x 8'**	**MO**				**0.00**	**0.00**	**207.88**	**207.88**
01500 0209	Temporary Office Trailer 20' x 8' without Hookup	MO	0	N/A	1.000	0.00	0.00	168.39	168.39
01500 0213	Temporary Construction Office Air Conditioner	MO	0	N/A	1.000	0.00	0.00	39.48	39.48
99 04 0102	**Temporary Office 32' x 8'**	**MO**				**0.00**	**0.00**	**248.52**	**248.52**
01500 0210	Temporary Office Trailer 32' x 8' without Hookup	MO	0	N/A	1.000	0.00	0.00	209.04	209.04
01500 0213	Temporary Construction Office Air Conditioner	MO	0	N/A	1.000	0.00	0.00	39.48	39.48
99 04 0103	**Temporary Office 50' x 10'**	**MO**				**0.00**	**0.00**	**364.65**	**364.65**
01500 0211	Temporary Office Trailer 50' x 10' without Hookup	MO	0	N/A	1.000	0.00	0.00	325.17	325.17
01500 0213	Temporary Construction Office Air Conditioner	MO	0	N/A	1.000	0.00	0.00	39.48	39.48
99 04 0104	**Temporary Office 50' x 12'**	**MO**				**0.00**	**0.00**	**418.07**	**418.07**
01500 0212	Temporary Office Trailer 50' x 12' without Hookup	MO	0	N/A	1.000	0.00	0.00	378.59	378.59
01500 0213	Temporary Construction Office Air Conditioner	MO	0	N/A	1.000	0.00	0.00	39.48	39.48

99 04 02 Temporary Storage Facilities		Unit	Hourly Output	Crew	Factor	Labor	Equip.	Matl.	Total
99 04 0201	**Temporary Storage Trailer 16' x 8'**	**MO**				**0.00**	**0.00**	**81.29**	**81.29**
01500 0214	Temporary Construction Storage Van 16' x 8'	MO	0	N/A	1.000	0.00	0.00	81.29	81.29
99 04 0202	**Temporary Storage Trailer 28' x 10'**	**MO**				**0.00**	**0.00**	**110.33**	**110.33**
01500 0215	Temporary Construction Storage Van 28' x 10'	MO	0	N/A	1.000	0.00	0.00	110.33	110.33

99 04 03 Security Fencing		Unit	Hourly Output	Crew	Factor	Labor	Equip.	Matl.	Total
99 04 0301	**Security Fencing 5' Chain-Link**	**LF**				**1.89**	**0.00**	**4.61**	**6.50**
01500 0201	Construction Fence Chain Link 5'	LF	50	ULABC	1.000	1.89	0.00	4.61	6.50

99 04 Prime Contractor Field Expenses

99 04 03 Security Fencing

		Unit	Hourly Output	Crew	Factor	Labor	Equip.	Matl.	Total
99 04 0302	**Security Fencing 6' Chain-Link**	**LF**				**2.10**	**0.00**	**4.77**	**6.87**
01500 0202	Construction Fence Chain Link 6'	LF	45	ULABC	1.000	2.10	0.00	4.77	6.87

99 04 04 Construction Signs

		Unit	Hourly Output	Crew	Factor	Labor	Equip.	Matl.	Total
99 04 0401	**Construction Signs**	**SF**				**0.00**	**0.00**	**11.74**	**11.74**
01500 0102	Temporary Construction Sign	SF	0	N/A	1.000	0.00	0.00	11.74	11.74

99 04 05 Portable Toilets

		Unit	Hourly Output	Crew	Factor	Labor	Equip.	Matl.	Total
99 04 0501	**Portable Toilets - Chemical**	**MO**				**0.00**	**0.00**	**83.23**	**83.23**
01500 0216	Toilet Portable Chemical	MO	0	N/A	1.000	0.00	0.00	83.23	83.23

99 04 06 Temporary Heat

		Unit	Hourly Output	Crew	Factor	Labor	Equip.	Matl.	Total
99 04 0601	**Temporary Heat - Buildings (12 Hours/Day)**	**CSF*W**				**0.00**	**0.00**	**1.49**	**1.49**
01500 0203	Temporary Construction Heat (12 Hours/Day)	CSF	0	N/A	1.000	0.00	0.00	1.49	1.49
99 04 0602	**Temporary Heat - Buildings (24 Hours/Day)**	**CSF*W**				**0.00**	**0.00**	**2.22**	**2.22**
01500 0204	Temporary Construction Heat/Week (24 Hours/Day)	CSF	0	N/A	1.000	0.00	0.00	2.22	2.22

99 04 07 Temporary Lighting

		Unit	Hourly Output	Crew	Factor	Labor	Equip.	Matl.	Total
99 04 0701	**Temporary Lighting - Minimum**	**CSF**				**10.48**	**0.00**	**2.36**	**12.84**
01500 0205	Temporary Construction Lighting with Service - Minimum	CSF	10	EELEC	1.000	10.48	0.00	2.36	12.84
99 04 0702	**Temporary Lighting - Maximum**	**CSF**				**20.97**	**0.00**	**5.13**	**26.10**
01500 0206	Temporary Construction Lighting with Service - Maximum	CSF	5	EELEC	1.000	20.97	0.00	5.13	26.10

99 04 08 Temporary Electrical Power

		Unit	Hourly Output	Crew	Factor	Labor	Equip.	Matl.	Total
99 04 0801	**Temporary Electric Power - Average**	**CSF**				**0.00**	**0.00**	**91.16**	**91.16**
01500 0207	Temporary Power and Lighting per Month Average	CSF	0	N/A	1.000	0.00	0.00	91.16	91.16
99 04 0802	**Temporary Electric Power - Maximum**	**CSF**				**0.00**	**0.00**	**127.75**	**127.75**
01500 0208	Temporary Power and Lighting per Month Maximum	CSF	0	N/A	1.000	0.00	0.00	127.75	127.75

99 04 11 Photographs

		Unit	Hourly Output	Crew	Factor	Labor	Equip.	Matl.	Total
99 04 1101	**Construction Photographs**	**SET**				**320.00**	**0.00**	**0.00**	**320.00**
01050 0101	Field Office Clerk	WK	0.025	XYBBA	1.000	320.00	0.00	0.00	320.00
99 04 1102	**Time Lapse Video Equipment & Tapes, Per Month**	**MO**				**5.15**	**0.00**	**358.91**	**364.06**
01953 2113	Time Lapse Recorder, S-VHS/VHS Compatible, High Performance	EA	1.63	EELEA	0.163	31.56	0.00	2,200	364.06

99 04 Prime Contractor Field Expenses

99 04 12 Surveying Expenses	Unit	Hourly Output	Crew	Factor	Labor	Equip.	Matl.	Total
99 04 1201 Surveying - 2-man Crew	**DAY**				**656.95**	**217.78**	**0.00**	**874.73**
01050 1001 2-person Crew	DAY	0.125	XYABC	1.000	656.95	217.78	0.00	874.73
99 04 1202 Surveying - 3-man Crew	**DAY**				**920.95**	**217.78**	**0.00**	**1,139**
01050 1002 3-person Crew	DAY	0.125	XYABD	1.000	920.95	217.78	0.00	1,139
99 04 1203 Surveying - 4-man Crew	**DAY**				**1,185**	**217.78**	**0.00**	**1,403**
01050 1003 4-person Crew	DAY	0.125	XYABE	1.000	1,185	217.78	0.00	1,403
99 04 1204 Hand Held GPS Unit, Battery Powered	**EA**				**0.00**	**0.00**	**551.63**	**551.63**
01953 3301 Hand Held GPS Unit, Battery Powered	EA	0	N/A	1.000	0.00	0.00	551.63	551.63
99 04 1205 Portable GPS Set with Mapping, 5 cm Accuracy	**MO**				**0.00**	**0.00**	**638.72**	**638.72**
01953 3302 Portable, Differential GPS Surveying Set with Map Output	MO	0	N/A	1.000	0.00	0.00	638.72	638.72
99 04 1206 Data Automation System (Computer)	**MO**				**0.00**	**0.00**	**232.26**	**232.26**
01953 3303 Pentium Computer with Windows	MO	0	N/A	1.000	0.00	0.00	232.26	232.26

99 04 13 Watchmen	Unit	Hourly Output	Crew	Factor	Labor	Equip.	Matl.	Total
99 04 1301 Uniformed Watchman - Minimum	**HR**				**8.90**	**0.00**	**0.00**	**8.90**
01050 0113 Uniformed Watchman - Minimum	HR	1	XYBBN	1.000	8.90	0.00	0.00	8.90
99 04 1302 Uniformed Watchman - Maximum	**HR**				**16.13**	**0.00**	**0.00**	**16.13**
01050 0114 Uniformed Watchman - Maximum	HR	1	XYBBO	1.000	16.13	0.00	0.00	16.13

99 04 14 Continuous Cleanup	Unit	Hourly Output	Crew	Factor	Labor	Equip.	Matl.	Total
99 04 1401 Continuous Cleanup	**MWK**				**1,068**	**0.00**	**0.00**	**1,068**
01050 0105 General Purpose Laborer	WK	0.025	XYBBF	1.000	1,068	0.00	0.00	1,068

99 11 01 Site Project Manager	Unit	Hourly Output	Crew	Factor	Labor	Equip.	Matl.	Total
99 11 0101 Site Project Manager - Minimum Cost	**MWK**				**1,424**	**0.00**	**0.00**	**1,424**
01050 0106 Project Manager - Minimum Cost	WK	0.025	XYBBG	1.000	1,424	0.00	0.00	1,424
99 11 0102 Site Project Manager - Average Cost	**MWK**				**1,596**	**0.00**	**0.00**	**1,596**
01050 0107 Project Manager - Average Cost	WK	0.025	XYBBH	1.000	1,596	0.00	0.00	1,596
99 11 0103 Site Project Manager - Maximum Cost	**MWK**				**1,803**	**0.00**	**0.00**	**1,803**
01050 0108 Project Manager - Maximum Cost	WK	0.025	XYBBI	1.000	1,803	0.00	0.00	1,803

99 11 02 Superintendent	Unit	Hourly Output	Crew	Factor	Labor	Equip.	Matl.	Total
99 11 0201 Superintendent - Minimum Cost	**MWK**				**1,375**	**0.00**	**0.00**	**1,375**
01050 0109 Field Superintendent - Minimum Cost	WK	0.025	XYBBJ	1.000	1,375	0.00	0.00	1,375

99 11 Major Subcontractor Field Office Staff

99 11 02 Superintendent

Code	Description	Unit	Hourly Output	Crew	Factor	Labor	Equip.	Matl.	Total
99 11 0202	**Superintendent - Average Cost**	**MWK**				**1,500**	**0.00**	**0.00**	**1,500**
01050 0110	Field Superintendent - Average Cost	WK	0.025	XYBBK	1.000	1,500	0.00	0.00	1,500
99 11 0203	**Superintendent - Maximum Cost**	**MWK**				**1,725**	**0.00**	**0.00**	**1,725**
01050 0111	Field Superintendent - Maximum Cost	WK	0.025	XYBBL	1.000	1,725	0.00	0.00	1,725

99 11 03 Clerk

Code	Description	Unit	Hourly Output	Crew	Factor	Labor	Equip.	Matl.	Total
99 11 0301	**Clerk - Average Cost**	**MWK**				**320.00**	**0.00**	**0.00**	**320.00**
01050 0101	Field Office Clerk	WK	0.025	XYBBA	1.000	320.00	0.00	0.00	320.00

99 11 04 Field Engineer

Code	Description	Unit	Hourly Output	Crew	Factor	Labor	Equip.	Matl.	Total
99 11 0401	**Field Engineer - Minimum Cost**	**MWK**				**765.00**	**0.00**	**0.00**	**765.00**
01050 0102	Field Engineer (Minimum Cost)	WK	0.025	XYBBC	1.000	765.00	0.00	0.00	765.00
99 11 0402	**Field Engineer - Average Cost**	**MWK**				**995.00**	**0.00**	**0.00**	**995.00**
01050 0103	Field Engineer (Average Cost)	WK	0.025	XYBBD	1.000	995.00	0.00	0.00	995.00
99 11 0403	**Field Engineer - Maximum Cost**	**MWK**				**1,150**	**0.00**	**0.00**	**1,150**
01050 0104	Field Engineer - Maximum Cost	WK	0.025	XYBBE	1.000	1,150	0.00	0.00	1,150

99 11 05 Timekeeper

Code	Description	Unit	Hourly Output	Crew	Factor	Labor	Equip.	Matl.	Total
99 11 0501	**Timekeeper - Average Cost**	**MWK**				**890.00**	**0.00**	**0.00**	**890.00**
01050 0112	Timekeeper - Average Cost	WK	0.025	XYBBM	1.000	890.00	0.00	0.00	890.00

99 11 06 General Purpose Laborer

Code	Description	Unit	Hourly Output	Crew	Factor	Labor	Equip.	Matl.	Total
99 11 0601	**General-purpose Laborer**	**MWK**				**1,068**	**0.00**	**0.00**	**1,068**
01050 0105	General Purpose Laborer	WK	0.025	XYBBF	1.000	1,068	0.00	0.00	1,068

99 11 07 Safety Engineer

Code	Description	Unit	Hourly Output	Crew	Factor	Labor	Equip.	Matl.	Total
99 11 0701	**Safety Engineer - Minimum Cost**	**MWK**				**1,545**	**0.00**	**0.00**	**1,545**
01050 0121	Safety Engineer - Minimum Cost	WK	0.025	XYBBQ	1.000	1,545	0.00	0.00	1,545
99 11 0702	**Safety Engineer - Average Cost**	**MWK**				**1,545**	**0.00**	**0.00**	**1,545**
01050 0122	Safety Engineer - Average Cost	WK	0.025	XYBBR	1.000	1,545	0.00	0.00	1,545
99 11 0703	**Safety Engineer - Maximum Cost**	**MWK**				**1,545**	**0.00**	**0.00**	**1,545**
01050 0123	Safety Engineer - Maximum Cost	WK	0.025	XYBBS	1.000	1,545	0.00	0.00	1,545

99 11 08 Quality Control

Code	Description	Unit	Hourly Output	Crew	Factor	Labor	Equip.	Matl.	Total
99 11 0801	**Quality Control - Minimum Cost**	**MWK**				**1,224**	**0.00**	**0.00**	**1,224**
01050 0124	Quality Control - Minimum Cost	WK	0.025	XYBBT	1.000	1,224	0.00	0.00	1,224
99 11 0802	**Quality Control - Average Cost**	**MWK**				**1,402**	**0.00**	**0.00**	**1,402**
01050 0125	Quality Control - Average Cost	WK	0.025	XYBBU	1.000	1,402	0.00	0.00	1,402

99 11 Major Subcontractor Field Office Staff

99 11 08 Quality Control	Unit	Hourly Output	Crew	Factor	Labor	Equip.	Matl.	Total
99 11 0803 Quality Control - Maximum Cost	**MWK**				**1,291**	**0.00**	**0.00**	**1,291**
01050 0126 Quality Control - Maximum Cost	WK	0.025	XYBBV	1.000	1,291	0.00	0.00	1,291

99 14 01 Temporary Offices	Unit	Hourly Output	Crew	Factor	Labor	Equip.	Matl.	Total
99 14 0101 Temporary Office 20' x 8'	**MO**				**0.00**	**0.00**	**207.88**	**207.88**
01500 0209 Temporary Office Trailer 20' x 8' without Hookup	MO	0	N/A	1.000	0.00	0.00	168.39	168.39
01500 0213 Temporary Construction Office Air Conditioner	MO	0	N/A	1.000	0.00	0.00	39.48	39.48
99 14 0102 Temporary Office 32' x 8'	**MO**				**0.00**	**0.00**	**248.52**	**248.52**
01500 0210 Temporary Office Trailer 32' x 8' without Hookup	MO	0	N/A	1.000	0.00	0.00	209.04	209.04
01500 0213 Temporary Construction Office Air Conditioner	MO	0	N/A	1.000	0.00	0.00	39.48	39.48
99 14 0103 Temporary Office 50' x 10'	**MO**				**0.00**	**0.00**	**364.65**	**364.65**
01500 0211 Temporary Office Trailer 50' x 10' without Hookup	MO	0	N/A	1.000	0.00	0.00	325.17	325.17
01500 0213 Temporary Construction Office Air Conditioner	MO	0	N/A	1.000	0.00	0.00	39.48	39.48
99 14 0104 Temporary Office 50' x 12'	**MO**				**0.00**	**0.00**	**418.07**	**418.07**
01500 0212 Temporary Office Trailer 50' x 12' without Hookup	MO	0	N/A	1.000	0.00	0.00	378.59	378.59
01500 0213 Temporary Construction Office Air Conditioner	MO	0	N/A	1.000	0.00	0.00	39.48	39.48

99 14 02 Temporary Storage Facilities	Unit	Hourly Output	Crew	Factor	Labor	Equip.	Matl.	Total
99 14 0201 Temporary Storage Trailer 16' x 8'	**MO**				**0.00**	**0.00**	**81.29**	**81.29**
01500 0214 Temporary Construction Storage Van 16' x 8'	MO	0	N/A	1.000	0.00	0.00	81.29	81.29
99 14 0202 Temporary Storage Trailer 28' x 10'	**MO**				**0.00**	**0.00**	**110.33**	**110.33**
01500 0215 Temporary Construction Storage Van 28' x 10'	MO	0	N/A	1.000	0.00	0.00	110.33	110.33

99 14 03 Security Fencing	Unit	Hourly Output	Crew	Factor	Labor	Equip.	Matl.	Total
99 14 0301 Security Fencing 5' Chain-Link	**LF**				**1.89**	**0.00**	**4.61**	**6.50**
01500 0201 Construction Fence Chain Link 5'	LF	50	ULABC	1.000	1.89	0.00	4.61	6.50
99 14 0302 Security Fencing 6' Chain-Link	**LF**				**2.10**	**0.00**	**4.77**	**6.87**
01500 0202 Construction Fence Chain Link 6'	LF	45	ULABC	1.000	2.10	0.00	4.77	6.87

99 14 04 Construction Signs	Unit	Hourly Output	Crew	Factor	Labor	Equip.	Matl.	Total
99 14 0401 Construction Signs	**SF**				**0.00**	**0.00**	**11.74**	**11.74**
01500 0102 Temporary Construction Sign	SF	0	N/A	1.000	0.00	0.00	11.74	11.74

99 14 05 Portable Toilets	Unit	Hourly Output	Crew	Factor	Labor	Equip.	Matl.	Total
99 14 0501 Portable Toilets - Chemical	**MO**				**0.00**	**0.00**	**83.23**	**83.23**
01500 0216 Toilet Portable Chemical	MO	0	N/A	1.000	0.00	0.00	83.23	83.23

99 14 06 Temporary Heat	Unit	Hourly Output	Crew	Factor	Labor	Equip.	Matl.	Total

99 14 Major Subcontractor Field Expenses

99 14 06 Temporary Heat	Unit	Hourly Output	Crew	Factor	Labor	Equip.	Matl.	Total
99 14 0601 Temporary Heat - Buildings (12 Hours/Day)	**CSF*W**				**0.00**	**0.00**	**1.49**	**1.49**
01500 0203 Temporary Construction Heat (12 Hours/Day)	CSF	0	N/A	1.000	0.00	0.00	1.49	1.49
99 14 0602 Temporary Heat - Buildings (24 Hours/Day)	**CSF*W**				**0.00**	**0.00**	**2.22**	**2.22**
01500 0204 Temporary Construction Heat/Week (24 Hours/Day)	CSF	0	N/A	1.000	0.00	0.00	2.22	2.22

99 14 07 Temporary Lighting	Unit	Hourly Output	Crew	Factor	Labor	Equip.	Matl.	Total
99 14 0701 Temporary Lighting - Minimum	**CSF**				**10.48**	**0.00**	**2.36**	**12.84**
01500 0205 Temporary Construction Lighting with Service - Minimum	CSF	10	EELEC	1.000	10.48	0.00	2.36	12.84
99 14 0702 Temporary Lighting - Maximum	**CSF**				**20.97**	**0.00**	**5.13**	**26.10**
01500 0206 Temporary Construction Lighting with Service - Maximum	CSF	5	EELEC	1.000	20.97	0.00	5.13	26.10

99 14 08 Temporary Electrical Power	Unit	Hourly Output	Crew	Factor	Labor	Equip.	Matl.	Total
99 14 0801 Temporary Electric Power - Average	**CSF**				**0.00**	**0.00**	**91.16**	**91.16**
01500 0207 Temporary Power and Lighting per Month Average	CSF	0	N/A	1.000	0.00	0.00	91.16	91.16
99 14 0802 Temporary Electric Power - Maximum	**CSF**				**0.00**	**0.00**	**127.75**	**127.75**
01500 0208 Temporary Power and Lighting per Month Maximum	CSF	0	N/A	1.000	0.00	0.00	127.75	127.75

99 14 11 Photographs	Unit	Hourly Output	Crew	Factor	Labor	Equip.	Matl.	Total
99 14 1101 Construction Photographs	**SET**				**320.00**	**0.00**	**0.00**	**320.00**
01050 0101 Field Office Clerk	WK	0.025	XYBBA	1.000	320.00	0.00	0.00	320.00
99 14 1102 Time Lapse Video Equipment & Tapes, Per Month	**MO**				**5.15**	**0.00**	**358.91**	**364.06**
01953 2113 Time Lapse Recorder, S-VHS/VHS Compatible, High Performance	EA	1.63	EELEA	0.163	31.56	0.00	2,200	364.06

99 14 12 Surveying Expenses	Unit	Hourly Output	Crew	Factor	Labor	Equip.	Matl.	Total
99 14 1201 Surveying - 2-man Crew	**DAY**				**656.95**	**217.78**	**0.00**	**874.73**
01050 1001 2-person Crew	DAY	0.125	XYABC	1.000	656.95	217.78	0.00	874.73
99 14 1202 Surveying - 3-man Crew	**DAY**				**920.95**	**217.78**	**0.00**	**1,139**
01050 1002 3-person Crew	DAY	0.125	XYABD	1.000	920.95	217.78	0.00	1,139
99 14 1203 Surveying - 4-man Crew	**DAY**				**1,185**	**217.78**	**0.00**	**1,403**
01050 1003 4-person Crew	DAY	0.125	XYABE	1.000	1,185	217.78	0.00	1,403

99 14 13 Watchmen	Unit	Hourly Output	Crew	Factor	Labor	Equip.	Matl.	Total

99 14 Major Subcontractor Field Expenses

99 14 13 Watchmen	Unit	Hourly Output	Crew	Factor	Labor	Equip.	Matl.	Total
99 14 1301 Uniformed Watchman - Minimum	**HR**				**8.90**	**0.00**	**0.00**	**8.90**
01050 0113 Uniformed Watchman - Minimum	HR	1	XYBBN	1.000	8.90	0.00	0.00	8.90
99 14 1302 Uniformed Watchman - Maximum	**HR**				**16.13**	**0.00**	**0.00**	**16.13**
01050 0114 Uniformed Watchman - Maximum	HR	1	XYBBO	1.000	16.13	0.00	0.00	16.13

99 14 14 Continuous Cleanup	Unit	Hourly Output	Crew	Factor	Labor	Equip.	Matl.	Total
99 14 1401 Continuous Cleanup	**MWK**				**1,068**	**0.00**	**0.00**	**1,068**
01050 0105 General Purpose Laborer	WK	0.025	XYBBF	1.000	1,068	0.00	0.00	1,068

99 21 01 Site Project Manager	Unit	Hourly Output	Crew	Factor	Labor	Equip.	Matl.	Total
99 21 0101 Site Project Manager - Minimum Cost	**MWK**				**1,424**	**0.00**	**0.00**	**1,424**
01050 0106 Project Manager - Minimum Cost	WK	0.025	XYBBG	1.000	1,424	0.00	0.00	1,424
99 21 0102 Site Project Manager - Average Cost	**MWK**				**1,596**	**0.00**	**0.00**	**1,596**
01050 0107 Project Manager - Average Cost	WK	0.025	XYBBH	1.000	1,596	0.00	0.00	1,596
99 21 0103 Site Project Manager - Maximum Cost	**MWK**				**1,803**	**0.00**	**0.00**	**1,803**
01050 0108 Project Manager - Maximum Cost	WK	0.025	XYBBI	1.000	1,803	0.00	0.00	1,803

99 21 02 Superintendent	Unit	Hourly Output	Crew	Factor	Labor	Equip.	Matl.	Total
99 21 0201 Superintendent - Minimum Cost	**MWK**				**1,375**	**0.00**	**0.00**	**1,375**
01050 0109 Field Superintendent - Minimum Cost	WK	0.025	XYBBJ	1.000	1,375	0.00	0.00	1,375
99 21 0202 Superintendent - Average Cost	**MWK**				**1,500**	**0.00**	**0.00**	**1,500**
01050 0110 Field Superintendent - Average Cost	WK	0.025	XYBBK	1.000	1,500	0.00	0.00	1,500
99 21 0203 Superintendent - Maximum Cost	**MWK**				**1,725**	**0.00**	**0.00**	**1,725**
01050 0111 Field Superintendent - Maximum Cost	WK	0.025	XYBBL	1.000	1,725	0.00	0.00	1,725

99 21 03 Clerk	Unit	Hourly Output	Crew	Factor	Labor	Equip.	Matl.	Total
99 21 0301 Clerk - Average Cost	**MWK**				**320.00**	**0.00**	**0.00**	**320.00**
01050 0101 Field Office Clerk	WK	0.025	XYBBA	1.000	320.00	0.00	0.00	320.00

99 21 04 Field Engineer	Unit	Hourly Output	Crew	Factor	Labor	Equip.	Matl.	Total
99 21 0401 Field Engineer - Minimum Cost	**MWK**				**765.00**	**0.00**	**0.00**	**765.00**
01050 0102 Field Engineer (Minimum Cost)	WK	0.025	XYBBC	1.000	765.00	0.00	0.00	765.00
99 21 0402 Field Engineer - Average Cost	**MWK**				**995.00**	**0.00**	**0.00**	**995.00**
01050 0103 Field Engineer (Average Cost)	WK	0.025	XYBBD	1.000	995.00	0.00	0.00	995.00
99 21 0403 Field Engineer - Maximum Cost	**MWK**				**1,150**	**0.00**	**0.00**	**1,150**
01050 0104 Field Engineer - Maximum Cost	WK	0.025	XYBBE	1.000	1,150	0.00	0.00	1,150

99 21 05 Timekeeper	Unit	Hourly Output	Crew	Factor	Labor	Equip.	Matl.	Total

99 21 Minor Subcontractor Field Office Staff

99 21 05 Timekeeper	Unit	Hourly Output	Crew	Factor	Labor	Equip.	Matl.	Total
99 21 0501 Timekeeper - Average Cost	**MWK**				**890.00**	**0.00**	**0.00**	**890.00**
01050 0112 Timekeeper - Average Cost	WK	0.025	XYBBM	1.000	890.00	0.00	0.00	890.00

99 21 06 General Purpose Laborer	Unit	Hourly Output	Crew	Factor	Labor	Equip.	Matl.	Total
99 21 0601 General-purpose Laborer	**MWK**				**1,068**	**0.00**	**0.00**	**1,068**
01050 0105 General Purpose Laborer	WK	0.025	XYBBF	1.000	1,068	0.00	0.00	1,068

99 21 07 Safety Engineer	Unit	Hourly Output	Crew	Factor	Labor	Equip.	Matl.	Total
99 21 0701 Safety Engineer - Minimum Cost	**MWK**				**1,545**	**0.00**	**0.00**	**1,545**
01050 0121 Safety Engineer - Minimum Cost	WK	0.025	XYBBQ	1.000	1,545	0.00	0.00	1,545
99 21 0702 Safety Engineer - Average Cost	**MWK**				**1,545**	**0.00**	**0.00**	**1,545**
01050 0122 Safety Engineer - Average Cost	WK	0.025	XYBBR	1.000	1,545	0.00	0.00	1,545
99 21 0703 Safety Engineer - Maximum Cost	**MWK**				**1,545**	**0.00**	**0.00**	**1,545**
01050 0123 Safety Engineer - Maximum Cost	WK	0.025	XYBBS	1.000	1,545	0.00	0.00	1,545

99 21 08 Quality Control	Unit	Hourly Output	Crew	Factor	Labor	Equip.	Matl.	Total
99 21 0801 Quality Control - Minimum Cost	**MWK**				**1,224**	**0.00**	**0.00**	**1,224**
01050 0124 Quality Control - Minimum Cost	WK	0.025	XYBBT	1.000	1,224	0.00	0.00	1,224
99 21 0802 Quality Control - Average Cost	**MWK**				**1,402**	**0.00**	**0.00**	**1,402**
01050 0125 Quality Control - Average Cost	WK	0.025	XYBBU	1.000	1,402	0.00	0.00	1,402
99 21 0803 Quality Control - Maximum Cost	**MWK**				**1,291**	**0.00**	**0.00**	**1,291**
01050 0126 Quality Control - Maximum Cost	WK	0.025	XYBBV	1.000	1,291	0.00	0.00	1,291

99 24 01 Temporary Offices	Unit	Hourly Output	Crew	Factor	Labor	Equip.	Matl.	Total
99 24 0101 Temporary Office 20' x 8'	**MO**				**0.00**	**0.00**	**207.88**	**207.88**
01500 0209 Temporary Office Trailer 20' x 8' without Hookup	MO	0	N/A	1.000	0.00	0.00	168.39	168.39
01500 0213 Temporary Construction Office Air Conditioner	MO	0	N/A	1.000	0.00	0.00	39.48	39.48
99 24 0102 Temporary Office 32' x 8'	**MO**				**0.00**	**0.00**	**248.52**	**248.52**
01500 0210 Temporary Office Trailer 32' x 8' without Hookup	MO	0	N/A	1.000	0.00	0.00	209.04	209.04
01500 0213 Temporary Construction Office Air Conditioner	MO	0	N/A	1.000	0.00	0.00	39.48	39.48
99 24 0103 Temporary Office 50' x 10'	**MO**				**0.00**	**0.00**	**364.65**	**364.65**
01500 0211 Temporary Office Trailer 50' x 10' without Hookup	MO	0	N/A	1.000	0.00	0.00	325.17	325.17
01500 0213 Temporary Construction Office Air Conditioner	MO	0	N/A	1.000	0.00	0.00	39.48	39.48
99 24 0104 Temporary Office 50' x 12'	**MO**				**0.00**	**0.00**	**418.07**	**418.07**
01500 0212 Temporary Office Trailer 50' x 12' without Hookup	MO	0	N/A	1.000	0.00	0.00	378.59	378.59
01500 0213 Temporary Construction Office Air Conditioner	MO	0	N/A	1.000	0.00	0.00	39.48	39.48

99 24 02 Temporary Storage Facilities	Unit	Hourly Output	Crew	Factor	Labor	Equip.	Matl.	Total

99 24 Minor Subcontractor Field Expenses

99 24 02	Temporary Storage Facilities	Unit	Hourly Output	Crew	Factor	Labor	Equip.	Matl.	Total
99 24 0201	**Temporary Storage Trailer 16' x 8'**	**MO**				**0.00**	**0.00**	**81.29**	**81.29**
01500 0214	Temporary Construction Storage Van 16' x 8'	MO	0	N/A	1.000	0.00	0.00	81.29	81.29
99 24 0202	**Temporary Storage Trailer 28' x 10'**	**MO**				**0.00**	**0.00**	**110.33**	**110.33**
01500 0215	Temporary Construction Storage Van 28' x 10'	MO	0	N/A	1.000	0.00	0.00	110.33	110.33

99 24 03	Security Fencing	Unit	Hourly Output	Crew	Factor	Labor	Equip.	Matl.	Total
99 24 0301	**Security Fencing 5' Chain-Link**	**LF**				**1.89**	**0.00**	**4.61**	**6.50**
01500 0201	Construction Fence Chain Link 5'	LF	50	ULABC	1.000	1.89	0.00	4.61	6.50
99 24 0302	**Security Fencing 6' Chain-Link**	**LF**				**2.10**	**0.00**	**4.77**	**6.87**
01500 0202	Construction Fence Chain Link 6'	LF	45	ULABC	1.000	2.10	0.00	4.77	6.87

99 24 04	Construction Signs	Unit	Hourly Output	Crew	Factor	Labor	Equip.	Matl.	Total
99 24 0401	**Construction Signs**	**SF**				**0.00**	**0.00**	**11.74**	**11.74**
01500 0102	Temporary Construction Sign	SF	0	N/A	1.000	0.00	0.00	11.74	11.74

99 24 05	Portable Toilets	Unit	Hourly Output	Crew	Factor	Labor	Equip.	Matl.	Total
99 24 0501	**Portable Toilets - Chemical**	**MO**				**0.00**	**0.00**	**83.23**	**83.23**
01500 0216	Toilet Portable Chemical	MO	0	N/A	1.000	0.00	0.00	83.23	83.23

99 24 06	Temporary Heat	Unit	Hourly Output	Crew	Factor	Labor	Equip.	Matl.	Total
99 24 0601	**Temporary Heat - Buildings (12 Hours/Day)**	**CSF*W**				**0.00**	**0.00**	**1.49**	**1.49**
01500 0203	Temporary Construction Heat (12 Hours/Day)	CSF	0	N/A	1.000	0.00	0.00	1.49	1.49
99 24 0602	**Temporary Heat - Buildings (24 Hours/Day)**	**CSF*W**				**0.00**	**0.00**	**2.22**	**2.22**
01500 0204	Temporary Construction Heat/Week (24 Hours/Day)	CSF	0	N/A	1.000	0.00	0.00	2.22	2.22

99 24 07	Temporary Lighting	Unit	Hourly Output	Crew	Factor	Labor	Equip.	Matl.	Total
99 24 0701	**Temporary Lighting - Minimum**	**CSF**				**10.48**	**0.00**	**2.36**	**12.84**
01500 0205	Temporary Construction Lighting with Service - Minimum	CSF	10	EELEC	1.000	10.48	0.00	2.36	12.84
99 24 0702	**Temporary Lighting - Maximum**	**CSF**				**20.97**	**0.00**	**5.13**	**26.10**
01500 0206	Temporary Construction Lighting with Service - Maximum	CSF	5	EELEC	1.000	20.97	0.00	5.13	26.10

99 24 08	Temporary Electrical Power	Unit	Hourly Output	Crew	Factor	Labor	Equip.	Matl.	Total
99 24 0801	**Temporary Electric Power - Average**	**CSF**				**0.00**	**0.00**	**91.16**	**91.16**
01500 0207	Temporary Power and Lighting per Month Average	CSF	0	N/A	1.000	0.00	0.00	91.16	91.16

99 24 Minor Subcontractor Field Expenses

99 24 08 Temporary Electrical Power		Unit	Hourly Output	Crew	Factor	Labor	Equip.	Matl.	Total
99 24 0802	**Temporary Electric Power - Maximum**	**CSF**				**0.00**	**0.00**	**127.75**	**127.75**
01500 0208	Temporary Power and Lighting per Month Maximum	CSF	0	N/A	1.000	0.00	0.00	127.75	127.75

99 24 11 Photographs		Unit	Hourly Output	Crew	Factor	Labor	Equip.	Matl.	Total
99 24 1101	**Construction Photographs**	**SET**				**320.00**	**0.00**	**0.00**	**320.00**
01050 0101	Field Office Clerk	WK	0.025	XYBBA	1.000	320.00	0.00	0.00	320.00

99 24 12 Surveying Expenses		Unit	Hourly Output	Crew	Factor	Labor	Equip.	Matl.	Total
99 24 1201	**Surveying - 2-man Crew**	**DAY**				**656.95**	**217.78**	**0.00**	**874.73**
01050 1001	2-person Crew	DAY	0.125	XYABC	1.000	656.95	217.78	0.00	874.73
99 24 1202	**Surveying - 3-man Crew**	**DAY**				**920.95**	**217.78**	**0.00**	**1,139**
01050 1002	3-person Crew	DAY	0.125	XYABD	1.000	920.95	217.78	0.00	1,139
99 24 1203	**Surveying - 4-man Crew**	**DAY**				**1,185**	**217.78**	**0.00**	**1,403**
01050 1003	4-person Crew	DAY	0.125	XYABE	1.000	1,185	217.78	0.00	1,403

99 24 13 Watchmen		Unit	Hourly Output	Crew	Factor	Labor	Equip.	Matl.	Total
99 24 1301	**Uniformed Watchman - Minimum**	**HR**				**8.90**	**0.00**	**0.00**	**8.90**
01050 0113	Uniformed Watchman - Minimum	HR	1	XYBBN	1.000	8.90	0.00	0.00	8.90
99 24 1302	**Uniformed Watchman - Maximum**	**HR**				**16.13**	**0.00**	**0.00**	**16.13**
01050 0114	Uniformed Watchman - Maximum	HR	1	XYBBO	1.000	16.13	0.00	0.00	16.13

99 24 14 Continuous Cleanup		Unit	Hourly Output	Crew	Factor	Labor	Equip.	Matl.	Total
99 24 1401	**Continuous Cleanup**	**MWK**				**1,068**	**0.00**	**0.00**	**1,068**
01050 0105	General Purpose Laborer	WK	0.025	XYBBF	1.000	1,068	0.00	0.00	1,068

Reed Construction Data/RSMeans . . . a tradition of excellence in Construction Cost Information and Services since 1942.

Table of Contents

Book Selection Guide

The following table provides definitive information on the content of each cost data publication. The number of lines of data provided in each unit price or assemblies division, as well as the number of reference tables and crews is listed for each book. The presence of other elements such as an historical cost index, city cost indexes, square foot models or cross-referenced index is also indicated. You can use the table to help select the Means' book that has the quantity and type of information you most need in your work.

Unit Cost Divisions	Building Construction Costs	Mechanical	Electrical	Repair & Remodel.	Square Foot	Site Work Landsc.	Assemblies	Interior	Concrete Masonry	Open Shop	Heavy Construc.	Light Commercial	Facil. Construc.	Plumbing	Western Construction Costs	Residential
1	1104	822	907	1013		1045		847	1014	1102	1051	752	1528	874	1102	710
2	2505	1468	460	1450		7794		547	1387	2467	4978	692	4305	1584	2490	772
3	1401	113	100	726		1259		201	1782	1396	1404	229	1312	82	1399	248
4	835	18	0	647		657		574	1056	811	594	407	1058	0	821	334
5	1820	239	193	957		768		927	691	1788	1023	805	1802	316	1802	752
6	1489	82	78	1475		452		1406	318	1481	608	1626	1596	47	1826	1757
7	1264	158	74	1241		468		489	408	1264	349	956	1318	168	1264	752
8	1839	28	0	1907		313		1673	645	1821	51	1227	2034	0	1841	1193
9	1595	47	0	1424		233		1688	360	1547	159	1337	1809	47	1586	1216
10	861	47	25	498		193		700	170	863	0	394	906	233	861	217
11	1018	322	169	502		137		813	44	925	107	218	1173	291	925	104
12	307	0	0	47		210		1416	27	298	0	62	1435	0	298	64
13	1153	1005	375	494		388		884	75	1136	279	484	1783	928	1119	193
14	345	36	0	258		36		292	0	344	30	12	343	35	343	6
15	1987	13138	636	1779		1569		1174	59	1996	1766	1217	10865	9695	2017	826
16	1277	470	10063	1002		739		1108	55	1265	760	1081	9834	415	1226	552
17	427	354	427	0		0		0	0	427	0	0	427	356	427	0
Totals	21227	18347	13507	15420		16261		14739	8091	20961	13159	11499	43528	15071	21347	9696

Assembly Divisions	Building Construction Costs	Mechanical	Elec-trical	Repair & Remodel.	Square Foot	Site Work Landsc.	Assemblies	Interior	Concrete Masonry	Open Shop	Heavy Construc.	Light Commercial	Facil. Construc.	Plumbing	Western Construction Costs	Asm Div	Resi-dential
A		19	0	192	150	540	612	0	550		542	149	24	0		1	374
B		0	0	809	2480	0	5590	333	1914		0	2024	144	0		2	217
C		0	0	635	862	0	1220	1568	145		0	767	238	0		3	588
D		1031	780	693	1823	0	2439	753	0		0	1310	1027	896		4	871
E		0	0	85	255	0	292	5	0		0	255	5	0		5	393
F		0	0	0	123	0	126	0	0		0	123	3	0		6	358
G		465	172	332	111	1856	584	0	482		432	110	113	560		7	299
																8	760
																9	80
																10	0
																11	0
																12	0
Totals		1515	952	2746	5804	2396	10863	2659	3091		974	4738	1554	1456			3940

Reference Section	Building Construction Costs	Mechanical	Electrical	Repair & Remodel.	Square Foot	Site Work Landsc.	Assemblies	Interior	Concrete Masonry	Open Shop	Heavy Construc.	Light Commercial	Facil. Construc.	Plumbing	Western Construction Costs	Residential
Tables	130	45	85	69	4	80	219	46	71	130	61	57	104	51	131	42
Models					102							43				32
Crews	410	410	410	391		410		410	410	393	410	393	391	410	410	393
City Cost Indexes	yes	yes	yes	yes	yes	yes	yes	yes	yes	yes	yes	yes	yes	yes	yes	yes
Historical Cost Indexes	yes	yes	yes	yes	yes	yes	yes	yes	yes	yes	yes	yes	yes	yes	yes	no

Annual Cost Guides

For more information visit Means Web Site at www.rsmeans.com

Means Building Construction Cost Data 2005

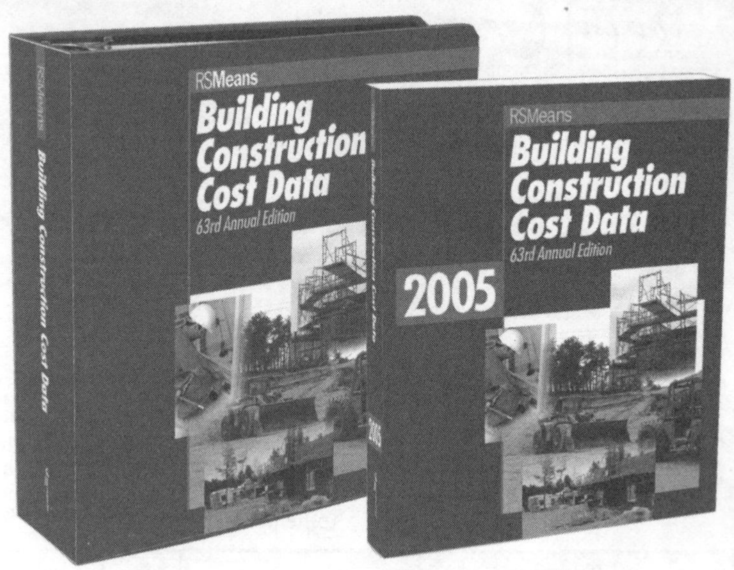

Available in Both Softbound and Looseleaf Editions

The "Bible" of the industry comes in the standard softcover edition or the looseleaf edition.

Many customers enjoy the convenience and flexibility of the looseleaf binder, which increases the usefulness of *Means Building Construction Cost Data 2005* by making it easy to add and remove pages. You can insert your own cost information pages, so everything is in one place. Copying pages for faxing is easier also. Whichever edition you prefer, softbound or the convenient looseleaf edition, you'll be eligible to receive *The Change Notice* FREE. Current subscribers receive *The Change Notice* via e-mail.

$115.95 per copy, Softbound
Catalog No. 60015

$144.95 per copy, Looseleaf
Catalog No. 61015

Means Building Construction Cost Data 2005

Offers you unchallenged unit price reliability in an easy-to-use arrangement. Whether used for complete, finished estimates or for periodic checks, it supplies more cost facts better and faster than any comparable source. Over 23,000 unit prices for 2005. The City Cost Indexes cover over 930 areas, for indexing to any project location in North America. Order and get *The Change Notice* FREE. You'll have year-long access to the Means Estimating **HOTLINE** FREE with your subscription. Expert assistance when using Means data is just a phone call away.

$115.95 per copy
Over 700 pages, illustrated, available Oct. 2004
Catalog No. 60015

Means Building Construction Cost Data 2005

Metric Version

The Federal Government has stated that all federal construction projects must now use metric documentation. The *Metric Version* of *Means Building Construction Cost Data 2005* is presented in metric measurements covering all construction areas. Don't miss out on these billion dollar opportunities. Make the switch to metric today.

$115.95 per copy
Over 700 pages, illus., available Dec. 2004
Catalog No. 63015

For more information
visit Means Web Site
at www.rsmeans.com

Annual Cost Guides

Means Mechanical Cost Data 2005

• HVAC • Controls

Total unit and systems price guidance for mechanical construction...materials, parts, fittings, and complete labor cost information. Includes prices for piping, heating, air conditioning, ventilation, and all related construction.

Plus new 2005 unit costs for:
- Over 2500 installed HVAC/controls assemblies
- "On Site" Location Factors for over 930 cities and towns in the U.S. and Canada
- Crews, labor and equipment

$115.95 per copy
Over 600 pages, illustrated, available Oct. 2004
Catalog No. 60025

Means Plumbing Cost Data 2005

Comprehensive unit prices and assemblies for plumbing, irrigation systems, commercial and residential fire protection, point-of-use water heaters, and the latest approved materials. This publication and its companion, *Means Mechanical Cost Data*, provide full-range cost estimating coverage for all the mechanical trades.

$115.95 per copy
Over 550 pages, illustrated, available Oct. 2004
Catalog No. 60215

Means Electrical Cost Data 2005

Pricing information for every part of electrical cost planning: More than 15,000 unit and systems costs with design tables; clear specifications and drawings; engineering guides and illustrated estimating procedures; complete labor-hour and materials costs for better scheduling and procurement; the latest electrical products and construction methods.
- A Variety of Special Electrical Systems including Cathodic Protection
- Costs for maintenance, demolition, HVAC/ mechanical, specialties, equipment, and more

$115.95 per copy
Over 450 pages, illustrated, available Oct. 2004
Catalog No. 60035

Means Electrical Change Order Cost Data 2005

You are provided with electrical unit prices exclusively for pricing change orders based on the recent, direct experience of contractors and suppliers. Analyze and check your own change order estimates against the experience others have had doing the same work. It also covers productivity analysis and change order cost justifications. With useful information for calculating the effects of change orders and dealing with their administration.

$115.95 per copy
Over 450 pages, available Oct. 2004
Catalog No. 60235

Means Facilities Maintenance & Repair Cost Data 2005

Published in a looseleaf format, *Means Facilities Maintenance & Repair Cost Data* gives you a complete system to manage and plan your facility repair and maintenance costs and budget efficiently. Guidelines for auditing a facility and developing an annual maintenance plan. Budgeting is included, along with reference tables on cost and management and information on frequency and productivity of maintenance operations.

The only nationally recognized source of maintenance and repair costs. Developed in cooperation with the Army Corps of Engineers.

$252.95 per copy
Over 600 pages, illustrated, available Dec. 2004
Catalog No. 60305

Means Square Foot Costs 2005

It's Accurate and Easy To Use!

- **Updated 2005 price information,** based on nationwide figures from suppliers, estimators, labor experts and contractors.
- "How-to-Use" Sections, with **clear examples** of commercial, residential, industrial, and institutional structures.
- Realistic graphics, offering true-to-life illustrations of building projects.
- Extensive information on using square foot cost data, including **sample estimates** and **alternate pricing methods.**

$126.95 per copy
Over 450 pages, illustrated, available Nov. 2004
Catalog No. 60055

Annual Cost Guides

For more information visit Means Web Site at www.rsmeans.com

Means Repair & Remodeling Cost Data 2005

Commercial/Residential

You can use this valuable tool to estimate commercial and residential renovation and remodeling.

Includes: New costs for hundreds of unique methods, materials and conditions that only come up in repair and remodeling. PLUS:

- Unit costs for over 16,000 construction components
- Installed costs for over 90 assemblies
- Costs for 300+ construction crews
- Over 930 "On Site" localization factors for the U.S. and Canada.

$99.95 per copy
Over 650 pages, illustrated, available Nov. 2004
Catalog No. 60045

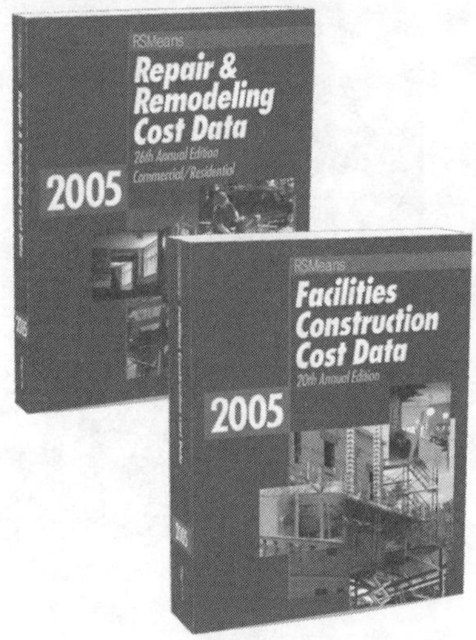

Means Facilities Construction Cost Data 2005

For the maintenance and construction of commercial, industrial, municipal, and institutional properties. Costs are shown for new and remodeling construction and are broken down into materials, labor, equipment, overhead, and profit. Special emphasis is given to sections on mechanical, electrical, furnishings, site work, building maintenance, finish work, and demolition. More than 45,000 unit costs, plus assemblies and reference sections are included.

$279.95 per copy
Over 1200 pages, illustrated, available Nov. 2004
Catalog No. 60205

Means Residential Cost Data 2005

Contains square foot costs for 30 basic home models with the look of today, plus hundreds of custom additions and modifications you can quote right off the page. With costs for the 100 residential systems you're most likely to use in the year ahead. Complete with blank estimating forms, sample estimates and step-by-step instructions.

$99.95 per copy
Over 600 pages, illustrated, available Oct. 2004
Catalog No. 60175

Means Light Commercial Cost Data 2005

Specifically addresses the light commercial market, which is an increasingly specialized niche in the industry. Aids you, the owner/designer/contractor, in preparing all types of estimates, from budgets to detailed bids. Includes new advances in methods and materials. Assemblies section allows you to evaluate alternatives in the early stages of design/planning.

Over 13,000 unit costs for 2005 ensure you have the prices you need...when you need them.

$99.95 per copy
Over 650 pages, illustrated, available Nov. 2004
Catalog No. 60185

Means Assemblies Cost Data 2005

Means Assemblies Cost Data 2005 takes the guesswork out of preliminary or conceptual estimates. Now you don't have to try to calculate the assembled cost by working up individual components costs. We've done all the work for you.

Presents detailed illustrations, descriptions, specifications and costs for every conceivable building assembly—240 types in all—arranged in the easy-to-use UNIFORMAT II system. Each illustrated "assembled" cost includes a complete grouping of materials and associated installation costs including the installing contractor's overhead and profit.

$189.95 per copy
Over 600 pages, illustrated, available Oct. 2004
Catalog No. 60065

Means Site Work & Landscape Cost Data 2005

Means Site Work & Landscape Cost Data 2005 is organized to assist you in all your estimating needs. Hundreds of fact-filled pages help you make accurate cost estimates efficiently.

Updated for 2005!

- Demolition features—including ceilings, doors, electrical, flooring, HVAC, millwork, plumbing, roofing, walls and windows
- State-of-the-art segmental retaining walls
- Flywheel trenching costs and details
- Updated Wells section
- Landscape materials, flowers, shrubs and trees

$115.95 per copy
Over 600 pages, illustrated, available Nov. 2004
Catalog No. 60285

Annual Cost Guides

Means Open Shop Building Construction Cost Data 2005

The latest costs for accurate budgeting and estimating of new commercial and residential construction... renovation work... change orders... cost engineering. *Means Open Shop BCCD* will assist you to...

- Develop benchmark prices for change orders
- Plug gaps in preliminary estimates, budgets
- Estimate complex projects
- Substantiate invoices on contracts
- Price ADA-related renovations

$115.95 per copy
Over 700 pages, illustrated, available Dec. 2004
Catalog No. 60155

Means Heavy Construction Cost Data 2005

A comprehensive guide to heavy construction costs. Includes costs for highly specialized projects such as tunnels, dams, highways, airports, and waterways. Information on different labor rates, equipment, and material costs is included. Has unit price costs, systems costs, and numerous reference tables for costs and design. Valuable not only to contractors and civil engineers, but also to government agencies and city/town engineers.

$115.95 per copy
Over 450 pages, illustrated, available Nov. 2004
Catalog No. 60165

Means Building Construction Cost Data 2005

Western Edition

This regional edition provides more precise cost information for western North America. Labor rates are based on union rates from 13 western states and western Canada. Included are western practices and materials not found in our national edition: tilt-up concrete walls, glu-lam structural systems, specialized timber construction, seismic restraints, landscape and irrigation systems.

$115.95 per copy
Over 650 pages, illustrated, available Dec. 2004
Catalog No. 60225

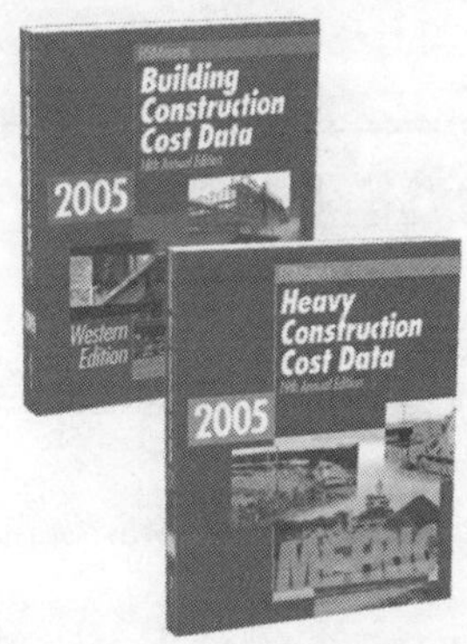

Means Heavy Construction Cost Data 2005

Metric Version

Make sure you have the Means industry standard metric costs for the federal, state, municipal and private marketplace. With thousands of up-to-date metric unit prices in tables by CSI standard divisions. Supplies you with assemblies costs using the metric standard for reliable cost projections in the design stage of your project. Helps you determine sizes, material amounts, and has tips for handling metric estimates.

$115.95 per copy
Over 450 pages, illustrated, available Dec. 2004
Catalog No. 63165

Means Construction Cost Indexes 2005

Who knows what 2005 holds? What materials and labor costs will change unexpectedly? By how much?

- Breakdowns for 316 major cities.
- National averages for 30 key cities.
- Expanded five major city indexes.
- Historical construction cost indexes.

$252.95 per year
$63.00 individual quarters
Catalog No. 60145 A,B,C,D

Means Interior Cost Data 2005

Provides you with prices and guidance needed to make accurate interior work estimates. Contains costs on materials, equipment, hardware, custom installations, furnishings, labor costs . . . every cost factor for new and remodel commercial and industrial interior construction, including updated information on office furnishings, plus more than 50 reference tables. For contractors, facility managers, owners.

$115.95 per copy
Over 600 pages, illustrated, available Oct. 2004
Catalog No. 60095

Means Concrete & Masonry Cost Data 2005

Provides you with cost facts for virtually all concrete/masonry estimating needs, from complicated formwork to various sizes and face finishes of brick and block, all in great detail. The comprehensive unit cost section contains more than 8,500 selected entries. Also contains an assemblies cost section, and a detailed reference section which supplements the cost data.

$105.95 per copy
Over 450 pages, illustrated, available Dec. 2004
Catalog No. 60115

Means Labor Rates for the Construction Industry 2005

Complete information for estimating labor costs, making comparisons and negotiating wage rates by trade for over 300 cities (United States and Canada). With 46 construction trades listed by local union number in each city, and historical wage rates included for comparison. No similar book is available through the trade.

Each city chart lists the county and is alphabetically arranged with handy visual flip tabs for quick reference.

$253.95 per copy
Over 300 pages, available Dec. 2004
Catalog No. 60125

Reference Books

Building Security: Strategies & Costs
By David Owen

Unauthorized systems access, terrorism, hurricanes and tornadoes, sabotage, vandalism, fire, explosions, and other threats... All are considerations as building owners, facility managers, and design and construction professionals seek ways to meet security needs.

This comprehensive resource will help you evaluate your facility's security needs — and design and budget for the materials and devices needed to fulfill them. The text and cost data will help you to:

- Identify threats, probability of occurrence, and the potential losses— to determine and address your real vulnerabilities.
- Perform a detailed risk assessment of an existing facility, and prioritize and budget for security enhancement.
- Evaluate and price security systems and construction solutions, so you can make cost-effective choices.

Includes over 130 pages of Means Cost Data for installation of security systems and materials, plus a review of more than 50 security devices and construction solutions—how they work, and how they compare.

$89.95 per copy
Over 400 pages, Hardcover
Catalog No. 37339

Green Building: Project Planning & Cost Estimating
By RSMeans and Contributing Authors

Written by a team of leading experts in sustainable design, this new book is a complete guide to planning and estimating green building projects, a growing trend in building design and construction – commercial, industrial, institutional and residential. It explains:

- All the different criteria for "green-ness"
- What criteria your building needs to meet to get a LEED, Energy Star, or other recognized rating for green buildings
- How the project team works differently on a green versus a traditional building project
- How to select and specify green products
- How to evaluate the cost and value of green products versus conventional ones — not only for their first (installation) cost, but their cost over time (in maintenance and operation).

Features an extensive Green Building Cost Data section, which details the available products, how they are specified, and how much they cost.

$89.95 per copy
350 pages, illustrated, Hardcover
Catalog No. 67338

Life Cycle Costing for Facilities
By Alphonse Dell'Isola and Dr. Steven Kirk

Guidance for achieving higher quality design and construction projects at lower costs!

Facility designers and owners are frustrated with cost-cutting efforts that yield the cheapest product, but sacrifice quality. Life Cycle Costing, properly done, enables them to achieve both—high quality, incorporating innovative design, and costs that meet their budgets.

The authors show how LCC can work for a broad variety of projects – from several types of buildings, to roads and bridges, to HVAC and electrical upgrades, to materials and equipment procurement. Case studies include:

- Health care and nursing facilities
- College campus and high schools
- Office buildings, courthouses, and banks
- Exterior walls, elevators, lighting, HVAC, and more

The book's extensive cost section provides maintenance and replacement costs for facility elements.

$99.95 per copy
450 pages, Hardcover
Catalog No. 67341

Reference Books

For more information visit Means Web Site at www.rsmeans.com

Interior Home Improvement Costs, New 9th Edition

Estimates for the most popular remodeling and repair projects—from small, do-it-yourself jobs—to major renovations and new construction. Includes: Kitchens & Baths; New Living Space from Your Attic, Basement or Garage; New Floors, Paint & Wallpaper; Tearing Out or Building New Walls; Closets, Stairs & Fireplaces; New Energy-Saving Improvements, Home Theatres, and More!

$24.95 per copy
250 pages, illustrated, Softcover
Catalog No. 67308E

Exterior Home Improvement Costs New 9th Edition

Estimates for the most popular remodeling and repair projects—from small, do-it-yourself jobs, to major renovations and new construction. Includes: Curb Appeal Projects—Landscaping, Patios, Porches, Driveways and Walkways; New Windows and Doors; Decks, Greenhouses, and Sunrooms; Room Additions and Garages; Roofing, Siding, and Painting; "Green" Improvements to Save Energy & Water

$24.95 per copy
Over 275 pages, illustrated, softcover
Catalog No. 67309E

Builder's Essentials: Plan Reading & Material Takeoff

By Wayne J. DelPico

For Residential and Light Commercial Construction

A valuable tool for understanding plans and specs, and accurately calculating material quantities.
Step-by-step instructions and takeoff procedures based on a full set of working drawings.

$35.95 per copy
Over 420 pages, Softcover
Catalog No. 67307

Planning & Managing Interior Projects 2nd Edition

By Carol E. Farren, CFM

Addresses changes in technology and business, guiding you through commercial design and construction from initial client meetings to post-project administration. Includes: evaluating space requirements, alternative work models, telecommunications and data management, and environmental issues.

$69.95 per copy
Over 400 pages, illustrated, Hardcover
Catalog No. 67245A

Builder's Essentials: Best Business Practices for Builders & Remodelers: An Easy-to-Use Checklist System

By Thomas N. Frisby

A comprehensive guide covering all aspects of running a construction business, with more than 40 user–friendly checklists. This book provides expert guidance on: increasing your revenue and keeping more of your profit; planning for long-term growth; keeping good employees and managing subcontractors.

$29.95 per copy
Over 220 pages, Softcover
Catalog No. 67329

Builder's Essentials: Framing & Rough Carpentry 2nd Edition

By Scot Simpson

A complete training manual for apprentice and experienced carpenters. Develop and improve your skills with "framer-friendly," easy-to-follow instructions, and step-by-step illustrations. Learn proven techniques for framing walls, floors, roofs, stairs, doors, and windows. Updated guidance on standards, building codes, safety requirements, and more. Also available in Spanish!

$24.95 per copy
Over 150 pages, Softcover
Catalog No. 67298A Spanish Catalog No. 67298AS

How to Estimate with Means Data & CostWorks

By RSMeans and Saleh Mubarak

New 2nd Edition!

Learn estimating techniques using Means cost data. Includes an instructional version of Means CostWorks CD–ROM with Sample Building Plans.
The step-by-step guide takes you through all the major construction items. Over 300 sample estimating problems are included.

$59.95 per copy
Over 190 pages, Softcover
Catalog No. 67324A

Unit Price Estimating Methods New 3rd Edition

This new edition includes up-to-date cost data and estimating examples, updated to reflect changes to the CSI numbering system and new features of Means cost data. It describes the most productive, universally accepted ways to estimate, and uses checklists and forms to illustrate shortcuts and timesavers. A model estimate demonstrates procedures. A new chapter explores computer estimating alternatives.

$59.95 per copy
Over 350 pages, illustrated, Hardcover
Catalog No. 67303A

Means Landscape Estimating Methods 4th Edition

By Sylvia H. Fee

This revised edition offers expert guidance for preparing accurate estimates for new landscape construction and grounds maintenance. Includes a complete project estimate featuring the latest equipment and methods, and **two chapters on Life Cycle Costing, and Landscape Maintenance Estimating.**

$62.95 per copy
Over 300 pages, illustrated, Hardcover
Catalog No. 67295B

Means Environmental Remediation Estimating Methods, 2nd Edition

By Richard R. Rast

Guidelines for estimating 50 standard remediation technologies Use it to prepare preliminary budgets, develop estimates, compare costs and solutions, estimate liability, review quotes, negotiate settlements.

A valuable support tool for ***Means Environmental Remediation Unit Price*** and ***Assemblies*** books.

$99.95 per copy
Over 750 pages, illustrated, Hardcover
Catalog No. 64777A

For more information
visit Means Web Site
at www.rsmeans.com

Reference Books

Means Illustrated Construction Dictionary, Condensed Edition, 2nd Edition

By RSMeans

Recognized in the industry as the best resource of its kind, this has been further enhanced with updates to existing terms and the addition of hundreds of new terms and illustrations . . . in keeping with the most recent developments in the industry.
The best portable dictionary for office or field use—an essential tool for contractors, architects, insurance and real estate personnel, homeowners, and anyone who needs quick, clear definitions for construction terms.

$59.95 per copy
Over 500 pages
Catalog No. 67282A

Means Repair and Remodeling Estimating

New 4th Edition
By Edward B. Wetherill & RSMeans

Focuses on the unique problems of estimating renovations of existing structures. It helps you determine the true costs of remodeling through careful evaluation of architectural details and a site visit.
New section on disaster restoration costs.

$69.95 per copy
Over 450 pages, illustrated, Hardcover
Catalog No. 67265B

Facilities Planning & Relocation

New, lower price and user-friendly format.
By David D. Owen

A complete system for planning space needs and managing relocations. Includes step-by-step manual, over 50 forms, and extensive reference section on materials and furnishings.

$89.95 per copy
Over 450 pages, Softcover
Catalog No. 67301

Means Square Foot & Assemblies Estimating Methods

3rd Edition!

Develop realistic Square Foot and Assemblies Costs for budgeting and construction funding. The new edition features updated guidance on square foot and assemblies estimating using UNIFORMAT II. An essential reference for anyone who performs conceptual estimates.

$69.95 per copy
Over 300 pages, illustrated, Hardcover
Catalog No. 67145B

Means Electrical Estimating Methods 3rd Edition

Expanded new edition includes sample estimates and cost information in keeping with the latest version of the CSI MasterFormat and UNIFORMAT II. Complete coverage of Fiber Optic and Uninterruptible Power Supply electrical systems, broken down by components and explained in detail. Includes a new chapter on computerized estimating methods. A practical companion to *Means Electrical Cost Data.*

$64.95 per copy
Over 325 pages, Hardcover
Catalog No. 67230A

Means Mechanical Estimating Methods 3rd Edition

This guide assists you in making a review of plans, specs, and bid packages with suggestions for takeoff procedures, listings, substitutions and pre-bid scheduling. Includes suggestions for budgeting labor and equipment usage. Compares materials and construction methods to allow you to select the best option.

$64.95 per copy
Over 350 pages, illustrated, Hardcover
Catalog No. 67294A

Means Spanish/English Construction Dictionary

By RSMeans, The International Conference of Building Officials (ICBO), and Rolf Jensen & Associates (RJA)

Designed to facilitate communication among Spanish- and English-speaking construction personnel—improving performance and job-site safety. Features the most common words and phrases used in the construction industry, with easy-to-follow pronunciations. Includes extensive building systems and tools illustrations.

$22.95 per copy
250 pages, illustrated, Softcover
Catalog No. 67327

Project Scheduling & Management for Construction

New 3rd Edition
By David R. Pierce, Jr.

A comprehensive yet easy-to-follow guide to construction project scheduling and control—from vital project management principles through the latest scheduling, tracking, and controlling techniques. The author is a leading authority on scheduling with years of field and teaching experience at leading academic institutions. Spend a few hours with this book and come away with a solid understanding of this essential management topic.

$64.95 per copy
Over 300 pages, illustrated, Hardcover
Catalog No. 67247B

Reference Books

Concrete Repair and Maintenance Illustrated
By Peter H. Emmons
$69.95 per copy
Catalog No. 67146

Superintending for Contractors:
How to Bring Jobs in On-Time, On-Budget
By Paul J. Cook
$35.95 per copy
Catalog No. 67233

HVAC Systems Evaluation
By Harold R. Colen, PE
$84.95 per copy
Catalog No. 67281

Basics for Builders: How to Survive and Prosper in Construction
By Thomas N. Frisby
$34.95 per copy
Catalog No. 67273

Successful Interior Projects Through Effective Contract Documents
By Joel Downey & Patricia K. Gilbert
Now $24.98 per copy, limited quantity
Catalog No. 67313

Building Spec Homes Profitably
By Kenneth V. Johnson
$29.95 per copy
Catalog No. 67312

Estimating for Contractors
How to Make Estimates that Win Jobs
By Paul J. Cook
$35.95 per copy
Catalog No. 67160

Successful Estimating Methods:
From Concept to Bid
By John D. Bledsoe, PhD, PE
$32.48 per copy
Catalog No. 67287

Total Productive Facilities Management
By Richard W. Sievert, Jr.
$29.98 per copy
Over 270 pages, illustrated, Hardcover
Catalog No. 67321

Means Productivity Standards for Construction
Expanded Edition *(Formerly Man-Hour Standards)*
$49.98 per copy
Over 800 pages, Hardcover
Catalog No. 67236A

Means Plumbing Estimating Methods, 3rd Edition
By Joseph Galeno and Sheldon Greene
Now $29.98 per copy
Catalog No. 67283B

For more information
visit Means Web Site
at www.rsmeans.com

Reference Books

Preventive Maintenance Guidelines for School Facilities

By John C. Maciha

A complete PM program for K-12 schools that ensures sustained security, safety, property integrity, user satisfaction, and reasonable ongoing expenditures.

Includes schedules for weekly, monthly, semiannual, and annual maintenance with hard copy and electronic forms.

$149.95 per copy
Over 225 pages, Hardcover
Catalog No. 67326

Preventive Maintenance for Higher Education Facilities

By Applied Management Engineering, Inc.

An easy-to-use system to help facilities professionals establish the value of PM, and to develop and budget for an appropriate PM program for their college or university. Features interactive campus building models typical of those found in different-sized higher education facilities, and PM checklists linked to each piece of equipment or system in hard copy and electronic format.

$149.95 per copy
150 pages, Hardcover
Catalog No. 67337

Historic Preservation: Project Planning & Estimating

By Swanke Hayden Connell Architects

Expert guidance on managing historic restoration, rehabilitation, and preservation building projects and determining and controlling their costs. Includes:

- How to determine whether a structure qualifies as historic
- Where to obtain funding and other assistance
- How to evaluate and repair more than 75 historic building materials

$99.95 per copy
Over 675 pages, Hardcover
Catalog No. 67323

Means Illustrated Construction Dictionary, 3rd Edition

Long regarded as the Industry's finest, the Means *Illustrated Construction Dictionary* is now even better.
With the addition of over 1,000 new terms and hundreds of new illustrations, it is the clear choice for the most comprehensive and current information. **The companion CD-ROM that comes with this new edition adds many extra features: larger graphics, expanded definitions, and links to both CSI MasterFormat numbers and product information.**

$99.95 per copy
Over 790 pages, Illustrated, Hardcover
Catalog No. 67292A

Designing & Building with the IBC, 2nd Edition

By Rolf Jensen & Associates, Inc.

This updated comprehensive guide helps building professionals make the transition to the 2003 *International Building Code®*. Includes a side-by-side code comparison of the IBC 2003 to the IBC 2000 and the three primary model codes, a quick-find index, and professional code commentary. With illustrations, abbreviations key, and an extensive Resource section.

$99.95 per copy
Over 875 pages
Catalog No. 67328A

Residential & Light Commercial Construction Standards, 2nd Edition

By RSMeans and Contributing Authors

New, updated second edition of this unique collection of industry standards that define quality construction. For contractors, subcontractors, owners, developers, architects, engineers, attorneys, and insurance personnel, this book provides authoritative requirements and recommendations compiled from the nation's leading professional associations, industry publications, and building code organizations.

$59.95 per copy
600 pages, illustrated, Softcover
Catalog No. 67322A

Means Estimating Handbook, 2nd Edition

By RSMeans

Updated new Second Edition answers virtually any estimating technical question - all organized by CSI MasterFormat. This comprehensive reference covers the full spectrum of technical data required to estimate construction costs. The book includes information on sizing, productivity, equipment requirements, code-mandated specifications, design standards and engineering factors.

$99.95 per copy
Over 900 pages, Hardcover
Catalog No. 67276A

Seminars

Means CostWorks Training

This one-day seminar course has been designed with the intention of assisting both new and existing users to become more familiar with CostWorks program. The class is broken into two unique sections: (1) A one-half day presentation on the function of each icon; and each student will be shown how to use the software to develop a cost estimate. (2) Hands-on estimating exercises that will ensure that each student thoroughly understands how to use CostWorks. You must bring your own laptop computer to this course.

CostWorks Benefits/Features:
- Estimate in your own spreadsheet format
- Power of Means National Database
- Database automatically regionalized
- Save time with keyword searches
- Save time by establishing common estimate items in "Bookmark" files
- Customize your spreadsheet template
- Hot key to Product Manufacturers' listings and specs
- Merge capability for networking environments
- View crews and assembly components
- AutoSave capability
- Enhanced sorting capability

Unit Price Estimating

This interactive two-day seminar teaches attendees how to interpret project information and process it into final, detailed estimates with the greatest accuracy level.

The single most important credential an estimator can take to the job is the ability to visualize construction in the mind's eye, and thereby estimate accurately.

Some Of What You'll Learn:
- Interpreting the design in terms of cost
- The most detailed, time tested methodology for accurate "pricing"
- Key cost drivers—material, labor, equipment, staging and subcontracts
- Understanding direct and indirect costs for accurate job cost accounting and change order management

Who Should Attend: Corporate and government estimators and purchasers, architects, engineers…and others needing to produce accurate project estimates.

Square Foot and Assemblies Cost Estimating

This two-day course teaches attendees how to quickly deliver accurate square foot estimates using limited budget and design information.

Some Of What You'll Learn:
- How square foot costing gets the estimate done faster
- Taking advantage of a "systems" or "assemblies" format
- The Means "building assemblies/square foot cost approach"
- How to create a very reliable preliminary and systems estimate using bare-bones design information

Who Should Attend: Facilities managers, facilities engineers, estimators, planners, developers, construction finance professionals…and others needing to make quick, accurate construction cost estimates at commercial, government, educational and medical facilities.

Repair and Remodeling Estimating

This two-day seminar emphasizes all the underlying considerations unique to repair/remodeling estimating and presents the correct methods for generating accurate, reliable R&R project costs using the unit price and assemblies methods.

Some Of What You'll Learn:
- Estimating considerations—like labor-hours, building code compliance, working within existing structures, purchasing materials in smaller quantities, unforeseen deficiencies
- Identifies problems and provides solutions to estimating building alterations
- Rules for factoring in minimum labor costs, accurate productivity estimates and allowances for project contingencies
- R&R estimating examples are calculated using unite prices and assemblies data

Who Should Attend: Facilities managers, plant engineers, architects, contractors, estimators, builders…and others who are concerned with the proper preparation and/or evaluation of repair and remodeling estimates.

Mechanical and Electrical Estimating

This two-day course teaches attendees how to prepare more accurate and complete mechanical/electrical estimates, avoiding the pitfalls of omission and double-counting, while understanding the composition and rationale within the Means Mechanical/Electrical database.

Some Of What You'll Learn:
- The unique way mechanical and electrical systems are interrelated
- M&E estimates, conceptual, planning, budgeting and bidding stages
- Order of magnitude, square foot, assemblies and unit price estimating
- Comparative cost analysis of equipment and design alternatives

Who Should Attend: Architects, engineers, facilities managers, mechanical and electrical contractors…and others needing a highly reliable method for developing, understanding and evaluating mechanical and electrical contracts.

Plan Reading and Material Takeoff

This two-day program teaches attendees to read and understand construction documents and to use them in the preparation of material takeoffs.

Some of What You'll Learn:
- Skills necessary to read and understand typical contract documents—blueprints and specifications
- Details and symbols used by architects and engineers
- Construction specifications' importance in conjunction with blueprints
- Accurate takeoff of construction materials and industry-accepted takeoff methods

Who Should Attend: Facilities managers, construction supervisors, office managers…and other responsible for the execution and administration of a construction project including government, medical, commercial, educational or retail facilities.

Facilities Maintenance and Repair Estimating

This two-day course teaches attendees how to plan, budget, and estimate the cost of ongoing and preventive maintenance and repair for existing buildings and grounds.

Some Of What You'll Learn:
- The most financially favorable maintenance, repair and replacement scheduling and estimating
- Auditing and value engineering facilities
- Preventive planning and facilities upgrading
- Determining both in-house and contract-out service costs; annual, asset-protecting M&R plan

Who Should Attend: Facility managers, maintenance supervisors, buildings and grounds superintendents, plant managers, planners, estimators…and others involved in facilities planning and budgeting.

Scheduling and Project Management

This two-day course teaches attendees the most current and proven scheduling and management techniques needed to bring projects in on time and on budget.

Some Of What You'll Learn:
- Crucial phases of planning and scheduling
- How to establish project priorities, develop realistic schedules and management techniques
- Critical Path and Precedence Methods
- Special emphasis on cost control

Who Should Attend: Construction project managers, supervisors, engineers, estimators, contractors…and others who want to improve their project planning, scheduling and management skills.

Advanced Project Management

This two-day seminar will teach you how to effectively manage and control the entire design-build process and allow you to take home tangible skills that will be immediately applicable on existing projects.

Some Of What You'll Learn:
- Value engineering, bonding, fast-tracking and bid package creation
- How estimates and schedules can be integrated to provide advanced project management tools
- Cost engineering, quality control, productivity measurement and improvement
- Front loading a project and predicting its cash flow

Who Should Attend: Owners, project managers, architectural and engineering managers, construction managers, contractors…and anyone else who is responsible for the timely design and completion of construction projects.

Seminars

2005 Means Seminar Schedule

Location	Dates
Las Vegas, NV	March
Washington, DC	April
Phoenix, AZ	April
Denver, CO	May
San Francisco, CA	June
Philadelphia, PA	June
Washington, DC	September
Dallas, TX	September
Las Vegas, NV	October
Orlando, FL	November
Atlantic City, NJ	November
San Diego, CA	December

Note: Call for exact dates and details.

Registration Information

Register Early... Save up to $100! Register 30 days before the start date of a seminar and save $100 off your total fee. *Note: This discount can be applied only once per order. It cannot be applied to team discount registrations or any other special offer.*

How to Register Register by phone today! Means toll-free number for making reservations is: **1-800-334-3509.**

Individual Seminar Registration Fee $895. Individual CostWorks Training Registration Fee $349. To register by mail, complete the registration form and return with your full fee to: Seminar Division, Reed Construction Data, RSMeans Seminars, 63 Smiths Lane, Kingston, MA 02364.

Federal Government Pricing All Federal Government employees save 25% off regular seminar price. Other promotional discounts cannot be combined with Federal Government discount.

Team Discount Program Two to four seminar registrations: Call for pricing.

Multiple Course Discounts When signing up for two or more courses, call for pricing.

Refund Policy Cancellations will be accepted up to ten days prior to the seminar start. There are no refunds for cancellations received later than ten working days prior to the first day of the seminar. A $150 processing fee will be applied for all cancellations. Written notice of cancellation is required . Substitutions can be made at anytime before the session starts. **No-shows are subject to the full seminar fee.**

AACE Approved Courses The RSMeans Construction Estimating and Management Seminars described and offered to you here have each been approved for 14 hours (1.4 recertification credits) of credit by the AACE International Certification Board toward meeting the continuing education requirements for re-certification as a Certified Cost Engineer/Certified Cost Consultant.

AIA Continuing Education We are registered with the AIA Continuing Education System (AIA/CES) and are committed to developing quality learning activities in accordance with the CES criteria. RSMeans seminars meet the AIA/CES criteria for Quality Level 2. AIA members will receive (14) learning units (LUs) for each two-day RSMeans Course.

NASBA CPE Sponsor Credits We are part of the National Registry of CPE Sponsors. Attendees may be eligable for (16) CPE credits.

Daily Course Schedule The first day of each seminar session begins at 8:30 A.M. and ends at 4:30 P.M. The second day is 8:00 A.M.–4:00 P.M. Participants are urged to bring a hand-held calculator since many actual problems will be worked out in each session.

Continental Breakfast Your registration includes the cost of a continental breakfast, a morning coffee break, and an afternoon break. These informal segments will allow you to discuss topics of mutual interest with other members of the seminar. (You are free to make your own lunch and dinner arrangements.)

Hotel/Transportation Arrangements RSMeans has arranged to hold a block of rooms at each hotel hosting a seminar. To take advantage of special group rates when making your reservation, be sure to mention that you are attending the Means Seminar. You are, of course, free to stay at the lodging place of your choice. **(Hotel reservations and transportation arrangements should be made directly by seminar attendees.)**

Important Class sizes are limited, so please register as soon as possible.

Note: *Pricing subject to change.*

Registration Form

Call 1-800-334-3509 to register or FAX 1-800-632-6732. Visit our Web site www.rsmeans.com

Please register the following people for the Means Construction Seminars as shown here. Full payment or deposit is enclosed, and we understand that we must make our own hotel reservations if overnight stays are necessary.

☐ Full payment of $ ______________________ enclosed.

☐ Bill me

Name of Registrant(s)
(To appear on certificate of completion)

____

____

____

P.O. #: ____

GOVERNMENT AGENCIES MUST SUPPLY PURCHASE ORDER NUMBER

Firm Name ____

Address ____

City/State/Zip ____

Telephone No. fax No. ____

E-mail Address ____

Charge our registration(s) to: ☐ MasterCard ☐ VISA ☐ American Express ☐ Discover

Account No. ______________________ Exp. Date ____________

Cardholder's Signature ____

Seminar Name City Dates

____

____

Please mail check to: Seminar Division, Reed Construction Data, RSMeans Seminars, 63 Smiths Lane, P.O.Box 800, Kingston, MA 02364 USA

MeansData™

CONSTRUCTION COSTS FOR SOFTWARE APPLICATIONS

Your construction estimating software is only as good as your cost data.

A proven construction cost database is a mandatory part of any estimating package. The following list of softwa providers can offer you MeansData™ as an added feature for their estimating systems. See the table below for w types of products and services they offer (match their numbers). Visit online at **www.rsmeans.com/demosource/** for more information and free demos. Or call their numbers listed below.

1. **3D International**
713-871-7000
venegas@3di.com

2. **4Clicks-Solutions, LLC**
719-574-7721
mbrown@4clicks-solutions.com

3. **Aepco, Inc.**
301-670-4642
blueworks@aepco.com

4. **Applied Flow Technology**
800-589-4943
info@aft.com

5. **ArenaSoft Estimating**
888-370-8806
info@arenasoft.com

6. **ARES Corporation**
925-299-6700
sales@arescorporation.com

7. **BSD - Building Systems Design, Inc.**
888-273-7638
bsd@bsdsoftlink.com

8. **CMS - Computerized Micro Solutions**
800-255-7407
cms@proest.com

9. **Corecon Technologies, Inc.**
714-895-7222
sales@corecon.com

10. **CorVet Systems**
301-622-9069
sales@corvetsys.com

11. **Discover Software**
727-559-0161
sales@discoversoftware.com

12. **Estimating Systems, Inc.**
800-967-8572
esipulsar@adelphia.net

13. **MAESTRO Estimator Schwaab Technology Solutions, Inc.**
281-578-3039
Stefan@schwaabtech.com

14. **Magellan K-12**
936-447-1744
sam.wilson@magellan-K12.com

15. **MC2 - Management Computer**
800-225-5622
vkeys@mc2-ice.com

16. **Maximus Asset Solutions**
800-659-9001
assetsolutions@maximus.com

17. **Prime Time**
610-964-8200
dick@primetime.com

18. **Prism Computer Corp.**
800-774-7622
famis@prismcc.com

19. **Quest Solutions, Inc.**
800-452-2342
info@questsolutions.com

20. **RIB Software (Americas), Inc.**
800-945-7093
san@rib-software.com

21. **Shaw Beneco Enterprises, Inc.**
877-719-4748
inquire@beneco.com

22. **Timberline Software Corp.**
800-628-6583
product.info@timberline.com

23. **TMA SYSTEMS, Inc.**
800-862-1130
sales@tmasys.com

24. **US Cost, Inc.**
800-372-4003
sales@uscost.com

25. **Vanderweil Facility Advisors**
617-451-5100
info@VFA.com

26. **Vertigraph, Inc.**
800-989-4243
info-request@vertigraph.com

27. **WinEstimator, Inc.**
800-950-2374
sales@winest.com

TYPE	1	2	3	4	5	6	7	8	9	10	11	12	13	14	15	16	17	18	19	20	21	22	23	24	25	26	27
BID					●		●	●		●	●				●		●		●	●		●				●	●
Estimating		●			●	●	●	●	●	●	●	●	●		●		●		●	●	●	●		●		●	●
DOC/JOC/SABER		●			●		●			●		●				●					●	●					●
IDIQ		●								●		●				●					●						●
Asset Mgmt.											●			●		●		●							●		●
Facility Mgmt.	●		●								●		●	●		●	●	●		●			●		●		
Project Mgmt.	●	●				●			●	●	●		●	●			●			●	●	●					
TAKE-OFF					●			●	●	●	●				●		●		●	●		●		●		●	●
EARTHWORK									●		●				●		●		●							●	
Pipe Flow				●							●						●										
HVAC/Plumbing					●			●		●	●						●										
Roofing					●			●		●	●						●										
Design	●				●						●						●				●	●					●
Other Offers/Links:																											
Accounting/HR		●			●						●									●	●	●					
Scheduling					●						●		●							●		●		●		●	●
CAD											●			●								●					●
PDA																						●		●			●
Lt. Versions		●						●			●						●					●		●			●
Consulting	●	●			●			●		●	●		●			●	●	●		●		●		●	●		●
Training		●			●		●	●	●	●	●	●	●			●	●	●	●		●	●	●	●	●	●	●

Reseller applications now being accepted. Call Carol Polio Ext. 5107.

FOR MORE INFORMATION

CALL 1-800-448-8182, EXT. 5107 OR FAX 1-800-632-6732

Builder's Essentials: Estimating Building Costs for the Residential & Light Commercial Contractor

By Wayne J. DelPico

Step-by-step estimating methods for residential and light commercial contractors. Includes a detailed look at every construction specialty—explaining all the components, takeoff units, and labor needed for well-organized, complete estimates covers:

Correctly interpreting plans and specifications.
Developing accurate and complete labor and material costs.
Understanding direct and indirect overhead costs... and accounting for time-sensitive costs.
Using historical cost data to generate new project budgets.

Plus hard-to-find, professional guidance on what to consider so you can allocate the right amount for profit and contingencies.

$29.95 per copy
Over 400 pages, illustrated, softcover
Catalog No. 67343

Building & Renovating Schools

This all-inclusive guide covers every step of the school construction process—from initial planning, needs assessment, and design, right through moving into the new facility. A must-have resource for anyone concerned with new school construction or renovation, including architects and engineers, contractors and project managers, facility managers, school administrators and school board members, building committees, community leaders, and anyone else who wants to ensure that the project meets the schools' needs in a cost-effective, timely manner. With square foot cost models for elementary, middle, and high school facilities and real-life case studies of recently completed school projects.

The contributors to this book – architects, construction project managers, contractors, and estimators who specialize in school construction – provide start-to-finish, expert guidance on the process.

$99.95 per copy
Over 425 pages
Catalog No. 67342

Means ADA Compliance Pricing Guide, New Second Edition

By Adaptive Environments and RSMeans

Completely updated and revised to the new 2004 Americans with Disabilities Act Accessible Guidelines, this book features more than 70 of the most commonly needed modifications for ADA compliance—their design requirements, suggestions, and final cost. Projects range from installing ramps and walkways, widening doorways and entryways, and installing and refitting elevators, to relocating light switches and signage, and remodeling bathrooms and kitchens. Also provided are:

Detailed cost estimates for budgeting modification projects, including estimates for each of 260 alternates.
An assembly estimate for every project, with detailed cost breakdown including materials, labor hours, and contractor's overhead.
3,000 Additional ADA compliance-related unit cost line items.
Costs that are easily adjusted to over 900 cities and towns.

Over 350 pages
$79.99 per copy
Catalog No. 67310A

ORDER TOLL FREE 1-800-334-3509
OR FAX 1-800-632-6732.

2005 Order Form

Qty.	Book No.	COST ESTIMATING BOOKS	Unit Price	Total
	60065	Assemblies Cost Data 2005	$189.95	
	60015	Building Construction Cost Data 2005	115.95	
	61015	Building Const. Cost Data–Looseleaf Ed. 2005	144.95	
	63015	Building Const. Cost Data–Metric Version 2005	115.95	
	60225	Building Const. Cost Data–Western Ed. 2005	115.95	
	60115	Concrete & Masonry Cost Data 2005	105.95	
	60145	Construction Cost Indexes 2005	251.95	
	60145A	Construction Cost Index–January 2005	63.00	
	60145B	Construction Cost Index–April 2005	63.00	
	60145C	Construction Cost Index–July 2005	63.00	
	60145D	Construction Cost Index–October 2005	63.00	
	60345	Contr. Pricing Guide: Resid. R & R Costs 2005	39.95	
	60335	Contr. Pricing Guide: Resid. Detailed 2005	39.95	
	60325	Contr. Pricing Guide: Resid. Sq. Ft. 2005	39.95	
	64025	ECHOS Assemblies Cost Book 2005	189.95	
	64015	ECHOS Unit Cost Book 2005	126.95	
	54005	ECHOS (Combo set of both books)	251.95	
	60235	Electrical Change Order Cost Data 2005	115.95	
	60035	Electrical Cost Data 2005	115.95	
	60205	Facilities Construction Cost Data 2005	279.95	
	60305	Facilities Maintenance & Repair Cost Data 2005	252.95	
	60165	Heavy Construction Cost Data 2005	115.95	
	63165	Heavy Const. Cost Data–Metric Version 2005	115.95	
	60095	Interior Cost Data 2005	115.95	
	60125	Labor Rates for the Const. Industry 2005	253.95	
	60185	Light Commercial Cost Data 2005	99.95	
	60025	Mechanical Cost Data 2005	115.95	
	60155	Open Shop Building Const. Cost Data 2005	115.95	
	60215	Plumbing Cost Data 2005	115.95	
	60045	Repair and Remodeling Cost Data 2005	99.95	
	60175	Residential Cost Data 2005	99.95	
	60285	Site Work & Landscape Cost Data 2005	115.95	
	60055	Square Foot Costs 2005	126.95	
		REFERENCE BOOKS		
	67147A	ADA in Practice	59.98	
	67310A	ADA Compliance Pricing Guide, 2nd Ed.	79.95	
	67273	Basics for Builders: How to Survive and Prosper	34.95	
	67330	Bldrs Essentials: Adv. Framing Methods	24.95	
	67329	Bldrs Essentials: Best Bus. Practices for Bldrs	29.95	
	67298A	Bldrs Essentials: Framing/Carpentry 2nd Ed.	24.95	
	67298AS	Bldrs Essentials: Framing/Carpentry Spanish	24.95	
	67307	Bldrs Essentials: Plan Reading & Takeoff	35.95	
	67261A	Bldg. Prof. Guide to Contract Documents 3rd Ed.	64.95	
	67342	Building & Renovating Schools	99.95	
	67339	Building Security: Strategies & Costs	89.95	
	67312	Building Spec Homes Profitably	29.95	
	67146	Concrete Repair & Maintenance Illustrated	69.95	
	67314	Cost Planning & Est. for Facil. Maint.	89.95	
	67317A	Cyberplaces: The Internet Guide 2nd Ed.	59.95	
	67328A	Designing & Building with the IBC, 2nd Ed.	99.95	
	67230B	Electrical Estimating Methods 3rd Ed.	64.95	

Qty.	Book No.	REFERENCE BOOKS (Cont.)	Unit Price	Total
	64777A	Environmental Remediation Est. Methods 2nd Ed.	$ 99.95	
	67160	Estimating for Contractors	35.95	
	67276A	Estimating Handbook 2nd Ed.	99.95	
	67249	Facilities Maintenance Management	86.95	
	67246	Facilities Maintenance Standards	79.95	
	67318	Facilities Operations & Engineering Reference	109.95	
	67301	Facilities Planning & Relocation	89.95	
	67231	Forms for Building Const. Professional	47.48	
	67260	Fundamentals of the Construction Process	34.98	
	67323	Historic Preservation: Proj. Planning & Est.	99.95	
	67308E	Home Improvement Costs–Int. Projects 9th Ed.	29.95	
	67309E	Home Improvement Costs–Ext. Projects 9th Ed.	29.95	
	67324A	How to Est. w/Means Data & CostWorks 2nd Ed.	59.95	
	67304	How to Estimate with Metric Units	9.98	
	67306	HVAC: Design Criteria, Options, Select. 2nd Ed.	84.95	
	67281	HVAC Systems Evaluation	84.95	
	67282A	Illustrated Construction Dictionary, Condensed	59.95	
	67292A	Illustrated Construction Dictionary, w/CD-ROM	99.95	
	67295B	Landscape Estimating 4th Ed.	62.95	
	67341	Life Cycle Costing	99.95	
	67302	Managing Construction Purchasing	19.98	
	67294A	Mechanical Estimating 3rd Ed.	64.95	
	67245A	Planning and Managing Interior Projects 2nd Ed.	69.95	
	67283A	Plumbing Estimating Methods 2nd Ed.	59.95	
	67326	Preventive Maint. Guidelines for School Facil.	149.95	
	67337	Preventive Maint. for Higher Education Facilities	149.95	
	67236A	Productivity Standards for Constr.–3rd Ed.	49.98	
	67247B	Project Scheduling & Management for Constr. 3rd Ed.	64.95	
	67265B	Repair & Remodeling Estimating 4th Ed.	69.95	
	67322A	Resi. & Light Commercial Const. Stds. 2nd Ed.	59.95	
	67327	Spanish/English Construction Dictionary	22.95	
	67145A	Sq. Ft. & Assem. Estimating Methods 3rd Ed.	69.95	
	67287	Successful Estimating Methods	32.48	
	67313	Successful Interior Projects	24.98	
	67233	Superintending for Contractors	35.95	
	67321	Total Productive Facilities Management	29.98	
	67284	Understanding Building Automation Systems	29.98	
	67303A	Unit Price Estimating Methods 3rd Ed.	59.95	
	67319	Value Engineering: Practical Applications	79.95	

MA residents add 5% state sales tax	
Shipping & Handling**	
Total (U.S. Funds)*	

Prices are subject to change and are for U.S. delivery only. *Canadian customers may call for current prices. **Shipping & handling charges: Add 7% of total order for check and credit card payments. Add 9% of total order for invoiced orders.

Send Order To: **ADDV-1001**

Name (Please Print) ____________________

Company ____________________

☐ **Company**

☐ **Home** Address ____________________

____

City/State/Zip ____________________

Phone # ____________ P.O. # ____________

(Must accompany all orders being billed)

Mail To: **RSMeans** P.O. Box 800, Kingston, MA 02364-0800